Particle Physics Reference Library

Stephen Myers • Herwig Schopper
Editors

Particle Physics Reference Library

Volume 3: Accelerators and Colliders

Editors
Stephen Myers
ADAM SA
Geneva, Switzerland

Herwig Schopper
CERN
Geneva, Switzerland

ISBN 978-3-030-34247-0 ISBN 978-3-030-34245-6 (eBook)
https://doi.org/10.1007/978-3-030-34245-6

This Springer imprint is published by the registered company Springer Nature Switzerland AG.
The registered company address is: Gewerbestrasse 11, 6330 Cham, Switzerland

Preface

For many years the series Landolt-Börnstein—Group I Elementary Particles, Nuclei and Atoms, Volume 21A (Physics and Methods Theory and Experiments, 2008), Vol. 21B1 (Elementary Particles Detectors for Particles and Radiation. Part 1: Principles and Methods, 2011), Vol. 21B2 (Elementary Particles Detectors for Particles and Radiation. Part 2: Systems and Applications), and Vol. 21C (Elementary Particles Accelerators and Colliders, 2013), has served as a major reference work in the field of high-energy physics.

When, not long after the publication of the last volume, open access became a reality for HEP journals in 2014, discussions between Springer and CERN intensified to find a solution for the "Labö" which would make the content available in the same spirit to readers worldwide. This was helped by the fact that many researchers in the field expressed similar views and their readiness to contribute.

Eventually, in 2016, at the initiative of Springer, CERN and the original Labö volume editors agreed in tackling the issue by proposing to the contributing authors a new OA edition of their work. From these discussions a compromise emerged along the following lines: transfer as much as possible of the original material into open access; add some new material reflecting new developments and important discoveries, such as the Higgs boson; and adapt to the conditions due to the change from copyright to a CC BY 4.0 license.

Some authors were no longer available for making such changes, having either retired or, in some cases, deceased. In most such cases, it was possible to find colleagues willing to take care of the necessary revisions. A few manuscripts could not be updated and are therefore not included in the present edition.

We consider that this new edition essentially fulfills the main goal that motivated us in the first place—there are some gaps compared to the original edition, as explained, as there are some entirely new contributions. Many contributions have been only minimally revised in order to make the original status of the field available as historical testimony. Others are in the form of the original contribution being supplemented with a detailed appendix relating recent developments in the field. However, a substantial fraction of contributions has been thoroughly revisited by their authors resulting in true new editions of their original material.

We would like to express our appreciation and gratitude to the contributing authors, to the colleagues at CERN involved in the project, and to the publisher, who has helped making this very special endeavor possible.

Vienna, Austria
Geneva, Switzerland
Geneva, Switzerland
July 2019

Christian Fabjan
Stephen Myers
Herwig Schopper

Contents

About the Editors

Stephen Myers was born in Belfast, Northern Ireland, and worked at CERN, Geneva, from 1972 until 2015. He was responsible for the commissioning and the energy upgrade of the CERN Large Electron-Positron Collider (LEP). In 2008, he was nominated CERN Director of Accelerators and Technology until January 2014; during this time, he directed the repair of the CERN Large Hadron Collider (LHC) after the serious accident and steered the operation of the collider in 2010, 2011, and 2012. On July 4, 2012, the data from the collider allowed the discovery of a "Higgs" boson for which Peter Higgs and Francois Engelert received the Nobel Physics Prize in 2013.

He is currently executive chair of a Geneva-based company (ADAM SA), which is developing a linear accelerator for proton therapy of cancer, and non-executive Director of the parent company Advanced Oncotherapy (AVO).

Herwig Franz Schopper joined as a research associate at CERN since 1966 and returned in 1970 as leader of the Nuclear Physics Division and went on to become a member of the directorate responsible for the coordination of CERN's experimental program. He was chairman of the ISR Committee at CERN from 1973 to 1976 and was elected as member of the Scientific Policy Committee in 1979. Following Léon Van Hove and John Adams' years as Director-General for research and executive Director-General, Schopper became the sole Director-General of CERN in 1981.

Schopper's years as CERN's Director-General saw the construction and installation of the Large Electron-Positron Collider (LEP) and the first tests of four detectors for the LEP experiments. Several facilities (including ISR, BEBC, and EHS) had to be closed to free up resources for LEP.

Chapter 1
Accelerators, Colliders and Their Application

E. Wilson and B. J. Holzer

1.1 Why Build Accelerators?

Accelerators are modern, high precision tools with applications in a broad spectrum that ranges from material treatment, isotope production for nuclear physics and medicine, probe analysis in industry and research, to the production of high energy particle beams in physics and astronomy. At present about 35,000 accelerators exist world-wide, the majority of them being used for industrial and medical applications. Originally however the design of accelerators arose from the request in basic physics research, namely to study the basic constituents of matter.

The first accelerators were inspired by the early experiments in nuclear physics. In the early years of the twentieth century Rutherford discovered that by using alpha particles from radioactive disintegration and detecting the pattern of particles scattered by atoms one might deduce that the nucleus was a tiny but massive central element in the atom. Alpha particles from disintegration can only be of energies of 10 MeV; comparable with the nuclear binding forces. Higher energies were needed and a more reliable and steady supply to ease the tedium of counting occasional flashes of light on the scintillation screen that was Rutherford's detector. De Broglie had shown that there was an inverse relationship between the momentum, and hence the energy of a particle and the wavelength of its representation in quantum mechanics.

$$\lambda = \frac{h}{p}$$

E. Wilson · B. J. Holzer (✉)
CERN (European Organization for Nuclear Research), Meyrin, Genève, Switzerland
e-mail: Bernhard.Holzer@cern.ch

S. Myers, H. Schopper (eds.), *Particle Physics Reference Library* ,
https://doi.org/10.1007/978-3-030-34245-6_1

where h represents Planck's constant and p the particles' momentum which relates to its energy via the well-known equation of special relativity, $E^2 = p^2c^2 + m^2c^4$. The limit to the scale of detail that experiments can reveal is set by the length of the wave which is scattered: rather as the wave breaking in the beach can only be deflected by islands larger than itself. It was argued correctly that higher energy particle, having the property of shorter wavelengths could better reveal the structure of the nuclei that Rutherford has detected.

Such arguments led to the invention of the first accelerators and have sustained the development of particle accelerators of higher and higher energy over the best part of the last 100 years. At first, physicists used accelerators to probe the structure of the nucleus, but went on to use higher energy accelerators to search for structure in the "fundamental" particles—protons, neutrons and electrons they discovered. Inevitably higher energies implied larger accelerators, for it was quickly discovered that the best way to accelerate repetitively was to keep particle in a circular path whose radius was itself proportional to energy, limited by the strength of the magnetic field one might use to do the bending.

As energies were raised physicists found new and interesting particles to fit into the pattern of those that their theories might predict. Einstein's

$$E = mc^2$$

tells us that only high energies will create the more massive particles. The latest and largest accelerator, LHC, flagship of the whole community, was designed to search for the Higgs Boson and the successful discovery of this missing puzzle piece in 2013 allowed us to complete the Standard Model of Particle Physics.

As we write, this machine is carrying on the search for physics beyond the standard model, seeking to disclose the nature of dark matter and dark energy.

As more powerful accelerators have been developed for high energy particle physics, advances in the field have been exploited in a whole range of smaller accelerators for other applications. From the time of the first cyclotrons they have been used for producing isotopes and for treating cancer. The development of compact high-frequency linac structures triggered the manufacture of hundreds of small electron linacs producing X-rays for cancer treatment in hospitals around the developed and, latterly, the developing world. Electron rings of a few GeV, specially designed to produce beams of synchrotron radiation have become popular. Each facility serves scores of experiments to investigate the structure of complex molecules—particularly the proteins of today's biomedical studies. Proton accelerators of about 1 GeV produce pulsed beams of neutrons by spallation which are used principally to study the structure of materials. In addition thousands of lower energy accelerators are used in industry for sterilisation and ion implantation in the fabrication of sophisticated CPU chips for computers.

1.2 Types and Evolution of Accelerators

The development of accelerators to ever higher energy is marked by a number of milestones. Each of these marks the invention of a new type of accelerator or the invention of a new principle of transverse or longitudinal focusing which enables a higher energy to be reached for a lower unit cost. The best way to describe this evolution and introduce the different types of accelerator is to follow the road charted by these milestones. Each is described in one of the sections which follow.

1.2.1 Early Accelerators

The nineteenth century had produced a number of electrostatic high-voltage generators. They were unpredictable in performance and electrical breakdown became a serious problem above a few tens of kV. Early accelerators were simply two electrodes enclosed in an evacuated tube with external connections to such high voltage source. A proton or electron source close to one electrode at a potential of V (or $-V$ for electrons) provided the particles which were then accelerated towards the second electrode at earth potential. They emerged or were observed through a small hole in the earthed electrode. The energy acquired by each particle with charge, e Coulombs, was just $e_* V$ Joules or, in the units commonly used for accelerated beams, V electron-Volts. An electron Volt is then just 1.6×10^{-19} Joules. If the particle is a fully stripped ion of an atom with atomic number A and charge Z then the energy is ZV/A electron Volts per nucleon.

The first high-voltage generator to approach 1 MeV was built by Cockcroft and Walton [1–3] in the 1930s to accelerate particles for their fission experiments. Their combination of diodes and capacitors, also known as rectifier circuit, is still used today to apply high voltage to the ion or proton source at the beginning of many linacs and synchrotrons although these are gradually being replaced by radio frequency quadrupoles.

The early 1930s also saw the invention by R.J. Van der Graaf [4] of an electrostatic generator which used a moving belt to carry charge into the high voltage terminal until it reaches a potential of several MV (Fig. 1.1). Van der Graaf accelerators have proved a useful source of low energy particles to this day but are inevitably limited by problems of voltage breakdown. Voltages up to 27 MV have been reached, putting the device in a discharge suppressing gas atmosphere (e.g. SF_6). Although it is possible in theory to chain together several electrostatic accelerators, each with its cathode connected to the anode of the next, each stage increases the potential between the ends of the device and between the ends and ground and eventually electrical breakdown discharges the high voltage terminals.

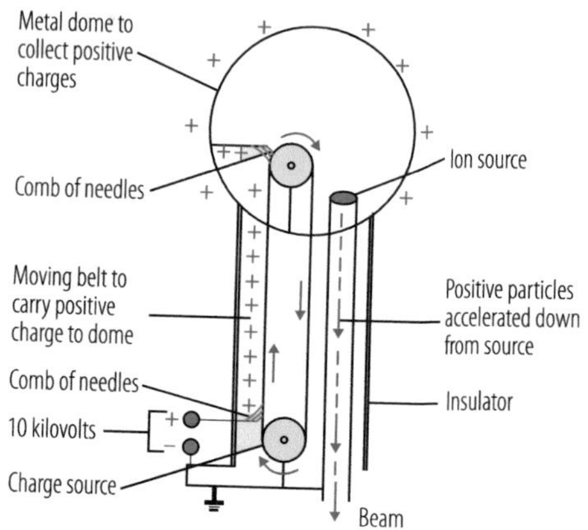

Fig. 1.1 Van der Graaf accelerator

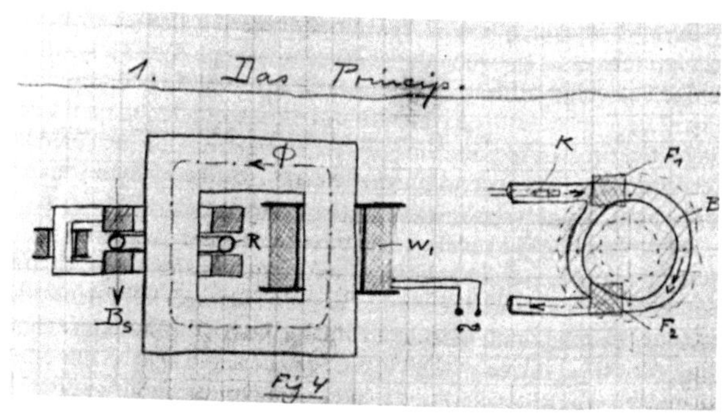

Fig. 1.2 Wideröe's sketch of the ray transformer

1.2.2 *The Ray Transformer*

The earliest idea of how to overcome the limitations of electrostatic acceleration involved using the time varying property of magnetic fields and came from the inventive mind of Rolf Wideröe.

Beginning his studies at Karlsruhe Technical University in 1923, he wondered if electrons in an evacuated ring would flow in the same way as the electrons in copper if they replaced the secondary winding of a transformer· His notebooks of that time contain sketches of a device he called a "ray transformer"; the first circular accelerator and the precursor of the "betatron" [5].

These sketches show a beam tube, in the form of an annulus, *R*, placed in the gap between the parallel poles or faces of a small electromagnet (on the left in Fig. 1.2). This magnet is in the form of a "C" and the field between the poles, B_z, guides particles in a circular orbit in the mid plane between the poles. A circular hole is cut in each pole through which the yoke of the transformer passes linking the beam tube. The primary winding of the transformer, labelled W_1, is powered with alternative voltage from the mains. The beam tube is placed where one would normally expect the secondary winding. The beam within it carries the induced secondary current.

Unlike almost all accelerators that followed, the ray transformer relied entirely upon the inductive effect of a varying magnetic field. It is the rate of change of flux, ϕ, in the yoke which establishes an accelerating potential difference around the beam's path. The windings, that of the C-magnet and of the primary of the transformer, W_1, give independent control of the guide field and accelerating flux.

Wideröe calculated that electrons circulating in a ring of only 10 or 20 cm diameter could reach several MeV within one quarter wave of the AC excitation of the transformer. He had to use Einstein's newly discovered theory of special relativity to correctly describe the motion of particles close to the speed of light. He also found an important principle which ensures that the beam radius does not change as it accelerates. To ensure constant radius during acceleration the total flux linking the beam including that generated by both sets of the coils, B_a, must be twice that generated by the left hand coil pair which produces the field keeping the beam in a circular orbit, B_g.

$$\dot{B}_a = 2\dot{B}_g$$

Unfortunately, Wideröe was dissuaded from building the ray transformer by difficulties with surface fields and by his professor, who wrongly assumed the beam would be lost because of gas scattering. However, his Ray Transformer and the 2 to 1 ratio, now known as the Wideröe principle, were important discoveries which were put into practice 15 years later when D.W. Kerst and R. Serber [6] built a series of betatrons.

Wideröe went on to develop a second basic acceleration method to overcome the electrostatic limitation: the drift tube linac.

1.2.3 Repetitive Acceleration

There are two broad classes of accelerator characterized by the way they achieve repetitive acceleration and which overcome the insulation problems of the electrostatic machines. The simplest concept is that of the linear accelerator. Particles pass though cavities excited by radio frequency generators. They arrive on the threshold of each cavity with the energy they have already received and gain a further increment in energy from the electric field in the cavity which points in their direction of motion. Each cavity performs the function of the gap between the anode and cathode of an electrostatic accelerator but, unlike the electrostatic case, the increments of energy may be added together without developing a huge voltage to ground in any part of the apparatus. Of course, there is a limit to the voltage (energy increment) each cavity can apply and the length of the device becomes very long for energies above 1 GeV. Nevertheless, a linac has become the only way of accelerating highly relativistic electrons which radiate a large fraction of their energy when bent into a circular path.

As alternative concept, circular machines, like cyclotrons and synchrotrons use the same set of accelerating cavities over and over again as the particles make complete turns around the accelerator, being guided and focused by the magnet structure of the ring which is thus constraining their orbit. On each turn an increment of energy is added and, once accelerated, particles may be allowed to circulate indefinitely at their top energy. Two circulating beams of say protons and antiprotons or electrons and positrons can be sustained in the same ring and, colliding at experiments around the circumference, create new particles up to a mass (centre of mass energy as it is called) that is the sum of the two energies. Colliders are today the preferred configuration for a high-energy machine. Earlier, new particles were sought in the debris from a particles collision with a nucleon in a fixed target but such collisions are limited to a smaller centre of mass energy—which rises only as the square root of the accelerated beam energy.

1.2.4 Linear Accelerators

Although disappointed by the rejection of his ray-transformer as a subject for his PhD, Wideröe was led to the idea of a linear accelerator by a paper by G. Ising [7] who tried to overcome the voltage breakdown problem of a single stage of acceleration by placing a series of hollow cylindrical electrodes one after another in a straight line to form what today we would call a 'drift tube linac' or linear accelerator. Wideröe realised that an oscillating potential applied to one drift tube flanked by two others which are earthed, accelerates at both gaps provided the oscillator's phase changes by 180° during the flight time between gaps.

In 1927 he built a three-tube model which accelerated sodium ions. At the wavelengths that radio transmitters generated at that time a particle travelling near the velocity of light would travel hundreds of meters in the time it would take for the r.f. to swing by half a sine wave. This would make the length of a drift tube impractically large. Sodium ions, being rather heavy compared with protons or electrons, travelled much slower than the velocity of light and this helped keep the apparatus down to table-top proportions. Although he realised that one might extend such a series of tubes indefinitely he did not take the idea any further as he was due to start his professional employment designing high voltage circuit breakers. Between 1931 and 1934, D. Sloan and E.O. Lawrence at Berkeley took up Wideröe's idea and constructed linacs with as many as 30 drift tubes to accelerate mercury ions but, these were never used for research.

Much later, in the mid-1940s, and when suitable high-power high-frequency oscillators had become available to meet the needs of war-time radar, L.W. Alvarez (1946) started to build the first serious proton linac at the Radiation Laboratory of the University of California. Figure 1.3 shows an Alvarez linac. A series of drift tubes are mounted within a copper-lined cylinder excited by a radio transmitter. As in Wideröe's linac, particles gain energy from the accelerating potential differences between the ends of the drift tube, but now the phase shift between drift tube

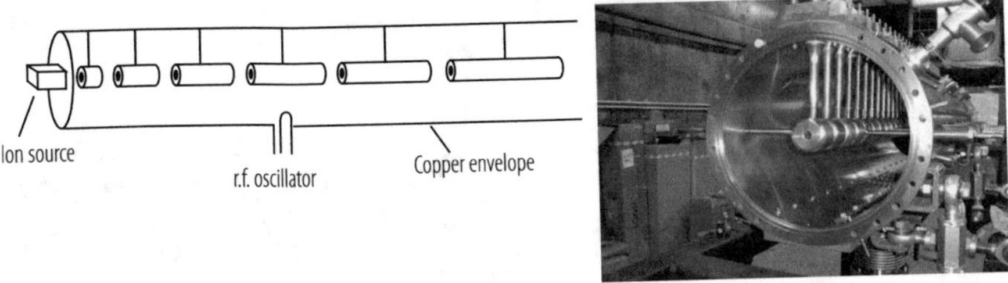

Fig. 1.3 left: The concept of the drift tube linac (from [8]); right: CERN's Linac 1

gaps is 360°. Each gap appears to the particle to be an identical field gradient which accelerates particles from left to right. The particles are protected from the decelerating phase while inside the metallic drift tubes. Although the particle gains energy steadily as it passes each gap, the total voltage between parts of the assembly and ground does not become larger along the length of the device as it would for an electrostatic machine.

The distances between gaps, or the lengths of the tubes, increase as the particle is accelerated since it travels an ever increasing distance during one swing of the radio frequency oscillation. At low energy, we would expect this distance to increase with the velocity or the root of the kinetic energy but when the energy is large we find the length of the drift tubes and their spacing no longer increases—a practical demonstration of special relativity. The Alvarez structure is still widely used, especially for non-relativistic proton and ion beams.

It was well known at the time that waves might be propagated along a much simpler smooth waveguide and that some of the modes have an accelerating electric field in the direction of propagation. Closer examination however shows that the stumbling block is that the phase velocity of these modes in a wave guide is always greater than that of light and hence the particle sees a field which sometimes accelerates and then decelerates as the wave overtakes the particle. It was later found that the phase velocity could be reduced by a series of iris diaphragms in the pipe. Such a structure (Fig. 1.4) is very popular in electron linacs and also in storage rings in which the particle is close to the velocity of light and cavities need not be tuned to follow the acceleration cycle.

These diaphragm-loaded linac structures have been commonly used as injectors for circular accelerators to accelerate electrons and protons to energies in the range 10 to 1000 MeV. As compact high frequency structures they have also been widely used to accelerate electrons to, typically 10 MeV, as a source of X-rays for cancer therapy. An early and very successful adventure in the electron linac development was the "two-mile long" Stanford Linear Accelerator at SLAC in California which has been the work horse for a number of ground breaking fixed target experiments and circulating beam storage ring projects at 20 to 50 GeV. With the help of two semi-circular arcs it was used to bring beams of electrons and positrons into head-on collision in the Stanford Linear Collider Project. This project, is forerunner for

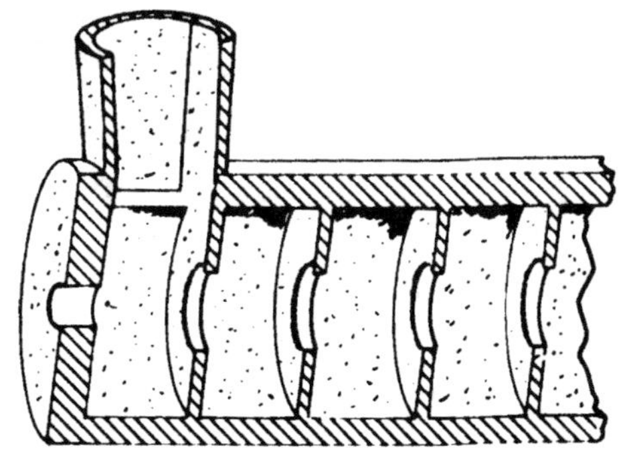

Fig. 1.4 Iris loaded structure (from [9]). The 'chimney' is the input waveguide

today's projected Linear Colliders in which linear accelerators accelerate positrons and electrons to energies approaching 1 TeV to collide them head on in a bid to overcome the very considerable energy lost by an electron to synchrotron radiation in circular lepton rings at high energy.

1.2.5 Cyclotrons

Unlike a linac, whose length must be extended to reach a higher energy, the cyclotron, as it is called, is a relatively compact accelerator in which the energy is only limited by the diameter and field strength of the magnet. The cyclotron idea first occurred to E.O. Lawrence who, reading through Wideröe's thesis, ruminated on the possibility of using a magnetic field to recirculate the beam through two of drift tubes. The cyclotron idea was published in 1930 [10] and another colleague, M.S. Livingston, who was also later to contribute much to the field, was given the job of making a working model as his doctoral thesis.

In Fig. 1.5 we see the two 'Dee's' which comprise the positive and negative electrodes of the accelerating system between the poles of the magnet. These are like two halves of a closed cylinder divided along its diameter. A radio-frequency generator excites them with an alternating field of constant frequency. The potential difference between the 'Dee's' accelerates the ions as they pass the gap between the two halves of the structure. The fundamental trick is that the field oscillates at the particle's circulation frequency and hence the sign of the potential difference at each gap is always in the accelerating direction.

As long as cyclotrons accelerate ions to modest energies, classical rather than relativistic mechanics still applies. In Fig. 1.6 we see the balance between centripetal acceleration of motion in a circle and the force exerted by the vertical magnetic field,

$$evB = \frac{mv^2}{\rho}, \text{ if } v \ll c, \tag{1.1}$$

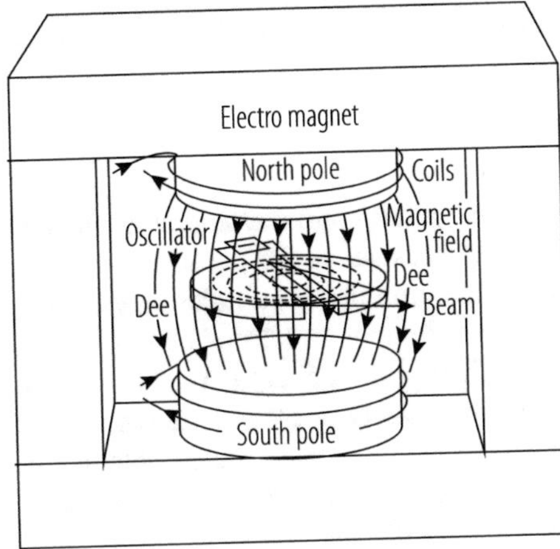

Fig. 1.5 The principle of the cyclotron

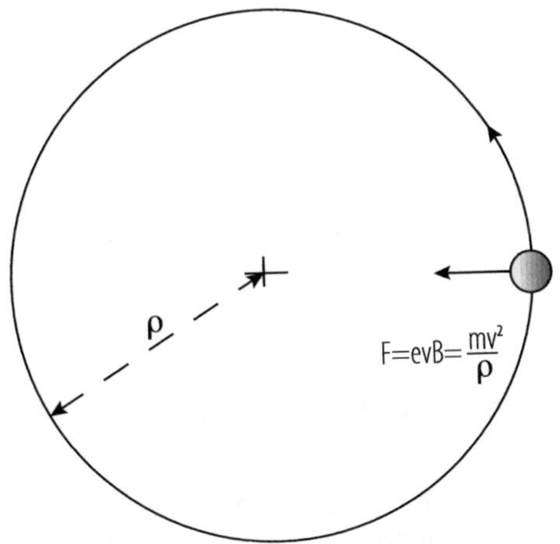

Fig. 1.6 Balance of forces in a cyclotron

and, rearranging, we can define the magnetic rigidity—the reluctance of the beam to be bent in a curve:

$$B\varrho = \frac{mv}{e}, \text{ if } v \ll c. \tag{1.2}$$

In the relativistic regime if we replace the classical momentum, *mv,* by the relativistic momentum, $p = \gamma mv$, with γ being the Lorentz factor, we obtain the equation, valid in the relativistic regime:

$$B\varrho = \frac{p}{e}. \tag{1.3}$$

By good fortune the radius of the orbit in a cyclotron is proportional to the velocity and the frequency of revolution this being the inverse of the time of revolution—just the length of one turn divided by the particle's velocity

$$f = \frac{v}{2\pi\varrho} = \frac{v}{2\pi} \cdot \frac{eB}{mv}. \tag{1.4}$$

has a numerator and denominator which are both proportional to v. This frequency remains constant as the particle is accelerated in the low energy, classical, regime. Thus, the circulating particles stay in synchronisation with the oscillating RF field and a continuous stream of ions injected in the centre will follow a spiral path to reach their highest energy at the rim of the poles.

Unfortunately, the synchronism between r.f. voltage and revolution frequency breaks down as the particles velocity begins to approach that of light and the relativistic mass in the above equation is no longer constant. This happens over 30 MeV for protons and at double this energy for deuterons. Electrons are much too light and relativistic to be accelerated in a cyclotron to any significant energy. For them other acceleration concepts are more adequate, like the disk loaded travelling wave linac or the betatron that both were described before.

The possible remedy of making the field stronger at the edge of the poles would have preserved synchronism and continuous beams but, as we shall see, was in conflict with the need to have a negative radial gradient to the field to provide vertical weak focusing. As a consequence a more powerful concept had to be developed to achieve highest particle beam energies: The synchrotron.

1.2.6 The Synchrotron

Meanwhile, in the 1940s, still higher energies were needed to pursue the aims of physics and the stage was set for the discovery of the synchrotron principle which opened the way to the series of circular accelerators and storage rings which have served particles physics up to the present day. It was Australian physicist Mark Oliphant who synthesized three old ideas into a new concept—the synchrotron. The ideas were: accelerating between the gaps of resonators, varying the frequency, and pulsing the magnet. In 1943 he described his invention in a memo to the UK Atomic Energy Directorate (see [11]).

> Particles should be constrained to move in a circle of constant radius thus enabling the use of an annular ring of magnetic field ... which would be varied in such a way that the radius of curvature remains constant as the particles gain energy through successive accelerations by an alternating electric field applied between coaxial hollow electrodes.

Unlike the cyclotron, the synchrotron accelerates the beam as a series of discrete pulses or "bunches" as they are called. Each short pulse is injected at low field and then the field rises in proportion to the momentum of particles as they are accelerated. This ensures that the radius of the orbit remains constant. In contrast to cyclotrons and betatrons, the synchrotron needs no massive poles to support a

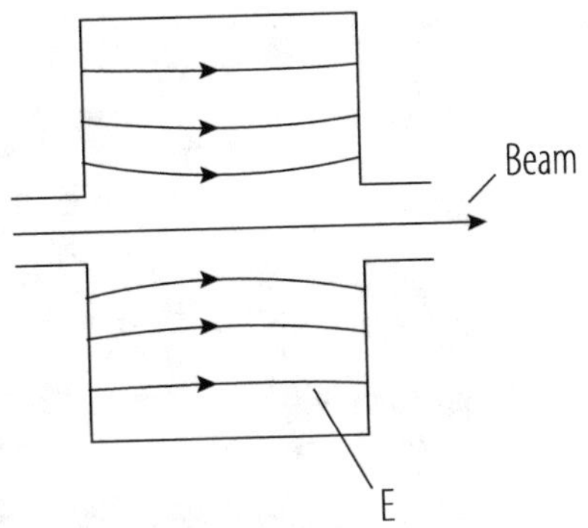

Fig. 1.7 A simple accelerating cavity

magnetic field within the beam's circular orbit. The guide field is instead provided by a slender ring of individual magnets. The fact that the machine is pulsed and the frequency must be controlled to track the increasing speed of particles is a complication, but it solves the difficulty that isochronous cyclotron builders had encountered in accelerating relativistic particles.

Instead of the Dees of a cyclotron acceleration is provided in a synchrotron by fields within a hollow cylindrical resonator or "pillbox" cavity, Fig. 1.7, excited by a radio transmitter. A particle passes from left to right as it completes each turn of the synchrotron receiving another increment in energy at each revolution.

The early synchrotrons, like the cyclotron before them, relied on a slight negative radial gradient in the vertical magnet field to produce field lines which belly outwards from the magnet gap. A small radial field component deflects any particles which head off towards the poles back to the median plane. Unfortunately, this field shape has the opposite (defocusing) effect horizontally but, up to a certain, rather weak, gradient strength focusing is assured by a slight imbalance between the central force and the centrifugal acceleration. The gradient cannot be too large—hence the term "weak focusing".

Oliphant was the first to start building a proton synchrotron (at Birmingham University) but he was overtaken by Stan Livingston's 3 GeV Cosmotron at Brookhaven National Laboratory and later by the 6 GeV Bevatron at Berkeley.

Due to the weak focusing forces in these first synchrotrons, the particles' excursions, both horizontally and vertically are large and the magnet pole width and gap correspondingly so. Strong focusing changed this. It was invented at the Cosmotron, which was actually the first proton synchrotron to operate, whose weak focusing 'C' shaped magnet was open to the outside. The top energy of the Cosmotron was limited by the extra fall-off in field caused by the effect of saturation. Stan Livingston and E.D. Courant wanted to compensate this by re-installing some of the C magnets with their return yokes towards the outside. They were afraid of the variations in gradient around the ring but were surprised to calculate that the focusing seemed to improve as the strength of the alternating component of the gradient increased. Courant, Livingston, and H.S. Snyder [12, 13] were able to explain this retrospectively with an optical analogy of alternating focusing by equal

Fig. 1.8 The CERN 25 GeV proton synchrotron

convex and concave lenses which will transport rays which pass through the centres of defocusing lenses.

Alternating gradient or strong focusing greatly reduces the beam's excursions and so the cross section of the magnet gap by more than an order of magnitude. Its discovery enabled Brookhaven and CERN to build the next generation of proton synchrotrons, AGS and PS, to reach 30 GeV—five times the energy of the Bevatron—yet use beam pipes of only a few centimetres height and width.

This was to lead to huge economies in the cost per unit length of the magnet system. Figure 1.8 shows how this was applied to the first of the two synchrotrons, AGS and PS that used this focusing system. From then on all synchrotrons and, later, storage ring colliders use this scheme. The history of synchrotrons has been always to seek methods of improving focusing and economizing on magnet aperture. The only other step function in their development to higher energies has been the use of superconducting magnets whose higher fields reduce the circumference of the machine by a factor between 3 and 5.

1.2.7 Phase Stability

When the first synchrotrons were built it was by no means obvious that the circulating beam and the accelerating voltage would remain in step. There were those who thought that any slight mistiming of the sine wave of accelerating voltage in the cavity might build up over many turns until particles would begin to arrive within the negative, decelerating, phase of the sine wave and be left behind. Even if one succeeded in achieving synchronism for the ideal, *synchronous particle*, others of slightly different energy would not have the same velocity and take a different time to circulate around the machine. Would not these particles gradually get out of step until they were lost? After all, particles had to make many hundred thousand turns before reaching full energy and while transverse focusing was understood there was no apparent focusing available in the longitudinal direction. Fortunately the comforting principle of phase stability, which prevents this happening, was soon to be independently discovered by V. I. Veksler in Moscow in 1944 [14] and McMillan in Berkeley in 1945 [15], opening the way to the construction of the first synchrotrons. We shall return to this later.

When it came to the next generation of synchrotrons, interest focused on colliding two opposing beams of particles. It had been known for some time that the energy available in the centre of mass from a collision of particles, one in the beam with energy E and the other of mass m_0 in a fixed target, only increased with the square root of the accelerators energy, $\sqrt{m_0 E}$. Two particles of the same mass and energy E colliding head on made available all their energy in the centre of mass, $2E$. The difficulty was making the two bunches of particles of sufficient density to have a significant probability of collision or, in technical jargon a high enough luminosity. Once this problem was solved a series of colliders: ISR, Sp$\bar{\text{p}}$S, LEP, Tevatron, HERA and finally LHC followed. Some of these (ISR, HERA and LHC) were two separate rings which intersected to collide particles at several points around the circumference. Others (Sp$\bar{\text{p}}$S and LEP) collided protons with antiprotons and electrons with their anti-particles: positrons. These exploited the fact that beams of particles and antiparticles will circulate on identical trajectories, but in opposite directions, in a single ring of bending and focusing magnets.

At present several studies are ongoing, to pave the way to even higher energies, mainly increasing the size of the machine and using super conducting magnets with higher critical field, to gain more bending and focusing fields in the lattice. One example, the Future Circular Collider study, FCC, under the guidance of CERN, is studying a 100 km proton storage ring to achieve centre of mass energies of up to 100 TeV. The R & D effort of accelerators of this dimension and complexity, in any case, has to be done by a truly international, in other words worldwide effort.

References

1. J.D. Cockcroft, E.T.S. Walton: Proc. Roy. Soc. A 129 (1930) 477-489.
2. J.D. Cockcroft, E.T.S. Walton: Proc. Roy. Soc. A136 (1932) 619-630.
3. J.D. Cockcroft, E.T.S. Walton: Proc. Roy. Soc. A 137 (1932) 229-242.
4. R.J. Van der Graaf: Phys. Rev. 38 (1931) 1919.
5. R. Wideröe.: ETH Library, Zürich, Hs 903 (1923-28) 633-638.
6. D.W. Kerst, R. Serber: Phys. Rev. 60 (1941) 53-58.
7. G. Ising: Arkiv för matematik o. fysik 18 (1924) 1-4.
8. J.J. Livingood: *Principles of cyclic particle accelerators*, van Nostrand (1961).

9. P. Lapostolle, A. Septier: *Linear Accelerators*, North Holland (1971).
10. E.O. Lawrence, N.E. Edelfsen: Science 72 (1930) 376-7.
11. M. Oliphant: *The genesis of the Nuffield Cyclotron and the Proton Synchroton*, Publ. Department of Physics, University of Birmingham.
12. E.D. Courant, M.S. Livingston, H.S. Snyder: Phys. Rev. 88 (1952) 1190-1196.
13. E.D. Courant, H.S. Snyder: Annals of Physics 3 (1958) 1-48.
14. V.I. Veksler: Comptes rendues (Doklady) de l'Academie des Sciences de l'URSS 43 (1944) 329-341.
15. E.M. MacMillan: Phys. Rev. 68 (1945) 143.

Chapter 2
Beam Dynamics

E. Wilson and B. J. Holzer

2.1 Linear Transverse Beam Dynamics

Now let us look in detail at the motion of particles in the transverse coordinates of the coordinate system defined in Fig. 2.1.

2.1.1 *Co-ordinate System*

The guide field of a synchrotron is usually vertically directed, causing the particle to follow a curved path in the horizontal plane (Fig. 2.1). The force guiding the particle in a circle is horizontal and is given by:

$$\boldsymbol{F} = e \cdot \boldsymbol{v} \times \boldsymbol{B}, \tag{2.1}$$

where:

$\boldsymbol{v}$ is the velocity of the charged particle in the direction tangential to its path,
$\boldsymbol{B}$ is the magnetic guide field.

The guide field inside a dipole magnet is uniform and the ideal motion of the particle is simply a circle of (local) radius of curvature, $\rho(s)$. The trajectory of an ideal particle (ideal in energy and without any amplitude) that is defined by the arrangement of the dipole magnets is called *design orbit*. The machine is usually designed with this orbit at the centre of its vacuum chamber. Now there is no such

E. Wilson · B. J. Holzer (✉)
CERN (European Organization for Nuclear Research), Meyrin, Geneva, Switzerland
e-mail: Bernhard.holzer@cern.ch

S. Myers, H. Schopper (eds.), *Particle Physics Reference Library*,
https://doi.org/10.1007/978-3-030-34245-6_2

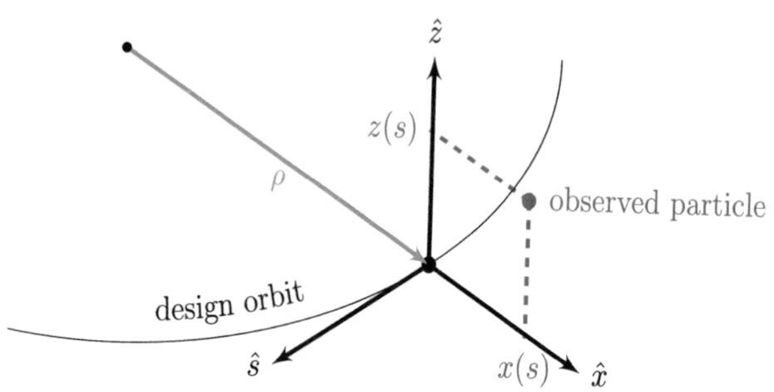

Fig. 2.1 Charged particle orbit in magnetic field

thing as an ideal particle. Still, we shall suppose that it is possible to find an orbit or curved path for the non-ideal particle which closes on itself around the synchrotron, which we call the *closed* or *equilibrium orbit* and it should be close enough to the ideal design orbit.

2.1.2 Displacement and Divergence

A beam of particles enters the machine as a bundle of trajectories spread about the ideal orbit. At any instant a particle may be displaced horizontally by x and vertically by z from the ideal position and may also have divergence angles horizontally and vertically:

$$x' = dx/ds, \quad \text{and } z' = dz/ds. \tag{2.2}$$

The divergence would cause particles to leave the vacuum pipe except for the carefully shaped field which restores them back towards the beam centre so that they oscillate about the ideal orbit. The design of the restoring fields determines the transverse excursions of the beam and the size of the cross section of the magnets and is therefore of crucial importance to the cost of a project.

2.1.3 Bending Magnets and Magnetic Rigidity

The design of a synchrotron; the diameter of the ring and its sheer size and cost for a given energy is driven by the fact that bending particle trajectories depends on a magnetic rigidity. The rigidity increases with momentum and is a function of the bending field which, for room temperature magnets, saturates at about 2 T.

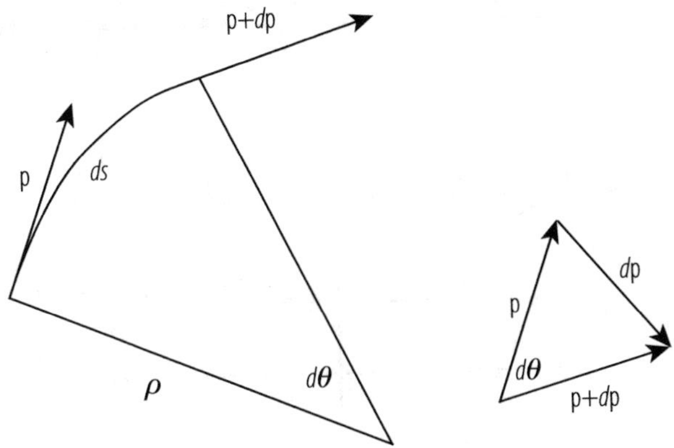

Fig. 2.2 Vector diagram showing differential changes in momentum for a particle trajectory

We will now briefly derive an expression for the magnetic rigidity of a relativistic. A particle has a relativistic momentum vector $\boldsymbol{p}$ and travels perpendicular to a field $\boldsymbol{B}$ which is into the plane of the diagram (Fig. 2.2).

We write the Lorentz force on the particle on its circular path as

$$\boldsymbol{F}_{Lorentz} = e * (\boldsymbol{v} \times \boldsymbol{B})$$

Assuming an idealized homogeneous dipole magnet along the particle orbit, having pure vertical field lines, the condition for a perfect circular orbit is defined as equality between this Lorentz force and the centrifugal force.

$$\boldsymbol{F}_{centrifugal} = \frac{\gamma m v^2}{\rho}$$

This yields the following condition for the idealized ring:

$$B\rho = \frac{p}{e}$$

where we are referring to protons and have accordingly set q = e. We conclude that the beam rigidity Bρ, given by the magnetic field and the size of the machine, defines the momentum of a particle that can be carried in the storage ring, or in other words, it ultimately defines, for a given particle energy, the magnetic field of the dipole magnets and the size of the storage ring.

We really should use the units Newton-second for p and express e in Coulombs to give $(B\rho)$ in Tesla·metres. However, in charged particle dynamics we often talk in a careless way about the 'momentum' pc. This actually has the dimensions of an energy and is expressed in units of GeV.

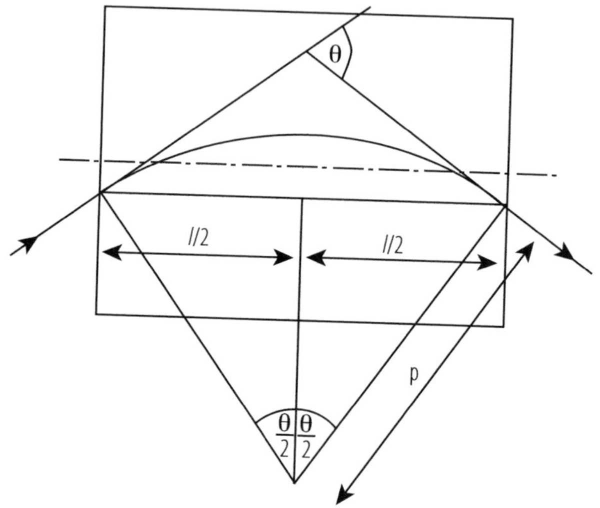

Fig. 2.3 Geometry of a particle trajectory in a bending magnet of length ℓ and deflecting angle θ

A useful rule of thumb formula based on these units is:

$$B\rho\ [\mathrm{T}\cdot\mathrm{m}\,] = 3.3356\ pc\ [\mathrm{GeV}\,]\,. \tag{2.3}$$

2.1.4 Particle Trajectory in a Dipole Bending Magnet

The trajectory of a particle in a bending magnet or dipole of length ℓ is shown in Fig. 2.3. Usually the magnet is placed symmetrically about the arc of the particle's path. One may see from the geometry that:

$$\sin(\theta/2) = \frac{\ell}{2\varrho} = \frac{\ell B}{2\,(B\varrho)}, \tag{2.4}$$

and, if $\theta << \pi/2$

$$\theta \approx \frac{\ell B}{(B\varrho)}. \tag{2.5}$$

So the bending angle provided by a dipole magnet is given by the ration of its integrated field strength and the beam rigidity.

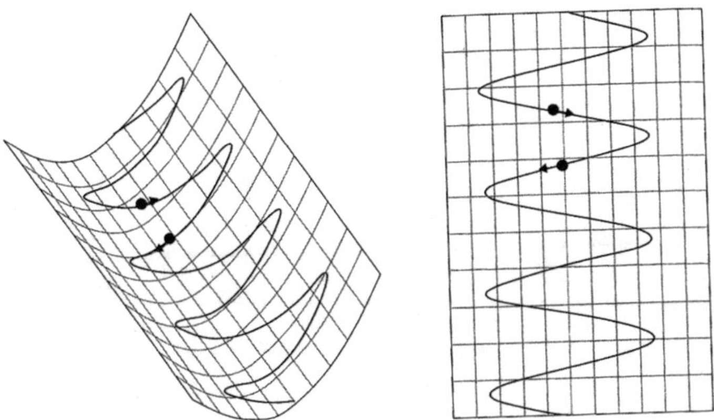

Fig. 2.4 Two views of a sphere rolling down a gutter as it is focused by the walls

2.1.5 *Weak Focusing*

We have mentioned that cyclotrons and early synchrotrons relied on weak focusing to constrain circulating particles within the vacuum chamber. In order to provide this, the vertical guide field has a slight negative gradient in the radial direction around the rim of the accelerator. The field lines belly out from the outer gap of the magnet. It can be shown, by applying $\nabla \times \overline{\boldsymbol{B}} = \boldsymbol{0}$, that there will be horizontal field components in this region. These produce vertically directed forces on errant particles causing them to oscillate about the median plane in a potential well (Fig. 2.4).

The motion is analogous to a small sphere rolling down a slightly inclined gutter with constant speed. Figure 2.4 shows two views of this motion and from the bottom view we recognise the motion as a sine wave. Note too that the sphere makes four complete oscillations along the gutter. In the language of accelerators, its motion has a wave number or "tune", $Q = 4$.

To complete the analogy of a weak focusing synchrotron we imagine that we bend the gutter into a circle rather like the brim of a hat. We provide the necessary instrumentation to measure the displacement of the sphere from the centre of the gutter each time it passes a given mark on the brim and we also have a means to measure its transverse velocity. With the aid of a computer, we might convert this information into the divergence angle, which is used as vertical axis in Fig. 2.5:

$$x' = \frac{dx}{ds} = \frac{v_\perp}{v_\parallel}. \tag{2.6}$$

We can make a ring shaped gutter out of a slightly different length of gutter than is shown so that Q is not an integer. We can plot a point for each arrival of

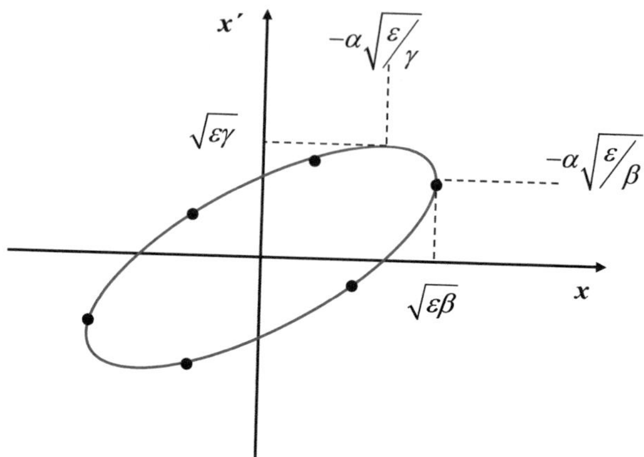

Fig. 2.5 The elliptical locus of a particle's history in phase space as it circulates in a synchrotron

the sphere in a diagram of x' against x which we call a 'phase space diagram' of transverse motion. The sphere has a large transverse velocity as it crosses the axis of the gutter and has almost zero transverse velocity as it reaches its maximum displacement.

If we plot these 'observations' they will be an ellipse (Fig. 2.5) and the phase of the oscillator will advance by Q evolutions each time the particle returns. Of course, only the fractional part of Q may be deduced from our observations since our measurements do not reveal what happens round the rest of the hat's brim.

Now let us use the analogy to define some of the transverse dynamical quantities of a particle beam. The area of the ellipse is a measure of how much the particle departs from the ideal trajectory, represented in the diagram by the origin.

$$\text{Area} = \pi\varepsilon \ [\text{mm} \cdot \text{rad}] . \tag{2.7}$$

In accelerator notation we use ε, the product of the semi-axes of the ellipse as a measure of the area called the emittance. The emittance is usually quoted in units of π mm·mradians. Thus if the semi-axes are 1 mm and 1 mrad the emittance will be 1 mm·mradian but the area will be π mm·mradian. The maximum excursion in displacement, the major axis, of the ellipse is defined as:

$$\hat{x} = \sqrt{\varepsilon\beta}, \tag{2.8}$$

At locations where the beta function reaches an extremum, i.e. $\alpha = 0$, we obtain hence

$$\hat{x}' = \sqrt{\varepsilon/\beta}. \tag{2.9}$$

We shall see that β (later to be called the envelope or betatron function) is a property of the gutter, not the beam. In the synchrotron it varies around the ring and is the envelope function we have plotted in Fig. 2.10 and again in Fig. 2.11. By analogy, the "brim of the hat" which represents the alternating gradient focusing system shown in this figure will vary its width and curvature around the crown and β will follow this variation in some way.

2.1.6 Alternating Gradient Focusing

In Chap. 1 we described a major break-through in the design of synchrotrons: the discovery of alternating gradient (AG) focusing (see [1] for an excellent summary of the dynamics of AG focussing). This allowed designers to use much stronger focusing systems with considerable savings in the space needed for the beam cross section.

The principle is shown in Fig. 2.6 which depicts an optical system in which each lens is concave in one plane while convex in the other and they alternate. It is possible, even with lenses of equal strength, to find a ray which is always on axis at the D lenses in the horizontal plane and therefore only sees the F lenses. To appear like Fig. 2.6 the spacing of the lenses would have to be $2f$. If the ray is also central in the lenses which are vertically defocusing, the same condition will apply simultaneously in the vertical plane. At least one particular particle or trajectory corresponding to this ray will never be defocused and be contained indefinitely.

The alternating gradient idea will work even when the rays in the D lenses do not pass exactly at their centre and the lenses are not spaced by precisely $2f$. In fact it is sufficient for the lens strengths and spacing to be chosen to ensure that the particle trajectories *tend* to be closer to the axis in D lenses than in F lenses as shown in Fig. 2.10.

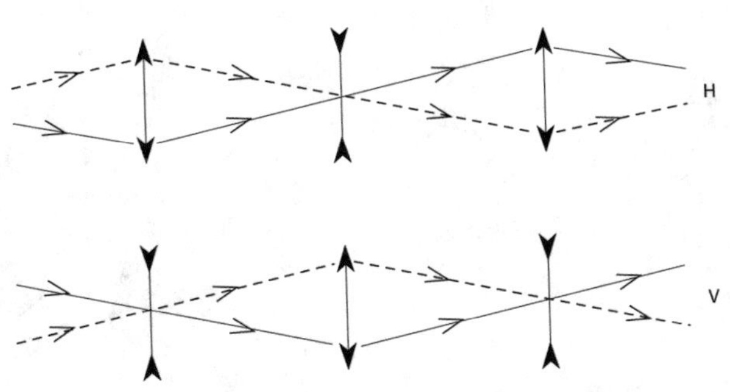

Fig. 2.6 Optical analogy with an alternating pattern of lenses

2.1.7 Quadrupole Magnets

The first alternating gradient synchrotorns used alternating magnetic lenses formed by bending magnets having the same vertical guide field but a radial gradient of alternating sign. In a modern synchrotron the functions of guiding and focussing the beam are separated. The dipole magnets which do the guiding have no gradient. The principal focusing elements are quite a different kind of magnet with four poles which produce gradient but no bending. The poles of these quadrupole magnets are truncated rectangular hyperbolae and alternate in polarity around the aperture circle which just touches the poles.

Figure 2.7 shows a particle's view of the fields and forces in the aperture of a quadrupole as it passes through normal to the plane of the paper. The field shape is such that it is zero on the axis of the device but its strength rises linearly with distance from the axis. This can be seen from a superficial examination of Fig. 2.7 if we remember that the product of field and length of any field line joining the poles is a constant. Symmetry tells us that the field is vertical in the median plane (and purely horizontal in the vertical plane of asymmetry). The field must be downwards on the left of the axis if it is upwards on the right.

The horizontal focusing force, $-evB_z$, has an inward direction on both sides and, like the restoring force on a weight suspended from a spring, rises linearly with displacement, x. The strength of the quadrupole is characterised by its gradient dB_z/dx normalised with respect to magnetic rigidity:

$$k = \frac{1}{(B\varrho)} \frac{dB_z}{dx}. \tag{2.10}$$

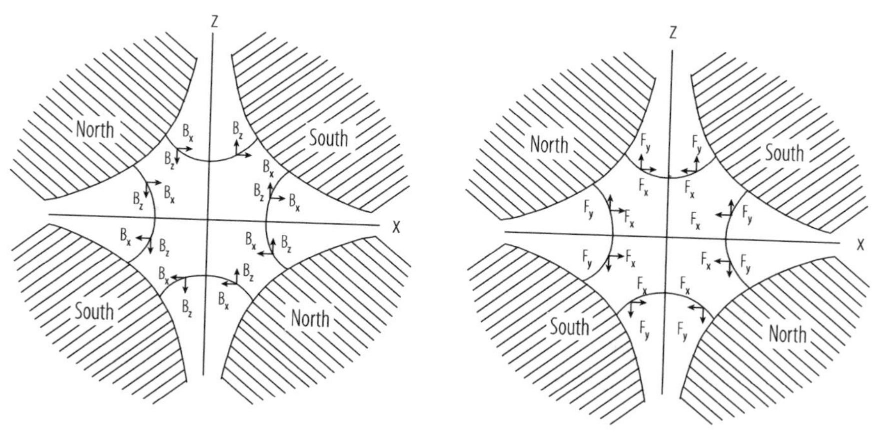

Fig. 2.7 Components of field and force in a magnetic quadrupole. Positive ions approach the reader on paths parallel to the s axis (orthogonal to x and z) [2]

The angular deflection given to a particle passing through a short quadrupole of length, ℓ and strength k, at a displacement x is therefore:

$$\Delta x' = \theta = \ell B/\left(B\rho\right) = \ell B' x/\left(B\rho\right) = \ell k x. \tag{2.11}$$

The use of x' to indicate the divergence angle of a trajectory is defined in Fig. 2.5. Compare this with a converging lens in optics:

$$\Delta x' = -x/f \tag{2.12}$$

and we see that the focal length of a horizontally focusing quadrupole is

$$f = -1/\left(k\ell\right) \tag{2.13}$$

The particular quadrupole shown in Fig. 2.7 would focus positive particles coming out of the paper or negative particles going into the paper in the horizontal plane. A closer examination reveals that such a quadrupole deflects particles with a vertical displacement away from the axis—vertical displacements are defocused. This can be seen if Fig. 2.7 is rotated through 90°.

2.1.8 *The Equation of Motion*

Earlier we derived an expression for the change in divergence of a particle passing through the quadrupole. A horizontally focusing quadrupole (which is at the same time vertically defocusing) has a negative k.

We first look at the vertical plane. The angular deflection given to a particle passing through a short quadrupole of length ds and strength k at a displacement z is therefore:

$$dz' = -kzds. \tag{2.14}$$

From this we can deduce a differential equation for the motion

$$z'' + k(s)z = 0. \tag{2.15}$$

Here we would like to make a clear statement: While inside a lattice element, say a quadrupole lens, the normalised gradient k is constant and we get a equation that we know from Hook's law in classical mechanics, (see Eq. 2.12), the situation now is more general. We allow $k(s)$ to change, while our particles are running through the accelerator. The corresponding equation (2.15) is called Hill's Equation, a second order linear equation with a periodic coefficient, $k(s)$ which describes the distribution of focusing strength around the ring. The above form of Hill's equation applies to the motion in the vertical plane while in the horizontal plane the effect of the dipole magnets has to be included:

$$x'' + \left[\frac{1}{\rho^2(s)} - k(s)\right]x = 0. \tag{2.16}$$

Here the sign in front of $k(s)$ is reversed so that the quadrupole focuses. The extra focusing term $1/\rho^2$ due to the curvature of the orbit can be significant in small rings. In the old constant gradient synchrotrons, this weak focusing term was the only form of horizontal focusing.

We see in Fig. 2.10, the pattern of one cell of a simple synchrotron lattice—a pattern which is repeated many times around the circumference as may be seen in Fig. 2.11 which shows—in addition to the focusing and defocusing lenses also the bending magnets—bending magnets. Within this pattern of dipole and quadrupole focusing and defocusing (F and D), particles make betatron oscillations within the envelopes described by β_x and β_z, or more precisely, the square roots of these quantities (here we use the variable y to represent either the horizontal or the vertical coordinate, x or z)

$$y = \sqrt{\varepsilon\beta(s)}\,\sin\left(\phi(s) + \phi_0\right). \tag{2.17}$$

If one tries to verify that this is the solution of Hills Equation an important and necessary condition emerges:

$$\phi' = 1/\beta \tag{2.18}$$

From which we see that $2\pi\beta$ is the local wavelength of the transverse oscillations.

2.1.9 *Matrix Description*

Usually in alternating gradient (AG) machines, the ring is a repetitive pattern of focusing fields that we call the "lattice". Each lattice element may be expressed by a matrix and whole sections of the ring which transport the beam from place to place may be represented as the product matrix of the single element matrices involved, which makes the description of particle trajectories very simple and very elegant at the same time. Any linear differential equation, like Hill's Equation, has solutions which can be traced from one point, s_1, to another, s_2, by a 2×2 matrix, the transport matrix:

$$\begin{pmatrix} y(s_2) \\ y'(s_2) \end{pmatrix} = \begin{pmatrix} a & b \\ c & d \end{pmatrix}\begin{pmatrix} y(s_1) \\ y'(s_1) \end{pmatrix} = M_{21}\begin{pmatrix} y(s_1) \\ y'(s_1) \end{pmatrix}. \tag{2.19}$$

The transport matrix M_{21} has a rather simple form for each focusing quadrupole that the particle encounters and for the drift length between quadrupoles and it is easy to compute the four elements numerically once we define the length and focusing strength. We can trace particles by simply forming the product of these elementary matrices. But there is also a general relation between the elements a, b, c, d and the amplitude and phase of transverse motion between any two points. Each term in M_{21} can be expressed as a function of $\beta(s)$ and $\phi(s)$. The functions of $\beta(s)$ and $\phi(s)$ may be calculated by comparing the numerical result of multiplying the individual matrices for quadrupoles and drift lengths with what we know must be the general form of each element.

As a first step, we derive the general form of a periodic transport matrix.

To simplify the notation we drop the explicit dependence of β and ϕ on s from the expressions—we will just have to remember that they vary with s. We also introduce a new quantity:

$$w = \sqrt{\beta}. \tag{2.20}$$

just to avoid too many terms in what follows.

In this new notation we can write the solution of the Hill Equation:

$$y = \varepsilon^{1/2} w \cos(\varphi + \phi_0). \tag{2.21}$$

Taking the derivative and substituting $\varphi' = 1/\beta = 1/w^2$ we have:

$$y' = \varepsilon^{1/2} w' \cos(\varphi + \phi_0) - \frac{\varepsilon^{\frac{1}{2}}}{w} \sin(\varphi + \phi_0). \tag{2.22}$$

Next we substitute these explicit expressions for y and y' in both sides of the matrix equation. We do this first with the initial condition $\varphi_0 = 0$, this is the so-called 'cosine' solution, and then we do it again for the 'sine' solution with $\varphi_0 = \pi/2$. This is exactly equivalent to tracing the paraxial and central rays through an optical lens. We write $\phi_2 - \phi_1 = \phi$ for each case. Each of the two solutions give us two equations for y and y' and thus we obtain four simultaneous equations which can be solved for a, b, c, d in terms of w, w', and φ. The result is the most general form of the transport matrix between the positions s_1 and s_2 :

$$M_{12} = \begin{pmatrix} \frac{w_2}{w_1}\cos\varphi - w_2 w_1' \sin\varphi & w_1 w_2 \sin\varphi \\ -\frac{1+w_1 w_1' w_2 w_2'}{w_1 w_2}\sin\varphi - \left(\frac{w_1'}{w_2'} - \frac{w_2}{w_1}\right)\cos\varphi & \frac{w_1}{w_2}\cos\varphi + w_1 w_2' \sin\varphi \end{pmatrix}. \tag{2.23}$$

This rather formidable looking expression simplifies a lot, if we refer to a full circle, in other words, if we restrict M to apply between two identical points in successive turns or cells of a periodic structure. Then $w_2 = w_1$, $w_2' = w_1'$, and φ to become μ, the phase advance per cell. The matrix for one period is now:

$$M = \begin{pmatrix} \cos\mu - ww'\sin\mu & w^2\sin\mu \\ -\frac{1+w^2w'^2}{w^2}\sin\mu & \cos\mu + ww'\sin\mu \end{pmatrix}. \tag{2.24}$$

Next we invent some new functions of β:

$$\begin{aligned} \beta &= w^2, \\ \alpha &= -ww' = -\tfrac{\beta'}{2}, \\ \gamma &= \tfrac{1+(ww')^2}{w^2} = \tfrac{1+\alpha^2}{\beta}. \end{aligned} \tag{2.25}$$

These functions (which are not the same as the parameters used in special relativity!) are a complete and compact description of the dynamics. The matrix now becomes even simpler:

$$M = \begin{pmatrix} \cos\mu + \alpha\sin\mu & \beta\sin\mu \\ -\gamma\sin\mu & \cos\mu - \alpha\sin\mu \end{pmatrix} = \begin{pmatrix} a & b \\ c & d \end{pmatrix}. \tag{2.26}$$

This is the Twiss matrix. It is the basic matrix for periodic lattices and should be memorized.

We can imagine that if we can only find an independent way of computing the numerical values of the four elements we can solve and find:

$$\begin{aligned} &\cos\mu = (\mathrm{Tr}\,\mathbf{M})/2 = (a+d)/2, \\ &\beta = b/\sin\mu > 0, \\ &\alpha = (a-b)/(2\sin\mu), \\ &\gamma = -c/\sin\mu. \end{aligned} \tag{2.27}$$

These Twiss parameters, μ, β, α, and γ, are therefore rigorously determined by the overall effect of the focusing properties of the lattice elements. Still, they vary around the ring and apply to the point chosen in the period as a starting and finishing point. We shall see that each individual component, quadrupole, dipole, or drift space in the ring has its own matrix and this provides the independent method of calculation. We must first choose the starting point, the location, s, where we wish to know β and the other Twiss parameters. By starting there and multiplying the element matrices together for one turn we are able to find a, b, c, d numerically for that location. We can then apply the above four equations to find the Twiss matrix. If the machine has a natural symmetry in which there are a number of identical periods, it is sufficient to do the multiplication up to the corresponding point in the next period. The values of α, β, and γ would be the same if we went on for the whole ring. By choosing different starting points we can trace $\beta(s)$ and $\alpha(s)$. We now give the matrices for the three basic lattice elements.

2.1.10 Transport Matrices for Lattice Components

An empty space or drift length is the simplest of the lattice component matrices. Figure 2.8(a) shows the analogy between a particle trajectory and a diverging ray in optics. The angle of the ray and the divergence of the trajectory are related:

$$\theta = \tan^{-1}\left(x'\right). \tag{2.28}$$

The effect of a drift length in phase space is a simple horizontal translation from (x, x') to $(x+\ell x', x')$ and can therefore be written as a matrix:

$$\begin{pmatrix} x_2 \\ x_2' \end{pmatrix} = \begin{pmatrix} 1 & \ell \\ 0 & 1 \end{pmatrix} \begin{pmatrix} x_1 \\ x_1' \end{pmatrix}. \tag{2.29}$$

The next case is that of a thin quadrupole magnet of infinitely small length but finite integrated gradient:

$$\ell k = \frac{1}{(B\rho)} \frac{\partial B_z}{\partial x}. \tag{2.30}$$

The optical analogy of a thin quadrupole with a converging lens is illustrated in Fig. 2.8(b). A ray, diverging from the focal point arrives at the lens at a displacement, x, and is turned parallel by a deflection:

$$\theta \approx \frac{1}{f} x. \tag{2.31}$$

This deflection will be the same for any ray at displacement x irrespective of its divergence. This behaviour can be expressed by a simple matrix, the thin lens matrix:

$$\begin{pmatrix} x_2 \\ x_2' \end{pmatrix} = \begin{pmatrix} 1 & 0 \\ -1/f & 1 \end{pmatrix} \begin{pmatrix} x_1 \\ x_1' \end{pmatrix}. \tag{2.32}$$

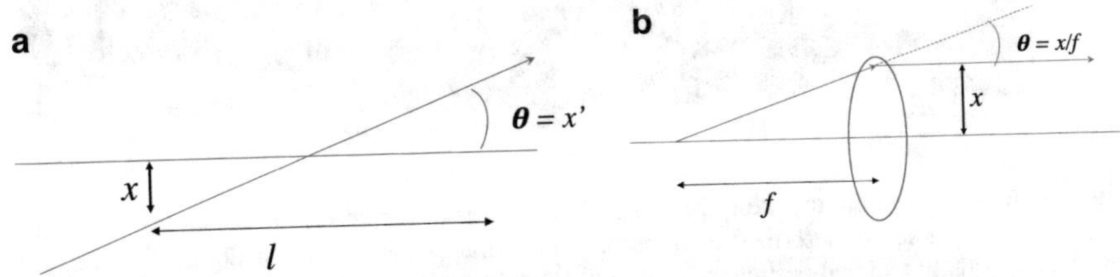

Fig. 2.8 The effect of a drift—**(a)**, left side—and a focusing quadrupole lens—**(b)** right side—on a particle trajectory. The mathematical expressions are given in Eqs. (2.29) and (2.32)

A particle arriving at a quadrupole lens at a displacement x obeys Hill's equation

$$x'' + kx = 0. \tag{2.33}$$

Hence the small deflection θ is just:

$$\Delta x' = -kx\ell. \tag{2.34}$$

Comparing quadrupoles with optical lenses we remember that $\ell k = 1/f$ and is the power of the lens and that the matrix, for a thin lens, can be written:

$$\begin{pmatrix} 1 & 0 \\ -k\ell & 1 \end{pmatrix}. \tag{2.35}$$

Under the influence of these focusing and defocusing fields, a particle trajectory will finally look like a more or less zig-zag shaped curve; which for the example of eight regular cells it is shown in Fig. 2.9.

Quadrupoles are sometimes not short compared to their focal length. One must therefore use the matrices for a long quadrupole when one comes to compute the final machine:

$$\begin{aligned} M_F &= \begin{pmatrix} \cos \ell\sqrt{k} & \frac{1}{\sqrt{k}} \sin \ell\sqrt{k} \\ -\sqrt{k} \sin \ell\sqrt{k} & \cos \ell\sqrt{k} \end{pmatrix}, \text{ and} \\ M_D &= \begin{pmatrix} \cosh \ell\sqrt{k} & \frac{1}{\sqrt{k}} \sinh \ell\sqrt{k} \\ -\sqrt{k} \sinh \ell\sqrt{k} & \cosh \ell\sqrt{k} \end{pmatrix}. \end{aligned} \tag{2.36}$$

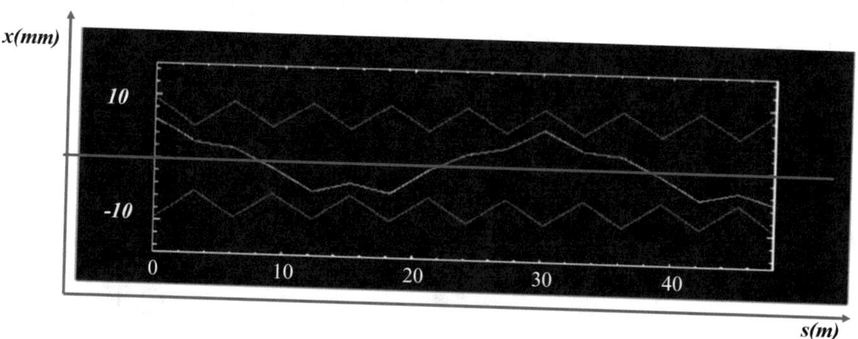

Fig. 2.9 A single particle trajectory in a ring: At each part of the lattice the amplitude and angle, (x, x′) of the particle are described by a matrix transformation, according to Eq. (2.32). The blue line corresponds to an ideal particle, with $x = x' = 0$ and so refers to the ideal orbit

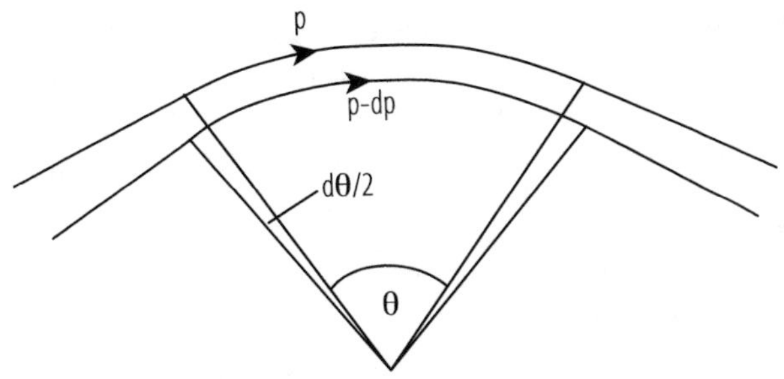

Fig. 2.10 The focusing effect of trajectory length in a pure sector dipole magnet

We can compare this with the solutions of Hill's equations within F and D quadrupoles:

$$\begin{aligned} z &= \cos\sqrt{k}\ell\ z_0 + \frac{1}{\sqrt{k}}\sin\sqrt{k}\ell\ z_0', \\ x &= \cosh\sqrt{k}\ell\ x_0 + \frac{1}{\sqrt{k}}\sinh\sqrt{k}\ell\ x_0'. \end{aligned} \tag{2.37}$$

We have so far ignored the bending that takes place in dipole magnets and these may be thought of as drift lengths in a first approximation. An exact calculation should include the focusing effect of their ends. A pure sector magnet, whose ends are normal to the beam will give more deflection to a ray which passes at a displacement x away from the centre of curvature (Fig. 2.10). This particle will have a longer trajectory in the magnet. The effect is exactly like a lens which focuses horizontally but not vertically. The matrices for a sector magnet are:

$$\begin{aligned} M_{\mathrm{H}} &= \begin{pmatrix} \cos\theta & \rho\sin\theta \\ -(1/\rho)\sin\theta & \cos\theta \end{pmatrix}, \\ M_{\mathrm{V}} &= \begin{pmatrix} 1 & \rho\theta \\ 0 & 1 \end{pmatrix}. \end{aligned} \tag{2.38}$$

Some bending magnets are not sector magnets as in Fig. 2.9, but have end faces which are parallel. It is easier to stack laminations this way than on a curve. The entry and exit angles are therefore, $\theta/2$, and the horizontal focusing effect is reduced but there is an additional focusing effect for a particle whose trajectory is displaced vertically. In the computer model one may convert a pure sector magnet into a parallel faced magnet by simply adding two thin lenses at each face. They are horizontally defocusing and vertically focusing and their strength is:

$$k\ell = -\frac{\tan\ (\theta/2)}{\rho}. \tag{2.39}$$

Unlike early lattice designers we have computers to help when we come to multiply these elements together to form the matrix for a ring or a period of the lattice [3–5]. A lattice program such as MAD [6] does all the matrix multiplication to obtain (a, b, c, d) from each specified point, s, and back again. It prints out β and φ and other lattice variables in each plane, and we can plot the result to find the beam envelope around the machine. This is the way machines are designed. Lengths, gradients, and numbers of FODO normal periods are varied to match the desired beam sizes and Q values.

In Fig. 2.8 we saw the trajectory of a particle, oscillating in a pattern of alternating focussing and defocusing quadrupoles (FODO). The trajectories in general all lie within an envelope which has the general features of the optical model in Fig. 2.6. If we were to repeat the observation of the displacement and divergence of a particle on successive turns we would find the elliptical locus of its motion (Fig. 2.5). The aspect ratio of this ellipse would depend upon where in the ring we choose to make the observation. The ellipse would be squat near D lenses and elongated near F's. The figure would appear just the same if we were to plot it between what are F quadrupoles in the vertical plane. Of course, the whole pattern of quadruples and the envelope is shifted by the distance between adjacent quadrupoles because F-quadrupoles in one plane are D in the other (et vice versa).

2.1.11 The Betatron Envelopes

To recapitulate, a modern synchrotron consists of pure bending magnets and quadrupole magnets or lenses which provide focusing. These are interspersed among the bending magnets of the ring in a pattern called the lattice. By suitable choice of strength and spacing of the lenses the envelope function $\beta(s)$ can be made periodic in such a way that it is large at all F quadrupoles and small at all D's. Symmetry will ensure this is true also in the vertical plane. Particles oscillating within this envelope will always tend to be further off axis in F quadrupoles than in D quadrupoles and there will therefore be a net focusing action. We have already seen that β is the aspect ratio of the phase space ellipse (see also [7, 8]).

In Fig. 2.11 we see an example of such a magnet pattern which is one cell, or about 1% of the circumference, of the 400 GeV SPS at CERN. Although the SPS is now considered a rather old fashioned machine its simplicity leads us to use it as an example. The focusing structure is FODO and in this pattern half of the quadrupoles (F) focus, while the other half, defocus (D) the beam. Bending magnets, which in a first approximation do no focussing are represented together with other non-focussing elements by the letter "O". The envelope of these oscillations follows a function $\beta(s)$ which has waists near each defocusing magnet and has a maximum at the centres of F quadrupoles. Since F quadrupoles in the horizontal plane are D quadrupoles vertically, and vice versa, the two functions $\beta_h(s)$ and $\beta_v(s)$ are one half-cell out of register in the two transverse planes. The function β has the dimensions of length but the units bear no relation at this stage to physical beam

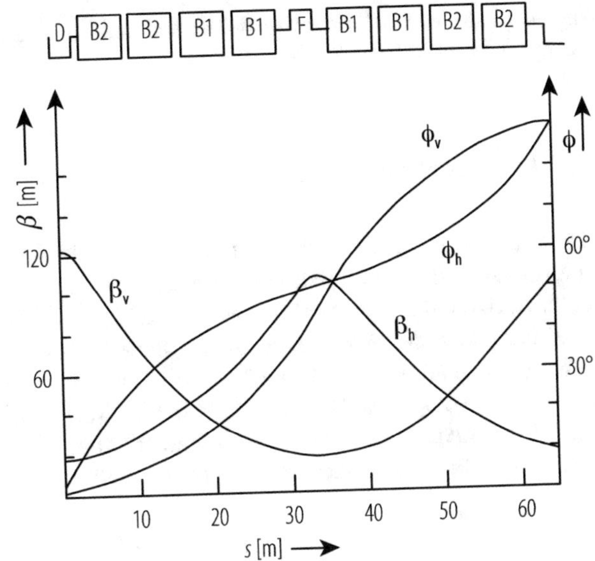

Fig. 2.11 One cell of the CERN SPS representing 1/108 of the circumference. The pattern of dipole (B) magnets and quadrupole (*F* and *D*) lenses is shown above

size. The reader should be persuaded that particles do not follow the $\beta(s)$ curves but oscillate within them in a form of modified sinusoidal motion whose phase advance is described by $\phi(s)$. The phase change per cell in the example shown is close to $\pi/2$ but the rate of phase advance is modulated throughout the cell.

2.2 Coupling

Until now we have considered the motion in the vertical and horizontal direction to be orthogonal and independent. This is the ideal case. Now we look at what happens when there is a skew quadrupole or solenoidal field in the machine which couples horizontal motion into vertical and vice versa. This is rather a special case affecting mainly electron synchrotrons and the reader may choose to skip to Sect. 2.3 and leave coupling to a second reading.

In a fully coupled machine the betatron oscillations in the two transverse directions are like two harmonic oscillators which transfer energy from one to the other with a frequency which is just the difference between their Q's. They act like coupled pendula. In this way all the horizontal "emittance" can add to the vertical emittance and the beam exceeds the available vertical aperture.

The phenomenon is particularly important in electron rings. The electron beam would damp to zero emittance were it not for quantum emission in the horizontal plane exciting betatron oscillations. There is no comparable excitation in the vertical plane and only coupling of the horizontal oscillations into the vertical plane gives the beam any vertical dimensions. Vertical emittance, and the magnet gap needed to accept it, is directly proportional to coupling.

2.2.1 Coupling Fields

There are two principal configurations of field which excite coupling. The first we shall consider is a skew quadrupole, i.e. a quadrupole whose poles lie symmetrically in the horizontal and vertical planes (Fig. 2.12).

A particle with horizontal position x, experiences not a B_z as would be the case in a normal quadrupole and which would change its x', but a B_x which together with the paraxial velocity deflects vertically in the direction of $\boldsymbol{v} \times \boldsymbol{B}$. Of course once the particle has acquired a vertical displacement z after a number of turns it experiences a vertical field, for in a skew quadrupole the field is:

$$\begin{aligned} B_x / (B\rho) &= kx, \\ B_z / (B\rho) &= -kz. \end{aligned} \tag{2.40}$$

Thus a horizontal displacement couples into the vertical plane leading to a vertical divergence and displacement. The vertical displacement goes on to couple back into the horizontal plane modifying the horizontal displacement and divergence—and so it proceeds transferring transverse momentum back and forth from one plane to the other.

A solenoid is the other field configuration that can couple the two planes but this kind of coupling is less important in synchrotrons and we leave it to the reader to consult a more exhaustive treatment of coupling in [9].

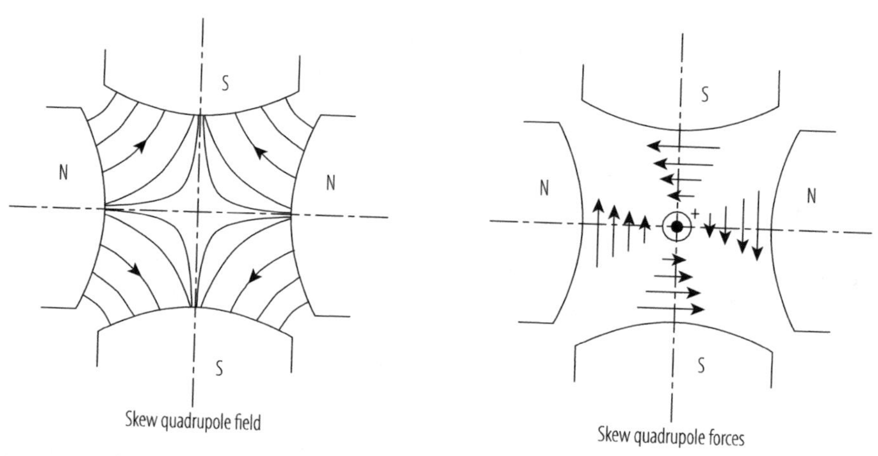

Fig. 2.12 The magnetic field and force in a skew quadrupole

2.2.2 *Qualitative Treatment of Coupling*

In our treatment the theory is deliberately simplified to reveal the physical mechanisms at work. We assume that the coupling is driven by a single skew quadrupole at the centre of one of the existing lattice machine quadrupoles where β is maximum and its derivative zero. We ignore the changes in betatron phase of one plane with respect to the other within a single turn.

The skew quadrupole gradient is normalized:

$$k = \frac{1}{(B\rho)}\left(\frac{\partial B_x}{\partial x}\right)_{z=0}, \tag{2.41}$$
$$l = \text{length of the quadrupole.}$$

Figure 2.13 on the left shows the betatron motion in the horizontal plane. We have normalized the elliptical phase space trajectory into a circle at the location of the skew quadrupole by multiplying the divergence by β_x. On the right we have done the same for the vertical plane. The angular kick Δp, on passing the skew quadrupole is calculated from a similar diagram for the vertical plane and

$$\Delta p_x = \beta_x k\ell w \cos Q_V\theta, \tag{2.42}$$

where $w = \sqrt{\varepsilon_V \beta_z}$ is the radius of the circle for vertical motion, and $u = \sqrt{\varepsilon_H \beta_x}$ is the radius horizontally.

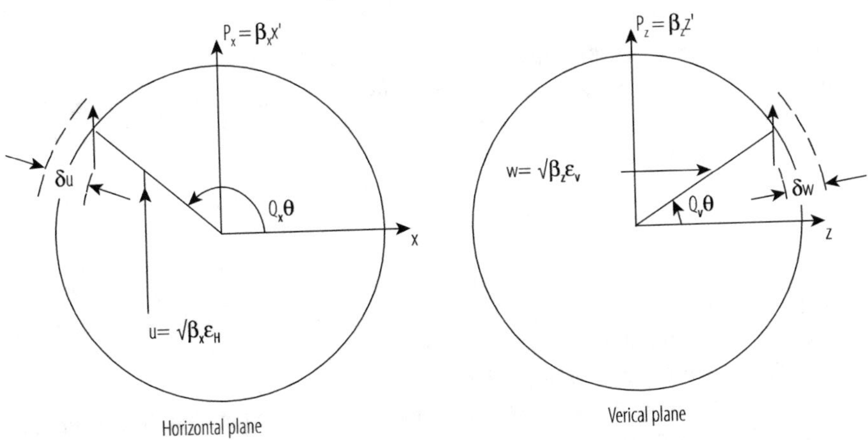

Fig. 2.13 Phase space diagram

The kick, projected as an amplitude increment becomes:

$$\delta u = w\beta_x k\ell \sin Q_{\mathrm{H}}\theta \cos Q_{\mathrm{V}}\theta. \tag{2.43}$$

When we use:

$$\sin A \cos B = \frac{1}{2}\sin(A - B) + \frac{1}{2}\sin(A + B)$$

and ignore the second, high frequency term; we obtain the coupled equations for a single passage:

$$\begin{aligned} \frac{\delta w}{w} &= -\sqrt{\frac{\varepsilon_{\mathrm{H}}}{\varepsilon_{\mathrm{V}}}}\sqrt{\frac{\beta_x\beta_z}{2}}k\ell \sin(Q_{\mathrm{H}} - Q_{\mathrm{V}})\,\theta, \\ \frac{\delta u}{u} &= \sqrt{\frac{\varepsilon_{\mathrm{V}}}{\varepsilon_{\mathrm{H}}}}\sqrt{\frac{\beta_x\beta_z}{2}}k\ell \sin(Q_{\mathrm{H}} - Q_{\mathrm{V}})\,\theta. \end{aligned} \tag{2.44}$$

These are incremental equations which we must sum over the n turns as the coupling enhances u at the expense of w.

Figure 2.14 shows diagrammatically the coupled motion. The vertical betatron amplitude decrease from $w + \Delta w$ to w in one quarter period of the slow oscillation which takes $1/4|Q_{\mathrm{H}} - Q_{\mathrm{v}}|$ turns. The mean value of the cosine is taken as $2/\pi$. We then arrive at the expressions for the maximum excursions in amplitude:

$$\begin{aligned} \frac{\Delta w}{w} &= \sqrt{\frac{\varepsilon_{\mathrm{H}}}{\varepsilon_{\mathrm{V}}}}\frac{\sqrt{\beta_x\beta_z}}{4\pi|Q_{\mathrm{H}}-Q_{\mathrm{v}}|}k\ell, \\ \frac{\Delta u}{u} &= \sqrt{\frac{\varepsilon_{\mathrm{V}}}{\varepsilon_{\mathrm{H}}}}\frac{\sqrt{\beta_x\beta_z}}{4\pi|Q_{\mathrm{H}}-Q_{\mathrm{v}}|}k\ell. \end{aligned} \tag{2.45}$$

We now move from the phase plane into real space. Some machines were designed to have a rectangular "vacuum chamber" which would accept particles which simultaneously have large horizontal and vertical "emittances". In this sense emittance is defined for a single particle

$$\varepsilon_{\mathrm{H}} = \pi u^2/\beta_x,\, \varepsilon_{\mathrm{V}} = \pi w^2/\beta_y. \tag{2.46}$$

In the presence of coupling, the particle motion is a series of Lissajous figures filling the rectangular cross-section but always touching it somewhere on each turn (Fig. 2.15). It is inevitable therefore that if coupling increases either amplitude by $\Delta u/u$ or $\Delta w/w$, some fraction of particles will be lost.

A rigorous treatment of coupling is too lengthy to include here but the reader may consult [9–11] for a complete description. This lengthier treatment leads to a model in which the modes of betatron oscillations are no longer about the vertical and horizontal planes but about two orthogonal principal planes inclined with respect to the vertical and horizontal frame.

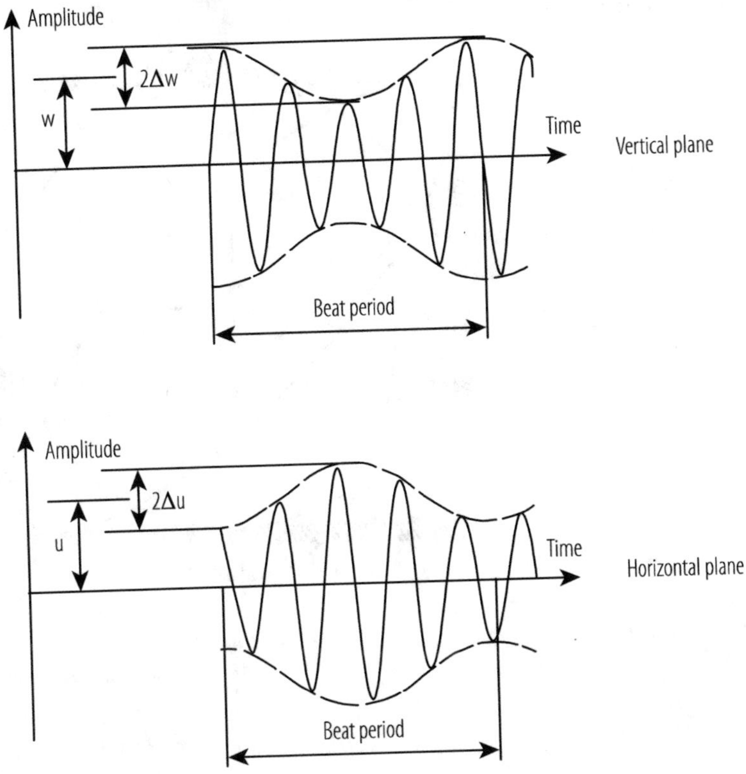

Fig. 2.14 Coupled betatron oscillations for $1/|Q_H - Q_V|$ turns

2.3 Liouville's Theorem

Now let us return to the 'mainstream' of transverse dynamics. Liouville's theorem is a conservation law that applies to the area occupied by a number of particles plotted in phase space.

We should think of a beam of particles as a cloud of points within a closed contour in a transverse phase space diagram (Fig. 2.16). Liouville's theorem tells us that this area within the contour is conserved. The contour is usually, but not always, an ellipse. In Fig. 2.5 we came across such an elliptical contour—the locus of a particle's motion plotted in phase space (x,x′) and we call its area, the emittance. We could also think of it as a limiting contour enclosing all the particles in the beam which we would again call the emittance—not of the particle but of the beam as a particle ensemble.

We express beam emittance in units of π mm·milliradians. According to Liouville the emittance area will be conserved as the beam passes down a transport line or circulates in a synchrotron whatever magnetic focusing or bending operation we do on the beam—provided that only conservative forces are taken into account.

Fig. 2.15 Particle lost due to coupling

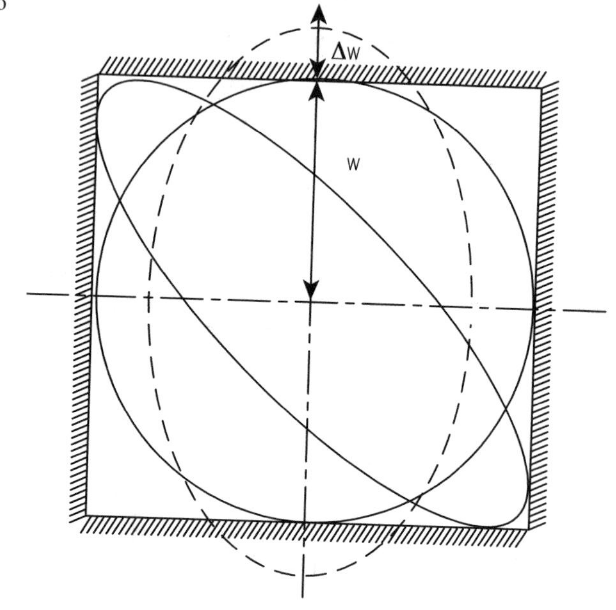

Fig. 2.16 Liouville's theorem applies to this contour

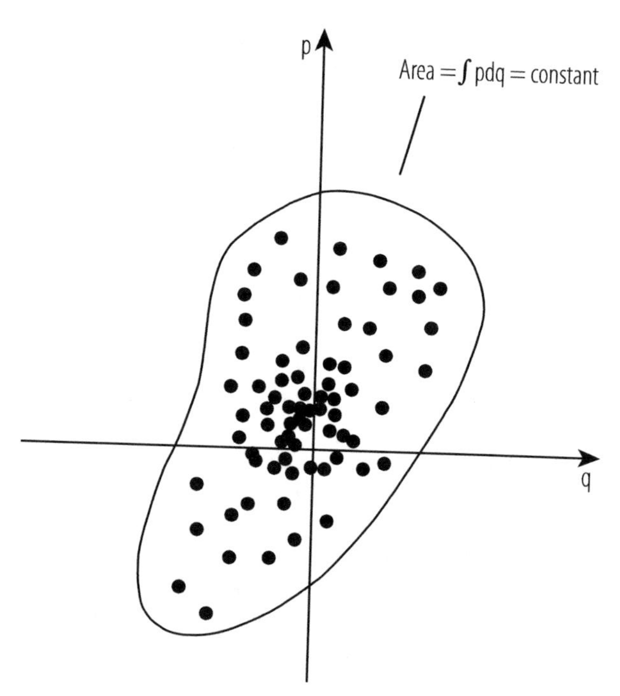

Even though the ellipse may appear to have many shapes around the accelerator its phase space area will not change (Fig. 2.17). The aspect ratio of the ellipse will change however. At a narrow waist, near a *D* quadrupole (a) in Fig. 2.17, its

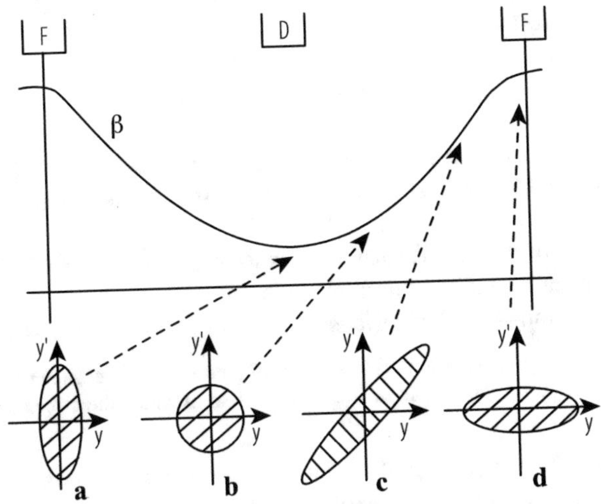

Fig. 2.17 How the conserved phase space appears at different points in a FODO cell

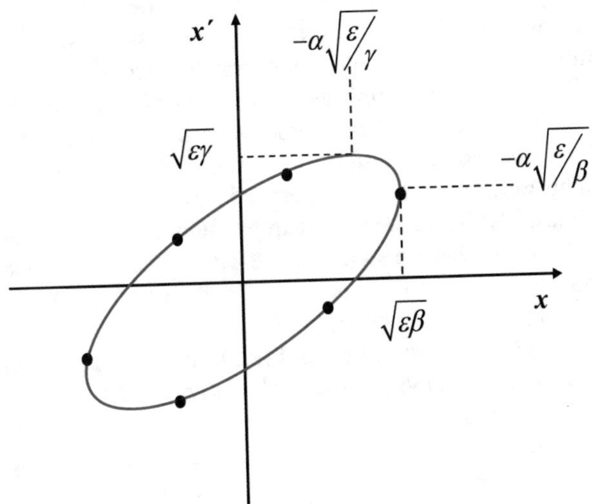

Fig. 2.18 The parameters of a phase-space ellipse containing an emittance ε at a certain point in the lattice. The shape and orientation of the ellipse are determined by the Twiss parameters at the given location

divergence will be large, while in an *F* quadrupole (d) where the betatron function is maximum, its divergence will be small. The beam is also seen at a broad waist or maximum in the beta function and a place where the beam is diverging.

In Fig. 2.18 we see how the various features of the ellipse are related to the Twiss parameters. The equation of the ellipse, often called the Courant and Snyder invariant, has the form

$$\gamma(s)y^2 + 2\alpha(s)yy' + \beta(s)y'^2 = \varepsilon. \tag{2.47}$$

Here y is used to mean either of the transverse displacements, x or z. It is straight forward determine the relation between the shape and orientation of the (x,x′) ellipse and the Twiss parameters α, β, γ as indicated in Fig. 2.18.

The invariance of the (x,x′) space area, as we move to different points in the ring is an alternative statement of Liouville's theorem.

A word of caution—another, stricter, version of Liouville's theorem states that:

> In the vicinity of a particle, the particle density in phase space is constant if the particles move in an external magnetic field or in a general field in which the forces do not depend upon velocity.

This rules out the application of Liouville's theorem to situations in which space charge forces within the beam play a role or when there is a velocity dependent effect such as when particles emit synchrotron light. However we may apply Liouville to proton beams which do not normally emit synchrotron light and to electrons travelling for a few turns in a synchrotron. This is usually too short a time for electrons to emit enough synchrotron light energy to affect their transverse motion.

Liouville's theorem does not apply as a proton beam is accelerated. Observations tell us this is not the case. The beam appears to shrink. This is because the co-ordinates we have used so far, y and y', are not 'canonical' in the sense defined by Hamiltonian in his mechanics, which is part and parcel of Liouville's mathematical theory of dynamics. We should therefore express emittance in Hamilton's canonical phase space and relate this carefully to the co-ordinates, displacement, y, and divergence, y', which we have been using so far. We can then define an emittance which is conserved even as we accelerate.

We shall have to be particularly careful not to confuse Twiss parameters, and the parameters of special relativity: In special relativity we use β as the ratio of the particles velocity and the speed of light and the Lorentz factor γ describes the total energy divided by the rest energy. The reader will have to examine the context to be sure. For those who have not met Hamiltonian mechanics, it is sufficient to know that the canonical co-ordinates of relativistic mechanics are:

$$p = \frac{m_0 \dot{y}}{\sqrt{1 - v^2/c^2}}, \; q = y. \tag{2.48}$$

Here q or y is a general transverse co-ordinate, p its conjugate momentum and we define β and γ when used in the context of special relativity to be:

$$\begin{aligned} \beta &= v/c, \\ \gamma &= 1/\sqrt{1-\beta^2}, \\ m_0 &= \text{rest mass}, \\ c &= \text{velocity of light}. \end{aligned} \tag{2.49}$$

We may find the relationship between canonical momentum and divergence from the substitution:

$$p_y = m_0 \frac{dy}{dt} \gamma = m_0 \frac{ds}{dt} \frac{dy}{ds} \gamma = mc\,(\beta\gamma)\, y'. \tag{2.50}$$

Writing Liouville's Theorem expressed in canonical coordinates we can use the above expression to define a conserved quantity and relate it to the area in (y, y') space

$$\int p_y dy = m_0 c\,(\beta\gamma) \int y' dy = p_0 \int y' dy \tag{2.51}$$

where p_0 is the momentum in the direction of motion of the particle.

This invariant is the emittance, ε, of our transverse phase space multiplied by the relativistic $\beta\gamma$ which is proportional to momentum. Accelerator physicists often call this the invariant or 'normalised' emittance:

$$\varepsilon^* = \beta\gamma\varepsilon \;\; [\pi\ \mathrm{mm} \cdot \mathrm{mrad}] \tag{2.52}$$

This normalised emittance, ε_*, is conserved as acceleration proceeds in a synchrotron and the physical emittance within the right-hand side of the equation must fall inversely with momentum if the whole term is to be conserved. Close to the velocity of light this implies that it is inversely proportional to energy.

$$Emittance = \pi\varepsilon = \int y' dy = \pi\varepsilon^* / (\beta\gamma) \propto 1/p_0. \tag{2.53}$$

We therefore expect the beam dimensions to shrink as (Fig. 2.19) a phenomenon called 'adiabatic damping'.

2.3.1 *Chains of Accelerators*

As a consequence of the adiabatic shrinking, the beam emittance is largest at low energy, and so is the beam dimension. Proton accelerators need their full aperture at injection and it is then that their design is most critical. For this reason it is economic to split a single large ring into a chain of accelerators—the smaller radius rings having a large aperture while the higher energy rings with large radius can have smaller apertures. In these chains of proton accelerators, such as the CERN accelerator complex, Linac – Booster – PS – SPS, the invariant emittance, determined by the parameters of the beam as it leaves the ion source at the beginning of the linac, may be conserved to several hundred GeV. Of course one must guard against mismatches between machines or non-linear fields which dilate the emittance.

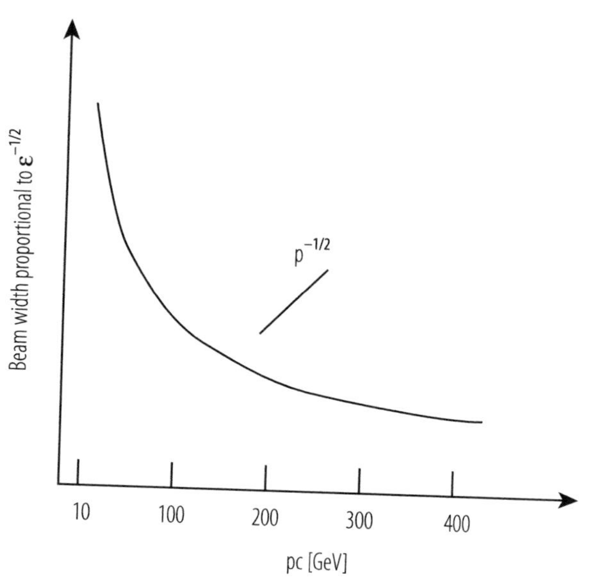

Fig. 2.19 Adiabatic shrinking of the beam as function of the beam momentum during acceleration

2.3.2 *Exceptions to Liouville's Theorem*

The invariance of normalised emittance of a proton beam and the shrinking of its physical emittance with energy is quite the opposite of what happens in an electron machine. Liouville's theorem only applies to conservative systems, where particles are guided by external fields and not to electron machines where particles emit some of their own energy. Electrons, being lighter than protons and hence more relativistic, emit quanta of radiation as they are accelerated. This quantised emission causes particles to jump around in momentum, leading to changes in the trajectories amplitude and angle. These changes couple into both planes of transverse phase space. At the same time, there is a steady tendency for particles near the edge of the emittance to lose transverse energy and fall back towards the centre. In an electron machine the emittance is determined not by the Liouville but by the equilibrium between these two effects. In fact, it grows with E^2.

Consider a number of protons which have the maximum amplitude present in the beam. They follow trajectories at the perimeter of the ellipse but at any instant have a random distribution of initial phases ϕ_0. If we were able to measure y and y' for each and plot them in phase space, they would lie around the ellipse of area $\pi\varepsilon$ and their co-ordinates would lie in the range of

$$\begin{aligned} -\sqrt{\beta\varepsilon} \leq y \leq \sqrt{\beta\varepsilon}, \\ -\sqrt{\varepsilon\gamma} \leq y' \leq \sqrt{\varepsilon\gamma}. \end{aligned} \tag{2.54}$$

Particles in a beam are usually distributed in a population which appears Gaussian when projected on a vertical or horizontal plane. In a proton machine the emittance boundary used to be conventionally chosen to be that of a proton with amplitude 2σ. This would include about 90% (strictly 87%) of a Gaussian beam where σ is the standard distribution. In an electron machine a 2σ boundary would be too close to the beam and an aperture stop placed at this distance would rather rapidly absorb most of the beam as particles redistribute themselves, moving temporarily into the tails due to quantum emission and damping. The safe physical boundary for electrons depends on the lifetime required but is in the region of 6σ to 10σ. The emittance which is normally quoted for an electron beam corresponds to an electron with the amplitude of σ in the Gaussian projection. We are then free to choose how many σ's we must allow. There is consequently a factor 4 between emittance defined by electron and proton experts.

2.4 Momentum Dependent Transverse Motion

In the previous chapters, we have studied the motion of a particle as it swings from side to side about the ideal orbit around the synchrotron: the transverse motion. Nothing is perfect, however, and so we cannot assume that each and every proton in a large ensemble of up to 10^{11} particles will have exactly the ideal momentum. Instead we expect a certain momentum spread in the beam and therefore we have to study how transverse motion depends on small departures, $\Delta p/p_0$, from the synchronous momentum p_0.

2.4.1 *Dispersion*

The central closed orbit of a synchrotron is matched to an ideal (synchronous) momentum p_0. A particle of this momentum and of zero betatron amplitude will pass down the centre of each quadrupole, be bent by exactly 2π by the bending magnets in one turn of the ring and remain synchronous with the r.f. frequency. Its path is called the central (or synchronous) momentum closed orbit. In Fig. 2.8 this ideal orbit is the horizontal axis and we see particles executing betatron oscillations about it but these oscillations do not replicate every turn. The synchronous orbit, however, closes on itself so that x and x' remain zero.

We now consider a closed orbit which is distorted in the horizontal plane by non-ideal bends in the dipole. Figure 2.20 shows a particle with a lower momentum $\Delta p/p < 0$ and which is bent horizontally more in each dipole of a FODO lattice. We could argue that the total deflection, being more than 2π would cause it to spiral inwards and hit the vacuum chamber wall. On the other hand there is a closed orbit for this lower momentum in which the extra bending forces are compensated by extra focusing forces as the orbit is displaced inwards in the F quadrupoles and less so in the defocusing in the D's (Fig. 2.20). We may describe the shape of this new closed orbit for a particle of unit $\Delta p/p$ by a *dispersion* function $D(s)$. The

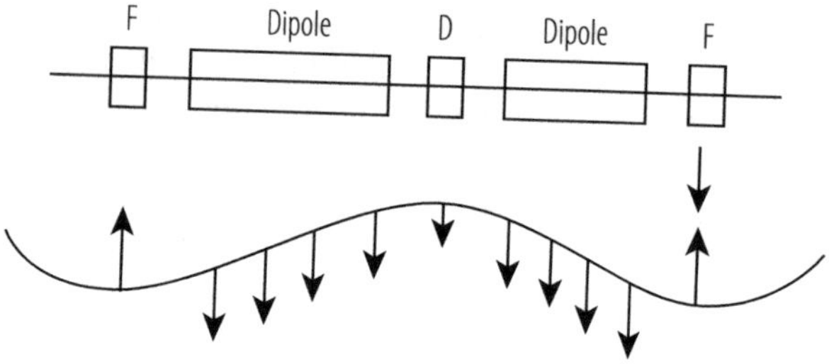

F Dipole D Dipole F

Fig. 2.20 The extra inward force given to a low momentum particle by the dipoles is balanced by the focusing the quadrupoles and defines a dispersion function

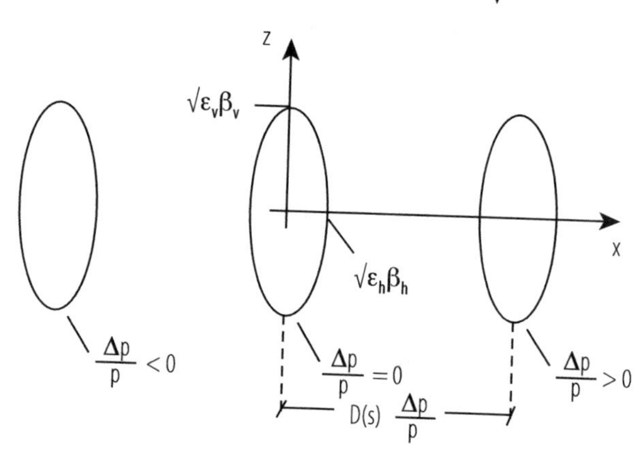

Fig. 2.21 The beam cross sections in real space for beams of three different momenta at a point where the dispersion function is large

displacement of the closed orbit is:

$$x(s) = D(s)\frac{\Delta p}{p_0}. \tag{2.55}$$

In Fig. 2.21 we see how the effect of dispersion for off momentum orbits adds to the betatron motion to widen the beam cross section. The betatron motion of each of the three particles: $\Delta p/p < 0$, $\Delta p/p = 0$, and $\Delta p/p > 0$, is within an ellipse in physical (x, z) space. The ellipses for each momentum are separated by a distance $D(s)\Delta p/p$. The semi-aperture required will be:

$$a_{\mathrm{V}} = \sqrt{\beta_{\mathrm{V}}\varepsilon_{\mathrm{V}}}, a_{\mathrm{H}} = \sqrt{\beta_{\mathrm{H}}\varepsilon_{\mathrm{H}}} + D(s)\frac{\Delta p}{p}. \tag{2.56}$$

2.4.2 *Chromaticity*

This effect is equivalent to the chromatic aberration in a lens. It is defined as a quantity Q':

$$\Delta Q = Q'\frac{\Delta p}{p}. \tag{2.57}$$

The chromaticity [12] arises because the focusing strength of a quadrupole has $(B\rho)$ in the denominator and is therefore inversely proportional to momentum:

$$k = \frac{1}{(B\rho)} \frac{dB_z}{dx}. \tag{2.58}$$

A small spread in momentum in the beam, $\pm \Delta p/p$, causes a spread in focusing strength:

$$\frac{\Delta k}{k} = \mp \frac{\Delta p}{p}. \tag{2.59}$$

Integrated over all focusing (and defocusing) elements in the ring, we obtain a change in the tune of the machine

$$\Delta Q = \frac{1}{4\pi} \int \beta(s)\delta k(s) ds. \tag{2.60}$$

This enables us to calculate Q':

$$\Delta Q = \frac{1}{4\pi} \int \beta(s)\delta k(s) ds = \left[\frac{-1}{4\pi} \int \beta(s) k(s) ds \right] \frac{\Delta p}{p}. \tag{2.61}$$

The quantity in square brackets is the chromaticity Q'. To be clear, this is called the natural chromaticity. For most alternating gradient machines, its value is about $-1.3Q$. Of course there are two Q values relating to horizontal and vertical oscillations and therefore two chromaticities. Chromaticity may be corrected with sextupole magnets (see Chap. 3, and Sects. 6.1 and 8.1).

2.5 Longitudinal Motion

2.5.1 *Stability of the Lagging Particle*

Suppose two particles are well below the velocity of light. A particle A, that arrives at the right moment to in the RF resonator and thus will obtain exactly the right acceleration voltage. We call this particle "*synchronous*" (see Fig. 2.22). A second particle, B, arrives late, and so receives an extra energy increment which will cause it to speed up and overtake the synchronous particle, A. In so doing, its energy defect, ΔE, grows and, provided the amplitude is not too large, its trajectory will follow an ellipse in phase space. This describes this motion up and down the r.f. wave (Fig. 2.22) and may remind some readers of the representation of a simple harmonic oscillator, or pendulum. When plotted in a phase space diagram of velocity versus longitudinal displacement we indeed obtain a shape that is elliptical. The trajectory

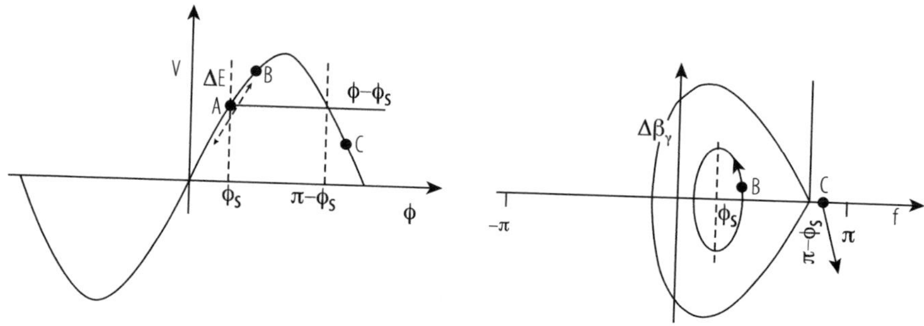

Fig. 2.22 The limiting trajectory for a particle in a 'moving' or accelerating bucket when the stable phase is not zero

of this longitudinal movement in phase space is closed and over many turns the average deviation from the synchronous energy is zero. This *phase stability* depends upon the fact that δE is positive when $\phi - \phi_s$ is small and positive [13, 14].

When a particle reaches the non-linear part of the r.f. wave and over the top of the wave, it will still be restored and oscillate about the stable phase provided it does not reach and pass the point where it receives less incremental energy than the synchronous particle. On this non-linear part of the curve the motion is no longer an ellipse but is distorted into a fish-shape but its trajectory is still closed and stable. However, if a particle, C, oscillates with such large amplitude that it falls below the synchronous voltage, an increase in ϕ will cause a negative ΔE which in turn causes ϕ to move further away from the synchrotron particle. This particle is clearly unstable and will be continuously decelerated. There is a particle which, starting at $\phi = \pi - \phi_s$, would trace out a limiting fish-shaped trajectory which is the boundary or *separatrix* between stable and unstable motion. The region within this separatrix is called the r.f. bucket and is shown in the lower half of Fig. 2.22. Formulae for the calculation of the parameters of moving buckets are to be found in [15].

Let us look more carefully at the argument that a particle, arriving late because of its lower energy, would see a higher RF voltage from the rising waveform and, accelerated to a higher velocity, would catch up with the synchronous particle. Dispersion may make the situation more complicated. Giving the errant particle more energy will speed it up but may also send it on an orbit of larger radius.

The path length that the particle, B, must travel around the machine, or more correctly, the change in path length with momentum, must depend upon the dispersion function. The closed orbit will have a mean radius:

$$R = R_0 + \overline{D}\frac{\Delta p}{p}. \tag{2.62}$$

Close to the velocity of light where acceleration can increase momentum but not velocity, the longer path length will more than cancel the small effect of velocity and the particle, instead of catching up with its synchronous partner, will arrive even later than it did on the previous turn. This seems to defeat the whole idea of phase stability. Depending on how the synchrotron is designed and which particles it accelerates, there can be a certain energy where our initial ideas of phase stability break down. This is called the transition energy. Fortunately there is also a way of ensuring stability above transition.

2.5.2 *Transition Energy*

The rigorous argument to resolve the question of velocity versus path length is to examine how the revolution time (or its reciprocal, the revolution frequency) varies as the particle is given extra acceleration. The revolution frequency is:

$$f = \frac{\beta c}{2\pi R}, \ (\beta = v/c)\,. \tag{2.63}$$

This revolution frequency, f, depends on two momentum dependent variables, the relativistic $\beta=v/c$ and R, the mean radius. The penultimate equation gives the change in the radius. The momentum dependence of β is determined by:

$$p = \frac{E_0 \beta}{\sqrt{1-\beta^2}}\,. \tag{2.64}$$

The rate of "catching up" depends upon a "slip factor", η, which is defined as logarithmic differential of frequency as a function of momentum. The procedure of partial derivatives tells us there must be two terms. Hence:

$$\eta_{\rm rf} = \frac{\Delta f/f}{\Delta p/p} = \frac{p}{\beta}\frac{d\beta}{dp} - \frac{p}{R}\frac{dR}{dp} = \frac{1}{\gamma^2} - \frac{\overline{D}}{R_0}\,. \tag{2.65}$$

The first term on the right-hand side describes the increase in speed with p and the other (negative), how the path to be completed increases with p.

The second term is energy independent while the first term shrinks as acceleration proceeds. At low energy this is largest and η is >0. But, since $\gamma = E/E_0$, the first term becomes smaller than the second at high energy so that η changes sign from positive to negative. During the acceleration process there is a certain energy, the transition energy, at which η is momentarily zero. At transition, the value of γ satisfies:

$$\frac{1}{\gamma_{\rm tr}^2} = \frac{\overline{D}}{R}\,. \tag{2.66}$$

In proton synchrotron design this condition tends to be encountered mid–way through the acceleration cycle and can only be avoided with some ingenuity in the design of the lattice. This was a worry to the designers of the PS and AGS, the

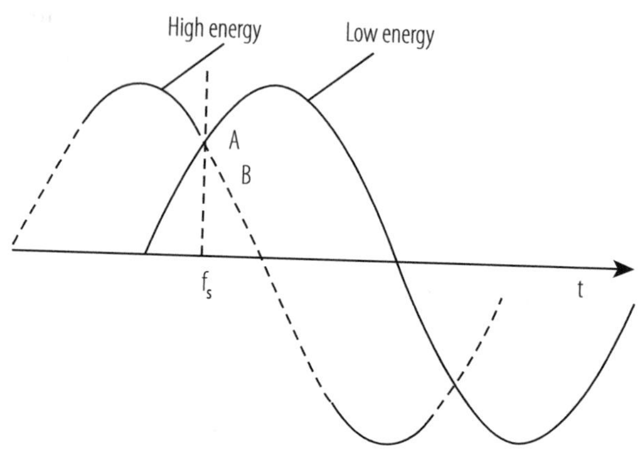

Fig. 2.23 Shows how changing the phase of the RF voltage waveform can give the lagging particle, B, less energy rather than more and lead to stability above transition

first proton synchrotrons of high enough energy to encounter this problem during acceleration but it was then realised that one could, almost instantaneously, change the phase of the voltage wave in the RF cavities to be falling rather than rising at the moment of the synchronous particles arrival (see Fig. 2.23). With such a reversed slope, particles arriving late are given less than their ration of energy and take a inner circular path—a short cut—to arrive earlier next time.

Electron machines are fortunate in that due to the small rest mass their Lorentz factor γ, being 2000 times higher, ensures that the first term may be neglected and such machines operate always above transition.

2.5.3 Synchrotron Motion

If we consider the motion of a particle on the linear part of the voltage wave of an r.f. cavity it is not difficult to imagine that it approximates rather closely to a harmonic oscillator. Unlike to the situation in the transverse plane, however, the motion becomes more complicated when the oscillation amplitude is larger, and the particle feels the non-linear part of the sinusoidal RF wave, or even more, for part of its motion it finds itself over the crest of the wave. But first let us focus on a small amplitude solution.

It is not hard to deduce from special relativity that the momentum may be written

$$p = m_0 c\,(\beta\gamma)\,. \tag{2.67}$$

The quantity $\Delta(\beta\gamma)$ serves as the momentum co-ordinate in longitudinal phase space. The other co-ordinate is the particle's arrival phase, ϕ, with respect to the zero crossing of the r.f. voltage at the cavity. Let us consider the simplest case of a small oscillation in a stationary bucket, $\phi_s = 0$ (when the particle is not being accelerated).

A particle with a small phase error will describe an ellipse in phase space which one may write parametrically as

$$\begin{aligned}\Delta\left(\beta\gamma\right) &= \widehat{\Delta(\beta\gamma)}\sin 2\pi f_{s}t,\\ \phi &= \hat{\phi}\cos 2\pi f_{s}t,\end{aligned} \tag{2.68}$$

where f_s is the frequency of execution of these oscillations in phase which we call the synchrotron frequency.

To reveal the differential equation behind this motion we must first remember that the angular frequency $2\pi f$ of an oscillator is nothing other than the rate of change of phase, $\dot{\phi}$ or to be exact $-\dot{\phi}$. (The negative sign stems from the fact that ϕ is a phase lag.) We may therefore relate the rate of change in arrival phase to the difference in revolution frequency of the particle, compared to that of the synchronous particle.

$$\dot{\phi} = -2\pi h\left[f\left(\Delta\beta\gamma\right) - f(0)\right] = -2\pi h\Delta f. \tag{2.69}$$

We have multiplied by, h, the harmonic number of the r.f. since ϕ is the phase angle of the r.f. swing while $f(\Delta\beta\gamma)$ is the revolution frequency. Here we can use the definition of the slip factor η and then simply use some standard relativistic relations to end up with Δf as a function of ΔE, the energy defect with respect to the synchronous particle:

$$\Delta f = \eta f\frac{\Delta p}{p} = \eta f\frac{\Delta\left(\beta\gamma\right)}{\left(\beta\gamma\right)} = \frac{\eta f}{\beta^{2}}\frac{\Delta\gamma}{\gamma} = \frac{\eta f}{E_{0}\beta^{2}\gamma}\Delta E. \tag{2.70}$$

where E_0 the total energy (including its rest mass) of the synchronous particle.

We differentiate once more to obtain a second order differential equation which we hope to resemble a simple oscillator.

$$\ddot{\phi} = -\frac{2\pi h\eta f}{E_{0}\beta^{2}\gamma}\left(\Delta\dot{E}\right). \tag{2.71}$$

The extra energy given per turn to a particle whose arrival phase is ϕ will be

$$\Delta E = eV_{0}\left(\sin\phi - \sin\phi_{s}\right), \tag{2.72}$$

and the rate of change of energy will be this times, f, the revolution frequency. So we can write

$$\ddot{\phi} = -\frac{2\pi eV_{0}h\eta f^{2}}{E_{0}\beta^{2}\gamma}\left(\sin\phi - \sin\phi_{s}\right). \tag{2.73}$$

This is a fundamental and exact description of the motion provided the parameters should change slowly (the adiabatic assumption). We can simply integrate to find its solution numerically but to see an analytic solution for small amplitudes we set $\phi_s = 0$ and $\phi \approx \sin\phi$:

$$\ddot{\phi} + \frac{2\pi e V_0 h\eta f^2}{E_0\beta^2\gamma}\phi = 0. \tag{2.74}$$

The frequency of these synchrotron oscillations in longitudinal phase space is

$$f_s = \sqrt{\frac{|\eta|\, he V_0}{2\pi\, E_0\beta^2\gamma}}\, f, \tag{2.75}$$

or writing $f_{rf} = hf$ we could equally express

$$f_s = \sqrt{\frac{|\eta|\, e V_0}{2\pi\, E_0\beta^2\gamma h}}\, f_{rf}. \tag{2.76}$$

In analogy to the transverse plane, we define a synchrotron tune, Q_s, as the number of such oscillations per revolution of the machine. This is analogous to Q in transverse phase space.

$$Q_s = \frac{f_s}{f} = \sqrt{\frac{|\eta|\, eh V_0 \cos\phi_s}{2\pi\, E_0\beta^2\gamma}}. \tag{2.77}$$

Usually Q_s is less than 10% of the revolution frequency. It drops down to zero at γ transition where η is zero and then rises again. In large proton machines it can be in the region 0 to 100 Hz and, but for the vacuum, one might hear it!

Close to γ_{tr} we cannot strictly assume that β, γ, η, and f vary slowly in comparison with the synchrotron oscillation which this equation describes. Hence we should use a more exact form of the equation of motion and approximate only when it seems that this is justified:

$$\frac{d}{dt}\left[\frac{E_0\beta^2\gamma\dot{\phi}}{2\pi\eta h f^2}\right] + eV_0\,(\sin\phi - \sin\phi_s) = 0. \tag{2.78}$$

In a stationary bucket, when $\phi_s = 0$, this exact differential equation for large amplitude motion is identical to that for a rigid pendulum:

$$\ell\frac{d^2\theta}{dt^2} + g\sin\theta = 0. \tag{2.79}$$

There is an extra term, $\sin\phi_s$, on the right hand side of the synchrotron equation which is not there in the pendulum case but it could be introduced too for the pendulum by using a magnetic ‘bob’ and biasing its equilibrium position to one side by attaching a weight on a cantilever at right angles to the rod of the pendulum.

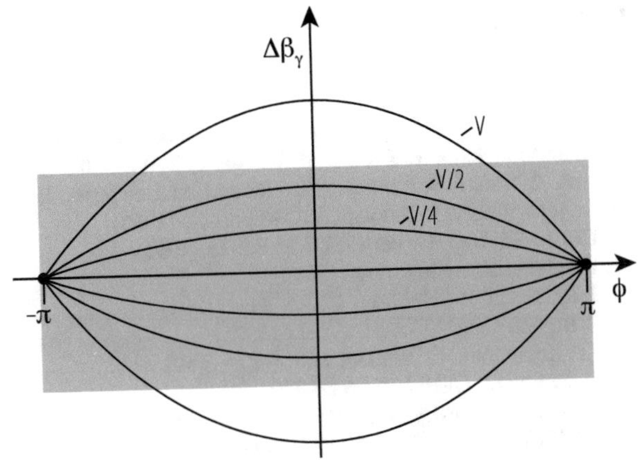

Fig. 2.24 Adiabatic trapping of coasting beam in growing stationary bucket

In fact the unbiased pendulum corresponds to synchrotron motion when there is no acceleration—we say the bucket is stationary. In Fig. 2.22 we saw how particles close to the edge of the stable area of the bucket follow a fish-shaped trajectory when $\phi_S = 0$; before acceleration starts or when the beam is held at the same energy in collider mode (see Fig. 2.24).

In order to accelerate ϕ_S must be made finite, in which case the figure changes somewhat. The stable area becomes smaller and shaped like a fish—or rather a series of fish chasing each other's tails. Small amplitude motion will still be sinusoidal but the ellipse will be centred on the stable phase ϕ_S and not on $\phi = 0$.

2.5.4 *Stationary Buckets*

The size of the bucket depends on how close the stable phase, ϕ_S is to the crest of the sine-wave. It shrinks to zero if $\phi_S = 90°$. There is a special case if ϕ_S is zero. This is often the case as a beam injected into a synchrotron before acceleration has started or in a collider where the r.f. simply holds the bunches together. The bucket is then said to be 'stationary' stretching over all phases from $-\pi$ to π. Its height is the range of energies $2\Delta E$ which the r.f. wave can constrain and this turns out to be dependent on $\sqrt{V}$ for a given ϕ_S. If V is reduced, the more energetic particles spill out of the bucket.

Very often the particles are injected as a continuous ribbon without any longitudinal structure crosshatched in Fig. 2.24. Usually acceleration has not yet started, the magnetic field B is constant, and ϕ_S is zero. If V is increased slowly, the height of the stationary bucket grows, and more and more of the energy spread in the beam, ΔE, is trapped (Fig. 2.24). This is called "adiabatic trapping".

References

1. E.D. Courant, H.S. Snyder: Annals of Physics 3 (1958) 1–48.
2. J.J. Livingood: *Principles of cyclic particle accelerators*, van Nostrand (1961).
3. R. Servranckx, K.L. Brown: SLAC Report 270 UC-288 (1984).
4. A.A. Garren, A.S. Kenney, E.D. Courant, M.J. Syphers: Fermilab report FN40 (1985).
5. L. Schachinger, R. Talman: SSC Report-52 (1985).
6. F.C. Iselin, H.G. Grote: CERN/SL/90-13 (1991).
7. P. Schmüser: CERN 87-10 (1987).
8. J. Rossbach, P. Schmüser: CERN 94-1 (1992).
9. P. Bryant: CERN ISR-MA/75-28 (1975).
10. J.P. Koutchouk: CERN ISR-OP/80-27 (1980).
11. G. Guignard: CERN 76-06 (1976).
12. S. Guiducci: CERN 94-01 (1992).
13. B.W. Montague: CERN 77-13 (1977).
14. J. Le Duff: CERN 94-01 (1992).
15. C. Bovet, R. Gouiran, I. Gumowski, H. Reich: CERN/MPS-SI/Int.DL/70/4 (1970).

Chapter 3
Non-linear Dynamics in Accelerators

Werner Herr and Etienne Forest

3.1 Introduction

Non-linear effects in accelerator physics are important both during the design stage and for successful operation of accelerators. Since both of these aspects are closely related, they will be treated together in this overview. Some of the most important aspects are well described by methods established in other areas of physics and mathematics. Given the scope of this handbook, the treatment will be focused on the problems in accelerators used for particle physics experiments. Although the main emphasis will be on accelerator physics issues, some of the aspects of more general interest will be discussed. In particular to demonstrate that in recent years a framework has been built to handle the complex problems in a consistent form, technically superior and conceptually simpler than the traditional techniques. The need to understand the stability of particle beams has substantially contributed to the development of new techniques and is an important source of examples which can be verified experimentally. Unfortunately the documentation of these developments is often poor or even unpublished, in many cases only available as lectures or conference proceedings.

This article is neither rigorous nor a complete treatment of the topic, but rather an introduction to a limited set of contemporary tools and methods we consider useful in accelerator theory.

W. Herr (✉)
CERN (European Organization for Nuclear Research), Meyrin, Genève, Switzerland
e-mail: werner.herr@cern.ch

E. Forest
KEK, High Energy Research Organisation, Tsukuba, Japan

S. Myers, H. Schopper (eds.), *Particle Physics Reference Library*,
https://doi.org/10.1007/978-3-030-34245-6_3

3.1.1 Motivation

The most reliable tools to study (i.e. description of the machine) are simulations (e.g. tracking codes).

- Particle Tracking is a numerical solution of the (nonlinear) Initial Value Problem. It is a "integrator" of the equation of motion and a vast amount of tracking codes are available, together with analysis tools (Examples: Lyapunov, Chirikov, chaos detection, frequency analysis, ...)
- It is unfortunate that theoretical and computational tools exist side by side without an undertaking how they can be integrated.
- It should be undertaken to find an approach to link simulations with theoretical analysis, would allow a better understanding of the physics in realistic machines.
- A particularly promising approach is based on finite maps [1].

3.1.2 Single Particle Dynamics

The concepts developed here are used to describe single particle transverse dynamics in rings, i.e. circular accelerators or storage rings. This is not a restriction for the application of the presented tools and methods. In the case of linear betatron motion the theory is rather complete and the standard treatment [2] suffices to describe the dynamics. In parallel with this theory the well known concepts such as closed orbit and Twiss parameters are introduced and emerged automatically from the Courant-Snyder formalism [2]. The formalism and applications are found in many textbooks (e.g. [3–5]).

In many new accelerators or storage rings (e.g. LHC) the description of the machine with a linear formalism becomes insufficient and the linear theory must be extended to treat non-linear effects. The stability and confinement of the particles is not given a priori and should rather emerge from the analysis. Non-linear effects are a main source of performance limitations in such machines. A reliable treatment is required and the progress in recent years allows to evaluate the consequences. Very useful overview and details can be found in [6–8].

3.1.3 Layout of the Treatment

Following a summary of the sources of non-linearities in circular machine, the basic methods to evaluate the consequences of non-linear behaviour are discussed. Since the traditional approach has caused misconception and the simplifications led to wrong conclusions, more recent and contemporary tools are introduced to treat these problems. An attempt is made to provide the physical picture behind these tools rather than a rigorous mathematical description and we shall show how the new concepts are a natural extension of the Courant-Snyder formalism to non-linear dynamics. An extensive treatment of these tools and many examples can be found in [7]. In the last part we summarize the most important physical phenomena caused by the non-linearities in an accelerator.

3.2 Variables

For what follows one should always use canonical variables!

In Cartesian coordinates:

$$R = (X, P_X, Y, P_Y, Z, P_Z, t) \tag{3.1}$$

If the energy is constant (i.e. $P_Z = \text{const.}$), we use:

$$(X, P_X, Y, P_Y, Z, t) \tag{3.2}$$

This system is rather inconvenient, what we want is the description of the particle in the neighbourhood of the reference orbit/trajectory:

$$R_d = (X, P_X, Y, P_Y, Z, t) \tag{3.3}$$

which are considered now the deviations from the reference and which are zero for a particle on the reference trajectory

It is very important that it is the reference not the design trajectory! (so far it is a straight line along the Z-direction)

3.2.1 *Trace Space and Phase Space*

A confusion often arises about the terms Phase Space $(x, p_x, \ldots)$ or Trace Space $(x, x', \ldots)$

It is not laziness nor stupidity to use one or the other:

- Beam dynamics is strictly correct only with $(x, p_x, \ldots)$, (see later chapter) but in general quantities cannot be measured easily
- Beam dynamics with $(x, x', \ldots)$ needs special precaution, but quantities based on these coordinates are much easier to measure
- Some quantities are different (e.g. emittance)

It comes back to a remark made at the beginning, i.e. that we shall use rings for our arguments. In single pass machine, e.g. linac, beam lines, spectrometers, the beam is not circulating over many turns and several hours, therefore there is no interest in stability issues. Instead for most of these applications what counts is the coordinates and angles at a given position (x, x', y, y'), e.g. at the end of a beam line or a small spot one an electron microscope. When "accelerator physicists" talk about concepts such as tune, resonances, β-functions, equilibrium emittances etc., all these are irrelevant for single pass machine. There is no need to study iterating systems. In these cases the use of the trace space is fully adequate, in fact preferred because the quantities can be measured. In the end, the mathematical tools are very different from the ones discussed in this article.

3.2.2 *Curved Coordinate System*

For a "curved" trajectory, in general not circular, with a local radius of curvature $\rho(s)$ in the horizontal (X–Z plane), we have to transform to a new coordinate system (x, y, s) (co-moving frame) with:

$$\begin{aligned} X &= (x+\rho)\cos\left(\tfrac{s}{\rho}\right) - \rho \\ Y &= y \\ Z &= (x+\rho)\sin\left(\tfrac{s}{\rho}\right) \end{aligned} \tag{3.4}$$

The new canonical momenta become:

$$\begin{aligned} p_x &= P_X\cos\left(\tfrac{s}{\rho}\right) + P_Z\sin\left(\tfrac{s}{\rho}\right) \\ p_y &= P_Y \\ p_s &= P_Z\left(1+\tfrac{x}{\rho}\right)\cos\left(\tfrac{s}{\rho}\right) - P_X\left(1+\tfrac{x}{\rho}\right)\sin\left(\tfrac{s}{\rho}\right) \end{aligned} \tag{3.5}$$

3.3 Sources of Non-linearities

Any object creating non-linear electromagnetic fields on the trajectory of the beam can strongly influence the beam dynamics. They can be generated by the environment or by the beam itself.

3.3.1 *Non-linear Machine Elements*

Non-linear elements can be introduced into the machine on purpose or can be the result of field imperfections. Both types can have adverse effects on the beam stability and must be taken into account.

3.3.1.1 Unwanted Non-linear Machine Elements

The largest fraction of machine elements are either dipole or quadrupole magnets. In the ideal case, these types of magnets have pure dipolar or quadrupolar fields and behave approximately as linear machine elements. Any systematic or random deviation from this linear field introduces non-linear fields into the machine lattice. These effects can dominate the aperture required and limit the stable region of the beam. The definition of tolerances on these imperfections is an important part of any accelerator design.

Normally magnets are long enough that a 2-dimensional field representation is sufficient. The components of the magnetic field can be derived from the potential and in cylindrical coordinates $(r, \Theta, s = 0)$ can be written as:

$$B_r(r,\Theta) = \sum_{n=1}^{\infty}(B_n \sin(n\Theta) + A_n \cos(n\Theta)) \left(\frac{r}{R_{ref}}\right)^{n-1}, \tag{3.6}$$

$$B_\Theta(r,\Theta) = \sum_{n=1}^{\infty}(B_n \cos(n\Theta) - A_n \sin(n\Theta)) \left(\frac{r}{R_{ref}}\right)^{n-1}, \tag{3.7}$$

where R_{ref} is a reference radius and B_n and A_n are constants. Written in Cartesian coordinates we have:

$$B(z) = \sum_{n=1}^{\infty}(B_n + iA_n) \left(\frac{r}{R_{ref}}\right)^{n-1} \tag{3.8}$$

where $z = x + iy = re^{i\Theta}$. The terms n correspond to $2n$-pole magnets and the B_n and A_n are the normal and skew multipole coefficients. The beam dynamics set limits on the allowed multipole components of the installed magnets.

3.3.1.2 Wanted Non-linear Machine Elements

In most accelerators the momentum dependent focusing of the lattice (chromaticity) needs to be corrected with sextupoles [3, 4]. Sextupoles introduce non-linear fields into the lattice that are larger than the intrinsic non-linearities of the so-called linear elements (dipoles and quadrupoles). In a strictly periodic machine the correction can be done close to the origin and the required sextupole strengths can be kept small. For colliding beam accelerators usually special insertions are foreseen to host the experiments where the dispersion is kept small and the β-function is reduced to a minimum. The required sextupole correction is strong and can lead to a reduction of the dynamic aperture, i.e. the region of stability of the beam. In most accelerators the sextupoles are the dominant source of non-linearity. To minimize this effect is an important issue in any design of an accelerator.

Another source of non-linearities can be octupoles used to generate amplitude dependent detuning to provide Landau damping in case of instabilities.

3.3.2 Beam–Beam Effects and Space Charge

A strong source of non-linearities are the fields generated by the beam itself. They can cause significant perturbations on the same beam (space charge effects) or on the opposing beam (beam-beam effects) in the case of a colliding beam facility.

As an example, for the simplest case of round beams with the line density n and the beam size σ the field components can be written as:

$$E_r = -\frac{ne}{4\pi\epsilon_0} \cdot \frac{\partial}{\partial r} \int_0^\infty \frac{\exp(-\frac{r^2}{(2\sigma^2+q)})}{(2\sigma^2+q)} dq, \tag{3.9}$$

and

$$B_\Phi = -\frac{ne\beta c\mu_0}{4\pi} \cdot \frac{\partial}{\partial r} \int_0^\infty \frac{\exp(-\frac{r^2}{(2\sigma^2+q)})}{(2\sigma^2+q)} dq. \tag{3.10}$$

In colliding beams with high density and small beams these fields are the dominating source of non-linearities. The full treatment of beam-beam effects is complicated due to mutual interactions between the two beams and a self-consistent treatment is required. in the presence of all other magnets in the ring.

3.4 Map Based Techniques

In the standard approach to single particle dynamics in rings, the equations of motion are introduced together with an ansatz to solve these equations. In the case of linear motion, this ansatz is due to Courant-Snyder [2]. However, this treatment must assume that the motion of a particle in the ring is stable and confined. For a non-linear system this is a priori not known and the attempt to find a complete description of the particle motion must fail.

The starting point for the treatment of the linear dynamics in synchrotrons is based on solving a linear differential equation of the Hill type.

$$\frac{d^2x(s)}{ds^2} + \underbrace{\left(a_0 + 2\sum_{n=1}^{\infty} a_n \cdot \cos(2ns)\right)}_{K(s)} x(s) = 0 .$$

Each element at position s acts as a source of forces, i.e. we must write for the forces $K \quad \rightarrow \quad K(s)$ which is assumed to be a periodic function, i.e. $K(s + C) = K(s)_{ring}$

The solution of this Boundary Value Problem must be periodic too!

It is therefore not applicable in the general case (e.g. Linacs, Beamlines, FFAG, Recirculators, ...), much better to treat it as an Initial Value Problem.

In a more useful approach we do not attempt to solve such an overall equation but rather consider the fundamental objects of an accelerators, i.e. the machine elements themselves. These elements, e.g. magnets or other beam elements, are the basic building blocks of the machine. All elements have a well defined action on a particle which can be described independent of other elements or concepts such as closed orbit or β-functions. Mathematically, they provide a "map" from one face of a building block to the other, i.e. a description of how the particles move inside and between elements. In this context, a map can be anything from linear matrices to high order integration routines.

A map based technique is also the basis for the treatment of particle dynamics as an Initial value Problem (IVP).

It follow immediately that for a linear, 1st order equation of the type

$$\frac{dx(s)}{ds} = K(s)\, x(s) \qquad \text{(and initial values at } s_0\text{)}$$

the solution can always be written as:

$$\begin{aligned} x(s) &= a \cdot x(s_0) + b \cdot x'(s_0) \\ x'(s) &= c \cdot x(s_0) + d \cdot x'(s_0) \end{aligned} \Longrightarrow \begin{pmatrix} x \\ x' \end{pmatrix}_s = \overbrace{\begin{pmatrix} a & b \\ c & d \end{pmatrix}}^{A} \begin{pmatrix} x \\ x' \end{pmatrix}_{s_0}$$

where the function $K(s)$ does not have to be periodic. Furthermore, the determinant of the matrix A is always 1. Therefore it is an advantage to use maps (matrices) for a linear systems from the start, without trying to solve a differential equation.

The collection of all machine elements make up the ring pr beam line and it is the combination of the associated maps which is necessary for the description and analysis of the physical phenomena in the accelerator ring or beam line.

For a circular machine the most interesting map is the one which describes the motion once around the machine, the so-called One-Turn-Map. It contains all necessary information on stability, existence of closed orbit, and optical parameters. The reader is assumed to be familiar with this concept in the case of linear beam dynamics (Chap. 2) where all maps are matrices and the Courant-Snyder analysis of the corresponding one-turn-map produces the desired information such as e.g. closed orbit or Twiss parameters.

It should therefore be the goal to generalize this concept to non-linear dynamics. The computation of a reliable one-turn-map and the analysis of its properties will provide all relevant information.

Given that the non-linear maps can be rather complex objects, the analysis of the one-turn-map should be separated from the calculation of the map itself.

3.5 Linear Normal Forms

3.5.1 *Sequence of Maps*

Starting from a position s_0 and combining all matrices to get the matrix to position $s_0 + L$ (shown for 1D only):

$$\begin{pmatrix} x \\ x' \end{pmatrix}_{s_0 + L} = \underbrace{M_N \circ M_{N-1} \circ \ldots \circ M_1}_{M(s_0,L)} \circ \begin{pmatrix} x \\ x' \end{pmatrix}_{s_0} \tag{3.11}$$

For a ring with circumference C one obtains the One-Turn-Matrix (OTM) at s_0

$$\begin{pmatrix} x \\ x' \end{pmatrix}_{s_0 + C} = \underbrace{\begin{pmatrix} m_{11} & m_{12} \\ m_{21} & m_{22} \end{pmatrix}}_{M_{OTM}} \circ \begin{pmatrix} x \\ x' \end{pmatrix}_{s_0} \tag{3.12}$$

Without proof, the scalar product:

$$\begin{pmatrix} x \\ x' \end{pmatrix}_{s_0} \cdot M_{OTM} \begin{pmatrix} x \\ x' \end{pmatrix}_{s_0} = \text{const.} = \text{J} \tag{3.13}$$

is a constant of the motion: invariant of the One Turn Map.

With this approach we have a strong argument that the construction of the One Turn Map is based on the properties of each element in the machine. It is entirely independent of the purpose of the machine and their global properties. It is not restricted to rings or in general to circular machine.

Once the One Turn Map is constructed, it can be analysed, but this analysis does not depend on how it was constructed.

As a paradigm: the construction of a map (being for a circular machine or not) and its analysis are conceptual and computational separated undertakings.

3.5.2 *Analysis of the One Turn Map*

The key for the analysis is that matrices can be transformed into **Normal Forms.** Starting with the One-Turn-Matrix, and try to find a (invertible) transformation A such that:

$$AMA^{-1} = R \qquad (\text{or}: \qquad A^{-1}RA = M)$$

- The matrix R is:
  - A "Normal Form", (or at least a very simplified form of the matrix)
  - For example (most important case): R becomes a pure rotation
- The matrix R describes the same dynamics as M, but:
  - All coordinates are transformed by A
  - This transformation A "analyses" the complexity of the motion, it contains the structure of the phase space

$$M = A \circ R \circ A^{-1} \quad \text{or}: \quad R = A^{-1} \circ M \circ A$$

The motion on an ellipse becomes a motion on a circle (i.e. a rotation): R is the simple part of the map and its shape is dumped into the matrix A. R can be obtained by the evaluation of the Eigenvectors and Eigenvalues.

One finds for the two components of the original map:

$$A = \begin{pmatrix} \sqrt{\beta(s)} & 0 \\ -\frac{\alpha(s)}{\sqrt{\beta(s)}} & \frac{1}{\sqrt{\beta(s)}} \end{pmatrix} \quad \text{and} \quad R = \begin{pmatrix} \cos(\Delta\mu) & \sin(\Delta\mu) \\ -\sin(\Delta\mu) & \cos(\Delta\mu) \end{pmatrix} \tag{3.14}$$

Please note that the normal form analysis gives the eigenvectors (3.14) without any physical picture related to their interpretation. The formulation using α and β is due to Courant and Snyder. Amongst other advantages it can be used to "normalise" the position x: the normalised position x_n is the "non-normalized" divided by $\sqrt{\beta}$. The variation of the normalised position x_n is then smaller than in the non-normalized case. This is also better suited for analytical calculation, e.g. involving perturbation theory.

The Normal Form transformation together with this choice gives the required information:

- μ_x is the "tune" $Q_x \cdot 2\pi$ (now we can talk about phase advance!)
- $\beta, \alpha, \ldots$ are the optical parameters and describe the ellipse
- The closed orbit (an invariant, identical coordinates after one turn!):
- $M_{OTM} \circ (x,\ x')_{co} \equiv (x,\ x')_{co}$

3.5.3 Action-Angle Variables

More appropriate for studies of beam dynamics is the use of Action-Angle variables.

Once the particles "travel" on a circle, the motion is better described by the canonical variables action J_x and angle Ψ_x:

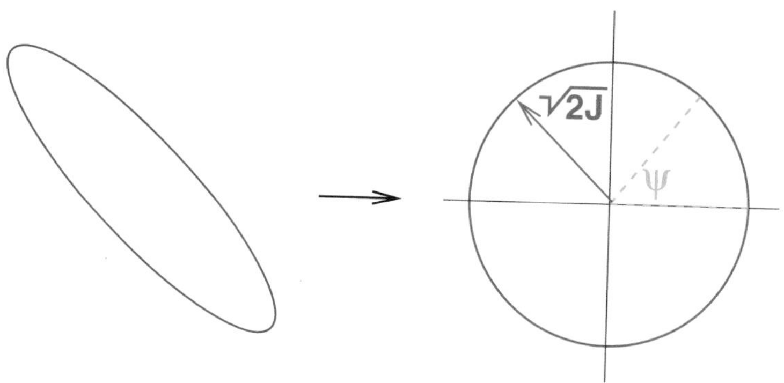

with the definitions and the choice is (3.14):

$$\begin{aligned} x &= \sqrt{2J_x\beta_x}\ \cos(\Psi_x) \\ p_x &= -\sqrt{\frac{2J_x}{\beta_x}}\ (\sin(\Psi_x) + \alpha_x \cos(\Psi_x)) \\ J_x &= \frac{1}{2}(\gamma_x x^2\ +\ 2\alpha_x x p_x\ +\ \beta_x p_x^2) \end{aligned} \tag{3.15}$$

- the angular position along the ring Ψ becomes the independent variable!
- The trajectory of a particle is now independent of the position s!
- The constant radius of the circle $\sqrt{2J}$ defines the action J (invariant of motion)

3.5.4 Beam Emittance

A sad and dismal story in accelerator physics is the definition of the emittance. Most foolish in this context is to relate emittance to single particles. This is true in particular when we have a beam line which is not periodic. In that case the Courant-Snyder parameters can be determined from the beam. These parameters are related to the moments of the beam, e.g. the beam size is directly related to the second order moment $< x^2 >$. Using the expression above for the action and angle, we can write for this expression:

$$< x^2 > \ = \ < 2J_x\beta_x \cdot \cos^2(\Psi_x) > \ = \ 2\beta_x < J_x \cdot \cos^2(\Psi_x) > . \tag{3.16}$$

The average of $\cos^2$ can immediately be evaluated as 0.5 and defining the emittance as:

$$\epsilon_x = < J_x >, \tag{3.17}$$

we write

$$< x^2 > = \beta_x \cdot \epsilon_x . \tag{3.18}$$

Using a similar procedure (details and derivation in e.g. [3], and to a much lesser extent in [1]) one can determine the moments

$$< p_x^2 > = \gamma_x \cdot \epsilon_x , \tag{3.19}$$

and

$$< x \cdot p_x > = -\alpha_x \cdot \epsilon_x . \tag{3.20}$$

Using these expressions, the emittance becomes readily

$$\epsilon_x = \sqrt{< x^2 >< p_x^2 > - < x \cdot p_x >^2} \tag{3.21}$$

Therefore, once the emittance is measured, the Courant-Snyder parameters are determined by Eqs. (3.18), (3.19), and (3.20).

Since other definitions often refer to the treatment by Courant and Snyder, here a quote from Courant himself in [9]:

Interlude 1

The invariant J is simply related to the area enclosed by the ellipse:

$$\text{Area enclosed} = 2\pi J. \tag{3.22}$$

In accelerator and storage ring terminology there is a quantity called the *emittance* which is closely related to this invariant. The emittance, however, is a property of a distribution of particles, not a single particle. Consider a Gaussian distribution in amplitudes. Then the (rms) emittance, ϵ, is given by:

$$(x_{rms})^2 = \beta_x(s) \cdot \epsilon_x . \tag{3.23}$$

In terms of the action variable, J, this can be rewritten

$$\epsilon_x = < J > . \tag{3.24}$$

where the bracket indicates an average over the distribution in J.

Other definitions based on handwaving arguments or those approximately valid only in special cases, should be discarded, in particular those relying on presumed distributions, e.g. Gaussian.

3.6 Techniques and Tools to Evaluate and Correct Non-linear Effects

The key to a more modern approach shown in this section is to avoid the prejudices about the stability and other properties of the ring. Instead, we must describe the machine in terms of the objects it consists of with all their properties, including the non-linear elements. The analysis will reveal the properties of the particles such as e.g. stability. In the simplest case, the ring is made of individual machine elements such as magnets which have an existence on their own, i.e. the interaction of a particle with a given element is independent of the motion in the rest of the machine. Also for the study of non-linear effects, the description of elements should be independent of concepts such as tune, chromaticity and closed orbit. To successfully study single particle dynamics, one must be able to describe the action of the machine element on the particle as well as the machine element.

3.6.1 Particle Tracking

The ring being a collection of maps, a particle tracking code, i.e. an integrator of the equation of motion, is the most reliable map for the analysis of the machine. Of course, this requires an appropriate description of the non-linear maps in the code. It is not the purpose of this article to describe the details of tracking codes and the underlying philosophy, such details can be found in the literature (see e.g. [6]). Here we review and demonstrate the basic principles and analysis techniques.

3.6.1.1 Symplecticity

If we define a map through $\vec{z}_2 = M_{12}(\vec{z}_1)$ as a propagator from a location "1" to a location "2" in the ring, we have to consider that not all possible maps are allowed. The required property of the map is called "symplecticity" and in the simplest case where M_{12} is a matrix, the symplecticity condition can be written as:

$$M \Rightarrow M^T \cdot S \cdot M = S \quad \text{where} \quad S = \begin{pmatrix} 0 & 1 & 0 & 0 \\ -1 & 0 & 0 & 0 \\ 0 & 0 & 0 & 1 \\ 0 & 0 & -1 & 0 \end{pmatrix} \tag{3.25}$$

The physical meaning of this condition is that the map is area preserving in the phase space. The condition can easily be derived from a Hamiltonian treatment, closely related to Liouville's theorem.

3.6.2 *Approximations and Tools*

The concept of symplecticity is vital for the treatment of Hamiltonian systems. This is true in particular when the stability of a system is investigated using particle tracking. However, in practice it is difficult to accomplish for a given exact problem. As an example we may have the exact fields and potentials of electromagnetic elements. For a single pass system a (slightly) non-symplectic integrator may be sufficient, but for an iterative system the results are meaningless.

To track particles using the exact model may result in a non-symplectic tracking, i.e. the underlying model is correct, but the resulting physics is wrong.

It is much better to approximate the model to the extend that the tracking is symplectic. One might compromise on the exactness of the final result, but the correct physics is ensured.

As a typical example one might observe possible chaotic motion during the tracking procedure. However, there is always a non-negligible probability that this interpretation of the results may be wrong. To conclude that it is not a consequence of non-symplecticity of the procedure or a numerical artifact it is necessary to identify the physical mechanism leading to this observation.

This may not be possible to achieve using the exact model as input to a (possibly) non-symplectic procedure. Involving approximations to the definition of the problem should reveal the correct physics at the expense of a (hopefully) small error. Staying exact, the physics may be wrong.

As a result, care must be taken to positively identify the underlying process.

This procedure should be based on a approximations as close as possible to the exact problem, but allowing a symplectic evaluation.

An example for this will be shown in Sect. 3.6.3.4.

3.6.3 *Taylor and Power Maps*

A non-linear element cannot be represented in the form of a linear matrix and more complicated maps have to be introduced [5]. In principle, any well behaved, non-linear function can be developed as a Taylor series. This expansion can be truncated at the desired precision.

Another option is the representation as Lie transformations [8, 10]. Both types are discussed in this section.

3.6.3.1 Taylor Maps

A Taylor map can be written using higher order matrices and in the case of two dimensions we have:

$$z_j(s_2) = \sum_{k=1}^{4} R_{jk} z_k(s_1) + \sum_{k=1}^{4}\sum_{l=1}^{4} T_{jkl} z_k(s_1) z_l(s_1) \tag{3.26}$$

(where z_j, $j = 1, \ldots, 4$, stand for x, x', y, y'). Let us call the collection: $A_2 = (R, T)$ the second order map A_2. Higher orders can be defined as needed, e.g. for the 3rd order map $A_3 = (R, T, U)$ we add a third order matrix:

$$+ \sum_{k=1}^{4}\sum_{l=1}^{4}\sum_{m=1}^{4} U_{jklm} z_k(s_1) z_l(s_1) z_m(s_1) \tag{3.27}$$

Since Taylor expansions are not matrices, to provide a symplectic map, it is the associated Jacobian matrix J which must fulfill the symplecticity condition:

$$J_{ik} = \frac{\partial z_i(s_2)}{\partial z_k(s_1)} \quad \text{and} \quad J \text{ must fulfill}: \; J^t \cdot S \cdot J = S \tag{3.28}$$

However, in general $J_{ik} \neq$ const and for a truncated Taylor map it can be difficult to fulfill this condition for all z. As a consequence, the number of independent coefficients in the Taylor expansion is reduced and the complete, symplectic Taylor map requires more coefficients than necessary [7].

The explicit maps for a sextupole is:

$$\begin{aligned}
x_2 &= x_1 + Lx_1' - k_2\left(\tfrac{L^2}{4}(x_1^2 - y_1^2) + \tfrac{L^3}{12}(x_1x_1' - y_1y_1') + \tfrac{L^4}{24}(x_1'^2 - y_1'^2)\right)\\
x_2' &= x_1' \qquad\quad - k_2\left(\tfrac{L}{2}(x_1^2 - y_1^2) + \tfrac{L^2}{4}(x_1x_1' - y_1y_1') + \tfrac{L^3}{6}(x_1'^2 - y_1'^2)\right)\\
y_2 &= y_1 + Ly_1' + k_2\left(\tfrac{L^2}{4}x_1y_1 + \tfrac{L^3}{12}(x_1y_1' + y_1x_1') + \tfrac{L^4}{24}(x_1'y_1')\right)\\
y_2' &= y_1' \qquad\quad + k_2\left(\tfrac{L}{2}x_1y_1 + \tfrac{L^2}{4}(x_1y_1' + y_1x_1') + \tfrac{L^3}{6}(x_1'y_1')\right)
\end{aligned} \tag{3.29}$$

Writing the explicit form of the Jacobian matrix:

$$J_{ik} = \begin{pmatrix}
\frac{\partial x_2}{\partial x_1} & \frac{\partial x_2}{\partial x_1'} & \frac{\partial x_2}{\partial y_1} & \frac{\partial x_2}{\partial y_1'}\\
\frac{\partial x_2'}{\partial x_1} & \frac{\partial x_2'}{\partial x_1'} & \frac{\partial x_2'}{\partial y_1} & \frac{\partial x_2'}{\partial y_1'}\\
\frac{\partial y_2}{\partial x_1} & \frac{\partial y_2}{\partial x_1'} & \frac{\partial y_2}{\partial y_1} & \frac{\partial y_2}{\partial y_1'}\\
\frac{\partial y_2'}{\partial x_1} & \frac{\partial y_2'}{\partial x_1'} & \frac{\partial y_2'}{\partial y_1} & \frac{\partial y_2'}{\partial y_1'}
\end{pmatrix} \quad \rightarrow k_2 = 0 \quad \begin{pmatrix}
1 & L & 0 & 0\\
0 & 1 & 0 & 0\\
0 & 0 & 1 & L\\
0 & 0 & 0 & 1
\end{pmatrix} \tag{3.30}$$

For $k_2 \neq 0$ coefficients depend on initial values, e.g.:

$$\frac{\partial y_2}{\partial y_1} = 1 + k_2 \left(\frac{L^2}{4} x_1 + \frac{L^3}{12 x_1}' \right) \rightarrow \text{Power series are not symplectic, cannot be used}$$

The non-symplecticity can be recovered in the case of elements with L = 0. It becomes small (probably small enough) when the length is small.

As a result, the model is approximated by a small amount, but the symplecticity (and therefore the physics) is ensured. An exact model but compromised integration can fabricate non-existing features and conceal important underlying physics.

The situation is rather different in the case of single pass machines. The long term stability (and therefore symplecticity) is not an issue and the Taylor expansion around the closed orbit is what is really needed. Techniques like the one described in Sect. 3.7.6 provide exactly this in an advanced and flexible formalism.

3.6.3.2 Thick and Thin Lenses

All elements in a ring have a finite length and therefore should be treated as "thick lenses". However, in general a solution for the motion in a thick element does not exist. It has become a standard technique to avoid using approximate formulae to track through thick lenses and rather perform exact tracking through thin lenses. This approximation is improved by breaking the thick element into several thin elements which is equivalent to a numerical integration. A major advantage of this technique is that "thin lens tracking" is automatically symplectic. In this context it becomes important to understand the implied approximations and how they influence the desired results. We proceed by an analysis of these approximations and show how "symplectic integration" techniques can be applied to this problem.

We demonstrate the approximation using a quadrupole. Although an exact solution of the motion through a quadrupole exists, it is a useful demonstration since it can be shown that all concepts developed here apply also to arbitrary non-linear elements.

Let us assume the transfer map (matrix) for a thick, linearized quadrupole of length L and strength K:

$$M_{s \rightarrow s+L} = \begin{pmatrix} \cos(L \cdot \sqrt{K}) & \frac{1}{\sqrt{K}} \cdot \sin(L \cdot \sqrt{K}) \\ -\sqrt{K} \cdot \sin(L \cdot \sqrt{K}) & \cos(L \cdot \sqrt{K}) \end{pmatrix} \tag{3.31}$$

This map is exact and can be expanded as a Taylor series for a "small" length L:

$$M_{s \rightarrow s+L} = L^0 \cdot \begin{pmatrix} 1 & 0 \\ 0 & 1 \end{pmatrix} + L^1 \cdot \begin{pmatrix} 0 & 1 \\ -K & 0 \end{pmatrix} + L^2 \cdot \begin{pmatrix} -\frac{1}{2}K & 0 \\ 0 & -\frac{1}{2}K \end{pmatrix} + \ldots \tag{3.32}$$

If we keep only terms up to first order in L we get:

$$M_{s\to s+L} = L^0 \cdot \begin{pmatrix} 1 & 0 \\ 0 & 1 \end{pmatrix} + L^1 \cdot \begin{pmatrix} 0 & 1 \\ -K & 0 \end{pmatrix} + O(L^2) \tag{3.33}$$

$$M_{s\to s+L} = \begin{pmatrix} 1 & L \\ -K \cdot L & 1 \end{pmatrix} + O(L^2) \tag{3.34}$$

This map is precise to order $O(L^1)$, but since we have det $M \neq 1$, this truncated expansion is not symplectic.

3.6.3.3 Symplectic Matrices and Symplectic Integration

However, the map (3.34) can be made symplectic by adding a term $-\mathbf{K^2L^2}$. This term is of order $O(L^2)$, i.e. does not deteriorate the approximation because the inaccuracy is of the same order.

$$M_{s\to s+L} = \begin{pmatrix} 1 & L \\ -K \cdot L & 1 - \mathbf{KL^2} \end{pmatrix} \tag{3.35}$$

Following the same procedure we can compute a symplectic approximation precise to order $O(L^2)$ from (3.32) using:

$$M_{s\to s+L} = \begin{pmatrix} 1 - \frac{1}{2}KL^2 & L \\ -K \cdot L & 1 - \frac{1}{2}KL^2 \end{pmatrix} \Rightarrow \begin{pmatrix} 1 - \frac{1}{2}KL^2 & L - \frac{\mathbf{1}}{\mathbf{4}}\mathbf{KL^3} \\ -K \cdot L & 1 - \frac{1}{2}KL^2 \end{pmatrix} \tag{3.36}$$

It can be shown that this "symplectification" corresponds to the approximation of a quadrupole by a single kick in the centre between two drift spaces of length $L/2$:

$$\begin{pmatrix} 1 & \frac{1}{2}L \\ 0 & 1 \end{pmatrix} \begin{pmatrix} 1 & 0 \\ -K \cdot L & 1 \end{pmatrix} \begin{pmatrix} 1 & \frac{1}{2}L \\ 0 & 1 \end{pmatrix} = \begin{pmatrix} 1 - \frac{1}{2}KL^2 & L - \frac{1}{4}KL^3 \\ -K \cdot L & 1 - \frac{1}{2}KL^2 \end{pmatrix} \tag{3.37}$$

It may be mentioned that the previous approximation to 1st order corresponds to a kick at the **end** of a quadrupole, preceded by a drift space of length L. Both cases are illustrated in Fig. 3.1.

One can try to further improve the approximation by adding 3 kicks like in Fig. 3.2 where the distance between kicks and the kick strengths are optimized to obtain the highest order. The thin lens approximation in Fig. 3.2 with the constants:

$$a \approx 0.675602, b \approx -0.175602, \alpha \approx 1.351204, \beta \approx -1.702410 \tag{3.38}$$

provides an $O(L^4)$ integrator [11].

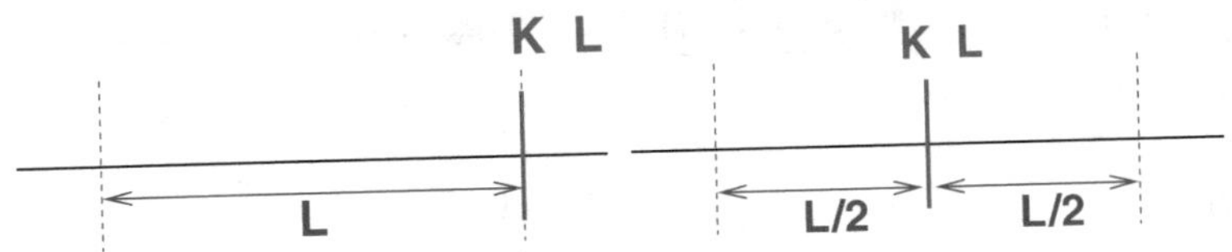

Fig. 3.1 Schematic representation of a symplectic kick of first order (left) and second order (right)

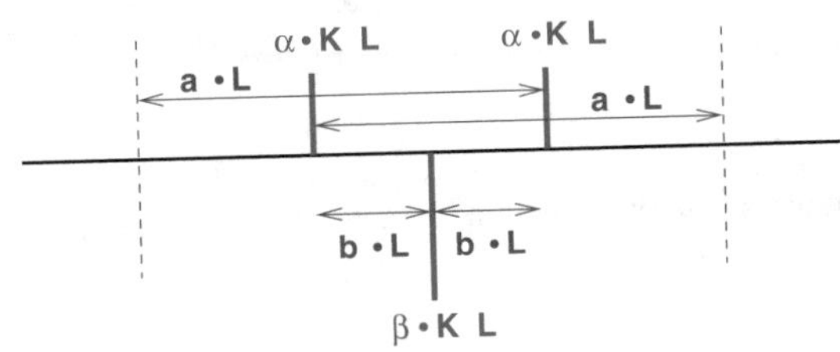

Fig. 3.2 Schematic representation of a symplectic integration with thin lenses of fourth order. The figure shows the size of drifts and thin lens kicks

This process is a Symplectic Integration [12] and is a formal procedure to construct higher order integrators from lower order ones. From a 2nd order scheme (1 kick) $S_2(t)$ we construct a 4th order scheme (3 kicks $= 3 \times 1$ kick) like: $S_4(t) = S_2(x_1 t) \circ S_2(x_0 t) \circ S_2(x_1 t)$ with:

$$x_0 = \frac{-2^{1/3}}{2 - 2^{1/3}} \approx -1.702410 \quad x_1 = \frac{1}{2 - 2^{1/3}} \approx 1.351204 \tag{3.39}$$

In general: If $S_{2k}(t)$ is a symmetric integrator of order $2k$, then we obtain a symmetric integrator of order $2k + 2$ by: $S_{2k+2}(t) = S_{2k}(x_1 t) \circ S_{2k}(x_0 t) \circ S_{2k}(x_1 t)$ with:

$$x_0 = \frac{-\sqrt[2k+1]{2}}{2 - \sqrt[2k+1]{2}} \quad x_1 = \frac{1}{2 - \sqrt[2k+1]{2}} \tag{3.40}$$

Higher order integrators can be obtained in a similar way in an iterative procedure. A very explicit example of the iterative construction of a higher order map from a lower order can be found in [7].

This method can be applied to any other non-linear map and we obtain the same integrators. The proof of this statement and the systematic extension can be done in the form of Lie operators [12].

It should be noted that higher order integrators require maps which drift backwards (3.38) as shown in Fig. 3.2 right. This has two profound consequences. First, a straightforward "physical" interpretation of thin lens models representing drifts and individual small "magnets" (a la MAD) makes no sense and prohibits the

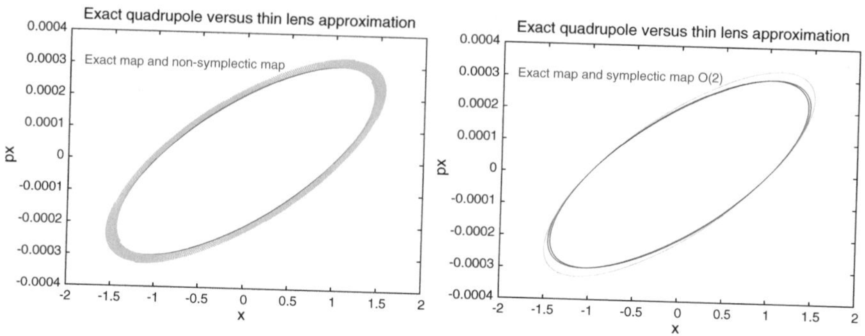

Fig. 3.3 Poincare section for tracking through a quadrupole. Comparison between exact solution, non-symplectic (left) and symplectic (right) tracking. Shown are symplectic integrators of order 1 and 2

use of high order integrators. Secondly, models which require self-consistent time tracking or s tracking (e.g. space charge calculations) must use integrators for which $s(t)$ is monotonic in the magnets.

3.6.3.4 Comparison Symplectic Versus Non-symplectic Integration

A demonstration of a non-symplectic tracking is shown in Fig. 3.3. A particle is tracked through a quadrupole and the poincare section is shown. A quadrupole is chosen because it allows a comparison with the exact solution. The non-symplecticity causes the particle to spiral outwards. As comparison to the exact tracking is shown. In Fig. 3.3 (right) the symplectic integrators of order 1 and 2 as derived above are used instead. The trajectory is now constant and the difference to the exact solution is small. Although the model is approximated but symplectic, the underlying physics (i.e. constant energy in this case) is correct at the expense of a small discrepancy with respect to the exact solution.

3.7 Hamiltonian Treatment of Electro-Magnetic Fields

A frequently asked question is why one should not just use Newton's laws and the Lorentz force. Some of the main reasons are:

- Newton requires rectangular coordinates and time, trajectories with e.g. "curvature" or "torsion" need to introduce "reaction forces". (For example: LHC has locally non-planar (cork-screw) "design" orbits!).
- For linear dynamics done by ad hoc introduction of new coordinate frame.

- With Hamiltonian it is free: The formalism is "coordinate invariant", i.e. the equations have the same form in every coordinate system.
- The basic equations ensure that the phase space is conserved

3.7.1 Lagrangian of Electro-Magnetic Fields

3.7.1.1 Lagrangian and Hamiltonian

It is common practice to use q for the coordinates when Hamiltonian and Lagrangian formalisms are used. This is deplorable because q is also used for particle charge.

The motion of a particle is usually described in classical mechanics using the Langrange functional:

$$L(\, q_1(t), \ldots q_n(t),\; \dot{q}_1(t), \ldots \dot{q}_n(t), t\,) \quad \text{short}: \quad L(q_i, \dot{q}_i, t) \tag{3.41}$$

where $q_1(t), \ldots q_n(t)$ are generalized coordinates and $\dot{q}_1(t), \ldots \dot{q}_n(t)$ the corresponding generalized velocities. Here q_i can stand for any coordinate and any particle, and n can be a very large number.

The integral

$$S = \int L(\, q_i(t), \dot{q}_i(t), t\,)\, dt. \tag{3.42}$$

defines the action S.

The action S is used with the Hamiltonian principle: a system moves along a path such that the action S becomes stationary, i.e. $\delta S = 0$

Is fulfilled when:

$$\frac{d}{dt}\frac{\partial L}{\partial \dot{q}_i} - \frac{\partial L}{\partial q_i} = 0 \quad \text{(Euler} - \text{Lagrange equation)} \tag{3.43}$$

It is unfortunate that the term action is used in different contexts and must not be confused with the action-angle variables defined earlier. The action above is a functional rather than a variable.

Without proof or derivation it should be stated that $L = T - V =$ kinetic energy−potential energy.

Given the Lagrangian, the Hamiltonian can be derived as:

$$H(\vec{q}, \vec{p}, t) = \sum_i [p_i \dot{q}_i - L(\vec{q}, \vec{\dot{q}}, t)]. \tag{3.44}$$

The coordinates q_i are identical to those in the Lagrangian (3.41), whereas the conjugate momenta p_i are derived from L as:

$$p_i \quad = \quad \frac{\partial L}{\partial \dot{q}_i}. \tag{3.45}$$

3.7.2 Hamiltonian with Electro-Magnetic Fields

Readers only interested in the final result can skip Eqs. (3.46)–(3.54).

A key for the correct Hamiltonian is the relativistic treatment. An intuitive derivation is presented here, a simpler and elegant derivation should be based on 4-vectors [13]. The action S must be a relativistic invariant and becomes (now using coordinates x and velocities v):

$$S = \int L(\, x_i(t), v_i(t), t\,)\; \gamma \cdot \mathrm{d}\tau. \tag{3.46}$$

since the proper time τ is Lorentz invariant, and therefore also $\gamma \cdot L$.

The Lagrangian for a free particle is usually a function of the velocity (see classical formula of the kinematic term), but must not depend on its position.

The only Lorentz invariant with the velocity is [13]:

$$U^{\mu} U_{\mu} \quad = \quad c^2 \tag{3.47}$$

where U is the four-velocity.

For the Lagrangian of a (relativistic) free particle we must write

$$L_{free} \; = \; -mc^2\sqrt{1 \; - \; \beta_r^2} \; = \; -mc^2\sqrt{1 \; - \; (\frac{v}{c})^2} \; = \; -\frac{mc^2}{\gamma} \tag{3.48}$$

Using for the electromagnetic Lagrangian a form (without derivation, any textbook):

$$L \; = \; \frac{e}{c}\, v \cdot \vec{A} \; - \; e\phi \tag{3.49}$$

Combining (3.48) and (3.49) we obtain the complete Lagrangian:

$$L \; = \; -\frac{mc^2}{\gamma} \; + \; \frac{e}{c} \cdot \vec{v} \cdot \vec{A} \; - \; e \cdot \phi \tag{3.50}$$

thus the conjugate momentum is derived as:

$$\vec{P} = \frac{\partial L}{\partial v_i} = \vec{p} + \frac{e}{c}\vec{A} \qquad \left(\text{or} \quad \vec{P} = \vec{p} - \frac{q}{c}\vec{A}\right) \tag{3.51}$$

where $\vec{p}$ is the ordinary kinetic momentum.

A consequence is that the canonical momentum cannot be written as:

$$P_x = mc\gamma\beta_x \tag{3.52}$$

Using the conjugate momentum the Hamiltonian takes the simple form:

$$H = \vec{P} \cdot \vec{v} - L \tag{3.53}$$

The Hamiltonian must be a function of the conjugate variables P and x and after a bit of algebra one can eliminate $\vec{v}$ using:

$$\vec{v} = \frac{c\vec{P} - e\vec{A}}{\sqrt{(\vec{P} - \frac{e\vec{A}}{c})^2 + m^2c^2}} \tag{3.54}$$

With (3.50) and (3.54) we write for the Hamiltonian for a (ultra relativistic, i.e. $\gamma \gg 1,\ \beta \approx 1$) particle in an electro-magnetic field is given by:

$$H(\vec{x}, \vec{p}, t) = c\sqrt{(\vec{P} - e\vec{A}(\vec{x}, t))^2 + m^2c^2} + e\Phi(\vec{x}, t) \tag{3.55}$$

where $\vec{A}(\vec{x}, t)$, $\Phi(\vec{x}, t)$ are the vector and scalar potentials.

Interlude 2
A short interlude, *one may want to skip to Eq. (3.60)*

Equation (3.55) is the total energy E of the particle where the difference is the potential energy $e\phi$ and the new conjugate momentum $\vec{P} = (\vec{p} - \frac{e}{c}\vec{A})$, replacing $\vec{p}$.

From the classical expression

$$E^2 = p^2c^2 + (mc^2)^2 \tag{3.56}$$

one can re-write

$$(W - e\phi)^2 - (c\vec{P} - e\vec{A})^2 = (mc^2)^2 \tag{3.57}$$

(continued)

The expression $(mc^2)^2$ *is the invariant mass [13], i.e.*

$$p_\mu p^\mu = (mc)^2 \tag{3.58}$$

with the 4-vector for the momentum [13]:

$$p^\mu = (\frac{E}{c}, \vec{p}) = \left(\frac{1}{c}(W - e\phi),\ \vec{P} - (\frac{e}{c}\vec{A})\right) \tag{3.59}$$

The changes are a consequence using 4-vectors in the presence of electromagnetic fields (potentials).

An interesting consequence of (3.51) is that the momentum is linked to the fields $(\vec{A})$ and the angle x' cannot easily be derived from the total momentum and the conjugate momentum. That is using $(x,\ x')$ as coordinate are strictly speaking not valid in the presence of electromagnetic fields.

In this context using $(x,\ x')$ or (x, p_x) is not equivalent. A general, strong statement that $(x,\ x')$ is used in accelerator physics is at best bizarre.

3.7.3 Hamiltonian Used for Accelerator Physics

In a more convenient (and useful) form, using canonical variables x and p_x, p_y and the design path length s as independent variable (bending field B_0 in y-plane) and no electric fields (for details of the derivation see [14]):

$$H = \underbrace{\overbrace{-(1 + \frac{x}{\rho})}^{\text{due to t} \to \text{s}}} \cdot \overbrace{\sqrt{(1+\delta)^2 - p_x^2 - p_y^2}}^{kinematic} + \underbrace{\overbrace{\frac{x}{\rho} + \frac{x^2}{2\rho^2}}^{\text{due to t} \to s}} - \overbrace{\frac{A_s(x,y)}{B_0\rho}}^{normalized} \tag{3.60}$$

where $p = \sqrt{E^2/c^2 - m^2c^2}$ total momentum, $\delta = (p - p_0)/p_0$ is relative momentum deviation and $A_s(x, y)$ (normalized) longitudinal (along s) component of the vector potential. Only transverse field and no electric fields are considered.

After square root expansion and sorting the A_s contributions:

$$H = \overbrace{\frac{p_x^2 + p_y^2}{2(1+\delta)}}^{kinematic} - \overbrace{\underbrace{\frac{x\delta}{\rho}}_{bending} + \underbrace{\frac{x^2}{2\rho^2}}_{focusing}}^{dipole} + \overbrace{\frac{k_1}{2}(x^2 - y^2)}^{quadrupole} + \overbrace{\frac{k_2}{6}(x^3 - 3xy^2)}^{sextupole} + \cdots \tag{3.61}$$

$$\text{using}: \quad k_n = k_n^{(n)} = \frac{1}{B\rho}\frac{\partial^n B_y}{\partial x^n} \qquad \left(k_n^{(s)} = \frac{1}{B\rho}\frac{\partial^n B_x}{\partial x^n}\right) \tag{3.62}$$

- The Hamiltonian describes the motion of a particle through an element
- Each element has a component in the Hamiltonian
- Basis to extend the linear to a nonlinear formalism

A short list of Hamiltonians of some machine elements (3D)
In general for multipoles of order n:

$$H_n = \frac{1}{1+n} Re\left[(k_n + ik_n^{(s)})(x + iy)^{n+1}\right] + \frac{p_x^2 + p_y^2}{2(1+\delta)} \tag{3.63}$$

We get for some important types (normal components k_n only):

$$\text{drift space}: \; H = -\sqrt{(1+\delta)^2 - p_x^2 - p_y^2} \approx \frac{p_x^2 + p_y^2}{2(1+\delta)} \tag{3.64}$$

$$\text{dipole}: \; H = -\frac{-x\delta}{\rho} + \frac{x^2}{2\rho^2} + \frac{p_x^2 + p_y^2}{2(1+\delta)} \tag{3.65}$$

$$\text{quadrupole}: \; H = \frac{1}{2}k_1(x^2 - y^2) + \frac{p_x^2 + p_y^2}{2(1+\delta)} \tag{3.66}$$

$$\text{sextupole}: \; H = \frac{1}{3}k_2(x^3 - 3xy^2) + \frac{p_x^2 + p_y^2}{2(1+\delta)} \tag{3.67}$$

$$\text{octupole}: \; H = \frac{1}{4}k_3(x^4 - 6x^2y^2 + y^4) + \frac{p_x^2 + p_y^2}{2(1+\delta)} \tag{3.68}$$

Interlude 3
A few remarks are required after this list of Hamiltonian for particular elements.

- *Unlike said in many introductory textbooks and lectures, a multipole of order n is not required to drive a nth order resonance—nothing could be more wrong!!*
- *In leading order perturbation theory, only elements with an even order (and larger than 2) in the Hamiltonian can produce an amplitude dependent tune shift and tune spread.*

3.7.3.1 Lie Maps and Transformations

In this chapter we would like to introduce Lie algebraic tools and Lie transformations [15–17]. We use the symbol $z_i = (x_i, p_i)$ where x and p stand for canonically conjugate position and momentum. We let $f(z)$ and $g(z)$ be any function of x, p and can define the Poisson bracket for a differential operator [18]:

$$[f, g] = \sum_{i=1}^{n} \left(\frac{\partial f}{\partial x_i}\frac{\partial g}{\partial p_i} - \frac{\partial f}{\partial p_i}\frac{\partial g}{\partial x_i} \right) \tag{3.69}$$

Assuming that the motion of a dynamic system is defined by a Hamiltonian H, we can now write for the equations of motion [18]:

$$[x_i, H] = \frac{\partial H}{\partial p_i} = \frac{dx_i}{dt} \tag{3.70}$$

$$[p_i, H] = -\frac{\partial H}{\partial x_i} = \frac{dp_i}{dt} \tag{3.71}$$

If H does not explicitly depend on time then:

$$[f, H] = 0 \tag{3.72}$$

implies that f is an invariant of the motion. To proceed, we can define a Lie operator $: f :$ via the notation:

$$: f : g = [f, g] \tag{3.73}$$

where $: f :$ is an operator acting on the function g.

We can define powers as:

$$(: f :)^2 g =: f : (: f : g) = [f, [f, g]] \quad \text{etc.} \tag{3.74}$$

One can collect a set of useful formulae for calculations:
Some common special (very useful) cases for f:

$$\begin{aligned}
&: x : \; = \frac{\partial}{\partial p} && : p : \; = -\frac{\partial}{\partial x} \\
&: x :^2 \; = \overbrace{: x :: x :}^{\text{applied twice}} = \frac{\partial^2}{\partial p^2} && : p :^2 \; = \overbrace{: p :: p :}^{\text{applied twice}} = \frac{\partial^2}{\partial x^2} \\
&: xp : \; = p\frac{\partial}{\partial p} - x\frac{\partial}{\partial x} && : x :: p : \; = \; : p :: x : \; = -\frac{\partial^2}{\partial x \partial p} \\
&: x^2 : \; = 2x\frac{\partial}{\partial p} && : p^2 : \; = -2p\frac{\partial}{\partial x} \\
&: x^n : \; = n \cdot x^{n-1}\frac{\partial}{\partial p} && : p^n : \; = -n \cdot p^{n-1}\frac{\partial}{\partial x}
\end{aligned} \tag{3.75}$$

Once powers of the Lie operators are defined, they can be used to formulated an exponential form:

$$e^{:f:} = \sum_{i=0}^{\infty} \frac{1}{i!}(: f :)^i \tag{3.76}$$

This expression is call a "Lie transformation".

Give the Hamiltonian H of an element, the generator f is this Hamiltonian multiplied by the length L of the element.

To evaluate a simple example, for the case $H = -p^2/2$ using the exponential form and (3.75):

$$\begin{aligned}
e^{: -Lp^2/2 :} x &= x - \frac{1}{2}L : p^2 : x + \frac{1}{8}L^2(: p^2 :)^2 x + .. \\
&= x + Lp
\end{aligned} \tag{3.77}$$

$$\begin{aligned}
e^{: -Lp^2/2 :} p &= p - \frac{1}{2}L : p^2 : p + \ldots \\
&= p
\end{aligned} \tag{3.78}$$

One can easily verify that for 1D and $\delta = 0$ this is the transformation of a drift space of length L (if $p \approx x'$) as introduced previously. The function $f(x, p) = -Lp^2/2$ is the generator of this transformation.

Interlude 4

The exact Hamiltonian in two transverse dimensions and with a relative momentum deviation δ *is (full Hamiltonian with* $\vec{A}(\vec{x}, t) = 0$*):*

$$H = -\sqrt{(1+\delta)^2 - p_x^2 - p_y^2} \quad \longrightarrow \quad f_{drift} = L \cdot H$$

The exact map for a drift space is now:

$$x^{new} = x + L \cdot \frac{p_x}{\sqrt{(1+\delta)^2 - p_x^2 - p_y^2}}$$

$$p_x^{new} = p_x$$

$$y^{new} = y + L \cdot \frac{p_y}{\sqrt{(1+\delta)^2 - p_x^2 - p_y^2}}$$

$$p_y^{new} = p_y$$

In 2D and with $\delta \neq 0$ *it is more complicated than Eq. (3.78). In practice the map can (often) be simplified to the well known form.*

More general, acting on the phase space coordinates:

$$e^{:f:}(x, p)_1 = (x, p)_2 \tag{3.79}$$

is the Lie transformation which describes how to go from one point to another.

While a Lie operator propagates variables over an infinitesimal distance, the Lie transformation propagates over a finite distance.

To illustrate this technique with some simple examples, it can be shown easily, using the formulae above, that the transformation:

$$e^{:-\frac{1}{2f}x^2:} \tag{3.80}$$

corresponds to the map of a thin quadrupole with focusing length f, i.e.

$$x_2 = x_1$$

$$p_2 = p_1 - \frac{1}{f}x_1$$

A transformation of the form:

$$e^{:-\frac{1}{2}L(k^2x^2+p^2):} \tag{3.81}$$

corresponds to the map of a thick quadrupole with length L and strength k:

$$x_2 = x_1\cos(kL) + \frac{p_1}{k}\sin(kL) \tag{3.82}$$

$$p_2 = -kx_1\sin(kL) + p_1\cos(kL) \tag{3.83}$$

The linear map using Twiss parameters in Lie representation (we shall call it $: f_2 :$ from now on) is always of the form:

$$e^{:f_2:} \quad \text{with :} \quad f_2(x) = -\frac{\mu}{2}(\gamma x^2 + 2\alpha xp + \beta p^2) \tag{3.84}$$

In case of a general non-linear function f(x), i.e. with a (thin lens) kick like:

$$x_2 = x_1 \tag{3.85}$$

$$p_2 = p_1 + f(x_1) \tag{3.86}$$

the corresponding Lie operator can be written as:

$$e^{:h:} = e^{:\int_0^x f(u)du:} \quad \text{or} \quad e^{:F:} \quad \text{with} \quad F = \int_0^x f(u)du. \tag{3.87}$$

An important property of the Lie transformation is that the one turn map is the exponential of the effective Hamiltonian and the circumference C:

$$M_{ring} = e^{:-CH_{eff}:}. \tag{3.88}$$

The main advantages of Lie transformations are that the exponential form is always symplectic and that a formalism exists for the concatenation of transformations. An overview of this formalism and many examples can be found in [7]. As for the Lie operator, one can collect a set of useful formulae. Another neat package with useful formulae:

With a constant and f, g, h arbitrary functions:

$$:a: \; = 0 \quad \longrightarrow \quad e^{:a:} = 1$$

$$:f:a = 0 \quad \longrightarrow \quad e^{:f:}a = a$$

$$e^{:f:}[g,h] = [e^{:f:}g,\ e^{:f:}h]$$

$$e^{:f:}(g \cdot h) = e^{:f:}g\ \cdot\ e^{:f:}h$$

and very important:

$$\begin{aligned} M\, g(x) &= e^{:f:}g(x) = g(e^{:f:}x) \qquad \text{e.g.} \quad e^{:f:}x^2 = (e^{:f:}x)^2 \\ M^{-1}\, g(x) &= (e^{:f:})^{-1} g(x) = e^{-:f:}g(x) \qquad \text{note :} \quad \frac{1}{e^{:f:}} \neq (e^{:f:})^{-1} \end{aligned} \tag{3.89}$$

3.7.3.2 Concatenation of Lie Transformations

The concatenation is very easy when f and g commute (i.e. $[f,g] = [g,f] = 0$) and we have:

$$e^{:h:} = e^{:f:}e^{:g:} = e^{:f+g:} \tag{3.90}$$

The generators of the transformations can just be added.

To combine two transformations in the general case (i.e. $[f,g] \neq 0$) we can use the Baker–Campbell–Hausdorff formula (BCH) which in our convention can be written as:

$$\begin{aligned} h = f &+ g + \tfrac{1}{2}:f:g + \tfrac{1}{12}:f:^2 g + \tfrac{1}{12}:g:^2 f \\ &+ \tfrac{1}{24}:f::g:^2 f - \tfrac{1}{720}:g:^4 f \\ &- \tfrac{1}{720}:f:^4 g + \tfrac{1}{360}:g::f:^3 g + \ldots \end{aligned} \tag{3.91}$$

In many practical cases, non-linear perturbations are localized and small compared to the rest of the (often linear) ring, i.e. one of f or g is much smaller, e.g. f corresponds to one turn, g to a small, local distortion.

In that case we can sum up the BCH formula to first order in the perturbation g and get:

$$e^{:h:} = e^{:f:}e^{:g:} = \exp\left[:f + \left(\frac{:f:}{1-e^{-:f:}}\right) g + O(g^2):\right] \tag{3.92}$$

When g is small compared to f, the first order is a good approximation.

For example, we may have a full ring $e^{:f_2:}$ with a small (local) distortion, e.g. a multipole $e^{:g:}$ with $g = kx^n$ then the expression:

$$e^{:h:} = e^{:f_2:}e^{:kx^n:}, \tag{3.93}$$

allows the evaluation of the invariant h for a single multipole of order n in this case.

In the case that f_2, f_3, f_4, are 2nd, 3rd, 4th order polynomials (Dragt-Finn factorization [19]):

$$e^{:f:} = e^{:f_2:} e^{:f_3:} e^{:f_4:}, \tag{3.94}$$

each term is symplectic and the truncation at any order does not violate symplecticity.

One may argue that this method is clumsy when we do the analysis of a linear system. The reader is invited to prove this by concatenating by hand a drift space and a thin quadrupole lens. However, the central point of this method is that the technique works whether we do linear or non-linear beam dynamics and provides a formal procedure. Lie transformations are the natural extension of the linear matrix formalism to a non-linear formalism. There is no need to move from one method to another as required in the traditional treatment.

In the case an element is described by a Hamiltonian H, the Lie map of an element of length L and the Hamiltonian H is:

$$e^{-L:H:} = \sum_{i=0}^{\infty} \frac{1}{i!}(-L:H:)^i \tag{3.95}$$

For example, the Hamiltonian for a thick sextupole is:

$$H = \frac{1}{3}k(x^3 - 3xy^2) + \frac{1}{2}(p_x^2 + p_y^2) \tag{3.96}$$

To find the transformation we search for:

$$e^{-L:H:}x \quad \text{and} \quad e^{-L:H:}p_x \quad \text{i.e. for} \tag{3.97}$$

$$e^{-L:H:}x = \sum_{i=0}^{\infty} \frac{-L^i}{i!}(:H:)^i x \tag{3.98}$$

We can compute:

$$:H:^i x \quad \text{for each } i \tag{3.99}$$

to get:

$$:H:^1 x = -p_x, \tag{3.100}$$

$$:H:^2 x = -k(x^2 - y^2), \tag{3.101}$$

$$: H :^3 x = 2k(xp_x - yp_y), \tag{3.102}$$

$$\dots . \tag{3.103}$$

Putting the terms together one obtains:

$$e^{-L : H :} x = x + p_x L - \frac{1}{2} k L^2 (x^2 - y^2) - \frac{1}{3} k L^3 (xp_x - yp_y) + \dots \tag{3.104}$$

3.7.4 Analysis Techniques: Poincare Surface of Section

Under normal circumstances it is not required to examine the complete time development of a particle trajectory around the machine. Given the experimental fact that the trajectory can be measured only at a finite number of positions around the machine, it is only useful to sample the trajectory periodically at a fixed position. The plot of the rate of change of the phase space variables at the beginning (or end) of each period is the appropriate method and also known as Poincare Surface of Section [20]. An example of such a plot is shown in Fig. 3.4 where the one-dimensional phase space is plotted for a completely linear machine (Fig. 3.4, left) and close to a 5th order resonance in the presence of a single non-linear element (in this case a sextupole) in the machine (Fig. 3.4, right).

It shows very clearly the distortion of the phase space due to the non-linearity, the appearance of resonance islands and chaotic behaviour between the islands. From this plot is immediately clear that the region of stability is strongly reduced in the

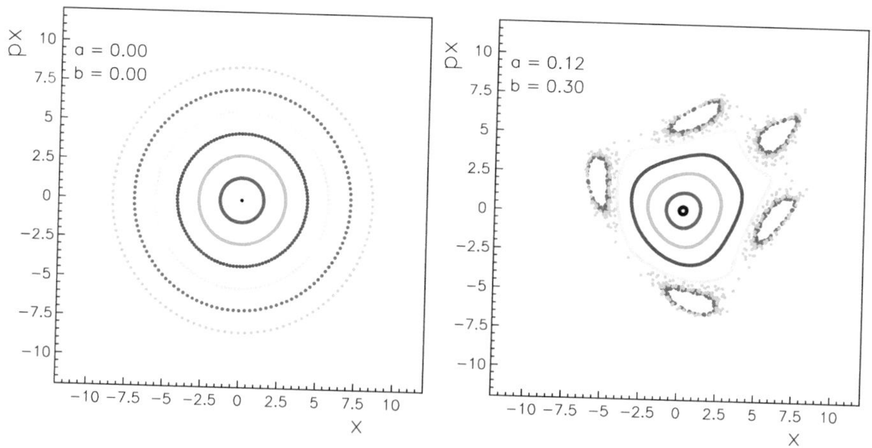

Fig. 3.4 Poincare surface of section of a particle near the 5th order resonances. Left without non-linear elements, right with one sextupole

presence of the non-linear element. The main features we can observe in Fig. 3.4 are that particles can:

- Move on closed curves
- Lie on islands, i.e. jump from one island to the next from turn to turn
- Move on chaotic trajectories

The introduction of these techniques by Poincare mark a paradigm shift from the old classical treatment to a more modern approach. The question of long term stability of a dynamic system is not answered by getting the solution to the differential equation of motion, but by the determination of the properties of the surface where the motion is mapped out. Independent how this surface of section is obtained, i.e. by analytical or numerical methods, its analysis is the key to understand the stability.

3.7.5 *Analysis Techniques: Normal Forms*

The idea behind this technique is that maps can be transformed into Normal Forms. This tool can be used to:

- Study invariants of the motion and the effective Hamiltonian
- Extract non-linear tune shifts (detuning)
- Perform resonance analysis

In the following we demonstrate the use of normal forms away from resonances. The treatment of the beam dynamics close to resonances is beyond the scope of this review and can be found in the literature (see e.g. [6, 7]).

3.7.5.1 Normal Form Transformation: Linear Case

The strategy is to make a transformation to get a simpler form of the map M, e.g. a pure rotation $R(\Delta\mu)$ as schematically shown in Fig. 3.5 using a transformation like:

$$M = U \circ R(\Delta\mu) \circ U^{-1} \quad \text{or :} \quad R(\Delta\mu) = U^{-1} \circ M \circ U \tag{3.105}$$

with

$$U = \begin{pmatrix} \sqrt{\beta(s)} & 0 \\ -\frac{\alpha(s)}{\sqrt{\beta(s)}} & \frac{1}{\sqrt{\beta(s)}} \end{pmatrix} \quad \text{and} \quad R = \begin{pmatrix} \cos(\Delta\mu) & \sin(\Delta\mu) \\ -\sin(\Delta\mu) & \cos(\Delta\mu) \end{pmatrix} \tag{3.106}$$

This transformation corresponds to the Courant-Snyder analysis in the linear case and directly provides the phase advance and optical parameters. The optical

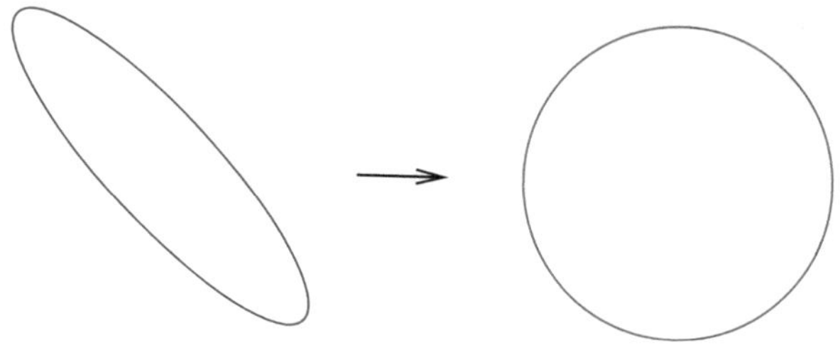

Fig. 3.5 Normal form transformation in the linear case, related to the Courant-Snyder analysis

parameters emerge automatically from the normal form analysis of the one-turn-map.

Although not required in the linear case, we demonstrate how this normal form transformation is performed using the Lie formalism. Starting from the general expression:

$$R(\Delta\mu) = U^{-1} \circ M \circ U \tag{3.107}$$

we know that a linear map M in Lie representation is always:

$$e^{:\, f_2 \,:} \quad \text{with :} \quad f_2 = -\frac{\mu}{2}(\gamma x^2 + 2\alpha x p_x + \beta p_x^2) \tag{3.108}$$

therefore:

$$\begin{aligned} R(\Delta\mu) &= U^{-1} \circ e^{:\, f_2(x) \,:} \circ U \\ &= e^{U^{-1} \,:\, f_2 \,:\, U} = e^{:\, U^{-1} f_2 \,:} \end{aligned} \tag{3.109}$$

and (with $U^{-1} f_2$) f_2 expressed in the new variables X, P_x it assumes the form:

$$f_2 = -\frac{\mu}{2}(X^2 + P_x^2) \quad \text{because :} \quad \begin{pmatrix} X \\ P_x \end{pmatrix} = U^{-1}\begin{pmatrix} x \\ p_x \end{pmatrix} \tag{3.110}$$

i.e. with the transformation U^{-1} the rotation $:\ f_2\ :$ becomes a circle in the transformed coordinates. We transform to action and angle variables J and Φ, related to the variables X and P_x through the transformations:

$$X = \sqrt{2J\beta}\sin\Phi, \qquad P_x = \sqrt{\frac{2J}{\beta}}\cos\Phi \tag{3.111}$$

With this transformation we get a simple representation for the linear transfer map f_2:

$$f_2 = -\mu J \quad \text{and}: \quad R(\Delta\mu) = e^{: -\Delta\mu J :} \tag{3.112}$$

3.7.5.2 Normal Form Transformation: Non-linear Case

In the more general, non-linear case the transformation is more complicated and one must expect that the rotation angle becomes amplitude dependent (see e.g. [6]). A schematic view of this scheme is shown in Fig. 3.6 where the transformation leads to the desired rotation, however the rotation frequency (phase advance) is now amplitude dependent.

We demonstrate the power by a simple example in one dimension, but the treatment is similar for more complex cases. In particular, it demonstrates that this analysis using the algorithm based on Lie transforms leads easily to the desired result. A very detailed discussion of this method is found in [6].

From the general map we have made a transformation such that the transformed map can be expressed in the form $e^{:h_2:}$ where the function h_2 is now a function only of J_x, J_y, and δ and it is the effective Hamiltonian.

In the non-linear case and away from resonances we can get the map in a similar form:

$$N = e^{: h_{eff}(J_x, J_y, \delta) :} \tag{3.113}$$

where the effective Hamiltonian h_{eff} depends only on J_x, J_x, and δ.

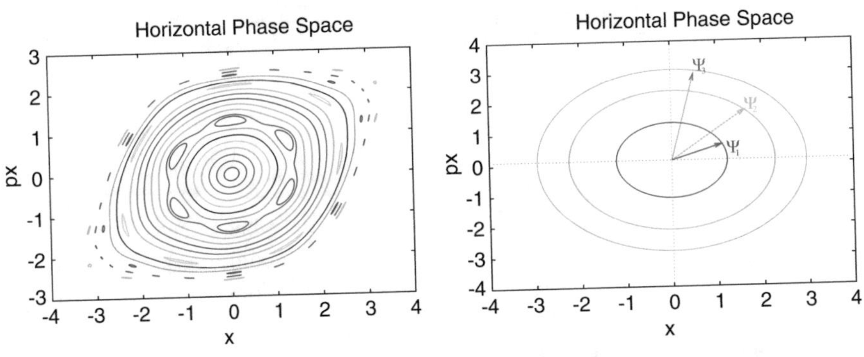

Fig. 3.6 Normal form transformation in the non-linear case, leading to amplitude dependent phase advance. The transformation was done for non-resonant amplitudes

If the map for h_{eff} corresponds to a one-turn-map, we can write for the tunes:

$$Q_x(J_x, J_y, \delta) = \frac{1}{2\pi}\frac{\partial h_{eff}}{\partial J_x} \tag{3.114}$$

$$Q_y(J_x, J_y, \delta) = \frac{1}{2\pi}\frac{\partial h_{eff}}{\partial J_y} \tag{3.115}$$

and the change of path length:

$$\Delta s = -\frac{\partial h_{eff}}{\partial \delta} = \alpha_c \delta \tag{3.116}$$

In the non-linear case, particles with different J_x, J_y, δ have different tunes. Their dependence on J_x, J_y is the amplitude detuning, the dependence on δ are the chromaticities.

The effective Hamiltonian can always be written (here to 3rd order) in a form:

$$h_{eff} = +\mu_x J_x + \mu_y J_y + \frac{1}{2}\alpha_c \delta^2 \tag{3.117}$$

$$+ c_{x1} J_x \delta + c_{y1} J_y \delta + c_3 \delta^3 \tag{3.118}$$

$$+ c_{xx} J_x^2 + c_{xy} J_x J_y + c_{yy} J_y^2 + c_{x2} J_x \delta^2 + c_{y2} J_y \delta^2 + c_4 \delta^4 \tag{3.119}$$

and then tune depends on action J and momentum deviation δ:

$$Q_x(J_x, J_y, \delta) = \frac{1}{2\pi}\frac{\partial h_{eff}}{\partial J_x} = \frac{1}{2\pi}\left(\mu_x + \overbrace{2c_{xx}J_x + c_{xy}J_y}^{detuning} + \overbrace{c_{x1}\delta + c_{x2}\delta^2}^{chromaticity}\right) \tag{3.120}$$

$$Q_y(J_x, J_y, \delta) = \frac{1}{2\pi}\frac{\partial h_{eff}}{\partial J_y} = \frac{1}{2\pi}\left(\mu_y + \overbrace{2c_{yy}J_y + c_{xy}J_x}^{detuning} + \overbrace{c_{y1}\delta + c_{y2}\delta^2}^{chromaticity}\right) \tag{3.121}$$

The meaning of the different contributions are:

- μ_x, μ_y: linear phase advance or $2\pi\cdot$ i.e. the tunes for rings
- $\frac{1}{2}\alpha_c, c_3, c_4$: linear and nonlinear “momentum compaction”
- c_{x1}, c_{y1}: first order chromaticities
- c_{x2}, c_{y2}: second order chromaticities

– c_{xx}, c_{xy}, c_{yy}: detuning with amplitude

The coefficients are the various aberrations of the optics.

As a first example one can look at the effect of a single (thin) sextupole. The map is:

$$\mathcal{M} = e^{-:\mu J_x + \mu J_y + \frac{1}{2}\alpha_c \delta^2 :} e^{: k(x^3 - 3xy^2) + \frac{p_x^2+p_y^2}{2(1+\delta)} :} \tag{3.122}$$

we get for h_{eff} (see e.g. [6, 7]):

$$h_{eff} = \mu_x J_x + \mu_y J_y + \frac{1}{2}\alpha_c \delta^2 - kD^3\delta^3 - 3k\beta_x J_x D\delta + 3k\beta_y J_y D\delta$$

Then it follows:

$$Q_x(J_x, J_y, \delta) = \frac{1}{2\pi}\frac{\partial h_{eff}}{\partial J_x} = \frac{1}{2\pi}(\mu_x - 3k\beta_x D\delta) \tag{3.123}$$

$$Q_y(J_x, J_y, \delta) = \frac{1}{2\pi}\frac{\partial h_{eff}}{\partial J_y} = \frac{1}{2\pi}(\mu_y + 3k\beta_y D\delta) \tag{3.124}$$

Since it was developed to first order only, there is no non-linear detuning with amplitude.

As a second example one can use a linear rotation followed by an octupole, the Hamiltonian is:

$$H = \frac{\mu}{2}(x^2 + p_x^2) + \delta(s - s_0)\frac{x^4}{4} = \mu J + \delta(s - s_0)\frac{x^4}{4} \quad \text{with}: \ J = \frac{(x^2 + p_x^2)}{2} \tag{3.125}$$

The first part of the Hamiltonian corresponds to the generator of a linear rotation and the second part to the localized octupole.

The map, written in Lie representation becomes:

$$M = e^{(-\frac{\mu}{2} : x^2 + p_x^2 :)} e^{: \frac{x^4}{4} :} = e^{: -\mu J :} e^{: \frac{x^4}{4} :} = R\, e^{: \frac{x^4}{4} :} \tag{3.126}$$

The purpose is now to find a generator F for a transformation

$$e^{-:F:} M e^{:F:} = e^{-:F:} e^{: \frac{x^4}{4} :} e^{:F:} \tag{3.127}$$

such that the exponents of the map depend only on J and not on x.

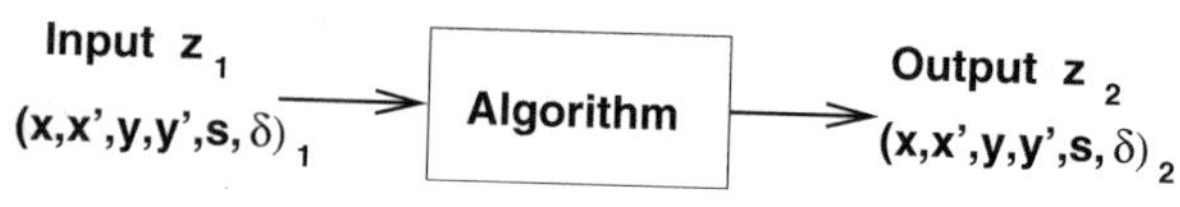

Fig. 3.7 Schematic view of tracking through a complex element

Without going through the algebra (advanced tools exist for this purpose, see e.g. [6]) we quote the result and with

$$F = -\frac{1}{64}\{-5x^4+3p_x^4+6x^2p_x^2+x^3p_x(8\cot(\mu)+4\cot(2\mu))+xp_x^3(8\cot(\mu)-4\cot(2\mu))\} \tag{3.128}$$

we can write the map:

$$M = e^{-:F:}\,e^{:-\mu J+\frac{3}{8}J^2:}\,e^{:F:} \tag{3.129}$$

the term $\frac{3}{8}J^2$ implies a tune shift with amplitude for an octupole.

3.7.6 *Truncated Power Series Algebra Based on Automatic Differentiation*

It was argued that an appropriate technique to evaluate the behaviour of complex, non-linear systems is by numerically tracking through the individual elements. Schematically this is shown in Fig. 3.7 and the tracking through a complicated system relates the output **numerically** to the input. When the algorithm depicted in Fig. 3.7 represents the full turn in a ring, we obtained the most reliable one-turn-map through this tracking procedure, assuming we have chosen an appropriate representation of the maps for the individual elements.

3.7.6.1 Automatic Differentiation: Concept

This procedure may not be fully satisfactory in all cases and one might like to get an analytical expression for the one-turn-map or equivalent. Could we imagine something that relates the output algebraically to the input? This might for example be a Taylor series of the type:

$$z_2 = \sum C_j z_1^j = \sum d_j f^{(n)} z_1^j \tag{3.130}$$

Then we have an analytic map (for all z_1).

To understand why this could be useful, we can study the paraxial behaviour. In Fig. 3.8 we show schematically the trajectories of particles close to the ideal orbit. The red line refers to the ideal trajectory while the other lines show the motion of

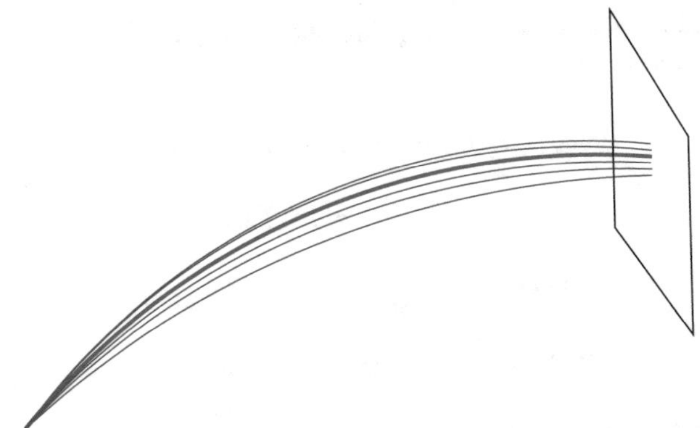

Fig. 3.8 Pictorial view of a paraxial analysis. Red line represents the ideal trajectory

individual particles with small deviations from the ideal path. The idea is that if we understand how small deviations behave, we understand the system much better.

If we now remember the definition of the Taylor series:

$$f(x + \Delta x) = f(x) + \sum_{n=1}^{\infty} \frac{\Delta x^n}{n!} f^{(n)}(x) \tag{3.131}$$

we immediately realize that the coefficients determine the behaviour of small deviations Δx from the ideal orbit x. Therefore the Taylor expansion does a paraxial analysis of the system and the main question is how to get these coefficients without extra work?

The problem is getting the derivatives $f^{(n)}(a)$ of $f(x)$ at a:

$$f'(a) = \lim_{\epsilon \to 0} \frac{f(a + \epsilon) - f(a)}{\epsilon} \tag{3.132}$$

Numerically this corresponds to the need to subtract almost equal numbers and divide by a small number. For higher orders f'', f'''.., one must expect numerical problems. An elegant solution to this problem is the use of Differential Algebra (DA) [21].

3.7.6.2 Automatic Differentiation: The Algebra

Here we demonstrate the concept, for more details the literature should be consulted [6, 7, 21].

1. Define a pair (q_0, q_1), where q_0, q_1 are real numbers
2. Define operations on such pairs like:

$$(q_0, q_1) + (r_0, r_1) = (q_0 + r_0, q_1 + r_1) \tag{3.133}$$

$$c \cdot (q_0, q_1) = (c \cdot q_0, c \cdot q_1) \tag{3.134}$$

$$(q_0, q_1) \cdot (r_0, r_1) = (q_0 \cdot r_0, q_0 \cdot r_1 + q_1 \cdot r_0) \tag{3.135}$$

3. We define the ordering like:

$$(q_0, q_1) < (r_0, r_1) \quad \text{if} \quad q_0 < r_0 \quad \text{or} \quad (q_0 = r_0 \quad \text{and} \quad q_1 < r_1) \tag{3.136}$$

$$(q_0, q_1) > (r_0, r_1) \quad \text{if} \quad q_0 > r_0 \quad \text{or} \quad (q_0 = r_0 \quad \text{and} \quad q_1 > r_1) \tag{3.137}$$

4. This implies that:

$$(0, 0) < (0, 1) < (r, 0) \quad (\text{for any } r) \tag{3.138}$$

This means that (0,1) is between 0 and ANY real number, i.e. it is infinitely small, corresponding to the "ϵ" in standard calculus.

Therefore we call this special pair "differential unit" $d = (0, 1)$.

With our rules we can further see that:

$$(1, 0) \cdot (q_0, q_1) = (q_0, q_1) \quad \text{and} \quad (q_0, q_1)^{-1} = \left(\frac{1}{q_0}, -\frac{q_1}{q_0^2}\right) \tag{3.139}$$

In general the inverse of a function $f(q_0, q_1)$ can de derived like:

$$f((q_0, q_1)) \cdot (r_0, r_1) = (1, 0) \tag{3.140}$$

using the multiplication rules. The inverse is then (r_0, r_1). For example:

$$(q_0, q_1)^2 \cdot (r_0, r_1) = (1, 0) \tag{3.141}$$

gives for the inverse:

$$(r_0, r_1) = \left(\frac{1}{q_0^2}, \frac{-2q_1}{q_0^3}\right) \tag{3.142}$$

3.7.6.3 Automatic Differentiation: The Application

Of course $(q, 0)$ is just the real number q and we define the "real" and the "differential part":

$$q_0 = R(q_0, q_1) \text{ and } q_1 = D(q_0, q_1) \tag{3.143}$$

For a function $f(x)$ we have (without proof, see e.g. [21]):

$$D[f(x+d)] = D[f((x, 0) + (0, 1))] = f'(x) \tag{3.144}$$

We use an example instead to demonstrate this with the function:

$$f(x) = x^2 + \frac{1}{x} \tag{3.145}$$

Using school calculus we have for the derivative:

$$f'(x) = 2x - \frac{1}{x^2} \text{ and for x } = \text{ 2 we get : } f(2) = \frac{9}{2}, \quad f'(2) = \frac{15}{4} \tag{3.146}$$

We now apply Automatic Differentiation instead. For the variable x in (3.145) we substitute $\text{x} \rightarrow (x, 1) = (2, 1)$ and using our rules:

$$\begin{aligned} f[(2, 1)] = (x, 1)^2 + (x, 1)^{-1} &= (2, 1)^2 + (2, 1)^{-1} \\ = (4, 4) + (\frac{1}{2}, -\frac{1}{4}) &= (\frac{9}{2}, \frac{15}{4}) = (f(2), f'(2)) \end{aligned}$$

we arrive at a vector containing the differentials at $x = 2$. The computation of derivatives becomes an algebraic problem, without need for small numbers. No numerical difficulties are expected and the differential is exact.

3.7.6.4 Automatic Differentiation: Higher Orders

To obtain higher orders, we need higher derivatives, i.e. larger dimension for our vectors:

1. The pair $(q_0, 1)$, becomes a vector of length N and with equivalent rules:

$$(q_0, 1) \Rightarrow (q_0, 1, 0, 0, \ldots, 0) \tag{3.147}$$

$$(q_0, q_1, q_2, \ldots q_N) + (r_0, r_1, r_2, \ldots r_N) = (s_0, s_1, s_2, \ldots s_N) \tag{3.148}$$

$$c \cdot (q_0, q_1, q_2, \ldots q_N) = (c \cdot q_0, c \cdot q_1, c \cdot q_2, \ldots c \cdot q_N) \tag{3.149}$$

$$(q_0, q_1, q_2, \ldots q_N) \cdot (r_0, r_1, r_2, \ldots r_N) = (s_0, s_1, s_2, \ldots s_N) \tag{3.150}$$

with:

$$s_i = \sum_{k=0}^{i} \frac{i!}{k!(i-k)!} q_k r_{i-k} \tag{3.151}$$

If we had started with:

$$x = (a, 1, 0, 0, 0 \ldots) \tag{3.152}$$

we would get:

$$f(x) = (\, f(a),\ f'(a),\ f''(a),\ f'''(a), \ldots\ f^{(n)}(a)\,) \tag{3.153}$$

Some special cases are:

$$(x, 0, 0, 0, ..)^n = (x^n, 0, 0, 0, ..) \tag{3.154}$$

$$(0, 1, 0, 0, ..)^n = (0, 0, 0, .., \overbrace{n!}^{n+1}, 0, 0, ..) \tag{3.155}$$

$$(x, 1, 0, 0, ..)^2 = (x^2, 2x, 2, 0, ..) \tag{3.156}$$

$$(x, 1, 0, 0, ..)^3 = (x^3, 3x^2, 6x, 6, ..) \tag{3.157}$$

As another exercise one can consider the function $f(x) = x^{-3}$

$$f(x) \rightarrow f(x, 1, 0, 0, ..) = \underbrace{(x, 1, 0, 0, ..)^{-3} = (f_0, f' f'', f''', ..)}_{\text{next : multiply both sides with } (x, 1, 0, 0)^3}$$

$$(1, 0, 0, ..) = (x, 1, 0, 0, ..)^3 \cdot (f_0, f', f'', f''', \ldots)$$

$$(1, 0, 0, ..) = (x^3, 3x^2, 6x, 6, ..) \cdot (f_0, f', f'', f''', \ldots) \quad \text{using (3.157)}$$

$$(1, 0, 0, ..) = (\underbrace{x^3 \cdot f_0}_{q_0 = 1}, \underbrace{3x^2 \cdot f_0 + x^3 \cdot f'}_{q_1 = 0}, \underbrace{6x \cdot f_0 + 2 \cdot 3x^2 \cdot f' + x^3 \cdot f''}_{q_2 = 0}, \ldots)$$

This can easily be solved by forward substitution:

$$\begin{aligned}
1 &= x^3 \cdot f_0 && \rightarrow \quad f_0 = x^{-3} \\
0 &= 3x^2 \cdot f_0 + x^3 \cdot f' && \rightarrow \quad f' = -3x^{-4} \\
0 &= 6x \cdot f_0 + 2 \cdot 3x^2 \cdot f' + x^3 \cdot f'' && \rightarrow \quad f'' = 12x^{-5} \\
&\ldots
\end{aligned} \tag{3.158}$$

Using the same procedure for $f(x) = x^{-1}$ one obtains:

$$(x, 1, 0, 0, ..)^{-1} = (\frac{1}{x}. - \frac{1}{x^2}, \frac{2}{x^3}, \ldots) \tag{3.159}$$

For the function we have used before (3.145) :

$$f(x) = x^2 + \frac{1}{x} \tag{3.160}$$

and using (adding!) the expressions (3.156) and (3.159) one has:

$$(f_0, f', f'', f''') = (x^2 + \frac{1}{x}, \ 2x - \frac{1}{x^2}, \ 2 + \frac{2}{x^3}, \ ..) \tag{3.161}$$

3.7.6.5 Automatic Differentiation: More Variables

It can be extended to more variables x, y and a function $f(x, y)$:

$$x = (a, 1, 0, 0, 0 \ldots) \tag{3.162}$$

$$y = (b, 0, 1, 0, 0 \ldots) \tag{3.163}$$

and get (with more complicated multiplication rules):

$$f((x + dx), \ y + dy)) = \left(f, \frac{\partial f}{\partial x}, \frac{\partial f}{\partial y}, \frac{\partial^2 f}{\partial x^2}, \frac{\partial^2 f}{\partial x \partial y}, \ldots\right)(x, y) \tag{3.164}$$

3.7.6.6 Differential Algebra: Applications to Accelerators

Of course it is not the purpose of these tools to compute analytical expressions for the derivatives, the examples were used to demonstrate the techniques. The application of these techniques (i.e. Truncated Power Series Algebra [6, 21]) is schematically shown in Fig. 3.9. Given an algorithm, which may be a complex simulation program with several thousands of lines of code, we can use the techniques to "teach" the code how to compute the derivatives automatically.

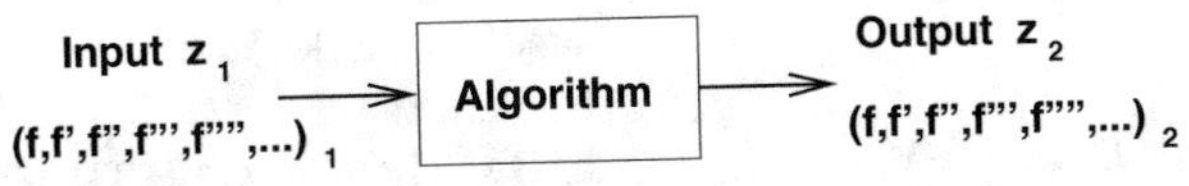

Input z 1 (f,f',f'',f''',f'''',...) 1 → Algorithm → Output z 2 (f,f',f'',f''',f'''',...) 2

Fig. 3.9 Schematic view of application of Truncated Power Series Algebra

When we push $f(x) = (a, 1, 0, 0, 0 \ldots)$ through the algorithm, using our rules, we get all derivatives around a, i.e. we get the Taylor coefficients and can construct the map!

What is needed is to replace the standard operations performed by the computer on real numbers by the algebra defined above. The maps are provided with the desired accuracy and to any order.

Given a Taylor series to high accuracy, the wanted information about stability of the system, global behaviour and optical parameters can be derived more easily. It should be stressed again that the origin is the underlying tracking code, just acting on different data types with different operations.

3.7.6.7 Differential Algebra: Simple Example

A simple example is shown below where the original "tracking code" is shown in the left column (DATEST1) and the corresponding modified code in the right column (DATEST2). The operation is rather trivial to demonstrate the procedure more easily. The code is written in standard FORTRAN 95 which allows operator overloading, but an object oriented language such as C++ or Python are obviously well suited for this purpose. Standard FORTRAN-95 is however more flexible overloading arbitrary operations. The DA-package used for demonstration only is loaded by the command *use myownda* in the code. To make the program perform the wanted operation we have to make two small modifications:

1. Replace the types *real* by the type *mytaylor* (defined in the package).
2. Add the "differential unit" (0, 1), the monomial in this implementation to the variable.

```
PROGRAM DATEST1
use my_own_da
real(8) x,z, dx
my_order=3
dx=0.0
x=3.141592653_8/6.0_8+dx
call track(x, z)
call print(z,6)
END PROGRAM DATEST1

SUBROUTINE TRACK(a, b)
use my_own_da
real(8) a,b
b = sin(a)
END SUBROUTINE TRACK
```

```
PROGRAM DATEST2
use my_own_da
type(my_taylor) x,z, dx
my_order=3
dx=1.0d0.mono.1    ! this is our (0,1)
x=3.141592653_8/6.0_8+dx
call track(x, z)
call print(z,6)
END PROGRAM DATEST2

SUBROUTINE TRACK(a, b)
use my_own_da
type(my_taylor) a,b
b = sin(a)
END SUBROUTINE TRACK
```

Running these two programs we get the results in the two columns below. In the left column we get the expected result from the real calculation of the expression

$\sin(\pi/6)) = 0.5$, while in the right column we get additional numbers sorted according to the array index.

```
(0,0) 0.50000000E+00
```

```
(0,0) 0.50000000E+00
(1,0) 0.86602540E+00
(0,1) 0.00000000E+00
(2,0) -0.25000000E+00
(0,2) 0.00000000E+00
(1,1) 0.00000000E+00
(3,0) -0.14433756E+00
(0,3) 0.00000000E+00
(2,1) 0.00000000E+00
(1,2) 0.00000000E+00
```

The inspection shows that these numbers are the coefficients of the Taylor expansion of $\sin(x)$ around $x = \pi/6$:

$$\sin(\frac{\pi}{6}+\Delta x) = \sin(\frac{\pi}{6})+\cos(\frac{\pi}{6})\Delta x^1 - \frac{1}{2}\sin(\frac{\pi}{6})\Delta x^2 - \frac{1}{6}\cos(\frac{\pi}{6})\Delta x^3 \tag{3.165}$$

We have indeed obtained the derivatives of our "algorithm" through the tracking code.

Some examples related to the analysis of accelerator physics lattices.

In example 1 a lattice with 8 FODO cells is constructed and the quadrupole is implemented as a thin lens "kick" in the center of the element. Note that the example is implemented in the horizontal and the longitudinal planes. For the second example an octupole kick is added to demonstrate the correct computation of the non-linear effect, i.e. the detuning with amplitude.

The procedure is:

1. Track through the lattice and get Taylor coefficients
2. Produce a map from the coefficients
3. Perform a Normal Form Analysis on the map

```
program ex1

use my_own_da
use my_analysis
type(my_taylor) z(3)
type(normalform) NORMAL
type(my_map) M,id

real(dp) L,DL,k1,k3,fix(3)
```

```
do j   = 1,8
z(1)   = z(1)+DL/2*z(2)
z(2)   = z(2)-kf*DL*z(1)/(1 + z(3))
z(1)   = z(1)+DL/2*z(2)

z(1)   =z(1)+LC*z(2)

z(1)      = z(1)+DL/2*z(2)
z(2)      = z(2)-kd*DL*z(1)/(1 + z(3))
```

```
! set up initial parameters
my_order=4 ! maximum order
4
fix=0.0 ! fixed point
id=1
z=fix+id

! set up lattice parameters
LC=62.5     ! half cell length
DL=3.0      ! quadrupole length
kf= 0.00295278     ! strength
kd=-0.00295278     ! strength
```

```
z(1)      = z(1)+DL/2*z(2)

z(1)      =z(1)+LC*z(2)
enddo

call print(z(1),6)
call print(z(2),6)

M=z ! overloads coefficient with the map
normal=m ! overloads map with normal
form

write(6,*) normal%tune, normal%dtune_da
end program ex1
```

```
(0,0,0) 0.9369211296691E-01
(0,0,1) -0.9649503806747E-01

(1,0,0) 0.9083165810508E-01
(0,1,0) 0.1667704101367E+03

(1,0,1) 0.1238115392391E+01
(0,1,1) -0.3527698956093E+02
(1,0,2) -0.1567062442887E+01
(0,1,2) 0.3478356898518E+02
(1,0,3) 0.1896009493384E+01
(0,1,3) -0.3429014840944E+02

(1,0,0) -0.5139797664004E-02
(0,1,0) 0.1572511594903E+01
(1,0,1) 0.1027959532801E-01
(0,1,1) -0.5648018984066E+00
(1,0,2) -0.1541939299201E-01
(0,1,2) 0.5570922019106E+00
(1,0,3) 0.2055919065602E-01
(0,1,3) -0.5493825054146E+01
```

From the elements in the Taylor expansion, the result for the matrix per cell:

$$\Delta x_f = 0.09083\Delta x_i + 166.77\Delta p_i$$

$$\Delta p_f = -0.00514\Delta x_i + 1.5725\Delta p_i$$

The output from the normal form analysis are (per cell!):
Tune = (0,0,0) = 0.093692
Chromaticity = (0,0,1) = -0.096495

```
program ex2

use my_own_da
```

```
do j   = 1,8
z(1)   = z(1)+DL/2*z(2)
z(2)   = z(2)-kf*DL*z(1)/(1 + z(3))
z(1)   = z(1)+DL/2*z(2)
```

```
use my_analysis
type(my_taylor) z(3)
type(normalform) NORMAL
type(my_map) M,id

real(dp) L,DL,k1,k3,fix(3)

! set up initial parameters
my_order=4 ! maximum order 4
fix=0.0 ! fixed point
id=1
z=fix+id

! set up lattice parameters
LC=62.5    ! half cell length
DL=3.0   ! quadrupole length
kf= 0.00295278    ! strength
kd=-0.00295278    ! strength
```

```
z(2)   =z(2)*k3*z(1)**3/1+z(3)  ! add
octupole kick
z(1)   =z(1)+LC*z(2)

z(1)      = z(1)+DL/2*z(2)
z(2)      = z(2)-kd*DL*z(1)/(1 + z(3))
z(1)      = z(1)+DL/2*z(2)

z(1)      =z(1)+LC*z(2)
enddo
call print(z(1),6)
call print(z(2),6)

M=z  !  overloads coefficient with the map
normal=m  ! overloads map with normal
form

write(6,*) normal%tune, normal%dtune_da
end program ex2
```

```
(0,0,0) 0.9369211296691E-01
(0,0,1) -0.9649503806747E-01

(2,0,0) 0.5383744464902E+02
(0,2,0) 0.5383744464902E+02
(0,0,2) 0.1009289258270E+00
(2,0,1) 0.2116575633218E+02
..........
(1,0,0) 0.9083165810508E-01
(0,1,0) 0.1667704101367E+03
(1,0,1) 0.1238115392391E+01
(0,1,1)-0.3527698956093E+02
(3,0,0)-0.1578216232118E+01
(2,1,0)-0.1429958442579E+02
(1,2,0)-0.4318760015031E+02

..........
(1,0,0)-0.5139797664004E-02
(0,1,0) 0.1572511594903E+01
(1,0,1) 0.1027959532801E-01
(0,1,1)-0.5648018984066E+00
(3,0,0)-0.1505298087837E-01
```

From the elements in the Taylor expansion, the result for the matrix per cell:

$$\Delta x_f = 0.09083\Delta x_i + 166.77\Delta p_i$$

$$\Delta p_f = -0.00514\Delta x_i + 1.5725\Delta p_i$$

The output from the normal form analysis are (per cell!):
Tune = (0,0,0) = 0.093692
Chromaticity = (0,0,1) = -0.096495

The added octupole kick results in a detuning with amplitude of dQ/dJ = 53.837

Defined assignments:

M = z, constructs a map M using the coefficients z
NORMAL = M, computes a normal form NORMAL using the map M

In FORTRAN95 derived "type" plays the role of "structures" in C, and NORMAL contains:

NORMAL%tune is the tune Q
NORMAL%dtune_da is the detuning with amplitude $\frac{dQ}{da}$

NORMAL%R, NORMAL%A, NORMAL%A**-1 are the matrices such that: $\text{M} = \text{A R A}^{-1}$

from the normal form transformation one obtains α, β, γ ...

Below a comparison is shown in Fig. 3.10 using the lattice function (e.g. β) obtained with this procedure and the optical functions from the corresponding MAD-X output.

One finds that $\beta_{max} \approx 300\,m$, $\beta_{min} \approx 170\,m$ and perfect agreement.

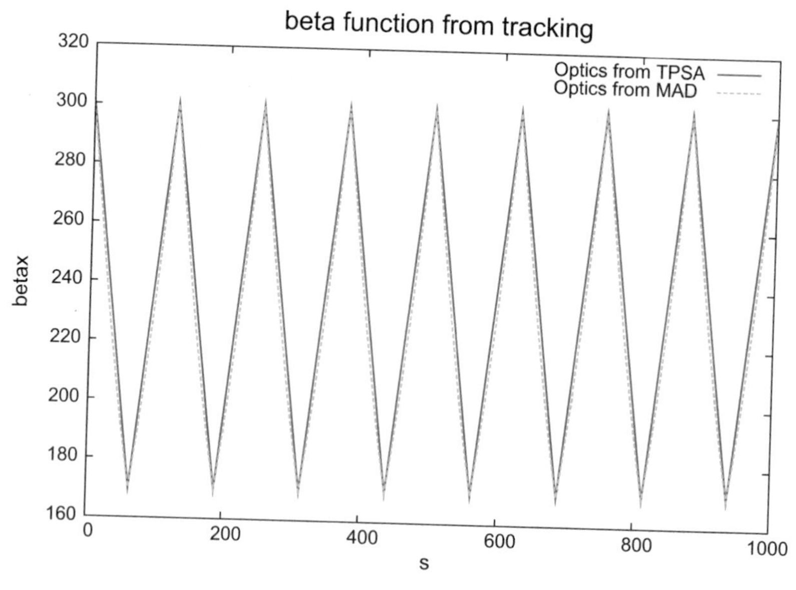

Fig. 3.10 Comparison: β-function from the model and the corresponding result from MAD-X

3.8 Beam Dynamics with Non-linearities

Following the overview of the evaluation and analysis tools, it is now possible to analyse and classify the behaviour of particles in the presence of non-linearities. The tools presented beforehand allow a better physical insight to the mechanisms leading to the various phenomena, the most important ones being:

- Amplitude detuning
- Excitation of non-linear resonances
- Reduction of dynamic aperture and chaotic behaviour.

This list is necessarily incomplete but will serve to demonstrate the most important aspects.

To demonstrate these aspects, we take a realistic case and show how the effects emerge automatically.

3.8.1 Amplitude Detuning

It was discussed in Sect. 3.7.5 that the one-turn-map can be transformed into a simpler map where the rotation is separated. A consequence of the non-linearities was that the rotation frequency becomes amplitude dependent to perform this transformation. Therefore the amplitude detuning is directly obtained from this normal form transformation.

3.8.1.1 Amplitude Detuning due to Non-linearities in Machine Elements

Non-linear elements cause an amplitude dependent phase advance. The computational procedure to derive this detuning was demonstrated in the discussion on normal for transformations in the case of an octupole Eqs. (3.125) and (3.129). This formalism is valid for any non-linear element.

Numerous other examples can be found in [6] and [5].

3.8.1.2 Amplitude Detuning due to Beam–Beam Effects

For the demonstration we use the example of a beam-beam interaction because it is a very complex non-linear problem and of large practical importance [7, 22].

In this simplest case of one beam-beam interaction we can factorize the machine in a linear transfer map $e^{:f_2:}$ and the beam-beam interaction $e^{:F:}$, i.e.:

$$e^{:f_2:} \cdot e^{:F:} = e^{:h:} \tag{3.166}$$

with

$$f_2 = -\frac{\mu}{2}(\frac{x^2}{\beta} + \beta p_x^2) \tag{3.167}$$

where μ is the overall phase, i.e. the tune Q multiplied by 2π, and β is the β-function at the interaction point. We assume the waist of the β-function at the collision point ($\alpha = 0$). The function $F(x)$ corresponds to the beam-beam potential (3.87):

$$F(x) = \int_0^x f(u)\mathrm{d}u \tag{3.168}$$

For a round Gaussian beam we use for $f(x)$ the well known expression:

$$f(x) = \frac{2Nr_0}{\gamma x}(1 - e^{\frac{-x^2}{2\sigma^2}}) \tag{3.169}$$

Here N is the number of particles per bunch, r_0 the classical particle radius, γ the relativistic parameter and σ the transverse beam size.

For the analysis we examine the invariant h which determines the one-turn-map (OTM) written as a Lie transformation $e^{:h:}$. The invariant h is the effective Hamiltonian for this problem.

As usual we transform to action and angle variables J and Φ, related to the variables x and p_x through the transformations:

$$x = \sqrt{2J\beta}\sin\Phi, \qquad p_x = \sqrt{\frac{2J}{\beta}}\cos\Phi \tag{3.170}$$

With this transformation we get a simple representation for the linear transfer map f_2:

$$f_2 = -\mu J \tag{3.171}$$

The function $F(x)$ we write as Fourier series:

$$F(x) \Rightarrow \sum_{n=-\infty}^{\infty} c_n(J)e^{in\Phi} \text{ with } c_n(J) = \frac{1}{2\pi}\int_0^{2\pi} e^{-in\Phi}F(x)\mathrm{d}\Phi \tag{3.172}$$

For the evaluation of (3.172) see [7]. We take some useful properties of Lie operators (e.g. [6, 7]):

$$:f_2: g(J) = 0, \qquad :f_2: e^{in\Phi} = in\mu e^{in\Phi}, \qquad g(:f_2:)e^{in\Phi} = g(in\mu)e^{in\Phi} \tag{3.173}$$

and the CBH-formula for the concatenation of the maps (3.92):

$$e^{:f_2:}e^{:F:} = e^{:h:} = \exp\left[: f_2 + \left(\frac{:f_2:}{1 - e^{-:f_2:}}\right) F + O(F^2) :\right] \tag{3.174}$$

which gives immediately for h:

$$h = -\mu J + \sum_n c_n(J) \frac{in\mu}{1 - e^{-in\mu}} e^{in\Phi} = -\mu J + \sum_n c_n(J) \frac{n\mu}{2\sin(\frac{n\mu}{2})} e^{(in\Phi + i\frac{n\mu}{2})} \tag{3.175}$$

Equation (3.175) is the beam-beam perturbed invariant to first order in the perturbation using (3.92).

From (3.175) we observe that for $\nu = \frac{\mu}{2\pi} = \frac{p}{n}$ resonances appear for all integers p and n when $c_n(J) \neq 0$.

Away from resonances a normal form transformation gives:

$$h = -\mu J + c_0(J) = const. \tag{3.176}$$

and the oscillating term disappears. The first term is the linear rotation and the second term gives the amplitude dependent tune shift (see (3.114)):

$$\Delta\mu(J) = -\frac{1}{2\pi}\frac{dc_0(J)}{dJ} \tag{3.177}$$

The computation of this tuneshift from the equation above can be found in the literature [7, 23].

3.8.1.3 Phase Space Structure

To demonstrate how this technique can be used to reconstruct the phase space structure in the presence of non-linearities, we continue with the very non-linear problem of the beam-beam interaction treated above. To test our result, we compare the invariant h to the results of a particle tracking program.

The model we use in the program is rather simple:

- linear transfer between interactions
- beam-beam kick for round beams
- compute action $J = \frac{\beta^*}{2\sigma^2}(\frac{x^2}{\beta^*} + p_x^2\beta^*)$
- compute phase $\Phi = \arctan(\frac{p_x}{x})$
- compare J with h as a function of the phase Φ

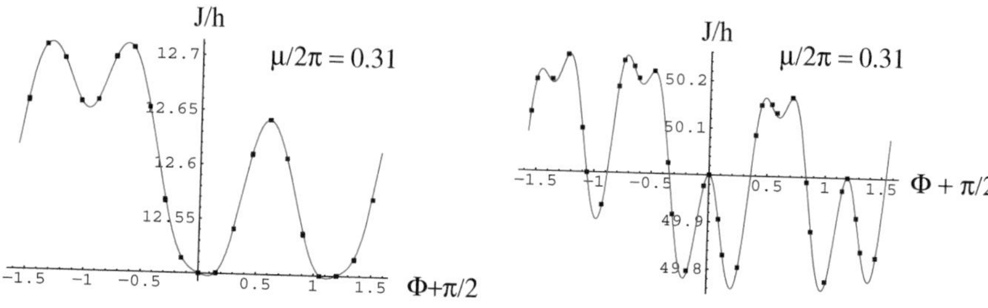

Fig. 3.11 Comparison: numerical and analytical model for one interaction point. Shown for $5\sigma_x$ (left) and $10\sigma_x$ (right). Full symbols from numerical model and solid lines from invariant (3.175)

The evaluation of the invariant (3.175) is done numerically with Mathematica. The comparison between the tracking results and the invariant h from the analytical calculation is shown in Fig. 3.11 in the (J,Φ) space. One interaction point is used in this comparison and the particles are tracked for 1024 turns. The symbols are the results from the tracking and the solid lines are the invariants computed as above. The two figures are computed for amplitudes of 5 σ and 10 σ. The agreement between the models is excellent. The analytic calculation was done up to the order $N = 40$. Using a lower number, the analytic model can reproduce the envelope of the tracking results, but not the details. The results can easily be generalized to more interaction points [22]. Close to resonances these tools can reproduce the envelope of the phase space structure [22].

3.8.2 *Non-linear Resonances*

Non-linear resonances can be excited in the presence of non-linear fields and play a vital role for the long term stability of the particles.

3.8.2.1 Resonance Condition in One Dimension

For the special case of the beam-beam perturbed invariant (3.175) we have seen that the expansion (3.175) diverges when the resonance condition for the phase advance is fulfilled, i.e.:

$$\nu = \frac{\mu}{2\pi} = \frac{p}{n} \tag{3.178}$$

The formal treatment would imply to use the n-turn map with the n-turn effective Hamiltonian or other techniques. This is beyond the scope of this handbook and can be found in the literature [6, 7]. We should like to discuss the consequences of resonant behaviour and possible applications in this section.

3.8.2.2 Driving Terms

The treatment of the resonance map is still not fully understood and a standard treatment using first order perturbation theory leads to a few wrong conclusions. In particular it is believed that a resonance cannot be excited unless a driving term for the resonance is explicitly present in the Hamiltonian. This implies that the related map must contain the term for a resonance in leading order to reproduce the resonance. This regularly leads to the conclusion that 3rd order resonances are driven by sextupoles, 4th order are driven by octupoles etc. This is only a consequence of the perturbation theory which is often not carried beyond leading order, and e.g. a sextupole can potentially drive resonances of any order. Such a treatment is valid only for special operational conditions such as resonant extraction where strong resonant effects can be well described by a perturbation theory. A detailed discussion of this misconception is given in [6]. A correct evaluation must be carried out to the necessary orders and the tools presented here allow such a treatment in an easier way.

3.8.3 *Chromaticity and Chromaticity Correction*

For reasons explained earlier, sextupoles are required to correct the chromaticities. In large machines and in particular in colliders with insertions, these sextupoles dominate over the non-linear effects of so-called linear elements.

3.8.4 *Dynamic Aperture*

Often in the context of the discussion of non-linear resonance phenomena the concept of *dynamic aperture* in introduced. This is the maximum stable oscillation amplitude in the transverse (x, y)-space due to non-linear fields. It must be distinguished from the physical aperture of the vacuum chamber or other physical restrictions such as collimators.

One of the most important tasks in the analysis of non-linear effects is to provide answers to the questions:

- Determination of the dynamic aperture
- Maximising the dynamic aperture

The computation of the dynamic aperture is a very difficult task since no mathematical methods are available to calculate it analytically except for the trivial cases. Following the concepts described earlier, the theory is much more complete from the simulation point of view. Therefore the standard approach to compute the dynamic aperture is done by numerical tracking of particles.

The same techniques can be employed to maximise the dynamic aperture, in the ideal case beyond the limits of the physical aperture. Usually one can define tolerances for the allowed multipole components of the magnets or the optimized parameters for colliding beams when the dominant non-linear effect comes from beam-beam interactions.

3.8.4.1 Long Term Stability and Chaotic Behaviour

In accelerators such as particle colliders, the beams have to remain stable for many hours and we may be asked to answer the question about stability for as many as 10^9 turns in the machine. This important question cannot be answered by perturbative techniques. In the discussion of Poincare surface-of-section we have tasted the complexity of the phase space topology and the final question is whether particles eventually reach the entire region of the available phase space.

It was proven by Kolmogorov, Arnol'd and Moser (KAM theorem) that for weakly perturbed systems invariant surfaces exist in the neighbourhood of integrable ones. Poincare gave a first hint that stochastic behaviour may be generated in non-linear systems. In fact, higher order resonances change the topology of the phase space and lead to the formation of island chains on an increasingly fine scale. Satisfactory insight to the fine structure of the phase space can only be gained with numerical computation. Although the motion near resonances may be stochastic, the trajectories are constrained by nearby KAM surfaces (at least in one degree of freedom) and the motion remains confined.

3.8.4.2 Practical Implications

In numerical simulations where particles are tracked for millions of turns we would like to determine the region of stability, i.e. dynamic aperture. Since we cannot track ad infinitum, we have to specify criteria whether a particle is stable or not. A straightforward method is to test the particle amplitudes against well defined apertures and declare a particle lost when the aperture is reached. A sufficient number of turns, usually determined by careful testing, is required with this method.

Usually this means to find the particle survival time as a function of the initial amplitude. In general the survival time decreases as the amplitude increases and should reach an asymptotic value at some amplitude. The latter can be identified as the dynamic aperture.

Other methods rely on the assumption that a particle that is unstable in the long term, exhibits features such as a certain amount of chaotic motion.

Typical methods to detect and quantify chaotic motion are:

- Frequency Map Analysis [24, 25].
- Lyapunov exponent [26].
- Chirikov criterion [27].

In all cases care must be taken to avoid numerical problems due to the computation techniques when a simulation over many turns is performed.

References

1. W. Herr, **Mathematical and Numerical Methods for Nonlinear Dynamics** Proc. CERN Accelerator School: **Advanced Accelerator Physics** (2013), published as CERN Yellow Report CERN-2014-009, arXiv:1601.07311.
2. E. Courant and H. Snyder, **Theory of the Alternating Gradient Synchrotron**, Ann. Phys. 3 (1958) 1.
3. A. Wolski, **Beam Dynamics in High Energy Particle Accelerators**, Imperial College Press (2014).
4. H. Wiedemann, **Particle Accelerator Physics—basic Principles and Linear Beam Dynamics**, Springer-Verlag (1993).
5. A. Chao and M. Tigner, **Handbook of Accelerator Physics and Engineering**, World Scientific (1998).
6. E. Forest, **Beam Dynamics**, Harwood Academic Publishers (1998).
7. A. Chao, **Lecture Notes on Topics in Accelerator Physics**, SLAC (2001).
8. A. Dragt, **Lie Methods for Nonlinear Dynamics with Applications to Accelerator Physics**, in preparation, Univ. of Maryland (Nov. 2019). https://www.physics.umd.edu/dsat/dsatliemethods.html.
9. E. Courant and R. Ruth, **Stability in Dynamical Systems**, Summer School on High Energy Particle Accelerators, Upton, New York, July 6–16, 1983.
10. A. Dragt and E. Forest, **Computation of Nonlinear Behaviour of Hamiltonian Systems using Lie Algebraic Methods**, J.Math.Phys. **24**, 2734 (1983).
11. E. Forest and R. Ruth, Physica D43, (1990) 105.
12. H. Yoshida, Phys. Lett A150 (1990) 262.
13. W. Herr, **Relativity**, lecture at CAS-CERN Accelerator School on "Introduction to accelerator Physics", Budapest, Hungary (2016).
14. S. Sheehy, **Motion of Particles in Electro-magnetic Fields**, lecture at CAS-CERN Accelerator School on "Introduction to accelerator Physics", Budapest, Hungary (2016).
15. A. Dragt, AIP Proc. 87, Phys. High Energy Accelerators, Fermilab, (1981) 147.
16. A. Dragt et al., Ann. Rev. Nucl. Part. Sci 38 (1988) 455.
17. A. Dragt et al, Phys. Rev. A45, (1992) 2572.
18. H. Goldstein, **Classical Mechanics**, Addison Wesley, (2001).
19. A. Dragt and J. Finn, J. Math. Phys., **17**, 2215 (1976); A. Dragt et al., Ann. Rev. Nucl. Part. Sci., **38**, 455 (1988).
20. H. Poincare, **Les Methods Nouvelles de la Mecanique Celeste**, Gauthier-Villars, Paris (1892).
21. M. Berz, Particle Accelerators 24, (1989) 109.
22. W. Herr, D. Kaltchev, **Effect of phase advance between interaction points in the LHC on the beam-beam interaction**, LHC Project Report 1082, unpublished, (2008).
23. T. Pieloni, **A Study of Beam-Beam Effects in Hadron Colliders with a Large Number of Bunches**, PhD thesis Nr. 4211, EPFL Lausanne, (2008).
24. H.S. Dumas, J. Laskar, Phys. Rev. Lett. 70 (1989) 2975.
25. J. Laskar, D. Robin, Particle Accelerators 54 (1996)183.
26. G. Benettin et al., Phys. Rev. A14 (1976) 2338.
27. B.V. Chirikov, At. Energ. 6, (1959) 630.

Chapter 4
Impedance and Collective Effects

E. Metral, G. Rumolo, and W. Herr

As the beam intensity increases, the beam can no longer be considered as a collection of non-interacting single particles: in addition to the "single-particle phenomena", "collective effects" become significant. At low intensity a beam of charged particles moves around an accelerator under the Lorentz force produced by the "external" electromagnetic fields (from the guiding and focusing magnets, RF cavities, etc.). However, the charged particles also interact with themselves (leading to space charge effects) and with their environment, inducing charges and currents in the surrounding structures, which create electromagnetic fields called wake fields. In the ultra-relativistic limit, causality dictates that there can be no electromagnetic field in front of the beam, which explains the term "wake". It is often useful to examine the frequency content of the wake field (a time domain quantity) by performing a Fourier transformation on it. This leads to the concept of impedance (a frequency domain quantity), which is a complex function of frequency. The charged particles can also interact with other charged particles present in the accelerator (leading to two-stream effects, and in particular to electron cloud effects in positron/hadron machines) and with the counter-rotating beam in a collider (leading to beam–beam effects). As the beam intensity increases, all these "perturbations" should be properly quantified and the motion of the charged particles will eventually still be governed by the Lorentz force but using the total electromagnetic fields, which are the sum of the external and perturbation fields. Note that in some cases a perturbative treatment is not sufficient and the problem has to be solved self consistently. These perturbations can lead to both incoherent (i.e. of

Coordinated by E. Metral

E. Metral (✉) · G. Rumolo · W. Herr
CERN (European Organization for Nuclear Research), Geneva, Switzerland
e-mail: Elias.Metral@cern.ch; Giovanni.Rumolo@cern.ch; Werner.Herr@cern.ch

S. Myers, H. Schopper (eds.), *Particle Physics Reference Library*,
https://doi.org/10.1007/978-3-030-34245-6_4

a single particle) and coherent (i.e. of the centre of mass) effects, in the longitudinal and in one or both transverse directions, leading to beam quality degradation or even partial or total beam losses. Fortunately, stabilising mechanisms exist, such as Landau damping, electronic feedback systems and linear coupling between the transverse planes (as in the case of a transverse coherent instability, one plane is usually more critical than the other).

Beam instabilities cover a wide range of effects in particle accelerators and they have been the subjects of intense research for several decades. As the machines performance was pushed new mechanisms were revealed and nowadays the challenge consists in studying the interplays between all these intricate phenomena, as it is very often not possible to treat the different effects separately [1, 2]. This field is still very active as can be revealed by the recent (and future) international workshops devoted to this subject [3–5].

This chapter is structured as follows: space charge is discussed in Sect. 4.1, wake fields (and related impedances) in Sect. 4.2, the induced coherent instabilities in Sect. 4.3 and the Landau damping mechanism in Sect. 4.4. The two-stream effects are analyzed in Sect. 4.5, concentrating mainly on electron cloud, while beam–beam effects are reviewed in Sect. 4.6, before concluding in Sect. 4.7 by the numerical modelling of collective effects.

4.1 Space Charge

E. Metral

4.1.1 Direct Space Charge

Two space charge effects are distinguished: the direct space charge and the indirect (or image) one ([6–9], and references therein). The direct space charge comes from the interaction between the particles of a single beam, without interaction with the surrounding vacuum chamber. Consider two particles with the same charge (for instance protons) in vacuum. They will feel two forces: the Coulomb repulsion (as they have the same charge) and the magnetic attraction (as they represent currents moving in the same direction, leading to an azimuthal magnetic field). Let's assume that particle 1 is moving with speed $v_1 = \beta_1 c$ with respect to the laboratory frame, with β the relativistic velocity factor and c the speed of light. In its rest frame, particle 1 produces only an electrostatic field, which can be computed, and applying the relativistic transformation of the electromagnetic fields between the rest and laboratory frames, the magnetic contribution can be obtained. Note that there is no magnetic contribution in the longitudinal plane ($B_s = 0$), which leads to the longitudinal Lorentz force $F_s = eE_s$, where e is the elementary charge, s the azimuthal coordinate and E_s the longitudinal electric field. The transverse (horizontal and vertical) Lorentz force on particle 2, moving with speed $v_2 = \beta_2 c$ with respect to the laboratory frame, is written

$$F_{x,y} = eE_{x,y} \begin{cases} (1 - \beta_1\beta_2), & \text{if 2 moves in same direction as 1} \\ (1 + \beta_1\beta_2), & \text{if 2 moves in opposite direction as 1} \end{cases}. \tag{4.1}$$

The first case corresponds to the space charge case where both particles move in the same direction, while the second corresponds to the beam–beam case (see Sect. 4.6) where the particles move in opposite direction. In both cases, the first term comes from the electric field while the second comes from the magnetic one. The main difference between the two regimes is that for the space charge case there is a partial compensation of the two forces, while for the beam–beam case the two forces add. The space charge force is maximum at low energy and vanishes at high energy, while the beam–beam force is maximum at high energy. Considering the space charge regime and assuming the same speed for both beams, the Lorentz force simplifies to

$$F_{x,y} = eE_{x,y}\left(1 - \beta^2\right) = e\frac{E_{x,y}}{\gamma^2}, \quad F_s = eE_s, \tag{4.2}$$

where γ is the relativistic mass factor. Assuming a circular beam pipe with radius b (which is important only for the computation of the longitudinal force) and applying Gauss's law, the electromagnetic fields can be computed for a bunch with Gaussian radial density (with rms $\sigma_x = \sigma_y = \sigma$) using the cylindrical coordinates (r, θ, s). The associated Lorentz forces are given by

$$F_r = \frac{e}{\gamma^2}E_r = \frac{e\lambda(z)}{2\pi\varepsilon_0\gamma^2}\left(\frac{1 - e^{-\frac{r^2}{2\sigma^2}}}{r}\right), \quad F_s = -\frac{e}{2\pi\varepsilon_0\gamma^2}\frac{d\lambda(z)}{dz}\int_{r'=r}^{b}\frac{1 - e^{-\frac{r'^2}{2\sigma^2}}}{r'}dr', \tag{4.3}$$

where $\lambda(z)$ is the longitudinal line density, $z = s - vt$ with t being the time, and ε_0 the vacuum permittivity. A first observation is that the space charge forces are highly nonlinear. Another important observation is that the radial force is proportional to the longitudinal density while the longitudinal one is proportional to the derivative of the longitudinal density. Linearizing both forces (for very small amplitudes where $r << \sigma$) leads to

$$F_r \approx \frac{e\lambda(z)}{2\pi\varepsilon_0\gamma^2}\frac{r}{2\sigma^2}, \quad F_s \approx -\frac{e}{4\pi\varepsilon_0\gamma^2}\frac{d\lambda(z)}{dz}\left[1 + 2\ln\left(\frac{b}{\sqrt{2}\sigma}\right)\right]. \tag{4.4}$$

This means that the transverse space charge force is linear for small amplitudes and defocusing. Due to the additional space charge force, e.g. the horizontal betatron tune will no longer be the unperturbed tune Q_{x0} but will be $Q_x = Q_{x0} + \Delta Q_x$, where ΔQ_x is the horizontal incoherent betatron tune shift. Similarly, the new synchrotron tune will be $Q_s = Q_{s0} + \Delta Q_s$, where ΔQ_s is the incoherent synchrotron tune shift. The betatron and synchrotron linearized incoherent space charge tune shifts are given by

$$\Delta Q_x^{\rm Lin} = -\frac{N_b r_p}{4\pi\beta\gamma^2\varepsilon_{x,\rm rms}^{\rm norm} B}, \qquad \Delta Q_s^{\rm Lin} = +\frac{\eta N_b e^2 R^2}{8\pi\sqrt{2\pi}\varepsilon_0 E_{\rm total}\beta^2\gamma^2\sigma_z^3 Q_{s0}}\left[1 + 2\ln\left(\frac{b}{\sqrt{2}\sigma}\right)\right], \tag{4.5}$$

where N_b is the number of protons in the bunch, r_p the classical proton radius, $\varepsilon_{x,\rm rms}^{\rm norm} = \beta\gamma\varepsilon_{x,\rm rms}$ the normalized rms horizontal emittance, with $\varepsilon_{x,\rm rms} = \sigma^2/\beta_x$ at a place of zero dispersion with β_x the horizontal betatron function, $B = \sqrt{2\pi}\sigma_z/2\pi R$ is the bunching factor (assuming a Gaussian longitudinal distribution with rms σ_z) with R the average machine radius, $\eta = \gamma_{\rm tr}^{-2} - \gamma^{-2}$ the slip factor (where $\gamma_{\rm tr}$ stands for γ at transition energy) and $E_{\rm total}$ is the total particles' energy. It is shown from Eq. (4.5) that the transverse betatron tune shift is always negative, revealing that the space charge force is always defocusing. Note that for the case of an ion with mass number A and charge state Z, the transverse tune shift has to be multiplied by Z^2/A.

Contrary to the transverse case, the longitudinal space charge is defocusing below transition (as $\eta < 0$) and focusing above (as $\eta > 0$). One can therefore already anticipate some longitudinal mismatch issues when crossing transition with high-intensity bunches, i.e. the bunch length will not be in equilibrium anymore and will oscillate inside the RF buckets. Such a case is depicted in Fig. 4.1(left) for the particular case of the high-intensity bunch in the CERN PS machine, which is sent to the nTOF (neutron Time-Of-Flight) experiment [10]. The computed quadrupolar oscillation is induced when transition is crossed and is a consequence of the longitudinal mismatch: below transition space charge is defocusing which reduces the bucket height and increases the bunch length, while above transition space charge is focusing which increases the bucket height and decreases the bunch length.

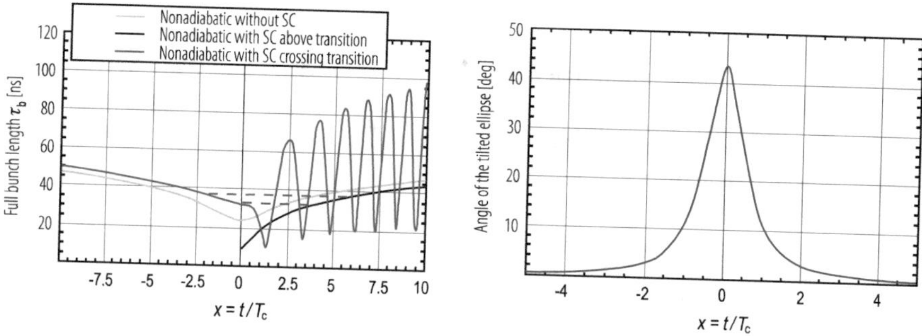

Fig. 4.1 (Left) Computation of the evolution of the full bunch length vs. time for the case of the nonadiabatic theory (the adiabatic theory is not valid anymore close to transition) with and without space charge, applied to the PS nTOF bunch [10]. T_c is the nonadiabatic time, equal to ~2 ms in the present case. (Right) Evolution of the angle of the tilted ellipse around transition (without space charge)

Therefore, there is an intensity-dependent step in the equilibrium bunch length at transition, which leads to a longitudinal mismatch and subsequent quadrupolar oscillations. If these bunch shape oscillations are not damped they will eventually result in filamentation and longitudinal emittance blow-up. It's worth mentioning that in presence of significant space charge, the minimum of bunch length is not reached right at transition anymore, but after about one nonadiabatic time T_c, i.e. after ~2 ms in the present case [11, 12]. The same kind of mechanism appears with the inductive part of the longitudinal machine impedance (see Sect. 4.2). The only difference is that in this case, the equilibrium bunch length is shorter below transition and longer above transition.

If transition crossing cannot be avoided, the γ_t jump is the only (known) method to overcome all the intensity limitations. It consists in an artificial increase of the transition crossing speed by means of fast pulsed quadrupoles. The idea is that quadrupoles at nonzero dispersion locations can be used to adjust the momentum compaction factor. The change in momentum compaction (called γ_t jump) depends on the unperturbed and perturbed dispersion functions at the kick-quadrupole locations. These schemes were pioneered by the CERN PS group [13–16]. Such a γ_t jump scheme makes it possible to keep the beam at a safe distance from transition, except for the very short time during which the transition region is crossed at a speed increased by one or two orders of magnitude. Looking at Fig. 4.1(left) clearly reveals why an asymmetric jump was proposed in the past [14] to damp the longitudinal quadrupolar oscillations arising from the space charge induced mismatch: the idea is to jump rapidly from an equilibrium bunch length below transition to the same value above. The amplitude of the jump is defined by the time needed to go to the same equilibrium bunch length above transition. The minimum amplitude of the jump corresponds to the case represented with the dashed blue line starting right at transition. However, in this case the initial longitudinal phase space ellipse is tilted (see Fig. 4.1(right)), while the final one is almost not, which is not ideal. One might want therefore to start the jump earlier, when the longitudinal phase space is almost not tilted, for instance at $x \approx -2$, which requires a larger jump (see the dashed orange line in Fig. 4.1(left)).

Coming back to the transverse space charge, in the case of an elliptical beam (instead of a round one), one has to replace $2\sigma_x{}^2$ by $\sigma_x(\sigma_x + \sigma_y)$ and $2\sigma_y{}^2$ by $\sigma_y(\sigma_x + \sigma_y)$. Furthermore, due to the nonlinear nature of the space charge, the tune shifts of the different particles will not be the same, which will lead to a tune spread: plotted in the tune diagram it is called a tune footprint. The latter has to be accommodated in the tune diagram, without crossing harmful resonance lines, which might lead to emittance growth and/or beam losses. The exact tune footprint depends on the distribution and to illustrate this effect we consider in the following a round beam with quasi-parabolic distribution function, whose particle density extends up to ~3.2σ [17, 18]. The corresponding horizontal 2D (i.e. neglecting the longitudinal distribution) space charge force is plotted in Fig. 4.2(left), and the tune footprint in Fig. 4.2(right). The unperturbed (low-intensity) working point is in the top right corner, the small-amplitude particles have the largest tune shifts while the largest amplitude particles have the smallest tune shifts. If the longitudinal

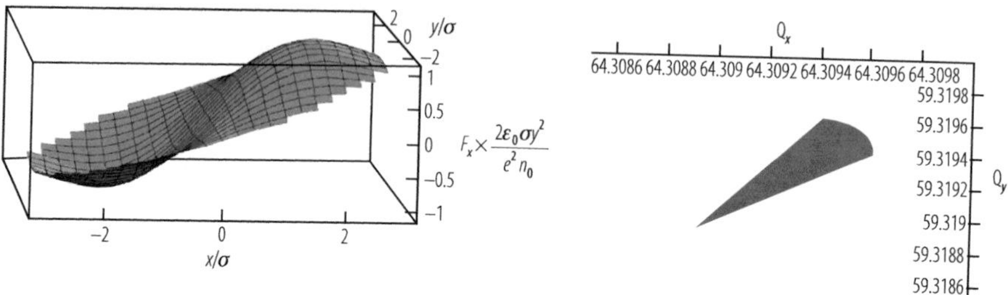

Fig. 4.2 Horizontal 2D (i.e. neglecting the longitudinal distribution) space charge force (left) and tune footprint for the case of the CERN LHC at injection, assuming the tunes in collision (64.31, 59.32). The parameter n_0 is the constant term in the particle density [18]

distribution is taken into account, the longitudinal variation (due to synchrotron oscillations) of the transverse space-charge force fills the gap until the low-intensity working point. However, it is interesting to plot it like this to clearly see the region occupied by the large synchrotron amplitude particles, because the interaction with a nonlinear resonance will depend on the overlapping position. Several possibilities exist with core emittance blow-up, creation of tails and/or beam losses. In particular, if the resonance interacts with the small amplitude particles, there could be a regime of loss-free (core-)emittance blow-up, while if the resonance interacts with the particles with large synchrotron amplitudes (i.e. if the resonance line is in the gap between the 2D tune footprint and the low-intensity working point) there could be a regime with continuous loss due to the trapping–detrapping mechanisms, as observed both in the PS [19] and at SIS18 [20].

Finally, another space charge mechanism, which could be important in high-intensity synchrotrons with unsplit transverse tunes (i.e. having the same integer) is the Montague resonance which can lead to emittance transfer from one plane to the other and might lead to losses if the beam fills the aperture [21, 22].

4.1.2 *Indirect Space Charge*

In the case of a beam off-axis in a perfectly conducting circular beam pipe (with radius b), a coherent (or dipolar, i.e. of the centre of mass) force arises, which can be found by using the method of the images (to satisfy the boundary condition on a perfect conductor, i.e. of a vanishing tangential electrical field). The electric field is always assumed to be non-penetrating. However, for the magnetic field, the situation is more complicated as it may or may not penetrate the vacuum chamber: the high-frequency components, called "ac" will not penetrate, while the low-frequency ones, called "dc" will penetrate and form images on the magnet pole faces (if there are some; otherwise they will go to infinity and will not act back on the beam). In the case of a non-penetrating "ac" magnetic field, one finally obtains (keeping only the linear terms, i.e. $\overline{x} \ll b$ and $\overline{y} \ll b$, where $\overline{x}$ and $\overline{y}$ are the transverse displacements of the centre of mass):

$$F_x = \Lambda_c \overline{x}, \quad F_y = \Lambda_c \overline{y}, \quad \Lambda_c = \frac{\lambda e}{2\pi \varepsilon_0 \gamma^2 b^2}. \tag{4.6}$$

It can be seen that the transverse coherent space charge force of Eq. (4.6) is similar to the transverse incoherent space charge force of Eq. (4.4, left): $2\sigma^2$ has been replaced by b^2.

The same analysis can be performed in the case of two infinite (horizontal) parallel plates spaced by $2h$ and the results are the following (assuming that the transverse beam sizes are much smaller than h, assuming only the "ac" magnetic part and keeping only the linear terms)

$$F_x = \Lambda_c \left(\frac{\pi^2}{24} \overline{x} - \frac{\pi^2}{24} x \right), \quad F_y = \Lambda_c \left(\frac{\pi^2}{12} \overline{y} + \frac{\pi^2}{24} y \right). \tag{4.7}$$

Therefore, compared to the circular case, the coherent force is smaller by $\pi^2/24 \approx 0.4$ in the horizontal plane and $\pi^2/12 \approx 0.8$ in the vertical one. Furthermore, there is a second incoherent (or quadrupolar, as it is linear with the particle position) term with opposite sign in both planes. The coefficients are linked to the Laslett coefficients usually used in the literature [23], and they are the same as the ones obtained by Yokoya [24] in the case of a resistive beam pipe under some assumptions (see Sect. 4.2). General formulae exist for the "real" tune shifts of coasting or bunched beams in pipes with different geometries, considering both the "ac" and "dc" magnetic parts and can be found for instance in Refs. [6, 7].

4.2 Wake Fields and Impedances

E. Metral

If the wall of the beam pipe is perfectly conducting and smooth, as it was the case in the previous section, a ring of negative charges is formed on the walls of the beam pipe where the electric field ends, and these induced charges travel at the same pace with the particles, creating the so-called "image" (or induced) current, which leads to real tune shifts. However, if the wall of the beam pipe is not perfectly conducting or contains discontinuities, the movement of the induced charges will be slowed down, thus leaving electromagnetic fields (which are proportional to the beam intensity) mainly behind: this is why these electromagnetic fields are called wake fields. The latter will create complex tune shifts leading to instabilities (see Sect. 4.3). What needs to be computed are the wake fields at the distance $z = s - vt$ behind the source particle (which is at position $s_{\mathrm{source}} = vt$; with this convention, one has $z < 0$) and their effects on the test or witness particles that compose the beam. The computation of these wake fields is quite involved and two fundamental approximations are introduced:

1. *The rigid-beam approximation*: The beam traverses a piece of equipment rigidly, i.e. the wake field perturbation does not affect the motion of the beam during the traversal of the impedance. The distance z of the test particle behind some source particle does not change.
2. *The impulse approximation*: As the test particle moves at the fixed velocity $v = \beta c$ through a piece of equipment, the important quantity is the impulse (and not the force) given by

$$\Delta \boldsymbol{p}(x, y, z) = \int_{-\infty}^{+\infty} dt \boldsymbol{F}(x, y, s = z + \beta ct, t) = \int_{-\infty}^{+\infty} dte (\boldsymbol{E} + \boldsymbol{v} \times \boldsymbol{B}), \quad (4.8)$$

where vectors are designated by boldtype letters. Starting from the four Maxwell equations for a particle in the beam, it can be shown that for a constant β (which does not need to be 1) [9]

$$\nabla \times \Delta \boldsymbol{p}(x, y, z) = 0, \quad (4.9)$$

which is known as Panofsky-Wenzel theorem. This relation is very general, as no boundary conditions have been imposed. Only the two fundamental approximations have been made. Another important relation can be obtained when $\beta = 1$, taking the divergence of the impulse, which is

$$\nabla_\perp \cdot \Delta \boldsymbol{p}_\perp = 0. \quad (4.10)$$

Considering the case of a cylindrically symmetric chamber (using the cylindrical coordinates r, θ, z), yields the following three equations from Panofsky-Wenzel theorem

$$\frac{1}{r}\frac{\partial \Delta p_z}{\partial \theta} = \frac{\partial \Delta p_\theta}{\partial z}, \quad \frac{\partial \Delta p_r}{\partial z} = \frac{\partial \Delta p_z}{\partial r}, \quad \frac{\partial (r \Delta p_\theta)}{\partial r} = \frac{\partial \Delta p_r}{\partial \theta}. \quad (4.11)$$

The fourth relation when $\beta = 1$ writes

$$\frac{\partial (r \Delta p_r)}{\partial r} = -\frac{\partial \Delta p_\theta}{\partial \theta}. \quad (4.12)$$

Consider now as a source charge density a macro-particle of charge $Q = N_b e$ moving along the pipe (in the s-direction) with an offset $r = a$ in the $\vartheta = 0$ direction and with velocity $v = \beta c$

$$\begin{aligned} \rho(r, \vartheta, s; t) &= \frac{Q}{a}\delta(r-a)\,\delta_p(\vartheta)\,\delta(s - vt) \\ &= \sum_{m=0}^{\infty} \frac{Q_m \cos(m\vartheta)}{\pi a^{m+1}(1+\delta_{m0})}\delta(r-a)\,\delta(s-vt) = \sum_{m=0}^{\infty} \rho_m, \end{aligned} \quad (4.13)$$

with $\boldsymbol{J}_m = \rho_m \boldsymbol{v}$, where $Q_m = Qa^m$ and δ is the Dirac function. In this case the whole solution can be written as, for $m \geq 0$ and $\beta = 1$ (with q the charge of the test particle and L the length of the structure)

$$\begin{aligned} v\Delta p_s(r,\theta,z) &= \int_0^L F_s ds = -qQa^m r^m \cos m\theta \, W_m'(z) \\ v\Delta p_r(r,\theta,z) &= \int_0^L F_r ds = -qQa^m m r^{m-1} \cos m\theta \, W_m(z), \\ v\Delta p_\theta(r,\theta,z) &= \int_0^L F_\theta ds = qQa^m m r^{m-1} \sin m\theta \, W_m(z). \end{aligned} \quad (4.14)$$

The function $W_m(z)$ is called the transverse wake function (whose unit is $\mathrm{VC}^{-1}m^{-2m}$) and $W_m{}'(z)$ is called the longitudinal wake function (whose unit is $\mathrm{VC}^{-1}m^{-2m+1}$) of azimuthal mode m. They describe the shock response of the vacuum chamber environment to a δ-function beam which carries a mth moment. The integrals (on the left) are called wake potentials. The longitudinal wake function for $m = 0$ and transverse wake function for $m = 1$ are therefore given by

$$\begin{aligned} W_0'(z) &= -\frac{1}{qQ}\int_0^L F_s ds = -\frac{1}{Q}\int_0^L E_s ds, \\ W_1(z) &= -\frac{1}{qQa}\int_0^L F_x ds = -\frac{1}{Qa}\int_0^L \left(E_x - vB_y\right) ds. \end{aligned} \quad (4.15)$$

The Fourier transform of the wake function is called the impedance. The idea of representing the accelerator environment by an impedance was introduced by Vaccaro [25] and Sessler [26]. As the conductivity, permittivity and permeability of a material depend in general on frequency, it is usually better (or easier) to treat the problem in the frequency domain, i.e. compute the impedance instead of the wake function. It is also easier to treat the case $\beta \neq 1$. Then, an inverse Fourier transform is applied to obtain the wake function in the time domain. The different relations linking the wake functions and the impedances are given by (with $k = \omega/v$, $\omega = 2\pi f$ with f the frequency, and j the imaginary unit)

$$\begin{aligned} Z_m^{\parallel}(\omega) &= \int_{-\infty}^{\infty} W_m'(z) e^{jkz} \frac{dz}{v} = \int_{-\infty}^{\infty} W_m'(t) e^{jks} e^{-j\omega t} dt, \\ Z_m^{\perp}(\omega) &= -j\int_{-\infty}^{\infty} W_m(z) e^{jkz} \frac{dz}{v} = -j\int_{-\infty}^{\infty} W_m(t) e^{jks} e^{-j\omega t} dt. \\ W_m'(z) &= \frac{1}{2\pi}\int_{-\infty}^{\infty} Z_m^{\parallel}(\omega)\, e^{-jkz} d\omega = \frac{1}{2\pi}\int_{-\infty}^{\infty} Z_m^{\parallel}(\omega)\, e^{-jks} e^{j\omega t} d\omega, \\ W_m(z) &= \frac{j}{2\pi}\int_{-\infty}^{\infty} Z_m^{\perp}(\omega)\, e^{-jkz} d\omega = \frac{j}{2\pi}\int_{-\infty}^{\infty} Z_m^{\perp}(\omega)\, e^{-jks} e^{j\omega t} d\omega. \end{aligned} \quad (4.16)$$

The unit of the longitudinal impedance $Z_m^{\parallel}(\omega)$ is Ωm^{-2m} while the unit of the transverse impedance $Z_m^{\perp}(\omega)$ is Ωm^{-2m+1}. Furthermore, two important properties of impedances can be derived. The first is a consequence of the fact that the wake function is real, which leads to

$$\left[Z_m^{\parallel}(\omega)\right]^* = Z_m^{\parallel}(-\omega), \quad -\left[Z_m^{\perp}(\omega)\right]^* = Z_m^{\perp}(-\omega), \tag{4.17}$$

where * stands for the complex conjugate. The second is a consequence of Panofsky-Wenzel theorem

$$Z_m^{\parallel}(\omega) = k Z_m^{\perp}(\omega). \tag{4.18}$$

Another interesting property of the impedances is the directional symmetry (Lorentz reciprocity theorem): the same impedance is obtained from both sides if the entrance and exit are the same.

A more general definition of the impedances (still for a cylindrically symmetric structure) is the following

$$Z_m^{\parallel}(\omega) = -\frac{1}{Q_m^2}\int dV E_m^{\parallel} J_m^*, \quad Z_m^{\perp}(\omega) = -\frac{1}{kQ_m^2}\int dV E_m^{\parallel} J_m^*, \tag{4.19}$$

where $dV = rdrd\vartheta ds$. For the previous ring-shape source it yields

$$\begin{aligned} Z_0^{\parallel}(\omega) &= -\tfrac{1}{Q_0}\textstyle\int_0^L ds E_s(r=a)\, e^{jks}, \\ Z_1^{\perp}(\omega) &= -\tfrac{L}{k\pi a Q_1}\textstyle\int_0^{2\pi} d\vartheta\, E_s(r=a,\vartheta,s)\cos\vartheta\, e^{jks}. \end{aligned} \tag{4.20}$$

The situation is more involved in the case of non axi-symmetric structures (due in particular to the presence of the quadrupolar wake field, already discussed in Sect. 4.1) and for $\beta \neq 1$, as in this case some electromagnetic fields also appear in front of the source particle. In the case of axi-symmetric structures, a current density with some azimuthal Fourier component creates electromagnetic fields with the same azimuthal Fourier component. In the case of non axi-symmetric structures, a generalized notion of impedances was introduced by Tsutsui [27], where a current density with some azimuthal Fourier component may create an electromagnetic field with various different azimuthal Fourier components. If the source particle 1 and test particle 2 have the same charge q, and in the ultra-relativistic case, the transverse wake potentials can be written (taking into account only the linear terms with respect to the source and test particles and neglecting the constant, coupling and high order terms) [28]

$$\begin{aligned} \textstyle\int_0^L F_x ds &= -q^2\left[x_1 W_x^{\text{driving}}(z) - x_2 W^{\text{detuning}}(z)\right], \\ \textstyle\int_0^L F_y ds &= -q^2\left[y_1 W_y^{\text{driving}}(z) + y_2 W^{\text{detuning}}(z)\right], \end{aligned} \tag{4.21}$$

where the driving term is used here instead of dipolar and detuning instead of quadrupolar (or incoherent). In the frequency domain, Eq. (4.21) leads to the following generalized impedances

$$Z_x\,[\Omega] = x_1 Z_x^{\text{driving}} - x_2 Z^{\text{detuning}}, \\ Z_y\,[\Omega] = y_1 Z_x^{\text{driving}} + y_2 Z^{\text{detuning}}. \tag{4.22}$$

Note that in the case $\beta \neq 1$, another quadrupolar term is found [29].

From Eqs. (4.21) and (4.22), the procedure to simulate or measure the driving and detuning contributions can be deduced. In the time domain, using some time-domain electromagnetic codes like for instance CST Particle Studio [30], the driving and detuning contributions can be disentangled. A first simulation with $x_2 = 0$ gives the dipolar part while a second one with $x_1 = 0$ provides the quadrupolar part. It should be noted that if the simulation is done with $x_1 = x_2$, only the sum of the dipolar and quadrupolar parts is obtained. The situation is more involved in the frequency domain, which is used for instance for impedance measurements on a bench [31]. Two measurement techniques can be used to disentangle the transverse driving and detuning impedances, which are both important for the beam dynamics (this can also be simulated with codes like Ansoft-HFSS [32]). The first uses two wires excited in opposite phase (to simulate a dipole), which yields the transverse driving impedance only. The second consists in measuring the longitudinal impedance, as a function of frequency, for different transverse offsets using a single displaced wire. The sum of the transverse driving and detuning impedances is then deduced applying the Panofsky-Wenzel theorem in the case of top/bottom and left/right symmetry [33]. Subtracting finally the transverse driving impedance from the sum of the transverse driving and detuning impedances obtained from the one-wire measurement yields the detuning impedance only. If there is no top/bottom or left/right symmetry the situation is more involved [34].

Both longitudinal and transverse resistive-wall impedances were already calculated 40 years ago by Laslett, Neil and Sessler [35]. However, a new physical regime was revealed by the CERN LHC collimators. A small aperture paired with a large wall thickness asks for a different physical picture of the transverse resistive-wall effect from the classical one. The first unstable betatron line in the LHC is around 8 kHz, where the skin depth for graphite (whose measured isotropic DC resistivity is 10 μΩm) is 1.8 cm. It is smaller than the collimator thickness of 2.5 cm. Hence one could think that the resistive thick-wall formula would be about right. In fact it is not. The resistive impedance is about two orders of magnitude lower at this frequency, as can be seen on Fig. 4.3. A number of papers have been published on this subject using the field matching technique starting from the Maxwell equations and assuming a circular geometry [36–40]. New results have been also obtained for flat chambers, extending the (constant) Yokoya factors to frequency and material dependent ones [41], as was already found with some simplified kicker impedance models [42, 43]. Note that the material resistivity may vary with the magnetic field

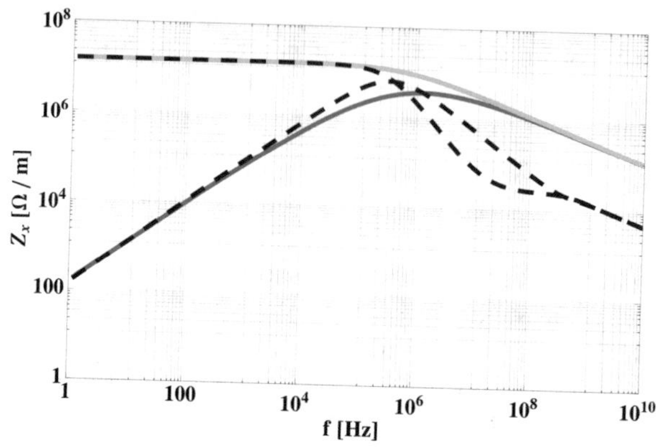

Fig. 4.3 Horizontal (driving) impedance of a cylindrical one-meter long LHC collimator (even if in reality the LHC collimators are composed of two parallel plates), with $b = 2$ mm and $\rho = 10$ μΩm. The real part is in red while the imaginary part is in green (note that in the classical thick-wall regime, the real and imaginary parts are equal). The dashed curves correspond to the case with a copper coating of 5 μm [1]

through the magneto-resistance and the surface impedance can also increase due to the anomalous skin effect ([44] and references therein).

In the case of a cavity, an equivalent RLC circuit can be used with R_s the longitudinal shunt impedance, C the capacity and L the inductance. In a real cavity, these three parameters cannot be separated easily and some other related parameters are used, which can be measured directly such as the resonance (angular) frequency $\omega_r = 1/\sqrt{LC}$, the quality factor $Q = R_s\sqrt{C/L} = R_s/(L\omega_r) = R_s C\omega_r$ and the damping rate $\alpha = \omega_r/(2Q)$. When $Q = 1$, the resonator impedance is called "broad-band", and this model was extensively used in the past in many analytical computations. The longitudinal and transverse impedances and wake functions (with $R_\perp$ the transverse shunt impedance) are given by [45] (see Fig. 4.4):

$$Z_m^{\parallel}(\omega) = \frac{R_s}{1+jQ\left(\frac{\omega}{\omega_r}-\frac{\omega_r}{\omega}\right)},\; W_m^{\parallel}(t) = \frac{\omega_r R_s}{Q}\mathrm{e}^{-\alpha t}\left[\cos\left(\overline{\omega}_r t\right) - \frac{\alpha}{\overline{\omega}_r}\sin\left(\overline{\omega}_r t\right)\right]$$
$$Z_m^{\perp}(\omega) = \frac{\omega_r}{\omega}\frac{R_\perp}{1+jQ\left(\frac{\omega}{\omega_r}-\frac{\omega_r}{\omega}\right)},\; W_m^{\perp}(t) = \frac{\omega_r^2 R_\perp}{Q\overline{\omega}_r}\,\mathrm{e}^{-\alpha t}\sin\left(\overline{\omega}_r t\right),\; \overline{\omega}_r = \omega_r\sqrt{1-\frac{1}{4Q^2}}. \tag{4.23}$$

Finally, all the transverse impedances (dipolar or driving and quadrupolar or detuning) should be weighted by the betatron function at the location of the impedances, as this is what matters for the effect on the beam, i.e. for the beam dynamics. Furthermore, all the weighted impedances can be summed and lumped at a single location around the ring (as the betatron phase advance does not play a role [46]) such that the transverse impedance-induced instabilities can be studied by considering only one interaction per turn, as it was confirmed in the past by performing macro-particle tracking simulations and comparing the cases of

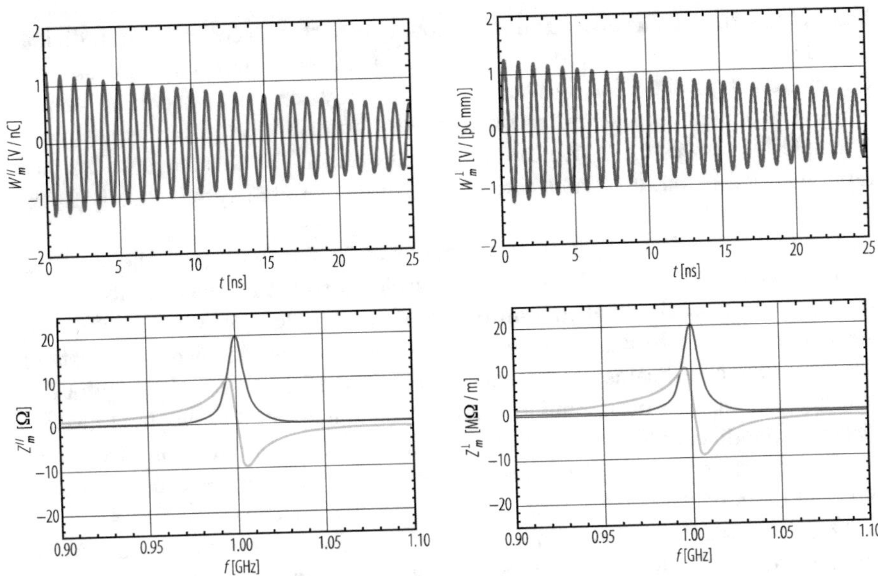

Fig. 4.4 Longitudinal and transverse impedances and wake functions, in the case of resonator impedances ($f_r = 1$ GHz, $Q = 100$, $R_s = 20\ \Omega$, and $R_\perp = 20\ \text{M}\Omega/\text{m}$)

distributed kickers in the CERN SPS with the corresponding lumped impedance: exactly the same result was obtained [47].

4.3 Coherent Instabilities

E. Metral

The wake fields can influence the motion of trailing particles, in the longitudinal and in one or both transverse directions, leading to energy loss, beam instabilities, or producing undesirable secondary effects such as excessive heating of sensitive components at or near the chamber wall. Therefore, in practice the elements of the vacuum chamber should be designed to minimise the self-generated electromagnetic fields. For example, chambers with different cross-sections should be connected with tapered transitions; bellows need to be separated from the beam by shielding; plates should be grounded or terminated to avoid reflections; high-resistivity materials should be coated with a thin layer of very good conductor (such as copper) when possible; etc.

Two approaches are usually used to deal with collective instabilities. One starts from the single-particle equation while the other solves the Vlasov equation, which is nothing else but an expression for the Liouville conservation of phase-space density seen by a stationary observer. In the second approach, the motion of the beam is described by a superposition of modes, rather than a collection of individual particles. The detailed methods of analysis in the two approaches are different, the particle representation is usually conveniently treated in the time domain, while in the mode representation the frequency domain is more convenient, but in principle they necessarily give the same final results. The advantage of the mode representation is that it offers a formalism that can be used systematically to treat the instability problem.

The first formalism was used by Courant and Sessler to describe the transverse coupled-bunch instabilities [48]. In most accelerators, the RF acceleration mechanism generates an azimuthal non-uniformity of the particle density and consequently the work of Laslett, Neil and Sessler for continuous beams [35] is not applicable in the case of bunched beams. Courant and Sessler studied the case of rigid (point-like) bunches, i.e. bunches oscillating as rigid units, and they showed that the transverse electromagnetic coupling of bunches of particles with each other can lead (due to the imperfectly conducting vacuum chamber walls) to a coherent instability. The physical basis of the instability is that in a resistive vacuum tank, fields due to a particle decay only very slowly in time after the particle has left (this leads to a long-range interaction). The decay can be so slow that when a bunch returns after one (or more) revolutions it is subject to its own residual wake field which, depending upon its phase relative to the wake field, can lead to damped or anti-damped transverse motion. For M equi-populated equi-spaced bunches, M coupled-bunch mode numbers exist ($n = 0, 1, \ldots, M - 1$), characterized by the integer number of waves of the coherent motion around the ring. Therefore the coupled-bunch mode number resembles the azimuthal mode number for coasting beams, except that for coasting beams there is an infinite number of modes. The bunch-to-bunch phase shift $\Delta\phi$ is related to the coupled-bunch mode number n by $\Delta\phi = 2\pi n/M$.

Pellegrini [49] and, independently, Sands [50, 51] then showed that short-range wake fields (i.e. fields that provide an interaction between the particles of a bunch but have a negligible effect on subsequent passages of the bunch or of other bunches in the beam) together with the internal circulation of the particles in a bunch can cause internal coherent modes within the bunch to become unstable. The important point here is that the betatron phase advance per unit of time (or betatron frequency) of a particle depends on its instantaneous momentum deviation (from the ideal momentum) in first order through the chromaticity and the slip factor. Considering a non-zero chromaticity couples the betatron and synchrotron motions, since the betatron frequency varies around a synchrotron orbit. The betatron phase varies linearly along the bunch (from the head) and attains its maximum value at the tail. The total betatron phase shift between head and tail is the physical origin of the head tail instability. The head and the tail of the bunch oscillate therefore with a phase difference, which reduces to rigid-bunch oscillations only in the limit of zero chromaticity. A new (within-bunch) mode number $m = \ldots, -1, 0, 1, \ldots$, also called head–tail (or azimuthal) mode number, was introduced. This mode describes the number of betatron wavelengths (with sign) per synchrotron period.

The work of Courant and Sessler, or Pellegrini and Sands, was done for particular impedances and oscillation modes. Using the Vlasov formalism, Sacherer unified

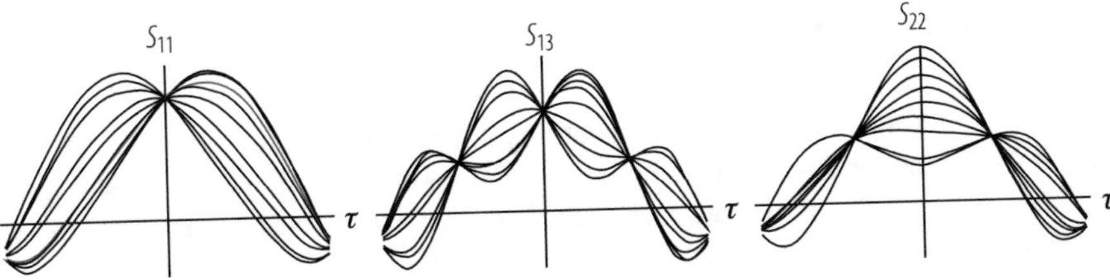

Fig. 4.5 Signal at the pick-up electrode for three different modes shown for several superimposed turns (the red line corresponds to one particular turn), for the case of the parabolic amplitude distribution and a constant inductive impedance (exhibiting therefore no growing oscillations!)

the two previous approaches, introducing a third mode number $q \equiv |m| + 2k$ (with $0 \leq k < +\infty$), called radial mode number, which comes from the distribution of synchrotron oscillation amplitudes [52, 53]. It can be obtained by superimposing several traces of the directly observable average displacement along the bunch at a particular pick-up. The number of nodes is the mode number q (see Figs. 4.5 and 4.8). The advantage of this formalism is that it is valid for generic impedances and any high order head–tail modes. This approach starts from a distribution of particles (split into two different parts, a stationary distribution and a perturbation), on which Liouville theorem is applied. After linearization of the Vlasov equation, one ends up with Sacherer's integral equation or Laclare's eigenvalue problem to be solved [53]. Because there are two degrees of freedom (phase and amplitude), the general solution is a twofold infinity of coherent modes of oscillation (m, q). At sufficiently low intensity, only the most coherent mode (largest value for the coherent tune shift) is generally considered, leading to the classical Sacherer's formulae in both transverse and longitudinal planes. Note that contrary to the space charge case, these tune shifts are now complex, the imaginary part being linked to the instability growth rate. For protons a parabolic density distribution is generally assumed and the corresponding oscillation modes are sinusoidal (or close to it). For electrons, the distribution is usually Gaussian, and the oscillation modes are described in this case by Hermite polynomials. In reality, the oscillation modes depend both on the distribution function and the impedance, and can only be found numerically by solving the (infinite) eigenvalue problem. However, the mode frequencies are usually not very sensitive to the accuracy of the eigenfunctions. Similar results are obtained for the longitudinal plane.

It is worth mentioning that the CERN ISR suffered from a beam instability brought about by beams having different revolution frequencies. They could be in the same vacuum chamber or coupled by the beam–beam effect. The name of "overlap knock-out" [54] has been given to this phenomenon by which the stack is subjected to transverse kicks from the bunches. This produces blow-up of the stacked beam when the longitudinal frequency spectrum of the bunches overlaps with the betatron frequency spectrum of the coasting stacked beam. Similar problems limit the energy range of RHIC and proton lead in LHC [55].

4.3.1 Longitudinal

The most fundamental longitudinal instability encountered in circular accelerators is called the Robinson instability. The (Radio-Frequency) RF frequency accelerating cavities in a circular accelerator are tuned so that the resonant frequency of the fundamental mode is very close to an integral multiple of the revolution frequency of the beam. This necessarily means that the wake field excited by the beam in the cavities contains a major frequency component near a multiple of the revolution frequency. The exact value of the resonant frequency relative to the multiple of the revolution frequency is of critical importance for the stability of the beam. Above the transition energy, the beam will be unstable if the resonant frequency is slightly above it and stable if slightly below. This is the opposite below transition. This instability mechanism was first analyzed by Robinson [56]. Physically, the Robinson instability comes from the fact that the revolution frequency of an off-momentum beam is not given by the on-momentum revolution frequency, but by a quantity slightly different, depending on both the slip factor and the energy deviation.

Let's assume in the following that the Robinson criterion is met. A bunch is longitudinally stable if the longitudinal profile observed at a wall-current monitor is constant turn after turn and it is unstable if the longitudinal profile is not constant turn after turn. In the case of instability, the way the longitudinal profile oscillates gives some information about the type of instabilities. This was studied in detail by Laclare [53], who explained theoretically such pictures of "longitudinal (single-bunch) instability" starting from the single-particle longitudinal signal at a pick-up electrode (assuming infinite bandwidth). The current signal induced by the test particle is a series of impulses delivered on each passage

$$s_z\left(t,\vartheta\right)=e\sum_{k=-\infty}^{k=+\infty}\delta\left(t-\tau-\frac{\vartheta}{\Omega_0}-\frac{2k\pi}{\Omega_0}\right),\tag{4.24}$$

where τ is the time interval between the passage of the synchronous particle and the test particle, for a fixed observer at azimuthal position ϑ and Ω_0 is the angular revolution frequency. In the frequency domain, the single-particle spectrum is therefore a line spectrum at (angular) frequencies $\omega_{pm}=p\Omega_0+m\omega_{s0}$, where ω_{s0} is the small-amplitude synchrotron frequency. Around every harmonic of the revolution frequency $p\Omega_0$, there is an infinite number of synchrotron satellites m (it is different from the one used in Sect. 4.2!), whose spectral amplitude is given by the Bessel function $J_m\left(p\Omega_0\hat{\tau}\right)$, where $\hat{\tau}$ is the synchrotron amplitude. The spectrum is centred at the origin and because the argument of the Bessel functions is proportional to $\hat{\tau}$, the width of the spectrum behaves like $\hat{\tau}^{-1}$. Applying the Vlasov equation, linearizing it, and studying the effect of the impedance on the unperturbed distribution leads to the potential-well effect: a new fixed point is reached, with a new synchronous phase, a new effective voltage, a new synchrotron frequency, a new bunch length and a new momentum spread, which all depend on intensity. Studying a perturbation on top of the new stationary distribution, one ends up at low intensity, i.e. considering independently the modes m (which is valid up to a certain intensity), with the following eigenvalue system

$$\Delta\omega_{cmq}\sigma_{mq}(l) = \sum_{p=-\infty}^{p=+\infty} K_{lp}^{m}\sigma_{mq}(p), \text{ with } \Delta\omega_{cmq} = \omega_{cmq} - m\omega_s, \tag{4.25}$$

where

$$K_{lp}^{m} = -\frac{2\pi I_b m\omega_s}{\Omega_0^2 \hat{V}_{\mathrm{T}} h \cos\phi_s} j \frac{Z_l(p)}{p} \int_{\hat{\tau}=0}^{\hat{\tau}=+\infty} \frac{dg_0}{d\hat{\tau}} J_m\left(p\Omega_0\hat{\tau}\right) J_m\left(l\Omega_0\hat{\tau}\right) d\hat{\tau}. \tag{4.26}$$

Here, $\Delta\omega_{cmq}$ is the coherent complex synchrotron frequency shift to be determined, $I_b = N_b e\Omega_0/2\pi$ is the bunch current, ω_s, $\hat{V}_{\mathrm{T}}$ and ϕ_s are the new synchrotron frequency, total voltage and synchronous phase (taking into account the potential-well distortion), $Z_l = Z_0^{\|}$ is the longitudinal impedance, g_0 is the longitudinal amplitude density function and h the RF harmonic number. The procedure to obtain first order exact solutions, with realistic modes and a general interaction, thus consists of finding the eigenvalues and eigenvectors of the infinite complex matrix whose elements are given by Eq. (4.26). The result is an infinite number of modes mq of oscillation. To each mode, one can associate a coherent frequency shift $\Delta\omega_{cmq} = \omega_{cmq} - m\omega_s$ (which is the qth eigenvalue), a coherent spectrum $\sigma_{mq}(p)$ (which is the qth eigenvector) and a perturbation distribution $g_{mq}\left(\hat{\tau}\right)$. For numerical reasons, the matrix needs to be truncated, and thus only a finite frequency domain is explored. For the case of the parabolic amplitude distribution and a constant inductive impedance (which leads to real tune shifts only and therefore no instability), the signal at the pick-up electrode shown for several superimposed turns is depicted on Fig. 4.5. In the case of a complex impedance, the real part will lead in addition to a growing amplitude with an associated instability rise-time. The spectrum of mode mq is peaked at $f_q \approx (q + 1)/(2\tau_b)$ and extends $\sim \pm\tau_b^{-1}$, where τ_b is the full bunch length (in second). It can be seen from Fig. 4.5 that there are q nodes on these "standing-wave" patterns. The longitudinal signal at the pick-up electrode is given by

$$S_{mq}\left(t, \vartheta\right) = S_{z0}\left(t, \vartheta\right) + \Delta S_{zmq}\left(t, \vartheta\right), \tag{4.27}$$

$$S_{z0}\left(t, \vartheta\right) = 2\pi I_b \sum_{p=-\infty}^{p=+\infty} \sigma_0(p) e^{jp\Omega_0 t} e^{-jp\vartheta}, \quad \sigma_0(p) = \int_{\hat{\tau}=0}^{\hat{\tau}=+\infty} J_0\left(p\Omega_0\hat{\tau}\right) g_0\left(\hat{\tau}\right)\hat{\tau}d\hat{\tau}, \tag{4.28}$$

$$\Delta S_{zmq}\left(t, \vartheta\right) = 2\pi I_b \sum_{p=-\infty}^{p=+\infty} \sigma_{mq}(p) e^{j(p\Omega_0 + m\omega_s + \Delta\omega_{cmq})t} e^{-jp\vartheta}. \tag{4.29}$$

Finding the eigenvalues and eigenvectors of a complex matrix by computer can be difficult in some cases, and a simple approximate formula for the eigenvalues is useful in practice to have a rough estimate. This is known as Sacherer's (longitudinal) formula [52]. Sacherer's formula is also valid for coupled-bunch instability with M equally-populated equally-spaced bunches, assuming multi-bunch modes with only one type of internal motion. In the case of gaps between bunch trains, a time-domain approach is usually better suited.

As the bunch intensity increases, the different longitudinal modes can no longer be treated separately and the situation is more involved. In the longitudinal plane, the microwave instability for coasting beams is well understood. It leads to a stability diagram, which is a graphical representation of the solution of the dispersion relation (taking into account the momentum spread) depicting curves of constant growth rates, and especially a threshold contour in the complex plane of the driving impedance (see Sect. 4.4) [57]. When the real part of the driving impedance is much greater than the modulus of the imaginary part, a simple approximation, known as the Keil-Schnell (or circle) stability criterion, may be used to estimate the threshold curve [58]. For bunched beams, it has been proposed by Boussard to use the coasting-beam formalism with local values of bunch current and momentum spread [59]. A first approach to explain this instability, without coasting-beam approximations, has been suggested by Sacherer through Longitudinal Mode-Coupling (LMC) [60]. The equivalence between LMC Instabilities (LMCI) and microwave instabilities has been pointed out by Sacherer and Laclare [53] in the case of broad-band driving resonator impedances, when the bunch length is much greater than the inverse of twice the resonance frequency. Furthermore, due to the potential well-distortion, a bunch is more stable below transition than above [53, 61, 62]. Typical pictures of LMCI are shown in Fig. 4.6 [63] and a comparison with macroparticle tracking simulations, which revealed a good agreement, is discussed in Ref. [64]. Experimentally, the most evident signature of the LMCI is the intensity-dependent longitudinal beam emittance blow-up to remain just below the threshold [65], as revealed also by macroparticle tracking simulations (see Fig. 4.7 [64]).

4.3.2 *Transverse*

A similar analysis as the one done for the longitudinal plane can be done in the transverse plane [53, 63]. Following the same procedure, the horizontal coherent oscillations (over several turns) of a "water-bag" bunch (i.e. with constant longitudinal amplitude density) interacting with a constant inductive impedance are shown in Fig. 4.8.

The main difference with the longitudinal plane is that there is no effect of the stationary distribution and the bunch spectrum is now centered at the chromatic frequency $f_\xi = Q_{x0} f_0 \xi / \eta$, where f_0 is the revolution frequency and ξ is the relative chromaticity. The sign of the chromatic frequency is very important and to avoid the head–tail instability (of mode 0) it should be slightly positive, meaning that

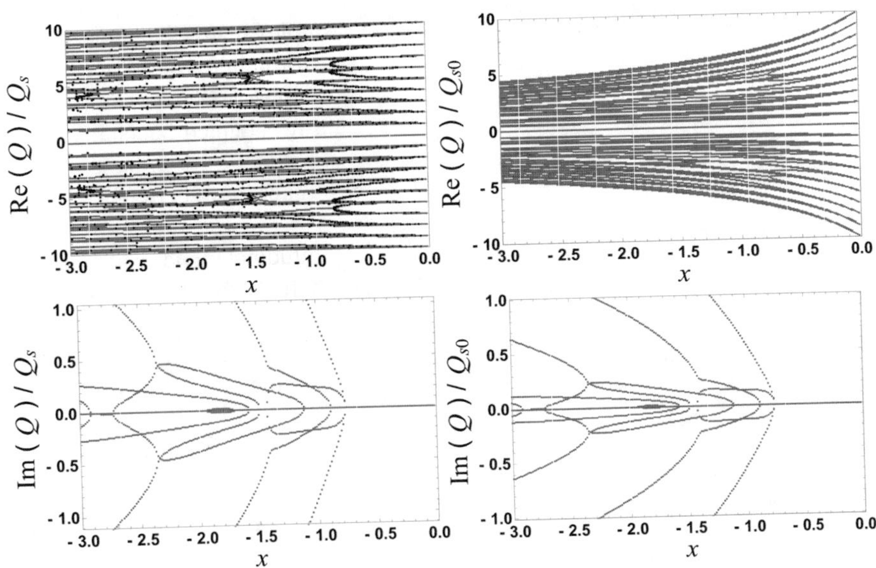

Fig. 4.6 (Left) Comparison between GALACLIC Vlasov Solver [63] (in red) and Laclare's approach [53] (in black) of the normalised mode-frequency shifts vs. the normalised parameter x (proportional to the bunch intensity [63]), in the case of a broad-band resonator impedance (with a quality factor of 1 and a resonance frequency f_r such that $f_r\ \tau_b = 2.8$), above transition, without taking into account the potential-well distortion (this is why the intensity-dependent synchrotron tune Q_s is used) and for a "Parabolic Amplitude Density" (PAD) longitudinal distribution [53]. (Right) Similar plot from GALACLIC only, taking into account the potential-well distortion (this is why the low-intensity synchrotron tune Q_{s0} is used)

the chromaticity should be negative below transition and positive above. Sacherer's formula is also valid for coupled-bunch instability with M equally-populated equally-spaced bunches, assuming multi-bunch modes with only one type of internal motion (i.e. the same head–tail mode number). This analysis was extended in Ref. [66] to include also the coupling between the modes (and the possibility to have different head–tail modes in the different bunches). In the case of gaps between bunch trains, a time-domain approach is usually better suited.

At low intensity (i.e. below a certain intensity threshold), the standing-wave patterns (head–tail modes) are treated independently. This leads to instabilities where the head and the tail of the bunch exchange their roles (due to synchrotron oscillation) several times during the rise-time of the instability. The (approximate) complex transverse coherent betatron frequency shift of bunched-beam modes is given by Sacherer's formula for round pipes [52]. For flat chambers a quadrupolar effect (see Sect. 4.2) has to be added to obtain the real part of the coherent tune shift, which explains why the horizontal coherent tune shift is zero in horizontally flat chambers (of good conductors). As an example, a head–tail instability with mode $q = 10$ is shown in Fig. 4.9(left). It is worth mentioning that there is also a head–tail instability in the longitudinal plane. The longitudinal head–tail instability,

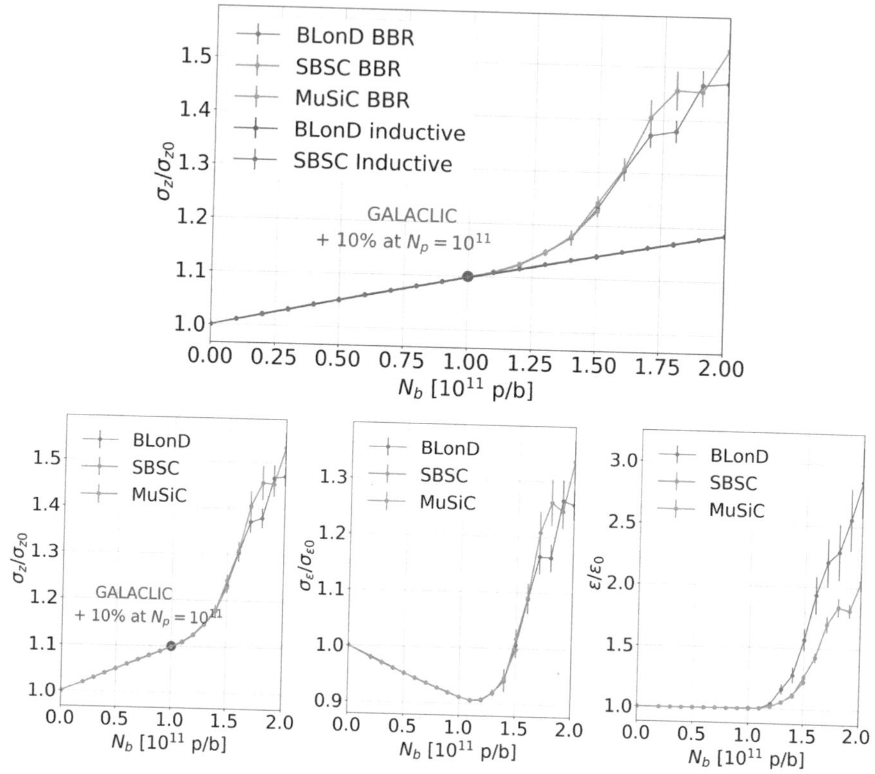

Fig. 4.7 Simulation results from BLonD, SBSC and MuSIC codes [64] corresponding to the case mentioned above: (upper) evolution of the normalised rms bunch length vs. bunch intensity for the cases of broad-band resonator and constant inductive impedances; (lower) evolution of the normalised rms bunch length, energy spread and longitudinal emittance vs. bunch intensity for the case of the broad-band resonator impedance. Courtesy of M. Migliorati [64]

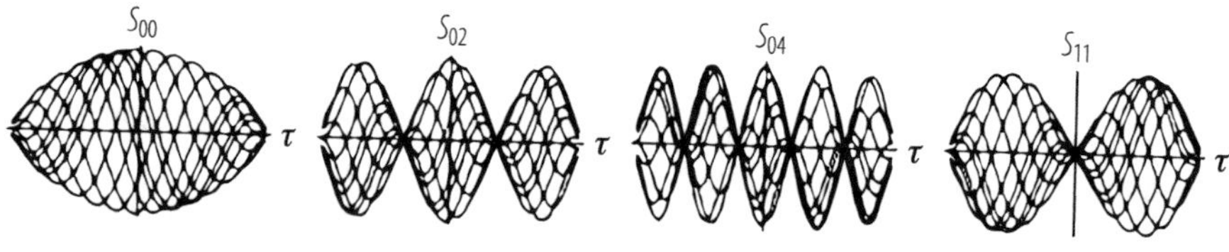

Fig. 4.8 Transverse signal at the pick-up electrode for four different modes shown for several superimposed turns, for the case of the "water-bag" bunch (i.e. with constant longitudinal amplitude density) and a constant inductive impedance. In the present example, the total phase shift between the head and the tail is given $\chi = \omega_\xi \tau_b = 10$ (see below)

first suggested by Hereward [67] and possibly observed at the CERN SPS [68] results from the fact that the slip factor is not strictly a constant: it depends on the instantaneous energy error just as the betatron frequency does. The longitudinal beam distribution then acquires a head–tail phase, and instability may arise as a result.

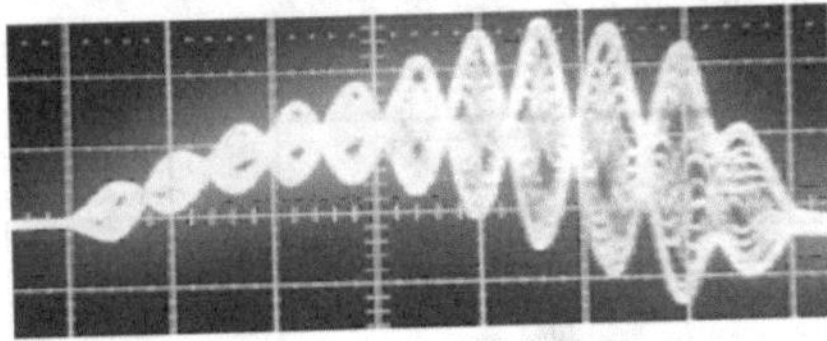

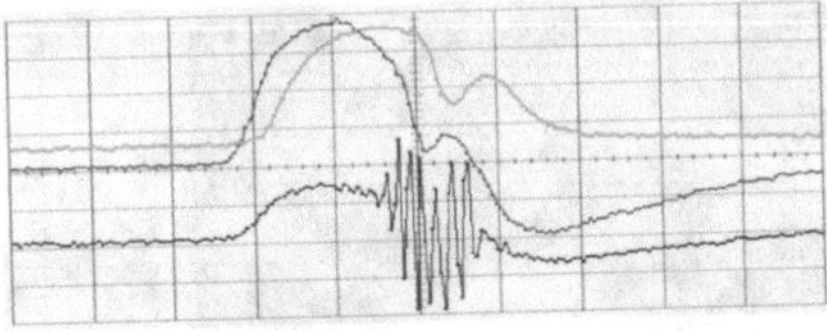

Fig. 4.9 (Left) Signal from a radial beam position monitor during 20 consecutive turns observed in the CERN PS at 1.4 GeV kinetic energy in 1999. Time scale: 20 ns/div. (Right) Fast instability observed in the CERN PS near transition (~6 GeV total energy) in 2000. Single-turn signals from a wide-band pick-up. From top to bottom: $\sum$, Δx, and Δy. Time scale: 10 ns/div. The head of the bunch is stable and only the tail is unstable in the vertical plane. The particles lost at the tail of the bunch can be seen from the hollow in the bunch profile

As the bunch intensity increases, the different head–tail modes can no longer be treated separately. In this regime, the wake fields couple the modes together and a wave pattern travelling along the bunch is created: this is the Transverse Mode Coupling Instability (TMCI). The TMCI for circular accelerators has been first described by Kohaupt [69] in terms of coupling of Sacherer's head–tail modes. This extended to the transverse motion, the theory proposed by Sacherer to explain the longitudinal microwave instability through coupling of the longitudinal coherent bunch modes. The TMCI is the manifestation in synchrotrons of the Beam Break-Up (BBU) mechanism observed in linacs. The only difference comes from the synchrotron oscillation, which stabilises the beam in synchrotrons below a threshold intensity by swapping the head and the tail continuously. In fact, several analytical formalisms exist for fast (compared to the synchrotron period) instabilities, but the same formula is obtained (within a factor smaller than two) from five, seemingly diverse, formalisms in the case of a broad-band resonator impedance in the "long-bunch" regime [70], as recently confirmed in Ref. [71]: (i) Coasting-beam approach with peak values, (ii) Fast blow-up, (iii) BBU (for 0 chromaticity), (iv) Post head–tail, and (v) TMCI with 2 modes in the "long-bunch" regime (for 0 chromaticity). Two regimes are indeed possible for the TMCI according to whether the total bunch length is larger or smaller than the inverse of twice the resonance frequency of the impedance. The simple (approximate) formula reveals the scaling with the different parameters. In particular it can be seen that the instability does not disappear at high energy but saturates like the slip factor (what is important is not the energy but the distance from the transition energy) [72]. This means that the TMCI intensity threshold can be raised by changing the transition energy, i.e. by modifying the optics. Furthermore, the intensity threshold increases with the resonance frequency (as high-order head–tail modes will couple), with longitudinal emittance and with chromaticity. Note that the coherent synchrobetatron resonances, important in large machines, are not discussed here. This was checked with the MOSES Vlasov solver [73], which is a program computing the coherent bunched-beam modes. Below is a comparison between the MOSES code and the HEADTAIL code [74], which is a tracking code simulating single-bunch phenomena, in the case of a LHC-type single bunch at SPS injection [75]. As can be seen from Fig. 4.10, a very good

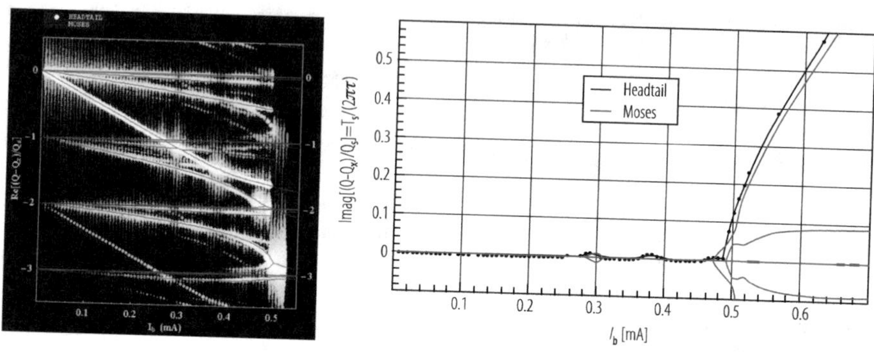

Fig. 4.10 Comparison between MOSES (in red) and HEADTAIL (in white) in the case of a broad-band resonator (Courtesy of Benoit Salvant [75]). Evolution of the real (left) and imaginary (right) parts of the shifts of the transverse modes (with respect to the unperturbed betatron tune), normalized by the synchrotron tune, vs. bunch intensity

agreement between the two was found. For a general impedance (i.e. not a resonator impedance) the situation is more involved and MOSES cannot be used: one should rely on HEADTAIL simulations or on the recently developed Vlasov solvers such as NHT [76] or DELPHI [77]. In the case of flat chambers, the intensity threshold is higher in one plane than in the other and linear coupling can be used to raise the TMCI intensity threshold [78]. Note finally that with many bunches the TMCI intensity threshold can be considerably reduced [66].

It is worth mentioning also all the work done for the TMCI in LEP, as Chin's work (with MOSES) came later. It was proposed to cure the TMCI with a reactive feedback that would prevent the zero mode frequency from changing with increasing beam intensity [79]. In [80, 81] a theory of reactive feedback has been developed in the two-particle approach and with the Vlasov equation. Theory has revealed that the reactive feedback can really appreciably increase the TMCI intensity threshold, which was confirmed by simulation [82, 83]. On the contrary, the resistive feedback was found to be "completely" ineffective as a cure for the TMCI [81]. An action of a feedback on the TMCI intensity threshold was later examined experimentally at PEP [84]. It was confirmed that a reactive feedback is indeed capable to increase the TMCI intensity threshold. But it turned out unexpectedly that a resistive feedback can also increase the TMCI intensity threshold and even more effectively [84]. In [85], an attempt was made to develop an advanced transverse feedback theory capable to elucidate the conditions at which the resistive or reactive or some intermediate feedback can cure the TMCI. Positive chromaticity above transition helps, but depending on the coupling impedance, beam stability may require a large value of the chromaticity either unattainable or which reduces the beam lifetime. It was proposed to have a negative chromaticity (what is usually avoided), where the zero mode is unstable (by head–tail instability) and all the other modes are damped, and stabilise this mode by a resistive feedback, keeping the higher order modes stable. In this case, the TMCI intensity threshold could be increased by a factor 3–5 [85]. In the last few years, several Vlasov solvers

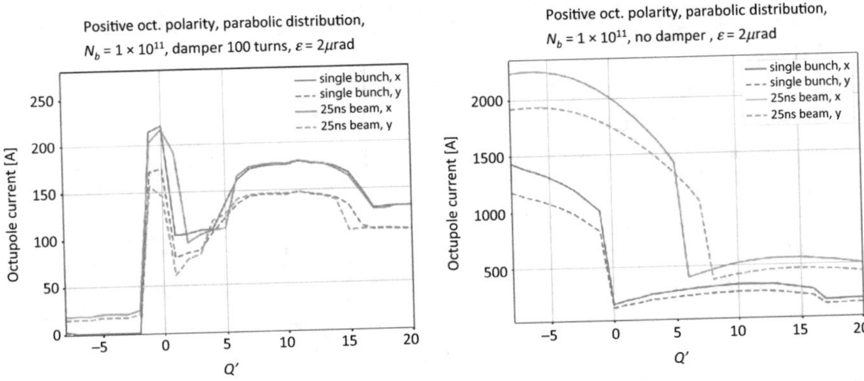

Fig. 4.11 Required Landau octupole current to stabilise the 2018 CERN LHC beam vs. chromaticity: (left) with resistive transverse damper and (right) without resistive transverse damper. Courtsey of N. Mounet [87]

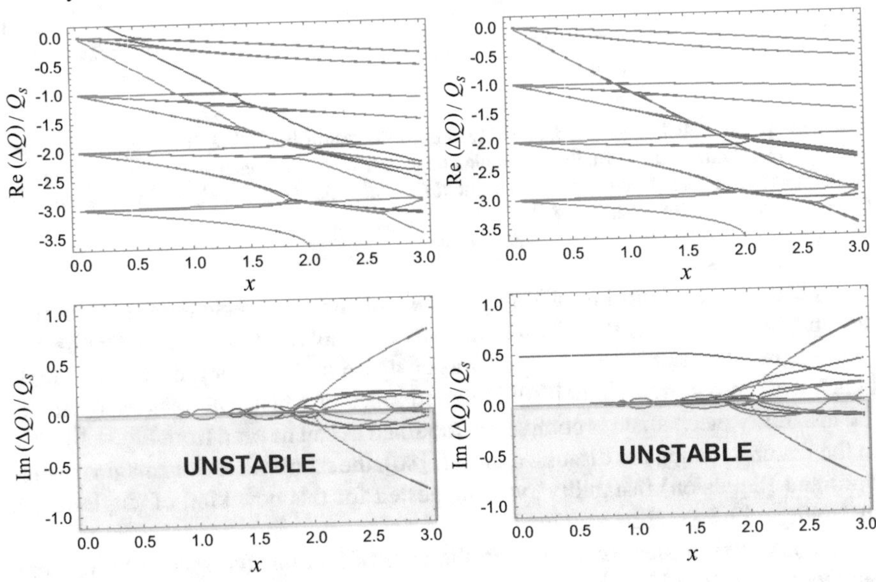

Fig. 4.12 Usual TMCI plots (for $f_r \tau_b = 2.8$, i.e. in the "long-bunch" regime) showing the real and imaginary parts of the normalised complex tune shift vs. the normalised parameter x (which is proportional to the bunch intensity [63]) without (in blue) and with (in red) a transverse damper: (left) reactive and (right) resistive [86]

were developed to take into account the effect of a transverse damper, such as NHT [76], DELPHI [77] and GALACTIC [86]. An example of DELPHI for the case of the LHC in 2018 is shown in Fig. 4.11, where the beneficial effect of the transverse resistive damper (on the required Landau octupole current needed to stabilise the beam) can be clearly seen. A comparison between a reactive and a resistive damper is shown in Figs. 4.12 and 4.13 using GALACTIC [86] (and a comparison between GALACTIC and Laclare's approach [53] is discussed in Ref. [63]).

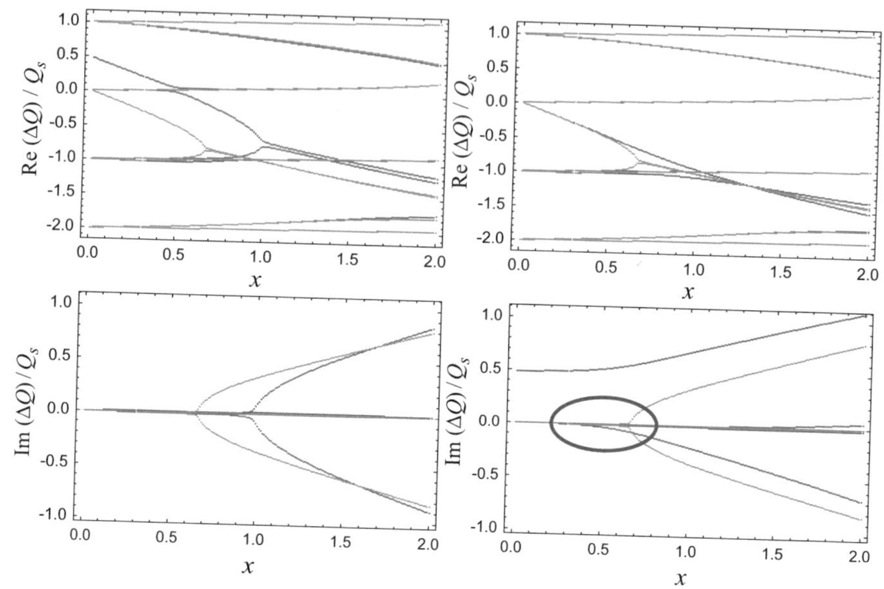

Fig. 4.13 Usual TMCI plots (for $f_r \tau_b = 0.8$, i.e. in the "short-bunch" regime) showing the real and imaginary parts of the normalised complex tune shift vs. the normalised parameter x (which is proportional to the bunch intensity [63]) without (in blue) and with (in red) a transverse damper: (left) reactive and (right) resistive [86]

As can be seen from Fig. 4.13(right) the resistive transverse damper exhibits a destabilising effect below the TMCI intensity threshold. This destabilising effect of (perfect) resistive transverse dampers was analysed in detail for the case of a single bunch with zero chromaticity [86]: in the presence of a resistive transverse damper the instability mechanism is completely modified as can be seen from Fig. 4.14. Due to the features, which are discussed in Ref. [86], the name "ISR (for Imaginary tune Split and Repulsion) instability" was suggested for this new kind of single-bunch instability with zero chromaticity.

It is also worth mentioning that in the case of hadrons (compared to leptons), another ingredient which should be taken into account while studying the transverse instabilities is space charge. This has been a subject of discussion for the last two decades as space charge was believed initially to be mainly beneficial as e.g. for the previous case of the CERN SPS TMCI predicted in the absence of space charge. It was recently found that space charge is actually destabilising in such a case ("long-bunch" regime) [88–90], while it is beneficial in the "short-bunch" regime [88, 89]. This is clearly revealed in Figs. 4.15 and 4.16, but still some work is needed to fully understand what happens.

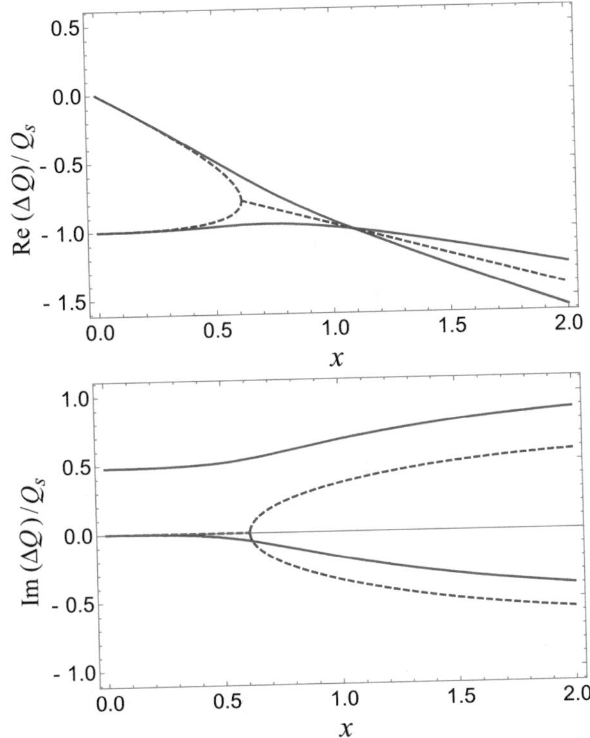

Fig. 4.14 Usual TMCI plots showing the real and imaginary parts of the normalised complex tune shift vs. the normalised parameter x (which is proportional to the bunch intensity [63]) without (in blue) and with (in red) a resistive transverse damper [86]

4.4 Landau Damping

E. Metral

Several stabilising mechanisms exist which can prevent the previous instabilities from developing. One of them is Landau damping, which is a general process that arises when one considers a whole collection of particles or other systems, which have a spectrum of resonant frequencies, and interact in some way. In accelerators we are usually concerned with an interaction of a kind that may make the beam unstable (wake fields), and we want to find out whether or not (and how) the spread of resonant frequencies will stabilise it. If the particles have a spread in their natural frequencies, the motion of the particles can lose its coherency. In order to understand the physical origin of this effect, let us first consider a simple harmonic oscillator, which oscillates in the x-direction with its natural frequency [8]. Let this oscillator be driven, starting at time $t = 0$, by a sinusoidal force. The equation of motion is

$$\ddot{x} + \omega_x^2 x = f \cos(\omega_c t), \tag{4.30}$$

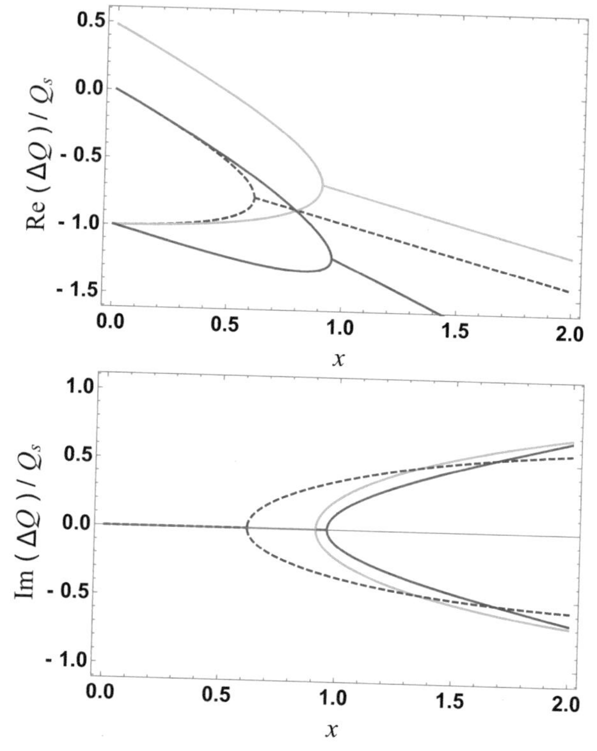

Fig. 4.15 Simplified model/example of Ref. [86], describing the mode-coupling in the "short-bunch" regime, i.e. the mode-coupling between modes 0 and −1, extended here to take into account also space charge, using the parameters mentioned above: (dashed blue) with impedance only, (green) with impedance and a reactive transverse damper, (red) with impedance and space charge [88]. The normalised parameter x is proportional to the bunch intensity [63]

where a dot stands for derivative with respect to time and with $x(0) = 0$ and $\dot{x}(0) = 0$. The solution is

$$x\,(t > 0) = -\frac{f}{\omega_c^2 - \omega_x^2}\,[\cos{(\omega_c t)} - \cos{(\omega_x t)}] = \frac{f}{2\omega_{x0}} \sin{(\omega_{x0} t)}\,\frac{\sin[(\omega_c - \omega_x)t/2]}{(\omega_c - \omega_x)/2}. \tag{4.31}$$

Consider now an ensemble of oscillators (each oscillator represents a single particle in the beam) which do not interact with each other and have a spectrum of natural frequency ω_x with a distribution $\rho_x(\omega_x)$ normalised to unity. Let's assume first that the origin of the betatron frequency spread is not specified: an externally given beam frequency spectrum is supposed. Now starting at time $t = 0$, subject this ensemble of particles to the driving force f cos $(\omega_c t)$ with all particles starting with initial conditions $x(0) = 0$ and $\dot{x}(0) = 0$. We are interested in the ensemble average

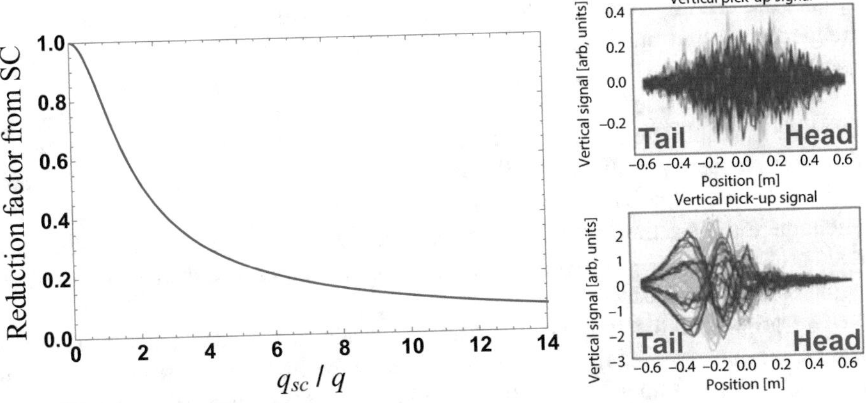

Fig. 4.16 (Left) reduction factor from a simplified model with space charge of the TMCI intensity threshold, as a function of the ratio between the space charge parameter q_{sc} and the radial mode number q, in the case of the "long-bunch" regime, as e.g. for the CERN SPS at injection [88]. (Right) simulated stable bunch without space charge (top) and unstable bunch with space charge (bottom) for the case $q_{sc}/q = 13.5$ with a bunch intensity a factor 3 lower than the TMCI intensity threshold without space charge [91] (Courtesy of A. Oeftiger)

of the response, which is given by superposition by

$$\overline{x}(t) = \frac{f}{2\omega_{x0}} \left[\cos(\omega_c t) \, \text{P.V.} \int_{-\infty}^{+\infty} \frac{\rho_x(\omega_x)}{\omega_x - \omega_c} d\omega_x + \pi \rho_x(\omega_c) \sin(\omega_c t) \right], \tag{4.32}$$

where P.V. stands for Principal Value. The sinus term has a definite sign relative to the driving force, because $\rho_x(\omega_c)$ is always positive. In particular, $\dot{\overline{x}}$ is always in phase with the force, indicating that work is being done on the system, which always reacts to the force "resistively". The Landau damping effect is to be distinguished from a "decoherence (also called phase-mixing, or filamentation) effect" that occurs when the beam has nonzero initial conditions. Had we included an initial offset, we would have introduced two additional terms into the ensemble response, which do not participate in the dynamic interaction of the beam particles and are not interesting for our purposes here. In this decoherence effect, individual particles continue to execute oscillations of constant amplitude, but the total beam response $\overline{x}$ decreases with time. As mentioned above, work is continuously being done on the system. However, the amplitude of $\overline{x}$, as given before, does not increase with time. Where did the energy go? The system absorbs energy from the driving force indefinitely while holding the ensemble beam response within bounds. The stored energy is incoherent in the sense that the energy is contained in the individual particles, but it is not to be regarded as heat in the system. This is because the stored energy is not distributed more or less uniformly in all particles, but is selectively stored in particles with continuously narrowing range of frequencies around the driving frequency. If one observes two particles, one with the exciting frequency ω_c and one with a frequency slightly different, at the beginning, they oscillate "coherently" (same amplitude and same phase). However, after a while the particle with the exciting frequency, being resonantly driven, continues to increase in amplitude as time increases, whereas the other particle with a slightly different frequency realizes that its frequency is not the same as the driving one and the "beating" phenomenon is observed for this particle. If one considers the phenomenon for a time t, the number of particles which still oscillate coherently decreases with time as $1/t$, while their amplitude increases as t, the net contribution being constant with time.

The origins of the frequency spread that leads to Landau damping have not been taken into account till now. The case where the frequency spread comes from the longitudinal momentum spread of the beam is straightforward (for a coasting beam), because the longitudinal momentum is a constant, which just affects the coefficients in the equations of motion of the transverse oscillations, and hence their frequencies. It can be dealt with the same method as in the previous sections, i.e. it is the distribution function which is important. The same result applies also if one considers a tune spread that is due to a non-linearity (e.g. from octupole lenses) in the other plane. However, this result is no longer valid if the non-linearity is in the plane of coherent motion. In this case, the steady-state is more involved because the coherent motion is then a small addition to the large incoherent amplitudes that make the frequency spread, and it is inconsistent to assume that it can be treated as a linear superposition [92]. One needs to consider "second order" non-linear terms and the final result is that in this case it is not the distribution function which matters but its derivative. Using the Vlasov formalism, this result is recovered more straightforwardly.

4.4.1 Transverse

Considering the case of a beam having the same normalized rms beam size $\sigma = \sqrt{\varepsilon}$ in both transverse planes, the Landau damping mechanism from octupoles of coherent instabilities, e.g. in the horizontal plane, is discussed from the following dispersion relation [17, 93]

$$1 = -\Delta Q_{\mathrm{coh}}^{x} \int\limits_{J_x=0}^{+\infty} dJ_x \int\limits_{J_y=0}^{+\infty} dJ_y \frac{J_x \frac{\partial f(J_x, J_y)}{\partial J_x}}{Q_c - Q_x\left(J_x, J_y\right) - m Q_s}, \tag{4.33}$$

with

$$Q_x\left(J_x, J_y\right) = Q_0 + a_0 J_x + b_0 J_y. \tag{4.34}$$

Here, Q_c is the coherent betatron tune to be determined, $J_{x,y}$ are the action variables in the horizontal and vertical plane respectively, with $f(J_x, J_y)$ the distribution function, ΔQ^x_{coh} is the horizontal coherent tune shift, $Q_x(J_x, J_y)$ is the horizontal tune in the presence of octupoles, m is the head–tail mode number, and Q_s is the small-amplitude synchrotron tune (the longitudinal spread is neglected).

The nth order distribution function is assumed to be

$$f\left(J_x, J_y\right) = a\left(1 - \frac{J_x + J_y}{b}\right)^n, \tag{4.35}$$

where a and b are constants to be determined by normalization, and which corresponds to a profile extending up to $\sqrt{2\,(n+3)}\sigma$. The dispersion relation of Eq. (4.33) can be re-written as

$$\Delta Q^x_{\text{coh}} = -\frac{a_0}{nab} I_n^{-1}\left(c, q\right), \tag{4.36}$$

with

$$I_n\left(c, q\right) = \int\limits_{J_x=0}^{1} dJ_x \int\limits_{J_y=0}^{1-J_x} dJ_y \frac{J_x\left(1 - J_x - J_y\right)^{n-1}}{q + J_x + cJ_y}, \tag{4.37}$$

$$q = \frac{Q_c - Q_0 - mQ_s}{-ba_0}, \text{ and } c = \frac{b_0}{a_0}. \tag{4.38}$$

It is convenient to write Eq. (4.36) in this way, with the left-hand-side (l.h.s) containing information about the beam intensity and the impedance and the right-hand-side (r.h.s) containing information about the beam frequency spectrum only. In the absence of frequency spread, the r.h.s. of Eq. (4.36) is equal to $Q_c - Q_0 - mQ_s$, which is thus given by ΔQ^x_{coh} (i.e. the l.h.s). Calculation of the l.h.s is now straightforward (following Sect. 4.3): for a given impedance (and transverse damper), one only needs to calculate the complex mode frequency shift, in the absence of Landau damping. Without frequency spread, the condition for the beam to be stable is thus simply $\text{Im}\left(\Delta Q^x_{\text{coh}}\right) \geq 0$ (oscillations of the form $e^{j\omega t}$ are considered). Once its l.h.s is obtained, Eq. (4.36) can be used to determine the coherent betatron tune Q_c in the presence of Landau damping when the beam is at the edge of instability (i.e. Q_c real). However, the exact value of Q_c is not a very useful piece of information. The more useful question to ask is under what conditions the beam becomes unstable regardless of the exact value of Q_c under these conditions, and Eq. (4.36) can be used in a reversed manner to address this question. To do so, one considers the real parameter $Q_c - Q_0 - mQ_s$ (stability limit) and observes the locus traced out in the complex plane by the r.h.s of Eq. (4.36), as $Q_c - Q_0 - mQ_s$ is scanned form $-\infty$ to $+\infty$. This locus defines a "stability boundary diagram". The l.h.s of Eq. (4.36), a complex quantity, is then plotted in this plane as a single point. If this point lies on the locus, it means the solution of Q_c for Eq. (4.36) is real, and this $Q_c - Q_0 - mQ_s$

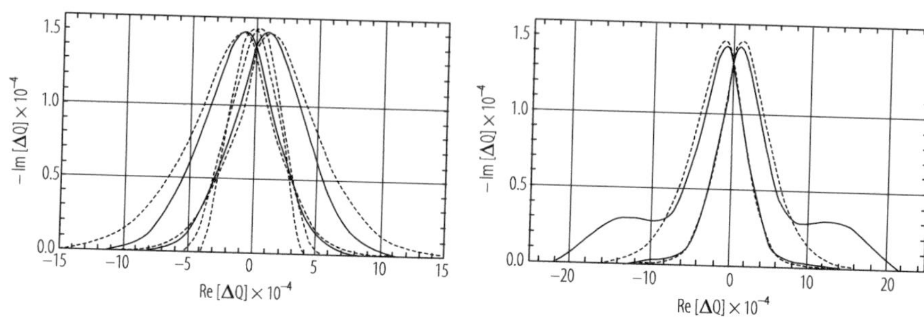

Fig. 4.17 Stability diagrams (for both positive and negative detunings a_0) for the LHC at top energy (7 TeV) with maximum available octupole strength: (Left) for the 2nd order (dashed curves), the 15th order (full curves), and the Gaussian (dotted curves) distribution; (Right) for the Gaussian distribution (dotted curve) and a distribution with more populated tails than the Gaussian (full curve)

is such that the beam is just at the edge of instability. If it lies on the inside of the locus (the side which contains the origin), the beam is stable. If it lies on the outside of the locus, the beam is unstable. The stability diagrams for the 2nd order, 15th order and Gaussian distribution functions are plotted in Fig. 4.17 for the case of the LHC at top energy (7 TeV) with maximum available octupole strength ($\varepsilon = 0.5$ nm, $|a_0| = 270440$ and $c = -0.65$).

The case of a distribution extending up to 6σ (as the 15th order distribution) but with more populated tails than the Gaussian distribution has also been considered and revealed a significant enhancement of the stable region compared to the Gaussian case ([93], see also Fig. 4.17(right)). This may be the case in reality in proton machines due to diffusive mechanisms.

It is worth reminding that Landau damping of coherent instabilities and maximization of the dynamic aperture are partly conflicting requirements. On the one hand, a spread of the betatron frequencies is needed for the stability of the beam coherent motion, which requires nonlinearities to be effective at small amplitude. On the other hand, the nonlinearities of the lattice must be minimized at large amplitude to guarantee the stability of the single-particle motion. A trade-off between Landau damping and dynamic aperture is therefore usually necessary [87].

Despite the destabilising effect of a resistive transverse damper in the case of a single bunch with zero chromaticity (as discussed in Sect. 4.3.2) below the TMCI intensity threshold (in the case of the "short-bunch" regime) without transverse damper, a transverse damper helps to reduce the amount of tune spread which would be needed to stabilise the bunch above the TMCI intensity threshold, as it can be seen in Fig. 4.18.

Note that linear coupling between the transverse planes can also influence the Landau damping mechanism [95], leading to a sharing of the Landau damping between the transverse planes, which can have a beneficial effect (i.e. stabilising the other plane, as it was used in the CERN PS for many years [96]) or a detrimental effect (i.e. destabilising one or two planes by loss of Landau damping, as it was

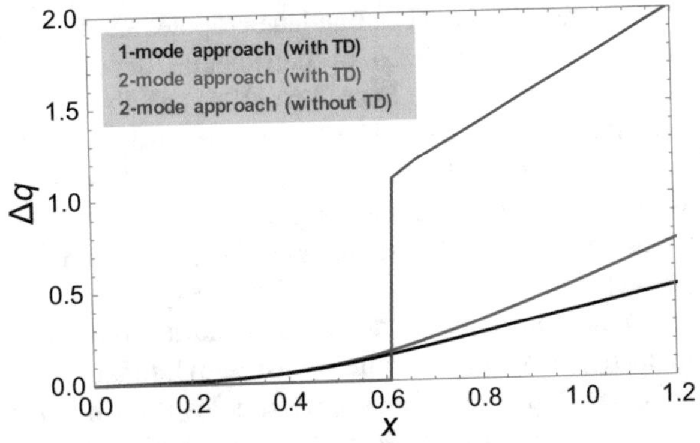

Fig. 4.18 Required normalised (to the synchrotron tune Q_s) tune spread Δq to stabilise the bunch in both cases of instabilities without and with Transverse Damper (TD), corresponding to the cases of Fig. 4.14 [94]. The normalised parameter x is proportional to the bunch intensity [63]

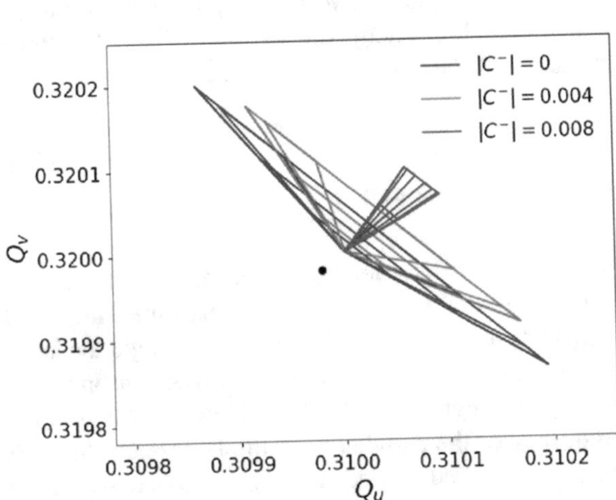

Fig. 4.19 Reduction of the tune footprint (and associated projections on the transverse tunes axes, responsible for Landau damping) vs. linear coupling (described here by the "closest tune approach" $|C^-|$) [98]. Courtesy of L.R. Carver

believed to be the case with the Batman instability of the HERA proton ring [97]: due to the features discussed in Ref. [97], the name "coupled head–tail instability" was suggested for this instability in the HERA proton ring). Recently, linear coupling was also observed to be detrimental in the CERN LHC [98], as revealed by both measurements and macroparticle simulations (see Fig. 4.19). This required a careful measurement and correction of linear coupling all along the LHC cycle to avoid to use much more Landau octupoles current than foreseen. One has also to remember that linear coupling modifies also the transverse emittances [99, 100].

In the case of additional space-charge nonlinearities, the stability diagram will be shifted and beam stability can be obtained or lost, depending on the coherent tune. The influence of space-charge nonlinearities on the Landau damping mechanism of transverse coherent instabilities has first been studied by Möhl and Schönauer for coasting and rigid bunched beams [101, 102]. It was studied in detail in the past years for higher-order head–tail modes from both theory [103–106] and numerical simulations [107].

The interplay between Landau octupoles and beam–beam long-range interactions can be either beneficial or detrimental depending on the sign of the Landau octupoles current (see Sect. 4.6) [108] and this effect has to be careful taken into account in the CERN LHC to be able to push its performance.

Another destabilising effect currently under investigation at the CERN LHC (and which could explain some long latencies observed in the past, of the order of few minutes or even tens of minutes) is the effect of noise, whose detrimental effect was predicted in 2012 [109] and confirmed experimentally in 2018 [110].

Some work is being done to try and use the nonlinear optics as a path to high intensity, providing "infinite (transverse) Landau damping" [111], or electron lenses [112] or Radio Frequency Quadrupoles (or similarly second order chromaticity) [113–115]. The latter two methods are believed to be more efficient than Landau octupoles at high energy due to the adiabatic damping and the associated significant reduction of the transverse beam sizes.

4.4.2 Longitudinal

When the bunch is very small inside the RF bucket, the motion of the particles is linear and all the particles have the same (unperturbed, maximum) synchrotron frequency ω_{s0}. By increasing the bunch length the incoherent synchrotron frequency spread S is increased (the maximum synchrotron frequency spread is obtained when the bunch length is equal to the RF bucket length as in this case the synchrotron frequency of the particles with the largest amplitude is equal to 0: the synchrotron frequency spread S is equal to ω_{s0} in this case). In the presence of an impedance, the coherent synchrotron frequency of the dipole mode ω_{c11}, which is equal to the low-intensity synchrotron frequency ω_{s0} without synchrotron frequency spread (due to the compensation between the incoherent and coherent tune shifts), moves closer and closer to the incoherent band (stable region). The two possible cases are represented in Fig. 4.20 (using the rigid-bunch approximation), which is similar to what was obtained by Besnier (who considered a parabolic distribution function, which introduces some pathologies in the stability diagram due to its sharp edge) ([116], and references therein): the case of a capacitive impedance below transition or inductive impedance above transition corresponds to $U > 0$ (the coherent synchrotron frequency shift of the dipole mode has been written $\Delta\omega_{c11} = U - jV$) and the incoherent synchrotron frequency shift (due to the potential-well distortion) is $\Delta\omega_s^i < 0$ (and thus $\omega_s < \omega_{s0}$), and the case of a capacitive impedance above transition or inductive impedance below transition corresponds to $U < 0$ and $\Delta\omega_s^i > 0$ (and thus $\omega_s > \omega_{s0}$). Motions $\propto e^{j\omega t}$ are considered, which means that the beam is unstable when $V > 0$ (V is called the instability growth rate). The usual case where the resistive part of the impedance is small compared to the imaginary part is assumed, i.e. $V << |U|$. Beam stability is obtained when ω_{c11} enters into the incoherent band. In both cases, the stability limit is reached for $k = 4$, i.e. $S = 4|U|$, which is Sacherer's stability criterion for the dipole mode.

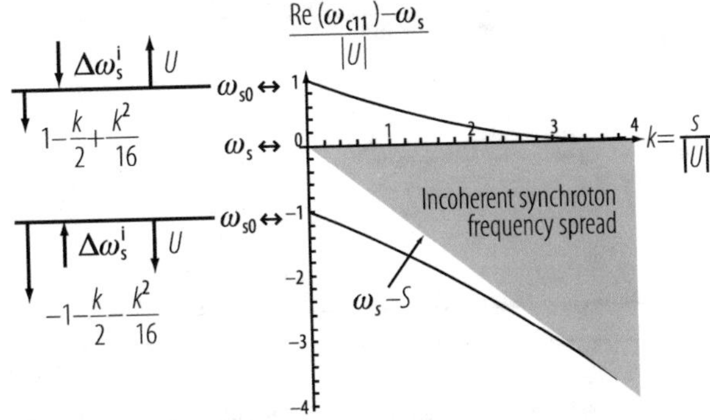

Fig. 4.20 Evolution of the coherent synchrotron frequency for the dipole mode with respect to the incoherent frequency spread (using the rigid-bunch approximation)

4.5 Two-Stream Effects (Electron Cloud and Ions)

G. Rumolo

4.5.1 *Electron Cloud Build-Up in Positron/Hadron Machines*

The term "electron cloud" is used for describing an accumulation of electrons inside the beam chamber of a circular accelerator, in which bunched beams of positively charged particles are accelerated or stored. The electron cloud can affect the accelerator operation by causing beam tune shift, emittance growth and coherent instabilities, as well as increase of the vacuum pressure and interference with beam diagnostics devices. Most of these effects eventually lead to beam quality degradation and loss. Electrons can be initially produced in the vacuum chamber by a number of processes. These electrons are called primary because, although some times their number can be sufficiently high to affect the circulating beam, they are usually only the seed for an avalanche process (see Fig. 4.21). In general, the primary electrons rapidly multiply via a beam-induced multipacting mechanism, which involves acceleration of the electrons in the beam field and secondary emission from their impact on the chamber wall. In the following, we first give an overview on the electron cloud formation (or build-up) process. We therefore list the main primary generation mechanisms and then discuss how the thus generated electrons can multipact in the presence of a train of bunches. The other important stages of the build-up of an electron cloud are its equilibrium and successive decay in the gap behind a bunch train. In the second part, we briefly discuss dynamics and consequences of the beam instabilities caused by the electron cloud that has formed in a beam pipe. In conclusion, we will try to give an up-to-date list of the possible techniques for electron cloud mitigation or suppression.

Primary electrons are the electrons generated during the passage of a bunch. They can be photo-electrons from synchrotron radiation in bending regions (mainly for positron beams) or secondary electrons desorbed from beam particles lost at the

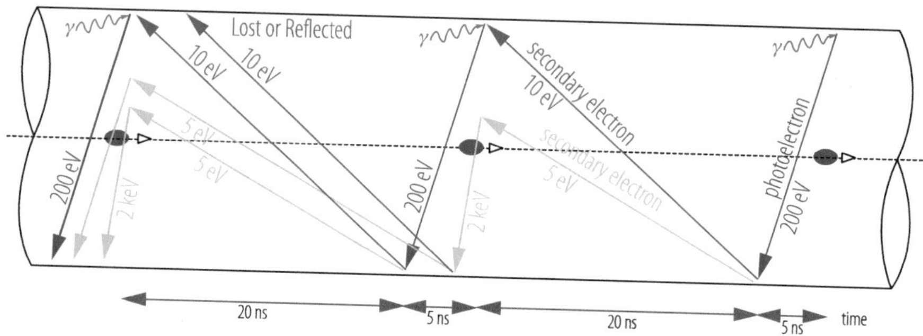

Fig. 4.21 Schematic of electron cloud build-up in the LHC beam pipe during multiple bunch passages, via photo-emission (due to synchrotron radiation) and secondary emission (Courtesy of F. Ruggiero). Note that the LHC is the 1st proton storage ring for which synchrotron radiation becomes a noticeable effect

walls (especially for ion beams). In this case they are emitted from the chamber wall. Primary electrons can also be created within the volume swept by the beam if the production mechanism is ionization of the residual gas. The location where the electrons are created can determine the energy gain of the primary electrons in the beam potential. The number of electrons created per unit length by synchrotron radiation or by beam loss during one bunch passage can be comparable to the average line density of beam particles, in which case these processes can alone give rise to amounts of electrons critical for the beam stability. The rate of photoemission (number of photoelectrons created per unit length) can be estimated as the product of the photo-electron yield Y_γ by the photoemission rate dN_γ/ds:

$$\frac{dN_{e\gamma}}{ds} = Y_\gamma \frac{dN_\gamma}{ds} = Y_\gamma \frac{5\alpha\gamma}{2\sqrt{3}\rho}, \tag{4.39}$$

where α denotes the fine structure constant and ρ the curvature radius of the beam in the dipole. For many materials, the photo-emission yield can be correctly approximated as being about 0.1 over a fairly large photon energy range, e.g. between a few eV and a few tens of keV. The azimuthal distribution of absorbed photons around the chamber wall and, thus, the launch positions of the emitted primary photo-electrons depend on the reflective properties of the chamber wall. The first simulation of an electron cloud build-up for short bunches was written by K. Ohmi. It served to explain coupled-bunch instabilities observed with positron beams at the KEK photon factory [117]. Ohmi's pioneering study considered only photo-emission at the chamber wall as a source of electrons, though a little later his initial code was extended to include secondary emission by electrons as well.

Ionization from scattering of individual charged beam particles against molecules of the residual gas occurs with typical cross sections of 1–2 Mbarn for most of the gas species that can be found in a beam chamber. However, a lower cross section of about 0.2 Mbarn applies to the lighter species, like H_2 [118]. These numbers refer to singly charged particles at ultrarelativistic energies. For fully ionized atoms, the cross section scales roughly with the square of the atomic number, i.e. like Z^2. Additionally, it increases by several orders of magnitude towards lower beam energies. If the beam density is sufficiently high, as it will be in certain sections of the next generation of linear colliders, ionization by the collective electric field of the bunch replaces single-particle scattering ionization as the dominant ionization process [119]. When this happens, the beam completely and instantly ionizes the residual gas in its neighbourhood.

Protons or ions impacting on the wall can be responsible for the generation of a large number of electrons. The secondary-electron yield from ion impact is approximately proportional to the projectile stopping power and inversely proportional to the cosine of the angle of incidence [120]. Since the stopping power is in turn proportional to the square of the charge number divided by the mass number, this value, usually very high because of the shallow angles at which losses typically occur, can be further amplified by one or two orders of magnitude for heavy ion beams. Production of this type of electrons also occurs with a high rate at the collimators, where significant beam loss routinely occurs by design.

The mechanism responsible for an exponential growth of the number of electrons is beam induced multipacting. The primary electrons are accelerated by the electric field of a passing bunch to such high energies that they produce, on average, more than one secondary electron when they again hit the wall of the vacuum chamber. The Secondary Emission Yield (SEY) of the chamber material is by definition the number of secondary electrons produced on average by an electron impact. It is obviously a function of the impinging electron energy, its angle of incidence, and the chamber history. For a round chamber of radius h and a short bunch, the resonance condition for beam-induced multipacting from electrons produced at the pipe walls takes the simple form [121]

$$N_b r_e L_b = h^2, \tag{4.40}$$

where r_e denotes the classical electron radius and L_b the bunch spacing in units of length. However, the condition of Eq. (4.40) is by far too stringent. Most secondary electrons have low energy and tend to stay in the vacuum chamber for a long time after a bunch passage. The survival of low energy electrons is made even more likely by the fact that their probability of being elastically backscattered at the chamber walls is almost one [122]. Moreover, a different multipacting regime, called "trailing edge multipacting", also exists in presence of long bunches. The electrons produced on the falling edge of the bunch can gain energy as they cross the pipe section and cause multipacting just during the passage of the second part of the bunch. This is a special type of single-bunch multipacting, causing electron clouds that can be either significantly cleared before the arrival of the next long bunch, or accumulate further in a mixed single-bunch and multi-bunch process.

Many simulation codes have been developed over the years to study numerically the process of electron cloud build-up and explore different beam/machine parameter ranges (e.g. [123–125]). Due to the variety of possible regimes and processes involved, simple considerations have usually turned out to be insufficient to predict the build-up thresholds and the change in electron cloud caused by some specific parameter change. It was a gratifying confirmation of the predictive power of the early simulations that the expected electron clouds were indeed observed at both B factories (PEP-II and KEKB), at the Large Hadron Collider injectors PS and SPS when operated with LHC-type beams, and finally in the LHC itself when operation with trains of closely spaced bunches started. In these rings, the electron cloud was seen to cause tune shift and emittance growth along the bunches of a train, both coupled and single-bunch instabilities and a degradation of certain beam diagnostics signals [126–128]. In the SPS and PS, significant beam loss could be observed at the end of a train if no countermeasures were put in place. As the electron cloud is potentially one of the main bottlenecks of the SPS after the upgrade of the LHC injector chain in the framework of the LHC Injectors Upgrade (LIU) project [129], dedicated electron diagnostics devices are installed in the machine to measure the electron flux and spatial electron distribution on surfaces with different coatings, as well as its variation with the time of exposure, i.e. what we call "beam-induced machine scrubbing" [129]. Since 2011, the electron cloud has been also observed routinely in the LHC, causing beam instability and emittance growth at the end of the multi-bunch trains but also additional heat load on the cold beam screens of the arcs as well as of the matching and final focusing quadrupoles [130]. While the effects linked to electron cloud have quickly disappeared for beams with 50 ns bunch spacing thanks to a relatively rapid beam induced machine scrubbing, running with 25 ns beams has proved to be rather challenging in this machine. With this type of beams, even after extensive machine scrubbing, the undesired effects of the electron cloud have remained visible on the beam and the machine equipment. In particular, the 25 ns beam needs to be stabilised with high values of chromaticity in both planes and large octupole settings. Besides, the large heat load on the cold beam screens still remains very close to the capacity of the cryogenic system in nominal operating conditions [131, 132]. While these effects have not prevented running LHC close to the nominal conditions from 2015 to 2018, they could still be a significant showstopper for future operation with double beam current in the High Luminosity LHC era [133]. Concerning lepton machines, the electron cloud is typically associated with a reduction of specific luminosity in e-p colliders and is expected to be one of the main limiting factors for the damping rings of future linear collider projects. The Cornell Electron Storage Ring (CESR) was reconfigured in 2008 as a Test Accelerator (CesrTA) for a program of electron cloud research with lepton beams. With its new local diagnostics for measurement of cloud density and improved instrumentation for the characterization of the beam dynamics of high intensity bunch trains interacting with the cloud, this test facility provided for many years both a benchmark case for the existing simulation codes and testing the effectiveness of several types of countermeasures [134]. Since the processes of secondary electron emission and elastic reflection at the walls play a fundamental role in causing beam induced multipacting, we now shortly describe their key parameters. The true secondary yield for perpendicular incidence, δ, can be expressed by a universal function [135]

$$\delta(x) = \delta_{\max} \frac{sx}{s - 1 + x^2}, \qquad (4.41)$$

where $x = E/E_{\max}$, with E the energy of the incident electron and $E_{\max}$ the energy at which the yield assumes the maximum value, and s is a fit parameter that was measured to be about equal to 1.35 for LHC Cu samples [135]. The two variable parameters in Eq. (4.41) are $\delta_{\max}$, the maximum yield, which typically assumes values between about 1.0 and 3.0 for conductive materials (but it can be higher for dielectrics), and $E_{\max}$. For non-normal incidence of the primary electron these two parameters are usually both increased by a factor depending on the cosine of the incidence angle [136]. Elastic reflection of electrons is mostly important at low energies, i.e. below about 20 eV. The measured electron reflection probability [122] can be parametrized as

$$\delta_{\mathrm{el}}(E) = \left(\frac{\sqrt{E} - \sqrt{E + E_0}}{\sqrt{E} + \sqrt{E + E_0}} \right)^2, \qquad (4.42)$$

with only one fit parameter, E_0. Equation (4.42) implies that the reflection probability approaches one in the limit of vanishing electron energy, even if presently no general consensus has been reached around this point, which is still very controversial, as it is extremely difficult to measure the secondary emission yield at very low energy.

The electron cloud build-up saturates when the electron losses balance the electron generation rate. This can happen either at low bunch charges, when the average neutralization density is reached, or at high bunch currents, when the electrons rapidly accumulate until the kinetic energy of the newly emitted ones becomes too low to let them penetrate into the space charge field of the cloud. Simulations have demonstrated a complex behaviour of the electron cloud equilibrium, which strongly depends on the combination between bunch length, charge, spacing and on the chamber radius. First of all, the saturation phase is generally characterized by an oscillating behaviour of the electron cloud density over the bunch spacing and the amplitude of this oscillation can be very large. Furthermore, in some cases the steady-state value of the electron cloud density has been found not to be monotonically increasing with the bunch intensity. For instance, a beam with 50 ns spaced bunches in the SPS is predicted to hit its highest electron cloud equilibrium density for bunch populations of about 10^{11} p, while this value decreases both for lower and higher intensities.

The electron density decays after the passage of a bunch train (or in the gap between bunch trains) and two different regimes can be distinguished during this phase. In the first one, right after the train passage, the cloud decays quickly due to the space charge effects and the reminiscent energy distribution from the last bunch passage. In the second one, only low energy electrons will be left, which move slowly and exhibit a dissipation rate depending on their probability of being elastically reflected at the chamber surface. Due to the second part of the decay evolution, the electron clearing time between bunch trains can become painfully long. Three effects are suspected to be responsible for the long memory and lifetime of the electron cloud. First, nonuniform fields, such as quadrupoles or sextupoles, may act as magnetic bottles and trap electrons for an indefinite time period [125, 137]. Second, if the probability of elastic reflection really approaches one in the limit of zero electron energy, as is suggested by measurements [122], low-energetic electrons could survive nearly forever, bouncing back and forth between the chamber walls, independently of the magnetic field. Third, slow ions produced by residual gas ionization have been also suspected to be long lived in the beam chamber and, therefore, possibly cause (or help) electron trapping and long survival time.

4.5.2 The Electron Cloud Instability

When a positron/hadron beam interacts electromagnetically with the electron cloud that has formed in the beam chamber, a coherent oscillation of both electrons and beam particles can grow from any small initial perturbation of the beam distribution, e.g. from the statistical fluctuations due to the finite number of beam particles. This instability can be considered as a two-stream instability of the same type as studied in plasma physics. Such instabilities can be very fast, since in the new generation of high intensity rings operating with many closely spaced bunches, the density of electrons can become quickly very large. Even machines operating with bunches spaced by hundreds of ns can actually suffer from electron cloud, because of the long survival time of low energy electrons in the beam pipe. Electron clouds can cause single-bunch instabilities as well as multi-bunch dipole mode instabilities. The multi-bunch instability appears when the electron cloud can carry a sufficiently long memory as to couple subsequent bunches. The single-bunch phenomenon, instead, is driven by a pinched electron cloud, which, over one single passage of the bunch through it, is able to transfer information from an offset bunch head to the bunch tail. Obviously, although this second type of instability is caused by a single-bunch mechanism, it can only occur in multi-bunch operation, since the electron cloud requires a train of several bunches to build up. For single-bunch instabilities caused by multi-bunch built electron clouds, electrons usually only perform a low number of oscillations while the bunch is passing (typically between fractions of unit and few units), and the bunch effectively interacts with a pre-existing cloud produced by the preceding bunches and filling almost uniformly the beam pipe prior to the bunch arrival. The number of electrons does not change appreciably during one bunch passage. In reality, another possible head tail effect resulting into a different type of two-stream instability was observed in some machines operating with long bunches. In this case, the instability is intimately related to an electron cloud from “trailing-edge multipacting”, described in the previous section. The electrons multiplication happening over the falling edge of the bunch can reach levels as to render the beam unstable. Even coasting beams are not immune from electron cloud problems. The electrons produced from residual gas ionization remain trapped in the transverse potential of the uniform beam and tend to accumulate to very high central density values. The electrons created at the chamber walls (e.g. from beam loss) are accelerated and decelerated in the beam field and eventually hit the chamber with the same energy with which they were emitted. Multipacting can play here a role, since electrons can gain energy and create secondaries when hitting the wall, if the beam line density is perturbed. The interaction of the coasting beam with the electrons can make noise evolve into an unstable coupled oscillation, called e–p instability, which was widely studied already at the beginning of the 70s [138, 139]. While the single-bunch instability described earlier in this section can be treated separately from the build-up of the electron cloud that causes it, in all other cases the two processes are coupled together and need to be solved with a joint model.

Electron cloud instabilities for short bunches have been observed in form of emittance growth and beam loss at the KEKB LER, at the CERN PS, SPS and LHC, and at the PEP-II LER. At the KEKB LER a blow-up of the vertical beam size was already observed at the early commissioning time [140]. This blow-up was not accompanied by any coherent beam motion, which could be easily suppressed by transverse feedback and chromaticity, and the blow-up was only seen in multi-bunch operation with a narrow bunch spacing. The single-bunch two-stream instability provided a plausible explanation of the observed beam blow-up [141]. This explanation has since been reinforced by the simultaneous observation of a tune shift along the bunch train, which appears for the same bunches exhibiting vertical size blow-up. Also the experimental evidence that the installation of solenoids around the ring could increase the instability threshold in regular operation shows the relation between electron cloud and the instability. At the CERN SPS the electron cloud has been observed since the ring has been regularly operated with LHC-type bunch trains [142] and it has been held responsible for strong transverse instabilities. In the horizontal plane a low order coupled bunch instability develops within a few tens of turns after injection. In the vertical plane, a single-bunch head tail instability rises on a much shorter time. The reason of the different behavior in the two transverse planes is ascribed to the confinement of the electron cloud mostly in dipole regions, which can limit the intra-bunch electron pinching in the horizontal plane and therefore inhibit the single-bunch mechanism for instability. The horizontal (coupled-bunch) instability is cured by means of a transverse feedback system. Similar to the situation at the KEKB LER, running the SPS at high positive chromaticity can cure the vertical instability [143]. Upstream from the SPS, when the nominal LHC beam was generated by the PS machine, one of the standard signatures of the electron cloud was observed shortly before extraction: a baseline drift in electrostatic devices. However, the beam resided too shortly in the machine to become unstable, even if a dedicated experiment proved the onset of an electron cloud instability on the short bunches, if they are kept in the machine for a sufficiently long time. In the LHC, transverse beam instabilities affecting the last bunches of long 25 ns trains in both transverse planes have been systematically observed at injection [144]. The reason why the instability appears in both planes is that the integrated central density of electrons causing the instability comes mainly from the electron cloud in quadrupole magnets and therefore affects equally both planes. This instability is controlled by means of high chromaticity and high octupole strength. Besides, electron cloud instabilities have been observed also at high energy (6.5 TeV) but mainly in the vertical plane and for values of bunch currents lower than nominal. This has been explained as due to the onset of a central stripe in all the dipoles (appearing when the bunch intensity decays), which leads to an integrated electron density capable of making a 6.5 TeV beam unstable [145]. Concerning long bunches, a great deal of evidence indicates that the primary instability limiting the performance of the LANL-PSR is an e-p instability [146]. Growth of the electron cloud results from multipacting of the electrons on the walls of the vacuum chamber during passage of the trailing edge of the proton beam, when the electrons can receive a net acceleration toward the wall. The instability was controlled by various measures to enhance Landau damping and transverse feedback. In coasting beams, an e-p instability was first observed in the LBNL-Bevatron [147] and CERN-ISR [139]. While in the Bevatron this instability was combated with active feedback and beam bunching, in the ISR additional pumping was installed to improve the vacuum from 0.1 to 0.01 nTorr and the number of clearing electrodes was increased to sweep away the electrons.

Several analytical approaches have been used to study the electron cloud instability, including few-particles models and an attempt to apply the TMCI theory to the electron cloud wake field, modeled as a broadband resonator. However the most widespread and comprehensive approach makes use of particle tracking simulations with localized electron cloud kicks. The numerical modeling of the interaction between an electron cloud and a particle bunch is discussed in Sect. 4.7. Simulation codes have had the merit to reveal interesting features of the electron cloud instability, which distinguish it from other types of conventional instabilities. For example, the electron cloud wake field is not only a function of the distance between source and probe particles, but it depends on the locations of the two separately. This translates into an impedance with a double frequency dependence [148]. Another interesting finding was that, for constant beam emittances (transverse and longitudinal) and constant bunch length, the electron cloud instability threshold decreases with the beam energy [149]. The reason of this anomalous behaviour is that, although the beam becomes stiffer at higher energies, its transverse sizes become smaller and the pinching effect on the electron cloud is amplified.

Concerning the multi-bunch instability, the usual approach is to calculate the bunch-to-bunch wake field with an electron cloud build-up code (which correctly models the electron cloud dynamics in the space between two bunches) and then apply the multi-bunch analytical formula to assess the threshold for the beam stability. Simulation codes with bunches modeled as single macroparticles, valid for machines operating with short bunches, have been also developed to study numerically the multi-bunch instabilities due to electron cloud in a more self-consistent manner.

A number of simulation tools have been developed over the years in order to study the electron cloud single-bunch instability for short bunches via direct particle tracking. The simulations of electron cloud build-up are generally treated separately, since they make use of a weak-strong approach, in which the beam is rigid and is approximated by bunches with static transverse and longitudinal Gaussian distributions, while the electrons are macroparticles. Build-up simulations need to be run prior to the instability simulations, because they provide the necessary input on the transverse distribution of the electron cloud density at saturation just before the arrival of a bunch. A variety of simulation codes are presently available for this purpose [150]. Fully self-consistent computations, in which the cloud generation over a bunch train around the ring as well as the resulting bunch instabilities, are treated by a single program are still under development. The existing simulation programs to study the electron cloud instabilities model the interaction of a single bunch with an electron cloud on successive turns. The cloud is always assumed to be generated by the preceding bunches, and can be considered initially uniform or the distribution is imported from a build-up code. The electrons give rise to a head-to-tail wake field, which amplifies any initial small deformation in the bunch offset, e.g. due to the finite number of macroparticles in the simulation. All simulation tools that have been developed for this study are essentially of the strong–strong type, since the purpose is to investigate how the bunch particles are affected by the electron cloud via the continuous interaction. In particular, electrons are always modeled as macroparticles either concentrated at one or several locations along the ring or uniformly smeared along the axis of the machine. The bunch consists of macroparticles or of microbunches with a fixed transverse size. The bunch is then subdivided into slices, which interact in sequence with the electrons of the cloud, creating the distortion of the initially uniform cloud distribution that can affect the body and tail of the bunch. The electric fields of the electrons and of the beam acting mutually on each other are calculated by means of a Particle-In-Cell (PIC) algorithm. The transformation of the 6D phase-space vectors of the beam particles between two kick points is achieved using the appropriate transport matrices or nonlinear tracking. The field of the electron cloud acting on itself can be optionally included, but in general does not seem to play a significant role in this type of mechanisms and hence it is neglected. For the purpose of studying the interplay of electron cloud instability with other mechanisms, the simulation codes contain synchrotron motion, chromaticity and usually additional options to model the action of an independent impedance source beside the electron cloud, as well as space charge and detuning with amplitude.

4.5.3 Mitigation and Suppression

There are at least three possible actions to reduce, or even suppress, the electron cloud: (i) reducing the production rate of primary electrons or confining their motion to a region where they are not likely to do any harm; (ii) eliminating the possibility of multipacting by lowering the SEY via surface treatment; (iii) alleviating the effect on the beam or on the diagnostics. In most cases a combined approach is desirable, that's why most machines affected by electron cloud problems have usually chosen to implement more than one of these mitigating techniques. The primary production of photoelectrons needs to be reduced, because it may be so high that the electron cloud could reach saturation within a few bunch passages even without any multipacting. This is the case at KEKB, the photon factory, if no countermeasures were taken. The obvious solution is an antechamber to intercept most of the synchrotron radiation, or also photon absorbers (as those designed for the CLIC Damping Rings). For dipole fields, a saw-tooth pattern impressed on the chamber wall, as was implemented for the LHC (actually on the beam screen that forms the inner part of the chamber and serves to protect the cold bore of the magnets from synchrotron radiation), is used for effectively reducing the photon reflectivity thanks to the perpendicular impact. Weak solenoids of the order of 50 G are a possibility in field-free regions, which was successfully implemented in the straight sections of KEKB and in RHIC. The solenoids do not really affect the photoemission process, but they keep the photoelectrons close to the wall and, thus, strongly mitigate the subsequent beam–electron interaction. Since the gas ionization rate is linearly proportional to the vacuum pressure in the beam chamber, the number of electrons created by gas ionization can only be reduced by significant factors improving the vacuum. If field ionization is important, however, a possible cure would be lengthening the bunches, though this will mainly be a concern for future projects such as linear colliders or X-ray FELs operating with positrons. Electrons generated by beam loss can be controlled if the localization of the losses is known with good precision. For example, electrons produced by the beam losses at a collimator can be controlled by solenoids or clearing electrodes. A large number of electrons is also generated at the injection stripping foils, for accelerators employing charge-exchange injection. At the SNS, a 10-kV clearing voltage is applied to channel the electrons liberated at the stripping foil onto a collector plate that is monitored by a TV camera, while solenoids are used along the collimator straights.

The reduction of the SEY of the inner wall of the beam chamber can be achieved in different manners. First of all, a serendipitous feature of the electron cloud build up in an accelerator's chamber is that, while the electrons hit the beam chamber with high energy and multiply, they also 'scrub' the surface by first removing layers of impurities responsible for high SEY values and eventually graphitising the surface with a further reduction of the SEY from that of the pure metal [151]. This means that, if a method is found to run an accelerator with electron cloud and stable beam, e.g. by stabilising the beam against the electron cloud through appropriate machine settings, it will be the electron cloud itself to gradually lower the SEY of the inner wall of the chamber and eventually turn itself off. This obviously relies on that the final SEY reachable through scrubbing is below the value that sets off the electron cloud build up in the operational configuration. Besides, it may take a significant amount of time to reach this condition, because the electron flux is decreasing while scrubbing and the electron doses required to perform further SEY reduction steps are also exponentially increasing when moving to SEYs below 1.3–1.4. The technique described here is what we call 'beam induced machine scrubbing' and machine like the SPS and LHC fully rely on it to run successfully with 25 ns spaced beams. While beam induced scrubbing is an important option for already built machines, coatings with intrinsically low SEY materials can be envisaged at the design stage to limit the creation of an electron cloud in future machines. A well established method to reduce multipacting is coating with TiN, a material whose secondary emission yield becomes quickly low after some conditioning (through illumination under synchrotron light). The thickness of the coating must be of the order of a μm, such as not to alter the resistive impedance seen by the beam. A more favorable getter material made from TiZrV, called Non-Evaporable Getter (NEG), was developed at CERN. This getter material is characterized by its greater structural stability than TiN, its pumping capability and its low activation temperature. The warm sections of the LHC, about 10% of the circumference, have been coated with NEG. The NEG coating was also tested at several light source insertion devices, where circumstantial evidence suggests an increase in the effective impedance, presumably due to a larger surface roughness and low conductivity. The additional contributions to the ring impedance from the surface roughness and low conductivity impedance of the coating layer is of no concern for the longer proton bunches in the LHC, but could significantly affect the stability of the short positron beams in the Damping Rings of a future linear collider. From 2007 on, new efforts have been put on the search for coating materials that do not require high temperature activation and do not suffer from aging. In particular, amorphous carbon (a-C) thin films, deposited with d.c. magnetron sputtering, have shown to possess all these qualities. Besides, their maximum secondary emission yields, measured in the lab, reach values even below one and the films are also stable against mechanical stress. Testing of a-C coating in accelerator environments (SPS and Cesr-TA) has demonstrated all these features. Another method to reduce the secondary emission yield of a surface is to use a naturally rough material. Here the SEY reduction is a geometrical effect due to the high probability of quick re-absorption of the electrons emitted with low energy.

Multipacting can also be suppressed by solenoids, though one should pay attention to the possibility of exciting undesired cyclotron resonances. Electric clearing fields are an efficient cure, as shown both in simulations and measurements of electron cloud in the CERN PS. They were already used to cure electron-proton instabilities for the coasting proton beams in the CERN ISR during the early 70s. At the SNS operating with long proton bunches all BPMs can be biased with a clearing voltage of 1 kV. To be effective for the multipacting experienced by short bunches with close spacing, the clearing electrodes must be mounted all around the ring, in distances of a few tens of cm and voltages of the order 1 kV are probably required. The impedance introduced by many such devices could be prohibitive, as it appeared to be the case in the DAΦNE positron ring with the very first clearing electrode design. Other options for a practical implementation of electric clearing fields may be splitting the beam pipe into a top and bottom half, isolated from each other and held at different potential. Biasing the two jaws of a collimator against each other is a similar idea. Recently, there is a growing interest towards the suppression of multipacting by means of grooves on the chamber wall. This technique, first tested in simulations, has proven to be efficient in KEKB and Cesr-TA. Similarly to a rough surface, but in a controlled way and on a macroscopic scale, these grooves essentially act as electron traps. Angle and depth of the grooves are key parameters and specifications are different in dipole or field-free regions. Proper tailoring of the bunch filling patterns (bunch spacing, bunch trains and bunch charges) is yet another way of achieving an acceptable electron density. Examples include the actual bunch spacing chosen for PEP-II and KEKB operation, which are twice or three times the design spacing, and satellite bunches proposed for the LHC [152]. Gaps within or between trains can lower the density and reset the cloud at least to some extent. Extensive studies of the electron cloud formation as a function of the bunch filling patterns were also carried out at RHIC, in which the optimization could be found using the maps approach to quickly scan the build-up for different configurations.

The electron cloud causes a large variety of undesired effects. Common stabilising measures can be taken against the resulting instabilities, which include transverse bunch-to-bunch feedback, increased chromaticity, Landau-damping octupoles, intra-bunch head–tail feedback, and linear coupling. All these measures are anyway necessary when a machine is operating in beam-induced scrubbing mode. Degradation of diagnostics signals due to impacting electrons can be also overcome with local solenoid windings.

4.6 Beam–Beam Effects

W. Herr

4.6.1 Introduction

The problem of the beam–beam interaction is the subject of many studies since the introduction of the first particle colliders [153]. It has been and will be one of the most important limits to the performance and therefore attracts the interest at the design stage of a new colliding beams facility. A particle beam is a collection of a large number of charges and represents an electromagnetic potential for other charges. It will therefore exert forces on itself and other beams. The forces are most important for high density beams, i.e. high intensity and small beam sizes, which are the key to high luminosity.

The electromagnetic forces from particle beams are very non-linear and result in a wide spectrum of consequences for the beam dynamics. Furthermore, as a result of the interaction, the charge distribution creating the disturbing fields can change as well. This has to be taken into account in the evaluation of beam–

beam effects and in general a self-consistent treatment is required. Although we now have a good qualitative understanding of the various phenomena, a complete theory does not exist and exact predictions are still difficult. Numerical techniques such as computer simulations have been used with great success to improve the picture on some aspects of the beam–beam interaction while for other problems the available models are not fully satisfactory in their predictive power [154].

4.6.2 Beam–Beam Force

In the rest frame of a beam we have only electrostatic fields and to find the forces on other moving charges, we have to transform the fields into the moving frame and to calculate the Lorentz forces (see [153, 155–160] and references therein).

The fields are obtained by integrating over the charge distributions. The forces can be defocusing or focusing since the test particle can have the same or opposite charge with respect to the beam producing the forces.

The distribution of particles producing the fields can follow various functions, leading to different fields and forces. It is not always possible to integrate the distribution to arrive at an analytical expression for the forces in which case either an approximation or numerical methods have to be used. This is in particular true for hadron beams, which usually do not experience significant synchrotron radiation and damping. For $e^- e^+$ colliders the distribution functions are most likely Gaussian with truncated tails.

In the two-dimensional case of a beam with bi-Gaussian beam density distributions in the transverse planes, i.e. $\rho(x, y) = \rho_x(x)\, \rho_y(y)$ with r.m.s. of σ_x and σ_y

$$\rho_u(u) = \frac{1}{\sigma_u \sqrt{2\pi}} \exp\left(-\frac{u^2}{2\sigma_u^2}\right) \quad \text{where } u = x, y \tag{4.43}$$

one can give the two-dimensional potential $U(x, y, \sigma_x, \sigma_y)$ as a closed expression

$$U\left(x, y, \sigma_x, \sigma_y\right) = \frac{ne}{4\pi\varepsilon_0} \int_0^\infty \frac{\exp\left(-\frac{x^2}{2\sigma_x^2+q} - \frac{y^2}{2\sigma_y^2+q}\right)}{\sqrt{\left(2\sigma_x^2 + q\right)\left(2\sigma_y^2 + q\right)}} dq \tag{4.44}$$

where n is the line density of particles in the beam, e is the elementary charge and ε_0 the permittivity of free space [159]. From the potential one can derive the transverse fields $\vec{E}$ by taking the gradient $\vec{E} = -\nabla U\left(x, y, \sigma_x, \sigma_y\right)$.

4.6.2.1 Elliptical Beams

For the above case of bi-Gaussian distributions (i.e. elliptical beams with $\sigma_x \neq \sigma_y$) the fields can be derived and for the case of $\sigma_x > \sigma_y$ we have [160]

$$E_x = \frac{ne}{2\varepsilon_0\sqrt{2\pi\left(\sigma_x^2-\sigma_y^2\right)}}\,\mathrm{Im}\times\left[\mathrm{erf}\left(\frac{x+iy}{\sqrt{2\left(\sigma_x^2-\sigma_y^2\right)}}\right)-\exp\left(-\frac{x^2}{2\sigma_x^2}+\frac{y^2}{2\sigma_y^2}\right)\mathrm{erf}\left(\frac{x\frac{\sigma_y}{\sigma_x}+iy\frac{\sigma_x}{\sigma_y}}{\sqrt{2\left(\sigma_x^2-\sigma_y^2\right)}}\right)\right] \quad (4.45)$$

$$E_y = \frac{ne}{2\varepsilon_0\sqrt{2\pi\left(\sigma_x^2-\sigma_y^2\right)}}\,\mathrm{Re}\times\left[\mathrm{erf}\left(\frac{x+iy}{\sqrt{2\left(\sigma_x^2-\sigma_y^2\right)}}\right)-\exp\left(-\frac{x^2}{2\sigma_x^2}+\frac{y^2}{2\sigma_y^2}\right)\mathrm{erf}\left(\frac{x\frac{\sigma_y}{\sigma_x}+iy\frac{\sigma_x}{\sigma_y}}{\sqrt{2\left(\sigma_x^2-\sigma_y^2\right)}}\right)\right] \quad (4.46)$$

The function erf(t) is the complex error function

$$\mathrm{erf}(t) = \exp\left(-t^2\right)\left[1+\frac{2i}{\sqrt{\pi}}\int_0^t \exp\left(z^2\right)dz\right] \quad (4.47)$$

The magnetic field components follow from

$$B_y = -\beta\frac{E_x}{c} \text{ and } B_x = \beta\frac{E_y}{c} \quad (4.48)$$

The Lorentz force acting on a particle with charge q is finally

$$\vec{F} = q\left(\vec{E}+\vec{v}\times\vec{B}\right) \quad (4.49)$$

4.6.2.2 Round Beams

With the simplifying assumption of round beams ($\sigma_x = \sigma_y = \sigma$), one can re-write (4.49) in cylindrical coordinates

$$\vec{F} = q\left(E_r+\beta c B_\phi\right)\times\vec{r} \quad (4.50)$$

From (4.44) and with $r^2 = x^2 + y^2$ one can immediately write the fields from (4.50) as

$$E_r = -\frac{ne}{4\pi\varepsilon_0}\frac{\delta}{\delta_r}\int_0^\infty \frac{\exp\left(-\frac{r^2}{2\sigma^2+q}\right)}{2\sigma^2+q}dq \tag{4.51}$$

and

$$B_\phi = -\frac{ne\beta c\mu_0}{4\pi}\frac{\delta}{\delta_r}\int_0^\infty \frac{\exp\left(-\frac{r^2}{2\sigma^2+q}\right)}{2\sigma^2+q}dq \tag{4.52}$$

δ stands for derivative (Eqs. 4.51 and 4.52) and μ_0 is the permeability of free space (Eq. 4.52).

We find from (4.51) and (4.52) that the force (4.50) has only a radial component. The expressions (4.51) and (4.52) can easily be evaluated when the derivative is done first and $1/(2\sigma^2 + q)$ is used as integration variable. We can now express the radial force in a closed form (using $\varepsilon_0\mu_0 = c^{-2}$)

$$F_r(r) = -\frac{ne^2\left(1+\beta^2\right)}{2\pi\varepsilon_0}\frac{1}{r}\left[1-\exp\left(-\frac{r^2}{2\sigma^2}\right)\right] \tag{4.53}$$

and for the Cartesian components in the two transverse planes we get

$$F_x(r) = -\frac{ne^2\left(1+\beta^2\right)}{2\pi\varepsilon_0}\frac{x}{r^2}\left[1-\exp\left(-\frac{r^2}{2\sigma^2}\right)\right] \tag{4.54}$$

and

$$F_y(r) = -\frac{ne^2\left(1+\beta^2\right)}{2\pi\varepsilon_0}\frac{y}{r^2}\left[1-\exp\left(-\frac{r^2}{2\sigma^2}\right)\right] \tag{4.55}$$

The forces (4.54) and (4.55) are computed when the charges of the test particle and the opposing beam have opposite signs. For equally charged beams the forces change signs. For small amplitudes the force is approximately linear and a particle crossing a beam at small amplitudes will experience a linear field. This results in a change of the tune like in a quadrupole. At larger amplitudes (i.e. above ~1σ) the force deviates strongly from this linear behaviour. Particles at larger amplitudes will also experience a tune change, however this tune change will depend on the amplitude. Already from the analytical form (4.55) one can see that the beam–beam force includes higher multipoles.

4.6.3 *Incoherent Effects: Single Particle Effects*

The force we have derived is the force of a beam on a single test particle. It can be used to study single particle or incoherent effects. For that we treat a particle crossing a beam like it was moving through a static electromagnetic lens. We have to expect all effects that are known from resonance and non-linear theory such as

- Unstable and/or irregular motion
- Beam blow up or bad lifetime

4.6.3.1 Beam–Beam Parameter

We can derive the linear tune shift of a small amplitude particle crossing a round beam of a finite length. We use the force to calculate the kick it receives from the opposing beam, i.e. the change of the slope of the particle trajectory. Starting from the two-dimensional force and multiplying with the longitudinal distribution which depends on both position *s* and time *t*, and assuming a Gaussian shape with a width of σ_s

$$F_r(r,s,t) = -\frac{Ne^2(1+\beta^2)}{\sqrt{(2\pi)^3}\varepsilon_0\sigma_s}\frac{1}{r}\left[1-\exp\left(-\frac{r^2}{2\sigma^2}\right)\right]\exp\left[-\frac{(s+v\,t)^2}{2\sigma_s^2}\right]$$

Now N is the total number of particles. We make use of Newton's law and integrate over the collision to get the radial deflection

$$\Delta r' = \frac{1}{mc\beta\gamma}\int_{-\infty}^{\infty} F_r(r,s,t)\; dt$$

The radial kick $\Delta r'$ a particle with a radial distance r from the opposing beam centre receives is then

$$\Delta r' = -\frac{2Nr_0}{\gamma}\frac{1}{r}\left[1-\exp\left(-\frac{r^2}{2\sigma^2}\right)\right] \tag{4.56}$$

where I have re-written the constants and use the classical particle radius

$$r_0 = \frac{e^2}{4\pi\varepsilon_0 mc^2} \tag{4.57}$$

where m is the mass of the particle. After the integration along the bunch length, N is the total number of particles. For small amplitudes r one can derive the asymptotic limit

$$\Delta r'\big|_{r\to 0} = -\frac{N r_0 r}{\gamma \sigma^2} = -r\ f \tag{4.58}$$

This limit is the slope of the force at $r = 0$ and the force becomes linear with a focal length as the proportionality factor.

It is well known how the focal length relates to a tune change and one can derive a quantity ξ which is known as the linear beam–beam parameter

$$\xi = \frac{N r_0 \beta^*}{4\pi \gamma \sigma^2} \tag{4.59}$$

r_0 is the classical particle radius, (e.g.: r_e, r_p) and β^* is the optical amplitude function (β-function) at the interaction point.

For small values of ξ and a tune far enough away from linear resonances this parameter is equal to the linear tune shift ΔQ.

The beam–beam parameter can be generalized for the case of non-round beams and becomes

$$\xi_{x,y} = \frac{N r_0 \beta^*_{x,y}}{2\pi \gamma \sigma_{x,y} \left(\sigma_x + \sigma_y\right)} \tag{4.60}$$

The beam–beam parameter is often used to quantify the strength of the beam–beam interaction, however it does not reflect the non-linear nature.

4.6.3.2 Non-linear Effects

Since the beam–beam forces are strongly non-linear, the study of beam–beam effects encompasses the entire field of non-linear dynamics (see earlier chapter) as well as collective effects. First, we briefly discuss the immediate effect of the non-linearity of the beam–beam force on a single particle. It manifests as an amplitude dependent tune shift and for a beam with many particles as a tune spread. The instantaneous tune shift of a particle when it crosses the other beam is related to the derivative of the force with respect to the amplitude $\delta F/\delta x$. For a particle performing an oscillation with a given amplitude the tune shift is calculated by averaging the slopes of the force over the range (i.e. the phases) of the particle's oscillation amplitudes. An elegant calculation can be done using the Hamiltonian formalism [156] developed for non-linear dynamics and as demonstrated in the chapter on non-linear dynamics using the Lie formalism. We get the formula for the non-linear detuning with the amplitude J

$$\Delta Q(J) = \xi \frac{2}{J}\left[1 - I_0\left(\frac{J}{2}\right) \mathrm{e}^{-\frac{J}{2}}\right] \tag{4.61}$$

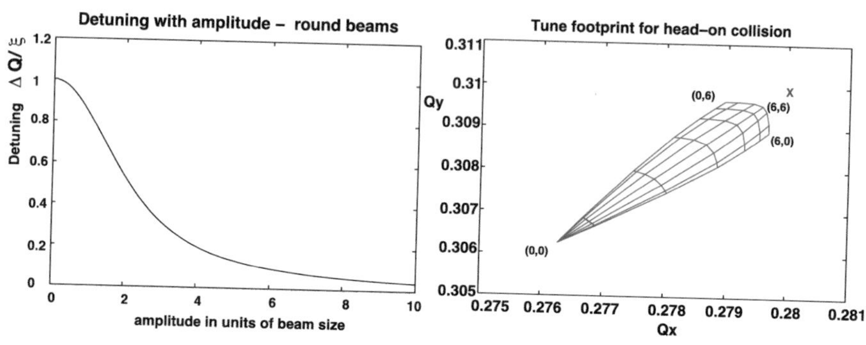

Fig. 4.22 Tune shift (non-linear detuning) as a function of the amplitude (left) and 2-dimensional tune footprint (right)

where $I_0(x)$ is the modified Bessel function and $J = \varepsilon\beta/2\sigma^2$ in the usual units. Here ε is the particle "emittance" and not the beam emittance.

In the 2-dimensional case, the tune shifts (ΔQ_x, ΔQ_y) of a particle with amplitudes x and y depend on both, horizontal and vertical amplitudes. The detuning must be computed and presented in a 2-dimensional form, i.e. the amplitude (x, y) is mapped into the tune space (Q_x, Q_y) or alternatively to the 2-dimensional tune change (ΔQ_x, ΔQ_y). Such a presentation is usually called a "tune footprint" and an example is shown in Fig. 4.22(right) and it maps the amplitudes into the tune space and each "knot" of the mesh corresponds to a pair of amplitudes. Amplitudes between 0 and 6σ in both planes are used. The cross indicates the original, unperturbed tunes without the beam–beam interaction.

The maximum tune spread for a single head-on collision is equal to the tune shift of a particle with small amplitudes and for small tune shifts equal to the beam–beam parameter ξ. In the simple case of a single head-on collision the parameter ξ is therefore a measure for the tune spread in the beam.

4.6.3.3 Beam Stability

When the beam–beam interaction becomes too strong, the beam can become unstable or the beam dynamics is strongly distorted. One can distinguish different types of distortions and a few examples are

- Non-linear motion can become stochastic and can result in a reduction of the dynamic aperture and particle loss and bad lifetime.
- Distortion of beam optics: dynamic beta (LEP) [161].
- Vertical blow-up above the so-called beam–beam limit.

Since the beam–beam force is very non-linear, the motion can become "chaotic". This often leads to a reduction of the available dynamic aperture. The dynamic aperture is the maximum amplitude where the beam remains stable. Particles outside the dynamic aperture are eventually lost. The dynamic aperture is usually evaluated by tracking particles with a computer program through the machine where they experience the fields from the machine elements and other effects such as wake fields or the beam–beam interaction.

Since the beam–beam interaction is basically a very non-linear lens in the machine, it distorts the optical properties and it may create a noticeable beating of the β-function around the whole machine and at the location of the beam–beam interaction itself. This can be approximated by inserting a quadrupole which produces the same tune shift at the position of the beam–beam interaction. The r.m.s. beam size at the collision point is now proportional to $\sqrt{\beta_p^*}$, where $\beta_p^*$is the perturbed β-function which can be significantly different from the unperturbed β-function $\beta*$. This in turn changes the strength of the beam–beam interaction and the parameters have to be found in a self-consistent form. This is called the dynamic beta effect. This is a first deviation from our assumption that the beams are static non-linear lenses. A strong dynamic beta effect was found in LEP [161] due to its very large tune shift parameters.

Another effect that can be observed in particular in $e^+ e^-$ colliders is the blow up of the emittance which naturally limits the reachable beam–beam tune shifts.

4.6.3.4 Beam–Beam Limit

In $e^+ e^-$ colliders the beam sizes are usually an equilibrium between the damping due to the synchrotron radiation and heating mechanisms such as quantum excitation, intra-beam scattering and very importantly, the beam–beam effect. This leads to a behaviour that is not observed in a hadron collider. When the luminosity is plotted as a function of the beam intensity, it should increase approximately as the current squared [162], in agreement with

$$\mathcal{L} = \frac{N^2 k f}{4\pi \sigma_x \sigma_y} \tag{4.62}$$

Here k is the number of bunches per beam and f the revolution frequency [162]. At the same time the beam–beam parameter ξ should increase linearly with the beam intensity according to (4.60)

$$\xi_y = \frac{N r_e \beta_y}{2\pi \gamma \sigma_y \left(\sigma_x + \sigma_y\right)} \tag{4.63}$$

In all $e^+ e^-$ colliders the observation can be made that above a certain current, the luminosity increases approximately proportional to the current, or at least much less than with the second power. Another observation is that at the same value of the intensity the beam–beam parameter ξ saturates. This limiting value of ξ is commonly known as the beam–beam limit.

When we re-write the luminosity as

$$\mathcal{L} = \frac{N^2 k f}{4\pi \sigma_x \sigma_y} = \frac{N k f}{4\pi \sigma_x} \frac{N}{\sigma_y} \tag{4.64}$$

we get an idea of what is happening. In $e^+ e^-$ colliders the horizontal beam size σ_x is usually much larger than the vertical beam size σ_y and changes very little. In order for the luminosity to increase proportionally to the intensity *N*, the factor N/σ_y must be constant. This implies that with increasing current the vertical beam size increases in proportion above the beam–beam limit. This has been observed in all $e^+ e^-$ colliders and since the vertical beam size is usually small, this emittance growth can be very substantial before the life time of the beam is affected or beam losses are observed [163].

The dynamics of machines with high synchrotron radiation is dominated by the damping properties and the beam–beam limit is not a universal constant nor can it be predicted. Simulation of beams with many particles can provide an idea of the order of magnitude [164, 165].

4.6.4 Studies of Head-on Collisions at the LHC

The layout of experimental regions in the LHC is shown in Fig. 4.23. The beams travel in separate vacuum chambers and cross in the experimental areas where they share a common beam pipe. In these common regions the beams experience head-on collisions as well as a large number of long range beam–beam encounters [166]. This arrangement together with the bunch filling scheme of the LHC as shown in Fig. 4.24 [166, 167] leads to very different collision pattern for different bunches, often referred to as “PACMAN” bunches. The number of both, head-on as well as long range encounters, can be very different for different bunches in the bunch trains and lead to a different integrated beam–beam effect [167]. This was always a worry in the LHC design and the effects have been observed in an early stage of the commissioning. Strategies have been provided to minimize these effects, e.g. different planes for the crossing angles [166, 167].

4.6.4.1 PACMAN Bunches

The bunches in the LHC do not form a continuous train of equidistant bunches spaced by 25 ns, but some empty space must be provided to allow for the rise time of kickers (Fig. 4.24). These gaps and the number of bunches per train are determined by requirements from the LHC injectors. The whole LHC bunch pattern is composed of 39 smaller trains (each with 72 bunches) separated by gaps of various length followed by a large abort gap for the dump kicker. Due to the symmetry,

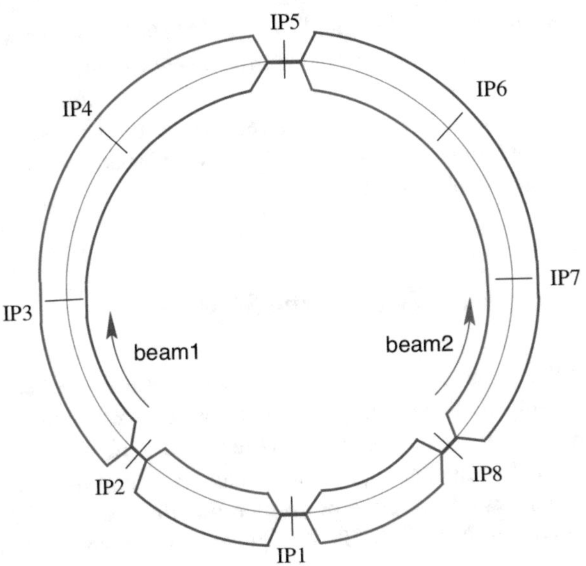

Fig. 4.23 Layout of the experimental collision points in the LHC [166]

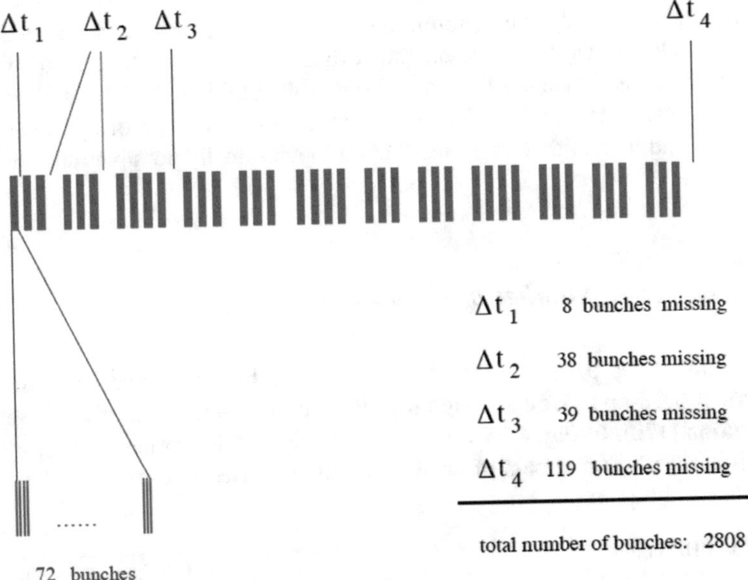

Fig. 4.24 Bunch filling scheme of the nominal LHC

bunches normally meet other bunches at the head-on collision point. For the long-range interactions this is no longer the case. Bunches at the beginning and at the end of a small train will encounter a hole and as a result experience fewer long-range interactions than bunches from the middle of a train [168]. Bunches with fewer long-range interactions have a very different integrated beam–beam effect and a different dynamics must be expected. In particular they will have a different tune and occupy a different area in the working diagram, therefore may be susceptible to resonances which can be avoided for nominal bunches. The overall space needed in the working diagram is therefore largely increased [166, 168].

4.6.5 Head-on Beam–Beam Tune Shift

The nominal LHC parameters have been chosen to reach the design luminosity of 10^{34} cm^{-2}s^{-1} [166]. The main parameters relevant for beam–beam effects are summarized in Table 4.1. At a very early stage of the LHC operation it was tested whether the nominal beam–beam parameters can be achieved. After this has been successfully demonstrated, we have performed a dedicated experiment to test the achievable beam–beam tune shift. To that purpose we have filled the LHC with single bunches per beam, colliding in IP1 and IP5 (see Fig. 4.23). We have used bunch intensities of ~1.9×10^{11} p/b, i.e. well above the nominal and the emittances have been reduced below 1.20 μm in both planes. It was shown that such bunches can be collided in both interaction points without significant losses or emittance increase [169] and we have demonstrated that a beam–beam tune shift of 0.017 for a single interaction and an integrated tune shift of 0.034 for both collision was possible. These tune shifts have been obtained in the absence of any long range encounters and it should be expected that the operationally possible tune shifts are lower.

4.6.6 Effect of Number of Head-on Collisions

Due to the filling pattern in the LHC, different bunches experience different numbers of head-on as well as long range interactions. Details are given in another contribution [170]. In Fig. 4.25 we show as illustration the losses of bunches with very different (0–3) numbers of head-on collisions. The data was taken during a

Table 4.1 LHC nominal parameters and achieved during operation and experiments in 2010/2011

Parameter	Nominal	Achieved
Intensity (p/bunch)	1.15×10^{11}	2.3×10^{11}
Emittance	3.75 μm	≤2.00 μm
β_*	0.55 m	1.5 m
ξ/IP	0.0035	0.0170
Bunch spacing	25 ns	50 ns
Bunches/beam	2808	1380

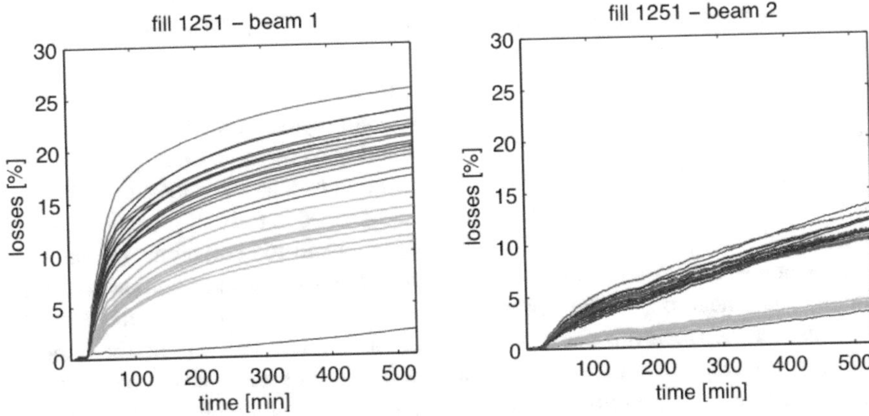

Fig. 4.25 Losses of bunches with different number of head-on collisions [170]. Numerology: blue (3 collisions), red (2 collisions), green (1 collision), black (no collision)

regular operational fill of 10 h duration. The correlation between losses and number of head-on collisions is apparent and a more detailed analysis is found in [170]. The transverse emittances during normal operation are larger (~2.5 μm) than in the head-on test. In a second experiment we increased the bunch intensity further to ~2.3×10^{11} p/b with emittances of ~1.80 μm. Although the tune shift was slightly lower than in the previous experiment (0.015), the lifetime was worse. We interprete these results as losses of particles at large amplitudes. This is supported by the observation that the strongest losses occur at the very beginning of a fill (Fig. 4.25).

4.6.7 *Crossing Angle and Long Range Interactions*

To reach the highest luminosity, it is desirable to operate a collider with as many bunches as possible since the luminosity is proportional to their number (4.62) [162].

In a single ring collider such as the SPS, Tevatron or LEP, the operation with k bunches leads to $2k$ collision points. When k is a large number, most of them are unwanted and must be avoided to reduce the perturbation due to the beam–beam effects. Various schemes have been used to avoid these unwanted "parasitic" interactions. In the SPS, Tevatron and in LEP so-called Pretzel schemes were used. When the bunches are equidistant, this is the most promising method. When two beams of opposite charge travel in the same beam pipe, they can be moved onto separate orbits using electrostatic separators. In a well-defined configuration the two beams cross when the beams are separated. To avoid a separation around the whole machine, the bunches can be arranged in so-called trains of bunches following each other closely. In that case a separation with electrostatic separators is only needed around the interaction regions. Such a scheme was used in LEP in the second phase [171].

Contrary to the majority of the colliders, the LHC collides particles of the same type which therefore must travel in separate beam pipes. At the collision points of the LHC the two beams are brought together and into collision. During that process it is unavoidable that the beams travel in a common vacuum chamber for more than 120 m. In the LHC the distance between the bunches is only 25 ns and therefore the bunches will meet in this region. In order to avoid the collisions, the bunches collide at a small crossing angle of 285 μrad. While two bunches collide at a small angle (quasi head-on) at the centre, the other bunches are kept separated by the crossing angle. However, since they travel in a common beam pipe, the bunches still feel the electromagnetic forces from the bunches of the opposite beam. When the separation is large enough, these so-called long-range interactions should be weak.

4.6.7.1 Long-Range Beam–Beam Effects

Although the long-range interactions distort the beams much less than a head-on interaction, their large number and some particular properties require careful studies:

- They break the symmetry between planes.
- While the effect of head-on collisions is strongest for small amplitude particles, they mostly affect particles at large amplitudes.
- The tune shift caused by long-range interactions has opposite sign in the plane of separation compared to the head-on tune shift.
- They cause changes of the closed orbit [153].
- They largely enhance so-called PACMAN effects [168].

4.6.7.2 Opposite Sign Tune Shift

The opposite sign of the tune shift can easily be understood when we average the oscillation of a small amplitude particle as it samples the focusing force of the beam–beam interaction. When the separation is larger than ~1.5σ, the focusing (slope of the force as a function of the amplitude) changes the sign and the resulting tune shift assumes the opposite sign.

To some extend this property could be used to partially compensate long-range interactions when a configuration is used where the beams are separated in the horizontal plane in one interaction region and in the vertical plane in another one.

4.6.7.3 Strength of Long-Range Interactions

Assuming a separation d in the horizontal plane, the kicks in the two planes can be written as

$$\Delta x' = -\frac{2Nr_0}{\gamma}\frac{(x+d)}{r^2}\left(1 - \mathrm{e}^{-\frac{r^2}{2\sigma^2}}\right) \quad (4.65)$$

with $r^2 = (x+d)^2 + y^2$. The equivalent formula for the plane orthogonal to the separation is

$$\Delta y' = -\frac{2Nr_0}{\gamma}\frac{y}{r^2}\left(1 - \mathrm{e}^{-\frac{r^2}{2\sigma^2}}\right) \quad (4.66)$$

The effect of long-range interactions must strongly depend on the separation. The calculation shows that the tune spread ΔQ_{lr} from long-range interactions alone follows an approximate scaling (for large enough separation, i.e. above ~6σ)

$$\Delta Q_{lr} \propto -\frac{N}{d^2} \quad (4.67)$$

where N is the bunch intensity and d the separation. Small changes in the separation can therefore result in significant differences. Since the symmetry between the two planes is broken, the resulting footprint shows no symmetry. In fact, the tune shifts have different signs for x and y, as expected.

4.6.7.4 Footprint for Long-Range Interactions

Contrary to the head-on interaction where the small amplitude particles are mostly affected, now the large amplitude particles experience the strongest long-range beam–beam perturbations. This is rather intuitive since the large amplitude particles are the ones which can come closest to the opposing beam as they perform their oscillations. We must therefore expect a totally different tune footprint. Such a footprint for only long-range interactions is shown in Fig. 4.26.

4.6.8 Studies of Long Range Interactions in the LHC

To study the effect of long range beam–beam interactions we have performed a dedicated experiment [172]. The LHC was set up with single trains of 36 bunches per beam, spaced by 50 ns. The bunch intensities were ~1.2×10^{11} p/b and the normalized emittances around 2.5 μm. The trains collided in IP1 and IP5, leading to a maximum of 16 long range encounters per interaction point for nominal bunches. First, the crossing angle (vertical plane) in IP1 was decreased in small steps and the

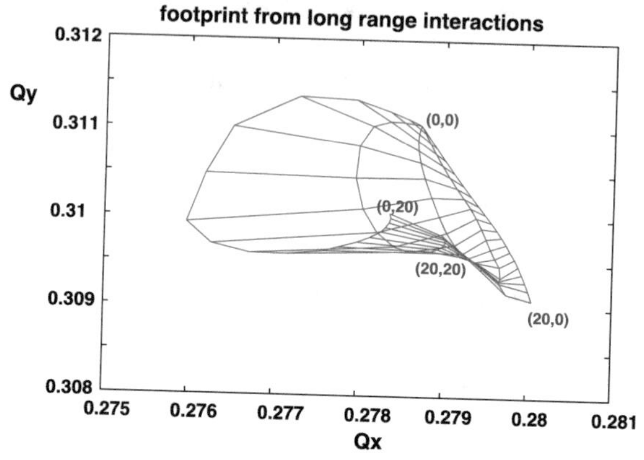

Fig. 4.26 Tune footprint for long-range interactions only. Vertical separation and amplitudes between 0 and 20σ

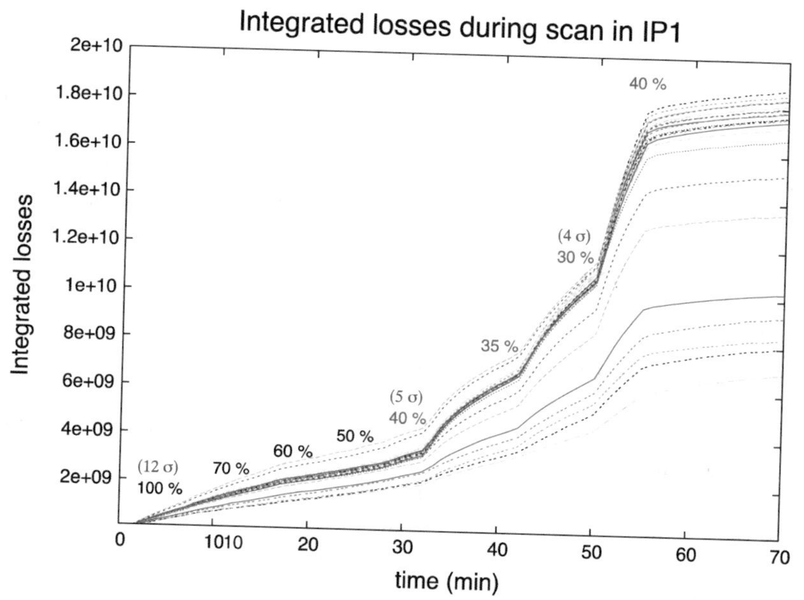

Fig. 4.27 Integrated losses of all bunches as a function of time during scan of beam separation in IP1. Numbers show percentage of full crossing angle

losses of each bunch recorded. The details of this procedure are described in [173] and the results are shown in Fig. 4.27 where the integrated losses for the 36 bunches in beam 1 are shown as a function of time and the relative change of the crossing angle is given in percentage of the nominal (100% $\equiv$ 240 μrad). The nominal value corresponds to a separation of approximately 12σ at the parasitic encounters. From Fig. 4.27 we observe significantly increased losses for some bunches when the separation is reduced to about 40%, i.e. around 5σ. Not all bunches are equally

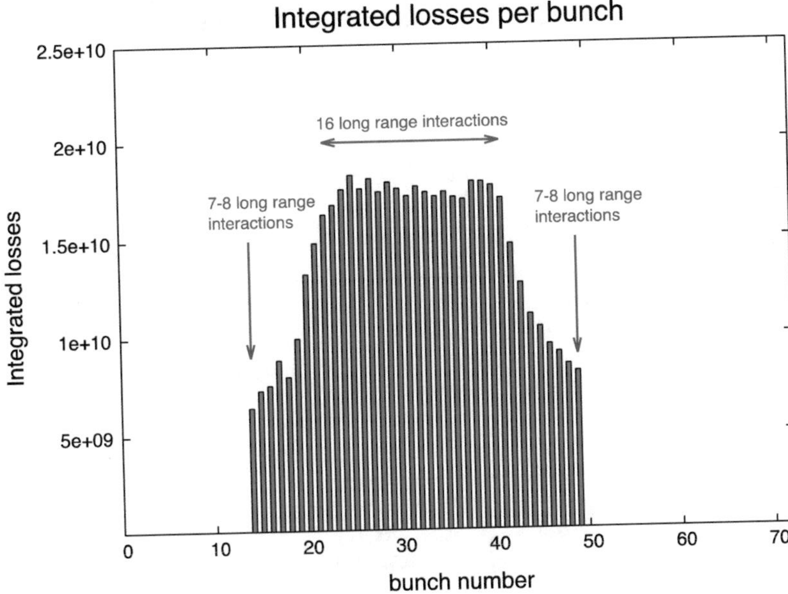

Fig. 4.28 Integrated losses of all bunches along a train of 36 bunches, after reducing the crossing angle in IP1

affected. At a smaller separation of 30% all bunches experience significant losses (~4σ). Returning to a separation of 40% reduces the losses significantly, suggesting that mainly particles at large amplitudes have been lost during the scan due to a reduced dynamic aperture. Such a behaviour is expected [174]. The different behaviour is interpreted as a "PACMAN" effect and should depend on the number of long range encounters, which varies along the train. This is demonstrated in Fig. 4.28 where we show the integrated losses for the 36 bunches in the train at the end of the experiment. The maximum loss is clearly observed for the bunches in the centre of the train with the maximum number of long range interactions (16) and the losses decrease as the number of parasitic encounters decrease. The smallest loss is found for bunches with the minimum number of interactions, i.e. bunches at the beginning and end of the train [166, 167]. This is a very clear demonstration of the expected different behaviour, depending on the number of interactions.

In the second part of the experiment we kept the separation at 40% in IP1 and started to reduce the crossing angle in the collision point IP5, opposite in azimuth to IP1. Due to this geometry, the same pairs of bunches meet at the interaction points, but the long range separation is in the orthogonal plane. This alternating crossing scheme was designed to compensate first order effects from long range interactions [166]. The Fig. 4.29 shows the evolution of the luminosity in IP1 as we performed the scan in IP5. The numbers indicate again the relative change of separation, this time the horizontal crossing angle in IP5. The luminosity seems to show that the lifetime is best when the separation and crossing angles are equal for

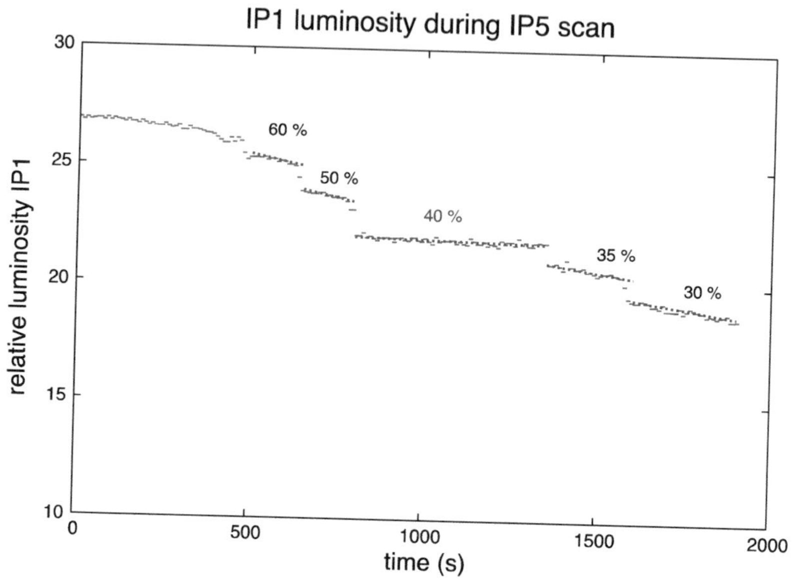

Fig. 4.29 Luminosity in IP1 as a function of time during scan of beam separation in IP5

the two collision points. It is worse for smaller as well as for larger separation. This is the expected behaviour for a passive compensation due to alternating crossing planes, although further studies are required to conclude.

4.6.8.1 Dynamic Aperture Reduction Due to Long-Range Interactions

For too small separation, the tune spread induced by long-range interactions can become very large and resonances cannot be avoided any more. The motion can become irregular and as a result particles at large amplitudes may be lost.

To evaluate the dynamic aperture in the presence of beam–beam interactions, a simulation of the complete machine is necessary and the interplay between the beam–beam perturbation and possible machine imperfections is important [174].

For the present LHC parameters we consider the minimum crossing angle to be 285 μrad.

4.6.8.2 Beam–Beam Induced Orbit Effects

When two beams do not collide exactly head-on, the force has a constant contribution which can easily be seen when the kick $\Delta x'$ from (4.65), for sufficiently large separation, is developed in a series

$$\Delta x' = \frac{const}{d}\left[1 - \frac{x}{d} + O\left(\frac{x^2}{d^2}\right)\right] \tag{4.68}$$

A constant contribution, i.e. more precisely an amplitude independent contribution, changes the orbit of the bunch as a whole. When the beam–beam effect is strong enough, i.e. for high intensity and/or small separation, the orbit effects are large enough to be observed.

When the orbit of a beam changes, the separation between the beams will change as well, which in turn will lead to a slightly different beam–beam effect and so on. The orbit effects must therefore be computed in a self-consistent way [175], in particular when the effects are sizeable. The closed orbit of an accelerator can usually be corrected, however an additional effect which is present in some form in many colliders, sets a limit to the correction possibilities. A particularly important example is the LHC and therefore it will be used to illustrate this feature.

We have to expect a slightly different orbit from bunch to bunch. The bunches in the middle of a train have all interactions and therefore the same orbit while the bunches at the beginning and end of a train show a structure which exhibits the decreasing number of long-range interactions. The orbit spread is approximately 10–15% of the beam size. Since the orbits of the two beams are not the same, it is impossible to make all bunches collide exactly head-on. A significant fraction will collide with an offset. Although the immediate effect on the luminosity is small [162], collisions at an offset can potentially affect the dynamics and are undesirable. The LHC design should try to minimize these offsets [168, 176]. A further consequence of the LHC filling and collision scheme is that not all bunches experience all head-on collisions [176]. Some of the bunches will collide only in 2 instead of the 4 nominal interaction points, leading to further bunch-to-bunch differences. In Fig. 4.30 we show a prediction for the vertical offsets in IP1 [166, 167]. The offsets should vary along the bunch train. Although the orbit measurement in the LHC is not able to resolve these effects, the vertex centroid can be measured bunch by bunch in the experiment (Fig. 4.31).

4.6.9 Coherent Beam–Beam Effects

So far, we have mainly studied how the beam–beam interaction affects the single particle behaviour and treated the beam–beam interaction as a static lens. In the literature, this is often called a "weak–strong" model: a "weak" beam (a single particle) is perturbed by a "strong" beam (not affected by the weak beam). When the beam–beam perturbation is important, the model of an unperturbed, strong beam is not valid anymore since its parameters change under the influence of the other beam and vice versa. When this is the case, we talk about so-called "strong–strong" conditions. The first example of such a "strong–strong" situation was the orbit effect

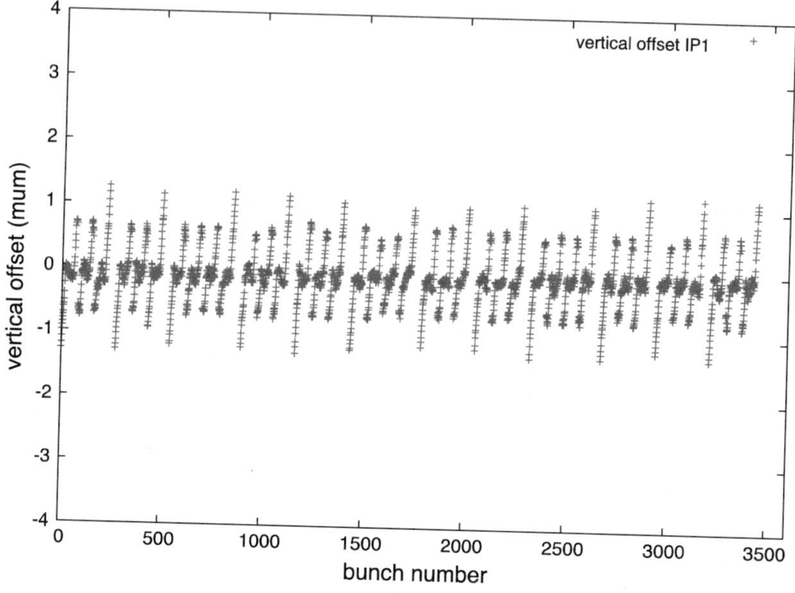

Fig. 4.30 Computed orbit offsets in IP1 along the bunch train [166, 167]

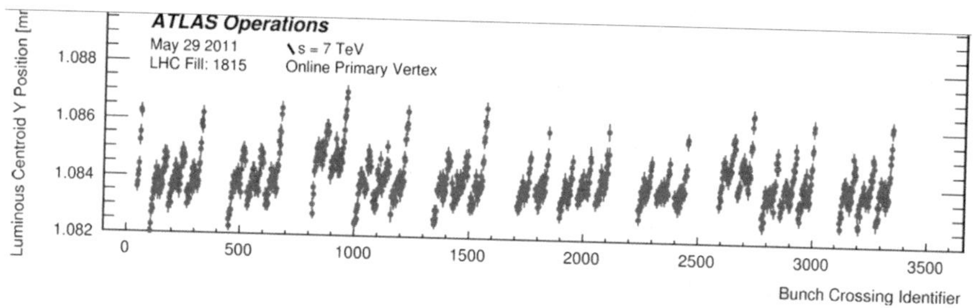

Fig. 4.31 Measured orbit offsets in IP1 along the bunch train [177, 178]

where the beams mutually changed their closed orbits. These closed orbits had to be found in a self-consistent way. This represents a static strong–strong effect.

In the next step, we investigate dynamic effects under the strong–strong condition [179].

When we consider the coherent motion of bunches, the collective behaviour of all particles in a bunch is studied. A coherent motion requires an organized behaviour of all particles in a bunch. A typical example are oscillations of the centre of mass of the bunches, so-called dipole oscillations. Such oscillations can be driven by external forces such as impedances and may be unstable. At the collision of two counter-rotating bunches not only the individual particles receive a kick from the opposing beam, but the bunch as an entity gets a coherent kick. This coherent kick of separated beams can excite coherent dipole oscillations. Its strength depends on the distance between the bunch centres at the collision point. It can be computed

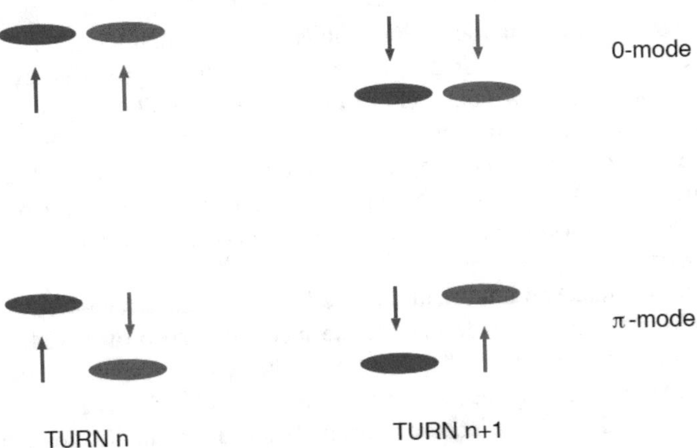

Fig. 4.32 Basic dipole modes of two bunches. Relative position of the bunches at the interaction point at two consecutive turns

by adding the individual contributions of all particles. For small distances, it can be shown [153, 180] that it is just one half of the incoherent kick a single particle would receive at the same distance. For distances large enough the incoherent and coherent kicks become the same.

4.6.9.1 Coherent Beam–Beam Modes

To understand the dynamics of dipole oscillations we first study the simplest case with one bunch in each beam. When the bunches meet turn after turn at the collision point, their oscillation can either be exactly in phase (0 degree phase difference) or out of phase (180 degrees or π phase difference). Any other oscillation can be constructed from these basic modes. The modes are sketched very schematically in Fig. 4.32. The relative positions of the bunches as observed at the interaction point are shown for two consecutive turns n and $n + 1$. The first mode is called the 0-mode (or sometimes called σ-mode) and the second the π-mode. In the first mode, the distance between the bunches does not change turn by turn and therefore there is no net force driving an oscillation. This mode must oscillate with the unperturbed frequency (tune) Q_0. For the second mode, the net force difference between two turns is a maximum and the tune becomes $Q_0 + \Delta Q_{coh}$. The sign of ΔQ_{coh} depends whether the two beams have equal charge (defocusing case) or opposite charge (focusing case). The calculation of ΔQ_{coh} is non-trivial: when the bunches are considered as rigid objects, the tune shift can be computed easily using the coherent kick but is underestimated [181]. The correct calculation must allow for changes of the density distribution during the collision and moreover, must allow a deviation from a Gaussian function. The computation requires to solve the Vlasov-equation of two coupled beams [182–185].

Self-consistent multi-particle simulations form a complementary study, but require large computing resources. Furthermore, the fields produced by the beams must be computed in a self-consistent form before every collision [186].

The 0-mode is found at the unperturbed tune as expected. The π-mode is shifted by 1.2–1.3ξ. The precise value depends on the ratio of the horizontal and vertical beam sizes [183]. We have seen before the incoherent tune spread (footprint) the individual particles occupy and we know that it spans the interval $[0.0, 1.0]\xi$, starting at the 0-mode.

Here one can make an important observation: under the strong–strong condition the π-mode is a discrete mode outside the incoherent spectrum [184, 185]. This has dramatic consequences for the stability of the beams. A coherent mode that is outside an incoherent frequency spectrum cannot be stabilized by Landau damping. Under these conditions the coherent beam–beam effect could drive the dipole oscillation to large amplitudes and may result in the loss of the beam. Observations of the coherent beam–beam effects have been made at PETRA [181]. Beam–beam modes have been observed with high intensity coasting beams in the ISR [187], and recently in a bunched hadron collider at RHIC [188].

Coherent beam–beam modes can be driven by head-on collisions with a small offset or by long-range interactions. In the first case and for small oscillations, the problem can be linearized and the theoretical treatment is simplified. The forces from long-range interactions are very nonlinear but the numerical evaluation is feasible. Since the coherent shift must have the opposite sign for long-range interactions, the situation is very different. In particular the π-mode from long-range interactions alone would appear on the opposite side of the 0-mode in the frequency spectrum [185, 186]. Both, the incoherent and the coherent spectra include both types of interactions.

4.6.10 Compensation of Beam–Beam Effects

For the case the beam–beam effects limit the performance of a collider, several schemes have been proposed to compensate all or part of the detrimental effects. The basic principle is to design correction devices which act as non-linear “lenses” to counteract the distortions from the non-linear beam–beam “lens”. For both head-on and long-range effects schemes have been proposed

- Head on effects:
  - Electron lenses
  - Linear lens to shift tunes
  - Non-linear lens to decrease tune spread
- Long-range effects:
  - At large distance: beam–beam force changes like 1/r
  - Same force as a wire!

4.6.10.1 Electron Lenses

The basic principle of a compensation of proton–proton (or antiproton) collisions with an "electron lens" implies that the proton (antiproton) beam travels through a counter-rotating high current electron beam ("electron lens") [189, 190]. The negative electron space charge can reduce the effect from the collision with the other proton beam.

An electron beam with a size much larger than the proton beam can be used to shift the tune of the proton beam ("linear lens"). When the current in the electron bunches can be varied fast enough, the tune shift can be different for the different proton bunches, thus correcting PACMAN tune shifts.

When the electron charge distribution is chosen to be the same as the counter-rotating proton beam, the non-linear focusing of this proton beam can be compensated ("non-linear lens"). When it is correctly applied, the tune spread in the beam can be strongly reduced.

Such lenses have been constructed at the Tevatron at Fermilab [190] and experiments are in progress.

4.6.10.2 Electrostatic Wire

To compensate the tune spread from long-range interactions, one needs a non-linear lens that resembles a separated beam. At large enough separation, the long-range force changes approximately with $1/r$ and this can be simulated by a wire parallel to the beam [191].

In order to compensate PACMAN effects, the wires have to be pulsed according to the bunch filling scheme. Tests are in progress at the SPS to study the feasibility of such a compensation for the LHC.

4.6.10.3 Möbius Scheme

The beam profiles of $e^+ e^-$ colliders are usually flat, i.e. the vertical beam size is much smaller than the horizontal beam size. Some studies indicate that the collision of round beams, even for $e^+ e^-$ colliders, show more promise for higher luminosity since larger beam–beam parameters can be achieved. Round beams can always be produced by strong coupling between horizontal and vertical planes. A more elegant way is the so-called Möbius lattice [192, 193]. In this lattice, the horizontal and vertical betatron oscillations are exchanged by an insertion. A horizontal oscillation in one turn becomes a vertical oscillation in the next turn and vice versa. Tests with such a scheme have been done at CESR at Cornell [193].

4.7 Numerical Modelling

G. Rumolo

Collective effects can be studied analytically, either through the perturbation formalism applied to the Vlasov equation or by means of few (typically two or three) particles models with the basic ingredients such as to reproduce the essential features of the phenomenon under study. Both ways are usually based on simplified approaches, in which some assumptions are necessary to make the models analytically solvable and lead to limited sets of equations relatively easy to interpret and handy to use.

The analytically solvable two or three particles models can be refined further to more realistic models, in which more than just few particles are assumed to represent the full particle beam. However, since the number of coupled differential equations to be solved grows proportionally with the number of particles used in the model, the resulting set of equations will rapidly become unmanageable as we increase the number of macroparticles (which are used to approximate the beam with a reduced number of particles), unless it is fed into a numerical simulation to be run on a computer. By using computers, the number of macroparticles necessary to model a beam can be pushed up to several millions, which is very useful to study the details of all possible internal oscillation modes of a bunch (or train of bunches), and also incoherent effects like emittance growth. Although ideally we would like to develop simulation programs that take into account the highest possible number of effects, in practice the existing codes narrow down their models to one or few effects, whose consequences in the beam dynamics are interesting to single out. For example, to study the effects of electron clouds, the beam will be made to interact with a given electron cloud at some locations around the accelerator ring, but in general other possible interactions with impedances, or the concurrent effects of space charge, beam–beam, Intra Beam Scattering, will be neglected. Although the study of two or more effects simultaneously is technically possible in most cases, at the present state of art of the simulations, it is generally preferred to limit the study of such interplays, because the combined models are difficult to control and tend to break down. Pushing further on this line, not only different effects can be decoupled in simulations, but also different regimes can be studied separately in the beam dynamics. For instance, for some problems only a partial description of the beam will be sufficient, so that transverse problems can be treated separately from longitudinal problems as well as single-bunch/multi-turn effects can be studied ignoring that these bunches are parts of long trains. In some cases, single-bunch/single- or multi-turn effects can also be modelled to generate driving terms to be used in reduced studies extending over longer time scales.

To perform a simulation, we will therefore have to define our beam as an ensemble of macroparticles, identified through arrays containing the phase space coordinates of each macro-particle (2–6-dimensional). This beam is first initialized and then transported across selected points of the accelerator ring using the appropriate transformation matrices. At each of these points, the interaction with the desired collective effect will be applied (e.g. the beam's own space charge field, an electron cloud, a wake field). It is clear, therefore, that a numerical simulation requires in the first place the knowledge of the driving term to be applied at each interaction point. That is why in the following we separate the general simulation into the solution of an electromagnetic problem, in which the collective interaction is modelled and the resulting excitation on the beam is calculated (at least, its non-self-consistent part), and the beam tracking part, in which the evolution of a beam is studied under the effect of this excitation. Note that most of numerical simulations including collective effects are based on time domain models, as these are best suited to describe the usually non-stationary beam evolution under the effect of collective interactions.

4.7.1 The Electromagnetic Problem

The first step to set up a numerical simulation including a collective effect consists of identifying the source of the self-induced perturbation acting on the beam and modelling it in a way that can be subsequently used. We usually distinguish three different types of collective interactions, which can take place with: (i) space charge (see Sect. 4.1); (ii) wake fields from an accelerator component or part of the resistive beam pipe (see Sect. 4.2); (iii) another "beam" of charged particles. This secondary beam can be either a counter-rotating beam in a collider (see Sect. 4.6), or a static cloud formed by the accumulation of particles, usually of opposite charge, around the primary beam (see Sect. 4.5).

If the source of the perturbation is space charge, then two different approaches are possible to compute its effect. Analytical formulae are available for the electromagnetic fields of coasting beams with ellipsoidal or Gaussian transverse sizes, as well as of ellipsoidal or Gaussian bunches (in all dimensions). The additional kicks given by these electromagnetic fields can be therefore calculated and applied to the beam macroparticles in a finite number of locations along the ring (even if the space charge interaction is in reality continuous). When doing that, self-consistency requires that the sizes of the bunch are updated at every kick point. Another possible approach consists of using the macroparticle distributions at each selected kick point to calculate self-consistently the electric field with a Poisson solver and use it to calculate the electromagnetic kicks on the macroparticles. It is worth noting that the same approach can be used for beam–beam problems, because the shape of the required electric field is the same, even if the coefficients need to be adapted (electric and magnetic forces tend to cancel at ultra-relativistic energies for space charge, while they add up for beam–beam).

If the source of the perturbation is a wake field from an accelerator component or resistive wall, the shape of the relative wake function has to be calculated beforehand. This is done analytically for some specific cases (e.g. resistive wall, step or tapered transitions), but in general dedicated electromagnetic codes can be used for this purpose. Some of them work in time domain, and can provide the wake potentials for given source bunches (usually chosen to be short enough as to simulate ideal pulse excitations and thus provide directly the wake functions). Other work in frequency domain and output impedances, which need then to be back-transformed into time domain to obtain the wake functions.

If the source of the perturbation is for example an electron cloud, then the electron distribution of the cloud is usually calculated beforehand by means of an electron cloud build-up code, and then its interaction with a coming bunch is calculated. Programs tracking simultaneously electrons and beam particles are presently under development or test, due to the massive memory and CPU requirements to solve this type of problems. On the other hand, generation and tracking of the ions can be included in multi-bunch beam tracking programs to calculate the effect of ions on bunched electron beams in a fully self-consistent manner. This is due to the fact that, while ions do not move significantly during the passage of an electron beam and allow modelling the bunches as charged disks, electrons can even perform several oscillations during the passage of a bunch, which requires a much more detailed modelling of the bunch.

4.7.2 Beam Dynamics

Beam dynamics tracking codes simulate the motion of beam particles inside an accelerator by transporting them across a number of discrete points by means of transformation matrices. In each of these points, additional kicks can be added, modelling either nonlinear components and errors of the external fields or the collective interactions. In the previous subsection, we have outlined the procedure to calculate the excitation to be applied to the beam to compute its evolution when it feels one or more collective interactions. To model the effects of space charge, wake fields and electron clouds, it is certainly necessary to describe the beam as an ensemble of macroparticles but beside that, its longitudinal structure needs also to be detailed. In particular, to model coupled bunch instabilities the relative positions of the macroparticles across the different bunches are necessary to determine the total effect of the wake acting on each of them. For single-bunch effects, a possible technique is to subdivide the bunch into several slices, so that the macroparticles of each slice can feel the integrated effect of the wakes left behind by the preceding slices (or the space charge from its own and the neighbouring slices, or the electron cloud as was deformed by the previous slices). A possible scheme of numerical simulation of a single bunch under the effect of a longitudinal wake field is illustrated in Fig. 4.33.

The bunch is first divided into N slices and a kick must be applied to each macroparticle within a given slice at a certain kick point. The kick depends on the longitudinal wake function and the charge distribution of the preceding slices. In the longitudinal plane, particles within a slice feel also the effect of the same slice to which they belong, because the bunch suffers a net energy loss. After all

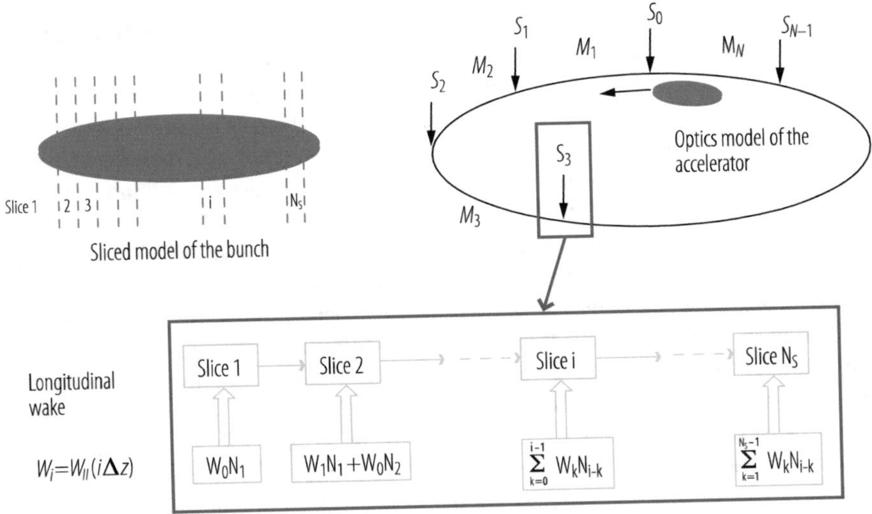

Fig. 4.33 Example of numerical simulation with collective effects: schematic view of the interaction of a single bunch with a longitudinal wake field at several locations in an accelerator ring

the particles in the bunch feel the effect of the wake at the kick point, they are subsequently transported to the next kick point in the accelerator optical model. Since synchrotron motion plays a key role in most of the effects under study, it is essential that particles are made to execute their synchrotron oscillations and move across slices from turn to turn. This means that collective effects dealing with single-bunch problems need to have at least one model of synchrotron motion built in, and that the bunch binning has to be regularly updated. Since the synchrotron motion is slow enough, and in reality the RF cavities do really kick the beam particles once or few times per turn, the longitudinal coordinates and the bunch slicing are usually not updated more frequently than once per turn. However, the update of the longitudinal coordinates from kick to kick point within one single turn, based on the only drift from momentum spread, could become significant especially in space charge simulations.

The simulation scheme with transverse wakes is basically the same as the one displayed in Fig. 4.33, except that particles inside one slice do not feel the effect of the same slice (as the transverse wakes are zero in the origin, for ultra-relativistic particles) and dipolar and quadrupolar contributions can be separated, making the wake kicks depending not only on the position of the source slice but also on the position of the witness macroparticles. The simulation scheme with the electron cloud is again similar to the one shown in Fig. 4.33, but the fundamental difference is that there is a mutual action between beam and electron cloud, so that, while macroparticles within a slice feel the effect of the electron cloud, the electron cloud itself is also deformed by the action of the passing slice.

The modelling described in the previous subsections has been frequently applied to explain collective instabilities observed in running machines, as well as to predict instability thresholds (both in existing and future machines) and develop strategies to circumvent limitations from collective effects. For instance, a detailed impedance model of the SPS comprises the contributions from several accelerator components and is used for deriving single-bunch wake fields, which are the driving terms for HEADTAIL simulations. The kicks given to the beam particles by the different wake fields can be then either applied at the real locations in which the sources are situated, or weighted by the beta functions, summed up and applied in a single location using a one-kick approximation. These simulations can be used for predicting at which intensity transverse mode coupling occurs and the effects of chromaticity on this threshold value [194]. This is very important to extrapolate the beam stability limits in different conditions of operation, e.g. with a different optics or to the upgraded machine, which will be in principle enabled to receive higher intensity bunches. The mode shift can be plotted as a function of the bunch intensity, because the main modes are detectable from the Fourier analysis of the centroid motion. A typical plot of mode shift provided by simulations is displayed in Fig. 4.10.

References

1. E. Métral et al., *Beam Instabilities in Hadron Synchrotrons*, IEEE Transactions on Nuclear Science, Vol. 63, No. 2, 50 p, April 2016 (invitation for the 50th anniversary of the PAC conference).
2. E. Métral (Issue Editor), ICFA Beam Dynamics Newsletter No. 69 devoted to the Collective Effects in Particle Accelerators, 310 p, December 2016.
3. E. Métral and V.G. Vaccaro (chairs), ICFA Mini-Workshop on “Electromagnetic Wake Fields and Impedances in Particle Accelerators”, Erice (Sicily, Italy), 2014: https://indico.cern.ch/event/287930/
4. M.R. Masullo, S. Petracca and G. Rumolo (chairs), ICFA Mini-Workshop on “Impedances and Beam Instabilities in Particle Accelerators”, Benevento (Italy), 2017: https://agenda.infn.it/event/12603/
5. E. Métral, G. Rumolo and T. Pieloni (chairs), ICFA Mini-Workshop on “Mitigation of Coherent Beam Instabilities in Particle Accelerators”, Zermatt (Switzerland), 2019: https://indico.cern.ch/event/775147/
6. B. Zotter: *Betatron Frequency Shifts due to Image and Self Fields*, CERN Accelerator School: General Accelerator Physics, CERN 85-19, Vol. I, p. 253, 1985.
7. K. Schindl: *Space Charge*, CERN-PS-99-012-DI, 1999.
8. A.W. Chao: *Physics of Collective Beam Instabilities in High Energy Accelerators*, New York: Wiley, 371 p, 1993.
9. K.Y. Ng: *Physics of Intensity Dependent Beam Instabilities*, World Scientific, 776 p, 2006.
10. S. Andriamonje, et al.: *Neutron TOF facility (PS213): Technical Design Report*, CERN-INTC-2000-004, 2000.
11. E. Métral, *Some effects near transition*, Proceedings of the CAS-CERN Accelerator School on Intensity Limitations in Particle Beams on November 2015, CERN Yellow Reports: School Proceedings, CERN-2017-006-SP.
12. E. Métral and G. Rumolo, USPAS course on “Collective Effects in Beam Dynamics” in Albuquerque, New Mexico, USA, June 22–26, 2009: http://emetral.web.cern.ch/emetral/

13. W. Hardt, D. Möhl: *Q-jump at Transition*, CERN ISR-300/GS/69-16, 1969.
14. D. Möhl: *Compensation of Space Charge Effects at Transition by a Asymmetric Q-jump – A Theoretical Study*, CERN ISR-300/GS/69-62, 1969.
15. W. Hardt: Proc. 9th Intern. Conf. High Energy Accelerators, Washington DC, 1974.
16. E. Métral and D. Möhl, *Transition Crossing*, Fifty years of the CERN Proton Synchrotron - Volume I (editors: S. Gilardoni and D. Manglunki), CERN–2011–004, 16 June 2011.
17. J.S. Berg, F. Ruggiero: *Landau Damping with Two-Dimensional Betatron Tune Spread*, CERN SL-AP-96-71 (AP), December 1996.
18. E. Métral, F. Ruggiero: *Stability Diagrams for Landau Damping with Two-Dimensional Betatron Tune Spread from both Octupoles and Nonlinear Space Charge Applied to the LHC at Injection*, Proc. EPAC2004, Lucerne, Switzerland, 5–9 July 2004.
19. G. Franchetti, et al.: *Space Charge and Octupole Driven Resonance Trapping Observed at the CERN Proton Synchrotron*, Phys. Rev. ST Accel. Beams 6, 124201 (2003).
20. G. Franchetti: *Simulations of the Space Charge and Beam Loss in the SIS18 Experiment*, Proc. ICFA HB2010, Morschach, Switzerland, Sept. 27–Oct. 1, 2010.
21. B.W. Montague: *Fourth-Order Coupling Resonance Excited by Space-Charge Forces in a Synchrotron*, CERN 68-38, 1968.
22. E. Métral: *Space Charge Experiment at the CERN Proton Synchrotron*, Proc. ICFA HB2004, Bensheim, Germany, 18–22/10/2004.
23. L.J. Laslett: *On Intensity Limitations Imposed by Space Charge Effects in Circular Particle Accelerators*, in: Proc. Summer Study on Storage Rings, Accelerators and Instrumentation at Super High Energies, BNL Report 7534, p. 324, 1963.
24. K. Yokoya: *Resistive Wall Impedance of Beam Pipes of General Cross Section*, Part. Accel., Vol. 41, 3–4, p. 221, 1993.
25. V.G. Vaccaro, *Longitudinal Instability of a Coasting Beam Above Transition, due to the Action of Lumped Discontinuities*, CERN Rep. ISRRF/66-35, Tech. Rep., 1966.
26. A. Sessler and V. Vaccaro, CERN Report ISR-RF/67-2 (1967).
27. H. Tsutsui: *On single Wire Technique for Transverse Coupling Impedance Measurement*, CERN-SL-Note-2002-034 AP, 2002.
28. A. Burov, V. Danilov: *Suppression of Transverse Bunch Instabilities by Asymmetries in the Chamber Geometry*, Phys. Rev. Lett. 82, 2286–2289 (1999).
29. N. Mounet, E. Métral: *Impedances of Two Dimensional Multilayer Cylindrical and Flat Chambers in the Non-Ultrarelativistic case*, Proc. ICFA HB2010, Morschach, Switzerland, Sept. 27–Oct. 1, 2010.
30. CST Particle Studio. http://www.cst.com/Content/Products/PS/Overview.aspx
31. F. Caspers: *Bench Measurements*, in: Handbook of Accelerator Physics and Engineering, 2nd printing, edited by A.W. Chao and M. Tigner, p. 574.
32. Ansoft – HFSS. http://www.ansoft.com/products/hf/hfss/
33. E. Métral, et al.: *Kicker Impedance Measurements for the Future Multiturn Extraction of the CERN Proton Synchrotron*, Proc. EPAC2006, Edinburgh, UK, 26–30 June 2006.
34. H.A. Day: *Measurements and Simulations of Impedance Reduction Techniques in Particle Accelerators*, PHD Thesis, University of Manchester (England), 2013.
35. L.J. Laslett, V.K. Neil, A.M. Sessler: *Transverse Resistive Instabilities of Intense Coasting Beams in Particle Accelerators*, R.S.I 36(4), 436–448, 1965.
36. B. Zotter: *New Results on the Impedance of Resistive Metal Walls of Finite Thickness*, CERN-AB-2005-043, 2005.
37. E. Métral: *Transverse Resistive-Wall Impedance from Very Low to Very High Frequencies*, CERN-AB-2005-084, 08 August 2005.
38. A.M. Al-Khateeb, et al.: *Transverse Resistive Wall Impedances and Shielding Effectiveness for Beam Pipes of Arbitrary Wall Thickness*, Phys. Rev. ST Accel. Beams 10, 064401 (2007).
39. F. Roncarolo, et al.: *Comparison Between Laboratory Measurements, Simulations, and Analytical Predictions of the Transverse Wall Impedance at Low Frequencies*, Phys. Rev. ST Accel. Beams 12, 084401 (2009).

40. N. Mounet, E. Métral: *Impedances of an Infinitely Long and Axisymmetric Multilayer Beam Pipe: Matrix Formalism and Multimode Analysis*, Proc. 1st Intern. Particle Accelerator Conf., Kyoto, Japan, 23–28 May 2010.
41. N. Mounet, E. Métral: *Generalized Form Factors for the Beam Coupling Impedances in a Flat Chamber*, Proc. 1st Intern. Particle Accelerator Conf., Kyoto, Japan, 23–28 May 2010.
42. B. Salvant, et al.: *Quadrupolar Transverse Impedance of Simple Models of Kickers*, Proc. 1st Intern. Particle Accelerator Conf., Kyoto, Japan, 23–28 May 2010.
43. C. Zannini, et al.: *Electromagnetic Simulations of Simple Models of Ferrite Loaded Kickers*, Proc. 1st Intern. Particle Accelerator Conf., Kyoto, Japan, 23–28 May 2010.
44. E. Métral: *Beam Screen Issues (with 20 T dipole magnets instead of 8.3 T)*, Proc. High-Energy LHC workshop, Malta, 14–16 October 2010.
45. A. Hofmann: *Beam Instabilities*, CERN Accelerator School, Rhodes, Greece, 20 Sept.–1 Oct., 1993, p. 307.
46. N. Biancacci, B. Salvant: *Dependence of Impedance vs. Phase Advance in HEADTAIL*, https://indico.cern.ch/event/164172/contributions/1420072/attachments/196173/275289/19-12-2011_Impedance-Vs-PhaseAdvance.pdf
47. N. Biancacci, *HEADTAIL + MAD-X*, https://indico.cern.ch/event/178920/contributions/1446485/attachments/235706/329825/HDTL_lattice_def.pdf
48. E.D. Courant, A.M. Sessler: *Transverse Coherent Resistive Instabilities of Azimuthally Bunched Beams in Particle Accelerators*, R.S.I, 37(11), 1579–1588, 1966.
49. C. Pellegrini, Nuovo Cimento, 64°, 477, 1969.
50. M. Sands: *The Head-Tail Effect: an Instability Mechanism in Storage Rings*, SLAC-TN-69-8, 1969.
51. M. Sands: *Head-Tail Effect II: From a Resistive-Wall Wake*, SLAC-TN-69-10, 1969.
52. F. Sacherer: *Single beam collective phenomena-transverse (bunched beams)*, Proc. First Course Intern. School on Accelerators, Erice, 1976: Theoretical Aspects of the Behaviour of Beams in Accelerators and Storage Rings, CERN 77-13, p. 210, 1977.
53. J.L. Laclare: *Bunched Beam Coherent Instabilities*, CERN Accelerator School, Oxford, CERN 87-03, p. 264, 1987.
54. J.P. Gourber, H.G. Hereward, S. Myers: *Overlap Knock-Out Effects in the CERN Intersecting Storage Rings (ISR)*, IEEE Trans. Nucl. Sci. NS-24(3), June 1977.
55. S. Myers: private communication (2011).
56. K.W. Robinson: Cambridge Electron Accel. Report CEAL-1010 (1964).
57. J.L. Laclare: *Introduction to Coherent Instabilities – Coasting Beam Case*, CERN Accelerator School: General Accelerator Physics, CERN 85-19, Vol. II, p. 377, 1985.
58. E. Keil, W. Schnell: CERN Report ISR-TH-RF/69-48, 1969.
59. D. Boussard: *Observation of Microwave Longitudinal Instabilities in the CPS*, CERN Report LabII/RF/Int./75-2, 1975.
60. F.J. Sacherer: *Bunch Lengthening and Microwave Instability*, IEEE Trans. Nucl. Sci. NS-24(3), June 1977.
61. K. Ng: *Potential-Well Distortion and Mode-Mixing Instability in Proton Machines*, FERMILAB-FN-630, 1995.
62. E. Métral: *Stability Criterion for the Longitudinal Mode-Coupling Instability in the Presence of Both Space-Charge and Resonator Impedances*, CERN/PS 2001-063 (AE), 2001.
63. E. Métral, *GALACTIC and GALACLIC: Two Vlasov Solvers for the Transverse and Longitudinal Planes*, Proceedings of IPAC2019, Melbourne, Australia, May 19–24, 2019.
64. E. Métral and M. Migliorati, *Longitudinal Mode-Coupling Instability: GALACLIC Vlasov Solver vs. Macroparticle Tracking Simulations*, Proceedings of IPAC2019, Melbourne, Australia, May 19–24, 2019.
65. E. Shaposhnikova: *Signatures of Microwave Instability*, CERN-SL-99-008 HRF, 1999.
66. J.S. Berg, R.D. Ruth: *Transverse Instabilities for Multiple Nonrigid Bunches in a Storage Ring*, Phys. Rev. E 52(3), September 1995.
67. H. Hereward: Rutherford Lab. Reports RL-74-062, EPIC/MC/48 (1974), and RL-75-021, EPIC/MC/70 (1975). See also B. Chen, A. W. Chao: SSCL Report 606 (1992).

68. D. Boussard, T. Linnecar: Proc. 2nd Euro. Part. Accel. Conf, Nice, 1990, p. 1560.
69. R.D. Kohaupt: *Head Tail Turbulence and the Transverse PETRA Instability*, DESY 80/22, 1980.
70. E. Métral: *Stability Criteria for High-Intensity Single-Bunch Beams in Synchrotrons*, Proc. 8th EPAC, Paris, France, 3–7 June 2002.
71. A. Burov and T. Zolkin, *TMCI with Resonator Wakes*, arXiv:1806.07521v2 [physics.acc-ph].
72. G. Rumolo et al., *Simulation Study on the Energy Dependence of the TMCI Threshold in the CERN-SPS*, Proceedings of EPAC 2006, Edinburgh, Scotland.
73. Y.H. Chin: *User's guide for new MOSES version 2.0: Mode-coupling Single bunch instability in an Electron Storage ring*, CERN/LEP-TH/88-05, 1988.
74. G. Rumolo, F. Zimmermann: *Practical user guide for HEADTAIL*, CERN-SL-Note-2002-036-AP, 2002.
75. B. Salvant: *Impedance Model of the CERN SPS and Aspects of LHC Single-Bunch Stability*, PHD Thesis, Ecole Polytechnique Fédérale de Lausanne (Switzerland), 2010.
76. A. Burov, *Nested Head-Tail Vlasov Solver*, Phys. Rev. ST Accel. Beams 17, 021007 (2014).
77. N. Mounet, *DELPHI: An Analytic Vlasov Solver for Impedance-Driven Modes*, CERN internal HSC meeting, May 07, 2014.
78. E. Métral, G. Rumolo: *Simulation Study on the Beneficial Effect of Linear Coupling for the Transverse Mode Coupling Instability in the CERN Super Proton Synchrotron*, Proc. EPAC 2006, Edinburgh, UK, 26–30 June 2006.
79. S. Myers: LEP Note 436, 1983.
80. R. Ruth: CERN LEP-TH/83-22, 1983.
81. R. Ruth: Proc. 12th Intern. Conf. High Energy Accelerators, Fermilab, Batavia, August 11–16, 1983, p. 389.
82. S. Myers: J. Vancraeynest: CERN LEP-RF/84-13, 1984.
83. S. Myers: CERN LEP-RF/85-22, 1985.
84. S. Myers: Proc. IEEE PAC, Washington D.C., March 16, 1987, p. 503–507.
85. M. Karliner, K. Popov: *Theory of a Feedback to Cure Transverse Mode Coupling Instability*, Nucl. Instrum. Meth. A 537 (2005), p. 481–500.
86. E. Métral *et al.*, *Destabilising Effect of the LHC Transverse Damper*, in *Proc. IPAC'18*, Vancouver, BC, Canada, Apr–May 2018, paper THPAF048, pp. 3076-3079. https://doi.org/10.18429/JACoW-IPAC2018-THPAF048
87. E. Métral et al., *Impedance Models, Operational Experience and Expected Limitations*, International Review of the HL-LHC Collimation System, CERN, 11-12/02/2019.
88. E. Métral, *A Two-Mode Model to Study the Effect of Space Charge on TMCI in the "Long-Bunch" regime*, Proceedings of IPAC2019, Melbourne, Australia, May 19–24, 2019.
89. E. Métral et al., *Space Charge and Transverse Instabilities at the CERN SPS and LHC*, Proceedings of ICAP'18, Key West, Florida, USA, October 20–24, 2018.
90. A. Burov, "Convective Instabilities of Bunched Beams with Space Charge", arXiv:1807.04887v4 [physics.acc-ph].
91. E. Métral, *Summary of my 3 IPAC19 papers*, CERN LIU-SPS BD WG meeting, 17/01/2019.
92. H.G. Hereward: *Landau Damping by Non-Linearity*, CERN/MPS/DL 69-11, 1969.
93. E. Métral, A. Verdier: *Stability Diagrams for Landau Damping with a Beam Collimated at an Arbitrary Number of Sigmas*, CERN-AB-2004-019-ABP.
94. E. Métral, *Landau Damping for TMCI: Without vs. With Transverse Damper (TD)*, CERN HSC internal meeting, 25/03/2019, https://indico.cern.ch/event/807899/contributions/3362767/attachments/1816203/2971974/LDforTMCI_EM_25-03-2019_2.pdf
95. E. Métral: Theory of Coupled Landau Damping, Part. Accelerators 62(3–4), p. 259, 1999.
96. E. Métral et al., Simulation Study of the Horizontal Head-Tail Instability Observed at Injection of the CERN Proton Synchrotron, Proceedings of PAC07, Albuquerque, New Mexico, USA, June 25–29, 2007.
97. E. Métral, et al.: Destabilising Effect of Linear Coupling in the HERA Proton Ring, Proc. 8th EPAC, Paris, France, 3–7 June 2002.

98. L.R. Carver et al., Transverse Beam Instabilities in the Presence of Linear Coupling in the Large Hadron Collider, PRAB 21, 044401 (2018).
99. E. Métral: Simple Theory of Emittance Sharing and Exchange due to Linear Betatron Coupling, CERN/PS 2001-066 (AE), December 2001.
100. A. Franchi, E. Métral, R. Tomás: *Emittance Sharing and Exchange Driven by Linear Betatron Coupling in Circular Accelerators*, Phys. Rev. ST Accel. Beams 10, 064003 (2007).
101. D. Möhl, H. Schönauer: Landau Damping by Non-Linear Space-Charge Forces and Octupoles, Proc. IX Intern. Conf. High Energy Acc., Stanford, 1974 (AEC, Washington, D.C., 1974), [CONF 740522] p. 380.
102. D. Möhl: On Landau Damping of Dipole Modes by Non-Linear Space Charge and Octupoles, CERN/PS 95-08 (DI), 1995.
103. M. Blaskiewicz: Phys. Rev. ST Accel. Beams 1, 044201 (1998).
104. E. Métral, F. Ruggiero: Stability Diagrams for Landau Damping with Two-Dimensional Betatron Tune Spread from both Octupoles and Nonlinear Space Charge, CERN-AB-2004-025 (ABP), 2004.
105. A. Burov: Phys. Rev. ST Accel. Beams 12, 044202 (2009).
106. V. Balbekov: Phys. Rev. ST Accel. Beams 12, 124402 (2009).
107. V. Kornilov: Head-Tail Bunch Dynamics with Space Charge, Proc. ICFA HB2010, Morschach, Switzerland, Sept. 27–Oct. 1, 2010.
108. X. Buffat et al., *Transverse Mode Coupling Instability of Colliding Beams*, Phys. Rev. ST Accel. Beams 17, 041002 (2014).
109. X. Buffat, *Transverse Beams Stability Studies at the Large Hadron Collider*, Ph.D. dissertation, EPFL, 2015.
110. S.V. Furuseth et al., MD3288: Instability latency with controlled noise, CERN internal note, under publication.
111. S. Nagaitsev, et al.: Nonlinear Optics as a Path to High-Intensity Circular Machines, Proc. ICFA HB2010, Morschach, Switzerland, Sept. 27–Oct. 1, 2010.
112. V. Shiltsev et al., *Landau Damping of Beam Instabilities by Electron Lenses*, Phys. Rev. Lett. 119, 134802 – Published 27 September 2017.
113. M. Schenk et al., *Analysis of Transverse Beam Stabilization with Radio Frequency Quadrupoles*, PRAB 20, 104402 (2017).
114. M. Schenk et al., *Experimental Stabilization of Transverse Collective Instabilities in the LHC with Second Order Chromaticity*, PRAB 21, 084401 (2018).
115. M. Schenk et al., *Vlasov Description of the Effects of Nonlinear Chromaticity on Transverse Coherent Beam Instabilities*, PRAB 21, 084402 (2018).
116. E. Métral: *Stability of the Longitudinal Bunched-Beam Coherent Dipole Mode*, Proc. APAC'04, Gyeongju, Korea, 22–26 March 2004.
117. K. Ohmi: Phys. Rev. Lett. 75, 1526 (1995).
118. *Tables and Graphs of electron-Interaction Cross Sections from 10 eV to 100 GeV Derived from the LLNL Evaluated Electron Data Library (EDLL), Z=1-100*, UCRL-50400, 31 (1991).
119. T. Raubenheimer, P. Chen: *Ions in the Linacs of Future Linear Colliders*, Proc. LINAC'92, Ottawa, Canada (1992).
120. P. Thieberger, et al: Phys. Rev. A 61, 042901 (2000).
121. O. Grobner: *Bunch Induced Multipacting*, 10th Intern. Conf. High Energy Accelerators, Protvino, July (1977).
122. R. Cimino, et al.: Phys. Rev. Lett. 93, 014801 (2004).
123. M. Furman, G. Lambertson: *The Electron-Cloud Instability in PEP-II*, Proc. EPAC'96, Sitges (1996).
124. F. Zimmermann: LHC Project Report 95 (1997).
125. L. Wang, et al.: PRST-AB 5, 124402 (2002).
126. H. Fukuma, et al.: *Observation of Vertical Beam Blow-up in KEKB Low Energy Ring*, Proc. EPAC'00, Vienna (2000).

127. G. Arduini, et al.: *Transverse Behaviour of the LHC Proton Beam in the SPS: An Update*, PAC2001, Chicago (2001).
128. K. Cornelis: *The Electron Cloud Instability in the SPS*, ECLOUD'02, Geneva (2002).
129. C. Yin Valgren, et al.: *Amorphous Carbon Coatings for Mitigation of Electron Cloud in the CERN SPS*, Proc. 1st Intern. Particle Accelerator Conf., Kyoto, Japan, 23–28 May 2010.
130. G. Rumolo et al., *Electron Cloud observation in the LHC*, Proc. IPAC'11, San Sebastian, Spain (2011).
131. G. Iadarola, *Electron cloud studies for CERN particle accelerators and simulation code development*, PhD thesis, CERN-THESIS-2014-047 (2014).
132. L. Mether et al., *Electron cloud in 2016: cloudy or clear?*, Proc. 7th Evian Workshop, Evian, France, 2016, CERN-ACC-2017-094 (2017).
133. G. Arduini et al., *How to Maximize the HL-LHC performance*, Proc. of the Review of the LHC and Injector Upgrade Plans Workshop (RLIUP), 29–31 October 2013 Archamps, France, CERN-2014-006 (2013)
134. M. Palmer, et al.: *Electron Cloud at Low Emittance in CesrTA*, Proc. 1st Intern. Particle Accelerator Conf., Kyoto, Japan, 23–28 May 2010.
135. V. Baglin, et al.: LHC Project Report 472, (2002).
136. M. Furman, G. Lambertson: *The Electron-Cloud Instability in the Arcs of the PEP-II Positron Ring*, Proc. MBI97 Tsukuba (1997).
137. P. Channell, A. Jason: LA AOT-3 TN 93-11, private communication (1994).
138. P.R. Zenkevich, D.G. Koskarev: *Coupling Resonances of the Transverse Oscillations of Two Circular Beams*, ITEF-841, (1970), translated as UCRL-Trans. 1451 (1971).
139. E. Keil, B. Zotter: CERN-ISR-TH/71-58 (1971).
140. K. Oide, et al., in: Proc. International Workshop on Performance Improvement of Electron-Positron Collider Particle Factories, KEKB Report No. 99-24 (2000).
141. K. Ohmi, F. Zimmermann: Phys. Rev. Lett. 85, 3821 (2000)
142. G. Arduini, et al., in: Proc. Workshop LEP-SPS Performance Chamonix X (2000), pp. 119–122.
143. G. Arduini, et al., in: Proc. PAC 2003, 12–16 May, Portland, USA (2003).
144. K. Li et al., *Beams during the cycle: Quality and behaviour*, LHC Performance Workshop 2017, Chamonix, France (2017).
145. A. Romano et al., *Electron cloud instabilities triggered by low bunch intensity at the Large Hadron Collider*, presently submitted to PRAB.
146. D. Neuffer, et al: *Observations of a fast transverse instability in the PSR*, Nucl. Instrum. Meth. A 321, pp. 1–12 (1992).
147. A. Grunder, G.R. Lambertson: Report UCRL 20691 (1971).
148. G. Rumolo, in: Proc. CARE-HHH-APD Workshop, CERN (2004).
149. G. Rumolo, et al.: Phys. Rev. Lett. 100, 144801 (2008).
150. Proc. ECLOUD02, CERN, Geneva, 2002, edited by G. Rumolo and F. Zimmermann as CERN Yellow Report No. CERN-2002-001 (2002).
151. R. Cimino et al., Phys. Rev. Lett., 109, 064801 (2012).
152. F. Ruggiero, X.L. Zhu: *Collective Instabilities in the LHC: Electron Cloud and Satellite Bunches*, Proc. Workshop Instabilities of High Intensity Hadron Beams in Rings, Upton, NY, USA, 28 Jun–1 Jul 1999, pp. 40–48.
153. W. Herr: *Beam-Beam Interactions*, Proceedings of CERN Accelerator School, Zeuthen 2003, CERN-2006-002 (2006).
154. Proceedings of the ICFA-Mini-Workshop on Beam-Beam effects in Hadron Colliders, CERN, Geneva, Switzerland, 18–22 April 2013, edited by W. Herr and G. Papotti, CERN-2014-004, https://doi.org/10.5170/CERN-2014-004
155. A. Chao: *The Beam-Beam Instability*, SLAC-PUB-3179 (1983).
156. L. Evans, J. Gareyte: *Beam-Beam Effects*, CERN Accelerator School, Oxford 1985, CERN 87-03 (1987).
157. A. Zholents: *Beam-Beam Effects in Electron-Positron Storage Rings*, Joint US-CERN School on Particle Accelerators, in Springer, Lecture Notes in Physics, Vol. 400 (1992).

158. E. Keil: *Beam-Beam Dynamics*, CERN Accelerator School, Rhodes 1993, CERN 95-06 (1995).
159. S. Kheifets, PETRA-Kurzmitteilung 119 (DESY 1976).
160. M. Basetti and G.A. Erskine, *Closed Expression for the Electrical Field of a 2-Dimensional Gaussian Charge*, CERN-ISR-TH/80-06 (1980).
161. D. Brandt, et al.: *Is LEP Beam-Beam Limited at its Highest Energy?* In Proceedings of the Particle Accelerator Conference 1999, New York, (1999) p. 3005.
162. W. Herr: *Concept of Luminosity*, Proceedings of CERN Accelerator School, Zeuthen 2003, CERN-2006-002 (2006).
163. J.T. Seeman: *Observations of the Beam-Beam Interaction*, Lecture Notes in Physics, Vol. 247, Springer, (1986).
164. S. Myers: IEEE Trans. Nucl. Sci. NS-28, 2503 (1981).
165. S. Myers: *Review of beam-Beam Simulations*, Lecture Notes in Physics, Vol. 247, Springer, (1986).
166. W. Herr, *Features and Implications of Different LHC Crossing Schemes*, CERN LHC Project Report 628 (2003).
167. W. Herr, *Dynamic Behaviour of Nominal and PACMAN Bunches for Different LHC Crossing Schemes*, CERN LHC Project Report 856 (2005).
168. W. Herr, *Effects of PACMAN Bunches in the LHC*, CERN LHC Project Report 39 (1996).
169. W. Herr et al., *Head-On Beam-Beam Tune Shifts with High Brightness Beams in the LHC*, CERN-ATS-Note-2011-029 (2011).
170. G. Papotti et al., *Observations of Bunch to Bunch Differences due to Beam-Beam Effects*, Proceedings of IPAC2011, San Sebastián, Spain.
171. B. Goddard, et al.: *Bunch Trains for LEP*, Particle Accelerators 57, 237 (1998).
172. W. Herr et al., *Observations of beam-beam effects at high intensities in the LHC*, In Proceedings of the Particle Accelerator Conference 2011, San Sebastion (2011).
173. W. Herr et al., *Head-on beam-beam interactions with high intensities and long range beam-beam studies in the LHC*, CERN-ATS-Note-2011-058 (2011).
174. W. Herr, D. Kaltchev et al., *Large Scale Beam-beam Simulations for the CERN LHC using distributed computing*, Proc. EPAC 2006, Edinburgh (2006).
175. H. Grote, W. Herr: *Self-Consistent Orbits with Beam-Beam Effects in the LHC*, In Proceedings of the 2001 Workshop Beam-Beam Effects, FNAL, 25-27/6/2001.
176. W. Herr, *Consequences of Periodicity and Symmetry for the Beam-Beam Effects in the LHC*, CERN LHC Project Report 49 (1996).
177. W. Kozanecki and J. Cogan, private communication (2011).
178. R. Bartoldus, *Online determination of the LHC Luminous Region with the ATLAS High Level trigger*, TIPP 2011, Int. Conf. on Tech. and Instr. in Particle Physics, Chicago (2011).
179. Y. Alexahin, et al., *Coherent Beam-Beam Effects in the LHC*, Proc. HEACC 2001, Tsukuba, Japan, CERN LHC Project Report 466 (2001).
180. K. Hirata: Nucl. Instrum. Meth. A 269, 7 (1988).
181. A. Piwinski: *Observation of Beam-Beam Effects in PETRA*, IEEE Trans. Nucl. Sci. NS-26, 4268 (1979).
182. R.E. Meller, R.H. Siemann: IEEE Trans. Nucl. Sci. NS-28, 2431 (1981).
183. K. Yokoya, et al., *Tune Shift of Coherent Beam-Beam Oscillations*, Part. Acc. 27, 181 (1990).
184. Y. Alexahin, *On the Landau Damping and Decoherence of Transverse Dipole Oscillations in Colliding Beams*, Part. Acc. 59, 43 (1996).
185. Y. Alexahin, *A Study of the Coherent Beam-Beam Effect in the Framework of the Vlasov Perturbation Theory*, Nucl. Instrum. Meth. A 380, 253 (2002).
186. W. Herr et al., *A Hybrid Fast Multipole Method Applied to Beam-Beam Collisions in the Strong-Strong Regime*, Phys. Rev. ST Accel. Beams 4, 054402 (2001).
187. J.P. Koutchouk, *ISR Performance Report - A Numerical Estimate of the Coherent Beam-Beam Effect in the ISR*, CERN ISR-OP/JPK-bm (1982).
188. W. Fischer, et al., *Observation of Coherent Beam-Beam Modes in RHIC*, BNL C-AD/AP/75 (2002).

189. V.D. Shiltsev, et al., *Compensation of Beam-Beam Effects in the Tevatron Collider with Electron Beams*, In Proceedings of the Particle Accelerator Conference 1999, New York, p. 3728.
190. V.D. Shiltsev, et al., *Considerations on Compensation of Beam-Beam Effects in the Tevatron with Electron Beams*, Phys. Rev. ST Accel. Beams 2, 071001 (1999).
191. J.P. Koutchouk, et al., *Correction of the Long-Range Beam-Beam Effect in LHC using Electro-Magnetic Lenses*, In Proceedings of the Particle Accelerator Conference 1999, New York, p. 1681.
192. R. Talman, *A Proposed Möbius Accelerator*, Phys. Rev. Lett. 74, 1590 (1995).
193. S. Henderson et al., *Investigation of the Möbius Accelerator at CESR*, CBN 99-5 (1999).
194. LHC Injectors Upgrade Technical Design Report – Volume I: Protons, edited by J. Coupard et al., CERN-ACC-2014-0337

Chapter 5
Interactions of Beams with Surroundings

M. Brugger, H. Burkhardt, B. Goddard, F. Cerutti, and R. G. Alia

With the exceptions of Synchrotron Radiation sources, beams of accelerated particles are generally designed to interact either with one another (in the case of colliders) or with a specific target (for the operation of Fixed Target experiments, the production of secondary beams and for medical applications). However, in addition to the desired interactions there are unwanted interactions of the high energy particles which can produce undesirable side effects. These interactions can arise from the unavoidable presence of residual gas in the accelerator vacuum chamber, or from the impact of particles lost from the beam on aperture limits around the accelerator, as well as the final beam dump. The wanted collisions of the beams in a collider to produce potentially interesting High Energy Physics events also reduces the density of the circulating beam and can produce high fluxes of secondary particles.

All of these unwanted interactions affect the performance of an accelerator, in desorption of gas from the vacuum system, in reduced lifetime of the circulating beam, in reduction of the collider luminosity and in background for the detectors. In this chapter the basic physical phenomena of particle interactions with matter are described, together with the techniques used to simulate the interaction with matter. The different types of particle interactions with the surroundings are elaborated in the context of their adverse effects on the accelerator performance and the mitigation measures. A full description of the effects and mitigation measures associated with the vacuum systems of accelerators is given separately in Chap. 8.

M. Brugger (✉) · H. Burkhardt (✉) · B. Goddard (✉) · F. Cerutti · R. G. Alia
CERN (European Organization for Nuclear Research), Meyrin, Genève, Switzerland
e-mail: Markus.Brugger@cern.ch; Helmut.Burkhardt@cern.ch; Brennan.Goddard@cern.ch

S. Myers, H. Schopper (eds.), *Particle Physics Reference Library*,
https://doi.org/10.1007/978-3-030-34245-6_5

5.1 The Interactions of High Energy Particles with Matter

Modern accelerators use leptons or hadrons (including ions) and have beam energies spanning the MeV to TeV range. Therefore, the capability of modelling particle interactions and showers from these energy ranges down to thermal energies is essential during all stages of the lifetime cycle of an accelerator, from the accelerator design through operation to its final decommissioning.

Before briefly discussing general aspects of hadronic and electromagnetic showers, a short overview is given of important ingredients for accelerator applications:

- energy deposition for the design of accelerator components and elements (e.g., collimators, magnets);
- particle fluences as a function of energy, angle and position (e.g., detector or radiation damage and radiation to electronics studies);
- distribution of particle interactions, inelastic interaction density (e.g., for tracking and loss pattern studies);
- residual nuclei production and generation of radioactive isotopes by beam interactions (e.g., radiation protection aspects like air activation or equipment handling).

To allow for calculations of related quantities, the underlying physical processes must not only be well understood and described in models, but also included in calculation codes able to yield allow reliable estimates within a reasonable time.

5.1.1 Basic Physical Processes in Radiation Transport Through Matter

Hadron and electromagnetic showers are very complex phenomena, whose description in terms of basic physical interactions requires a detailed and complex modelling.

As soon as the energy of a primary hadron beam exceeds a few tens of MeV, inelastic interactions start playing a major role and generate secondary particles that will have enough energy to trigger further interactions giving rise to hadronic showers. Furthermore, whenever the beam energy is high enough that significant pion production can occur, an increasing fraction of the energy will be transferred from the hadronic to the electromagnetic part due to meson decay (e.g., π^0 decaying into a gamma pair). The pion production threshold for nucleons interacting with stationary nucleons is around 290 MeV.

Therefore, energetic hadronic showers are always accompanied by significant electromagnetic showers, where the latter ones tend to develop independently without further hadronic particle production (with the exception of electro- and photo-nuclear interactions of lower importance for hadron accelerators, which however have to be considered for lepton accelerators).

While electromagnetic interactions can be described in one coherent theory (QED), the same does not apply to hadronic nuclear interactions.

The development of hadron initiated showers is determined both by atomic processes (e.g. ionization, multiple Coulomb Scattering, etc.), which take place very frequently, as well as relatively rare nuclear interactions (both elastic and nonelastic). Electromagnetic showers are determined by the same atomic processes plus additional ones (e.g., Bremsstrahlung, pair production, Compton scattering, etc.) which are specific for electrons, positrons or photons. Nuclear interactions coming from the electromagnetic component usually play a minor role, and whenever the interest is not in the small fraction of hadrons produced by electromagnetic particles they can be safely neglected.

Concerning particle production, energetic (shower) particles are concentrated mainly around the primary beam axis, regardless of their identity. Their ionization as well as the electromagnetic cascades define the core of the energy deposition distribution. At the same time, neutrons (since these are the only neutral hadrons with a long enough lifetime) will dominate at energies where charged particle ranges become shorter than the respective interaction length. In this sense, the energy deposition associated with low energy neutron interactions constitute the peripheral tails of the energy deposition distribution. Most of the interactions are due to particles (mainly neutrons) of moderate energy.

Pions can be only produced by shower particle interactions, so that they are often considered as the real indicator of a high energy cascade. Neutrons, and to a lesser extent protons, are copiously produced also in the final stages of nuclear interactions (e.g., evaporation) down to projectile energies that are comparable to their nuclear binding energy.

As previously mentioned, to focus on the relevant processes one usually distinguishes between continuous and discrete (or explicit) processes. This distinction reflects a real physical distinction, between processes which occur very frequently with mean free paths much shorter than particle ranges in matter, and others that, however, are often the dominating ones in determining the shower development.

The most important discrete processes are:

- inelastic nuclear interaction;
- decay;
- elastic nuclear interaction;
- delta-ray production;
- bremsstrahlung;
- annihilation;
- photoelectric effect;
- Compton scattering;
- pair production;
- coherent (Rayleigh) scattering.

In addition to these processes, nuclear interactions with a much lower rate can occur also for photons (as well as with a reduced rate of about 1/137 for electrons and positrons). Bremsstrahlung radiation can also be produced by "heavy" charged particles even though it is significantly suppressed. Furthermore, charged particles (light and heavy ones) can produce electron-positron pairs.

5.1.2 *Simulation Tools*

In all life cycle stages of an accelerator the use of simulation, notably Monte-Carlo codes, became fundamental. Thanks to the variety of such codes over different particle physics applications and the associated extensive benchmarking with experimental data, the modelling has reached an unprecedented accuracy. Furthermore, most codes allow the user to simulate all aspects of a high energy particle cascade in one and the same run: from the first interaction over the transport and re-interactions (hadronic and electromagnetic) of the produced secondaries, to detailed nuclear fragmentation, the calculation of radioactive decays and even of the electromagnetic shower from such delayed decays.

In the following we give a brief overview of the most used multi-purpose codes around accelerator applications.

5.1.2.1 FLUKA

FLUKA is a general-purpose particle interaction and transport code with roots in radiation protection and respective design and detector studies for high energy accelerators [1, 2]. It comprises all features needed in this area of application, such as detailed nuclear interaction models, full coupling between hadronic and electromagnetic processes and numerous variance reduction options.

The module for hadronic interactions is called PEANUT and consists of a phenomenological description (Dual Parton Model-based Glauber Gribov cascade) of high energy interactions, through a generalized intranuclear cascade, pre-equilibrium emissions as well as evaporation, fragmentation, fission and de-excitation by gamma emission. Interactions of ions are simulated through interfaces with different codes depending on the energy range (DPMJET3 above 5 GeV/n and rQMD-2.4 between 0.125 and 5 GeV/n, while the embedded Boltzmann Master Equation model is applied below 0.125 GeV/n).

The transport of neutrons with energies below 20 MeV is performed by a multi-group algorithm based on evaluated cross section data (ENDF/B, JEFF, JENDL etc.) binned into 260 energy groups, 31 of which in the thermal region. For a few isotopes point-wise cross sections can be optionally used. The detailed implementation of electromagnetic processes in the energy range between 1 keV and 1 PeV is fully coupled with the models for hadronic interactions.

5.1.2.2 GEANT4

GEANT4 is an object-oriented toolkit originally designed to simulate detector responses of modern particle and nuclear physics experiments [3, 4]. It consists of a kernel which provides the framework for particle transport, including tracking, geometry description, material specifications, management of events and interfaces to external graphics systems.

The kernel also provides interfaces to physics processes. In this regard the flexibility of GEANT4 is unique as it allows the user to select freely the physics models which serve best the particular application needs. This freedom comes with high responsibility as the user must ensure that the most adequate models are used for a given problem. Implementations of interaction models exist over an extended range of energies, from optical photons and thermal neutrons to high energy interactions as required for the simulation of accelerator and cosmic ray experiments. In many cases complementary or alternative modelling approaches are offered from which the user can choose.

Descriptions of intranuclear cascades include implementations of the Binary and the Bertini cascade models (the latter significantly reworked and not at all linked to the original Bertini model). Both are valid for interactions of nucleons and charged mesons, the former for energies below 3 GeV and the latter below 10 GeV. The Intranuclear Cascade of Liège (INCL) is also an usable option. At higher energies (up to 10 TeV), three models are available: a high-energy parameterized model (using fits to experimental data), a quark-gluon string model and the Fritiof fragmentation model, both based on string excitations and decay into hadrons. Nuclear de-excitation models include abrasion-ablation and Fermi-breakup models. Furthermore, heavy ion interactions can also be simulated if the appropriate packages are linked.

The package for electromagnetic physics comprises the standard physics processes as well as extensions to energies below 1 keV, including emissions of X-rays, optical photon transport, etc.

5.1.2.3 MARS15

The MARS15 code system [5, 6] is a set of Monte Carlo programs for the simulation of hadronic and electromagnetic cascades which is used for shielding, accelerator design and detector studies. Correspondingly, it covers a wide energy range: 100 keV–100 TeV for muons, charged hadrons and heavy ions, 1 keV–100 TeV for electromagnetic showers and down to 0.00215 eV for neutrons.

Hadronic interactions above 5 GeV can be simulated with either an inclusive or an exclusive event generator. While the former is CPU-efficient (especially at high energy) and based on a wealth of experimental data on inclusive interaction spectra, the latter provides final states on a single interaction level and preserves correlations. In the exclusive mode, the cascade-exciton model code CEM03, the Quark-Gluon String Model code LAQGSM03 and the DPMJET3 code are implemented, including models for a detailed calculation of nuclide production via evaporation, fission and fragmentation processes.

Interfaced to MARS, the MCNP4C code handles all interactions of neutrons with energies below 14 MeV. Produced secondaries other than neutrons are directed back to the MARS15 modules for further transport.

5.1.2.4 MCNP

MCNP6 [7] originates from the Monte Carlo N-Particle transport (MCNP)-family of neutron interaction and transport codes and, therefore, features one of the most comprehensive and detailed description of the related physical processes. The extension to other particle types, including ions and electromagnetic particles, allowed an expansion of the areas of application from purely neutronics to, among others, accelerator shielding design, medical physics and space radiation.

The neutron interaction and transport modules use standard evaluated data libraries mixed with physics models where such libraries are not available. The transport is continuous in energy and includes all features necessary for reactor simulations, including burn-up, depletion and transmutation. Different generalized intranuclear cascade codes can be linked to explore different physics implementations, such as CEM03, INCL4 and ISABEL. They either contain fission-evaporation models or can be coupled to such models (i.e., ABLA) allowing detailed predictions for radio-nuclide production. While the intranuclear cascade codes are limited to interaction energies below a few GeV, a link to the Quark-Gluon String Model code LAQGSM03 extends this energy range to about 800 GeV. The latter code also allows the simulation of ion interactions.

5.1.2.5 PHITS

The Particle and Heavy-Ion Transport code System PHITS (see [8] and [9] and references therein) was among the first general-purpose codes to simulate the transport and interactions of heavy ions in a wide energy range, from 10 MeV/nucleon to 100 GeV/nucleon. It is based on the high-energy hadron transport code NMTC/JAM which was extended to heavy ions by incorporating the JAERI Quantum Molecular Dynamics code JQMD.

Below energies of a few GeV hadron-nucleus interactions in PHITS are described through the production and decay of resonances while at higher energies (up to 200 GeV) inelastic hadron-nucleus collisions proceed via the formation and decay of so-called strings which eventually hadronize through the creation of (di)quark-anti(di)quark pairs. Both are embedded into an intranuclear cascade calculation. Nucleus-nucleus interactions are simulated, within a molecular dynamics framework, based on effective interactions between the two self-binding system of nucleons.

The generalized evaporation model GEM treats the fragmentation and de-excitation of the spectator nuclei and includes 66 different ejectiles (up to Mg) as well as fission processes. The production of radioactive nuclides, both from projectile and target nuclei, thus follows directly from the microscopic interaction models.

The transport of low energy neutrons employs cross sections from evaluated nuclear data libraries such as JENDL below 20 MeV. Electromagnetic interactions are simulated based on the EGS5 code in the energy range between 1 keV and 1 TeV.

Due to its capability to transport nuclei PHITS is frequently applied in ion-therapy and space radiation studies. The code is also used for general radiation transport simulations, such as in the design of spallation neutron sources.

5.1.2.6 Simulation Uncertainties

Depending on the complexity of Monte-Carlo simulations, the size of geometries and the energy range involved, calculations can carry considerable uncertainties of various sources which are in many cases difficult to evaluate. While statistical uncertainties are generally below a few percent thanks to the available computing power, systematic errors are in most cases very difficult, or even impossible, to predict in an accurate way.

The main sources of error are:

- Error due to the physics modelling: e.g., in the uncertainty in cross sections, especially at modern accelerators operating at very high energies or applications sensitive to other uncertainties in the modelling used in the simulation code. One can usually expect up to few 10% uncertainty on integral quantities, while for multi differential quantities the uncertainty can be much worse.
- Further uncertainties due to the assumptions used in the description of the geometry and of the materials under study. Usually it is difficult to quantify this uncertainty and experience shows that a factor of 2 can be taken as a safe limit for general calculations. For special cases, even in case of rather complex geometries, but when the design is implemented to a very detailed level, the latter can be reduced to about 10–20%, but rarely significantly below.
- Typical for accelerator applications, additional errors appear when having beams grazing at small angles to surfaces, where either the surface roughness is not taken into account, small misalignments can have large effects, or where one is interested in scattering effects over large distances. Therefore, especially for the latter, a safety factor of 2–3 has to be considered.

Only a detailed case-by-case analysis and careful evaluation with monitoring data can reduce the above to very low levels. Such studies are continuously done by code developers, as well as core code users and nicely show the possible high-accuracy reached with modern Monte-Carlo codes (see for instance [10]).

5.1.3 Practical Shielding Considerations

In most of the cases around high-energy hadron accelerator operation, the particle cascades originate from beam interactions due to four basic source terms:

- beam–beam interactions (at and close to the experiments);
- beam–residual gas interactions (all along the accelerator);
- direct losses: beam cleaning at collimator locations or beam dump;
- spurious losses (random locations around the accelerator).

The emerging secondary particle cascade is then again defining a multi particle-type and energy spectrum, interacting with the various materials around the accelerator. Therefore, in order to reduce the radiation levels and the corresponding fluences, shielding material can be employed to initiate and absorb showers. If the shielding material is thick, the cascades continue until most of the charged particles and photons have been absorbed, except for neutrons and secondary photons.

The following aspects have to be considered in the shielding conceptual design:

5.1.3.1 Radiation Attenuation

The propagation of high-energy hadrons features an exponential decrease due to nuclear reactions, while their energy loss, in case they are charged, is mostly due to ionization. The latter accounts for the majority of the energy loss in electromagnetic showers and for about 2/3 of the energy deposited in hadronic showers. Most of the other 1/3 of the hadronic energy is carried away by neutrons. Being neutral these are not affected by ionization losses, are decoupled from the rest of the shower, are subject to elastic and inelastic collisions and are considerably more penetrating than the charged component. Hadronic showers above 100 MeV progress through a variety of hadronic and nuclear processes that result in secondaries that are predominantly pions (π^+, π^-,π^0), followed in importance by nucleons (protons, neutrons), strange mesons/baryons and photons. The π^0 component, appearing whenever hadron energies are above the pion production threshold, is particularly important, since it decays immediately to two photons producing electromagnetic showers and shifting the shower energy from the hadronic to the electromagnetic sector. Hadrons above few tens of MeV undergo nuclear reactions that increase the neutron multiplicity. The spallation process breaks the nucleus into a few large fragments. Additional neutrons may also be produced just during this process, as well as by the subsequent evaporation from these excited fragments.

Since the fragments have large mass (M) and total charge (Z) and ionize, they will be stopped quickly in dense shield materials.

5.1.3.2 Shielding of Electromagnetic Showers

As mentioned earlier, electromagnetic showers in the multi-GeV range develop through successive bremsstrahlung and pair production. The particles involved are photons, electrons and positrons that in dense materials have typical radiation lengths in the centimeter range. As a result, high-energy electromagnetic showers are halted by a typical concrete shielding wall having thickness of the order of 1 m or by a tungsten layer of a few centimeters. It has to be noted that secondary neutrons are also generated through photo-production, thus associated shielding issues have to be correctly considered.

5.1.3.3 Shielding of Neutrons

The elastic scattering cross section of neutrons on nuclei is large at all energies and has a role in attenuating them. The scattered neutron will lose energy on every elastic collision, especially if the target nucleus is light and so carries away a larger fraction of the energy as it recoils. Therefore, the presence of hydrogen is most effective in reducing the neutron kinetic energy. Special care must be taken with respect to:

- neutrons streaming through ventilation channels or other penetrations, where intermediate and low-energy neutrons readily scatter from the wall. Propagation down the channel is thus possible via a series of 'reflections' even if the channel is not in the original direction of the neutron;
- the thermalization of neutrons, in the presence of certain elements with high capture cross-sections (e.g., Boron or Cobalt) that effectively clean the neutron field but can induce relevant secondary effects (e.g., electronic damage or activation issues).

Monte Carlo calculations aimed at shielding design optimization may require dedicated biasing techniques to overcome the associated statistical convergence challenges.

5.2 Lifetimes, Intensity and Luminosity

Particles which circulate on stable orbits in an accelerator within the geometrical and energy acceptance can get lost as a result of collisions with other particles. This leads to a decrease of the number of particles N circulating with time t.

The beam intensity lifetime τ is the inverse of the total loss rate dN/dt, normalized to the actual intensity N (at the time t)

$$\frac{1}{\tau} = -\frac{1}{N}\frac{dN}{dt}\,. \tag{5.1}$$

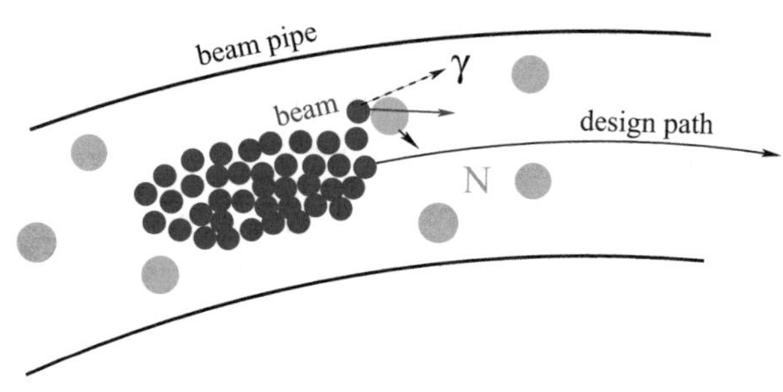

Fig. 5.1 Schematic view of beam collisions with the nuclei (N) of the rest-gas

A constant lifetime corresponds to an exponential decay of the intensity with time t proportional to $\exp(-t/\tau)$.

The intensity and beam lifetime can be monitored by measuring the beam current I which is directly proportional to the intensity, $I = Nef$, where e is the charge of the particle and f the revolution frequency in a ring [11].

We can distinguish between several types of collision processes.

- Beam-gas scattering: collisions of beam particles with particles from the residual gas in the evacuated beam pipe, sketched in Fig. 5.1.
- Thermal photon scattering: inverse Compton scattering of electrons or positrons with photons from black-body radiation in the beam pipe; relevant for beam energies $> 10\,\text{GeV}$
- Intra-beam scattering: collisions or close encounters with other particles in the same beam; most relevant at lower energies and for very dense beams
- Quantum lifetime: particles leaving the acceptance by quantum fluctuations in synchrotron radiation
- Colliding beams: particle collisions or close encounters when beams cross in colliders

Estimates for the first two types which involve collisions with particles outside the beam will be given below.

The losses from the different loss mechanism have to be added. This implies, that the corresponding lifetime contributions add reciprocally

$$\frac{1}{\tau} = \sum_i \frac{1}{\tau_i}$$

An example of observed lifetimes in LEP is listed in Table 5.1, where losses from intrabeam scattering were negligible.

The single beam lifetime is observed before beams are brought into collisions. In LEP this was dominated by the thermal photon scattering. LEP was generally operated with sufficient aperture and over-voltage from the RF-acceleration system, so that losses from the quantum lifetime were negligible. More details on lifetimes observed in LEP with a discussion of quantum lifetime and small, but non-negligible effects which made lifetimes longer than originally anticipated can be found in [12].

Table 5.1 Example of the beam lifetime in LEP, fill 4163 from 14-Sep-1997 at Eb = 91.5 GeV

Component	Lifetime τ in hours
Thermal Compton	50
Beam Gas, 0.3 nTorr CO	160
Combined, single beam	38
e^+e^- collisions, $\sigma = 0.21$ barn	8.6
Total	7

5.2.1 Beam-Gas

From ideal gas theory, the density ρ_m in terms of molecules or atoms per unit volume is

$$\rho_m = \frac{p}{kT} \,. \tag{5.2}$$

Multiplying ρ_m with the cross section σ for beam-gas collisions, gives us the collision probability per unit length

$$P_{coll} = \sigma \, \rho_m \,.$$

Further multiplication with the velocity of the beam particles $v = \beta c$ gives us the collision rate per unit time

$$\beta c \, \sigma \, \rho_m = \frac{1}{\tau} \tag{5.3}$$

which corresponds to the inverse lifetime, if σ is the cross section for collisions which lead to a loss of the beam particles. This will be the case for inelastic scattering processes and for elastic scattering in which the scattering angle is larger than the angular acceptance.

For numerical estimates in this section we take $\rho_m = 3.26 \times 10^{13}$ molecules/m^3 which corresponds to a pressure of $p = 1$ Torr $= 1.33 \times 10^{-7}$ Pa at room temperature ($T = 296.15$ K $= 23\,°$C) and can be considered as typical number for good vacuum conditions. At high energy we have $\beta \approx 1$. For a cross section of $\sigma = 1$ b (one barn, where 1b $= 10^{-28}$ m^2), we obtain a beam-gas lifetime $\tau = 284$ h.

The main beam-gas scattering processes are shown in Fig. 5.2.

eN Scattering Relevant for Electron Rings

The elastic cross section for eN scattering scales strongly with energy (with $1/\gamma^2$) and scattering angle $1/\theta^4$. Elastic scattering is mostly relevant as a halo production process for lower energy rings and becomes negligible for lifetime estimates for high energy electron rings.

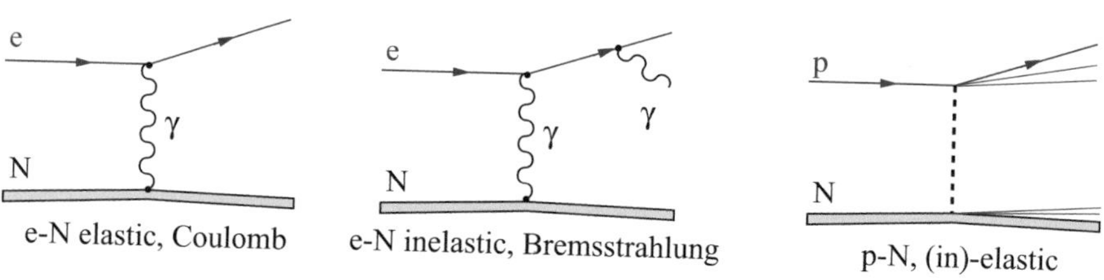

Fig. 5.2 Beam-gas scattering processes; elastic and inelastic eN scattering relevant for e^+, e^- machines is shown on the left and pN scattering relevant for proton machines on the right

At high energy, the dominating beam-gas process for electron rings is the inelastic scattering or bremsstrahlung in which the incident electron interacts with the field of the residual gas nucleus and radiates a photon.

The high energy cross section for eN scattering can be written in good approximation in dependently of the electron energy as [13]

$$\sigma_{\mathrm{eN}} = 4\alpha\, r_e^2\, Z(Z+1)\, \log(287/\sqrt{Z}) \left(-\frac{4}{3} \log k_{\min} - \frac{5}{6} + \frac{4}{3} k_{\min} - \frac{k_{\min}^2}{2} \right) , \tag{5.4}$$

where $k_{\min}$ is the fractional energy loss or minimum photon energy in units of the electron energy, α the fine-structure constant $(1/137)$ and Z the atomic number (or number of protons). We can see that the cross section scales with $Z(Z+1)$. Numerical values obtained from Eq. 5.4 for $k_{\min} = 0.01$ are shown in Table 5.4.

pN Scattering Relevant for Electron Rings

At high energies ($p_{\mathrm{lab}} > 10\,\mathrm{GeV}$), the pN cross section is mostly inelastic ($> 80\%$). It depends only weakly on the proton energy and scales approximately with the atomic mass $\propto A^{2/3}$ [14], as can be expected for the cross-section of a sphere. Numerical values are listed in Table 5.2. We can see that pN cross sections are much smaller ($\sim 10\times$) than eN cross sections. Good beam-gas lifetimes in proton machines can be several 10^3 h compared to typically 10^2 h in high energy electron machines.

5.2.2 *Thermal Photons*

Even a perfectly evacuated beam pipe remains "filled" with photons from black body radiation which can be relevant as source of backgrounds and reduction of beam lifetime as first pointed out by V. Telnov in 1987 [15]. The photon density from black-body radiation is

Table 5.2 Numerical values for σ_{eN} for an energy loss of at least 1% and for σ_{pN}, the pN cross section at high energy ($p_{lab} = 0.01$–10 TeV)

Gas	σ_{eN} b	σ_{pN} b
H_2	0.28	0.08
He	0.39	0.19
CH_4	3.02	0.43
H_2O	4.38	0.40
N_2	6.47	0.56
CO	6.56	0.56
CO_2	10.7	0.87
Ar	17.8	0.60

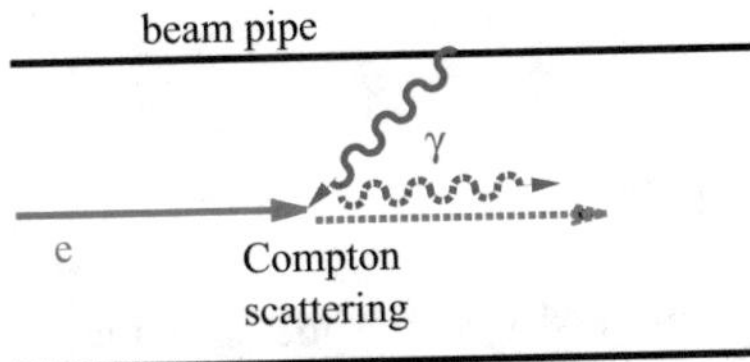

Fig. 5.3 Schematic view of the inverse Compton scattering with thermal photons. A high energy beam particle collides and loses energy to a low energy photon radiated from the beam pipe by black body radiation

$$\rho_\gamma = 8\pi \left(\frac{kT}{hc}\right)^3 \underbrace{\int_0^\infty \frac{x^2}{e^x - 1} dx}_{=2.404} \tag{5.5}$$

where T is the absolute temperature, and k,h,c the Boltzmann, Planck constants and the speed of light. For a beam pipe at room temperature (23 °C), we get a photon density of $\rho_\gamma = 5.3 \times 10^{14}\,\mathrm{m}^{-3}$ which is an order of magnitude higher than the typical residual gas molecular densities ρ_m, considered in this chapter for beam gas estimates (Fig. 5.3).

The lifetime from thermal photon scattering is

$$\tau_t = \underbrace{\frac{1}{\rho_\gamma\, c\, \sigma_C}}_{\sim 26\,\mathrm{h}} \frac{1}{f_{loss}} \tag{5.6}$$

where ρ_γ is the photon density, σ_C the Compton cross section (at high energy ∼ 0.665 barn) and f_{loss} the fraction of the $e^\pm$ lost after collision. Numerical values calculated using the program described in [16] are given in Table 5.3. We can see that thermal photon scattering at room temperature becomes only relevant for electron beam energies above 10 GeV.

Table 5.3 Thermal photon scattering

E_b (GeV)	f_{loss} (%)	τ_t (h)
10	0.3	9000
45.6	19	144
100	39	72
250	61	49

Fraction of beam particles lost and lifetime, for various electron beam energies E_b. At room temperature and for an energy acceptance of 2%

5.2.3 Luminosity Lifetime

In analogy to Eq. 5.1, the luminosity L lifetime is defined as

$$\frac{1}{\tau_L} = -\frac{1}{L}\frac{dL}{dt}. \tag{5.7}$$

The luminosity for colliding beams depends on the product of the colliding beam intensities, or N^2 in case of equal beam intensities. At constant beam sizes, the luminosity decreases as $dN^2/dt = 2dN/dt$, so that the luminosity lifetime is half of the intensity lifetime $\tau_L = \tau/2$. For operation at the beam-beam limit as is typical for high luminosity e^+e^- storage rings, beam sizes increase with intensity such that the beam-beam parameter is constant, and $L \propto N$ and therefore $\tau_L = \tau$. In proton machines, beam sizes tend to increase with time due to intrabeam scattering, noise and vibrations, resulting in luminosity lifetimes $\tau_L < \tau/2$.

5.3 Experimental Conditions

The most important performance parameters for colliders for particle physics are

- the beam energy; higher beam energies allow to study smaller distances and to produce new heavier particles;
- high luminosity to obtain sufficient collisions rates to observe new processes or to improve the measurement precisions.

It is also essential to provide good experimental conditions for the particle detectors installed around the collision regions. Criteria for good experimental conditions are

- low backgrounds;
- good knowledge and stability of beam parameters;

- minimize the risk of damage of the detectors by beam loss and irradiation;
- minimize the size of the beam pipe in the detector region to allow for the installation of vertex detectors as close as possible to the interaction region;
- maximize the space available for the detector and the solid angle coverage down to low angles close to the beam.

These are rather conflicting requirements. Minimizing the size of the beam pipe for installation of sensitive vertex detectors close to the beam pipe will increase the background rates hitting the detector region and increase the risk of damage by beam loss. The requirements for maximum space and solid angle coverage for the detectors are also in conflict with the requirements of the accelerator to maximize luminosity by

- allowing for space close to the interaction point for final focus quadrupoles;
- reducing the β-function at the interaction point, which increases the beam size in the final focus quadrupoles with the risk to create local aperture limits and losses and which limits the space and low angle coverage available for particle detection;
- installation of beam-separators close to the interaction region to allow to fill the machine with many bunches.

It is essential to consider the accelerator and detector requirements together, both during the design stage and also later in the optimization of the running parameters.

Experience shows that it typically takes several years to commission and optimize the performance of a new accelerator. During these first years, it will often be possible to increase the luminosity without compromising on the experimental conditions for the detectors. Detailed simulation and continuous monitoring of the background conditions are important to identify potential limitations, to guide further optimization and to identify the potential for upgrades towards higher luminosities or smaller beam pipes.

Different types of backgrounds and their mitigation with examples from LEP and LHC are now discussed.

5.3.1 Sources of Detector Background and Detector Performance

It is possible to distinguish between two main types of machine induced backgrounds

- backgrounds induced by losses of beam particles;
- background from synchrotron radiation, relevant for high energy e^+, e^- beams.

Table 5.4 Lifetimes τ and beam loss rates dN/dt in LEP2 and the (nominal) LHC, compared to the bunch crossing rates f_c

	N_{tot}	τ (h)	$-dN/dt$ (Hz)	f_c (Hz)	$-\frac{dN}{dt}/f_c$ (Hz)
LEP2	3.2e12	5	1.8×10^8	4.5×10^3	4×10^4
LHC	6.5e14	10	1.8×10^{10}	3.2×10^7	6×10^2

Machine induced backgrounds by particle loss are relevant for all (circular and linear) colliders. Even under good conditions, millions of particles will be lost per second, exceeding by several orders of magnitude the beam crossing rates, see Table 5.4. A minimum requirement is that only a very small fraction of these particles gets lost close to the detector, such that the background rate in the interaction region is small compared to the bunch crossing rates.

To achieve this one has to assure

- good vacuum conditions in the region around the detectors, in order to minimize local losses by beam-gas scattering in the detector region;
- that there is no aperture limitation which would concentrate losses close to the detectors.

the latter imposes limits on the minimum β in the interaction region and hence the maximum luminosity. A standard method to reduce backgrounds from particle losses is to use aperture limiting collimators to remove high amplitude halo particles. These should be placed far from the experiments, to minimize the probability that secondary particles scattered off the collimators reach the experiments.

For beam energies above about 50 GeV, the production of secondary muons in electromagnetic showers has to be taken into account. High energy muons are hard to shield. Muon production and shielding is taken into account in the design studies for high energy linear colliders [17, 18].

The final focus quadrupoles placed around the interaction regions of colliders generate a high local chromaticity which can lead to a concentration of losses of off-momentum particles into the detectors. LEP2 was equipped with momentum collimators in the dispersion suppressors around all experimental sections to reduce the flux of off-momentum particle generated by e^+e^- collisions, bremsstrahlung in the residual gas and thermal photon scattering.

5.3.2 *Synchrotron Radiation Background*

The energy spectrum of the synchrotron radiation photons radiated by a high energy electron (or positron or proton) travelling on a circular orbit of radius ρ

is characterised by the critical energy

$$E_c = \frac{3}{2}\,\hbar c\,\frac{\gamma^3}{\rho}\,. \tag{5.8}$$

The number of photons radiated in a bending magnet of length L and bending radius ρ is ($\alpha = e^2/\,4\pi\epsilon_0\hbar c$ is the fine-structure constant and γ the Lorentz factor of the particle) and the energy loss U_0 per beam particle and turn are

$$N_{\gamma,L} = \frac{5\alpha\gamma L}{2\sqrt{3}\rho}\,, \qquad U_0 = \frac{4\pi\alpha\hbar c\,\gamma^4}{3\rho} \tag{5.9}$$

The photon energy increases with γ^3 and the energy loss in synchrotron radiation over one turn as γ^4. A practical limit was reached with electrons at LEP at beam energies around 100 GeV, corresponding to a Lorentz factor $\gamma \approx 2 \times 10^5$ when 3% of the particle energy was lost on a single turn, while for the 1836 times heavier protons synchrotron radiation only becomes noticeable at TeV energies. Numerical values for the main synchrotron radiation parameters for several e^+e^- colliders, the LHC with protons at 7 TeV and the proposed FCC colliders[23, 24] are shown in Table 5.5.

The normalised quadrupole power spectrum (for flat beams as typical for e^+, e^- colliders) is

$$s_d(k) = \frac{9\sqrt{3}}{8\pi}k\int_k^\infty K_{5/3}(s)\,ds\,, \quad k = \frac{E}{E_{cr}}\,.$$

Table 5.5 Synchrotron radiation parameters; p is the beam-momentum, ρ the bending radius of the main dipole magnets

Machine	p (GeV/c)	γ	ρ (m)	E_c	N_γ	U_0	N_{tot}	P_{tot}
DaΦne	0.51	998	97.7	457 eV	66	9.3 keV	2.1×10^{13}	98 kW
PEP-II HER	9.0	17,613	163	9.9 keV	1166	3.57 MeV	4.5×10^{13}	3.4 MW
PETRA	23.4	45,792	192	148 keV	3031	138 MeV	1.0×10^{12}	3.0 MW
TRISTAN	32.	62,622	244	298 keV	4144	380 MeV	8.8×10^{11}	5.3 MW
LEP1	45.6	89,237	3026	70 keV	5906	126 MeV	1.9×10^{12}	0.4 MW
LEP2	94.5	184,932	3026	619 keV	12,239	2.3 GeV	3.2×10^{12}	13 MW
FCC-ee,Z	45.6	89,237	10,190	22 keV	5973	36 MeV	2.8×10^{15}	50 MW
FCC-ee,tt	182.5	357,144	10,190	1323 keV	23,636	9.19 GeV	1.1×10^{13}	50 MW
LHC	7000	7460.5	2784	44.1 eV	494	6.7 keV	6.5×10^{14}	7.8 kW
FCC-hh	50,000	53,289	10,572	4.24 keV	3733	4.6 MeV	1.1×10^{15}	2.5 MW

N_γ, U_0 are the number of photons radiated and the energy loss per beam particle and turn. N_{tot}, P_{tot} are the number of beam particles and the power radiated in the accelerator

Half of the power is radiated below the critical photon energy.

Quadrupoles can be considered as bending magnets which increase in strength with the distance from the magnet axis. The equations given above also hold for quadrupoles for Gaussian beams if we take as ρ the bending radius of the quadrupole at 1σ offset from the beam axis [19]. The normalised quadrupole power spectrum is

$$s_q(k) = \frac{9\sqrt{3}}{8\pi} k \int_0^\infty \left(1 - \mathrm{erf}(k/\sqrt{2}s)\right) K_{5/3}(s)\, ds\,, \quad k = \frac{E}{E_{\mathrm{cr},1\sigma}}\,,$$

and is shown in Fig. 5.4 together with the spectrum for a dipole.

Averaged over the ring, the synchrotron power radiated in the quadrupoles remains usually very small compared to the power radiated in the main dipole magnets. Due to the vicinity and strength of the quadrupoles around the experiments, it is mandatory to include these in background estimates and also important to consider the possibility of non-Gaussian tails which increase the synchrotron radiation from quadrupoles.

Background estimates for synchrotron radiation depend critically on design details: the magnet lattice, beam pipe geometry, materials and beam parameters. Estimates were often done using dedicated "home-grown" programs with ad-hoc interfaces between separated simulations of the accelerator components, synchrotron radiation generation and simulation of the interactions in the detectors. More recently it has become feasible with the programs BDSIM and MDISim, both based on GEANT4, to perform more flexible integrated simulations which include all relevant components and processes [21, 22].

We will now shortly look at LEP2 which had the strongest synchrotron radiation of all colliders and still tolerable background levels for the detectors. The amount of synchrotron radiation in LEP was huge, particularly at LEP2 energies: about 6×10^{20} photons were emitted per second and a power of 18 MW lost to synchrotron radiation. The experiments had to be very well screened using a sophisticated collimation system with about 100 movable collimators and in addition fixed masks close the experiments [20]. The typical layout of the collimators in a straight section,

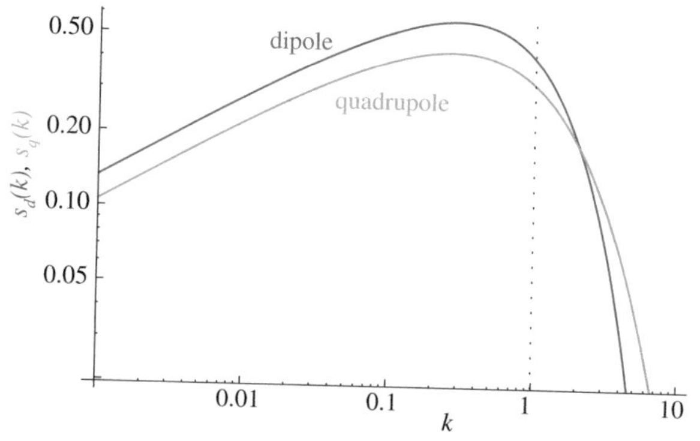

Fig. 5.4 Normalised power spectra for synchrotron radiation in a dipole and a quadrupole

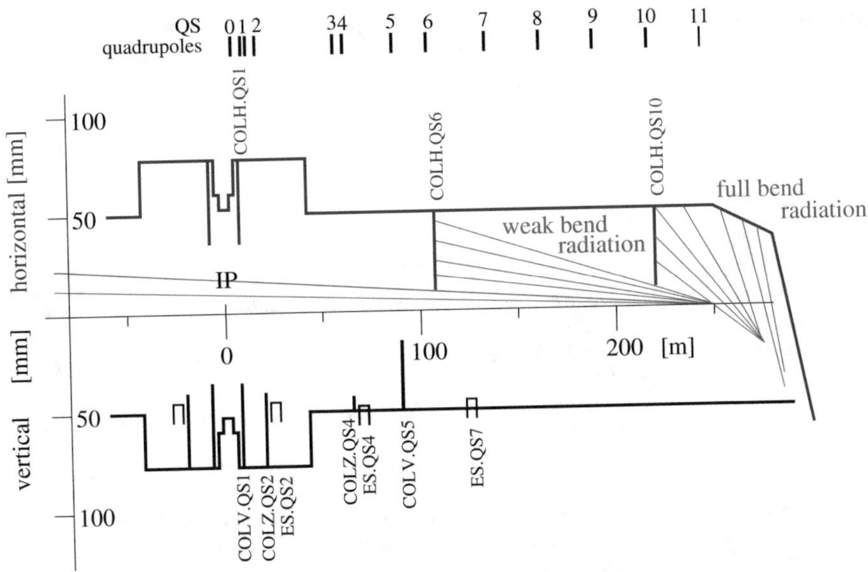

Fig. 5.5 Schematic layout of a straight section at an interaction point (IP) of LEP in the horizontal (top) and vertical (bottom) planes. Shown are the locations of the quadrupoles (QS), electrostatic separators (ES) and collimators (COLH, COLV, COLZ). The solid lines mark the inner vacuum chamber radii

where only scattered synchrotron light could reach the detectors, is shown in Fig. 5.5. The synchrotron radiation spectrum is broad and photons down to about 20 keV can leave the beam pipe. The lower energy X-ray radiation can undergo low angle (multiple) reflection. The strong radiation from the main dipoles of LEP was intercepted close to the arcs, with collimators located between 100 and 220 m away from the interaction point, before the photons could be scattered at low angle towards the experiments. To reduce the radiation shining into the straight sections further, the first dipoles in the arcs had only 10% of the field of the normal arc dipoles.

Local masks were installed about 2.4 m from the interaction points to improve the shielding of the experiments from the increased synchrotron radiation at LEP2. The collimators and masks close to the interaction point were however also a source of scattered background particles. The surface material and inclination of the masks was optimized to minimize the scattering towards the experiment: the masks were made of tungsten and the surface coated with silver and copper layers to reduce the emission of fluorescence photons.

The background photons observed in the detectors originated mainly from synchrotron radiation in the last quadrupoles and was backscattered into the experiment from local collimators. The bunch crossing rate in LEP was about 45 kHz and typically only a few background photons were recorded per bunch crossing in the large wire chambers of the LEP detectors. There was no problem with detector occupancy, but the currents drawn in the gas-chambers were reported to be not too far from the tolerable limit. The experience gained with LEP has been essential for the design studies for a possible future circular lepton collider at CERN [23, 25].

References

1. G. Battistoni et al., *Overview of the FLUKA code*, Annals of Nuclear Energy 82 (2015) 10–18.
2. T.T. Bohlen et al., *The FLUKA Code: Developments and Challenges for High Energy and Medical Applications*, Nuclear Data Sheets 120 (2014) 211–214.
3. S. Agostinelli et al., *GEANT4-A simulation toolkit*, Nuclear Instruments and Methods in Physics Research A 506 (2003) 250–303.
4. J. Allison et al., *Recent developments in GEANT4*, Nuclear Instruments and Methods in Physics Research A 835 (2016) 186–225.
5. N.V. Mokhov and C.C.James, *The MARS Code System User's Guide, Version 15*, (2016), https://mars.fnal.gov
6. N.V. Mokhov and S.I. Striganov, *MARS15 Overview*, Proceedings of the Hadronic Shower Simulation Workshop 2006, Fermilab 6–8 September 2006, M. Albrow, R. Raja eds., AIP Conference Proceeding 896 (2007).
7. C.J. Werner ed., *MCNP User's Manual, Version 6.2*, Los Alamos National Laboratory report, LA-UR-17-29981 (2017).
8. T. Sato et al., *Features of Particle and Heavy Ion Transport code System (PHITS) version 3.02*, J. Nucl. Sci. Technol. 55 (2018) 684–690.
9. Y. Iwamoto et al., *Benchmark study of the recent version of the PHITS code*, J. Nucl. Sci. Technol. 54 (2017) 617–635.
10. A. Lechner et al., *Validation of energy deposition simulations for proton and heavy ion losses in the CERN Large Hadron Collider*, submitted to Physical Review Accelerators and Beams (2019), https://doi.org/10.1103/PhysRevAccelBeams.22.071003.
11. A. Peters, H. Schmickler, K. Witternburg (Editors), "Proceedings of Workshop on DC Current Transformers and Beam-Lifetime Evaluations, Lyon 2004", CARE-Note-2004-023-HHH
12. H. Burkhardt, "Beam Lifetime and Beam Tails in LEP", Proc. e^+e^- factories, KEK 1999, CERN-SL-99-061-AP (1999)
13. Y.-S. Tsai, "Pair production and Bremsstrahlung of charged leptons," *Rev. Mod. Phys.* **46** (1974) 815–851.
14. J. Letaw, R. Silberberg, and C. Tsao, "Proton-nucleus inelastic cross sections: an empirical formula for E>10 MeV", The Astrophys. Journ.Suppl.Ser. 51, (1983), pp 271–276
15. V. I. Telnov, "Scattering of electrons on thermal radiation photons in electron - positron storage rings", *Nucl. Instrum. Meth.* A260 (1987) 304,
16. H. Burkhardt, "Monte Carlo Simulation of Beam Particles and Thermal Photons", July, 1993. CERN SL Note 93-73 (OP).
17. I. Agapov, H. Burkhardt, D. Schulte, A. Latina, G. A. Blair, S. Malton, and J. Resta-López, "Tracking studies of the Compact Linear Collider collimation system," *Phys. Rev. ST Accel. Beams* **12** (Aug, 2009) 081001. 10.1103/PhysRevSTAB.12.081001
18. H. Burkhardt, G. A. Blair, and L. Deacon, "Muon Backgrounds in CLIC", Proc. IPAC 2010
19. E. Keil, "Synchrotron Radiation from a Large electron-Positron Storage Ring", CERN-ISR-LTD-76-23, June 1976
20. G. von Holtey, A. Ball, *et al.*, "Study of beam-induced particle backgrounds at the LEP detectors", *Nucl. Instrum. Meth.* A403 (1998) 205–246, https://doi.org/10.1016/S0168-9002(97)01094-2.

21. L. Nevay et al., BDSIM: An Accelerator Tracking Code with Particle-Matter Interactions. https://urldefense.proofpoint.com/v2/url?u=https-3A__arxiv.org_abs_1808.10745v1&d=DwIFaQ&c=vh6FgFnduejNhPPD0fl_yRaSfZy8CWbWnIf4XJhSqx8&r=XRE98stjh0DZlXWNYx-jk11s49AfQ9rwMw8TKownaLA&m=nT9ubpnYZV6ASvZRvylWgRRpic9Q8sGOtaj0hVsLltE&s=sVWR0RTCWGrRe0NT2jXWKQJUXrvTIg_zydlInZnlwOs&e=
22. H. Burkhardt and M. Boscolo. Tools for Flexible Optimisation of IR Designs with Application to FCC. In *Proc IPAC 2015*, https://doi.org/10.18429/JACoW-IPAC2015-TUPTY031.
23. A. Abada et al. FCC-ee: The Lepton Collider. https://doi.org/10.1140/epjst/e2019-900045-4. [Eur. Phys. J. ST228,no.2,261(2019)].
24. A. Abada et al. FCC-hh: The Hadron Collider. *Eur. Phys. J. ST*, 228(4):755–1107, https://doi.org/10.1140/epjst/e2019-900087-0.
25. M. Boscolo, H. Burkhardt, and M. Sullivan. Machine detector interface studies: Layout and synchrotron radiation estimate in the future circular collider interaction region. *Phys. Rev. Accel. Beams*, 20(1):011008, 2017. https://doi.org/10.1103/PhysRevAccelBeams.20.011008.

Chapter 6
Design and Principles of Synchrotrons and Circular Colliders

B. J. Holzer, B. Goddard, Werner Herr, Bruno Muratori, L. Rivkin, M. E. Biagini, J. M. Jowett, K. Hanke, W. Fischer, F. Caspers, and D. Möhl

6.1 Beam Optics and Lattice Design in High Energy Particle Accelerators

B. J. Holzer

Lattice design in the context we will describe it here is the design and optimization of the principle elements—the lattice cells—of a circular accelerator, and it includes the dedicated variation of the accelerator elements (as for example position and strength of the magnets in the machine) to obtain well defined and predictable parameters of the stored particle beam. It is therefore closely related to the theory of linear beam optics that has been described in Chap. 2 [1].

Coordinated by K. Hanke

B. J. Holzer · B. Goddard · W. Herr · B. Muratori · M. E. Biagini · J. M. Jowett · K. Hanke (✉) · F. Caspers · D. Möhl
CERN (European Organization for Nuclear Research)Meyrin, Genève, Switzerland
e-mail: Bernhard.holzer@cern.ch; Brennan.Goddard@cern.ch; werner.herr@cern.ch; bruno.muratori@stfc.ac.uk; John.Jowett@cern.ch; Klaus.Hanke@cern.ch; Fritz.Caspers@cern.ch

L. Rivkin
Paul Scherrer Institut, Villigen, Switzerland
e-mail: Leonid.Rivkin@psi.ch

W. Fischer
Brookhaven National Laboratory, New York, NY, USA
e-mail: Wolfram.Fischer@bnl.gov

S. Myers, H. Schopper (eds.), *Particle Physics Reference Library*,
https://doi.org/10.1007/978-3-030-34245-6_6

6.1.1 Geometry of the Ring

For the bending force as well as for the focusing of a particle beam, magnetic fields are applied in an accelerator. In principle, electrostatic fields would also be possible but at high momenta (i.e. if the particle velocity is close to the speed of light) the usage of magnetic fields is much more efficient. In its most general form, the force acting on the particles is given by the Lorentz-force

$$\boldsymbol{F} = q\,(\boldsymbol{E} + \boldsymbol{v} \times \boldsymbol{B}) \tag{6.1}$$

In high energy accelerators, the velocity v is close to the speed of light and so represents a nice amplification factor whenever we apply a magnetic field. As a consequence, it is much more convenient to use magnetic fields for bending and focusing the particles. Neglecting the $\boldsymbol{E}$ component therefore in Eq. (6.1), the condition for a circular orbit is defined as the equality of the Lorentz force and the centrifugal force:

$$qvB = \frac{mv^2}{\varrho} \tag{6.2}$$

In a constant transverse magnetic field B, the particle will see a constant deflecting force and the trajectory will be a part of a circle, whose bending radius ρ is determined by the particle momentum $p = mv$ and the external B field.

$$\rho = \frac{p}{qB} \tag{6.3}$$

The term $B\rho$ is called beam rigidity. Inside each dipole magnet in a storage ring the bending angle—sketched out in Fig. 6.1—is given by the integrated field strength via

$$\alpha = \frac{\int Bds}{B\rho} \tag{6.4}$$

Requiring a bending angle of 2π for a full circle, we get the condition for the magnetic dipole fields in the ring. In the case of the LHC e.g. for a momentum of $p = 7000$ GeV/c a number of 1232 dipole magnets are needed each having a length of ~15 m with a B-field of 8.3 T. As a general rule in high energy rings, about 66% (2/3) of the circumference of the machine should be foreseen to install dipole magnets, as they define the maximum particle momentum that can be carried by the machine. This basic dipole structure is completed with focusing elements, beam diagnostic tools etc. and forms the arcs of the ring. They are connected by long

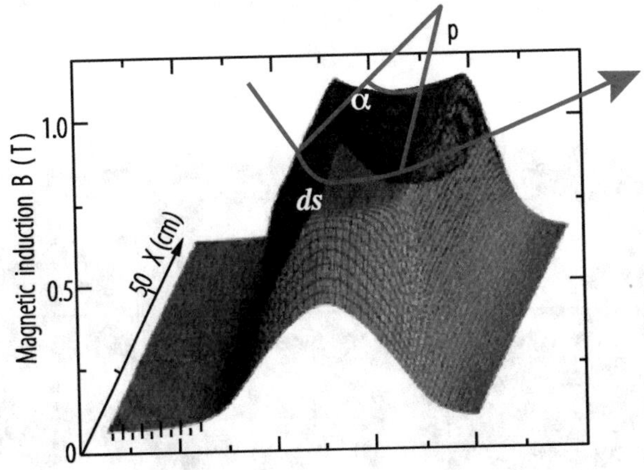

Fig. 6.1 *B*-field in a storage ring dipole magnet and schematic particle orbit

straight sections, so-called insertions, where the optics are modified to establish conditions needed e.g. for particle injection or extraction and the installation of the radio-frequency resonators for the particle acceleration. In the case of collider rings so-called mini-beta insertions are included, where the beam dimensions are reduced considerably to increase the particle collision rate and where space is needed for the installation of the particle detectors.

The lattice and correspondingly the beam optics therefore are split in different characteristic parts: arc structures that are used to guide the particle beam and define the geometry of the ring; they establish a regular pattern of focusing elements. And the straight sections, that are optimised for the installation of a manifold of technical devices, including the high-energy physics detectors.

6.1.2 Lattice Design

An example of such a high-energy lattice and the corresponding beam optics is shown in Fig. 6.2. In the upper part of the figure the regular pattern of the beta function is plotted in red and green for the two transverse planes. As a consequence of the periodic structure of the lattice, the beta function—and so the beam size—reaches a maximum value in the centre of the focusing, and a minimum in the centre of the defocusing quadrupoles. The lower part of the figure shows the horizontal and vertical dispersion function. The lattice of the complete machine is designed on the basis of small periodic lattice structures—called cells—that repeat many times in the ring. One of the most widespread lattice cells used in high-energy rings is the

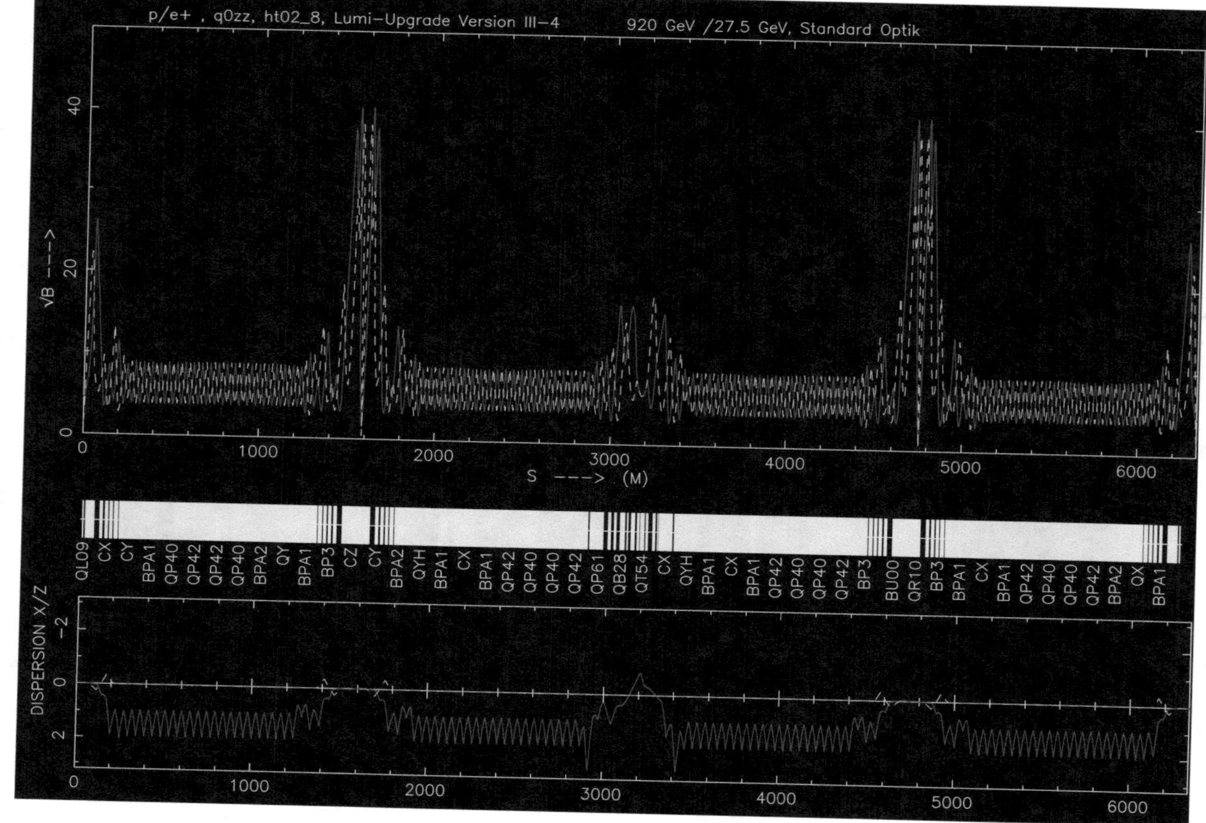

Fig. 6.2 Lattice of a high energy storage ring with periodic arc structures and straight sections

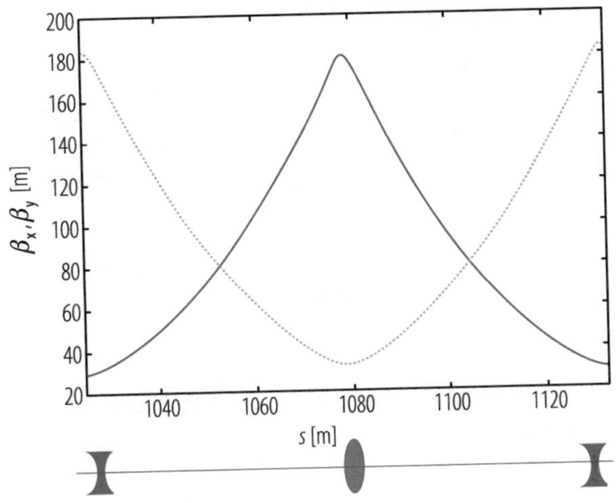

Fig. 6.3 Basic element of a high-energy storage ring: the FODO cell

so-called FODO cell: A magnet structure consisting of focusing and defocusing quadrupole lenses in alternating order. In between the focusing elements the dipole magnets are located and any other machine elements like orbit corrector dipoles, multipole correction coils or diagnostic instruments can be installed.

In Fig. 6.3 the optical solution of such a FODO cell is plotted: The graph shows the β-function in the two transverse planes (red curve for the horizontal, green curve for the vertical plane). In the lower part of the plot the position of the magnet lenses, the lattice, is indicated schematically. In first order the optical properties of such a lattice are determined only by the parameters of the focusing (F) and de-focusing (D) quadrupole lenses. In between these two quadrupole magnets only lattice elements are installed that have zero ("O") or negligible influence on the transverse particle dynamics. Hence the acronym *FODO* for such a structure. Due to the symmetry of the cell the solution for the β function is periodic (in general such a FODO cell is the smallest periodic structure in a storage ring) and it reaches its extreme values at the position of the quadrupole lenses. As a consequence, at these locations in the arc, the beam will reach its maximum dimension $\sigma = \sqrt{\varepsilon\beta}$, and the aperture need will be highest.

Accordingly, the "Twiss" parameter α, which is the derivative of β is generally zero in the middle of the FODO quadrupoles. Based on the thin lens approximation a number of scaling laws and rules can be established to understand the properties of such a FODO structure [2]: How do we arrange the strength and position of the quadrupole lenses in the lattice to obtain a certain beta-function? How does the cell length influence the phase advance of the particle trajectories? How do we guarantee that, turn by turn, a stable particle oscillation is obtained?

In the following we briefly summarise these rules.

- Stability of the motion: the strengths of the focusing (and defocusing) elements in the lattice have to be such that the particle oscillation does not increase. This condition—the stability criterion for a periodic structure in a lattice—is obtained in a FODO if the focal length of the magnets is larger than a quarter of the cell length:

$$f = \frac{1}{kl} = \frac{L_{cell}}{4}. \tag{6.5}$$

- The beta function—and so the beam size—is determined by the phase advance of the cell and its length:

$$\beta_{\max,\min} = \frac{1 \pm \sin(\varphi_{cell}/2)}{\sin \varphi_{cell}} L_{cell}. \tag{6.6}$$

- A similar scaling law is obtained for the dispersion:

$$D_{\max,\min} = \frac{L_{cell}^2}{4\rho} \frac{1 \pm \frac{1}{2}\sin(\varphi_{cell}/2)}{\sin^2(\varphi_{cell}/2)}. \tag{6.7}$$

In general, small values for the β functions as well as for the dispersion are desired. It will be the intention of the lattice designer to minimise the beam size, and so to optimise the aperture need of the beam. In addition the β-function indicates the sensitivity of the beam with respect to external fields and field errors. A change in a quadrupole field e.g. will shift the tune of the beam by

$$\Delta Q = \frac{1}{4\pi} \int \Delta k(s)\beta(s)ds. \tag{6.8}$$

The effect is proportional to the size of the applied change in quadrupole field, Δk but also to the value of the beta function at this position. Therefore, the phase advance of the FODO cell has to be chosen to obtain smallest values for β in both transverse planes, which leads in the case of protons or heavy ions to an optimum phase advance of 90° per cell. It will be no surprise that the focusing structure of typical high energy proton rings like SPS, Tevatron, HERA-p and LHC were optimised for this value.

In addition to the main building blocks, the dipoles and quadrupole magnets, the FODO will be equipped with a number of correction magnets for orbit correction, compensation of higher harmonic field errors of the main magnets, and sextupoles

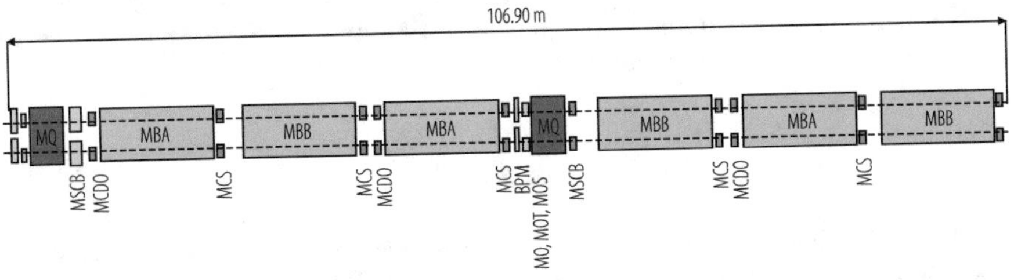

Fig. 6.4 FODO cell of LHC. In addition to the two main quadrupoles and six dipole magnets, diagnostic instruments and multipole compensation coils are included in the arc lattice

for chromaticity compensation of the machine. The FODO cell of LHC, including these corrector magnets is illustrated in Fig. 6.4.

Six dipoles and two main quadrupoles are forming the basic structure of the cell; they are complemented by orbit correction dipoles, trim quadrupoles that are used for fine tuning of the working point and multipole correction coils to compensate higher order field distortions up to 12 pole [3].

Among the higher order correction coils mentioned above the sextupoles play the most critical role in the arc structure, as they are indispensable to compensate the chromatic errors in the lattice. Chromaticity is an optical error that describes the distortion of the focusing properties in a lattice under the presence of momentum spread of the particle beam. In general a sextupole magnet will be installed to support each quadrupole in the arc. At least two sextupole families are required, one for each transverse plane. In some cases several families per plane are installed to improve the region of stability in the transverse plane (the so-called dynamic aperture of the storage ring). They have to be strong enough to correct the chromaticity created in the arc cells as well as in the insertion sections. The mechanism of chromaticity correction is based on the combination of the dispersion function that sorts the particles according to their momentum and the nonlinear field of a sextupole magnet:

$$B_z = \frac{1}{2}\tilde{g}\left(x^2 - z^2\right), \tag{6.9}$$

where

$$\tilde{g} = \frac{d^2 B_z}{dx^2} \tag{6.10}$$

describes the sextupole "gradient".

Normalizing the sextupole field to the beam rigidity we write the contribution of each sextupole to the chromaticity as

$$\Delta Q = \frac{1}{4\pi} \int k_{sext} D\beta dl \tag{6.11}$$

and it depends indeed on the value of both, beta function and dispersion. Therefore the sextupole magnets that are needed to compensate the natural chromaticity in the ring will be located in the lattice at places where at the same time the dispersion and the beta function are large, i.e. close to the corresponding quadrupole lenses.

6.2 Lattice Insertions

B. J. Holzer

The arc structure of a storage ring is usually built out of regular patterns like FODO cells that are repeated periodically and determine the geometry of the machine. Straight sections are inserted to combine these arcs and provide the space required for beam injection, extraction, or dispersion free lattice parts to install e.g. RF systems. Finally space is needed to establish the conditions that are required for the collisions of the two counter rotating beams. As an example of the general layout of a storage ring we refer again to the LHC lattice. Eight straight sections connect eight arcs: four of them are used for beam injection, extraction and collimation, the remaining four are optimised to house the high-energy detectors (IR1, 5, 2, 8 in Fig. 6.5). Here the storage ring lattice has to provide the free space needed for the installation of a large modern particle detector and the beam optics has to be modified to provide the strong focusing needed at the collision point.

6.2.1 Low Beta Insertions

The most important “insertion” for a particle collider ring is the so-called mini beta structure: The key issue of a collider is its luminosity [4] that defines the rate of produced collision events (particles or particle reactions of interest) in the machine. Its value is defined by the machine lattice and under the assumption of equal beam properties in the two colliding beams it is given by the stored currents in the two beams, I_{p1}, I_{p2}, the revolution frequency f_0, the number of stored bunches, n_b, and most of all by the transverse size of the two beams, σ_x^* and σ_y^*. In the simplest case we get:

$$L = \frac{1}{4\pi e^2 f_0 n_b} \frac{I_{p1} I_{p2}}{\sigma_x^* \sigma_y^*}. \tag{6.12}$$

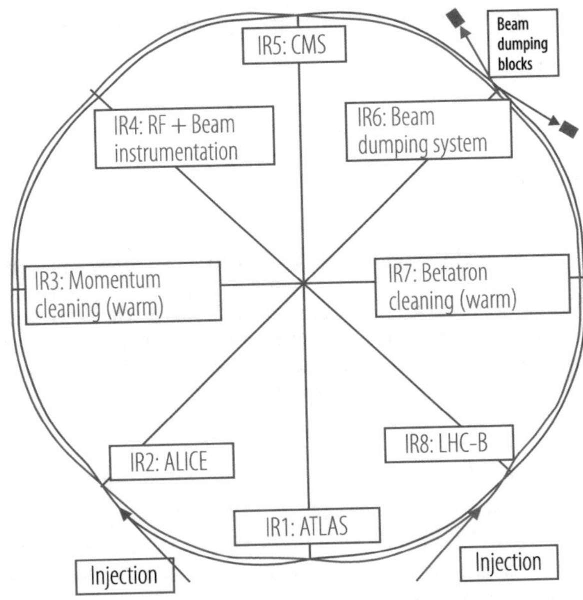

Fig. 6.5 Lattice geometry of the LHC

A more general formula that includes geometric and optical reduction factors is presented in Sect. 6.4 [4]. At the interaction point "IP", the intention of the lattice designer will be to reduce the beta function as much as possible in order to obtain the smallest possible beam. The main limiting factor comes from a basic principle which is valid for any system of particles under the influence of conservative forces ("Liouville's Theorem"): Under conservative forces, the density of the particle's phase space volume is constant. Applying this law to a particle beam in an accelerator it means that the beam dimension and divergence are not independent of each other. Namely for the design of symmetric drift space in a storage ring we can deduce a rule for the beta function: Starting from a waist ($\alpha^* = 0$ at the collision point) the beta function develops as

$$\beta(s) = \beta^* + \frac{s^2}{\beta^*}. \tag{6.13}$$

The star refers to the value at the waist (e.g. the interaction point "IP"). This relation is a direct consequence of Liouville's theorem and therefore of fundamental nature. As a consequence the behaviour of β in a symmetric drift cannot be changed and has a strong impact on the design of a storage ring: Small beta functions at the collision point and a large distance to the first focusing element lead to high values of the beta function and correspondingly to large beam dimensions at the first focusing element in front and after the IP.

The preparation of the beam optics for the installation of modern high-energy detectors therefore needs special treatment in the lattice design to provide the large space needed for the detector hardware. An illustrative example is shown in

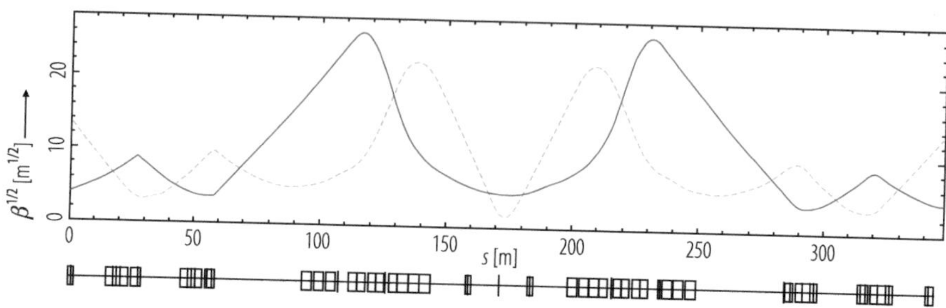

Fig. 6.6 Layout of a mini beta insertion scheme. The example shows a low beta insertion based on a quadrupole doublet. The vertical beta function (green line) starting with smaller values at the IP shows a stronger increase than the beta in the horizontal plane. Accordingly the doublet quadrupoles are powered in QD-QF polarity

Fig. 6.6: a long symmetric drift space that holds the experiment is centred around the interaction point of the colliding beams. Depending on the respective value of beta at the IP the beta functions increase in the horizontal (red) and vertical (green) plane and are focused back using a couple of strong, large aperture and high quality quadrupole lenses. Depending on the particular situation (namely the ratio of the two $\beta*$ values in the two planes a quadrupole doublet or triplet arrangement will be the adequate choice for these mini beta quadrupoles. Additional independent quadrupole magnets (i.e. individually powered magnets) will be needed to create a smooth transition of the optics from the IP to the periodic solution of the FODO cells in the arc. In general eight parameters have to be optimised: the β and α values in the two planes, the dispersion and its derivative and the phase advance of the complete mini beta system. As a consequence such a mini beta insertion will have to be equipped with at least eight individually powered quadrupole magnets to fulfil this requirement.

It has been pointed out in the previous chapter that the emittance of a particle beam is not constant during acceleration but depends on the energy of the particle beam. In the case of a proton or ion beam the adiabatic shrinking is the dominant effect and the emittance follows the rule $\varepsilon \propto 1/\beta\gamma$ where β and γ are the relativistic parameters. As a consequence the emittance in a proton storage ring is highest at injection energy and the beam optics has to be optimised to limit the beta function at any place in the machine to values that guarantee sufficient aperture. At high energy (the so-called flat-top) the emittance is small enough that the mini beta concept can be used to full extend and only here the $\beta*$ can be reduced to the small values that are required to deliver the design luminosity values. The lattice of the mini beta insertion therefore has to be optimised in a way, that two quite different beam optics can be established by corresponding adjustment of the quadrupole gradients: A low energy optics for injection and the early steps of the acceleration and a true mini beta optics that will be used for the collider run at high energy.

The procedure to pass from the injection optics to the luminosity case is often called "beta squeeze" and is a critical situation as optics, orbits and global beam parameters like tune and chromaticity have to be maintained constant and well

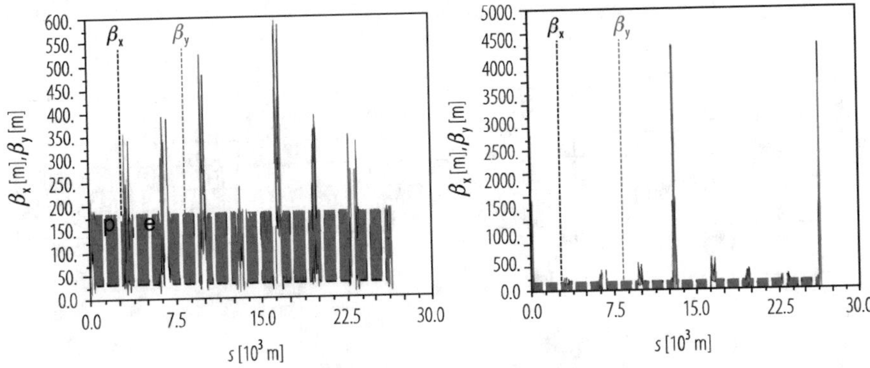

Fig. 6.7 (Left) Beam optics for the LHC: 450 GeV injection optics optimised for small values of beta to gain highest aperture in the machine. (Right) Low beta optics for the LHC luminosity operation: due to the small values at the IP the beta function reaches large values in the low beta quadrupole lenses. (Note the different scale of the vertical axis)

controlled during the changing quadrupole settings. Several intermediate steps might be needed to guarantee a smooth transition between the two operation modes. In the case of the LHC the 450 GeV injection case and the 7 TeV luminosity optics are compared in Fig. 6.7.

6.2.2 *Injection and Extraction Insertions*

In addition to the mini beta insertions where the beams are optimised for highest collision rates, additional insertions are needed in the storage ring for beam injection and extraction. In these cases the same rules are valid as for the mini beta insertions but in general the consequences are more relaxed. Additional hardware that has to be installed for the injection process (fast kicker magnets and septum dipoles to inject the new beam) is much smaller than the detectors at the collision points. Still, however, some modifications of the lattice will be needed and the optics will have to be re-matched to establish the required space. A special additional feature should be mentioned here: the new beam that is being injected has to match perfectly in energy and in phase space to the optical parameters of the storage ring or synchrotron. At the end of the beam transfer line as well as in the storage ring the focusing fields have to be optimised to obtain the same values of the Twiss functions α and β in both transverse planes. As in the case of the mini beta insertions additional individually powered quadrupole magnets are needed. As an example the beam optics of the SPS-LHC transfer-line is plotted in Fig. 6.8. At the beginning and the end of the lattice structure—indicated by red markers in the figure—the beta function is modified to match the optics from the SPS to the FODO channel of the transfer line and from

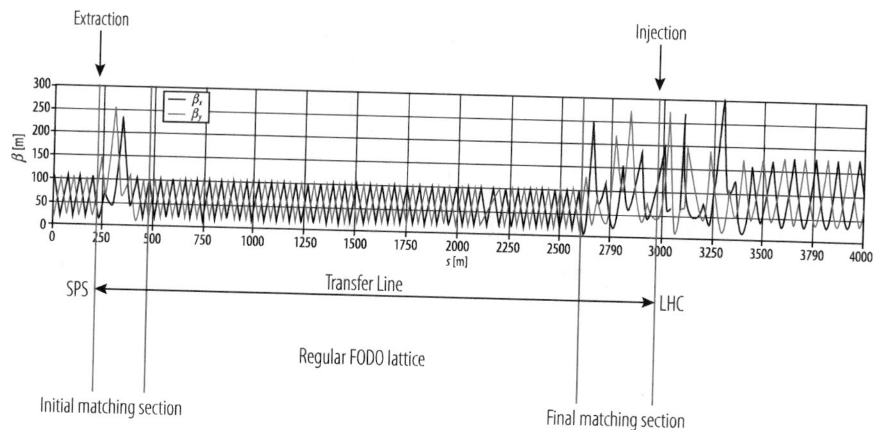

Fig. 6.8 Transfer line between the SPS and the LHC. Two matching sections have to be introduced to adopt the beam optics from the SPS to the transfer line and to the LHC

the FODO to the LHC insertion at IR2 and IR8 where the injection elements are located.

6.2.3 Dispersion Suppressors

The dispersion function $D(s)$ has already been introduced in Sects. 2.4 and 6.1. It describes the trajectory in the case of a momentum deviation of the particle and is the consequence of the corresponding error in the bending strength of the dipole magnets. In the arc structure with its regular pattern of dipole magnets, dispersive effects cannot be avoided (but they should be minimised) and the additional amplitude due to the dispersion has to be considered if we are talking about particle trajectories or beam sizes. In linear approximation and for a small momentum spread $\Delta p/p$ in the beam, the amplitude of a particle oscillation is obtained by

$$x(s) = x_\beta(s) + D(s)\frac{\Delta p}{p_0}, \tag{6.14}$$

where x_β describes the solution of the homogeneous differential equation (the usual betatron oscillations of the particle) and the second term—the dispersion term—corresponds to the additional oscillation amplitude for particles with a relative momentum error $\Delta p/p_0$. At the interaction point where the smallest beam sizes are required to obtain the highest luminosity, we intend to suppress the dispersion and as the collision point is generally located in a straight section of the accelerator, techniques have been developed to obtain dispersion free sections inside the lattice. The insertions that are used to reduce the dispersion function from its periodic value in the arc to zero are called dispersion suppressors [2, 5, 6].

It has to be mentioned in this context that especially in the case of synchrotron light sources a variety of lattice types has been developed with the goal to achieve small or even zero dispersion in the ring or in parts of it. However, these lattices are optimised for the purpose of high brilliant synchrotron radiation and are not ideal for high-energy particle accelerators, where FODO cells are usually the most appropriate choice.

Referring to high energy colliders we will concentrate therefore on the interaction region, i.e. a straight section of a ring where two counter rotating beams collide in a dispersion free part of the storage ring. A non-vanishing dispersion dilutes the luminosity of the machine and leads to additional stop bands in the working diagram of the accelerator ("synchro-betatron resonances"), that are driven by the beam-beam interaction. Therefore sections are inserted in our magnet lattice that are designed to reduce the function $D(s)$ to zero. Three main techniques are widely used: the quadrupole based dispersion suppressor, the missing bend scheme and the half bend scheme. We will not present all of them in detail but instead restrict ourselves to the basic idea behind it.

6.2.3.1 The "Straightforward" Way: Dispersion Suppression Using Quadrupole Magnets

Let us assume here that a periodic lattice is given in the arc (see Fig. 6.2) and that this FODO structure simply is continued through the straight section—but with vanishing dispersion. Given an optical solution in the arc cells, as for example shown in Fig. 6.9, we have to guarantee that starting from the periodic solution

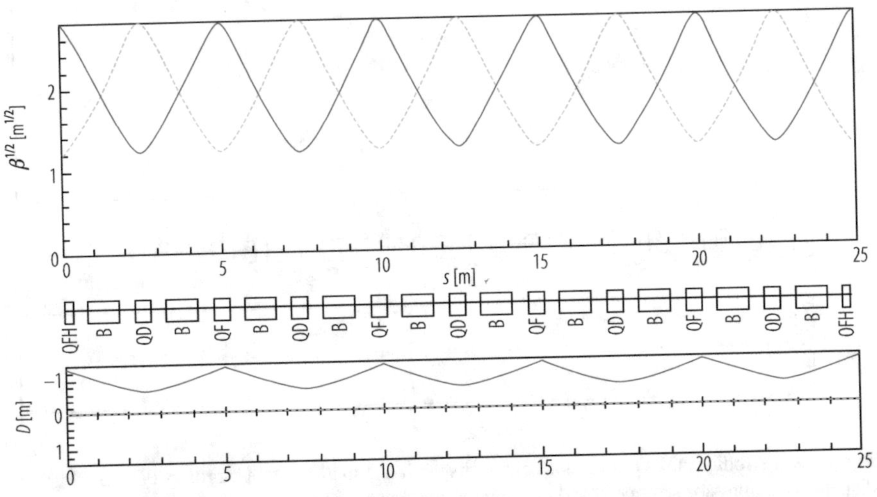

Fig. 6.9 Periodic FODO and horizontal dispersion function in a regular FODO structure

of the optical parameters $\alpha(s)$, $\beta(s)$ and $D(s)$ we obtain a situation at the end of the suppressor where we get $D(s) = D'(s) = 0$ and the values for α and β unchanged.

The boundary conditions after the suppressor section

$$D(s) = D'(s) = 0,$$
$$\beta_x(s) = \beta_{x\ \mathrm{arc}}, \alpha_x(s) = \alpha_{x\ \mathrm{arc}}, \tag{6.15}$$
$$\beta_y(s) = \beta_{y\ \mathrm{arc}}, \alpha_x(s) = \alpha_{y\ \mathrm{arc}},$$

can be fulfilled by introducing six additional quadrupole lenses whose strengths have to be matched individually in an adequate way. This can be done by using one of the beam optics codes that are available today in every accelerator laboratory. An example is shown in Fig. 6.10, starting from a FODO structure with a phase advance of $\varphi \approx 70^\circ$ per cell.

The advantages of this scheme are:

- it works for any phase advance of the arc structure;
- matching works also for different optical parameters α and β before and after the dispersion suppressor as—within a certain range—the quadrupoles can be used to match the Twiss functions to different values;
- the ring geometry is unchanged as the number and location of dipole magnets in the ring is unchanged.

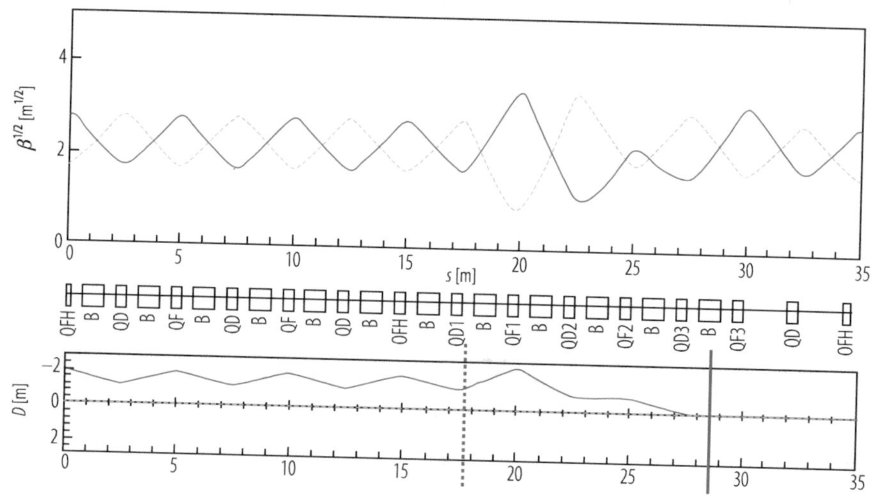

Fig. 6.10 Periodic FODO and horizontal dispersion function in a regular FODO structure dispersion suppressor scheme based on individually powered quadrupole lenses

On the other hand there are a number of disadvantages that have to be mentioned:

- as the strength of the additional quadrupole magnets have to be matched individually the scheme needs additional power supplies and quadrupole magnet types which can be an expensive requirement;
- the required quadrupole fields are in general stronger than in the arc;
- the β function reaches higher values (sometimes *really* high values) which leads to higher beam sensitivity and larger aperture needs.

There are alternative ways to suppress the dispersion, which do not need individually powered quadrupole lenses but instead change the strength of the dipole magnets at the end of the arc structure.

6.2.3.2 The "Clever" Way: Half Bend Schemes

This dispersion suppressing scheme is made up of *n* additional FODO cells that are added to the periodic arc structure but where the bending strength of the dipole magnets is reduced. As before we split the lattice into three parts: the periodic structure of the FODO cells in the arc, the lattice insertion where the dispersion is suppressed, followed by a dispersion free section which can be another FODO structure without bending magnets or a mini beta insertion.

Starting from the dispersion free straight section the basic idea of this scheme is to create with a special arrangement of dipole magnets inside the dispersion suppressor—exactly the dispersion that corresponds to the periodic solution of the arc FODO cells. The solution will depend on the phase advance of the cells as well as on the strength of the bending magnets inside the suppressor magnets.

As explained before in the beam optics chapter, the matrix for a periodic part of the lattice (namely one single cell in our case) can be expressed as

$$M_{cell} = \begin{pmatrix} C & S & D \\ C' & S' & D' \\ 0 & 0 & 1 \end{pmatrix} = \begin{pmatrix} \cos\phi_c & \beta_C \sin\phi_c & D \\ -\frac{1}{\beta_c}\sin\phi_C & \cos\phi_c & D' \\ 0 & 0 & 1 \end{pmatrix}, \tag{6.16}$$

where the index "*c*" reflects the solution of a cell, ϕ_c denotes the phase advance for a single cell and the elements *D* and *D'* correspond to its periodic dispersion.

As usual the dispersion elements are obtained by

$$D(l) = S(l)\int_0^l \frac{C\left(\tilde{s}\right)}{\varrho\left(\tilde{s}\right)}d\tilde{s} - C(l)\int_0^l \frac{S\left(\tilde{s}\right)}{\varrho\left(\tilde{s}\right)}d\tilde{s}. \tag{6.17}$$

The functions $C(s)$ and $S(s)$ are the cosine and sine like matrix elements of the lattice element in the sense that e.g. $C(s) = M[1,1]$, and the integral is executed over one complete cell.

In the dispersion suppressor section, the dispersion $D(s)$ starts with the value D_0 the end of the arc cell and is reduced to zero. Or turning it around and thinking from right to left: the dispersion has to be created inside the suppressor part by proper arrangement of the dipole magnets, starting from $D = D' = 0$ in the straight section to reach the values that correspond to the periodic dispersion of the arc cells. Solving the equation above by integrating over a certain number of cells will determine the bending strength $1/\rho$ and the number n of cells in the suppressor part that is needed to fulfill the boundary condition and get the values of the dispersion in the following periodic arc cell.

For a given phase advance φ_c per cell two conditions for the dispersion matching are obtained that combine the number of suppressor cells, n, and the strength of the suppressor dipoles, δ_{supr}:

$$\left.\begin{array}{r} 2\delta_{\text{supr}}\sin^2\left(\frac{n\phi_c}{2}\right) = \delta_{\text{arc}} \\ \sin\left(n\phi_c\right) = 0 \end{array}\right\} \delta_{\text{supr}} = \frac{1}{2}\delta_{\text{arc}}. \tag{6.18}$$

If the phase advance per cell in the arc fulfills the condition $\sin(n\phi_c) = 0$, the strength of the dipoles in the suppressor region is just half the strength of the arc dipoles. In other words the phase advance has to fulfill the condition

$$n\phi_c = k\pi, \quad k = 1, 3, \ldots. \tag{6.19}$$

There are a number of possible phase advances that fulfill that relation, but clearly not every arbitrary phase is allowed. Possible constellations would be for example, $\phi_c = 90^\circ$, $n = 2$ cells, or, $\phi_c = 60^\circ$, $n = 3$ cells in the suppressor.

Figure 6.11 shows such a half bend dispersion suppressor, starting from a FODO structure with 60° phase advance per cell. The focusing strength of the FODO cells before and after the suppressor are identical, with the exception that—clearly—the FODO cells on the right are "empty", i.e. they have no bending magnets.

It is evident that unlike to the suppressor scheme with quadrupole lenses now the beta function is unchanged in the suppressor region.

Again this scheme has advantages:

- no additional quadrupole lenses are needed and no individual power supplies;
- in first order the β functions are unchanged; aperture needs and beam sensitivity are not increased;

and disadvantages:

- it works only for certain values of the phase advance in the structure and therefore restricts the free choice of the optics in the arc;
- special dipole magnets are needed (having half the strength of the arc types);
- the geometry of the ring is changed.

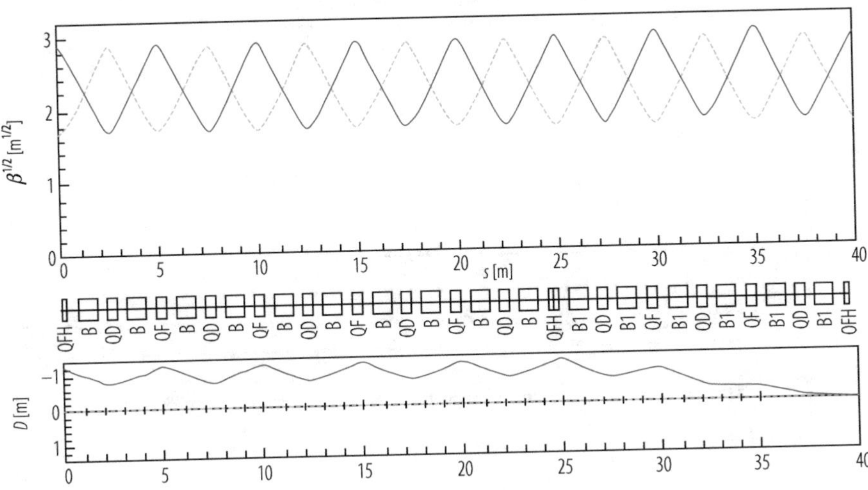

Fig. 6.11 Dispersion suppressor based on the half bend scheme

It has to be mentioned here, that in theses equations the phase advance of the suppressor part is equal to the one of the arc structure—which is not completely true as the weak focusing term $1/\varrho^2$ in the arc FODO differs from the term $1/(2\rho)^2$ in the half bend scheme. As, however, the impact of the weak focusing on the beam optics can be neglected in many practical cases Eq. (6.18) is *nearly* correct.

The application of such a scheme is very elegant, but as it has a strong impact on the beam optics and geometry it has to be embedded in the accelerator design at an early stage.

6.2.3.3 The "Missing Bend" Dispersion Suppressor Scheme

A similar approach is used in the case of the missing bend dispersion suppressor: It consists of a number of n cells without dipole magnets at the end of the arc, followed by m cells that are identical to the arc cells. The matching condition for this *missing bend* scheme with respect to the phase advance is

$$\frac{2n+m}{2}\phi_c = (2k+1)\,\frac{\pi}{2}. \tag{6.20}$$

For the number m of the required cells after the empty cells we get:

$$\sin\frac{m\phi_c}{2} = \frac{1}{2}, \qquad k = 0, 2, \ldots, \quad \text{or} \quad \sin\frac{m\phi_c}{2} = -\frac{1}{2}, \qquad k = 1, 3, \ldots. \tag{6.21}$$

The following example is based on $\phi_c = 60^\circ$, where the conditions above are fulfilled for $m = n = 1$, Fig. 6.12.

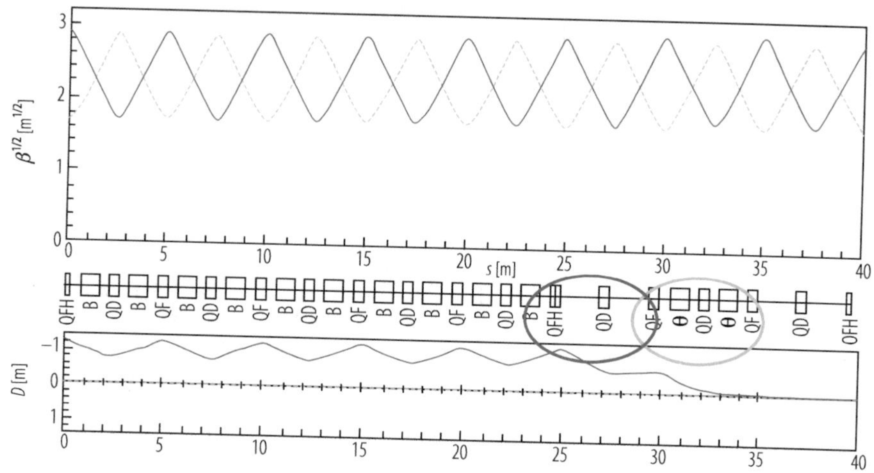

Fig. 6.12 Dispersion suppressor based on the missing bend scheme. The FODO cell without dipoles and the following standard cell are indicated by blue and green markers in the plot

There are more scenarios for a variety of phase relations in the arc and the corresponding bending strength needed to reduce $D(s)$, see [2, 3]. In general, one will combine one of the two schemes (missing or half bend suppressor) with a certain number of individual quadrupole lenses to guarantee the flexibility of the system with respect to phases changes in the lattice and to keep the size of the beta-function moderate.

6.3 Injection and Extraction Techniques

B. Goddard

Transfer of a beam between accelerators or onto external dumps, targets and measurement devices is a specialized topic and requires dedicated systems for injection and extraction [7], as well as beam transfer lines. Injection is the final process of the transfer of beam between one accelerator and another, either from a linear to a circular accelerator or between circular accelerators. Extraction is the removal of beam from an accelerator, either for the transfer to another accelerator or to deposit the beam on a target, dump or measurement system. Both injection and extraction systems need to be designed to transfer beam with minimum beam loss, to achieve the desired beam parameters and usually to minimize the dilution of the beam emittance.

Single-turn injection and extraction methods are rather straightforward for both lepton and hadron machines. They generally involve a kicker system to deflect the beam onto or away from the closed orbit, a septum (or series of septa for higher energy beams) to deflect the beam into or out of the accelerator aperture, and frequently also a closed orbit bump to approach the septum and reduce the required kicker strength. For these single-turn methods, the beam losses can be very low, and the emittance dilution associated with the injection or extraction can be very small, defined by the delivery precision, the optics mismatch, the kicker flat top ripple and septum stability. For both injection and extraction, the circulating beam can be adversely affected by septum stray fields penetrating into the circulating beam region and by the kicker field rise time which can overlap temporally with circulating bunches. Injecting a bunched beam into another accelerator also requires that the momentum spread and phase be matched to the RF bucket, and that the RF system can accept the transient beam loading which arises from the sudden change in beam intensity.

Multiple-turn injection is used to fill the circumference of a receiving accelerator and to accumulate bunch intensity. A wide variety of multiple-turn injection and extraction schemes exist, and these can be very different for lepton and hadron machines. Lepton injection schemes can take advantage of synchrotron radiation damping to achieve high beam brightness, while for hadron machines space charge effects dominate, especially at low energy. High brightness proton injection can make use of phase-space "painting" to precisely tailor the transverse and longitudinal distributions, particularly with H^- charge exchange injection or slip stacking; while resonant multiple-turn extraction schemes have been developed to provide quasi-continuous particle fluxes for periods which range from milliseconds to hours. The additional hardware systems required for these more advanced injection and extraction techniques include multiple RF systems, programmed fast closed-orbit bumps, stripping foils and non-linear lattice elements.

Overall, injection and extraction techniques share many similarities and hardware requirements [8]: one important difference between them is that extraction is usually at higher beam rigidity, which implies less effect from space charge and also stronger and hence longer deflecting systems, which can have a significant effect on lattice and insertion design [9–11].

6.3.1 Fast Injection

Fast injection [12–14] is typically used to fill another machine with bunch-to-bucket transfer, or to fill a collider over several injections with 'boxcar' stacking, where bunches or trains of bunches are added sequentially like boxcars (wagons) to a train. The system design depends critically on the aperture needed for the beam, and the kicker rise time, fall time and flat top duration. Very fast kicker rise times are often required to maximize the amount of beam which can be injected, especially in machines with small circumferences, since the kicker rise and fall times must be significantly shorter than the revolution time.

6.3.2 Slip-Stacking Injection

In slip-stacking [15], two trains of bunches are merged to increase the bunch intensity, using separate RF systems. A first train of bunches is injected on the closed orbit and captured by the first RF system. This train of bunches is then decelerated, and as a result circulates on a different orbit. A second batch is then injected on the closed orbit and captured by the second RF system. The two trains of bunches have slightly different energies and can be made to move relative to each other in phase. When the phase difference reaches zero, both sets of bunches are captured together and merged, by a rapid change of the RF frequency. The accelerator needs enough momentum aperture to accept both beams, and sophisticated RF control to make the manipulations. The final longitudinal emittance is the sum of the two individual emittances multiplied by an unavoidable blowup factor, typically around 1.5.

6.3.3 H^- Charge-Exchange Injection

High brightness, low energy proton machines frequently make use of H^- charge exchange injection [16]. In this technique, a linac accelerates H^- ions which are then merged with the circulating proton beam in a dipole magnet, Fig. 6.13, before the two loosely-attached electrons are stripped away in a foil which is almost transparent to the circulating beam.

This technique allows the accumulation of high brightness beams, since unlike other methods it allows injection into the already occupied phase space area. Transverse particle distributions can be controlled using phase space painting, to ameliorate space charge effects, reduce beam losses and increase accumulated intensity. The stripping is achieved with thin foils of carbon or diamond-like carbon, with a thickness typically in the micron range, which is a compromise between obtaining high stripping efficiency and minimizing the beam losses and emittance growth from scattering.

Fast painting bumpers or kickers in both planes are used to displace the circulating beam access with respect to the foil, and the waveform of the bumper

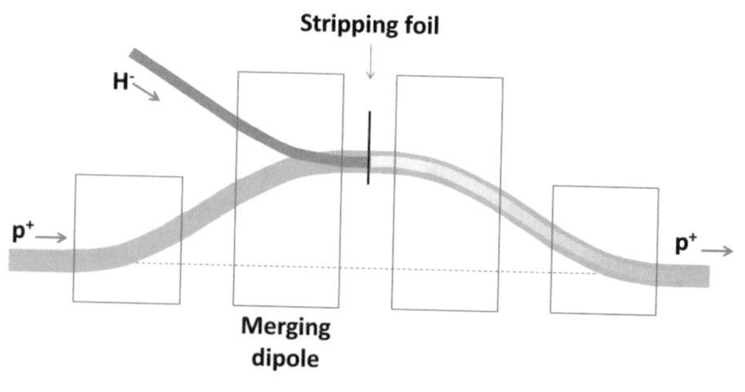

Fig. 6.13 Merging H^- and p^+ beams in H^- charge exchange injection

field can be varied to achieve the desired phase space density distribution. This is the process of phase space painting, where the small emittance LINAC beam is the brush and the large acceptance of the receiving machine is the canvas.

In addition to beam loss from scattering at the foil, another significant source of beam loss can be the field-stripping in the third chicane magnet of excited H^0. In ISIS [17] the injection is made on the ramp, and the dispersion at the foil provides some of the transverse phase space painting. For SNS [18], where the average beam power is over 1 MW, the uncontrolled beam losses must be kept extremely low and the accumulation is made over 1160 turns.

The use of stripping foils is disadvantageous for several reasons, in particular the associated uncontrolled beam losses, but also due to the simple mechanical and radiological difficulties in handling such fragile objects. A foil-free method of H^- stripping using a high-powered laser to resonantly excite neutral H^0 before field stripping in a dipole has been proposed and demonstrated in principle, and is promising for very high energy H^- injection systems [19].

6.3.4 Lepton Accumulation Injection

Injection of leptons can take advantage of the strong damping which is present from synchrotron radiation to accumulate intensity. This is very commonly used at Synchrotron Radiation rings, where top-up operation [20] consists of frequently injecting small amounts of beam to replace beam losses and keep the beam and synchrotron radiation intensities stable in a very small range.

In betatron injection, Fig. 6.14, the injected bunch or train is injected with an orbit offset with respect to the circulating beam, which is moved towards the injection septum with a fast closed-orbit bump. The offset between the injected beam and the circulating beam must be large enough to accommodate the injection septum. The particles of the newly-injected bunches then perform damped betatron oscillations

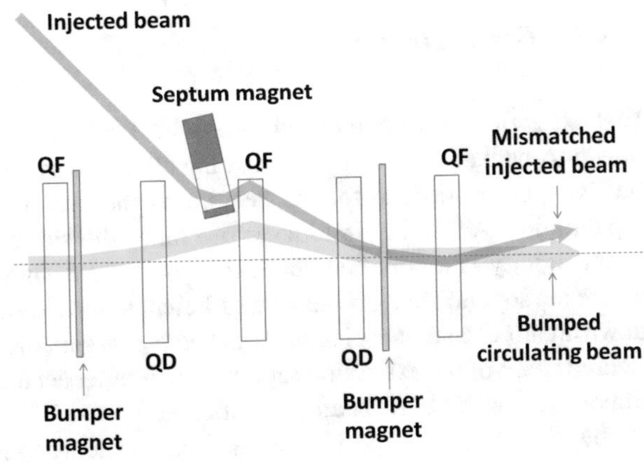

Fig. 6.14 Betatron injection. The injected beam is mismatched and performs betatron oscillations until damped by emission of synchrotron radiation

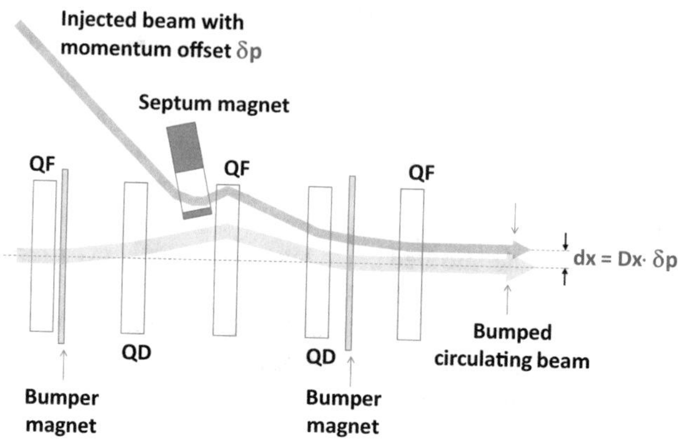

Fig. 6.15 Synchrotron injection. The injected beam has a momentum offset, and the injection trajectory is matched to the local dispersion orbit. The beam then performs oscillations about the closed orbit determined by the dispersion function, as the momentum changes with the synchrotron oscillations

around the closed orbit, until they merge with the already circulating beam. This technique has the disadvantage that the betatron amplitude may be large in regions of the accelerator where the β-function is large. In the alternative synchrotron injection [21], Fig. 6.15, the new particles are injected with a momentum offset δp and a position offset X into a region with dispersion D, such that $X = \delta p \times D$. The particles are injected onto the matched betatron orbit for their momentum, and thus only perform synchrotron oscillations around the stored particles, with the transverse offsets following the dispersion function. For LEP a combination of betatron and synchrotron injection was preferred [22], since the dispersion in the long straight sections was very small and the background to the experiments could be significantly improved.

6.3.5 *Fast Extraction*

Fast extraction is typically used to provide beam to a higher energy machine with bunch-to-bucket transfer. As for fast injection, the system design depends critically on the aperture needed for the beam, and the kicker rise time, fall time and flat top duration. Achieving fast kicker rise time with sufficient deflection angle at high beam rigidity is a common challenge, as is the design of the extraction insertion where the septum strength must be sufficient to provide enough clearance at the next downstream accelerator element. As beam energies increase, protection from mis-steered beam of the extraction septum and of other accelerator components becomes important; for the LHC beam extraction system at 7 TeV [23], the synchronization of the kicker system and protection from asynchronous kicker firing is a critical

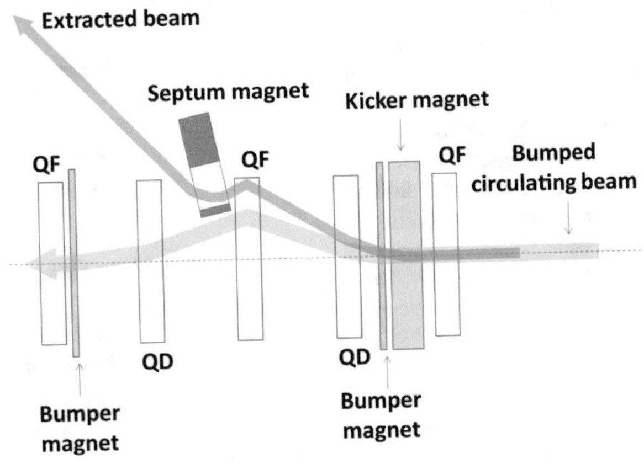

Fig. 6.16 Schematic of fast extraction system with kicker, septum and orbit bumpers. For higher energy machines, protection devices to intercept and dilute any mis-kicked extracted beam are placed in front of the septum and downstream QF quadrupole

system design feature. Closed orbit bumps can be used to move the beam closer to the septum, to reduce the required kick strength, Fig. 6.16.

6.3.6 *Resonant Extraction*

Many rate-limited applications such as physics experiments, test beams or medical treatment beams require a slow flux of particles with as uniform a time structure as possible. Resonant extraction using the third integer is the most common method of providing such uniform spills. In this 'slow' extraction [24], a triangular stable area in phase space (usually horizontal) is defined by exciting sextupole elements, and by moving the machine tune close to the third integer resonance. Before the start of the extraction process, particles remain stable if their single-particle emittances are smaller than the area of the stable triangle.

The beam is extracted by driving some particles unstable in a controlled way. The unstable particle amplitudes increase rapidly, following the outward-going separatrix every three turns, and the particles eventually move into the high-field region of a very thin electrostatic septum and are extracted, Fig. 6.17. The rate of extraction is controlled either by modulating the excitation process or by controlling the stable area. Several techniques for driving the particles unstable are possible:

i) the stable area can be reduced by increasing the resonance (sextupole) strength or by moving the tune closer to the third integer. Increasing the resonance strength reduces the stable area, but the smallest amplitude particles cannot be extracted, and changing the resonance affects the machine optics. Crossing the resonant tune offers the advantage that all of the beam can be extracted; however, the optics is still perturbed and in addition the position of the extracted beam in phase space changes as particles are extracted;

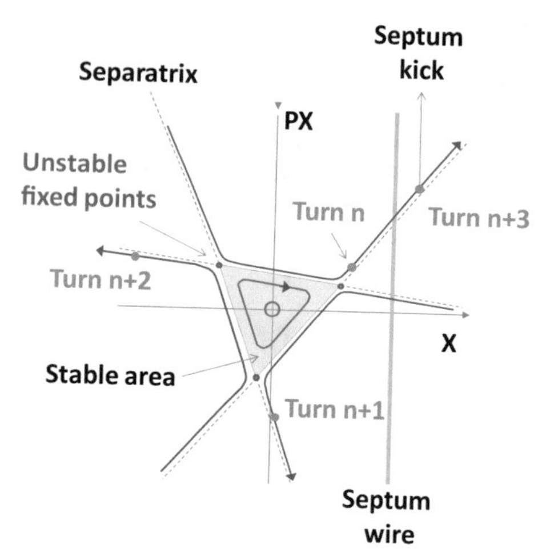

Fig. 6.17 Resonant extraction in normalised phased space. The amplitudes of particles outside the stable area grow rapidly, following the outward-going separatrix lines every three turns until they reach the electrostatic septum

ii) the particle amplitudes can be increased by use of a transverse excitation. The stable area is kept fixed and the particle amplitudes increased, as in RF-knockout [25] where a high-frequency damper is used near the betatron resonance frequency to excite the beam. The machine optics is not changed and this method allows very fine control of the spill flux, suitable for medical machines;
iii) the particles can be accelerated into the resonance where the chromaticity couples the momentum and the tune. A betatron core can be used [26] to accelerate the beam smoothly through the resonance. As the momentum of the beam changes, this is coupled via the chromaticity into a tune change. This method provides stability and insensitivity to power supply ripple. An alternative method (Constant Optics Slow Extraction) is to change the strengths of all machine elements to achieve the same effect, where the beam momentum remains fixed but the accelerator momentum changes [27];
iv) RF noise can be applied to gradually diffuse particles longitudinally, which through the chromaticity are brought into resonance. This stochastic extraction [28] allows extremely long and uniform spills, and again has the advantage of leaving the machine lattice functions unchanged.

It should be noted that extraction can also be made using the second order resonance, where octupole fields are used to define a stable area in phase space. The amplitude growth with time is much faster, and the beam can be extracted in several hundred turns.

The use of a physical septum means that losses and activation are key performance aspects for slow extraction. Several interesting techniques exist to reduce beam losses at extraction [29], including the use of scatterers to reduce the particle density at the septum, multipoles to manipulate the separatrix density and techniques to reduce the angular spread of the beam and reduce the effective septum width.

6.3.7 Continuous Transfer Extraction

A frequent requirement in an accelerator complex is to fill a large circumference machine with the contents of a smaller machine. One way of doing this is boxcar stacking; another technique is continuous transfer [30], where the beam in the first machine is extracted over a number of turns, like peeling the skin from an orange in a continuous strip. The machine tune is brought near to the appropriate integer n, where the beam will be extracted in $n+1$ turns. A fast closed bump is then applied to the circulating beam with kickers to move the beam partly across a septum, such that a fraction of the beam is cut and extracted. The machine tune rotates the beam in phase space such that subsequent slices are extracted—when the nth turn is extracted, the bump amplitude is increased to extract the remaining central part. This process is of use where the injector can service other machines or experiments while the receiving machine is accelerating the beam, since it minimises the time spent filling. The disadvantage of the technique is that large beam losses occur at the septum, with the transfer efficiency typically 85%. The transfer can be made with a bunched beam, leaving space for the kicker rise time, but this means that the receiving machine will need to capture a beam with strong intensity modulation. Another feature of this extraction is that the extracted slices all have different emittances, as the slices in phase space are all different.

6.3.8 Resonant Continuous Transfer Extraction

To reduce the beam losses from continuous transfer, a hybrid technique has been developed and deployed called Multi-Turn Extraction [31] where non-linear resonances are excited which define stable areas in phase space. These are populated by the controlled crossing of a resonance, and the islands are then separated by varying the multipole strength to provide a physical separation at the septum, to reduce or avoid transverse losses. The beam needs to be bunched with a gap to avoid losses during the kicker rise time. In addition to the lower losses, another advantage of this technique is that the extracted islands all have the same emittance.

6.3.9 Other Injection and Extraction Techniques

More exotic injection and extraction techniques also exist as working systems or concepts. These include radio-frequency stacking [32], pion-decay injection into muon storage rings [33] and combined cooling and stacking [34]. Charge exchange extraction [35] is used in cyclotrons, with a stripping foil, to convert for example H^- to p+, or H_2^+ to H_2^{2+} so that the beam is then deflected out of the accelerator. Finally, very high energy particle extraction can be envisaged with a bent crystal replacing the septum [36].

6.4 Concept of Luminosity

Werner Herr · Bernhard Holzer · Bruno Muratori

6.4.1 Introduction

In particle physics experiments the energy available for the production of new effects is the most important parameter. Besides the energy the number of useful interactions (events), is important. The quantity that measures the ability of a particle accelerator to produce the required number of interactions is called the luminosity (see Chap. 2) and is the proportionality factor between the number of events per second dR/dt and the cross section σ_p:

$$\frac{dR}{dt} = \mathcal{L} \cdot \sigma_p \tag{6.22}$$

The unit of the luminosity is therefore cm^{-2} s^{-1}.

Here we will derive a general expression for the luminosity and give formulae for basic cases. Additional complications such as crossing angle and offset collisions are added to the calculation. Other effects such as the hourglass effect are estimated from the generalized expression.

In the final section we will discuss the measurement and calibration of the luminosity for both e^+e^- as well as hadron colliders.

6.4.2 Computation of Luminosity

In the case of two colliding bunches, both serve as "target" as well as "incoming" beam at the same time. A schematic picture is shown in Fig. 6.18. The overlap integral which is proportional to the luminosity L can be written as [37]:

$$\mathcal{L} \propto K N_1 N_2 \cdot \iiiint_{-\infty}^{+\infty} \rho_1 (x, y, s, -s_0)\, \rho_2 (x, y, s, s_0)\, \mathrm{d}x \mathrm{d}y \mathrm{d}s \mathrm{d}s_0 \tag{6.23}$$

Here $\rho_1(x, y, s, s_0)$ and $\rho_2(x, y, s, s_0)$ are the time dependent beam density distribution functions and N_1 and N_2 the number of particles per bunch. We assume, that the two bunches meet at $s_0 = 0$ and $s_0 = c \cdot t$ is used as the "time" variable. Because the beams are moving against each other, we have to introduce the kinematic factor [38]:

$$K = \sqrt{(\vec{v}_1 - \vec{v}_2)^2 - (\vec{v}_1 \times \vec{v}_2)^2 / c^2} \tag{6.24}$$

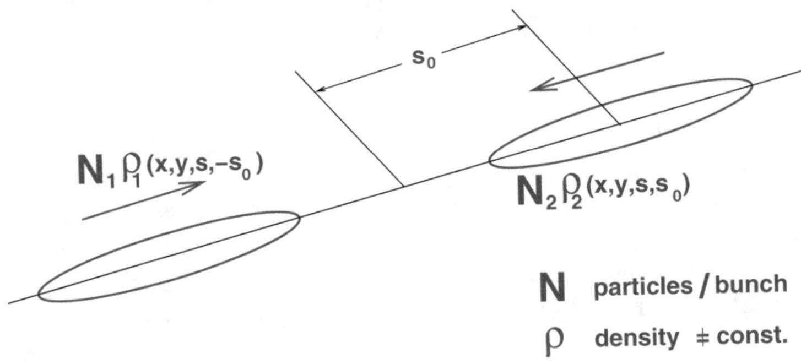

Fig. 6.18 Schematic view of a colliding beam interaction

This factor is needed to make the luminosity and therefore the cross section relativistically invariant.

For the calculation we assume Gaussian profiles in all dimensions of the form:

$$\rho_{iz}(z) = \frac{1}{\sigma_z\sqrt{2\pi}} \exp\left(-\frac{z^2}{2\sigma_z^2}\right) \quad \text{where} \quad i = 1, 2, \quad z = x, y \tag{6.25}$$

in the transverse planes and

$$\rho_s\,(s \pm s_0) = \frac{1}{\sigma_s\sqrt{2\pi}} \exp\left(-\frac{(s \pm s_0)^2}{2\sigma_s^2}\right) \tag{6.26}$$

in the longitudinal plane.

We further assume that the distributions are independent in the three coordinates and can be factorized. The integral (6.23) can then be evaluated. For the general case of: $\sigma_{1x} \neq \sigma_{2x}$, $\sigma_{1y} \neq \sigma_{2y}$, but assuming approximately equal bunch lengths $\sigma_{1s} \approx \sigma_{2s}$ we get the formula:

$$\mathcal{L} = \frac{N_1 N_2 f_c}{2\pi\sqrt{\sigma_{1x}^2 + \sigma_{2x}^2}\sqrt{\sigma_{2y}^2 + \sigma_{2y}^2}} \tag{6.27}$$

Where N_1 and N_2 are the bunch intensities and f_c the repetition rate. In the case of a circular collider with N_b bunches and a revolution frequency of f_{rev}, we have $f_c = f_{rev} \cdot N_b$.

6.4.3 Luminosity with Correction Factors

The Eq. (6.26) requires correction factors when the beam do not fully overlap (crossing angle and offset), the beam size varies in the longitudinal plane (hour glass effect) or in the case of non-Gaussian beams.

6.4.3.1 Effect of Crossing Angle and Transverse Offset

Here we give the correction to the luminosity calculation in the case where two bunches do not collide exactly head-on, but with a crossing angle and/or transverse offset. In that case the luminosity is reduced and we must apply a correction factor to compute the correct value. For simplicity we assume crossing angle and offset in the horizontal (x) plane, but this is not a restriction. The integration (6.23) can be carried out by rotating the coordinate systems of the two beams each by half the crossing angle [37] and can be simplified introducing the factors:

$$A = \frac{\sin^2\frac{\phi}{2}}{\sigma_x^2} + \frac{\cos^2\frac{\phi}{2}}{\sigma_s^2}, \quad B = \frac{(d_2-d_1)\sin(\phi/2)}{2\sigma_x^2}, \quad W = e^{-\frac{1}{4\sigma_x^2}(d_2-d_1)^2} \tag{6.28}$$

$$S = \frac{1}{\sqrt{1+\left(\frac{\sigma_s}{\sigma_x}\tan\frac{\phi}{2}\right)^2}} \approx \frac{1}{\sqrt{1+\left(\frac{\sigma_s}{\sigma_x}\frac{\phi}{2}\right)^2}} \tag{6.29}$$

where $\Phi/2$ is half the crossing angle and d_1 and d_2 are the transverse offsets of the two beams (Fig. 6.19).

We can re-write the luminosity with three correction factors:

$$\mathcal{L} = \frac{N_1 N_2 f N_b}{4\pi\sigma_x\sigma_y}\frac{N_1 N_2 f N_b}{4\pi\sigma_x\sigma_y} \cdot W \cdot e^{\frac{B^2}{A}} \cdot S \tag{6.30}$$

This factorization enlightens the different contributions and allows straightforward calculations. The last factor S is the luminosity reduction factor for a crossing angle. One factor W reduces the luminosity in the presence of beam offsets and the factor $e^{\frac{B^2}{A}}$ is only present when we have a crossing angle and offsets simultaneously in the same plane. The formulae for the luminosity under very general conditions can be found in [39]. A popular interpretation of this result is to consider it a correction to the beam size and to introduce an "effective beam size" like:

$$\sigma_{eff} = \sigma/\sqrt{1+\left(\frac{\sigma_s}{\sigma}\frac{\phi}{2}\right)^2} \tag{6.31}$$

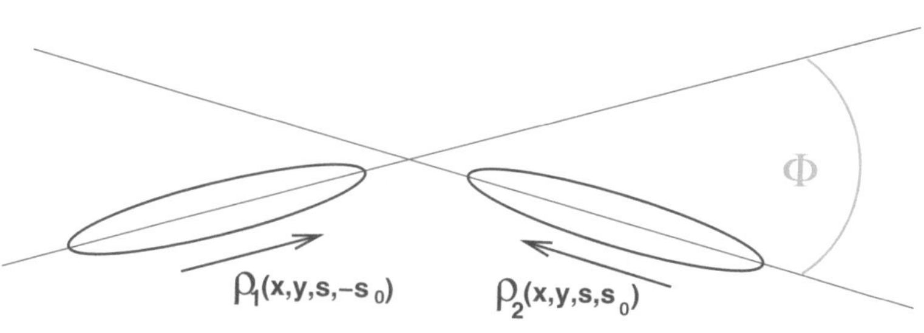

Fig. 6.19 Schematic view of a colliding beam interaction at a crossing angle

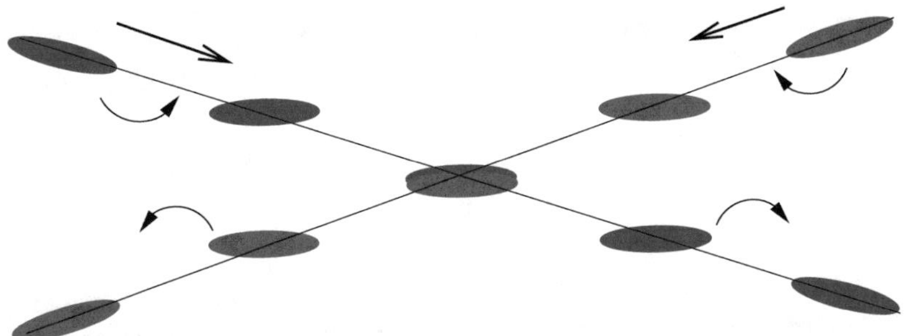

Fig. 6.20 Scheme of crab crossing with transversely deflecting cavities

This equation is valid when $\sigma_z \gg \sigma$. The effective beam size can then be used in the standard formula for the beam size in the crossing plane. This concept of an effective beam size is interesting because it also applies to the calculation of beam-beam effects of bunched beams with a crossing angle [40, 41].

In the case of flat beams, (i.e. $\sigma_z \ll \sigma_z$) a more general expression has to be used, (see e.g. [39]).

To avoid the loss of luminosity, the use of crab cavities is an option, where the bunches are deflected transversely before and after the collision Fig. 6.20.

6.4.3.2 Hour Glass Effect

In a low-β region the β-function varies with the distance s to the minimum like:

$$\beta(s) = \beta^* \left(1 + \left(\frac{s}{\beta^*}\right)^2\right) \tag{6.32}$$

For very small $\beta*$ comparable to the bunch length, the β-function is not a constant along the longitudinal dimension of the bunch. It cannot be considered a constant in Eq. (6.23). It follows a parabola and rises very fast and can become very large for small $\beta*$.

In our formulae we have to replace σ by $\sigma(\text{s})$ and get a more general expression for the luminosity (assuming equal parameters in both beams, the most general expression can be found in [39]):

$$\frac{\mathcal{L}(\sigma_s)}{\mathcal{L}(0)} = \int_{-\infty}^{+\infty} \frac{1}{\sqrt{\pi}} \frac{e^{-u^2}}{\sqrt{\left[1 + \left(\frac{u}{u_x}\right)^2\right] \cdot \left[1 + \left(\frac{u}{u_y}\right)^2\right]}} \mathrm{d}u \tag{6.33}$$

Using the expressions: $u_x = \beta_x^*/\sigma_s$ and $u_y = \beta_y^*/\sigma_s$
For the case of round beams it can be simplified and the integral becomes:

$$\frac{\mathcal{L}(\sigma_s)}{\mathcal{L}(0)} = \int_{-\infty}^{+\infty} \frac{1}{\sqrt{\pi}} \frac{e^{-u^2}}{\left[1 + \left(\frac{u}{u_x}\right)^2\right]} du = \sqrt{\pi} \cdot u_x \cdot e^{u_x^2} \cdot \text{erfc}(u_x) \tag{6.34}$$

Here erfc(u) is the complex error function. The hourglass effect depends strongly on the relative value of $\beta*$ and the bunch length σ_s. For small $\beta*$ the effect becomes relevant since the beam size varies rapidly along the longitudinal bunch direction, i.e. when s becomes comparable to the bunch length in Eq. (6.32). A loss of luminosity according to Eq. (6.34) is the consequence.

6.4.3.3 Crabbed Waist Scheme

In the case of a large crossing angle, the collision point of particles is displaced. Schematically this is shown in Figs. 6.21 and 6.22.

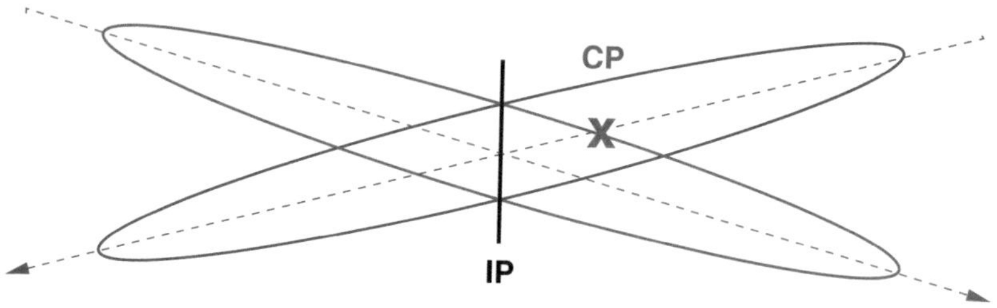

Fig. 6.21 Collision with large crossing angle and longitudinally displaced collison point

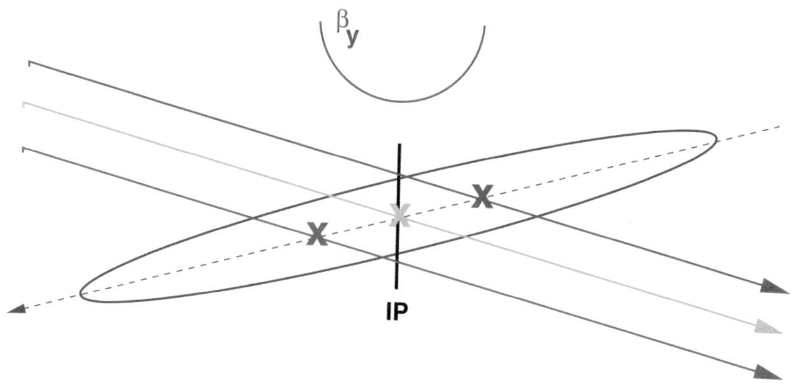

Fig. 6.22 Collision with large crossing angle and longitudinally displaced collison point. Shown for three particles with different amplitudes

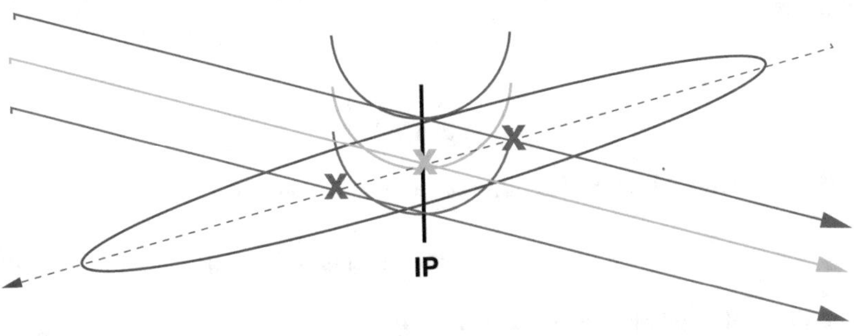

Fig. 6.23 Collisions with different vertical β-functions

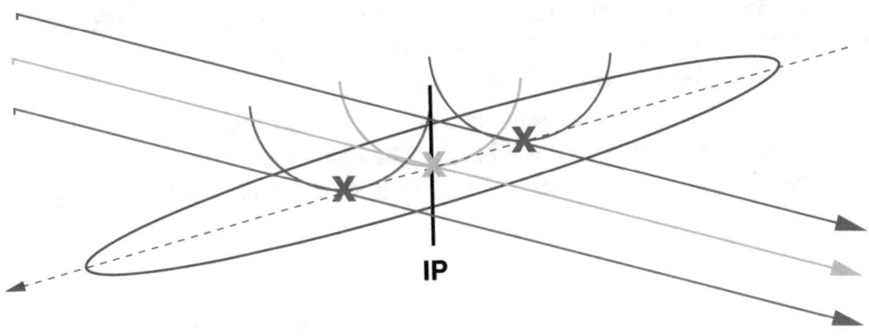

Fig. 6.24 All collisions at a minimum of the vertical β-functions using a crabbed waist scheme

One possible consequence can be the coupling between the transverse and and longitudinal plane. Such a coupling is particularly bad for flat beams since the vertical beam size will increase significantly.

In Fig. 6.23 the vertical β-function is indicated and the result of this effect is that the particles collide at positions with different vertical β-functions.

This can be mitigated [42] by making the vertical waist (β_y^{min}) amplitude dependent in the horizontal plane Fig. 6.24. All particle collide now at the minimum of the vertical β-function.

It should be emphasized that the main purpose of such a scheme is not to reduce a geometrical loss but to reduce the coupling. Therefore it is of interest only for flat beams.

This scheme is established using two sextupoles.

6.4.4 *Integrated Luminosity and Event Pile Up*

The maximum luminosity, and therefore the instantaneous number of interactions per second, is very important, but the final figure of merit is the so-called integrated luminosity:

$$\mathcal{L}_{\text{int}} = \int_0^T \mathcal{L}(t')\,dt' \tag{6.35}$$

because it directly relates to the number of observed events:

$$\mathcal{L}_{\text{int}} \cdot \sigma_p = \text{number of events of interest} \tag{6.36}$$

The integral is taken over the sensitive time, i.e. excluding possible dead time. The unit of the integrated luminosity is cm^{-2} and often expressed in inverse barn ($1\ \text{barn}^{-1} = 10^{24}\ \text{cm}^{-2}$).

Another important parameter for a beam with high luminosity and bunched beams are the number of collisions per bunch crossing, the so-called pile up. In particular for collisions with a large cross section this can become a problem. In the case of the LHC, bunch crossings occur every 25 ns and the expected pile up is more than 20 for proton-proton collisions. The challenge is to maximise the useful luminosity while keeping the pile up to a level that can be handled by the particle detectors.

6.4.5 Measurement and Calibration of Luminosity

To obtain the exact integrated luminosity, it has to be recorded continuously. It is rather straightforward to obtain a counting rate directly proportional to the total interaction rate dR/dt. This relative signal has to be calibrated to deliver the absolute luminosity. We have already seen some effects that affect the absolute luminosity and therefore to a large extent the luminosity measurement. In particular the crossing angle and the luminous region are of importance since they have immediate implications for the geometrical acceptance of the instruments.

In principle one can determine the absolute luminosity when all relevant beam parameters are known, i.e. the bunch intensities, beam sizes (r.m.s. in case of unknown beam profiles) and the exact geometry. However the precise measurement of beam sizes is a challenge, in particular for hadron colliders when a non-destructive measurement is required. When the energy spread in the beams is large (e.g. some e^+e^- colliders), a residual dispersion at the interaction point increases significantly the beam size and must be included.

There exist other methods which relate the counting rate to well known processes which can be used for calibration. We shall discuss several methods for both, lepton and hadron colliders.

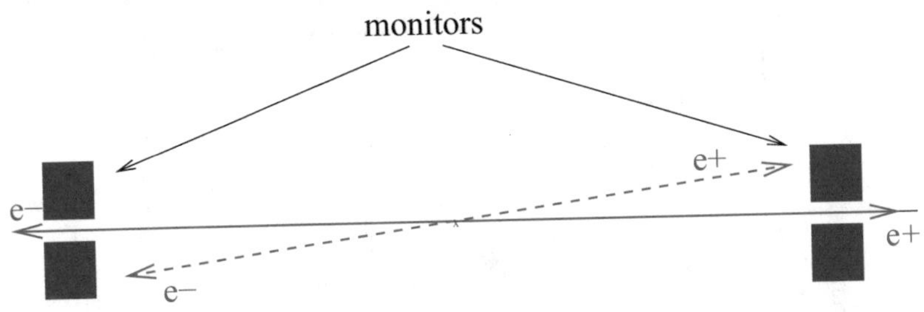

Fig. 6.25 Principle of luminosity measurement using Bhabha scattering for e^+e^- colliders

6.4.6 *Absolute Luminosity: Lepton Colliders*

Once the relative luminosity is known, a very precise method is to compare the counting rate to well known and calculable processes. In case of e^+e^- colliders these are electromagnetic processes such as elastic scattering (Bhabha scattering). The principle is shown in Fig. 6.25. Particle detectors are used to measure the trajectories at very small angles and with a coincidence of particles on both sides of the interaction point. For a precise measurement one has to go to very small angles since the elastic cross section σ_{el} has a strong dependence on the scattering angle ($\sigma_{el} \propto \Theta^{-3}$).

Furthermore, the cross section diminishes rapidly with increasing energy ($\sigma_{el} \propto \frac{1}{E^2}$) and the result may be small counting rates. At LEP energies with $\mathcal{L} = 10^{30}$ cm^{-2} s^{-1} one can expect only about 25 Hz for the counting rate. Background from other processes can become problematic when the signal is small.

6.4.7 *Absolute Luminosity: Hadron Colliders*

For hadron colliders two types of calibration have become part of regular operation, the measurement of the beam size by scanning the beam and the calibration with the cross section for small angle scattering. The determination of the bunch intensities is usually easier, although non-trivial in the case of a collider with several thousand bunches.

6.4.7.1 Measurement by Profile Monitors and Beam Displacement

Typical profile measurement devices are wire scanners where a thin wire is moved through the beam and the interaction of the beam with the wire gives the signal. For high intensity hadron beams this has however limitations. Non-destructive devices

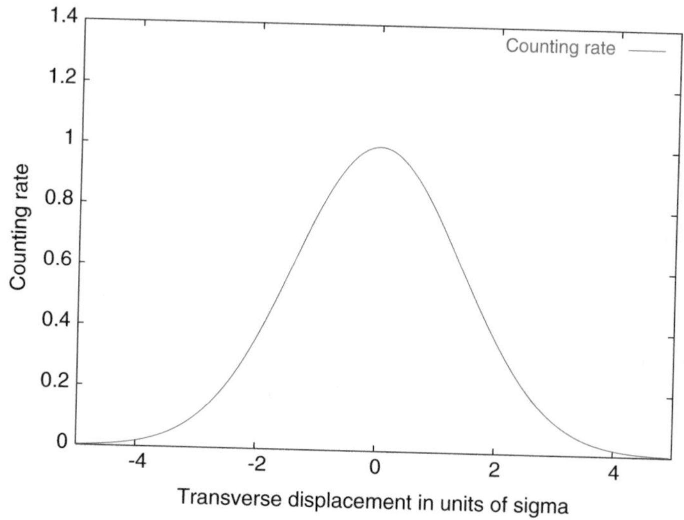

Fig. 6.26 Principle of luminosity measurement using transverse beam displacement

such as synchrotron light monitors are available but the emitted light from hadrons is often not sufficient for a precise measurement.

An alternative is to measure the beam size by displacing the two beams against each other. The relative luminosity reduction due to this offset can be measured and is described by the formula (6.28) developped earlier:

$$\mathcal{L}(d)/\mathcal{L}_0 \;\; W = e^{-\frac{d^2}{4\sigma^2}} \tag{6.37}$$

where d is the separation between the beams and the measurement of the luminosity ratio is a direct measurement of W. This method was already used in the CERN Intersection Storage Rings (ISR) and known as "van der Meer scan".

The expected counting rate of such a scan is shown in Fig. 6.26. A fit to the above formula gives the beam size. A drawback of this method is the distortion of the beam optics in case of very strong beam-beam interactions [40]. This effect has to be evaluated carefully.

6.4.7.2 Absolute Measurement with Optical Theorem

This method is similar to the measurement of Bhabha scattering for e^+e^- colliders but requires dedicated experiments and often special machine conditions.

The total elastic and inelastic counting rate is related to the luminosity and the total cross section (elastic and inelastic) by the expression:

$$\sigma_{tot} \cdot \mathcal{L} = N_{inel} + N_{el} \quad \text{(Total counting rate)} \tag{6.38}$$

The key to this method is that the total cross section is related to the elastic cross section for small values of the momentum transfer t by the so-called optical theorem [43]:

$$\lim_{t\to 0}\frac{d\sigma_{el}}{dt} = \left(1+\rho^2\right)\frac{\sigma_{tot}^2}{16\pi} = \frac{1}{\mathcal{L}}\frac{dN_{el}}{dt}\bigg|_{t=0} \tag{6.39}$$

Therefore the luminosity can in principle be calculated directly from experimental rates through:

$$\mathcal{L} = \frac{\left(1+\rho^2\right)}{16\pi}\frac{(N_{inel}+N_{el})^2}{(dN_{el}/dt)_{t=0}} \tag{6.40}$$

All counting rates, the total number of events $N_{inel} + N_{el}$ and the differential elastic counting rate dN_{el}/dt at small t have to be measured with high precision. This requires a very good detector coverage of the whole space (4π) for the inelastic rate and the possibility to measure to very small values of t.

A slightly modified version of the above uses the Coulomb scattering amplitude which can be precisely calculated. The elastic scattering amplitude is a superposition of the strong (f_s) and Coulomb (f_c) amplitudes, the latter dominates at small t. We can re-write the differential elastic cross section $\frac{d\sigma_{el}}{dt}$:

$$\lim_{t\to 0}\frac{d\sigma_{el}}{dt} = \frac{1}{\mathcal{L}}\frac{dN_{el}}{dt}\bigg|_{t=0} = \pi\,|f_c+f_s|^2 \simeq \pi\,\left|\frac{2\alpha_{em}}{-t} + \frac{\sigma_{tot}}{4\pi}(\rho+i)\,e^{B\frac{t}{2}}\right|^2 \simeq \frac{4\pi\alpha_{em}^2}{t^2}\bigg|_{|t|\to 0} \tag{6.41}$$

If the differential cross section is measured over a large enough range, the unknown parameters σ_{tot}, ρ, B and $\mathcal{L}$ can be determined by a fit. A measurement [44–46] together with some crude fits is shown in Fig. 6.27 to demonstrate the principle. The advantage of this method is that it can be performed measuring only elastic scattering without the need of a full coverage to measure N_{inel}. It is therefore a good way to measure the luminosity (and total cross section σ_{tot} and interference parameter ρ!) although the previous method is of more practical importance for regular use.

The measurement of the Coulomb amplitude usually requires dedicated experiments with detectors very close to the beam (e.g. with so-called Roman Pots) and therefore special parameters such as reduced intensity and zero crossing angle. Furthermore, in order to measure very small angle scattering, one has to reduce the divergence in the beam itself ($\sigma' = \sqrt{\epsilon/\beta}$). For that purpose special running conditions with a **high** $\beta*$ at the collision point are often needed ($\beta* > 1000$ m) [45]. The precision of such a measurement is however as good as a few percent.

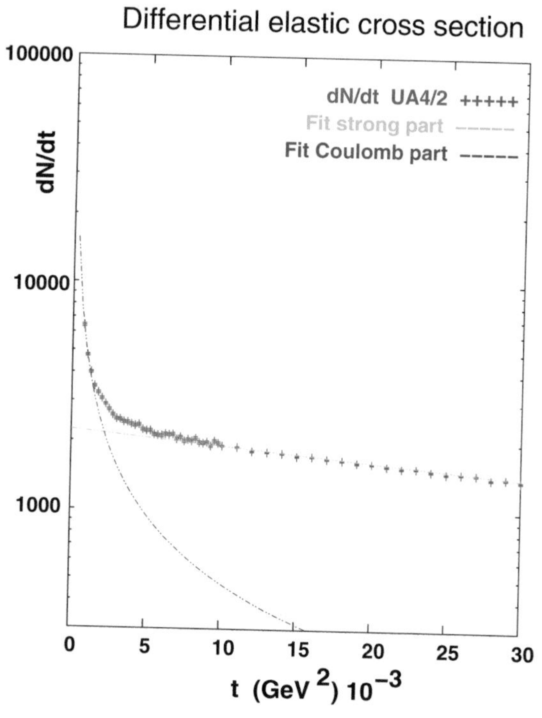

Fig. 6.27 Principle of luminosity measurement using optical theorem in proton proton (antiproton) collisions

6.4.8 *Luminosity in Linear Colliders*

In linear colliders the beams collide only once and to get a high luminosity a very small beam size and therefore small β at the collision point are required.

This implies additional effects such a beam disruption and an enhanced luminosity due to the so-called pinch effect.

Due to very strong field of the quadrupoles of the final focusing, significant synchrotron radiation is produced.

6.4.8.1 Disruption and Luminosity Enhancement Factor

The basic formula for the luminosity of a linear collider is shown in Eq. (6.42).

$$\mathcal{L} = \frac{N^2 \, f_{rep} \, n_b}{4\pi \overline{\sigma_x} \; \overline{\sigma_y}} \quad \to \quad \mathcal{L} = \frac{H_D \cdot N^2 \, f_{rep} \, n_b}{4\pi \, \sigma_x \; \sigma_y} \tag{6.42}$$

The revolution frequency has to be reaplced by the repetition rate f_{rep} of the colliding bunches.

The luminosity is increased by the enhancement factor H_D which takes into account the reduction of the nominal beam size by the disruptive field (pinch effects).

This enhancement foctor is related to the beam disruption parameter

$$D_{x,y} = \frac{2r_e N\sigma_z}{\gamma\sigma_{x,y}\left(\sigma_x + \sigma_y\right)} \tag{6.43}$$

For weak disruption ($D \ll 1$) and round beams the enhancement factor can be written as:

$$H_D = 1 + \frac{2}{3\sqrt{\pi}}D + \mathcal{O}\left(\mathcal{D}^2\right) \tag{6.44}$$

When the disruption is strong or for flat beams, computer simulations are necessary.

6.4.8.2 Beamstrahlung

The strong synchrotron radiation (beamstrahlung) has two main effects:

- Spread of the centre of mass energy.
- Pair creation and background in the detectors.

It is parametrized by the parameter Y which can be written as the mean field strength in the rest frame, normalized to the critical field B_c:

$$Y = \frac{< E + B >}{B_c} \approx \frac{5}{6}\frac{r_e^2\gamma N}{\alpha\sigma_z\left(\sigma_x + \sigma_y\right)} \tag{6.45}$$

$$B_c = \frac{m^2c^3}{e\hbar} \approx 4.4 \times 10^{13}G \tag{6.46}$$

6.5 Synchrotron Radiation and Damping

L. Rivkin

6.5.1 *Basic Properties of Synchrotron Radiation*

Charged particles radiate when they are deflected in the magnetic field [47] (transverse acceleration) [see also Sect. 11.1 for a more detailed treatment]. In the ultra-relativistic case, when the particle speed is very close to the speed of light, $\beta \approx c$, most of the radiation is emitted in the forward direction [48] into a cone centred on the tangent to the trajectory and with an opening angle of $1/\gamma$, where γ is the Lorentz factor (since for a few GeV electron or a few TeV proton, $\gamma \approx 1000$, the photon emission angles are within a milliradian of the tangent to the trajectory).

The power emitted by a particle is proportional to the square of its energy E and to the square of the deflecting magnetic field B:

$$P_{SR} \propto E^2 B^2, \tag{6.47}$$

and in terms of Lorentz factor γ and the local bending radius ρ can be written as follows:

$$P_{SR} = \frac{2}{3}\alpha\, \hbar c^2\, \frac{\gamma^4}{\rho^2}, \tag{6.48}$$

where α is the fine-structure constant and the Plank's constant is given in a convenient conversion constant:

$$\alpha = \frac{1}{137} \quad \text{and} \quad \hbar c = 197 \text{ MeV fm}. \tag{6.49}$$

The emitted power is a very steep function of both the particle energy and particle mass, being proportional to the fourth power of γ.

Integrating the above expression around the machine we obtain the amount of energy lost per turn:

$$U_0 = \frac{4\pi}{3}\alpha\, \hbar c\, \frac{\gamma^4}{\rho}. \tag{6.50}$$

The emitted radiation spectrum consists of harmonics of the revolution frequency and peaks near the so-called critical frequency or critical photon energy. It is defined such that exactly half of the radiated power is emitted below it:

$$\varepsilon_c = \frac{2}{3}\, \hbar c\, \frac{\gamma^3}{\rho}. \tag{6.51}$$

On the average a particle then emits $n_c \approx 2\pi\alpha\gamma$ photons per turn.

6.5.2 *Radiation Damping*

In a storage ring the steady loss of energy to synchrotron radiation is compensated in the RF cavities, where the particle receives each turn the average amount of energy lost. The energy lost per turn is normally a small fraction of the total particle energy, typically of the order of one part per thousand.

Transverse Oscillations
Since the radiation is emitted along the tangent to the trajectory, only the amplitude of the momentum changes. As the RF cavities increase the longitudinal component of the momentum only, the transverse component is damped exponentially with the damping rate of the order of U_0 per revolution time. A typical transverse damping time corresponds simply to the number of turns it would take to lose the amount of energy equal to the particle energy. The damping times are very fast, in case of a few GeV electron ring being on the order of a few milliseconds.

$$A_{\perp} = A_0 e^{-\frac{t}{\tau}}, \text{ where } \frac{1}{\tau} = \frac{U_0}{2ET_0}. \tag{6.52}$$

In a given storage ring the damping time is inversely proportional to the cube of the particle energy.

Longitudinal or Synchrotron Oscillations
Synchrotron oscillations are damped because the energy loss per turn is a quadratic function of the particle's energy. The damping rate is typically twice the rate for transverse oscillations.

Damping Partition Numbers and Robinson Theorem
For particles that emit synchrotron radiation the dynamics is characterized by the damping of particle oscillations in all three degrees of freedom. In fact, the total amount of damping (Robinson theorem [49]), i.e. the sum of the damping decrements depends only on the particle energy and the emitted synchrotron radiation power:

$$\frac{1}{\tau_x} + \frac{1}{\tau_y} + \frac{1}{\tau_\varepsilon} = \frac{2U_0}{ET_0} = \frac{U_0}{2ET_0}\left(J_x + J_y + J_\varepsilon\right) \tag{6.53}$$

where we have introduced the usual notation of *damping partition numbers* that show how the total amount of damping in the system is distributed among the three degrees of freedom. A typical set of the damping partition numbers is (1,1,2) and their sum is, according to the Robinson theorem, a constant.

$$J_x + J_y + J_\varepsilon = 4. \tag{6.54}$$

Adjustment of Damping Rates
The partition numbers can differ from the above values, while their sum remains a constant. In fact, under certain circumstances, the motion can become "anti-damped", i.e. the damping time can become negative, leading to an exponential growth of the oscillations amplitudes. From a more detailed analysis of damping

rates [50] the damping time can be written as

$$\frac{1}{\tau_\varepsilon} = \frac{U_0}{2ET_0}\,(2+\mathcal{D})\,, \quad \text{and} \quad \frac{1}{\tau_x} = \frac{U_0}{2ET_0}\,(1-\mathcal{D})\,, \quad \text{where } \mathcal{D} \equiv \frac{\oint \frac{D}{\rho}\left(2k+\frac{1}{\rho^2}\right)ds}{\oint \frac{ds}{\rho^2}}. \tag{6.55}$$

The constant introduced above is an integral of the dispersion function $\mathcal{D}$ and the magnetic guide field functions, i.e. bending radius and gradient around the ring and is independent of the particle energy. It deviates substantially from zero only when a particle encounters combined function elements, i.e. where the product of the field gradient and the curvature is non-zero. The damping partition numbers then are:

$$J_x = 1 - \mathcal{D},\ J_\varepsilon = 2 + \mathcal{D},\ J_x + J_\varepsilon = 3. \tag{6.56}$$

The vertical damping partition number is usually unchanged as the vertical dispersion is zero in storage rings that are built in one (horizontal) plane.

The amount of damping can be repartitioned between the horizontal and energy-time oscillations by altering the value of the $\mathcal{D}$ constant [50]. This can be achieved by either using combined function magnetic elements in the lattice, or by introducing a special combined function wiggler magnet (so-called Robinson wiggler). Values of horizontal partition number as high as 2.5 have been obtained that way. Values of $\mathcal{D} > 1$ lead to anti-damping of horizontal betatron oscillations, while for $\mathcal{D} < -2$ the synchrotron oscillations become unstable.

6.6 Computer Codes for Beam Dynamics

Werner Herr

6.6.1 *Introduction*

The design and operation of an accelerator today is unthinkable without the help of computer codes, the reason being large, complex structures (like in the case of big accelerators and colliders, e.g. LHC) or complications in the beam dynamics of small or special purpose machines (e.g. FFAG). Their complexity does not allow the computation with pencil and paper. Here we address only the codes for beam dynamics, i.e. special codes for the design of accelerators components such as magnets or RF equipment will not be treated but can be found in the literature. The main fields where beam dynamics codes are essential are:

- Determination of parameters and the design of beam lines and accelerators
- Evaluation of performance
- Control, machine protection and operation

Different classes of codes are used in these fields which also resemble the life cycle of an accelerator.

Given the scope of this handbook and the rapid evolution of computer codes and software techniques, we do not attempt to provide a list of existing codes, but rather will describe the main features, techniques and applications of the different types of codes. Details and access to existing codes can be found in computer code libraries on the internet. A supported library is provided by the Los Alamos Accelerator Code Group (LAACG) [51], another one supported by Astec (UK) [52]. It contains links to popular and frequently used codes from many laboratories and institutions.

6.6.2 Classes of Beam Dynamics Codes

The different classes of codes can be divided according to their application:

- General purpose optics codes
- Beam dynamics of single particles
- Beam dynamics of multi particles

Optics codes are used mainly in the initial design phase of an accelerator, rings as well as beam lines and linear accelerators. The evaluation of the performance (stability etc.) is done using codes to simulate the beam dynamics of single particles as well as ensembles of particles and their interaction with the environment or other particles in the beam(s).

6.6.3 Optics Codes

A large group of computer codes for beam dynamics are used to design the lattice of an accelerator or beam line and to compute and optimize the optical parameters. The range of available codes extends from small codes for pedagogical purpose to large general purpose programs. Such codes can have easily 100,000 lines of codes or more. The accelerator physics is described in the existing literature [53] and in this handbook. The main applications of general purpose optics codes are:

- Determination of main parameters and the computation of linear and non-linear optics. This implies to find periodic solutions for the optical parameters and the closed orbit.
- Parameter matching (optical/geometrical) and lattice optimization, i.e. the properties of elements are varied until the optical functions assume their desired values.
- Simulation of imperfections and algorithms for their corrections.
- Simulation of synchrotron radiation and evaluation of radiation integrals to derive estimates for parameters (e.g. equilibrium emittances) in lepton machines.

The result should be a consistent set of parameters fulfilling the design requirements. They are the basis for the design of machine elements.

Depending on the complexity of the problem, different techniques are in use for optics codes. The majority of these codes rely on the description of machine elements using maps, which can be of higher order for non-linear elements. In the simplest case for the description of linear machines the maps become matrices and are therefore often referred to as "matrix codes" [54]). The concatenation of the matrices provide a matrix for the entire ring and its analysis gives the optical parameters, closed orbit etc.

Another technique is to follow the particles through the accelerator, i.e. integrating the equation of motion in the electromagnetic fields of the machine elements. The analysis of the results of these "tracking programs" provides the required parameters and information about the stability of the machine (for some details see [54]).

Dealing with complex machines, other considerations may become important such as e.g.:

- Definition of an input language which can be used by other programs. This input language defines the sequence of elements, i.e. the ring or a beam line, as well as the properties of the elements such as e.g. their types (dipole, quadrupole,..), lengths and strengths.
- For large machines with a large number of elements the interface to a data base may be required. Large machines such as the LHC or future colliders have several thousand elements.
- An interface to the control system for on-line modelling is desirable

6.6.4 Single Particle Tracking Codes

To evaluate the performance of accelerators, in particular multi pass, i.e. circular machines, one has to deal with complex iterative processes. The standard perturbation theories can fail to correctly describe the behaviour beyond leading orders. Single particle tracking codes are successfully used when analytical methods fail to describe the effect of non-linear forces on the stability of the particles. Many tracking codes have been developed together with the necessary tools to analyse the results and from the simulation point of view the treatment of non-linear effects is well established. Conceptually, in a tracking code the equation of motion of a particle in an accelerator element is solved and the phase space coordinates of the particle are followed through all elements of the accelerator or beam line. To obtain the desired information, it may be necessary to repeat this process for up to 10^7 turns which require appropriate algorithms and techniques to avoid numerical problems. Similar problems exist and some of these techniques have been developped for celestial mechanics. In order to draw conclusions from the tracking data it is necessary to provide tools to allow a qualitative and quantitative understanding of the results [55]. The outcome of the analysis allows to answer the most important questions for the design of a machine such as:

- Stability of particle motion
- Dynamic aperture
- Specifications for the properties of machine elements
- Optimization or the particle stability

In general the results of these studies are used in an iterative procedure to improve and optimize the design of the machine.

6.6.4.1 Techniques

A requirement for all techniques employed for particle tracking is that the associated maps must be symplectic. To solve the equation of motion, most programs use explicit canonical integration techniques, e.g.:

- Thin lens tracking (most common since they are automatically symplectic and fast)
- Ray tracing (accuracy by slicing into large number of steps, but time consuming)
- Symplectic integration (see [54] and references therein)

6.6.4.2 Analysis of Tracking Data

Some of the analysis techniques are discussed in the chapter on non-linear dynamics in this handbook in more detail and some are mentioned here for completeness:

- Taylor maps using Truncated Power Series Algebra (TPSA, [54])
- Lie algebraic maps [54, 56]
- Normal form analysis

The results of the analysis include non-linear resonances and distortion, non-linear tuneshift with amplitude and an evaluation of the long term stability. In all cases the interpretation of the results requires a careful analysis of the range where the data is meaningful to avoid wrong conclusions. Typical problems are numerical effects which can lead to unphysical features.

6.6.5 *Multi Particle Tracking Codes*

Multi particle tracking codes are used when we are concerned with the behaviour of an ensemble of particle. The calculations largely rely on techniques developped for single particle dynamics. Typical applications are the simulation of:

- Space charge effects, mutual interaction of particles within the same beam.

- Collective instabilities and interaction with environment (impedance)
- Beam-beam effects in case of particle colliders, i.e. the interactions with the fields produced by the counter-rotating beam.
- Electron cloud effects, i.e. secondary electron production by synchrotron radiation

A key issue for multi particle simulation codes is the evaluation of the electromagnetic fields produced by the beams or the environment. New techniques and the availability of parallel computing facilities have allowed vast progress in this field in the last 20 years.

6.6.6 Machine Protection

For large energy and high intensity machines the protection of the machine elements becomes an important part of the design. Simulation codes have to include the interaction of particles with matter.

6.7 Electron-Positron Circular Colliders

M. E. Biagini · J. M. Jowett

Electron-positron (e^+e^-) collider rings have been a mainstay of both discovery and precision physics for half a century: discovery, since the simple initial state can create any particle coupled to the electromagnetic field; precision, from the combination of high luminosity and large cross-sections at a rich spectrum of resonances up to $\sqrt{s} \simeq 200\,\text{GeV}$. While the fundamentals of these machines have remained in essence the same, the technology has matured to the point where luminosities of the latest "factories" exceed what was thought possible in the 1970s and early 1980s by 2–3 orders of magnitude.

These colliders are based on the principle of the synchrotron (Sect. 1.2.6) although the name is barely appropriate for those which enjoy the advantage of full-energy injection. Beams are necessarily bunched by an RF system, which must provide sufficient voltage to compensate the energy lost by synchrotron radiation.

6.7.1 Physics of Electron-Positron Rings

Consider an ideal storage ring constructed with bending and focussing magnets such that a particle of charge e and *constant* momentum p_0 could circulate on a stable closed orbit, O_{xy}, in transverse phase space (x, p_x, y, p_y), with local radius of curvature $\rho(s)$. The orbits and optical functions ($\beta_{x,y}(s)$, dispersion $D_{x,y}(s)$, etc., Chap. 2) of such *hypothetical, non-radiating* particles are a construct useful in the description of $e^{\pm}$ dynamics. Real, radiating, e^{+} of energy $E = \sqrt{p^2c^2 + m^2c^4} \simeq pc \simeq p_0c$ can circulate in a phase-space neighbourhood of O_{xy} provided RF cavities of a proper frequency and sufficient voltage are added to compensate the average radiative energy loss and provide longitudinal phase stability (e^{-} can circulate in the opposite direction). In a semi-classical picture [49, 53, 57, 58], $e^{\pm}$ emit photons at random times according to the classical synchrotron radiation spectrum [47, 48] and make stochastic transitions between betatron trajectories corresponding to their instantaneous momenta. This picture can be understood [59] by recognising that a storage ring differs from an atom in that changes, $\Delta n = n\,u/E$, in orbital quantum number, n, corresponding to typical photon emissions of energy u, satisfy $n \gg \Delta n \gg 1$.

There is no deterministic closed orbit but the full 6D *central orbit*, O_{xyz} of a bunch of many electrons normally coincides with the attractive stable orbit calculated by averaging over photon emissions to include only the classical deterministic part of the synchrotron radiation (this includes the stable phase with respect to the RF system). If the domain of attraction of this orbit is large enough, the beam can have a good lifetime (Eq. 6.60 below). Because of the energy variation round the ring (localised RF cavities giving "energy-sawtooth"), the transverse projection of O_{xyz} does not coincide with O_{xy}. Figure 6.28 shows an example.

Neglecting intensity-dependent phenomena, the equilibrium dimensions of the beam are macroscopic quantum effects determined by the balance between radiation damping (the dependence of the classical radiation lost in magnetic fields on the energy, [49, 57, 58] and Sect. 6.5), and the quantum fluctuations (discrete photon nature) of the synchrotron radiation [57, 58]. Generally, the effects are linear enough that the core of the distribution is gaussian in each normal mode coordinate.

The mean-square fractional energy spread in the beam is

$$\frac{\sigma_E^2}{E^2} = \frac{55}{32\sqrt{3}}\frac{\hbar}{mc}\left(\frac{E_0}{mc^2}\right)^2 \frac{\oint |G^3|\,ds}{J_z \oint G^2 ds} \simeq \frac{1}{2}\gamma^2\frac{\lambda_e}{\rho_0}, \tag{6.57}$$

where $G = eB/p_0c = \rho^{-1}$ is the inverse of the local bending radius of O_{xy}, $\oint \cdots ds$ denotes an integral around O_{xy}, J_z is the longitudinal damping partition number (Sect. 6.5), $\lambda_e = \hbar/mc$ is the reduced Compton wavelength of the electron and the last equality holds to the extent that $G(s)$ is zero or has a constant value $1/\rho_0$ (isomagnetic ring).

Economic arguments, balancing construction cost against power consumption, are sometimes invoked to derive a scaling of radius with energy squared but this only applies for the highest energy rings with a few bunches (see [60] for the scaling of design parameters). More generally, the chromaticity correction and dynamic aperture constraints (Sect. 3.4) in collider rings require $6\sigma_E/E \lesssim 1\%$, so imposing a minimum radius $\rho/\mathrm{m} \approx 0.26(E_0/\mathrm{GeV})^2$. The spread in centre-of-mass energies of collisions $\sigma_{\sqrt{s}} = \sqrt{2}\sigma_E$ (if $D_x = 0$ at the collision point) should also be kept small.

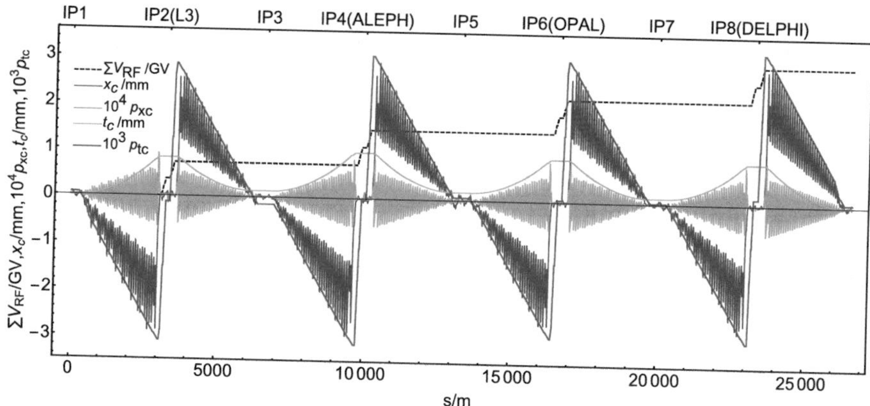

Fig. 6.28 The cumulative RF voltage (black dashed line) around the ring and four components of the ideal six-dimensional closed orbit of the e$^+$ beam in CERN's LEP collider, at a central beam energy of 94 GeV in an optics used in 1998. The fractional deviation of the beam energy on the closed orbit p_{tc} (red) exhibits the "energy sawtooth" due to the energy lost by synchrotron radiation in the eight arcs and its replenishment by RF systems located around the experimental interaction points. The conjugate time-lag coordinate t_c (green) reflects the corresponding path-length changes. The horizontal closed orbit $x_c = D_x p_{tc} + x_{cB}$ (blue) is a combination of the local dispersion orbit and a forced betatron oscillation and its conjugate p_{xc}. Without radiation, these four orbit components would be zero. LEP was a single ring collider with e$^+$ and e$^-$ beams of similar intensity circulating in the same beam pipe. In normal operation, the average x_c for the two beams, which were approximately equal and opposite, $x_c^+ \simeq -x_c^-$, was measured and corrected to the central trajectory. At higher energies, these orbits could be separated by a few cm near the RF systems

The equilibrium horizontal emittance for flat rings without betatron coupling is

$$\varepsilon_x = \frac{55}{32\sqrt{3}} \frac{\hbar}{mc} \left(\frac{E_0}{mc^2} \right)^2 \frac{\oint \left| \mathcal{H} G^3 \right| ds}{J_x \oint G^2 ds}, \tag{6.58}$$

where $\mathcal{H} = \beta_x {D_x}^2 + 2\alpha_x D_x D_x' + \gamma_x D'^2_x$ is a quadratic form constructed from the dispersion and betatron matrix (Sect. 2.1). Together, Eqs. (6.57) and (6.58) give the mean-square equilibrium beam size at any point in the ring

$$\sigma_x^2 = \left\langle \left(x_\beta + D_x \left(E - E_0 \right) / E_0 \right)^2 \right\rangle = \beta_x \varepsilon_x + D_x^2 (\sigma_E / E)^2. \tag{6.59}$$

The vertical emittance is usually smaller and due to some coupling of horizontal betatron motion into the vertical and vertical dispersion from orbit errors or other vertical bends. More general formalisms [53, 61] describe the radiation-generated emittances for the eigenmodes of linear oscillations about general six-dimensional central orbits.

A true equilibrium (strictly, stationary) state does not exist because the quantum fluctuations lead to loss from the tails of the beam with lifetimes for the three modes given by

$$\tau_{q,u} = \frac{1}{2}\tau_u \frac{e^{\xi_u}}{\xi_u}, \quad \text{where} \quad \xi_u = \frac{A_u^2}{2\sigma_u^2}, \quad \text{for} \quad u = x, y, z, \quad \xi_u \gtrsim 20, \tag{6.60}$$

where the τ_u are the radiation damping times and A_u are appropriate acceptances [53, 57, 58]. For the synchrotron mode, A_z is the RF bucket half-height

$$\left(\frac{A_z}{E_0}\right)^2 = \frac{2U_0}{\pi\,|\eta|\,h_{\mathrm{RF}}E_0}\left[\sqrt{(eV_{\mathrm{RF}}/U_0)^2 - 1} - \arccos\left(U_0/eV_{\mathrm{RF}}\right)\right]. \tag{6.61}$$

For adequate lifetime at small intensity, the mechanical and dynamic apertures and RF voltage must be large enough.

The bunch length is given by $\sigma_z = c|\eta|\sigma_E/(\omega_s E_0)$ where ω_s is the angular synchrotron frequency and $\eta \simeq \alpha_c$ the frequency slip factor ([53], Sects. 2.5.2 and 2.5.3).

6.7.2 Design of Colliders

Colliders are designed from the interaction point outwards. The classical design is based on head-on collisions of flat beams. However a number of other configurations have been explored and the most promising among them is described in the following section. In the classical scheme, luminosity (Sect. 6.4) is maximised by achieving very flat beams, $\kappa = \varepsilon_y/\varepsilon_x \ll 1$; we consider only beams of equal energy, size and single bunch population, N_b, colliding head-on, with $\sigma_y^* \ll \beta_y^*$; for generalisations see [53]. The beam-beam effect (Sect. 4.6.1) generally imposes maximum attainable values on the horizontal and vertical beam-beam parameters

$$\xi_{x,y} = \frac{r_e N_b \beta_{x,y}^*}{2\pi\left(E_0/mc^2\right)\sigma_{x,y}^*\left(\sigma_x^* + \sigma_y^*\right)}, \tag{6.62}$$

where r_e is the electron classical radius, $\beta_{x,y}^*$ and $\sigma_{x,y,z}^*$ are the optical functions and beam sizes at the collision point. Typically, one finds $\max\xi_y = 0.03 - 0.1$ with the highest values attained when the machine is very well corrected (favourable tunes, central orbits close to design, minimised vertical dispersion) and when radiation damping is strong. Then the luminosity (Sect. 6.4) can be expressed as

$$L = \frac{f_c N_b}{2r_e}\left(\frac{E_0}{mc^2}\right)\frac{(1+\kappa)\,\xi_y}{\beta_y^*}, \tag{6.63}$$

where f_c is the frequency at which identical bunches collide; in the simplest case $f_c = k_b f_0$ where k_b is the number of bunches per beam.

The number of bunches in a single-ring collider is limited by the possibilities for separating the opposing beams at unwanted encounters, e.g., by local or long-range ("pretzel scheme") electrostatic orbit bumps [53]. Collective effects limiting the single-bunch intensity (bunch-lengthening, transverse mode-coupling, see Chap. 4) are a major concern. In recent double-ring colliders, many more bunches can be stored. A crossing angle at the collision point separates the beams at encounters in the adjacent common section of the beam pipe. In recent years, the highest luminosity collider designs have adopted a new scheme described in the following section.

Multi-bunch collective effects and other limits related to total beam current (e.g., component heating by wakefields or synchrotron radiation, beam-loading, electron-cloud, ion-trapping [53]) tend to dominate. The impedance and surface properties of the vacuum chamber are critical.

Integrated luminosity can be further maximised in moderate energy rings for which a full-energy injector is available by topping up the intensity of the stored beam rather than dumping and refilling. The static magnetic configuration (no ramp and squeeze cycle) simplifies operation dramatically.

The arcs of collider rings are usually composed of FODO cells whose length and phase advance determine the emittance through Eq. (6.58). To minimise radiation power, the bending magnets are made as long as possible. In the highest energy rings, the quadrupoles must also be lengthened.

Low-β insertions (Sect. 6.2.1) provide small values of β_y^* at the interaction point(s) of the experiment(s) in long straight sections. These can also accommodate the accelerating cavities of the RF system, beam instrumentation and wiggler magnets and are connected to the arcs via dispersion suppressors.

Wiggler magnets modify the radiation damping, bunch length and/or emittance by contributing additional terms [53] with large $|G|$ to the integrals in Eqs. (6.57) and (6.58), so providing additional flexibility to maximise performance (e.g., at lower energy).

Sextupoles incorporated in the arcs must correct the large chromatic aberrations generated in the low-β quadrupoles while preserving adequate dynamic aperture (Sect. 3.4.4).

Many variations on this classical e^+e^- collider design are possible with new interaction region concepts showing promise (Sect. 6.4) in overcoming the need for ever-increasing beam current and ever-shorter bunches.

At higher intensities, phenomena such as the Touschek effect and intra-beam scattering [53], sometimes in combination with non-linear single particle dynamics or beam-beam effects, can reduce the lifetime below the values implied by Eq. (6.60); see Chap. 3 and Sect. 4.6.

6.7.3 *Large Piwinski Angle and Crab Waist Collision Scheme*

The need for precision measurements of rare decay modes with small cross sections at e^+e^- factories has driven requirements on peak luminosity to unprecedented levels. Conventional collision schemes, see Eq. (6.63), are based on pushing up the beam currents, lowering the β_y^*, and increasing the beam emittance so as not to exceed the beam-beam tune-shift limits. Passing from single to double ring colliders allowed the number of bunches to be increased considerably. However in order to avoid luminosity reduction due to parasitic (or long-range) bunch encounters near the collision point, beams had to be collided with a small horizontal crossing angle rather than head-on. However, this approach has come to a dead end since high currents result in high power losses, beam instabilities and increased power consumption.

Because of the parabolic variation of $\beta_y(s) = \beta_y^* + s^2/\beta_y^*$ in the vicinity of the interaction point (IP), the longitudinal region in which individual particle collisions occur will include places where the effective $\beta_y(s) >> \beta_y^*$ at the IP, and will therefore contribute less to the luminosity. This so-called *hour-glass* effect imposes a condition on the bunch-length: $\sigma_z \lesssim \beta_y(s)$. Unfortunately, shortening the bunch length is costly since it requires high voltage in the RF cavities, can excite collective instabilities, induce higher-order mode (HOM) heating in the beam pipe, and lead to coherent synchrotron radiation emission, which in turn deteriorates the bunch shape. On the other hand, increasing the bunch current leads to coupled bunch instabilities, HOM heating of the beam pipe, and higher wall-plug power.

A solution to these problems came with the idea of a new collision scheme, called "*Large Piwinski Angle and Crab Waist Sextupoles*" (LPA&CW), by P. Raimondi in 2006 [62]. This scheme has two main ingredients:

1. A large horizontal crossing angle at the IP, combined with very small horizontal beam size, resulting in a large Piwinski angle;
2. a pair of sextupoles, each placed on one side of the IP at a specific betatron phase from it.

The Piwinski angle is defined as:

$$\Phi = \frac{\sigma_z \, tan\,(\theta)}{\sigma_x} \approx \theta \frac{\sigma_z}{\sigma_x}. \tag{6.64}$$

Consider two bunches with RMS beam size σ_x and bunch length σ_x, colliding at a horizontal crossing angle 2θ. For flat beams colliding at a small crossing angle $\theta \ll 1$ and large Piwinski angle $\Phi \gg 1$, the luminosity L and the tune-shifts scale as [63]:

$$L \quad \propto \quad \frac{N\xi_y}{\beta_y^*} \tag{6.65}$$

$$\xi_y \quad \propto \quad \frac{N}{2\theta\sigma_z}\sqrt{\beta_y^*/\epsilon_y} \tag{6.66}$$

$$\xi_x \quad \propto \quad \frac{N}{(2\theta\sigma_z)^2} \tag{6.67}$$

In the LPA scheme the Piwinski angle is increased by decreasing σ_x and increasing θ. The most relevant consequence is that the overlap area of the two colliding beams is now reduced, since it is proportional to σ_x/θ. As a plus, as can be seen from Eq. (6.67), the horizontal tune shift in this case drops like $(2\theta\sigma_z)^2$, so the beam-beam interaction can be considered as one-dimensional and only the vertical plane is relevant.

Now, the vertical β_y^* function at the IP can be decreased, as much as the focussing magnet technology allows, to be comparable to the overlap area size that, in this case, is smaller than the bunch length. In this case that is much smaller the bunch length, so relaxing the problems of HOM heating, coherent synchrotron radiation and excessive power consumption:

$$\beta_y^* \approx \frac{\sigma_x}{2\theta} \ll \sigma_z \tag{6.68}$$

This scheme has several advantages:

- a smaller spot size at the IP, leading to higher luminosity,
- a reduction of the vertical tune-shift parameter,
- the mitigation of synchro-betatron resonances.

Long range beam-beam interactions no longer limit the maximum achievable luminosity when the distance between bunches is short. These parasitic crossings become negligible because of the larger crossing angle and the smaller horizontal beam size. The separation at each encounter is larger in terms of σ_x.

However the large Piwinski angle itself may introduce new beam-beam resonances which can limit the maximum achievable tune shifts. The second ingredient of the LPA&CW scheme, the pair of *Crab Waist sextupoles*, is designed to solve this problem. The CW transformation causes the horizontal oscillations to modulate the vertical motion modulation and thereby suppresses the betatron and synchro-betatron resonances. The CW scheme is realised by installing a couple of sextupole magnets on the two sides of the IP, preferably in a high β and zero dispersion region. To provide the exact compensation the sextupoles be at π horizontal and a $\pi/2$ vertical betatron phase advance from the IP.

The CW transformation can be described by the Hamiltonian:

$$H = H_0 + \frac{1}{2\theta}xp_y^2 \tag{6.69}$$

where H_0 is the Hamiltonian of the particle's motion without the CW, x is the horizontal particle coordinate and p_y the vertical momentum. The effect of the CW

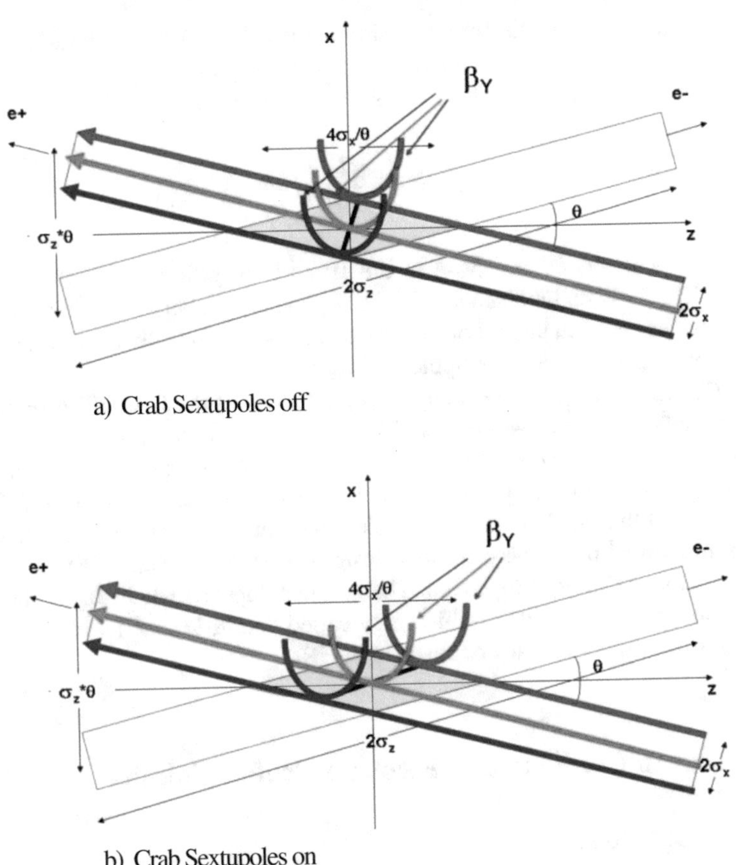

Fig. 6.29 Crab Waist collision scheme

transformation is a vertical betatron function *twist* according to:

$$\beta_y = \beta_y^* + \frac{(s - x/2\theta)^2}{\beta_y^*} \tag{6.70}$$

In this case, the β_y waist of one beam is twisted to be oriented along the central trajectory of the other beam. As a consequence, all particles, independently of their x position, collide at the minimum β_y spot of both beams, with an increase of few percent in the geometric luminosity due to the β_y redistribution along the overlapping beams area. A sketch of this shown in Fig. 6.29.

However the main CW effect is to suppress the betatron and synchro-betatron resonances which would arise due to the vertical motion modulation induced by the horizontal oscillations. This increases the space for the working betatron tunes of the collider. Moreover beam-beam simulations showed that beam tails are very much reduced and the beam-beam blow-up is also suppressed.

The CW sextupoles strength should satisfy the following condition:

$$K = \frac{1}{2\theta}\frac{1}{\beta_y^*\beta_y}\sqrt{\frac{\beta_x^*}{\beta_y}} \tag{6.71}$$

where starred β values are those at the IP and the others are at the sextupole location. The CW sextupoles can reduce the dynamic aperture if there are other non-linearities between them. For this reason they should ideally be installed before the chromaticity correction sextupoles.

The (LPA&CW) collision scheme was first tested at the DAΦNE Φ-Factory in Frascati (Italy) in 2008 [64], by modifying the interaction region to increase the crossing angle, decrease both the β^* and allocate space for the sextupoles. The result was a boost in luminosity of about a factor of 4 and measurement of the beam profile showed that the bunches kept their Gaussian shape. This scheme has since then been adopted by all new collider designs worldwide (SuperKEKB, FCC-ee, various τ-charm Factory proposals). The collider SuperKEKB in Japan, however, has adopted the LPA scheme (which they called "*nano-beams*") without the CW sextupoles because of the lack of space in the IR.

6.8 Hadron Colliders and Electron-Proton Colliders

K. Hanke · B. J. Holzer

6.8.1 Principles of Hadron Colliders

Hadron colliders are discovery machines which provide high centre-of-mass energy and cover a wide energy range. Contrary to electron-positron colliders, where the energy and quantum state of the initial particles is precisely known, the input conditions are less well defined in the case of proton-proton or proton-antiproton collisions. In fact, such collisions are, unlike e^+e^- annihilation, collisions of quarks and antiquarks the momenta of which are distributed according to the structure function of the hadron and are hence not precisely defined. As far as the analysis of the events is concerned, hadronic collisions result in a much larger number of tracks in the detector than in the case of e^+e^- annihilation, providing an additional challenge. From a machine physics point of view hadron machines have the enormous advantage that the particle beam energy is not limited by synchrotron radiation. This is because the proton mass is 2000 times higher than the one of the electron, and the energy loss due to synchrotron radiation scales with m^{-4}. What is limiting the achievable beam energy in a hadron collider is the magnetic field to be provided by the dipole magnets in order to bend the particle beam on a circular trajectory. The radius of curvature of a particle in a dipole field is given by

$$\frac{1}{\rho}\left[m^{-1}\right] = \frac{eB}{p}p = 0.2998\frac{B\ [T]}{p\,[GeV/c]}$$

where ρ is the radius of curvature, B is the magnetic field, e is the elementary charge and p is the particle momentum. The quantity Bρ is called the rigidity [65, 66]. The beam rigidity determines the B-field required to bend the beam on a circular trajectory with given bending radius. The maximum achievable B-field being limited to about 2 Tesla for normal conducting magnets, today's high-energy hadron colliders use superconducting bending magnets.

The quasi absence of synchrotron radiation leads to another feature of hadron colliders, which is the fact that there is no synchrotron radiation damping and the transverse emittance is hence determined and preserved throughout the injector chain. Emittance blow up, e.g. via injection mismatch, is therefore critical.

6.8.2 *Proton-Antiproton Colliders*

A machine with one single vacuum chamber, e.g. the Super Proton Synchrotron SPS ("SppbarS in this operation mode) in the 1980s or the Tevatron can accomplish the acceleration of protons and antiprotons.

During the years 1981–1987, the CERN SPS was operated as a proton-antiproton collider, providing high energy collisions for two major experiments located in adjacent sextants of the accelerator. This operation was first with three dense bunches of protons in collision with three rather weak bunches of antiprotons, with no separation of the beams at the unused crossing points. After increasing the antiproton production rate, six bunches per beam were used. The SPS has normal conducting bending magnets and a circumference of 6.9 km. The beam energy provided by the SPS as proton-antiproton collider was 315 GeV [67, 68]. The SppS was the first hadron collider operating with bunched beams. Before the commissioning of the machine it was debated if it was possible to collide proton and antiproton bunches, or if the beams would become unstable due to the presence of the beam-beam interaction without damping as in e^+e^- colliders. Its success demonstrated the feasibility of high energy hadron colliders [69].

Higher energy was achieved by the Tevatron, using superconducting magnets with a maximum B field of 4.5 T. The circumference of the machine is 6.28 km; comparable to that of the SPS. In the final stage of operations ("run II"), beams are injected at 150 GeV and accelerated to 980 GeV. The bunches (36+36) circulate in the same aperture, the protons clockwise and the antiprotons anticlockwise. The machine has a lattice with four dipoles followed by a quadrupole, with a total of 772+2 dipoles and 90+90 quadrupoles, plus a number of corrector magnets.

6.8.3 Proton-Proton Colliders

Proton-proton colliders require a dedicated magnet design with two separate vacuum chambers for the two equally charged beams. The first proton-proton collider was the ISR (Intersecting Storage Rings) at CERN. It consisted of two rings of 943 m length which were intersecting at eight points. Out of these eight intersection points six were used for experiments. The ISR was operated between 1970 and 1984. The top energy achieved for protons was 31.4 GeV/c. The ISR allowed not only proton-proton collisions, but stored and collided later also deuterons, alpha particles and antiprotons. The ISR pioneered a number of techniques which were beneficial which paved the path for future high energy colliders like the SPS.

The proton-proton collider with the highest energy ever built is the Large Hadron Collider (LHC) at CERN [70]. It uses superconducting magnets with two separate vacuum chambers for the two equally charged beams. The design field is 8.36 T, and the machine circumference 26.659 km which yields a design beam energy of 7 TeV. Higher field levels are being studied in the frame of possible further energy upgrades.

A machine with an even higher beam energy of up to 50 TeV is presently being studied by an international collaboration. The Future Circular Collider (FCC) has a hadron-hadron option (FCC-hh) with a beam energy of 50 TeV [71]. The latest design features a machine circumference of 97.75 km with a maximum dipole field of 15.7 T. The size of the machine is a compromise of civil engineering constraints and dipole feasibility.

6.8.4 Electron-Proton Colliders

Collisions between electrons and protons are used to study the inner structure of the proton e.g. the quark gluon distribution underneath the valence quarks. The electrons are used as a point like probe to determine the inner structures in the target. This deep inelastic scattering studies were performed in the beginning using an accelerated electron beam colliding on a fixed target. Due to kinematic considerations however a much higher resolution is obtained if two accelerated beams are brought into collision.

Due to the different nature and beam dynamics of the two particles an electron-proton collider cannot be built as a single ring machine: It consists of two storage rings of equal circumference, one being optimised for the acceleration and storage of electrons, the other for a high energy proton beam. The design of these two rings looks quite different and completely different effects determine the performance limitations of the rings. Figure 6.30 shows the two storage rings of the HERA collider:

HERA was built as a 6.3 km long double ring collider with beam energy of 27.5 GeV for the electron beam, and 920 GeV for the proton beam [72]. The fundamental layout was based on four arcs and four straight sections where

Fig. 6.30 View of the two independent storage rings for electron and proton acceleration in HERA. The super conducting proton lattice is placed on top of the conventional electron ring

the high-energy detectors were located. The proton machine was designed as a superconducting magnet lattice in the arcs to achieve the highest possible beam rigidity (or particle energy). The electron storage ring was built in conventional magnet technology: here the limiting factor was the synchrotron radiation emitted by the electrons which was too strong to justify super conducting magnet technology. Basic limits for the achievable beam energy therefore were in the case of the protons the magnetic field of the bending magnets (B = 5.1 T) and for the electron ring the available RF power that was needed to compensate the synchrotron radiation losses. Both rings had been built on top of each other to guarantee an equal revolution time of the circulating particle bunches.

The interaction region of such a two ring collider deserves special attention: While the two beams are brought into collision in a common vacuum system and magnet lattice, they have to be separated after the IP and guided into their respective magnet lattices. Especially in the case of the electron beams the separation has to be performed fast enough, as the strong focusing fields of proton mini beta magnets can only be applied after a full separation of the beams.

Two mini beta insertions therefore have to be installed and combined with an effective beam separation scheme. In the case of HERA the separation has been achieved by using the different momenta of the beams: The mini beta quadrupoles of the electron beam have been placed offset with respect to their magnetic axis and acted as combined function magnets. Consequently the electron beam was bent due to its smaller beam rigidity to the inner side of the ring and at a distance $s^* = 20$ m the first proton magnet could be installed. A schematic view of this nested interaction region is shown in Fig. 6.31.

The advantage of this scheme is its compactness as beam separation and focusing are obtained at the same time. Special care however is needed as the electron quadrupoles of the mini beat section will have an effect on the proton beam that depends on the corresponding energy of the electron beam. This dynamic influence

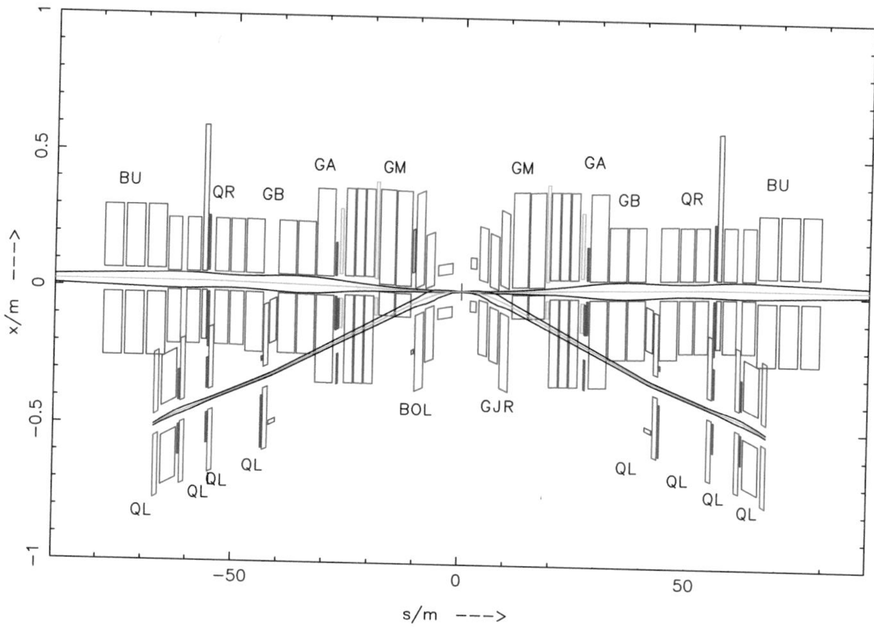

Fig. 6.31 Layout of the HERA interaction region: The inner triplet of the electron lattice is combined with the beam separation scheme and embedded inside the doublet quadrupoles of the proton mini beta insertion

on the optics and orbit of the protons therefore has to be compensated during the acceleration of the electrons as well as during the beta squeeze.

The luminosity formula for such a double ring collider is given by

$$L = \frac{1}{2\pi e^2 f_0} * \frac{\sum_i (I_{pi} * I_{ei})}{\sqrt{\sigma_{xp}^2 + \sigma_{xe}^2} * \sqrt{\sigma_{yp}^2 + \sigma_{ye}^2}}$$

It depends on the product of the single bunch currents I_{pi} and I_{ei} and the sum of this contribution over the overall number *i* of colliding bunches in the rings. As the beams are guided in different magnet lattices the beam sizes σ are independent of each other. Nevertheless the beams have to be matched, i.e. the beam sizes of the two beams at the IP have to be equal in both planes: $\sigma_{xp} = \sigma_{xe}$ and $\sigma_{yp} = \sigma_{ye}$. This condition deserves special attention as the beam emittances of protons and electrons are quite different and independent beam optics have to be established to achieve matched beam sizes.

Another special feature of an electron proton collider is the synchrotron radiation that is emitted by the electron beam. Usually this effect is present in the arc structure where the dipole fields bend the beam on the design orbit. Due to the separation fields needed in the interaction region the synchrotron radiation is also emitted close to the IP and special care is needed to shield the high energy detector from the

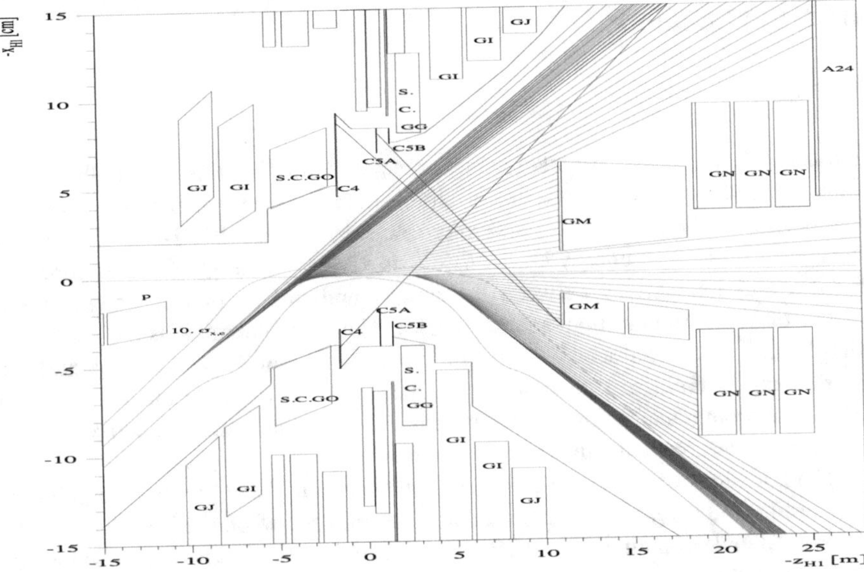

Fig. 6.32 Synchrotron radiation emitted during beam separation in the HERA interaction region. The plot shows schematically the interaction region lattice, the beam dimension and the direction and density of the synchrotron light

emitted photons. For the HERA collider this problem has been studied in detail (see Fig. 6.32) and a combination of absorbers and movable collimation masks have been used to avoid hits from direct or back scattered photons into the detector parts.

The performance limitations of such an e-p-collider are given by the singe bunch intensity of the protons (limited by the particle source), the overall current of the electrons (limited by RF power or beam instabilities), the number of bunches that can be stored in the machine (limited by technical reasons of the injection elements) and the usual limits of the mini beta insertions that have been mentioned above.

The main parameters of HERA are summarised in Table 6.1:

Table 6.1 Summary of the main parameters of HERA

	Electrons	Protons
Energy	27.5 GeV	920 GeV
Beam current	58 mA	140 mA
Particles per bunch	4×10^{10}	1×10^{11}
Number of bunches	189	180
Beta function at IP x/y	0.63 m/0.26 m	2.45 m/0.18 m
Hor. emittance	20 nm	5.1 nm
Emittance ratio $\varepsilon_y/\varepsilon_x$	0.18	1.0
Beam size (IP) σ_x/σ_y	112/30 μm	112/30 μm
Luminosity	7×10^{31} cm^{-2} s^{-1}	

6.9 Ion Colliders[1]

W. Fischer · J. M. Jowett

Ion colliders are research tools for high-energy nuclear physics. The collisions of fully stripped high-energy ions, that is, atomic nuclei, create matter of a temperature and density that existed in the first microseconds after the Big Bang. The matter created in these high-energy ion collisions is known as the Quark Gluon Plasma (QGP), and interactions between the quarks and gluons is the subject of the theory of quantum chromodynamics (QCD). The basic interactions are studied in simpler collisions such as e^+e^- or pp but heavy-ion collisions allow the study of more complex collective phenomena in QCD. The collisions in ion colliders can create hadronic matter at much higher densities and temperatures than fixed target experiments although at a much lower luminosity.

The collisions of heavy ions in RHIC and the LHC have yielded a number of new results and revealed phenomena that were unexpected on the basis of previous theoretical understanding. The QGP generated in the heavy ion collisions in RHIC was expected to be weakly interacting, but found to be strongly interacting like an almost perfect liquid [73, 74]. Hadronic jets created in the collisions have a rather short mean free path in the QGP leading to a phenomenon termed "jet quenching" [74], and the largest ever measured vorticity was seen in heavy ion collisions [75]. The collisions also created the heaviest artificially made antimatter nuclei, anti-helium-4 [76, 77]. The higher energies in the LHC create many more hard probes and heavy bound states such as charmonium (J/ψ) or bottomonium (Υ) and, in the highest-energy p-Pb collisions, toponium. Z and W bosons, particles that do not interact with the QGP via the strong interaction, were never before seen in heavy ion collisions. The ALICE experiment also reported the highest temperatures directly measured in the laboratory [78].

The colliding nuclei also have high electric charges ($Z \sim 80$). Together with the powerful Lorentz-compression at high energies, these generate enormous electro-magnetic fields outside the nuclear radius. As first shown by Fermi, Weizsäcker and Williams, these fields can be represented as a beam of high energy quasi-real photons, leading to so-called ultraperipheral photonuclear and photon-photon collisions. Besides their intrinsic interest, the high cross-sections for these processes have consequences for the operation of the collider. The ATLAS experiment at the LHC has published the first evidence for light-on-light elastic scattering, a long-predicted fundamental process of nonlinear quantum electrodynamics, transcending Maxwell's equations.

---

[1]This section has been authored by Brookhaven Science Associates, LLC under Contract No. DE-SC0012704 with the U.S. Department of Energy. The United States Government retains and the publisher, by accepting the article for publication, acknowledges that the United States Government retains a non-exclusive, paid-up, irrevocable, world-wide license to publish or reproduce the published form of this manuscript, or allow others to do so, for United States Government purposes.

The first ion collider was the CERN Intersecting Storage Rings (ISR), which briefly collided light ions [79, 80] in the late 1970s. The BNL Relativistic Heavy Ion Collider (RHIC) has been in operation since 2000 and collided a number of species at numerous energies. The CERN Large Hadron Collider (LHC) started its Run 1 heavy ion program in 2010 and has provided mainly p-p, p-Pb and Pb-Pb at increasing luminosity with a substantial increase in energy in Run 2 (2015–2018). Both RHIC and the LHC have an expected operating time exceeding 20 years. Further upgrades to the LHC, its injector complex and its experiments, foreseen in the shutdown after Run 2, should allow the integrated luminosity in Runs 3 and 4 (up to 2029) to exceed Runs 1 and 2 by an order of magnitude. Table 6.2 shows all species combinations and energy ranges demonstrated to date for the ISR, RHIC and LHC. All three machines also collide protons. In RHIC the protons are spin-polarized, making the machine the only collider of spin-polarized protons ever built. The LHC is the highest energy proton-proton and heavy-ion collider ever built. Critically, proton-proton collisions at the same energy per nucleon provide reference data for heavy ion collisions. In the following, we will limit our comments to the ion operation in RHIC and the LHC.

Ion colliders differ from proton or antiproton colliders in a number of ways: the preparation of the ions in the source and the pre-injector chain is limited by other effects than for protons; frequent changes in the collision energy and particle species, including asymmetric species, are typical; and the interaction of

Table 6.2 Ion species and energies achieved in ISR, RHIC and LHC as of 2017

Machine	Species	Energies [GeV/nucleon]
ISR	α–α	13.3–15.7
	p–α	26.6–31.4 (p), 13.3–15.7 (α)
	d–d	13.3–15.7
	p–d	26.6–31.4 (p), 13.3–15.7 (d)
	p–p	13.5–31.2
RHIC	U–U	96.4
	Au–Au	3.85–100
	Cu–Au	100
	Cu–Cu	11.2–100
	h–Au	103.5 (h)–100 (Au)
	d–Au	9.9–100
	p↑–Au	103.9 (p)–98.6 (Au)
	p↑–Al	103.9 (p)–98.7 (Al)
	p↑–p↑	31.2–255
LHC	Pb–Pb	1380–2511 (2563 briefly)
	Xe–Xe	2721
	p–Pb	4000–6500 (p), 1577–2563 (Pb)
	p–p	3500–6500

p, d, h and α denote the nuclei of the hydrogen, deuterium, helium-3 and helium-4 atoms respectively. All three machines also collide proton beams, which are spin-polarized in RHIC. The quoted energy is the sum of rest and kinetic energy per nucleon

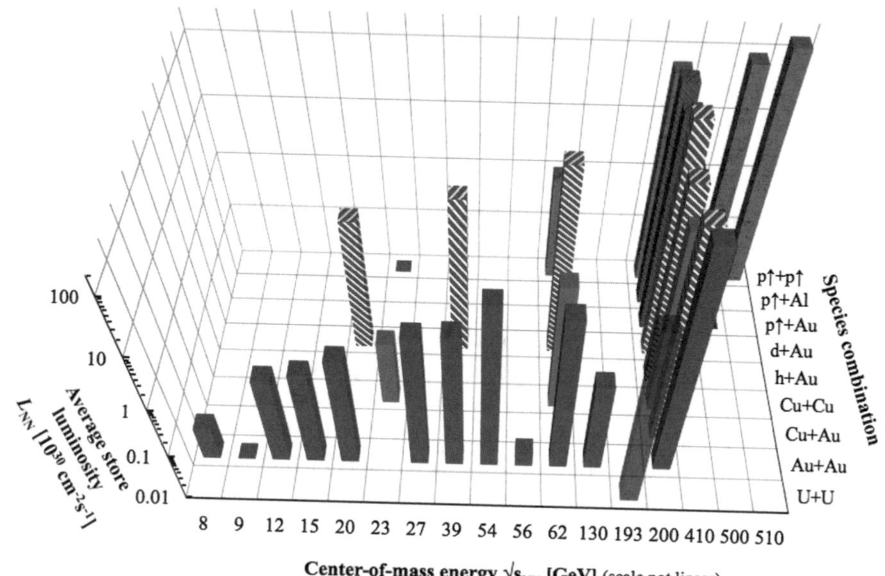

Fig. 6.33 Achieved nucleon-pair luminosity L_{NN}, averaged over a store, for all species combination and energies in RHIC

ions with each other and accelerator components is different from protons. This has implications for collision products, collimation, the beam dump, and intercepting instrumentation devices such as profile monitors. Thus, the performance limitations of heavy-ion colliders are also different from proton-proton colliders.

Figure 6.33 shows the achieved nucleon-pair luminosities L_{NN}, averaged over a store, for all species combinations and energies in RHIC. The plot demonstrates the flexibility of RHIC in colliding different species combinations (all of them at or near the center of mass energy $\sqrt{s_{NN}} = 100$ GeV), energy scans for a number of species combinations (Au+Au, Cu+Cu, d+Au p↑+p↑), and a luminosity that is strongly decreasing with the collision energy.

In the preparation for the collider use, the charge state Z of the ions is successively increased. A high charge state Z increases the bending and acceleration efficiency, but also increases the effects of space charge and intrabeam scattering (IBS). The direct space charge tune shift ΔQ, typically limited to values of less than 0.5, is given by [81]

$$\Delta Q = -\frac{\lambda R}{2\varepsilon_n \beta \gamma^2}\frac{r_0 Z^2}{A},$$

where λ is the particle line density, R the machine circumference, ε_n the normalized emittance, β and γ the relativistic factors, r_0 the classical proton radius, and A the

Table 6.3 Preparation of the heavy ions for RHIC and LHC

RHIC (Au)			LHC (Pb)		
	Charge state Z	Ion energy [eV/nucleon]		Charge state Z	Ion energy [eV/nucleon]
LION[a]	1+	150	ECR	27+	2.5 k
EBIS[b]	32+	0.9 M	LINAC3	54+	4.2 M
Booster	77+	101 M	LEIR	54+	72.2 M
AGS	79+	8.8 G	PS	82+	5.9 G
RHIC	79+	99 G	SPS	82+	177 G
			LHC	82+	2.51 T

For each accelerator, the kinetic energy of the ions is given at extraction, and the charge state in the following transfer line
[a]Laser Ion Source; [b]Electron Beam Ion Source

mass number. IBS growth rates $1/T_{x,y,s}$ scale like [81]

$$\frac{1}{T_{x,y,s}} \propto \frac{Z^4}{A^2} \frac{N_b}{\gamma \varepsilon_x \varepsilon_y \varepsilon_s}$$

where N_b is the bunch intensity, and $\varepsilon_{x,y,s}$ are the normalized emittances. High charge states also reduce the electron stripping probability, and electron stripping at higher energies is generally more efficient.

Table 6.3 shows the charge states and energies in the RHIC and LHC injector chains for the Au and Pb respectively, the heavy ion species most often used in these machines. For RHIC singly charge ions are generated in a hollow cathode or laser ion source (LION) [82], and transferred into an Electron Beam Ion Source (EBIS) [83]. With EBIS, beams of almost any element can be prepared for RHIC including uranium and spin-polarized ^{3}He. After increasing the charge state to $Z = +32$ the ions are accelerated through an RFQ and short linac, and injected into the Booster. After acceleration in the Booster, all but two electrons are stripped before injection into the AGS, and the ions are further accelerated. To increase the intensity of the ion bunches, bunches are merged in both the Booster and AGS. The last two electrons are stripped in the transfer line from the AGS to RHIC. In RHIC all ions except protons have to cross the transition energy, when bunches become short and peak currents high. In addition, the longitudinal motion is frozen for a short period, and the short bunches can trigger the creation of an electron cloud [84]. This situation makes the beams vulnerable to instabilities [85], which limited the bunch intensity for a number of years [84].

At CERN an ECR ion source is used, followed by an RFQ and the heavy ion LINAC3 [86, 87]. After passing a carbon foil that strips electrons, the ions are then accumulated in Low Energy Ion Ring (LEIR) [88]. During the 71-turn injection and before acceleration the ions are cooled with an electron beam, with a transverse cooling time of 0.2 s. To minimize dynamic vacuum effects from charge-change processes the LEIR vacuum system is designed for a dynamic pressure of less than 10^{-12} mbar. From LEIR the ions are injected into the PS where the bunches are split to obtain the bunch spacing needed for the LHC. After acceleration in the PS the last remaining electrons are stripped before injection into the SPS. In the SPS at injection space charge and intrabeam scattering were a concern, and an emittance growth of about 20% is observed at injection. Acceleration in the SPS requires a special fixed frequency acceleration scheme since the main 200 MHz RF system does not have the frequency range required to accelerate heavy ions with a constant harmonic number. The SPS acceleration scheme takes advantage of the fact than the ion bunch train only fills a fraction of the circumference allowing for an adjustment of the RF phase during the time without beam [89].

The luminosity is given by

$$L = (\beta\gamma)\frac{f_{rev}}{4\pi}k_c\frac{N_{b1}N_{b2}}{\varepsilon_n\beta*}H$$

where f_{rev} is the revolution frequency, k_c the number of bunch-bunch collisions per turn, N_{b1} and N_{b2} the bunch intensities in the two beams respectively, and $\beta*$ the lattice envelope function at the interaction point. The factor H accounts for the hourglass effect and crossing angles, and is smaller than and of order 1. The luminosity is limited by different effects in RHIC and the LHC.

In RHIC bunches of fully stripped heavy ions like Au^{79+} with the same number of charges as proton bunches have IBS growth rates an order of magnitude larger. In RHIC at injection IBS leads to bunch lengthening, and at store to particle loss out of the RF buckets and an increase in the transverse emittance. Longitudinal and transverse bunched beam stochastic cooling at store has been implemented [90] to counteract IBS. This and an increase in the bunch intensity have significantly increased the average store luminosity (Fig. 6.34). Table 6.4 shows the latest RHIC parameters for Au–Au operation.

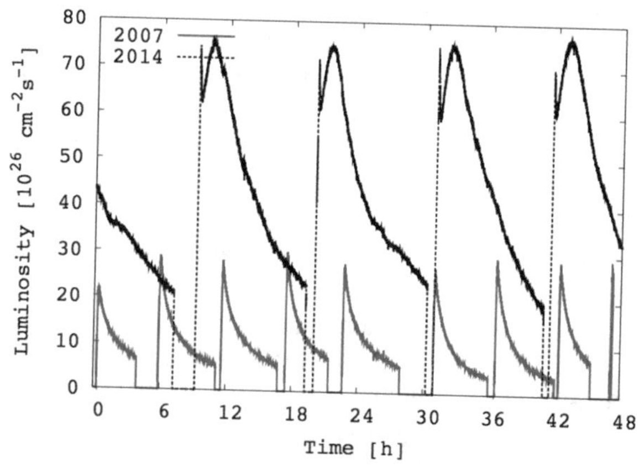

Fig. 6.34 RHIC instantaneous Au+Au luminosity in 2007 with longitudinal stochastic cooling in the Yellow ring only, and in 2014 with 3D cooing in both rings. The increase in the initial luminosity is due to an increase in the bunch intensity

Table 6.4 Main operating parameters achieved for the most commonly heavy ions in RHIC and LHC as of 2017

Parameter	Unit	RHIC	LHC
Circumference C	km	3.8	26.7
Ion species		$^{197}Au^{79+}$	$^{208}Pb^{82+}$
Maximum energy	GeV/nucleon	100	2511
Bunch intensity	10^9	2.0	0.20
Number of colliding bunches		111	492
Peak luminosity	10^{26} cm^{-2}s^{-1}	155	36
Average store luminosity	10^{26} cm^{-2}s^{-1}	87	17

Other effects that have limited the heavy ion performance in the past include: the availability of high intensity bunches from the injector chain, instabilities at transition [91] driven by the machine impedance and electron clouds (RHIC is the only superconducting accelerator that crosses the transition energy), dynamic pressure increases including pressure instabilities caused by electron clouds [84], beam loading in the storage RF system (bunches are accelerated with $h = 360$ and transferred into a $h = 7{\times}360$ system at store), and chromatic lattice aberrations at $\beta* < 70$ cm.

The LHC heavy ion operation started in 2010 and the luminosity is principally limited by two effects [92, 93]. Firstly, secondary beams generated in collision and having a *Z*/*A* ratio different from the primary beam will be lost in the dispersion suppressor, a location with superconducting magnets with a limited ability to absorb heat [94]. Secondly, the collimation efficiency for ions is lower than for protons leading again to losses in uncontrolled regions [95]. Expected LHC ion parameters are shown in Table 6.4. The two most important processes for the generation of secondary beams in collisions are Bound-Free Pair Production (BFPP),

$$^{208}\mathrm{Pb}^{82+} + {}^{208}\mathrm{Pb}^{82+} \xrightarrow{\gamma} {}^{208}\mathrm{Pb}^{82+} + {}^{208}\mathrm{Pb}^{81+} + \mathrm{e}^{+}, \tag{6.72}$$

with a cross section of 281 barn; and Electromagnetic Dissociation (EMD) with a total cross section of 226 barn, about half of which is from the 1-neutron reaction

$$^{208}\mathrm{Pb}^{82+} + {}^{208}\mathrm{Pb}^{82+} \xrightarrow{\gamma} {}^{208}\mathrm{Pb}^{82+} + {}^{207}\mathrm{Pb}^{81+} + \mathrm{n}. \tag{6.73}$$

Beam losses due to BFPP were observed in RHIC with $^{63}Cu^{29+}$ ions [96] and effective mitigation measures have now been implemented at the LHC [97, 98]. These have allowed Pb-Pb luminosities far beyond the design value from 2015 onwards. Collimation of heavy ions is fundamentally different from protons. Protons are scattered at a primary collimator and collected at a secondary collimator. Heavy ions undergo nuclear fragmentation and electromagnetic dissociation in the primary collimator. The fragments created have a wide range of *Z*/*A* ratios that are not collected by the secondary collimators. Measurements of collimation efficiency were done in the CERN SPS and compared with detailed simulations to obtain reliable estimates of the heavy ion collimation efficiency in the LHC [95].

The LHC also collided Xe nuclei in 2017 [98] and may collide other species in future, generally with a view to increasing the nucleon-nucleon luminosity.

In the collision of asymmetric species the 2-in-1 magnet design of the LHC requires that the magnetic fields in two rings are the same (the two RHIC rings are independent and can have different fields). For p–Pb operation it is then necessary to have different revolution frequencies at injection and during the energy ramp [99]. Lead beams in the LHC at design energy have noticeable synchrotron radiation damping times (6 h and 13 h longitudinally and transversally) that are of the same order as the IBS emittance growth times (8 h and 13 h longitudinally and transversally) [92].

6.10 Beam Cooling

F. Caspers · D. Möhl

6.10.1 Introduction

Beam cooling aims at reducing the size and the energy spread of a particle beam circulating in a storage ring or in an ion trap. This reduction of size should not be accompanied by beam loss; the goal is to increase the particle density [100]. Since the beam size varies with the focusing properties of the storage ring, it is useful to introduce normalized measures of size and density. Such quantities are the (horizontal, vertical and longitudinal) emittances and the phase-space density. For our present purpose they may be regarded as the (squares of the) horizontal and vertical beam diameters, the energy spread, and the density, normalized by the focusing strength and the size of the ring to make them independent of the storage ring properties. Phase-space density is then a general figure of merit of a particle beam, and cooling improves this figure of merit. The terms beam temperature and beam cooling have been taken over from the kinetic theory of gases. For visualization one may imagine a beam of particles going around in a storage ring. Particles will oscillate around the beam centre in much the same way that particles of a hot gas bounce back and forth between the walls of a container. The larger the mean square of the velocity of these oscillations in a beam, the larger the beam size. The mean square velocity spread is used to define the beam temperature in analogy to the temperature of the gas which is determined by the kinetic energy of the molecules.

There are several basic motivations for the application and development of different beam cooling techniques:

- Collection and accumulation of rare particles, e.g. antiprotons or short lived particles such as muons.
- Improvement of interaction rate and resolution, e.g. collision experiments with antiprotons or with ions; increase in luminosity. For fixed target experiments: sharply collimated and/or highly mono-energetic beams for precision experiments.
- Preservation of beam quality, mitigation and suppression of beam blow-up.
- Preparation of crystalline beams.

Several cooling techniques are operational or have been discussed:

- Radiation cooling (often referred to as radiation damping); linked to energy loss of particles via synchrotron radiation (used in virtually all modern electron synchrotrons).
- Stochastic cooling (works well for "hot" beams to get them "tempered).
- Electron cooling (most suitable for "tempered" beams to get them "cold").
- Laser cooling (essentially for ions where two level transitions of electrons can be excited).
- Ionization- and friction-cooling (mainly discussed in the context of muon cooling).
- Resistive cooling; used to cool charged particles in a trap where the kinetic energy of the particle is dissipated in the resistive losses of a resonant circuit.
- Coherent electron cooling, a kind of blend from stochastic cooling at very high frequencies and electron cooling (under development at BNL theses days)

The use of the terms cooling and damping is not always well distinguished and unambiguous in the literature. Even in the context of stochastic cooling the authors were using the term damping in the early days. A similar observation can be made for radiation damping and cooling. One may consider defining any action on individual particles as "cooling" and any action on groups of particles as damping. Examples are the feedback systems in circular machines which are commonly referred to as dampers and which prevent emittance blow up, while a very similar feedback system just having a much higher electronic gain can work as (stochastic) cooler and reduce the emittance. However typically such damper systems have a much smaller bandwidth and lower operation frequency as compared to stochastic cooling hardware.

6.10.2 Beam Cooling Techniques

6.10.2.1 Radiation Cooling

Back in 1956, A.A. Kolomenski and A.N. Lebedev [101] pointed out that the 'synchrotron light' emitted by an electron moving on a curved orbit can have a

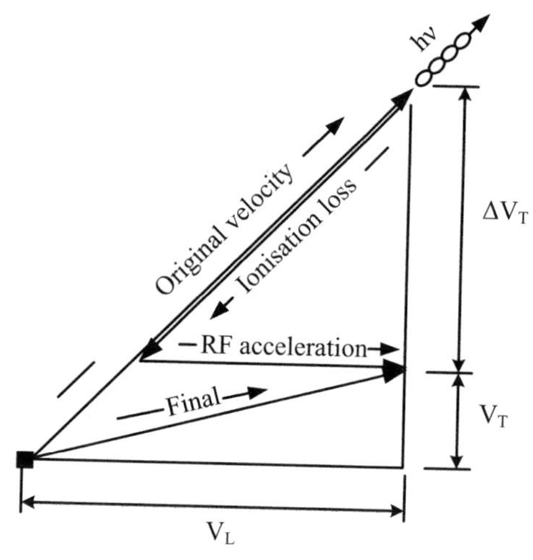

Fig. 6.35 The principle of transverse cooling by synchrotron radiation (transverse velocities exaggerated)

damping effect on the motion of the particle. This is because the radiation is sharply peaked in the forward direction. The continuous emission of synchrotron radiation leads to a friction force opposite to the direction of the motion. For a particle moving on the design orbit, the energy loss is restored and the friction force is on average compensated by the RF-system. For a real particle the residual friction force tends to damp the deviation from the design orbit (Fig. 6.35). This cooling force is counteracted by the 'radiation excitation': synchrotron light is really emitted in discrete quanta and these many small kicks tend to heat the particle. The final emittances result from the equilibrium of radiation damping and excitation. We will see that a similar interplay between a specific cooling and heating mechanism is characteristic also for the other cooling methods.

The theory of cooling by synchrotron radiation is in a mature state. Following up on Sands' classical treatment on "the physics of electron storage rings", radiation cooling has found its place in text books. The immense success of modern electron–positron machines, both 'synchrotron light facilities' (e.g. ESRF, ALS, APS, BESSY, SPRING8) and colliders (e.g. LEP, PEP II, KEKB) would not have been possible without the full understanding of radiation effects. Virtually all these machines depend critically on radiation cooling to attain the minute emittances necessary in their application. Linear e^+e^--collider schemes (like CLIC, TESLA, NLC, JLC) too, have to rely on 'damping rings' in their injector chain to produce the ultra-high phase-space density required. For historical reasons the reduction of beam emittance due to the emission of synchrotron radiation (typically from leptons) is usually referred to as radiation damping, although the term "cooling "might be more consistent.

The cooling rates as well as the final beam size and momentum spread depend on the lattice functions in regions where the orbit is curved. The art is then to 'arrange' these functions such that the desired beam property results. The strategy for 'low emittance lattices' is well developed and 'third-generation machines' providing beams of extremely high brightness have come into operation. To enhance the cooling, wiggler magnets are used, producing a succession of left and right bends. This increases the radiation and thereby the damping rates. The heating can be kept small by placing the wiggler at locations where the focusing functions of the ring are appropriate to make the particle motion insensitive to kicks. More details for cooling by synchrotron radiation are given in Sect. 6.5.

Radiation cooling and lattice properties of the storage ring are thus intimately linked and by smart design, orders of magnitude in the equilibrium emittances have been gained. This may serve as example for other cooling techniques for which the art of 'low emittance lattices' is only now emerging.

6.10.2.2 Microwave Stochastic Cooling

For (anti-)protons and heavier ions, radiation damping is almost negligible at the energies currently accessible in accelerators except for the LHC. One of the 'artificial' cooling methods devised for these heavy particles is stochastic cooling by a broadband feedback system (Fig. 6.36). The name "stochastic damping" was coined by Simon van der Meer who invented this method in 1968 (first published in 1972) [102] to underline the statistical basis of the method. First successful tests and observations were done at the CERN ISR (Intersecting Storage Rings) [102] followed by a dedicated "Initial Cooling Experiment" ICE [103]. In 1984 Simon van der Meer shared the Nobel Prize [104] in physics with Carlo Rubbia for his contribution to the observation of the intermediate vector boson. Microwave stochastic cooling was considered a key ingredient for reaching sufficient phase space density of the precious and rare antiprotons to produce a small number of W- and Z-Bosons in the CERN Super Proton-Antiproton Synchrotron (SppS) experiment in 1982. At the core of stochastic cooling is the observation, that the phase-space density can be increased by a system that acts to reduce the deviation of small sections, called samples, of the beam. By measuring and correcting the statistical fluctuations ('Schottky noise') of the sample averages, the *spreads* in the corresponding beam properties are gradually reduced. Stochastic cooling may thus be viewed as a 'sampling procedure' where samples are continuously taken from the beam and the average of each sample is corrected. The basic principle of (transverse) stochastic cooling is sketched in Fig. 6.36.

A somewhat different picture is based on the behavior of a test particle. At each passage it receives its own 'coherent' kick plus the 'incoherent' random kicks due to all other sample members. The sample length T_s (response-time) is given by the bandwidth W of the system through $T_s \approx 1/2W$ and the number N_s of particles per sample is proportional to T_s. Hence a large bandwidth is important to work with small samples. Present day cooling rings have a revolution time between a fraction of a μs (e.g. CERN AD) up to about 20 μs (Relativistic Heavy Ion Collider (RHIC), Fermilab bunched beam cooling systems). The sample length T_s amounts usually to less than 1 ns which corresponds to a cooling system bandwidth of 500 MHz in this case assuming the generalized Nyquist criterion for band-limited signals under

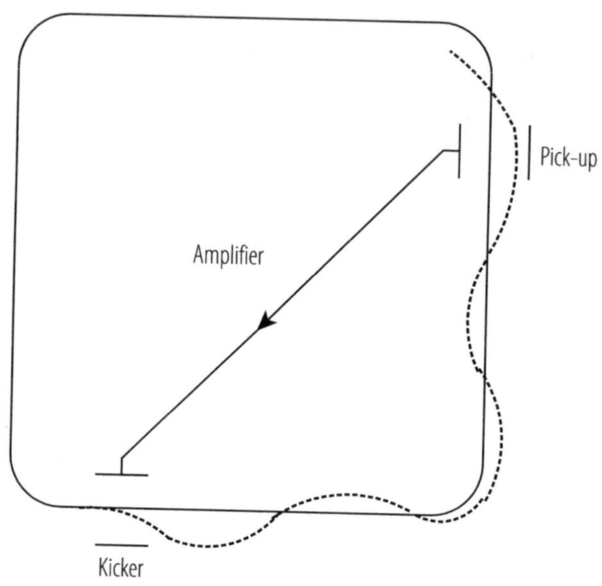

Fig. 6.36 The basic set up for (horizontal) stochastic cooling

ideal assumptions. Thus each sample contains only a small fraction of the total beam population circulating in the machine. Another important ingredient is 'mixing', i.e. the renewal of the sample population due to the spread of the particle revolution frequencies.

Based on the 'sampling' and/or the 'test particle picture' one derives in a few steps [105] a simplified relation for the cooling rate $1/\tau$ of the transverse emittance ε with $(1/\tau = (1/\varepsilon)d\varepsilon/dt)$ or for longitudinal phase space the momentum deviation $(1/\tau = (1/\Delta p)d\Delta p/dt)$:

$$\frac{1}{\tau} = \frac{W}{N}\Big[\underbrace{2g\left(1-\tilde{M}^{-2}\right)}_{\text{<coherent effect>}} - \underbrace{g^2\left(M+U/Z^2\right)}_{\text{<incoherent effect>}}\Big]. \tag{6.74}$$

The parameters appearing in Eq. (6.74) have the following significance:

N	number of particles in the coasting beam	
W	cooling system bandwidth	
g	gain parameter (fraction of sample error corrected per turn)	$(g < 1)$
M	desired mixing factor (mixing on the way kicker–pick-up = good mixing)	$(M > 1)$
$\tilde{M}$	undesired mixing factor (slippage on the way pick-up–kicker = bad mixing)	$(\tilde{M} > 1)$
U	noise to signal power ratio (for single charged particles)	$(U > 0)$
Z	charge number of beam particles ($\leq$ atomic number of the ion!)	$(Z \geq 1)$

There is an optimum value of g for which Eq. (6.74) has a maximum. As to the other parameters, N and Z are properties of the beam, W is a property of the cooling system and M, $\tilde{M}$ and U depend on the interplay of cooling system-, beam- and storage ring characteristics. The term in the bracket can at best be 1 but is more like 1/10 to 1/100 in real systems, depending on how well the mixing and noise problems are solved. The ideal cooling rate W/N can be interpreted as the maximum rate at which information on single particles can be acquired. Note that the gain parameter g (fractional sample error correction) should not be confounded with the electronic gain of cooling system which is typically 120 db or 12 orders of magnitude in power.

Lattice parameters are especially important for the achievement of 'good' values of M, $\tilde{M}$ and U, maximising the bracket in Eq. (6.74). In addition to the struggle for large bandwidth, the advance in stochastic cooling is intimately linked to progress in dealing with the noise and mixing factors. In summary it can be said that present-day systems are working with a bandwidth of around 1 GHz for an individual cooling system with the possibility of extensions up to nearly 10 GHz by using several cooling bands in the same ring. Limitations on W are discussed in [106].

Turning to the mixing dilemma discussed at length in [107], we note that stochastic cooling only works if after each correction the samples (at least partly) re-randomise (desired mixing), and at the same time a particle on its way from pick-up to kicker does not slip too much with respect to its own signal (undesired mixing). The mixing rates $1/M$ and $1/\tilde{M}$ are related to the fraction of the sample length by which a particle with the typical momentum deviation slips with respect to the nominal particle. Here M refers to the way from kicker to pick-up ('K to P'), and $\tilde{M}$ to the way pick-up to kicker ('P to K'). Both depend on the flight-time dispersion which in turn is given by the local 'off-momentum factors',

$$\eta_{\mathrm{kp}} = \left(\frac{dT}{T} / \frac{dp}{p} \right)_{\mathrm{kp}}, \tag{6.75}$$

and the similar quantity η_{pk} respectively. For a regular lattice the beam paths 'K to P' and 'P to K' consist of a number of identical cells and one has

$$\eta_{\mathrm{kp}} \approx \eta_{\mathrm{pk}} \approx \eta = \left| \gamma_{\mathrm{tr}}^{-2} - \gamma^{-2} \right|, \tag{6.76}$$

i.e. the local η-factors are close to the off-momentum factor for the whole ring. In this situation the ratio $\tilde{M}/M$ is simply given by the corresponding path lengths (T_{pk} and T_{kp}). Then, e.g. in the case of the CERN AD (antiproton decelerator) where the cooling loop cuts diagonally across the ring, one has $\tilde{M} \approx M$ instead of the desired $\tilde{M} \gg 1$, $M = 1$. The usual compromise is to accept imperfect mixing, letting both $\tilde{M}$ and M be in the range of 3–5, say. The price to pay is a slower cooling rate, for example $1/\tau \leq 0.28W/N$ in the case of $\tilde{M} = M$ instead of $1/\tau \leq W/N$ for perfect mixing.

'Optimum mixing lattices' (also referred to as 'split ring designs') have been proposed for the 10 GeV 'SuperLEAR' ring [108] (which was, however, never built). The idea is to make the path P to K isochronous ($\eta_{\text{pk}} = 0$) and the path K to P strongly flight-time dispersive ($\eta_{\text{kp}} \gg 0$). These lattice properties have to be reconciled with the many other requirements of the storage ring. The next generation of stochastic cooling rings will use such split rings lattices. They were discussed for RIKEN in Japan [109] and are under construction for GSI and FAIR in Germany [110, 111]. It should be mentioned that the condition $\eta_{\text{pk}} = 0$, $\eta_{\text{kp}} \gg 0$ can increase the cooling rate for transverse and for longitudinal 'Palmer-Hereward' cooling where the transverse displacement concurrent with the betatron amplitude and the momentum error of the particles is used. For momentum cooling by the filter ('Thorndahl') method, the split ring design brings less improvement since here the time of flight over a full revolution is used as a measure of momentum. A storage ring with $\eta_{\text{pk}} = 0$ and $\eta \approx 1$–2% is under construction for GSI and FAIR in Germany, meeting best conditions for both transverse cooling and filter momentum cooling of antiprotons [112].

Regarding the situation at GSI it should be mentioned that a first successful experiment was performed at the ESR to measure the nuclear radius of the radioactive nucleus 56Ni. To this purpose stochastic precooling and subsequent electron cooling were used in order to accumulate enough intensity for a sufficient S/N in a scattering experiment with an internal hydrogen target [113].

This is not the end of the mixing dilemma: during momentum cooling, as $\Delta p/p$ decreases, the M-factors increase (c.f. Eq. 6.75) and the mixing situation tends to degrade. One can in principle stay close to the optimum by changing η ('dynamic transition tuning') as cooling proceeds. Similar considerations hold for machines with variable working energy where, through a change of η, good mixing can be maintained. Again these improvements might be incorporated in the next generation of cooling rings (e.g. at FAIR [112]).

As for the noise, from Eq. (6.74) it is clear that a balanced design aims at $U/Z^2 \ll M$. The noise to signal (power-)ratio depends on the technology of the pre-amplifier and other 'low level components' on the one hand and on the sensitivity of the pick-up device on the other hand. There has been great progress in the design of the pick-up and kicker structures and the other components of the cooling loop. These components developed in different labs (e.g. BNL [114], CERN [107], Fermilab [115], Forschungszentrum Jülich (FZJ) [116], GSI [112, 117]) are in fact formidable 'high-fidelity (HiFi) systems' with an unprecedented combination of high sensitivity, low noise, great bandwidth, large amplification, very linear phase response, and excellent compatibility with the ultra-high vacuum of the storage ring.

A more detailed discussion of stochastic cooling hardware progress over the last 30 years can be found in [118]. Regarding pick-up and kicker structures we have seen printed versions arriving in the late 1980s of the classical $\lambda/4$ strip-line couplers which are referred to as printed loop or printed slotline couplers which are normally used as “phased arrays” [119, 120]. As for travelling wave structures starting from the TEM type slotted line version of Faltin [121] McGinnis developed a related device [122] not based on a TEM line, but essentially a waveguide directional coupler with slots masks for the coupling. Those waveguide type slot array couplers have the advantage (in contrast to the Faltin version) that they can operate efficiently also for highly relativistic beams and they exhibit a very high longitudinal and transverse sensitivity over a bandwidth of several 100 MHz in the GHz region. As a particular development the kicker structure for the BNL RHIC bunched beam stochastic cooling system [123–125] is worth mentioning. It consists of an array of cavities which are cut in length and can be opened by a mechanical plunging mechanism in order to let the injected beam pass without aperture limitations.

Another travelling wave structure is the perforated structure which was originally proposed in 2011 [126] and later developed for HIRFL-CSRe stochastic cooling. A large number of small slots in the electrode provides distributed inductive loading, slowing down the phase velocity of the travelling wave structure for the low beta beams. This device is very broadband and operates from low frequencies onwards as a forward coupler. Even for 2.76 m long electrodes used in HIFRL-CSRe, it can be used from a few MHz to 1.2 GHz [127].

Another very promising recent development for pick-ups and kickers are “slot ring” structures [128]. These structures were originally developed for the High Energy Storage Ring (HESR) of the FAIR project at GSI, Germany and successfully tested at the Nuclotron (JINR, Russia) for longitudinal cooling and at COSY (FZJ) for longitudinal and transverse cooling. Slot ring couplers have a fixed aperture and can be used for all three cooling planes simultaneously [128].

In CERN’s anti-proton decelerator AD stochastic cooling is employed at 3.57 GeV/c and 2 GeV/c in both transverse planes and for the longitudinal plane (filter cooling) [129]. The current system uses a set of two kickers and pick-ups, each combining one transverse plane and the longitudinal cooling, with a total of 4.8 kW installed power. It is undergoing a consolidation and upgrade [130] which is including a notch filter with optical delay lines. Cooling times of 15–20 s reduce transverse emittances to 3–4 π mm rad and Dp/p to $\pm 0.3 \times 10^{-3}$ at 3.57 GeV/c and to $\pm 0.08 \times 10^{-3}$ at 2 GeV/c at intensities of 5×10^7 antiprotons. The system uses a bandwidth of one octave between 850 MHz and 1.7 GHz. This is the actual status in early 2019.

In parallel, CLASS A solid state amplifiers [131] (kicker driver) gradually took over from TWT (travelling wave tube) units, although TWTs are still in operation for stochastic cooling e.g. at Fermilab where they work reliably. Notch filters, required for Thorndahl type longitudinal cooling (filter cooling) are implemented since about 1990 with good success in optical fibre technology [125, 132].

Optical signal transmission across the ring (Fermilab de-buncher) has been realized with a laser beam in an evacuated metal pipe (no signal fluctuation from temperature effects of air and humidity on the laser beam). The driving force to select this method of signal transmission was the very tight requirement in terms of transmission delay and delay stability. Anything slower than speed of light would not have permitted timely arrival of the correction signal at the kicker. In 2017 very fast (around 99% speed of light) hollow optical fibres were applied successfully for analog and wideband signal transmission across the ring at COSY (FZ-Jülich, Germany) [133].

Front end amplifiers showed slow but steady progress and these days we can easily get an uncooled 1–2 GHz or 2–4 GHz device with a noise temperature of 30 K.

Examples of remarkable recent progress in the field of microwave stochastic cooling are

- bunched beam stochastic cooling at RHIC and Fermilab [134–137],
- the impressive improvements of the performance and the interplay of all stochastic cooling and stacking systems at Fermilab together with elaborate beam handling methods such as "slip stacking" [135, 136].

It should be noted that there have been unsuccessful attempts to get bunched beam stochastic cooling operational in large machines despite the fact that one of the first evidence on stochastic cooling at all, in ICE [137] already worked with a bunched beam. However the bunch length was very large. Attempts which failed were in the frame of the SPS p-pbar program at CERN [124] and later (around 1990) also in the Tevatron [138, 139]. Bunched beam cooling is of course hampered by the higher particle density in the bunch. In fact in Eq. (6.74) the number N for the coasting beam has to be replaced by $N_b/B_f = N_b \cdot 1.4 \cdot 2\pi R/l_b$ (with R radius and l_b length of bunch) for a rough estimate [137, 140]. In addition to those expected effects the direct (coherent) bunch signal (proportional to N at low frequencies) tends to mask the very weak Schottky signals required for cooling [137, 140]. This is one of the reasons to place the cooling bands towards high frequencies. In addition a subtle but important difficulty is related to the presence of unexpected and rather strong coherent signals in the bunched beam spectrum which lead to saturation of the front end amplifiers via intermodulation [139, 140]. A theoretical treatment of these persisting "turbulence islands" in the bunch was given by Blaskiewiecz [141]. Just in the recent years this problem became mastered at BNL (gold ions) [135] and also in the Fermilab recycler [140]. However, bunched beam stochastic cooling has always been working reasonably well in small machine like the CERN AC [107], LEAR [142], Fermilab de-buncher and accumulator [143] and others since the relative intensity of those coherent signals was less violent compared to large machines. However for the small machines there was little interest in bunched beam stochastic cooling.

New applications of stochastic cooling may include:

- fast cooling and stacking of low intensity radioactive ion beams with cooling times of 100 ms or less as discussed for RIKEN [109] and under construction at GSI [110],
- fast optical stochastic cooling [144–146] (e.g. of intense muon beams but also for bunched beam cooling in large rings) for which a bandwidth of 10^{12}–10^{13} Hz and a new pick-up, kicker and amplifier technology, and new lattice designs have been contemplated.

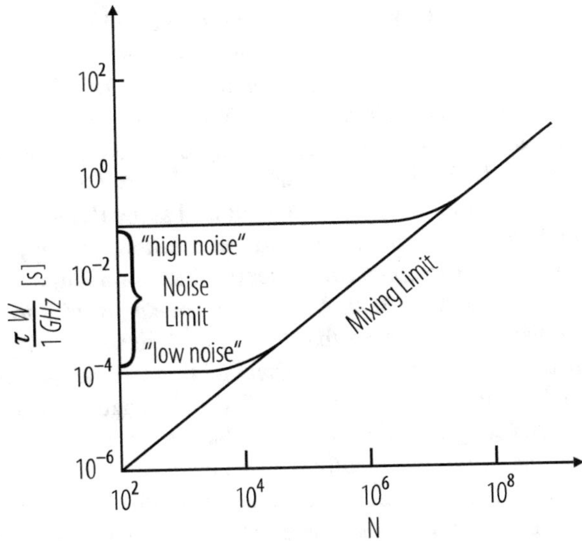

Fig. 6.37 Sketch of cooling time vs. intensity (for the 'mixing limit' $\tau = 10\ N/W$ is taken in the figure)

Let us have a quick look at these developments. The challenge of fast low-intensity cooling can be discussed with reference to Fig. 6.37 [108], which illustrates the optimum cooling time vs. intensity N. For large N the cooling time increases linearly with N with the slope $1/W\left[M/\left(1-\tilde{M}^{-2}\right)^2\right]$. This is the mixing and bandwidth limit. For small N, cooling time levels off to a constant 'noise limited' value reached for $U/Z^2 \gg M$ (note that $U \propto 1/N$).

The art is to shift the levelling off to small N by improving the signal to noise ratio. Theoretically, short cooling times are then possible (e.g. 10 ms for $N = 10^5$ Sn^{50+} ions and a few 100 MHz bandwidth as discussed for RIKEN). However, other difficulties like the broadband power needed for such a rapid emittance decrease, and the residual RF-structure after debunching may pose new problems for fast cooling and stacking.

Optical stochastic cooling (OSC) proposed by Mikhailchenko, Zholents and Zolotorev [144, 145] in 1993 is an extension of certain basic concepts of microwave stochastic cooling into the optical frequency range using different pickup and kicker mechanism and structures. It is a potentially very promising technique but has not been tested in practice so far. Challenges may be amongst other items the stability and linearity of the optical signal transmission chain as well as of the circulating hadron beam. Maybe we shall soon see important steps towards this technology at BNL in a forthcoming "coherent electron cooling experiment".

At BNL stochastic cooling has been implemented at top energy in the RHIC [147]. The bunch cores have full length 5 ns and are spaced by 100 ns. The root mean square Schottky voltage is typically 10% of the coherent voltage generated by the average bunch shape and multi-kilovolt kicker voltages are required for optimal longitudinal cooling. Several novel technologies were required to meet the challenges. The kicker voltage was obtained by periodically extending the pickup signal, passing it through narrow band filters spaced by 200 MHz (1/5 ns) and driving individual cavities. Taming coherent lines while meeting timing requirements was a serious challenge [148]. When all was said and done the cooling system increased the integrated luminosity of uranium-uranium collisions by a factor of five and typically doubled the gold-gold luminosity.

At Fermilab, OSC is being pursued at the IOTA facility [149]. In OSC a particle emits electromagnetic radiation in the first (pickup) wiggler. Then, the radiation amplified in an optical amplifier (OA) makes a longitudinal kick to the same particle in the second (kicker). A magnetic chicane is used to make space for the OA and to delay a particle so that to compensate for a delay of its radiation in the OA resulting in simultaneous arrival of the particle and its amplified radiation to the kicker wiggler. The chosen optical wavelength is 800 nm, resulting in bandwidths approaching 10^{14} Hz. In the proposed test, the use of 100-MeV ($\gamma = 200$) electrons instead of protons greatly reduces the cost of the experiment but does not limit its generality and applicability to hadron colliders. Conceptual design of the system is complete, with engineering design of the wiggler and optical hardware underway.

Already in late 1970s the need for "stochastic stacking" has been realized [150]. In the "old" CERN AA (antiproton accumulator) [151] early stacking methods were tested and applied in routine operation. In the CERN AAC (antiproton accumulator complex) [152] the antiprotons (pbar or p) coming from the AC (collector ring) were transferred to the inner ring (AA = accumulator). There dedicated stack tail and stack core systems took over the antiprotons after they have passed a pre-cooling system in the AA and were transferred to another orbit by means of RF manipulations. At Fermilab [138] stacking is done in the accumulator ring and later also in the recycler. For the future stacking with stochastic cooling is planned in the frame of the FAIR project [110]. Stochastic stacking of rare radioactive ions has been considered during the planning phase of RIKEN [110] upgrades between 1900 and about 2000 and for FAIR [110].

At Fermilab huge progress has been made since the year 2000 [153], this includes stacking with stochastic cooling was done in three separate machines. The debuncher [154], which accepted ~1.5×10^8 antiprotons every 2.1 s, used the McGinnis waveguide directional couplers in eight bands over the frequency range 4–8 GHz for a factor of 10 reduction in longitudinal and transverse size. A key piece was the implementation of ramping the amplifier gain down during the cycle, to counter act noise to signal for the momentum bands in the notch filters. A 6 dB decrease in gain resulted in a 12% decrease in the 95% momentum width after 2 s of cooling. The Accumulator [154] accepted the same ~1.5×10^8 antiprotons and used the Palmer method to build a 'stack'. Peak performance reached 2.6×10^{11} antiprotons in an hour, with regular transfers to the Recycler to mitigate the known decrease in performance with larger stacks. The Recycler, using a combination of stochastic and electron cooling [155], reached intensities of greater than 4×10^{12} regularly, with peak intensity of 6.1×10^{12} and delivering over 4×10^{13} per week to the collider program.

6.10.2.3 Electron Cooling

The concept of cooling a "hot" beam of ions by mixing it over a short distance in a circular machine with a cold electron beam had been developed by Budker [156] in 1966. It was first tested in 1974 with 68 MeV protons at the NAP-M storage ring at in Novosibirsk. The notions of 'beam temperature' and 'beam cooling' were introduced and become lucid in the context of electron cooling, which is readily viewed as temperature relaxation in the mixture of a hot ion beam with a co-moving cold electron 'fluid'. The equilibrium emittances, obtainable when other 'heating mechanisms' are negligible, can easily be estimated from this analogy, assuming equalisation of the temperatures ($(M\Delta v^2)/_{\text{ion}} \to (m\Delta v^2)/_{\text{electron}}$). For a simple estimate of the cooling time, another resemblance, namely the analogy with slowing down of swift particles in matter, can be helpful. A nice presentation of this subject is given in Jackson's book [157]: the energy loss in matter is due to the interaction with the shell electrons and in first approximation these electrons are regarded as free rather than bound. Results for this case can be directly applied to the 'stopping of the heavy particles in the co-moving electron plasma'. The calculations are performed assuming 'binary collisions' involving only one ion and one electron at a time.

Using this approximation the cooling time can be written as

$$\frac{1}{\tau} \approx \frac{1}{k}\frac{q^2}{A}\eta_c L_C r_e r_p \frac{j}{e}\frac{1}{\beta^4\gamma^5\theta^3}, \tag{6.77}$$

where

$k = 0.6$: for a Gaussian distribution (not realistic),
$k = 0.16$: for a flattened distribution,
q: ion charge number,
A: ion mass number,
η_c: length of cooling section/circumference,
$L_C \approx 10$: Coulomb logarithm (log of max/min impact parameter),
$r_e \approx 2.8 \times 10^{-13}$ cm: classical electron radius,
$r_p \approx 1.5 \times 10^{-16}$ cm: classical proton radius,
j (A/cm^2): electron beam current density,
$e \approx 1.6 \times 10^{-19}$ C: elementary charge
$\theta = \left(\theta_e^2 + \theta_i^2\right)^{1/2} = \left(\frac{T_e}{m_e c^2} + \frac{T_i}{m_i c^2}\right)$: r.m.s. angle between electron and ion beams,
β, γ: relativistic factors.

The cooling rate $(1/\tau)$ thus obtained exhibits the dependence on the main beam and storage ring parameters [158]. Notable is the dependence on both the electron and the ion (both longitudinal and transverse) velocity spreads: $\tau \propto \theta^3 \propto \left(\left|\Delta \boldsymbol{v}_{e_{\text{rms}}}\right|^3 + \left|\Delta \boldsymbol{v}_{i_{\text{rms}}}\right|^3\right)$. This indicates an 'ion spread dominated regime', where cooling gets faster as the ions cool down until it saturates for $\left|\Delta \boldsymbol{v}_{i_{\text{rms}}}\right| < \left|\Delta \boldsymbol{v}_{e_{\text{rms}}}\right|$ ('electron dominated regime'). Remarkable also is the strong energy dependence predicted in this model: $\tau \propto \beta^4\gamma^5$, with all other parameters (including the electron current density j) kept constant [159].

Neglected in the simple theory are the 'flattened distribution', the 'magnetisation' and the 'electron space-charge' effects, all three (also) discovered and explained at Novosibirsk [159, 160]. In essence the flattened distribution effect takes into account that (due to the acceleration) the electron velocity spread is not isotropic but contracted (by $[E_{\text{cathode}}/E_{\text{final}}]^{1/2}$) in the longitudinal direction. The magnetisation effect is due to the spiraling (Larmor-) motion of the electrons in the magnetic field of the solenoid that is used to guide the electron beam. Then for electron-ion encounters with long 'collision times' (impact parameter $\gg$ Larmor radius), the transverse electron velocity spread averages to zero. Finally the electron space-charge induces a potential that leads to a parabolic velocity profile $v(r)$ over the beam whereas the ions exhibit a linear dependence $v(x)$ and $v(y)$ given by the storage ring lattice. Hence the difficulty arises to match the ion and electron velocities. Flattening and magnetisation can have a beneficial outcome, whereas space-charge has a hampering influence on the cooling process. All three effects complicate the theory, spoil the hope for simple analytical formulae and obscure the comparison between measurements at different machines, and even different situations at the same cooler. As an example the cooling assembly used in the low energy antiproton ring (LEAR) is sketched in Fig. 6.38.

The electrons are produced in a gun and directed into the cooling region where they overlap the ion beam over a length 1 m. At the end of the cooling section the electrons are steered away from the ions into a collector where their energy is recuperated. On their whole way from the cathode of the gun to the collector the electrons are usually immersed in a longitudinal magnetic guiding field. This field is constant over the full length or stronger in the gun region. In the latter case the transverse electron temperature in the overlap region decreases (due to "magnetic expansion") at the expense of the longitudinal temperature. This can reduce the cooling time in situations where the electron temperature dominates.

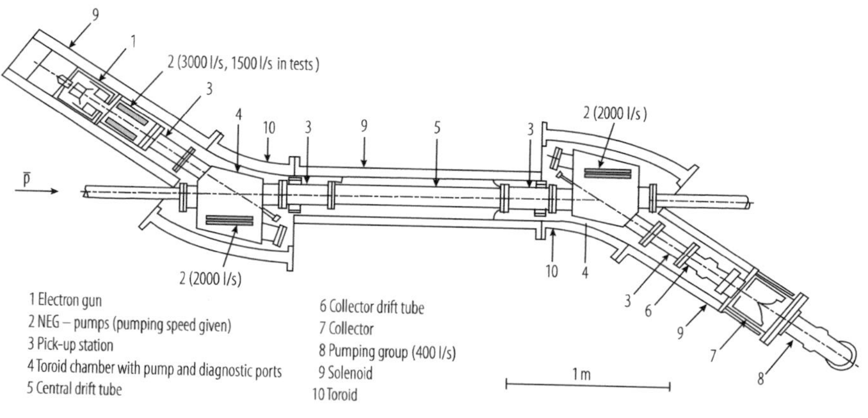

Fig. 6.38 An electron cooling assembly (LEAR electron cooler) from [158]

An important technical problem is electron beam power consumption. To reduce direct losses of the beam power the recuperation (recovering) method is used. It assumes biasing of the collector to negative potential slightly above the cathode potential. Then the power consumption is defined mainly by product of the beam current by the difference of the collector and gun potentials.

In the two toroidal sections, adjacent to the overlap region, the solenoid to create the longitudinal field is curved to guided the electron beam parallel to the ions at the entrance and away from them at the exit. Also in the toroidal regions the solenoid has a larger diameter to permit the penetration of the ion beam.

Many papers deal with the 'exact and general theory' [161] and computer programs like BETACOOL [162] try to include all the subtle effects. Numerous also are the experimental results from 11 (or so) present and past cooling rings. It is not easy to compare the data from different experiments because the cooling in each plane depends in a complicated way on the emittances in all three directions both of the ion and the electron beam. Moreover different quantities are used to measure/define 'cooling strength' (examples: cooling of a large injected beam, response of a cold beam to a 'kick' or to a transverse or an energy displacement, equilibrium with heating by noise).

In the context of the accumulation of lead ions for the future Large Hadron Collider (LHC) [163], a program of experiments [164] was performed at the LEAR ring to determine optimum lattice functions [165]. Results indicate rather small optimum betatron functions (3–5 m instead of the expected 10 m) and large dispersion ($D = 2$–3 m instead of the expected 0–1 m). The dependence on dispersion is not fully reproduced by simple analytical formulae. There are other old questions: e.g. the (dis)advantage of magnetic expansion, the dependence of the cooling time on the charge of the ion, the (dis)advantage of neutralising the electron beam, the enigma of the stability of the cooled beam [166], the puzzle of the anomalously fast recombination of certain ions with cooling electrons [167], the (dis)advantage of a hollow electron beam [168].

Considerations so far concern electron cooling at 'low energies' ($T_e = 2$–300 keV) where cooling rings have flourished since the 1980s. More recently medium energy cooling ($T_e = 1$–10 MeV) has re-gained a lot of interest [167–169]. Clearly the higher energy requires new technology and extrapolation to a new range of parameters. At Fermilab high energy e-cooling (with up to 5 MeV electrons using electrostatic acceleration for cooling of 8 GeV bunched antiprotons in the recycler) has been successfully developed and implemented. The generation and recirculation of the 4.3 MeV and 0.5 Ampere electron beam and its adaptation to the antiproton beam over a cooling length of 20 m are remarkable achievements. Finally the idea of 'very high energy electron cooling' ($T_e \geq 50$ MeV) has been revived as this might improve the luminosity of RHIC [168, 170]. At this energy the electron beam could circulate in a small 'low-emittance storage ring' with strong radiation damping. An attractive alternative is a scheme [170], in which the low-emittance beam after acceleration is re-decelerated after the passage through the cooling section to recuperate its energy.

In summary: 45 years after its invention, the field of electron cooling continues to expand with exciting old and new questions to be answered. Bunched beam cooling is no longer a magic barrier and even a merger between electron cooling and stochastic cooling i.e. the "coherent electron cooling" [171] appears at the horizon. In the concept of coherent electron information of the particle distribution of the hadron beam to be cooled is sampled by the electron beam, amplified and further downstream fed back onto the hadron beam.

6.10.2.4 Laser Cooling

Due to the pioneering work of the Heidelberg (TSR) [172] and Aarhus (ASTRID) [173] groups in the 1990s, laser cooling in storage rings has evolved into a very powerful technique. Longitudinal cooling times as short as a few milliseconds and momentum spreads as small as 10^{-6} are reported. These bright perspectives are somewhat mitigated by two specific attributes [174]: laser cooling takes place (mainly) in the longitudinal plane and it works (only) for special ions that have a closed transition between a stable (or meta-stable) lower state and a short-lived higher state. The transition is excited by laser light, and the return to the lower state occurs through spontaneous re-emission (Fig. 6.39). 'Unclosed' transitions, where the de-excitation to more than one level is possible, are not suited because ions decaying to the 'wrong' states are lost for further cooling cycles. This limits the number of ion candidates (although extended schemes with additional lasers to 'pump back' from the unwanted states could enlarge the number of ion species susceptible to cooling). Up to now, a few singly charged ions (like Li^{1+}, Be^{1+} or Mg^{1+}) have been used with 'normal' transitions accessible to laser frequencies. Transitions between fine structure, or even hyperfine levels of highly-charged heavy ions have also been considered, but in that case the cooling force is less pronounced and not so much superior to the electron cooling force which increases with charge (like $Q^{1.5}$ or even Q^2).

The laser irradiates the circulating ions co-linearly over the length of a straight section of the storage ring [174]. The absorption is very sharply resonant at the transition frequency. Then the Doppler shift ($\omega = (1 \pm v/c)\gamma\omega_{\text{laser}}$) seen by

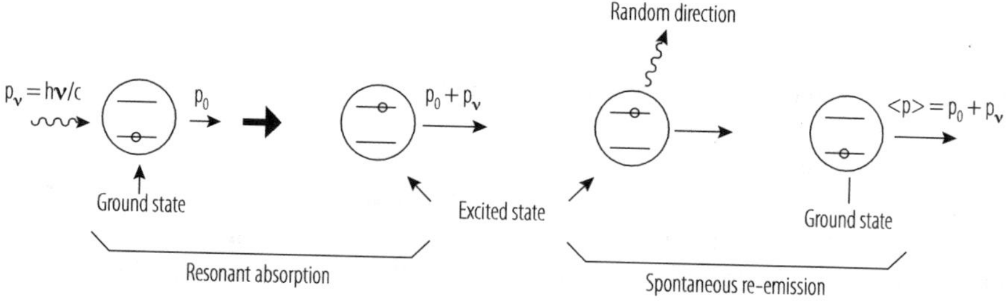

Fig. 6.39 Sketch of Laser–ion interaction

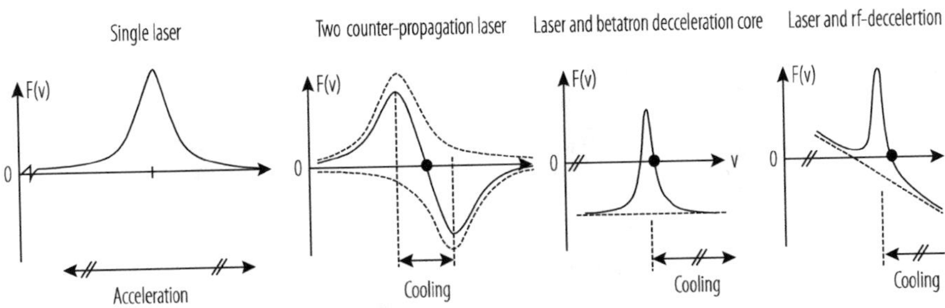

Fig. 6.40 Force $F(v)$ due to a single laser and different schemes for cooling to a fixed velocity

the ion makes the interaction strongly dependent on its velocity. This leads to a sharp resonance of the absorption as a function of the velocity (Fig. 6.40). The corresponding recoil (friction) force accelerates/decelerates the ions with a maximum rate at the resonant momentum. To obtain cooling to a fixed momentum, a second force f(v) is necessary. It can be provided by a second (counter-propagating) laser or by a betatron core or by an RF-cavity, which decelerate the ions 'towards the resonance of the first laser' (Fig. 6.40). The interaction with the laser photons (and hence the cooling) takes place in the direction of the laser beam (longitudinal plane of the ions). De-excitation proceeds by re-emission of photons in all directions and this leads to heating of the ions in all three planes.

Through transverse-to-longitudinal coupling, part of the cooling can be transferred to the horizontal and vertical planes. Intra-beam scattering [175], dispersion [176] and special coupling cavities [177] have been considered for this purpose. Transfer by scattering and by dispersion has been demonstrated at the cooling rings, although the transverse cooling thus obtained was weak, a fact explainable by the weakness of the coupling.

The main motivation for laser cooling has been the goal of achieving ultra-cold *crystalline beams* [178] where the ions are held in place because the Coulomb repulsion overrides the energy of their thermal motion. A second application, cooling of low-charge states of heavy ions, was proposed [179] in order to prepare high-density drive beams for inertial confinement fusion. Several years ago a study [180] on the use of laser cooling of ions for the LHC was published. All these applications for the moment meet with difficulties: crystallisation, in full three-dimensional beauty, is hampered by the lattice properties of (present) storage rings and by the relative weakness of transverse cooling. Cooling for fusion is not fast enough [181] to 'compress' the high-intensity large-momentum-spread beam during the few milliseconds lifetime given by intra-beam charge exchange between the ions. And, finally laser cooling of highly charged ions for colliders meets with the competition of electron cooling and also with the restrictions on the choice of suitable ion species and states [180]. The investigations on laser cooling to obtain crystalline beams continue [182] and a special storage ring (S-LSR) with lattice properties apt to reach this goal [183, 184] has been built at Kyoto university.

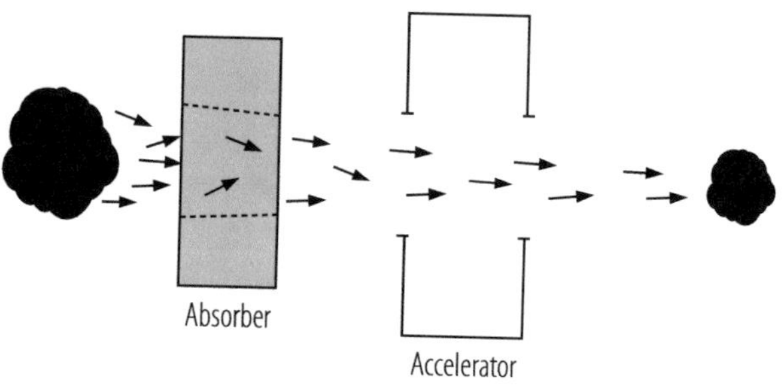

Fig. 6.41 Basic setup for (transverse) ionisation cooling (adapted from [186])

In conclusion: laser cooling in storage rings has led to very interesting and important results concerning the physics of cooling and cooling rings and, also atomic and laser physics. However, 'accelerator applications' like for electron or stochastic cooling are not realistic for the near future. The goal to obtain crystalline beams in special storage rings is under intense investigation.

6.10.2.5 Ionisation Cooling

Excellent reviews of ionisation cooling are given in papers by Skrinsky [185] and Neuffer [186]. The basic setup (Fig. 6.41) consists of a block of material (absorber) in which the particles lose energy, followed by an accelerating gap (RF-cavity) where the energy loss is restored. Losses in the absorber reduce both the longitudinal and the transverse momentum of the particle. The RF-cavity (ideally) only restores the longitudinal component and the net result is transverse cooling (Fig. 6.41). There is an obvious resemblance to radiation damping (Fig. 6.35), in which energy loss by synchrotron radiation followed by RF-acceleration results in cooling. Longitudinal ionisation cooling is also possible, especially in the range where the loss increases with energy (i.e. above the energy where the minimum of dE/ds occurs). At the expense of horizontal cooling, the longitudinal effect can be enhanced by using a wedge-shaped absorber in a region where the orbits exhibit dispersion with energy.

The statistical fluctuations ('straggling') of the loss and the angular (multiple) scattering introduce heating of the longitudinal and transverse emittances. The ratio of ionisation loss due to angular scattering favours light absorber material. Equilibrium emittances depend strongly on the lattice functions at the position of the absorber and the cavity. As in the case of radiation damping, the sum of the cooling rates (also in the case of a wedge absorber) is invariant with a value $J_x + J_y + J_E \approx 2 + J_E \approx 2$ for the 'damping partition numbers', instead of $J_x + J_y + J_E = 4$ for radiation damping. The quantity J_E depends on the slope of the dE/ds vs. E curve and is about constant and roughly equal to 0.12 for light materials above the minimum of dE/ds, but is strongly negative below. In terms of the partition numbers, the three emittance damping rates can be expressed by the energy loss ΔE_μ of the muons in the absorber and the length Δs of the basic cell (Fig. 6.41) as:

$$\frac{1}{\varepsilon_i}\frac{d\varepsilon_i}{ds} = J_i \frac{1}{E_\mu}\frac{\Delta E_\mu}{\Delta s}. \tag{6.78}$$

A large number of cells or traversals through a cell is necessary to obtain appreciable emittance reduction.

Almost by a miracle, the muon mass falls into a narrow 'window' where ionisation cooling within the short life of the particle looks possible (although not easy). For electrons as well as for protons and heavier particles, the method is not practical, because the effect of bremsstrahlung (for e's) and non-elastic processes in the absorber (for p's), leads to unacceptable loss.

With the revival of interest for muon colliders and, related to that, neutrino factories [187], large collaborations (including more than 15 institutes, [188]) is undertaking a demonstration experiment. The ISIS accelerator at the Rutherford lab. is chosen for this task. Neutrino factory and muon collider proposals have to rely critically on muon cooling: typically 50 m to several 100 m long channels with solenoidal focussing (superconducting solenoids) are foreseen to reduce the phase-space of the muons emerging from pion decay. Liquid hydrogen absorbers, each 0.5–1 m in length, alternate with high-field accelerating cavities.

The variant selected by MICE is a 'single particle experiment' where one muon at a time is traced. Fast spectrometers, capable of resolving 1 muon per 25 ns, record/compare the three position coordinates and the three velocity components of the muon at the entrance and the exit of a short cooling section. Typically such a test-section should lead to 10% emittance reduction. The emittance pattern is 'painted' by a scatterer or a steering magnets changing the entrance conditions of the particle at random (scatterer) or in a programmed manner. A large number of muons are necessary to establish the six-dimensional phase-space reduction with sufficient statistics.

Apart from the spectrometers, other challenges can be identified: long term mechanical stability, muon decay and birth, contamination with other particles and non-linarites in focussing which deform the emittance pattern. In the coming years we will see a large effort on muon cooling scenarios and tests.

6.10.2.6 Cooling of Particles in Traps

In many experiments utilizing ion traps, the ions must first be cooled in order to perform high precision measurements. Cooling refers here to the reduction of kinetic energy of confined particles. A detailed review of cooling traps is given in [189] and the implementation of several cooling methods into a big project is described in [190].

With adequate modifications, most of methods discussed above for storage rings stochastic- [191], electron- [192], or laser cooling [193] can also be applied to traps.

Some (like stochastic cooling) are more difficult others (e.g. laser cooling) are much more powerful in the traps.

There are however several techniques which are especially adapted to or even working only in the environment of particles confined in a trap which is frequently cryogenically cooled. A gross classification is to divide them into lossy and lossless methods in terms of conservation of number of particles. A lossy technique (which in the strict definition of [100] would not be classified as "phase density cooling") is evaporative cooling. In this case, just as in the evaporation of water, the more energetic molecules leave the trap and the temperature of the condensate is thereby strongly reduced. Experiments at the forefront of physics making Bose-Einstein condensates at a temperature of a tiny fraction of a degree have thus become possible [194].

An example of a widely used lossless process is resistive cooling: the trap electrodes are connected to an external circuit to dissipate energy from the ions through induced currents [189, 195] (Fig. 6.42).

In other words the particle's kinetic energy is dampened by I^2R losses in a resistive circuit [195, 196]. Idealistically speaking, the resistor or the losses in a resonant circuit and absorb the particle's energy to create a thermal equilibrium when there is no other heating source involved. Since the resistor has a specific physical temperature, it generates Johnson noise that in turn stochastically drives the trapped particles. Resistive cooling was first applied by H. Dehmelt and collaborators in 1975 [195].

To estimate the cooling time, a simple single particle model is used, where-by it is harmonically bound between two capacitor plates [195]. Due to this model, the energy is dampened with a time constant τ calculated by:

$$\tau = \frac{4mz_0}{q^2R}. \tag{6.79}$$

Here $2z_0$ is the separation of the capacitor plates (the electrodes of the trap) and R stands for the real part of the impedance from the attached external circuit, q is the charge and m the mass of the trapped particles. From Eq. (6.79) [189] one can easily conclude that light, highly charged particles are efficiently cooled. The cooling rate can be further improved by developing a high resistance in the external circuit.

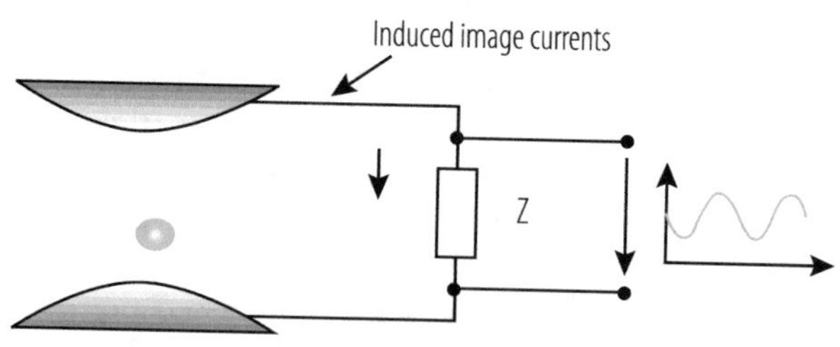

Fig. 6.42 Principle of resistive cooling of a trapped ion

In general, the external circuit which is often in vacuum and at cryo-temperature includes a low-noise amplifier to couple the induced current signal to room temperature, and thus enable plasma diagnostics. The input noise temperature of the amplifier is, depending on its coupling to the resonant circuit, closely related to the minimum achievable temperature of the particles in the trap. Most frequently, the impedance Z (with the real part R) shown in Fig. 6.42 is implemented as inductance L so that the circuit becomes resonant at the oscillation frequency of the ions. This is to tune out (compensate at resonance) the parasitic capacitance of the electrodes. This inductance may be implemented as discrete solenoid coil made of copper or superconducting wire. The quality factor $Q = R/\omega L$ of the tuned circuit has to be large to guarantee efficient resistive cooling A high Q in turn means to incorporate a low loss network. As a caveat it should be mentioned that extremely high Q values (above say 10^5) may be problematic if the bandwidth of the resonance becomes smaller than the width of the particle spectrum. For further reading we refer to [197], where Shockley presents the basic equations for trapped and charged particles in a Penning trap.

There exist a large number of other cooling techniques used in traps, such as collisional cooling, RF- and optical sideband cooling (resolved and unresolved sideband methods) and sympathetic cooling. The list of examples given here is certainly not exhaustive and a detailed description can be found in review articles [189]. As for the term RF cooling (which may be confounded with RF-sideband cooling) [189] it should be pointed out that this refers to a reduction of temperature or vibration amplitude of a microscopic cantilevered bar in vacuum using an RF resonant circuit which is fed externally with a few milli-Watt of RF power [198].

Acknowledgements The authors would like to thank Michael Blaskiewicz, Paul Derwent, Christina Dimopoulou, Wolfgang Hofle, Fritz Nolden, Rolf Stassen and Junxia Wu for their contribution to this chapter and Jens Vigen for reading and correcting the manuscript.

References

1. E. Wilson, B.J. Holzer: *Beam Optics*, this handbook, Chapter 2.
2. B.J. Holzer: *Lattice Design in High Energy Accelerators*, Proc. CERN Accelerator School, CERN-2006-002.
3. LHC Design Report, CERN-2004-003.
4. W. Herr, B. Holzer, B. Muratori: *Concept of Luminosity*, this handbook, Sect. 6.4.
5. K. Steffen: *Periodic dispersion suppressors II*, DESY-HERA 83/02, 1983.
6. R. Brinkmann: *Insertions*, Proc. CERN Accelerator School, CERN-1987-010.
7. Proceedings of CERN Accelerator School: Beam Injection, Extraction and Transfer, 2017, Erice, CERN-2018-008-SP, 2018
8. V. Mertens. Kicker and septa technology, Chapter 5.5.
9. Y. Yamazaki ed., “Accelerator Technical Design Report for J-PARC”, KEK Report 2002-13; JAERI-Tech 2003-044, 2003.
10. J. Wei et al., Evolution of the Spallation Neutron Source Ring Lattice, Proc 20th ICFA Advanced beam dynamics workshop, Chicago IL, 2002,

11. W. Bartmann, B. Goddard and C. Hessler, a doublet-based injection-extraction straight for PS2, Proceedings of IPAC'10, Kyoto, Japan, pp. 3945-3947, 2010.
12. O.D. Kazachkovskii, Particle injection into a ring accelerator, Atomic Energy Vol. 84, Number 5, 345-350, 1998.
13. A. Hilaire, V. Mertens and E. Weiss, Beam transfer to and Injection into LHC, Proc. 6th European Particle Accelerator Conference, Stockholm, Sweden, pp. 2117-2119, 1998.
14. J. Uythoven et al., Fast injection into the PS2, Proceedings of PAC09, Vancouver, BC, Canada, 2009.
15. K. Seiya et al., Slip Stacking, Proc. CARE-HHH-APD Workshop on Finalizing the Roadmap for the Upgrade of the CERN and GSI Accelerator Complex, CERN, Geneva, Switzerland, pp. 66-68, 2007.
16. G.I. Budker et al., Experiments on charge-exchange injection of protons into a storage ring, Atomnaya Énergiya, Vol. 19, No. 6, pp. 507–510, December, 1965.
17. J.W.G. Thomason et al., Injection Upgrades for the ISIS Synchrotron, Proceedings of IPAC'10, Kyoto, Japan, pp 705-707, 2010.
18. J. Beebe-Wang et al., Transverse phase space painting for SNS accumulator ring injection, Proc. 1999 Particle Accelerator Conference, New York, pp. 1743-1745, 1999.
19. V. Danilov et al., Three-step H- charge exchange injection with a narrow-band laser, Phys. Rev. ST Accel. Beams 6, 2003.
20. T.S. Ueng et al., Topping Up Experiments at SRRC, Proceedings of the 5th European Particle Accelerator Conference, 10 to 14 June 1996.
21. S. Myers, A Possible New Injection and Accumulation Scheme for LEP, CERN LEP Note 334, 1981.
22. P. Collier, Synchrotron phase space injection into LEP, Proc. 16th IEEE Particle Accelerator Conference, Dallas, TX, USA, pp. 551-553, 1995.
23. B. Goddard and V. Mertens, Moving the beam into and out of the LHC, Chapter 4.4, in 'The Large Hadron Collider: a marvel of technology', L. Evans (ed), EPFL press, Laussane, p 114-133, 1999.
24. M.Q. Barton, Beam extraction from synchrotrons, Proc. VIIIth Int. Conference on High Energy Accelerators, CERN, Geneva, p. 85-8, 1971.
25. A. Miyamoto et al., Study of slow beam extraction through the third order resonance with transverse phase space manipulation by a mono-frequency RFKO, Proc 2005 Particle Accelerator Conference, Knoxville, Tennessee, pp 1892-1894, 2005.
26. L. Badano and S. Rossi, Characteristics of a betatron core for extraction in a proton-ion medical synchrotron, CERN-PS-97-019-DI, 1997.
27. M. Fraser et al., Improvements to the SPS Slow Extraction for High Intensity Operation, CERN-ACC-NOTE-2019-0010, 2019.
28. S. van der Meer, Stochastic extraction, a low-ripple version of resonant extraction, CERN, CERN/PS/AA 78-6, 1978.
29. M. Fraser et al., SPS Slow Extraction Losses and Activation: Challenges and Possibilities for Improvement, CERN-ACC-2017-123, 10.18429/JACoW-IPAC2017-MOPIK045 2017.
30. C. Bovet et al., The fast shaving ejection for beam transfer from the CPS to the CERN 300 GeV machine, Proc 1973 Particle Accelerator Conference, San Francisco, California, p. 438, 1973.
31. R. Cappi and M. Giovannozzi, Multiturn extraction and injection by means of adiabatic capture in stable islands of phase space, Phys. Rev. ST Accel. Beams 7, 2004.
32. K. Johnsen, CERN Intersecting Storage Rings (ISR), Proc. Nat. Acad. Sci. USA, Vol. 70, No. 2, pp. 619-626, 1973.
33. D. Neuffer, Design of muon storage rings for neutrino oscillation experiments, IEEE Transactions on Nuclear Science, Vol. NS-28, No. 3, 1981.
34. S. van der Meer, Stochastic cooling in the CERN antiproton accumulators, IEEE Transactions on Nuclear Science, Vol. NS-28, No. 3, 1981.
35. G.N. Vialov, Y.C. Oganessian and G.N. Flerov, Method for heavy ion beam extraction from a cyclotron with azimuthal variation of the magnetic field, JINR Preprint 1884, Dubna, 1964.

36. V. Biryukov, Computer Simulation of Crystal Extraction of Protons from a Large-Hadron-Collider Beam, Phys. Rev. Lett. 74, 2471–2474, 1995.
37. W. Herr, *Concept of Luminosity*, Proceedings of CERN Accelerator School, Zeuthen 2003, CERN-2006-002 (2006).
38. C. Moller, K. Danske Vidensk. Selsk. Mat.-Fys. Medd., **23**, 1 (1945).
39. A. Chao and M. Tigner, **Handbook of Accelerator Physics and Engineering**, World Scientific (1998).
40. W. Herr, *Beam-beam interactions*, Proceedings of CERN Accelerator School, Zeuthen 2003, CERN-2006-002 (2006).
41. J.E. Augustin, *Space charge effects in e^+e^-storage rings with beams crossing at an angle*, Orsay, Note interne 35-69 (1969).
42. P. Raimondi, 2nd SuperB Workshop, Frascati, Italy, (2006).
43. Review of Particle Physics, Vol. **15**, Number 1-4, 213 (2000).
44. C. Augier et al., Physics Letters, **B 315**, 503 (1993).
45. C. Augier et al., Physics Letters, **B 316**, 448 (1993).
46. C. Augier et al., Physics Letters, **B 344**, 451 (1993).
47. J. D. Jackson: Classical Electrodynamics, John Wiley & Sons, 1998.
48. A. Hofmann: The Physics of Synchrotron Radiation, Cambridge, 2004.
49. K.W. Robinson: Phys. Rev. 111 (1958) 373.
50. H. Wiedemann: Particle Accelerator Physics, Springer, 2007.
51. Los Alamos Accelerator Code Group, http://laacg1.lanl.gov/laacg/componl.html
52. ASTEC, https://projects.astec.ac.uk/Plone/Codes/optics/
53. A. Chao and M. Tigner, **Handbook of Accelerator Physics and Engineering**, World Scientific (1998).
54. W. Herr and E. Forest, **Non-linear dynamics**, This handbook.
55. E. Forest, **Beam Dynamics**, Harwood Academic publishers (1998).
56. A. Dragt and J. Finn, J. Math. Phys., **17**, 2215 (1976); A. Dragt et al., Ann. Rev. Nucl. Part. Sci., **38**, 455 (1988).
57. M. Sands: SLAC Report 121 (1969).
58. J.M. Jowett: AIP Conf. Proc. 153 (1985) 934.
59. J.D. Jackson: Rev. Mod. Phys. 48 (1976) 417.
60. A. Chao: AIP Conf. Proc. 153 (1985) 103.
61. A.W. Chao: J. Appl. Phys. 50 (1979) 595; DOI:10.1063/1.326070.
62. P. Raimondi: *Status on SuperB effort,* Second SuperB Workshop, LNF, Frascati (Italy), March 2006.
63. P. Raimondi, D. Shatilov, M. Zobov: *Beam-beam issues for colliding schemes with large Piwinski angle and crabbed waist,* Preprint physics/0702033.
64. M. Zobov et al: *Test of crab-waist collisions at DAΦNE Φ-Factory,* Phys. Rev. Lett. 104 (2010).
65. D. A. Edwards, M. J. Syphers, An introduction to the physics of high energy accelerators, John Wiley & Sons, Inc. New York, NY, 1993
66. J. Rossbach, P. Schmüser, Basic course on accelerator optics, CERN accelerator school, fifth general accelerator physics course, CERN 94-01 vol. 1, 1994
67. B. de Raad, The CERN SPS proton-antiproton collider, IEEE Transactions on Nuclear Science, Vol. NS-32, No. 5, October 1985
68. R. Bailey, E. Brouzet, K. Cornelis, L. Evans. A. Faugier, V. Hatton, J. Miles, R. Schmidt, D. Thomas, High luminosity performance of the SPS proton-antiproton collider, conf. proc. Particle Accelerator Conference 1989, Chicago, 1989
69. R. Schmidt, Beam-beam observations in the SPS proton antiproton collider, Particle Accelerators, 1995, Vol. 50, pp. 47-60
70. LHC design report, CERN 2004-003 vol. 1, 2004
71. A. Chance, D. Boutin, B. Dalena, B. Holzer, A. S. Langner, D. Schulte, Updates on the Optics of the Future Hadron-Hadron Collider FCC-hh, conf. proc. IPAC2017, Copenhagen, 2017.
72. HERA Design Report, DESY HERA-1981-010

73. N. Auerbach and S. Chlomo, Phys. Rev. Lett. 103, 172501 (2009).
74. B.V. Jacak and B. Müller, Science 337, pp. 310-314 (2012).
75. The STAR collaboration, Nature, Vol. 548, pp. 62-65 (2017).
76. The STAR collaboration, Nature 473, pp. 353-35 (2011).
77. B. Dönigus, Nucl. Phys. A, 904-905, pp. 547c-550c (2013).
78. M. Wilde et al., Nucl. Phys. A 904-905, pp. 547c-550c (2013).
79. P. Asbo-Hansen, et al.: IEEE Trans. Nucl. Sci. 24 (1997) 1557.
80. M. Boutheon, et al.: IEEE Trans. Nucl. Sci. 28 (1981) 2049.
81. A.W. Chao, M. Tigner (eds.): Handbook of accelerator physics, World Scientific (1999).
82. M. Okamura, HIAT2015, pp. 274-276 (2015).
83. J.G. Alessi et al., PAC2011, pp. 1966-1968 (2011).
84. W. Fischer et al., Phys. Rev. ST Accel. Beams 11, 041002 (2008).
85. M. Blaskiewicz et al. PAC2001, pp. 3026-3028 (2003).
86. M. Benedict, P. Collier, V. Mertens, J. Poole and K. Schindl (eds.), CERN-2004-003-V3 (2004).
87. D. Manglunki et al., IPAC 2012, pp. 3752-3754 (2012).
88. M. Chanel, Nucl. Instrum. Methods A 532, pp. 137-143 (2014).
89. D. Boussard, J.M. Brennan, T. Linnecar, PAC1995, pp. 1506-1508 (1995).
90. M. Blaskiewicz, J.M. Brennan, K. Mernick: Phys. Rev. Lett. 105 (2010) 094801.
91. C. Montag, et al.: Phys. Rev. ST Accel. Beams 5 (2002) 084401.
92. O. Brüning, et al. (eds.): LHC Design Report, CERN-2004-003 (2004).
93. J.M. Jowett, et al.: Proc. EPAC2004, Luzern, Switzerland (2004) 578.
94. R. Bruce, et al.: Phys. Rev. ST Accel. Beams 12 (2009) 071002.
95. R. Bruce, et al.: Phys. Rev. ST Accel. Beams 12 (2009) 011001.
96. R. Bruce, et al.: Phys. Rev. Lett. 99 (2007) 144801.
97. J.M. Jowett, M. Schaumann et al; IPAC2016, Busan, Korea (2016) 1493.
98. J. M. Jowett, IPAC2018, Vancouver, Canada (2018) 584.
99. J.M. Jowett, C. Carli: Proc. EPAC2006, Edinburgh, Scotland (2006) 550.
100. A. Sessler, "Comment on the word "cooling" as used in Beam Physics," in *International Workshop on Beam Cooling and Related Topics - COOL05*, Galena, Illinois (USA), 2006.
101. A. A. Kolomenski and A. N. Lebedev, "The effect of radiation on the motion of relativistic electrons in synchrotrons," in *Proc. CERN Symposium*, Geneva, 1956.
102. P. Bramham et al., "Stochastic cooling of a stored proton beam," *Nucl. Instr. Meth.*, vol. 125, pp. 201-202, 1975.
103. G. Carron et al., "Experiments on stochastic cooling in ICE (Initial Cooling Experiment)," *IEEE Trans. Nucl. Sci. NS-26,* p. 3456, 1979.
104. S. van der Meer, "Stochastic cooling and the accumulation of antiprotons," *Rev. Mod. Phys.,* vol. 57, p. 689, 1985.
105. D. Möhl, "Stochastic cooling," in *Proc. of CERN Accelerator School*, Rhodes, 1993.
106. F. Caspers, "Techniques of stochastic cooling," in *Workshop on Beam Cooling and Related Topics*, Bad Honnef, 2001.
107. D. Möhl, "The status of stochastic cooling," *Nucl. Instrum. Meth. A,* vol. 391, p. 164, 1997.
108. R. Giannini, P. Lefevre and D. Möhl, "Super LEAR, conceptual machine design," *Nucl. Phys. A,* vol. 558, p. 519, 1993.
109. T. Katayama and H. Tsutsui, "Electron-RI collider and internal target operation of RIKEN storage ring project," *Nucl. Instrum. Methods Phys. Res., A,* vol. 532, pp. 157-171, 2004.
110. FAIR Technical Design Team, "Fair Baseline Technical Report, vols. 1-6," GSI, Darmstadt, 2007.
111. S. Wunderlich et al., "Progress of the RF-system developments for stochastic cooling at the FAIR Collector Ring," in *International Workshop on Beam Cooling and Related Topics (COOL'15)*, Newport News, 2015.
112. C. Dimopoulou, "Stochastic cooling for FAIR," *ICFA Beam Dynamics Newsletter,* no. 64, pp. 106-120, 2014.

113. F. Nolden et al., "Radioactive beam accumulation for a storage ring experiment with an internal target," in *4th Int. Particle Accelerator Conf. (IPAC'13)*, Shanghai, China, 2013.
114. J. M. Brennan and M. Blaskiewicz, "Bunched beam stochastic cooling project for RHIC," *AIP Conf. Proc.*, vol. 821, pp. 185-189, 2006.
115. R. J. Pasquinelli, "Stochastic cooling technology at Fermilab," *Nucl. Instrum. Methods Phys. Res., A*, vol. 532, pp. 313-320, 2004.
116. R. Stassen et al., "Recent developments for the HESR stochastic cooling system," in *International Workshop on Beam Cooling and Related Topics (COOL'07)*, Bad Kreuznach, 2007.
117. C. Peschke et al., "Prototype pick-up module for CR stochastic cooling at FAIR," in *Workshop, COOL'09*, Lanzhou, 2009.
118. J. Marriner, "Stochastic cooling overview," *Nucl. Instrum. Methods Phys. Res., A*, vol. 532, pp. 11-18, 2004.
119. F. Caspers, "Planar slotline pick-ups and kickers for stochastic cooling," in *1987 IEEE Particle Accelerator Conference (PAC 87)*, Washington, 1987.
120. J. Petter, D. McGinnis and J. Marriner, "Novel stochastic cooling pickups/kickers," in *IEEE Particle Accelerator Conference (PAC 89)*, Chicago, 1989.
121. L. Faltin, "Slot type pickup and kicker for stochastic beam cooling," *Nucl. Instrum. Meth.*, vol. 148, pp. 449-455, 1978.
122. D. McGinnis, "The 4-GHz to 8-GHz stochastic cooling upgrade for the Fermilab debuncher," in *Particle Accelerator Conference (PAC 99)*, New York, 1999.
123. G. Carron, F. Caspers and L. Thorndahl, "Development of power amplifier modules for the ACOL stochastic cooling systems," CERN, Geneva, 1985.
124. D. Boussard, "Frontiers of particle beams," in *Joint US - CERN School on Particle Accelerators: Topical Course on Frontiers of Particle Beams*, Vols. Lecture Notes in Physics, vol 296, M. Month and S. Turner, Eds., South Padre Island, Springer, 1986, pp. 269-296.
125. M. Blaskiewicz, J. M. Brennan and K. Mernick, "Three-dimensional stochastic cooling in the Relativistic Heavy Ion Collider," *Phys. Rev. Lett.*, vol. 105, p. 094801, 2010.
126. F. Caspers, "A novel type of forward coupler slotted stripline pickup electrode for non-relativistic particle beams," CERN, Geneva, 2011.
127. J. X. Wu et al., "A novel type of forward coupler slotted stripline pickup electrode for CSRe stochastic cooling," in *4th International Particle Accelerator Conference*, Shanghai, 2013.
128. H. Stockhorst et al., "Status of stochastic cooling predictions at the HESR," in *2nd International Particle Accelerator Conference (IPAC 2011)*, San Sebastian, 2011.
129. C. Carli and F. Caspers, "Stochastic cooling at the CERN Antiproton Decelerator," in *7th European Particle Accelerator Conference (EPAC 2000)*, 2000.
130. T. Eriksson et al., "AD status and consolidation plans," in *International Workshop on Beam Cooling and Releated Topics (COOL13)*, Mürren, 2013.
131. R. J. Pasquinelli, "Bulk acoustic wave (BAW) devices for stochastic cooling notch filters," in *IEEE Particle Accelerator Conference (PAC 91)*, San Francisco, 1991.
132. W. Maier et al., "The novel optical notch filter for stochastic cooling at the ESR," in *International Workshop on Beam Cooling and Related Topics (COOL'13)*, Mürren, 2013.
133. R. Stassen et al., "The HESR stochastic cooling system, design, construction and test experiments in COSY," in *11th International Workshop on Beam Cooling and Related Topics (COOL2017)*, Bonn, 2017.
134. D. R. Broemmelsiek et al., "Bunched beam stochastic cooling in the Fermilab Recycler Ring," in *Particle Accelerator Conference (PAC 05)*, Knoxville, 2005.
135. P. F. Derwent et al., "Performance and upgrades of the Fermilab Accumulator Stacktail Stochastic Cooling," in *International Workshop on Beam Cooling and Related Topics (COOL 05)*, 2005.
136. V. A. Lebedev, "Improvements to the stacktail and debuncher cooling systems," in *International workshop on beam cooling and related topics (COOL 09)*, Lanzhou, 2009.
137. H. Herr and D. Möhl, "Bunched beam stochastic cooling," in *Workshop on the Cooling of High-Energy Beams*, Madison, 1978.

138. R. J. Pasquinelli, "Twenty-five years of stochastic cooling experience at Fermilab," in *International Workshop on Beam Cooling and Related Topics (COOL09)* , Lanzhou, 2009.
139. G. Jackson et al., "Bunched beam stochastic cooling in the Fermilab Tevatron Collider," in *International Conference on Particle Accelerators (PAC'93)*, Washington, 1993.
140. F. Caspers and D. Möhl, "Stochastic cooling in hadron colliders," in *17th International Conference on High-Energy Accelerators*, Dubna, 1998.
141. M. Blaskiewicz et al., "Longitudinal solitons in RHIC," in *2003 Particle Accelerator Conference (PAC 2003)*, 2003.
142. J. Bosser et al., "LEAR MD report: bunched beam Schottky spectrum," CERN, Geneva, 1994.
143. J. Marriner et al., "Bunched beam cooling in the FNAL Antiproton Accumulator," in *2nd European Particle Accelerator Conference*, Nice, 1990.
144. A. A. Mikhailichenko and M. S. Zolotorev, "Optical stochastic cooling," *Phys. Rev. Lett,* vol. 71, p. 4146–4149, 1993.
145. A. Zholents and M. Zolotorev, "Transit time method of optical stochastic cooling," *Phys. Rev. E,* vol. 50, p. 3087, 1994.
146. W. A. Franklin et al., "Optical stochastic cooling experiment at the MIT-Bates South Hall Ring," in *Workshop on Beam Cooling and Related Topics (COOL 07)*, Bad Kreuzbach, 2007.
147. M. Blaskiewicz, J. M. Brennan and F. Severino, "Operational stochastic cooling in the Relativistic Heavy-Ion Collider," *Phys. Rev. Lett.,* vol. 100, p. 174802, 2008.
148. J. M. Brennan and M. Blaskiewicz, "Stochastic cooling in RHIC," in *Particle Accelerator Conference (PAC 09)*, Geneva, 2009.
149. S. Antipov et al., "IOTA (Integrable Optics Test Accelerator): facility and experimental beam physics program," *JINST,* vol. 12, p. T03002, 2017.
150. S. van der Meer, "Stochastic stacking in the Antiproton Accumulator," CERN, Geneva, 1978.
151. B. Autin et al., "Design study of a proton-antiproton colliding beam facility," CERN, Geneva, 1978.
152. E. J. N. Wilson, "Design study of an antiproton collector for the antiproton accumulator (ACOL)," 1983.
153. V. Lebedev and V. Shiltsev, Accelerator physics at the Tevatron Collider, New York: Springer, 2014.
154. R. J. Pasquinelli, "Implementation of stochastic cooling hardware at Fermilab's Tevatron Collider," *JINST,* vol. 6, p. T08002, 2011.
155. S. Nagaitsev, L. Prost and A. Shemyakin, "Fermilab 4.3 MeV electron cooler," *JINST 10,* vol. 10, no. 01, p. T01001, 2015.
156. G. L. Budker, "Experiences sur l'obtention d'un faisceau intense de protons par la methode d'injection par echange de charges," in *Symposium International sur les anneaux de collisions a electrons et positrons, Saclay, 26–30 septembre 1966*, Saclay, 1967.
157. J. D. Jackson, Classical electrodynamics, New Work: John Wiley & Sons, 1975.
158. H. Poth, "Electron Cooling," in *CAS - CERN Accelerator School: Accelerator Physics, Oxford, UK, 16 - 27 Sep 1985*, Vols. CERN-1987-003-V-2, Geneva, CERN, 1987, pp. 534-569.
159. Y. Derbenev and I. Meshkov, "Studies on electron cooling of heavy particle beams made by the VAPP-NAP group at the Nuclear Physics Institute of the Siberian branch of the USSR Academy of Science at Novosibirsk," CERN, Geneva, 1977.
160. G. Budker et al., "Experimental studies of electron cooling," *Part. Accel.,* vol. 7, pp. 197-211, 1976.
161. H. Poth, "Electron cooling: theory, experiment, application," *Phys. Rep.,* vol. 196, p. 135, 1990.
162. A. Y. Larentev and I. N. Meshkov, "The computation of electron cooling process in a storage ring," in *International Conference on Crystal Beams "Ettore Majorana"*, Erice, 1996.
163. M. Chanel, "Ion accumulation for LHC," in *Workshop on Beam Cooling and Related Topics*, Bad Honnef, 2001.

164. J. Bosser et al., "Experimental investigation of electron cooling and stacking of lead ions in a low-energy accumulation ring," *Part. Acc.,* vol. 63, p. 171, 1999.
165. G. Tranquille, "Optimum parameters for electron cooling," CERN, Geneva, 2001.
166. J. Bosser et al., "Stability of cooled beams," *Nucl. Instrum. Methods Phys. Res. A,* vol. 441, pp. 1-8, 2000.
167. S. Nagaitsev et al., "Antiproton cooling in the Fermilab Recycler Ring," in *International Workshop on Beam Cooling and Related Topics (COOL05)*, 2005.
168. A. V. Fedotov, "Progress of high-energy electron cooling for RHIC," in *International Workshop on Beam Cooling and Related Topics (COOL07)*, Bad Kreuznach, 2007.
169. D. Reistad, "Calculations on high-energy electron cooling in the HESR," in *International Workshop on Beam Cooling and Related Topics (COOL07)*, Bad Kreuznach, 2007.
170. I. Ben-Zvi, "R&D towards cooling of the RHIC Collider," *Nucl. Instrum. Meth. A,* vol. 532, pp. 177-183, 2004.
171. Y. Derbenev, "Coherent electron cooling," *Phys. Rev. Lett.,* vol. 102, p. 114801, 2009.
172. W. Petrich et al., "Laser cooling at the Heidelberg Test Storage Ring (TSR)," in *Workshop on Electron Cooling and new Cooling Techniques*, Legnaro, 1990.
173. N. Kjærgaard et al., "Recent results from laser cooling experiments in ASTRID – real-time imaging of ion beams," *Nucl. Instrum. Meth. A,* vol. 441, p. 196, 2000.
174. E. Bonderup, "Laser cooling," in *CAS - CERN Accelerator School: 5th Advanced Accelerator Physics Course, Rhodes, Greece, 20 Sep - 1 Oct 1993*, Vols. CERN Report 95-06, S. Turner, Ed., Rhodes, CERN, 1993, pp. 731-747.
175. H. J. Miesner et al., "Efficient, indirect transverse laser cooling of a fast stored ion beam," *Phys. Rev. Let.,* vol. 77, p. 623, 1996.
176. I. Lauer et al., "Transverse laser cooling of a fast stored ion beam through dispersive coupling," *Phys.Rev. Lett.,* vol. 81, p. 2052, 1998.
177. H. Okamoto, A. M. Sessler and D. Möhl, " Three-Dimensional laser cooling of stored and circulating ion beams by means of a coupling cavity," *Phys. Rev. Lett.,* vol. 72, p. 3977, 1994.
178. D. Maletic and A. Ruggiero, "Crystalline beams and related issues," in *31st INFN Eloisatron Workshop*, Erice, 1996.
179. D. Habs and R. Grimm, "Crystalline Ion Beams," *Ann. Rev. Nucl. Sci.,* vol. 45, pp. 391-428, 1995.
180. N. Madsen and J. S. Nielsen, "Laser-cooling for light ion accumulation," in *7th European Particle Accelerator Conference (EPAC 2000)*, Vienna, 2000.
181. G. Plass, "The status of the HIDIF study," *Nucl. Instrum. Meth. A,* vol. 415, p. 204, 1998.
182. A. Noda et al., "Present status and recent activity on laser cooling at S-LSR," in *International Workshop On Beam Cooling And Related Topics (COOL07)*, Bad Kreuznach, 2007.
183. A. Noda et al., "Laser cooling for 3-D crystalline state at S-LSR," in *International Workshop on Beam Cooling and Related Topics (COOL05)*, 2005.
184. M. Ikegami et al., "Heavy ion storage ring without linear dispersion," *Phys. Rev. Accel. Beams,* vol. 7, p. 120101, 2004.
185. A. N. Skrinsky, "Ionization cooling and muon collider," *Nucl. Instrum. Meth. A,* vol. 391, pp. 188-195, 1997.
186. D. N. D. Neuffer, "Introduction to muon cooling," *Nucl. Instrum. Meth. A,* vol. 532, pp. 26-31, 2004.
187. D. Kaplan, "Muon-cooling research and development," *Nucl. Instrum. Meth. A,* vol. 532, pp. 241-248, 2004.
188. A. Blondel and G. Hanson, "MICE status report September 2010," 2010.
189. W. M. Itano et al., "Cooling methods in ion traps," *Phys. Scr.,* vol. T59, pp. 106-120, 1995.
190. F. Herfurth et al., "Highly charged ions at rest: The HITRAP project at GSI," Wako, 2005.
191. N. Beverini et al., "Stochastic cooling in Penning traps," *Phys. Rev. A,* vol. 38, no. 1, pp. 107-114, 1988.
192. D. S. Hall and G. Gabrielse, "Electron cooling of protons in a nested Penning trap," *Phys. Rev. Lett.,* vol. 77, no. 10, p. 1962, 1997.

193. S. Chu, "Nobel Lecture: The manipulation of neutral particles," *Rev. Mod. Phys.*, vol. 70, p. 685, 1996.
194. E. A. Cornell and Wieman, "Nobel lecture: Bose-Einstein condensation in a dilute gas: the first 70 years and some recent experiments," *Int. J. Mod. Phys. B,* vol. 16, no. 30, pp. 4503-4536, 2002.
195. H. G. Dehmelt, "Radiofrequency spectroscopy of stored ions I: Storage," *Adv. At. Mol. Phys.*, vol. 3, pp. 53-72, 1968.
196. D. J. Wineland and H. G. Demelt, "Principles of the stored ion calorimeter," *J. Appl. Phys.*, vol. 46, p. 919, 1975.
197. W. J. Shockley, "Currents to conductors induced by a moving point charge," *Appl. Phys.*, vol. 9, p. 635, 1938.
198. K. R. Brown et al., "Passive cooling of a micromechanical oscillator with a resonant electric circuit," *Phys. Rev. Lett.*, vol. 99, p. 137205, 2007.
199. S. van der Meer, "Stochastic damping of betatron oscillations," CERN, Geneva, 1972.

Chapter 7
Design and Principles of Linear Accelerators and Colliders

J. Seeman, D. Schulte, J. P. Delahaye, M. Ross, S. Stapnes, A. Grudiev, A. Yamamoto, A. Latina, A. Seryi, R. Tomás García, S. Guiducci, Y. Papaphilippou, S. A. Bogacz, and G. A. Krafft

7.1 General Introduction on Linear Accelerators

J. Seeman

Linear accelerators (linacs) use alternating radiofrequency (RF) electromagnetic fields to accelerate charged particles in a straight line. Linacs were invented about 95 years ago and have seen many significant technical innovations since. A

Coordinated by J. Seeman, J. P. Delahaye.

J. Seeman · M. Ross
SLAC National Accelerator Laboratory, Stanford University, Menlo Park, CA, USA
e-mail: seeman@slac.stanford.edu; mcrec@slac.stanford.edu

D. Schulte · J. P. Delahaye (✉) · S. Stapnes · A. Grudiev · A. Latina · R. Tomás García · Y. Papaphilippou
CERN (European Organization for Nuclear Research) Meyrin, Geneva, Switzerland
e-mail: daniel.schulte@cern.ch; Jean-Pierre.Delahaye@cern.ch; steinar.stapnes@cern.ch; alexej.grudiev@cern.ch; andrea.latina@cern.ch; rogelio.tomas@cern.ch; ioannis.papaphilippou@cern.ch

A. Yamamoto
KEK, Tsukuba, Ibaraki, Japan

CERN (European Organization for Nuclear Research) Meyrin, Geneva, Switzerland
e-mail: akira.yamamoto@cern.ch

A. Seryi · S. A. Bogacz · G. A. Krafft
Center for Advanced Studies of Accelerators, Jefferson Lab, Newport News, VA, USA
e-mail: seryi@jlab.org; bogacz@jlab.org; krafft@jlab.org

S. Guiducci
National Laboratory of Frascati/National Institute for Nuclear Physics, Frascati, Roma, Italy
e-mail: susanna.guiducci@lnf.infn.it

S. Myers, H. Schopper (eds.), *Particle Physics Reference Library*,
https://doi.org/10.1007/978-3-030-34245-6_7

wide range of particle beams have been accelerated with linacs including beams of electrons, positrons, protons, antiprotons, and heavy ions. Linac parameter possibilities include pulsed versus continuous wave, low and high beam powers, low and high repetition rates, low transverse emittance beams, short bunches with small energy spreads, and accelerated multiple bunches in a single pulse. The number of linacs around the world has grown tremendously with thousands of linacs in present use, many for medical therapy, in industry, and for research and development in a broad spectrum of scientific fields. Researchers have developed accelerators for scientific tools in their own right, being awarded several Nobel prizes. Moreover, linacs and particle accelerators in general have enabled many discovery level science experiments in related fields, resulting in many Nobel prizes as well.

In this chapter the various types, near term uses, and future directions of linacs are discussed. There are many standard types of linac structures, several are shown in Figs. 7.1 and in Figs. 7.6 and 7.7 in Sect. 7.4. A complete linac system includes an RF power source, the microwave power waveguide distribution, the accelerating structure itself, a power load, a control system, a vacuum system, survey-alignment, a cooling system, and beam diagnostics such as beam position monitors and profile monitors. Examples of present operating linacs from around the world, linacs under construction, and proposed large scale linacs are shown in Table 7.1 [3–11]. There are many constraints to design a successful linac [12, 13] with the basic being to

Fig. 7.1 Examples of linac structures: Cu linac 3-GHz (upper left), Cu cells 12-GHz (upper centre), Drift Tube Linac 202-MHz (upper right), and Super-conducting 9-cell cavity 1.3-GHz (lower centre)

Table 7.1 Several existing and near future (F) Linacs used for scientific studies in the world

Location	Length (m)	E (GeV)	G (MV/m)	Part.	RF Freq. (MHz)	Type
DAPHNE, IT	60	0.51	17	e−/e+	2856	NC-Cu-constant grad
Linac, KEK	608	8.0	15	e−/e+	2856	NC-Cu-constant grad
SLC/LCLS, SLAC	3000	54.0	18	e−/e+	2856	NC-Cu-constant grad
SwissFEL	620	5.8	28		5712	NC-Cu-constant grad
CLS-II, SLAC (F)	500	4.1	16	e−	1300	SC-CW-9-cell cavities
J-PARC, JP	400	0.18/0.4	2	p	324/972	RFQ/DTL/SDTL/ACS
SNS, ORNL	493	0.97	10	p	402/805	RFQ/DTL/CCL/SC
Spring8FEL, JP	400	8.0	35	e−	5712	NC-Cu-constant imp
E-XFEL, DESY	1700	17.5	23.6	e−	1300	SC-Nb-9-cell cavities
ILC, World (F)	2 × 11300	250	35	e−/e+	1300	SC-Nb-9-cell cavities
CLIC, World (F)	2 × 21000	1500	100	e−/e+	12,000	NC-travelling wave

choose an RF structure such that electromagnetic fields and the beam particles are in phase as the beam traverses each cell in the RF cavity. Much of the design work has gone into maximizing the accelerating gradients, avoiding arcing, RF discharges, and multipacting, to minimize the construction costs, and to make efficient use of overall AC power. Proton and ion beams are often made in drift tube linacs (DTL) with gradients of 2–8 MeV/m using RF frequencies of 30–400 MHz. Electron linacs are typically made of either copper structures with 15–75 MeV/m at 3–12 GHz or superconducting structures with 10–30 MeV/m near 1.3 GHz. Small proton or ion linacs are used for medical therapy and patient diagnostics. Larger proton linacs are injectors for large particle colliders or proton drivers for neutron or neutrino production. Small electron linacs are used for electron or gamma ray medical therapy and in industry. Large electron linacs are often injectors into GeV energy storage rings for synchrotron radiation sources and e−/e+ colliders and as injectors into FEL undulators.

An active area of present research is the conditioning of beams in linacs to make them useful for high energy physics, basic energy sciences, nuclear physics, material sciences, and technology security. These beam parameters include the calculation and control of longitudinal and transverse wakefields (Sect. 7.5), low emittance generation and preservation (Sect. 7.7), incoherent and coherent synchrotron radiation effects, electron cloud effects, ion effects, energy recovery through recirculation (Sect. 7.8), multibunch effects (Sect. 7.3), high luminosity requirements (Sect. 7.2), final focus systems (Sect. 7.5), 2D and 3D emittance exchanges, and beam-undulator interactions in x-ray FELs.

The use of linacs in energy frontier e+e− colliders is essential to avoid excessive synchrotron radiation in ultrahigh energy beams. The first linear collider used the SLAC linac called the SLC (Sect. 7.2) which provided frontier particle physics results as well as establishing a basis to build upon for a future linear collider design. Recent linear collider studies (Sect. 7.3) have concentrated on the ILC (1.3 GHz Superconducting) [4] and CLIC (12 GHz normal conducting) [7]. Years of studies and experiments have illustrated the approaching viability of these two collider technologies.

Very attractive schemes like for example the energy recovery linacs [14] as described in Sect. 7.8 are being developed. Recent avenues of study for far future linacs are in the area of excited plasmas and dielectrics as structures [15, 16]. Examples are electron beam driven plasma, wakefield accelerator PWFA that recently produced 40 GeV/m acceleration for electrons and 0.23 GeV/m for positrons [17–20] and laser driven plasma wakefield accelerator demonstrating up to 4.2 GeV in a 9 cm plasma [21, 22] as described in Chap. 12. Another active research area is direct laser driven accelerating nanostructures in silicon [23, 24]. New ongoing studies in these technologies will illuminate possible future uses.

7.2 High Luminosity Issues and Beam-Beam Effects

D. Schulte

In linear colliders, the colliding beams have extremely small transverse dimensions $\sigma_{x,y}$ to reach high luminosity. Each beam exerts a strong electro-magnetic force on the other beam, which is focusing in case of electron-positron collisions. This disruption can shrink the beam size significantly during collision, the so-called pinch effect [25–27]. This increases the luminosity but the bending of the particles' trajectories stimulates them to radiate so-called beamstrahlung photons, a process similar to synchrotron radiation [28–32]. Consequently, not all collisions take place at the nominal centre-of-mass energy. Hence, one needs to choose the beam parameters in order to limit the beamstrahlung and to achieve an acceptable luminosity spectrum for the experiment. High luminosity with low beamstrahlung is usually achieved by using flat beams $\sigma_x \gg \sigma_y$) as shown below. Approximate formulae are used, since full analytic treatment of the beam-beam interaction is for most parameters not possible. Simulation codes are used for precise numerical predictions, in particular CAIN and GUINEA-PIG [33–35].

It should be noted that one usually needs a horizontal crossing angle θ_c between the two beam lines at the interaction point. This separation of incoming and outgoing beam allows to efficiently extract collision debris and hence to avoid large beam losses after the collision. For short distances in-between bunches in each beam pulse, the crossing angle also reduces the impact of parasitic crossings of incoming and outgoing bunches. The luminosity reduction due to the crossing angle is avoided by using a so-called crab-crossing scheme. Before the collision a transverse deflecting cavity introduces a rotation around the vertical axis, such that the beams are aligned to the longitudinal axis of the laboratory system at the collision rather than the direction of motion. Hence the two bunches fully overlap during the collision while moving together horizontally, like crabs. In this case the crossing angle hardly affects the beam-beam interaction and can be neglected in the further considerations.

In high energy linear colliders like ILC and CLIC developed in Sect. 7.3, the luminosity is limited by beamstrahlung, the achievable vertical beam size and the efficiency of the main linac RF. This is seen by expressing it as a function of the number of particles per bunch N, the number of bunches per beam pulse n_b, the repetition rate of beam pulses f_r and the luminosity enhancement factor H_D, which tends to be in the range of 1–2:

$$L = H_D \frac{N^2}{4\pi\sigma_x\sigma_y} n_b f_r . \tag{7.1}$$

Ignoring the usually small variations of H_D, one obtains the simple dependence:

$$L \propto \frac{N}{\sigma_x}\frac{1}{\sigma_y}\eta P_{\text{wall}} . \tag{7.2}$$

The first factor N/σ_x is a measure of the beamstrahlung, the second σ_y depends strongly on the beam quality and the efficiency η of transforming the wall plug power P_{wall} into beam power is dominated by the RF to beam transfer efficiency of the main linac.

The beamstrahlung can conveniently be described with the beamstrahlung parameter Y, the ratio of the average critical energy $\hbar\omega_c$ to the beam energy E:

$$Y = \frac{2}{3}\frac{\hbar\omega_c}{E} = \frac{5}{6}\frac{N r_e^2 \gamma}{\alpha\left(\sigma_x + \sigma_y\right)\sigma_z}. \tag{7.3}$$

Here, α is the fine structure constant, r_e the classical electron radius and γ the relativistic factor of the beam. In the classical limit $Y \gg 1$, which is applicable to the ILC or CLIC at 500 GeV, the number of beamstrahlung photons emitted per beam particle n_γ and their average energy E_γ can be approximated as

$$n_\gamma \approx 2.1\alpha \frac{N r_e}{\sigma_x + \sigma_y} \frac{E_\gamma}{E} \approx 0.385 \frac{N r_e^2 \gamma}{\alpha\left(\sigma_x + \sigma_y\right)\sigma_z}. \tag{7.4}$$

Hence, one uses $\sigma_x \gg \sigma_y$ to maximise luminosity ($\propto N/(\sigma_x\sigma_y)$) while limiting the beamstrahlung ($\propto N/(\sigma_x + \sigma_y) \approx N/\sigma_x$). Typically one aims for $n_\gamma \leq 1 - 2$ to maximise luminosity while maintaining the degradation of the luminosity spectrum due to beamstrahlung comparable to the degradation due to initial state radiation. Hence, the machine is designed such that the optimum value of N/σ_x can be reached.

The vertical beam size depends on the vertical beta-function and emittance at the interaction point $\sigma_y = \sqrt{\beta_y \varepsilon_y / \gamma}$. Hence the vertical emittance is minimised as much as possible, with limits arising from the lattices designs and dynamic and static imperfections in the beam transport system. In addition one aims to minimise the beta-function. However, a beta-function smaller than the bunch length leads to a rapidly increasing beam size just before and after the collision point still during the collision with the other bunch. Ignoring beam-beam forces, the optimum choice is $\beta_y = \sigma_z/4$ due to this so-called hourglass effect. The luminosity would only be 20% larger than for the more relaxed value of $\beta_y = \sigma_z$. The beam-beam force strongly modifies the collision and impacts the optimum choice of vertical beta-function and the longitudinal position of beam waist [36]. Pinching of the beams is more effective for larger vertical beta-functions, For ILC and CLIC parameters the luminosity enhancement factor is strongly reduced if the beta-function is pushed below the bunch length resulting in an optimum choice of about $\beta_y = \sigma_z$. It is also advantageous to focus the beams slightly before the collision point as this further improves the luminosity enhancement.

A small value of σ_y also has a strong impact on the beam-beam collision dynamics and tolerances. The beam-beam jitter must be significantly smaller than σ_y, but the disruption can tighten the tolerance even more. The strength of the pinch effect can be described using the disruption parameters D_x and D_y:

$$D_{x,y} = \frac{2Nr_e\sigma_z}{\gamma\sigma_{x,y}\left(\sigma_x + \sigma_y\right)}. \tag{7.5}$$

If $D_{x,\,y} \ll 1$ each beam acts as a thin lens on the other beam with a focal length $f_{x,\,y} = \sigma_z/D_x$ close to its centre. If $D_x \gg 1$ the beam particles oscillate in the field of the other beam; ILC and CLIC have $D_x < 1$ and $D_y \gg 1$. For $D_y \geq 15 - 20$ the beam-beam interaction becomes unstable. In this case very small beam-beam offsets lead to a large loss of luminosity.

In order to increase the wall plug to beam power efficiency, η, two different strategies exist. The first is to use superconducting structures to minimise the RF losses in the structure itself. The second it to use normal conducting structures but to increase the structure impedance and the beam current as much as possible to maximise the power transfer to the beam. A limit arises from single and multi-bunch beam instabilities, see Sect. 7.5.

At multi-TeV energies, the beamstrahlung parameter is much larger, $Y \gg 1$, which slightly changes the functional dependence of the luminosity on the number of beamstrahlung photons to

$$L \propto n_\gamma^{\frac{3}{2}} \frac{\sqrt{\gamma}}{\sqrt{\sigma_z}} \frac{1}{\sigma_y} P_b. \tag{7.6}$$

However, the fundamental considerations remain unchanged.

The beam-beam interaction is an important source of background. The disrupted beam and the beamstrahlung photons need to be extracted from the detector without large losses, this requires typically an exit hole of a few milliradian. In addition secondary particles are produced.

The collision of beamstrahlung photons and beam particles produces low energy electron-positron pairs, a process called incoherent pair production. The number of these particles per bunch crossing can be of the order of 10^5–10^6. They are an important source of background in the innermost layer of the vertex detector and define a lower limit for its radius. Most of the particles go into the forward region of the detector.

In a similar fashion, hadrons can also be produced in the collision, with a rate ranging typically from one event every few bunch crossing to a few events per bunch crossing. Most of the tracks of these events go into the forward region but they can impact the jet reconstruction.

At high beam energies, if $Y \gg 1$, beamstrahlung photons and the virtual photons accompanying the beam particles can turn into electron-positron pairs due to the strong beam fields, a process referred to as coherent pair production [37]. The number of pairs can be a significant fraction of the number of beam particles. These particles can lead to background in the forward region, depending on the detector design.

7.3 CLIC & ILC

J. P. Delahaye · M. Ross · S. Stapnes

7.3.1 Introduction

The case for an e+e− collider to explore the physics opened up by the discovery of the Higgs boson is widely accepted. More generally, with the completion of the Standard Model the energy scales needed to explore it in great detail are well established. Two alternative technologies for linear e+e− colliders are being pursued, with different potential energy reach and performance considerations:

- The International Linear Collider (ILC) [38] being proposed in Japan with an initial energy of 250 GeV has the potential to study the Higgs sector in great detail. The ILC is based on beam acceleration by RF Super-Conducting cavities and is prepared as an international project lead by Japan. The ILC Technical Design Report (TDR) [39] was published in 2012 focusing on a 500 GeV machine. Recently a 250 GeV initial stage has been defined [40].
- The Compact Linear Collider (CLIC) [41] study is exploring the possibility of a Linear Collider with a Multi-TeV energy range through the development of Two Beam Acceleration, a novel technology. Normal conducting 12 GHz X-band accelerating structures are used. The study is carried out an international collaboration hosted by CERN. The CLIC Conceptual Design Report (CDR) was published in 2012 [42]. In 2014 an initial stage at 380 GeV was defined and is currently the main focus of the study [43].

These two studies, with the basic parameters shown in Table 7.2, aim to devise appropriate facilities to complement the LHC in the e+e− area. A collaboration between CLIC and ILC that takes advantage of the overlapping portions of the two schemes has proven to be extremely fruitful.

The main challenge for both concepts is reliable and power efficient acceleration, hence the focus on RF technology developments. Secondly, the required luminosity of about 1.5×10^{34} cm^{-2} s^{-1} is a major challenge. It requires collisions of powerful beams with extremely small beam dimensions (a few nm in the vertical plane). Cost is another key factor, both ILC and CLIC aiming for project-costs comparable to LHC [44].

7.3.2 ILC Design

The design of the International Linear Collider (ILC) is based on 1.3 GHz Super-Conducting RF technology (SCRF). The configuration of the linac power and utility

Table 7.2 Basic parameters for ILC and CLIC

	ILC	CLIC
Centre-of-mass energy	250 GeV (upgradable to 1 TeV)	380 GeV (upgradable to 3 TeV)
Total luminosity (cm^{-2} s^{-1})	1.4×10^{34}	1.5×10^{34}
Total site length (km)	20	11
Loaded accel. gradient (MV/m)	31.5 (35)	72 (100)
Main linac technol. & RF frequency	Super-conduct @ 1.3 GHz	Normal-conduct @ 12 GHz
Beam power/beam (MW)	5	3
Bunch charge ($10^9 e^{+/-}$)	20	5.2
Bunch separation (ns)	554	0.5
Beam pulse duration (μs)	722	0.176
Repetition rate (Hz)	5	50
Hor./vert. norm. emitt ($10^{-6}/10^{-9}$)	5/35	0.95/30
Hor./vert. IP beam size (nm)	520/8	150/3
Beamstrahlung photon/electron	1.9	1.5
Total power consumption (MW)	130	200

infrastructure is based on klystron sources and waveguide distribution. A train of 1312 bunches is accelerated during a ~1.6 ms macro-pulse, corresponding to the beam-pulse duration plus cavity fill-time, at a repetition rate of 5 Hz. The average gradient foreseen is 31.5 MV/m but on-going R&D opens for the possibility to increase to 35 MV/m with higher Q cavities (see Sect. 7.3.4). The cryo-modules that make up the main linacs are 12.65 m long. There are two types: a module with nine 1.3 GHz nine-cell cavities and a module with eight nine-cell cavities and one superconducting quadrupole package located at the centre of the module.

The RF power is provided by 10 MW multi-beam klystrons each driven by a 120 kV Marx modulator. The 10 MW klystrons has achieved the ILC specifications and is now a well-established technology with several vendors worldwide.

The long time scale of the 722 μs macro-pulse, with 554 ns between bunches, provides the time needed for effective intra-train trajectory, energy and interaction region collision feedback resulting in very relaxed mechanical vibration tolerances. Accelerating cavity positioning tolerances are also relaxed due to the large 70 mm diameter clear aperture of the accelerating cavities.

The ILC overall layout is shown on Fig. 7.2. The figure shows two centrally-positioned detectors and the electron and positron damping rings. It also shows the mid-linac undulator-based positron source.

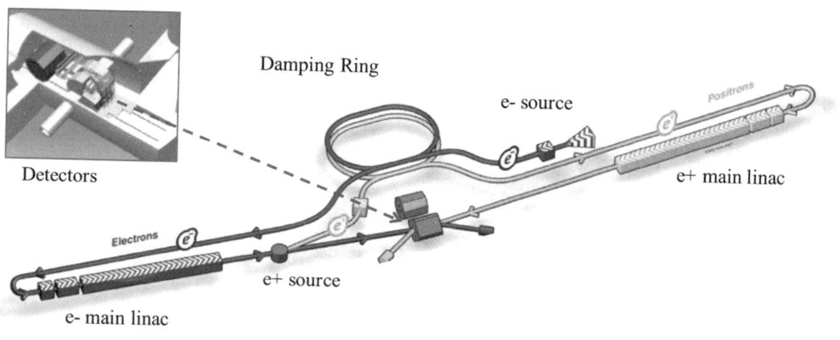

Fig. 7.2 ILC overall layout with the central region expanded

7.3.3 CLIC Design

The CLIC design is based on a novel Two Beam Acceleration (TBA) scheme where a high intensity drive beam running all along the linacs accelerating the main beams. The CLIC main linacs are made of normal-conducting structures resonating at RF frequency of 12 GHz with average accelerating fields of 72/100 MV/m resulting from an overall cost/performance optimisation at 380 GeV/3 TeV, respectively. The accelerating field results from a cost optimisation primarily being a trade-off between the linac extension and the required RF power. The X-band frequency of 12 GHz also results from a trade-off between the required RF power, scaling with inverse square of RF frequency, and the corresponding wake-fields which limit the charge per bunch, therefore the luminosity.

The RF power which is necessary to feed the main linac accelerating structures with high field is efficiently generated by the TBA scheme where the energy of a high intensity drive beam is converted into RF power by specially designed Power Extraction and Transfer Structures (PETS). The 100 A drive beam is generated from a 148 μs long train of bunches accelerated by 20 MW 1 GHz klystrons in a 2.4 GeV normal conducting linac at low intensity and low frequency working in fully loaded mode. For the initial stage of CLIC at 380 GeV one single drive beam at 2 GeV is needed with shortened bunch train, while at 3 TeV two will be needed. The drive-beam trains are compressed in a delay loop and two combiner rings thus multiplying the beam intensity and frequency by a factor 24 and providing series of trains with the required 100 A current and 12 GHz bunch repetition frequency. Each train is used to power one 878 m long sector of the main linac. Upgrade in energy by adding sectors powered by additional drive beam generated by the same drive beam generation complex is particularly cost effective.

The overall layout is shown in the left part of Fig. 7.3 whereas the principle of the two beam scheme is displayed on the right.

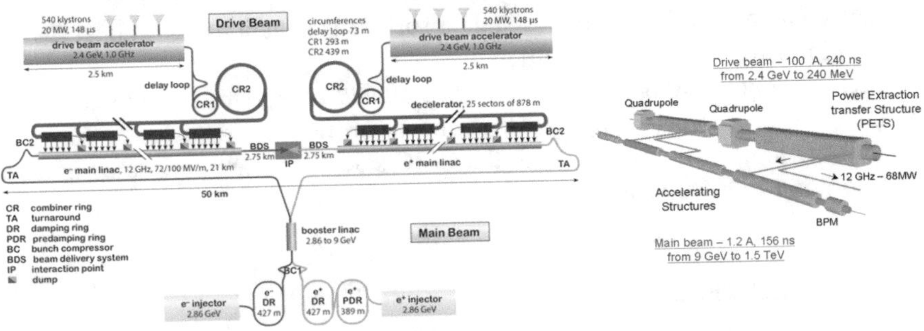

Fig. 7.3 CLIC two-beam scheme (left) and overall layout (right). In the case of a 380 GeV initial phase only one drive-beam is needed and the layout is simplified accordingly

Fig. 7.4 The E-XFEL linac (©European XFEL/Heiner Müller-Elsner)

7.3.4 *On-Going or Recent R&D*

7.3.4.1 ILC Specific

High quality ILC-style SCRF modules are built in all three regions, Americas, Asia and Europe. Furthermore, linacs based on SCRF technology have been/are being constructed, the largest being the 1.7 km European XFEL [45] now in operation with more than 100 cryo-modules (Fig. 7.4). In the US the LCLS II with around 40 cryo-modules at SLAC [46] is being constructed aiming for first beam in 2020. These large scale projects have firmly established the industrial capabilities for SCRF module production.

For ILC slightly higher gradients are needed than for these projects and recent R&D and results for high Q cavities provide promises of gradient and/or efficiency gains [47].

Since 2012 when the Japanese particle physics community expressed the wish to host the ILC [48] site specific studies have been pursued with high priority. These include civil engineering studies including surface installations, as well as studies of local infrastructure and capabilities.

R&D is continuing on various non-linac related subsystem technologies, such as the positron source, damping rings, and beam delivery/final focus. Many of these R&D topics are common with the CLIC studies and are performed in close collaboration with CLIC teams.

7.3.4.2 CLIC Specific

The feasibility of the novel two-beam scheme has been addressed in the CLIC Test Facility (CTF3) which consists of a complex of accelerators for drive beam generation and experimental studies [49]. The drive beam is used to test the Two Beam Acceleration scheme accelerating a probe beam with a gradient well above 100 MV/m. The stability of the drive-beam itself has been another major verification study in CTF3, as well as studies of prototype RF structures, quadrupoles, instrumentation, vacuum, beam alignment and stabilisation (Fig. 7.5).

The accelerating gradient of 100 MV/m with the specified breakdown rate of 3×10^{-7}/pulse/m has been demonstrated in test-stands where prototype test-structures are conditioned to the required power, pulse-lengths and breakdown rate. The requirement for the breakdown rate—these are discharges on the structure surface with the potential of disturbing the beam—is set to cause less than 1% luminosity loss in a 3 TeV machine.

Fig. 7.5 The CTF3 test facility at CERN, which has demonstrated CLIC's novel two-beam acceleration technology (image credit: Maximilien Brice – CERN)

Technological developments are pursued for all critical elements of the machine. Of particular relevance are novel methods of alignment in the micron range and stabilisation in the nano-meter range. Power reductions studies with high efficiency klystrons and permanent magnets are important R&D activities. Civil engineering and infrastructure studies have been done to establish the cost and schedule of the project implementation.

Also in the case of the normal conducting technology XFELs linacs provide important industrial lessons, so far using S or C-band technology. X-band technology is now widely considered for future compact linac installations [50].

7.3.5 *Common Issues and Prospects*

Apart from the linacs based on different RF-technologies, ILC and CLIC have similar technology challenges for several sub-systems. This is especially so for the beam delivery system, the machine detector interface and the civil engineering & conventional facilities. To take advantage of the overlapping aspects of the two studies, common working groups have been set-up and actively address common issues for both studies including beam dynamics, low beam emittance generation, positron generation, beam delivery system as well as cost and schedule. Issues of low emittance beam generation, electron cloud collective instabilities, emittance conservation studies, and beam optics for the interaction region are being tested in test facilities supported by linear collider groups, notably in the CESR-Test Accelerator at Cornell, FACET and SLAC, and the Accelerator Test Facility (ATF2) at KEK [51–53].

The 250 GeV ILC project is currently being evaluated for implementation in Japan. During 2018 one is expecting that Japan can conclude this evaluation which will determine if the project will move forward towards realisation. Such a machine could start operation in the early 2030s. The CLIC collaboration will submit a Project Implementation Plan by the end of year, describing a project that could come into operation after completion of the LHC programme in the mid 2030'ies.

7.4 Accelerating Structures Design and Efficiency

A. Grudiev · A. Yamamoto

In linear accelerators, beam is accelerated by accelerating structures made of a chain of cavities (cells) fed with RF power establishing an electromagnetic field from which part of the energy is transferred to the beam.

The efficiency of a cavity to produce an accelerating field with given RF power is defined by the shunt impedance R. This is equivalent to Ohm's law where the resistance is the proportionality factor between the square of the voltage and the power loss as given by:

$$V_{\rm acc}^2 = RP, \tag{7.7}$$

where $V_{\rm acc}$ is the accelerating voltage per cavity, R is the shunt impedance per cavity, and P is the RF power loss per cavity [1].

At a given power loss P, the accelerating voltage can be maximised by optimum shape of the cavity and through the use of low-loss cavity-surface material. Therefore the shunt impedance can be divided into two factors, R/Q and Q, as follows:

$$R = \{R/Q\}\, Q, \tag{7.8}$$

where R/Q is the so-called "cavity shape factor", only depending on the cavity shape, and Q is so-called "quality factor of the cavity resonator", mainly depending on the conductivity of the cavity wall. This brings:

$$V_{\rm acc}^2 = \{R/Q\}\, QP. \tag{7.9}$$

The R/Q value can be calculated using several available cavity codes or can be measured, and compared with a simple cavity that can be evaluated analytically. Superconducting cavities have a typical value of $R/Q = 100\ \Omega$ per cell. The shunt impedance R and the shape factor R/Q value with normal-conducting cavities must be optimized to reduce the RF power.

The quality factor Q is proportional to the ratio of the stored energy and the RF power loss,

$$Q = \omega U/P, \tag{7.10}$$

where $\omega = 2\pi f$ is the angular frequency, U is the stored energy, and P is the RF power loss dissipated in the cavity wall. A typical value for Q is 1×10^{10} for niobium superconducting cavities at 1.3 GHz and 1.8 K [1].

Assuming one cell length, l, is 0.5 wavelength, we can calculate the RF power loss per meter:

$$P/l = \left(V_{\rm acc}^2/l\right) / \{(R/Q)\, Q\} = E_{\rm acc}^2 l / \{(R/Q)\, Q\}. \tag{7.11}$$

Historically, normal conducting structures have been in use since the very beginning of RF acceleration covering the whole range of applications from very low frequency drift tube linac structures up to very high frequency travelling wave accelerating structures. Typically, normal conducting structures handle high peak power to provide high gradient and/or to support high beam loading but the pulse length is limited by the ohmic heating of the copper walls. To overcome this limitation at least in some cases, superconducting accelerating structures are being developed for several decades, in the frequency range from few hundreds of megahertz to a few gigahertz, with improving performances as described below in Sect. 7.4.2.

7.4.1 Normal Conducting Accelerating Structures

In normal conducting (NC) linacs, acceleration of charged particles using RF power is typically done in a chain of cavities (cells) which are strongly coupled and where the electromagnetic wave propagates through the cells from the input to the output of the structure. This allows a single RF source to feed many cells via single input coupler thus minimizing the feeding waveguide network. The chain of cells forms a periodic structure which, in the simplest case of a disk-loaded circular waveguide, is shown in Fig. 7.6a where the input and output couplers allow to feed the structure with RF power and extract the remaining RF power out. The property of an electromagnetic wave propagating in an infinitely long periodic structure of period d is described by dispersion curves $\omega(k_z)$, so called the Brillouin diagram as shown in Fig. 7.6b by the thick solid line. If the structure is excited at a frequency f_0 inside the passband (shown in gray in Fig. 7.6b), then the wave propagates along the structure with an RF phase advance per cell: $0 < \varphi_0 < \pi$. A travelling-wave accelerating structure is a structure where the wave is matched at both input and output ends. Since most of the normal conducting lepton linacs are based on this type of accelerating structures we will restrict ourselves to this case.

The following synchronism condition is fundamental for acceleration in periodic structures and must be satisfied in order that all cells contribute in phase to beam acceleration:

$$v_{ph} = v_p, \tag{7.12}$$

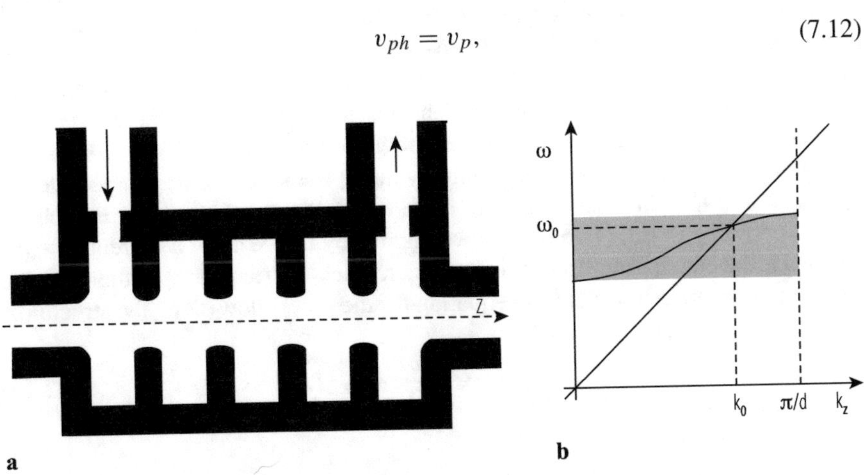

Fig. 7.6 Schematic geometry of a travelling wave accelerating structure with input and output coupler cells is shown in (**a**). Brillouin diagram for a periodic structure of period d is shown in (**b**), where $\omega_0 = 2\pi f_0$ is the operating frequency, $k_0 = \varphi_0/d$ is the propagation constant, and $\omega_0 = \omega(k_0)$

where v_{ph} is the phase velocity of the wave excited at operating frequency and v_p corresponds to the velocity of the charged particle. Ultra-relativistic case with $v_p = c$, the speed of light, is considered below. In this case, the beam line with a slope c (strait line in Fig. 7.6b as $\omega = v_p k_z$) must intersect the dispersion curve at the operating point: (f_0; φ_0) on the Brillouin diagram in order to satisfy the synchronism condition (7.12). The slope of the dispersion curve provides another important parameter of the wave propagating in the structure, the so called group velocity:

$$v_g = \frac{\partial \omega}{\partial k_z}, \tag{7.13}$$

which can also be expressed using the cell stored energy U and power flow through the cell iris aperture P_z:

$$v_g = \frac{P_z d}{U}. \tag{7.14}$$

The stronger the coupling between the cells the higher the group velocity and the faster energy propagates along the structure. Combining Eqs. (7.9, 7.10 and 7.14) yields expression for the power flow along the travelling-wave structure which is needed to maintain an accelerating gradient $E_{\mathrm{acc}} = V_{\mathrm{acc}}/d$:

$$P_z = v_g \frac{E_{\mathrm{acc}}^2}{\omega R'/Q}, \tag{7.15}$$

where $R' = R/d$ is the shunt impedance per meter length. For a given working point (f_0; φ_0), the group velocity, the Q-factor and the R-upon-Q fully describe the accelerating properties of the cell. In so-called constant impedance, the geometry of all cells is identical and the three above parameters are identical in all cells. In practice, the so-called constant gradient structures are used. In these structures, the geometry of the cells is tapered in order to maintain $E_{\mathrm{acc}}(z) \approx$ *const*. This is achieved by reducing the group velocity along the structure to compensate the reduction in power flow along the structure which is caused by two terms: ohmic losses according to Eq. (7.15) and power gained by the beam of a current $I = qf_b$, where q is the bunch charge and f_b is the bunch repetition frequency. In this case, the energy conservation law yields an equation for the power flow along the structure:

$$\frac{dP_z}{dz} = -\frac{P_z \omega}{v_g Q} - E_{\mathrm{acc}} I. \tag{7.16}$$

The distribution of accelerating gradient $E_{acc}(z)$ along the structure is obtained by solving Eqs. (7.15 and 7.16) for a given input power P_{in}. Integrating it over the structure length L gives the overall structure energy gain:

$$V_{AS} = \int_0^L E_{acc}(z)dz. \tag{7.17}$$

Then the steady-state RF-to-beam efficiency is defined as following:

$$\eta_0 = \frac{V_{AS}I}{P_{in}}. \tag{7.18}$$

For linacs operating in pulsed mode, the structure must be filled on each pulse before beam is injected. Filling time of the structure is defined as

$$t_f = \int_0^L \frac{dz}{v_g(z)}. \tag{7.19}$$

In this case, RF-to-beam efficiency, η, is reduced by the ratio of the bunch train length $t_b = N_b/f_b$, where N_b is the number of bunches and the RF pulse length $t_p = t_b + t_f$:

$$\eta = \eta_0 \frac{t_b}{t_p}. \tag{7.20}$$

A few examples of normal-conducting travelling wave cavity parameters are provided in Table 7.3.

Table 7.3 Examples of normal-conducting travelling wave cavities

	SLC [54]	CTF3 [55]	CLIC-ML [56]
Frequency (GHz)	2.9	3	12
Average gradient (MV/m)	17	7	100
Average Q	13,000	12,500	5640
Current (A)	Two bunches e+e−	4	1
Repetition rate (Hz)	180	50	50
Pulse width (μs)	0.82	1.6	0.24
RF-to-beam efficiency (%)	2	90	28
# cell/cavity unit	85 + 2	32 + 2	26 + 2
Status	In operation	In operation	R&D

7.4.2 Superconducting Accelerating Structures

Linear accelerators have benefitted greatly through the use of superconducting radio-frequency (SCRF) cavity technology [1, 57]. This technology, when applied in standing-wave RF operation, provides the following important advantages:

- Small RF surface resistance and large quality factor, Q, resulting *long pulse operation*, with a range of 1 ms, and much higher duty factor in beam acceleration.
- Lower operational frequency with enlarged beam-apertures in the range of ~70 mm diameter, (1.3 GHz), resulting in large acceptance and providing practical solutions for *very intense* beams.

Two salient characteristics of superconducting cavities are (1) the average accelerating field gradient E_{acc} and (2) the intrinsic quality factor Q. Quality factor Q is a universal figure of merit for resonators and is defined in the usual manner as the ratio of the energy, U, stored in the cavity to the power, P_c, lost in one RF period. Q depends on the microwave surface resistance of the metal. In general, one would like to have as high accelerating field and as high Q as possible.

The strongest incentive to use superconducting cavities in an accelerator is that continuous wave (CW) mode or high duty factor (>1%) operation is practical. For CW operation power dissipation in the walls of a copper structure is substantial and often not possible. Here superconductivity comes to the rescue. The microwave surface resistance of a superconductor is typically five orders of magnitude lower than that of copper, and therefore the Q value is five orders of magnitude higher [58]. The above advantages may be of benefit even though superconducting technology requires low temperature (1.8 K) cryogenic system operation, resulting additional power consumption, discussed below.

Figure 7.7a shows a schematic cavity shape of a normal-conducting multi-cell cavity (top), which represents larger impedance to the beam due to the small beam

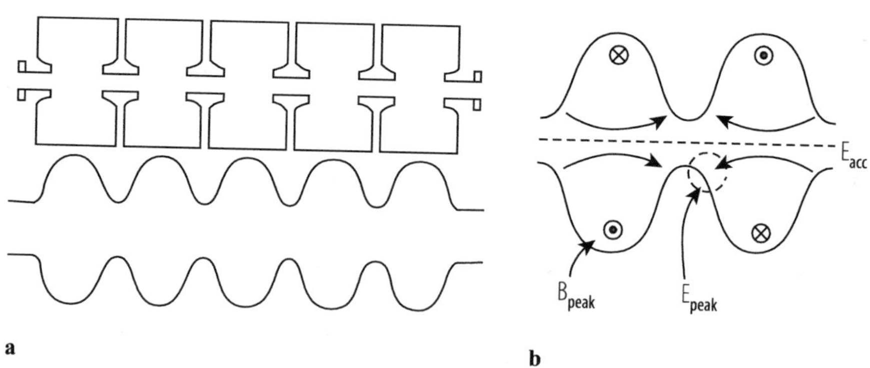

Fig. 7.7 (**a**) A comparison of cylindrical shaped normal-conducting RF cavity (top) and elliptical shaped superconducting RF cavity (bottom), and (**b**) electric and magnetic field profile in the elliptical cavity structure [1, 2]

holes and nose cones, and a schematic shape of a superconducting multi-cell cavity (bottom). The corresponding electromagnetic field profile inside an elliptical cavity is shown in Fig. 7.7b [1, 2].

The RF loss is proportional to the square of the surface current and proportional to the resistivity. For a superconducting cavity, the microwave surface resistivity is several orders of magnitude smaller than that of Cu, although it is not zero. The shape of a superconducting cavity is optimised for properties such as: (1) reduced excitation of higher order harmonics by the beam, (2) reduced surface magnetic field to enhance the critical limit of the resultant superconducting to normal-conducting phase transition, (3) reduced surface electric field to suppress field emission, and (4) reduced multipacting behaviour [59–61]. The large iris opening and elliptical shape result from these considerations.

Following Eq. (7.11), an RF power loss per meter of ~0.1 W/m is generated in an accelerating cavity gradient of 1 MV/m, and it is proportionally increased with .square of the frequency.

The above RF power loss dissipated into the 1.8 K cryogenic fluid needs to be converted to an AC power load at room temperature using a total cryogenic efficiency estimate, η_{cr}, which includes the 'Carnot efficiency' and the 'thermodynamic efficiency as follows:

$$\eta_{cr} = \eta_c \eta_d = (P_{1.8K}/P_{300K-i})\,(P_{300K-i}/P_{300K-r})\,, \tag{7.21}$$

where η_c is the 'Carnot efficiency', η_d is the 'thermodynamic efficiency' for the compressor work at 300 K [2]. Assuming η_c is ~1/(300/1.8) and η_d is ~0.2 (typical for large scale refrigerators), the total cryogenics efficiency η_{cr} can be ~1/800. Including this effect in the estimate of power loss, the relative RF power loss saving factor of superconducting cavities, (surface resistance ~10^{-5} lower than normal-conducting cavities), may be approximately two orders of magnitude better in AC power consumption performance for a CW operation. Therefore, superconducting cavity technology enables the nearly entire (>99%) RF power to be transmitted to the beam from the power source,

However, in case of the pulsed operation, which is the usual mode of normal-conducting cavity operation, general power loss should be evaluated including a duty factor, and is given by

$$(\omega U/Q) \times (\text{pulse length}) \times (\text{repetition rate}) = (\omega U/Q) \times (\text{duty factor})\,. \tag{7.22}$$

The pulse duration is typically in the μsec range in case of normal conducting cavity operation, and the duty factor is normally about three orders of magnitude smaller for normal-conducting cavity operation, compared with the superconducting cavity operation. It results in the general power balance between the normal-conducting cavity and superconducting cavity operation become in similar level, with including the total cryogenic efficiency for the superconducting cavity operation. On the other hand, it should be noted that a pulse duration in the level of msec,

Table 7.4 Summary of superconducting cavities in operation, and planned [65, 66]

	FLASH	European-XFEL	ILC-ML
Gradient (MV/m)	18–35	23.8	31.5
Q	5×10^9 to 1×10^{10}	1×10^{10}	1×10^{10}
Current (mA)	1–9	5	6
Repetition rate (Hz)	5	10	5
Pulse width (μs)	800	650	730
# cell/cavity unit	9	9	9
Status	In operation	In operation	Planned

thus three order magnitude longer than that of the normal-conducting cavity, would be much helpful in other linear collider sub-system such as detectors and feedback system. A superconducting cavity system may be expected to operate with more than two times better efficiency, in CW operation than that of the normal conducting cavity system. However, the power consumption in pulsed operation is less different in either case.

The superconducting cavity intended for use in high-energy accelerators was designed for operation at 1.3 GHz with a cell length of 115.4 mm. The 9-cell elliptical cavity was designed and developed for the FLASH/TESLA Test Facility program at DESY [62], and has become a standard for further programs such as the European XFEL program [63] and for the ILC project [64]. Table 7.4 gives a summary of superconducting cavity operation in FLASH, planned operation at European-XFEL, and planned for ILC [65, 66]. Superconducting cavity technology is expected to advance substantially through further optimization of cavity materials, shapes (TESLA, Low-Loss, Re-entrant), and cost-effective fabrication techniques [57, 59–61, 65–68].

7.5 Wakefields and Emittance Preservation

A. Latina

Wakefields induced by particles in high impedance environment interact with the following particles and can therefore affect the beam quality. They can be described in either the time domain, using the wake-potential W, or in the frequency domain, using the impedance Z. Analytical approximations to describe wakefields due to resistive walls or geometry variations in periodic accelerating structures exist. In case of complex geometries, however, the analytical computation of wake-potentials and impedances must be performed numerically. In the following paragraphs, we provide models for short- and long- range wakefields and describe their impact on the beam. The symbols used in the following paragraphs are defined in Table 7.5. The unit 1/m in the wakefield functions indicates that the effect is normalized to the length of the generating element. The unit 1/mm (and its second power) relates to the

Table 7.5 Symbols and constants used in the text

Symbol	Description	Notes	Symbol	Description	Notes
e	Absolute electron charge	1.6×10^{-19} C	L_0	Length of the linac	(m)
c	Speed of light	299,792,458 m/s	k_β	Average lattice focusing strength	(1/m)
N	Bunch population	Number of particles	Z	Impedance	(Ω)
q	Bunch charge	$q = Ne$	$W_{\parallel,0}$	Wake: longitudinal monopole	(V/C/m)
σ_z	r.m.s. bunch length	(m)	$W_{\perp,1}$	Wake: transverse dipole	(V/C/m/mm)
E	Bunch energy	(GeV)	$W_{\perp,2}$	Wake: transverse quadrupole	(V/C/m/mm^2)
l_b	Bunch-to-bunch distance	(m)	V	Cavity voltage	$V = GL$
L	Length of a cavity	(m)	G	Cavity accelerating gradient	(V/m)

transverse offset of the exciting charge. For example, the transverse and longitudinal momentum kicks experienced by a charge q following at a distance z a charge Q, due to the transverse dipole and the longitudinal monopole modes, are respectively $\Delta\vec{p}_\perp = -qQL_{structure}\vec{r}_\perp W_{\perp,1}(z)$, and $\Delta\vec{p}_\parallel = -qQL_{structure}W_{\parallel,0}(z)$, where $L_{structure}$ is the length of the structure and $\vec{r}_\perp$ is the (small) transverse offset of the exciting charge.

7.5.1 Short-Range Wakefields

The wake-functions of a periodic accelerating structure have been parameterized by a number of authors. Here we present a convenient formula for the longitudinal and the transverse component of the wake-potential, $W_\parallel$, provided by Bane et al. [69]:

$$W_\parallel = \frac{Zc}{\pi a^2}\exp\left(-\sqrt{\frac{s}{s_0}}\right), \text{ and } s_0 \approx 0.41\frac{a^{1.8}g^{1.6}}{d^{2.4}}, \tag{7.23}$$

$$W_{\perp,1} = 4\frac{Zc}{\pi a^4}s_0\left(1-\left(1+\sqrt{\frac{s}{s_0}}\right)\right)\exp\left(-\sqrt{\frac{s}{s_0}}\right), \text{ and } s_0 \approx 0.169\frac{a^{1.79}g^{0.38}}{d^{1.17}},$$

where s is the distance from the source charge, a is the radius of the iris aperture, g is the interior cell width and d is the cell period (i.e. $g = d - h$ where h is the disc thickness); Z is the impedance of the medium, typically 377 Ω for an evacuated accelerating structure.

7.5.2 Long-Range Wakefields

The long-range wakefields are usually characterized by a set of cavity modes, obtained numerically. Three numbers (c_m, Q_m, k_m) are necessary to describe a mode. Following Eq. 2.88, in [70], the wake-function for each mode m, is

$$W_{\perp,m}(s) = c_m\frac{R}{Q_m}\exp\left(\frac{k_m z}{2Q_m}\right)\sin(k_m s), \tag{7.24}$$

where c_m is the amplitude of the mode in V/C/m/mm^m, Q_m is the quality factor, k_m is the wake number, and s is the distance from the source to the witness particle and is negative for all particles affected by the wake. Note that $W_{\perp,m}(s)$ is a decaying exponential as expected. The total wake-potential is the sum of all modes.

7.5.3 *Single-Bunch Wakefield-Induced Effects*

7.5.3.1 Beam Loading

The electromagnetic interaction between the bunch tail and the wakes induced by the bunch head, causes the tail to radiate and lose energy. This decelerating effect is called *beam loading*. The energy loss experienced by each particle is estimated by adding to the self-generated wake the decelerating voltage due to the upstream generated wakefields:

$$\Delta E = -eL\sum_i \left[\frac{1}{2}\left|q_i\right| W_{\parallel}(0) + \sum_{\forall j/z_i<z_j} \left\{\left|q_j\right| W_{\parallel}\left(z_{ij}\right)\right\}\right], \tag{7.25}$$

where $z_{ij} = z_j - z_i$ is the distance between the ith and the jth macroparticles; the inner summation runs over all particles preceding q_i, i.e. with $z_i < z_j$. Notice that the energy loss due to the self-generated wakefield, in $z = 0$, is *half* the energy loss given by the upstream generated wakefield. This is the *fundamental theorem of beam loading* [71].

7.5.3.2 Wake-Induced Energy Spread

Since the decelerating voltage in Eq. (7.2) varies along the bunch, an RMS energy spread arises. An estimate of this wakefield-induced energy spread can be obtained considering $W_{\parallel}$ in Eq. (7.23) and a bunch modeled with two macro-particles located at $z = 0$ and $z = 2\sigma_z$ respectively,[1] each with charge $q/2$. The decelerating voltage experienced by the two particles is

$$\begin{aligned}\Delta V_1 &= \tfrac{1}{2}\tfrac{qL}{2} W_{\parallel}(0),\\ \Delta V_2 &= \tfrac{1}{2}\tfrac{qL}{2} W_{\parallel}(0) + \tfrac{qL}{2} W_{\parallel}(2\sigma_z) = \tfrac{1}{2}\tfrac{qL}{2} W_{\parallel}(0)\left(1 + 2\mathrm{e}^{-\Delta}\right), \text{ with } \Delta = \sqrt{2\sigma_z/s_0}.\end{aligned} \tag{7.26}$$

The two particles experience two different energy losses: ΔE_1 and ΔE_2 (where $\Delta E = -e\Delta V$), this introduces energy spread within the bunch:

$$\delta E = e\frac{qL}{2} W_{\parallel}(2\sigma_z). \tag{7.27}$$

[1] This gives the overall distribution an RMS length of σ_z.

7.5.3.3 Energy Spread Compensation

To compensate for the wake-induced energy spread, one can adjust the RF phase offset ϕ_{RF} (at the cost of a slight reduction of the acceleration rate), so that the change in energy gain equals the change in wakefield deceleration. In the approximation $V \gg qLW_{\parallel}(0)$, $\sigma_z \ll s_0$, and $\sigma_z \ll \lambda$ the result is [71]

$$\phi_{RF} = \frac{qLW_{\parallel}(0)}{8\pi V}\frac{\lambda}{\sigma_z}, \tag{7.28}$$

where λ is the wavelength of the accelerating mode and V its voltage. The residual energy spread after compensation is found from the convolution of the bunch with the longitudinal wakefield and the acceleration RF

$$\frac{\Delta E}{E} \approx \frac{1}{4}\frac{qW_{\parallel}(0)}{G}. \tag{7.29}$$

7.5.3.4 Single-Bunch Beam Break-up

If the beam is traversing off-center an acceleration cavity, the bunch head can excite a transverse dipole wakefield $W_{\perp,1}$ that causes transverse deflection of the tail. This deflection affects the tails' betatron motion and can lead to a transverse beam break-up. Using a two-particle bunch model, the oscillation amplitude of the bunch tail relative to the head, at the linac end, is characterized by the dimensionless growth BBU parameter [71, 72]:

$$T_{\mathrm{BBU}} = -\frac{eqW_{\perp,1}(2\sigma_z)}{4k_\beta}\frac{L_0}{L}\frac{2}{\sqrt{E_i E_j}}. \tag{7.30}$$

Here we assume a lattice design where $k_\beta \approx const$ and $\beta \approx \gamma^2$. Equation (7.6) holds also in case of no acceleration, with $1/E$ replacing $2/\sqrt{E_i E_j}$. The BBU parameter can be interpreted as the following: if a beam is injected with a certain betatron oscillation, the transverse wake-functions cause an oscillation of the tail that increases by a factor T_{BBU} long the linac. BBU instability can be mitigated by using BNS damping.

7.5.3.5 Single-Bunch BNS Damping

The defocusing effect of $W_{\perp,1}$ can be compensated by increasing the focusing strength of the tail particles, from k_β to $k_\beta + \Delta k_\beta$. To do this, RF quadrupoles with rapidly varying field can be used, or the bunches can be offset with respect to the crest of the RF wave so that the tail acquires less energy than the head. Using a two-particles model, the equation of the motion for the trailing particle is [71, 72]

$$x_2''(s) + \left(k_\beta^2 + \Delta k_\beta^2\right) x_2(s) = -\frac{eq W_{\perp,1}(2\sigma_z)}{2E} x_1(s). \tag{7.31}$$

BNS damping is achieved if the "auto-phasing" condition is met,

$$\left(1 + \frac{\Delta k_\beta}{k_\beta}\right)^2 = 1 + \frac{2T_{\mathrm{BBU}}}{k_\beta L_0}. \tag{7.32}$$

BNS damping should be applied at low energies, where the instability is stronger. In this regime, the energy reducing effect of the longitudinal wakefield actually helps to maximize BNS damping.[2]

7.5.4 Multi-Bunch Wakefield-Induced Effects

7.5.4.1 Multi-Bunch Beam Break-Up

Multi-bunch BBU leads to an amplification of the incoming trajectory jitter to cause trailing bunches to be strongly deflected transversely. Each bunch can be assimilated to a point charge. For n equally-charged, equally-spaced bunches, each bunch represented by a single macroparticle with a charge Ne, the equation of motion is

$$x_n'' + \frac{dE}{ds}\frac{1}{E}x_n' + k_\beta^2 x_n = -\frac{eq}{E}\sum_{i=1}^{n-1} W_{\perp,1}((n-1)\, l_B)\, x_i. \tag{7.33}$$

A difference from the single-bunch BBU is that $W_\perp$ is now dominated by one or few resonators having large shunt impedance Q_m (see Eq. (7.1)).

7.5.4.2 Control of Multi-Bunch BBU

The multi-bunch BBU is mitigated by minimizing the long-range transverse wakefield in the structure design. Assuming the Daisy chain model, the criterion for little or no blow-up is

$$\left|\frac{eq W_{\perp,1}(l_B)}{ek_\beta}\frac{L_0}{L}\right| \frac{2}{\sqrt{E_i E_j}} < 1, \tag{7.34}$$

where $W_{\perp,1}(l_B)$ is the wakefield at the following bunch (see Sect. 4.3 in [73]).

[2]Toward the end of the linac, at high beam energies, the beam break-up effect becomes small, and the bunch should be moved ahead of the crest to reduce the energy spread in the beam.

7.6 Focusing at Interaction Point

A. Seryi · R. Tomás García

7.6.1 Final Focus Design

The main task of a Final Focus (FF) system is to focus the beams to the small sizes required at the interaction point (IP) of a Collider. To achieve this, the FF forms a large and almost parallel beam at the entrance to the final doublet (FD), which contains two or more strong quadrupole lenses. However, even for a beam with a minor energy spread of a fraction of a percent, the focused beam size will be diluted by the chromaticity of these strong lenses. The design of a FF is therefore driven primarily by the necessity of compensating the chromaticity of the FD.

There are two primary approaches for chromaticity compensation—the non-local scheme, implemented particularly at FFTB [74] and B-factories [75, 76] and the local compensation Scheme [77] at ATF2 [78, 79]. Further developments in optics with smaller vertical beam size and larger chromaticity are also being investigated at ATF2 [80, 81] to explore the feasibility of the local compensation scheme at different chromaticity levels.

In the non-local FF, the chromaticity is compensated in dedicated sections by sextupole magnets placed at maxima of dispersion and beta-functions. The geometric aberrations generated by the sextupoles are cancelled when used in pairs with a minus identity transformation between them.

The non-local FF is built from separated optics blocks with strictly defined functions, and its design and analysis is relatively simple. The major drawback of the non-local FF rests in its required length for multi-TeV colliders. This can be partly mitigated by allowing a smaller peak dispersion function and adding extra sextupoles to the design [82, 83].

Local compensation of chromaticity is achieved by interleaving a pair of sextupole magnets with the quadrupoles of the final doublet, see Fig. 7.8. The dispersion throughout the FD is created by upstream bends, and is designed to cancel at the IP. Geometric aberrations, generated by FD sextupoles, are cancelled by two or more sextupoles located upstream. Sextupoles placed in FD generate second order dispersion, which, however, can be compensated simultaneously with x and y chromaticities provided that half of the total horizontal chromaticity of the whole FF is generated upstream. The second order aberrations are cancelled when the x and y pairs of sextupoles are separated by transfer matrices M with block-diagonal structure {A 0; 0 B} where A = {f 0; c − 1/f}, provided the optics is flexible enough to adjust the coefficients and provide compensation of third and fourth order aberrations. The FF with local compensation requires fewer bends, and allows the design of a 3 TeV CM FF system with about half a kilometre length. The recipe for the design of such final focus is described in [84].

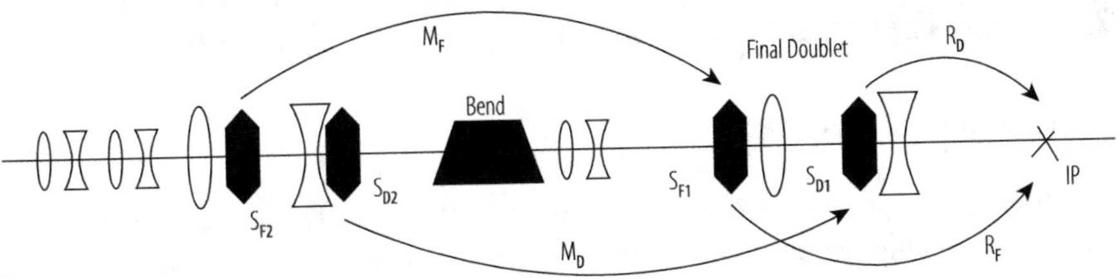

Fig. 7.8 Optical layout of the final focus with local chromaticity correction. The final doublet consists of two quadrupoles (represented by lenses) and two sextupoles (represented by hexagons) to locally cancel the chromatic aberrations. A replica of the final doublet is placed upstream to cancel the sextupolar geometrical aberrations

Synchrotron radiation in the FD sets a lower limit to the achievable IP rms spot size [85, 86] that depends on the FFS optics parameters and the beam emittance. This effect drives the length of the FD quadrupoles for high energy colliders.

7.6.2 *Final Focus Optimization*

The transfer map between the start of the FFS and the IP is given by $x_{IP} = X_{jklmn}$ $x^j\, p_x{}^k\, y^l\, p_y{}^m\, \delta^n$, where X_{jklm} are the map coefficients that can be extracted from MAD-X [87] and PTC [88] and the sum over repeated indexes applies. The standard quadratic deviation of the particle distribution at the IP is expressed as a function of the X_{jklm} coefficients and the entry beam sigmas as given in [89]. This allows for a semi-analytical optimization of any lattice parameters (like the strength of non-linear elements) so as to minimize the IP beam size.

7.6.3 *Final Focus tuning*

The unavoidable misalignments and field errors of the different components of the FFS result in an emittance dilution at the IP. FFS tuning refers to the process of bringing the machine to nominal performance and to maintain it in the presence of dynamic errors. The initial set-up procedure involves steering the beam through the centre of critical apertures and magnetic elements with known higher-order fields. Pre-computed knobs use orbit bumps at the sextupoles to orthogonally control all the different IP particle distribution correlations. These knobs are iteratively scanned until the minimum IP beam size is reached. Finally either single magnets strengths or higher order knobs [90, 91] can be scanned to minimize the higher-order aberrations.

7.7 Low Emittance Generation

S. Guiducci · Y. Papaphilippou

The high luminosity of a linear collider depends strongly on the generation of ultra-low emittance high-intensity bunches, with remarkable stability. Conventional electron sources and positron production schemes provide beams with several orders of magnitude larger emittances, than the ones needed. The required cooling mechanism is generated by the natural synchrotron radiation damping of the beam when circulating in rings.

The requested performance of the damping rings (DRs) is driven by the collider's principal parameters, the upstream or downstream systems' requirements, and especially the main linac RF. The parameters driving the design of the ILC [92] and CLIC are shown in Table 7.6. The technological choice of super-conducting over copper main linac RF cavities, clearly diversifies the DR design, although a number of approaches and challenges remain common. In the one flavour of DRs as CLIC, the bunch trains are relatively short with even shorter bunch spacing and with a high repetition rate. The ILC bunch train is ~220 km long and needs to be compressed and stored in a 3.2 km-long ring. In order to achieve the high luminosity, the ILC is based on bunches with high bunch charge and small emittances, whereas CLIC targets small bunch charges with much lower emittances. Modern X-ray storage rings in operation or construction phase are rapidly approaching these regimes, targeting ultra-low transverse emittances. Especially for the vertical emittance, requiring challenging alignment tolerances and stringent control of the optics and orbit, X-ray rings in operation have approached the quantum limit of vertical emittance, i.e. values below 1 pm [93].

For CLIC, the large input emittance for the positron beam and the high repetition rate necessitates a two-stage beam damping, with a pre-damping ring [94]. For ILC, there is no pre-damping stage and the DRs need a large acceptance for the injected beams, especially for positrons.

Most of the design challenges of the DRs are driven by the extremely high bunch density and the associated collective effects. In this respect, the DR parameters (Table 7.7) are carefully chosen and optimised in order to mitigate these effects [95, 96].

Table 7.6 CLIC versus ILC parameters driving the DRs design

Parameters	ILC	CLIC
Bunch population (10^9)	20	4.1
Bunch spacing (ns)	554	0.5
Number of bunches/train	1312	312
Number of trains	1	1
Repetition rate (Hz)	5	50
Ex. H/V/L norm. emittances (μm, nm, keV m)	(5.5, 20, 33)	(0.5, 5, 6)

Table 7.7 ILC and CLIC DRs design parameters

Damping ring parameters	ILC	CLIC
Energy (GeV)	5.0	2.86
Circumference (m)	3.238	359.4
Energy loss/turn (MeV)	4.5	5.8
RF voltage (MV)	14	6.5
Compaction factor (10^{-4})	3.3	1.2
Damping time x/s (ms)	24/12	1.2/0.6
Number of arc cells/wigglers	150/54	90/40
Dipole/wiggler field (T)	0.23/2.2	0.69–2.3/3.5

In the case of CLIC, the steady state emittance is dominated by Intra-Beam Scattering (IBS). The ring energy [97] and lattice design [96], (racetrack shape with TME arc cells with variable field dipoles [98] and long straight sections filled with super-conducting wiggler [99] FODOs) is optimized for reducing IBS. The larger emittance specification of ILC, allows for higher ring energy, thus relaxing collective effects.

Due to the very small beam size especially in the vertical plane, IBS is large. In order to mitigate its value within manageable limits, the ring is made as compact as possible, and the longitudinal beam size has to be increased [10P2, 100].

The key systems to allow damping the beam to ultra-low horizontal emittance in a compact ring during the short time between two machine pulses like in CLIC are high-field super-conducting damping wigglers with short period. Prototypes have been built and tested in a synchrotron light source, including novel cooling concepts [101]. Higher field mock-ups based on Nb_3Sn technology are under development and tests [102].

The combination of high bunch density and short bunch spacing triggers two stream instabilities for both ILC and CLIC DRs. In the e^--ring, the fast ion instability can be avoided with low vacuum pressure, partial ring filling and bunch-by-bunch transverse feedback [103]. In order to mitigate the electron cloud build up and avoid the instability to occur in the e^+-ring, the secondary electron yield (SEY) of the vacuum chambers has to be limited to below 1.2–1.3 and the photo-emission yield (PEY) has to be very low, from a few down to 0.1% [104]. The low SEY can be achieved with chamber coatings, as TiN, NEG or amorphous carbon [105, 106], whereas the low PEY necessitates an efficient photon absorption scheme.

The e-cloud mitigation, low emittance generation, fast kicker technology and the associated diagnostics are studied in dedicated test facilities (CESR-TA, ATF), synchrotron light sources as well as at various laboratories around the world.

The very high peak and average current of CLIC presents a big challenge due to the transient beam loading, especially for a high frequency RF system. Concepts of RF design including low-level RF feedback have been developed extrapolated from the design of e^+/e^- ring colliders [107].

As the beam stability requirement is quite stringent and typically 10% of the beam size, tight jitter tolerances for the rings extraction kickers are imposed, down to a few10^{-4}. Especially for the bunch-by-bunch extraction scheme of ILC, the kicker rise time of a few ns, is extremely challenging. An ILC extraction experiment using a prototype strip-line kicker was carried out at KEK-ATF [108]. It achieved multi-bunch beam extraction with 5.6 ns bunch spacing. The angle jitter of a single bunch beam was reduced to 3.5×10^{-4}, using a double kicker system. For CLIC, a stripline with an ultra-stable inductive adder as power source is being currently tested at ALBA synchrotron [109].

7.8 Recirculated Linacs and Energy Recovery

S. A. Bogacz · G. A. Krafft

Linear accelerators provide superb beam quality as defined by the sources, but they are current-limited due to the high cost of their RF drive. Circular accelerators, by contrast, offer high average beam current and exhibit high electrical efficiency (and associated cost reductions) because of the limited required investment in RF power. However, effects such as quantum excitation (incoherent synchrotron radiation) or space charge limit their beam quality. On the other hand, rings typically only come to equilibrium after many hundreds or thousands of turns, which significantly degrades beam quality (emittance, momentum spread, etc.).

Recirculated Linear Accelerators (RLAs) have several advantages that support electron beam parameters outside of the scope of the traditional ring accelerators or linacs [110]. Synchrotron radiation effects, as in electron storage rings, do not limit beam emittances and pulse lengths emerging from an RLA. The beam is circulated only a modest number of times, so the impacts of the degrading effects are then limited, and the beam quality may be much higher than the equilibrium beam quality inherent for rings.

Going a step further, in the Energy-Recovered Linac (ERL), the full-energy beam is returned to the accelerating structure out of phase with respect to the accelerating field after being used (collides, produces synchrotron radiation, drives a specific reaction, etc.). The beam is decelerated, returning the RF power in the beam to the accelerating structure, making this RF power available to accelerate a subsequent beam. The resulting system retains the excellent beam quality of a conventional linac—and, because of the recovery of significant levels of RF power—can provide high average beam current at excellent electrical efficiency and lower associated cost. The ERL concept is quite attractive because it provides linac-quality/brightness beam at storage ring beam powers. With the advent of operational ERLs, it may be possible to push average currents to levels approaching the best lepton storage rings in existence as of 2019. Furthermore, the production of high beam power with reduced RF drive represents improved electrical efficiency, introducing ‘green technology’ with significant cost reductions. Energy recovery also provides an additional environmental advantage, as the total stored energy in the system is deposited at very low (injection) energy, providing mitigation of serious environmental/safety issues. Finally, ERLs, like linacs, offer a flexible time structure, allowing operation with single bunches, CW bunch trains, and virtually every combination between these options. As in other linac-based systems, ERLs can easily manipulate various portions of phase space independently of other portions; they are fully six-dimensional systems, supporting transverse matching to desired spot sizes, longitudinal matching to desired bunch length/energy spread ratios (via transverse/longitudinal coupling), and any (or all) of horizontal/vertical transverse/longitudinal phase space exchanges.

Next-generation light source or collider applications requiring the following elements should generally be well-suited to deploying a recirculated and/or energy-recovered linac: CW or other high duty factor operation, high beam average current, low delivered beam energy spread, and low delivered beam emittance. CW beam acceleration with high accelerating gradients (>10–20 MV/m) generally requires deploying a multi-pass RLA consisting of superconducting accelerator structures. GeV-scale RLAs at 100 mA average current would ordinarily require at least 100 MW of installed RF power merely to accelerate the beam load. Beam energy recovery allows substantial reduction of the RF beam loading of the cavities.

In applying this idea with a back-to-front beam recirculation, as illustrated in Fig. 7.9, the beam recirculation path length is chosen to be an integral number of RF wavelengths, plus approximately one-half of the RF wavelength. Because the beam sees accelerating phase on the lower accelerating beam passes through the linac, after a phase shift of 180 degrees, energy is delivered back to the (S)RF cavities by higher beam passes, and transferred directly to the accelerating beams without the need for additional power from other RF sources [111]. To the extent that the average beam load from the accelerating passes completely cancels the beam load from the decelerating beam passes, there is no limit to the average current that may be accelerated due to RF source capacity. Because the beam transit time through the recirculated linac is much smaller than the radiation-induced emittance growth times in the bending arcs, the beam longitudinal and transverse emittances can be much smaller in energy-recovered linacs than in storage ring accelerators that operate at the same energy. It should be noted that energy recovery is also an important element in the design of high average current electrostatic accelerators.

Beam energy recovery was first proposed as a way to construct high-luminosity colliders for high energy physics [112]. Although never realized in this application, energy-recovered accelerators have been built as electron cooling drivers and high-power free-electron laser drivers [113–115].

Many proposed applications benefit from the advantages of energy-recovered linacs. For example, Cornell University is investigating the energy-recovered linac as an undulator driver yielding superior, high average brilliance X-ray sources as an upgrade to their conventional synchrotron light facility [116]. Similar programs exist at Argonne and Daresbury Laboratories [117], and in Japan [118, 119]. Brookhaven National Laboratory and CERN are investigating the use of high average current

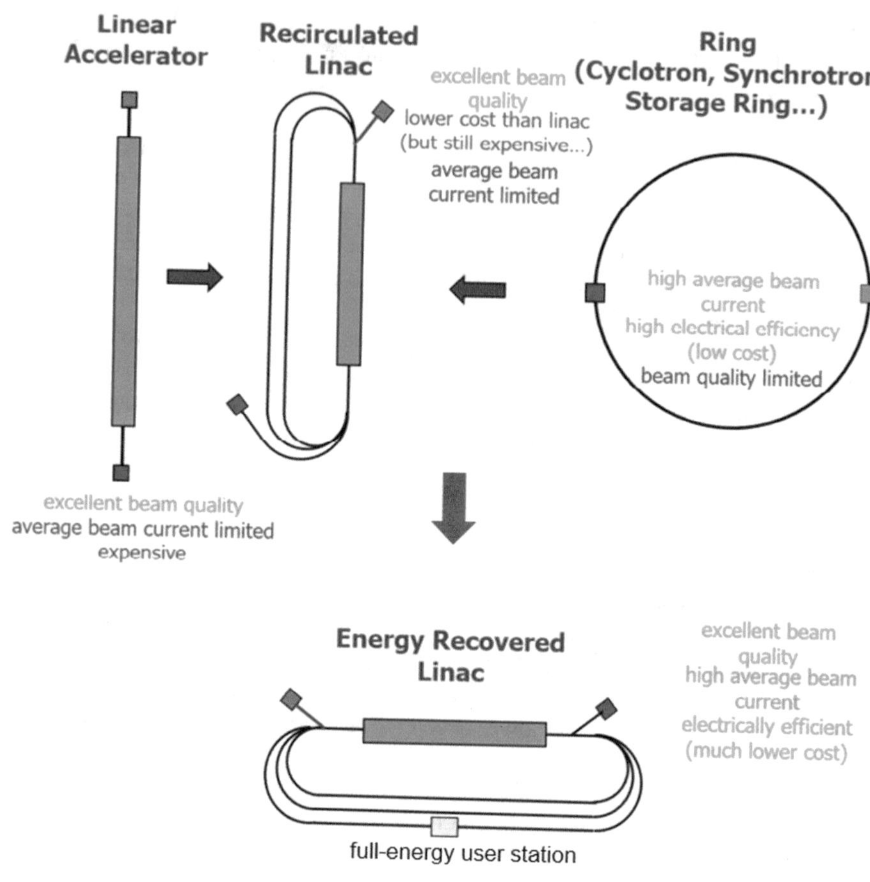

Fig. 7.9 Evolution of accelerator architectures driven by cost/performance optimization—the ERL architecture emerges as the final step—'the best of both worlds' (linacs, rings)

energy-recovery linacs as electron sources for high-luminosity electron-ion colliders [120]. Advanced RF-coolers have been proposed based on energy-recovered linacs.

Several important design issues for recirculated linacs are the high average current gun and injector, linac design, recirculating arc design, and beam stability. Historically, DC guns have been applied at recirculated and energy-recovered linacs, largely because they easily support CW beam. The desire to support higher beam quality in the injector has led to efforts to develop CW RF-based electron guns. Linac design philosophy has largely followed the example of the CEBAF accelerator where 'graded gradient' first-pass optics are designed to yield constant phase advance, and the higher beam passes are less focused because of the higher energy of higher beam passes. For a variety of reasons, superconducting recirculated linacs prior to 1990 used recirculation arcs that were isochronous, leading to FODO-type systems as in the large recirculation arcs of CEBAF, or the large energy acceptance design first deployed at Bates Laboratory [121, 122]. On the other hand, normal-conducting devices have operated with non-isochronous recirculation and microtron phase stability [123] as an inherent design choice [124]. Recirculator optics based on two-bend and three-bend achromatic optics have been deployed. With the advent of beam energies where synchrotron radiation is an important part of the beam quality from the recirculator, arcs are being suitably designed and implemented to minimize emittance and energy spread growth, as at synchrotron radiation sources.

An additional multi-bunch beam instability potentially affects recirculated and energy-recovered linacs: multi-pass beam breakup (BBU) [125, 126]. In this instability, deflections of the beam generated by high order modes in the accelerating cavities can drive unstable feedback if the deflections translate after recirculation into position offsets that act to further excite the high order mode. This instability was first observed in the first superconducting recirculator at Illinois, and restricted the operating current in the early Stanford superconducting recirculator. When the much-larger-scale CEBAF accelerator was built, the solution to the instability problem was provided by building the linac from cavities that were known to have good HOM damping. This approach has continued to the present, where cavities have been designed and tested that are expected to support 100 mA to 1 A currents in recirculating linac arrangements.

Successful operations of high-power FEL drivers stimulated community interest in ERL technology, and there followed numerous proposals and a number of actual systems. These included: FEL drivers: the JLab IR Upgrade FEL [127], ALICE [128], and the JLab UV Demo FEL [129]; test facilities: CEBAF-ER [130], the KEK cERL [131], bERLinPro [132], ER@CEBAF [133], and PERLE [134, 135]; and systems associated with nuclear physics facilities: the BNL test ERL, an electron cooler concept [136], MESA [137], and most recently, conversion of the Darmstadt S-DALINAC to an ERL [138].

Having described the motivation for ERLs and how they operate, we briefly survey the contemporary ERL landscape, review progress to date, and detail challenges confronting upcoming generations of these machines. We also provide an overview of applications (servicing FELs, high-energy electron cooling systems, inverse Compton-driven gamma sources, internal-target experiments, accelerator science/technology test platforms).

7.8.1 Novosibirsk ERL

The Novosibirsk ERL—developed, built and commissioned at BINP in 2003—was the first multi-pass ERL operating in CW mode [139]. It initially reached a top energy of 12 MeV, with high average current of 30 mA [140]. The ERL is fed by a 300 kV electrostatic gun with a thermionic cathode ($Q \sim 1$ nC, $\tau = 1$ ns, $f_{rep} = 10$ kHz–50 MHz), followed by one bunching and two accelerating cavities for effective bunch compression. The facility uses a normal-conducting accelerating system at 180 MHz, with average power up to 0.5 kW (peak power of about 1 MW). The ERL drives three separate FELs (NovoFEL facility) [141] operating in a spectral range of 90–240 microns. The ERL had recently (2015) been upgraded to 42 MeV, based on four-pass energy recovery to drive a short wave FELs in a spectral range of 8–15 microns [142].

7.8.2 S-DALINAC

The S-DALINAC energy-recovery linac is a superconducting electron accelerator operated at Technical University Darmstadt since 1991. The ERL was recently upgraded (2015–2016) and as of 2019 it is capable of operating as a one-pass, or two-pass ERL with maximum energies of approximately 34 or 68 MeV, respectively [143]. The ERL provides beam for Compton scattering of laser beams on intense electron beams to generate quasi-monochromatic, energy-tunable, fully polarized gamma-ray beams for photonuclear reactions [144]. The most recent upgrade [145] started in 2017, and would enable an increase of the maximum achievable energy close to its design value of 130 MeV. A newly-added beamline features a path-length adjustment system capable of changing the phase of the beam by a full RF cycle.

7.8.3 MESA

MESA is a recirculating superconducting accelerator under construction at Johannes Gutenberg-Universität Mainz. The facility [146] uses a superconducting accelerating system based on the TESLA operation frequency of 1.3 GHz. The 2-pass recirculating linac has been configured to operate in two different modes: the external beam (EB) mode, where a 150 μA polarized electron beam at 155 MeV is dumped after being used at the experiment, and in energy-recovery mode (ERL) with an unpolarized beam of 1 mA at 105 MeV [147]. At an upcoming later construction stage, MESA's maximum achievable beam current (in the ERL-mode) will be upgraded to 10 mA (unpolarized).

7.8.4 Compact ERL

The compact ERL (cERL) at KEK is a test accelerator to develop ERL technologies for high average beam current operation with high-quality beam performance [148]. The cERL consists of a photoinjector, a main linac for energy recovery, a recirculation loop, and a beam dump. To achieve energy-recovery operation with high average beam current, collimator tuning to reduce unwanted beam loss has been very important. After fine beam tuning and collimator tuning, stable CW operation with 0.9 mA average beam current was achieved. Upgrade efforts are under way to increase CW beam current to 10 mA through improved instrumentation. The

ERL transports a short electron bunch with a high repetition frequency of 1.3 GHz. Coherent, high-intensity THz radiation from such short electron bunches has many unique applications in material science.

7.8.5 *bERLinPro*

As of 2019, the Helmholtz-Zentrum Berlin is constructing the Energy-Recovery Linac Prototype bERLinPro, an SRF-based demonstration facility for the science and technology of ERLs for future high power, high brilliance electron beam applications [149]. bERLinPro is designed to accelerate a high current (100 mA, 50 MeV), high brilliance (normalized emittances below 1 mm·mrad) CW electron beam. The ERL as a prototype to demonstrate low normalized beam emittance of 1 mm·mrad at 100 mA and short pulses of about 2 ps. The high-brilliance beam will originate from a gun configured with (1.4 λ)/2 cell SRF cavity with a normal-conducting, high quantum efficiency. This injector is planned to support 6 mA beam current and up to 3.5 MeV beam kinetic energy.

7.8.6 *CBETA*

Cornell-BNL-ERL-Test-Accelerator (CBETA) is a test ERL [150], featuring four accelerating passes through the superconducting linac with a single Fixed Field Alternating Linear Gradient (FFA-LG) return beamline built of the Halbach-type permanent magnets. The CBETA ERL accelerates electrons from 42 to 150 MeV, with a 6 MeV injector. The novelty is that four electron beams, with energies of 42, 78, 114, and 150 MeV, are merged by spreader beamlines into single-arc FFA-LG beamlines. The electron beams from the main linac cryomodule pass through the FFA-LG arc and are adiabatically merged into a single straight line. This is the first 4-pass superconducting ERL and the first single permanent magnet return line. It promises to deliver unprecedentedly high beam current with simultaneously small emittance. A collaboration between Cornell and Brookhaven National Laboratory has constructed the CBETA facility on the Cornell campus [151], and commissioning is ongoing as of the summer of 2019. A DC photo-emitter electron source, a high-power SRF injector linac, a high-current SRF linac for energy recovery, and a permanent-magnet return loop have been assembled to the 4-turn SRF ERL. CBETA provides essential R&D for the EIC. Furthermore, the high-brightness beam with 150 MeV and up to 40 mA will have applications beyond EIC cooling and basic accelerator research, for industry, nuclear physics, and X-ray science.

7.8.7 PERLE

PERLE (Powerful ERL for Experiments) is a novel ERL test facility [152] which has been designed to validate choices for a 60 GeV ERL foreseen in the design of the LHeC [153] and the FCC-eh. Its main thrust is to probe high current, CW, multi-pass operation with superconducting cavities at 802 MHz (and perhaps other frequencies of interest). With very high transient beam power (~10 MW), PERLE offers an opportunity for controllable studies of every beam dynamic effect of interest in the next generation of ERL design; PERLE will become a 'stepping stone' between present state-of-the-art 1 MW ERLs and future 100 MW scale applications. PERLE design features a flexible momentum compaction lattice architecture in six vertically stacked return arcs, and a high-current, 5 MeV photo-injector. With only one pair of four-cavity cryomodules, 400 MeV beam energy will be reached in three recirculation passes, with beam currents of appoximately 20 mA. The beam will be decelerated in three consecutive passes back to the injection energy. Work on the engineering design of PERLE has just begun as of summer 2019.

In summary, the next-generation ERLs under study or in construction in 2019 take performance to the next (10 MW) level, and include bERLinPro, CBETA, and PERLE. These systems will provide the experience and knowledge base for fourth-generation objective systems such as electron-ion collider electron coolers, XFEL drivers, and high-energy colliders.

A more detailed comparison of present and next-generation systems is given in Table 7.8 [154]. Certain characteristics of the next generation are apparent: not only do projected beam powers climb by an order of magnitude as noted, but the number of passes, the power multiplier, and the dynamic range all increase, and both accelerated and recovered beams are transported in shared beamlines. All of these features are natural evolutionary consequences of system design optimization in the pursuit of higher performance with a lower cost.

Table 7.8 Comparison of selected present and future generation ERLs

	JLab ERLs	S-DALINAC	bERLinPro	CBETA	PERLE
Gun technology	DC	Thermionic	SRF	DC	DC
Total # passes	2	2	2	8	6
Recirculation architecture	Conventional	Conventional	Conventional	FFG	Conventional
Acceleration/recovery multipass transport	Linac only	Linac only	Linac only	Common	Common
RF frequency (GHz)	1.5	3	1.3	1.3	0.8
Nominal bunch charge (pC)	135	Very low	77	77	320
Design current (mA)	10	Very low	100	40	20
Total current in linac (mA)	9.1	Very low	200	320	120
Energy (GeV)	0.165	~0.0425	~0.05	~0.15	0.5–1
Beam power (MW)	1.25 (>>P_{RF})	Low (<P_{RF})	10 (>>P_{RF})	6 (>>P_{RF})	10–20 (>>P_{RF})
Energy at dump (MeV)	11	2.5	5	5	5
E_{full}/E_{dump}	15–20	17	10	30	100–200

References

1. D. Proch: *Quest for high gradient*, Proc. CERN Accelerator School, Superconductivity in Particle Accelerators, H. Rissen (ed.), CERN 96-03 (1996) 201.
2. W. Weingarten: *Superconducting cavities – Basics*, Proc. CERN Accelerator School, Superconductivity in Particle Accelerators, H. Rissen (ed.), CERN 96-03 (1996) 167.
3. R. Hardekopf, et al: Particle Accelerator Conf. (1999) 3597.
4. T. Raubenheimer, et al: Particle Accelerator Conf. (2007) 1944.
5. R. Brinkmann: Linear Accelerator Conf. (2008) 5.
6. H. Ao: *Status of J-PARC Linac Energy Upgrade*, Linear Accelerator Conf. (2010).
7. J. Delahaye, et al: Intern. Particle Accelerator Conf. (2010) 4769.
8. T. Sugimura, et al: Intern. Particle Accelerator Conf. (2010) 4290.
9. J.-Y. Raguin, et al: Linac Conf. (2012) 501.
10. T. Raubenheimer, et al: FEL Conf. (2015) 618.
11. F. Loehl, et al: Linac Conf. (2016) 22.
12. G. Loew: Handbook of Accelerator Physics and Engineering, World Scientific (1999) 26.
13. J. Potter: Handbook of Accelerator Physics and Engineering, World Scientific (1999) 513.
14. I. Bazarov: Particle Accelerator Conf. (2005) 382.
15. J. Simpson: Handbook of Accelerator Physics and Engineering, World Scientific (1999) 46.
16. B. O'Shea et al: Nature Comm. (2016) 7:12763.
17. B. Blue et al: Phys. Rev. Let. (2003) 90-21-214801-1.
18. M. Hogan, et al: Particle Accelerator Conf. (2007) 1910.
19. S. Gessner, et al: Nature Comm. 7 (2016) 11785.
20. C. Joshi, et al: Plasma Phys. Control Fusion (2018) 1.
21. A. Gonsalves, et al: Particle Accelerator Conf. (2007) 1911.
22. W. Leemans, et al: Phys. Rev. Let. 113 (2014) 245002.
23. E. Colby, et al: Particle Accelerator Conf. (2007) 3115.
24. J. England et al: Rev. Mod. Phys. (2014) 1337.
25. R. Hollebeek: Nucl. Instrum. Meth. 184 (1981) 33.
26. P. Chen, K. Yokoya: Phys. Rev. D 38 (1988a) 987.
27. R.J. Noble: Nucl. Instrum. Meth. A 256 (1987) 427.
28. M. Bell, J.S. Bell: Part. Accel. 24 (1988) 1.
29. R. Blankenbecler, S.D. Drell: Phys. Rev. Lett. 61 (1988) 2324.
30. P. Chen, K. Yokoya: Phys. Rev. Lett. 61 (1988b) 1101.
31. M. Jacob, T.T. Wu: Nucl. Phys. B 303 (1988) 389.
32. V.N. Baier, V.M. Katkov, V.M. Strakhovenko: Nucl. Phys. B 328 (1989) 387.
33. For CAIN see: P. Chen, et al.: CAIN: Conglomérat d'ABEL et d'Interactions Nonlinéaires, Nucl. Instrum. Meth. A 355 (1995) 107.
34. For GUINEA-PIG see: D. Schulte: *Electro-magnetic and hadronic background in the interaction region of the TESLA collider* (PhD thesis), DESY-TESLA-97-08 (1996).
35. For GUINEA-PIG++ see: D. Schulte, et al.: *GUINEA-PIG++ : An Upgraded Version of the Linear Collider Beam Beam Interaction Simulation Code GUINEA-PIG*, PAC07-THPMN010.
36. D. Schulte, *Beam-beam effects in linear colliders,* in CERN-2017-006-SP (CERN, Geneva, 2017).
37. P. Chen, V. Telnov: Phys. Rev. Lett. 63 (1990) 1976.
38. ILC WEB: http://www.linearcollider.org
39. ILC TDR: http://www.linearcollider.org/ILC/Publications/Technical-Design-Report
40. ILC 250 GeV: https://arxiv.org/abs/1711.00568
41. CLIC WEB: http://clic.cern
42. CLIC CDR: http://clic-study.web.cern.ch/content/conceptual-design-report
43. CLIC 380 GeV: https://arxiv.org/abs/1608.07537
44. LHC cost: http://cds.cern.ch/record/2255762/file

45. E-XFEL: http://accelconf.web.cern.ch/AccelConf/ipac2017/papers/moxaa1.pdf and https://www.xfel.eu/
46. LCLS II: http://accelconf.web.cern.ch/AccelConf/ipac2017/papers/tupab130.pdf and https://lcls.slac.stanford.edu/lcls-ii
47. SCRF R&D: A. Grassellino et al., "Unprecedented quality factors at accelerating gradients up to 45 MV/m in niobium superconducting resonators via low temperature nitrogen infusion", Supercond. Sci. Technol. 30 (2017) 094004 (https://doi.org/10.1088/1361-6668/aa7afe)
48. ILC Japan 2012: http://www.jahep.org/office/doc/201202_hecsubc_report.pdf
49. CTF3 results: http://accelconf.web.cern.ch/AccelConf/ipac2017/papers/tuzb1.pdf
50. X-band applications: http://cerncourier.com/cws/article/cern/52358
51. CESRTA results: http://accelconf.web.cern.ch/AccelConf/IPAC10/papers/tuymh02.pdf
52. FACET studies: Latina, Andrea et al., "Experimental demonstration of a global dispersion-free steering correction at the new linac test facility at SLAC", Physical Review Special Topics - Accelerators and Beams. 17. 042803 (https://doi.org/10.1103/PhysRevSTAB.17.042803) (2014)
53. ATF2 results: https://agenda.linearcollider.org/event/7014/contributions/36882/attachments/30069/44951/ATF2_okugi_20160601.pdf
54. R.B. Neal, et al., The Stanford two-miles linear accelerator, New York: W.A. Benjamin, 1968.
55. E. Jensen, CTF3 Drive Beam accelerating structures, Proc. LINAC2002, 2002, Gyeongju, Korea, CERN/PS 2002-068(RF), and CLIC note 538
56. A. Grudiev, W. Wuensch, DESIGN OF THE CLIC MAIN LINAC ACCELERATING STRUCTURE FOR CLIC CONCEPTUAL DESIGN REPORT, Proceedings of Linear Accelerator Conference LINAC2010, Tsukuba, Japan, 2010.
57. H. Padamsee: *RF superconductivity*, Wiley-VCH Verlag (2009).
58. H. Padamsee, J. Knobloch, Tom Hays: *RF superconductivity for accelerators*, John Wiley & Sons Inc. (1998).
59. H. Padamsee: *Designing superconducting cavities for accelerators*, Proc. CERN Accelerator School, S. Russenschuck, G. Vandoni (eds.), CERN-2004-008, (2004).
60. J. Sekutowicz: *Design of a low loss SRF cavity for the ILC*, PAC'05, Knoxville, (2005) 3342.
61. R. Geng: *Review for new shapes for high gradients*, Physica C 441 (2006) 145.
62. FLASH/TESLA facility project: http://flash.desy.de/tesla/tesla_documentation/
63. European XFEL project: http://www.xfel.eu/
64. ILC project: http://www.linearcollider.org/
65. A. Yamamoto: *Global R&D effort for the ILC LINAC technology*, Proc. EPAC-08, MOY-BGM01, Genova, (2008).
66. L. Evans, S. Michizono, and A. Yamamoto: *International Linear Collider (ILC) – Overview* "KASOKUKI", Journal of Particle Accelerator Society of Japan, **14**, No. 4 (2017) 194-200.
67. A. Yamamoto: *Superconducting RF Cavity Development for the International Linear Collider*, IEEE Trans. Appl. Superconductivity. **19** (3) (2009) 1387-1393.
68. A. Yamamoto and K. Yokoya: *Linear Colliders*, Review of Accelerator Science and Technology, Vol. 7 (2014) 1-22.
69. K.L.F. Bane, A. Mosnier, A. Novokhatsky, K. Yokoya: *Calculation of the Short-Range longitudinal wakeelds in the NLC linac*, in: Proc. ICAP 1998, Monterey, CA, November 1998.
70. A.W. Chao: Physics of Collective Beam Instabilities in High Energy Accelerators, Wiley-Interscience, 1st edition, January 1993.
71. Th.P. Wangler: RF Linear Accelerators. Wiley-VCH, 2nd edition, March 2008.
72. M.G. Minty, F. Zimmermann: Measurement and Control of Charged Particle Beams, Springer, 1st edition, August 2003.
73. A.W. Chao: Handbook of Accelerator Physics and Engineering, Second Edition, World Scientific Publ., 2013.
74. V. Balakin, et al.: Phys. Rev. Lett. 74 (1995) 2479-2482.
75. M.S. Zisman: *The PEP-II Project*, LBL-34556 CBP Note-036.
76. E. Kikutani, et al.: *KEKB Accelerator Papers*, KEK Preprint 2001-157, December 2001.

77. P. Raimondi, A. Seryi: Phys. Rev. Lett. 86 (2001) 3779-3782.
78. P. Bambade, et al. (ATF Collaboration): Phys. Rev. ST Accel. Beams 13 (2010) 042801.
79. G.R. White et al.: Phys. Rev. Lett. 112, 034802 (2014).
80. E. Marin, et al.: Phys. Rev. ST Accel. Beams 17, 021002 (2014)
81. M. Patecki, et al.: Phys. Rev. Accel. Beams 19, 101001 (2016).
82. R. Brinkmann, Report No. DESYM-90-14, 1990.
83. H. Garcia Morales and R. Tomas Garcia: Phys. Rev. ST Accel. Beams, 17, 101001 (2014).
84. A. Seryi et al.: *A recipe for linear collider final focus system design,* PAC 2003.
85. K. Oide: Phys. Rev. Lett. 61, 1713 (1988).
86. O. R. Blanco, et al.: Phys. Rev. Accel. Beams 19, 021002 (2016).
87. http://mad.web.cern.ch/mad/
88. E. Forest, F. Schmidt, E. McIntosh: KEK Report 2002–3.
89. R. Tomas: Phys. Rev. ST Accel. Beams 9, 081001 (2006).
90. N.J. Walker, J. Irwin, M. Woodley: *Global tuning knobs for the SLC final focus*, PAC 93, and *Third-Order Corrections to the SLC Final Focus*, SLAC-PUB 6206, May 1993
91. T. Okugi et al.: Phys. Rev. ST. Accel. Beams 17, 023501 (2014).
92. Adolphsen, C., et al. (eds.): "The International Linear Collider Technical Design Report - Vol. 3 – Accelerator Baseline Design, (2013), arXiv: 1306.6328 [physics.acc-ph] ILC-REPORT-2013-040.
93. Low Emittance Rings workshop series, 2010, 2011, 2013, 2014, 2015, 2016, 2018.
94. Antoniou, F. and Papaphilippou, Y.: "Analytical considerations for linear and nonlinear optimization of TME cells. Application to the CLIC pre-damping rings", Phys. Rev. ST Accel. Beams 17, 064002, 2014.
95. Guiducci, S., et al.: " 10G1 Updates to the International Linear Collider Damping Rings Baseline Design", (2011), Proceedings of IPAC2011, San Sebastián, Spain
96. Papadopoulou, S., Papaphilippou, Y., Antoniou, F.: "Emittance reduction with variable bending magnet strengths: Analytical optics considerations", Phys. Rev. ST Accel. Beams, submitted, 2018.
97. Papaphilippou, Y, et al.: "Parameter scan for the CLIC damping rings", EPAC'08, Genova, 2008.
98. Dominguez Martinez, M. A., et al.: "Design of a dipole with longitudinally variable field using permanent magnets for CLIC damping rings", IEEE Trans. on Applied Superconductivity 28, 3, 2018.
99. Schoerling, D., et al.: "Design and System Integration of the Superconducting Wiggler Magnets for the CLIC Damping Rings", Phys. Rev. ST Accel. Beams 15, 042401, 2012.
100. Antoniou, F., Optics design of Intrabeam Scattering dominated damping rings, Ph.D. thesis, National Technical University of Athens, 2013.
101. Bernhard, A., et al.: "A CLIC Damping Wiggler Prototype at ANKA: Commissioning and Preparations for a Beam Dynamics Experimental Program", IPAC'16, Busan, 2016.
102. Garcia Fajardo, L., et al.: "A CLIC Damping Wiggler Prototype at ANKA: Commissioning and Preparations for a Beam Dynamics Experimental Program", Proceedings, 24th International Conference on Magnet Technology (MT-24): Seoul, Korea, October 18-23, 2015, IEEE Trans. Appl. Supercond. 26, 4100506, 2016.
103. Kim, E.-S., Ohmi, K.: "Simulations on the Fast-Ion Instability in the International Linear Collider Damping Rings", Jpn. J. Appl. Phys. 48 (2009) 086501.
104. Rumolo, G., et al.: "Electron Cloud Build Up and Instability in the CLIC Damping Rings", EPAC'08, Genova, 2008.
105. Shaposhnikova, E., et al.: "Experimental studies of carbon coatings as possible means of suppressing beam induced electron multi-pacting in the CERN SPS", PAC'09, Vancouver, 2009.
106. Palmer M., et al.: "Electron Cloud at Low Emittance in CesrTA", IPAC'10, Kyoto, 2010.
107. Grudiev, A.: "Conceptual Design of the CLIC Damping Ring RF System", TUPPR026, Proceedings of IPAC2012, New Orleans, Louisiana, USA, 1870-1872, 2012.

108. Naito, T., et al.: "Multi-bunch Beam Extraction by using Strip-line Kicker at KEK-ATF", IPAC'10, Kyoto, 2010.
109. Belver-Aguilar, C., et al.: "The Stripline Kicker Prototype for the CLIC Damping Rings at ALBA: Installation, Commissioning and Beam Characterisation", IPAC'18, Vancouver, 2018.
110. L. Merminga, D.R. Douglas and G.A. Krafft, 'High-Current Energy-Recovering Electron Linacs', Ann. Rev. Nuc. Part. Sci., 53, 387-429 (2003)
111. T. Smith *et al.*, "Development of the SCA/FEL for Use In Biomedical and Materials Science Experiments", NIM-A, **259**, 26-30 (1987).
112. M. Tigner, 'A Possible Apparatus for Electron Clashing-Beam Experiments', Nuovo Cimento, **37**, 1228 (1965)
113. G.R. Neil *et al.*, 'Sustained Kilowatt Lasing in a Free-Electron Laser with Same-Cell Energy Recovery', Phys. Rev. Lett. 84, 662-665 24 January 2000a
114. R. Hajima, *et al.*,' Optics and beam transport in energy-recovery linacs', *Nucl. Instr. and Meth. A***507** 115 (2003)
115. E.J. Minehara *et al.*, 'JAERI superconducting RF linac-based free-electron laser-facility', NIM-A, **445** pp. 183-186 (2000)
116. S.M. Gruner, *et al.*, *Review of Scientific Instruments***73**, p. 1402 (2002)
117. S. Saveliev, *et al.*, *Proc. IPAC2010*, p. 2350, 2010
118. K. Umemori, *et al.*, *Proc. SRF-2009*, p. 896, 2009
119. T. Sakanaka, *et al.*, *Proc. IPAC2010,* p. 2338, 2010
120. V.N. Litvinenko, *Proc. IPAC2010,* p. 2364, 2010
121. J. Flanz, S. Kowalski and C. Sargent, IEEE Trans. Nucl. Sci. **NS-28** 2847 (1981)
122. J. Flanz, *Proc. 1989 Particle Accelerator Conference* p. 1349 (1989)
123. V.I. Veksler, *Proc. USSR Acad. Sci.***43** (1944) 346; *J. Phys. USSR***9** 153 (1945)
124. H. Herminghaus, *et al.*, *IEEE Trans. Nucl. Sci.***NS-30** 3274 (1983)
125. R.E. Rand, '*Recirculating Electron Accelerators'*, Harwood Academic Publishing, (1984)
126. G.A. Krafft and J.J. Bisognano, *Proc. 1987 Part. Acc. Conf.* p. 1356 (1987)
127. G.R. Neil *et al.*, 'Sustained Kilowatt Lasing in a Free-Electron Laser with Same-Cell Energy Recovery' Phys. Rev. Lett. 84, 662-665 24 (2000b)
128. F. Jackson *et al.*, 'The Status of the ALICE R&D Facility at STFC Daresbury Laboratory', Proc. IPAC2011, TUODA03, pp. 934-936, 2011
129. R. Legg *et al.*, 'Operation and Commissioning of the Jefferson Lab UV FEL Using an SRF Driver ERL', Proc. IPAC2011, THP172, pp. 2432-2434, 2011
130. S.A. Bogacz *et al.*, 'CEBAF Energy Recovery Experiment', Proc. PAC 2003, IEEE, pp 195, (2003)
131. T. Kasuga, 'Future Light Source Based On Energy Recovery Linac in Japan', Proc. APAC 2007, TUPMA046, pp. 172-174, 2007
132. J. Knobloch *et al.*, 'Status of the bERLinPro Energy Recovery Project', Proc. IPAC2012, MOPPP015, pp. 601-603, 2012
133. F. Méot et al., 'ER@CEBAF - A High Energy, Multi-pass Energy Recovery Experiment at CEBAF', Proc. of IPAC2016, TUOBA02, 2016
134. D Angal-Kalinin *et al.*, 'PERLE. Powerful Energy Recovery Linac for Experiments. Conceptual Design Report', J. Phys. G: Nucl. Part. Phys. **45** 065003 (2018)
135. S. A. Bogacz *et al.*, 'PERLE - Lattice Design and Beam Dynamics Studies', Proc. IPAC2018, THPMK105, 2018
136. V. Litvinenko *et al.*, 'High Current Energy Recovery Linac at BNL', Proc. PAC2005, RPPT032, pp. 2242-2244, 2005
137. K. Aulenbacher, 'The MESA accelerator', *Workshop to Explore Physics with Intense, Polarized Electron Beams at 50-300 MEV*, R. Milner *et al.*, ed., AIP Conf. Proc. 1563 (2013)
138. M. Arnold *et al.*, 'ERL Mode of S-DALINAC: Design and Status', Proc. ERL 2017, CERN (2017a)
139. N. A. Vinokurov *et al.*, Preprint BINP, (1978) 78-88; Proc. of the 6th Soviet Union Accelerator Conference, VII (1978) 233-236

140. N.G Gavrilov *et al.*, 'Project of CW Racetrack Microtron-Recuperator for Free-Electron Lasers' Nucl. Instr. and Meth. A304 (1991) 228-229
141. G. N. Kulipanov *et al.*, 'Novosibirsk Free-Electron Laser – Facility Description and Recent Experiments', IEEE Trans. on Terahertz Science and Technology, 5(5) (2015) 798–809
142. Y.V. Getmanov *et al.*, 'Full spatial coherent multi-turn ERL x-ray source (MARS) based on two Linacs', Bristol: IOP Publishing – Journal of Physics: Conference Series, (2013) 4
143. N. Pietralla, Nuclear Physics News, Vol. 28, No. 2, (2018) 4
144. C. Kremer *et al.*, Phys. Rev. Lett. 117, 172503 (2016)
145. M. Arnold *et al.*, 'Construction of a Third Recirculation for the S-DALINAC' *Proc. LINAC 2016*, pp. 168-170, 2017b
146. D. Simon, K. Aulenbacher, R. Heine and F. Schlander, 'Lattice and beam dynamics of the energy recovery mode of the Mainz energy recovering superconducting accelerator MESA' *Proc. IPAC2015*, 2015
147. F. Hug, K. Aulenbacher, R. Heine, B. Ledroit and D. Simon, 'MESA - an ERL project for particle physics experiments' *Proc. LINAC2016*, 2016
148. N. Nakamura *et al.*, 'Present Status of the Compact ERL at KEK' *Proc. of IPAC2014, MOPRO110*, 2014
149. M. Abo-Bakr *et al.*, 'Status Report of the Berlin Energy Recovery Linac Project BERLinPro' *Proc. IPAC2018,* THPMF034, 2018
150. R.J. Michnoff *et al.*, 'CBETA – Novel Superconducting ERL' *Proc. IPAC2019*, TUPGW102, 2019
151. C. Gulliford *et al.*, 'CBETA Beam Commissioning Results' *Proc. IPAC2019,* MOPRB076, 2019
152. G. Arduini *et al.*, 'PERLE: Powerful ERL for Experiments - Conceptual Design Report' accepted for publication in Journal of Physics G (2017)
153. D. Pellegrini, A. Latina, D. Schulte and S.A. Bogacz, 'Beam-dynamics Driven Design of the LHeC Energy Recovery Linac', PRST-AB, **18**, 121004 (2015)
154. D. Douglas, C. Tennant, S. Benson and S.A. Bogacz, 'Why PERLE? – Historical Context and Technological Motivation', JLAB-TN-18-014 (2018)

Chapter 8
Accelerator Engineering and Technology: Accelerator Technology

F. Bordry, L. Bottura, A. Milanese, D. Tommasini, E. Jensen, Ph. Lebrun, L. Tavian, J. P. Burnet, M. Cerqueira Bastos, V. Baglin, J. M. Jimenez, R. Jones, T. Lefevre, H. Schmickler, M. J. Barnes, J. Borburgh, V. Mertens, R. W. Aβmann, S. Redaelli, and D. Missiaen

8.1 Magnets, Normal and Superconducting

L. Bottura · A. Milanese · D. Tommasini

8.1.1 Introduction

Magnets are at the core of both circular and linear accelerators. The main function of a magnet is to guide the charged particle beam by virtue of the *Lorentz force*, given by the following expression:

$$\boldsymbol{F} = q\, \boldsymbol{v} \times \boldsymbol{B}, \tag{8.1}$$

where q is the electrical charge of the particle, $\boldsymbol{v}$ its velocity, and $\boldsymbol{B}$ the magnetic field induction. The trajectory of a particle in the field depends hence on the particle

---

Coordinated by F. Bordry

F. Bordry (✉) · L. Bottura · A. Milanese · D. Tommasini · E. Jensen · P. Lebrun · L. Tavian · J. P. Burnet · M. C. Bastos · V. Baglin · J. M. Jimenez · R. Jones · T. Lefevre · H. Schmickler · M. J. Barnes · J. Borburgh · V. Mertens · S. Redaelli · D. Missiaen
CERN (European Organization for Nuclear Research)Meyrin, Genève, Switzerland
e-mail: Frederick.Bordry@cern.ch; Davide.Tommasini@cern.ch; Erk.Jensen@cern.ch; Laurent.Jean.Tavian@cern.ch; Vincent.Baglin@cern.ch; Rhodri.Jones@cern.ch; Jan.Borburgh@cern.ch; Dominique.Missiaen@cern.ch

R. W. Aβmann
DESY, Hamburg, Germany
e-mail: ralph.assmann@desy.de

S. Myers, H. Schopper (eds.), *Particle Physics Reference Library*,
https://doi.org/10.1007/978-3-030-34245-6_8

velocity and on the space distribution of the field. The simplest case is that of a uniform magnetic field with a single component and velocity $\boldsymbol{v}$ normal to it, in which case the particle trajectory is a circle. A uniform field has thus a pure *bending* effect on a charged particle, and the magnet that generates it is generally referred to as a *dipole*.

By equating the Lorentz force to the centripetal force, we obtain the bending radius ρ of the motion of a particle of charge q under the action of a magnetic field B perpendicular to the motion:

$$\frac{1}{\rho} = \frac{qB}{pc}, \tag{8.2}$$

By expressing the momentum p in practical units [GeV/c], we can write:

$$B\rho \ [\mathrm{Tm}] = \frac{10^9}{c}\frac{p}{Z} = 3.3356\,\frac{p\ [\mathrm{GeV/c}]}{Z}, \tag{8.3}$$

where Z is the charge number of the particle, with $q = Ze$.

The product $B\rho$ is known as *magnetic rigidity* and provides the link between dipole strength and length based on the momentum of a charged particle in a circular accelerator. Note how the formula shows clearly the trade-off between the bending magnetic field B and the size of the machine (related to ρ).

Besides bending magnets, a number of other field shapes are required to focus and control the beam. Most important are magnets that generate a pure gradient field, i.e. a field that is zero on the axis of the magnet and grows linearly with distance. This type of magnet is referred to as a *quadrupole* and is used to focus the particles on the central trajectory of the accelerator. The strength of a quadrupoles is customarily quoted in terms of the field *gradient* G, in units of [T/m]. A *normalised quadrupole strength* for a quadrupole of length l is defined as the ratio of the integrated quadrupole gradient to the beam rigidity, or: $K = Gl/(B\rho)$. The angular deflection α (in radians) of a particle passing at a distance x from the centre of a quadrupole can be computed using the normalised quadrupole strength as:

$$\alpha \ [\mathrm{rad}] = Kx, \tag{8.4}$$

which shows that a particle on the quadrupole axis ($x = 0$) has a straight trajectory, while a particle off-axis receives a kick proportional to its distance from the centre, i.e. the expected focussing effect. Higher order gradient fields, such as sextupoles, or octupoles, behave similarly and provide further non-linear means to control, correct and stabilize the dynamics of the motion of the particles, as described elsewhere in this handbook.

The accelerator magnets considered here have typically slender, long *apertures* (the space available for the beam), where the magnetic field has components only in the plane of the magnet cross section. In this plane, 2-D configuration, the most compact representation of the magnetic field shape in the magnet aperture is provided by the complex formalism [1] and its multipole expansion. Defining the complex variable $z = x + iy$, where the plane (x,y) is that of the magnet cross section, the function $B_y + iB_x$ of the two non-zero components of the magnetic field is expanded in series:

$$B_y + i B_x = \sum_{n=1}^{\infty} (B_n + i A_n) z^{n-1}. \tag{8.5}$$

The coefficients B_n and A_n of the series expansion are the *multipoles* of the field, and determine the shape of the field lines. As an example, a magnet in which only the term B_1 is non-zero, corresponds to a magnetic field:

$$B_x = 0;\ B_y = B_1,$$

i.e. a perfect dipole field (constant in amplitude and direction) oriented in y direction. If the y direction is taken perpendicular to the plane of the accelerator (e.g. vertical), this is usually called a *normal* dipole, which provides bending in the plane of the accelerator (e.g. horizontal). A magnet in which only A_1 is non-zero results in a perfect *skew* dipole field:

$$B_x = A_1;\ B_y = 0,$$

which is in the plane of the accelerator (e.g. horizontal) and provides bending perpendicular to it (e.g. vertical). Multipoles B_2 and A_2 correspond to magnets generating a pure normal and skew quadrupole field. Higher order gradients (sextupole, octupole, etc.) are obtained by simple analogy in the continuation of the series.

The explicit expressions of the field components corresponding to the first four multipoles, and a sketch of the corresponding field lines are reported in Table 8.1. It is useful to remark that the coefficients B_n and A_n appearing in Table 8.1 have units of $[\mathrm{T/m}^{n-1}]$ and are hence the *generalized normal and skew gradient* of order n. Specifically, B_2 corresponds to the normal quadrupole gradient G discussed earlier. To complete this short review of field configuration, we use the complex expansion to evaluate the module of the field B in the case of a pure normal multipole (chosen for convenience, the case of a pure skew multipole field yields identical results). Simple algebra, writing that $z = Re^{i\theta}$ where R is the module and θ is the argument of z, gives the following result:

$$B = B_n R^{n-1}, \tag{8.6}$$

which shows that the field strength in a pure multipole field of order n is proportional to the generalized gradient and grows with the power $n-1$ of the distance from the magnet centre. This extends the case of the dipole ($n = 1$, constant field), and

Table 8.1 Multipole coefficients, field components and field lines for pure multipole magnets

Magnet type	n	Normal ($B_n \neq 0$)		Skew ($A_n \neq 0$)	
Dipole	1	$B_x = 0$ $B_y = B_1$	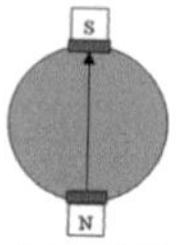	$B_x = A_1$ $B_y = 0$	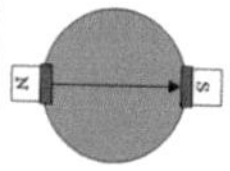
Quadrupole	2	$B_x = B_2 y$ $B_y = B_2 x$	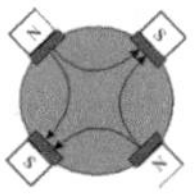	$B_x = A_2 x$ $B_y = -A_2 y$	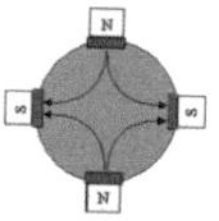
Sextupole	3	$B_x = B_3 2xy$ $B_y = B_3(x^2 - y^2)$	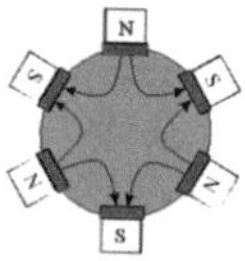	$B_x = A_3(x^2 - y^2)$ $B_y = -A_3 2xy$	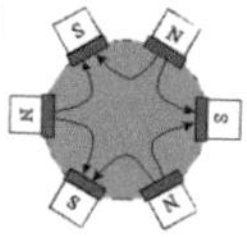
Octupole	4	$B_x = B_4(3x^2 y - y^3)$ $B_y = B_4(x^3 - 3xy^2)$	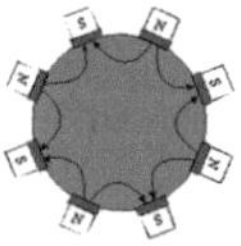	$B_x = A_4(x^3 - 3xy^2)$ $B_y = -A_4(3x^2 y - y^3)$	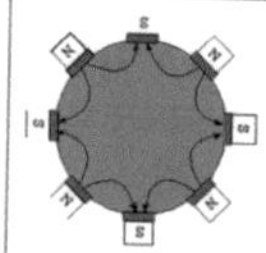

quadrupole ($n = 2$, linear field), to higher order multipoles such as the sextupole ($n = 3$, quadratic field profile), octupole ($n = 4$, cubic field profile), and so on.

Besides its compact form, the complex notation is useful because there is a direct relation between multipoles (order and strength) and beam properties. This is why accelerator magnets are often characterised using their *harmonic content* in terms of the B_n and A_n coefficients of the field expansion. Indeed, the pure multipolar fields discussed so far can only be approximated to a suitable degree in real magnets. The field generated by a magnet contains then all multipoles, normal and skew, i.e. a dense harmonic spectrum. Symmetries cause cancellation effects, resulting in low (ideally zero) *non-allowed* multipoles when compared to the multipoles *allowed* by the magnet symmetry. Selected multipoles can be further reduced by using design features such as optimization of the coil and iron geometry, or corrections such as passive and active magnetic shims.

We define the *field quality* of an accelerator magnet as the relative difference between the field produced and the ideal field distribution, usually a pure multipole, in the region of interest for the beam, which is generally referred to as the *good field region*. Depending on the shape of the good field region, it may be convenient to quote field quality as an overall homogeneity (i.e. $\Delta B/B$, typically done for magnets with a rectangular or elliptic aperture), or providing the spectrum of multipoles other than the one corresponding to the main magnet function (ratio of A_n and B_n to the main field strength, typically done for magnets with round bore). Whichever the form, the field quality of accelerator magnets is generally requested to be in the range of few 10^{-4}. To maintain practical orders of magnitude, field errors are then quoted in relative units of 1×10^{-4} of the main field, or simply *units*.

Gradient magnets (i.e. quadrupole and higher order) are also characterised by a *magnetic axis*, which is usually taken as the locus of the points in the magnet aperture where the field is zero. Magnets are aligned with respect to their axis (or an average of the locus when it deviates from a straight line) to the specified beam trajectory to avoid unwanted *feed-down* effects. Typical alignment tolerances in circular machines range from few tens of μm in synchrotron light sources to fractions of mm in large colliders (e.g. the LHC). Linear colliders are more demanding, with typical tolerances at the sub-μm level.

Dipoles and quadrupoles are the main elements of the linear optics in modern synchrotrons. Depending on the beam specifications, any residual field and alignment imperfections, as well as drift in magnet properties, may require active correction to ensure stable and efficient operation. This is done using *corrector magnets* that are powered using information established from previous knowledge on the main magnets, or parameters measured on the beam, or both. Corrector magnets are often designed to generate a single multipole, so to act on the beam as an orthogonal knob, thus making the correction easier to execute.

8.1.2 Normal Conducting Magnets

"Normal conducting", and alternatively "resistive", "warm" or "conventional" magnets, are electro-magnets in which the magnetic field is generated by conductors like copper or aluminium, which oppose an electrical resistance to the flow of current. The magnetic field induction provided in the physical aperture of these magnets rarely exceeds 1.7 to 2.0 T, such that the working point of the ferromagnetic yoke remains below saturation. In these conditions, the yoke provides a closure of the magnetic path with small use of magneto-motive force, and its pole profile determines the magnetic field quality.

The integral form of the static part of the last Maxwell equation, the Ampere's law, provides a simple analytical expression for the relationship between magnetic field and magneto-motive force in most of magnet configurations used in particle accelerators.

As an example, we illustrate in Fig. 8.1 a non-saturated C-type dipole magnet, made of two coils of $N/2$ turns each, connected in series and supplied by a current I.

$$NI = \oint H dl = H_{\text{iron}} l_{\text{iron}} + H_{\text{air}} l_{\text{air}} = \frac{B}{\mu_0 \mu_r} l_{\text{iron}} + \frac{B}{\mu_0} l_{\text{air}}. \tag{8.7}$$

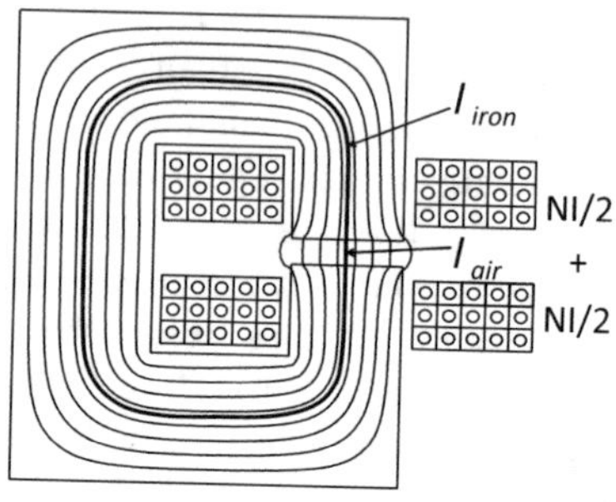

Fig. 8.1 C-dipole magnetic circuit, of physical vertical aperture l_{air}, supplied with a total magnetomotive force *NI*

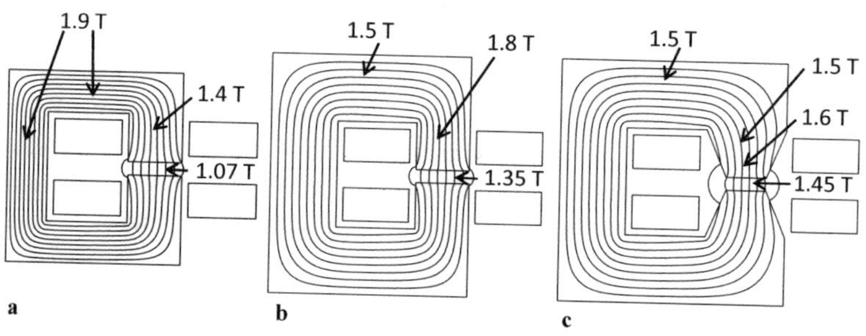

Fig. 8.2 C-dipole magnetic circuit of physical vertical aperture $l_{\mathrm{air}} = 10$ mm, supplied with $N_{\mathrm{tot}}I = 12{,}000$ A

If $\mu_r \gg l_{\mathrm{iron}}/l_{\mathrm{air}}$ we can neglect the magneto-motive force "used" in the iron and obtain:

$$B \approx \mu_0 NI/l_{\mathrm{air}}. \tag{8.8}$$

If part of the iron is saturated, its permeability will be lower and part of the *ampere-turns NI* will be used to magnetize the iron as discussed later.

In case the magnetic field exceeds a value of typically about 1.5 T along the path corresponding to l_{iron} the magneto-motive force used in the iron may become no longer negligible with respect to that used in the air. As the iron yoke gathers also the stray field, the field induction in the iron poles is always higher than the one between poles. To reduce the iron portion working at fields above 1.5 T, the iron pole can be tapered. This allows designing iron-dominated magnets capable of producing magnetic fields intensities in their physical aperture rather close to the saturation limit of the iron, i.e. up to about 1.7–2.0 T. A quantitative example of the effect of saturated iron in dipole magnet is given in Fig. 8.2.

8.1.2.1 Magnetic Design

Transfer function (the ratio of the magnetic field intensity in the magnet aperture to the supply current) and inductance can be computed starting from the Ampere's law and considering the relationship between magnet inductance (L), current (I) and energy (E) as $E = ½·LI^2$.

In practical cases the theoretical transfer function of an ideal magnetic circuit is reduced by an "efficiency" η, typically of the order of $\eta = 0.95 \ldots 0.98$, which depends on the length, stacking factor and working conditions of the magnetic yoke. The formulas in Table 8.2 provide an analytical formulation of the inductance values for different magnet configurations.

The inductance depends on how the pole geometry is trimmed (shims, tapered poles, chamfers) and on saturation. For quadrupole and sextupole magnets different tapering of the poles can strongly modify the inductance. For such magnets these simplified formulas can cover only standard designs.

The field homogeneity typically required by an accelerator magnet within its *good field region* is of the order of few parts in 10^{-4}. Transfer line and corrector magnets may be specified with lower homogeneity.

Field quality in a given volume is determined by several factors:

- the size of the magnet aperture with respect to the good field region
- the shape of the iron poles
- manufacture and assembly tolerances
- the position of active conductors (coils), in particular in window-frame magnets
- the ferromagnetic properties at the working conditions of the steel used for the yoke
- dynamic effects

Optimizing field quality is achieved considering all above aspects. In particular magnets operating below 2 T, as the ones treated in this chapter, are also described as iron dominated magnets because the shape of the magnetic field induction is dominated by the shape of the ferromagnetic poles. At the interface between the magnet aperture and the poles, i.e. between steel and air, the component of the magnetic field induction $B_\perp$ perpendicular to the interface surface is the same on both media. The tangential component of the magnetic field H_t also remains the same in case no surface currents are present: this corresponds to a change of the tangential components B_t of the field induction by the ratio of the magnetic permeability between the two media. As a result, in case of infinite permeability the direction of the magnetic field induction in the air at the exit of a magnet pole is always perpendicular to the pole surface.

An example of trimming field quality with pole shims in a dipole magnet is shown in Fig. 8.3.

Table 8.2 Basic magnets: approximate pole shape, transfer function and inductance

Magnet type	Descriptions	Pole shape	Transfer function	Inductance
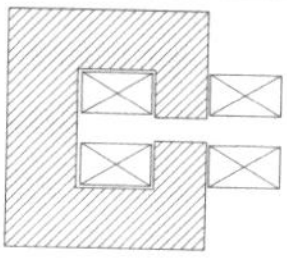	C-type dipole w: pole width g: distance between poles NI: total ampereturns l: yoke longitudinal length	$y = \pm g/2$	$B = \eta\mu_0 NI/g$	$L = \eta\mu_0 N^2 A/g$ A~$(w + 1.2g)$ $(l + g)$
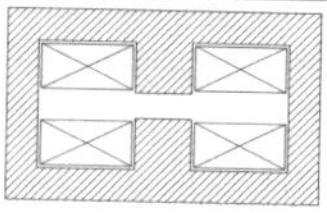	H-type dipole w: pole width g: distance between poles NI: total amperturns l: yoke longitudinal length	$y = \pm g/2$	$B = \eta\mu_0 NI/g$	$L = \eta\mu_0 N^2 A/g$ A~$(w + 1.2g)$ $(l + g)$
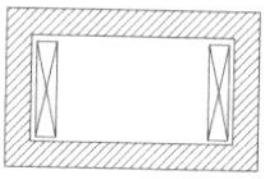	Window-frame d: aperture between coils t: coil width g: physical vertical aperture NI: total ampereturns l: yoke longitudinal length	$y = \pm g/2$	$B = \eta\mu_0 NI/g$	$L = \eta\mu_0 N^2 A/g$ A~$(d + 2/3t)$ $(l + g)$
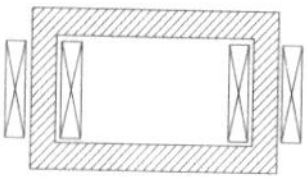	Window-frame d: aperture between coils t: coil width g: physical vertical aperture NI: ampereturns on one leg l: yoke longitudinal length	$y = \pm g/2$	$B = \eta\mu_0 NI/g$	$L = \eta 2\mu_0 N^2$ A/gA~$(d + 2/3t)$ $(l + g)$
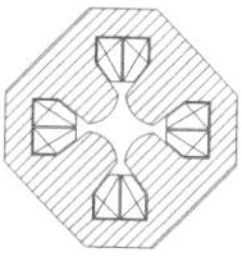	Quadrupole R: radius at pole tip d: distance centre-inner coil NI: ampere-turns per pole l: yoke longitudinal length	$2xy = R^2$	$\lvert B\rvert(r) = Gr$ $G = \eta 2\mu_0 NI/R^2$	$L = 8\pi\mu_0 N^2$ $l_m\sqrt{d/R}$ $l_m = (l + 2/3R)$
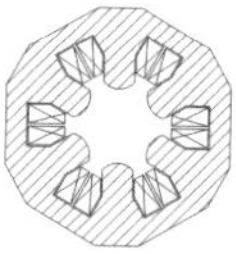	Sextupole R: radius at pole tip d: distance centre-inner coil NI: ampere-turns per pole l: yoke longitudinal length	$3x^2y - y^3 = R^3$	$\lvert B\rvert(r) = Sr^2$ $S = B''/2$ $S = \eta 3\mu_0 NI/R^3$	$L = 12\pi\mu_0 N^2$ $l_m\sqrt{d/R}$ $l_m = (l + 1/2R)$

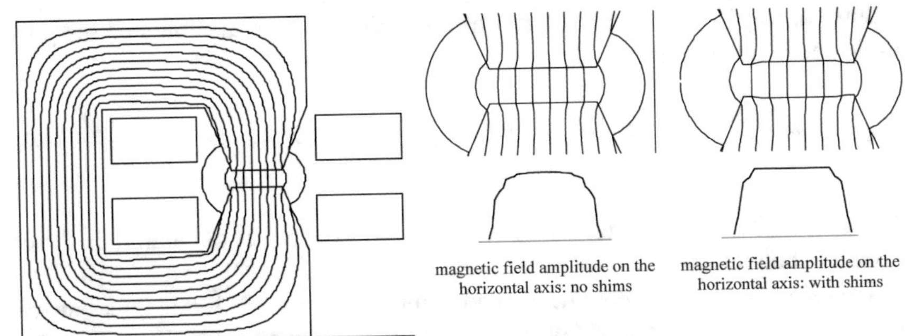

Fig. 8.3 C-dipole with pole shims. The shims extend the good field region

8.1.2.2 Coils

The coils generate the magneto-motive force necessary to produce the required magnetic field induction in the magnet physical aperture. Coils produce losses due to their electrical resistance.

The power dissipated in a conductor of electrical resistivity ρ, volume V, effective current density J_{rms}, is:

$$P = \rho V J_{\mathrm{rms}}^2 = \rho \frac{l}{S} I_{\mathrm{rms}}^2 = R I_{\mathrm{rms}}^2, \tag{8.9}$$

where S is the conductor section, l its length, $I_{\mathrm{rms}} = S J_{\mathrm{rms}}$ the effective current, R the electrical resistance. In case the current is not uniformly distributed in the conductor section, in particular for fast transients where the penetration depth is smaller than the conductor size, an effective section has to be considered.

We recall the resistivity of copper and aluminium as a function of the temperature T:

$$\rho_{\mathrm{Cu}}\,[\Omega \mathrm{m}] = 1.72\,(1 + 0.0039\,(T - 20)) \cdot 10^{-8},$$

$$\rho_{\mathrm{Al}}\,[\Omega \mathrm{m}] = 2.65\,(1 + 0.0040\,(T - 20)) \cdot 10^{-8}. \tag{8.10}$$

The effective current density J_{rms} determines coil size, power consumption and cooling: the designer shall consider the right balance between requirements, technological limits, investment and operation cost.

Typical values of J_{rms} are around 5 A/mm^2 for water cooled magnets, and around 1 A/mm^2 for air cooled magnets. Air cooled coils with favourable configurations (large heat exchange surface with respect to their section) can be operated at higher current densities.

Introducing l_{av} as the average coil turn length, it is easy to show that the power dissipated in a magnet is:

$$P_{\text{dipole}} = \rho \frac{B_{rms}g}{\eta\mu_0} J_{\text{rms}} l_{\text{av}};\ P_{\text{quadrupole}} = 2\rho \frac{G_{\text{rms}} R^2}{\eta\mu_0} J_{\text{rms}} l_{\text{av}};\ P_{\text{sextupole}} = \rho \frac{B''_{\text{rms}} R^3}{\eta\mu_0} J_{\text{rms}} l_{\text{av}}. \tag{8.11}$$

In case of water-cooled magnets, heat is removed by water circulating in the coil (hollow) conductors.

The choice of cooling parameters and number of circuits is based on a few main principles: set the water flow corresponding to the allowed temperature drop for a given power to be removed, having a moderate turbulent flow to provide an efficient cooling, keeping the water velocity within reasonable limits to avoid erosion-corrosion and impingement of the cooling pipes and of the junctions, keeping the pressure drop across the circuit within reasonable limits (typically within 10–15 bars). Properties of water cooling circuits are provided in Table 8.3

Finally, we remark that coils are submitted to forces: their own weight and the electromagnetic forces produced by the interaction between magnetic field and current.

Table 8.3 Criteria and formula for the determination of water cooling circuits for circular conduits

Parameter	Fundament	Formula (in practical units)
Cooling flow	1 kcal = 4186 J increases the temperature of 1 kg of water by 1 °C. Q is the cooling flow, P the dissipated power, ΔT the allowed temperature drop increase	$Q\left[\frac{\text{liter}}{\text{min}}\right] \sim 14.3 \frac{P\ [\text{kW}]}{\Delta T\ [\text{K}]}$
Water velocity	$Q = vA$, where Q is the flow, v the water velocity and A the section of the pipe. In this and the next ones d is the diameter of the conduit.	$v\left[\frac{\text{m}}{\text{s}}\right] = \frac{1000}{15\pi d^2} \cdot Q\left[\frac{\text{l}}{\text{min}}\right]$
Turbolent flow	Reynolds number > 2000, where $Re = dv/\nu$, v the fluid velocity and ν the kinematic viscosity	$\text{Re} \sim 1400 \cdot d\ [\text{mm}] \cdot v\left[\frac{\text{m}}{\text{s}}\right] > 2000$ valid for water at ~ 40 °C
Water velocity limit	Limited by erosion-corrosion and impingement that might start already at $v > 1.5$ m/s in copper pipes, tee pieces and elbow fittings. Velocities up to 10 m/s can still be considered in particular cases, depending on water characteristics, temperature, sizing and layout of pipes and junctions	$v < 3\ \frac{m}{s}$
Pressure drop	The pressure drop ΔP of a smooth pipe with length L can be computed as a function of the cooling flow Q from the Blasius law.	$\Delta P\ [\text{bar}] \sim 60 \cdot L\ [\text{m}] \cdot \frac{Q\left[\frac{\text{liter}}{\text{min}}\right]^{1.75}}{d[\text{mm}]^{4.75}}$

The interaction between a moving charge and the magnetic field is described by the so called Lorentz force, which in the macroscopic form for a wire carrying a current I is referred as *Laplace force*:

$$\boldsymbol{F} = I\,\boldsymbol{l} \times \boldsymbol{B}, \tag{8.12}$$

where the length l is oriented towards the direction of the current flow. For example 1.5 m of straight coil immerged in an average magnetic field component perpendicular to the coil of 0.5 T, carrying a total current of 60,000 ampere-turns, is subjected to a force of $F = 60{,}000 \cdot 1.5 \cdot 0.5 = 45$ kN.

8.1.2.3 Yoke

The magnet yoke has the function of directing and shaping the magnetic field generated by the coils. While magnets operated in persistent mode can be built either with solid or with laminated steel, the yokes of cycled magnets are composed of laminations electrically insulated from each other to reduce the eddy currents generated by the change of magnetic field with time. This electrical insulation can be inorganic (oxidation, phosphating, Carlite) or organic (epoxy). Epoxy coating in a B-stage form can be used to glue laminations together, a technique widely used for small to medium magnets, possibly reinforced by welded bars on the yoke periphery.

The magnetic properties of steel depend on the chemical composition and on the temperature/mechanical history of the material. Important parameters for accelerator magnets are the coercive field H_c and the saturation induction. The coercive field has an impact on the reproducibility of the magnetic field at low currents. A typical requirement for the steel used in accelerator magnets is $H_c < 80$ A/m. Tighter constraints ($H_c < 20$ A/m) apply when the operation covers a large field factor starting from low field inductions (few hundred gauss). The saturation induction is highest with low carbon steel (carbon content in the final state <0.006%). It is common to specify points along the normal magnetization curve, with the condition that the magnetic induction B shall exceed specification values at given field levels H.

To increase its electrical resistivity and at the same time narrow the hysteresis cycle, laminated steel used in cycled magnets usually contains 2 . . . 3% of silicon.

With 3% of Si the electrical resistivity increases from $\rho = 2 \cdot 10^{-7}$ Ωm to $\rho = 5 \cdot 10^{-7}$ Ωm.

The work performed during the hysteresis cycle and the eddy currents produce losses in cycled magnets. An estimate of hysteresis losses can be obtained by the Steinmetz law:

$$P\left[\frac{\mathrm{W}}{\mathrm{kg}}\right] = \eta \cdot f \cdot B^{1.6}, \tag{8.13}$$

where f is the frequency and $\eta = 0.01 \ldots 0.1$, about 0.02 for silicon steel, and of eddy current losses, for silicon steel, by:

$$P\left[\frac{\mathrm{W}}{\mathrm{kg}}\right] = 0.05 \cdot \left(d_{\mathrm{lam}} \cdot \frac{f}{10} \cdot B\right)^2, \tag{8.14}$$

where d_{lam} is the lamination thickness in mm.

8.1.2.4 Costs

We can distinguish

- *fixed costs*: design, coil tooling (winding, molding), yoke tooling (punching, stacking), quality assurance (including tools for specific measurements/checks, as magnetic measurements if requested);
- *unitary costs*: main materials (conductor, insulation, steel), manufacture of parts (coil, laminations, yoke), final assembly, ancillaries (connectors, interlocks, hoses), tests (mechanical, electrical, magnetic);
- *other systems*: cooling, power converters, controls and interlocks, electrical distribution. These parameters have to be taken into account at the magnet design phase: for example for cycled magnets a low inductance can minimize the voltage levels, however the corresponding higher current would require larger supply cables from the power converters to the magnets.
- *running costs*: electric power, maintenance over the life of the project.

A compromise between capital and operational cost is typically found with magnets operating with:

- current densities of about 5 A/mm^2: higher current densities correspond to smaller coils and consequently smaller and cheaper yokes, lower current densities correspond to lower power consumption (less electricity, smaller cooling plant) but to larger magnets;
- field induction levels in the region between 1.2 T and 1.7 T: a given required integrated strength can be provided by short magnet with high field induction, long magnets with low field induction or a compromise between the two. Since, below saturation, the pole width size depends essentially on the good field region size and not on the field induction level, the highest possible field and the corresponding lowest magnetic length represent in most cases a cost-optimized yoke design.

8.1.2.5 Undulators, Wigglers, Permanent Magnets

Wigglers and *undulators* produce a periodic field variation along the beam trajectory causing relativistic charged particles to wiggle emitting electromagnetic radiation with special properties, in particular with a small angle $\alpha = 1/\gamma$ where γ is the relativistic factor.

To a first approximation, these magnets produce a series of dipole fields with alternated directions, of period λ. This is typically obtained with conventional electromagnets when the period is relatively large allowing sufficient space for the coils, and with permanent magnets for shorter periods. Superconducting windings, in general cryo-cooled, are used in case the required field exceeds 2 T and/or for small periods where a high current density is needed.

The difference between wigglers and undulators is in the nature of the radiation produced by the particle. When the amplitude of the beam excursion expressed in meters is small with respect to the angle of the synchrotron radiation emission expressed in radians, the device is called undulator: the emitted radiation is concentrated in a small opening angle and the radiation produced by the different periods interferes coherently producing sharp peaks at harmonics of a fundamental wavelength. Wigglers on the contrary produce particle displacements of larger amplitude: the emitted radiation is similar to the continuous spectrum generated by bending magnets, with in addition the effect coming from the incoherent superposition of radiation from individual poles.

It is useful to introduce the *deflection parameter* $K = \delta_0/\alpha$ as the ratio between the maximum trajectory deflection δ_0 (in meters) and the emission angle α (in radians). For electrons:

$$K = \frac{eB_{0\lambda}}{2\pi mc} = 93.4 \cdot B_0 \cdot \lambda. \tag{8.15}$$

In case $K < 1$ the device is an undulator, in case $K >> 1$ the device is a wiggler.

As anticipated, these magnets are often built with the use of permanent magnets.

Two types of high performance permanent magnet materials, both composed of rare earth elements, are available: Neodymium-Iron-Boron (NdFeB) and Samarium-Cobalt (in the form $SmCo_5$ or Sm_2Co_{17}, also referred as SmCo 1:5 and SmCo 2:17).

NdFeB materials show the highest remanent induction, up to B_r~$1.4T$, and the highest energy product up to $BH_{\max}$~50 MGOe, they are ductile, but they require coating to avoid corrosion and have a relatively low stability versus temperature. Their relative change of remanent field induction with temperature (temperature coefficient) is $\Delta B_r/B_r \sim -0.11\%$ per °C: field induction decreases when temperature increases.

SmCo magnets show a lower remanent induction, up to B_r~1.1 T, are brittle, but they are corrosion and radiation resistant. Furthermore, their temperature coefficient is about −0.03%, lower than that of NdFeB.

The use of permanent magnets in particle accelerators is not limited to wigglers and undulators. For example, the 3.3 km long recycler ring at FNAL, commissioned in May 1999, stands as a pioneer, as it is solely composed of permanent magnets dipoles and quadrupoles [2]. More recently, permanent magnets are used in compact quadrupoles for LINAC4 at CERN [3], in the main bending magnets of the ESRF-

EBS light source [4], and even as spectrometers as for example in the nTOF experiment at CERN [5].

8.1.2.6 Solenoids

Solenoids are made by electrical conductors wound in the form of a helix. The magnetic field induction inside an ideal solenoid is parallel to the longitudinal axis and its intensity is $B = \mu_0 NI/l$ where NI is the total number of ampere-turns and l the solenoid length. By introducing the coil thickness t, the formula can be written as $B = \mu_0 Jt$, where J is the current density.

Solenoids can be built as ironless magnets or can have an external ferromagnetic yoke, used for shielding and to increase the magnetic field uniformity particularly at the solenoid extremities.

The design and construction of solenoids, thanks to their use in many electrical and electro-mechanical devices, is well assessed since more than a century. A comprehensive treatment of solenoid electromagnets was compiled by C. Underhill already in 1910. The treatment issued by Montgomery in 1969 is still a reference [6] nowadays, in spite of new materials now available in particular for wire dielectric insulation and for the containment of stresses.

In solenoids the electromagnetic forces produced by the interaction between the magnetic field and the currents in the coils can reach extremely high values capable of breaking the wires or even, especially for pulsed magnets, leading to an explosion of the device. Their design should consider conductor characteristics and reinforcements, winding tension during manufacture and the containment structure.

8.1.3 Superconducting Magnets

Superconducting magnet technology has been instrumental to the realization of the largest particle accelerators on Earth. Table 8.4 reports the main characteristics of the four large scale hadron accelerators built and operated since the beginning of superconducting magnet technology for accelerators. In parallel, superconductivity has fostered the construction of high field and large volume detector magnets that have become commonplace in high energy physics. The first such detector magnet was installed at Argonne National Laboratory, and operated in the mid 1960s as an instrument in the Zero Gradient Synchrotron (ZGS) [11]. Atlas [12] and CMS [13] at the LHC are the latest and most impressive example of superconducting detector magnets.

The prime difference between superconducting and normal conducting magnets is in the way the magnetic field induction is generated. While in normal conducting magnets the field is dominated by the magnetization of the iron yoke, in their superconducting "siblings" the field is generated by a suitable distribution of currents, properly arranged around the beam aperture. This is possible because a

Table 8.4 Characteristics of the four major superconducting hadron accelerators

		Tevatron [7]	HERA [8]	RHIC [9]	LHC [10]
Maximum beam energy	[GeV]	980	820[a]	250[b]100/n[c]	7000
Injection energy	[GeV]	151	45	12	450
Ring length	[km]	6.3	6.3	3.8	26.7
Dipole field induction	[T]	4.3	4.7	3.5	8.3
Aperture	[mm]	76	75	80	56
Configuration	[mm]	Single bore	Single bore	Single boreTwo rings	Twin bore
Operating temperature	[K]	4.2	4.5	4.3–4.6	1.9
First beam		7–1983	4–1991	6–2000	9–2008

[a]Energy of the proton beam, colliding with the 27.5 GeV electron beam
[b]Particle energy for proton beams
[c]Particle energy per nucleon, for ion beams (Au)

superconducting material can carry large currents with no loss, and ampere-turns become *cheap*, thus opening the way to magnetic fields much above saturation of ferromagnetic materials. Very schematically, superconducting magnets for large scale accelerators consist of a coil wound with highly compacted cables, tightly packed around the bore that delimits a vacuum chamber hosting the beam. The coil shape is optimized to maximize the bore field and achieve acceptable field quality, as described later. The large forces that are experienced by the coil (several tens to hundreds of tons/m) cannot be reacted on the winding alone, that has the characteristic shape of a slender racetrack. The force is hence transferred to a structure that guarantees mechanical stability and rigidity. The iron yoke that surrounds this assembly closes the magnetic circuit, yields to a marginal gain of magnetic field in the bore, and shields the surrounding from stray fields. Finally, the magnet is enclosed in a cryostat that provides the thermal barrier features necessary to cool the magnet to the operating temperature, which is in the cryogenic range (1.9 to 4.5 K for accelerators built to date). Various implementations of this basic concept can be seen in Fig. 8.4 that shows the cross sections of the superconducting dipoles of the four large superconducting hadron accelerators listed in Table 8.4.

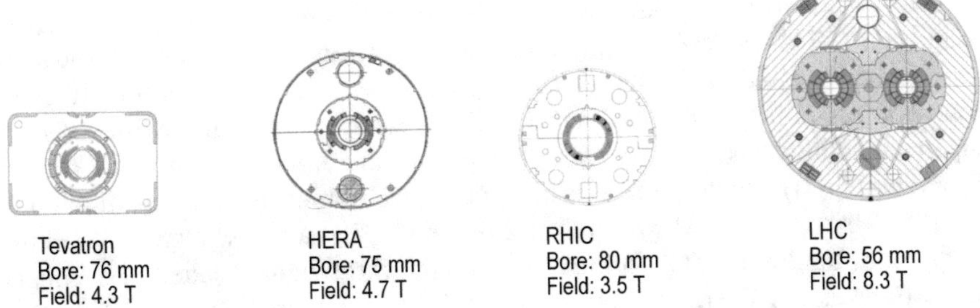

Fig. 8.4 Cross section (to scale) of the dipoles of the four major superconducting hadron accelerators built to date

Superconducting magnet technology relies heavily on the ability to produce technical superconducting materials in the form of high current cables. Beyond considerations on the magnetic field that are specific to the arrangements described above, the design of superconducting magnets must take into account issues related to the superconductor, such as stability, quench protection, magnetization and AC loss. The magnet powering requires warm-to-cold transitions capable to transport the large currents in the range of few to few tens of kA with minimal thermal loss. The construction of the magnet, and in particular the insulation, coil winding, assembly in the mechanical structure and yoke, and the placement in a cryostat necessary for thermal management, all have aspects specific to superconducting magnet technology. Finally, cooling by helium is a science by itself. In the following sections we will discuss some of the main principles, without entering into detail, and provide extensive references on the above matters. The bibliography in the references directs the reader to excellent books on superconducting magnet engineering and science, especially [14–18].

8.1.3.1 Superconducting Materials

A superconductor is such only if it operates below the *critical surface*, a combination of temperature T, magnetic field induction B and current density J that delimits the boundary between the superconducting and normal-conducting phases. This surface is best expressed using a function $J_C(B,T)$, the *critical current density* which is the main engineering characteristic of a superconductor. A good example is the critical current density for the highly optimised Nb-47%Ti alloy used for the production of the LHC magnets, shown in Fig. 8.5. When cooled to 4.2 K, this superconductor can carry a current density up to 3000 A/mm^2 in a background field of 5 T. Indeed, this is the order of magnitude of current density that is of interest to make the design of a superconducting accelerator competitive. As we see from Fig. 8.5, higher fields or higher temperatures result in a reduction of the critical current density, while a decrease of any yields an increase in J_C.

Table 8.5 reports a summary of the critical temperature T_C and critical field B_C for the technical superconducting materials that have found practical applications over the past 50 years, as well as materials that are expected to come into use in a few years. The materials are generally classified as *low-temperature* superconductors (LTS) and *high-temperature* superconductors (HTS). This classification was originally based on the temperature of the superconducting transition T_C, but now it refers rather to the different mechanisms that explain the existence of a superconducting phase. According to this classification, the alloy of Niobium and Titanium (Nb-Ti), the inter-metallic compounds of Niobium and Tin or Aluminium (Nb_3Sn, Nb_3Al) and of Magnesium and Boron (MgB_2) are LTS materials. On the other hand, the Perovskites formed by Bismuth, Strontium, Calcium and Copper oxide (conventionally referred to as BSCCO) and a Rare Earth (e.g. Yttrium), Barium and Copper oxide (referred to as REBCO) are HTS materials. The details of the material composition and production route influence the values of T_C and

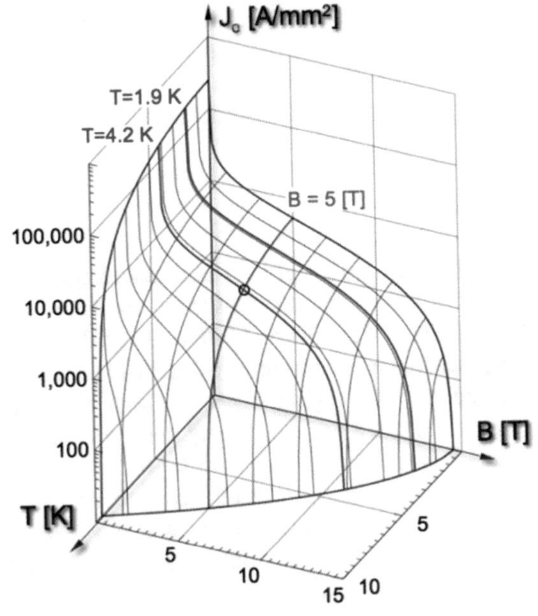

Fig. 8.5 Critical current density of Nb-47%Ti representative of the LHC production. The superconductor carries 3000 A/mm^2 at a temperature of 4.2 K and in a background field of 5 T (marked point). The LHC dipoles operate at 1.9 K, taking advantage of the increase of J_C at lower temperature to reach a nominal field of 8.33 T

Table 8.5 Material reference and typical values of critical parameters for technical superconductors

	LTS				HTS	
Material	Nb-Ti	Nb_3Sn	Nb_3Al	MgB_2	YBCO	BSCCO
Discovery	1961 [19]	1954 [20]	1958 [21]	2001 [22]	1987 [23]	1988 [24]
T_C [K]	9.2	18.2	19.1	39	≈93	95[a], 108[b]
B_C [T]	14.5	≈30	33	36...74	120[c], 250[d]	≈200

[a]BSCCO-2212
[b]BSCCO-2223
[c]B parallel to *c*-axis
[d]B parallel to *ab*-axes

B_C. For this reason the values quoted in the table should be regarded as indicative, especially for developmental materials such as MgB_2, BSCCO and REBCO.

Note for completeness that many other elements and compounds become superconducting when cooled to sufficiently low temperatures and, in some cases, when submitted to large pressure. The family of superconductors is hence still developing. The recent discovery of a new class of HTS, the iron based superconductors [25] (also named IBS, FeSC, or ferropnictides), bears the promise of understanding the theory of high-temperature superconductivity, as well as the potential development of a new class of technical materials for high field applications. At the time of writing, the search for the highest possible T_C has reached the maximum verified value of 203 K in a hydrogen sulphide under a pressure of the order of 155 GPa [26].

Though exciting and promising, these recent developments are still far from being ripe for applications. In the following sections we will focus on the main characteristics of the technical superconductors of Table 8.5, which are at present the only ones available in sufficient quantity and quality for magnet applications.

8.1.3.1.1 Nb-Ti

The alloy of Nb-47%Ti [19] is undoubtedly the most successful practical superconductor, and has been used in all superconducting particle accelerators built to date. Nb-Ti is a ductile and tough material, easily available in long lengths (a few km piece length) in the form of multi-filamentary wires where the superconductor is dispersed in a copper matrix of high purity and low electrical resistivity. Nb-Ti at 4.2 K has a critical current density of about 1500 A/mm^2 at 7.5 T. Cooling it to 1.9 K shifts this point up to 10.5 T. This field range represents the upper (quench) limit for the use of Nb-Ti in accelerator magnets. Standard industrial production yields filament size of a few μm (5 to 10), which is beneficial to reduce the field perturbations induced by persistent currents (see later). Smaller filaments (1 to 3 μm) have been produced to reduce magnetization and losses, but this R&D products are not industrial standards. Homogeneity of the production is at the level of few % for key parameters such as critical current, magnetization, wire composition and geometry, which demonstrates the maturity of the technology. Figure 8.6 shows multi-filamentary Nb-Ti strands used in the LHC.

8.1.3.1.2 Nb_3Sn

The inter-metallic compound Nb_3Sn [20] is the second LTS material that founds its way from material research to large-scale applications. Nb_3Sn is a brittle and fragile compound, which is why after an initial success in high field solenoids of the 1960's, attaining record fields of 10 T, it was quickly replaced by the ductile Nb-Ti for more modest field values. All manufacturing routes involve assembly of

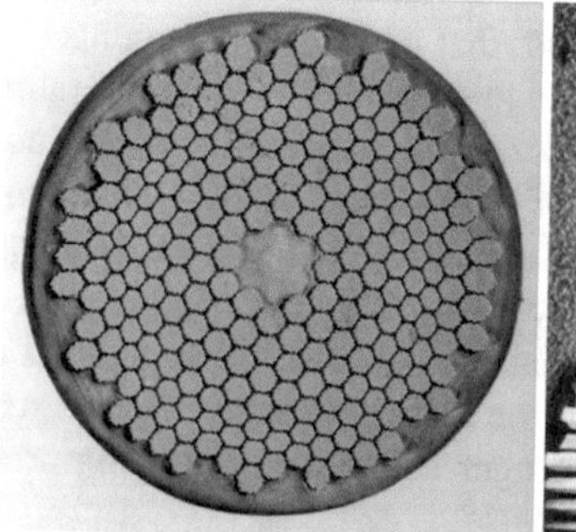

Fig. 8.6 One of the multi-filamentary Nb-Ti strands used in the LHC. The strand has a diameter of approximately 1 mm, and each Nb-Ti filament (shown in the detail micrograph) has a diameter of 7 μm. The matrix is pure copper

the precursor elements Nb and Sn into large size billets that are extruded and/or drawn to the final diameter wire. The Nb_3Sn is then formed by a chemical reaction induced by a heat treatment to temperatures in the range of 650 °C for durations of few tens to few hundreds hours. Various manufacturing routes have been established industrially, resulting in wires with different properties of critical current density and filament size. Most common fabrication methods are based on the so-called "bronze-route", or "internal-tin", which distinguish themselves by the way that the relatively mobile Sn is made available for reaction with Nb. The interest in Nb_3Sn comes from the improved T_C and B_C with respect to Nb-Ti, resulting in record critical current density exceeding 1500 A/mm^2 at 15 T and 4.2 K. This potential is commonly exploited in commercial solenoids built for NMR spectroscopy or other laboratory applications, but to date found no accelerator applications due to the technical difficulties associated with the heat treatment and the handling of a fragile magnet coil. In addition, high J_C Nb_3Sn has presently larger filament size than Nb-Ti, in the range of 50 to 100 μm. This material is nonetheless very relevant to extend the reach of present accelerators such as the LHC beyond the limit of Nb-Ti (e.g. the High-Luminosity upgrade of the LHC and the R&D on magnet technology for a Future Circular Collider), or to build compact accelerators that could be of interest for industrial and medical applications (e.g. high-field compact cyclotrons for proton therapy).

8.1.3.1.3 MgB_2

The most recent of the technical superconductors [22], the inter-metallic compound MgB_2 is a relatively inexpensive material obtained from readily available precursor elements. Superconducting MgB_2 wire can be produced through the powder-in-tube (PIT) process. Variants of this process exist, depending on whether the MgB_2 is formed at the end of the wire processing from Mg and B precursors (in-situ variant), or rather powders of pre-reacted MgB_2 are sintered in the finished wire (ex-situ variant). In both cases, a heat treatment is required in the range of 600 °C to 1000 °C. As for Nb_3Sn, the wires and tapes of MgB_2 are fragile and require careful handling to limit deformation. In spite of a high critical field, which in thin films has reached values above 70 T, the bulk material becomes irreversible at much smaller applied fields. MgB_2 is hence of interest in the range of low to medium field applications (presently up to a maximum 5 T range), but for operating temperatures up to 20 K, i.e. well above boiling helium conditions. The main technical realizations presently based on MgB_2 are open MRI systems at modest magnetic field (0.5 T), and cables for power transmission planned for the High Luminosity upgrade of the LHC (the so called SC links) or electric energy distribution. This material is to date still in the developmental state, with specific interest in enlarging the field range for magnet applications. While the main motivation remains helium-free MRI magnets, MgB_2 could be an option for accelerator magnets subjected to radiation loads that require high operating temperature and energy margin (see later).

8.1.3.1.4 Cuprate Superconductors

These are the two HTS materials that seem to bear the largest potential for future use in accelerator magnets, mainly because of the high critical field at low operating temperature that makes them the ideal candidates to break the 20 T barrier. Both BSCCO and REBCO are ceramic compounds obtained respectively by chemical synthesis of B, Sr, Ca, Cu and O, or Y, Ba, Cu and O. As for all other HTS materials, their production is a complex process that may involve assisted crystal growth, at high temperature, and under very well controlled chemical conditions. BSCCO and REBCO still have unresolved engineering issues, they are fragile, and have comparatively high material and production costs. One of the main limitations to critical current in HTS materials comes from the effect of grain boundaries. Mis-aligned grain boundaries are a barrier to the free flow of the super-currents. In principle, this means that HTS conductors would require solving the tantalizing task of growing single crystals of km length. Different ways have been found to deal with this limitation in BSCCO and REBCO. BSCCO is manufactured using a Powder in Tube (PIT) technique, filling Ag tubes with a precursor BSCCO oxide powder. Two different techniques are used to produce the 2212 and 2223 variants of the BSCCO compound. In the case of BSCCO-2212, the tubes are stacked and drawn to wires that can be wound, cable and finally reacted, much like Nb_3Sn, albeit at higher temperature (900 °C) and under controlled oxygen atmosphere. BSCCO-2212 undergoes a melt process, critical to produce the connected grain structure required for high critical current. For BSCCO-2223 the common form is tape. An initial liquid-mediated reaction of the precursor is used to form the 2223 compound. In this case the required grain alignment cannot be obtained upon heat treatment, but can be induced by mechanical deformation. This is why reacted BSCCO-2223 is mechanically deformed to the final tape dimension, and undergoes a final sintering heat treatment. REBCO is also manufactured in the form of tape. A robust substrate such as stainless steel or hastelloy, is polished and prepared with a series of buffer layers that imprint a crystal texture to a thin layer of REBCO superconductor. This is the crucial step to induce aligned crystal growth of the superconducting phase. The superconductor growth can be achieved by various chemical or physical deposition techniques. The tape is then capped with a sealing Ag layer, finally adding Cu as a stabilizer. When cooled down to 4.2 K, REBCO and BSCCO attain exceptional current densities at high-field, surpassing the performance of LTS materials at fields of 10 to 15 T. In spite of the early phase of development, the HTS BSCCO-2223 already found a large-scale application in the current leads of the LHC, where the gain in operating efficiency and margin has offset the additional investment.

8.1.3.2 Superconducting Cables

Wires and tapes manufactured with the LTS and HTS materials listed above carry currents in the range of few hundreds of A, and are appropriate to wind small magnets, where the magnet inductance and stored energy are not an issue (see later

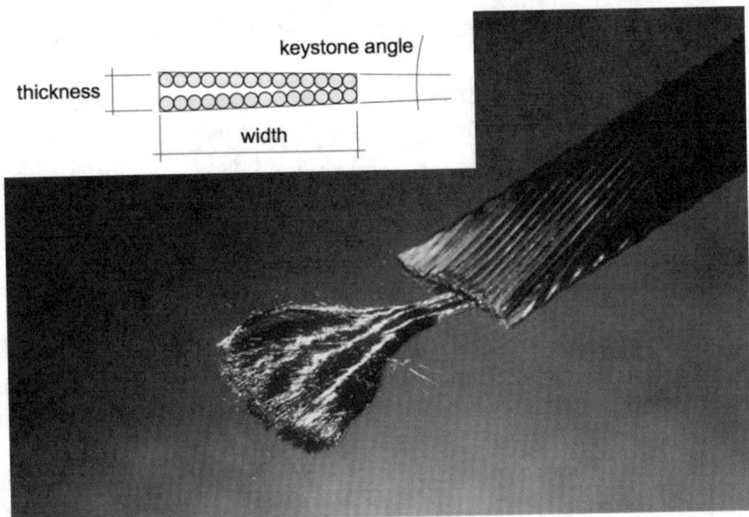

Fig. 8.7 The Rutherford cable for the inner layer of the LHC dipoles, showing the Nb-Ti filaments in a few etched strands. The schematic in the inset shows the definition of the quantities reported in Table 8.6

discussion on protection). On the other hand, the large-scale dipole and quadrupole magnets of an accelerator are connected in km-long strings, and stored energy that can reach hundreds of MJ. To decrease their inductance and limit the operating voltage, it is mandatory to use cables made of several wires in parallel, able to carry much larger currents, typically in the range of 10 kA. An additional benefit of a cable is to provide parallel paths for the current in case of local wire defects. Such cables must insure good current distribution through transposition, combined with precisely controlled dimensions necessary to obtain coils of accurate geometry, as well as good winding characteristics. These properties are the characteristic of the flat cable invented at the Rutherford Laboratory in England [27]. A typical Rutherford cable, shown in Fig. 8.7, is composed of fully transposed twisted wires (Nb-Ti in Fig. 8.7). The transposition length, also referred to as *twist pitch*, is usually kept short, a few cm. To improve winding properties the cable is slightly keystoned, i.e. the cable width is not constant from side to side. The angle formed by the planes of the cable upper and lower faces is called the keystone angle, usually in the range of 1 to 2 degrees. A summary of cable characteristics for the major superconducting accelerator projects is reported in Table 8.6.

The concept of Rutherford cables can be easily applied to LTS materials that come in the form of round wires and has been recently extended to round BSCCO-2212 HTS wires. The rectangular geometry of the cable provides high strand packing and a flexible cable for winding magnet coils of various geometries. The cabling process is invariably associated with large deformations at the edges of the cable, where the wires are plastically deformed. This is necessary to achieve mechanical stability of the cable, but can lead to a degradation of the critical

Table 8.6 Main characteristics of the superconducting cables used to wind the dipoles for the four superconducting colliders

Name	Strand diameter [mm]	Thickness [mm]	Width [mm]	Twist pitch [mm]	Keystone angle [deg]
Tevatron	0.68	1.257	7.75	66	2.06
Hera	0.90	1.471	9.97	95	2.22
RHIC	0.65	1.163	9.67	94	1.21
LHC inner	1.07	1.895	15.06	115	1.24
LHC outer	0.83	1.476	15.06	100	0.89

current of the superconductor. Optimization of this delicate balance between limited wire deformation and desired cable compaction is done empirically, and the I_C degradation in an optimized cable is in the range of a few %.

The electro-dynamic and mechanical requirements for a cable are essentially independent of the superconducting material, at least to first order, and apply both to LTS and HTS. Cabling of HTS materials, however, and in particular those only available in the form of tapes, is far by being standard practice. Several configurations are under development, from compact stacks and Roebel bars, to twisted assemblies of tapes, or wound tapes around cores. Indeed, the matter of HTS cabling is an R&D that will need to be resolved before these materials can become a viable superconductor for accelerator magnets.

8.1.3.3 Stability and Margins, Quench and Protection

We have already remarked that a superconductor is such only when it operates below its critical surface. Once in normal conducting state, e.g. because of a sudden temperature increase caused by internal mechanical energy release or a beam loss, the superconductor generates resistive power, causing a thermal runaway, i.e. an unstable behaviour. It is for this reason that the operating point of current density J_{op}, field B_{op} and temperature T_{op} are chosen by design well inside the allowable envelope, i.e. with proper *margins* that ensure *stability* at the operating point. The typical metrics used for operating margins are:

- Critical current margin i, expressed as the operating fraction of the critical current density $i = J_{op}/J_C(B_{op},T_{op})$, where the critical current is evaluated at the operating field and temperature;
- Margin along the loadline f, expressed as the ratio of operating to critical current $f = J_{op}/J_C(t/J_C,T_{op})$ where the critical current is evaluated at the intersection of the magnet loadline, i.e. the straight line with slope $t = B/J$ and the critical surface;
- Temperature margin ΔT, given by the difference in temperature from operating conditions T_{op} to *current sharing* conditions T_{CS}, evaluated at the operating field and current density $\Delta T = T_{CS}(J_{op},B_{op}) - T_{op}$. The current sharing temperature

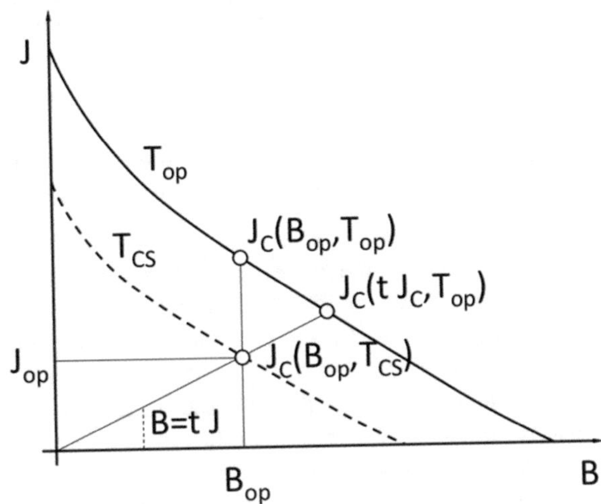

Fig. 8.8 Definition of the operating margins discussed in the text

is defined as the temperature at which the operating current density equals the critical current, or $J_{op} = J_C(B_{op},T_{CS})$.

The margins defined above are shown graphically in Fig. 8.8. Representative values for the design of the large-scale Nb-Ti accelerator dipoles listed earlier are $i \approx 0.5, f \approx 0.8$, and $\Delta T = 0.5 \ldots 1.5$ K.

An additional quantity that measures the stability of the operating point is the *energy margin*, i.e. the quantity of heat necessary to drive the superconductor normal. The energy margin depends on the time and space structure of the heat deposition. A lower bound of the energy margin is given by the enthalpy difference between operating and current sharing conditions $\Delta H = H(T_{CS}) - H(T_{op})$. A robust magnet design is such that the energy margin is larger than the expected amplitude of perturbation over the whole spectrum of operating conditions and characteristic times, which is the basic idea behind all stabilization strategies adopted.

In spite of good design, an irreversible transition to the normal conducting state is always possible, resulting in a thermal runaway process that is referred to as a *quench*. Superconducting magnets in general, and more specifically the highly compact accelerator magnets, tend to have large stored magnetic energy density. Local dissipation of this energy has the potential to lead to material damage and cause loss of electrical insulation. For this reason all superconducting magnets must be protected against quench by detecting any irreversible resistive transition (quench detection electronics) and discharging the magnet. The peak temperature T_{hot} reached during a quench can be estimated by equating the Joule heat produced during the discharge to the enthalpy of the conductor, or $H(T_{hot}) - H(T_{op}) = \int \rho J^2 dt$, where ρ is the specific resistance of the superconductor composite in normal conducting state. We see from the above concept, borrowed from electrical blow-fuses design, that it is always advantageous to reduce the normal state resistance and the time of the discharge. The normal state resistance is decreased by backing the superconductor with a matrix of a material with good conductivity properties (e.g. copper, aluminium or silver). The discharge can be made faster by reducing the inductance of the magnet, which is the reason why large scale magnets are wound using cables with large operating current in the place of single wires. There are various possibilities to dump the magnetic energy, based on one or more of the following standard strategies:

- Energy extraction, in which the magnetic energy is extracted from the magnet and dissipated in an external circuit (e.g. a dump resistor or diode);
- Coupling to a secondary circuit, in which the magnet is coupled inductively to a secondary that absorbs and dissipates a part of the magnetic energy;
- Subdivision, a partition of the magnet in sections, with each section shunted by an alternative current path (resistance or diode) in case of quench;
- Active quench initiation, that relies most commonly on heater embedded in the winding pack, fired at the moment a quench is detected to spread the normal zone over the whole magnet mass and thus reduce the peak temperature. Alternative means to actively initiate quench have been revisited, based on over-current or fast field changes induced by capacitive discharge in the coil.

8.1.3.4 Magnetization, Coupling and AC Loss

An ideal superconductor (type-I) tends to exclude field variations from its bulk, i.e. a perfect diamagnetic behaviour. In practice, in the superconducting materials listed above (type-II) the magnetic field can penetrate the bulk, still resulting in partial diamagnetism. Macroscopically seen, a field change induces shielding currents, which, in a superconductor, do not decay. For this reason these currents are referred to as *persistent*. The magnetic moment per unit volume M associated with persistent currents is proportional to the current density J_C of the shielding currents, and the characteristic size D of the superconductor, i.e. $M \approx J_C D$. This magnetic moment can attain large values, perturb the field generated by the magnet, and lead to instabilities in case the magnetic energy inside the superconductor is dissipated in a process referred to as *flux-jump*. For this reason, the superconductor in wires and tapes is subdivided in fine filaments that have characteristic dimension in the range of 10 to 100 μm. Persistent currents and the associated magnetization produce a significant field perturbation in accelerator magnets, and must be subject to optimization and tight control.

Similar to the bulk behaviour described above, field variations also induce shielding currents between the superconducting filaments. These currents couple the filaments electromagnetically by finding a return path crossing the wire matrix. The amount of filament coupling depends on the resistivity of the matrix, which has to be low for good protection, and the geometry of the current loop. In the extreme case of wires and tapes with untwisted filaments, coupling currents could travel along long lengths (e.g. the km length in a magnet) and find a low cross-resistance. The net effect would be that the multi-filamentary matrix would respond

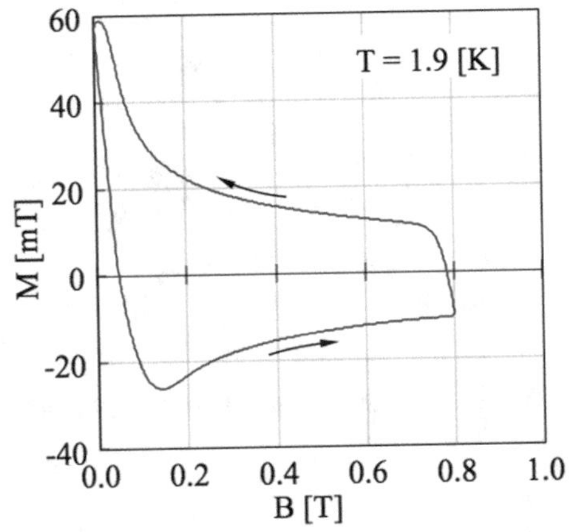

Fig. 8.9 Magnetization loops measured on an LHC Nb-Ti strand for the inner layer of the dipole magnets. The Nb-Ti strand has filaments of 7 μm geometric diameter. (Data by courtesy of S. Le Naour, CERN, Geneva)

to field changes as a single bulk filament, losing the advantage of fine subdivision. Decoupling of the filaments is achieved by shortening the current loop, twisting the wire with typical pitch in the range of few mm. This procedure cannot be applied in tapes that thus suffer from a higher degree of coupling. Similar reasoning applies to the superconducting cable, which explains why the strands in the cable must be transposed by twisting them. In addition, the cross resistance can be controlled in cables by applying resistive coating to the strands, or inserting resistive barriers (sheets, wraps) in the cable itself.

A typical value of magnetization due to persistent and coupling currents is shown in Fig. 8.9 for an LHC Nb-Ti strand. The magnetic moment has a hysteresis, and the area of the loop is proportional to the energy density dissipated during a powering cycle. This means that in superconducting magnets a field ramp is invariably associated with an energy loss, which is referred to as *AC loss*. When compared to other heat loads on the magnet, AC losses become relevant only at high ramp-rates, of the order of 1 T/s, which is of interest for fast cycled accelerators.

8.1.3.5 Magnetic Design of Superconducting Accelerator Magnets

Field calculations for superconducting magnets are very different from those described earlier for iron dominated normal-conducting magnets. The task in this case is to find the current distribution that generates the desired multipole magnetic field. Among the many possible solutions, the $\cos(n\theta)$ and *intercepting ellipses* of Beth [1] and Halbach [28] are most instructive examples and the reader should familiarise with their theory. These ideal current distributions are, however, not practical for winding coils with cables of the type described later. A good approximation of a coil cross section is obtained considering sectors of current shown in Fig. 8.10. The sectors have uniform current density J, a high degree of symmetry, but produce only an approximate multipolar field. The configuration

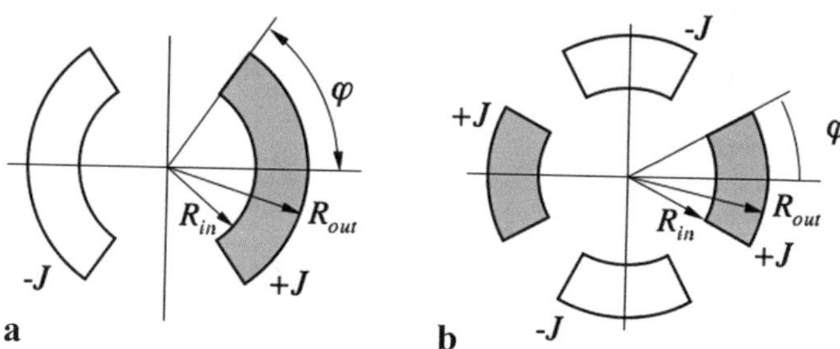

Fig. 8.10 Principle of sector coils that generate an approximate dipole field (**a**), and quadrupole field (**b**)

depicted in Fig. 8.10a generates an approximate dipole B_1, with higher order field errors. Because of symmetry, the only field errors produced (*allowed multipoles*) are normal multipoles of order $(2n + 1)$, i.e. B_3, B_5, $B_7 \ldots$ Similarly, the configuration of Fig. 8.10b produces an approximate quadrupole B_2 with normal higher order multipoles of order $2(2n + 1)$, i.e. B_6, B_{10}, $B_{14} \ldots$ The strength of field and field errors are reported in Table 8.7 that can be used as a starting point for an analytical design of multipole coils.

Examining the equations in Table 8.7a for the field of a sector coil dipole, we see that the strength of the field produced is directly proportional to the current density J and the coil width (R_{out}-R_{in}). In order to keep the coil cross section as small as practically feasible, it is always beneficial to maximise the coil current density, compatibly with mechanical limits (stress) and quench protection (heating rate in case of quench). A maximum current density results in the smallest possible coil, which is associated with minimum overall magnet dimension and cost. This is the simple and clear explanation for the push towards high current density in accelerator magnets.

Similar considerations also hold for a quadrupole, as we see from the expression of the field gradient in Table 8.7b. In this case, however, it is interesting to note that while the quadrupole gradient is proportional to the current density, as for the dipole, the dependence on the coil width is logarithmic. This limits the interest of increasing the coil size in the case of a quadrupole, and makes the role of current density even more prominent.

With respect to field quality, we note further that a choice of $\varphi = 60°$ in the sector dipole coil cancels the sextupole error b_3. The first non-zero multipole error is then the decapole b_5. For typical coil dimensions, the b_5 error is a few percent, i.e. much larger (two orders of magnitude) than acceptable field quality specification in an accelerator magnet. Better field quality can be obtained by segmenting the sectors using insulating wedges, and using two (or more) nested layers. This adds degrees of freedom in the coil geometry that can be used to improve the field homogeneity, at the cost of an increased complexity of the winding. In Fig. 8.11 we show the coil cross sections of the four large-scale superconducting world synchrotrons. It is evident from this how the coils have evolved in complex structures to follow

Table 8.7 Analytical formulae for the main field and field errors for the sector coil configurations shown in Fig. 8.10

	Dipole (a)
Main field	$B_1 = \frac{2\mu_0}{\pi} J\,(R_{out} - R_{in})\sin(\varphi)$
Field errors	$B_n = \frac{2\mu_0}{\pi} J \frac{R_{out}^{2-n} - R_{in}^{2-n}}{n(2-n)} \sin(n\varphi)$ $n = 3, 5, 7, \ldots, (2\,m+1)$
Force per coil quadrant	$F_x = \frac{\sqrt{3}\mu_0 J^2}{\pi}\left[\frac{2\pi-\sqrt{3}}{36} R_{out}^3 + \left(\frac{\sqrt{3}}{12}\ln\left(\frac{R_{out}}{R_{in}}\right) + \frac{4\pi+\sqrt{3}}{36}\right) R_{in}^3 - \frac{\pi}{6} R_{out} R_{in}^2\right]$ $F_y = \frac{\sqrt{3}\mu_0 J^2}{\pi}\left[\frac{1}{12} R_{out}^3 + \left(\frac{1}{4}\ln\left(\frac{R_{in}}{R_{out}}\right) - \frac{1}{12}\right) R_{in}^3\right]$ $F_z = \frac{3\mu_0 J^2}{\pi}\left[\frac{1}{6} R_{out}^4 - \frac{2}{3} R_{out} R_{in}^3 + \frac{1}{2} R_{in}^4\right]$
Energy per unit length	$\frac{E}{l} = \frac{\pi B_1^2 R_{in}^2}{\mu_0}\left\{1 + \frac{2}{3}\left(\frac{R_{out}}{R_{in}} - 1\right) + \frac{1}{6}\left(\frac{R_{out}}{R_{in}} - 1\right)^2\right\}$
	Quadrupole (b)
Main field	$G = B_2 = \frac{2\mu_0}{\pi} J\,\ln\left(\frac{R_{out}}{R_{in}}\right)\sin(2\varphi)$
Field errors	$B_n = \frac{4\mu_0}{\pi} J \frac{R_{out}^{2-n} - R_{in}^{2-n}}{n(2-n)} \sin(n\varphi)$ $n = 6, 10, 14, \ldots, 2(2\,m+1)$
Force per coil quadrant	$F_x = \frac{\sqrt{3}\mu_0 J^2}{6\pi}\left[\frac{1}{72}\frac{12R_{out}^4 - 36R_{in}^4}{R_{out}} + \left(\ln\left(\frac{R_{in}}{R_{out}}\right) + \frac{1}{3}\right) R_{in}^3\right]$ $F_y = \frac{\sqrt{3}\mu_0 J^2}{\pi}\left[\frac{5-2\sqrt{3}}{36} R_{out}^3 + \frac{1}{12}\frac{R_{in}^4}{R_{out}} \left(\frac{2-\sqrt{3}}{6}\ln\left(\frac{R_{in}}{R_{out}}\right) + \frac{\sqrt{3}-4}{18}\right) R_{in}^3\right]$ $F_y = \frac{3\mu_0 J^2}{4\pi}\left[\frac{1}{4} R_{out}^4 - \left(\ln\left(\frac{R_{out}}{R_{in}}\right) + \frac{1}{4}\right) R_{in}^4\right]$
Energy per unit length	$\frac{E}{l} = \frac{\pi B_2^2 R_{in}^4}{2\mu_0 \ln^2\left(\frac{R_{out}}{R_{in}}\right)}\left\{\frac{1}{8}\left[\left(\frac{R_{out}}{R_{in}}\right)^4 - 1\right] - \frac{1}{2}\ln\left(\frac{R_{out}}{R_{in}}\right)\right\}$

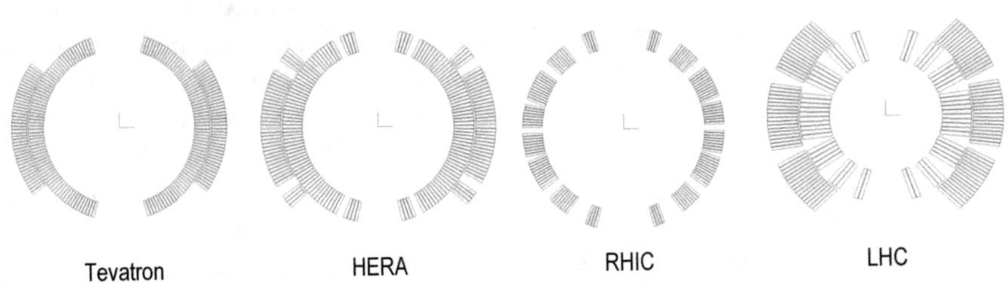

Fig. 8.11 Coil cross section, to scale, for the dipole magnets of the Tevatron, HERA, RHIC and LHC

the increased demand of field quality. Similar considerations, and optimization, are valid in the case of a quadrupole magnet.

Considering further field quality, it is important to recall that the magnetic moment associated with persistent and coupling currents produces field errors that are typically in the range of 10^{-4} to 10^{-3}. These field errors are not easy to predict and control in production, can exhibit large non-linearity and time dependence, and

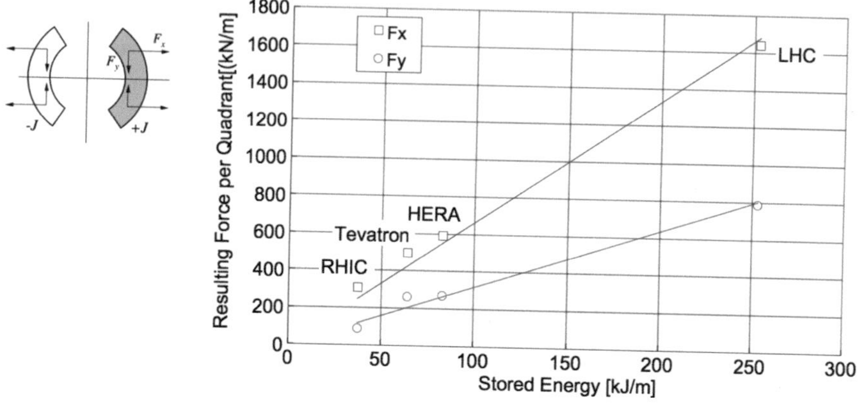

Fig. 8.12 Force on a coil quadrant (see inset), at nominal operating conditions, for the dipole magnets of the Tevatron, HERA, RHIC and LHC plotted vs. the magnetic energy

hence require due attention through direct measurement and appropriate corrections in accelerator operation.

The forces reported in Table 8.7 are intended as resultants on a coil quadrant (for a dipole) or octant (for a quadrupole). We see that the forces generally scale with the square of the current density, and hence with the square of the field both in a dipole and in a quadrupole (i.e. proportional to the magnetic pressure). The same holds for the magnetic energy per unit length, scaling with the square of the field in the bore. Practical values for the loads seen by the coils of accelerator dipoles are compiled in Fig. 8.12, which reports the Lorentz forces in the plane of the coil (referred to a coil quadrant) for the dipoles of the large scale accelerators discussed. We clearly see in Fig. 8.12 the progression in the level of electromagnetic forces and stored energy, from the modest field values of RHIC (3.5 T) to the state-of-the-art of the LHC (8.3 T). Large forces and stored energy are indeed the main engineering challenges of superconducting accelerator magnets, with increasing challenges in the mechanics (supporting structures and internal stress) and quench protection (quench detection and dump time) as the field is pushed to higher values.

Finally, the expressions of Table 8.7 do not take into account the presence of magnetic iron, that surrounds the coil and produces an additional contribution to the field and field errors. The magnitude of the iron contribution is usually small (in the range of 10 to 20%, with the exception of super-ferric magnets, described later) compared to the field generated by the coil current. When the iron is not saturated, its contribution can be approximated analytically using the method of images, which is simple in the case of a round iron cavity. A compact treatment of this method can be found in [17]. For complex geometries, or in the presence of saturation, it is mandatory to resort to computer codes to perform the appropriate calculations and optimizations.

8.1.3.6 Current Leads

The powering of superconducting magnets is done via current leads from room temperature, where the warm power cables are connected, to the cold mass at cryogenic temperature, where the magnets are operated. Current leads are often the dominant source of heat leak into the cryogenic environment because of thermal conduction under a large temperature gradient, as well as ohmic loss. The goal of a current lead design is the minimization of these losses, aiming at the optimum geometry which enables stable operation with a minimum heat in-leak.

Conventional current leads are made from metal, and are cooled either by conduction or by heat exchange to the cryogen (commonly helium) boil-off. The heat leak to cryogenic temperature for optimized conduction cooled leads at operating current is about 47 W/kA, while leads cooled by helium boil-off have a much more efficient value of 1.1 W/kA. These values are representative of the minimum heat leak that can be achieved. They are independent of the properties of the material chosen because of the proportionality relation between electrical conductivity (which governs Joule dissipation) and thermal conductivity (which rules heat in-leak) established by the Wiedemann-Franz law, to which most metals and alloys obey. The geometry of the lead that corresponds to the optimum performance (length and cross section) is however strongly dependent on the materials chosen.

The loss of a conventional lead can be further decreased replacing the cold part with High Temperature Superconducting (HTS) material, which is characterized by low thermal conductivity and zero electrical resistivity. The use of HTS leads was first pioneered on large scale at the LHC machine [29], where more than 1000 HTS leads operate at currents ranging from 600 A to 13,000 A and power the superconducting magnet circuits. The high temperature superconductor incorporated in the LHC leads is the Bi-2223 tape with a gold-doped silver matrix [30]. The HTS operates in a temperature range spanning from 50 K to 4.5 K, while a resistive heat exchanger, cooled by helium gas, provides the link between 50 K and the room temperature. In the LHC leads, the heat load into the helium bath is reduced by a factor 10 with respect to conventional self-cooled leads [29]. An additional heat load appears at intermediate temperature, which can however be removed with much better thermodynamic efficiency. The typical gain on the overall heat balance that can be achieved by proper use of high temperature superconductor in current leads is then a factor of 3 with respect to the optimized values quoted earlier.

8.1.3.7 Mechanics, Insulation, Cooling and Manufacturing Aspects

The performance of a superconducting magnet is invariably determined by proper consideration of the material physics and engineering aspects discussed earlier. Good material and magnet design, however, are not sufficient, and success relies heavily on a sound mechanical concept and the adapted manufacturing technology. The main issue in the mechanics of a superconducting magnet is how to support the electromagnetic force in situations of both large forces and large force density, at cryogenic conditions, maintaining the conductors in their nominal position, and avoiding excessive stress on superconducting cables, insulating materials and the structure itself. This simple task can be addressed using different strategies that depend on the level of field and electromagnetic force, on space and other operation constraints, on material selection, and, not last, on the specific choice of the magnet designer.

A few general lessons have been learned in the past 40 years of development, and are common practice in superconducting magnet engineering. It is important to constrain the winding pack in all directions, with a force that is greater than the Lorentz load, including an appropriate safety factor. This is done to reduce the energy inputs of mechanical origins that can trigger instabilities. For simple magnetic configurations (e.g. a solenoid) the coil itself can be the bearing structure. In the more complex situations encountered in accelerator magnets, the winding must always be in contact with a force-bearing surface. An initial load is applied on this surface, sometimes just a few MPa sufficient to remove the fluff. Caution must be used to avoid cracking or tearing at the interfaces, and in particular those bonded or glued during an impregnation process. In some cases, it may be of advantage to intentionally remove the bond between the windings and the surfaces that are not supporting the Lorentz load, especially at the surfaces that tend to separate. In this case the coil would be allowed to move as much as required, e.g. by field quality considerations. These principles may be difficult to apply in cases where the force distribution is complex, e.g. in high order field configurations such as nested multipole corrector magnets of particle accelerators. These magnets are then designed with larger operating margin to cope with the increased perturbation energy spectrum.

Coils can be *dry-wound*, in which case the conductor has free surfaces and can be permeated by the cryogen. Alternatively, they can be impregnated with a polymer resin that fills the coil spaces and once *cured* provides mechanical strength but prevents direct contact to the coolant. Common resins do not have sufficient mechanical strength to withstand thermal and mechanical stresses, and are *loaded* with fibers (e.g. glass). It is important to avoid volumes of unloaded resin, as these tend to crack and release energy that can lead to magnet quenches. Nb-Ti based conductors, ductile and strain tolerant, are well adapted to both techniques, while impregnation is favoured in the case of Nb_3Sn or HTS based conductors that are strain sensitive and require stress homogenization in the winding. Coil support is usually achieved using a stiff clamping system. For magnets working at moderate fields (up to about 5 T) a simple structure acting on the coil (referred to as *collars*), and locked by dowels or keys, may be adequate. This is the type of structure used for the Tevatron dipole (see Fig. 8.4), in which the collared coil assembly is enclosed in a cryostat (for thermal management) and centreed in the warm iron yoke by means of spring-loaded bolts. Higher fields require additional force transfer structures, for example the collared coil can be further clamped inside the magnetic yoke, thus increasing rigidity, as in the case of the RHIC, HERA and LHC dipoles, also shown in Fig. 8.4. In this case the collared coil assembly has a well defined outer surface that *mates* with the inner surface of the iron yoke. The iron yoke, assembled from packs, is held by an outer shell that takes part of the mechanical load.

Force transfer from cold to warm parts is usually kept to a minimum, and possibly reduced to the bare support of gravity loads. This is because any massive mechanical component also acts as a thermal bridge, affecting heat loads and cooling efficiency. Due to differential thermal contraction, any warm-to-cold transition is also subjected to relative movements, which shall be accomodated by adjusted kinematics or flexibility.

One of the recurrent issues in the manufacture of a superconducting magnet is the choice of electrical insulation between coil turns and between coil and ground. Most important is to include in the consideration of insulation all coil discontinuities (e.g. terminations, contact surfaces, coil heaters) as well as the instrumentation. The insulation of accelerator magnets is submitted to moderate dielectric stresses (in the range of few hundred V to a kV), but extremely high mechanical stresses under cryogenic temperatures (in the range of few ten to hundred MPa), and possibly in a radiation environment. Good coil insulation must rely in materials capable of retaining their dielectric and mechanical properties at cryogenic temperatures, such as polyimide tapes, cryogenic-grade epoxy resins and relevant glass-fiber composites. The dielectric strength of impregnated coils can be as good as for resistive magnets. In the case of insulation porous to the coolant, such as obtained by dry-winding the coil, the dielectric strength depends on the properties of the coolant itself. Liquid helium has a high breakdown strength, about 30 kV/mm which is one order of magnitude larger than that of dry air. Gaseous helium, on the contrary, has a much lower breakdown voltage, typically one tenth than that of air at the same pressure, while at sub-atmospheric pressures it decreases to a Paschen minimum of about 150 V [31]. In this case it is important to consider all possible operating conditions (e.g. the decrease of helium density during a magnet quench).

A further issue in the design of superconducting magnets is cooling. Any heat load from internal or external origin (e.g. particle energy deposition, heating at resistive splices, heat conduction and radiation to the cold mass, AC loss), of both steady state and transient nature, must be removed to a suitable cryogenic installation that provides the heat sink at the lowest temperature of the system. Magnets subjected to small heat loads (in the range of few mW) can be *indirectly* cooled by thermal conduction. This is the case of small-size magnets, operated at low current, and well shielded. The recent advance in cryo-coolers has made cryogen-free operation a convenient solution for this class of instrumentation magnets. At the other extreme, large cold masses, and high current cables, subjected to much larger heat load (few W to few tens of W) require the *direct* use of the coolant as a thermal vector. The magnet can then be cooled by immersion in a bath of liquid, either normal or superfluid helium, or by force-flow cooling. A further option in case of force-flow is to either cool the magnet as a whole, or to distribute the cooling channel within the coil using, as an example, internally cooled cables. An important aspect of any cooling method, either direct or indirect, is the temperature gradient that is established under the heat load between the superconductor and the heat sink. This temperature gradient affects the temperature margin discussed earlier. With the exception of internally cooled cables, the temperature gradient is directly proportional to the thermal resistance of the coil. Low thermal resistance insulation schemes can be obtained by proper choice of insulation thickness and overlap, as demonstrated by recent work [32] devoted to the upgrade of the quadrupoles for the inner triplet of the LHC.

The fabrication of a superconducting magnet resorts on the standard techniques, tools and instruments used in electrical and heavy industries, adapted to specificities of the materials used. Particular attention is devoted to preserve the physical properties of the conductor throughout the coil fabrication and handling, as these can be degraded by excessive strain during the winding or by heat treatments (e.g. curing of the resin). The case of Nb_3Sn magnets fabricated by the wind-and-react technique poses additional constraints on the structural and insulating materials. These need to withstand the high temperature heat treatment required for the formation of the superconducting phase (approximately 700 °C for a few hours to a few days). When compared to standard electrical equipment, superconducting coils have very tight manufacturing tolerances originating from demands of field quality (recall that the field is dominated by the current distribution) and to avoid movements during energization that could trigger quenches (recall the discussion on stability). Indeed, the level of accuracy demanded is often beyond the standard experience in electromechanical constructions. Which is why trained personnel and field experience are often critical to the successful performance of a superconducting magnet.

8.1.3.8 Super-Ferric Magnets

Super-ferric magnets are a special case of electromagnets, where the ferro-magnetic yoke, producing the dominant portion of the magnetic field in the useful aperture, is magnetized by superconducting coils. These electromagnets resemble the normal conducting magnets described earlier, apart for their cryogenic features. The prime interest of super-ferric magnets is to profit from the power advantage of superconductors (no ohmic loss). Provided the design is well optimized, the cost of cryogenic cooling is smaller than the cost of powering the resistive coils, resulting in a net gain. Part of this optimization is the trade-off between a *cold* or *warm* yoke, affecting the amount of required cooling power at cryogenic temperatures. In addition, super-ferric magnets provide operational flexibility. In particular, steady operation at nominal field is possible with no significant overhead. By contrast, high field operation can be very power-intensive, and cøstly, in normal conducting magnets. Finally, the high current density which can be achieved in superconducting coils is more than one order of magnitude larger than in normal conducting coils. This can bring additional benefits to the reduction of the overall magnet dimension, also thanks to a reduced yoke reluctance.

The limitation of super-ferric magnets is similar to normal conducting electro-magnets, namely that the field is limited by the saturation of the ferro-magnetic material of the yoke. In fact, super-ferric magnets for accelerators, such as the

Nuclotron [33] and FAIR [34] dipoles, achieve maximum field in the aperture of the order of 2 T. Super-ferric correctors are a good *niche* application, because the peak field at the pole tip remains modest. Recent examples of these magnets are the super-ferric high-order correctors of the new interaction regions of the High-Luminosity upgrade of the LHC [35].

8.2 RF Cavities

E. Jensen

The acceleration of charged particles is possible with electromagnetic fields thanks to the Lorentz force $\boldsymbol{F} = q(\boldsymbol{E} + \boldsymbol{v} \times \boldsymbol{B})$. Since the increase of particle energy is given by

$$dW = \boldsymbol{F} \cdot \boldsymbol{v}\, dt = q\,(\boldsymbol{E} \cdot \boldsymbol{v} + (\boldsymbol{v} \times \boldsymbol{B}) \cdot \boldsymbol{v})\ dt,$$

neither the magnetic field nor the transverse components of the electric field contribute to energy exchange with the particle. Only the longitudinal electric field component can be used to accelerate particles to higher energies; RF cavities for acceleration thus must provide a longitudinal electric field.

In addition to the necessity that the electric field must have a longitudinal component, a second condition is that the net electric field (or force) integrated through the RF cavity on the path of the particle trajectory (taking its finite speed into account) must not vanish. For substantial acceleration, this latter condition naturally leads to the need for a time-varying field, as can be concluded from the following line of thought: For field quantities constant in time, it follows from Maxwell's equations that the electric field is the gradient of a potential; to increase the kinetic energy of a charged particle, it will thus have to pass through a potential difference, which for technical (and safety) reasons will be limited to a few MV at best, which in turn will limit the possible energy gain to a few MeV just once, i.e. without the possibility to add stages. To reach larger energy gains, time-varying fields are necessary. RF cavities provide time-varying fields at high frequencies (radio frequency = RF, typically ranging from a few kHz to some 10 GHz).

However, when averaging over one period of the RF, the fields at any one location inside an RF cavity average out to zero; so if a particle travels a large distance through the time-varying fields in an RF cavity, it may experience both accelerating and decelerating fields, which will lead to a reduction of the net acceleration. For this reason, RF cavities are designed to concentrate the accelerating field over a relatively short distance (the *accelerating gap*). Cavities may have more than one gap, but in this case the distance between gaps must be adjusted to the particle velocity (see Sect. 8.2.6 below).

Since the particle beam is normally travelling in a vacuum pipe, the RF cavity must be compatible with this requirement as well. This can be done either by using

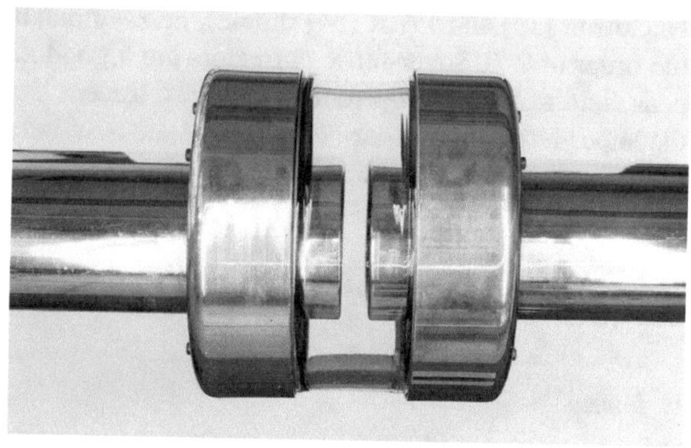

Fig. 8.13 Example of an accelerating gap with glass insulation. It is chosen for illustration; modern insulating gaps use opaque ceramics typically from SiO_2

an insulating, vacuum-tight tube inserted into the beam pipe at the location of the accelerating gap; this insulating tube is often itself referred to as the gap. Gaps are typically made of ceramic or glass (Fig. 8.13). Another possibility to make the cavity compatible with the vacuum requirements is to use the cavity itself as a vacuum vessel. In this case, the RF power has to be coupled into the cavity through vacuum tight feed-through. The design of high power RF couplers is a complex task and has almost become a discipline of its own.

Since RF electromagnetic fields radiate, RF cavities must be entirely closed on their outside with a well-conducting shield to prevent both power loss and electromagnetic interference. For this reason, cavities are normally fabricated from metal. This RF shield continues around the power coupler, the feeder cable and the RF amplifier. The beam pipe does not break this RF shield since due to its diameter it presents a hollow waveguide well below cut-off, through which electromagnetic field cannot propagate at the cavity operational frequency. A generic RF cavity (see Fig. 8.14) thus forms almost naturally an enclosed volume around the accelerating gap—the name "cavity" derives of course from this very property.

8.2.1 Parameters of a Cavity

If the metallic enclosure that forms the RF cavity were a perfect conductor and the cavity volume would not contain any lossy material, there would exist solutions to Maxwell's equations with non-vanishing fields even without any excitation. These so-called eigensolutions are also known as the cavity (oscillation) modes. Each mode is characterized by its (eigen-)frequency and its characteristic field distribution inside the cavity. If the cavity walls are made of a good rather than a perfect conductor, modes still exist and are useful to characterize the cavity, but their eigenfrequencies will become complex, describing damped oscillations, so each mode will be characterized by its frequency and its decay rate. If the field amplitudes of a mode decay as $\propto e^{-\alpha t}$, the stored energy decays as $\propto e^{-2\alpha t}$. The quality factor

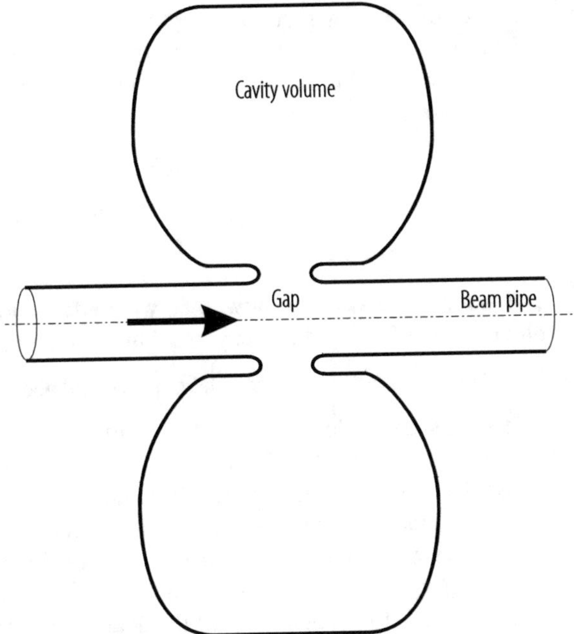

Fig. 8.14 A generic RF cavity. The arrow indicates the particle trajectory

Q is defined as

$$Q = \frac{\omega_n W}{-\frac{d}{dt} W} = \frac{\omega_n}{2\alpha}. \tag{8.16}$$

Here ω_n denotes the eigenfrequency and W the stored energy. The expression—dW/dt in the denominator of Eq. (8.15) describes the power that is lost into the cavity walls (or any other loss mechanism); it is equally the power that will have to be fed into the cavity in order to keep the stored energy at a constant value W. It is clear that the larger Q, the smaller will become the power necessary to compensate for cavity losses. In other words, one can design the cavity to be operated at or near one of its eigenfrequencies (often the lowest order mode, which is the one with the lowest eigenfrequency) and thus make use of the high Q by using the resonance phenomenon that will lead to large fields.

We define the "accelerating voltage" of a cavity (or more precisely of one cavity oscillation mode) as the integrated change of the kinetic energy of a traversing particle divided by its charge:

$$V_{\mathrm{acc}} = \frac{1}{q} \int_{-\infty}^{\infty} q\,(\boldsymbol{E} + \boldsymbol{v} \times \boldsymbol{B}) \cdot d\boldsymbol{s}, \tag{8.17}$$

where $\boldsymbol{ds}$ denotes integration along the particle trajectory, taking the fields at the actual position of the particle at the time of passage. With the fields varying at a single frequency ω and particles moving with the speed of light in the direction z, this expression simplifies to

$$V_{\text{acc}} = \int_{-\infty}^{\infty} \underline{\boldsymbol{E}}(z) e^{j\frac{\omega}{c}z} dz. \tag{8.18}$$

The underscore denotes now that we understand the field as the complex amplitude of the field of the cavity oscillation mode, with the real fields oscillating as $\text{Re}\left\{\underline{\boldsymbol{E}}(x, y, z)\, e^{j\omega t}\right\}$ in time. The exponential accounts for the movement of the particle with speed c through the cavity while the fields continue to oscillate. It is clear that the expression (8.17) is generally complex; the phase angle accounts for the phase difference between the RF field and the bunches of the passing beam; the complex amplitude is generally referred to as accelerating voltage. Since the energy W stored in the cavity is proportional to the square of the field (and thus the square of the accelerating voltage), it can be used to conveniently normalize the accelerating voltage; this leads to the definition of the quantity R-upon-Q:

$$\frac{R}{Q} = \frac{|V_{\text{acc}}|}{2\omega_0 W}. \tag{8.19}$$

The R-upon-Q thus simply quantifies how effectively the cavity converts stored energy into acceleration. Note that R-upon-Q is uniquely determined by the geometry of the cavity and not by the loss mechanism that leads to a finite Q. Multiplying the R-upon-Q with the quality factor Q, one obtains the *shunt impedance* R, which describes how effectively the cavity converts input power into acceleration voltage, as long as beam loading can be neglected:

$$R = \left(\frac{R}{Q}\right) Q = \frac{|V_{\text{acc}}|^2}{2P}. \tag{8.20}$$

Following this line of thought, the R-upon-Q may be considered a fundamental quantity and the shunt impedance R a derived quantity, in spite of the names that suggest otherwise. Note that there are a number of different definitions for these quantities in the technical literature. The definition given here is often used for synchrotrons, while in the definition used for linacs (Eq. 7.7), the factor 2 on the right hand side of Eqs. (8.18) and (8.19) is missing ("Linac-Ohms").

A cavity oscillation mode is conveniently described in an equivalent circuit as depicted in Fig. 8.15; driven by a current from the power source (or by the beam), the accelerating voltage develops across a parallel resonance circuit with resonance frequency ω_0 and quality factor Q. Losses appear in its resistive element R; the name "shunt impedance" now becomes obvious—it is "shunting" the gap.

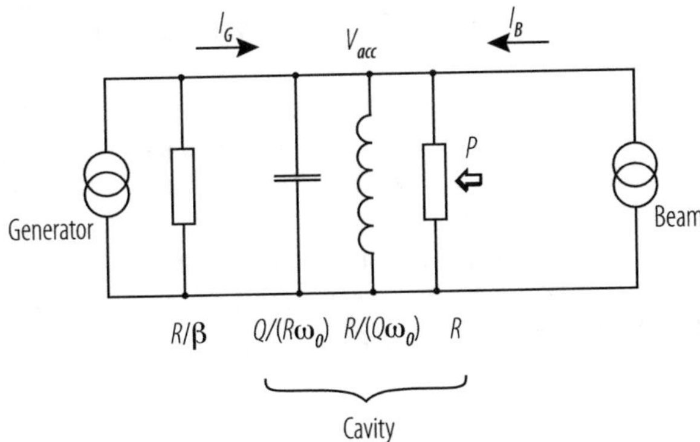

Fig. 8.15 Equivalent circuit of an RF cavity

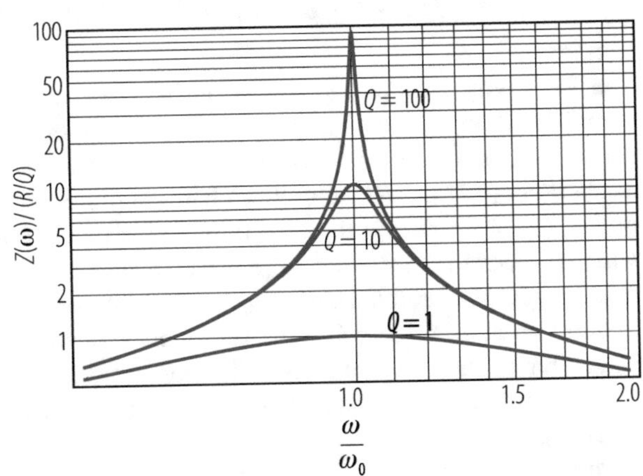

Fig. 8.16 Resonant behaviour of a cavity

When plotting the accelerating voltage versus frequency for different values of Q (Fig. 8.16), the resonance phenomenon becomes apparent that allows developing large voltages with modest powers. Consequently one trend in RF technology development has been to optimize the Q of cavities by design. Superconducting (SC) RF cavities are pushing this trend to the extreme; Q's in the order of 10^{10} are typical of SC cavities (Sect. 7.5.2). Also normal conducting cavities use high Q's to minimize the power losses; the technically obtained values depend on frequency and size and are typically in the range of some 10^4.

But high Q's also have disadvantages. As can be seen in Fig. 8.16, a high Q leads to a very sharp resonance or a very narrow bandwidth resonator, which has to be tuned very precisely and may become very delicate and sensitive to error (machining tolerances, temperature, pressure, vibrations . . .). A large stored energy will not allow for rapid changes of the field amplitude, frequency or phase. For a small ion synchrotron for example, one may wish to apply RF with non-sinusoidal form and/or with rapidly varying frequency to the beam; these requirements call for cavities which either have a large instantaneous bandwidth (i.e. a low Q) or cavities that can be rapidly tuned. In these cases moderate or even extremely small Q's can become optimum.

In addition to the fundamental mode, the field distribution of which is normally optimized for acceleration, other modes exist in the cavity, which can (and will) also interact with the particle beam. Even if not actively driven by an amplifier, these so-called higher order modes (HOM's) still present their impedance to the beam and may lead to instabilities and consequently have to be considered in the design. They are normally selectively coupled out and damped using external loads (HOM damper), thus reducing their Q.

8.2.2 *The RF Cavity as Part of the System*

The RF cavity is only one part of an RF system for a particle accelerator; the complete RF system typically consists of the following elements: (1) a master RF signal generator, controlled to have the correct frequency, phase and amplitude for acceleration, (2) the RF amplifier chain amplifies this signal to often very large power, (3) this power is then fed through the power coupler into the RF cavity, in which it leads to the desired large electromagnetic RF field, designed to optimally interact with the particle beam. As explained above, the cavity often uses resonance to build up large fields with relatively modest powers. In modern RF systems, the behaviour of the beam is constantly monitored and a multitude of feedback and feed-forward loops is constantly correcting the phase and amplitude of the RF; this latter is generally referred to as (4) low-level RF system (LLRF). Ancillary systems assure the correct tune of the resonance frequency (see below), correct vacuum and temperature conditions and interlocks for safety and protection.

8.2.3 *Ferrite Cavities*

In relatively small synchrotrons for protons and heavier ions, the speed of the particles is still changing substantially during acceleration. This requires the frequency to be swept over a large range during the acceleration cycles. For the CERN PS Booster e.g., which accelerates protons from to 50 MeV to 1.4 GeV, the proton velocity and consequently the revolution frequency varies by roughly a factor 3 in about 500 ms. In order to still take advantage of a resonance phenomenon, the resonance frequency must be varied simultaneously with the ramping of the magnetic field and the acceleration of the particles; in the PS Booster the corresponding frequency swing for the $h = 1$ system (where h denotes the harmonic number) is 0.6 MHz to 1.8 MHz, which was implemented as a ferrite cavity system with magnetic tuning [36].

A typical ferrite cavity is a double coaxial resonator with the accelerating gap in the middle, half of which is conceptually sketched in Fig. 8.17. The inner conductors of the coaxial are the beam pipes—the outer conductor is the cavity housing. Ferrite

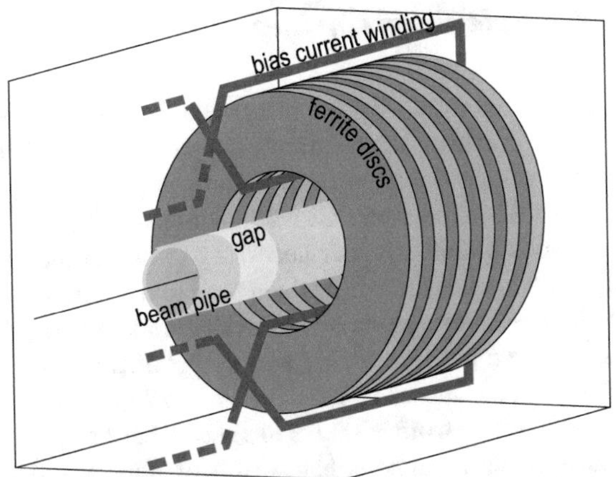

Fig. 8.17 Schematic of a ferrite cavity. The bias current allows magnetization of the ferrite rings which influences the resonant frequency

Fig. 8.18 The CERN PSB ferrite cavity with the top cover removed

rings between the two will present their magnetic permeability to the azimuthal RF magnetic field. A DC bias current allows controlling the magnetization of the magnetically soft ferrite material, which effectively changes the effective RF permeability and thus the resonance frequency of the cavity—a large bias current will magnetize the ferrites up to their saturation, which results in a small RF permeability and consequently a smaller inductivity in the equivalent circuit and thus a larger resonance frequency. The bias current windings form a figure of eight around the two separate blocks of ferrite rings, which ensures the DC circuit to be decoupled from the RF circuit.

Figure 8.18 shows the practical implementation at the example of one of the four CERN PS Booster cavities with the top cover removed. Air cooling ducts are visible on the left; the power amplifier is connected on the right.

8.2.4 Wide-Band Cavities

Instead of rapidly tuning the resonance frequency, a relatively new idea is to use a very low Q instead, which allows to operate in a large frequency range without any tuning at all. Cavities with this feature can for example be built using amorphous, nanocrystalline magnetic alloys like Finemet® or Metglas®. The price to pay in operation is a higher power necessary to obtain the same voltage, but this may be well worth it since there is no need for a tuning circuit or a tuning power supply. This technique has been pioneered by KEK in the nineties for the J-PARC facility [37]. An example for a wide-band cavity is that of LEIR (Low Energy Ion Ring) at CERN [38]. LEIR accelerates Pb ions from 4.2 to 72 MeV/n—the corresponding revolution frequency varies from 360 kHz to 1.42 MHz (2 octaves). The RF system is designed to have an instantaneous bandwidth of almost 4 octaves (350 kHz to 5 MHz), which allows operation with Pb ions at harmonics 1 and 2 simultaneously with the necessary large frequency swing and still could allow acceleration of other ion species if required. Due to the extremely low Q, the relatively modest accelerating voltage of 2 kV required a 60 kW amplifier for each of the two LEIR systems.

Figure 8.19 shows the longitudinal section of a LEIR cavity with 3 Finemet® cores on either side of the central ceramic gap. Figure 8.20 shows the large instantaneous bandwidth in excess of a decade measured for the complete system (blue curve). Comparison with Fig. 8.16 above indicates an effective Q in the order of 0.5.

Fig. 8.19 The LEIR wide-band cavity. 3 water-cooled Finemet® cores are placed symmetrically on each side of the central gap

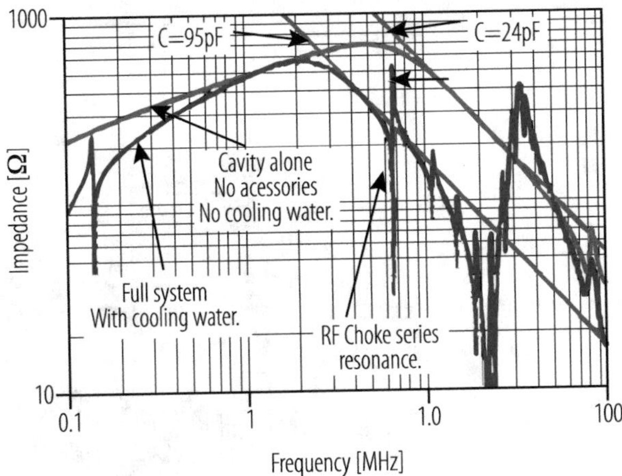

Fig. 8.20 The LEIR cavity gap impedance (blue curve). In the frequency range from 350 kHz to 5 MHz it varies from 275 Ω to 560 Ω

8.2.5 *Single-Gap Vacuum Cavities*

Once the particles are travelling close enough to the speed of light, their acceleration will not significantly change their speed anymore; electrons have reached 99.9% of the speed of light with a kinetic energy of 11 MeV, protons with 20 GeV—at larger energies their speed and thus the frequency varies less than 1‰, At larger energies, the operation frequency lies inside the natural bandwidth of the cavity, which makes rapid tuning of the resonance frequency unnecessary. In this case, vacuum cavities can be used, which do not require a ceramic gap and thus can potentially reach much larger accelerating voltages. Fixed frequency cavities are used in the CERN PS to form the bunch pattern for the LHC, which requires very short bunches (<4 ns) spaced at integer multiples of 25 ns. This bunching process requires not only relatively large RF voltages (hundreds of kV) at both 40 MHz and 80 MHz, but also fast switching of these voltages (20 μs).

As example of a single-gap fixed frequency cavity, a longitudinal section of the CERN PS 80 MHz cavity is sketched in Fig. 8.21. The gap on the left can be opened and closed by a pneumatically operated mechanical short-circuit, which allows to make the cavity "invisible" to the beam. Piston tuners entering from the left allow to set the fundamental mode resonance frequency to the operation frequency. The cavity is fabricated from stainless steel, galvanically copper plated on the inside, leading to a Q of 22,600. The R/Q of this cavity is 56 Ω. Figure 8.22 shows the cavity in the RF power test-stand. The final amplifier (visible in the foreground) is strongly coupled to the cavity, which allows the rapid filling of the cavity [39].

It is also part of a fast feed-back loop, which allows making the cavity almost invisible to the beam if the gap is mechanically open. This works as follows: the voltage present in the cavity is constantly monitored by a pick-up, the signal of which is fed into the amplifier chain with the proper phase. If the set-point value of the accelerating voltage is zero but the beam induces a voltage in the cavity

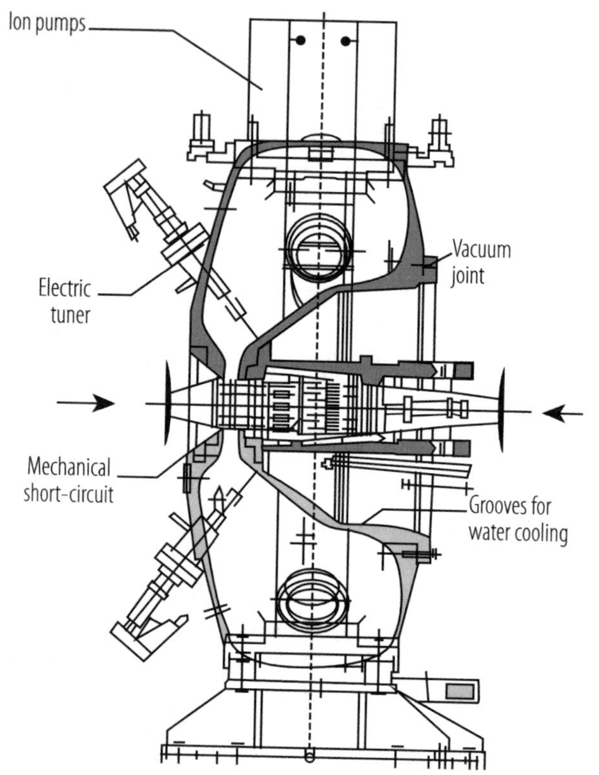

Fig. 8.21 CERN PS 80 MHz cavity, assembly drawing

(current source on the right in Fig. 8.15), the pick-up will detect this induced voltage and the feedback loop creates an equal voltage with opposite phase, such that the beam induced voltage is exactly compensated. The fast RF feedback loops in the PS 40 MHz and 80 MHz systems have a loop gain in excess of 40 dB, thus reducing the beam induced voltage by more than a factor 100. The loop also stabilizes the voltage precisely around any other set-point value.

8.2.6 Multi-Gap Cavities

In a normal-conducting, single-gap vacuum cavity, there exists a maximum value for the shunt impedance that can be obtained after full optimisation of the cavity. This value is in the order of a few MΩ, the exact value will depend on the frequency range. Limited by the available RF power and the cavity, this sets a upper limit to the accelerating voltage; for larger voltages one has to increase the number of RF systems and the power accordingly.

Fig. 8.22 CERN PS 80 MHz cavity in the RF test bunker. The final amplifier is visible in the foreground

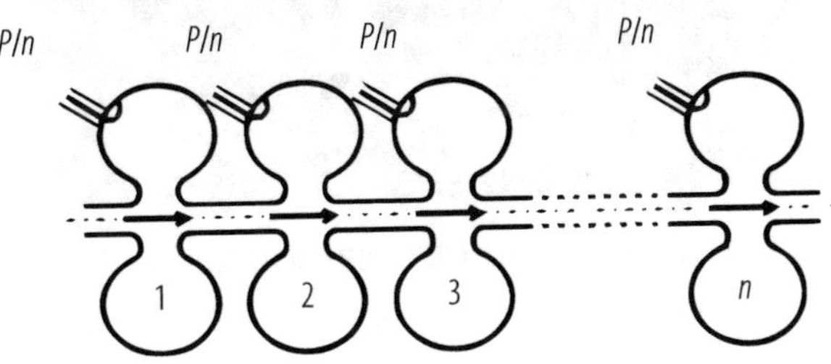

Fig. 8.23 Distributing a given power P to n cavities

To introduce multi-gap cavities let us see what happens if one just increases the number of gaps, keeping the total power constant: consider n single-gap cavities with a shunt impedance R, as sketched in Fig. 8.23. The available power is split in equal parts and evenly distributed to the n cavities. According to Eq. (8.19), each cavity will produce an accelerating voltage of $\sqrt{2R\,(P/n)}$, so with the correct phasing of the RF the total voltage will be just the sum, $V_{acc} = \sqrt{2(nR)P}$. If we now consider the assembly consisting of the n original cavities and the power splitter as a single cavity with n gaps, we notice that this new cavity has the shunt impedance nR; this is a significant increase. Consequently by just multiplying the number of gaps one can make much more efficient use of the available RF power to generate very large accelerating voltages.

Instead of using n individual power couplers and a large power splitter, much more elegant ways of distributing the available RF power to many gaps have been invented—one can combine the individual gaps in one vacuum vessel and one can in fact use this vacuum vessel itself as a distributed power splitter, which leads to standing wave or travelling wave cavities. In a travelling wave structure (Sect. 7.5.1), the RF power is fed via a power coupler into one end of the cavity, flowing

Fig. 8.24 A view inside the CERN SPS 200 MHz travelling wave cavity

(travelling) through the cavity, typically in the same direction as the beam, creating an accelerating voltage at every gap. An output coupler at the far end of the cavity is connected to a matched power load. By design it is assured that the phase velocity of the travelling wave is equal to the speed of the particles, i.e. the phase advance over a cell of length d is equal to $2\pi d/\lambda$ for particles travelling with the speed of light. If this condition is satisfied, one can imagine the particles "surfing" on this forward travelling wave.

As an example for a travelling wave structure, Fig. 8.24 shows a view inside one of the 4 SPS 200 MHz travelling wave cavities [40] with the end covers removed. The phase advance per cell is $\pi/2$, corresponding to a regular gap distances of 375 mm. The presently operating 4 SPS travelling wave cavities produce an unloaded accelerating voltage of 12 MV with a total power of about 4 MW at 200 MHz, but in the framework of the LHC injector upgrade project, the cavities will be shortened to allow for higher beam current and their number will be increased from 4 to 6.

Vacuum cavities can reach significantly higher voltages than cavities with a ceramic gap, which has a maximum hold-off voltage smaller than vacuum. Multi-gap accelerating structures, both travelling wave and standing wave, allow in addition to effectively convert available RF power to acceleration. For these reasons, multi-gap vacuum cavities are used where very large accelerating gradients are required, like in a future linear collider projects. Section 7.5 above is dedicated to the design of high gradient accelerating cavities for linear colliders.

8.2.7 Superconducting Cavities

Superconductivity denotes the effect of vanishing electrical resistivity in some materials at cryogenic temperatures. Superconductivity is applied in many present day high energy accelerators in coils to create very large (DC) magnetic fields (Sect. 8.1.3). If also the RF resistance were zero, one would be able to establish RF fields as cavity mode in a superconducting cavity without feeding any RF power, i.e. the shunt impedance would become infinite. The RF surface resistance is not exactly zero though, but it still can be made small enough to reach Q values in the order of 10^{10}, which makes RF superconductivity still extremely attractive.

It should be noted that even if the power lost in a superconducting cavity is much smaller than in normal-conducting cavities, this power has to be cooled at cryogenic temperatures, which is much less efficient. As a rule of thumb, an AC power for the refrigerator plant of about 1 kW is necessary to cool a lost RF power of 1 W lost at 2 K (Sect. 8.3.5).

The more serious limitation for superconducting RF cavities is however the maximum possible accelerating field: the RF magnetic field at the cavity surface of the accelerating mode is equal to a surface current density, which must stay below a critical value. For niobium, the best RF superconductor known today, this effect limits the accelerating gradient to below 50 MV/m. But already accelerating gradients above of 5 MV/m require extraordinary care, since also other effects like field emissions from impurities on the surface or multipactor induced quenches have to be dealt with. In order to obtain very large accelerating fields, a very complex technology has been developed over the last decades, primarily driven by the TESLA collaboration (and later the ILC Global Design Effort) with the aim to develop cavities for a linear collider. This R&D allowed pushing the practical values for reliably obtained accelerating gradients from a few MV/m to levels close to the above mentioned limit; the accelerating gradient considered practical for the ILC is 31.5 MV/m. Also the European XFEL, now under construction near Hamburg, uses this technology. The developed technology includes the forming and welding of niobium sheets, different abrasive and non-abrasive, chemical and electrochemical cleaning processes, a special technique with ultra-pure water applied under high pressure, and the strict application of clean-room methods.

Since Sect. 7.5.2 above is dedicated to superconducting accelerating structures for linacs, we will here give an example of a superconducting cavity using a different technology, which has been initially developed for LEP at CERN and is now also used for LHC. The cavities are fabricated from sheet metal copper—a process well understood—and eventually sputtered from the inside with a thin layer of Nb. The advantage of this technique is that the good thermal conductivity of copper will

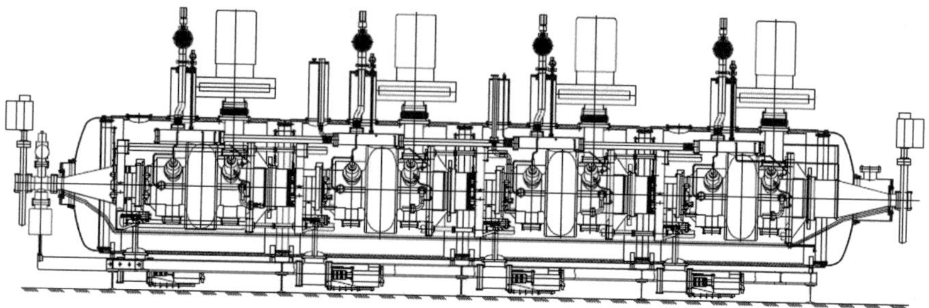

Fig. 8.25 Longitudinal section of LHC 400 MHz cryostat containing 4 single-cell cavities

rapidly equalize the temperature distribution if local heating occurs caused by a surface defect or an impurity; this retards the onset of a quench significantly and thus reduces the sensitivity to impurities. Due to the significant cost of niobium, this technique can also be very cost effective. The maximum accelerating fields however are still in the order of 10 MV/m today, well below what has been obtained with bulk Niobium cavities, which makes this technology less attractive for linacs, but very attractive for large synchrotrons.

The LHC 400 MHz cavities [41], sketched in Fig. 8.25 in a longitudinal section, are fabricated using this technique. One cryostat contains 4 individual cavities, each equipped with its helium tank, tuner, HOM dampers and variable power coupler. A total of 16 cavities (4 cryostats) are installed in LHC, 8 for each beam. They are operated at the relatively modest accelerating gradient of 5.5 MV/m resulting in a total accelerating voltage of 2 MV per cavity or 16 MV per beam.

8.2.8 *RF Cavities for Special Applications*

The use of RF cavities is not limited to particle acceleration in the direction of their motion. Cavities with longitudinal fields are also used for bunching, de-bunching and deceleration. Radio frequency quadrupoles (RFQ's), for example, serve at the same time transverse focussing, bunching and acceleration in a very effective way; most modern ion injectors make use of RFQ's.

Three examples for deceleration shall be mentioned here: antiprotons are decelerated for example in the CERN AD facility to enable studies of antiprotons at rest or for the creation of antimatter. Decelerating cavities knows as PETS (for Power Extraction and Transfer Structure) are used to slow down the CLIC drive beam in order to extract its kinetic energy as high frequency electromagnetic energy (12 GHz in the case of CLIC), which allows to use the drive beam as a high efficiency, high power microwave source [42]. And finally in energy recovery linacs (ERLs), one and the same cavity is used for both acceleration and deceleration to recover the beam energy (Sect. 7.9).

8.2.9 *Deflecting Cavities: Crab Cavities*

Another class of RF cavities employ the transverse, not the longitudinal component of the force to kick particles sideways. If a beam bunched at a frequency f travels through a deflecting cavity oscillating at a frequency $f/2$, every even bunch will be kicked in one direction, every odd bunched in the other. This can be used to separate the beam in two beams bunched at half the frequency. It can equally be used to combine two beams, interleaving the bunches and thus creating one beam with twice the current and bunched at double the frequency. This scheme is a basic building block of the drive beam recombination scheme in CLIC (see Sect. 7.4.3 above) and has been thoroughly tested in the CTF3 facility.

Probably the most important application of deflecting cavities is their use as so-called *crab cavities*. In a high energy collider like the LHC, very small bunches of counter-running beams are colliding in the interaction points. The beams are normally not colliding exactly head on, but with a small crossing angle in order to minimize long-range beam-beam effects. Due to this non-vanishing crossing angle, only some particles of one bunch will collide with some particles of the oncoming bunch (see Fig. 8.26, top). If the head and the tail of the bunches are kicked sideways in opposite directions in the plane of the collision, the bunches become tilted after some drift—they start to move sideways like crabs. If this tilt is equal to half the crossing angle at the interaction point, the bunches are again oriented exactly as in a head-on collision and the geometrical loss of luminosity is compensated (Fig. 8.26, bottom).

The plan to upgrade the LHC to reach larger luminosity contains a number of important elements: In addition to increasing the bunch intensity by an upgrade of the LHC injector chain, stronger final focus magnets (the inner triplet) are planned with larger apertures. These will allow working with smaller beam sizes at collision and larger crossing angles. In order to take full advantage of this upgrade, crab cavities are essential [43]. An important additional feature of the operation with crab cavities is the possibility to change the crab cavity voltage during the coast to compensate for the loss of particles (luminosity levelling).

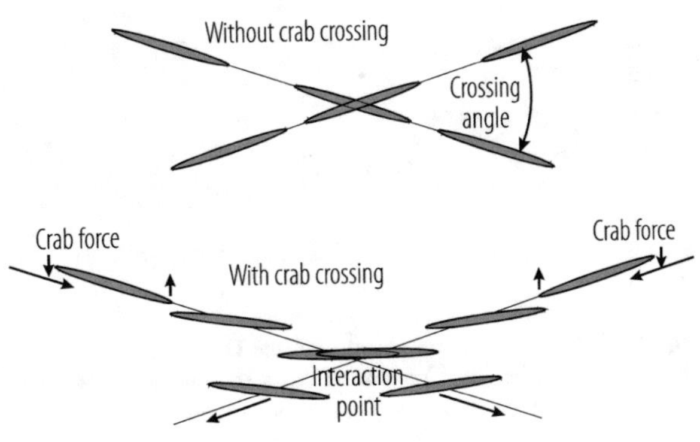

Fig. 8.26 With a finite crossing angle, only part of the colliding bunches interact without crab crossing (top). Crab cavities allow aligning the bunches such as to maximize interaction

For the construction of crab cavities for the LHC, RF engineers were confronted with an additional difficulty resulting from the proximity of the two beam pipes, which in most of the machine are spaced by only 194 mm. Conventional, elliptically shaped RF cavities have a radius of roughly half a wavelength, which in the case of LHC is not consistent at a frequency of 400 MHz. A special straight section near point 4 is provided for the 400 MHz accelerating cavities described in Sect. 8.2.7 above, but it is desirable that the crab cavities are located upstream and downstream of the high luminosity collision points.

In a rigorous R&D program over recent years, a collaboration between CERN, FNAL, JLAB, BNL, LBNL and ODU in the US, the STFC in the UK, KEK in Japan and industry has successfully designed, prototyped, fabricated and tested crab cavities compact enough to meet the demanding LHC constraints and specifications. The "Double Quarter-Wave" (DQW) cavity was optimized for vertical crossing; its inside geometry is sketched in Fig. 8.27a, indicating also the location of the second beam pipe. The "RF Dipole" (RFD) cavity with similar topology but different coupling for both fundamental and higher-order modes, was optimized for horizontal crossing and is sketched in Fig. 8.27b. Both designs are needed since the LHC high-luminosity interaction points IP1 (ATLAS) and IP5 (CMS) utilize different crossing planes. Each cavity produces a transverse kick voltage in the order of 4 MV; two cavities will be assembled in one cryomodule. A total of 16 cavities (8 cryomodules) will initially be installed in the LHC, two per beam per side and per IP.

Figure 8.27c shows a cut-open view of a 2-cavity cryomodule (illustrated by means of the DQW cavities). The cavities are located in individual helium vessels and equipped with tuners, power couplers and HOM couplers. The cryomodules are equipped equally with thermal and magnetic shields. A large cryogenic line (2 K) connected to a cryogenic service box can be seen on the left. The fundamental mode couplers are connected to two 400 MHz waveguides on the top. The beam tube for the second LHC beam passes through the cryomodule. The cryomodule shown was

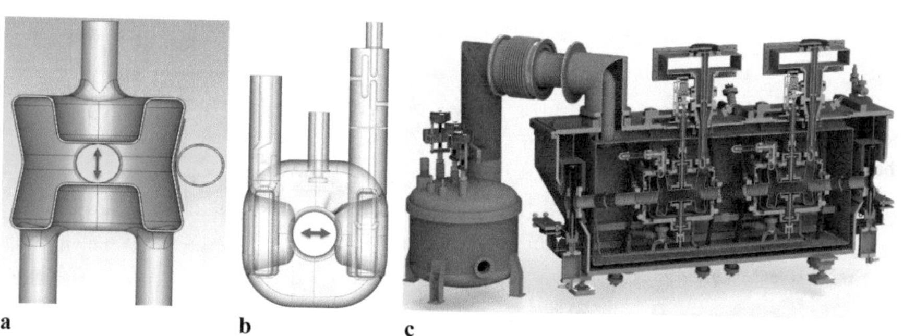

Fig. 8.27 Conceptual cross section of the DQW crab cavity, indicating the crossing plane and the second beam tube (**a**). Conceptual cross section of the RFD crab cavity indicating the crossing plane and the HOM couplers (**b**). Cut-open view of the 2-cavity cryomodule equipped with DQW cavities (**c**)

built for the test of the LHC crab cavities in the SPS for full validation in 2018, but it is conceptually compatible with the LHC requirements.

8.3 Cryogenics

Ph. Lebrun · L. Tavian

8.3.1 *Introduction*

Cryogenics has become a key technology for particle accelerators, primarily as ancillary to the developing use of superconducting magnets and RF cavities [44, 45]. In this class of applications, the superconductor must operate at a fraction of its critical temperature in order to preserve current-carrying capability at high field (magnets) or to limit a.c. losses (RF cavities), thus imposing the use of helium in the case of low-temperature superconductors. Additional important benefits of operating the accelerator beam pipes at low temperature are the achievement of high vacuum through cryo-pumping of all residual gas species except helium, and the reduction of wall resistance which controls image-current losses and transverse impedance. Following pioneering work at the TeVatron and first large-scale applications (HERA, LEP2), accelerators at the energy frontier (RHIC, LHC) use superconducting magnets, while high-intensity proton accelerators (SNS, ESS) and high-energy electron linacs (European X-FEL, ILC) are based on superconducting RF cavities, all requiring large helium cryogenic systems. A recent example of such a system is sorely described in [46]. On a smaller scale, cryogenics is also at work cooling compact superconducting cyclotrons for radionuclide production or particle therapy, as well as compact synchrotron sources for X-ray lithography.

8.3.2 *Cryogenic Fluids*

8.3.2.1 Thermophysical Properties

The simplest way of cooling equipment with a cryogenic fluid is to make use of its latent heat of vaporization, e.g. by immersion in a bath of boiling liquid. As a consequence, the useful temperature range of cryogenic fluids [47–49] is that in which there exists latent heat of vaporization, i.e. between the triple point and the critical point, with a particular interest in the normal boiling point, i.e. the saturation temperature at atmospheric pressure. This data are given in Table 8.8. In the following, we will concentrate on two cryogens: helium which is the only liquid at very low temperature and thus the coolant of low-temperature superconducting

Table 8.8 Characteristic temperature [K] of cryogenic fluids

Cryogen	Triple point	Normal boiling point	Critical point
Methane	90.7	111.6	190.5
Oxygen	54.4	90.2	154.6
Argon	83.8	87.3	150.9
Nitrogen	63.1	77.3	126.2
Neon	24.6	27.1	44.4
Hydrogen	13.8	20.4	33.2
Helium	2.2[a]	4.2	5.2

[a] λ point

Table 8.9 Properties of helium and nitrogen compared to water

Property	Helium	Nitrogen	Water
Normal boiling point [K]	4.2	77	373
Critical temperature [K]	5.2	126	647
Critical pressure [bar]	2.3	34	221
Liquid density[a] [kg/m^3]	125	808	960
Liquid/vapour density ratio	7.4	175	1600
Heat of vaporization[a] [kJ/kg]	20.4	199	2260
Liquid viscosity[a] [μPl]	3.3	152	278

[a] At normal boiling point

devices [50, 51], and nitrogen for its wide availability and ease of use for pre-cooling equipment and for thermal shielding.

To develop a feeling about properties of these cryogenic fluids, it is instructive to compare them with those of water (Table 8.9). In both cases, but particularly with helium, applications operate much closer to the critical point, i.e. in a domain where the difference between the liquid and vapour phases is much less marked: the ratio of liquid to vapour densities and the latent heat associated with the change of phase are much smaller. Due to the low values of its critical pressure and temperature, helium can also be used as a cryogenic coolant beyond the critical point, in the supercritical state. It is also interesting to note that, while liquid nitrogen resembles water as concerns density and viscosity, liquid helium is much lighter and less viscous. This latter property makes it a medium of choice for permeating small channels inside magnet windings and thus stabilizing the superconductor.

8.3.2.2 Liquid Boil-off

The factor of ten in latent heat of vaporization between helium and nitrogen, combined with the lower density of the former, induces a large difference in vaporization rates under the same applied heat load (Table 8.10). This illustrates the need for implementing much better insulation techniques in liquid helium vessels to achieve comparable holding times. Vaporization flow measurements during steady-

Table 8.10 Vaporization of cryogen at normal boiling point under 1 W applied heat load

Cryogen	[mg/s]	[l/h liquid]	[l/min gas NTP]
Helium	48	1.38	16.4
Nitrogen	5	0.02	0.24

state boil-off constitute a practical method for assessing the heat load of a cryostat holding a saturated cryogen bath.

8.3.2.3 Cryogen Usage for Equipment Cooldown

For both fluids, the specific heat of the vapour over the temperature range from liquid saturation to ambient is comparable to or larger than the latent heat of vaporization. This provides a valuable cooling potential at intermediate temperature, which can be used for thermal shielding or for pre-cooling of equipment from room temperature. The heat balance equation for cooling a mass of, say iron m_{Fe} of specific heat $C_{\mathrm{Fe}}(T)$ at temperature T by vaporizing a mass dm of cryogenic liquid at saturation temperature T_v, latent heat of vaporization L_v and vapour specific heat C_p (taken as constant), is assuming perfect heat exchange with the liquid and the vapour:

$$m_{\mathrm{Fe}} C_{\mathrm{Fe}}(T)\mathrm{d}T = \left[L_v + C_p\left(T - T_v\right)\right]\mathrm{d}m. \tag{8.21}$$

Hence the specific liquid cryogen requirement for cool-down from temperature T_0:

$$\frac{m}{m_{\mathrm{Fe}}} = \int_{T_0}^{T} \frac{C_{Fe}(T)\mathrm{d}T}{L_v + C_p\left(T - T_v\right)}. \tag{8.22}$$

Calculated values of specific liquid cryogen requirements for iron are given in Table 8.11, clearly demonstrating the interest of making use of the specific heat of helium vapour, as well as that of pre-cooling equipment with liquid nitrogen.

8.3.2.4 Phase Domain

Typical operating domains with cryogenic helium and nitrogen are shown in Figs. 8.28 and 8.29, superimposed on the phase diagrams of the substances. While

Table 8.11 Volume [l] of liquid cryogens required to cool down 1 kg of iron

Using	Latent heat only	Latent heat and specific heat of vapour
Liquid helium from 290 K to 4.2 K	29.5	0.75
Liquid helium from 77 K to 4.2 K	1.46	0.12
Liquid nitrogen from 290 K to 77 K	0.45	0.29

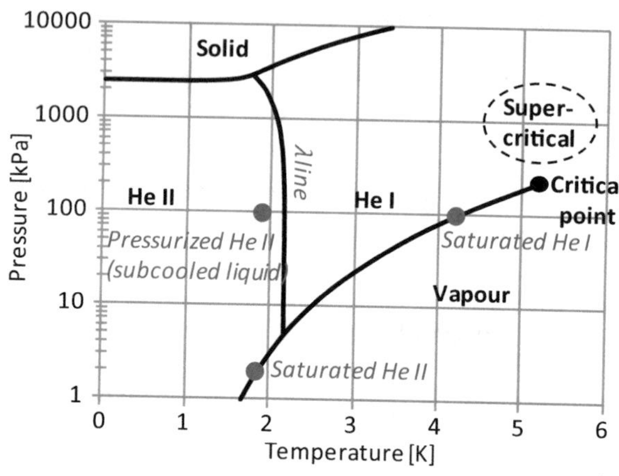

Fig. 8.28 Phase diagram of helium, showing typical operating domains

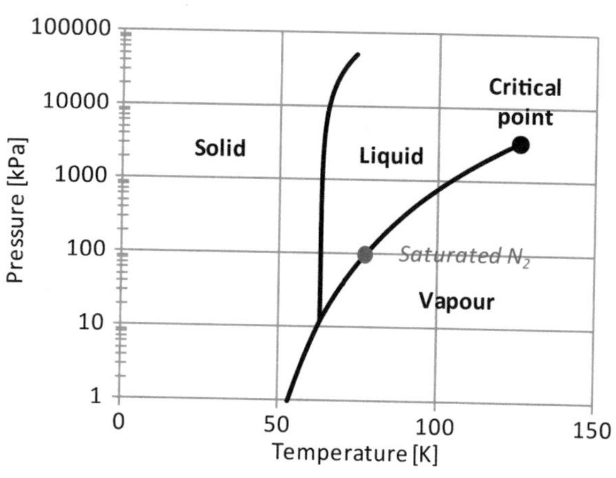

Fig. 8.29 Phase diagram of nitrogen, showing typical operating domains

nitrogen shows a classical phase diagram, that of helium shows several peculiarities. The solid phase of helium only exists under pressure and the normal liquid He I undergoes below 2.2 K a transition to another liquid phase, He II, instead of solidifying. There is no latent heat associated with this phase transition, but a peak in the specific heat, the shape of which gave the name "λ-line" to the phase boundary. He II exhibits superfluidity, a macroscopic quantum behaviour entailing very high thermal conductivity and very low viscosity which make it a coolant of choice for advanced superconducting devices [52, 53]. Besides the thermodynamic penalty of lower temperature, the use of He II imposes that at least part of the cryogenic circuits operate at sub-atmospheric pressure, thus requiring efficient compression of low-pressure vapour and creating risks of dielectric breakdown and contamination by air in-leaks.

While saturated He I provides fixed (saturation) temperature and high boiling heat transfer at moderate heat flux, it may develop instabilities in two-phase flow and is prone to boiling crisis above the peak nucleate boiling flux (about 1 W/cm^2).

The use of mono-phase supercritical helium in forced-flow systems avoids the problems of two-phase flow. However, the strongly varying properties of the fluid in the vicinity of the critical point may create other issues, such as density wave oscillations. More fundamentally, supercritical helium exhibits no latent heat, so that applied heat loads result in temperature increases which must be contained by high flow-rate or periodic re-cooling in extended systems.

8.3.3 *Materials at Low Temperatures*

Designing and building cryogenic apparatus require good knowledge of material properties at low temperatures [54–56], some of which may vary by orders of magnitude between ambient and cryogenic conditions, or even exhibit discontinuous behaviour. Of particular concern is the ductile-to-brittle transition undergone by metals and alloys of body-centered cubic and to a lesser extent, hexagonal close-packed structures: these materials should always be used above their ductile-to-brittle transition temperature.

Thermal contraction, shown in Fig. 8.30 for selected materials, range from very low to a few per mille (metals and alloys) up to a few percent (polymers). For practical purposes, it is important to note that most of the contraction occurs between ambient and 80 K.

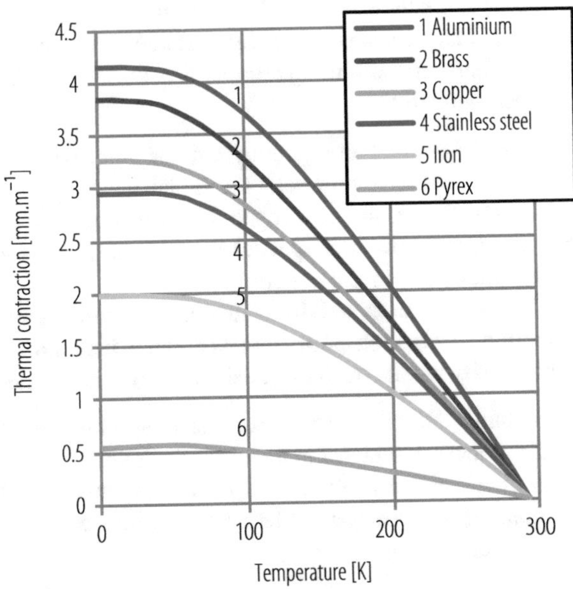

Fig. 8.30 Thermal contraction of selected materials at low temperature

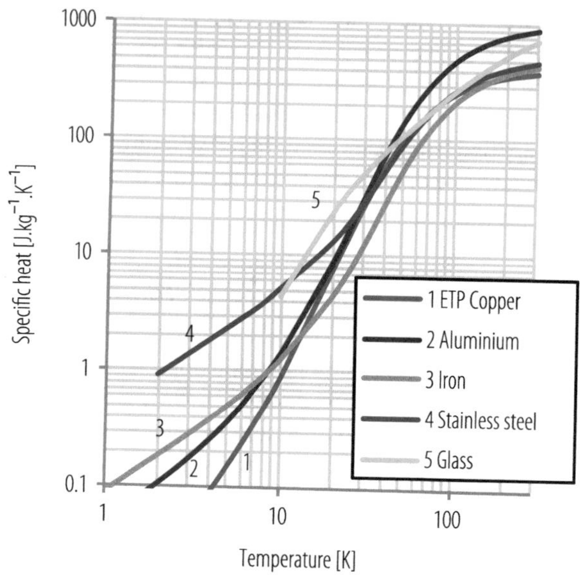

Fig. 8.31 Specific heat of selected materials at low temperature

Specific heat of materials always shows a marked drop at low temperatures, asymptotic to zero as one approaches absolute zero (Fig. 8.31). Solid state physics describes specific heat as the sum of several terms due to the contributions from:

- the crystal lattice, with a T^3-dependence (Debye law),
- the free electrons, with a T-dependence (Fermi gas model), and
- the phase transition undergone by the material, e.g. magnetic ordering or superconductivity.

Specific heat of compounds can be approximately predicted by the Kopp-Neumann additivity principle: the molar heat capacity of a compound is equal to the sum of the atomic heat capacities of its constituents. To be noted that, at liquid helium temperature, the specific heat of the fluid is several orders of magnitude higher than that of solid materials.

The thermal conductivity of technical materials spans across several orders of magnitude, from high-purity metals down to insulators (Fig. 8.32). The bulk thermal conductivity results from two basic mechanisms, namely:

- thermal conduction by electrons, scattered by lattice phonons and imperfections; this process dominates in metals and alloys,
- thermal conduction by lattice phonons, scattered by lattice imperfection or other phonons; this process, far less efficient than electronic conduction, dominates in non-metals.

It must be noted that, for pure metals, low temperature thermal conductivity is strongly influenced by the impurity level and microstructure (e.g. cold work).

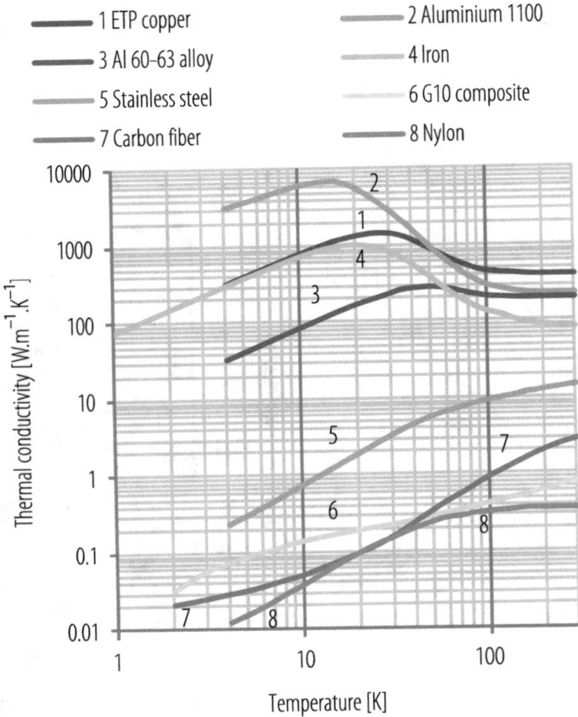

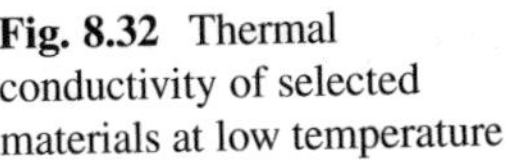
Fig. 8.32 Thermal conductivity of selected materials at low temperature

Electrical resistivity at low temperatures is a most interesting property both for engineering applications, assessment of material purity and microstructure, and understanding of its solid-state physics. Low temperature resistivity spans across several orders of magnitude, depending upon the type of material, degree of alloying and purity (Fig. 8.33). In all cases, the decrease of electrical resistivity upon cooldown shows a plateau, corresponding to a residual value ρ_0. Matthiessen's rule expresses total resistivity $\rho(T)$ as the sum of two terms,

$$\rho(T) = \rho_0 + \rho_i(T), \tag{8.23}$$

where ρ_0 is the residual resistivity due to electron scattering by impurities and lattice imperfections, and $\rho_i(T)$ is the intrinsic resistivity due to electron scattering by phonons, strongly temperature dependent.

A convenient way to characterize a pure metal or dilute alloy is to measure its residual resistivity ratio *RRR*:

$$RRR = \rho\,(273\ \mathrm{K})\,/\rho\,(4\ \mathrm{K})\,. \tag{8.24}$$

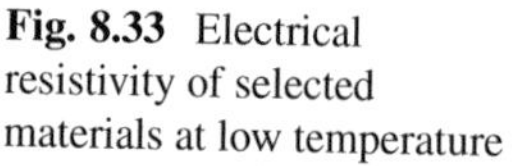

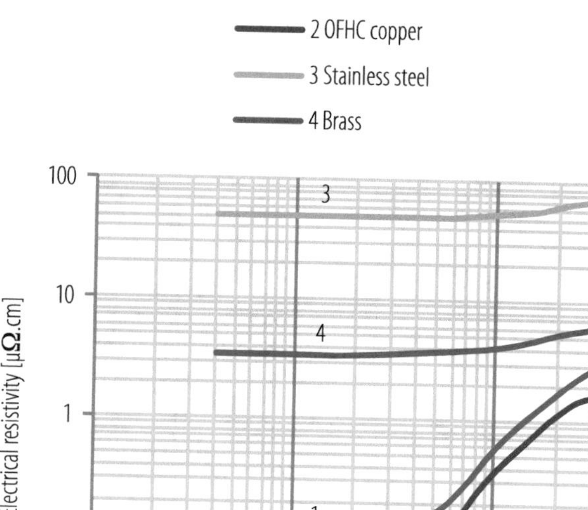

Fig. 8.33 Electrical resistivity of selected materials at low temperature

8.3.4 Heat Transfer and Thermal Design

With the exception of superfluid helium, the heat transfer processes at work in cryogenics are basically the same as for any engineering temperature range. The strong variation of thermal properties of materials and fluids at low temperature however has two consequences: the relative and absolute magnitudes of the processes may be very different from those at room temperature, and the equations become non-linear, thus requiring numerical integration. Cryogenic thermal design is the art of using these processes adequately, either for achieving thermal insulation (cryostats, transfer lines) or for improving thermal coupling between equipment and coolant (cool-down & warm-up, thermal stabilization, thermometry) [57, 58]. The diversity and complexity of convection processes cannot be treated here. Fortunately, in the majority of cases, the correlations established for fluids at higher temperature are fully applicable to the cryogenic domain [59], and reference is made to the abundant technical literature on the subject.

8.3.4.1 Solid Conduction

Heat conduction in solids is represented by Fourier's law, expressing proportionality of heat flux with thermal gradient

$$Q = k(T)S\mathrm{d}T/\mathrm{d}x. \tag{8.25}$$

This equation also defines the thermal conductivity $k(T)$ of the material, which varies with temperature. Conduction along a solid rod of length L, cross section S spanning a temperature range $[T_1, T_2]$, e.g. the support strut of a cryogenic vessel, is then given by the integral form

$$Q = \frac{S}{L}\int_{T_1}^{T_2} k(T)\mathrm{d}T, \tag{8.26}$$

where $\int_{T_1}^{T_2} k(T)\mathrm{d}T$ is called the thermal conductivity integral.

Thermal conductivity integrals of standard materials are tabulated in the literature [54]. A few examples are given in Table 8.12, showing the large differences between good and bad thermal conducting materials, the strong decrease of conductivity at low temperatures, particularly for pure metals, and the interest of thermal interception to reduce conductive heat in-leak in supports. As an example, the thermal conductivity integral of austenitic stainless steel from 80 K to vanishingly low temperature is nine times smaller than from 290 K, hence the benefit of providing a liquid nitrogen cooled heat sink on the supports of a liquid helium vessel.

8.3.4.2 Radiation

Blackbody radiation strongly and only depends on the temperature of the emitting body, with the maximum of the power spectrum given by Wien's law

$$\lambda_{max}T = 2898\ [\mu\mathrm{m}\cdot \mathrm{K}], \tag{8.27}$$

Table 8.12 Thermal conductivity integrals [W/m] of selected materials

From vanishingly low temperature up to	20 K	80 K	290 K
OFHC copper	11,000	60,600	152,000
DHP copper	395	5890	46,100
Aluminium 1100	2740	23,300	72,100
2024 aluminium alloy	160	2420	22,900
AISI 304 stainless steel	16.3	349	3060
G-10 glass-epoxy composite	2	18	153

and the total power radiated by an surface area A given by Stefan-Boltzmann's law

$$Q = \sigma\ A\ T^4, \tag{8.28}$$

with Stefan-Boltzmann's constant $\sigma \simeq 5.67\ 10^{-8}$ W m^{-2} K^{-4}. The dependence of the radiative heat flux on the fourth power of temperature makes a strong plea for radiation shielding of low-temperature vessels with one or several shields cooled by liquid nitrogen or cold helium vapour. Technical radiating surfaces are usually described as "gray" bodies, characterized by an emissivity ε smaller than 1:

$$Q = \varepsilon\ \sigma\ A\ T^4\,. \tag{8.29}$$

The emissivity ε strictly depends on the material, surface finish, radiation wavelength and angle of incidence. For materials of technical interest, measured average values are found in the literature [60], a subset of which is given in Table 8.13. As a general rule, emissivity decreases at low temperature, for good electrical conductors and for polished surfaces. As Table 8.13 shows, a simple way to obtain this combination of properties is to wrap cold equipment with aluminium foil. Conversely, radiative thermal coupling requires emissivity as close as possible to that of a blackbody, which can be achieved in practice by special paint or adequate surface treatment, e.g. anodizing of aluminium.

The net heat flux between two "gray" surfaces at temperature T_1 and T_2 is similarly given by

$$Q = E\ \sigma\ A\ \left(T_2{}^4 - T_1{}^4\right), \tag{8.30}$$

with the emissivity factor E being a function of the emissivities ε_1 and ε_2 of the surfaces, of the geometrical configuration and of the type of reflection (specular or

Table 8.13 Emissivity of some technical materials at low temperature

	Radiation from 290 K, surface at 77 K	Radiation from 77 K, surface at 4.2 K
Stainless steel, as found	0.34	0.12
Stainless steel, mechanically polished	0.12	0.07
Stainless steel, electro-polished	0.10	0.07
Stainless steel + aluminium foil	0.05	0.01
Aluminium, black anodized	0.95	0.75
Aluminium, as found	0.12	0.07
Aluminium, mechanically polished	0.10	0.06
Aluminium, electro-polished	0.08	0.04
Copper, as found	0.12	0.06
Copper, mechanically polished	0.06	0.02

diffuse) between the surfaces. Its precise determination can be quite tedious, apart from the few simple geometrical cases of flat plates, nested cylinders and nested spheres.

8.3.4.3 Gas Conduction

Since J. Dewar's invention (1898) of the cryogenic vessel which bears his name, evacuated envelopes provide the best insulation against heat transport in gaseous media. At low pressure, convection becomes negligible and only residual gas conduction is at work. This process operates in two distinct regimes, depending upon the value of the mean free path of gas molecules ℓ relative to the typical distance d between the cold and warm surfaces.

The mean free path of gas molecules, as predicted by kinetic theory, scales with the square root of temperature and inversely with pressure and the square root of molar mass. It therefore becomes large at low pressure, high temperature and for light gas species.

When ℓ is much less than d corresponding to higher pressure, the probability of interaction of a given molecule with others before it travels distance d is high (viscous regime), and heat diffuses as in any continuous medium:

$$Q = k(T)A\mathrm{d}T/\mathrm{d}x. \tag{8.31}$$

Note that the thermal conductivity $k(T)$ of the gas is independent of pressure.

When ℓ is much greater than d at low pressure, the molecular regime prevails and the heat transfer between two surfaces at temperatures T_1 and T_2 is given by Kennard's law:

$$Q = A\,\alpha(T)\,\Omega\,P\,\,(T_2 - T_1)\,, \tag{8.32}$$

where Ω is a parameter depending upon the gas species, and α is the "accommodation coefficient" representing the thermalization of molecules on the surfaces; its value depends on T_1, T_2, the gas species and the geometry of the facing surfaces. Note that the conductive heat flux in molecular regime is proportional to pressure P and independent of the spacing between the surfaces (and therefore not amenable to the concept of thermal conductivity). Typical values of heat flux by gas conduction at cryogenic temperature are given in Table 8.14.

8.3.4.4 Multilayer Insulation

Multi-layer insulation (MLI) is based on n multiple reflecting shields wrapped around the cryogenic piece of equipment to be insulated, with the aim of benefiting from the $n + 1$ reduction factor in radiative heat in-leak. In practice, this is implemented in the form of aluminium or aluminized polymer films, with low

Table 8.14 Typical values of heat flux [Wm^{-2}] to vanishingly low temperature between flat plates

Black-body radiation from 290 K	401
Black-body radiation from 80 K	2.3
Gas conduction (100 mPa helium) from 290 K	19
Gas conduction (1 mPa helium) from 290 K	0.19
Gas conduction (100 mPa helium) from 80 K	6.8
Gas conduction (1 mPa helium) from 80 K	0.07
MLI (30 layers) from 290 K, pressure <1 mPa	1 . . . 1.5
MLI (10 layers) from 80 K, pressure <1 mPa	0.05
MLI (10 layers) from 80 K, pressure 100 mPa	1 . . . 2

packing density achieved by crinkling or by insertion of a net-type spacer between layers. The wrapping can be made by winding the layers and spacer in situ, or by pre-fabricated blankets installed and held in place by insulating fasteners.

In all cases, MLI is a complex thermal system, involving the combination of radiation, solid-contact conduction, and residual-gas conduction between layers. As a result, increasing the number of layers, while beneficial for cutting radiation, usually results in increased packing with more contacts and trapped residual gas between layers, two effects which increase heat transfer. In view of the nonlinearity of these elementary processes, thermal optimization requires layer-to-layer modelling and efficient control of the critical parameters. In practice, performance is measured on test samples and measured data is available from an abundant literature. Typical values for some practical MLI systems are given in Table 8.14.

Of particular interest is the case of operation in degraded vacuum, where the heat in-leak by molecular conduction is directly proportional to the residual pressure. The presence of a multilayer system which segments the insulation space into many cells thermally in series, significantly contains the increase in heat in-leak to the low-temperature surface (Table 8.14). In this respect, the multilayer system is no longer used for its radiative properties, but for the reduction of molecular gas conduction. In the extreme case of complete loss of vacuum in a liquid helium vessel, MLI also efficiently limits the heat flux which would otherwise become very high due to condensation of air on the cold wall, thus alleviating the requirements for emergency discharge systems.

8.3.4.5 Vapour-Cooling of Necks and Supports

The enthalpy of cryogen vapour escaping from a liquid bath can be used to continuously intercept conduction heat along solid supports and necks connecting the cryogenic bath with the room temperature environment (Fig. 8.34).

Assuming perfect heat exchange between the escaping vapour and the solid, the energy balance equation reads

$$k(T)S\mathrm{d}T/\mathrm{d}x = Q_v + \dot{m}\,C_p(T)\;(T - T_v)\,, \tag{8.33}$$

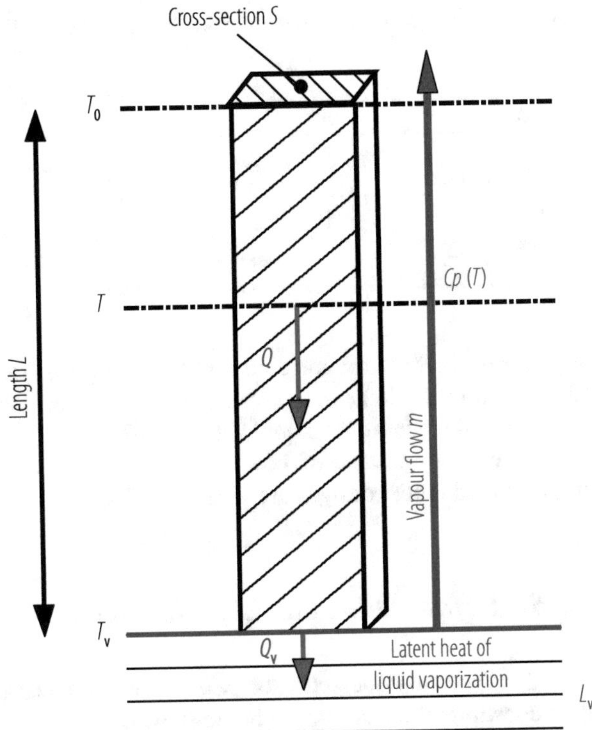

Fig. 8.34 Vapour cooling of necks and supports with perfect heat exchange

where Q_v is the heat reaching the liquid bath and $\dot{m}$ is the vapour mass flow-rate. In the particular case of self-sustained vapour cooling, i.e. when the vapour mass flow-rate $\dot{m}$ precisely equals the boil-off from the liquid bath,

$$Q_v = L_v\,\dot{m} \tag{8.34}$$

Combining Eqs. (8.32) and (8.33) and integrating yields the value of Q_v

$$Q_v = \frac{S}{L}\int_{T_v}^{T_0} \frac{k(T)}{1 + (T - T_v)\,\frac{C_p}{L_v}}\mathrm{d}T \tag{8.35}$$

The denominator of the integrand clearly acts as an attenuation term for the conduction integral. Numerical results for helium and a few materials of technical interest appear in Table 8.15. If properly used, the cooling power of the vapour brings an attenuation of one to two orders of magnitude in the conductive heat in-leak.

Vapour cooling can also be used for continuous interception of other heat loads than solid conduction. In cryogenic storage and transport vessels with vapour-cooled shields, it lowers shield temperature and thus reduces radiative heat in-leak to the

Table 8.15 Heat conduction attenuation between 290 K and 4 K by self-sustained helium cooling

Material	Purely conductive regime [W/m]	Self-sustained vapour cooling [W/m]	Attenuation factor
ETP copper	1620	128	13
OFHC copper	1520	110	14
Aluminium 1100	728	39.9	18
Nickel 99% pure	213	8.65	25
Constantan	51.6	1.94	27
AISI 300 stainless steel	30.6	0.92	33

liquid bath. In vapour-cooled current leads, a large fraction of the resistive power dissipation by Joule heating is taken by the vapour flow, in order to minimize the residual heat reaching the liquid bath [61, 62].

Worked-out example of how these diverse thermal insulation techniques are implemented in real designs are given in [63–65].

8.3.5 *Refrigeration and Liquefaction*

Refrigeration and liquefaction of gases are historically at the root of cryogenics, as they constitute the enabling technology which gave access to the low-temperature domain. They have developed over the years along several lines, to become a specialized subject which would deserve a thorough presentation. In the following, we shall briefly describe the basic thermodynamics, the cooling processes at work and the corresponding equipment in the case of helium. For more complete reviews, see [66, 67].

8.3.5.1 Thermodynamics of Refrigeration

A refrigerator is a machine raising heat Q_i from a low-temperature source T_i to a higher-temperature sink (usually room temperature) T_0, by absorbing mechanical work W_i; doing so, it rejects heat Q_0 (see Fig. 8.35). These quantities are related through the application of the first (Joule) and second (Clausius) principles of thermodynamics:

$$Q_0 = Q_i + W_i, \tag{8.36}$$

$$Q_0/T_0 \geq Q_i/T_i. \tag{8.37}$$

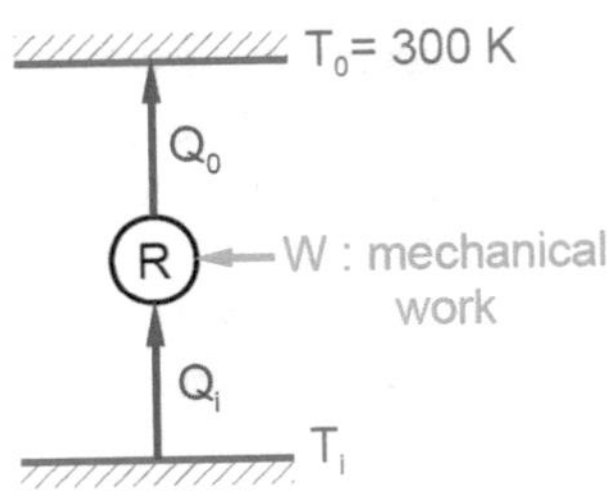

Fig. 8.35 Thermodynamic scheme of a refrigerator

In Eq. (8.36), the equality applies to the case of a reversible process. From the above

$$W_i \geq T_0 Q_i / T_i - Q_i . \tag{8.38}$$

This expression can be written in three different ways. Introducing the reversible entropy variation $\Delta S_i = Q_i/T_i$:

$$W_i \geq T_0 \Delta S_i - Q_i . \tag{8.39}$$

Another form isolates the group $Q_i(T_0/T_i - 1)$ as the proportionality factor between Q_i and W_i, i.e. the minimum specific refrigeration work,

$$W_i \geq Q_i \, (T_0 / T_i - 1) . \tag{8.40}$$

As Carnot has shown in 1824 [68], the minimum work can only be achieved through a cycle constituted of two isothermal and two adiabatic transforms (Carnot cycle). All other thermodynamic cycles entail higher refrigeration work for the same refrigeration duty.

A third form of Eq. (8.37) is

$$W_i \geq \Delta E_i . \tag{8.41}$$

This introduces the variation of “exergy” $\Delta E_i = Q_i(T_0/T_i - 1)$, a thermodynamic function representing the maximum mechanical work content (Gouy’s “énergie utilisable”) of a heat quantity Q_i at temperature T_i, given an environment at temperature T_0 [69].

Equation (8.39) enables to calculate the minimum mechanical power needed to extract 1 W at 4.5 K (saturated liquid helium temperature at 1.3 bar pressure, i.e. slightly above atmospheric) and reject it at 300 K (room temperature), yielding a value of 65.7 W. This is the power that would be absorbed by a refrigerator operating on a Carnot cycle between 4.5 K and 300 K. In practice, the best practical cryogenic helium refrigerators have an efficiency of about 30% with respect to a Carnot refrigerator, hence a specific refrigeration work of about 220 W/W.

Cryogenic refrigerators are often required to provide cooling duties at several temperatures or in several temperature ranges, e.g. for thermal shields or continuous heat interception. Equation (8.39) can then be applied to the cooling duty at every temperature and every elementary mechanical power W_i summed or integrated in the case of continuous cooling. This also allows comparison of different cooling duties in terms of required mechanical work.

8.3.5.2 Helium Refrigerators vs. Liquefiers

A 4.5 K helium refrigerator absorbs heat isothermally at this temperature, by re-condensing the cold helium vaporized at saturation (saturation pressure 1.3 bar). A liquefier also eventually condenses cold helium vapour at saturation, but starting from gaseous helium at 300 K which it must first pre-cool to 4.5 K (Fig. 8.36). From

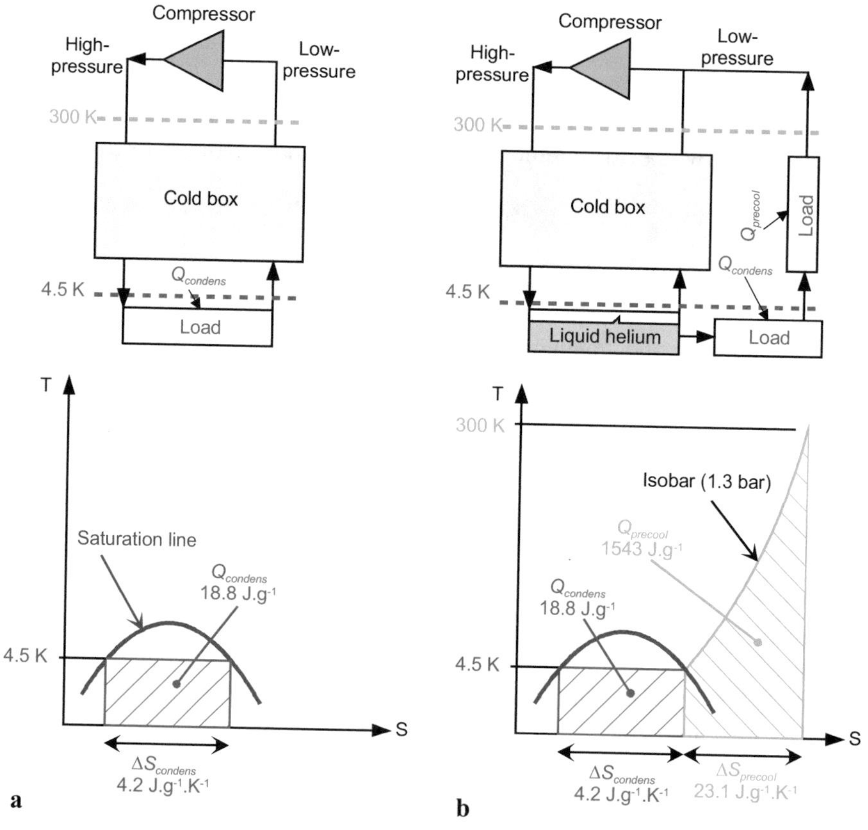

Fig. 8.36 A helium refrigerator (**a**) performs the duty by condensing cold helium vapour (red area). A helium liquefier (**b**) additionally needs to precool helium gas from room temperature (green area)

Eq. (8.38), the minimum mechanical power W_{liq} for helium liquefaction is:

$$W_{\mathrm{liq}} = W_{\mathrm{condens}} + W_{\mathrm{precool}}, \tag{8.42}$$

$$W_{\mathrm{liq}} = T_0 \Delta S_{\mathrm{condens}} - Q_{\mathrm{condens}} + T_0 \Delta S_{\mathrm{precool}} - Q_{\mathrm{precool}}. \tag{8.43}$$

The heat quantities Q_{condens} and Q_{precool} exchanged at constant pressure are—by definition—equal to the enthalpy variations $\Delta H_{\mathrm{condens}}$ and $\Delta H_{\mathrm{precool}}$. With $T_0 = 300$ K and the entropy and enthalpy differences taken from thermodynamic tables, one finds $W_{\mathrm{liq}} = 6628$ W per g/s of helium liquefied. Given the minimum specific mechanical work of 65.7 at 4.5 K, this yields an approximate equivalence of about 100 W at 4.5 K for 1 g/s liquefaction. More precisely, a liquefier producing 1 g/s liquid helium at 4.5 K will absorb the same power (and thus have similar size) as a refrigerator extracting about 100 W at 4.5 K, provided they both have the same efficiency with respect to the Carnot cycle. For machines with mixed refrigeration and liquefaction duties, this equivalence can be approximately verified by trading some liquefaction against refrigeration around the design point and *vice versa*. An example is given in reference [70].

8.3.5.3 Real Cycles and Refrigeration Equipment

So far we have only addressed cryogenic refrigeration and liquefaction through thermodynamics, i.e. through the exchanges of mass, heat and work at the boundaries of machines seen as "black boxes". We will now consider cycles, cooling methods and equipment of real refrigerators.

In order to minimize the specific mechanical work requirement (and hence the size and power consumption), an efficient refrigerator should try to approximate the Carnot cycle, which is represented by a rectangle on the temperature-entropy diagram: the two isotherms are horizontal lines, while the two isentropic transforms are vertical lines. To liquefy helium, the base of the rectangle should intercept the liquid-vapour dome (Fig. 8.37).

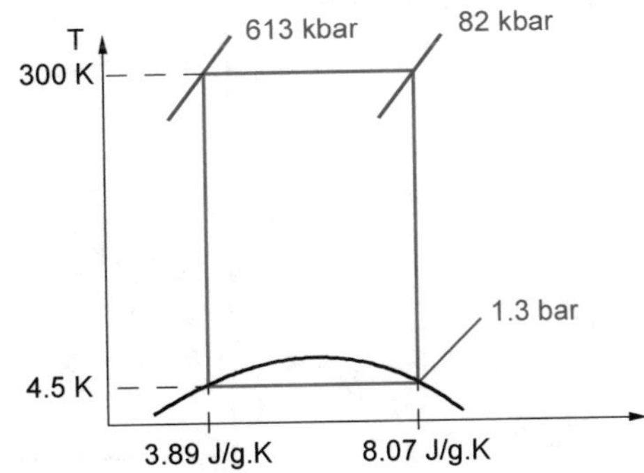

Fig. 8.37 A hypothetical Carnot cycle for helium liquefaction

However, superimposing this cycle on the temperature-entropy diagram of helium shows that one should operate at a high pressure of about 613 kbar (!), with a first isentropic compression from 1.3 bar to 82 kbar (!), followed by an isothermal compression. This is clearly impractical, and real helium cycles are elongated along isobar (or isochoric) lines, thus involving transforms which require heat exchange between the high- and low-pressure streams. This heat exchange can be performed in recuperative or regenerative heat exchangers, respectively for continuous or alternating flows. In the following, we focus on the continuous-flow cycles using recuperative heat exchangers which constitute the operating principles of large-capacity helium refrigerators and liquefiers.

Practical elementary cooling processes are shown on the temperature-entropy diagram in Fig. 8.38. Apart from the quasi-isobar cooling of the gas stream in a heat exchanger (segment AB_1), refrigeration can be produced by adiabatic (para-isentropic) expansion with extraction of mechanical work, usually in a gas turbine (segment AB_2'), and isenthalpic Joule-Thomson expansion in a valve or restriction (segment AB_3).

This latter process does not produce any cooling for ideal gases, the enthalpy of which is a sole function of temperature. For real gases, however, enthalpy depends both on temperature and pressure, so that isenthalpic expansion can produce warming or cooling, depending upon the slope of the isenthalps on the diagram in the region of interest. In order to cool the gas stream, Joule-Thomson expansion must start below a limit called the inversion temperature. The values of inversion temperature for cryogenic fluids (Table 8.16) show that while air can be cooled from room temperature by Joule-Thomson expansion (the risk of freezing the pressure reducer on the air bottle is well known to scuba divers), helium must first be pre-cooled down to below its inversion temperature of 43 K. The moderate downward slope of isenthalps on the temperature-entropy diagram indicates that in any case, Joule-Thomson expansion generates substantial entropy. Its relative inefficiency with respect to adiabatic expansion is however accepted in view of the simplicity of its implementation, particularly when it results in partial condensation of the

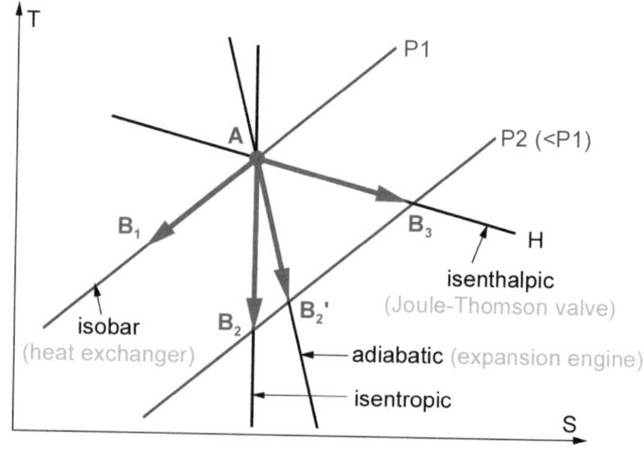

Fig. 8.38 Elementary cooling processes shown on temperature-entropy diagram

Table 8.16 Maximum values of Joule-Thomson inversion temperature

Cryogen	Maximum inversion temperature [K]
Helium	43
Hydrogen	202
Neon	260
Air	603
Nitrogen	623
Oxygen	761

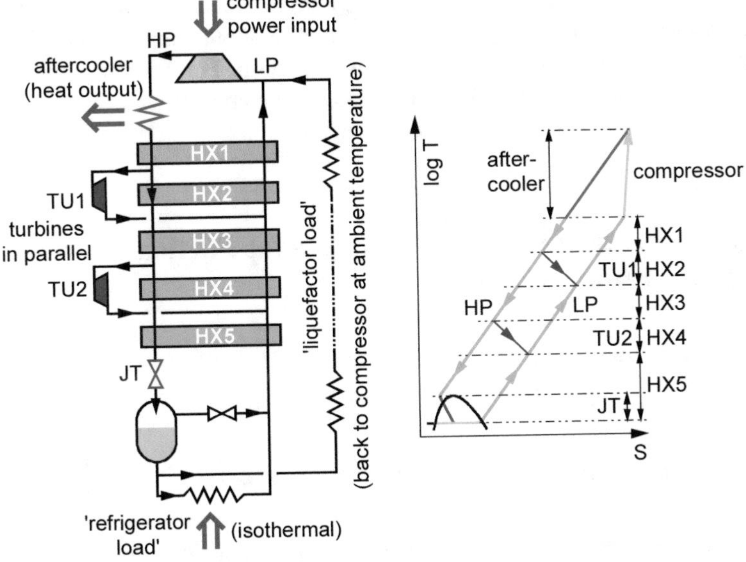

Fig. 8.39 Schematic example of two-pressure, two-stage Claude cycle: flow scheme (left) and T-S diagram (right)

stream entailing two-phase flow conditions which would be difficult to handle in an expansion turbine.

These elementary cooling processes are combined in practical cycles, a common example for helium refrigeration is provided by the Claude cycle and its refinements. A schematic two-pressure, two-stage Claude cycle is shown in Fig. 8.39: gaseous helium, compressed to high pressure (HP) in a lubricated screw compressor, is re-cooled to room temperature in water-coolers, dried and purified from oil aerosols down to the ppm level, before being sent to the HP side of the heat exchange line (HX1 to HX5) where it is refrigerated by heat exchange with the counter-flow of cold gas returning on the low pressure (LP) side. Part of the flow is tapped from the HP line and expanded in the turbines before escaping to the LP line. At the bottom of the heat exchange line, the remaining HP flow is expanded in a Joule-Thomson valve and partially liquefied.

Large-capacity helium refrigerators and liquefiers operate under this principle, however with many refinements aiming at meeting specific cooling duties and improving efficiency and flexibility of operation, such as three- and sometimes four-pressure cycles, liquid nitrogen pre-cooling of the helium stream, numerous heat exchangers, many turbines in series or parallel arrangements, Joule-Thomson expansion replaced by adiabatic expansion in a "wet" turbine, cold compressors to lower the refrigeration temperature below 4.5 K.

The capital cost of these complex machines is high, but scales less than linearly with refrigeration power, which favours large units. Operating costs are dominated by that of electrical energy, typically amounting to about ten percent of the capital cost per year in case of quasi-continuous operation. For overall economy, it is therefore very important to seek high efficiency, which is also easier to achieve on large units. For a review of these aspects, see [70].

8.4 High Precision Power Converters for Particle Accelerators

F. Bordry · J. P. Burnet · M. Cerqueira Bastos

8.4.1 *Introduction to Magnet Power Converters*

The trend in power electronics is evolving rapidly, mainly due to progress in power semiconductors. Since the nineties, the IGBT (Insulated Gate Bipolar Transistor) took the leadership in medium and large power, with the development of new converter topologies called switch-mode power converters [71, 72]. With the progress of electronics (analogue and digital) and measurement systems, the performance of the power converters made a step forward in precision and stability. This chapter will give an overview of the state of the art for high precision power converters for particle accelerators (conventional and superconducting magnet, RF klystrons, solenoids, . . .).

8.4.2 *Main Parameters of Magnet Power Converters*

The powering of particle accelerator magnets are mainly DC power converters and in most cases, the feedback control system regulates the magnet current. The choice of the power converter topology and thus the technology will be defined by the required performance, translated into a precise specification. First of all, the peak and the rms ratings of the current and voltage to power the magnet should be

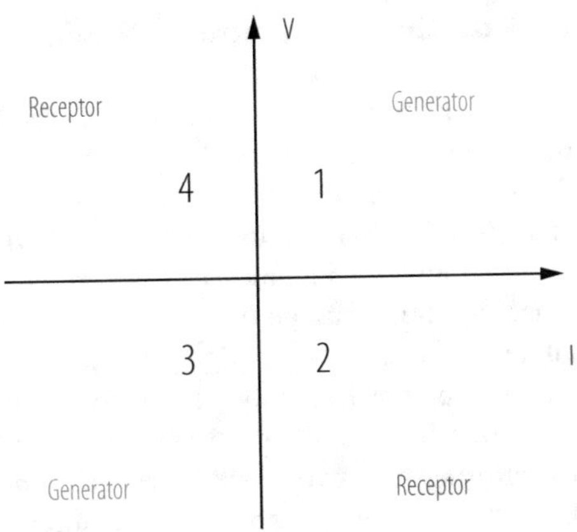

Fig. 8.40 Definition of the 4-quadrant operation

defined. Then, the operation in different quadrants has to be identified: shall the power converter be bipolar in current and/or in voltage? Shall the power converter operate as a generator and/or as a receptor? Figure 8.40 defines the 4-quadrants of operation of a power converter.

In quadrant 1 and 3, the power converter operates as a generator. In quadrant 2 and 4, the power converter operates as a receptor. In this mode, the magnet is giving back its stored energy during the ramping down of the current. The power converter can dissipate this energy, give it back to the mains, or store it locally making it available for the next cycle. The power converters can be classified in 3 categories: one-quadrant power converter; unipolar in current and voltage, two-quadrant power converter; unipolar in current but bipolar in voltage, and 4-quadrant power converter; bipolar in current and voltage.

The power converter can be controlled with different strategies: steady DC current control, variable current reference, or pulsed current.

Another important parameter for the design of the power converters is the voltage and current ripple. The power converter topology and the performance of the inner control loops define the voltage ripple. The current ripple is defined by the load transfer function (cables, magnet inductance . . .). To get good current ripple estimation, a good identification of the converter load is required.

To identify the optimal topology of the power converter, a complete list of parameters has to be reviewed between the accelerator physicists, the magnet designers and the power converter designers.

8.4.3 *Power Converter Topologies*

Three main families of power converters are used for particle accelerators: Thyristor-controlled rectifier, switch-mode power converter and discharged power converter. Each type will be described in the following paragraphs.

8.4.3.1 Thyristor Controlled Rectifier

The thyristor-controlled rectifier was the main topology used from the seventies up to the nineties. Many different topologies can be implemented with thyristor devices, but only the three main types will be described here. First, the simplest one is the 6-pulse thyristor rectifier with freewheeling diode, see Fig. 8.41.

The topology includes a transformer, a 6-pulse thyristor bridge with a free-wheeling diode and an output filter. The output voltage is controlled by the firing angle of the thyristor bridge. Due to the free-wheeling diode, the output voltage can only be positive, as well as the current. This topology is a one quadrant power converter. Second, the most used is the 12-pulse rectifier. This topology includes a transformer with two secondaries with a phase shift of ±15°, two 6-pulse thyristor bridges connected in series, and an output filter, see Fig. 8.42.

This topology allows positive and negative output voltages, but imposes a minimum current to control the thyristors. The first harmonic of the output voltage is 600 Hz which helps to reduce the current ripple compared to the previous topology. When the voltage is negative, the energy is given back to the grid. It is a two-quadrant power converter.

Last, a back to back thyristor rectifier is shown as an example of 4-quadrant power converter. In this case, two 6-pulse rectifiers are connected back to back, see Fig. 8.43.

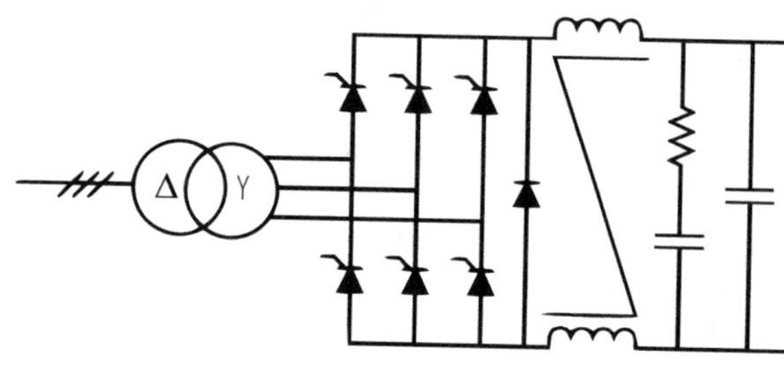

Fig. 8.41 6-pulse thyristor rectifier

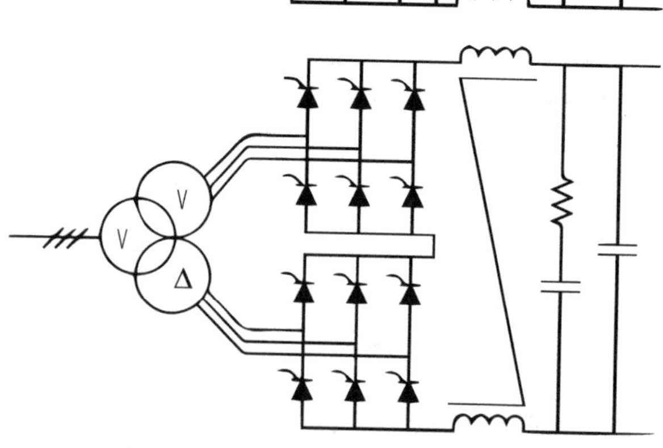

Fig. 8.42 12-pulses thyristor rectifier

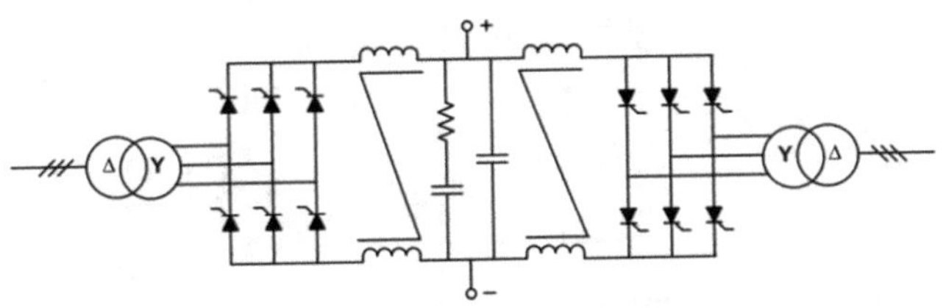

Fig. 8.43 Back-to-back thyristor rectifier with circulating current

This topology requires placing inductors between the rectifiers to limit the circulating current which is necessary to control the thyristor rectifiers. In this case, the current can be positive and negative in the load as well as the voltage.

Thyristor devices are known to be robust, with low losses, and easy to drive. The main drawbacks of this type of power converters are their high susceptibility to perturbations from the grid, they generate reactive power on the grid, and they have a slow dynamic response. It also requires 50Hz transformers which can be bulky. Nevertheless, this type of power converter is widely used for particle accelerators. It was for example, the main type for the LEP accelerator. Nowadays, the thyristor controlled rectifiers are mainly used for high power converters (above 500 kW) which requires very often a reactive power compensator on the grid.

8.4.3.2 Switch-Mode Power Converter

Switch-mode power converters incorporate power semiconductors which switch quickly between full-on and full-off states. The voltage regulation is provided by varying the ratio of on, off time, which is called duty cycle. Pulse-Width Modulation is the most familiar technique used to control the duty cycle. The main advantages are a better efficiency and a smaller size due to the reduction of the magnetics component size, at the cost of a greater complexity. Many types of switch-mode power converters exist and are too numerous to be listed in this chapter. Two main types will be presented, first, a topology with a 50Hz transformer and second, a topology with a high frequency transformer.

8.4.3.2.1 Switch-Mode Power Converter With 50 Hz Transformer

A transformer is always required to isolate the magnet from the grid. If the volume of the power converter is not an issue, 50 Hz transformers can be used as they are produced by industry for other applications at affordable prices. The most classical topology used for particle accelerators can be seen in Fig. 8.44. The 50 Hz transformer adapts the voltage for the rectifier, then an H-bridge comprising of IGBTs, control the voltage applied to the magnet. The switching frequency of the IGBT depends on the DC-link voltage and on the current to control. For power

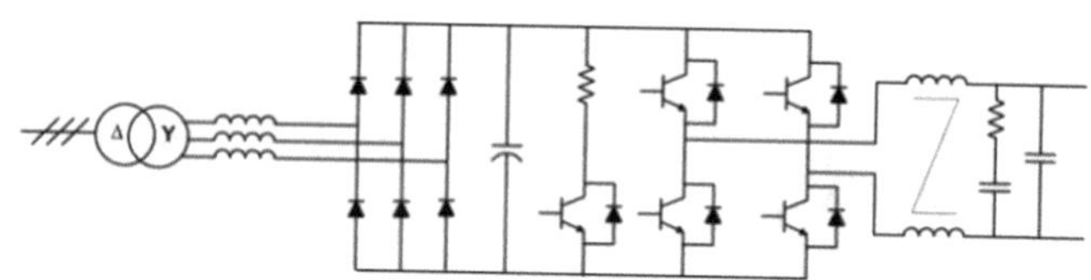

Fig. 8.44 Switch-mode power converter with 50 Hz transformer, connected to the grid at the left and connected to the magnet at the right

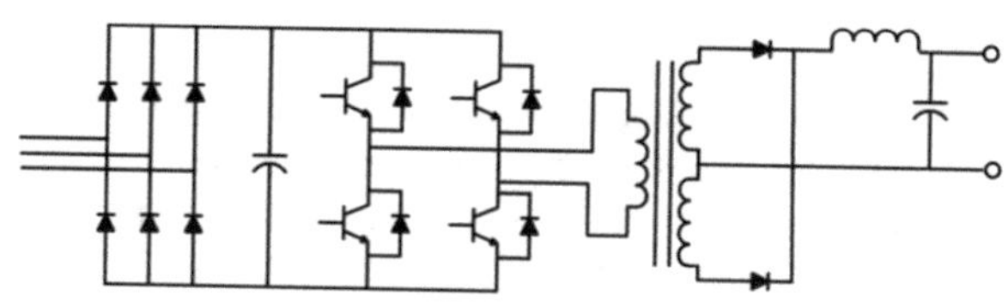

Fig. 8.45 Switch-mode power converter with high frequency transformer

converters above 10 kW, the classical switching frequency is in the range of 1 kHz to 20 kHz.

The H-bridge topology allows 4-quadrant operation. When the magnet gives back its energy, the return energy can be dissipated in a brake chopper or stored in the capacitor bank of the DC-link. Another possibility is to replace the diode rectifier by an active front end provided by IGBTs. In this case, the energy is returned to the grid. The main advantage of this topology is the use of the IGBT semiconductors which are widely used and produced at reasonable cost. They can be easily controlled with PWM (pulse-width modulation) technique at a switching frequency above few kHz; this helps to reduce the current ripple and improve the feedback loop performance. The IGBT rating ranges from 600 V to 6.5 kV and from 50 A to 3 kA, allowing a large scale of the requirements to be met.

8.4.3.2.2 Switch-Mode Power Converter with High Frequency Transformer

One of the main interests of using a high frequency transformer is to reduce the volume of the power converters. Ferrite cores are widely produced which allow the use of high frequency transformers at a competitive price. A classical topology is shown in Fig. 8.45.

The diode bridge is directly connected to the grid. An inverter drives a high frequency transformer. A diode bridge is connected at the output of the secondaries of the transformer to obtain a DC voltage to apply to the magnet. This type of power converter operates only in one-quadrant mode. 4-quadrant operation can be obtained by adding another stage but the return energy has to be dissipated at this level (Fig. 8.46) [73].

Energy recovery to the grid can be achieved, but with a much more sophisticated topology which will not be described here [71]. The switching frequency of the

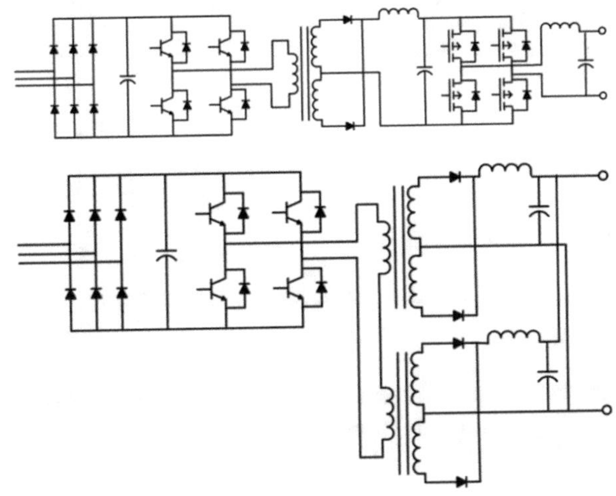

Fig. 8.46 4-quadrant switch-mode power converter with high frequency transformer

Fig. 8.47 Switch-mode power converter with 2 high frequency transformers

inverter is usually from 10 kHz to several 100 kHz. Due to the high frequency spectrum of the semiconductor's commutation, an important design constraint for the power converters is the EMC (Electro Magnetic compatibility) immunity and emission. Soft commutation of the inverter semiconductors is an elegant solution to reduce the switching losses and to improve the EMC. In the case of the CERN LHC, where power converters have to be installed underground with limited space, the switch-mode power converter was chosen for power converter up to 200 kW. In this case, many sub-converters have to be placed in parallel to reach this power level. The superconducting magnets require a high current (many kA) but few volts (less than 20 V); for this application, the solution was to drive many high-frequency transformers in series with one inverter. All the transformer secondaries are connected in parallel after the output filter, see Fig. 8.47.

To improve reliability by increasing redundancy, N+1 sub-converters are placed in parallel, where a sub-converter comprises one inverter with 8 transformers. In case of failure of one sub-converter, the other sub-converters compensate without any disturbance to the load. A modulator construction of the power converter can be done due to the high number of converters in parallel which eases the operation and maintenance of it.

For power converter below 1 kW, the preferred semiconductor is the MOSFET which has lower losses than the IGBT thereby allowing a higher switching frequency. Nevertheless, its voltage range is limited from 5 V to 500 V.

8.4.3.3 Fast Pulsed Power Converter

When the presence of the magnetic field in the magnet is required for a very short duration, discharged power converters are a very interesting technique to reduce power consumption. This is the case for example for beam transfer lines where

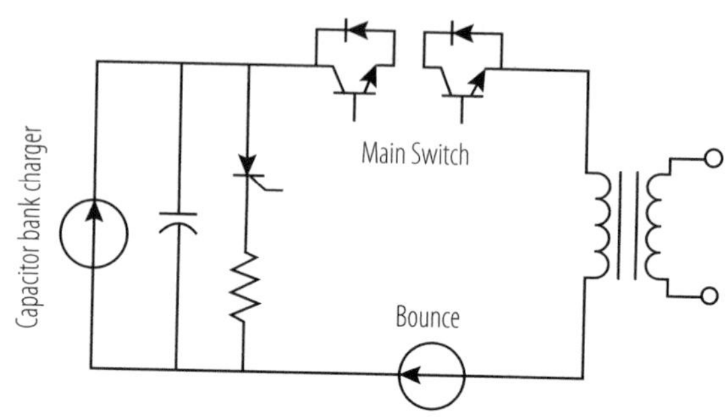

Fig. 8.48 Klystron modulator with pulse transformer

the beam can be present for few μs every second or minute. It is also the case for the Klystron Modulators in a Linac. An example of a discharged power converter topology is shown in Fig. 8.48.

In this case, a capacitor bank is charged to a nominal value (less than 15 kV) which can be low compared to the need of the load (from 60 kV to 140 kV). A main switch discharges the capacitor bank via a pulse transformer.

The voltage is applied to the load while the switch stays ON but generally for a short time (2.4 ms for LINAC4 at CERN), as the capacitor bank voltage decreases (few %), the droop voltage has to be compensated by a bouncer which is a passive resonant circuit. The main challenges are the design of the pulse transformer and of the main switch. In case of an arc in the Klystron, the energy deposit has to be limited to few joules (~20 J). This needs a fast turn OFF of the main switch. A redundancy policy has to be implemented to be sure to be able to open the circuit. For LINAC4, 4 IGCT (Integrated Gate-Commutated thyristor) are placed in series where only 3 are required.

8.4.3.4 High Power System with Local Energy Storage

For high power system (above 1 MW), Thyristor rectifiers are the preferred technology. The principal drawbacks are:

- the reactive power generated on the grid which needs to be compensated to stabilize the network voltage,
- the pulsed active power on the grid which requires a strong electrical network. To avoid the flow of pulsed power on the grid, local energy storage can be used.

One example is the new POPS system [74] designed for the CERN PS accelerator, see Fig. 8.49.

The principle is to store energy in capacitor banks and to exchange this energy with the magnets during the cycles. As the rating power is very high (60 MW peak), many switch-mode power converters are associated in series and in parallel. Only the losses of the system (magnets and converters) are taken from the electrical network (5 MW peak). 20 MJ are stored in the capacitor banks where 14 MJ can be

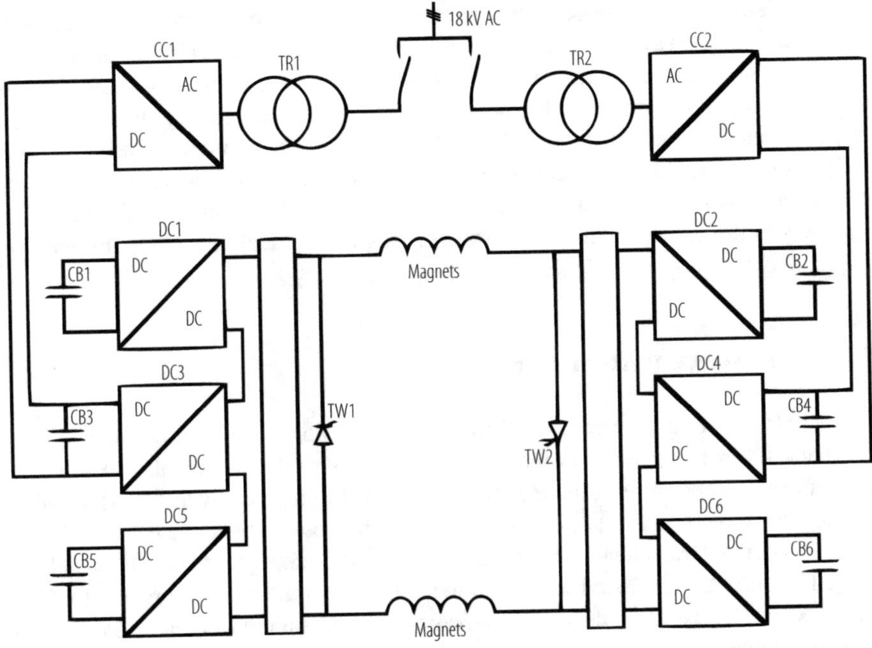

Fig. 8.49 Power system with capacitive energy storage

exchanged during each cycle with the magnets. A redundancy is also implemented; the system can operate without one capacitor bank or one DC/DC converter or one transformer. In this case, no reactive power is generated on the grid and the active power taken from the grid is reduced to the minimum. This new type of power system helps reduce the energy consumption of particle accelerators.

8.4.4 High Accuracy in Power Converters for Particle Accelerators

Magnets in particle accelerators are powered by electrical power converters. These can be seen as controlled current sources in which a feedback loop ensures that the output current follows the reference to the best possible accuracy. The accuracy that can be achieved is greatly determined by the current measurement transducer and, in the case of digitally controlled power converters, the control algorithm and the ADC employed in the feedback loop.

As energies reached by particle accelerators become greater, accuracy requirements for the magnet current increase proportionally. Accuracy requirements depend also on the powering strategy chosen for the accelerator. In a synchrotron, the main dipole and quadrupole circuits are normally powered in series to ensure synchronism and homogeneity in the magnetic field around the circumference. However this is not always true: in the case of the LHC the main circuits are divided into eight sectors due to the very high stored magnetic energy and constraints in the protection of the superconducting magnets. As a consequence, individual sector currents must be controlled with very high absolute accuracy in amplitude and time [75] to ensure tracking between all sectors. Requirements for accuracy in current control are also determined by the type of magnet and its function: the need for accuracy in the current control for corrector magnets is much less stringent than for the main quadrupole and dipole magnets.

8.4.4.1 Power Converter Control

In the past, power converter control in accelerators was usually based on analogue feedback loops using PID (Proportional/Integral/Derivative) control. In such systems, the reference value at the input of the loop is often given by a DAC (Digital to Analogue Converter) and the output current of the power converter measured by a precision current transducer which provides the feedback signal for the control loop [76]. The accuracy achieved with such techniques is limited by errors due to drift, linearity and temperature dependency. Moreover, adjustment of control parameters can be cumbersome, as it might require trimming of potentiometers or replacement of components.

Developments in digital electronics and in particular DSPs, PLDs and Micro-controllers, as well as the need for increased performance in power converter applications fuelled significant progress in digital control during the last decades. The first applications using digital control in power converters for accelerators were implemented in the late 1980s [77]. Amongst the advantages of using digital control are increased stability and reproducibility, less susceptibility to noise and thermal effects, easy implementation of different control methods (state-space, robust, fuzzy) as well as easy loop parameterization. On the other hand, the use of digital control increases system complexity and introduces new sources of error such as the ones resulting from ADC measurement uncertainty and limited resolution on arithmetic calculations, which might lead to arithmetic errors.

When high accuracy is required, one effective power converter control strategy is to have an external current loop controlling a power supply that works as a voltage source. In a digitally controlled system, the output current of the converter is measured by a current transducer connected to an ADC and then compared with a digital reference. The error is fed into a digital regulator and the result sent to a DAC that provides an analogue signal to control the voltage source [77].

This solution is implemented in the control of the LHC power converters at CERN. In this case, the control challenge is even more demanding due to the 8-sector powering strategy used for the main dipole and quadrupole circuits. This powering strategy requires not only an accurate control of the current of each power converter but also an accurate generation and synchronization of the current references sent to the converters along the 27 km circumference of the LHC. For this purpose each power converter in the LHC has a dedicated controls electronics which is actually an embedded microcontroller-based computer capable of performing full local state control, reference function generation and measurement acquisition as well as running a digital current regulation loop. Reference functions are synchronized using a timing network. Each digital controller is connected to a field bus (WorldFIP) and the timing network is used to synchronise the cycles of all segments of the field bus. The digital controller disciplines a phase-locked loop to align its clock to the start of each WorldFIP cycle guaranteeing synchronism of the references along the machine.

The digital control strategy implemented in the LHC power converters is based on an R-S-T algorithm. The canonical structure of an RST controller is presented in Fig. 8.50 [78].

This structure has two degrees of freedom, i.e. the digital filters R and S are designed in order to achieve the desired regulation performance and the digital filter T is designed to achive the desired tracking performance. The structure can be described by the following discrete equation [79]:

$$S\left(z^{-1}\right)u(t) + R\left(z^{-1}\right)y(t) = T\left(z^{-1}\right)r(t), \tag{8.44}$$

where $u(t)$ and $y(t)$ are the input and output of the plant, $r(t)$ the desired tracking trajectory, R, S and T are z^{-1} polynomials and t is the normalized discrete time.

The corresponding time domain expression is given by:

$$u(t) = -\sum\nolimits_{i=1}^{n_S} S_i u\,(t-i) - \sum\nolimits_{i=0}^{n_R} R_i y\,(t-i) + \sum\nolimits_{i=0}^{n_T} T_i r\,(t-i)\;[84]. \tag{8.45}$$

The RST controller makes it possible to obtain the desired tracking behaviour (following the reference) independent of the desired regulation behaviour (rejection of a disturbance) [80].

The application of this control strategy to the LHC power converter control resulted in excellent performance. Recent results proved that the tracking error on

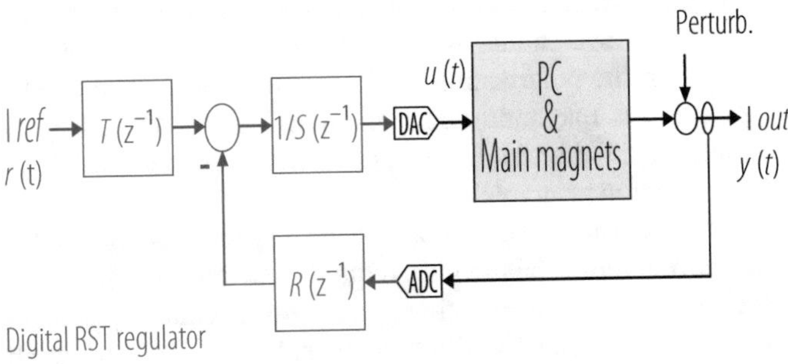

Fig. 8.50 Canonical structure of a digital RST controller

the LHC main power converters is in the order of one ppm and that tracking between different sectors is within a couple of ppm [78].

8.4.4.2 Current Measurement in Particle Accelerators

Traditionally, accurate current measurement devices for particle accelerators are associated with beam current measurement. Magnetic transducers and in particular current transformers have been for a long time the preferred transducers to measure beam currents. The requirements for beam current measurement have driven progress in transducer technology, culminating with the introduction of the DCCT (Direct-Current Current Transformer) for beam current measurement at CERN in late 1960s. The idea was to build a magnetic beam current transformer with frequency response extended down to DC to measure beam current in the ISR accelerator. The new transducer combined the zero flux detection principle used in flux-gate magnetometers since the 1930s, with the active transformer circuit as originally proposed by H.G. Hereward (and already used to measure the circulating beam in the CERN PS accelerator) [81]. Although the new transducer was not initially intended for power supply regulation applications, its advantages compared with previous DC instrument transformers (e.g. Kramer and Hingorani [82]) soon became obvious. The concept got picked up by industry and the new transducer was soon being used at accelerators such as DESY in Hamburg. In late 1970s, DCCTs were used for the first time in large quantities in the SPS project at CERN [83].

The use of DCCTs spread to other applications but it continued to be widely used in particle accelerators. In particular, the beginning of the twenty-first century saw important progress in DCCT technology with the development and deployment of the DCCTs for the main dipole and quadrupole power supplies of the LHC, at CERN [84]. Short term stability in the order of two part-per-million (ppm), yearly drifts better than fifteen ppm and linearity better than two ppm have been achieved.

8.4.4.2.1 Current Measurement Technologies

The most common current measurement technologies used in electrical power converters include resistive shunts and current sense resistors, Hall-Effect current transducers (based on the polarization of charges in an electrical conductor in the presence of an external magnetic field), Current Transformers, Rogowsky Coils (high current, high bandwidth applications), Active CTs and DCCTs (both based on the zero flux detection principle).

The choice of a current measuring device for a specific application depends on factors such as current range, bandwidth, required accuracy, required output signal, need for isolation, reliability, installation constraints, availability and cost. DCCTs provide isolated measurements for different current ranges and can reach very high accuracy, albeit with a higher cost.

8.4.4.2.2 DCCTs (Direct-Current Current Transformers)

A DCCT is a magnetic current transducer of the zero-flux type where a second harmonic or a peak detector is used in a feedback loop to generate a compensation current which is a fractional image of the current being measured, keeping zero flux in the magnetic cores. The working principle of a DCCT is illustrated in more detail in Fig. 8.51: two DC flux sensing cores are modulated by an oscillator. The resulting current peaks are unequal if there is a DC flux in the cores (originated by the DC current being measured). A balanced peak detector circuit will detect any unbalance and give a non zero output when a DC flux is present. The output of the peak detector is combined with the AC component measured by a third core which works as a normal transformer. A control loop is set up to generate the secondary current that will bring the total DC flux back to zero [84].

The secondary current is therefore a fractional image of the primary current, and it can be fed into a precision burden resistor to get a measurable voltage signal. This signal is amplified by an output amplifier to produce a 0 . . . 10 V output.

DCCTs have the potential of reaching short term stability and repeatability in the order of a few part-per-million. This requires not only a careful design of the magnetic part (magnetic "head") as well as the use of high quality burden resistors and very stable precision amplifiers, usually in a temperature controlled environment.

A DCCT "head" used in the LHC is shown in Fig. 8.52.

The head design and in particular the magnetic shielding are important to increase the sensitivity to the primary magnetic field while minimising the sensitivity to external magnetic fields, head centring and return bus-bar fields. Soft magnetic materials and in particular amorphous and nanocrystalline alloys are the

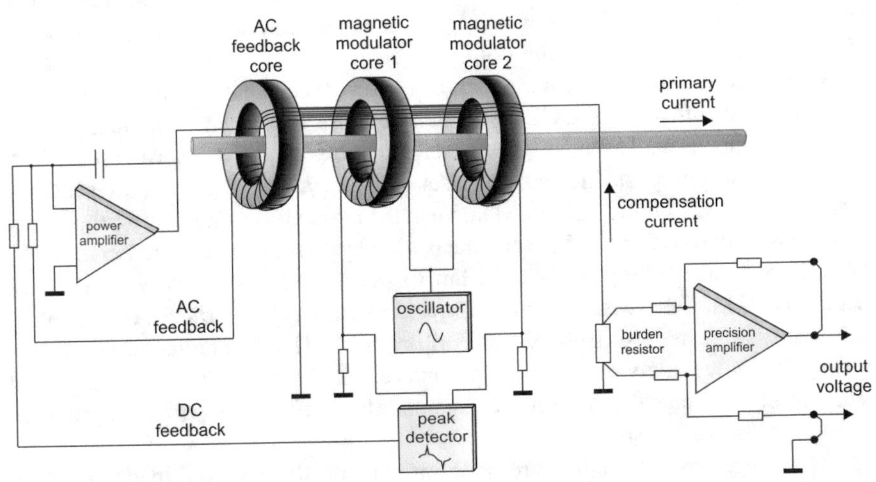

Fig. 8.51 The DCCT working principle

Fig. 8.52 A 13 kA DCCT "head", used in the LHC

most commonly used in the DCCT core and inner shielding design because of their magnetic properties such as high initial permeability and low coercivity. As for the outer layer of the shielding, a doughnut-shaped shell containing the cores and the inner shielding, ferromagnetic materials of lower permeability and high saturation are normally used. During the last couple of decades, improvements in the manufacturing process of amorphous alloys contributed to their widespread use and to the development of new materials hence making for a greater availability of candidate materials for magnetic sensors. More recent developments include progress in reducing coercivity and increasing saturation induction of amorphous and nanocrystalline materials [85]. These new breakthroughs have still to find their application in DCCT head design.

The current output of the DCCT is usually connected to a burden resistor, a 2-terminal or 4-terminal resistor depending on the required accuracy. The performance of this component is one of the dominant factors in the overall accuracy of the transducer. Well known effects that can cause resistance change and therefore can affect the performance of DCCTs are temperature coefficient, self heating and thermal settling as well as ageing. Less well know effects include power coefficient, humidity absorption and hysteresis under power cycling. The highest accuracy available in DCCT burden resistors known at the moment is offered by a proprietary Zeranin wire design from PM special measuring systems, which is used in the LHC main dipole and quadrupole DCCTs, but its price limits its use only to the most critical DCCTs. Otherwise there is only one resistor type on the market offering the performance needed: Bulk Metal Foil or "foil". This technique, pioneered by Vishay, but now widely spread, tightly bonds a rolled metal foil to a substrate and seeks to compensate the resulting consistent stress effects as part of the overall resistor performance [86].

The voltage across the burden resistor normally needs to be amplified to produce a voltage output adequate for subsequent ADC conversion. This task is done by a precision amplifier, usually a difference amplifier circuit making use of high precision network ratio resistors to establish the gain. Presently, the consumer electronics market offers a range of choice in precision amplifiers, including low offset, low drift amplifiers that can be used as the output amplifiers in the DCCT. However, this does not exempt the designer from careful design and implementation of the output circuit. Parameters such as offset and gain stability, noise, common mode rejection and output impedance depend strongly on circuit design and implementation.

8.4.4.2.3 ADCs (Analogue to Digital Converters)

With the advent of digital control, Analogue to Digital Converters have become essential components for ensuring the accuracy of the current measuring chain in the control loop of power converters. If in the past, most of the control was implemented using analogue electronics, in the last couple of decades digital control has almost completely taken over. The advantages of having a digital representation of the signal are numerous and not limited to power converter control: calibration, traditionally done by adjusting gain and offset potentiometers, is now performed using calibration constants which can be memorised and used to correct the output of the measuring device. If these constants are also kept in a database and updated whenever a calibration takes place, the task of following the behaviour of high precision devices becomes much simpler and less error prone. Another advantage is the possibility of using correction algorithms. A typical example is the use of a temperature sensor and a correction algorithm to correct for the temperature dependency of a measurement device.

ADC technology has significantly evolved in the last couple of decades fuelled mainly by the telecommunications industry. As a consequence, a wide range of solutions in AD conversion are available on the market with tradeoffs in resolution, speed and accuracy. In power converter control for accelerators, ADC speeds above the MHz are seldom required. As for resolution, requirements often range from 16 bits to 24 bits with accuracy following along. In this context Successive Approximation Register ADCs (SAR) have recently become of the most interesting technologies to follow as they are now competing with Delta Sigma ADCs and multi slope integrating ADCs in the high resolution, high accuracy segment. However, to reach effective resolution figures beyond 20 bits, they must operate in oversampling mode. Some ICs have built-in provisions for it, while for others it has to be implemented in external logic [87]. Therefore, Delta Sigma ADCs still remain the most commonly used solution in the segment. They do not require external components, their oversampled nature simplifies circuit design by relaxing the requirements of the analogue anti-aliasing filter and they are often simpler to drive than SARs. However, higher accuracy also requires more complex digital decimation filters which, for linear phase FIR filters, corresponds to increased latency. Filter latency limits the maximum loop bandwidth which means that high accuracy is normally associated with lower speed. The use of minimum phase FIR filters can substantially decrease latency.

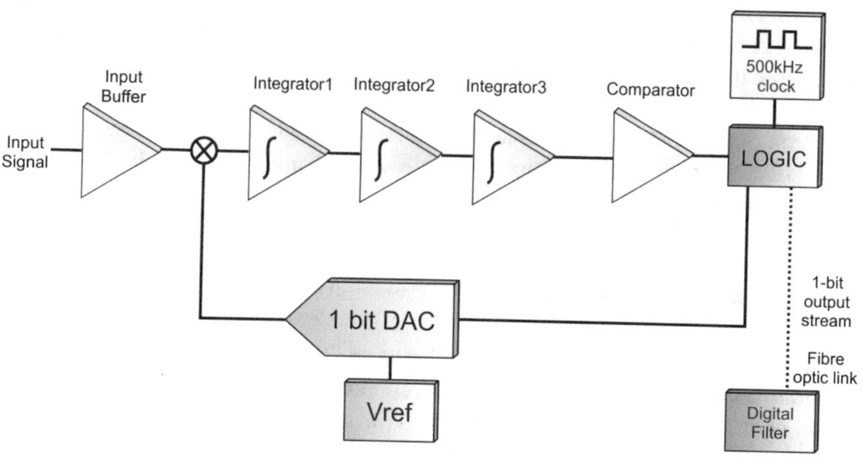

Fig. 8.53 The architecture of the Delta-Sigma Analogue to Digital converter used in the LHC main power converters

ADC accuracy depends greatly on the voltage reference employed. Precision external voltage references are preferred in applications demanding high accuracy as they have lower temperature coefficient, thermal hysteresis and long term drift than an on-chip voltage reference. High end Zener based high precision references usually use buried Zener technology. They can include internal temperature control or temperature compensation. However, for higher accuracy, they might need to be stabilised inside a temperature controlled oven. Reference annealing can also be used to accelerate the ageing process and reduce initial drift.

The main dipole and quadrupole power converters on the LHC use a Delta Sigma converter with a resolution of about 22 bits. The voltage reference is a buried zener type, previously submitted to a burn-in process for minimum drift. Sub-ppm accuracy is achieved albeit at very low sampling speeds (1 kHz) and in a temperature controlled environment.

Figure 8.53 shows the architecture of the Delta-Sigma Analogue to Digital converter used in the LHC [88].

8.4.4.2.4 Calibration

There can be different motivations for calibration of measurement devices in accelerator applications. The first is the requirement to keep long term drift within specified limits. Another motivation is to minimise the impact of replacements. When an operational equipment is replaced by a spare, the impact on the accuracy of the power converter depends on the difference between the measurement errors of the two devices. This difference can be kept within specified limits by means of periodic calibration. Finally, calibration might be necessary to guarantee good tracking between different transducers. Such is the case in the LHC at CERN, where the main dipole and quadrupole circuits are divided into eight sectors. In this case, calibration of the current measurement chain using the same reference ensures that all sectors track each other, allowing the beam to circulate around the machine without "seeing" any difference in the magnetic field between sectors.

The periodicity of calibrations must be set according to the accuracy requirements of the power converter. Some devices include automatic calibration mechanisms: a common practice is to use a multiplexer that connects the ADC input to a voltage reference for calibration. However, the references themselves might need periodic calibration, so human intervention might be unavoidable.

Calibration procedures usually involve the use of dedicated equipment which must be previously characterised in a laboratory using well known reference devices. The complexity of the necessary calibration infrastructure depends on the accuracy one is trying to achieve. Some of the methods employed to calibrate DCCTs and ADCs and required calibration equipment are described below.

DCCTs can be calibrated through different methods:

1. The reference DCCT method, where the output of the DCCT being calibrated is compared against the output of a "reference" DCCT measuring the same current.
2. The output stage method, which involves injecting a reference current in the burden resistor of the DCCT and measuring its output with a calibrated ADC or with a DVM. The value of the error is then memorised by the digital controller. This method has the disadvantage that it only calibrates the DCCT output stage (burden resistor + precision amplifier) and it requires a precision current source to generate the calibration current.
3. The calibration winding method, which involves injecting a reference current in an auxiliary winding in order to produce an Ampere-Turn value equivalent to the one produced by the primary current hence simulating real primary current. The output of the DCCT is measured with a calibrated ADC or with a DVM and the value of the error is memorised by the digital controller.

The latter method requires a precision current source to generate the calibration current. At CERN, a programmable current reference has been developed for the calibration of DCCTs equipped with calibration windings [89]. It can produce DC currents ranging from −5 A to 5 A with sub-ppm accuracy. In the LHC, in-situ calibration systems equipped with these devices are installed close to the main power converters. They are housed in temperature controlled racks and can be remotely controlled via an ethernet connection in order to calibrate one or several DCCTs in the main dipole and quadrupole power converters.

The remotely controlled calibration system used in the LHC is depicted in Fig. 8.54.

In applications where a reference current source is not available, DCCTs are usually calibrated using the reference DCCT method. However, since in-situ deployment of a reference device is not always easy, the calibration procedure often requires bringing the DCCT to a laboratory. As removal and transport of the magnetic heads is a difficult process, the calibration in the laboratory is normally

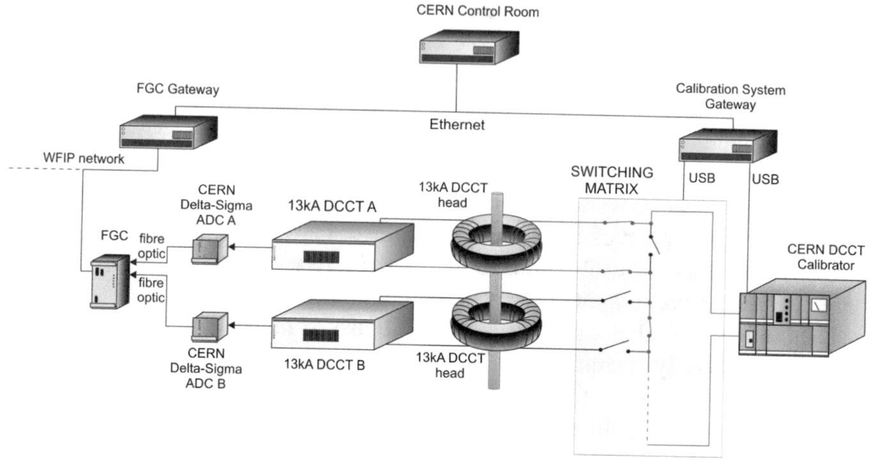

Fig. 8.54 The remotely controlled calibration system for the main power converters in the LHC

performed using a different magnetic head. This can introduce errors that must be accounted for.

ADCs can be calibrated using different methods depending on the requirements of the application. The most common is a three point calibration i.e. calibration at ± FS (Full Scale) and at zero. Calibration at FS can be performed using a voltage standard connected to the input of the ADC. The use of voltage standards requires constant monitoring and maintenance of these devices, including periodic calibration in an independent standards laboratory. In the LHC case, a set of voltage and resistor standards as well as forty unique 10 mA portable current standards [90] are kept in a controlled environment to be used as the basis for CERN's calibration infrastructure. Some units are used for field calibration whilst a set of standards is kept constantly in the laboratory, under permanent monitoring.

8.5 Ultra-High Vacuum

V. Baglin · J. M. Jimenez

8.5.1 Introduction

In particle accelerators, beams are travelling under vacuum primarily to reduce the beam-gas interactions i.e. the scattering of beam particles by the molecules of the residual gas. These interactions are responsible for machine performance limitations such as reduction of beam lifetime (nuclear scattering) and of luminosity (multiple coulomb scattering), intensity limitation by pressure instabilities (ionization) and, for positive beams only, electron (ionization) induced instabilities, e.g. beam blow up.

Beam-gas scattering can also increase the background to the detectors in the experimental areas (non-captured particles or nuclear cascade generated by the lost particles upstream the detectors) and the radiation dose rates in the accelerator tunnels. Thus leading to material activation, dose rates to intervention crews, premature degradation of tunnel infrastructures like cables and electronics and finally higher probability of electronic single events induced by neutrons which can destroy the electronics in the tunnel but also in the service galleries.

In addition, the design of an accelerator vacuum system must observe severe additional constraints which have to be considered at the design stage since retrofitting mitigation solutions is often impossible or very expensive. Among them, the vacuum system has to be designed to minimise beam impedance and radiofrequency higher-order-modes (HOM) generation as well as to optimise the beam aperture in particular in the magnets. It also must provide enough ports for the pumps and for the vacuum diagnostics and allow for bake-outs in order to achieve Ultra-High Vacuum (UHV) pressures ($<10^{-8}$ Pa). The impact of other constraints like integration, safety (material and personnel), operational issues (conditioning of RF and HV devices) and costs often lead to a compromise in performances of all systems of an accelerator. This explains why these issues must be addressed at the design stage [91].

For accelerators operating at cryogenic temperatures [92], the heat load induced by scattered beam particles and synchrotron radiations can also be an issue for the cryo-magnets since local heat loads can lead to a magnet quench i.e. a transition from the superconducting to the normal state. The heavy gases are the most dangerous because of their higher ionisation cross-sections. Thus, the beam-pipes shall be designed to intercept heat loads induced by synchrotron radiation, energy loss by nuclear scattering, image currents, energy dissipated during the development of electron clouds. In the LHC, these constraints required, for the first time in a particle accelerator, the use of a beam screen [93]. Increasing further the luminosity of a storage ring operating at cryogenic temperature, such as in the High-Luminosity LHC (HL-LHC), the very large production of collision debris must be intercepted at the level of the beam screen by heavy material, e.g. tungsten. This shielding is designed to offer a protection of the superconducting coil against radiation induced ageing and extract the beam induced heat load at an elevated temperature, e.g. 60 K, to optimise the Carnot efficiency [a].

8.5.2 *Vacuum Fundamentals*

Vacuum is defined as the absence of matter or a space empty of matter. These are idealistic definitions since a perfect vacuum is a theoretical limit. Realistically, whenever a pressure in an enclosed space is less than the pressure of the surrounding atmosphere, the enclosed space is defined to be under vacuum. Hence, vacuum is obtained by removing molecules of gas from an enclosed space to the necessary level required for a specific process or experiment. A gas in an enclosed space can be physically described by its volume, its temperature and the amount of molecules or the pressure.

8.5.2.1 Total, Partial and Vapor Pressures

The residual gas is usually composed of several types of molecules (ex: air, gas in vacuum systems). The total pressure, P_{tot}, is the sum of all the partial pressure, P_i, as shown by the Dalton law:

$$P_{tot} = \sum P_i = kT \sum n_i. \tag{8.46}$$

The saturated vapor pressure of gasses (Fig. 8.55) is of major importance for the accelerators operated at cryogenic temperatures. Above ambient temperature, only water condensation is a limitation justifying the use of bake-out to improve the outgassing. At 1.9 K, the operating temperature of the LHC cold mass, helium is of concern in case of leaks. In the 5–20 K range, the hydrogen is pumped onto the cold bore due to the beam screen's perforation which has a conductance designed to maintain the partial pressures below the 100 h life time limit. If operated between 20 and 40 K, then the scheme is less favorable and strong pressure oscillations can be expected in presence of temperature variations. This is one reason why this range of

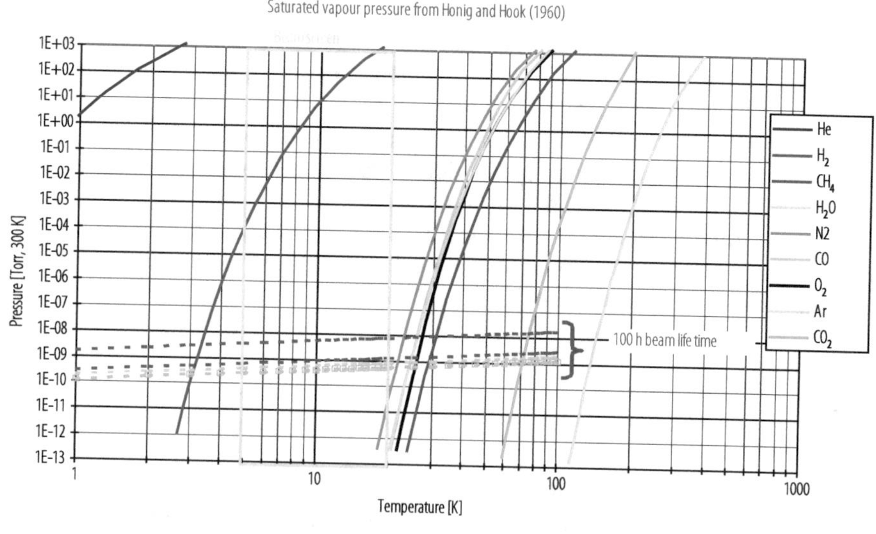

Fig. 8.55 Saturated vapour pressure [94]

temperature is avoided others are the cryogenic implications and magneto-resistance of the beam-pipe materials.

8.5.2.2 Gas Laws and Gas Densities

The vacuum system of an accelerator will behave as the gas it contains. The major Laws of interest to predict the behaviour of gasses are Boyle's Law, Charles's Law, Gay-Lussac Law and the General Gas Law which resulted from the combination of Boyle's and Charles' Laws.

The Avogadro's Law is very useful to convert pressures, P, into gas densities, n, for a given volume, V. Under the same conditions of pressure and temperature, equal volumes of all gases have the same number of molecules: called a mole.

$$PV = nk_B T, \tag{8.47}$$

with:

$$k_B = 1.38 \times 10^{-23} \left[\frac{\text{Nm}}{\text{K}} = \frac{\text{Pa}\cdot\text{m}^3}{\text{K}}\right] = 1.38 \times 10^{-22} \left[\frac{\text{mbar}\cdot\text{l}}{\text{K}}\right]. \tag{8.48}$$

For $T = T_{\text{RT}} = 296$ K (23 °C):

$$\frac{1}{k_B T_{\text{RT}}} = 2.45 \times 10^{20} \left[\text{Pa}\cdot\text{m}^3\right]^{-1} = 2.5 \times 10^{19} \,[\text{mbar}\cdot\text{l}]^{-1},$$

in other words, a volume of 1 l at room temperature and 1 mbar contains 2.5×10^{19} molecules.

8.5.2.3 Gas Flow, Mean Free Path, Throughput and Ultimate Pressure

After having pumped out the gas molecules from the volume (Fig. 8.56), the residual pressure in the beam-pipe is dominated by the surface desorption then, at much lower pressures, by the diffusion from the beam-pipe walls and the permeation through them. Modern pumping systems allow roughing the accelerator beam-pipes quickly enough to have only to consider the molecular regime, laminar and intermediate regimes being ruled out in few minutes. The molecular flow regime occurs when molecules are so far apart that they no longer have any influence on each other. Their motion is strictly random and their mean free path, i.e. the path length that a molecules traverse between two successive impacts with other molecules becomes larger than the beam-pipe dimensions. The mean free path for

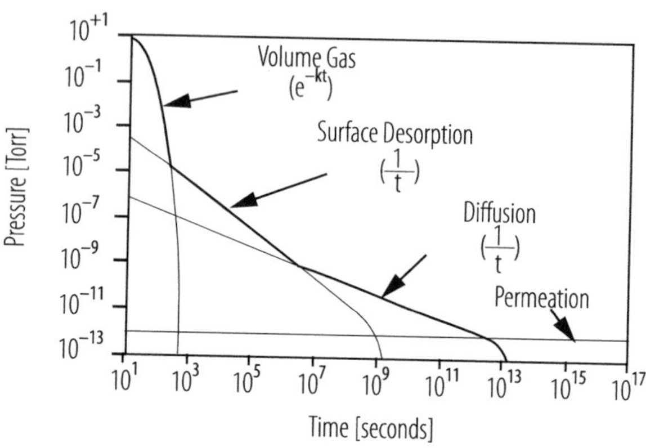

Fig. 8.56 Typical pump down curves observed in accelerator beampipes [b]

air at room temperature is given by:

$$\lambda_{\text{air}}\,[\text{m}] = \frac{6.67 \times 10^{-3}}{P\,[\text{Pa}]};\quad \lambda = 67\ \text{m at } 10^{-4}\text{Pa}\ \left(10^{-6}\text{mbar}\right), \tag{8.49}$$

In steady-state or equilibrium conditions, the throughput (quantity of gas per unit of time), Q, is conservative: the same at one end of a vacuum system as it is at the other.

$$Q = \frac{PV}{t} = P\frac{V}{t} = P \cdot S, \tag{8.50}$$

S being defined as the pumping speed [m^3/h], often in l/s.

In reality, Eq. (8.49) becomes Eq. (8.50) where, P_0, is defined as the ultimate pressure. In general, S varies in the range of three orders of magnitude ($\sim 1 \rightarrow 1000\ \text{l s}^{-1}$) while Q can extend over more than 10 orders of magnitudes: $\sim 10^{-5}$ mbar·l·s^{-1}·cm^{-2} for plastics $\rightarrow$ 10^{-15} mbar·l·s^{-1}·cm^{-2} for metals [95]. The right choice of materials and treatments is compulsory in the design of vacuum systems especially those for accelerators. In this respect the measurement of outgassing rates is a basic activity for an ultra-high vacuum expert.

Equation (8.50) shows that the ultimate pressure which is intrinsic to the system dominates at low pressure regimes. At a given stage, increasing the pumping speed will not lead to the expected pressure decrease.

$$P = \frac{Q}{S} + P_0, \tag{8.51}$$

8.5.2.4 Outgassing of Materials

The outgassing is the spontaneous evolution of gas from solids or liquids. To be compared with the degassing which is the deliberate removal of gas from a solid or a liquid and with the desorption which is the stimulated release of adsorbed chemical species from the surface of a solid or liquid.

The intrinsic outgassing rate is the quantity of gas leaving per unit time and per unit of exposed geometric surface, or per unit of mass, at a specified time after the start of the evacuation. The geometric surface is the visible surface without correction for roughness or open porosity. The measured outgassing rate is the net difference between the intrinsic outgassing rate and the rate of re-adsorption in the test chamber. The re-adsorption rate depends on test chamber and on method of measurement.

For unbaked metals, water is the main gas desorbed. The outgassing rate of water decreases following a $1/t$ law. The outgassing of unbaked metals is not an intrinsic value. However, measurements have shown that the water outgassing does not depend significantly on the nature of metals, on surface treatments and on operating temperature for temperatures lower than 110 °C. At present no methods, except heating (bake-out), exist to quickly remove water from unbaked metals. The fact that desorption rates depend exponentially on temperature allows a faster reduction of outgassing through heating (bake-out procedure). If a heat of adsorption of 15 kcal/mol and a bake-out temperature of 150 °C are assumed, the rate of gas removal will be increased by a factor of roughly 5000 relative to the value at room temperature.

The outgassing of baked systems is dominated by H_2 diffusing out of the metallic walls. The remaining outgassing species CH_4 and CO are believed to be related to the presence of H_2 and its recombination with the surface contaminants. These species will decrease if the H_2 level is reduced. Contrary to water outgassing, the value of the outgassing rate of hydrogen is stable at room temperature. When the thermal history is known, the outgassing of hydrogen is an intrinsic property of metals. The diffusion model predicts values for the hydrogen outgassing that are in agreement with experimental observations. For the vacuum-fired chambers (950 °C × 2 h, 10^{-3} Pa H_2), the outgassing rate is limited by the background signal induced by the gauges. The upper limit at room temperature is 10^{-14} mbar·l s^{-1}·cm^{-2}. In UHV systems, the outgassing of the beampipes can often be compared to the outgassing of the vacuum instrumentation. Therefore, the outgassing of the instrumentation has to be considered in the design of an accelerator, in particular close to the detectors in the experimental areas or in regions where non-evaporable getter (NEG) coatings are used.

8.5.2.5 Kinetic Theory of Gasses

The gas density velocity distribution (mass of gas m) at a temperature, T, in a stationary large volume and at equilibrium follows a Maxwell-Boltzmann distribution law:

$$\frac{dn}{dv} = 2n\sqrt{\frac{m^3}{2\pi k^3 T^3}}e^{-\frac{mv^2}{2kT}}v^2. \tag{8.52}$$

The average velocity of molecules is given by Eq. (8.52) in m/s with T in K and M the molar mass. The number of molecular hits on a wall per second is given by (8.53).

$$\overline{v} = \sqrt{\frac{8kT}{\pi m}} = 146\sqrt{\frac{T}{M}}, \tag{8.53}$$

$$\mu = \frac{n}{4}\sqrt{\frac{8kT}{\pi m}} = \frac{1}{4}n\overline{v}. \tag{8.54}$$

8.5.2.6 Conductance and Effective Pumping Speed

The conductance of an orifice of surface A, and of a tube of diameter D and length L, can be derived from the previous expression and are given for air at room temperature by Eqs. (8.54) and (8.55), respectively.

$$C_{\text{air, }20^\circ\text{C}}\ [\text{l/s}] = 11.6\ A\ \left[\text{cm}^2\right], \tag{8.55}$$

$$C_{\text{air,}20^\circ\text{C}}\ [\text{l/s}] = 12.1\ \frac{D[\text{cm}]^3}{L\,[\text{cm}]}, \tag{8.56}$$

to be noted that both equations show that the conductance varies with $\sqrt{T/M}$. The conductances add when in parallel ($C = \sum C_i$) and they add in the inverse when in series ($1/C = \sum (1/C_i)$). The flux through a conductance is given by: $Q = C(P_2 - P_1)$ [96, 97].

As any type of discrete pump is connected to the beampipe through a connecting piece with a conductance, C, the efficient pumping speed is given by:

$$S_{\text{eff}} = \frac{Q}{P} = \frac{C\ S}{C + S}, \tag{8.57}$$

in practice, $S_{\text{eff}} \sim S$ if $C >> S$, $S_{\text{eff}} \sim C$ if $C << S$, and $S_{\text{eff}} \sim S/2$ if $C \sim S$. It is therefore of great importance to optimise the conductance of the connecting pieces to benefit as much as possible of the pumping system.

8.5.3 *Vacuum Dynamics*

8.5.3.1 Synchrotron Radiation

When a particle beam circulates in a bending magnet or circulates off-axis in a quadrupole magnet, it might radiates photons and lose energy by synchrotron radiation (SR). The interaction of the synchrotron radiation with the vacuum chamber wall stimulates molecular gas desorption and dissipates heat load. Therefore, the vacuum chamber design must take into account these effects [98, 99].

The synchrotron radiation is due to the acceleration (centripetal or longitudinal) of the charged particles. Therefore, these charged particles loose energy which must be compensated by the RF system in order to keep the particles on the closed orbit. The dissipated power due to the centripetal acceleration is much larger than the longitudinal acceleration. The proportional factor is equal to the relativistic factor γ to the square. Due to the relativistic effects, the radiation is strongly emitted in the forward direction. The opening angle is roughly $1/\gamma$. The energy of the emitted photons varies from meV to MeV.

The synchrotron radiation spectrum is characterised by the critical energy ε_{C}. This energy separates the power spectrum in two equal amounts. However, 90% of the emitted photons have energy below the critical energy. The critical energy is given by Eq. (8.57) where h is the Planck constant, c the speed of light and ρ the curvature radius.

$$\varepsilon_C = \frac{3}{2}\frac{hc}{2\pi}\frac{\gamma^3}{\rho}, \tag{8.58}$$

A practical equation for engineers is given by Eq. (8.58) for electrons and protons.

$$\begin{aligned} \varepsilon_C\,[\mathrm{eV}] &= 2.218 \times 10^3 \frac{\mathrm{E}\ [\mathrm{GeV}]^3}{\rho[\mathrm{m}]} \quad \text{for electrons, and} \\ \varepsilon_C\,[\mathrm{eV}] &= 3.5837 \times 10^{-7} \frac{\mathrm{E}\ [\mathrm{GeV}]^3}{\rho[\mathrm{m}]} \quad \text{for protons.} \end{aligned} \tag{8.59}$$

At CERN, the critical energy of the Large Electron Positron Collider (LEP) was varying from 6 keV at injection till 660 keV at 104 GeV. For the Large Hadron Collider (LHC), the critical energy varies from 10 meV at injection till 44 eV at 7 TeV.

Practical equations for engineers of the dissipated power, P_0, and photon flux, Γ, as a function of beam energy, E, and beam current, I, are given by Eqs. (8.59) and (8.60).

$$P_0\,[\mathrm{W/m}] = \alpha \frac{\mathrm{E\,[GeV]}^4}{2\pi\rho[\mathrm{m}]^2}\mathrm{I\ [mA]}\,,\ \text{with}\ \alpha = 88.7\ \text{or}\ 7.79\times 10^{-12}\quad \text{for electrons or protons,} \tag{8.60}$$

$$\begin{aligned}&\dot{\Gamma}\left[\mathrm{ph/m/s}\right] = \beta\frac{\mathrm{E}\,[GeV]}{\rho[m]}\mathrm{I\ [mA]}\,,\quad \text{with}\\ &\beta = 1.288\times 10^{17}\ \text{or}\ 7.017\times 10^{13}\quad \text{for electrons or protons.}\end{aligned} \tag{8.61}$$

In accelerators with synchrotron radiation (SR), e.g. LEP, DIAMOND, SOLEIL, … , the heat load of some 10 kW/m requires the use of coolant and dedicated engineering design in order to evacuate the heat deposited on the beam-pipe wall. In the LHC, the heat load of 0.2 W/m is evacuated by the beam screen's cooling circuit maintained between 5 and 20 K while the cold bore is operating at 1.9 K.

The photo-desorption of neutral gas from the oxide layer is a 2 steps process where photo-electrons and secondary electrons desorb molecules from the near surface and excite strongly bounded molecules which subsequently desorb thermally. The desorb species are H_2, CH_4, H_2O, CO, O_2 and CO_2. The phenomenon is characterised by the photo-desorption yield η, which is the ratio of the desorb molecules to the number of incident photons. The photodesorption yield increases when decreasing the angle of incidence. Below 1 keV, the photo-desorption yield scales roughly like the critical energy and is almost constant above 1 keV. The desorption yield, which decreases under photon irradiation, can be expressed as a function of the initial desorption yield, η_0, initial and accumulated photon dose, D_0 and D, by Eq. (8.61) with $D_0 \sim 10^{21} \ldots 10^{22}$ ph/m and $a \sim 0.9 \ldots 1.3$.

$$\eta = \eta_0 \left(\frac{D}{D_0}\right)^{-a}. \tag{8.62}$$

At 3.75 keV, after a period of beam cleaning of $10^{24} \ldots 10^{25}$ ph/m accumulated, the photo-desorption yield is reduced from $10^{-2} \ldots 10^{-3}$ to $10^{-5} \ldots 10^{-6}$ molecules/ph for unbaked and baked Cu or baked stainless steel surfaces. The desorption yield of aluminium is a factor 10 higher, most probably owing to the fabrication process. The amount of gas desorbed during the process varies from 0.5 in the baked case to ~10 monolayers in the unbaked case. At large doses, the irradiated surfaces have a pumping speed of ~100 l/s/m for CO and CO_2. The pumping capacity is however very small (10^{-4} monolayer). Figure 8.57 shows typical curves measured during photon stimulated desorption studies.

Beside its pumping characteristics, Non Evaporable Getter (NEG) coating such as TiZrV is also used to reduce the desorption yield. The diffusion barrier created during the activation after the surface oxide layer has diffused into the bulk could be the origin the low intrinsic yields. At 4.5 keV critical energy, the intrinsic photodesorption yield is as low as 10^{-5} H_2/ph and even lower for other gas species. When the film is saturated with CO, the photo desorption yields remain constant. Photon induced pumping by the getter is also observed for CO at a rate of 10^{-5}

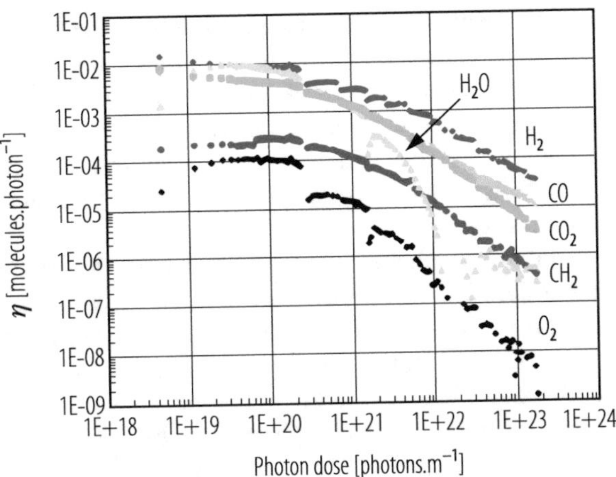

Fig. 8.57 Photo-desorption yields of unbaked stainless steel irradiated by 3.75 keV critical energy photons [100]

CO/ph [101]. A cleaning rate $a = 0.4$ (Eq. 8.61) was observed with 20 keV critical energy.

At cryogenic temperature, the photo-desorption yields are reduced by one order of magnitude as compared to room temperature. The primary desorption equals 10^{-4} molecules/ph for unbaked Cu held at 5 K when irradiated by SR of 200 eV critical energy. The cleaning rate a equals ~0.6. During irradiation, the previously strongly bounds molecules are weakly bounded (physisorbed/condensed) to the cold surface. In a closed geometry, these molecules can in turn be recycled into the gas phase by the subsequent irradiation. A recycling yield as large as 1 H_2/ph is reached at one monolayer. In this case, the pressure level inside the vessel can increase up to the saturated vapour pressure. However, the operating pressure, P_{op}, inside a cold bore can be controlled by inserting a perforated liner with a conductance c. In this case P_{op} is given by Eq. (8.62) [102].

$$\mathrm{P_{op}} = \frac{\eta \dot{\Gamma}}{c}. \tag{8.63}$$

8.5.3.2 Electron Cloud

Operation with trains of bunches of high intensity can lead to the formation of a cloud of electrons. This was and is the case of machines such as ISR, PSR, KEK-B, PEP-II, SPS, RHIC and LHC. This electron cloud affects the beam properties but also contribute to the vacuum dynamics. When an electron is in the vicinity of the beam potential, it can be attracted or repelled in the case of a positive or negative beam. In both cases, there can be beam conditions which produce photoelectrons, ionise the residual gas and/or allow the multiplication of secondary electrons with time. Once the energy given to electrons is above the binding energy of physisorbed and/or chemisorbed molecules (0.1 . . . 1 eV) the bombardment stimulate gas desorption from the vacuum chamber wall. Clearly, the highest gas load is observed in the multipacting regime. For instance, in the LHC, the proton bunch intensity is large enough to give a kick of ~100 eV to the stray electrons. At the wall, these electrons stimulate gas desorption but also produce secondary electrons. The secondary electrons are further accelerated to ~100 eV by the following bunch, 25 ns apart, leading to a multiplication of electrons in the vacuum system.

Similarly to photon stimulated molecular desorption, electron simulated molecular desorption is characterised by the electron desorption yield. The electron desorption yield is also decreased under electron bombardment according to Eq. (8.61). The electron desorption yield increases almost linearly with the electron energy. Table 8.17 shows the initial yield and cleaning rate for unbaked and baked copper when perpendicularly irradiated by 300 eV electrons. At an integrated dose of 5×10^{16} e/cm^2, the desorption yield of the baked Cu is about one order of magnitude lower than unbaked Cu. After an accumulated dose of 10^{18} e/cm^2, i.e. 1.6 mC/mm^2, 1 . . . 10 monolayers of gases have been desorbed, mainly H_2. The hydrogen electron desorption yield can be explained by a diffusion model ($a \sim 0.5$). The H atoms are produced by the dissociation of the hydroxides and diffuse under the electron bombardment.

Electron desorption yields of metallic surfaces are about the same. As a comparison to the baked case, after activation, the initial yields for NEG coatings are further reduced by 1 and 2 orders of magnitude for H_2 and CH_4, CO, respectively. Like the SR case, electron induced pumping is observed for CO at a rate of 10^{-3} CO/e [105]. At 4.2 K, the electron desorption yield of physisorbed/condensed molecules is linear up to a monolayer and then levels-off. Values up to 500 H_2/e were reported with 300 eV electrons [106].

The secondary electron yield (SEY) is another key parameter which defines the vacuum level in a beam tube. It is defined as the ratio of the number of produced electrons to the number of incident electrons. The SEY value determines the amount of stray electrons between each bunch and therefore the electron flux to the wall. Therefore, there are strong interests to reduce the SEY for the vacuum scientist. The maximum of the SEY curve δ_{max}, lies in the range 200 . . . 300 eV. Typical δ_{max} values for as received metallic surface are ~2. Baked Cu has 1.8

Table 8.17 Desorption yield parameters δ for unbaked and baked copper perpendicularly irradiated by 300 eV electrons [103, 104]

		H_2	CH_4	H_2O	CO	CO_2
Unbaked	η_0	2×10^{-1}	3×10^{-2}	1×10^{-1}	4×10^{-2}	5×10^{-2}
	D_0 [$\times 10^{14}$]	3	1	6	2	4
	a	0.5	0.6	0.7	0.5	0.5
Baked	η_0	4×10^{-3}	2×10^{-4}	–	1×10^{-3}	7×10^{-4}
	D_0 [$\times 10^{16}$]	5	5	–	5	5
	a	0.6	1.3	–	0.6	0.9

and atomically cleaned Cu has 1.5. Water condensation on atomically cleaned Cu resets δ_{max} to 2 which correspond to the as received state. Coating such as baked TiN, activated TiZrV or amorphous carbons are used to reduce δ_{max} close to 1. Due to graphitization of the carbon oxides, an accumulated electron dose of a few mC/mm^2 is also effective by beam scrubbing to reduce δ_{max} of a metallic surface (except Al) close to 1 [107]. These methods are used in recent machines to control the multipacting process. Thanks to their morphology, laser engineered structured surfaces can reduce δ_{max} below 1. However, their exploitation in future accelerator is still under study.

8.5.3.3 Vacuum Stability

In storage rings, the circulating particle beam ionise the residual gas. The produced ions are then repelled toward the vacuum chamber wall desorbing neutral gasses. As a consequence, the amount of beam-ionisation is increased leading to a higher gas load. Ultimately, this mechanism, which was first observed in the Intersecting Storage Rings (ISR), can conduct to a particle loss and a pressure run-away [108]. Modern accelerators operated with high beam intensities are designed to maintain vacuum stability (SNS, LHC).

The flux of molecules *Q*(*I*), inside the vacuum chamber is a function of the ion desorption yield, η, the gas pressure, *P*, the ionisation cross section, σ, and the thermal outgassing, Q_0. It is given by Eq. (8.63) where *I* is the beam intensity and *e* the electron charge.

$$Q(I) = \eta P \sigma I/e + Q_0. \tag{8.64}$$

With a linear pumping speed, *S*, the dynamic pressure is given by Eq. (8.64).

$$P(I) = \frac{Q(I)}{S} = \frac{Q_0}{S - \eta\sigma I/e}. \tag{8.65}$$

The stability limit is then given by Eq. (8.65) when the denominator of Eq. (8.64) equals zero. During operation, when the beam current approaches the critical current, the pressure increases drastically leading to strong particle losses and ultimately to beam dump.

$$(\eta I)_{\text{crit}} = \frac{eS}{\sigma}. \tag{8.66}$$

In practice, a treatment to reduce the ion desorption yield and the optimisation of the pumping speed at each position of the vacuum system is required to achieve vacuum stability. Ex-situ Argon glow discharge with in-situ bake-out was successfully applied in the ISR.

In the LHC, the ion impact energy is ~500 eV. For baked materials, ion desorption yields are a few 0.1 molecules/ion while minor beam conditioning is expected during operation. When using NEG coatings, the vacuum stability is ensured by lumped pumping to constraint CH_4. This allows to achieve a critical current larger than 4 A in the LHC experiments and in the long straight sections [109]. At 4.2 K and 5 keV, the ion desorption at a monolayer for a condensed gas is as large as 2000 and 2 for H_2 and CO_2 respectively. Due to the closed geometry of the beam tube and the possibility to physisorbed/condensed gas, the critical current of Eq. (8.65) is modified into Eq. (8.66). It is a function of the sticking probability α and the ion desorption yield at cryogenic temperature, η'.

$$I_{crit} = \frac{\alpha S}{(\eta + \eta')\,\sigma/e}. \tag{8.67}$$

For such a closed geometry when operating at cryogenic temperature, the critical current is less than 1 A for vacuum chambers of 1 m length. However, the perforation of the LHC beam screen inserted inside the cold bore guarantee a minimum pumping speed. Thus, the equilibrium coverage of the condensed gas is below a monolayer and the product $(\eta I)_{\text{crit}}$ is ~ 100 [98, 99].

8.5.3.4 Particle Losses

In heavy ions storage rings or accelerators, beam losses linked to charge transfer induce large pressure rises (RHIC, LEAR, GSI, LHC ...) which could trigger the beam dump. These pressure rises are due to heavy ions bombarding the vacuum chamber wall at grazing incidence. At high energy, the heavy ions lose energy in the solid by electronic losses (stopping power) which determines the desorption yield. The reported desorption yields are very large but decrease when increasing impact energy above the Bragg maximum (1 MeV/u for lead ions on Cu). At grazing incidence, the yields range from 10^2 to 10^5 molecules/ion as a function of surface treatment and impact energy. Possible mitigation techniques implemented in operating machines are the use of dedicated collimators, beam conditioning (2 orders of magnitude is gained after 100 h i.e. 10^{13} ions/cm^2), NEG coatings, etc. [110]. For machines operating at cryogenic temperatures, the dominant factor is the quench level of the cryomagnets. For these typical loss rates, pressure rises are often negligible.

8.5.4 Vacuum Engineering

The engineering of the vacuum system of a particle accelerator has to be started right from the beginning of the project. Indeed, if considered later on, the resulting integration and performance issues will lead to a significant increase of the cost of the vacuum system (complex shapes, more pumps, retrofitted modifications, etc.) and even into performance limitations. Experience has shown that the vacuum engineering shall proceed in parallel on the following topics: expertise provided to beam-related components (magnets, beam instrumentation, radio-frequency systems, etc.), engineering of vacuum related components (beampipes, bellows, pumping ports, etc.) and machine integration including cabling and services.

8.5.4.1 Vacuum Pumping

The pumping scheme is often decided at the early stages of the design since it affects the overall engineering of the accelerator. Indeed, the pumping scheme will define the reserved space for the pumping and in case of integration problems will lead to alternative solutions to preserve the expected performances or to compromise. The number and types of pumps will allow reaching the design operating pressure which is linked to the beam lifetime, a parameter in relation with the beam-gas scattering.

As the pumping speeds vary between a few litre-per-seconds to thousands of litre-per-seconds while the outgassing rates of materials in the vacuum system vary by up to 8 orders of magnitude, an appropriate selection of material and design is required to achieve the specified vacuum performances. Said differently, increasing the installed pumping speed in an accelerator by a factor 2 is already challenging and costly, whereas losing several orders of magnitude in outgassing rates could easily result from an inappropriate use of materials or bad vacuum engineering or cleaning of the materials.

8.5.4.1.1 Discrete Pumping

In particle accelerators, discrete pumping is the most commonly used solution and is often obtained by ion pumps combined with sublimation or non-evaporable getter (NEG) cartridge pumps, cryogenic and turbomolecular pumps.

Ion pumps [111] are widely used since they are very reliable and provide a large pumping speed, up to 800 l/s. In addition, the discharge current of an ion pump is directly proportional to the pressure, and thereby provides an indication of pressure for a limited cost. This signal is often used to trigger vacuum interlocks for machine protection. Ion pumps have also the advantage of pumping all gas species and once baked-out at 250–300 °C, the ultimate pressure is in the low 10^{-10} Pa. Special attention has to be taken while pumping noble gasses and water. Above a given quantity of gas pumped, part of the pumped noble gasses will be released as hydrogen (via dissociation of the water molecules) when pumping water.

Sublimation pumps [112] are often used to speed up the pump down to the UHV pressure range or as a complement to ion pumps at very low pressures ($<10^{-10}$ Pa). If used at high pressures, the pump's bodies must be water cooled to prevent the heating induced by the frequent activations (several per minute). Similarly to NEG pumps, the major limitation of the sublimation pumps is that the pumping speed of noble gasses and methane is very small.

NEG pumps are commercially available as different types: cartridges [113], strips and coatings. Only the first type is used as discrete pumping, strips and coatings are, by nature, used as distributed pumping. NEG cartridges offer a huge pumping speed for hydrogen, carbon mono- and dioxide. Once the NEG material is saturated, the pumping capacity can be recovered by heating via Joule effect, most of the commercial products include this option in the body of the NEG cartridge.

Turbo-molecular pumps are widely used by the industries and this has resulted in a significant decrease of prices and increase in reliability. Most vacuum suppliers provide a large variety of turbo-molecular pumps off-the-shelf. The improvement in performance and lifetime over the last decades has stimulated their use in particle accelerators. The major limitation comes from the radiation inherent in particle accelerators. Indeed, this requires decoupling the power supply and associated electronics from the body of the pump, a technical solution contrary to the needs of industry. Thus explaining why most of the suppliers are reluctant to develop and maintain such products in their catalogues.

Cryogenic pumps are also widely used by industry and similarly to turbo-molecular pumps, the prices have decreased. Cryogenic pumps are used in accelerators to cope with huge local outgassing, mainly on unbaked systems. Special operating precautions have to be taken to limit the back streaming of condensed gasses in case of power cuts or compressor failure.

8.5.4.1.2 Distributed Pumping

Distributed pumping has been used for decades in particle accelerators and is of special interest for conductance limited beam pipes. The most commonly used distributed pump was the ion pumps cells inserted in the dipole beampipes [114]; the dipole magnet field serving also for the ion pump cells. Later on and in particular in the large electron-positron collider (LEP), NEG strips were used. The NEG strips [115] were activated by the Joule effect.

More recently, the large hadron collider (LHC) could profit from the development of the NEG coatings [116], a technology developed and industrialized at CERN. These coatings provide many advantages such as a huge pumping speed, large pumping capacity for hydrogen and low yields for secondary electrons and stimulated desorption by electrons, photons and ions. The activation temperature has decreased; 180 °C is enough to obtain nominal pumping speed and capacity.

8.5.4.2 Vacuum Instrumentation

8.5.4.2.1 Pressure Sensors

Similar to the pumps, the evolution of the pressure sensors is stimulated by the needs of the industry and the tendency is to use pressure sensors with electronics on their heads. All types of pressure sensors are now available in catalogues covering the range of pressure from several atmospheres (10^6 Pa) down to UHV pressures (10^{-10} Pa). These products are cheap, reliable and easy to maintain.

However, their use in particle accelerators is also compromised by the inherent radiation. An alternative are the so-called passive gauges with the electronics decoupled from the pressure sensor. The existing sensors can be placed far away, up to 2 km, from their controllers. It should be noted that the installation of appropriate cables is an important issue to allow use of the sensors up to their design specifications. Special attention has to be taken while laying the cables, to avoid electromagnetic incompatibility and interference (EMI) perturbations; it is recommended to lay these cables in special cable trays or with instrumentation cables and to pay special attention to cable shielding and grounding loops.

The ion pumps are also often used as pressure indicators since their discharge current is proportional to the pressure. Modern controls electronics allow a fast conversion of the currents into pressures. The stability of the discharge current explains why these signals are used as pressure interlocks.

8.5.4.2.2 Residual Gas Analyzers (RGA)

The residual gas analysers are of primary importance to understand the beam-gas scattering since, as mentioned earlier, the ionisation cross-section varies with the gas species. Therefore, while operating an accelerator with beams, knowing the residual gas composition can allow calculating the beam lifetime and predicting potential limitations.

As the modern accelerators require more and more UHV conditions to operate, the RGAs must provide partial pressure resolutions in the low 10^{-11} Pa range to provide useful information. This type of residual gas analyser is available off-the-shelf, but in many cases, the electronics is attached to the head or placed at a few meters. This limitation still exists and the demand being very small, most of the suppliers do not plan to modify the position of the electronics thereby limiting the use of RGA to radiation free environment.

8.5.4.3 Vacuum Sectorisation

The sectorisation of the beam vacuum system results from the combination of various constraints, the major being: venting and bake-out requirements, conditioning requirements (RF and HV devices), protection of fragile and complex systems (experimental areas and ceramic chambers), decoupling of baked vacuum parts, which are at room temperature, from non-baked parts at cryogenic temperature, radiation issues, etc.

For UHV beam vacuum systems, all-metal gate valves are preferred in order to allow for bake-out at temperature above 250 °C. VITON-sealed valves are less expensive but limited in bake-out temperature. A special treatment of the VITON seals is recommended prior to their use nearby NEG coatings or pumps since minor outgassing of Fluor will degrade the pump characteristics.

Interlocking the sector valves is not an obvious task. Indeed, increasing the number of sensors will provide more pressure indications but might results in a degradation of the overall reliability. Protection at closure (pressure rise, leaks) is treated differently from the protection while recovering from a technical stop. In the latter case, parts of the accelerator beampipe vented or being pumped down must not be put in communication with upstream or downstream sectors which remained under vacuum.

8.5.4.4 Corrosion Issues

In vacuum systems, feed-throughs and bellows are particularly exposed to corrosion. Feed-throughs, particularly those of the ion pumps where high voltage is permanently present, are critical parts. Despite the design optimization made by many suppliers, a direct heating of the protective cover to reduce the relative humidity around the feed-through is the most efficient and commonly used technique to limit the corrosion.

The bellows are also vulnerable due to their thinness, often between 0.1 . . . 0.15 mm. Aluminum bellows are exposed to corrosion by nitric acid (HNO_3) which is generated by the combination of O_3 and NO_x. In presence of PVC material, radiation can lead to the creation of hydrochloric acid (HCl) which corrodes stainless steel materials. This corrosion is strongly penetrating, and once seen at the surface, it is often too late to mitigate the effects.

Relative humidity is the critical parameter and must be kept below 50%. However, accidental spillage can compromise locally the conditions and therefore, corrosion-resistant design are strongly recommended.

Corroded components like feed-throughs and bellows normally start to leak after a mechanical action or a venting to atmosphere. If several components are affected, this could result in a long sequence of pumping, leak searching, venting, repumping, etc., leading to a long downtime and often subsequent radiation dose rates to personnel doing the interventions.

8.5.4.5 Experimental Areas

Their design must include the following additional constraints: integration, reliability, availability, engineering and performances.

The installation of the vacuum system follows the closure of the detector. Therefore, the design has to be validated beforehand in order to prevent integration issues which could prevent the closure of the detector and lead to significant delay and increase of costs. Temporary supports and protection are required at each stage of the installation. Indeed, as compared to the size of the detectors, the beam pipes are small, fragile and need to be permanently supported and protected while moving the detector components.

The leak tightness and bake-out testing are compulsory at each step of the installation since all vacuum systems are later on encapsulated in the detector, thereby preventing any access, even for repair. Their reliability is critical.

The installation of the detectors dictates the speed and sequence of installation of the beam-pipes. To avoid delays, the vacuum components must be available (tested and ready for installation) in advance in order to adapt to the installation sequence.

The experimental beam-pipe must be thin and manufactured from light material in order to allow the particles created inside the beam-pipe to escape from it through the vacuum envelope. Since the detectors are extremely demanding in terms of operating pressure (normally expressed in gas densities) light materials are mandatory. Beryllium and aluminium materials are the most commonly used. Titanium is sometimes an alternative. These materials can be NEG coated by plasma discharge and baked-out in-situ. Beryllium being is extremely expensive and dangerous in case the beampipe breaks; beam-pipe design attempts have been made to replace the beryllium by composite chambers. Carbon-carbon materials have been used with metallic layers but the ultimate pressure is not yet in the range of the specification of the modern detectors. The engineering solutions for the bake-outs have also to be studied in details since the bake-out solutions (heaters, probes and cables) must fit within the limited space available between beam-pipes and the detector parts.

The excellent performances of the vacuum system in the experimental areas are a determinant factor for the performances of the detectors themselves. The most important issues to follow are the beam-pipe transparency, gas density and residual gas composition, impedance and high order mode trapping and precise alignment. The last point is difficult to achieve once the beam-pipes are installed in large detectors.

8.6 Beam Instrumentation and Diagnostics

R. Jones · T. Lefevre · H. Schmickler

Beam instrumentation and diagnostics [117, 118] combines the disciplines of accelerator physics with mechanical, electronic and software engineering, making it an extremely interesting field in which to work. The task of a beam instrumentation physicist or engineer is to design, build, maintain and improve the diagnostic equipment for the observation of particle beams with the precision required to tune, operate and improve the accelerators and their associated transfer lines. This chapter will give an overview of the instrumentation in use in modern accelerators, although the choice available today is so vast that inevitably it will not be possible to cover them all.

8.6.1 *Beam Position Measurement*

A beam position monitoring system can be found in every accelerator. Its role is to provide information on the position of the beam in the vacuum chamber. In linacs and transfer lines Beam Position Monitors (BPMs) are used to measure and correct beam trajectories, while in synchrotrons such monitors are distributed around the ring, providing the global orbit. Their location is usually chosen to be close to the main quadrupole magnets where the β-functions are largest and so any orbit distortion a maximum.

8.6.1.1 Pick-Ups

The measurement of beam position relies on processing information from pick-up electrodes located in the beam pipe.

Four pick-up families are commonly employed (see e.g. [118]):

- **Electrostatic**

  These consist of two electrodes insulated from and located on opposite sides of the vacuum chamber (either left & right for the measurement of horizontal position and/or up & down for the measurement of vertical position). As the beam passes inside the pick-up it charges up the electrodes, which act as capacitors. The amount of charge collected by each electrode will depend on the position of the beam. Hence, by comparing the charge on opposite electrodes, a beam position can be calculated. A so called "shoe-box" design (Fig. 8.58) is often used for linearity, while a button design (Fig. 8.59a, b) is favored for short bunches and its low cost, but needs correction for its non-linear response [119].
- **Electromagnetic**

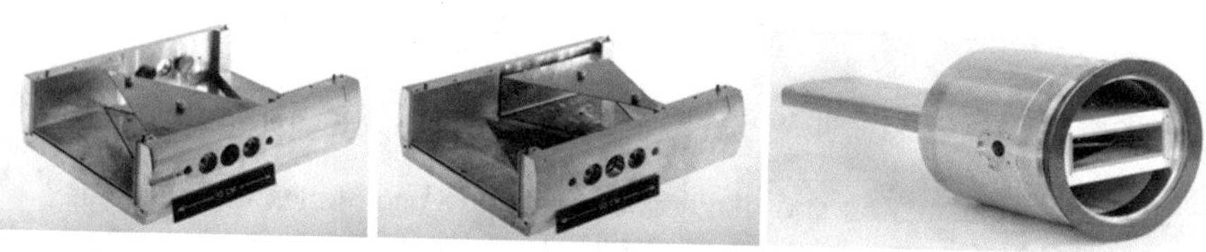

Fig. 8.58 Example of a horizontal shoe box pick-up from the CERN SPS showing right electrode, left electrode and the full assembly in its vacuum chamber

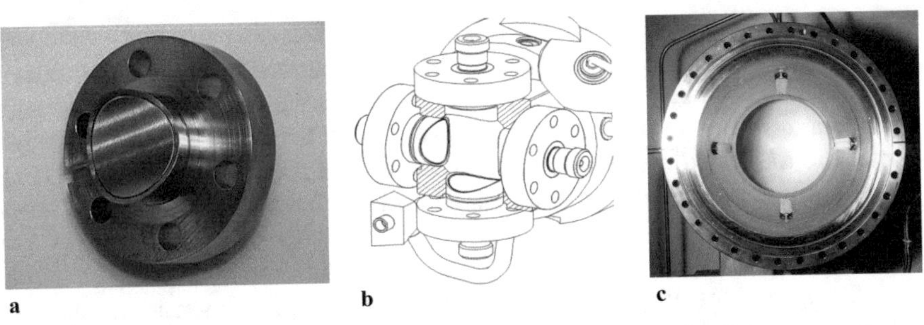

Fig. 8.59 (**a**) Example of a button electrode from the CERN LHC. (**b**) Schematic of the full LHC button pick-up assembly for dual plane measurement using four electrodes. (**c**) Example of a stripline pick-up from the CERN-SPS

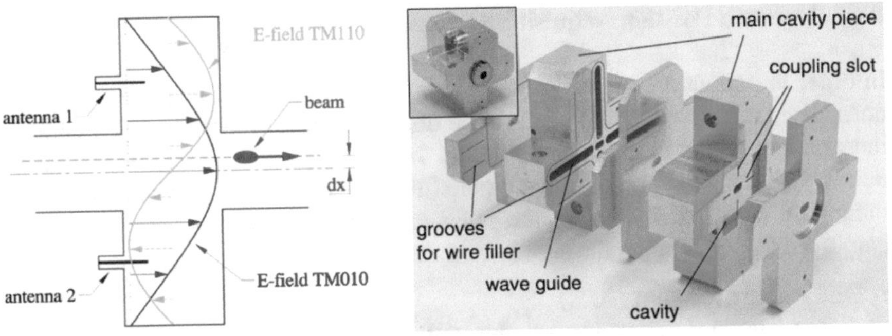

Fig. 8.60 Principle and example of a cavity pick-up (ATF2, KEK, Japan)

The most common type of electromagnetic pick-up is the stripline coupler (Fig. 8.59c). These consist of two strip electrodes located on opposite sides of the vacuum chamber, with an output port at each end. The particularity of this pick-up is that the beam induced signal is only produced at one end of each strip and depends on the direction of the beam. Commonly known as directional couplers, they are used to distinguish between counter rotating beams present in the same vacuum chamber. Due to their relatively high signal level they are also sometimes exploited instead of electrostatic monitors for low intensity beams. Striplines also provide beam signals with a fast time response, in excess of a few GHz, and are used to observe transverse beam instabilities [120].

A second type of electromagnetic monitor is the cavity pick-up (Fig. 8.60). These are constructed to exploit the fact that an off-centre beam excites a dipole mode (TM110) in the cavity, with the amplitude of excitation directly proportional to the off-axis displacement of the beam. This dipole mode has a slightly different frequency from the main monopole mode (TM010) of the cavity, which allows an excellent suppression of the intensity related signal. This in turn can lead to very good accuracy and nanometre scale resolution 121].

- **Resistive/Inductive**

  These monitors make direct use of the image current flowing on the wall of the vacuum chamber. A ceramic gap is used to force the image current through external resistors, in the case of the resistive (wall current) monitor, or through metallic rods fitted with transformers in the case of the inductive pick-up [122]. The position is calculated by comparing the output from suitably placed resistors or rods.
- **Magnetic**

  Such monitors are usually exploited for their relative insensitivity to stray particles. In this case the two electrodes are replaced by two loops orthogonal to the plane of measurement which couple to the magnetic field of the beam.

8.6.1.2 Beam Position Acquisition Systems

In order to extract an intensity independent position from all of these monitors a normalised difference signal needs to be obtained, i.e. a difference signal which is independent of the beam intensity [123]. In the case of the cavity this is simply done by dividing the output of the dipole (difference) cavity with that of a separate monopole (sum) cavity. In all the other cases one of the following methods is employed (A and B being the signal from each electrode):

- **Difference over sum (Δ/Σ)**

  The sum and difference are obtained either using a 0°/180° passive hybrid, a differential amplifier or calculated by software after digitising the signal from each electrode The resulting transfer function is highly linear with *Normalised Position* $= (A-B)/(A+B)$.
- **Logarithmic ratio**

  The two input signals are converted into their logarithmic counterparts and subtracted. In practice this is done using logarithmic amplifiers followed by a differential amplifier. The transfer function is an S-shaped curve which becomes highly non-linear when exceeding 70% of the normalised aperture: *Normalised Position* $= \text{Log}\,(A/B) = \text{Log}(A) - \text{Log}(B)$.
- **Amplitude to Phase or Time**

  The two input signals are converted into signals of equal amplitude but varying phase by combining them in a 90° passive hybrid. *Normalised Position* $= \varphi = 2\times\arctan(A/B)$. The transfer function is again an S-shaped curve but does not diverge for large excursions. In addition, the gradient is larger around zero, making it more sensitive towards the middle of the pick-up.

The type of normalisation to be used will depend on the choice of processing electronics, which in turn depends on the type of measurements to be performed (single pass, single bunch, average orbit etc.). In all cases the non-linearity is taken into account by calibration circuits and correction algorithms. Common acquisition electronics include:

- **Multiplexed**
  Each electrode is multiplexed in turn onto the same, single acquisition electronics chain. This eliminates channel to channel variations, giving very good accuracy. However, since the switching is generally quite slow such an acquisition only tends to be used in circulating machines where the average position is of importance.
- **Sigma/Delta**
  A passive hybrid is generally used to produce sum (Σ) and difference (Δ) from opposite electrodes. The resulting signals are then directly digitised, or downmixed to lower frequency before digitisation using homodyne detection (self-mixing the signal) or heterodyne detection (external mixing using a clock related to the accelerator radio-frequency). Once digitised, the position is computed as $K \times \Delta/\Sigma$ with K a scaling factor dependent on the pick-up geometry (typically close to ¼ of the aperture).
- **Individual Treatment**
  Each electrode is acquired separately but in parallel either by direct digitisation or logarithmic amplifiers. Position is calculated by difference over sum or logarithmic ratio normalisation by the central processing unit of the acquisition system.
- **Passive Normalisation**
  Such electronics convert the amplitude difference from opposite electrodes into a phase or time difference, which can then be detected using standard phase detection or time to digital conversion techniques. The position is extracted from this measurement using amplitude to phase or time normalisation as described above.

8.6.2 Beam Current and Intensity Measurement

The measurement of beam current or bunch intensity is one of the most basic measurements performed at any accelerator [124]. This is usually done by means of a Faraday Cup or Beam Current Transformer (BCT).

8.6.2.1 Faraday Cup

The Faraday cup technique relies on measuring the charge deposited in a metallic cup when intercepting the whole beam. This requires careful design, often using a bias voltage or magnetic field at the entrance of the cup, to avoid the secondary charges created by the interaction of the beam with the metal from escaping. They are usually used for relatively low current measurements due to problems related to the heat-load for higher intensity beams. Such devices act as an absolute calibration standard and are capable of providing fA resolution [125].

8.6.2.2 AC Beam Transformers

In order for a transformer to interact with the magnetic field of the beam it has to be placed over a ceramic gap in the vacuum chamber. The wall current accompanying the beam along the metallic wall of the vacuum chamber is deviated around the transformer using a low impedance radio-frequency bypass. A beam current, I_B, can be considered as the primary winding of the transformer, with the output voltage from the secondary windings given by $V \alpha L\, dI_B/dt$, L being the transformer inductance. An ideal transformer would give a differentiated response, with the integrated charge being zero, which is not of much use as a measuring device. In reality the secondary windings of the transformer have some capacitance and are terminated by some finite resistance. This leads to signals that closely resemble the beam intensity distribution, with the added inconvenience of a DC offset due to the transformer droop. This DC offset can be corrected for either electronically or by software treatment of directly digitised data. Bandwidths from 200 Hz to 1 GHz are now available, using ferromagnetic cores wound with high permeability metal tape to avoid eddy currents. Such transformers are capable of resolving the charges contained within single short bunches, acquired either by direct digitisation or using fast integrator electronics. Resistive or inductive wall current monitors can also be used to provide such signals, but again require DC offset compensation [126].

8.6.2.3 DC Beam Transformers

In storage rings and accelerators with cycle times of several seconds a DC beam Current Transformer (DCCT) can be used to measure the total current. Such an instrument has a bandwidth that extends from a few kHz down to DC (hence its name) and was developed for the CERN-ISR (Intersecting Storage Rings), the first machine to sustain beams for hours [127].

A DC transformer is based on a pair of matched, toroidal, ferromagnetic cores, which are driven into saturation by a modulation current at frequencies of up to a few kHz. The principle of operation makes use of the hysteresis loop of the toroid. If an equal but opposite modulation current is applied to both cores with the beam not present, then the voltage induced in the detection windings on each core will be equal but opposite. When, however, there is a beam current present, the starting point in the hysteresis loop for zero modulation current is offset due to the static magnetic field generated by the beam current. Since the modulation is opposite in each toroid, the time spent in saturation will be different for the two branches of the hysteresis loop. This results in the generation of voltage pulses at twice the modulation frequency when the induced voltage in the detection windings on each core is combined. The demodulation of this signal gives a train of pulses, with the width of each pulse being a direct measure of the beam current, i.e. by how much the hysteresis curves are offset.

In the "zero flux detector" implementation of the DC beam transformer, the result of the demodulation is fed back through the pair of transformers by a compensating

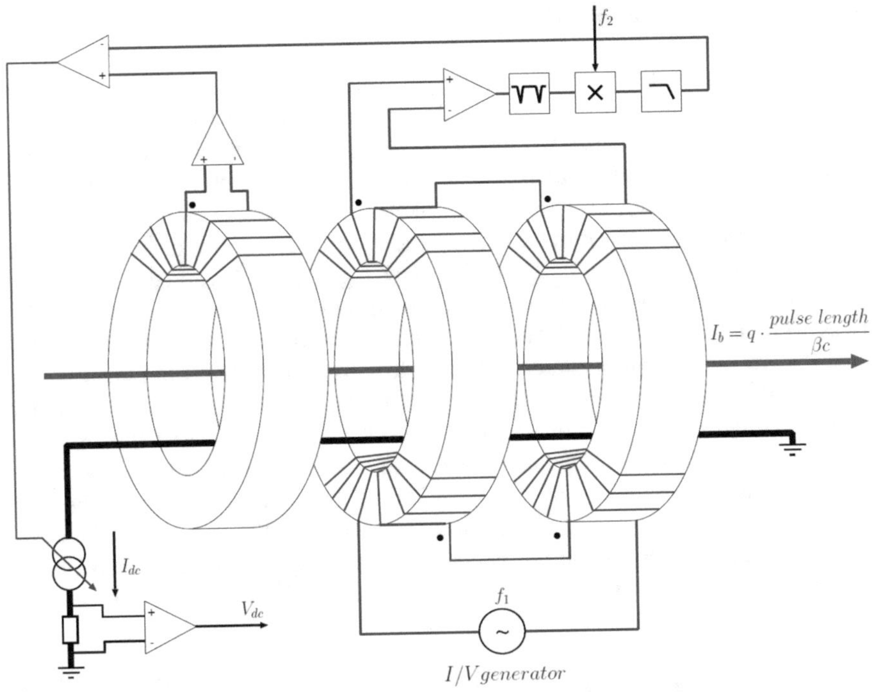

Fig. 8.61 Schematic of a Modern DCCT. The two cores on the right form the "zero flux" DCCT, while the left hand core is an AC transformer that extends the bandwidth of the system to higher frequency

current loop (Fig. 8.61). Once the compensation current and the beam current are identical the net static magnetic field seen by the toroids is zero (hence zero flux) and the output from the demodulator is also zero. The beam current is then simply obtained by measuring the voltage produced by this compensation current across a known resistor. For modern DC transformers such a zero flux detector is combined with an AC transformer. The AC transformer is used to feedback on fast changes which are not compensated by the DC transformer and hence significantly increases the bandwidth of the system, allowing measurement from DC to a few MHz.

8.6.3 Diagnostics of Transverse Beam Motion

The instrumentation used to look at transverse beam motion is very important for the efficient operation of any circular accelerator [128]. There are three main parameters that can be measured using such diagnostics: the betatron tune, chromaticity and coupling.

8.6.3.1 Tune Measurement

All betatron tune measurements are based on measuring the characteristic frequency of the transverse motion of the beam. In the simplest case the beam is given a single kick using a powerful stripline or magnetic kicker and allowed to oscillate freely. A position pick-up is then used to measure the resulting beam motion, with the betatron tune being the frequency which has the highest amplitude response in the power density spectrum obtained using Fourier Transform techniques. If the detection technique is sufficiently sensitive (see e.g. [129, 130]) and there is enough external excitation from other sources such as ground motion then the method also gives useful information without any specific, additional beam excitation (Fig. 8.62).

What is usually of more interest is to be able to track the tune evolution during the whole of the acceleration cycle. The simplest way of achieving this is to repeat the tune measurement at regular intervals, with the results displayed as a spectrograph. In order to minimise the frequency range over which power is put into the beam, a swept excitation frequency or "chirp excitation" is often used (so-called because if listened to at audio frequencies such a signal sounds like the chirp of a bird). The chirp range is set around the expected betatron tunes and the sweep time is determined depending on the requested time resolution and precision of the tune

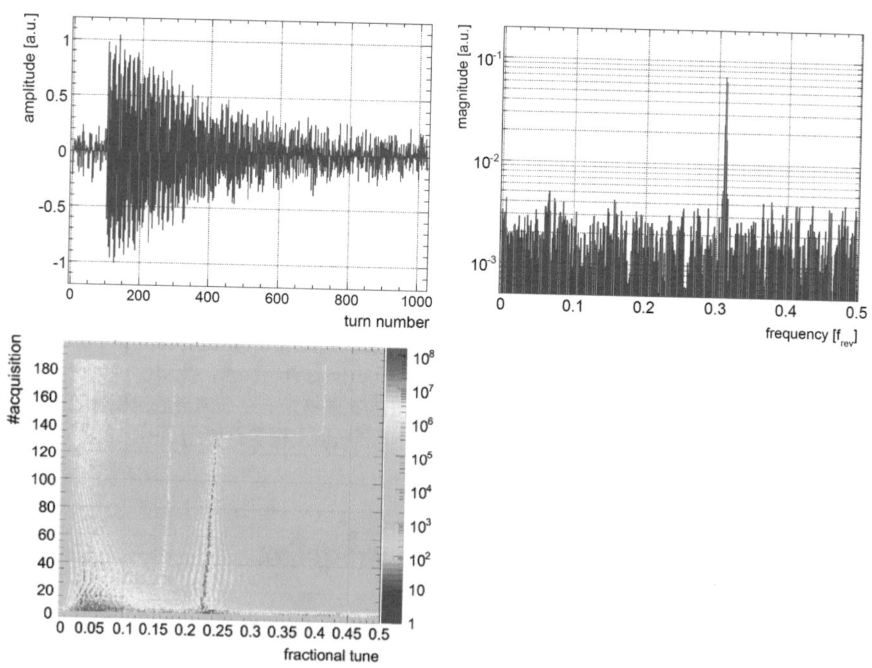

Fig. 8.62 Beam response to a single-turn kick excitation (top, left) and corresponding magnitude spectrum (top, right). Example of continuous tune measurement in the CERN Super Proton Synchrotron (left)

measurement. The advantage of this technique is that in addition to an amplitude response it also gives phase information. This makes it more sensitive than the single kick method and so allows smaller excitation amplitudes to be used.

To have a fully continuous measurement of the tune when no signal is visible due to external sources, a Phase Locked Loop (PLL) needs to be implemented. A voltage or numerically controlled oscillator (VCO or NCO) is used to put a sine wave excitation, $A\sin(\omega t)$, on the beam. The beam response to this signal is then observed using a pick-up and will be of the form $B\sin(\omega t+\phi)$, where ϕ is the phase difference between the excitation and the observed signal. A phase detector multiplies the excitation and observed signals together, giving a signal of the form $\frac{1}{2}AB\cos(-\phi) - \frac{1}{2}AB\cos(2\omega t+\phi)$ which has a DC component proportional to the cosine of the phase difference. This will therefore be zero when the phase difference is 90° and the amplitude response is a maximum, i.e. at the tune frequency. The aim of the PLL is to "lock-in" to this 90° phase difference between excitation and observed signal by correcting the VCO frequency until the DC component of the phase detector output is zero. Since the PLL will always try to maintain this 90° phase difference, the VCO frequency will track any tune changes, so giving a continuous tune measurement.

In practice things are not quite that simple. Many parameters have to be optimised in order to for the PLL to find, lock-in and subsequently track the tune peak. The beam spectra and dynamics also have to be well understood if the PLL is not to lock or jump to a spurious line, resonance, synchrotron sideband etc. In addition, for hadron machines, the continuous excitation will lead to some emittance blow-up. In order for this to be kept to a minimum the applied excitation has to be small and therefore the observation pick-up and following electronics very sensitive. This is less of a problem for lepton machines, where radiation damping takes care of any emittance blow-up caused by the excitation, making PLL systems much easier to implement on such machines.

8.6.3.2 Chromaticity Measurement

For any high energy synchrotron, the control of chromaticity is very important for beam stability and quality. If the chromaticity is of the wrong sign (positive below transition energy or negative above) then the beam quickly becomes unstable due to a head-tail instability. If the chromaticity is too large then the tune spread increases such that particles are inevitably lost as they hit resonance lines in tune space. The most common method of measuring the chromaticity of a circular machine is to measure the betatron tune as a function of the beam momentum and then to calculate the chromaticity from the resulting gradient. This is usually done by varying the RF frequency, keeping the magnetic field static. The equations of interest are:

$$\Delta Q = (\xi Q)\frac{\Delta p}{p} = Q'\frac{\Delta p}{p} = Q'\gamma_t^2\frac{\Delta R}{R} = Q'\left(\frac{-\gamma_t^2\gamma^2}{\gamma^2-\gamma_t^2}\right)\frac{\Delta f}{f}, \qquad (8.68)$$

where ΔQ is the change in tune, $\Delta p/p$ the momentum spread (or relative change in momentum), $\Delta R/R$ the relative change in average radius, $\Delta f/f$ the relative change in RF frequency and ξ the chromaticity. Please note that the chromaticity, ξ, is often expressed as $Q' = Q\xi$, where Q is the total betatron tune including the integer part. A typical chromaticity measurement therefore consists of performing a tune measurement for several different RF frequency settings. What is usually measured is the average change in closed orbit, due to dispersion, from which the relative change in radius, and hence momentum change can be calculated. The chromaticity is then simply the measured tune change divided by the applied momentum change.

In order to obtain a continuous chromaticity measurement the technique of RF modulation can be combined with PLL tune measurement [131].

8.6.3.3 Coupling Measurement

The control of coupling (the degree to which horizontal and vertical betatron motion is linked) is also important for circular accelerators. Excessive coupling will make tune and chromaticity measurements almost impossible, as the information from both planes are mixed-up in the observed signal. A direct measure for the total coupling coefficient $|C^-|$ can be obtained by the "closest tune approach" method. Both betatron tunes are measured during a linear quadrupole power converter ramp which crosses the values of the horizontal and vertical tunes. Any coupling present in the machine will introduce a forbidden band through which the tune cannot cross, instead jumping from one side to the other. The width of this band is given by the total coupling coefficient $|C^-|$.

In order to obtain a continuous measurement of coupling the PLL tune measurement can again be used, comparing the amplitude of the tune response in the plane of excitation with that of the other plane [132].

8.6.4 *Beam Profile Measurements*

Transverse beam distribution measurements are performed in all accelerators. As well as providing a direct measure of the beam quality, they can be used verify accelerator lattice functions and to calculate the emittance of particle beams. There are many ways to measure the beam size with the most commonly used techniques described in this section.

8.6.4.1 Secondary Emission Grids

Secondary Emission (SEM) Grids, also known as harps, consist of ribbons or wires which are placed in the beam [133]. See Fig. 8.63. As the beam intercepts the grid, secondary emission occurs leading to a current in each strip which is

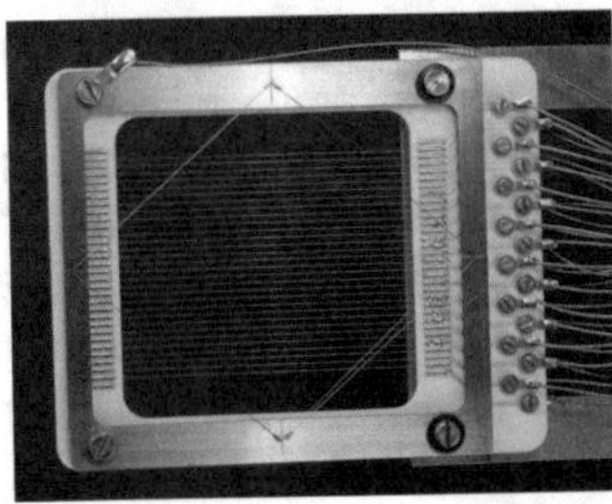

Fig. 8.63 Secondary emission grid and its insertion mechanism

proportional to the beam intensity at that location. By measuring this current for all strips a beam profile is obtained. SEM-grids are the most widely used means to measure the density profile of beams in transfer lines. In addition, sets of three, properly spaced grids (i.e. with the right phase advance between monitors), allow a direct determination of the complete emittance ellipse. What makes them popular is their simple and robust construction, the fact that there is little doubt about the measured distribution, and their high sensitivity, in particular at low energies and for ions. At higher energies they can be considered semi-transparent. Amongst their drawbacks are the limited spatial resolution (difficult to get the wire spacing much below 0.25 mm) and the rather high cost for the mechanisms and electronics.

8.6.4.2 Scintillator and Optical Transition Radiation Screens

Scintillator screens have been used for nearly a century and are the simplest and most convincing device when one has to thread a beam through any accelerator. The modern versions used in accelerator instrumentation applications [134] are typically based on an activated inorganic ceramic or crystal structures, such as chromium activated alumina (Al_2O_3) or Cerium doped YAG ($Y_3Al_5O_{12}$). These screens are directly inserted into the beam's path and can stand high intensities and large amounts of integrated charge. In its simplest form a graticuled screen is observed using a TV-camera. It can deliver a wealth of information to the eye of an experienced observer, but precautions need to be taken into account to extract quantitative measurements. Much can be done about that with modern means of rapid image treatment, but questions concerning the linearity of these screens at high beam densities remain.

Optical Transition Radiation (OTR) [135] screens are a cheap substitute for scintillator screens. OTR radiation is generated when a charged-particle beam transits the interface between two media with different dielectric constants (e.g. vacuum to metal or vice versa) [136]. Since this is a surface phenomenon, the screens can be made of very thin foils which reduces beam scattering and minimises heat deposition. The radiation produced when the beam enters the foil is emitted in a narrow cone around the angle of reflection for backward (vacuum to metal) OTR so that if the foil is placed at 45° to the beam direction, the radiation produced is at 90° to the beam direction. In addition, forward OTR (metal to vacuum) is also produced around the beam direction when the beam exits the foil. The angular distribution of the emitted radiation has a central hole and a peak located at $1/\gamma$. The higher the value of γ the sharper the peaks and the more light can be collected, which is why OTR is generally suited to lepton or high energy hadron machines. The imaging can again be performed using simple optics followed by a CCD camera. See Fig. 8.64. Recent studies have demonstrated that OTR imaging systems are capable of measuring a beam size below one micron [137, 138]. A non-invasive alternative to OTR screens has been developed using diffraction radiation from slits [139], which has also demonstrated good performance [140].

8.6.4.3 Wire Scanners

Of all the instruments used for measuring the emittance of circulating beams, wire-scanners are considered to be the most trustworthy. They come in two different types; rotative and linear. Rotative wire scanners can reach speeds of up to 20 ms^{-1} and consist of a thin wire (some tens of microns in diameter) mounted on a fork which is attached to a rotating motor, while linear scanners use motors which push/pull the wire across the beam. There are two ways of obtaining a beam profile with wire scanners; by measuring the secondary emission current in the wire as a function of wire position (similar to SEM-grid acquisition) or by measuring the flux of secondary particles created as the beam interacts with the wire. This latter technique is often used for high intensities, where the heating of the wire produces thermal emission which falsifies the secondary emission results. It relies on the use of radiation detectors, typically scintillators followed by photo-multipliers, placed downstream of the wire scanner to detect the γ-radiation and secondary particles produced when the wire intercepts the beam. To make the flux collected independent of the wire position may require the summation of the signals from two or more detectors positioned around the beam chamber. Fast wire scanners are nearly non-destructive over a wide range of energies. Their spatial resolution can reach the micrometer range [141] and, with fast gated electronics, the profiles of individual bunches can be observed. Their great sensitivity also allows them to be used for the study of beam halo.

For H^- and lepton accelerators the wire can be replaced by a laser beam, with the detection of stripped electrons [142] or Compton scattered [143] electrons respectively. This technique is ideally suited to high power machines, where a wire would not survive, or to machines where the transverse beam size is extremely small requiring a very thin ‘wire’ which is obtained by tightly focussing the laser beam.

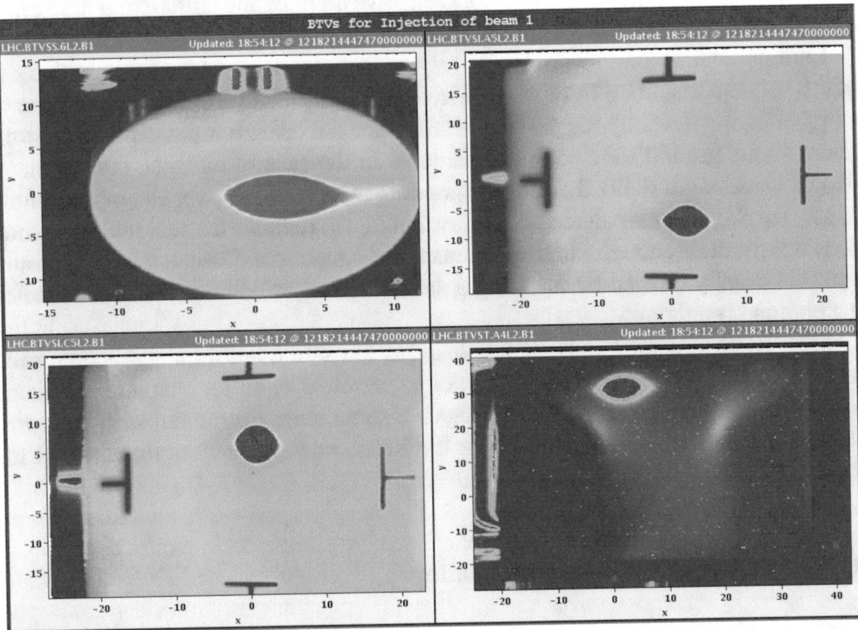

Fig. 8.64 (Left) Combined OTR and scintillator screen mechanism showing chamber in the out position and (bottom) an example of images obtained during the very first injection into the LHC

8.6.4.4 Residual Gas and Luminescence Monitors

Rest gas monitors are used in many high energy accelerators in order to reconstruct transverse beam distributions [144]. The signal results from the collection of either the ions or the electrons produced by the beam ionising the small amount of residual gas in the vacuum chamber. These ions or electrons are accelerated using a bias voltage of several kilovolts and typically collected on a micro channel plate

(MCP). The avalanche of electrons produced by the MCP are then collected on strip detectors or hit a phosphor screen to give an image of the beam profile that can be monitored using a CCD camera. Due to their rigidity, ions are less sensitive to the distorting effects of the space charge from the circulating beam, but their slow drift time, even with high bias voltages, means that they spend a long time in this beam field, making it difficult to analyse beam dimensions smaller than one millimetre. However, in order to use electrons to produce an image, a transverse magnetic field needs to be added around which the electrons spiral on their way to the MCP. This eliminates, to a large extent, the space charge effects of the beam and allows sharper images to be produced than with ions. This additional magnetic field is also seen by the beam and has to be compensated by two corrector magnets either side of the ionisation profile monitor. Direct detection of ionized particles using in-vacuum pixel sensors has recently been demonstrated as an alternative to MCP detection [145], providing a more sensitive monitor and eliminating the ageing effects associated with micro channel plates.

Luminescence monitors [144] also rely on the interaction of the beam with a gas in the vacuum chamber. In this case electrons are excited from the ground state to a higher energy level by the passing beam. Once the beam has passed the electrons return to the ground state and emit photons. In the case of nitrogen the dominant photon wavelength is 391.3 nm, corresponding to light at the lower end of the visible range, for which many detectors are available. In general, the residual gas alone does not produce enough photons for accurate imaging and hence a local pressure bump is usually created by injecting a small amount of additional gas to enhance the photon production.

Most users consider both the residual gas ionisation and luminescence profile monitors to be semi-quantitative and not to be relied upon for absolute emittance measurements, even after calibration against some other instrument such as a wire scanner. Their virtual transparency for the beam, however, makes them useful for the continuous on-line tracking of beam size.

8.6.4.5 Synchrotron Radiation Monitors

Synchrotron radiation monitors [146] are limited to highly relativistic particles and offer a completely non-destructive and continuous measurement of the 2-dimensional density distribution of the beam. These monitors make use of the light produced when highly relativistic particles are deflected by a magnetic field. They are therefore usually positioned to make use of parasitic light produced by a dipole magnet in the machine or behind a purpose built “undulator” magnet in which the beam is deflected several times to enhance the photon emission.

The most common way of measuring the beam size with synchrotron radiation is to directly image the extracted light using traditional optics and a camera. The spatial resolution for such systems is usually limited by diffraction and depth-of-field effects which can be overcome using interferometers [147]. If the beam is sufficiently relativistic then the photon emission extends into the hard X-ray region of the spectrum and X-ray detectors can be used [148], for which diffraction effects become less important.

8.6.5 *Beam Loss Monitoring*

Beam loss monitors (BLMs) have three main uses in particle accelerators: Damage prevention, diagnostics and machine optimisation [149]. The job of the BLM system is to establish the number of lost particles at a certain position within a specified time interval. Most BLM systems are mounted outside the vacuum chamber, so that the detector observes the shower caused by the lost particles interacting in the vacuum chamber walls or in the materials of the magnets. The number of detected particles and the signal from the BLM should be proportional to the number of lost particles. This proportionality depends on the position of the BLM with respect to the beam, the type of lost particles and the intervening material. It also, however, depends on the momentum of the lost particles, which may vary by a large amount during the acceleration cycle. One has to distinguish between two types of losses; fast losses, where a large amount of beam is lost in a short time (for circular machines) or distance (for LINACs and transport lines) and slow losses, where partial beam loss occurs over some time or distance. In storage-rings, the lifetime is defined by slow losses. Superconducting accelerators use the BLM system to prevent beam loss induced quenches. The fact that BLM systems have to cover both of these cases means that they are often required to function over a very large dynamic range, typically in the region of 10^4 to 10^6. BLMs can be classed into two main families: global—containing a few long detectors covering most of the accelerator; distributed—with many units of relatively small active area.

8.6.5.1 Global BLM Systems

The first global system was proposed by Panowsky for SLAC in 1963 [150] and was composed of a 3.5 km hollow coaxial cable filled with Ar (95%) + CO_2 (5%), mounted on the ceiling of the LINAC, about 2 m from the beam. When a beam loss occurs the secondary particles produced traverse the monitor and induce an electrical signal which propagates to both ends of the cable. Position sensitivity is achieved by comparing the time delay between the direct pulse from one end and the reflected pulse from the other. The time resolution achievable with such systems is about 30 ns (~8 m), which can be reduced to about 5 ns by using shorter cables. This principle of space resolution works for relativistic linear accelerators and transport lines with a bunch train much shorter than the machine. For particles travelling significantly slower than the signal in the cable the resolution of multiple hits in the cable becomes difficult. In this case, and for circular machines, it is necessary to split the cable. Each segment has to be read out separately, with a spatial resolution which becomes approximately equal to their length. A similar measurement can these days be achieved using Cherenkov or scintillation light created in long optical fibres [151]. The fast response of the Cherenkov signal is detected using photomultipliers at the ends of the fibre, allowing a longitudinal position resolution of 20cm to be achieved.

8.6.5.2 Distributed BLM Systems

By far the most common, distributed BLM systems are comprised of many individual beam loss monitors located throughout the accelerator, each acquired in parallel. The active medium in these systems is tailored to the type of loss to be measured. Ionisation chambers are commonly used due to their robustness and large dynamic range (see e.g. [152]). The chamber provides some medium with which the secondary particles created by the beam loss can interact, typically a gas such as nitrogen or argon. This interaction produces electron-ion pairs that are collected by a series of high voltage gaps along the length of the chamber. The resulting current is then measured and is proportional to the beam loss at the location of the monitor.

The drawback of the ionisation chamber is its relatively slow response time, dominated by the drift velocity of the ions created during a beam loss. In cases where a fast response is required scintillation counters are often employed. These come in plastic or liquid form and are read-out using a photomultiplier. The disadvantages of these systems are the susceptibility of plastic scintillators to radiation damage and long term drift of photomultiplier gain. It is also possible to directly use the photomultiplier as a detector by replacing the photocathode with an aluminium foil that works as a secondary electron emitter when irradiated. Known as an Aluminum Cathode Electron Multiplier (ACEM) it gives response times in the nanosecond regime [153]. Diamond detectors, consisting of a biased monocrystalline or polycrystalline diamond wafer with an active area of a few cm^2, have also recently been demonstrated to give nanosecond response times while being relatively radiation resistant [154].

In order to distinguish beam losses in accelerators where there are sources of background ionising radiation such as X-rays or synchrotron radiation, Cherenkov or PIN photodiode detectors can be used. Cherenkov detectors are useful as the threshold for light production is above the typical Compton-electron energies produced by such background radiation, while back to back PIN photodiodes can be used to distinguish between the hadronic shower created by beam losses and synchrotron radiation [155]. In the latter case only the charged particles will interact with both photodiodes, giving a coincidence signal, with the photons absorbed by the first diode. In contrast to the charge detection of most other BLM systems, PIN photodiode detection depends on counting coincidences, with the count rate proportional to the loss rate so long as the number of overlapping coincidences is small.

8.6.6 *Short Bunch Length Diagnostics*

Recent developments for Free Electron Laser (FEL) accelerators and studies for linear colliders in the TeV range are driving the typical bunch length of accelerated lepton beams down to the 10–100 femtosecond level. To measure the longitudinal behaviour of such beams is not trivial and a lot research has been carried out over the last decade to develop instruments that allow their characterisation. They broadly fall into three categories: direct beam observation; coherent radiation techniques; radio-frequency and electro-optic sampling techniques.

8.6.6.1 Direct Beam Observation

Direct Beam Observation, as its name suggests, relies on the direct measurement of longitudinal beam structure by means of fast detectors. Wall Current Monitors followed by fast sampling oscilloscopes are able to probe time structures down to well below the nanosecond, while streak cameras can reach some 200 femtoseconds [156]. The latter relies on analysing an optical replica of the longitudinal bunch distribution obtained by means of synchrotron, transition, diffraction or Cherenkov radiation.

8.6.6.2 Coherent Radiation

For wavelengths shorter than the bunch length, the particles within the bunch radiate incoherently, with the power emitted proportional to the number of particles. However, for wavelengths equal to or longer than the bunch length, the particles emit radiation in a coherent way with the emitted power dependent on the bunch length and scaling as the square of the number of particles.

$$S(\omega) = S_p(\omega)\,[\underbrace{N}_{\text{incoherent term}} + \underbrace{N(N-1)\,F(\omega)}_{\text{coherent term}}]\,,\quad F(\omega) = \left|\int_{-\infty}^{\infty} \rho(s) e^{-i\frac{\omega}{c}s} ds\right|^2 \tag{8.69}$$

where $S(\omega)$ represents the radiation spectrum, $S_p(\omega)$ the single particle spectrum, N the number of particles and $F(\omega)$ the longitudinal bunch form factor, which depends on the longitudinal particle distribution $\rho(s)$.

Measuring the power spectrum therefore allows the form factor to be calculated from which an indirect estimate of the bunch length is possible. This method is relatively simple to implement and has already demonstrated its capacity for the measurement of extremely short bunches [157].

8.6.6.3 Radio-Frequency and Electro-optic Sampling Techniques

These techniques rely on encoding the longitudinal bunch length information into spatial or optical information. Radio-frequency manipulation makes use of a transverse RF deflecting cavity [158] to kick the head of the bunch in one direction and the tail of the bunch in the opposite direction. By measuring the transverse beam size downstream, it is then possible to reconstruct the longitudinal bunch profile. This technique is the standard method for measurement of femtosecond bunches, with the drawbacks being its destructive nature and cost.

Electro-optic sampling [159] (Fig. 8.65) is an alternative to the RF technique and relies on converting the coulomb field of the bunch into an optical intensity variation by means of an electro-optic crystal. The crystal is placed close to the beam so that the charged particles in the bunch induce a polarization change of a laser beam passing through the crystal at the same time. Three main types of encoding are in common use; spectral, temporal and spatial. In spectral decoding the beam passes the electro-optic crystal at the same time as a chirped wavelength laser pulse. The longitudinal distribution of the bunch is hence encoded as an optical power variation at different wavelengths in the chirped laser pulse. A spectrometer is used to separate out the wavelengths and so reconstruct the longitudinal profile. Temporal decoding [160], works in much the same way as spectral decoding, with the spectrometer replaced by a non-linear crystal. A femtosecond laser pulse is mixed with the chirped laser pulse in this crystal to produce a frequency doubled replica of the longitudinal distribution. Spectral encoding can characterise bunch distributions down to the picoseconds level, while temporal encoding works down to

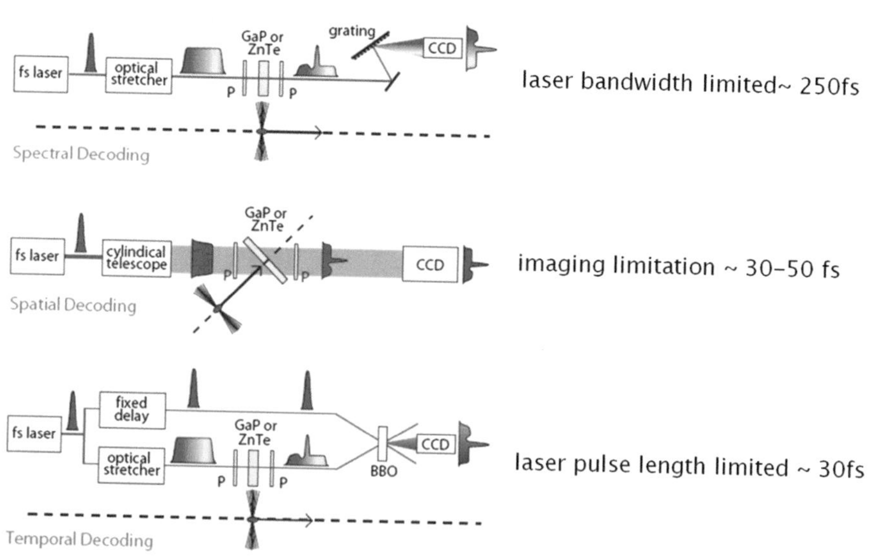

Fig. 8.65 Electro-optic sampling techniques for the measurement of short bunch lengths

some 100 femtoseconds. Spatial decoding [161] uses a short laser pulse that crosses the crystal at an angle such that when it interferes with the beam signal, the temporal profile of the electron bunch is encoded onto the transverse profile of the laser, i.e. mapping time into space.

Such techniques all rely on the use of very short laser pulses and precise laser to beam synchronization.

8.7 Injection and Extraction Related Hardware: Kickers and Septa

M. J. Barnes · J. Borburgh · V. Mertens

This chapter describes the main hardware elements involved in single-turn injection and fast extraction, as well as resonant slow extraction: fast pulsed systems (kickers) and magnetic and electrostatic septa. For a description of the beam processes involved reference is made to Sect. 6.3. For space reasons other associated components, such as stripping foils used for multi-turn injection, injection and extraction related protection systems, or specific beam instrumentation, are not discussed here. For certain of these components information can be found in the corresponding chapters of this volume.

Injection and extraction regions are particularly challenging areas in accelerators in several respects. The space allocated for these parts, which are usually located in straight sections, is naturally rather limited so that the space available for bending sections (in the case of a circular machine) or accelerating structures (in the case of a linear accelerator) can be maximized, and with it the achievable energy for a given circumference or length of the accelerator. As a result injection and extraction related elements are often required to be as physically short and electrically or magnetically strong as technically feasible (yet remaining reasonable in cost). Although magnetic kickers and septa are dipole magnets, following the same principles as ordinary bending elements, they have very distinct features and must often fulfil conflicting design requirements. They are typically purpose-built single elements or only produced in a small series. Many of these elements are installed under vacuum, with the resulting implications.

Figure 8.66 shows an example of fast single-turn injection in one plane [162, 163]. The injected beam passes through the homogeneous field region (aperture) of the septum: circulating beam is in the field-free region. The septum deflects the injected beam onto the closed orbit at the centre of the kicker magnet; the kicker magnet compensates the remaining angle. The septum and kicker are either side of a quadrupole (defocusing in the injection plane) which provides some of the required deflection and minimizes the required strength of the kicker magnet.

Magnetic septa usually have high current densities (in the order of 100 A/mm^2); a short interruption in the cooling can lead to their destruction. Electrostatic septa

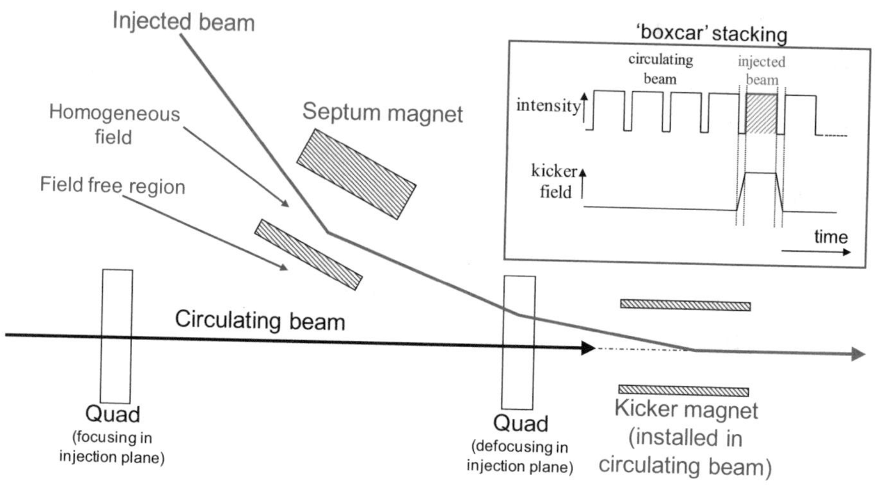

Fig. 8.66 Fast single-turn injection in one plane

are operated with very high fields and must also cope with the heat load from the impinging beam. To reduce the particle losses septa (coils, or foils/wires in case of electrostatic septa) must be as thin as possible; nevertheless, the characteristics of the beam processes involved mean that septa are usually among the most radioactive parts of an accelerator, together with collimators and dumps. This has to be taken into account during the design, for example, it must be possible to quickly assemble and release electrical, water and vacuum connections.

The rise time of the field produced by the kicker magnets must be very short (ranging from several tens of μs down to several ns), while the requirements for low ripple and good reproducibility of the flat-top amplitude are stringent (at least considering the pulse-type excitation). Redundancy can also be an important design criterion: if one of the devices fails the remaining devices should be sufficient, in combination with special protection elements, to handle the beam safely. For single-turn injection or extraction there is only one action on the beam and the timing needs to be precisely set up; although it may be feasible to compensate for small, non-ideal, deflection of the first and last bunches of a train [164], significant online corrections are not possible. Applications are often safety critical (beam abort systems) and imminent failures must be recognized in real time.

8.7.1 Fast Pulsed Systems (Kickers)

Apart from their use for injection and extraction purposes specific kickers are also used to excite the beam, for instance to measure the tune or probe the machine aperture. Kicker magnet yokes are frequently made out of NiZn ferrite because of its relatively high resistivity and magnetic permeability: with the NiZn ferrite field rise can track current rise to within ~1 ns [165], while slower kickers can use laminated steel yokes. Fast rise times are particularly required in smaller machines to optimize the filling factor (amount of beam which can be injected). Useable pulse durations are in the range of a few 100 ns to 100 μs (case of the CERN LHC beam dump kicker which must empty the whole 27 km circumference at once; here the fall time is not important as all beam is anyway extracted after one turn). Even faster rise times, in the range of a few ns, can be achieved with yoke-free stripline kickers, where the field is created by a pulse running along two longitudinal electrodes [166, 167].

The main design options for kicker systems include the type of magnet ("travelling wave" (also known as "transmission line") or "lumped inductance"), the installation of the magnet in- or outside of the machine vacuum, the type of aperture (window-frame or C-core), and the type of termination (impedance-matched or short-circuited) [162]. The actual choice depends on the demands and constraints of the application, but is also impacted by cost considerations. Applications where a considerable number of bunches must be placed at a precise position on the circumference of an accelerator, without disturbing the already circulating bunches, or where only a part of the circulating beam needs to be extracted, requires kickers with fast rise and fall time. Low flat-top ripple reduces the injection oscillations which could lead to emittance blow-up (in hadron machines), and ensures constant deflection angle for a train of extracted bunches. To achieve fast rise and fall times and low flat-top ripple, quasi-rectangular pulses are applied to "transmission line" kicker magnets. For space reasons most of the discussion below will be devoted to such systems.

A simplified block diagram of the power circuit is given in Fig. 8.67. Shortly prior to the required kicker pulse a resonant charging power supply (RCPS) charges either a high voltage coaxial cable (also known as a Pulse Forming Line (PFL)) or a Pulse Forming Network (PFN); the choice between a PFL and a PFN is discussed below. At the appropriate time the main switch (thyratron or semiconductor) is turned on (closed) and the pulse current is sent through a coaxial transmission line into the kicker magnet. In the schematic of Fig. 8.67, once the current pulse has passed through the magnet the energy is absorbed in a terminating resistor. If for any reason the pulse current must not be sent to the kicker magnet (for instance if the accelerator turns out to be no longer ready to receive beam or the kicker system exhibits an internal fault, e.g. a charging voltage not matching the beam energy), the dump switch is turned on and the PFN is thus discharged. The dump switch is also turned on when the pulse duration, in the kicker magnet, needs to

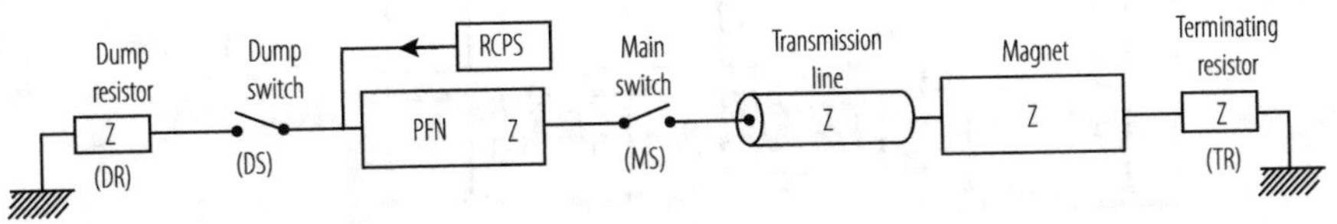

Fig. 8.67 Simplified block diagram of a typical transmission line kicker system

be reduced. An additional “clipper switch” can be used to connect the magnet side of the PFN to ground in order to improve the fall time. The main components of the electrical circuit are all matched to a characteristic impedance Z (typically in the range of a few Ohms to a few tens of Ohms) to avoid unwanted reflections and thus distortions of the pulse shape. The interplay of the different components is therefore important and a carefully chosen set of parameters and a coherent design are required to achieve optimum performance. Computer aided design tools (electric circuit simulation and finite element analysis software) are used extensively to predict and verify the performance achieved by the hardware, or to study how an existing system can be improved [162]. The main components of a kicker system are presented in more detail below.

8.7.1.1 Kicker Magnets

A transmission line magnet consists of typically between 5 and 30 cells (Fig. 8.68) to approximate the behaviour of a coaxial cable. The cells consist of either a sequence of metallic plates which are alternately connected to the high voltage conductor and ground, thus forming capacitors, or commercial capacitors. Ferrite cores are sandwiched between the high voltage plates (Fig. 8.69).

After general considerations, such as available space, deflection angle, rise time, flat-top ripple, and fault tolerance, the system design usually commences with the choice of the charging voltage (V) and the characteristic impedance (Z) of the system. The impedance is given by $Z = (L_c/C_c)^{1/2}$, where L_c and C_c are the inductance and capacitance of a cell, respectively (Fig. 8.68). The choice of Z depends on the required field rise time and the space available. The field rise time is given approximately by the sum of the current pulse rise time, at the terminating resistor, and the fill time of the magnet [162]. The fill time (t_m) of a kicker magnet, terminated in its characteristic impedance, is given by $t_m = L_m/Z$; to achieve a short fill time the series inductance of the magnet, L_m, should be small and the impedance rather high. In addition high impedance provides for a relatively high cut-off frequency [162] which in turn helps achieve a low flat-top ripple. In the case of a C-core kicker magnet the high voltage conductor usually consists of a single straight bar (single turn) which helps to ensure the low inductance. In

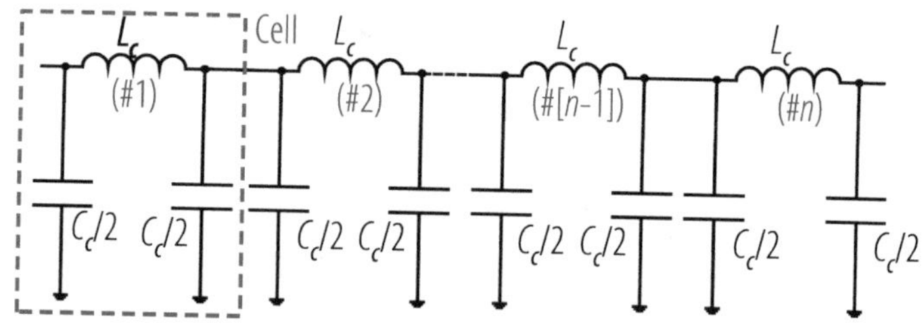

Fig. 8.68 Simplified equivalent electric circuit of a transmission line kicker magnet

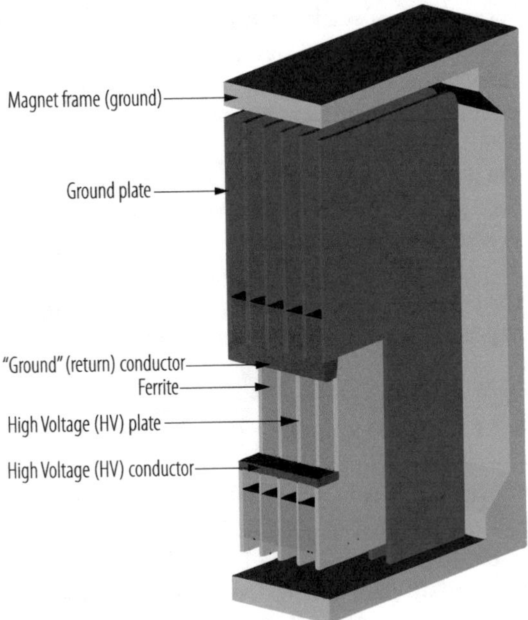

Fig. 8.69 Cut-away of a transmission line kicker magnet

practice impedances up to that of commercially available coaxial high voltage cables are used (50 Ω), within the boundaries dictated by the available length for the kicker magnets and the chosen PFN voltage. The lower the impedance needs to be the greater must be the cell capacitance. In magnets using vacuum as dielectric, high cell capacitance requires that either the plate surfaces are very large (which would be impractical for space reasons) or the separation between high voltage and ground plates is small (however the distance must not be made so small so as to risk electrical breakdown). Therefore in some cases capacitor media with higher permittivity must be used, e.g. ceramic capacitors, as for instance is done for the LHC injection kickers.

A design where the whole kicker magnet is under vacuum allows for the minimum size of magnetic aperture; for out-of-vacuum magnets the magnetic aperture must also enclose a vacuum chamber. For a larger aperture the current must be increased to ensure that, for a given impedance, the same magnetic flux is achieved, which in turn implies an increase of the PFN voltage [162]. Hence, for a given current, under-vacuum magnets can be of a shorter length than the equivalent out-of-vacuum magnet, but also have some disadvantages. The magnet materials must be compatible with the vacuum requirements and the whole construction must allow for bake-out and resulting thermal expansion. In addition a vacuum tank and pumps are required, and the repair in case of magnet or vacuum equipment failures is usually very time consuming.

The termination of the kicker magnet with a resistor (Fig. 8.67) avoids unwanted reflections and ensures a fast filling time of the magnet. The resistor is usually formed by a stack of high power ceramic-carbon resistor disks housed in a coaxial metallic enclosure, which is the return path of the current, to minimize the inductance and thus allow good matching over a wide range of frequencies. For cooling and insulation purposes the resistors are usually immersed in oil. Where available length for the kicker magnet has priority over rise time, a short circuit at the pulse output end of the magnet has the advantage of doubling the kick strength, for a given PFN voltage and system impedance. However the fill time of the magnet is also doubled; to establish full field in the magnet the pulse has to travel to the magnet output and then reflect back to the driven end of the magnet. In addition, for a magnet terminated in a short-circuit, during the field pulse the magnet also experiences both polarities of voltage [162].

8.7.1.2 Beam Coupling Impedance

Kicker magnets with ferrite or laminated steel yokes, directly exposed to the beam, can contribute a major fraction of the longitudinal and transverse beam coupling impedance of a machine [168]. The passage of the beam, without providing an appropriate path for the beam image current, can also cause resonances within the volume of the tank and lead to higher order mode losses. Since these effects can results in beam instabilities their reduction is becoming increasingly important with rising beam intensity. Electromagnetic coupling with high-intensity beams also leads to heating of the ferrites, sometimes beyond their Curie temperature.

Where needed and possible these issues are being addressed in existing applications by various retrofitting techniques, thus avoiding a costly redesign of the whole kicker system and possible implicit layout changes. The kicker magnet vacuum tank can be shielded from the beam by incorporating metallic transition pieces between the tank flanges and the kicker magnet [169] and between subsequent magnet modules in a common tank, thus minimising changes in aperture dimensions and avoiding some resonances.

To reduce the temperature rise of the ferrites, due to beam induced heating, indirect water cooling has been implemented with electrically insulating but thermally conducting Aluminium-Nitride plates in contact with the ferrites [170]. Particular precautions must be taken here because of the presence of the high voltage. To reduce the coupling of the beam with the ferrite yoke, without loss of aperture, two comb structures of silver paint stripes can be serigraphed onto a ferrite (Fig. 8.70, left) [171]. Each of the two sets of stripes is connected to a high voltage plate; the capacitive coupling between the two sets of stripes, which is enhanced by the permittivity of the ferrite, provides the path for the beam image current, without creating eddy current loops which would adversely impact the field rise time. Nevertheless the overlapping length of the two serigraphed combs should be optimized to avoid resonances at undesirable frequencies [172]. The serigraphy technique is less attractive for short cells because of both high voltage breakdown issues and the relatively small capacitive coupling between the two sets of stripes.

Fig. 8.70 Serigraphed ferrite yoke of the CERN SPS fast extraction kickers (left); ceramic tube with wires forming the beam screen of the LHC injection kickers (right)

In new designs, where the beam impedance is taken into account from the beginning of the design, or where the available aperture permits, a proper beam screen can be inserted between the magnet yoke and the beam. In the LHC injection kickers a 3 m long ceramic tube serves as support for up to 24 Nickel-Chrome (80/20) wires lodged in grooves around its inner circumference (Fig. 8.70, right). The tube also provides insulation between the wires and the ferrites and high voltage plates, but has no vacuum function. On one end the wires are directly connected to the beam pipe, on the other end through a capacitive coupling with a grounded metallization on the outside of the tube [173]. The LHC extraction kicker magnets, which have a rise time of approximately 3 μs, use a 1.4 m long ceramic vacuum chamber with a 3 μm thin uniform Ti layer on the inside which acts as beam screen [174]. To note that beam screening techniques have equally to be compatible with the particular demands from ultra-high vacuum, nearby high voltage and acceptably low secondary electron yield [175].

For certain applications, e.g. beam extraction, the problem of beam impedance and screening can be diminished or to some extent circumvented if the beam circulates most of the time outside of the kicker, and is only bumped into the magnetic aperture of an open C-core kicker shortly (few ms) before the kicker is used [176]. The advantage of such a configuration is that the kicker aperture can be smaller, because of the beam shrinkage during acceleration, than if the beam passes through its aperture right from injection. However the smaller aperture may increase beam coupling impedance again; detailed studies are required to find the best trade-off.

8.7.1.3 Pulse Generation and Forming

The simplest pulse forming circuit consists of a coaxial cable (a pulse forming line, PFL) charged to twice the required pulse voltage. A PFL provides fast and low-ripple pulses but requires low dispersion and attenuation, especially for longer pulses, to keep droop and "cable tail" within specification. Since semiconductor materials in the cable (used to improve the breakdown voltage) increase the attenuation and dispersion, SF6 pressurized polyethylene (PE) cables have historically been used for PFL voltages above 50 kV. The required length and cost of PFL cable, as well as the attenuation and dispersion, limits the practical application to pulse durations of less than approximately 3 μs.

Where long pulses with low droop are required, a PFN is used which acts as artificial coaxial cable. The PFN consists of a series of "cells" forming a delay line. There are many configurations for PFNs [177] but a commonly used design is where each cell consists of a series inductor (sometimes with a parallel damping resistor) and a capacitor connected to ground (Fig. 8.71, left). The precision of the mechanical and electrical dimensions is important for a good pulse shape, i.e. short rise and fall times with low over- and undershoot, and low ripple. To preserve the precise inductance and to avoid time consuming adjustment by hand, longer coils are often wound onto an insulating support (Fig. 8.71, right) [178]. To obtain good matching and thus low ripple, two parallel lines can be used (Fig. 8.71): in addition, to compensate for conduction losses along the coil, the values of the capacitors are selected prior to assembly. The first cell and last cell of a PFN are mechanically adjustable, to allow catering for as-built imperfections which cannot be simulated with precision. The whole assembly is normally immersed in a dielectric liquid (polydimethylsiloxane or ester oil) which permits a more compact design than with air insulation, and which serves also as coolant.

Prior to turning on the main switch, to send the pulse to the kicker magnet, the PFN is charged to the voltage V. The charging components are usually connected at the dump switch end of the PFN (Fig. 8.67). When the main switch closes a pulse with amplitude of $V/2$ is sent through the transmission line to the kicker magnet and a voltage pulse of $-V/2$ propagates from the main switch end of the PFN towards

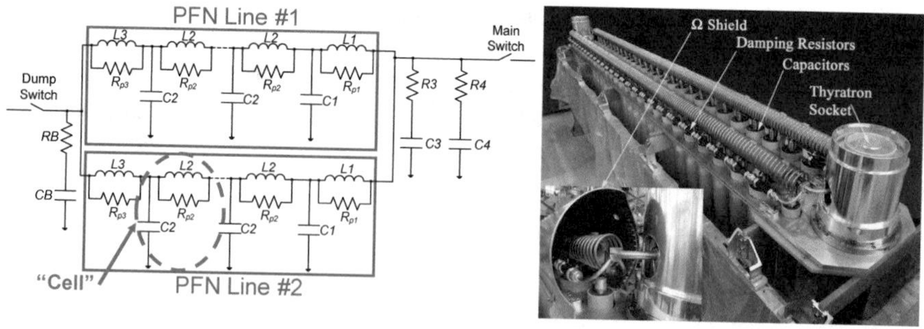

Fig. 8.71 PFN of the LHC injection kickers, consisting of two parallel 10 Ohm lines

the dump switch end. If the kicker magnet is terminated in its own characteristic impedance there is no reflection of the pulse which propagated through the magnet. However if the kicker magnet is short circuited at its output the pulse which propagated through the kicker magnet is reflected and travels backwards through the magnet, doubling the current and hence the field in the magnet; the fill time is in this case also increased by a factor 2.

8.7.1.4 Power Switching

Various technologies are used for power switching [177]. Thyratrons are still commonly used for fast, high power, switching applications. These deuterium filled gas tubes are characterized by a short turn-on time, high rate of rise of current and generally good reliability. They need however careful tuning and supervision of parameters such as reservoir voltage and grid bias voltages, as well as requiring occasional HV conditioning, to ensure optimum performance. Thyratron faults include failure to become conducting when required (so called "missing" pulses), or on the contrary becoming conducting before being triggered (so called "erratic" turn-on). High power thyratrons are also relatively expensive (up to several 10 k$ each) and have a limited lifetime which is strongly determined by the conducted charge (typically several 100 thousand pulses for very high current and long duration (tens of μs) pulses, to 100 million pulses for lower current and shorter pulses). To lower the risk of erratic turn-on the anode-cathode voltage is only applied to the thyratron a short time (typically few ms) before the kick is required. Although not many companies produce thyratrons anymore they are still widely used in accelerators.

For many years high power semiconductors have been used as switches in those kicker applications which do not require the high rate of rise of current of thyratrons. Semiconductors have the advantage of requiring little or no maintenance in comparison with gas tubes and they have a less critical dependence on the anode-cathode voltage, but generally have a lower rate of rise of current rating. An example where these switches are preferred are the LHC extraction (dump) kickers: the associated PFNs need to be charged over hours and the switches must therefore withstand the high voltage over extended periods without erratic turn-on. Semiconductors used for power switching in kicker systems include GTOs (Gate Turn-off Thyristors, which have been improved for fast switch-on; they are also called Fast High Current Thyristors (FHCTs)), IGBTs (Insulated Gate Bipolar Transistors), and MOSFETs (Metal-Oxide Semiconductor Field-Effect Transistors).

Although in principle semiconductors require no maintenance and have a very long lifetime when used in normal conditions, their application in accelerators is not completely straightforward. Since the typical breakdown voltage of a representative GTO is in the order of a few kV, the high voltages used in kicker systems requires the assembly of the semiconductors in series stacks (in practice up to 10–15 per stack, including redundancy) (Fig. 8.72). The distribution of the total voltage within the stack of series GTOs and the synchronous injection of a sufficiently high trigger current into the gates are important, to avoid excessive voltage across individual GTOs which could lead to unwanted conduction of the whole stack ("erratic") or destruction of the power semiconductors. It is noteworthy that such assemblies are often operated near or even beyond their normal specification limits; this is to achieve suitable high current with adequately fast rise time. In areas with a high background radiation (either from activated ambient materials or from beam losses) the radiation effects can be cumulative, with relatively slow deterioration of semiconductor performance, and/or sudden malfunction or failure [179]. Protection against particle-triggered breakdowns can be important. For installations in which the turn-on delay is critical, temperature stabilization of the semiconductors (usually through cooling of the surrounding air) is also an issue. High power GTOs used for the LHC extraction kicker systems (Fig. 8.72) have turn-on delay times in the order of 300 ns, a hold-off voltage of 2.5 kV per GTO for d.c. operation, a current rate of rise capability of 20 kA/μs, and conduct a current with an amplitude of 30 kA.

To provide redundancy in critical applications, two switches can be installed in parallel for each pulse generator, as for example in the LHC extraction kickers [174]. It should be noted, however, that putting switches in parallel has the disadvantage of increasing the risk of unwanted erratic turn-on. Other switch configurations include for example an ignitron in parallel with a thyratron, where the faster thyratron ensures the rapid and precise start of the pulse while the sturdier but somewhat slower ignitron takes over the bulk of the current [180]. Ignitrons are being phased-out from existing applications due to their mercury content, and resulting safety and environmental risk.

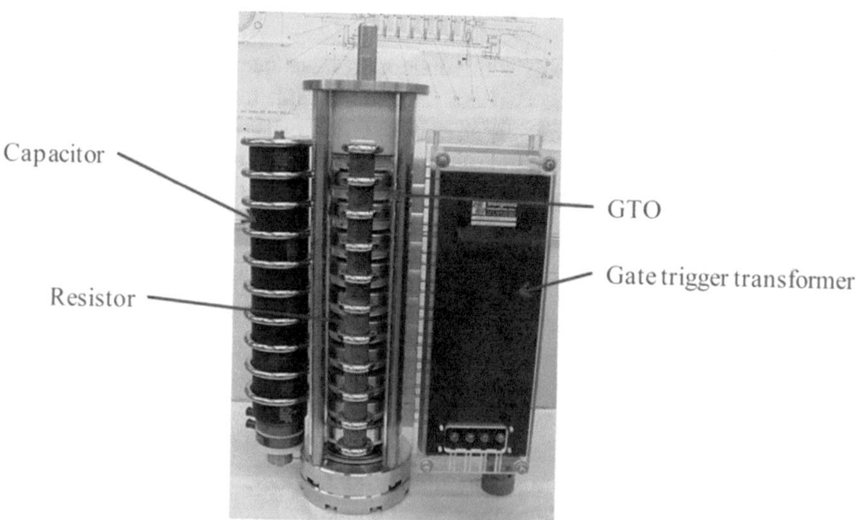

Fig. 8.72 Semiconductor switch stack of the LHC extraction kickers, together with voltage grading networks and gate trigger transformer

Topologies such as the inductive adder [181–183] and Marx generator [184, 185], utilizing fast and high-voltage semiconductor switches such as MOSFETs, are being actively pursued by CERN as possible alternatives to both thyratron and PFL technologies. The inductive adder and Marx generator use the series and parallel connection of power semiconductor switches to achieve high pulse power designs. The required energy is stored in capacitors. The MOSFET semiconductor switches have the ability to both turn on and turn off the current pulse.

Other types of switches for particular applications include so called "pseudo spark gaps", fast high-voltage MOSFETS and Fast Ionisation Dynistors (FID, also called Fast ionization Devices) [162]. With advances in semiconductor technology new, interesting, types of devices or switches with improved characteristics are expected to become commercially available.

8.7.1.5 Other Types of Circuits

In applications where only a few bunches need to be injected or extracted (e.g. LEP), fast deflection is most economically achieved by an oscillation system using half-sine wave pulses, where a capacitor is discharged into a purely inductive magnet terminated in a short circuit [186]. In this case no particular PFN is needed and the magnet does not need to be fitted with matching capacitors, permitting important economies. The charging voltage is half that for a travelling wave kicker system, with the same magnetic parameters, which often allows working with more economic air insulation in the pulse generator.

Capacitive discharge systems complemented by free-wheel circuits are mainly used for applications such as beam abort systems where a fast rise of the magnet pulse is required which can be followed by a long exponential decay (since all beam has been dumped after one turn), while the requirements of the flat-top are not very stringent. This is the case for instance for the LHC extraction kickers [174]. Here a capacitor is discharged into an inductive magnet with a stack of diodes in parallel. During the sinusoidal rise of the current the diodes are non-conducting, but when the magnet current pulse starts to reduce in magnitude the diodes conduct and the magnet current free-wheels in the low impedance circuit.

Where very fast rise times are required (order of few ns) stripline kickers are used [166, 167, 187]. These consist essentially of electrodes parallel to the beam direction, and contain no magnetic material. The deflection is provided by the electric and magnetic field between the plates. Generators with fast semiconductor (MOSFET) switches are typically used and sub-nanosecond jitter can be achieved. For an ultra-relativistic beam, the beam is deflected only if it is travelling from the terminated to the driven end of the striplines [188]. A schematic view of a stripline kicker system, for the tail clipper of the CTF3 facility, is given in Fig. 8.73.

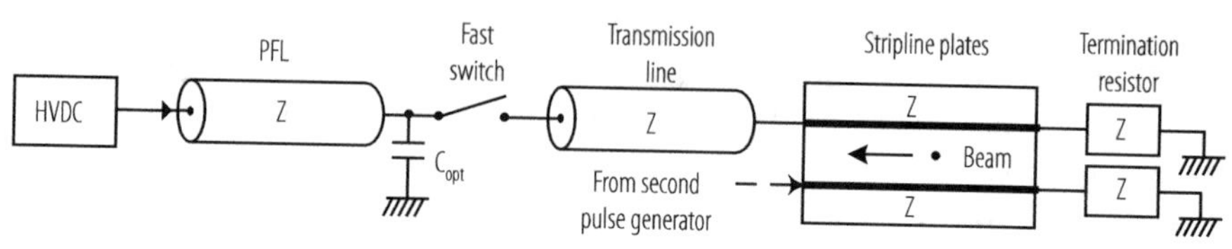

Fig. 8.73 Schematic view of a stripline kicker system

8.7.1.6 Electronics and Controls

The electronics and controls for a fast pulsed system is an important, complex and voluminous part of the overall system. They must be designed from the outset together with the power circuit. Typically they comprise systems for synchronization with the accelerators (mainly the radiofrequency and general timing system), systems to generate and distribute the timing across the system (including fine delays), and power triggers to drive the high voltage switches. A beam energy tracking system checks that the charging PFN/PFL voltage corresponds to the present energy of the accelerator. Fast interlocks, waveform acquisition and analysis hardware and software, as well as comprehensive software to monitor, follow-up and document the performance of the overall system complete the installation.

8.7.2 Electrostatic and Magnetic Septa

A septum, either electrostatic or magnetic, constitutes the separation between an area of ideally zero field, which is traversed by the circulating beam, from an area with high field in which the injected or extracted beam experiences a deflection. The septum should be as thin as possible to reduce particle losses and thus irradiation of the septum and the surroundings. If used in combination with a kicker (case of single-turn injection or extraction, Fig. 8.66) it serves also to keep the strength requirements for the kicker at an affordable level. Since the beam passes only once through the high field area of a septum its field homogeneity is not as critical as for normal bending magnets, however the stray field in the “field-free” region must be minimized, since the circulating beam experiences it at every passage. If the field homogeneity in the aperture is an issue, appropriate measures can help to improve it (e.g. deflectors, end plates).

8.7.2.1 Electrostatic Septa

Electrostatic septa [163, 189] for extracting beam provide an electric field in the direction of extraction, by applying a voltage between the septum and an electrode. Figure 8.74 left shows a schematic cross section of an electrostatic septum device.

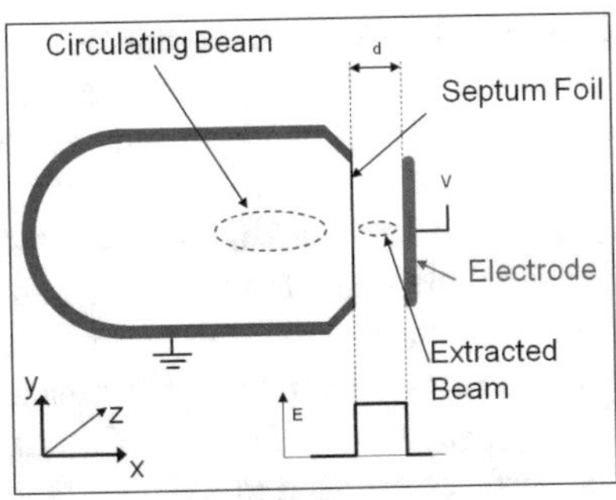

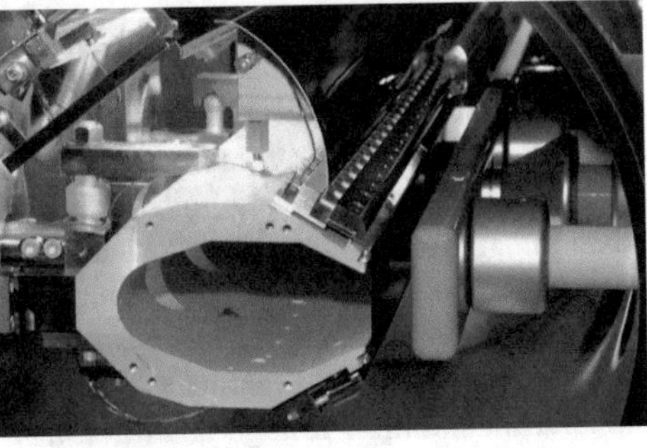

Fig. 8.74 Schematic of an electrostatic septum (left); CERN PS foil septum (right)

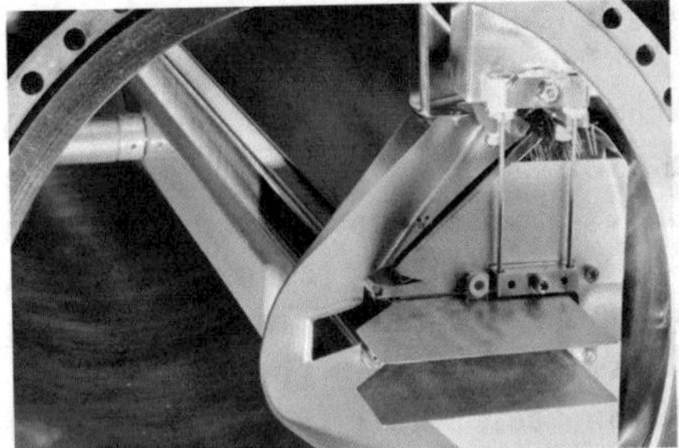

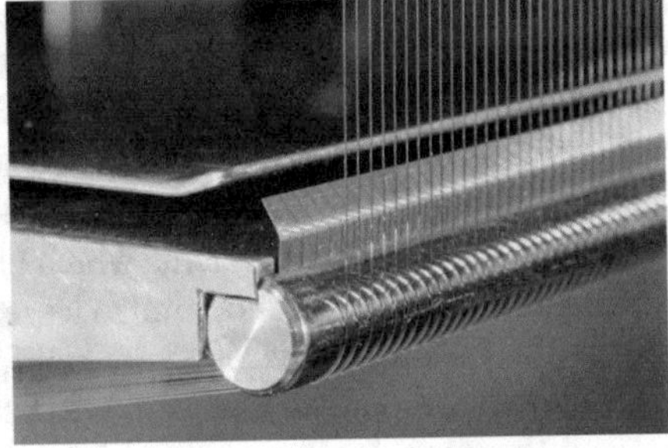

Fig. 8.75 Electrostatic wire septum for the CERN SPS (left); zoom on wires with bottom ion trap electrode (right)

The septum and its support structure are marked in blue; they form an electrode which is in most cases, for practical reasons, at ground potential. The opposite electrode is marked in red. In the lower part of the figure the electric field E is shown as it could be measured along the x-axis. The field in the gap between the septum and the electrode is homogeneous on the axis and equal to $E = V/d$, where V is the voltage applied to the electrode and d the distance (gap) between the septum and the electrode. As shown in Fig. 8.74 the orbiting beam passes through the hollow support structure which is a field free region.

During the process of slow extraction, which can take thousands to millions of turns, the beam is moved towards the septum, which is to be chosen very thin to have the least interaction with the beam. The part of the beam entering the gap between the septum and the electrode experiences an electric field which deflects it away from the circulating beam. The septum consists either of a series of wires (several 100 to over 1000, made for example out of tungsten-rhenium) or a set of foils (for instance of molybdenum), precision aligned on the support frame. Typical septum thicknesses range in the order of 50 to 100 μm, but can also be lower (10 μm). Figure 8.74 right shows an example of a foil septum in the CERN PS, Fig. 8.75 shows a photo of a wire septum used for slow extraction from the CERN SPS.

Beam impact leads to heating of the septum, and the thermal expansion can cause warping which increases the effective thickness seen by the beam (case of foil septa). Once the deformation becomes too important the device needs to be overhauled. Wires are typically spring-loaded such that damaged wires will be retracted from the beam and high voltage regions and not get stuck in an unwanted position. To further reduce deformations caused by thermal effects the septum support can be made out of Invar which is most relevant for high intensity or high energy extraction channels. The opposite electrode consists essentially of a bar made out of stainless steel or titanium and is at high voltage (up to over 300 kV). Careful preparation and high voltage conditioning of the surfaces exposed to the high field are essential for reliable operation.

To allow high electric fields electrostatic septa are installed under vacuum; the vacuum works as an insulator between septum and electrode. If the vacuum requirements are not too severe (order of 10^{-9} mbar) anodized aluminium can be used to eliminate the dark current on the cathode electrode, thus allowing operation with fields as high as 10 to 20 MV/m. To achieve a vacuum of up to 10^{-12} mbar septa may be designed to be bake-able at up to 300 °C. Electrostatic septa are usually fitted with remote positioning systems which allow alignment of the septum with respect to the circulating beam to minimize losses, optimize the trajectories, and adjust the gap width. This feature is also very helpful for HV conditioning purposes.

If the interaction with the beam is to be minimized, a wire septum is the preferred choice. However, it allows some of the field to penetrate into the circulating beam region, the degree of which depends on the wire diameter and spacing. Ions created by ionisation of the residual gas by the beam can cross through the wire array into the high field area and provoke HV breakdowns. To counteract this, clearing electrodes (so called "ion traps", visible in Fig. 8.75) are added which provide a small electric field (few kV) in the circulating beam area and capture the ions before they can escape. The price to pay for the reduced beam interaction with a wire septum is the additional complexity of the clearing electrodes and their associated power supplies and electronics. Foil septa do not need clearing electrodes; to avoid ionised residual gas entering the high field region longitudinally, aluminium screens are sometimes mounted at the entry and the exit of the extraction channel (septum).

Beam impact leads to activation of the septum (either foil or wires), and, due to the scattered particles, also of the septum support and surrounding material and equipment. The radiation level has an impact on the cool-down time needed before maintenance or an intervention can be carried out on the equipment. To reduce the amount of beam interacting with the septum, a passive diffuser can be installed upstream of and in line with the septum [189, 190]. A diffuser is typically composed of an array of wires with a thickness greater than the septum, to scatter away the beam that otherwise would hit the septum. By scattering this finite part of the beam, the beam density that hits the septum is reduced, reducing the overall activation of both diffuser and electrostatic septum. Relevant for the effectiveness of a diffuser is the angular spread of the beam at the septum, the distance between diffuser and septum (phase advance), as well as the alignment of the diffuser with respect to the septum [191]. To ease the alignment with the septum, the diffuser can be installed on

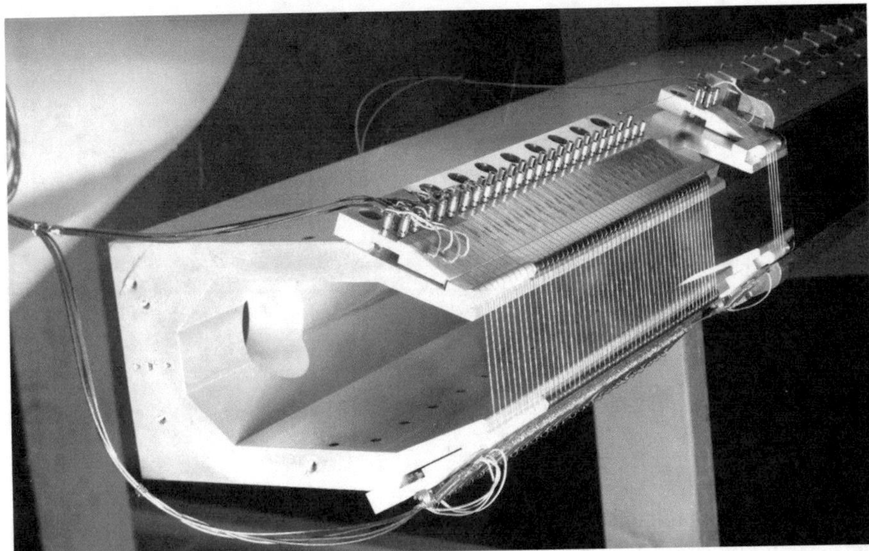

Fig. 8.76 A wire diffuser (left) on the common support in front of the foil septum (right) used for the CERN PS extraction towards SPS. The 4 wires directly in front of the foil are part of a beam observation system. The orbiting beam is circulating inside the hollow diffuser and septum support, while the extracted beam is passing through the high field of the septum outside the support (to the right)

the same support. In case a bigger distance is required between diffuser and septum, the diffuser is a separate device upstream of the septum (mainly for high-energy extractions). A diffuser was deployed in the CERN PS upstream of the beam slicing septum used to extract the so-called Continuous Transfer beam (CT) towards the SPS. Figure 8.76 shows the wires diffuser in front of the foil septum used for the CT extraction in the PS.

8.7.2.2 Magnetic Septa

There are several major types of magnetic septa [163]: direct-drive septa (pulsed or d.c.) [189], eddy-current septa, Lambertson (steel) septa, massless septa and superconducting septa. Figure 8.77 left shows a schematic cross section of a direct-drive magnetic septum. The coil, consisting of the septum conductor (also called "septum blade") and rear conductors, is marked in red. In the lower part of the figure the magnetic field B is shown as it could be measured on the axis O–X as indicated in the cross section. In this schematic drawing the field is homogeneous in the magnet gap, and dropping to zero in the septum conductor for the region with circulating beam. The stray (or leak) field of a septum magnet is kept to a minimum by using a magnetic yoke with a steel with a relative permeability as high as possible (thus limiting the use of the yoke near saturation), by assuring a uniform current density

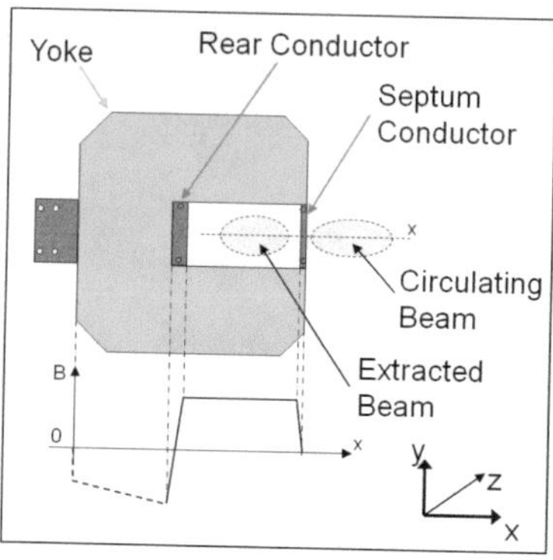

Fig. 8.77 Schematic drawing of a direct-drive magnetic septum (left); CERN SPS extraction septum (right)

in the septum conductor (thus minimising the impact of the cooling channels on the current uniformity), by reducing the space between the current carrying septum conductor and the magnet yoke, as well as by using a magnetic shield between septum conductor and orbiting beam area.

Figure 8.77 right shows a septum used in the extraction of the SPS. Such type of septa are often housed under vacuum (in the range 10^{-9} to 10^{-10} mbar) to avoid adding vacuum chambers which would increase the apparent septum thickness. The steel yoke is made from thin laminations; bake-out of up to 200 °C must thus be foreseen to permit reaching the desired vacuum levels. As another example Fig. 8.78 shows the extraction septum of the PS Booster, with 4 magnets in 2 vacuum vessels (for the 4 superimposed machine rings), with bake-out heating lamps switched on.

The coil is generally constructed as a single turn to minimize magnet self-inductance. This allows the magnet to be pulsed, thus reducing its power dissipation and hence cooling requirements. The septum thickness is typically in the range of a few mm. Powering is through a capacitor discharge providing a half sine wave pulse of a few ms length with peak currents of the order of 5 to 40 kA. To improve the flat-top a 3rd harmonic circuit and active filtering can be added [163]. As for electrostatic septa, the magnet is often equipped with a positioning system, to allow precise matching of the septum position with the circulating beam trajectory. The mechanical forces on the thin septum conductor can be quite significant (up to 10 kN) and lead to fatigue failure. The coil fixation is usually done with beryllium-copper springs at regular intervals along the magnet.

Slowly pulsed or d.c. septa typically use thicker conductors with cooling channels inside the individual conductors to remove the heat, or use edge cooling. In the latter case non-homogeneous temperature distributions may lead to imperfect current distributions causing important stray fields. The thermal design, hence the cooling circuitry of a d.c. septum, is very critical and has a significant impact on the robustness of the device. Generally, the cooling circuit is designed to use the cooling water in turbulent regime to profit from the improved heat exchange between the

Fig. 8.78 PS Booster extraction septa for the four rings, before closing the vacuum vessels

copper conductors and the cooling water. Particular attention needs to be paid to avoid cavitation inside the cooling circuit. The water quality (electrical resistance, but also filtering of microparticles) is essential to assure a reasonable lifetime of the septum magnet. A d.c. septum magnet is often used outside vacuum. In this case the coil and the magnet yoke can be split into an upper and a lower part allowing the magnet to be "clamped" over the vacuum chamber [192], permitting magnet and coil maintenance or repairs without breaking the accelerator vacuum (Fig. 8.79). The magnet is usually fitted with a multi-turn (series) coil, so as to reduce the current needed. Nevertheless required currents range between 0.5 and 4 kA and can consume up to 350 kW; efficient high-flow cooling and temperature surveillance are therefore essential.

In an eddy-current septum magnet [193–195] the septum blade is not part of the coil. Instead, the drive coil is situated around the back leg of the C-shaped yoke. Coil dimensions are therefore less critical, and can be designed for a lower current density, while the fixation as well as the robustness of the septum conductor are improved with respect to a direct-drive septum. The coil is constructed as a single turn, therefore minimizing the self-inductance of the magnet. Eddy-current septa are powered with a half or full sine wave current with a period of typically 50 μs. To avoid a leak field being created next to the magnet, the magnet opening is screened off with a copper plate, the so-called septum. When the magnet is pulsed, the magnetic field induces eddy currents in the septum, counteracting the fringe field

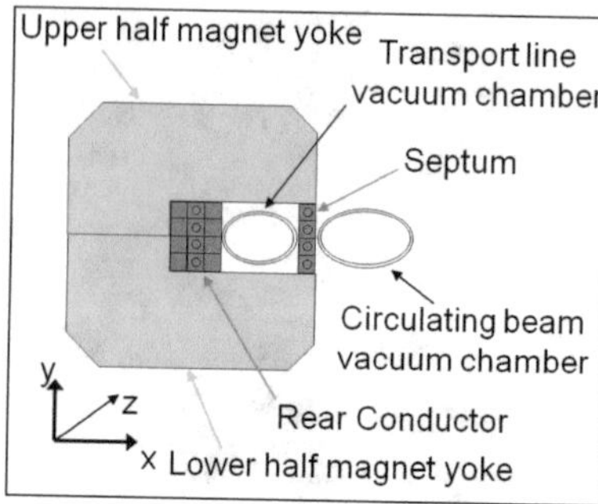

Fig. 8.79 Schematic of a d.c. septum (left) and a d.c. septum coil (cut open) showing the cooling channels (right)

created. These induced eddy currents in the septum and return box help to keep the field inside the magnet gap, thus improving the field homogeneity in the gap as well as reducing the stray (leak) field of the magnet. The septum conductor can be made somewhat thinner than for the direct-drive septum, but cooling circuits may still be needed at the edges to cool the septum. A schematic drawing of an eddy-current septum is shown in Fig. 8.80.

Without additional measures the typical maximum leakage field would be 10% of that in the gap. Using a magnetic screen between the septum and the circulating beam, and by installing a copper "return box" around the septum the fringe field can be reduced to 0.01% of the gap field at all times and places [196]. These magnets are often under vacuum to minimise the distance between circulating beam and extracted beam axis. The need for a fast pulse to make the eddy current shield effective, calls for very thin yoke steel laminations (typically 0.23 mm or less) which adds significantly to the gas load for the vacuum system, in particular for large aperture magnets.

Lambertson (steel) septa [197–200] can be constructed as d.c. or pulsed devices, mostly outside vacuum. The conductors are enclosed in a steel yoke, and are relatively far away from beam which allows a robust low current density design. The septum is formed by a thinner part of the yoke, between the magnet aperture and the circulating beam; additional steel is required to avoid saturation. Figure 8.81 left shows a schematic cross section of a Lambertson septum; Fig. 8.81 bottom row shows a cross section of a Lambertson septum used in the LHC injection. Here the two counter-rotating LHC beams circulate through the holes in the upper yoke part. Their vacuum tubes are clad with mu-metal to shield the circulating beams from the stray field.

A massless septum is a septum device that has a high field and a (near) zero field region, without a physical separation between them. The transition between the high field and zero field region is not instantaneous but gradual [201–203]. The septum thickness is defined as the space between the 2 regions, defined by chosen percentages of the main field (5–95% for example). Since the design of a

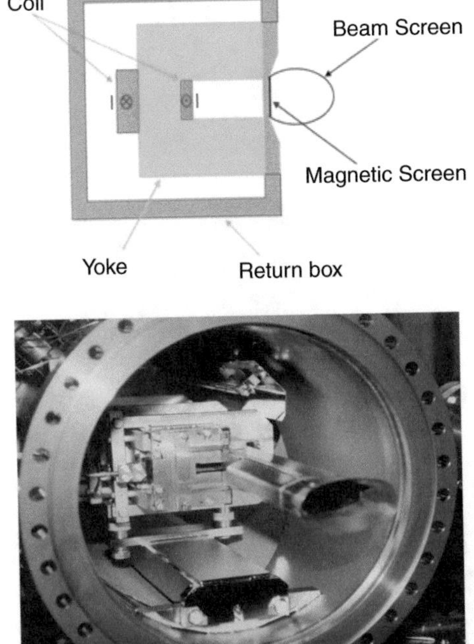

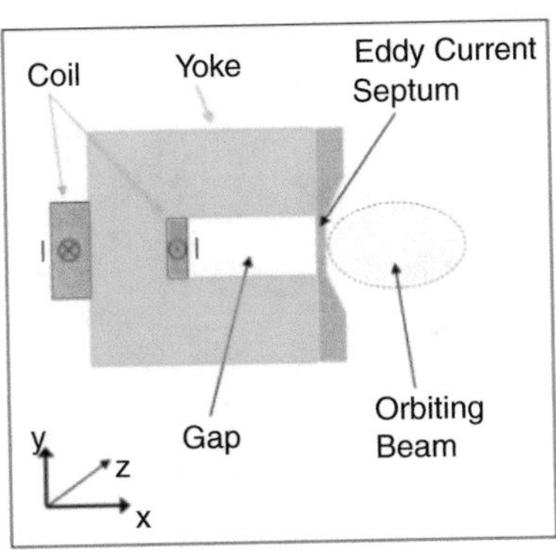

Fig. 8.80 Top: Schematic of an eddy current septum (left) with a return current box and magnetic shield (right). Photo shows the ESRF S3 eddy current septum constructed at CERN (1991)

massless septum is intimately linked to the performance of the magnetic material, the fields which can be achieved are relatively low (typically below 1 T), and the septum width is typically at least as big as the gap height. Massless septa require the injection or extraction to be designed to accommodate the particles that transit through the septum region. A fraction of these particles are only partially deflected, insufficiently deflected to enter the extraction channel. To deal with this, a dedicated absorber may need to be installed further downstream, to intercept these particles on a sufficiently robust device. Only few massless septa have been constructed and used in accelerators.

To limit electrical power consumption, superconducting septa are an attractive option [204], in particular for storage rings or accelerators with long cycle times. Although the direct-drive septum could be equipped with a superconducting coil, most superconducting septa are derived from superconducting cosine theta dipoles [205–207]. A novel concept under study uses a superconducting tube, used to shield a magnetic field (in principle generated by a superconducting magnet outside the tube), the tube becoming effectively the septum separation between the high field generated by the external magnet and the field-free region inside the tube [208]. This topology is often referred to as the SuShi (superconducting shield) septum. When

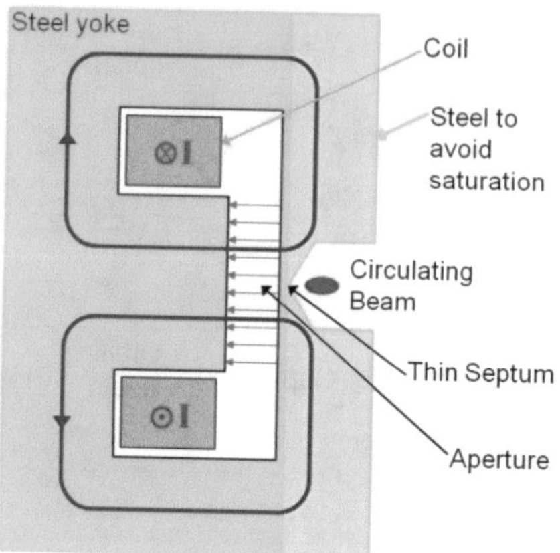

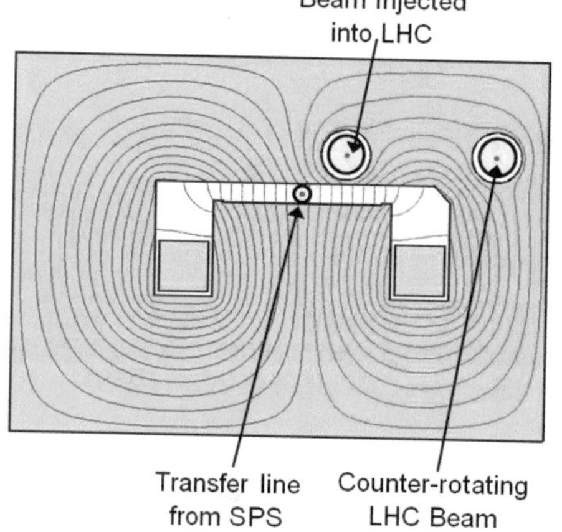

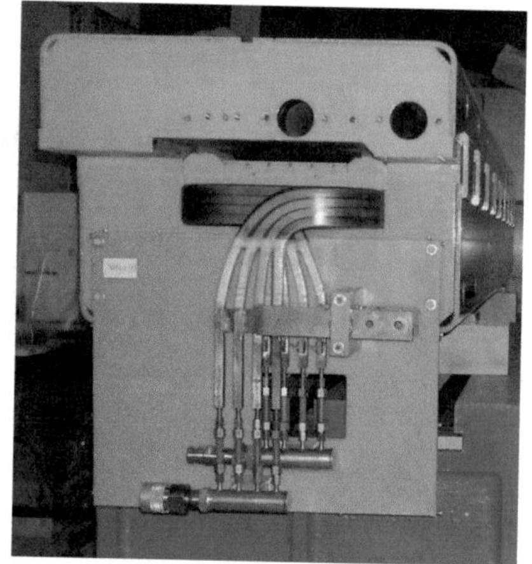

Fig. 8.81 Schematic of a Lambertson septum (left); a Lambertson septum used in the LHC injection (bottom row)

using a superconducting septum, care must be taken to protect the superconductors from beam impact, hence heat load, to avoid quenching the septum. Also because of the superconductors, the use of superconducting septa is more suitable for slowly varying fields (extraction towards dump lines of high-energy accelerators for example) of constant energy application (such as injection into colliders or storage rings).

To reduce activation of magnetic septa, caused by beam hitting the septum blade, dummy septa can be installed upstream of the septum, see Fig. 8.81 (left). A dummy septum is a passive element that is to intercept the particles that otherwise would hit the septum blade [209]. This is relevant for unbunched beam transfer. To be effective, the dummy septum needs to be perfectly aligned with the downstream septum, and therefore can be equipped with its own displacement system. The dummy septum being a more robust device than the septum magnet,

Fig. 8.82 Dummy septum TPS15 from the CERN PS extraction being assembled in the laboratory (left). The copper dummy septum blade can be seen through the vacuum flange. The carbon blade of the extraction protection element TPSG6 as used in the CERN SPS extraction towards the LHC (right)

it needs far less maintenance and moves the activation (partially) from the septum magnet to the dummy septum. This facilitates the maintenance activities related to the septum. Of similar technology are extraction protection elements upstream of septa magnets. Their role is to protect the septa from mis-steered beams by diluting the beam sufficiently that no permanent damage occurs in case a beam be accidentally steered onto the septum, see Fig. 8.82 (right). This is most relevant for high energy accelerators where a direct impact could lead to damage of the septa. The requirements for the extraction protection elements is related to the impact resistance to the beam, but most importantly to the protection of the septum downstream, in terms of sufficient dilution of the beam energy to avoid damage to the septum coil (either mechanical stress in the conductors) or shock and pressure waves in the septum cooling circuits.

8.8 Collimators

R. W. Aβmann · S. Redaelli

8.8.1 Introduction

Several definitions are introduced for the topic of collimators:

- **Collimators** are special accelerator devices that place scattering or absorbing blocks of materials around the beam. They can be fixed or movable with respect to the beam.
- The **collimator jaws** are the blocks of material that are placed close to the beam. The jaw material is characterized by its nuclear properties, its thermal conductivity, its electrical resistivity, mechanical properties (surface roughness and flatness) and vacuum properties (residual outgassing rates).
- A **collimation system** is an ensemble of collimators that is integrated into the accelerator layout to intercept stray particles and to protect the accelerator.
- The **impact parameter b** is the typical transverse offset from the edge of a collimator jaw for particles impacting the collimator. It is measured in m and can be as small as a few hundred nano-meters for the collimator jaws that are closest to the beam to intercept primary beam losses.
- Collimation is acting in the **normalized phase space**. With $z = x$ or y, the Twiss functions β_z and α_z, and the emittance ε_z we define the normalized coordinates z_n and z_n' as:

$$z_n = z/\sqrt{\varepsilon_z \beta_z} \tag{8.70}$$

$$z_n' = \frac{\alpha_z z + \beta_z z'}{\sqrt{\varepsilon_z \beta_z}} \tag{8.71}$$

It is noted that the **transverse betatron beam size** (Gaussian rms) for a beam with zero energy spread is given by:

$$\sigma_z = \sqrt{\varepsilon_z \beta_z} \tag{8.72}$$

- An unperturbed particle describes a circle in normalized phase space with **amplitude**:

$$a_z = \sqrt{z_n^2 + z_n'^2} \tag{8.73}$$

- Collimator **settings** (distance between jaw surface and beam center) are defined in normalized coordinates z_n with $z_n = n_1$ being the collimator family setting closest to the beam, $z_n = n_2$ the second closed setting, and so on. Several families usually define a hierarchy that must be respected. Often this results in stringent mechanical and operational tolerances for collimators.
- The **radiation length X_0** can be defined as the mean distance over which the energy of a relativistic electron is reduced to 1/e of its initial value by bremsstrahlung.

- The **nuclear interaction length λ_i** is the mean path length that a relativistic charged particle can traverse through matter before its energy is reduced to 1/e of its initial value.
- The **nuclear collision length λ_T** is the mean free path of a particle before nuclear reaction.

Collimator technology has seen significant advances over the last 40 years, driven by the requirements of more and more performing accelerators. The first collimation systems were conceived in the 1970's, see for example [210]. In the 1980's more elaborate collimation systems were designed and constructed for the e^+e^- colliders LEP [211] and SLC [212]. The 1990's saw the development of two-stage collimation theory [213–218], primarily for hadron beams with energies in the 100 GeV–10 TeV regime. The first two-stage collimation systems were constructed for HERA [219], Tevatron [220, 221] and RHIC [222]. The 2000's took modern collimator technology to linac-based, GeV range energy, high power facilities [223–231], where multi-stage collimation is required for ensuring hands-on maintenance and reducing environmental impact of high power accelerators. At the same time collimator technology was developed for linear collider concepts [232]. A major effort in the 2000's was spent at CERN to develop an ultra-efficient and robust collimation system for the LHC, in the end consisting of four stages and more than 100 fully movable collimators [213, 233, 234]. Recently, this design effort was further extended to cope with the challenges of the High-Luminosity upgrade of the LHC (HL-LHC) [235, 236], adding in-jaw orbit measurements [237] and conceiving improved designs for local protection of cold magnets with warm collimators [238, 239]. Recent R&D develops a new collimator design with low desorption for cryogenic accelerator regions [240].

8.8.2 Requirements for Modern Collimators

Modern collimators must support operation of accelerators with high power loss and beams that are often beyond the destruction limits of available materials [233]. At the same time collimators must be placed closer and closer to the beam and allowable tolerances have reached the micron regime [241]. A collimating slit is shown in Fig. 8.83 for the example of an LHC collimator. At the LHC, the closest jaws sit at distances as small as 1 mm from the circulating beam. Collimators become heavily cooled, high power devices, which are highly radioactive, must be good absorbers, extremely robust and work as precision tools under heavy beam loads. A few central requirements that drive the collimator design are introduced. The adopted solution must be closely targeted to the foreseen use case and requires a detailed analysis of accelerator physics and operational requirements before the design work is started.

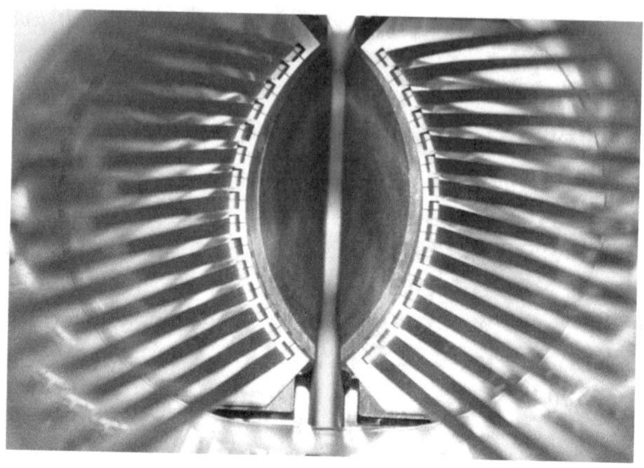

Fig. 8.83 Photograph of a collimating slit (LHC example) along the beam path. The collimator tank hosts two 1.2 m long parallel jaws that define a collimating slit. The beam passes through the middle. Collimator jaws, which are made of fiber-reinforced carbon (black material), intercept stray particles. RF contacts guide electro-magnetic image currents. The jaw material is tapered at its end to provide a smooth transition to the beam tank. More advanced designs implement orbit pick-up bottoms in the tapered part to measure the beam position [242]

8.8.2.1 High Power Loads

State-of-the-art accelerators have advanced into the regime of high beam brightness. This regime is characterized by high beam power that is compressed into small transverse beam size. The beam power can be characterized by considering the energy that is stored in one beam with N_p charged particles:

$$E_{stored} = pceN_p. \tag{8.74}$$

Here, c is the light velocity. We consider particles with charge $q = e$ and relativistic momentum p. The stored energy in the beams is compared in Fig. 8.84 for several accelerator facilities. It is seen that modern accelerators operate or are designed for beam momenta between a few GeV/c to a few TeV/c. The stored beam energies are in the range of 10 KJ to 500 MJ. Losses and power loads must be distinguished for different types of accelerators:

- The highest stored energies are achieved in proton storage rings (RHIC, Tevatron, HERA, LHC) where the beam is kept for many hours [219, 220, 222, 233]. Power loss can be 1–500 kW if 0.1% of the beam is lost over 1 s. Several future accelerators are under study and plan to exceed the design goal of the LHC: the High-Luminosity upgrade of the LHC [236] is an approved project that will increase by a factor 10 the LHC integrated luminosity starting in 2026 and HE-LHC [243] and FCC-hh [244] are studies for machines up to 50 TeV. Assuming similar loss scenarios for future accelerators, entails power losses easily exceeding the MW levels. The loss will most often appear fully

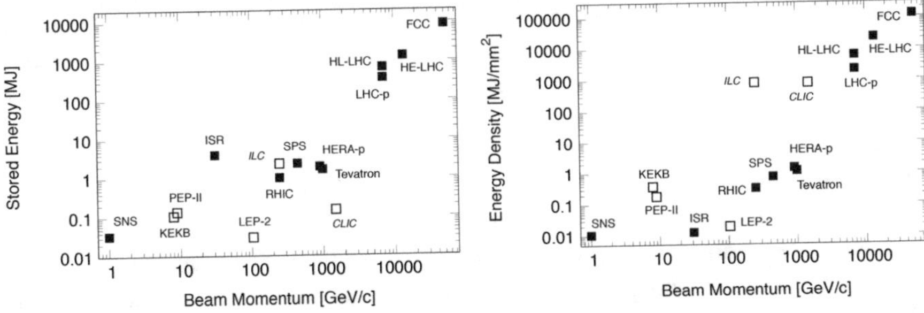

Fig. 8.84 The stored energy (left) and the density of stored energy (right) are shown versus the beam momentum for various accelerators. Filled symbols refer to proton, while open symbols indicate electron/positron accelerators. ILC and CLIC are design studies

at the location of the smallest ring aperture, by design a collimator. In modern proton colliders, these power losses must be controlled in the presence of superconducting magnets, placing high demands on the efficiency of collimators.

- Lepton storage rings (like LEP2) have additional power loss due to synchrotron radiation [211]. The power loss can reach several MW, for example in LEP2 where 3% of the beam energy was lost per turn. Synchrotron radiation losses are distributed over the length of the accelerator. Collimators intercept only part of the synchrotron radiation, mainly at strategic locations at the end of the arcs or around any experimental detectors.
- Linac-based accelerators (SNS, J-PARC, CLIC, ILC) do not store the beams but regenerate them at 5–100 Hz, using each beam pulse only once [223, 227, 232]. Power loss can range from a few kW at SNS to a few 10 kW at the ILC (for a 0.1% loss per beam pulse). Losses can appear concentrated at one collimator.

It is seen that modern collimators must usually be designed for an impacting power in the range from a few kW to 500 kW. Future projects must aim at designs capable to cope with multi-MW levels. Depending on the choice of jaw material and length only a fraction of the impacting beam power will remain in the collimator jaw, the rest being sprayed downstream. A proper design of thermal heat flow, collimator cooling and vacuum outgassing is essential.

8.8.2.2 Destructive Beam Densities

The transverse beam size σ_z at collimator locations has decreased over the years, either due to lower normalized emittances from injectors or due to the operation at high beam energies (adiabatic emittance damping). This increases the energy density in the beam:

$$\rho_E = E_{stored} / \left(\pi \sigma_x \sigma_y\right) \tag{8.75}$$

The stored energy densities are compared in Fig. 8.84 for several facilities, the beam size taken at typical collimator locations. It is noted that this parameter is directly proportional to relevant performance parameters like luminosity and therefore is usually maximized.

The stored energy densities range from 10 kJ/mm^2 to 4 GJ/mm^2 for operating accelerators but exceeds 100 GJ/mm^2 for future studies like the FCC-hh. Damage limits depend on the type and length of material that is hit by beam. Typical values for metals are around 50 kJ/mm^2. Very robust materials like fiber-reinforced carbon can survive an impacting proton energy density of around 5 MJ/mm^2 for 1 m long blocks. In many cases, collimators can only survive fractions of the collimated beam [233, 245]. A full beam impact must be avoided and collimators must be designed for maximum robustness, non-catastrophic failure in case of beam impact and in-situ handling of damaged surfaces.

8.8.2.3 Precision Tolerances

Precision requirements arise from various issues. Collimator settings are given in normalized distance to the beam center. Movable collimators are often designed to be placed as close as possible to the beam. Typical desired settings for half gaps are in the range of $5\sigma_z$ to $10\sigma_z$. As transverse beam sizes at collimators are reduced to sub-mm (as small as 0.14 mm in the LHC), collimator full gaps can be as small as 1 mm in existing accelerators. Maintaining the correct full gap over long collimators (e.g. 1 m long parallel jaws) introduces tolerances on jaw flatness, deformation during power impact, reproducibility, parallelism, angular alignment, etc. that are all in the order of 50 μm (μrad) or lower [234].

Other requirements arise in the case that several collimators are combined to form a multi-stage collimation system. The collimators then belong to different stages (families), where all collimators of a given stage sit at the same setting (in normalized coordinates). Most notably, in colliders where the performance is maximized by reducing the transverse spot sizes of colliding beams, limitations often occur from the aperture of the triplet quadrupole magnets, $n_{triplet}$, that are used to squeeze the beams [246, 247]. The system will only work correctly if the collimators fulfill the hierarchy requirement. For the example illustrated in Fig. 8.85 the following condition would apply:

$$n_1 < n_2 < n_3 < n_{triplet} < n_4 < n_{arc} \tag{8.76}$$

The smallest difference in this condition (for example $n_{triplet}$-n_3) can be a fraction of σ_z and then imposes additional tolerances for the design and operational control of the collimators.

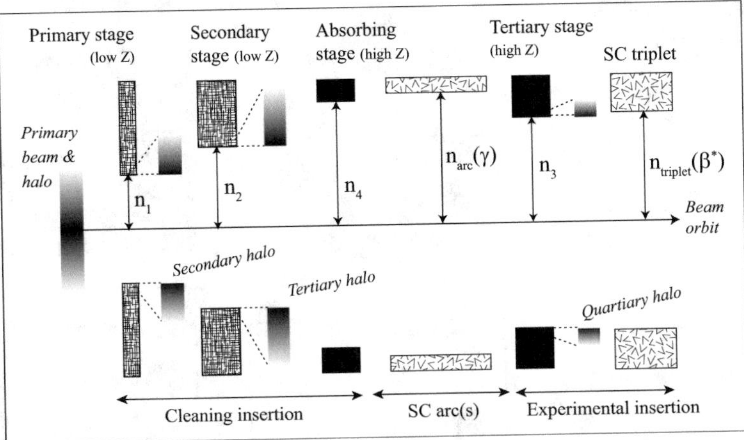

Fig. 8.85 Illustration of various collimator families that are put at different settings n_1, n_2, n_3, n_4 to form a multi-stage collimation system. The example illustrates the protection of superconducting magnets in the arcs and the experimental insertions. Robust collimators (low Z) are used for primary and secondary stages, while non-robust collimators (high Z) are used for the third and fourth stages

8.8.3 *Collimator Solutions*

Collimators can follow different concepts, from very simple to very advanced technology. A few typical concepts are illustrated in Fig. 8.86. Each concept has its use and one should always follow the simplest possible solution for the problem at hand. The different concepts are discussed:

– In the simplest case, collimators consist of fixed masks with either elliptical or rectangular shape. Such objects are best used either in cases of well-defined aperture bottlenecks where the beam energy, the beam size and the center at the collimator location are constant or in cases where broad showers or synchrotron radiation must be intercepted. Such collimators can easily be cooled and (in case required) heavily shielded as no movable parts need to be accessible. Collimator jaw materials can be shrink-fitted into a metallic pipe. These simple collimators are not adequate for high-efficiency collimation close to the beam. In addition, fixed-aperture devices are not suitable for accelerators that need collimation during dynamic changes of beam energy or optics, where multiple aperture bottlenecks occur, potentially at different ring locations. Some improvement can be implemented by making the whole assembly movable with respect to the beam, however, the collimating gap cannot be adjusted. Fixed collimators with very low desorption ("catchers") have recently been developed for handling losses from partly ionized beams, as foreseen for the FAIR project [240].
– A more advanced concept is an L-shaped, one-sided collimator design as used for the Tevatron and RHIC primary collimators [220, 222]. A single L-shaped

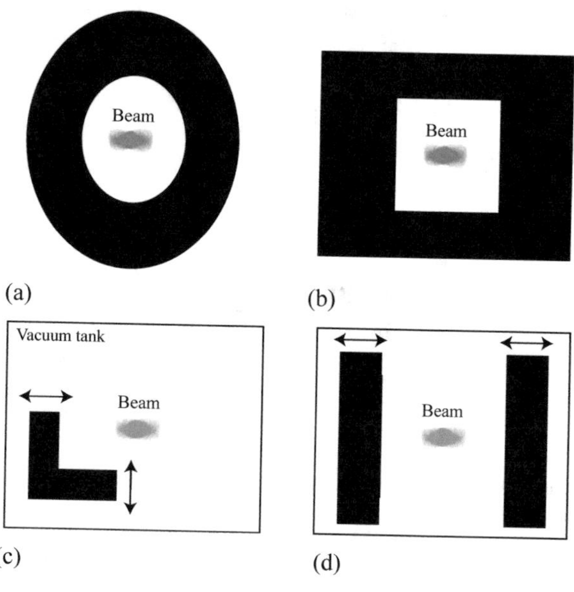

Fig. 8.86 Illustration of a few different collimator concepts, with fixed or moveable apertures, that have been used for various accelerators

collimator jaw is placed into a vacuum tank. The L-shaped design is very cost-effective as one device can be used to collimate the beam in both horizontal and vertical directions. The L-shaped jaw is fully movable and can be adjusted to the beam center and size. It is adequate for high-efficiency collimation close to the beam. While it is very cost-effective, it is compromised in terms of operational stability. There is no direct measure of the distance to the beam. Also, the beam centering is not constrained and the beam can wander off in one direction without a practical limit.

– The most advanced concept consists of a vacuum tank that hosts two parallel, fully movable jaws [234, 248]. The two jaws define a collimation gap that constrains the beam position. If the beam wanders off it will be collimated more deeply in one side. The design is operationally fail-safe and is adequate for high-efficiency collimation close to the beam. Each collimator can act only in one plane, depending on its azimuthal orientation.

8.8.3.1 An Advanced Two-Jaw Collimator Concept

The two-jaw collimator design has been modernized for the purposes of the LHC [234, 248, 249]. The final design offers fully movable jaws (position and angle to beam), choice of jaw materials, redundant position monitoring, a concept of spare surface (by moving the full tank) and a precise measurement of the collimation gap that is placed on the beam. The conceptual solution is shown in Fig. 8.87. The decision to support the collimator jaws from the bottom reduces the width

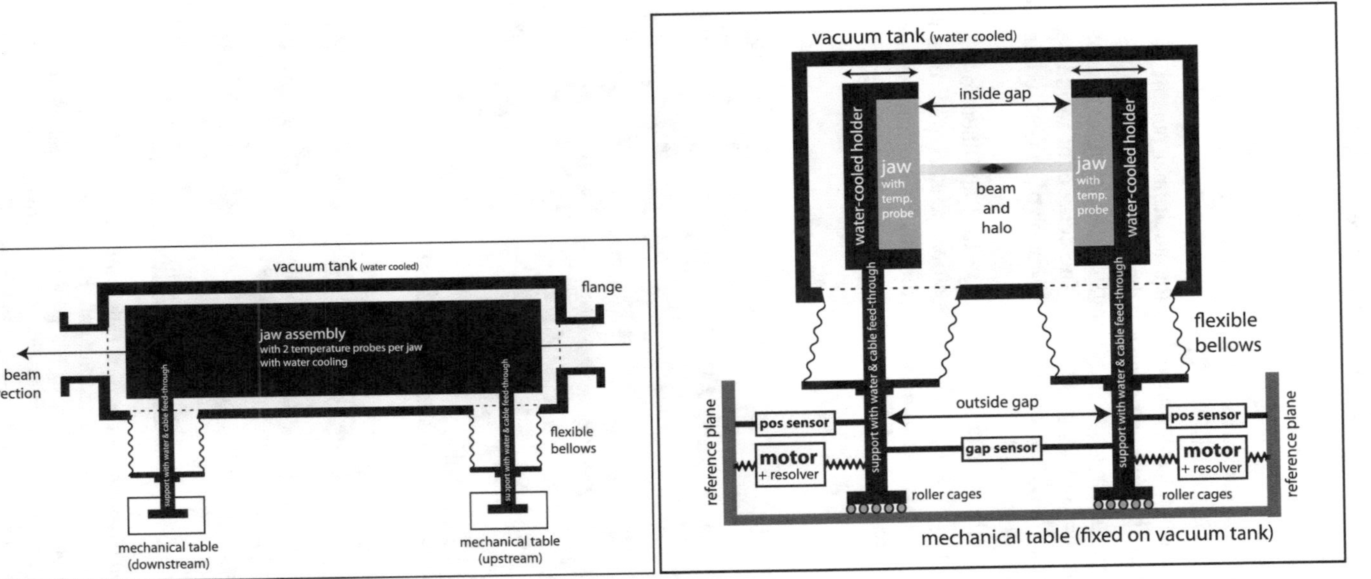

Fig. 8.87 Side view (left) and view along beam path (right) of the modernized two-jaw collimator design. Each jaw can be adjusted in distance and angle to the beam. The right view illustrates the possibility to measure the outside gap and jaw positions. Precision calibration during production allows inferring accurately the inside gap and jaw positions from the outside measurements

of the assembly and allows matching to tight space requirements in a two-beam pipe accelerator like the LHC. The second beam pipe can be passed besides the collimator vacuum tank. A further improvement on the initial design was achieved by adding at both jaw extremities beam position monitors for on-line measurements of the local beam orbit, as shown in Fig. 8.88 [236]. About twenty BPM collimators are already operational in the LHC [237].

The adopted mechanical solution offers the additional benefit of an accurate online measurement of the collimator gap and the jaw positions. It is possible to infer the inside gap from measurements of the outside gap, using precision calibration data that can be obtained during production. The concept of inside versus outside gap is illustrated in Fig. 8.87. The precision calibration of an LHC collimator during production is shown in Fig. 8.88.

Each end of the collimator has two motors [250] and therefore 2 degrees of freedom (DOF). Monitoring it with two position sensors plus one gap sensor provides important redundancy: three measurements are performed for two DOF. Self-consistency can be checked in real time and possible sensor problems can

a

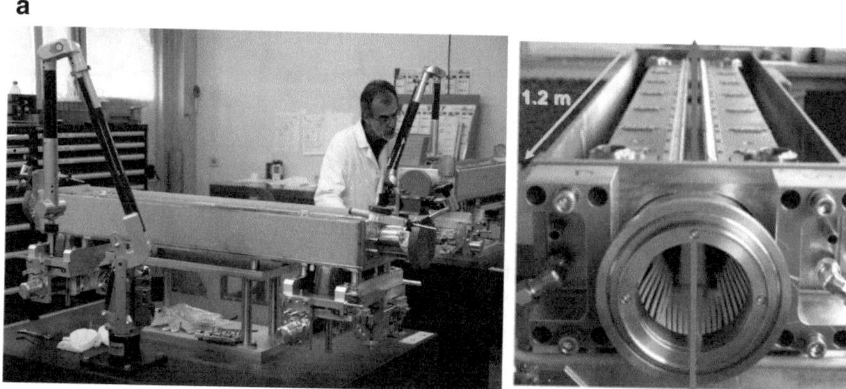

b

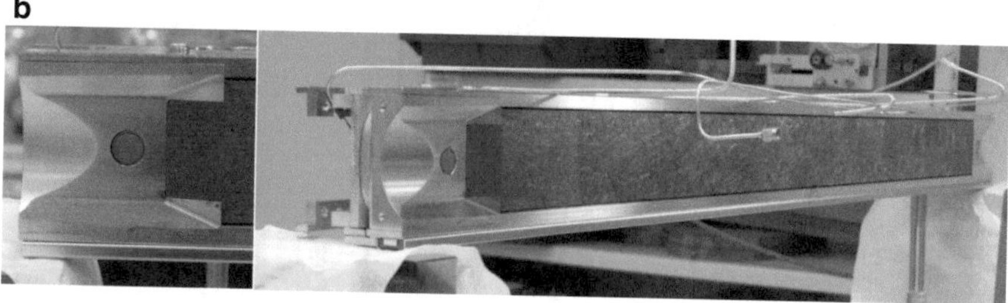

Fig. 8.88 (**a**) Photograph of collimator precision calibration during production (LHC example). The two jaws have been installed into a vacuum tank but the cover has not yet been welded, providing possibility for precise calibration of jaw positions versus mechanical stops. (**b**) New collimator jaw with integrated BPMs at each extremity. The active part if made of CFC. A detail of the BPM is given on the left-hand side. A variant of this design, made with a Glidcop support and tungsten inserts for the active jaw part, was used for adapting this concept to collimators made of higher-Z materials. From [236]

be detected and addressed. The LHC collimator controls system allows a synchronization to other accelerator systems like magnets and radio-frequency to the micro-second level. Motors are used in a dedicated tool to align in parallel several collimators, at frequencies up to 50 Hz, in a feedback loop that uses 100 Hz beam loss measurements [251, 252]. The recent addition of orbit measurements [242] improves further the on-line diagnostic capabilities, allowing an earlier detection of possibly critical shifts of the local beam orbit.

8.8.3.2 Mechanical Design, Cooling and Vacuum

The demanding requirements in modern high power accelerators were already shortly reviewed. These requirements translate into important constraints for the design of collimators. We summarize some of them:

- The design of **mechanical movement** must be precise, robust and reproducible. The lifetime of the collimator movement system should be a few 10,000 cycles for slowly ramped accelerators (like LHC) up to a few million cycles for rapidly cycled accelerators. The tolerances for mechanical plays and setting reproducibility can be as small as a few 10 μm.
- The use of traditional **grease** is often not permitted due to the presence of high beam losses and radiation. Grease could age and become sticky. Instead, advanced techniques like dry grease (graphite powder) or surface coatings must be used.
- **Thermal heat flow** must be carefully designed, from the jaw where heat is deposited to the cooling pipes that take it out. Heat loads from direct particle losses and electro-magnetic currents must be taken into account (W to 100 kW). Thermal expansion coefficients of materials must be well adapted to ensure minimal deformations during power gradients (few 10 μm for several kW) and good thermal contact for heat transfer [253].
- The impacting power loads often require powerful **cooling** of the jaw materials. High-pressure water flow (20 bar) must be integrated into the jaw design. Sometimes, it is required to also cool the vacuum tank and flanges.
- **Safety and robustness against beam shock impacts** (accidents) often requires that cooling pipes do not have brazed parts under vacuum.
- Many accelerators require **ultra-high vacuum** (~10^{-8} mbar) and the collimators must not disturb this. As heated materials show outgassing, additional constraints on maximum jaw temperatures and cooling must be respected.
- Intense beams induce electro-magnetic image currents on the surrounding materials: **machine impedance or wakefields** are induced [254–256]. As collimator materials are very close to the beam, their electro-magnetic properties must be well designed. Critical are good electrical resistivity (~μΩm) and smooth geometrical shapes. Tapering of jaws is used to avoid sharp edges close to the beam. RF fingers need to be used to guide image currents.
- The high **irradiation** at collimators [257] often requires special measures. Possibilities include quick plug-ins for cables and water pipes, absence of

shielding (to have easy access to devices), heavy shielding (to provide radiation-free passage), robotic survey techniques and remote handling. All active components, like motors, and sensors must be compatible with high radiation operation.

The use of modern programs is important for precise calculation of halo impact, heat deposition, heat transfer, deformations, electro-magnetic fields and currents [258–264]. The calculations must be used to verify the adequate performance of the chosen design before construction starts as well as to optimize the performance of operating system (see for example [265]).

8.8.3.3 Precision Actuation and Monitoring

The collimator is a mechanical system that (if movable) requires actuation and monitoring. In dependence of the design chosen, one or several motors must be used for moving the system. The motor of choice should be a stepping motor with a resolver mounted to count the steps performed. The mechanical system translates the rotary into a lateral movement. The absolute position of jaws in the collimator tank can be determined in presence of a precise calibration with well-known mechanical stops. Similar is true for collimator gaps.

The precise monitoring of jaw positions is often required for safety reasons. In this case independent position sensors should in real-time monitor the jaw positions. The measurements provide redundancy to the setting procedure that relies on stepping motors and mechanical stops. The sensors also monitor positions when the jaws are away from the mechanical stops. The addition of orbit measurements from in-jaw pickups [242] is complementary to mechanical jaw/gap measurements.

Actuation and monitoring is more complicated for large accelerators where issues of radiation resistance and signal transport over long cables arise. Specialized solutions must be adapted [250, 266].

8.8.3.4 Examples of Installed Collimators

Two examples of recently constructed and installed collimators are shown in Fig. 8.89. The selected examples from SNS and LHC illustrate the strong differences that are possible in collimator design, reflecting different requirements and boundary conditions.

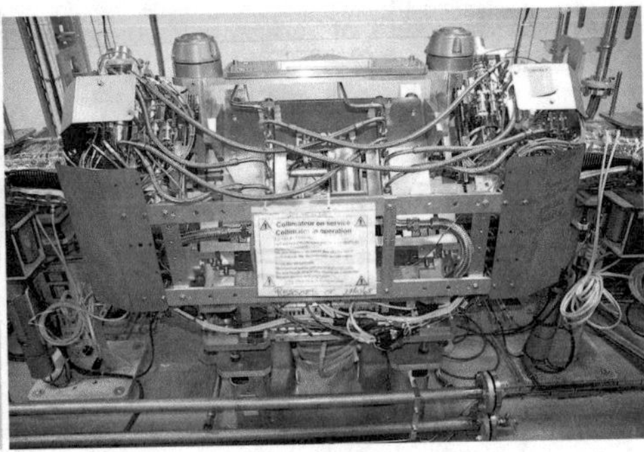

Fig. 8.89 Photographs of an installed collimator for the SNS linac (left, courtesy G. Murdoch) and an installed collimator for the LHC ring (right). The difference in design choices is clearly visible: SNS relies on heavily shielded, massive collimators (18 tons per collimator) while LHC implemented a non-shielded, cooled and heavily instrumented solution (0.5 tons per collimator). The SNS solution becomes highly activated inside but has little radiation outside. The LHC design avoids radiation hot spots but dilutes radiation over a larger volume

8.8.4 Choice of Collimator Jaw Material and Length

The purpose of a collimator is to intercept stray particles and its performance is determined by the choice of the collimator jaw material. The material must be adapted to the specific boundary conditions that the collimator should fulfill.

A summary of often-used collimator materials is listed in Table 8.18 [267]. The interaction with the beam particles is determined by the nuclear properties of the collimator material. Full descriptions of beam-matter interactions are published in literature and cannot be repeated here, but see for example [221]. We note a few general observations that can be important for collimator design:

- Energy loss of charged beam particles in matter (ionization and excitation) is described by the Bethe-Bloch equation. The average lost energy for relevant materials is in the range of 0.5 GeV/m to 5.8 GeV/m. Highly energetic particles can therefore traverse collimator blocks with small energy change. For example, a 7 TeV proton would lose 0.7 GeV in a 1m long carbon collimator, a relative energy loss of 10^{-4}.
- Multiple Coulomb-scattering (MCS) will induce a net deflection for charged particles by some angle θ_{MCS}:

$$\theta_{MCS} = \frac{13.6\, MeV}{\beta c p} z \sqrt{\frac{x}{X_0}} \left[1 + 0.038\ \ln\left(\frac{x}{X_0}\right)\right] \tag{8.77}$$

- Here, β_c is the velocity of the particle, p its momentum, z its charge number, x the length of material traversed and X_0 the radiation length of the material. The MCS-induced kick increases the normalized amplitude of the particle oscillation. This

Table 8.18 Overview of various commonly used collimator materials and their main nuclear physics parameters [267]

Element	Atomic number Z	Mass number of nucleus A	Nuclear collision length λ_T [cm]	Nuclear interaction length λ_i [cm]	Radiation length X_0 [cm]
Be	4	9	29.9	42.1	35.3
C	6	12	26.8	38.8	19.3
Al	13	27	25.8	39.7	8.9
Fe	26	56	10.4	16.8	1.8
Cu	29	63.5	9.4	15.3	1.4
W	74	184	5.7	9.9	0.35
Pb	82	207	10.1	17.6	0.56

effect is used in collimation system to intercept amplitude-increased (scattered) particles in a second stage downstream or in following turns. As seen from Table 8.18, it is advantageous to select high Z materials for maximizing MCS.

- Particles are considered "stopped" in the collimator material if they undergo an inelastic interaction and a secondary particle cascade is initiated. A short nuclear collision length and a long collimator jaw length are desirable for ensuring that particles are efficiently "stopped". This favors high Z materials. In case of high power beams, the deposited energy may become too large and low Z or low length solutions are required.
- As highly energetic particles traverse even long collimator blocks with small energy loss and reduced MCS kicks, it becomes more likely that their energy is dissipated through inelastic nuclear collisions. Some processes can become limiting for collimation performance. For example, the cross-section for single-diffractive scattering is as follows:

$$\sigma_{SD} = 0.68 \text{ mb } \ln(0.3pc) \tag{8.78}$$

It is seen, that the effect of single-diffractive scattering becomes stronger with higher beam energies. Protons that experience single-diffractive scattering can lose significant amounts of energy while escaping the material with small transverse kicks. An off-center proton can acquire an off-momentum component that might be important for their loss location.

- Ions must be treated specially [240, 268, 269]. Ion-specific processes like dissociation and fragmentation can dominate the processes that are used for efficient collimation of elementary particles like electrons, protons, etc. See for example some recent results from an LHC run [270].

It can be seen that long collimator jaws often seem beneficial. It is, however, important to take into account the edge feature of beam collimation. Most particles will impact close to the jaw edge. The typical impact parameters on primary collimators, for a circular machine like the LHC, can range from 100 nm to a few μm. Surface roughness and MCS will result in many particles exiting from the jaw before its full length has been traversed. It is seen that longer collimators jaws are often not useful for interception of primary beam particles. In addition, longer jaws also require a better design to control static and dynamic deformations to the required accuracy.

The optimum choices of collimator jaw length and material must be based on simulations with special programs [258–263, 271]. These programs take into account the beam properties (beam type, energy, impact parameters), the accelerator type (single-pass or multiple-pass) and the material properties (geometry, nuclear properties). Results from beam simulations are then put into programs for calculating energy deposition [264]. Figure 8.87 shows a FLUKA energy deposition calculation for 1m long carbon block.

A few common directions in collimator design are described:

- Short, high-Z collimators (for example W) are often used as primary scatterers, which increase particle amplitudes via MCS, while still limiting the energy deposited on the jaws. Length can range from a few 100 μm to a few cm.
- Long, high Z collimators (for example W) are often used as absorbers, which effectively stop the particles that hit with large impact parameters. Length can range from 10's of cm to several meters.
- Long, low Z collimators (for example C) are used as primary scatterers when energy deposition with short, high Z collimators would be too high for material survival. The use of special materials (for example fiber-reinforced graphite CFC) can maximize the robustness. Such solutions are important for high power beams. Length can range from 10's of cm to several meters.

It is noted that research on modern composite materials is ongoing [249, 272] and can provide new directions and solutions in the future. A key aspect for this development is to take into account effects related to specific aspect of materials' response to beam losses [273] while optimizing the machine impedance. Reviews of recent developments that are already being implemented for the upgrade of the LHC can be found in [236]. Promising results were achieved recently that identified valid solutions for a new generation of secondary collimators for the HL-LHC made of a novel Molybdenum-Graphite composite [274], possibly Mo-coated, as well as for metallic composites suitable for tertiary collimators, but about 15 times more robust against beam impact than the tungsten alloy used presently at the LHC [275, 276] (Fig. 8.90).

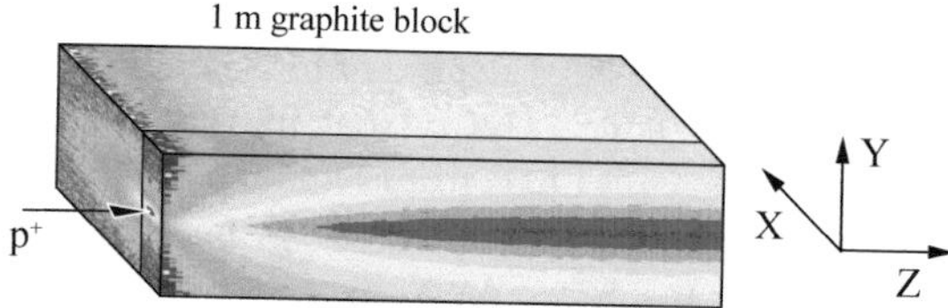

Fig. 8.90 Example of an energy deposition result as obtained with FLUKA [258, 262]. The impact of 7 TeV protons on a 1m long graphite block is considered. The development of the secondary showers from inelastic proton interactions is seen

8.8.5 Advanced Collimator Concepts

The traditional collimator design places materials close to the beam to intercept the beam halo. It is noted that several research efforts investigate alternative technological solutions for collimation:

- The **technique of bent crystals** investigates the channeling of stray particles (see for example [277] and references in there). Instead of amorphous scattering in a material block, particles enter into the channel of a bent crystal that guides them onto a different trajectory. The particles receive a net deflection (equivalent to the crystal bent angle). Large controlled deflections (~mrad) can be realized over a small distance (~mm). The crystals would replace primary collimators. Absorbers would be used to intercept the channeled beam halo. Research challenges include (1) the edge effects of crystals, (2) the design of an appropriate halo dump system and the (3) operational procedures for alignment of halo particles with the crystal channels [278]. Promising results were achieved in the LHC with a test system installed in the betatron cleaning insertion [279], demonstrating for the first time channeling of proton and heavy ion beam halos at 6.5 TeV. This technique is now being considered for the upgrade of LHC collimation system.
- The **technique of a hollow e-beam lens** [280] investigates the usage of a low energy, hollow electron beam for inducing a diffusion boundary for particles beyond some amplitude. Such a device could act as a scraper that cannot be destroyed by beam and could therefore act close to the beam core. The reduction of beam tails could lead to lower peak loss rates and higher feasible intensities. Research challenges include (1) the design of such a device with controllable diffusion rates and (2) the emittance preservation of the beam core. A well-advanced design for the LHC was produced and is being considered to enhance the performance of the LHC collimation system in view of the challenging requirements of the HL-LHC beams (see [281] and references in there).
- The **technique of non-linear collimation** [282] investigates the creation of non-linear fields with magnets, inducing a diffusive boundary for the beam. The goal is to replace primary collimators with such a device, while obtaining larger

impact parameters at the absorbing collimators. Research challenges include the emittance preservation of the beam core.

- The concept of a **rotatable collimator** [283] uses a circular jaw with many facets. Such a solution can be used for applications where occasional damage to the jaw surface cannot be avoided. After a damaging beam impact, the jaw is rotated and a fresh facet is presented to the beam. The collimator only needs to be exchanged once all facets have been used up. The testing of this concept is well advanced, thanks to beam tests carried out recently on a prototype developed at SLAC for the LHC upgrade [284]. Another, even more advanced variant of a circular jaw design uses liquid metal coating that is continually refreshed [285].

8.9 Geodesy and Alignment for Particle Accelerators

D. Missiaen

8.9.1 Introduction

Particle accelerators require very tight tolerances for the positioning of their components. These tolerances are coming mainly from optics requirements but can also be triggered by aperture and mechanical considerations as is the case for the LHC.

The task of the surveyors in this domain is to measure and align the position, the orientation, the shape and the size of big objects, such as electro-magnets and particle detectors with an accuracy never requested elsewhere. This activity is deeply linked to the geodesy.

8.9.2 Alignment Tolerances

The alignment precision requirements are the key values that will drive any survey study. The absolute accuracy in the vertical direction is the deviation to the theoretical plane of the collider, while it is the variation of its radius R with respect to the theoretical value in the transversal direction. The differential variations between several consecutive magnets represent the relative accuracy. This latter type of error has a more direct effect on the closed orbit of the particles.

8.9.3 Reference and Co-ordinate Systems

For the positioning of an object, one has to define a Reference system, a frame to which the position and the orientation of the object are referenced. A co-ordinates system is attached to this frame and defines the position of the object in units (m for distances, gon for angles) and ways of describing the position (Polar or Cartesian system).

At CERN, the reference system has evolved with the increase of the size of its installations. At the epoch of the PS (50's) and the ISR (60's), the surface of the earth could be considered as a plane without significant error. The XY plane adopted was the plane of the PS synchrotron and a polar co-ordinate system was used. The Z co-ordinate was the difference of height measured with respect to the XY plane.

With the extension towards the SPS (70's) with a total surface of 3 km by 3 km of CERN installations, the earth couldn't be considered any longer as a plane. A sphere, materializing the average sea level extended through the continents, was chosen as the reference surface and a new co-ordinate H, the altitude, was defined as the distance measured with respect to this surface (Fig. 8.91).

In the 80's, the size of the LEP, with its 27 km circumference, obliged to reconsider the reference surface for the earth as an ellipsoid of revolution. But, this surface is not accurate enough to take into account the anomalies of the vertical provoked by the neighboring Jura mountains and Geneva lake. An equipotential surface of gravity, called the Geoid, to which the force of gravity is perpendicular everywhere (Fig. 8.92) has been defined by means of zenithal camera and gravimetric measurements. The measurements taken with survey instruments are therefore linked to this Geoid. In the case of the LEP and of the LHC, the Geoid was calculated as a hyperbolic paraboloid tangent to a local ellipsoid in the CERN area.

The CERN reference system is therefore a local ellipsoid which fits the earth in CERN area, a geoid model has been calculated to take into account the deviation of vertical between the local vertical and the perpendicular to this ellipsoid. A Cartesian XYZ co-ordinate system has been defined; the XY plane being the PS synchrotron plane, and the Z is perpendicular to the XY plane. An H co-ordinate has been added to take into account the shape of the earth and the anomalies of the vertical due to the presence of mountains and valleys masses. This geoid model will

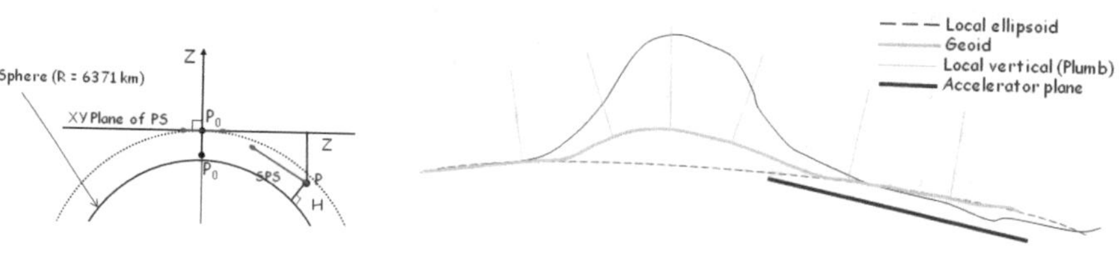

Fig. 8.91 The spherical model (left) and the Geoid (right)

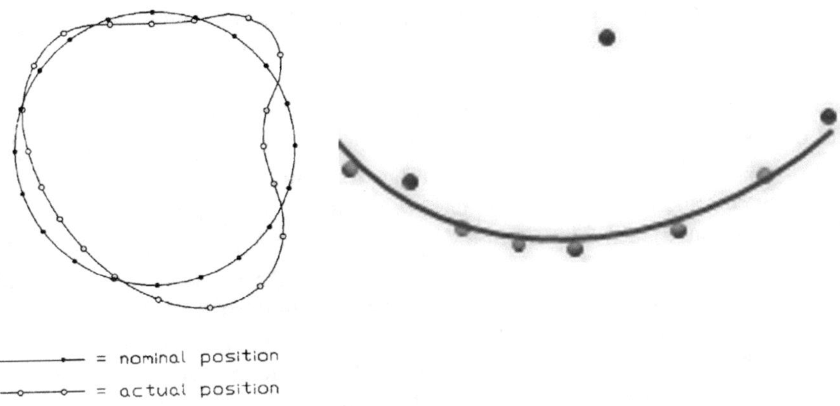

Fig. 8.92 Absolute and relative tolerance

have to be refined in the future in order to fit with the tight tolerances requested by the Compact Linear collider (CLIC) or by the Future Circular Collider (FCC).

8.9.4 Definition of the Beam Line on the Accelerator Site

A circulating particle beam is not influenced by gravity, therefore the physicists calculate the theoretical trajectory of a beam line in a local Cartesian co-ordinate system xyz. A specific software application (Beatch, MADx) providing the optics can also generate a 3D co-ordinate file in this local system. In order to obtain the co-ordinates in a global site co-ordinate system XYZ, the geodesists provides parameters for the geographical location of the accelerator with respect to the existing installations or geological/technical constraints. These parameters are typically the co-ordinates of a starting point and two angles, a slope and an orientation. They can also be three translations and three rotations, allowing the transformation from the local co-ordinate system one into the global one.

The 3D co-ordinates of the accelerator components will be used in particular for the 3D definition of the civil engineering works as well as for the accurate positioning of these components.

8.9.5 Geodetic Network

The absolute positioning of accelerator components, known in a XYZ co-ordinate system, is ensured by means of a geodetic network. The accuracy of this framework has, of course, a direct influence on the control of the absolute geometry of the accelerator to be built.

The first level of this network is a surface network. It is constituted of monuments solidly anchored to the earth by means of concrete works, forming a very well-defined basic framework and from which the links to national and international reference system can be established. It will also be used for regular checks and eventual extension of the project. The determination of the co-ordinates of these monuments is done by very accurate triangulation, trilateration, leveling measurements and nowadays when possible by Global Navigation Satellite System (GNSS) measurements. The accuracy of these network points has to be in the range of a few mm (1σ). It could be advisable to equip some network points with permanent GNSS receivers in order to have their position permanently recalculated together with permanent GNSS stations located in the vicinity of the site.

This surface network is transferred to an underground network by the means of several techniques among which one can mention:

- Angles and distances measurements with total stations
- Azimuthal orientation with gyro measurements
- Plumb lines measured by theodolites at the top and at the bottom of the shaft at the same time
- Nadiro-zenithal telescope
- Calibrated Electronic Distance Measurers (EDM) and optical levels for the altitude determination.

All the observations are processed together as a spatial block. Deflections of the vertical and meridians convergence are also taken into account. For a pit as deep as 140 m, the accuracy is estimated to be in the range of one mm in the XY plane and better than 2 mm in H (altitude), both values given at 1σ.

The underground network is constituted of tripods regularly spaced out along the accelerator tunnel, the distance between them being often directly linked to the lattice of the machine. The topographical traverse linking one access shaft to the adjacent one is realized by gyro-theodolites, theodolites and very accurate Electronic Distance Measurer, namely the Mekometer. Offset distances with respect to a nylon stretched wire are also frequently measured in order to improve the "smoothness" of the network. As an example, for a distance of 3.3 km between two points transferred from the surface network, as is the case for the LHC, the transverse deviation is estimated to be 4 mm (1σ) in the middle between these two points (Fig. 8.93).

In the vertical direction, the determination of the altitude of the points of the underground network is done by leveling measurements using optical or digital levels. The accuracy is of the order of 0.4 mm (1σ) per km. In order to have a stable reference plane, for calculation and future stability comparison, it is advisable to anchor references at a depth of 25 to 30 m under the tunnel level in stable rock. Two of these references installed in the SPS have given proof of their utility and therefore such a reference has been installed in the vicinity of each of the eight LHC shafts.

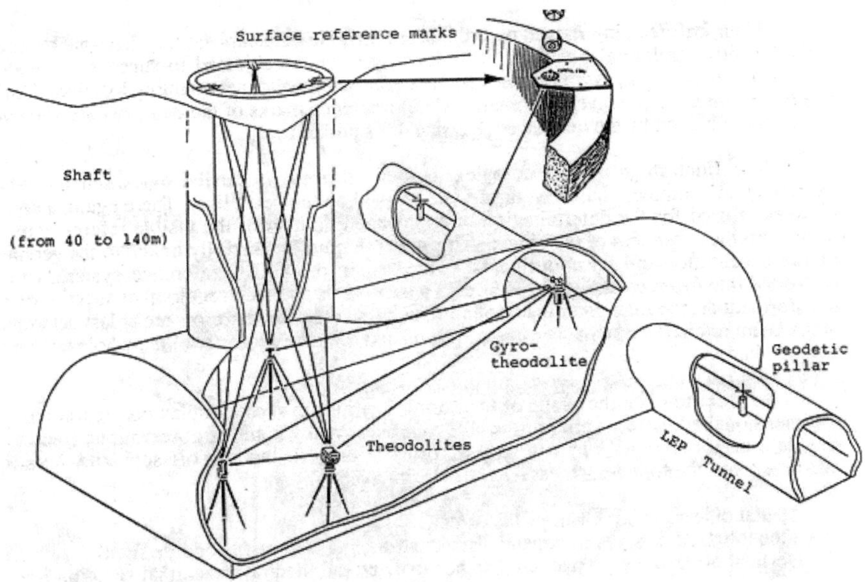

Fig. 8.93 From surface to the underground network

8.9.6 Tunnel Preliminary Works

The geometrical checks of the civil engineering works is done by the firm itself, often with the help of a consultant specialist in geodesy. They are done with respect to the geodetic surface network provided by CERN.

Once the civil engineering works are delivered to CERN, the beam line is marked on the tunnel floor, with respect to the underground network, as well as the longitudinal and transversal position of the components and their supporting systems. This information is also very useful for the installation of all the services from this early stage of installation.

8.9.7 The Alignment References

The "beam" points, entering and exiting each magnet, provided by the physicists are often not accessible in the tunnel at the time of the alignment. The surveyors, therefore, recommend the installation of "fiducials" or alignment references directly on the magnets. In order to align a component along its 6 degrees of freedom, at least two "fiducials" are requested and a reference surface for the measurement of the transverse inclination also called the roll (Fig. 8.94). At CERN, these "fiducials" are

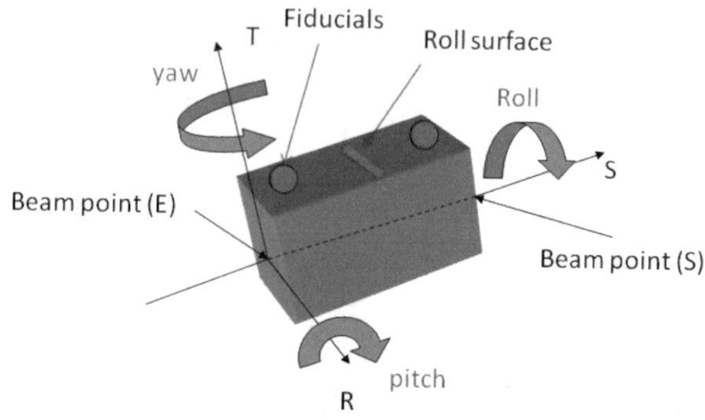

Fig. 8.94 Fiducials and local magnet co-ordinate system

Fig. 8.95 LHC fiducials and fiducialisation equipment

equipped with forced centering system. It allows not only the measurements of a Taylor-Hobson sphere located on top of it but also the installation of a theodolite or any specific device, considering then this point as part of the network (Fig. 8.95).

The "fiducialisation" is the operation during which the position of the fiducials is determined with respect to a reference axis (magnetic or mechanic). It can be done using:

- Magnetic measurements for most of the main magnetic components (quadrupoles and dipoles) which are realized by magnet people;
- Coordinate Measuring Machine (CMM) Mechanical measurements in a metrology laboratory;
- Laser tracker measurements when the size of the component is such that a metrological control is inadequate or impossible (Fig. 8.95).

For CLIC components, the fiducialisation has to be performed with an accuracy better than 10 microns. To achieve it, several techniques and instruments have been developed. Among them, one has to mention the micro-triangulation and the Frequency Scanner Interferometry (FSI). The principle of the micro-triangulation (Fig.

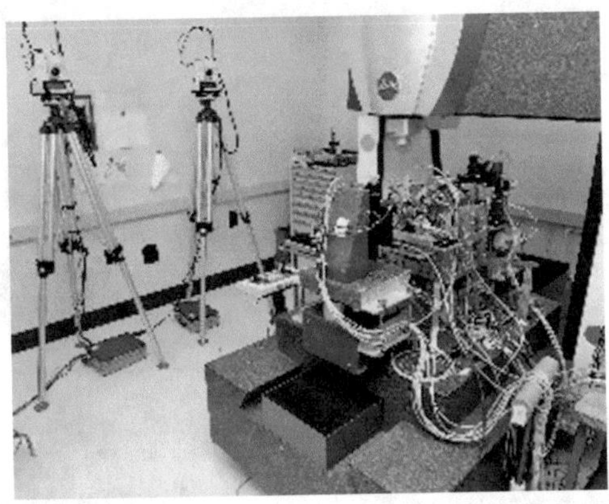

Fig. 8.96 Micro-triangulation

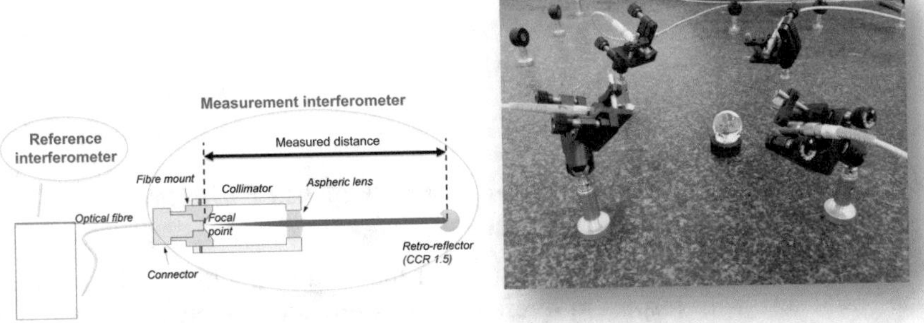

Fig. 8.97 FSI principle and instrumentation

8.96) is based on angle measurements taken automatically by several theodolites. This technique allows to measure at the same time the fiducials located on the magnet and the wire used for the determination of the magnetic axis. The FSI (Fig. 8.97) technique measures very accurate absolute distances between points located around the magnet, including the fiducials.

In the case of cryogenic components, the magnet is located inside a vacuum vessel. The fiducialisation having been done at warm or at cold temperature, it could be necessary to monitor the movement of the magnet with respect to the fiducials at different state. This determination could be done using the FSI technique or an optical measuring system called Brandeis Camera Angle Monitor (BCAM). It is a three points measuring system composed of a CCD camera measuring a light flashing through a lens (Fig. 8.98).

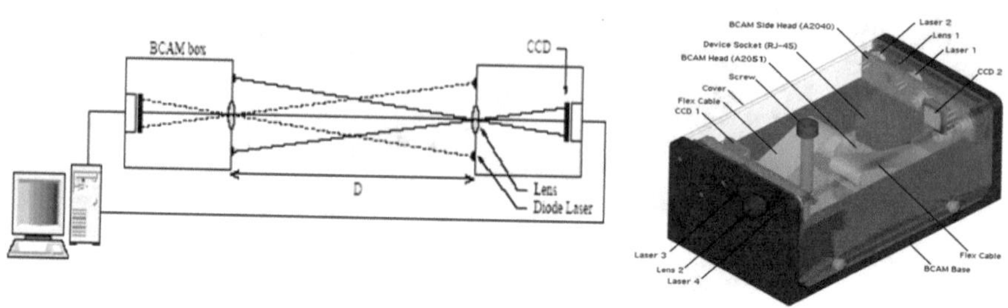

Fig. 8.98 BCAM

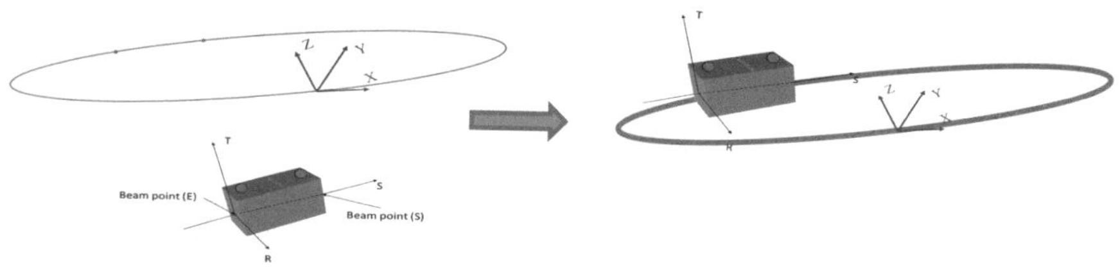

Fig. 8.99 Fiducials in the CCS

8.9.8 Determination of the Co-ordinates of the Fiducials

The combination of the theoretical position of the "beam" points and the position of the "fiducials" with respect to a reference axis provides the coordinates of the fiducials in the CERN Co-ordinates System (CCS) XYZ. The H co-ordinate is also generated (Fig. 8.99).

8.9.9 Alignment of Accelerator Components

The initial alignment is the phase during which the components are aligned at their absolute position, therefore with respect to the underground geodetic network. As the function of the quadrupoles magnets is to focus/defocus the particle beams, these are the most critical components in terms of alignment and consequently they are aligned first, followed by the dipole magnets. Optical or digital levels are used for the vertical positioning while total stations and wire offset measuring devices are used for the horizontal (Fig. 8.100). The roll angle is adjusted thanks to accurate inclinometers which are installed on the magnet reference surface. The adjustment of the components is realized thanks to three mechanical jacks, an auxiliary hydraulic device is sometimes needed when the necessary force to move the jack in the vertical direction is too important.

As the major requirement for the geometry of an accelerator is that the relative errors must be very small, a compulsory step is to check the alignment by measuring

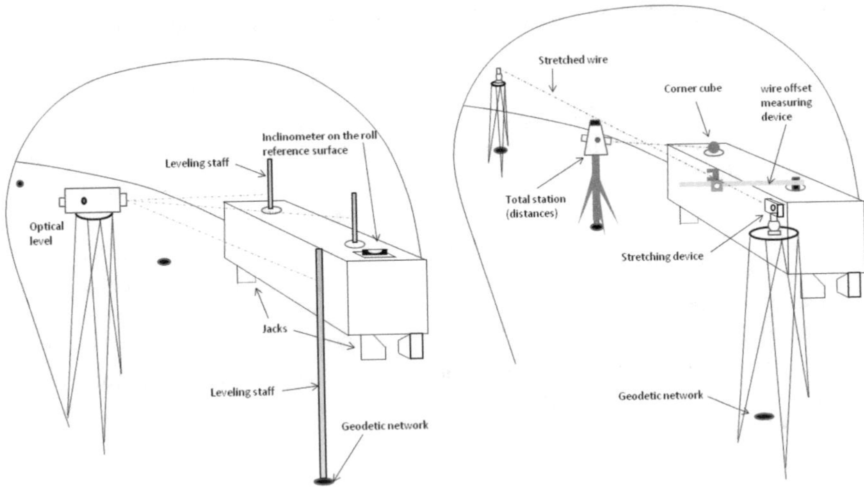

Fig. 8.100 Vertical (left) and horizontal (right) alignment of a magnet

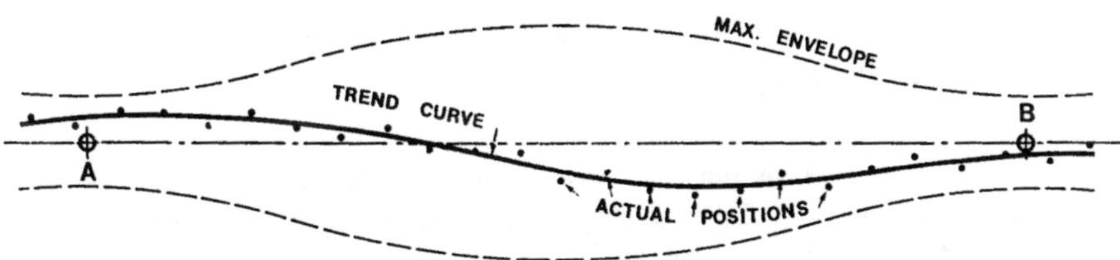

Fig. 8.101 Position of the components around the trend curve

and—if needed—improving the smoothness of the initial alignment. The magnets are finally positioned around an unknown mean trend curve contained within the envelope of maximum errors (Fig. 8.101). The measurements are done using only the "fiducials", the underground network is not used anymore. Digital levels and "ecartometers", a device developed at CERN to measure an offset with respect to a stretched wire installed directly on the "fiducials", are used for this smoothing operation. Some software has been developed in-house in order to determinate the trend curve. The principle is to fit, in a sliding window, a portion of an accelerator line with polynomials and to reject from the calculation a component located further than a defined tolerance. It can be compared to a carpenter's plane used for smoothing an irregular plank: depending on the length of the tool and on the adjustment of its blade, one can obtain different qualities of smoothness with more or less waves on the wood.

The polynomial degree of the curve depends on the redundancy, the overlap of measurements, and the betatron harmonics of the beam.

The relative position of the quadrupole magnets with respect to the trend curve is achieved with an accuracy better than 0.15 mm (1σ).

After this operation, the less critical components from the alignment point of view are positioned with respect to the "smoothed" components.

During the exploitation of the accelerator, when making successive maintenance surveys of these long and flexible figures, absolute comparisons would be a non sense and the differences between trend curves, corresponding to each survey, must be analytically eliminated. The state of the alignment is expressed by the statistical dispersion observed around the mean trend curve in these successive comparisons. In any case, and for successive measurements as well, the problem is the difference between these distorted curves and the theoretical geometry. Each "image" of the ideal line has the same likelihood of being "true", within the envelope of errors, but is nevertheless different. This difference has (globally) no physical meaning, but local discrepancies or distortions may be the signal of a move, either for a single element or for a group, depending on the deformations of the supporting structure (floor and tunnel) due to geo-mechanical forces, and/or to some constraints along the machine (vacuum, dissipated energy, etc.).

8.9.10 *Permanent Monitoring and Remote Alignment of Low Beta Quadrupoles*

The low beta quadrupoles magnets ensure the final focus of the particle beam before the collisions and for this purpose they are located in close vicinity to the experimental detectors. Their positioning accuracy (1σ) is requested to be as followed:

- 0.1 mm (1σ) for the positioning of one triplet of quadrupoles with respect to the triplet located on the other side.
- A few microns of stability of the position of one quadrupole inside its triplet

These tight tolerances, the absence of a direct line of sight through the detectors and the difficult environment (radiations, magnetic fields) have justified the permanent monitoring of the position of these critical components and the addition of dedicated survey galleries (Fig. 8.102).

In the vertical direction, the permanent monitoring is realized by Hydrostatic Leveling System (HLS). The principle is based on communicating vessels, the sensors measuring the distances to a water level by means of capacitive technology with a resolution below the micron. This high accuracy can only be obtained by taking into account the temperature gradients, the difference of pressure, the tidal effects and the influence of the deviation of the vertical. Both sides of the experiment are linked by a network of pipes (Fig. 8.103—below) running along the cavern and avoiding big changes in height which would generate temperature gradients.

In the horizontal plane, on each side of the experiment, a "short" (30 m to 45 m) stretched wire detected by biaxial sensor from Wire Positioning Systems (WPS), also using the capacitive technology, measures the relative position of

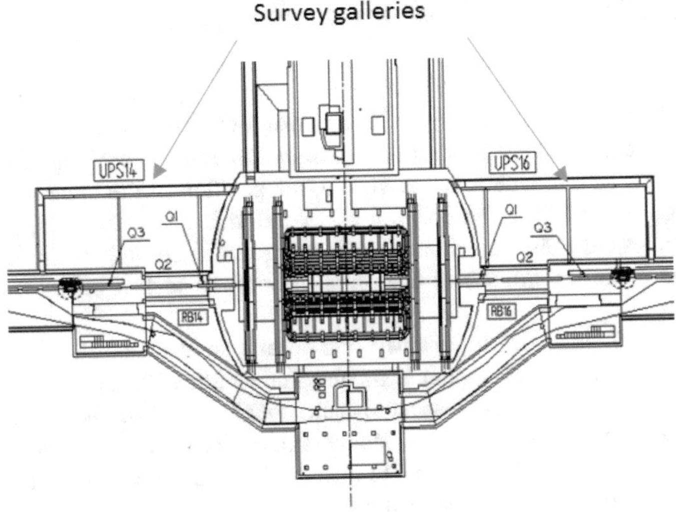

Fig. 8.102 Survey galleries

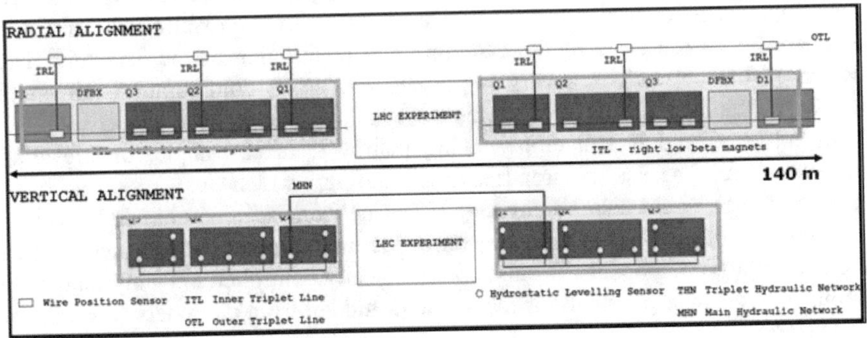

Fig. 8.103 The HLS and WPS around an LHC experiment

the quadrupoles. To link both sides, a "long" (120 m) wire has been installed in dedicated Survey galleries and goes through the experimental cavern (Fig. 8.103). The distances between the "long" and the "shorts" wires are measured by invar rods and proximity sensors in six locations to have redundant measurements. The accuracy obtained is in the range of several microns for the quadrupole magnets located along the same "short" wire and 0.1 mm between both sides of the experiment.

In addition, a motorized system has been installed to enable a remote readjustment for the improvement of the beam trajectory. This is not an active alignment system, the alignment is allowed only when the beam is off.

8.9.11 Alignment of Detector Components

Modern Experiments are complex structures made up of many individual modules, assembled concentrically, each one being design to detect a certain physics phenomena and can be likened to a set of Russian dolls. In the working position, only the outer skin of the experiment is visible, the position of the inner detectors has to be reconstructed from the position of the external ones. Their impressive size (several thousand tons, and thousands of m3) combined with their confinement in caverns make the alignment operation difficult in such a restrictive environment.

Furthermore, in order to obtain good results in the analysis of the events (reconstruction of particle paths), a precise knowledge of detector positions both with respect to the beam and respect to each other is required. The geodesists have therefore to ensure that the dimensional parameters of the experiment are fulfilled during fabrication and assembly and to give the final position of the whole detector with respect to the theoretical beam line.

The positioning of the experiment with respect to the beam line is done using a geodetic experiment network. This network of points is composed of forced centering sockets distributed in the whole cavern volume on walls and floor. It is used during all the steps of the assembly and positioning of the detectors. The design of the network is a very important step as its configuration influences the final accuracy of the measurements. It is measured, once the cavern has been delivered and is still empty, using mainly distances, angles and leveling measurements. The use of laser tracker technology, which uses a very accurate distance-meter, could also be envisaged. The network accuracy has to be within a few tenths of a mm (1σ) with respect to the underground network and later to the accelerator components. It must be periodically controlled knowing that some points may become physically inaccessible or hidden from the others (Figs. 8.104 and 8.105).

To obtain the final co-ordinates of each detector of the experiment, a succession of survey operations is needed from the first availability of an individual object through its final positioning. Each individual sub-detector has to be equipped with fiducial marks geometrically linked to the sensitive components. These references must be included in the design of the detectors with the right specifications and at the most strategic positions to allow their coordinate measurement in the future with the required accuracy. The surveyors have to be involved at an early stage of the projects.

One of the first in field operations, called the link geometry, is carried out in labs, workshop or assembly hall by means of micro-triangulation/trilateration. Fifteen years ago, the close range digital photogrammetry was added to these techniques and is now extensively used. Its principle relies on the reconstruction of an object simultaneously from several images taken from different positions and with the best possible perspective to ensure a suitable geometry of intersecting rays. The pictures are taken from free camera positions stationed in the object space and

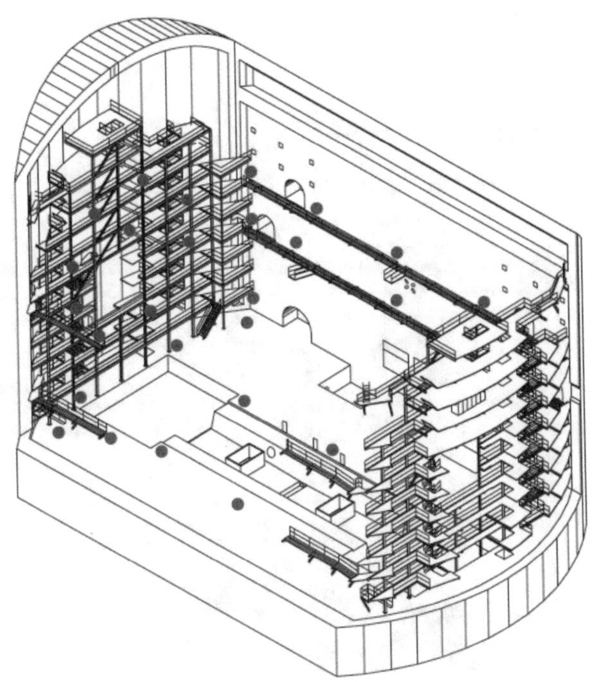

Fig. 8.104 Experimental network in the Atlas cavern

without a camera tripod. The photogrammetric network and the object coordinates are reconstructed from a bundle adjustment of rays. The valuable advantages are the non necessity to have stable stations, the short time of on-site intervention, nearly independent from the number of object points, coupled with a high measuring accuracy, even for voluminous objects.

After the survey involvement in all the phases of the detector construction: the preparatory work, the prototyping, the tests, the step-by-step adjustment and alignment of sub-detectors, the final stage of the complex geodetic process is carried out when the whole experimental equipment is installed on the beam line. From the experiment geodetic network, a control is done on a sufficient number of visible marks located on the outer skin of the object using techniques such as close range photogrammetry, leveling, angles, distances and offset measurements. From the obtained co-ordinates, successive 3D transformations will allow the calculation of each detector internal reference position both in the experimental local co-ordinate system and in the CERN co-ordinate system, the one in which the accelerator components, in particular the low beta quadrupoles, are also known.

The operation of the opening and the closure of all the parts of the detectors at the beginning and the end of each shut-down of an accelerator is quite delicate. During these two phases, the detectors shall remain within a few mm with respect to their nominal trajectory in order not to be damaged during the longitudinal movement. To limit the survey interventions, a positioning system based on the principle of the

Fig. 8.105 Fiducialisation of a CMS big wheel by photogrammetry

BCAM has been developed. The design of the lines of sight has been chosen either vertical or parallel to the movement of the detectors so that the BCAM can measure the most critical deviations during the movement of the detectors.

References

1. R.A. Beth: *Complex representation and computation of two-dimensional magnetic fields*, J. Appl. Phys. 37(7) (1966) 2568.
2. H.D.Glass: *Permanent magnets for beamlines and the recycler ring at Fermilab*, FERMILAB-CONF-98-267
3. D. Tommasini *et al.*, "Design, Manufacture and Measurements of Permanent Quadrupole Magnets for Linac4," IEEE Trans. Appl. Superc., v. 22, n. 33, Jun. 2012
4. C. Benabderrahmane, J. C. Biasci, J. F. Bouteille, J. Chavanne, L. Farvacque, L. Goirand, G. Le Bec, S. M. Luizzo, P. Raimondi, and F. Villar, Magnetf for the ESRF-EBS Project, Proceedings of the 7th International Particle Accelerator Conference, Busan, Korea (JACoW, Geneva, 2016), p. 1096.
5. P. Thonet, Use of Permanent Magnets in Multiple Projects at CERN, *IEEE Trans. Appl. Supercond. 26 (2016) 4101404*
6. B.D. Montgomery: Solenoid Magnet Design, Wiley-Interscience 1969, Krieger Publ. Co. 1980.
7. R.R. Wilson: *The Tevatron*, Fermilab Report TM-763, 1978.

8. R. Meinke: *Superconducting Magnet System for HERA*, IEEE Trans. Mag. 27(2) (1991) 1728.
9. M. Anerella, et al.: *The RHIC magnet system*, Nucl. Instrum. Meth. Phys. Res. Sect. A 499(2-3) (2003) 280-315.
10. L. Evans, P. Bryant (eds.): CERN 2004 LHC Design Report: The Main Ring Design Report, Vol. 1 CERN-2004-003, 2008 LHC Machine: J. Instrum. 3 (2008) S08001
11. M. Derrick, L.G. Hyman, E.G. Pewitt: *History of the Superconducting-Magnet Bubble Chambers*, Argonne National Laboratory Report ANL-HEP-CP-80-19 (1980) (unpublished).
12. H.H.J. ten Kate. *The ATLAS superconducting magnet system at the Large Hadron Collider*, Physica C 468 (15-20) (2008) 2137.
13. A. Herve (CMS Collaboration): *The CMS Detector Magnet*, IEEE Trans. Appl. Supercond. 10 (2000) 389.
14. P.J. Lee (ed.): Engineering Superconductivity, Wiley Interscience (2001), ISBN 0-471-41116-7.
15. M.N. Wilson: *Superconducting Magnets*, Oxford Univ. Press (1983) ISBN 0-019-854805-2.
16. Y. Iwasa: Case Studies in Superconducting Magnets, Plenum Press, New York (1994), ISBN 0-306-44881-5.
17. K.H. Mess, P. Schmuser, S. Wolf: Superconducting Accelerator Magnets, World Scientific, (1996) ISBN 981-02-2790-6.
18. F.M. Asner: High Field Superconducting Magnets, Oxford Univ. Press (1999) ISBN 0 19 851764 5.
19. J.K. Hulm, R.D. Blaugher: *Superconducting Solid Solution Alloys of the Transition Elements*, Phys. Rev. 123(5) (1961) 1569.
20. B.T. Matthias, et al.: *Superconductivity of Nb_3Sn*, Phys. Rev. 95(6) (1954) 1435.
21. E. Corenzwit: *Superconductivity of Nb_3Al*, J. Phys. Chem. Solids 9(1) (1959) 93.
22. J. Nagamatsu, N. Nakagawa, T. Muranaka, Y. Zenitani, J. Akimitsu: *Superconductivity at 39 K in magnesium diboride*. Nature 410(6824) (1 March 2001) 63. doi:https://doi.org/10.1038/35065039. PMID 11242039.
23. M.K. Wu, et al.: *Superconductivity at 93-K in a new Mixed-Phase Y-Ba-Cu-O Compound System at Ambient Pressure*, Phys. Rev. Lett. 58 (9) (1987) 908.
24. H. Maeda, Y. Tanaka, M. Fukutomi, T. Asano: *A New High-Tc Oxide Superconductor without a Rare Earth Element*, Jpn. J. Appl. Phys. 27(2) (1988) L361-4.
25. Kamihara, Yoichi; Hiramatsu, Hidenori; Hirano, Masahiro; Kawamura, Ryuto; Yanagi, Hiroshi; Kamiya, Toshio; Hosono, Hideo (2006). "Iron-Based Layered Superconductor: LaOFeP". J. Am. Chem. Soc. 128 (31): 10012–10013. doi:https://doi.org/10.1021/ja063355c. PMID 16881620.
26. A. P. Drozdov, M. I. Eremets, I. A. Troyan, V. Ksenofontov & S. I. Shylin. Conventional superconductivity at 203 Kelvin at high pressures in the sulfur hydride system. Nature 525, 73 (2015).
27. G.E. Gallagher-Daggitt: *Superconductor Cables for Pulsed Dipole Magnets*, Rutherford Laboratory Memorandum No. RHEL/M/A25 (1973) (unpublished).
28. K. Halbach: *Fields and First Order Perturbation Effects in Two-Dimensional Conductor Dominated Magnets*, Nucl. Instrum. Meth. 78 (1970) 185.
29. A. Ballarino: *Large capacity current leads*, Physica C 468 (2008) 2143.
30. A. Ballarino, L. Martini, S. Mathot, T. Taylor, R. Brambilla: IEEE Trans. Appl. Supercond. 17 (2007) 2121.
31. F. Paschen: *Ueber die zum Funkenübergang in Luft, Wasserstoff und Kohlensäure bei verschiedenen Drucken erforderliche Potentialdifferenz*, Annalen der Physik (1889) 273.
32. M. La China, D.Tommasini: *Comparative study of heat transfer from Nb-Ti and Nb_3Sn coils to He II*, Phys. Rev. ST Accel. Beams 11 (2008) 082401.
33. A.M. Baldin, et al., *Superconducting Fast Cycling Magnets of the Nuclotron*, IEEE Trans. Appl. Sup., 5(2), 875-877, 1995.
34. E. Fischer, et al., *Full Size Model Magnets for the FAIR SIS100 Synchrotron*, IEEE Trans. Appl. Sup., 18(2), 260-263, 2008.

35. M. Sorbi, et al., *Status of the Activity for the Construction of the HK-LHC Superconducting High Order Corrector Magnets at LASA Milano*, IEEE Trans. Appl. Sup., 28(3), 4100205, 2018.
36. A. Krusche, M. Paoluzzi: *The New Low Frequency Accelerating Systems for the CERN PS Booster*, 6th Eur. Part. Accel. Conf. (EPAC98), Stockholm, 1998.
37. C. Ohmori, et al.: *High Field-Gradient Cavities Loaded with Magnetic Alloys for Synchrotrons*, Proc. 1999 Part. Accel. Conf. (PAC99), New York, 1999.
38. M. Paoluzzi, et al.: *The LEIR RF System*, Proc. 2005 Part. Accel. Conf. (PAC 2005), Knoxville, 2005.
39. D. Grier, et al.: *The PS 80 MHz Cavities*", 6th Eur. Part. Accel. Conf. (EPAC98), Stockholm, 1998.
40. M. Benedikt, et al. (ed.): LHC Design Report, Vol. 3: The LHC Injector Chain, Chapter 16, CERN-2004-003, Geneva, 2004.
41. D. Boussard, T. Linnecar: *The LHC Superconducting RF System*, Cryogenic Engineering and Intern. Cryogenic Materials Conf. (CEC-ICMC'99), Montreal, 1999.
42. M. Aicheler *et al.* (ed.): *A Multi-TeV Linear Collider based on CLIC Technology: CLIC Conceptual Design Report,* CERN-2012-007, JAI-2012-001, KEK Report 2012-1, PSI-12-01, SLAC-R-985
43. R. Calaga: *Crab Cavities for the High-Luminosity LHC*, SRF2017, Lanzhou, China, THXA03
44. Ph. Lebrun: *Cryogenic systems for accelerators*, in: Frontiers of Accelerator Technology, World Scientific (1996) 681-700.
45. Ph. Lebrun: *Superconductivity and cryogenics for future high-energy accelerators*, Proc. ICEC21 Prague, Icaris (2006) 13-21.
46. O. Bruning, et al. (eds.): *LHC design report*, CERN 2004-003, Vol. I, Chapter 11.
47. R.D. Mc Carty: *Thermodynamic properties of helium 4 from 2 to 1500 K at pressures to 10^8 Pa*, J. Chem. Phys. Ref. Data 2 (1973) 923.
48. V. Arp: *HEPAK, Thermophysical properties of helium*, www.cryodata.com
49. R.D. Mc Carty: *GASPAK, Thermophysical properties of 36 fluids*, www.cryodata.com
50. J. Wilks, D.S. Betts: *An introduction to liquid helium*, Clarendon Press, Oxford (1987).
51. S.W. van Sciver: *Helium cryogenics*, 2nd ed., Springer, New York (2012).
52. F. Vinen: *The physics of superfluid helium*, CERN-2004-008, Geneva (2004) 363.
53. Ph. Lebrun, L. Tavian: *Cooling with superfluid helium*, CERN-2014-005, Geneva (2014) 453.
54. NIST Cryogenic Technologies Group: Cryogenic properties of materials, www.cryogenics.nist.gov
55. P. Duthil: *Material properties at low temperature*, CERN-2014-005, Geneva (2014) 77.
56. J.G. Weisend (ed.): *Handbook of cryogenic engineering*, Taylor & Francis, Philadelphia (1998).
57. G. Vandoni: *Heat transfer*, CERN-2004-008, Geneva (2004) 325.
58. B. Baudouy: *Heat transfer and cooling techniques at low temperature*, CERN-2014-005, Geneva (2014) 329.
59. M.C. Jones, V. Arp: *Review of hydrodynamics and heat transfer for large helium cooling systems*, in: Advances in refrigeration at the lowest temperatures, IIR-IIF Commission A1-2, Zürich (1978) 41.
60. W. Obert, et al.: *Emissivity measurements of metallic surfaces used in cryogenic applications*, Adv. Cryo. Eng. 27 (1982) 293.
61. Yu.L. Buyanov: *Current leads for use in cryogenic devices, principle of design and formulae for design calculations*, Cryogenics 25 (1985) 94.
62. A. Ballarino: *Current leads, links and buses*, CERN-2014-005, Geneva (2014) 547.
63. Ph. Lebrun: *Design of a cryostat for superconducting accelerator magnets: the LHC main dipole case*, CERN-2004-008, Geneva (2004) 348.
64. V. Parma: *Cryostat design*, CERN-2014-005, Geneva (2014) 353.
65. J.G. Weisend (ed.): *Cryostat design: case studies, principles and engineering*, Springer, New-York (2016).

66. U. Wagner: *Refrigeration*, CERN-2004-008, Geneva (2004) 295.
67. A. Alekseev: *Basics of low-temperature refrigeration*, CERN-2014-005, Geneva (2014) 111.
68. S. Carnot: *Réflexions sur la puissance motrice du feu et sur les machines propres à developper cette puissance*, Bachelier, Paris (1824) and Librairie philosophique Vrin, Paris (1978).
69. Z. Rant: *Exergie, ein neues Wort für "technische Arbeitsfähigkeit"*, Forsch.-Ing.-Wes., 22 (1956).
70. S. Claudet, et al.: *Economics of large helium cryogenic systems: experience from recent projects at CERN*, Adv. Cryo. Eng. 45B (2000) 1301.
71. F. Bordry: *Power Converters for Particle Accelerators*, Keynote presentation, 11th Eur. Conf. Power Electronics EPE 2005, Dresden, Germany, Sept. 2005.
72. F. Bordry: *Power converters: definitions, classification and converter topologies*, Specialised CERN Accelerator course "Power Converters" -Warrington, UK, May 2004.
73. A. Beuret, F. Bordry, J.P. Burnet, C. De Almeida Martins: *A 4-quadrant 300kW-peak high precision and bandwidth switch mode power converter for particle accelerator magnets supply*, 12th Eur. Conf. Power Electronics and Applications, Aalborg, Denmark, Sept. 2007.
74. C. Fahrni, A. Rufer, F. Bordry, J.P. Burnet: *A novel 60 MW Pulsed Power System based on Capacitive Energy Storage for Particle Accelerators*, EPE Journal 18(4) (Dec. 2008) 5.
75. G. Fernqvist, B. Halvarsson, J. Pett, J. Pickering: *A Novel Current Calibration System up to 20kA*, IEEE Trans. Instrum. Meas. 52(2) (April 2003) 445.
76. I. Barnett, G. Fernqvist, D. Hundzinger, J.-C. Perréard, J.G. Pett: *A strategy for controlling the LHC magnet currents*, 5th Eur. Part. Accel. Conf., Sitges, Barcelona, Spain, 10 - 14 Jun 1996, pp. 2317-2319.
77. J. Carwardine, F. Lenkszus: *Trends in the Use of Digital Technology for Control and Regulation of Power Supplies*, Intern. Conf. Accelerator and Large Experimental Physics Control Systems, 1999, Trieste, Italy.
78. H. Thiesen, M. Cerqueira Bastos, G. Hudson, Q. King, V. Montabonnet, D. Nisbet, S. Page: *High Precision Current Control for the LHC Main Power Converters Digital*, IPAC'10, Kyoto, Japan.
79. I.D. Landau: *The R-S-T digital controller design and applications*, Control Eng. Pract. 6 (1998) 155-165.
80. F. Bordry, H. Thiesen: *RST Digital Algorithm for controlling the LHC magnet current*, Electrical Power Techonology in European Physics Research EP2, Grenoble (France), Oct. 1998.
81. K. Unser: *Beam current transformer with DC to 200 MHz range*, IEEE Trans. Nucl. Sci. 16(3) (1969) 934-938.
82. C. Adamson, N.G. Hingorani: *New transductor type DC transformer particularly applicable to HV DC systems*, Proc. IEE 110(4) (April 1963) 739-750.
83. H. Appelo, M. Groenenboom, J. Lisser: *The zero-flux DC current transformer – a high precision bipolar wide-band measuring device*, IEEE Trans. Nucl. Sci. 24(3) (June 1977) 1810-1811.
84. G. Hudson, K. Bouwknegt: *4-13kA DC Current Transducers Enabling Accurate In-Situ Calibration for a New Particle Accelerator Project, LHC*, Eur. Conf. Power Electronics and Applications, 2005.
85. R. Hasegawa: *Advances in amorphous and nanocrystalline magnetic materials*, J. Magn. Magn. Mater. 304(2) (Sept. 2006) 187-191.
86. G. Fernqvist, P. Dreesen, G. Hudson, J. Pickering: *Characteristics of Burden Resistors for High precision DC Current Transducers*, Particle Accelerator Conference, June 2007, Albuquerque, New Mexico.
87. N. Beev, Analog-to-digital conversion beyond 20 bits: Applications, Architectures, State of the Art, Limitations, and Future Prospects, I2MTC 2018.
88. A. Belcher, J. Pett, J. Pickering: *Design and evaluation of a metrology class delta-sigma analogue to digital converter for the LHC project at CERN,* IEE NMC/BEMC 2001, 2nd National Measurement Conference, Harrogate, UK, p. 4.

89. G. Femqvist, B. Halvarsson, J. Pett: *The CERN Current Calibrator – a new type of instrument*, Conf. Precision Electromagnetic Measurements 2002.
90. G. Fernqvist, G. Hudson, J. Pickering, F. Power: *Design and Evaluation of a 10-mA DC Current Reference Standard*, IEEE Trans. Instrum. Meas. 52(2) (April 2003) 440.
91. C. Wyss (ed.): LEP Design Report, Vol. 3, CERN-AC-96-01-LEP-2, CERN (1996) 224 pages.
92. O.S. Brüning, P. Collier, P. Lebrun, S. Myers, R. Ostojic, J. Poole, P. Proudlock (eds.): LHC Design Report, Vol. 1: The LHC Main Ring, CERN-2004-003-V-1, CERN (2004) 548 pages.
93. O. Gröbner: *Overview of the LHC vacuum system*, Vacuum 60 (2001) 25-34.
94. V. Baglin: *Cold/sticky systems*, CAS - CERN Accelerator School and ALBA Synchrotron Light Facility: Course on Vacuum in Accelerators, Platja d'Aro, Spain, 16 - 24 May 2006, pp. 351-368.
95. Handbook of Vacuum Technology, new edition, New York, NY: Wiley, (2008) 1040 pages.
96. J.M. Lafferty: Foundations of Vacuum Science and Technology, New York, NY, Wiley, (1998) 728 pages.
97. A. Roth: Vacuum Technology, 3rd ed., Amsterdam : North-Holland, (1990) 554 pages.
98. CAS - CERN Accelerator School: Vacuum Technology, CERN 1999-05 19, CERN (1999).
99. CAS - CERN Accelerator School: Vacuum in Accelerators, CERN-2007-003, CERN (2007).
100. C. Herbeaux, et al.: J. Vac. Sci. Technol. A 17(2) (Mar/Apr 1999) 635.
101. O.B. Malyshev, et al.: Vacuum 75 (2004) 155.
102. V. Baglin, et al.: Proc. EPAC 2002, Paris, France.
103. N. Hilleret, et al.: Proc. EPAC 2002, Paris, France.
104. J. Gómez-Goñi, et al.: J. Vac. Sci. Technol. A 15(6) (Nov/Dec 1997) 3093.
105. O.B. Malyshev, et al.: J. Vac. Sci. Technol. A 28(8) (Sep/Oct 2010) 1215.
106. H. Tratnik, et al.: Vacuum 81 (2007) 731.
107. N. Hilleret, et al.: Proc. EPAC 2000, Vienna, Austria.
108. R. Calder: *Ion induced gas desorption problems in the ISR*, Vacuum 24 (1974) 437-443
109. A. Rossi, et al.: Proc. PAC 2001, Chicago, USA.
110. E. Mahner: Phys. Rev. ST Accel. Beams 11 (2008) 104801.
111. M. Audi, M. de Simon: Vacuum 37 (1987) 629.
112. A.K. Gupta, J.H. Leck, Vacuum 25 (1975) 362.
113. NEG cartridge pumps.
114. M.D. Malev, E.M. Trachtenberg: Vacuum 23 (1973) 403.
115. C. Benvenuti, F. Froncia: J. Vac. Sci. Technol. A 6 (1988) 2528.
116. C. Benvenuti, et al.: Vacuum 53 (1999) 317. Ref [a]: High-Luminosity Large Hadron Collider (HL-LHC) Technical Design Report V.0.1, CERN-2017-007-M Ref [b]: J. F. O'Hanlon, A user's guide to vacuum technology, Wiley, 2003.
117. J. Bosser (ed.): *Beam Instrumentation*, CERN-PE-ED 001-92, Rev. 1994.
118. D. Brandt (ed.): *Beam Diagnostics for Accelerators*, Proceedings of the CERN Accelerator School, Dourdan, CERN-2009-005 (2009).
119. A. Nosych et al: *Overview of the geometrical non-linear effects of button BPMs and methodology for their efficient suppression*, Proceedings of the International Beam Instrumentation Conference, Monterey, USA (2014) p. 298
120. T. Levens, K. Lasocha and T. Lefevre: *Recent developments for instability monitoring at the LHC*, Proceedings of the International Beam Instrumentation Conference, Barcelona, Spain (2016) p. 852
121. S. Walston *et al*: *Performance of a High Resolution Cavity Beam Position Monitor System*, Nuclear Instruments and Methods in Phys. Rev. A578, 2007, 1-22
122. M. Gasior: *An inductive pick-up for beam position and current measurements*, Proceedings of the Beam Diagnostics and Instrumentation for Particle Accelerators Conference, Mainz, Germany, (2003) p. 53
123. G. Vismara: *Signal Processing for Beam Position Monitors*, Proceedings of the Beam Instrumentation Workshop, Cambridge, MA, USA, (2000) p.36-60

124. R.C. Webber: *Charged particle beam current monitoring tutorial*, Proceedings of the Beam Instrumentation Workshop, Cambridge, MA, USA (2000)
125. K. L. Brown et G. W. Tautfest: *Faraday-Cup Monitors for High-Energy Electron Beams*, Review of Scientific Instruments 27 (1956) 696
126. M. Krupa and M. Gasior: *The wall current transformer – A new sensor for precise bunch-by-bunch intensity measurements in the LHC*, Proceedings of the International Beam Instrumentation Conference, Barcelona, Spain (2016) p. 568
127. K. Unser: IEEE Trans. Nucl. Sci. NS-16 (1969) p. 924-938
128. H. Schmickler: *Diagnostics and Control of the Time Evolution of Beam Parameters*, Proceedings of the Beam Diagnostics and Instrumentation for Particle Accelerators Conference, Frascati, Italy, (1997)
129. M. Gasior and R. Jones: *High Sensitivity Tune Measurement by Direct Diode Detection*, Proceedings of the Beam Diagnostics and Instrumentation for Particle Accelerators Conference, Lyon, France (2005), p. 310
130. M. Gasior: *Farady cup award - High Sensitivity Tune Measurement by Direct Diode Detection,* Proceedings of the Beam Instrumentation Workshop, Newport News, USA (2012) p. 1
131. A. Marusic: *Chromaticity Feedback at RHIC*, Proceedings of the International Particle Accelerator Conference, Kyoto, Japan (2010) p. 525
132. R. Jones, et al.: *Towards a robust phase locked loop tune feedback system*, Proceedings of the Beam Diagnostics and Instrumentation for Particle Accelerators Conference, Lyon, France (2005) p. 298
133. M. Plum: *Interceptive Beam Diagnostics-Signal Creation and Materials Interactions*, Proceedings of the Beam Instrumentation Workshop, Knoxville, TN, USA, (2004) p. 23-46
134. B. Walasek-Höhne et al: *Scintillating screen applications in accelerator beam diagnostics*, IEEE Transactions on Nuclear Science Vol. 59, No. 5 (2012) p. 2307
135. V. Ginsburg: Sov. Phys. JETP 6, 1079, (1958) and 10, 372 (1960)
136. J. Bosser, et al.: *Optical transition radiation proton beam profile monitor*, Nuclear Instrument And Methods In Phys. Res. A 238 (1985) 45
137. P. Karataev et al: *First observation of the point spread function of optical transition radiation*, Physical Review Letters 107 (2011) 174801
138. B. Bolzon et al: *Very high resolution optical transition radiation imaging system*, Physical Review special topics on Accelerator and Beams 18 (2015) 082803
139. P. Karataev *et al*, *"Beam-size measurement with Optical Diffraction Radiation at KEK Accelerator Test Facility"*, Physical Review Letters 93 (2004) 244802
140. A. Cianchi, M. Castellano, L. Catani, E. Chiadroni, K. Honkavaara, and G. Kube: *Non-intercepting electron beam size monitor using optical diffraction radiation interference*, Physical Review special topics on Accelerator and Beams 14 (2011) 102803
141. J.L. Sirvent: *Performance assessment of pre-series fast beam wire scanner prototypes for the upgrade of the CERN LHC injector complex*, Proceedings of the International Beam Instrumentation Conference, Grand Rapids, Michigan, USA (2017) p.338
142. T. Hofmann et al: *Demonstration of a laserwire emittance scanner for hydrogen ion beams at CERN*, Physical Review special topics on Accelerator and Beams 18 (2015) 122801
143. S.T. Boogert *et al*, "*Micron-scale laser-wire scanner for the KEK Accelerator Test Facility extraction line*", Physical Review Special Topics –Accelerators and Beams 13 (2010) 122801
144. P. Forck: *Minimal invasive beam profile monitors for high intense hadron beams*, Proceedings of the International Particle Accelerator Conference, Kyoto, Japan (2010) p. 1261
145. H. Sandberg et al: *First use of Timepix3 hybrid pixel detectors in ultra-high vacuum for beam profile measurements*, Journal of Instrumentation 14 (2019) C01013
146. A. Hofmann: *The Physics of Synchrotron Radiation*, Cambridge university press (2000)
147. T. Naito and T. Mitsuhashi: *Very small beam-size measurement by a reflective synchrotron radiation interferometer*, Physical Review Special Topics –Accelerators and Beams 9 (2006) 122802

148. S. Takano *et al*: *X-ray imaging of a small electron beam in a low-emittance synchrotron light source*, Nuclear Instruments and Methods in Phys. Rev. A 556 (2006) 357
149. R.E. Shafer: *A tutorial on beam loss monitoring*, Proceedings of the Beam Instrumentation Workshop New-York, USA (2002) p. 44
150. W. Panofsky: *The SLAC long ionisation chamber for machine protection,* SLAC Internal Report TN-63-57 (1963)
151. F. Wulf and M. Korfer: *Local beam loss and beam profile monitoring with optical fibers,* Proceedings of the Beam Diagnostics and Instrumentation for Particle Accelerators Conference, Basel, Switzerland (2009) p.411
152. E.B. Holzer et al: *Beam loss monitoring system for the LHC,* CERN-AB-2006-009, Proceedings of the IEEE Nuclear Science Symposium and Medical Imaging Conference, San Juan, Puerto Rico (2005) p.1052
153. S.S. Gilardoni et al: *Beam loss monitors comparison at the CERN proton synchrotron,* Proceedings of the International Particle Accelerator Conference, San Sebastián, Spain (2011), p. 1341
154. E. Griesmayer, et al: *A Fast CVD Diamond Beam Loss Monitor for LHC*, Proceedings of the Beam Diagnostics and Instrumentation for Particle Accelerators Conference, Hamburg, Germany, (2011) p. 143
155. K. Wittenburg: *The PIN-diode beam loss monitor system at HERA,* Proceedings of the Beam Instrumentation Workshop, Conference, Cambridge, Massachusetts, USA (2000), p. 3
156. C.P. Welsch, et al: *Longitudinal beam profile measurements at CTF3 using a streak camera*, Journal of Instrumentation 1 (2006) P09002
157. M. Castellano, et al: *Measurement of Coherent Diffraction Radiation and its Applications for Bunch Length Diagnostics in Particle Accelerators*, Physical Review E 63 (2001) 056501
158. J. Maxson et al: *Direct Measurement of sub-10fs Relativistic Electron Beams with Ultralow Emittance*, Physical Review Letters 118 (2017) 154802
159. I. Wilke et al: *Single-Shot Electron-Beam Bunch Length Measurements*, Physical Review Letters 88 (2002) 124801
160. G. Berden et al: *Benchmarking of Electro-Optic Monitors for Femtosecond Electron Bunches*, Physical Review Letters 99, (2007) 164801
161. A.L. Cavalieri et al: *Clocking Femtosecond X-rays*, Physical Review Letters 94, (2005) 114801
162. M.J. Barnes, et al.: *Injection and Extraction Magnets: Kicker Magnets*, Proc. CERN Accelerator School on Magnets, Bruges, Belgium, June 16–25, 2009.
163. M.J. Barnes, J. Borburgh, B. Goddard, M. Hourican: *Injection and Extraction Magnets: Septa*, Proc. CERN Accelerator School on Magnets, Bruges, Belgium, June 16–25, 2009.
164. W. Bartmann et al, *Impact of LHC and SPS Injection Kicker Rise Times on* LHC *Filling Schemes and Luminosity Reach,* Proc. 8th Intern. Particle Accelerator Conf. (IPAC17), Copenhagen, Denmark, May 14–19, 2017.
165. D. Fiander, K.D. Metzmacher, P.D. Pearce: *Kickers and septa at the PS complex*, CERN, KAON PDS Magnet Design Workshop, Vancouver, Canada, October 3–5, 1988, pp. 71–79.
166. T. Naito, et al.: *Development of strip-line kicker system for ILC damping ring*, Proc. 22nd Particle Accelerator Conf. (PAC'07), Albuquerque, New Mexico, USA, June 25–29, 2007, pp. 2772–2774.
167. D. Alesini, S. Guiducci, F. Marcellini, P. Raimondi: *Fast injection kickers for Daphne collider and ILC damping rings*, DAPHNE Technical Note, INFN - LNF, Accelerator Division. Note I-17, June 6, 2006.
168. M.J. Barnes, F. Caspers, T. Kroyer, E. Métral, F. Roncarolo, B. Salvant: *Measurement of longitudinal and transverse impedance of kicker magnets using the coaxial wire method*, Proc. 23rd Particle Accelerator Conf. (PAC'09), Vancouver, Canada, May 4–8, 2009.
169. F. Caspers, A. Mostacci, H. Tsutsui: *Impedance Evaluation of the SPS MKE Kicker with Transition Pieces between Tank and Kicker Module*, CERN CERN-SL-2000-071 (AP).

170. E.H.R. Gaxiola, J.A. Uythoven, M.A. Timmins: *Upgrade of the SPS Extraction Kickers for LHC and CNGS Operation*, Proc. 8th Eur. Particle Accelerator Conf. (EPAC'02), Paris, France, June 3–8, 2002.
171. F. Caspers, et al.: *The Fast Extraction Kicker System in SPS LSS6*, Proc. 10th Eur. Particle Accelerator Conf. (EPAC'06), Edinburgh, Scotland, June 26–30, 2006.
172. C. Zannini, *Electromagnetic Simulation of CERN Accelerator Components and Experimental Applications*, Thesis No. 5737 (2013), EPFL. https://cds.cern.ch/record/1561199
173. M.J. Barnes, F. Caspers, L. Ducimetière, N. Garrel, T. Kroyer: *An improved beam screen for the LHC injection kickers*, Proc. 22nd Particle Accelerator Conf. (PAC'07), Albuquerque, New Mexico, USA, June 25–29, 2007.
174. E. Carlier, F. Castronuovo, L. Ducimetière, E.B. Vossenberg: *A High Power Pulse System for the Beam Extraction from CERN's Large Hadron Collider*, Proc. 2008 IEEE Intern. Power Modulators and High Voltage Conf., May 27–31, 2008.
175. M.J. *Barnes* et al., *Operational Experience of the Upgraded LHC Injection Kicker Magnets During Run 2 and Future Plans,* Proc. 8th Intern. Particle Accelerator Conf. (IPAC17), Copenhagen, Denmark, May 14–19, 2017.
176. M.J. Barnes, B. Goddard: *Considerations on a New Fast Extraction Kicker Concept for SPS*, CERN sLHC Project Note 0018, June 2010.
177. K. Takayama, R.J. Briggs (eds.): Induction Accelerators, 2011, ISBN 978-3-642-13916-1 (hbk).
178. L. Ducimetière, U. Jansson, G.H. Schröder, E.B. Vossenberg, M.J. Barnes, G.D. Wait: *Design of the injection kicker magnet system for CERN's 14TeV proton collider LHC*, Proc. 10th Intern. Pulsed Power Conf., Albuquerque, New Mexico, USA, July 10–13, 1995.
179. V. Senaj, N. Voumard, M.J. Barnes, L. Ducimetière: *Optically isolated circuit for failure detection of a switch in an HV series connected stack*, Proc. 17th IEEE Intern. Pulsed Power Conf., Washington DC, USA, June 29 – July 2, 2009.
180. V. Senaj, L. Ducimetière, E. Vossenberg: *Upgrade of the Super Proton Synchrotron Vertical Beam Dump System*, Proc. 1st Intern. Particle Accelerator Conf. (IPAC10), Kyoto, Japan, May 23–28, 2010.
181. E.G. Cook, *Review of Solid-State Modulators*, Linac 2000, Proc. XX Int. Linear Accelerator Conf., Monterey, CA, USA, Aug. 21-25, 2000.
182. J. Holma, M.J. Barnes, *Measurements on a 20-Layer 12.5 kV Prototype Inductive Adder for the CLIC DR Kickers Magnet*, to be Publ. in Proc. 2017 IEEE Pulsed Power Conference, Brighton, U.K., June 18-22, 2017.
183. J. Holma, *A Pulse power modulator with extremely flat-top output pulses for the Compact Linear Collider at CERN*, Ph.D. Thesis, Aalto University publication series, Doctoral Dissertations 196/2015, Helsinki, Finland, 2015.
184. M.A. Kemp, A. Benwell, C. Burkhart, R. Larsen, D. MacNair, M. Nguyen, J. Olsen, *Status Update on the Second-Generation ILC Marx Modulator Prototype*, Power Modulator and High Voltage Conference (IPMHVC), 2010 IEEE International, Atlanta, GA, USA, 23-27 May, 2010.
185. L.M. Redondo, A. Kandratsyeu, M.J. Barnes, T. Fowler, *Design Strategies for a SiC Marx Generator for Kicker Magnet*, to be Publ. in Proc. 2017 IEEE Pulsed Power Conference, Brighton, U.K., June 18-22, 2017
186. G.H. Schröder, J. Bonthond, U. Jansson, H. Kuhn, M. Mayer, E.B. Vossenberg: *The Injection Kicker Systems of LEP*, Proc. Eur. Particle Accelerator Conf. (EPAC'88), Rome, Italy, June 7–11, 1988, Vol. 2, p. 1381, ISBN 9971-50-642-4, and CERN SPS/88-26 (ABT).
187. I. Rodriguez, F. Toral, M.J. Barnes, T. Fowler, G. Ravida: *Design, Manufacturing and Testing of the CTF3 Tail Clipper Kicker*, Proc. 1st Intern. Particle Accelerator Conf. (IPAC10), Kyoto, Japan, May 23–28, 2010.
188. A.W. Chao, M. Tigner (eds.): Handbook of Accelerator Physics and Engineering, World Scientific, 1998, ISBN 9180238584 (pbk).

189. M. Thivent: Développements Liés à la Construction des Deflecteurs Electrostatiques, CERN PS/PSR/Note 83-8 (in French).
190. A. Durand, *Scattering of protons in a wire array*, N.I.M. 127 (1975) 349-354
191. V. Nagaslaev et al., *Mars tracking simulations for the MU2E slow extracted proton beam*, Proc. 6th Intern. Particle Accelerator Conf. (IPAC15), Richmond, USA, May 3–8, 2015.
192. J. Borburgh, M. Crescenti, M. Hourican, T. Masson: *Design and Construction of the LEIR Extraction Septum*, IEEE Trans. Applied Superconductivity 16(2) (June 2006).
193. M.J. Barnes, et al.: *Development of an Eddy Current Septum for LINAC4*, Proc. 11th Eur. Particle Accelerator Conf. (EPAC'08), Genoa, Italy, June 23–27, 2008.
194. Z. Szoke et al., *Direct-drive and eddy current septa magnet designs for CERN's PSB extraction at 2 GeV*, IEEE trans. On appl. Superconductivity, vol. 26, No.4, June 2016.
195. M. Barnes et al., *Development of an eddy current septum for LINAC4*, Proc. 11th Europeans Particle Conf. (EPAC2008), Genoa, Italy, June 23-27, 2008.
196. P. Lebasque et al., *Eddy current septum magnets for booster injection and extraction, and storage ring injection at synchrotron Soleil*, Proc. 10th Europeans Particle Conf., Edinburgh, Scotland, June 26-30, 2006.
197. M. Sassowsky, et al.: *Steel Septum Magnets for the LHC Beam Injection and Extraction*, Proc. 8th Eur. Particle Accelerator Conf. (EPAC'02), Paris, France, June 3–8, 2002.
198. M. Hub, *Measuring machine and results of the magnetic measurements for the steel septum magnets for the ISR injection*, CERN-ISR/BT/71-19, CERN internal note, 1971.
199. J. Rank et al., *The extraction lambertson septum magnet of the SNS*, Proc. Particle Accelerator Conf. 05 (PAC'05), Knoxville, Tennessee, May 16-20, 2005
200. R. Muto et al., *Development of Lambertson magnet and septum magnets for splitting 30-GeV proton beam in Hadron experimental Facility at J-Parc*, IEEE trans. On appl. Superconductivity, vol. 26, No.4, June 2016
201. Y. Yonemura et al., *Beam extraction of the POP FFAG with a massless septum*, Proc. Particle Accelerator Conf. 03 (PAC'03), Portland, Oregon, May 12-16, 2003
202. Y. Iwashita et al., *Massless septum with hybrid magnet*, Proc. 11th Europeans Particle Conf. (EPAC2008), Genoa, Italy, June 23-27, 2008.
203. O. Payir et al., *Massless beam separation system for intense ion beams*, Proc. 6th Intern. Particle Accelerator Conf. (IPAC15), Richmond, USA, May 3–8, 2015.
204. P. Brindza et al., *Superconducting septum magnet design for Jefferson Lab hall A*, IEEE trans. On appl. Superconductivity, vol. 11, No.1, March 2001
205. F. Krienen, *The truncated double cosine theta superconducting septum magnet*, N.I.M. in Physics research sec. A, Vol. 283, Issue 1, pages 5-12, 20 October 1989
206. A. Yamamoto et al., *The superconducting inflector for the BNL g-2 experiment*, N.I.M. in Physics Research A 491 (2002) 23-40
207. K. Sugita, *Novel concept of truncated iron-yoked cosine theta magnets and design studies for FAIR septum magnets,* IEEE trans. On Appl. Superconductivity, vol. 22, NO. 3, June 2012
208. D. Barna, *High field septum magnet using a superconducting shield for Future Circular Collider*, Physical Review Accelerators and beam 20, 2017
209. H. Bartosik et al., *Proposal of a dummy septum to mitigate ring irradiation for the CERN PS multi-turn extraction*, Proc. 3rd Intern. Particle Accelerator Conf. (IPAC15), New Orleans, Louisiana, USA, May 20-25, 2012.
210. T. Risselada et al, "The ISR Collimation System". PAC 1979, San Francisco, NS-26, No. 3, 4131-3.
211. G. von Holtey, "Electron Beam Collimation at LEP Energies", Proc. PAC 1987 Washington, IEEE Cat. No. 87CH2387-9), Vol. 2, 1252-4.
212. D.R. Waltz, A. McFarlane, E. Lewandowski, J. Zabdyr, "Momentum Slits, Collimators and Masks in the SLC". SLAC-PUB-4965, C89-03-20.1. Apr 1989.
213. L. Burnod, J.B. Jeanneret, "Beam Losses and Collimation in the LHC: A Quantitative Approach", CERN SL/91-39 (EA), LHC Note 167 (1991).

214. M.A. Maslov, N.V. Mokhov, I.A. Yazynin, "The SSC Beam Scraper System", SSCL-484 June 1991.
215. P. Bryant and E. Klein, "The Design of Betatron and Momentum Collimation Systems". CERN/SL-42-90 (AP). Aug. 1992.
216. T. Trenkler and J.B. Jeanneret, "The Principles of Two-Stage Betatron and Momentum Collimation in Circular Accelerators". Particle Accelerators, 1995, Vol. 50, pp. 287-311.
217. D. Kaltchev, M.K. Craddock, R.V. Servranckx, J.B. Jeanneret, "Numerical optimization of collimator jaw orientations and locations in the LHC". CERN-LHC-PROJECT-REPORT-134. Sep 1997.
218. J.B. Jeanneret, "Optics of a two-stage collimation system". Physical Review Special Topics – Accelerators and Beams, Vol 1, 081001 (1998).
219. M. Seidel, "The Proton Collimation System of HERA", DESY 94-103 (1994).
220. M. Church, A.I. Drozhdin, A. Legan, N.V. Mokhov, R. Reilly. "Tevatron run-II beam collimation system". FERMILAB-CONF-99-059, Apr 1999. Given at PAC 99, New York, NY, 29 Mar - 2 Apr 1999.
221. N. Mokhov et al, "Beam Collimation at Hadron Colliders". ICFA Workshop on Beam Halo Dynamics, Diagnostics, and Collimation (HALO'03), Montauk, Long Island, NY, May 19-23, 2003. FERMILAB-Conf-03/220 July 2003.
222. A. Drees, R. Fliller, W. Fu, "RHIC loss limitations and collimation". AIP Conf. Proc. 773 (2005) 55-59.
223. N. Catalan-Lasheras et al, "Optimization of the collimation system for the Spallation Neutron Source accumulator ring". Physical Review Special Topics – Accelerators and Beams, Vol 4, 010101 (2001).
224. S. Cousineau, N. Catalan Lasheras, J. Holmes, D. Davino, H. Ludewig, "SNS beam-in-gap cleaning and collimation". AIP Conf. Proc. 642 (2003) 170-173.
225. N. Simos, H. Ludewig, D. Raparia, J. Brodowski, N. Catalan Lasheras, G. Murdoch, "SNS collimating system design: Performance and integration". AIP Conf. Proc. 693 (2004) 162-166.
226. M. Kinsho (JAERI, Tokai), "Lattice and collimation system for J-PARC". AIP Conf. Proc. 773 (2005) 45-49.
227. M.J. Shirakata, H. Oki, T. Oogoe, Y. Takeuchi, M. Yoshioka, "Beam collimator system in the J-PARC 3-50BT line". Proc. EPAC 06, 26-30 Jun 2006, Edinburgh, Scotland.
228. K. Yamamoto, M. Abe, H. Hanaue, A. Nakamura, Y. Takeuchi, Y. Hirooka, M. Okazaki, "Present status of beam collimation system of J-PARC RCS". Proc. EPAC 06, 26-30 Jun 2006, Edinburgh, Scotland.
229. M. Tomizawa, A. Molodozhentsev, M. Shirakata, "Design of dynamic collimator for J-PARC main ring". Proc. Particle Accelerator Conference (PAC 07), 25-29 Jun 2007, Albuquerque, New Mexico.
230. Y. Lee, M. Gandel, D. Kiselev, D. Reggiani, M. Seidel, S. Teichmann, "Simulation based optimization of a collimator system at the PSI proton accelerator facilities". IPAC-2010-THPEC088. May 2010.
231. S. Di Mitri, "Geometric efficiency of a two-stage fully absorbing collimation system in single-pass linacs". Physical Review Special Topics – Accelerators and Beams 13, 052801 (2010).
232. A. Drohzdin et al, "Comparison of the TESLA, NLC and CLIC Beam-Collimation System Performance". Proc. PAC03.
233. R.W. Assmann et al. "Requirements for the LHC collimation system". CERN-LHC-PROJECT-REPORT-599, 2002. EPAC02, La Vilette, Paris, France, 3-7 Jun 2002.
234. R. Assmann, "Collimation for the LHC High intensity beams". Proc. 46th ICFA Advanced Beam Dynamics Workshop on High-Intensity and High-Brightness Hadron Beams (HB2010). Sep. 27 – Oct 1 2010, Morschach, Switzerland.
235. S. Redaelli, "Beam cleaning and collimation systems," CERN Yellow Report CERN-2016-002, pp.403-437

236. G.Apollinari et al., “High-Luminosity Large Hadron Collider (HL-LHC): Technical Design Report V. 0.1.” CERN Yellow Reports: Monographs. CERN-2017-007-M, Geneva: CERN, 2017.
237. G. Valentino et al., “Final implementation, commissioning, and performance of embedded collimator beam position monitors in the Large Hadron Collider,” Phys. Rev. Accel. Beams 20 (2017) no.8, 081002
238. R. Bruce, A. Marsili, S. Redaelli, “Cleaning Performance with 11T Dipoles and Local Dispersion Suppressor Collimation at the LHC,” Proceedings, 5th International Particle Accelerator Conference (IPAC 2014): Dresden, Germany, June 15-20, 2014. DOI: 10.18429/JACoW-IPAC2014-MOPRO042
239. D. Mirarchi et al., “Cleaning Performance of the Collimation System of the High Luminosity Large Hadron Collider,” Proceedings, 7th International Particle Accelerator Conference (IPAC 2016): Busan, Korea, May 8-13, 2016, doi 10.18429/JACoW-IPAC2016-WEPMW007.
240. C. Omet, H. Kollmus, H. Reich-Sprenger, P.J. Spiller, “Ion Catcher System for the Stabilisation of the Dynamic Pressure in SIS18”. EPAC08-MOPC099. Jun 23, 2008.
241. R. Assmann et al, “The final collimation system for the LHC”. EPAC06. LHC-Project-Report-919.
242. D. Wollmann et al., “Beam feasibility study of a collimator with in-jaw beam position monitors,” Nucl.Instrum.Meth. A768 (2014) 62-68
243. The FCC collaboration (Abada, A. and others), “HE-LHC: The High-Energy Large Hadron Collider,” Eur.Phys.J.ST 228 (2019) no.5, 1109-1382
244. The FCC collaboration (Abada, A. and others), “FCC-hh: The Hadron Collider: Future Circular Collider Conceptual Design Report Volume 3,” Eur.Phys.J.ST 228 (2019) no.4, 755-1107
245. N.V. Mokhov, P.C. Czarapata, A.I. Drozhdin, D.A. Still, R.V. Samulyak. “Beam-induced damage to the Tevatron components and what has been done about it”. FERMILAB-CONF-06-415-AD, FERMILAB-APC, Nov 2006. Presented at HB2006, Tsukuba, Japan, 29 May - 2 Jun 2006.
246. R. Bruce, R. Assmann, S. Redaelli, “Calculations of safe collimator settings and β_* at the CERN Large Hadron Collider,” Phys.Rev.ST Accel.Beams 18 (2015) no.6, 061001
247. R. Bruce et al., “Reaching record-low $\beta_*\beta_*$ at the CERN Large Hadron Collider using a novel scheme of collimator settings and optics,” Nucl.Instrum.Meth. A848 (2017) 19-30
248. A.Bertarelli et al, “The Mechanical Design for the LHC Collimators”. EPAC04. LHC-Project-Report-786.
249. A. Bertarelli, A. Dallocchio, L. Gentini, N. Mariani, R. Perret, M. Timmins, “Mechanical Engineering and Design of the LHC Phase II Collimators”. IPAC-2010-TUPEB071. May 2010.
250. A. Masi and R. Losito, “LHC Collimator Lower Level Control System,” 15th IEEE NPSS Real Time Conference 2007.
251. G. Valentino et al., “Semiautomatic beam-based LHC collimator alignment,” Phys.Rev.ST Accel.Beams 15 (2012) 051002
252. G. Valentino et al., “Successive approximation algorithm for beam-position-monitor-based LHC collimator alignment,” Phys.Rev.ST Accel.Beams 17 (2014) no.2, 021005
253. A. Bertarelli, O. Aberle, R.W. Assmann, A. Dallocchio, T. Kurtyka, M. Magistris, M. Mayer, M. Santana-Leitner. “Permanent deformation of the LHC collimator jaws induced by shock beam impact: An analytical and numerical interpretation.” Proc. EPAC 06, Edinburgh, Scotland, 26-30 Jun 2006.
254. D. Onoprienko, M. Seidel, P. Tenenbaum, “Measurement of resistivity dominated collimator wakefield kicks at the SLC”. SLAC-PUB-10192. Jun 2002.
255. P. Tenenbaum et al, “Collimator Wakefield Calculations for ILC-TRC Report”. SLAC-TN-03-038. LCC-0101. Aug. 2002.
256. E. Metral et al, “Transverse Impedance of LHC Collimators”. PAC2007. CERN-LHC-PROJECT-Report-1015.

257. M. Brugger, S. Roesler. "Remanent dose rates around the collimators of the LHC beam cleaning insertions". Radiat. Prot. Dosim.115: 470-474, 2005.
258. A. Ferrari, T. Rancati, P.R. Sala, "FLUKA applications in high energy problems: From LHC to ICARUS and atmospheric showers". 3rd Workshop On Simulating Accelerator Radiation Environments (SARE3). 7-9 May 1997, Tsukuba, Japan.
259. R. Assmann, M. Brugger, M. Hayes, J.B. Jeanneret, F. Schmidt, I. Baishev, D. Kaltchev, "Tools for predicting cleaning efficiency in the LHC.". Proc. PAC03. CERN-LHC-PROJECT-REPORT-639.
260. S. Redaelli, R.W. Assmann, G. Robert-Demolaize (CERN), "LHC aperture and commissioning of the collimation system". LHC Project Workshop 14th Chamonix Workshop, 17-21 Jan 2005, Chamonix, Switzerland.
261. G. Robert-Demolaize, R. Assmann, S. Redaelli, F. Schmidt, "A new version of SixTrack with collimation and aperture interface". CERN-AB-2005-033, PAC-2005-FPAT081. Jun 2005. 3 pp.
262. M. Magistris, A. Ferrari, M. Santana-Leitner, K. Tsoulou, V. Vlachoudis. "Study for magnets and electronics protection in the LHC betatron-cleaning insertion". Nucl. Instrum. Meth. A562:989-992, 2006.
263. S. Redaelli (Ed.), "ICFA Mini-Workshop on Tracking for Collimation in Particle Accelerators," CERN-2018-011-CP, doi: 10.23732/CYRCP-2018-002.
264. E. Quaranta et al., "Modeling of beam-induced damage of the LHC tertiary collimators," Phys.Rev.Accel.Beams 20 (2017) no.9, 091002.
265. A. Lecner et al., "Validation of energy deposition simulations for proton and heavy ion losses in the CERN Large Hadron Collider," Phys.Rev.Accel.Beams 22 (2019) no.7, 071003
266. S. Redaelli et al., "Final Implementation and Performance of the LHC Collimator Control System," Proceedings, 23rd Conference, PAC'09, Vancouver, Canada, May 4-8, 2009.
267. Particle Data Group, S. Eidelman et al., Physics Letters B592, 1 (2004).
268. H. Braun et al, "Collimation of Heavy Ion Beams in LHC". EPAC2004. CERN-LHC-Project-Report-766.
269. R. Bruce, R.W. Assmann, G. Bellodi, C. Bracco, H.H. Braun, S.S. Gilardoni, Eva Barbara Holzer, J.M. Jowett, S. Redaelli, Th. Weiler, et al., "Measurements of Heavy Ion Beam Losses from Collimation". EPAC08-WEOAG02, CERN-LHC-PROJ.REP-1109, 2008.
270. N. Fuster-Martinez et al., "Performance of the Collimation System During the 2018 Lead Ion Run at the Large Hadron Collider," Proceedings, 10th International Particle Accelerator Conference (IPAC2019): Melbourne, Australia, May 19-24, 2019. DOI 10.18429/JACoW-IPAC2019-MOPRB050.
271. R. Bruce et al., "Simulations and measurements of beam loss patterns at the CERN Large Hadron Collider," Phys.Rev.ST Accel.Beams 17 (2014) 081004
272. J. Stadlmann, H. Kollmus, E. Mustafin, I. Petzenhauser, P. Spiller, I. Strasik, N. Tahir, C. Trautmann, L. Bozyk, M. Krause, et al., "Collimation and Material Science Studies COLMAT at GSI". IPAC-2010-THPEC079. May 2010.
273. A. Bertarelli, "Beam-Induced Damage Mechanisms and their Calculation," CERN Yellow Report CERN-2016-002
274. J. Guardia Valenzuela et al., "Development and properties of high thermal conductivity molybdenum carbide - graphite composites," Carbon 135 (2018) 72-84, DOI: https://doi.org/10.1016/j.carbon.2018.04.010
275. F. Carra et al., "Mechanical robustness of HL-LHC collimator designs," Proceedings, 10th International Particle Accelerator Conference (IPAC2019): Melbourne, Australia, May 19-24, 2019. DOI:10.18429/JACoW-IPAC2019-MOPTS091
276. G. Gobbi et al., "Novel LHC collimator materials: High-energy Hadron beam impact tests and nondestructive post-irradiation examination," Mech.Adv.Mat.Struct. (2019).
277. W. Scandale, "Crystal collimation as an option for the LHC". Proceedings 2nd International Conference on Charged and Neutral Particles Channeling Phenomena (Channeling 2006), Frascati, Rome, Italy, 3-7 Jul 2006.

278. R. Assmann, S. Redaelli, W. Scandale. "Optics study for a possible crystal-based collimation system for the LHC". CERN-LHC-PROJECT-REPORT-918, Jun 2006. 3pp. Proc. EPAC 06, Edinburgh, Scotland, 26-30 Jun 2006.
279. W. Scandale et al., Phys.Lett. B758 (2016) 129-133
280. J. Smith et al, "Prospects for Integrating a Hollow Electron Lens into the LHC Collimation System". Proceedings PAC09.
281. S. Redaelli et al., "Plans for Deployment of Hollow Electron Lenses at the LHC for Enhanced Beam Collimation," Proceedings, 6th International Particle Accelerator Conference (IPAC 2015): Richmond, Virginia, USA, May 3-8, 2015. DOI: 10.18429/JACoW-IPAC2015-WEBB1
282. J. Resta Lopez, R. Assmann, S. Redaelli, G. Robert-Demolaize, D. Schulte, F. Zimmermann, A. Faus-Golfe. "An alternative nonlinear collimation system for the LHC". CERN-LHC-PROJECT-REPORT-939, Jun 2006. 3pp. Proc. EPAC 06, Edinburgh, Scotland, 26-30 Jun 2006.
283. J.C. Smith, J.E. Doyle, L. Keller, S.A. Lundgren, Thomas W. Markiewicz, L. Lari. "Design of a Rotatable Copper Collimator for the LHC Phase II Collimation Upgrade." Proc. EPAC 08, Magazzini del Cotone, Genoa, Italy, 23-27 Jun 2008.
284. T. Markiewicz et al., "Design, construction, and beam tests of a rotatable collimator prototype for high-intensity and high-energy hadron accelerators," submitted to PRAB (2019).
285. J. Frisch, E. Doyle, K. Skarpaas, VIII, "Advanced collimator engineering for the NLC". SLAC-PUB-9417, PAC-2001-TPAH012. Aug 2002.
286. G.E. Fischer: *Iron Dominated Magnets*, AIP Conf. Proc. 153 (1987) 1120-1227.
287. P. Campbell: Permanent Magnet Materials and their Application, Cambridge Univ. Press, 1994.
288. B. Seeber (ed.): Handbook of Applied Superconductivity, UK Institute Physics (1998).
289. J.T. Tanabe: Iron Dominated Electromagnets, World Scientific, Singapore, 2005.
290. CERN Accelerator School *Magnets*, 16-25 June 2009, Bruges, Belgium, http://cas.web.cern.ch/cas/Belgium-2009/Lectures/Bruges-lectures.htm
291. R.L. Keizer: Dipole Septum Magnets, CERN PS/Int. 74-13, 1974.
292. J. Borburgh et al., *Modifications to the SPS LSS6 septa for LHC and the SPS septa diluters*, Proc. 10th Eur. Particle Accelerator Conf. (EPAC'06), Edinburgh, Scotland, June 26-30, 2002.
293. J. Gervaise and E.J.N Wilson, CERN, Geneva, High precision geodesy applied to CERN accelerator, CERN accelerator school on Applied Geodesy for Particle Accelerators, CERN, 14-18 April 1986
294. C. Lasseur, CERN, Geneva, Metrology for Experiments, CERN accelerator school on Applied Geodesy for Particle Accelerators, CERN, 14-18 April 1986
295. M. Mayoud, CERN, Geneva, Geodetic Metrology of Particle Accelerators and Physics Equipment, 1st International workshop on Accelerator Alignment, SLAC, July 31- Aug 2, 1989
296. M. Mayoud, CERN, Geneva, Specific Aspects of the Geodetic Metrology of Large Particle Accelerators, FIG XXIth Intern. Congress, Brighton, GB 1998.
297. J.C. Gayde, C. Humbertclaude, C. Lasseur, CERN, Geneva, Prospects of Close Range Digital Photogrammetry in large physics installations, IWAA97, Argonne, USA
298. M. Jones, CERN, Geneva, système de coordonnées et le référentiel géodésique CERN, internal presentation.
299. H. Mainaud Durand & al, CERN, Geneva, "Permanent monitoring of the LHC low beta triplets: latest results and perspectives", IWAA2010, DESY, 13-17 September 2010
300. J.-C. GAYDE, D. MERGELKUHL, M. RAYMOND, CERN, Geneva, Switzerland, M. DÖNSZELMANN, Radboud University, Nijmegen, The Netherlands, M. DAAKIR, Université Paris-Est, Champs-sur-Marne, France, V. BATUSOV, JINR, Dubna, Russia, "The AtlAs DEtector POsitioning system to control moving parts during atlas closure", IWAA 2016, ESRF, Grenoble

301. V.Vlachakis and al, CERN and ETH Zurich,Switzerland, "Recent development of micro-triangulation for magnet fiducialisation", IWAA 2016, ESRF, Grenoble
302. S. W. Kamugasa$_{*}$, CERN, Geneva, Switzerland & ETHZ, Zurich, Switzerland "FREQUENCY SCANNING INTERFEROMETRY FOR CLIC COMPONENT FIDUCIALISATION", IWAA 2016, ESRF, Grenoble

Chapter 9
Accelerator Operations

M. Lamont, J. Wenninger, R. Steinhagen, R. Tomás García, R. Garoby, R. W. Assmann, O. Brüning, M. Hostettler, and H. Damerau

9.1 Introduction

M. Lamont

The cost of building a particle accelerator is a major capital investment. Commissioning should be swift and the subsequent exploitation of a facility must provide an effective return. This return may be difficult to quantify unambiguously but generally acceptable measures of performance can be established. These measures might include: machine availability; integrated luminosity; protons on target; beam hours to users and so on.

The role of accelerator operations is to maximize the performance of an accelerator or accelerator complex by: minimizing downtime; maximizing the amount of beam delivered to the users; fully optimizing the quality of beam delivered to the users; and doing it all safely.

Accelerators and the control of particles beams have advanced considerably over recent years. There is deep understanding of particle dynamics, innovative measurement techniques, in-depth simulation and modelling, and widespread leverage of

Coordinated by M. Lamont.

M. Lamont (✉) · J. Wenninger · R. Steinhagen · R. Tomás García · R. Garoby · R. W. Assmann · O. Brüning · M. Hostettler · H. Damerau
CERN (European Organization for Nuclear Research), Meyrin, Genève, Switzerland
e-mail: Mike.Lamont@cern.ch

S. Myers, H. Schopper (eds.), *Particle Physics Reference Library*,
https://doi.org/10.1007/978-3-030-34245-6_9

twenty-first century technology. It is not possible to go into too much depth here, instead some fundamentals that have been established from experience are outlined. References are given to more detailed sources. Some key operational considerations are outlined below and addressed in more detail in this chapter.

- **Availability** Accelerators are complex and their operations demands interfacing to a wide number of systems. Some these may be regarded as technical services e.g. cooling; cryogenics; electricity distribution. Others will have a more direct relationship with beam based operation e.g. power converters; radio frequency systems; beam instrumentation. One main challenge is to maintain the facility in a operable state for the maximum amount of time. Preventative maintenance, consolidation, fault tracking and fast problem resolution are required. In larger complexes there can be a chain of dependencies from sources, linacs and so on. Availability will be the cross product of the whole chain and associated primary services. Mean time between failure of essential components have to be evaluated at the design stage and appropriate component reliability assured. An effective fault tracking system is essential to identify weaknesses and targets for improvement.
- **Reproducibility** One key operational driver is reproducibility. In terms of the magnetic machine this implies careful attention to the powering history via a well defined pre-cycling strategy in a collider or careful cycle configuration in a fast cycling machine. A on-line measurement of the dipole field in a dedicated reference magnet system may be required. Reproducibility and stability of orbit can be critical for a number of reasons including guaranteeing collimator hierarchy or available aperture. Reproducibility of machine settings is to be expected and must be guaranteed.
- **Control** The control system will act as the primary interface to the accelerator systems and provide the means to communicate with a sub-system components. It will also provide high level facilities for driving the accelerator through its duty cycle. Although sometimes taken for granted, poorly designed controls can have a debilitating effect on the operability of an accelerator. Careful evaluation of the requirements and sufficient resources for development are required. Flexibility is required for commissioning and machine studies; access to control functionality and measurements for non-standard development should be considered.
- **Instrumentation** Effective beam instrumentation underpins control, optimization and understanding. The importance of well specified, reliable, accurate systems with appropriate acquisition systems and software cannot be understated.
- **Optimization and stability** The high performance demanded of modern machines can demand operating within tight parameter envelopes. Appropriate flexibility and precision in parameter control should be anticipated. Feedback systems for control of the key beam parameters can become mandatory. These could range from orbit and tune to transverse feedback systems. Harnessing modern technology and techniques is vital and again, expert resources must be given over to ensuring appropriate solutions. Commissioning these systems early in the lifetime of a machine should be a priority.

- **Understanding** Accelerator operations offers almost infinite possibilities for empirical tweaking as a way around problems. There is no substitute for building up real understanding of the key properties of a machine. These might include aperture, optics, instabilities, beam losses, beam and luminosity lifetimes. Good control and instrumentation are the tools of this trade and their importance is again stressed.
- **Safety** High energy and high power machines bring with them a number of risks. These risks have to be properly understood and protected against. For a complex machine the process of understanding the risks and their, sometimes, subtle interplay can be a painstaking process. Failure to perform this process properly can prove costly.

9.2 Parameter Control

M. Lamont

The high level control system shall provide the following functionality:

- monitoring, recording and logging of accelerator status and process parameters;
- display of operator information regarding the accelerator status and beam parameters;
- operational settings management should provide facilities for the settings changes of all individual equipment systems; all settings changes shall be recorded, with simple-to-use roll-back possibilities;
- automatic process control and sequence control during all beam related modes of operation and covering all operational scenarios i.e. control within normal operating limits;
- prevention of automatic or manual control actions which might initiate a hazard;
- detection of the onset of a hazard and automatic hazard termination (e.g. dump the beam), or mitigation (e.g. establish control within safe operating limits).

In particular, control of the key beam parameters is vital to maintain good beam lifetime and minimize beam loss at all stages of operation. A general operational principle is to effect control at the appropriate level. Thus high level parameters such tune, chromaticity, synchrotron tune, bucket area, phase advance should used where appropriate. Software provides the appropriate translation to lower level hardware parameters such as current and voltage. The results of more complex beam dynamics analyses such as beta beating measurement and correction or the measurement and correction of non-linear effects have also to be incorporated in a sensible way into the machine settings.

9.2.1 Magnetic Elements

For a given machine configuration, optics programs are generally used off-line to generate the baseline settings in normalized strength for magnetic elements. These strengths should be imported into the settings management system and be amenable to adjustment as required. The required beam momentum is used to calculate the field or gradient required in a magnet or series of magnets.

For example, the required quadrupole gradient given a normalized quadrupole strength is calculated via Eq. (9.1).

$$k_2 = \frac{e}{p}\frac{\mathrm{d}B_z}{\mathrm{d}x} = \frac{1}{B\rho}\frac{\mathrm{d}B_z}{\mathrm{d}x}\,. \tag{9.1}$$

Given a required field or gradient in a magnet circuit the next challenge is to establish the required current to be supplied to the circuit. The two forms of Eq. (9.1) reflect the two main approaches to the challenge of conversion from required field or gradient to current.

The first approach depends on precise knowledge of the instantaneous total bending field in a synchrotron. This information can be obtained from a closed-loop measurement system, which generates and distributes a train of impulses ("B-train") representing the field in a reference unit powered in series with the machine, or by a mathematical field model ("synthetic B-train") possibly supplemented by off-line measurements. The real-time measurements of the main dipole field are then distributed to the power supply front-ends which perform a conversion from required gradient to current. Several major uncertainty sources apply in these cases, namely: temperature drifts, hysteresis behaviour in the iron, eddy current effects, material ageing [1].

An alternative, used at LEP and the LHC, is an off-line approach. Here look-up tables (gradient/field versus current) are pre-generated and used directly in the calculation of magnet currents, usually at the higher level of the control system. The LHC developed a semi-empirical model ("FiDeL" [2]) to generate the look-up tables and a description of the dynamic behaviour of multipole field errors. This model was based on a large database of test results which included all magnetic elements. The measurements were statistically analysed to extract model parameters. The required momentum is the taken as the driving parameter in the settings generation software; the current in all magnetic circuits is then given by pushing the required fields and gradients through the look-up tables.

9.2.2 Transverse Beam Parameters

Tune adjustment can be made using either the main quadrupoles circuits or dedicated trim quadrupole circuits. In the linear approximation it is straightforward to pre-calculate the coefficients relating strength changes to a given tune change. The key point here is to have on-line a rapid and easily available conversion between a beam parameter, magnet strength and ultimately power supply current. Once established the method can be used by on-line, off-line and real-time software.

The coefficients can either be established from standard definitions involving Twiss parameters or by pre-calculating the corresponding coefficients using, say, MADX, and use them with appropriate linear scaling to calculate the required changes in quadrupole strengths. The same arguments hold for all other commonly used operational parameters.

For chromaticity a matrix equation can be formed for the required change in sextupole strength given a desired change in chromaticity. Operationally the matrix coefficients can be pre-calculated either via the standard integrals or a machine model and then used on-line. Different sextupole families are often used in colliders to target different sources of chromaticity. Correction algorithms can be configured to weight different families as required.

Correction of the coupling can be important for machine performance and beam diagnostics. Sources of coupling include: tilted quadrupoles; solenoid fields; and vertical orbit deviations in sextupoles. Anticipating these sources at the design stage ensures that appropriately placed families of skew quadrupoles can be used to correct all sources of coupling. In principle the coupling generated by experimental solenoids and its correction can be pre-calculated. Correction of coupling from random sources in the machine can either be performed empirically or by compensating the coupling driving terms extracted from multi-turn BPM measurements. Knobs to correct the real and imaginary part of the coupling driving terms should be anticipated.

An effective method of measuring coupling is required. Methods range from measurements of the cross-plane tune amplitudes; the closest tune approach; measuring the cross-plane effects of beam excitation. More sophisticated measurement techniques can identified local sources of coupling and local correction may be possible if dedicated elements are available e.g. correction of coupling arising from tilts of the inner triplet magnets in the LHC. On-line methods using the transverse damper in AC-dipole mode to obtain spectral data around ring have provided automatic on-line correction of the real and imaginary parts of the coupling in the LHC.

The principles of orbit correction are described later in this chapter. Orbit dipole kicks should be treated as a parameter like any other. Thus

$$\theta_i = \left.\frac{\Delta B \Delta s}{B\rho}\right|_i \tag{9.2}$$

where θ_i is an orbit kick at a location s_i would slot into a settings management system in a consistent way, along with the use of orbit correctors in predefined local bumps and the like.

9.2.3 Generalization

The parameter space of a particle accelerator can be complex and large and might include the parameters such as tune, chromaticity and orbit but also others such as: Landau damping octupoles; bunch length control with wigglers; compensation of eddy current effects; higher order multipole correction with dedicated magnets etc. In a machine with an energy ramp these parameters and required corrections will be a function of time.

It should be easy to define combinations of parameters to give other higher level parameters (a "knob" in CERN parlance). These knobs can be then manipulated to control the ensemble. A simple example is a closed orbit bump. A more sophisticated example would be an optics correction which could include quadrupole strength adjustments and tune compensation.

Reliable settings management is mandatory. Generic tools for tracking all changes to settings should be deployed providing a full record of all settings changes. Archive and roll back facilities are essential. It is important that all parameter control be dealt within the same parameter and settings management system.

Besides the provision of tools for standard operations, the requirements of machine studies and commissioning should be borne in mind. Exploiting the machine in study mode often required non-standard measurements and non-standard control actions on accelerator hardware. An appropriate scripting environment (e.g. Python) with interfaces to control system functionality should allow the user to fully exploit the potential of the hardware and control system in a flexible and experimental way. Many of the tools thus developed can subsequently be incorporated into the standard operational environment. At the LHC this has included, for example, tools for beta beating measurement and correction, detection and logging of beam instabilities etc.

9.3 Orbit Correction

J. Wenninger

The main role of orbit correction is to centre the trajectory (in a transfer line) or the closed orbit (in a ring) inside the aperture or the magnetic elements, typically quadrupoles or wiggler magnets. This is usually achieved using global correction algorithms acting on all or a large part of the accelerator. In some cases local excursions may be desired, requiring the use of local 'bumps' of the orbit or trajectory.

9.3.1 Global Orbit Correction

Consider a storage ring with M beam position monitors (BPM) and N correctors. Orbit displacements $\vec{d}$ (M-component vector) arising from corrector kick angles $\vec{\theta}$ (N-component vector) are determined by the $M \times N$ linear *response matrix* **A**,

$$\mathbf{A}\vec{\theta} = \vec{d} \,, \tag{9.3}$$

$$A_{mn} = \frac{\sqrt{\beta_m \beta_n}}{2 \sin \pi \nu} \cos\left(|\phi_m - \phi_n| - \pi \nu\right) . \tag{9.4}$$

The elements of **A** may be obtained from the machine model or be determined experimentally by measuring the deviation at each BPM resulting from exciting each corrector individually.

The task of the orbit correction is to find a set of corrector kicks $\vec{\theta}$ that satisfy the following relation,

$$\vec{d} + \mathbf{A}\vec{\theta} = 0 \,. \tag{9.5}$$

In general the number of BPMs (M) and the number of correctors (N) are not identical and Eq. (9.5) is either over- ($M > N$) or under-constrained ($M < N$). In the former and most frequent case, Eq. (9.5) can not be solved exactly. Instead, an approximate solution must be found, and commonly used least square algorithms minimize the quadratic residual

$$S = \|\vec{d} + \mathbf{A}\vec{\theta}\|^2 \,. \tag{9.6}$$

9.3.2 SVD Algorithm

When $M \geq N$, the *Singular Value Decomposition* (SVD) [3] of matrix **A** has the form $\mathbf{A} = \mathbf{U}\mathbf{W}\mathbf{V}^t$,

$$A = \begin{bmatrix} u_1^{(1)} & u_1^{(2)} & \cdots & u_1^{(N)} \\ u_2^{(1)} & u_2^{(2)} & \cdots & u_2^{(N)} \\ \vdots & \vdots & & \vdots \\ u_M^{(1)} & u_M^{(2)} & \cdots & u_M^{(N)} \end{bmatrix} \begin{bmatrix} w_1 & & & \\ & w_2 & & \\ & & \ddots & \\ & & & w_N \end{bmatrix} \begin{bmatrix} v_1^{(1)} & v_2^{(1)} & \cdots & v_N^{(1)} \\ v_1^{(2)} & v_2^{(2)} & \cdots & v_N^{(2)} \\ \vdots & \vdots & & \vdots \\ v_1^{(N)} & v_2^{(N)} & \cdots & v_N^{(N)} \end{bmatrix} . \tag{9.7}$$

U is the $M \times N$ matrix whose column vectors $\vec{u}^{(\alpha)}$, ($\alpha = 1, \ldots, N$) form an orthonormal set, $\mathbf{U}^t\mathbf{U} = \mathbf{I}$. **W** is $N \times N$ diagonal matrix with non-negative elements. $\mathbf{V}^t$ is the transpose of the $N \times N$ matrix **V**, whose column vectors $\vec{v}^{(\alpha)}$, ($\alpha = 1, \ldots, N$) form an orthonormal set, $\mathbf{V}^t\mathbf{V} = \mathbf{V}\mathbf{V}^t = \mathbf{I}$.

From Eq.(9.7), it follows that ($\alpha = 1, \ldots, N$),

$$\mathbf{A}\vec{v}^{(\alpha)} = w_\alpha \vec{u}^{(\alpha)}, \quad \mathbf{A}^t \vec{u}^{(\alpha)} = w_\alpha \vec{v}^{(\alpha)}, \tag{9.8}$$

and

$$\mathbf{A}\mathbf{A}^t \vec{u}^{(\alpha)} = w_\alpha^2 \vec{u}^{(\alpha)}, \quad \mathbf{A}^t \mathbf{A}\vec{v}^{(\alpha)} = w_\alpha^2 \vec{v}^{(\alpha)}. \tag{9.9}$$

When none of the diagonal elements w_α vanish, the solution of Eq.(9.5) is $\vec{\theta} = -\mathbf{V}\mathbf{W}^{-1}\mathbf{U}^t\vec{d}$. $\vec{d}$ may be expanded in terms of eigenvectors $\vec{u}^{(\alpha)}$ [4],

$$\vec{d} = \sum_{\alpha=1}^{N} C_\alpha \vec{u}^{(\alpha)} + \vec{d}_0 , \tag{9.10}$$

where $C_\alpha = \vec{d} \cdot \vec{u}^{(\alpha)}$, while $\vec{d}_0$ corresponds to the uncorrectable part of the orbit. The corrector strength required for correction is

$$\vec{\theta} = -\sum_{\alpha=1}^{N} \frac{C_\alpha}{w_\alpha} \vec{v}^{(\alpha)} . \tag{9.11}$$

If a given $w_\alpha = 0$ indicating that the matrix is singular, one discards the corresponding term from Eq. (9.11). An example for an eigenvalue spectrum is given in Fig. 9.1 for LEP.

In practice one may want to limit the number of eigenvalues used for the correction to control the r.m.s. strength of the orbit correctors or to avoid small eigenvalues that are very sensitive to the accuracy of the model.

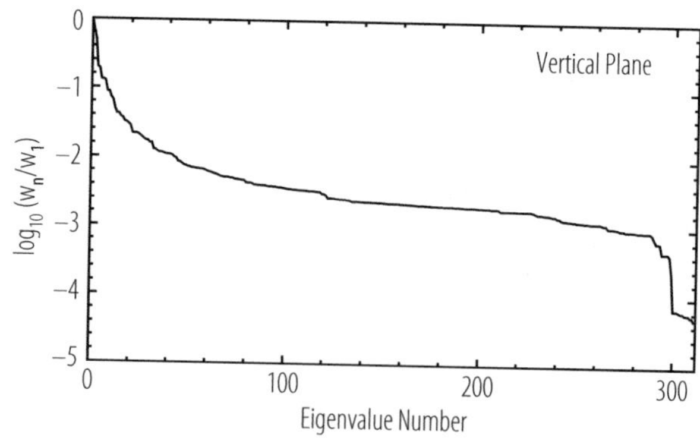

Fig. 9.1 Vertical orbit eigenvalue spectrum for LEP. The last four eigenvalues correspond to singular solutions in the low-beta sections around the interaction points

The SVD algorithm is ideally suited for feedback application since the correction can be cast in the simple form of a matrix multiplication once the SVD decomposition has been performed. This provides a fast and reliable correction procedure for realtime orbit feedback.

9.3.3 MICADO Algorithm

MICADO [5] is a least square correction algorithm based on Householder transformations. MICADO performs an iterative search for the most effective corrector and is, together with SVD, one of the most common orbit correction algorithms. For a non-singular matrix, a MICADO correction with all N correctors and an SVD correction with all N eigenvectors yield identical solutions. For corrections with a limited number of correctors or eigenvectors, and for singular matrices, the two algorithms converge differently.

A major difference between SVD and MICADO is the corrector strength distribution, MICADO using fewer but also much stronger kicks. The corrector strength r.m.s. can be easily controlled with SVD over the number of eigenvalues that are included in a correction. A correction of a small number of localized kicks is very effectively handled by MICADO, particularly when the response matrix is accurate, in which case MICADO can be used to identify the sources of the kicks. On the other hand, corrections based on few eigenvectors with the largest eigenvalues are similar to corrections of the main harmonics. Such a scheme spreads out the correction of a few kicks over the whole machine which can be an asset when the strength of correctors is limited. To compensate an isolated kick locally, a large number of eigenvectors must be included in the correction such that the linear combination forming $\vec{\theta}_c$ converges to a single nonzero corrector.

Singularities of the response matrix, associated to very small eigenvalues, are handled more easily with SVD, since it is sufficient to avoid using the corresponding eigenvectors in the corrections procedure. For the MICADO algorithm, it is necessary to regularize matrix **A** by removing redundant correctors.

9.3.4 Local Orbit Bumps

A local bump may be build from three correctors with deflections $\theta_{1,2,3}$ at locations 1, 2, 3. The deflections may be expressed in terms of the lattice parameters,

$$
\begin{aligned}
\frac{\theta_2}{\theta_1} &= -\sqrt{\frac{\beta_1}{\beta_2}\frac{\sin(\phi_3-\phi_1)}{\sin(\phi_3-\phi_2)}}\,,\\
\frac{\theta_3}{\theta_1} &= -\sqrt{\frac{\beta_1}{\beta_3}\frac{\sin(\phi_2-\phi_1)}{\sin(\phi_2-\phi_3)}}\,.
\end{aligned}
\qquad (9.12)
$$

At a target point t between 1 and 2, the position and angular displacements are

$$\begin{aligned} d_t &= \theta_1 \sqrt{\beta_1 \beta_t}\, \sin(\phi_t - \phi_1)\,, \\ d'_t &= \theta_1 \sqrt{\frac{\beta_1}{\beta_t}} \left[\cos(\phi_t - \phi_1) - \alpha_t \sin(\phi_t - \phi_1) \right]. \end{aligned} \tag{9.13}$$

At a point t between 2 and 3,

$$\begin{aligned} d_t &= \theta_3 \sqrt{\beta_3 \beta_t}\, \sin(\phi_3 - \phi_t)\,, \\ d'_t &= -\theta_3 \sqrt{\frac{\beta_3}{\beta_t}} \left[\cos(\phi_3 - \phi_t) + \alpha_t \sin(\phi_3 - \phi_t) \right]. \end{aligned} \tag{9.14}$$

To control both position d_t and angle d'_t at the source point, a four-magnet local bump is required. The four-magnet local bump with corrector locations 1, 2, 3, 4, where the source point t is located between correctors 2 and 3 is given in terms of optics functions by:

$$\begin{aligned} \theta_1 &= \frac{d_t\,(\cos(\phi_t - \phi_2) - \alpha_t \sin(\phi_t - \phi_2))}{\sqrt{\beta_t \beta_1}\, \sin(\phi_2 - \phi_1)} - \frac{d'_t \sqrt{\beta_t/\beta_1}\, \sin(\phi_t - \phi_2)}{\sin(\phi_2 - \phi_1)}\,, \\ \theta_2 &= \frac{-d_t\,(\cos(\phi_t - \phi_1) - \alpha_t \sin(\phi_t - \phi_1))}{\sqrt{\beta_t \beta_2}\, \sin(\phi_2 - \phi_1)} + \frac{d'_t \sqrt{\beta_t/\beta_2}\, \sin(\phi_t - \phi_1)}{\sin(\phi_2 - \phi_1)}\,, \\ \theta_3 &= \frac{-d_t\,(\cos(\phi_4 - \phi_t) - \alpha_t \sin(\phi_t - \phi_4))}{\sqrt{\beta_t \beta_3}\, \sin(\phi_4 - \phi_3)} + \frac{d'_t \sqrt{\beta_t/\beta_3}\, \sin(\phi_t - \phi_4)}{\sin(\phi_4 - \phi_3)}\,, \\ \theta_4 &= \frac{d_t\,(\cos(\phi_3 - \phi_t) - \alpha_t \sin(\phi_t - \phi_3))}{\sqrt{\beta_t \beta_4}\, \sin(\phi_4 - \phi_3)} - \frac{d'_t \sqrt{\beta_t/\beta_4}\, \sin(\phi_t - \phi_2)}{\sin(\phi_4 - \phi_3)}\,. \end{aligned} \tag{9.15}$$

9.3.5 Software

To be operationally useful the above algorithms must be fully integrated into application software. The software should typically provide the following functionality:

- an interface to orbit and trajectory acquisition system;
- the ability to compare measured orbits with saved references;
- the ability to calculate orbit corrections with a range of correction strategies (correction algorithm, number of correctors, number of eigenvalues) and send resulting correction to the machine;
- the ability to introduced a fully configurable local bump into the machine;

- facilities for dispersion measurement, harmonic analysis, energy offset analysis;
- facilities for threading, averaging and correction of the first N turns, injection point correction;
- multi-turn capture and analysis facilities.

9.4 Beam Feedback Systems

R. Steinhagen

The domain of control system design is too vast to be comprehensively and briefly covered and thus this section focuses only on the key aspects that are applicable to accelerator control. The inclined reader is referred for a more detailed introduction to [6]. One distinguishes two paradigms that are used to drive and stabilise any given beam parameter:

- Feed-Forward which relies on the knowledge of the transfer function between beam parameter, required current and corresponding effective magnetic field, and essentially consists of inverting the scalar or matrix relations discussed in Sect. 9.3, and for matrices most commonly done using a Singular-Value-Decomposition based approach[7]. While this scheme is often sufficient, its intrinsic weakness of being limited by inaccuracies of the magnet's transfer function, dynamic field effects such as e.g. the decay- and snapback phenomenon in superconducting magnets and random external perturbations may lead to a potentially out-of-tolerance systematic error $\Delta\epsilon = 1 - \epsilon$, which is illustrated as a difference between the reference and actual beam process variable in Fig. 9.2a. Still, this type of control is suitable for many beam parameters such as the beam momentum where the magnet and RF transfer functions involved are typically known on the 0.1% level.
- Feedback may be used in case the beam parameter model is insufficient (due to, for example, multiple dependencies or random perturbations) and at the same time the parameter deviations being accessible by beam instrumentation or diagnostics methods. In this case, the corresponding measurements can either be used to: (a) improve the knowledge of the beam parameter to magnetic field to current transfer function (feed-forward scheme), or (b) to 'feed back' part of the measured value into the reference that is send to the magnets or RF systems. In this case the difference $\Delta\epsilon = 1 - \epsilon$ drives a controller that iteratively optimises the actuator signal by minimising $\Delta\epsilon$ till the actual beam process variable matches the desired reference value as shown in Fig. 9.2b. Due to intrinsic limitations such as the bandwidth of the magnetic circuits and non-linear effects such as delays and rate-limits, the stabilisation is typically not instantaneous. To cope with these effects requires a more complex form of controller to optimise temporal convergence.

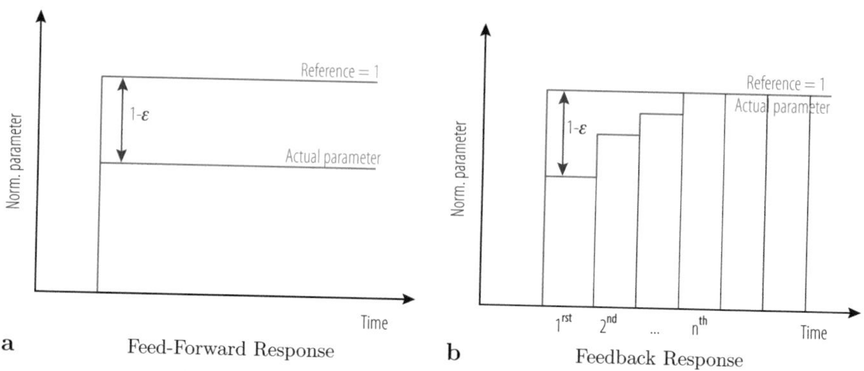

Fig. 9.2 Feed-forward and feedback signal in response to a step reference change: the actual value of process variable and residual errors 1-ϵ are indicated and vanishes to zero in case of an integral feedback

The strength of feedbacks is that they require, in comparison to feed-forward systems, only a rough process model while reducing the residual error through continuous adjustments or measure-and-correct iterations. For a steady-state scenario, the final parameter stability is ultimately limited by the noise and systematic error of the underlying beam parameter measurement. Since the robustness and stability of the beam instruments directly translates through this into the robustness and stability of the feedback loop, a good understanding of the underlying instrument, measurement and diagnostic principles is of paramount importance while designing beam-based feedback systems.

Two types of feedback are often distinguished: those acting within a given accelerator cycle (e.g. during injection, acceleration, store, dump) and those acting on a cycle-to-cycle basis where the disturbance measurements of the previous cycle are used to drive the actual state.[1] Both cases are similar and differ mainly on the time-scale in between measurements and corrections, the latter offers less strict requirements in terms of read-out speed and bandwidth of the involved beam instrumentation's acquisition system and is often initially favoured to a continuous read-out. The basic measurement, control principles and issues remain the same however.

9.4.1 Feedback Controller Design

A simple loop block diagram consisting of a single-input-single-output (SISO) beam response $G(s)$ and controller $D(s)$ is shown in Fig. 9.3. Here, r is the desired beam

[1]The cycle-to-cycle feedbacks are sometimes incorrectly referred to as "fill-to-fill feed-forward" though—in the strictest sense—they usually also rely on beam-based measurements.

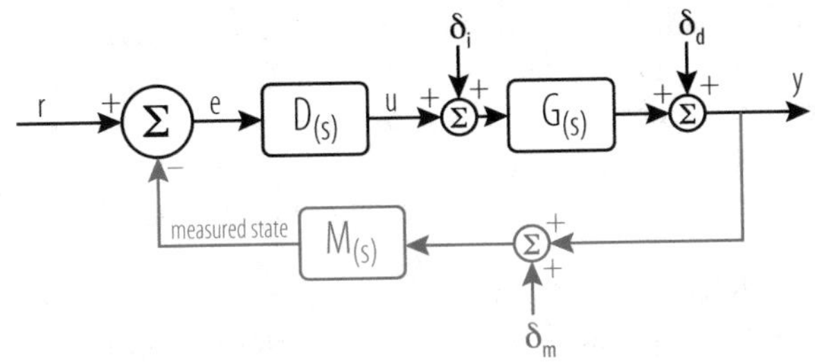

Fig. 9.3 First order closed loop block diagram

parameter reference, y the actual parameter state, e the feedback error being the difference between both, and u the output that is generated by the controller that minimise the error within the given constraints (e.g. response time, required power, etc.). One distinguishes typically at least three different error sources that can enter the feedback system: internal perturbations δ_i that are amplified by the same beam transfer function $G(s)$ as the to be stabilised parameter, external perturbations δ_d and measurement noise δ_m that while not directly affecting the actual state y may propagate to it depending on the choice of controller function $D(s)$. Depending on whether the system is implemented in the analogue or digital domain, it is convenient to transform the differential or difference equations of controller and beam process transfer functions using the Laplace[2] or Z-Transform. Both transforms translate complex convolutions in time domain to simple multiplications in the s- or z-domain. For the sake of simplicity, further discussion uses the Laplace transform but equally applies to Z-transformed digital systems.

For low-order beam parameters—such as orbit, tune, coupling, chromaticity and energy—the effects of the RF and individual corrector circuits are usually sufficiently linear and thus $G(s)$ is often approximated by a scalar K or matrix transfer function as described in Sect. 9.2 followed by a given time dependence. The time dependence of most electrical magnet circuits and RF systems is well approximated as first- (low-pass) and second-order (simple harmonic oscillator) systems with transfer functions given as:

$$G_{\text{first order}}(s) = \frac{K}{\tau s + 1}, \tag{9.16}$$

$$G_{\text{second order}}(s) = \frac{K\omega_0^2}{s^2 + 2\zeta\omega_0 \cdot s + \omega_0^2}, \tag{9.17}$$

[2]The Laplace transform of a function i$f(t)$ is given by $F(s) = \mathcal{L}\{f(t)\} = \int_0^\infty e^{-st} f(t)dt$ and is closely related to the Fourier transform. While not exact, in many cases replacing $s \rightarrow i\omega$ yields the Fourier transformed form. As a good approximation, Laplace-transformed functions can be translated to the Z-Transform using the bilinear transformation $s = \frac{2}{T_s}\frac{z-1}{z+1}$ with T_s being the sampling period.

with τ being the constant of the low-pass characteristic, ζ the damping factor and ω_0 the eigenfrequency of the system. In an ideal world the feed-forward ($M(s) = 0$) controller design would try to achieve a unity gain system transfer function

$$T_{\text{open-loop}}(s) := D(s)G(s) \rightarrow 1 \,, \tag{9.18}$$

which for a perfect actuator response corresponds to inverting Eqs. (9.16) and (9.17). However, in a real-world environment this can be only be achieved for reference changes with limited time-constants due to intrinsic limits on the power converter's slew-rate. Also, any scaling error in $D(s)$ or modelled $G(s)$ would yield a static error on the actual parameter output and discussed before.

If part of the actual parameter value is measured via a monitor transfer function $M(s)$, compared to the reference and the difference fed into the controller it is useful to define the following transfer functions that describe the stability and sensitivity to perturbations and noise

$$T(s) := \frac{y}{r} = \frac{D(s)G(s)}{1 + D(s)G(s)} \,, \tag{9.19}$$

$$S_d(s) := \frac{y}{\delta_d} = \frac{1}{1 + D(s)G(s)} \,, \tag{9.20}$$

$$S_i(s) := \frac{y}{\delta_i} = \frac{G(s)}{1 + D(s)G(s)} \,, \tag{9.21}$$

$$S_u(s) := \frac{u}{\delta_d} = \frac{D(s)}{1 + D(s)G(s)} \,, \tag{9.22}$$

where $T(s)$ is the nominal closed-loop (feed-back) transfer function, $S_d(s)$ the *nominal sensitivity* defining the loop disturbance rejection, $S_i(s)$ the *input-disturbance sensitivity* and $S_u(s)$ the *control sensitivity*. The state variables are indicated in Fig. 9.3.

It can be easily see, inserting Eq. (9.18) into 9.19, that using the same ideal open-loop controller relation described above in a closed-loop would yield a loop response error of 50% even for a perfect system. However, assuming only a perfect scalar controller $D(s) = D_0$ on can see that the loop response converges to one for larger controller gain D_0 and in particular becomes less dependent on relative errors of $G(s)$ which again illustrates the advantages of feed-back over feed-forward systems. However, at the same time one can show that the transfer function between actual beam parameter and measurement noise is equal to the nominal transfer function $T(s) = \frac{y}{\delta_m}$. Thus while simply increasing the scalar gain D_0 yields a perfect loop response to reference changes, it also causes the system to propagate any measurement error directly onto the actual beam parameter.

Thus, from the beam diagnostics point of view, in case the given instrument is to be used in beam-based feedbacks, the involved instrumentation performance is sometimes prematurely optimised—often including significant filtering in $M(s)$ to reduce the measurement noise—prior to closing the feedback loop. While this is suitable for systems that are used for verification or monitoring only, it has certain disadvantages when used in feedbacks e.g. through introducing additional sampling delays. Combining the filtering of beam instruments and closed-loop responses inside $D(s)$ is not just equivalent to this from a noise point of view but also improves the feedback response by minimising the total loop delay in these cases.

The discussed simple scalar controller are usually only used if the beam parameter response is much faster than the required reference changes (e.g. feedback that act on a cycle-to-cycle basis) but need to be extended for time dependent components for the other cases. Classic feedback designs typically rely on the discussion of denominator zeros in Eqs. (9.19) and (9.20) while keeping constraints such as required bandwidth, minimisation of overshoot, limits on the maximum possible excitation signal and robustness with respect to model and measurement errors. For ideal processes, this yields adequate controller designs but often falls short in providing a simple comprehensive method for estimating and modifying the loop sensitivity (robustness) in the presence of process uncertainties, non-linearities and noise.

Here, Youla's affine parameterisation method for optimal controllers is briefly introduced, which is based on the analytic process inversion, first introduced in [8]. For an open-loop stable process $G(s)$, the nominal closed-loop transfer function is stable if and only if $Q(s)$ is an arbitrary stable proper transfer function and $D(s)$ parameterised as:

$$D(s) = \frac{Q(s)}{1 - Q(s)G(s)} \,. \tag{9.23}$$

The stability of the closed loop system follows immediately out of the above definition if inserted into Eqs. (9.19)–(9.22). The sensitivity functions in the $Q(s)$ form are given as:

$$T(s) = Q(s)G(s) \,, \tag{9.24}$$

$$S_d(s) = 1 - Q(s)G(s) \,, \tag{9.25}$$

$$S_i(s) = (1 - Q(s)G(s))G(s) \,, \tag{9.26}$$

$$S_u(s) = Q(s) \,. \tag{9.27}$$

Assuming $G(s)$ is stable, the only requirement for closed loop stability is for $Q(s)$ to be stable. The strength of this method is the explicit controller design with respect to required closed loop performance, as visible in Eq. (9.24), and required stability (Eqs. (9.25)–(9.27)). Equations (9.24) and (9.25) are complementary and illustrate the intrinsic limiting trade-off of feedbacks that either have a good disturbance rejection or are robust with respect to noise. The ultimate limit is thus defined rather by the bandwidth and noise performance of the corrector circuits and beam measurements than by the feedback loop design itself.

9.4.1.1 First and Second Order Example

The design formalism can be demonstrated using a simple first order system $G_0(s) = \frac{K_0}{\tau \cdot s+1}$ with open-loop gain K_0 and time constant τ. A common controller design ansatz is to write $Q(s)$ as

$$Q(s) = F_Q(s) \cdot G_0^i(s) , \tag{9.28}$$

with $F_Q(s)$ a trade-off function and $G_0^i(s)$ the pseudo-inverse of the process. Since G_0 does not contain any unstable zeros, the pseudo-inverse equals the inverse and is given by $G_0^i(s) := [G_0(s)]^{-1} = \frac{\tau \cdot s+1}{K_0} \cdot Q(s)$. In order for $D(s)$ to be biproper, $F_Q(s)$ must have a degree of one and can be written as:

$$F_Q(s) = \frac{1}{\alpha s + 1} . \tag{9.29}$$

Inserting Eq. (9.28) into Youla's controller parameterisation equation (9.23) yields the following controller:

$$D(s) = \frac{\tau}{K_0 \alpha} + \frac{1}{K_0 \alpha s} = K_p + K_i \cdot \frac{1}{s} , \tag{9.30}$$

which shows a simple PI controller structure with proportional gains K_p and integral gain K_i. Inserting Eq. (9.28) into (9.24) yields

$$T_0(s) = F_Q(s) \tag{9.31}$$

that the closed loop response is essentially determined by the choice of trade-off function $F_Q(s)$ and that the closed loop bandwidth is proportional to the parameter $1/\alpha$. This can be used to tune the closed loop between: high disturbance rejection but high sensitivity to measurement noise (small α) and low noise sensitivity but low disturbance rejection (large α) depending on the operational scenario. The maximum possible closed loop bandwidth is limited by the excitation, as described by Eq. (9.27). In case of power converters, for example, the excitation is limited by the maximum available voltage.

One can derive a similar optimal controller for a second-order system (Eq. (9.17)) using a similar ansatz:

$$Q(s) = F_Q(s) \cdot G^i(s) = \frac{\omega_{cl}^2}{s^2 + 2\zeta_{cl}\omega_{cl}`s + \omega_{cl}^2} \cdot G^i(s) \tag{9.32}$$

yields the following controller PID structure

$$D(s) = K_p + K_i \cdot \frac{1}{s} + K_d \cdot \frac{s}{\tau_d s + 1} \tag{9.33}$$

with proportional gains K_p, integral gain K_i and integral gain K_d defined as:

$$K_p = \frac{4\zeta_{cl}\zeta_0\omega_0\omega_{cl} - \omega_0^2}{4K_0\zeta_{cl}^2}, \tag{9.34}$$

$$K_i = \frac{\omega_0^2\omega_{cl}}{2K_0\zeta_{cl}}, \tag{9.35}$$

$$K_d = \frac{4\zeta^2\omega_{cl}^2 - 4\zeta_0\omega_0\zeta_{cl} + \omega_0^2}{8K_0\zeta_{cl}^3\omega_{cl}}, \quad \text{with } \tau_d = \frac{1}{2\zeta_{cl}\omega_{cl}}. \tag{9.36}$$

There are several options but the most commonly used discrete 'velocity form' of above PID controller (without contracting sums) can be written as

$$\begin{aligned} u[n] = u[n-1] &+ K_p \cdot (e[n] - e[n-1]) + K_i \cdot T_s \cdot e[n] \\ &+ \frac{K_d}{T_s} \cdot (e[n] - 2e[n-1] + e[n-2]) \end{aligned} \tag{9.37}$$

with n being the sampling index, T_s the sampling frequency, $u[n]$ the controller output and $e[n]$ the measured error signal. The error signal $e[n]$ that is specifically used for the last differential part of the controller is nearly always low-pass filtered to suppress high-frequency noise that is common in discrete systems and that would otherwise be amplified by the K_d term. Compared to analogue feedbacks, digital feedbacks are more robust and their operation more reproducible compared to cases where temperature drifts or other external factors would affect analogue controller. Thus, in case the requested bandwidths are in the few Hz to MHz-range, digital feedbacks are the de-facto standard. Still, digital controllers are fundamentally limited by the intrinsic phase-lag caused by the sampling and dynamic range due to the finite number of ADC bits available at that frequency. Thus for feedbacks operating in the range of a few hundred to GHz range or where a high dynamic range is required, analogue feedbacks are still used. In any case, neither fully analogue nor fully digital representations are exact, and the final implementation is usually validated using beam-based optimisation techniques.

9.4.1.2 Non-linear Systems

The same method can be extended to open-loop unstable and multi-input-multi-output (MIMO) systems [8]. Real life feedbacks may contain significant delays λ (due to e.g. data transmission, data processing etc.) and non-linearities $G_{NL}(s)$, due to e.g. saturation and rate limits of the corrector circuits' power supplies. The modified process can be written, for example as:

$$G(s) = G_0(s) \cdot e^{-\lambda s} G_{NL}(s) \,. \tag{9.38}$$

Using the same pseudo-inverse $G_0^i(s)$ as for the above example and inserting Eq. (9.28) into (9.23) yields a controller parameterisation $D_{NL}(s)$ including a classic Smith-Predictor and anti-windup paths, discussed in more detail in [6, 9]. Inserting Eq. (9.28) including the delay and non-linearities into Eq. (9.24) yields the following closed loop transfer function:

$$T(s) = F_Q(s) \cdot e^{-\lambda s} G_{NL}(s) \,. \tag{9.39}$$

Similar to the linear case discussed above, the closed loop is essentially defined by the function $F_Q(s)$ that within limits can be chosen arbitrarily based on the required disturbance rejection and robustness during possibly different operational scenarios (gain-scheduling). Further information and a review on Youla's parameterisation can be found in [6, 10].

9.4.2 *Inter-Loop Dependencies*

Above described individual parameter control complexity is dwarfed by the challenge of operating parallel feedback loops on for example orbit, tune, chromaticity, coupling, radial position and transverse bunch-by-bunch motion. Even in a fully optimised scheme, some cross-talk is inevitable: the momentum modulation required to measure chromaticity induces tune and radial offsets that are seen by the tune, orbit and radial position feedbacks; transverse feedback, by design, minimises the very same beam oscillations required to measure the tune. If not addressed at an early design stage, a naïve one-by-one implementation of these feedback loops can lead to serious interferences, coupling and instabilities.

There are various classic de-coupling strategies such as: diagonalisation, e.g decoupling of horizontal and vertical planes; suppression of known cross-terms, i.e. allowing certain variations which are required for measurements; dead-bands to limit the operational ranges of one feedback in favour of another; time-scheduling between feedback actions, such as alternating tune measurements with transverse feedback operation; choosing different bandwidths for each loop.

An improved de-coupling strategy is to derive the dependent variable from the compensated feedback actuator control signal. In this case the tune feedback is operated at the maximum desired bandwidth, fully compensating radial modulation induced tune changes. Chromaticity is in turn derived and corrected from the amplitude of the actuator signal required to stabilise the tunes. Due to the finite bandwidth and gain of the feedback, the actuator signal does not typically contain the full modulation. An accurate chromaticity estimate needs to account for this and should be complemented by the demodulation of the residual tune frequency oscillation remaining on the beam. The required dispersion orbit variation and corresponding momentum mismatch need to be addressed differently. This is done by subtracting them dynamically from the orbit and radial-loop feedback reference targets. In machines running with transverse feedback systems, the tune can similarly be derived from its actuator signal while keeping beam oscillations and potential instabilities under control. Because of the various inter-loop dependencies, it is beneficial to implement the tune, chromaticity, coupling, orbit and radial-loop feedbacks in one global controller to minimise data exchange and synchronisation requirements.

9.5 Optics Measurement and Correction

R. Tomás García

9.5.1 *Introduction*

The unavoidable misalignments and field errors of the different components of an accelerator cause distortions of the machine optics with respect to the design. Optics errors deteriorate the accelerator performance and can even challenge the machine safety when operating with beam. Optics measurement and correction techniques are therefore fundamental to keep optics errors as low as possible or within specified tolerances [11]. The following sections review procedures and techniques to measure and correct various optics parameters, focusing on circular accelerators.

9.5.2 *Optics Measurement Techniques*

The tune is probably the most fundamental optics parameter as resonances need to be avoided for efficient accelerator operation. One- and two-dimensional resonance lines in the tune space have correspondances to Farey sequences [12]. The fractional part of the tune, Q, is given by the frequency of the beam oscillations when sampled turn-by-turn at any longitudinal location s,

$$z(N) = \sqrt{2J\beta(s)}\cos\left(2\pi QN + \phi(s) + \phi_0\right), \tag{9.40}$$

where z stands for horizontal or vertical position and J and ϕ_0 are the amplitude and phase invariants of the beam oscillations. Exciting the beam oscillations above the frequency noise floor of the Beam Position Monitors (BPMs) is crucial for the tune measurement. All optics measurements involve some kind of excitation as discussed below.

9.5.2.1 Quadrupole Strength Modulation

A change in the integrated strength of a quadrupole ΔKL yields a change in the tunes $\Delta Q_{x,y}$ that can be unambiguously used to determine the average $\beta_{x,y}$ functions over the quadrupole [13],

$$\overline{\beta}_{x,y} = \pm\frac{2}{\Delta KL}\Big(\cot(2\pi Q_{x,y})(1-\cos(2\pi\Delta Q_{x,y})) + \sin(2\pi\Delta Q_{x,y})\Big) \approx \pm 4\pi\frac{\Delta Q_{x,y}}{\Delta KL}\,, \tag{9.41}$$

where the $\pm$ sign refers to the horizontal and vertical planes, respectively. The approximation displayed to the right of Eq. (9.41) is applicable for $2\pi\Delta Q_{x,y} \ll 1$ and $Q_{x,y}$ far away from the integer and the half-integer. This technique is routinely used in many synchrotrons [14–18]. Hadron colliders typically operate with Q_x very close to Q_y. In this case a good correction of coupling is required prior to measurements with quadrupole strength modulation.

9.5.2.2 Closed Orbit Distortion

Exciting an orbit corrector with a deflection strength of $\Delta\theta$ yields a closed orbit distortion around the ring given by

$$\Delta x_{co}(s) = \Delta\theta\frac{\sqrt{\beta(s)\beta(s_0)}\cos\big(|\phi(s)-\phi(s_0)| - \pi Q\big)}{2\sin\pi Q}\,, \tag{9.42}$$

where s is the location of the BPM and s_0 is the location where the kick ($\Delta\theta$) is applied. The $\beta(s)$ and $\phi(s)$ functions can be obtained at the BPMs as the result of fitting to a collection of orbit distortions induced by, at least, three different orbit correctors per plane, as done in KEKB [19]. Closed orbit distortions are also the basis of another optics measurement and correction algorithm [20, 21] where quadrupole strengths and other machine parameters are fitted to reproduce a large ensemble of closed orbit acquisitions. This has demonstrated very effective in synchrotron light sources, however its application to large scale machines as the LHC requires unaffordable long times for measurements and computer calculations.

9.5.2.3 Betatron Oscillations, Free or Forced

The phase of the turn-by-turn free betatron oscillations, see Eq. (9.40), can be directly used to compute phase advances between pairs of nearby BPMs. The amplitude of the betatron oscillations can be used to measure β functions but it requires a good control of the BPM calibration errors [22–24]. Instead the phase advances between three or more BPMs can be used to obtain β and α functions as described in [25–27].

The Fast Fourier Transform (FFT) of the BPM signal, $z(N)$, offers a poor resolution on the phase measurement. Interpolated FTs like [28] or Singular Value Decomposition (SVD) algorithms like [29] feature a higher performance in terms of spectral resolution. The achievable resolution is usually limited by decoherence processes that damp the bunch centroid oscillations. Furthermore, beam decoherence also affects FT amplitudes and phases, which can be partially restored [30, 31, 42]. These limitations can be overcome by forcing betatron oscillations with the aid of an AC dipole with a frequency close to the machine tune. Moreover, if the AC dipole is ramped up and down adiabatically the beam emittance is preserved [32, 33].

The beam dynamics with forced oscillations features remarkable differences from that of free oscillations [34–37]. In presence of an AC dipole the measured β functions differ from the machine β functions. This difference is simply modelled as a quadrupole error of strength KL_{ac} in the location of the AC dipole [38], where KL_{ac} is given by

$$KL_{ac} = \pm 2\frac{\cos(2\pi Q_{x,y}) - \cos(2\pi Q_{ac})}{\beta_{ac}\sin(2\pi Q_{x,y})} \approx \pm 4\pi\frac{Q_{ac} - Q_{x,y}}{\beta_{ac}}, \tag{9.43}$$

where the $\pm$ sign refers to the horizontal and vertical planes, respectively. This equivalence allows to apply exactly the same analysis to all experimental data but using a modified reference model which includes the quadrupole error according to the AC dipole settings.

9.5.2.3.1 Transverse Momentum Reconstruction

It is convenient to study the transverse phase space using normalized coordinates, $\hat{z}(N) = z(N)/\sqrt{\beta}$. The turn-by-turn transverse normalized momentum, $\hat{z}'(N) = -\sqrt{2J}\sin\left(2\pi QN + \phi(s) + \phi_0\right)$, can be reconstructed by using the normalized signal from two BPMs, $\hat{z}_1(N)$ and $\hat{z}_2(N)$, as

$$\hat{z}'_1(N) = \left(\hat{z}_2(N) - \hat{z}_1(N)\cos(2\pi\Delta_{12})\right)/\sin(2\pi\Delta_{12}), \tag{9.44}$$

where Δ_{12} is the phase advance between the two BPMs. If non-linear elements are placed in between the two BPMs extra contributions to $\hat{z}'_1(N)$ appear as described in [39].

9.5.2.3.2 Coupling Measurement

Using the complex variable, $h_z(N) = \hat{z}(N) - i\hat{z}'(N)$, the turn-by-turn motion in presence of linear coupling is given by [40]

$$\begin{aligned} h_x(N) &= \sqrt{2J_x}e^{i\phi_x(N)} - i2f_{1001}\sqrt{2J_y}e^{i\phi_y(N)} - i2f_{1010}\sqrt{2J_y}e^{-i\phi_y(N)}, \\ h_y(N) &= \sqrt{2J_y}e^{i\phi_y(N)} - i2f^*_{1001}\sqrt{2J_x}e^{i\phi_x(N)} - i2f_{1010}\sqrt{2J_x}e^{-i\phi_x(N)}, \end{aligned} \tag{9.45}$$

where $\phi_{x,y}(N) = 2\pi N Q_{x,y} + \phi_{x0,y0}$ and f_{1001} and f_{1010} are the coupling difference and sum resonance terms, respectively. These terms are linearly related to the elements of the coupling matrix as described in [41]. All the monomials in the right-hand-side of Eqs. (9.45) correspond to a single spectral line of the complex spectrum with frequencies $\pm Q_x$ and $\pm Q_y$. By applying a complex FT to the $h_{x,y}$ variables as reconstructed from the BPMs it is possible to measure the amplitude and phases of the coupling terms. To achieve a measurement independent of BPM calibration and beam decoherence, the values obtained from the horizontal and vertical planes are geometrically averaged as described in [42]. The closest tune approach, ΔQ_{min}, can be computed from the skew quadrupolar fields as [13],

$$\Delta Q_{\text{min}} = \left| \frac{1}{2\pi} \oint \mathrm{d}s \; j(s) \sqrt{\beta_x \beta_y} e^{-i(\phi_x - \phi_y) + i(Q_x - Q_y)s/R} \right| , \tag{9.46}$$

where $j(s)$ is the skew quadrupolar gradient around the ring and R is the machine radius. ΔQ_{min} can also be computed from the difference resonance term f_{1001} around the ring by [43–45]

$$\Delta Q_{\text{min}} = \left| \frac{4(Q_x - Q_y)}{2\pi R} \oint \mathrm{d}s f_{1001} \mathrm{e}^{-i(\phi_x - \phi_y) + i(Q_x - Q_y)s/R} \right| \tag{9.47}$$

$$\approx \left| 4(Q_x - Q_y) \overline{f_{1001} \mathrm{e}^{-i(\phi_x - \phi_y)}} \right| \lesssim 4 \left| Q_x - Q_y \right| \overline{|f_{1001}|} , \tag{9.48}$$

where $\overline{x}$ represents the ring average of x. Linear coupling in combination with octupolar fields gives rise to an amplitude dependent closest tune approach [46, 47].

9.5.2.3.3 Measurement of Non-linearities

Betatron oscillations are also used to measure resonance driving terms f_{jklm} since these affect the motion as follows [40]

$$h_x(N) = \sqrt{2J_x} e^{i\phi_x(N)} - i2 \sum_{jklm} j f_{jklm} (2J_x)^{\frac{j+k-1}{2}} (2J_y)^{\frac{l+m}{2}} e^{i[(1-j+k)\phi_x(N) + (m-l)\phi_y(N)]} . \tag{9.49}$$

First sextupolar resonance driving terms measurements and applications were carried out in [42], where the effects of beam decoherence on these measurements are also described. First resonance terms measurements using AC dipoles to avoid beam decoherence are described in [39].

9.5.2.4 Dispersion Measurement

Dispersion is typically inferred from the orbit change induced by a shift in the RF frequency, see e.g. [13]. This measurement is affected by the BPM calibration errors. Alternatively it is possible to measure the ratio $D_x/\sqrt{\beta_x}$ (normalized dispersion) independently of BPM calibration errors if $\sqrt{\beta_x}$ is inferred from the amplitude of the beam oscillations [48].

9.5.3 *Optics Correction Techniques*

The most used correction approach consists in building a response matrix **R** of the available machine variables on a collection of observables. For instance,

$$\left(\Delta Q_x, \Delta Q_y, \frac{\Delta\vec{\beta}_x}{\beta_x}, \frac{\Delta\vec{\beta}_y}{\beta_y}, \Delta\vec{\phi}_x, \Delta\vec{\phi}_y, \Delta\frac{\vec{D}_x}{\sqrt{\beta_x}}\right)^T = \mathbf{R}\Delta\vec{k}^T\,, \tag{9.50}$$

$$\left(\Delta\Re\vec{f}_{1001}, \Delta\Im\vec{f}_{1001}, \Delta\Re\vec{f}_{1010}, \Delta\Im\vec{f}_{1010}, \Delta\vec{D}_y\right)^T = \mathbf{R}_s\Delta\vec{k}_s^T\,, \tag{9.51}$$

where $\vec{k}$ stands for quadrupole strengths but also horizontal orbit bumps at the sextupoles and $\vec{k}_s$ represents skew quadrupoles or vertical orbit bumps at sextupoles. **R** and $\mathbf{R}_s$ are pseudo-inverted and applied to the measured deviations of the optics parameters to compute the effective corrections. The correction should incorporate appropriate weights as illustrated in [49]. The success of this approach strongly depends on the configuration of errors and available correctors. Sextupolar resonance driving terms have been successfully corrected using an equivalent approach in [50]. For large localized errors this approach tends to distribute the correction over many correctors around the machine.

9.5.3.1 Segment-by-Segment Technique

A more local approach, the segment-by-segment technique [51, 52], consists in splitting the machine into a collection of independent beam lines by using as starting optics parameters the measured $\beta_{x,y}, \alpha_{x,y}, D_{x,y}, D'_{x,y}, f_{1001}, f_{1010}$. The comparison of the measurements and the propagated optics parameters along the segment will reveal discrepancies starting at the error location. The local error sources or the effective corrections can be computed by means of matching algorithms using only the machine variables in the segment.

9.6 Longitudinal Control and Manipulations

H. Damerau and R. Garoby

RF systems in synchrotrons are primarily installed for beam acceleration. However, they provide also the possibility to manipulate the longitudinal beam characteristics like bunch length, energy spread, distance between bunches, number of bunches, etc. [53]. The following section deals with the typical longitudinal beam manipulations in synchrotrons, when synchrotron radiation is negligible.

9.6.1 *Adiabaticity*

The longitudinal motion that we consider is conservative (i.e. there is no energy dissipation effect like synchrotron radiation, as well as no coupling of longitudinal and transverse motion). Liouville's theorem is then applicable, which states that the local density of particles in the longitudinal phase plane is always constant [54]. The time scale for the rate of change is given by the oscillation frequency of the individual particles at the centre of a bunch:

$$\omega_s \propto \left(\frac{hV \cdot \eta \cos \phi_s}{\gamma} \right)^{\frac{1}{2}}, \tag{9.52}$$

where h is the RF harmonic number, V the RF voltage, φ_s the stable phase and η the slip factor ($\eta = 1/\gamma^2 - 1/\gamma_t^2$). If the rate of change of the accelerator parameters is slow enough for the distribution of particles to be continuously at equilibrium in the longitudinal phase plane, longitudinal emittance is preserved [55]. Such a process is called "adiabatic". The degree of adiabaticity is assessed by the adiabaticity parameter ε. It is defined as the relative change of the synchrotron frequency, ω_s, during one period:

$$\varepsilon = \frac{1}{\omega_s^2} \left| \frac{d\omega_s}{dt} \right| . \tag{9.53}$$

Adiabatic processes can be reversed in time.

9.6.2 *Changing the Longitudinal Characteristics of the Bunches*

When a process is slow enough to be quasi-adiabatic ($\varepsilon < 0.1$), the particle distribution is completely determined by the instantaneous beam and accelerator parameters. The area occupied in the longitudinal phase plane (emittance) remaining constant, bunch length l_b (in ns) and energy spread ΔE_b (in eV) evolve like:

$$l_b \propto \frac{1}{\beta} \cdot \left(\frac{\eta}{\gamma \cdot hV \cos\varphi_s} \right)^{\frac{1}{4}} , \quad (9.54)$$

$$\Delta E_b \propto \beta \cdot \left(\frac{\gamma \cdot hV \cos\varphi_s}{\eta} \right)^{\frac{1}{4}} . \quad (9.55)$$

When the accelerator parameters (e.g. RF voltage or phase) are quickly varying, the process is non-adiabatic and tracking simulations are required to evaluate the final particle distribution. Although the local density of particles remains constant, the contour containing all particles is usually not a stable trajectory in the final state and the resulting emittance is therefore larger because of filamentation. Non-adiabatic beam manipulations allow getting bunch lengths and energy spreads which cannot be obtained with adiabatic processes ("bunch rotation"). They also give the possibility to blow-up the longitudinal emittance in a controlled fashion ("longitudinal controlled blow-up").

9.6.3 Bunch Rotation

A step increase of the RF voltage triggers a "bunch rotation" in the longitudinal phase plane during which the bunch length first decreases during 1/4 of a synchrotron period (Fig. 9.4a).

A step decrease has the opposite effect of first lengthening the bunch. To achieve the smallest possible bunch length, as required, for example, for transferring the beam to the following synchrotron equipped with a higher frequency RF system, this two effects can be combined [56, 57].

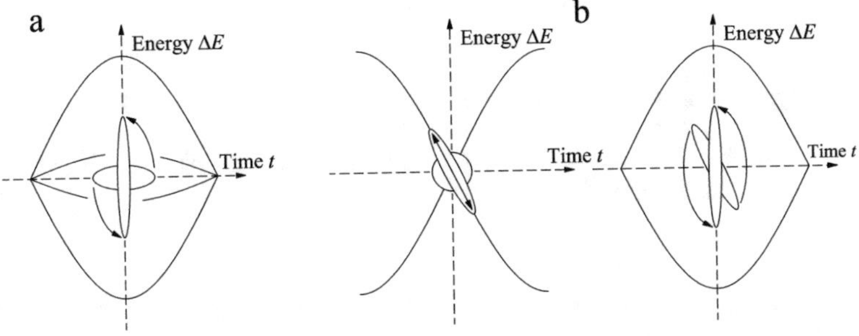

Fig. 9.4 Bunch rotation for bunch shortening. (**a**) 1/4 a synchrotron period in the longitudinal phase plane after a fast voltage step; (**b**) bunch stretching prior to rotation by phase jump

Because the focusing voltage is sinusoidal, the rotated bunch has a marked S-shape if, at any moment during the process, its length has exceeded $\simeq$ 1/3 of an RF period. If very short bunches have to be obtained, the maximum tolerable length is even smaller [53]. More sophisticated techniques making use of multiple RF harmonics have been developed to improve the results [58].

To avoid the step increase in RF voltage, a phase jump by π of the RF voltage can be applied instead [59]. The bunch is stretched at the unstable point between two buckets (Fig. 9.4b). Switching the RF voltage back to its initial phase triggers the rotation in the longitudinal phase plane.

9.6.4 Longitudinal Controlled Blow-Up

Increasing the longitudinal emittance is a convenient mitigation means against instabilities by keeping the beam below threshold. Such a blow-up should not generate tails in the distribution of particles. The most efficient and fast technique makes use of a phase-modulated high frequency (V_H, h_H) superimposed to the RF holding the beam $(h_{rf} \ll h_H)$ [60, 61]:

$$V_H = \hat{V}_H \sin(h_H \omega_R t + \alpha \sin \omega_M t + \vartheta_H) , \tag{9.56}$$

α being the peak phase modulation, ω_R the modulation frequency and ϑ_H a phase constant. This high frequency phase-modulated voltage perturbs motion in the longitudinal phase plane. Resonances can be induced which create a re-distribution of density in the bunch. Parameters are in practice determined with computer simulations and finely adjusted on the real accelerator. Typical ranges of values are shown in Table 9.1. A smaller harmonic ratio h_H/h_{rf} can also be used, although blow-up is then slower.

With RF voltage at a single harmonic controlled longitudinal blow-up is obtained by modulating either the voltage [62] or the phase [63] with noise, which introduces diffusion [64, 65]. To target specific parts of a bunch with the blow-up to shape its distribution, bandwidth limited noise can be applied.

Table 9.1 Typical ranges of blow-up parameters

$\hat{V}_H/\hat{V}_{rf}$	h_H/h_{rf}	α [rad]	ω_H/ω_s	Duration [s]
0.1–0.3	>10 for fast blow-up	0.8π to 1.2π	3–12	$\geq 20 \cdot (2\pi/\omega_s)$

9.6.5 Changing the Bunch Train

9.6.5.1 Iso-Adiabatic Rebunching (Debunching)

The simplest technique for bunching a continuous beam (e.g. a beam from a linac after it has debunched because of its energy spread) with a minimum emittance blow-up consists of applying an iso-adiabatic increase. With $\omega_s \propto V^{1/2}$ (Eq. (9.52)) the RF voltage function which keeps the adiabaticity ε constant ($\varepsilon < 0.1$) becomes (Fig. 9.5)

$$V(t) = \frac{V_I}{\left[1 - \left(1 - \sqrt{\frac{V_I}{V_F}}\right)\frac{t}{t_F}\right]^2}, \tag{9.57}$$

V_I being the initial RF voltage applied at the start of rebunching, and t_F the moment when the final RF voltage V_F is reached. The adiabaticity of the process is inversely proportional to its duration, $\varepsilon \propto 1/t_F$.

To minimize the emittance blow-up, V_I has to be low, typically such that the beam energy spread is significantly larger than the bucket height. However, for a given adiabaticity, the smaller V_I is, the smaller the initial synchrotron frequency and hence the slower the whole rebunching process.

Debunching is the inverse process to transform a bunched beam into a continuous beam. An iso-adiabatic voltage variation is also used, which is a time-reversed version of the one used for rebunching (Fig. 9.5).

9.6.5.2 Splitting (Merging)

Splitting is used to multiply the number of bunches by 2 or 3 and merging is the reverse process [66–68]. With respect to iso-adiabatic debunching-rebunching, these processes have the advantage of preserving gaps without beam and keeping the beam always under RF control. In theory as well as in practice, they can be quasi-adiabatic and almost without blow-up.

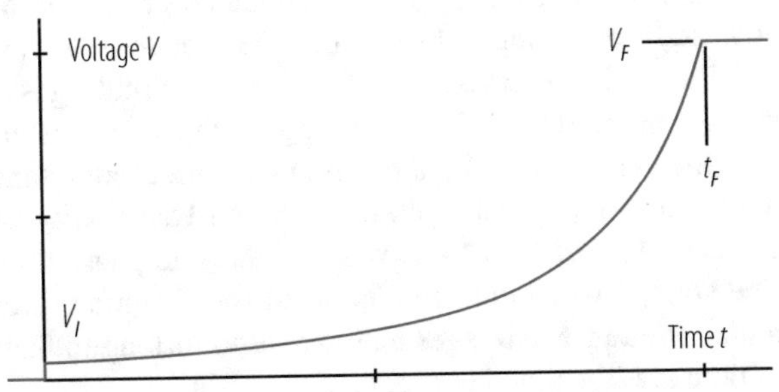

Fig. 9.5 Iso-adiabatic rebunching voltage

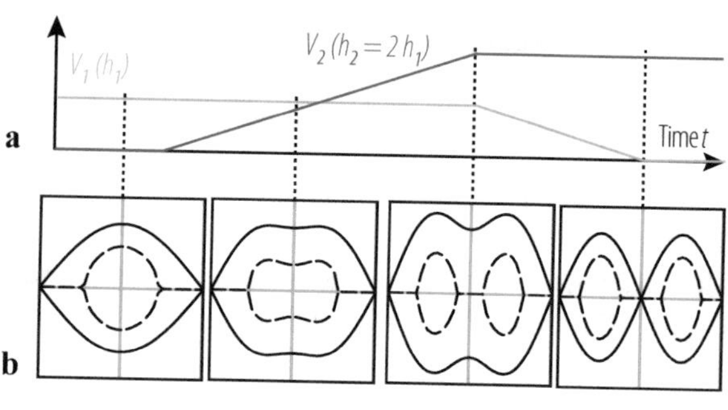

Fig. 9.6 Bunch splitting in two. (**a**) RF voltages vs. time; (**b**) longitudinal phase plane

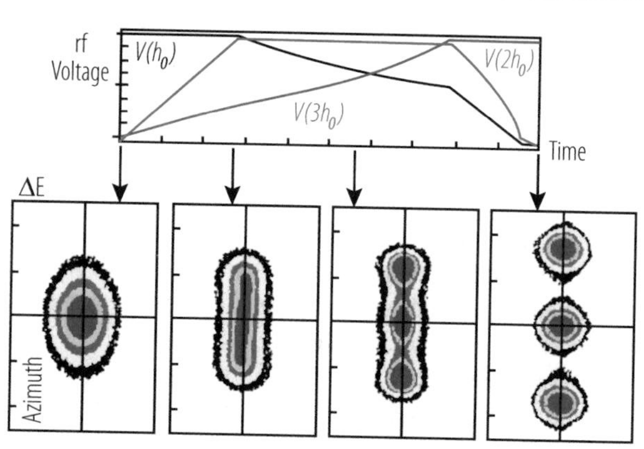

Fig. 9.7 Bunch splitting in three. (**a**) RF voltages vs. time; (**b**) longitudinal phase plane

Splitting bunches in 2 can be obtained using two RF systems with an harmonic ratio of 2 [67]. The bunch is initially held by the first system (V_1, h_1) while the second $(V_2, h_2 = 2h_1)$ is stopped. The unstable phase on the second harmonic is centred on the bunch. As V_2 is slowly increased and V_1 decreased the bunch lengthens and progressively splits in two as illustrated in Fig. 9.6.

Good results are consistently obtained when the voltage $V_1(h_1) = V_{1sep}$ is such that, at the moment when two separate bunches have just formed, the initial bunch would fill 1/3 of the bucket acceptance in the absence of second harmonic $(V_2(h_2) = 0)$. Linear voltage variations with a total duration larger than 5 synchrotron periods in the bucket (V_{1sep}, h_1) give satisfying results. Each final bunch has ideally 1/2 the emittance of the initial one. Practically, less than 10% additional longitudinal emittance growth is achieved. The longitudinal bunch distribution is preserved during the process. Splitting is also obtained when applying an RF voltage at $h_2 > 2h_1$, resulting in empty buckets in between split bunches [69].

Splitting bunches in three requires three simultaneous RF systems on three harmonics [68]. A stable phase on the third harmonic $(3h_0)$ coincides with the stable phase on first one (h_0) and with an unstable phase on the second harmonic $(2h_0)$. The voltage variations are computed for obtaining three equal bunches of 1/3 the initial emittance. Voltages and evolution in longitudinal phase space as a function of time are illustrated in Fig. 9.7.

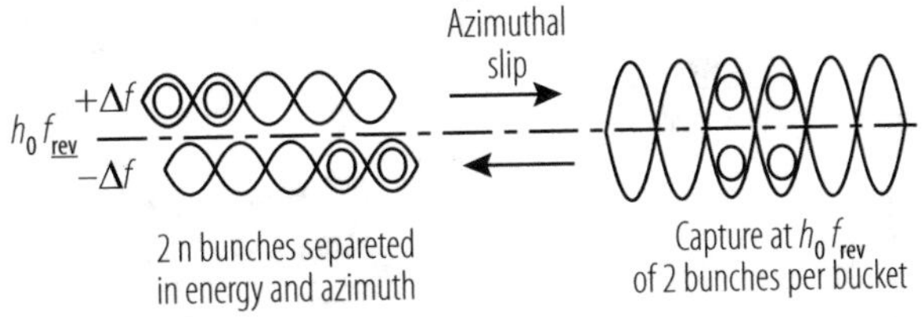

Fig. 9.8 Slip stacking: evolution in the longitudinal phase plane during the process

9.6.6 *Slip Stacking*

Slip stacking is a non-adiabatic technique for combining bunches two by two [70, 71]. It is fast, but it leads to large emittance blow-ups. The two beams have to be held by two slightly different RF frequencies. If the frequency difference is large enough ($\Delta\omega > 2\omega_s$, where ω_s is the synchrotron frequency in the centre of an unperturbed bucket of one family), two families of buckets coexist which drift towards each other because of their frequency difference (Fig. 9.8). Consequently, and provided the acceptance of the buckets is large enough (acceptance $> 2\times$ emittance), the bunches drift with them and slip past each other. When they are superimposed in azimuth, pairs of bunches can be captured in large buckets centred at the middle frequency.

Although improvements are possible, like reducing the frequency difference towards the end of the process, the longitudinal contour enclosing a pair of bunches in the final bucket always contains a large area without particles. Therefore the emittance after filamentation is much more than doubled and longitudinal density is accordingly reduced.

9.6.6.1 Batch Compression (Expansion)

Batch compression does not change the number of bunches but concentrates them in a reduced fraction of the accelerator circumference [72]. It can be quasi-adiabatic and consequently avoids longitudinal emittance blow-up.

The principle is slowly to increase the harmonic number of the RF controlling the beam as shown in Fig. 9.9. Starting from harmonic h_1, voltage is progressively increased on harmonic $h_2 > h_1$ and decreased on h_1, until harmonic h_2 finally holds the batch of bunches. The phase on h_2 with respect to h_1 must be such that bunches converge symmetrically towards the centre of the batch. This can be achieved for even (compression around unstable point between buckets) or odd number of bunches (central bucket does not move in phase).

Due to the presence of RF voltage at two harmonics simultaneously during each harmonic number step, the effective RF voltage is amplitude modulated at the

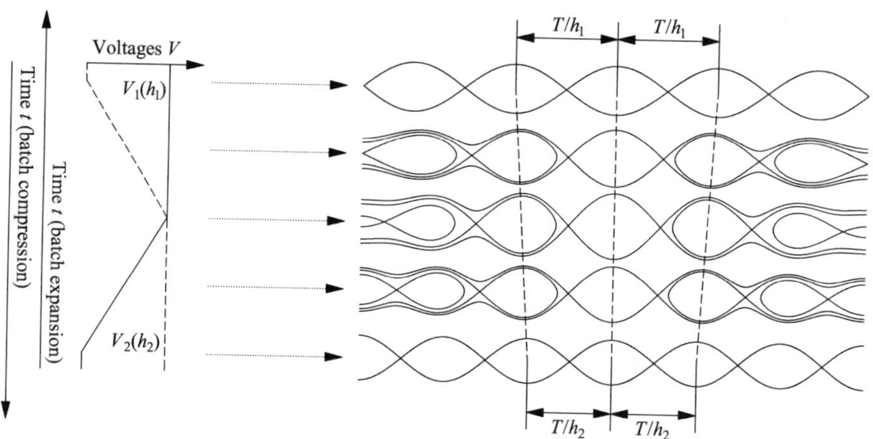

Fig. 9.9 Batch compression (top to bottom) or expansion (bottom to top)

difference harmonic $|h_1 - h_2|$,

$$\begin{aligned} V &= V_1 \sin h_1\omega_R t + V_2 \sin h_2\omega_R t \\ &= 2V_0 \cos\left(\frac{h_1 + h_2}{2}\omega_R t\right) \cdot \sin\left(\frac{h_1 - h_2}{2}\omega_R t\right) \quad \text{at} \quad V_0 = V_1 = V_2 \,. \end{aligned} \tag{9.58}$$

In case h_1 and h_2 have common dividers, the process repeats $\gcd(h_1, h_2)$ (gated common divider) times around the circumference, which allows to compress or expand multiple batches of bunches simultaneously.

The amount of compression achievable in a single step is limited by the acceptance of the buckets holding the edge bunches. A consequence is that multiple batch compression steps are necessary to reach large compression factors, and complicated manipulations of RF parameters are involved.

Non-linear voltage programs can be applied to improve the adiabaticity of the process [73].

9.7 Collimation

R. W. Assmann

9.7.1 Introduction

The concept of a collimator and its design are introduced in [74]. The design process for a collimation system and several important issues are discussed in [75]. In this chapter we focus on the performance and operational use of a collimation system. We quickly introduce a few central definitions:

- A collimation system is an ensemble of collimators that is integrated into the accelerator layout to intercept stray particles and to protect the accelerator.
- Collimation is acting in the normalized phase space. With $z = x$ or $z = y$, the Twiss functions β_z and α_z, and the emittance ϵ_z we define the normalized coordinates z_n and z_n' as:

$$z_n = \frac{z}{\sqrt{\epsilon_z \beta_z}}, \quad z_n' = \frac{\alpha_z z + \beta_z z'}{\sqrt{\epsilon_z \beta_z}}. \tag{9.59}$$

  It is noted that the transverse beam size (Gaussian rms) is given by $\sigma_z = \sqrt{\epsilon_z \beta_z}$.
- An unperturbed particle describes a circle in normalized phase space with amplitude:

$$a_z = \sqrt{z_n^2 + z_n'^2}. \tag{9.60}$$

- Collimator settings from the beam are defined in normalized coordinates (numbers of beam size σ_z) with n_1 being the collimator family setting closest to the beam, n_2 the second closest setting, and so on. Several families usually define a hierarchy that must be respected. Often this results in stringent tolerances on the positioning of collimators.

9.7.2 *Definition of Cleaning Efficiency and Performance*

A collimation system is designed to intercept stray particles with maximum efficiency and thus to protect critical regions of the accelerator. Critical regions can be experiments that must be protected against background from beam halo, super-conducting magnets that must be protected against quenches or hands-on maintenance equipment that must be protected against activation from beam losses. A system with 100% efficiency would absorb all impacting particles and power with zero leakage into critical zones. However, the various nuclear processes in the collimator jaw materials will always cause some particles to escape the system. The creation of secondary and tertiary halo in a two-stage collimation system is illustrated in Fig. 9.10.

9.7.2.1 Local Cleaning Inefficiency

For collimation it is convenient to define inefficiency or leakage [76]. We first introduce inefficiency and then connect it to efficiency. The inefficiency η_c of a collimation system with a primary collimation cut at n_1 is defined as the ratio

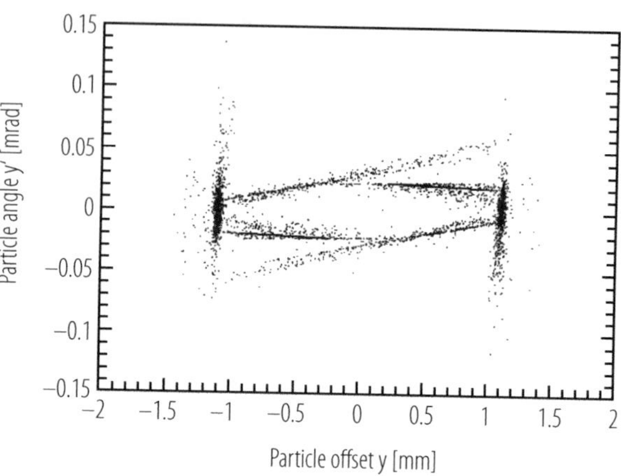

Fig. 9.10 Illustration of the secondary and tertiary halo created by a two-stage collimation system in vertical phase space (LHC example). The scattering along the jaw surface creates the characteristic lines of halo particles in phase space

between the number N_{leak} of particles that leak out and reach a normalized transverse amplitude a_z^{cut} and the number N_{impact} of impacting particles:

$$\eta_c = \frac{N_{leak}(a_z > a_z^{cut})}{N_{impact}} . \tag{9.61}$$

We require that $a_z^{cut} > n_1$. The value for a_z is given by the available machine aperture and is often around 10 sigma. Modern collimation systems can reach quite low inefficiencies with η_c in the range of 10^{-2} (1%) to 10^{-4} (0.01%). Efficiency η can then be defined as $\eta = 1 - \eta_c$ and is in the range of 99% to 99.99%.

Inefficiency is, however, not sufficient to characterize the performance of a collimation system. It is important to realize that the large amplitude particles are not lost at one location but are spread over some dilution length L_{dil}. For local losses we define a local cleaning inefficiency $\tilde{\eta}_c$ [76]:

$$\tilde{\eta}_c = \frac{\eta_c}{L_{dil}} . \tag{9.62}$$

Local cleaning inefficiency has a unit of 1/m. As the dilution is not uniform, simulations are used to predict the local cleaning inefficiency $\tilde{\eta}_c$ along the whole accelerator. This definition allows a direct comparison with measurements. Collimation systems must be designed to minimize local cleaning inefficiency. This is achieved by both minimizing global inefficiency (overall leakage) and maximizing dilution. It is crucial to work on both aspects for achieving best performance.

9.7.2.2 Performance Reach with Collimation

The overall performance of an accelerator is often limited by the peak residual loss that appears in one or few critical locations. It is therefore useful to define a maximum local cleaning inefficiency $\max[\tilde{\eta}_c]$ over all critical locations. For example, in a super-conducting storage ring $\max[\tilde{\eta}_c]$ describes the peak loss per m in super-conducting magnets. Alternatively, in a linac $\max[\tilde{\eta}_c]$ may describe the peak loss per m in the regions that must be protected for hands-on maintenance.

During the design phase one should define an allowable maximum beam loss rate R_{lim} for critical locations. The maximum allowed loss rate R_{loss} at the collimators is defined in particles per second. It depends on the allowable maximum beam loss rate R_{lim} for critical locations and the maximum local inefficiency of the system [76]:

$$R_{loss} = \frac{R_{lim}}{\max[\tilde{\eta}_c]} \,. \tag{9.63}$$

The loss rates R_{loss} can be converted into energy deposition rate P_{loss} using:

$$P_{loss} = \frac{R_{loss}}{(\mathrm{p/s})} \cdot \frac{E_b}{(\mathrm{GeV})} \cdot 1.6022 \times 10^{-10}\ \mathrm{W} \,. \tag{9.64}$$

Considering a stored beam and assuming that all leaked particles are lost at collimators we can relate $R_{loss} = \Delta N/\Delta T$ to the number of particles N_{max} and beam lifetime τ_{min}:

$$\tau_{min} = -\frac{\Delta T}{\ln\left(1 - \frac{R_{loss}\cdot\Delta T}{N_{max}}\right)} \approx \frac{N_{max}}{R_{loss}} \,. \tag{9.65}$$

For convenience we give the equation for calculating the energy deposition rate for any lifetime:

$$P_{loss} \approx N_{max} \cdot \frac{(\mathrm{h})}{\tau_{min}} \cdot \frac{E_b}{(\mathrm{GeV})} \cdot 4.45 \times 10^{-17}\ \mathrm{kW} \,. \tag{9.66}$$

The maximum achievable beam intensity can be expressed as a function of the maximum local cleaning inefficiency, the minimum beam lifetime that must be sustained and the limit of beam loss in critical regions [76]:

$$N_{max} = \frac{\tau_{min} \cdot R_{lim}}{\max[\tilde{\eta}_c]} \,. \tag{9.67}$$

Similar equations can be given for single-pass accelerators. This equation can be used during the design phase of an accelerator to specify the required collimation performance once beam intensity, minimum beam lifetime and loss limits have been determined. A proper design of a collimation system requires a well-defined target for cleaning efficiency.

9.7.3 Operational Settings and Tolerances

Modern collimation systems are often designed to form a multi-stage cleaning system. The collimators then belong to different stages (families), where all collimators of a given stage sit at the same setting (in normalized coordinates). The system will only work correctly if the collimators fulfil the hierarchy requirement [77, 78]. For the example illustrated in Fig. 9.11 the following condition must be fulfilled for the collimation half gaps:

$$n_1 < n_2 < n_3 < n_{\text{triplet}} < n_4 < n_{\text{arc}} \, . \tag{9.68}$$

Typical values for the half gaps n_1 to n_4 are in the range of 5 σ_z to 10 σ_z. The differences in normalized settings are called collimator retractions Δx:

$$\begin{aligned}
\Delta x_1 &= n_2 - n_1 \, , \\
\Delta x_2 &= n_3 - n_2 \, , \\
\Delta x_3 &= n_{\text{triplet}} - n_3 \, , \\
\Delta x_4 &= n_4 - n_{\text{triplet}} \, , \\
\Delta x_5 &= n_{\text{arc}} - n_4 \, .
\end{aligned} \tag{9.69}$$

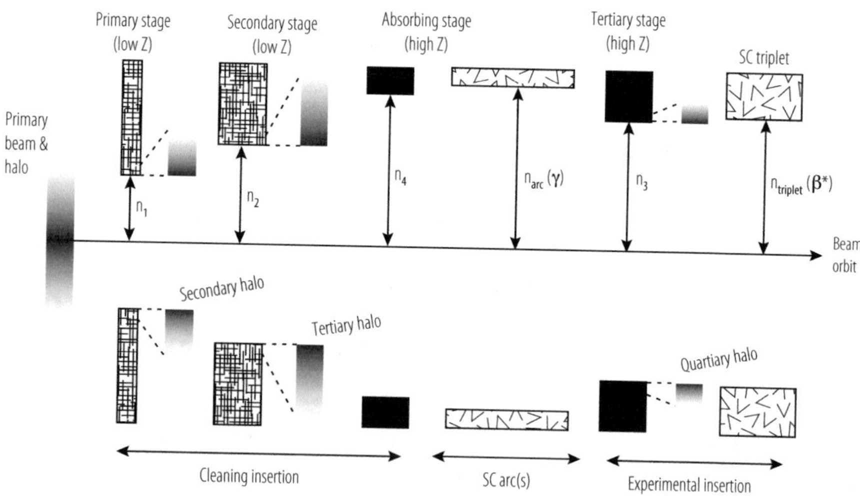

Fig. 9.11 Illustration of collimation hierarchy for the example of LHC collimation [79]. The system establishes global, multi-stage cleaning of halo particles. To be effective, the different collimator families must respect a strict hierarchy in normalized phase space

The retraction values can become quite small. For example, in LHC at 7 TeV retraction values can be smaller than 1 σ_z. At the same time emittance at high energy becomes very small (0.5 nm for the LHC at 7 TeV) and the transverse beam size σ_z can be as small as 140 μm. The smallest collimator retraction values can therefore be in the range of 100 μm, imposing strict operational tolerances.

Various imperfections can reduce the available collimator retraction [78]:

- Beam loss deposits energy on the collimator jaws. The resulting heating can lead to transient jaw deformations.
- If off-momentum beta beating is not or insufficiently corrected, the retraction becomes a function of particle momentum and can be different for particles inside a bunch. See discussion in [74].
- Inaccuracies in beam-based set-up for collimators in different families. See discussion in next section.
- Drifts in beam orbit around the ring since last beam-based set-up or during the beam cycle (for example during squeeze in IP beta function). The possible impact on retraction is illustrated in Fig. 9.12. Most critical is a zero orbit change at a primary collimator and a maximum change at a secondary collimator.
- Change in beta functions around the ring. The possible impact on retraction is illustrated in Fig. 9.12. Most critical is a reduction in beta function at primary collimators and an increase at secondary collimators.

A reduction in collimator retraction reduces the efficiency of the system (up to a factor 10 is possible [80]) and can render it operationally unstable. At some point a secondary collimator can start acting as a primary collimator and efficiency can suddenly be reduced by two orders of magnitude. A collimation system should always be quantified in terms of operational tolerances to make sure that it is appropriately designed and can deliver the required performance. Two sided collimators (as shown in the example of Fig. 9.12) are preferable for operational stability.

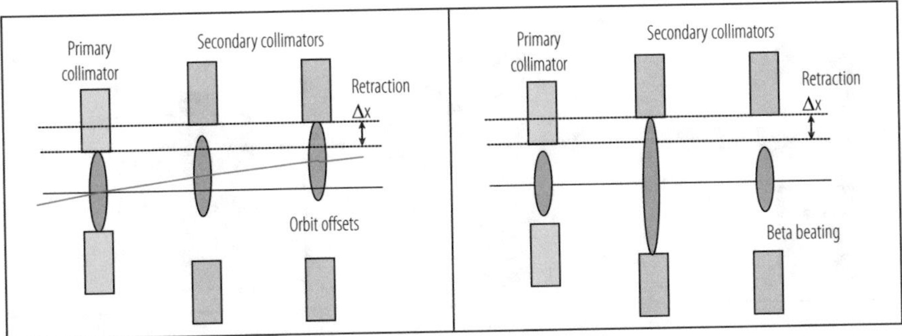

Fig. 9.12 Illustration of machine errors (top: orbit error; bottom: beta beating) that can affect the collimation hierarchy in normalized phase space and reduce cleaning efficiency

9.7.4 Beam-Based Set-Up of Collimation

The previous section explained the criticality of correct collimator hierarchy. It was shown that retraction values can be in the range of 100 μm. Collimators must be centred around the beam with an accuracy that is a fraction of the collimator retraction. Tolerances for collimator settings can then be in the range of a few 10's of μm. However, the exact beam position and size are not known a priori with this accuracy. Collimators are therefore set up in a beam-based process. This beam-based procedure differs for one-pass or stored beams.

- In a single pass accelerator or in transfer lines a collimator jaw is moved through the beam [81–83]. This process is illustrated in Fig. 9.13. Initially (the jaw is still out of the beam) there is a transmission of 100% of beam intensity while downstream beam loss monitors (BLM's) read zero (no showers from the collimator). When the jaw is cutting the beam in its centre then there is a transmission of about 50% and the BLM reads half of its maximum value. Finally, when the jaw intercepts the full beam, the transmission is almost zero and the BLM reads a maximum value, independent of the exact jaw position. This "collimator scan" method allows calibrating the beam centre and size. A related method establishes a collimation gap that is smaller than the beam size. This gap is then scanned across the beam. The transmission and beam loss signals are measured during the scan while recording the gap position. A precise determination of beam centre and size is possible.
- In a storage ring the effects of phase space mixing and amplitude conservation are used, as shown in Fig. 9.14. Any collimator (most often a primary collimator) can be used to define a betatron cut in normalized phase space [77, 78, 81]. This can be done with a single jaw. Assuming zero dispersion, the same phase space cut is present all around the ring after phase space mixing (particles oscillating around

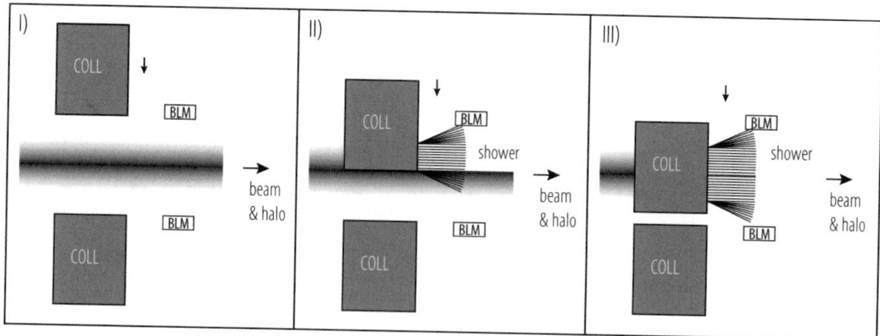

Fig. 9.13 Illustration of beam-based set-up of a collimator in a beam line with single pass beam. A collimator jaw is moved from out position (I) through a centre position (II) into a beam-intercepting position (III). The beam centre is inferred by (**a**) measurement of beam-induced showers with a beam-loss monitor and/or (**b**) by measurement of the not intercepted beam intensity

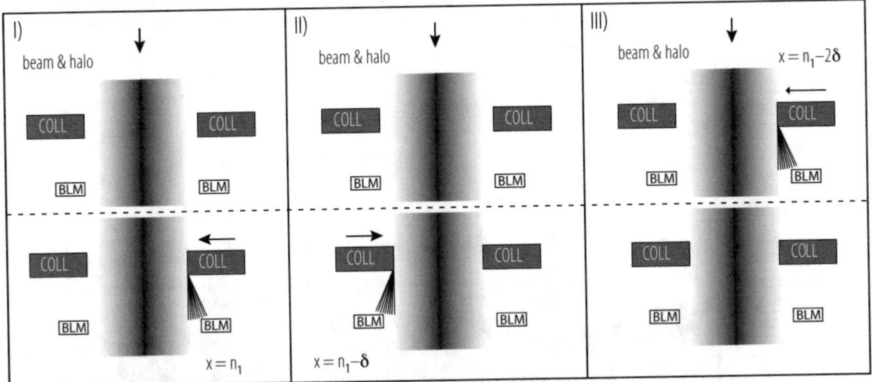

Fig. 9.14 Illustration of beam-based set-up of collimators in a ring with stored beam. At first a reference collimator is defined and one of its jaws is used to create an edge in the normalized beam shape (I). Phase space mixing establishes the same edge all around the ring, in positive and negative directions (in case of zero dispersion). Then the second jaw of the reference collimator is moved to the same normalized position by observing beam loss (II). Analogous, a jaw from any other collimator can be moved to the same defined beam cut (III)

the closed orbit, sweeping around the whole allowed phase space volume). The second jaw of the reference collimator can then be moved to the same cut. A sudden spike in beam loss measured downstream of the collimator is used to detect the halo edge. Successively all collimators around the ring are set up to the same cut in normalized phase space. In the end all jaws are centred on the beam and any beta variations have been calibrated. It is noted that the method is affected by systematic errors that must be taken into account. For example, each set-up will scrape the beam halo by a small additional amount δ. It is advisable to recheck the edge periodically with the reference collimator. Also, tilts in the collimator jaws can induce errors in the knowledge of the collimation gap which can limit the accuracy in determining local beam size.

Once collimators have been set up it is important to record all beam conditions, especially the orbit and optics of the accelerator. This information can then be used to re-establish the collimator set-up and hierarchy for extended periods of times. Collimator set-ups could be used and kept operational for up to 5 months without repeating a set-up for the LHC.

9.7.4.1 Measurement of Collimation Performance

Collimation performance can be measured if a distributed beam loss measurement system has been installed around the ring [84, 85]. Ideally beam loss is measured at all collimators [86, 87], all quadrupoles (here the beta functions are maximal) and other critical locations [88]. An example measurement of collimation performance

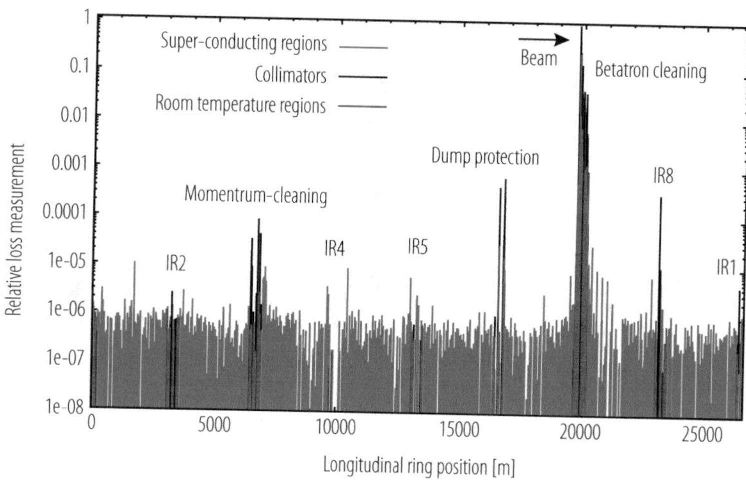

Fig. 9.15 Example for a measurement of collimation performance in the LHC at 450 GeV. The data shows peak integrated losses over 1.3 s. A beam loss is provoked for beam 1 in the horizontal plane. Losses are normalized to the peak loss in the ring. The peak loss appears as expected at the betatron collimators and falls off exponentially over the betatron cleaning insertion. Leakage around the ring is measured. The measurement resolution is limited by noise in the beam loss monitors (6 orders of magnitudes below the peak loss)

in the LHC at 450 GeV is shown in Fig. 9.15. The procedure for such a measurement is described below.

- The measurement should not disturb the orbit or the beta functions, as this would decrease the cleaning efficiency. Therefore one induces a strong diffusion process that rapidly increases the beam emittance. In a storage ring one can move the beam onto a resonance, for example the $\frac{1}{3}$ resonance.
- The integrated beam losses are monitored around the ring as the beam emittance is blown up, for example with a 1.3 s integration time as shown in Fig. 9.15. The data with the highest losses is selected.
- The loss data is normalized to the highest loss all around the ring, which by definition should occur at a collimator. Losses at different locations are distinguished by colour.

The example in Fig. 9.15 shows the results that can be achieved when generating a rapid horizontal emittance blow-up for one beam. The relative loss measurement shown is very similar to the local cleaning inefficiency as defined above, if we ignore differences in BLM response and realize that measured losses are per BLM and not per meter. The measured maximum "local cleaning inefficiency" in a critical region (super-conducting magnets) is about 2×10^{-5}, illustrating a very good performance. The collimators in the cleaning insertion and other areas in the ring intercept reliably all losses and leakage is very small.

9.8 Luminosity Optimization

O. Brüning and M. Hostettler

9.8.1 Introduction

The performance of a collider can be characterized by three main parameters:

- the centre of mass collision energy E_{CM} (in the following we will assume two beams with equal beam energies $\rightarrow E_{CM} = 2 \cdot E_{beam}$);
- the instantaneous luminosity specifying the rate at which certain events are generated in the beam collisions (number of events per second $= L(t) \cdot \sigma_{event}$ with σ_{event} being the cross section of the event of interest);
- the integrated luminosity specifying the total number of events that are produced over a time interval $t - t_0$.

The instantaneous luminosity is given by

$$L = \frac{f_{rev} \cdot n_b \cdot N_1 \cdot N_2}{2\pi \sqrt{(\sigma_{x,1}^2 + \sigma_{x,2}^2)} \cdot \sqrt{(\sigma_{y,1}^2 + \sigma_{y,2}^2)}} \cdot F \cdot H, \tag{9.70}$$

where f_{rev} is the revolution frequency, n_b the number of bunches colliding at the interaction point (IP), $N_{1,2}$ are the particles per bunch and $\sigma_{x,1,2}$ and $\sigma_{y,1,2}$ the horizontal and vertical beam sizes of the two colliding beams. F is the geometric luminosity reduction factor due to collisions with a transverse offset or crossing angle at the IP and H is the reduction factor for the hour glass effect that becomes relevant when the bunch length is comparable or larger than the beta functions at the IP ($\rightarrow$ the transverse beta function varies over the luminous region where the two beams interact with each other).

In the following we assume that all bunches of both beams have equal intensities ($N_1 = N_2 = N_b$) and the same size at the IP. The transverse beam sizes at the IP are given by

$$\sigma_{x,y} = \sqrt{(\beta_{x,y}^* \cdot \epsilon_{x,y}) + D_{x,y}^2 \cdot \delta_p^2}, \tag{9.71}$$

where δ_p is the relative momentum spread ($\delta_p = \frac{\Delta p}{p_0}$) of the particles within a bunch, $\beta_{x,y}^*$ and $D_{x,y}$ are the horizontal and vertical beta and dispersion functions at the IP and $\epsilon_{x,y}$ the horizontal and vertical emittances of the two beams.

Because the bunch intensities and beam sizes of a collider vary over time, the instantaneous luminosity is implicitly a function of time.

The integrated luminosity is defined by

$$\hat{L}(t - t_0) = \int_{t_0}^{t} L(\tau)d\tau, \tag{9.72}$$

where t_0 is an arbitrary starting point, $L(\tau)$ the instantaneous luminosity at a given time and $t - t_0$ the time period of interest.

Optimizing the luminosity of a collider aims essentially at two goals:

- maximize the total number of events over a given time interval → maximize the integrated luminosity;
- minimize the experimental background (e.g. events created by collisions of the beams with rest gas molecules).

The first goal can be achieved by three means: maximizing the instantaneous luminosity, maximizing the luminosity lifetime and minimizing the so called 'turnaround' time which specifies the time interval between the end of one physics fill and the start of the next one.

Then second goal can be achieved by minimizing the vacuum pressure near the IP (reduced rate of rest-gas collisions) and dedicated collimators and absorbers for removing synchrotron light and stray particles before the beams collide at the IP.

The following discussion concentrates on the optimization of the luminosity in circular colliders. The performance optimization of linear colliders will be discussed separately.

9.8.2 *Maximizing the Instantaneous Luminosity*

Maximizing the instantaneous luminosity implies (in order of priority):

- maximize the number of particles per bunch (enters quadratically into the luminosity);
- minimize the beam size at the interaction points (does not imply a 'cost' in terms of total beam power and impedance but might require special focusing quadrupoles near the experiment);
- maximize the number of bunches in the collider;
- optimize the overlap of the two beams at the IP (this essentially implies a precise control of the orbit and optics functions at the IP during operation).

Leaving aside potential single bunch intensity limits due to collective effects and instabilities, the instantaneous luminosity is limited by the strength of the non-linear beam-beam interaction that the particles experience when the bunches of both beams collide with each other at the IP. The strength of the beam-beam interaction can be characterized by the linear head-on beam-beam parameter which specifies the maximum tune shift due to the beam-beam interaction per IP that a particle at the centre of a bunch experiences when the two bunches collide without a crossing angle and transverse offset. The beam-beam parameter is given by

$$\xi_{x,y} = \frac{N_b \cdot r_p \cdot \beta^*_{x,y}}{2\pi \cdot \gamma \cdot \sigma_{x,y} \cdot (\sigma_x + \sigma_y)}, \tag{9.73}$$

where $\beta^*_{x,y}$ are the horizontal and vertical beta functions at the IP and r_p is the classical radius $r_p = e^2/(4\pi\epsilon_0 mc^2)$ for the colliding particles.

It is worthwhile underlining that for round beams with equal beam emittances in both planes the beam-beam force is independent of the transverse beta-functions at the IP and depends only on the normalized beam emittance. For such round beams the beam-beam parameter can be written

$$\xi = \frac{N_b r_p}{4\pi\epsilon\gamma}, \tag{9.74}$$

where $\epsilon\gamma$ is the normalized emittance ϵ_n in a Hadron storage ring.

The beam-beam force provides additional focusing for colliders with beams of opposite charge (e.g. particle anti-particle colliders like LEP) and an additional defocusing strength for beam collisions between particles of the same charge (e.g. particle-particle colliders like the LHC). In either case, the non-linear beam-beam force generates an amplitude growth of the particle oscillations within a bunch which eventually limits the maximum collider performance.

For large particle amplitudes ($>2\sigma$) the (de-)focusing due to the beam-beam force changes sign and there is a reduction of the effective tune shift which eventually becomes negligible for very large oscillation amplitudes and effectively averages to zero for head-on collisions. The beam-beam force therefore results in different tune shift values for different particle amplitudes. Figure 9.16 shows on

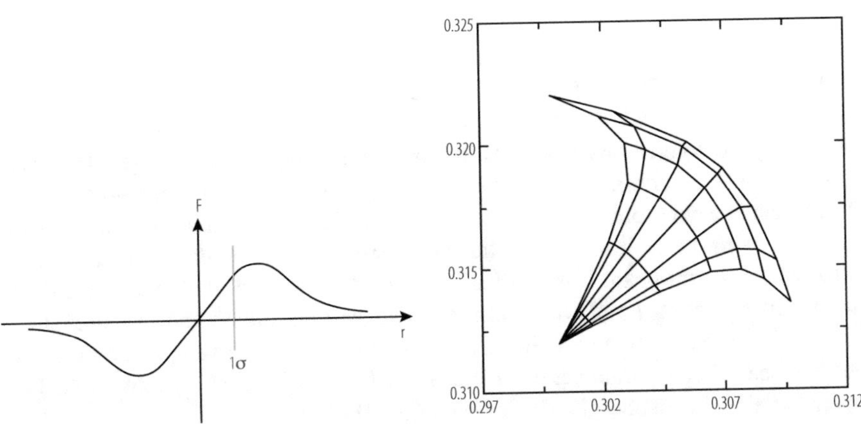

Fig. 9.16 Left: Schematic dependence of the beam-beam force on the particle oscillation amplitude at the IP. Right: Tune foot print for the nominal LHC collisions covering oscillation amplitudes from 0 to 6σ [89]

the left-hand side a schematic picture for the dependence of the beam-beam force on the particle oscillation amplitude and on the right-hand side as an example the resulting tune foot print for the nominal LHC collisions covering particle amplitudes from 0 to 6σ (the particles with zero amplitudes are located at the tip of the tune foot print). Note that the LHC tune foot print shown features the effect of both head-on and long range collisions.

With increasing instantaneous luminosity, the number of concurrent interactions increases. The number of collisions per bunch crossing, the pile-up, is given by

$$\mu = \frac{L}{f_{rev} n_b} \sigma_{tot} \tag{9.75}$$

where σ_{tot} is the total cross-section for the interaction of beam particles.

The maximum acceptable pile-up may be limited by the capabilities of the installed experimental detectors, e.g. due to a limited read-out rate or detector dead time. It may also be limited by the accelerator itself, e.g. due to heating or radiation limits. If this is the case, it imposes an additional limit for the instantaneous luminosity; increasing the luminosity per colliding bunch pair (through intensity or beam size at the IP) increases the pile-up proportionally. The luminosity of a collider operating at the pile-up limit can be optimized through luminosity levelling, which will be discussed in Sect. 9.8.7.

Similarly, the acceptable pile-up density in the luminous region around the IP may be limited e.g. by the detector resolution. Apart from reducing the overall pile-up, this can be mitigated by increasing the size of the luminous region or changing the bunch distribution.

9.8.3 Collider with Strong Synchrotron Radiation Damping

In circular colliders with strong synchrotron radiation the amplitude growth due to the beam-beam interaction is stabilized by the synchrotron radiation damping yielding an increased but stable beam size at the IP. As a result, the instantaneous luminosity no longer increases quadratically, but rather linearly, with the bunch current above a certain limit of the beam-beam force and the beam-beam tune shift reaches asymptotically a maximum value. This maximum value defines the so called beam-beam limit in a collider with strong radiation damping. The actual value of the beam-beam limit varies between different colliders and depends on the actual operating point (the horizontal and vertical tunes) and the strength of the synchrotron radiation damping. Figure 9.17 shows as an example the measured vertical beam-beam parameter in LEP as a function of the bunch current for the LEP operation with beam energies of 98 GeV and 101 GeV.

In an ideal storage ring without coupling and vertical dispersion, the vertical beam size shrinks to a minimum value which is in theory only limited by the quantum fluctuation of the synchrotron radiation. In the horizontal plane this effect

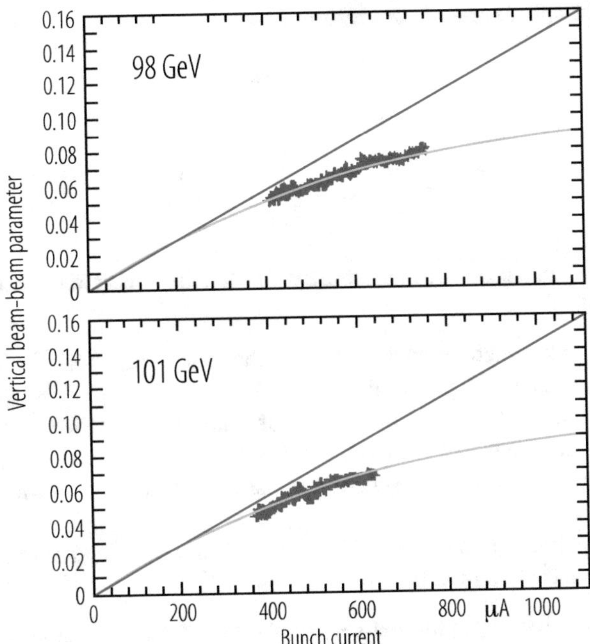

Fig. 9.17 Measured vertical beam-beam limit in LEP for operation with beam energies of 98 GeV and 101 GeV [90]. The beam-beam parameter has been derived from the measured luminosity assuming a constant value of β^*. The straight line indicates the expected linear scaling of the beam-beam parameter for operation below the beam-beam limit

is outweighed by the oscillation excitations due to the photon emission in regions with dispersion and the equilibrium horizontal beam size increases as a result with the beam energy. Provided the associated synchrotron radiation damping times are short compared to the typical store length of the collider, the beam sizes are therefore essentially determined by the synchrotron radiation yielding a flat beam with small vertical and large horizontal beam sizes. In practice, the minimum attainable vertical beam emittance is limited by the rms vertical dispersion and the coupling between the horizontal and vertical motion along the storage ring. Furthermore, the minimum vertical beam size at the IP is limited by the momentum spread within a bunch and the spurious vertical dispersion at the IP (see Eq. (9.71)). Colliders with strong synchrotron radiation are therefore normally optimized for the operation with flat beams that have a large aspect ratio between the horizontal and vertical beam sizes at the IP. The generation of non-equal betatron functions at the IP can be best provided by a doublet quadrupole configuration of the focusing elements next to the IP (→ non-equal β^* values in both planes for equal peak betatron function values inside the doublet magnets).

Assuming vanishing dispersion functions at the IP, the beam-beam parameters of a collider with flat beams and a large aspect ratio between the horizontal and vertical beam sizes are approximately proportional to

$$\xi_x \propto \frac{N_b}{\epsilon_x}, \tag{9.76}$$

$$\xi_y \propto N_b \cdot \frac{\sqrt{\beta_y^*}}{\sqrt{\epsilon_y}\sqrt{\epsilon_x}} \cdot \frac{1}{\sqrt{\beta_x^*}}. \tag{9.77}$$

Optimizing the instantaneous luminosity for flat beams at the beam-beam limit therefore implies:

- increasing the bunch intensity so that the horizontal beam-beam parameter approaches the beam-beam limit;
- minimizing the vertical dispersion (via, for example, dispersion free steering) and coupling along the whole the machine to get a small vertical emittance;
- minimizing the vertical beta function at the IP to reduce the vertical beam-beam parameter;
- adjusting the horizontal beta function at the IP for obtaining approximately equal beam-beam parameters in both planes.

For example, the LEP operation at the beam-beam limit for beam energies of 94.5 GeV, featured an aspect ratio of the horizontal and vertical beta functions at the IP of ca. 31 (1.25 m/0.04 m) and an aspect ratio of the horizontal and vertical beam sizes at the IP of ca. 51 (180 μm/3.5 μm) yielding for a bunch current of 780 μA (ca. 1.4×10^{11} particles per bunch) theoretical beam-beam parameters of $\xi_x \approx 0.04$ and $\xi_y \approx 0.06$. Depending on the operation configuration and beam energy the measured vertical beam-beam limit in LEP varied between $\xi_{beam-beam} = 0.05$ and 0.01 [90–93].

Another interesting feature of the operation with large beam-beam parameters is that the additional focusing due to the beam-beam interaction can also change the beta-functions at the IP. This dynamic β^* effect can result in a non-negligible second order perturbation of the optic functions at the IP.

The luminosity can be further increased by increasing the number of bunches in the machine. The maximum number of bunches in the collider is then either limited by collective effects (e.g. impedance and collective oscillations, electron-cloud effect for positron bunches or ion trapping for electron bunches) or by the maximum acceptable synchrotron radiation power or the simple fact of having one or two rings. If such considerations still permit a large number of bunches, it might be necessary to introduce a crossing angle at the IP in order to avoid unwanted collisions next to the IP (e.g. the KEKB factory). In this case, the luminosity optimization becomes more complex. The crossing angle decreases the luminosity and the beam-beam parameter at the IP via the geometric form factor F in Eq. (9.70) and introduces a dependence of the luminosity on the crossing angle, the bunch length and the transverse beam size in the plane of the crossing angle (see Sect. 6.4

for more details). The luminosity optimization in this configuration will be discussed in Sect. 9.8.5

9.8.4 Collider with Weak Synchrotron Radiation Damping

In circular colliders with weak synchrotron radiation damping (with damping times much larger than a typical store length—typically the case for hadron colliders where the relativistic γ factor is small) the amplitude growth of the particle oscillations within a bunch due to the beam-beam interaction is not stabilized and leads to emittance growth, the development of tails in the bunch distribution, particle losses and a reduction of the beam lifetime and possibly an increase of the experimental background. The above detrimental processes of the beam-beam interaction become more pronounced the more resonances are covered by the tune distribution for the particles within a bunch. The beam-beam limit in hadron colliders is therefore more commonly expressed by the total maximum acceptable tune spread within a bunch due to the beam-beam interactions rather than the maximum attainable beam-beam parameter for a single interaction. For hadron colliders the beam-beam limit is therefore the sum of all beam-beam related tune spreads. For purely head-on collisions it is proportional to the number of collisions a single bunch experiences during one revolution. For a collider that features parasitic long range beam-beam encounters it also includes the tune spread contributions from the long-range beam-beam interactions.

The maximum acceptable beam-beam related tune spread depends on the operation point (transverse tune values) and other sources for tune spread in the collider (e.g. tune spread due to magnetic octupole components) and typically varies for different Hadron colliders between values of $\Delta Q = 0.015$ and $\Delta Q = 0.03$ (e.g. RHIC and Tevatron). The 'precise' value of the quoted beam-beam limit in a Hadron collider depends on the acceptable luminosity lifetime and background rates for a given operation mode and is therefore somewhat less well defined as for the case of lepton colliders with strong synchrotron radiation. For example, a beam-beam driven tune spread of $\Delta Q = 0.03$ is in principle attainable in the Tevatron operation but the normal machine operation rather aims at a smaller beam-beam driven tune spread of $\Delta Q = 0.02$ (the quoted tune shift refers to the anti-proton beam—the beam-beam tune shift for the proton beam is ca. five times smaller) [94].

There are currently several studies under way for looking into possibilities of compensating part of the beam-beam generated tune spread either by the use of wires with DC or pulsed currents for the compensation of the tune spread arising from the long-range beam-beam interactions [95] or by the use of electron lenses for a compensation of the tune spread generated by the head-on beam-beam collisions [96, 97].

Without damping the beam sizes are mainly determined by the injector complex and the ability to preserve the injected emittances in the collider during injection and acceleration. Without natural equilibrium beam sizes, the luminosity with head-on collisions can be best optimized by deploying round beams (equal beam sizes and emittances in both planes). The generation of equal betatron functions at the IP can be best provided by a triplet quadrupole configuration of the focusing elements next to the IP (this implies equal β^* values in both planes for equal peak betatron functions for both planes inside the triplet magnets). The beam-beam parameter in this case is given by Eq. (9.74) and is entirely independent from the beta function values at the IP. Optimizing the instantaneous luminosity in this case therefore implies:

- increasing the beam brightness ($N_b/\epsilon_{x,y}$) until the beam-beam limit is reached;
- increasing the bunch population at constant brightness;
- minimizing the beta functions at the IP in both planes.

The second point is limited by the available aperture along the whole collider and by collective effects that might limit the maximum bunch intensity. The third point is limited by the available aperture of the triplet elements next to the IP (the maximum beta function in these elements is proportional to $1/\beta^*$), the 'hour glass' effect and eventual chromatic aberrations due to large betatron functions in the focusing elements. The hour glass effect depends strongly on the bunch length and is only relevant for a configuration with $\beta^* \leq \sigma_s$. Shortening the bunch length might be another strategy for maximizing the luminosity in this case but the minimum bunch length might itself be limited by other constrains such as the available RF voltage and limitations on the maximum acceptable Intra-Beam-Scattering (IBS) growth rates.

The maximum number of bunches in the collider is then either limited by collective effects (e.g. electron-cloud effect for proton bunches), the maximum acceptable stored beam power (e.g. machine protection issues) and the rate of particle production in case the collider uses anti-protons in the collisions (e.g. the Tevatron).

9.8.5 Luminosity Optimization in the Presence of a Crossing Angle

If the operation of the collider permits a large number of bunches, it might be necessary to introduce a crossing angle at the IP in order to avoid unwanted collisions of the two beams next to the IP (e.g. the KEK-B factory and the LHC). A crossing angle reduces the luminosity and the beam-beam parameter via the geometric form factor F in Eq. (9.70). The geometric form factor is a function of the crossing angle, the bunch length and the transverse beam size in the plane of the crossing angle and therefore crossing angle operation introduces additional parameter dependencies which make the luminosity optimization more complex.

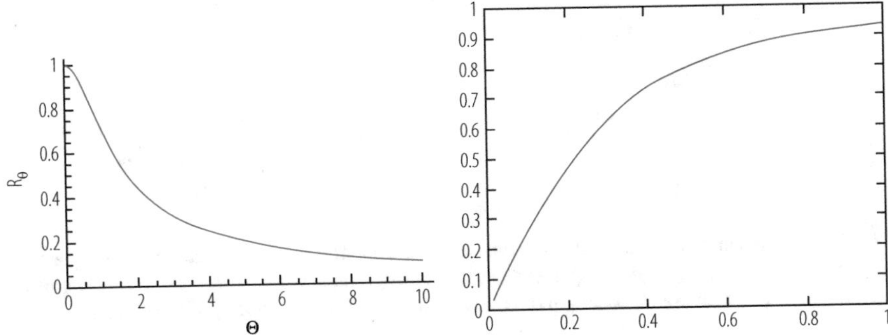

Fig. 9.18 Left: The geometric form factor as a function of the Piwinski angle. Right: the form factor for the nominal LHC parameters as a function of β^* for a constant bunch length and beam separation in terms of transverse rms beam sizes

The geometric luminosity reduction factor becomes relevant for large Piwinski angles ($\Theta \geq 1$) were the Piwinski angle is defined as:

$$\Theta = \frac{\sigma_s \phi}{2\sigma_\perp}. \tag{9.78}$$

ϕ is the full crossing angle and $\sigma_\perp$ the transverse beam size in the plane orthogonal to the crossing angle. In terms of the Piwinski angle, the geometric reduction factor can be expressed as

$$F = 1/\sqrt{1+\Theta^2}. \tag{9.79}$$

The left-hand side of Fig. 9.18 shows the geometric form factor as a function of the Piwinski angle and the right-hand side as a function of β^* for the data of the nominal LHC assuming a constant bunch length and beam separation in terms of rms beam sizes were the form factor is essentially a function of β^*. The right-hand side of Fig. 9.18 illustrates with the example of the LHC that a collider with crossing angle operation can rely on a performance increase though a reduction of β^* only up to a certain point. If the Piwinski angle becomes large, the desired performance gain is essentially directly lost via the geometric form factor.

The resulting loss in luminosity can partially be compensated either by the use of Crab cavities [98] which tilt the bunches longitudinally at the IP such that the bunches effectively still collide head-on (see Fig. 9.19) or by the implementation of a 'crabbed waist' crossing scheme [99] for very large crossing angles and long and flat bunches ($\sigma_z \phi_{x,y} >> \sigma_{x,y}$) which shifts the location of the minimum beta function in the plane orthogonal to the crossing angle along the bunch length with the help of dedicated crab sextupole magnets. The crab waist optimization requires in fact a crossing angle in the plane with the larger beam size dimension (the horizontal plane for a collider with strong synchrotron radiation damping) and features three separate

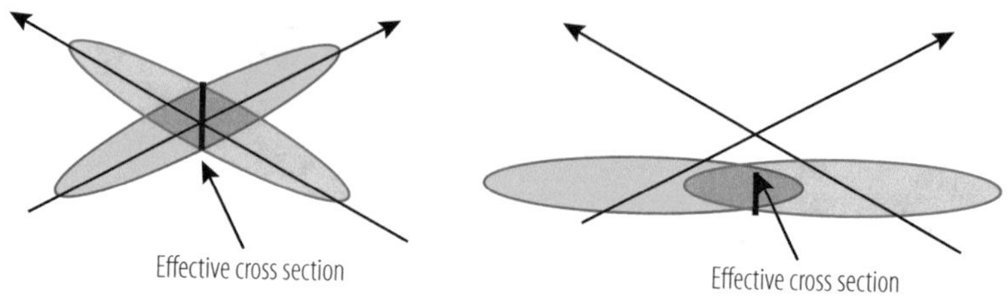

Fig. 9.19 Schematic illustration of the crab-crossing collision scheme for large crossing angles. Left: the beam interaction in the presence of a crossing angle without crab crossing. Right: the beam interaction with crab crossing and a maximized bunch overlap during the collision

ingredients: operation at large Piwinski angle, a vertical β^* value comparable to the luminous area and a shift of the vertical betatron function waist of one beam along the central trajectory of the other beam.

In the presence of two IPs with orthogonal crossing angle planes, small β^* values and round beams one can write the dependence of the luminosity and the head-on beam-beam parameter approximately as follows (if the collider does not feature two orthogonal crossing angle planes, the form factor will not reduce the beam-beam parameter equally in both planes):

$$L \propto \Delta Q_{bb} \cdot \frac{N_b}{\beta^*} , \tag{9.80}$$

$$\Delta Q_{bb} \propto \frac{N_b}{\epsilon} \cdot F(\beta^*) . \tag{9.81}$$

In this situation one can essentially distinguish four different strategies for optimizing the instantaneous luminosity (we limit the discussion here to the case of round beam operation):

- keep the geometric form factor (and thus β^*) and the beam-beam parameter constant and maximize the luminosity by increasing the bunch intensity at constant brightness—this strategy requires sufficient aperture in the collider for accommodating large beam emittances and the absence of any other bunch intensity limitations (e.g. collective effects and single bunch instabilities);
- keep the beam emittance constant and increase the bunch intensity inversely proportional to the geometric form factor when β^* is lowered. This strategy implies that the beam intensity in the collider is not limited by collective effects;
- keep the number of particles per bunch constant and vary the beam emittance proportionally to the geometric form factor when β^* is lowered;
- compensate the performance loss due to the geometric form factor (e.g. by the use of Crab cavities or crabbed waist operation) and reduce β^*.

9.8.6 *Maximizing the Integrated Luminosity*

Several effects can lead to an intensity decrease of the colliding beams over the length of a physics store:

- the loss of particles via the actual collisions (beam burn off);
- particle losses via collisions with the rest gas and photons in the vacuum system;
- loss of particles via resonances (e.g. driven by the beam-beam interaction);
- particle losses through the Touschek effect;
- particle losses through aperture restrictions (relevant for colliders where the beam distribution is an equilibrium distribution determined by the synchrotron radiation and were any part of the distribution (e.g. the tails of the Gaussian distribution) lost through aperture restrictions will be repopulated).

In addition, several effects can lead to an increase of the beam size via an emittance blow-up:

- resonances in the machine (e.g. driven by the beam-beam interaction);
- noise in machine equipment (e.g. kicker and RF elements);
- intra beam scattering (IBS).

All the above effects together result in a reduction of the luminosity with time and require the preparation of a new fill with fresh beams once the performance dropped too much. The integrated luminosity depends then on the luminosity lifetime, the run length and the 'turnaround' time required for preparing a new fill in the collider. The turnaround time is defined as the time interval between the end of one fill and the start of the next one. Assuming an exponential luminosity decay ($L(t) = L_0 \cdot e^{-t/\tau_L}$) the integrated luminosity per fill can be written as:

$$\hat{L}_{fill} = L_0 \cdot \left[1 - e^{-T_{run}/\tau_L}\right], \tag{9.82}$$

where τ_L is the luminosity lifetime, T_{run} the length of the physics operation and L_0 the initial instantaneous luminosity of the physics fill. The optimum run length is a function of the average turnaround time of the collider and a high integrated luminosity requires a short and reproducible turnaround time.

If the luminosity lifetime is not much longer than the average turnaround time the total integrated luminosity might be limited by the machine reliability and variations in the turnaround time. Furthermore, if the luminosity lifetime is too short, a fraction of the generated luminosity might be lost for the experiments if not all detector components can be fully operational from the very beginning of a physics fill when the operation still performs beam adjustments and measurements.

9.8.7 Luminosity levelling

For colliders limited by pile-up, the integrated luminosity can be significantly improved by implementing a luminosity levelling scheme that initially reduces the peak luminosity and levels it at a constant value that yields an acceptable pile-up throughout a physics fill [100, 101]. In case different detectors with different levels of acceptable pile-up are installed in a collider (e.g. at the LHC), it also allows satisfying the requirements of the low pile-up detectors without impairing the performance of the high luminosity experiments.

The effective luminosity lifetime increases when levelling the luminosity due to the slower burn off rate and, for certain means of levelling (e.g. offset, crossing angle), weaker head-on beam-beam effects. The luminosity evolution over a fill with luminosity levelling consists of a first part, where the levelling is used to reduce the instantaneous luminosity to a certain target value, possibly followed by a second part where the luminosity is left to decay without levelling (Fig. 9.20).

If the beam emittance is constant and beam losses are dominated by burn off while levelling, the beam current decays linearly and the levelling time is directly proportional to the initial beam current:

$$T_{\rm lvl} = \left(1 - \sqrt{\frac{L_{\rm lvl}}{L_{\rm peak}}}\right) \frac{n_b\, N_i}{n_{\rm IP}\sigma_{\rm tot} L_{\rm lvl}} \tag{9.83}$$

where n_b is the number of bunches with initial intensity N_i, $L_{\rm peak}$ is the theoretical peak luminosity, $L_{\rm lvl}$ is the levelled luminosity, $\sigma_{\rm tot}$ is the cross-section for the interaction of beam particles, and $n_{\rm IP}$ is the number of IPs.

In the following, the machine parameters commonly used to control the luminosity for levelling are discussed.

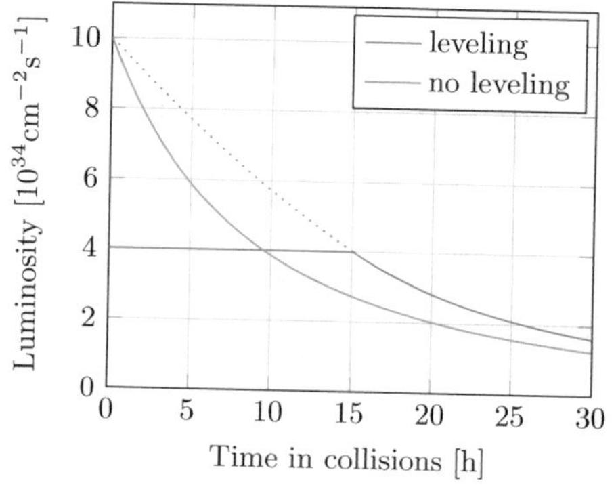

Fig. 9.20 Luminosity evolution in a burn-off dominated scenario with and without levelling. In the levelled case, the peak luminosity is initially levelled down by a factor of 0.4. The dotted line shows the "virtual" luminosity which is reduced to the target through levelling. The levelling time is $T_{\rm lvl} = 15$ h

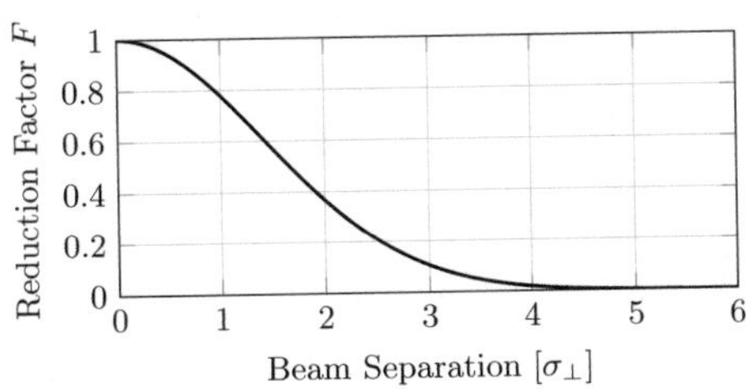

Fig. 9.21 Luminosity reduction factor as a function of the transverse separation

9.8.7.1 Levelling by Transverse Offset

A beam separation is introduced by local closed orbit bumps around the IPs. In the presence of a crossing angle, the separation is typically applied perpendicular to the crossing plane. This reduces the luminosity by a geometric factor given by

$$F = \exp\left(-\frac{d^2}{4\,\sigma_\perp}\right) \tag{9.84}$$

where d is the separation and $\sigma_\perp$ is the transverse beam size in the separation plane. This factor is shown in Fig. 9.21.

When levelling by offset, the pile-up density decreases along with the total pile-up. Since the required offsets are small (a few units of the transverse beam size) and the bumps are locally closed, this levelling approach is operationally easiest. However, it offsets the particles from the centre of the linear part of the beam-beam force, reduces the beam-beam tune spread and drives odd-order resonances, which can affect the beam stability. Furthermore, the luminosity becomes more sensitive to small variation of the offset, requiring a good orbit control around the IP.

Offset levelling has been used successfully e.g. at the LHC with levelling factors between 0.01 and 1. While this allowed for a great flexibility and worked well in general, instabilities were observed with separated beams close to the beam-beam limit [102].

9.8.7.2 Levelling by Crossing Angle

In the absence of crab-crossing, an increased crossing angle reduces the luminosity as explained in Sect. 9.8.5. Levelling by crossing angle only affects the total pile-up while the pile-up density remains constant (as the length of the luminous region changes with the crossing angle). While the crossing angle can be controlled through local closed orbit bumps, the resulting orbit excursions are significantly larger than for a transverse separation. In particular for small β^*, the maximum orbit excursion and hence the maximum crossing angle may be limited by the aperture of the triplet elements next to the IP.

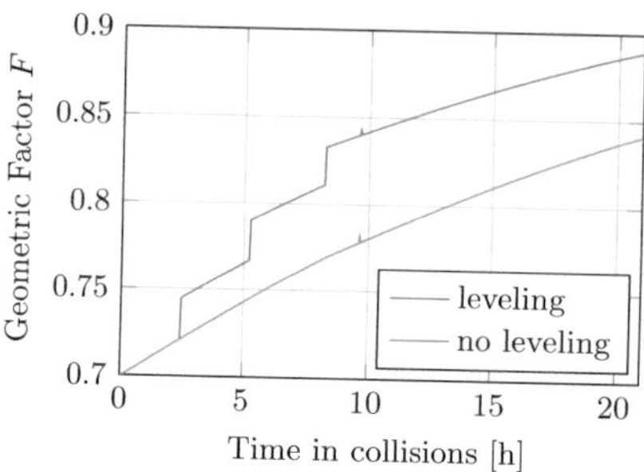

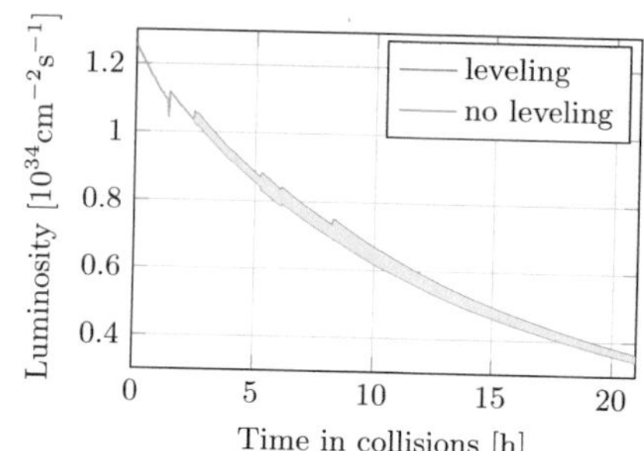

Fig. 9.22 Luminosity levelling by reducing the crossing angle in steps over the course of a fill at the LHC. Four crossing angle were used: 150 µrad (initial), 140 µrad, 130 µrad and 120 µrad (final). Left: the geometric reduction factor F, which increases as the crossing angle is decreased. Right: the resulting luminosity. The integrated gain due to the levelling (shaded area) was at the level of 5%

In a collider with crab cavities, a similar effect can be achieved by varying the crab cavity voltage instead of the external crossing angle. In this case, the bunches collide tilted at the IP, reducing the effective cross-section and hence the geometric factor.

Furthermore, levelling by crossing angle can also be used to improve the performance of a collider limited by beam-beam interactions. Since the bunch intensities decrease over the course of a fill, beam-beam forces (both head-on and parasitic) decrease and allow for a smaller crossing angle. By gradually reducing the crossing angle throughout a fill, the loss of luminosity due to the geometric factor can be minimized. Since reducing the crossing angle increases the length of the luminous region, the pile-up density remains constant during this operation. This approach has e.g. been used at the LHC as of 2017. An example is given in Fig. 9.22.

9.8.7.3 Levelling by β*

Adjusting the β^* allows controlling the luminosity while keeping the beams head-on, avoiding aperture restrictions of closed orbit bumps and providing a constant beam-beam tune spread. It implies changing the local optics around the IP, which will change the orbit and β functions locally. Therefore, care must be taken ensure smooth transitions between different β^* values and ensuring machine protection at all times. In particular, orbit excursions must be kept under control e.g. by using feedback systems and any collimators around the IP must be moved to match the new optics.

To allow for arbitrary values of β^* and independent levelling of all IPs, the corresponding IP optics would need to be matched online during collider operation. To reduce complexity, a discrete set of IP optics covering the range of β^* needed for operation can be pre-generated. If necessary, this can then be combined with levelling by separation or crossing angle for a fine-grained luminosity control between the foreseen β^* steps.

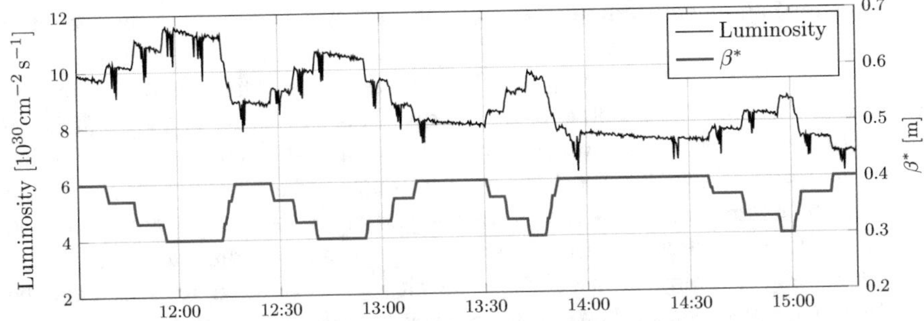

Fig. 9.23 β^* and luminosity during a β^* levelling test at the LHC. To demonstrate that the beams can be squeezed and un-squeezed at the IPs as necessary, multiple cycles between $\beta^* = 40\,\text{cm}$ and $\beta^* = 30\,\text{cm}$ were performed. The luminosity follows accordingly

levelling by β^* is a baseline concept for novel accelerators (e.g. HL-LHC, FCC-hh) and has been successfully demonstrated at the LHC. An example from a machine test is given in Fig. 9.23.

9.8.7.4 Alternative Methods and Combined Levelling Scenarios

In addition to the methods described above, the luminosity can also be reduced in a controlled way by creating a longitudinal separation through RF cogging. However, this is operationally challenging and intrinsically effects all IPs equally. In a collider with synchrotron radiation damping of the transverse and longitudinal emittances, controlled emittance blow-up (transverse or longitudinal) could be used likewise.

Also, two or more levelling methods described above can be combined to a full levelling scenario for a collider which yields the optimal integrated luminosity within the given constraints of pile-up, beam-beam effects and machine limitations (e.g. aperture). For example, the luminosity could initially be reduced to a target via a combination of β^* and offset levelling; once this is not effective anymore, the intensity and hence the beam-beam forces will have decreased, so the crossing angle could be reduced to increase the luminosity, gaining extra time at the levelling target. Once the crossing angle is small enough, the triplet aperture may allow for a further reduction of β^*.

9.9 Machine Protection

J. Wenninger

Two main machine protection domains are considered. Firstly the protection of accelerator equipment related to high power equipment. For example: the protection of powering equipment from overheating (magnets, power converters, high current cables); the protection of superconducting magnets from damage after a quench; and the safe operations of high power klystrons. Secondly there is the protection of equipment from high stored beam energy. This is related to the high beam power of high power proton accelerators such as ISIS, SNS and the PSI cyclotron, to the emission of synchrotron light by electron/positron accelerators and to the increase of energy stored in the beam (in particular for hadron colliders such as TEVATRON, HERA and LHC). Effective protection in this case requires an excellent understanding of accelerator physics and operation to anticipate the failures that could lead to damage. It includes sophisticated beam and equipment monitoring, a system to safely stop beam operation (e.g. dumping the beam) and an interlock system providing the means to integrate the various systems. Machine protection must be considered during design, construction and operation of the accelerator.

9.9.1 Definition of Risk

The risk factor may be defined as the product of the probability of a failure multiplied by the consequences of the failure (e.g. damage to equipment). The total risk is the product of the individual risks factors of the possible failures in play. Protection can be used to mitigate risk. In the accelerator domain it is clear that the higher the risk factor, the more important that protection becomes.

The cost of given failure scenario can be estimated. The consequences may be in terms of, for example: damage to equipment (repair requiring expenditure); downtime of the accelerator; or the radiation dose to personnel accessing equipment; or indeed a combination of the above.

The second factor entering into the risk is the probability that a given failure happens. One can imagine that such an estimate be measured in number of failures per year.

For operation with beam, a list of all possible failures that could lead to beam loss into equipment should be considered. This is not obvious to establish since there are a potentially large, but not innumerable, ways to lose the beam. However, the most likely failure modes and, in particular, the worst case failures (in terms of cost) and their probability must be considered.

9.9.2 Beam Losses and Consequences

Particle losses in a material lead to particle cascades. The maximum energy deposition can be deep in the material at the maxima of the hadronic and electromagnetic showers. The energy deposition leads to a temperature increase and stress in the material that can be vaporised, melt, deform or lose its mechanical properties, depending on the material and the beam impact.

There is already some risk of damage to sensitive equipment from an energy deposition of some 10 kJ for a beam impact on the time scale of a few milliseconds. The risk of damage to equipment from losses in the mega-Joule regime is enormous.

Equipment becomes activated due to beam losses. Rigorous radiation protection measures are required and acceptable exposure limits must be clearly established. For potential exposure to higher rates one possible approach is the ALARA principle which aims to keep the exposure of personnel to radiation "As Low As Reasonably Achievable". Cool-down times might be required to respect this principle with knock on costs to operational availability.

For accelerators with superconducting magnets there is a specific problem—even with beam loss much below the damage threshold superconducting magnets may quench due to the temperature increase induced by beam loss. In the case of a quench, beam operation is interrupted leading to downtime for recovery. To avoid beam induced quenches, beam losses are monitored and beam operation is aborted if a predefined threshold is exceeded. Since the damage threshold is well above the quench threshold, this strategy also protects the superconducting magnets from damage.

There is no simple expression for the energy deposition of high energy particles, since this depends on the particle type, the momentum, beam parameters and material parameters (atomic number, density, specific heat). Programs such as FLUKA [103], MARS [104] or [105] are used to estimate energy deposition as well as the activation of the exposed material.

9.9.3 *Time Constants for Beam Losses*

Potential beam losses scenarios can be broken down according to the time scale of the losses. Monitoring and reaction to losses must, of course, act at the appropriate time scale.

9.9.3.1 Ultra Fast Beam Losses

Sources of failures that lead to a beam loss within very short time, typically in the range of ns to μs are:

- failures of kicker magnets (during injection, at extraction, or during the use of special kicker magnets used for beam diagnostics);
- failure during beam transfer via transfer lines between different accelerators or from the accelerator to a target station.

9.9.3.2 Very Fast Beam Losses

Sources for failures that lead to a very fast beam loss (typically in the range of ms) are multi turn beam losses in circular accelerator. These can have a large number of possible causes, most of which will concern the powering system of the main magnets. Time constants could be in the order of some 10 turns.

9.9.3.3 Fast Beam Losses

Fast beam loss (some 10 ms to seconds) can have many different origins, for example failures in the magnet powering system; vacuum valves that close; trips of the RF acceleration system; beam instabilities; control system failures.

9.9.3.4 Slow Beam Losses

Slow beam losses (many seconds) can have many different origins, for example, high vacuum pressure; failures in the powering system for corrector magnets; operational error. The main difference between slow and fast losses is that the operation crew could still be involved in the decision how to continue operation (for example, to stop beam operation or to correct a parameter that is not set correctly).

9.9.4 Principles of Machine Protection

The key high level requirements for machine protection outline below.

- Protect the accelerator and ensure no, or minimal, damage. The highest priority to avoid any damage to accelerator equipment.
- Protect the beam and ensure maximum machine availability. Complex protection systems with many interlocks may reduce the availability of the machine. The number of false interlocks stopping operation must be minimized. A "false" interlock is defined as an interlock that stops operation although there is no risk (for example when a sensor involved in an interlock gives an incorrect reading).
- Provide the evidence and ensure complete availability of all post mortem data in case of failure, or false triggers of the machine protection system. When the protection systems stop operation (e.g. dump the beam or inhibit injection), clear diagnostics should be provided [106] to understand cause and consequences. This requires fast diagnostics based on synchronised transient recording of all the important parameters, as well as long term logging of parameters.

The principle components of a machine protection system are listed below.

- Systems to monitor the correct operation of all hardware.

- Beam instrumentation to measure and check that beam parameters are in the correct range.
- A beam dumping system, in general including a fast kicker magnet and absorber block.
- Collimators and beam absorbers to provide passive protection against beam loss.
- A beam interlock system that links the different protection systems. Its role is to ensure that the appropriate action is taken in case of problems (the beam is extracted from a synchrotron, injection is stopped etc.). The interlock system might include complex logic.

There are several principles that should be considered in the design of protection systems, although it might not be possible to follow all these principles in all cases.

- If the protection system does not work or is compromised, it should prevent continued operation. The design should be fail-safe. In case of a failure in the protection system, protection functionalities should not be compromised. As an example, if the cable that triggers the extraction kicker of the beam dumping system is disconnected, operation must be stopped and the beam dumped.
- Detection of internal faults: the protection system must monitor its status. In case of an internal fault, the fault should be reported. If the fault is critical, operation must be stopped.
- Remote testing of correction system behaviour should be possible. For example between two runs unit testing should allow verification of the correct status of the system.
- Critical systems should have built in redundancy of critical functionality (possibly diverse redundancy, with the same or similar functions executed by different systems).
- Critical processes for protection should not rely on complex software running under an operating system and requiring the availability of a computer network.
- It should not be possible to remotely change the most critical parameters of a given system. If parameters need to be changed, the changes must be controlled, logged and password protection should ensure that only authorised personnel can make the change.
- Safety, availability and reliability of the systems should be demonstrated. This is possible by using established methods to analyse critical systems and to predict failure rates.
- One should attempt to gain experience of operating the protection systems in good time before they become critical, to gain experience and to build up confidence. This could be done before beam operation, or during early beam operation when the beam intensity is deliberately kept relatively low.

9.9.5 Strategy for Protection

The best protection strategy is to prevent by design the occurrence of a potential failure. As an example, fast kicker magnets for diagnostics that could deflect the beams into the vacuum chamber should only be installed in high intensity machines if they are indispensable. If they are installed the available kick strength should be below the level that can kick the beam to the aperture.

Failure should be detected as early as possible, with priority at the hardware level of the system at the origin of the failure. If the failure detection is fast enough, beam operation may be stopped before the beam parameters are affected in a significant way. As an example, a failure of a magnet power converter should be detected as early as possible within the converter itself.

When a failure is detected, beam operation must be stopped. For synchrotrons and storage rings the beam is extracted by a fast kicker magnets into a beam dump block. Injection must be inhibited.

Since detecting failures at the hardware level is not always possible or fast enough, beam diagnostics is needed to detect abnormal beam behaviour. This requires reliable, and sometimes dedicated, beam instrumentation.

9.9.6 Active and Passive Protection

Active protection is based on detection of a failure or of the consequences of the failure on the beam. After detection the beam must be turned off and injection inhibited, or, the case of a storage ring, the beam must be dumped as soon as possible.

Passive protection there is a certain class of failures when active protection is not possible. For example, protection against misfiring of an injection or extraction kicker magnet might include a beam absorber or collimator to catch any mis-kicked beam. All possible beam trajectories in such case must be considered, and the absorber must be designed and positioned to absorb the beam energy without being damaged itself.

9.9.7 Interlock Management

In large complex accelerator with high beam currents it is unavoidable that there will be a high number of interlocks. In certain phases of operation (e.g. commissioning) when operating with limited beam power, the disabling of certain interlocks is certainly to be envisaged. However, it is vital to track any disable interlocks and ensure that they are re-enabled when required. In the LHC the disabling of a selected number of interlocks is possible for low intensity, low energy beams. When the beam energy or intensity increase above damage level, the interlocks are automatically enabled.

9.9.8 Beam Instrumentation for Machine Protection

Beam instrumentation plays an important role for machine protection, and is used to monitor beam parameters and stop beam operation if a parameter is outside a predefined range. If machine protection relies on the correct operation of beam instruments, failures in beam instrumentation sub-systems need to be considered.

9.9.8.1 Beam Loss Monitors—BLM

BLMs are used for monitoring beam losses and can clearly play an important role in machine protection. In simple terms the BLMs measure localized beam losses along the accelerator and stop beam operation in case losses the losses become too high. It is important that the monitors cover the entire accelerator and there is no region with potential for losses without BLM coverage.

The monitors can be fast, for example, the LHC BLMs operate on a 40 μs time-scale. The BLM system should be designed so that it can trigger a beam dump and stop operation before beam losses on this scale can damage equipment.

Failure cases for the system might include defective BLMs providing no or too low readings and therefore not providing a signal even in case of high beam losses; and the thresholds could be set incorrectly. The former can be caught via rigorous unit testing, the latter by strict and rigorous threshold management.

9.9.8.2 Beam Position Monitors—BPM

BPMs coupled with appropriate orbit or trajectory correction can ensure that the beam is in the correct place in respect to the aperture. This position would normal be around the centre of the beam pipe but there are many exceptions, for example during extraction where a closed orbit bump is applied to position the beam close to a septum magnet. BPMs can monitor the amplitude of such bumps and are effectively redundant monitors of the magnet current in the closed orbit dipoles.

One possible issues is the presence of constant offset in BPM readings which can be independent of the beam position. If a closed orbit feedback system is used, the feedback tries to correct an erroneous position. A closed-orbit bump can develop and, in the worst case, the beam touches the aperture. Even if the protection systems work correctly and the beam is dumped there is some risk—for example, the beam dump kicker might push the trajectories of part of the beam to the aperture.

9.9.8.3 Beam Current Monitors

Beam current monitors can be used to monitor transmission and lifetimes of beams. If the beam transmission between two locations of the accelerator complex is too low, implying beam loss, the obvious action is to stop operation and address the problem. If the beam lifetime in a synchrotron or storage ring is too low, one dumps the beam. The dangers of erroneous readings are fairly clear. Duplication of monitors can provide some level of redundancy.

9.9.9 Machine Protection at the LHC

A Livingston type plot (Fig. 9.24) shows the energy stored in the beam as a function of particle momentum.

Machine Protection is required during all phases of operation since the LHC is the first accelerator with the intensity of the injected beam already far above threshold for damage. Protection during the injection process is mandatory. It is striking that the energy stored in the nominal LHC beam at injection is about one order of magnitude higher than the stored energy in the beam for other accelerators. At 7 TeV fast beam loss with an intensity of about 5% of one single "nominal bunch" can damage equipment (e.g. superconducting coils).

The only component that can stand a loss of the full beam are the beam dump blocks—all other components would be severely damaged. The LHC beams must always be extracted onto the beam dump blocks at the end of a fill and in case of failure.

During powering at 7 TeV beam energy, about 10 GJ is stored in the superconducting magnets. Therefore quench protection and powering interlocks must be operational and fully debugged and tested long before the start of beam operation.

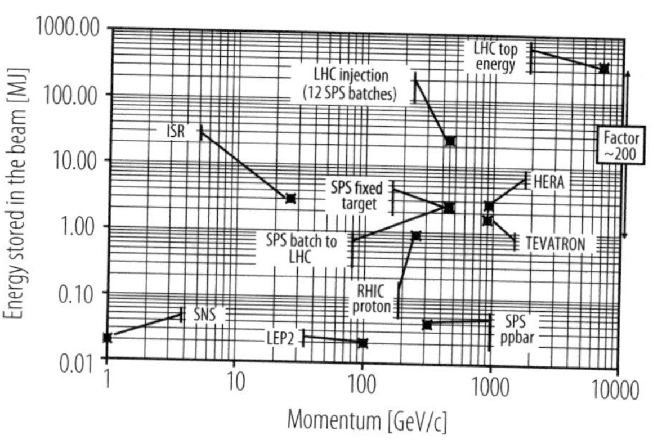

Fig. 9.24 Energy stored in the beams for different accelerators (based on a figure by R. Assmann)

References

1. M. Buzio, P. Galbraith, G. Golluccio, D. Giloteaux, S. Gilardoni, C. Petrone, L. Walckiers, A. Beaumont: *Development of upgraded magnetic instrumentation for CERN real-time reference field measurement systems*, CERN/ATS 2010-142.
2. M. Di Castro, D. Sernelius, L. Bottura, L. Deniau, N. Sammut, S. Sanfilippo, W. Venturini Delsolaro: *Parametric field modeling for the LHC main magnets in operating conditions*, Proc. IEEE Particle Accelerator Conf. 2007, 25–29 June 2007, Albuquerque, NM, pp. 1586–1588 (2007).
3. W.H. Press, et al.: Numerical Recipes, Cambridge U. Press (1988) p. 52
4. A. Friedman, E. Bozoki: Nucl. Instrum. Meth. A 344 (1994) 269.
5. B. Autin, Y. Marti: CERN report ISR MA/73-17, 1973.
6. Goodwin, Graebe, Salgado: Control System Design, Prentice Hall, 2000.
7. G. Golub, C. Reinsch: Handbook for automatic computation II, Linear Algebra, Springer, NY, 1971.
8. D.C. Youla, et al.: *Modern Wiener-Hopf Design of Optimal Controllers*, IEEE Trans. Automatic Control 21(1) (1976) 3–13 & 319–338.
9. O. Smith: Feedback Control Systems, McGraw-Hill, 1958.
10. B. Anderson: *From Youla-Kucera to Identification, Adaptive and Non-linear Control*, Automatica 34(12) (1998) 1485–1506.
11. R. Tomás, M. Aiba, A. Franchi, and U. Iriso, *Review of linear optics measurement and correction for charged particle accelerators*, Phys. Rev. Accel. Beams **20**, 054801 (2017).
12. R. Tomás, *From Farey sequences to resonance diagrams*, Phys. Rev. ST Accel. Beams **17**, 014001 (2014).
13. M. Minty, F. Zimmermann: Measurement and control of charged particle beams, Springer-Verlag (2003).
14. A. Hofmann and B. Zotter, *Measurement of the β-functions in the ISR*, Issued by: ISR-TH-AH-BZ-amb, Run: 640-641-642 (1975).
15. J. Borer A. Hofmann, J-P. Koutchouk, T. Risselada, and B. Zotter, Proceedings of the 1983 Particle Accelerator Conference, IEEE Transactions on Nuclear Science **30**, Issue 4, 2406–2408 (1983) and CERN/LEP/ISR/83-12 (1983).
16. G. Buur, P. Collier, K.D Lohmann, H. Schmickler: *Dynamic Tune and Chromaticity Measurements in LEP*, CERN SL 92-15 DI.
17. M. Böge, A. Streun, V. Scholtt: *Measurement and correction of imperfections in the SLS storage ring*, EPAC 2002.
18. F. Carlier and R. Tomás, *Accuracy & Feasibility of the β^* Measurement for LHC and HL-LHC using K-Modulation*, Phys. Rev. Accel. and Beams, **20**, 011005 (2017).
19. A. Morita, H. Koiso, Y. Ohnishi, K. Oide: *Measurement and correction of on- and off-momentum beta functions at KEKB*, Phys. Rev. ST Accel. Beams **10** (2007) 072801.
20. W.J. Corbett, M.J. Lee, and V. Ziemann, *A fast model-calibration procedure for storage rings*, Proceedings of the 1993 Particle Accelerator Conference, ISBN 0-7803-1203-1, 108–110 (1993).
21. J. Safranek: *Experimental determination of storage ring optics using orbit response measurements*, Nucl. Instrum. Meth. A 388 (1997) 27.
22. X. Shen, S. Y. Lee, M. Bai, S. White, G. Robert-Demolaize, Y. Luo, A. Marusic, and R. Tomás, Phys. Rev. ST Accel. Beams **16**, 111001 (2013).
23. M. Carlá, Z. Martí, G. Benedetti, and L. Nadolski, Proceedings of 6th International Particle Accelerator Conference, Richmond, VA, USA, 1686–1688 (2015).
24. A. Langner, et al.: , Phys. Rev. Accel. Beams **19**, 092803 (2016).
25. P. Castro-Garcia: *Luminosity and beta function measurement at the e^-e^+ collider ring LEP*, Ph.D. Thesis, CERN-SL-96-70-BI (1996).
26. A. Langner and R. Tomás: *Optics measurement algorithms and error analysis for the proton energy frontier*, Phys. Rev. ST Accel. Beams **18**, 031002 (2015).

27. A. Wegscheider, A. Langner, R. Tomas and A. Franchi: *Analytical N beam position monitor method*, Phys. Rev. Accel. Beams **20**, 111002 (2017).
28. R. Bartolini, F. Schmidt: *A Computer Code for Frequency Analysis of Non-Linear Betatron Motion*, CERN SL-Note-98-017-AP (1998).
29. J. Irwin, C.X. Wang, Y.T. Yan, K.L.F. Bane, Y. Cai, F.-J. Decker, M.G. Minty, G.V. Stupakov, F. Zimmermann: *Model-Independent Beam Dynamics Analysis*, Phys. Rev. Lett. **82**(8) (1999) 1684.
30. W. Guo et al., *A lattice correction approach through betatron phase advance*, in Proceedings of IPAC'16, Busan, Korea, 2016.
31. L. Malina et al., *Improving the precision of linear optics measurements based on turn-by-turn beam position monitor data after a pulsed excitation in lepton storage rings*, Phys. Rev. Accel. Beams **20**, 082802 (2017).
32. M. Bai, et al.: *Overcoming Intrinsic Spin Resonances with an rf Dipole*, Phys. Rev. Lett. **80**(21) (1998) 4673.
33. R. Tomás, *Adiabaticity of the ramping process of an ac dipole*, Phys. Rev. ST Accel. Beams **8**, 024401 (2005).
34. S. Peggs, C. Tang: *Nonlinear diagnostics using an AC Dipole*, BNL RHIC/AP/159 (1998).
35. R. Tomás: *Normal form of particle motion under the influence of an ac dipole*, Phys. Rev. ST Accel. Beams **5** (2002) 054001.
36. S. White, E. Maclean and R. Tomás: *Direct amplitude detuning measurement with ac dipole*, Phys. Rev. ST. Accel. Beams, **16**, 071002.
37. R. Tomás et al.: *Beam-beam amplitude detuning with forced oscillations*, Phys. Rev. Accel. and beams **20**, 101002 (2017).
38. R. Miyamoto, S.E. Kopp, A. Jansson, M.J. Syphers: *Parametrization of the driven betatron oscillation*, Phys. Rev. ST Accel. Beams **11** (2008) 084002.
39. R. Tomás, M. Bai, R. Calaga, W. Fischer, A. Franchi, and G. Rumolo, *Measurement of global and local resonance terms*, Phys. Rev. ST Accel. Beams **8**, issue 2, 024001 (2005).
40. R. Bartolini, F. Schmidt: *Normal Form via Tracking or Beam Data*, Part. Accel. 59 (1998) 93.
41. R. Calaga, R. Tomás, A. Franchi: *Betatron coupling: Merging Hamiltonian and matrix approaches*, Phys. Rev. ST Accel. Beams **8** (2005) 034001.
42. M. Benedikt, F. Schmidt, R. Tomás, P. Urschütz, A. Faus-Golfe: *Driving term experiments at CERN*, Phys. Rev. ST Accel. Beams **10** (2007) 034002.
43. Y. Alexahin and E. Gianfelice-Wendt: *Determination of linear optics functions from turn-by-turn data*, Journal of Instrumentation **6**, P10006 (2011).
44. T. H. B. Persson and R. Tomás: *Improved control of the betatron coupling in the Large Hadron Collider*, Phys. Rev. ST Accel. Beams **17**, 051004 (2014).
45. A. Franchi: *Studies and Measurements of Linear Coupling and Nonlinearities in Hadron Circular Accelerators*, PhD thesis, GSI DISS 2006-07 (2006).
46. E.H. Maclean, R. Tomás, F. Schmidt and T.H.B. Persson, *Measurement of LHC nonlinear observables using kicked beams*, Phys. Rev. ST. Accel. Beams, **17**, 081002.
47. R. Tomas, T.H.B Persson and E.H. Maclean, *Amplitude dependent closest tune approach*, Phys. Rev. Accel. Beams **19**, 071003 (2016)
48. M. Aiba, R. Calaga, A. Morita, R. Tomás, G. Vanbavinckhove: *Optics correction in the LHC*, Proc. EPAC 2008, Genoa, Italy.
49. T. Persson et al.: *HC optics commissioning: A journey towards 1% optics control*, Phys. Rev. Accel. Beams, **20**, 061002 (2017).
50. R. Bartolini, et al.: *Correction of multiple nonlinear resonances in storage rings*, Phys. Rev. ST Accel. Beams **11** (2008) 104002.
51. M. Aiba, et al.: *First β-beating measurement and optics analysis for the CERN Large Hadron Collider*, Phys. Rev. ST Accel. Beams **12** (2009) 081002.
52. R. Tomás, O. Brüning, M. Giovannozzi, P. Hagen, M. Lamont, F. Schmidt, G. Vanbavinckhove, M. Aiba, R. Calaga, R. Miyamoto: *CERN Large Hadron Collider optics model, measurements, and corrections*, Phys. Rev. ST Accel. Beams **13** (2010) 121004.

53. A.W. Chao, M. Tigner (eds.): Handbook of Accelerator Physics and Engineering, World Scientific (1999), p. 283.
54. M. Weiss: CERN 87-10, p.162.
55. C. G. Lilliequist, K. R. Symon: MURA-491.
56. M.R. Geiger: CERN-AR-Int-GS-61-6.
57. J. Griffin, et al.: Proc. PAC 83, p. 2630.
58. LHC Design Report, CERN-2004-03, Vol. 3, section 7.1.2.
59. R. Cappi, et al.: Proc. EPAC 94, p. 279.
60. V.V. Balandin, et al.: Part. Accel. 35 (1991) 114.
61. R. Cappi, R. Garoby, E. Chapochnikova: CERN/PS 92-40 (RF).
62. T. Toyama: NIM-A 447 (2000), p. 317.
63. S. V. Ivanov, O. P. Lebedev: At. Eng. (USA) 93, 6 (2002), p. 973.
64. H. G. Hereward, K. Johnsen: CERN 60–38.
65. E. Jones: CERN-AR-Int-SR-63-17, CERN-AR-Int-SR-64-6.
66. I. Bozsik: Proc. Computing in Acc. Design and Operation (1983), p. 128.
67. R. Garoby, S. Hancock: EPAC 94, p. 282.
68. R. Garoby, CERN/PS 98-048 (RF).
69. R. Garoby: Proc. Chamonix XI, 2001, p. 32.
70. F.E. Mills: BNL Report AADD 176 (1971).
71. D. Boussard, Y. Mizumachi: PAC 79, p. 3623.
72. R. Garoby: PAC 85, p. 2332.
73. H. Damerau: CERN-Proceedings-2017-002, p. 139.
74. R. Assmann: *Collimators*, this handbook, Sect. 8.8.
75. R. Assmann: *Beam Collimation*, Handbook for Accelerator Engineering, to be published.
76. R. Assmann: *Collimators and cleaning, could this limit the LHC performance?*, Proc. 12th Chamonix LHC Performance Workshop, 3-8 Mar 2003, Chamonix, France.
77. D. Wollmann, et al.: *First Cleaning with LHC Collimators*, IPAC-2010-TUOAMH01, May 2010.
78. R. Assmann: *Collimation for the LHC High intensity beams*, Proc. 46th ICFA Advanced Beam Dynamics Workshop High-Intensity and High-Brightness Hadron Beams (HB2010), Sep 27 - Oct 1 2010, Morschach, Switzerland.
79. R. Assmann, et al.: *The final collimation system for the LHC*. CERN-LHC-PROJECT-REPORT-919, Jun 2006.
80. C. Bracco: *Commissioning scenarios and tests for the LHC collimation system*, CERN-THESIS-2009-031.
81. T. Weiler, et al.: *Beam loss response measurements with an LHC prototype collimator in the SPS*, PAC07-TUPAN107, Jun 2007, 3 pp.
82. V. Kain, H. Burkhardt, B. Goddard, S. Redaelli: *Beam based alignment of the LHC transfer line collimators*, PAC-2005-MPPE048, CERN-LHC-PROJECT-REPORT-852, May 2005.
83. W. Bartmann, et al : *Beam Commissioning of the Injection Protection Systems of the LHC*, IPAC-2010-TUPEB067, May 2010.
84. B. Dehning, et al.: *The LHC Beam Loss Measurement System*, PAC07-FRPMN071, Jun 2007.
85. E.B. Holzer, et al.: *Lessons learnt from beam commissioning and early beam operation of the beam loss monitors (including outlook to 5-TeV)*, Proc. Chamonix 2010 Workshop LHC Performance, 25–29 Jan 2010, Chamonix, France.
86. S. Redaelli, G. Arduini, R. Assmann, G. Robert-Demolaize: *Comparison between measured and simulated beam loss patterns in the CERN SPS*. CERN-LHC-PROJECT-REPORT-938, Jun 2006.
87. R. Bruce, R. Assmann, G. Bellodi, C. Bracco, H.H. Braun, S. Gilardoni, J.M. Jowett, S. Redaelli, T. Weiler: *Ion and proton loss patterns at the SPS and LHC*, CERN-2008-005, 2008, 6 pp.
88. G. Robert-Demolaize, R.W. Assmann, C. Bracco, S. Redaelli, T. Weiler: *Critical beam losses during commissioning and initial operation of the LHC*. Proc. 3rd LHC Project Workshop: 15th Chamonix Workshop, 23–27 Jan 2006, Chamonix, Divonne-les-Bains, Switzerland.

89. O. Brüning, P. Collier, P. Lebrun, S. Myers, R. Ostojic, J. Poole, P. Proudlock: *LHC Design Report, Volume I: The LHC Main Ring*, CERN-2004-003, June 2004.
90. R. Assmann, K. Cornelis: *The beam-beam interaction in the presence of strong radiation damping*, CERN-SL-2000-046-OP.
91. E. Keil, R. Talman: Part. Accel. 14 (1983) 109.
92. W. Herr, D. Brandt, M. Meddahi, A. Verdier: Proc. 1999 Particle Accelerator Conf., New York, 1999.
93. R. Assmann, M. Lamont, S. Myers: *A brief history of the LEP collider*, Nucl. Phys. B Proc. Suppl. 109 (2002) 17–31.
94. V. Shiltsev, et al.: Phys. Rev. ST Accel. Beams 8 (2005) 101001.
95. J.-P. Koutchouk: *Correction of the Long-Range Beam-Beam Effect in LHC using Electro-Magnetic Lenses*, Proc. PAC2001 Chicago (2001), p. 1681, and/or the earlier LHC Project Note 223 (2000); U. Dorda, et al.: *Wire excitation experiments in the CERN SPS*, Proc. EPAC2008, Genoa (2008), p. 3176; R. Calaga, W. Fischer, G. Robert-Demolaize, and N. Milas: *Long-range beam-beam experiments in the Relativistic Heavy Ion Collider*, Phys. Rev. ST Accel. Beams 14 (2011) 091001.
96. V. Shiltsev, V. Danilov, D. Finley, A. Sery: *Considerations on compensation of beam-beam effects in the Tevatron with electron beams*, Phys. Rev. ST Accel. Beams 2(7) (1999) 071001; V. Shiltsev, Y. Alexahin, K. Bishofberger, V. Kamerdzhiev, V. Parkhomchuk, V. Reva, N. Solyak, D. Wildman, X.L. Zhang, F. Zimmermann: *Experimental studies of compensation of beam-beam effects with Tevatron electron lenses*, New J. Phys. 10(4) (2008) 043042.
97. W. Fischer, Z. Altinbas, M. Anerella, E. Beebe, M. Blaskiewicz, D. Bruno, W.C. Dawson, D.M. Gassner, X. Gu, R.C. Gupta, K. Hamdi, J. Hock, L.T. Hoff, A.K. Jain, R. Lambiase, Y. Luo, M. Mapes, A. Marone, T.A. Miller, M. Minty, C. Montag, M. Okamura, A.I. Pikin, S.R. Plate, D. Raparia, Y. Tan, C. Theisen, P. Thieberger, J. Tuozzolo, P. Wanderer, S.M. White, W. Zhang: *Construction progress of the RHIC electron lenses*, Proc. Intern. Particle Accelerator Conf. 2012, New Orleans, Louisiana, USA, (2012) 2125–2127; Y. Luo, W. Fischer, N.P. Abreu, A. Pikin, G. Robert-Demolaize: *Six-dimensional weak-strong simulation of head-on beam-beam compensation in the Relativistic Heavy Ion Collider*, Phys. Rev. ST Accel. Beams 15 (2012) 051004.
98. R.B. Palmer: *Energy Scaling, Crab Crossing and the Pair Problem*, invited talk at the DPF Summer Study Snowmass 88, SLAC-PUB-4707, Stanford 1988.
99. P. Raimondi: 2nd SuperB Workshop, Frascati, 2006.
100. F. Zimmermann: *Two Scenarios for the LHC Luminosity Upgrade*, Joint PAF/POFPA Meeting, CERN, 13 Feb 2007; W. Scandale, F. Zimmermann: *Scenarios for sLHC and vLHC*, Proc. Hadron Collider Physics Symposium, La Biodola, Italy, 20–26 May 2007, Nucl. Phys. B Proc. Suppl. 177–178 (2008) 207–211.
101. J.-P. Koutchouk, V. Shiltsev: Handbook of Accelerator Physics and Engineering, World Scientific, 1999.
102. T. Pieloni et al., *Observations of Two-beam Instabilities during the 2012 LHC Physics Run*, IPAC 2013, Shanghai, China, 2013.
103. A. Fasso, et al.: *The physics models of FLUKA: status and recent development*, CHEP 2003, LA Jolla, California, 2003.
104. http://www-ap.fnal.gov/MARS/
105. http://geant4.web.cern.ch/geant4/
106. E. Ciapala: *The LHC Post-mortem System*, CERN LHC-Project-Note 303, 2002, CERN, Geneva.

BY

Chapter 10
The Largest Accelerators and Colliders of Their Time

K. Hübner, S. Ivanov, R. Steerenberg, T. Roser, J. Seeman, K. Oide, Karl Hubert Mess, Peter Schmüser, R. Bailey, and J. Wenninger

10.1 Proton Accelerators and Colliders

K. Hübner · S. Ivanov · R. Steerenberg

10.1.1 CERN Proton Synchrotron (CPS)

The Study Group for a GeV-scale Proton Synchrotron was launched in 1952 at CERN. Initially, an up-scaled version of the 3 GeV Cosmotron was considered but soon a new design based on the newly discovered alternating-gradient principle and promising a proton energy of 30 GeV was adopted by the CERN Council in the same year. In order to limit cost the energy was subsequently limited to 25 GeV

K. Hübner · R. Steerenberg · K. Oide · K. H. Mess · R. Bailey (✉) · J. Wenninger
CERN (European Organization for Nuclear Research) Meyrin, Genève, Switzerland
e-mail: Rende.Steerenberg@cern.ch; roger.bailey@cern.ch; jorg.wenninger@cern.ch

S. Ivanov
Institute of High Energy Physics, Moscow, Russia
e-mail: sergey.ivanov@ihep.ru

T. Roser
Brookhaven National Laboratory, New York, NY, USA
e-mail: roser@bnl.gov

J. Seeman
SLAC National Accelerator Laboratory, Stanford University, Menlo Park, CA, USA
e-mail: seeman@slac.stanford.edu

P. Schmüser
DESY, Hamburg, Germany
e-mail: Peter.Schmueser@desy.de

S. Myers, H. Schopper (eds.), *Particle Physics Reference Library*,
https://doi.org/10.1007/978-3-030-34245-6_10

Table 10.1 Basic parameters of the CPS [5]

Accelerated particles	Protons, lead ions
Momentum protons/lead ions	26 GeV/c, 5.9 GeV/c nucleon
Circumference [m]	200 π
Magnetic lattice	Alternating-gradient focusing, combined-function
Focusing order	FOFDOD
Magnetic field index	$n = 288$
Number of main magnets	100
Bending magnetic field	0.1013 T (inj. at 1 GeV), 1.25 T (extr. at 26 GeV/c)
Betatron oscillations/turn	6.24 (h), 6.26 (v)
Rise time/flat top time	0.7 s/0.3 s (26 GeV/c)
Long straight sections	Number = 20, length = 3.0 m
RF system (tunable)	11 cavities, 2.8–9.55 MHz, 220 keV/turn total maximum
Auxiliary RF systems	13, 20, 40, 80, 200 MHz
Vacuum chamber	Stainless steel, 146 × 70 mm^2 in the bending magnets

and the project led by J.B. Adams was approved in 1953. The final parameters were fixed in 1954 and construction started in 1955. The CPS [1] became operational towards the end of 1959 reaching an energy of 28 GeV [2, 3]. It has turned out to be an extremely versatile facility [4] (Table 10.1).

Initially, the proton injector was a 50 MeV Alvarez-type linear accelerator (linac L1) operating at 200 MHz. In order to increase the intensity a four-ring synchrotron booster (PSB) was inserted in 1972 between L1 and CPS raising the kinetic injection energy to 0.8 GeV. Over the years, it had its top kinetic energy raised in steps to 1 GeV in 1985 and 1.4 GeV in 1999 to allow for the production of the LHC beams. As part of the LHC Injectors Upgrade (LIU) project [6] a further increase to 2 GeV is being prepared at present for first beam in 2020. In 1979, L1 was replaced by linac 2 (L2) of a more modern and robust design. The 50 MeV proton linac 2 is presently being replaced by the new linac 4 (L4) [7] that will provide H^- ions at 160 MeV to the charge exchange injection equipped PS Booster.

The CPS provided initially secondary beams by means of internal targets. Since 1967 a fast extraction system over one turn became available for fixed-target physics complemented in 1969 by a system providing slow-extracted beams spilling out particles over a large number of turns. The fast extraction was used to produce neutrino beams with protons and, between 1970 and 1983, 26 GeV/c protons for the Intersecting Storage Rings (ISR). At present, the fast extraction provides a 26 GeV/c proton beam (1.5×10^{13} per pulse) for the Antiproton Decelerator (AD) after compression of the four equidistant bunches that occupy half the CPS circumference. This manipulation by the radio-frequency accelerating system makes this proton bunch train so short that the secondary antiproton bunch train fits into the AD circumference being one quarter of that of the CPS. Further, a 20 GeV/c single high-intensity bunch containing up to 9×10^{12} protons is extracted after a non-adiabatic bunch shortening to produce neutrons from a lead-target. Since 2010, a large variety of LHC beams have been produced that over time have become much

Table 10.2 Pre- and post-LIU main LHC beam parameters at 26 GeV/c PS extraction

	Beam type	N [$\times 10^{11}$ p]	ε [mm mrad]	N_b/extr
2018	Standard	1.4	2.3	72
	BCMS	1.3	1.2	48
Post-LIU	Standard	2.6	1.85	72
	BCMS	2.6	1.45	48

brighter. As foreseen in the LHC design report [8] the PS produces single bunch beams varying in intensity from 0.05×10^{11} to 1.2×10^{11} protons. For the initial multi-bunch beams 12–72 bunches per extraction at 26 GeV/c with respective bunch spacings of 25, 50, 75 and 150 ns were routinely produced. The highest LHC bunch intensity of 1.7×10^{11} protons per bunch in a transverse emittance of 1.6 mm mrad and a bunch train length of 36 bunches was reached with the 50 ns bunch spacing that was initially used to limit electron cloud effects in the LHC. By mid-July 2015 the LHC requested the 25 ns bunch spacing with 1.15×10^{11} protons per bunch in a transverse emittance of 2.5 mm mrad and 72 bunches per extraction. Thanks to the versatile PS RF system an even brighter LHC beam, based on Bunch Merging, Compression and Splitting (BCMS), was established. Up to 1.3×10^{11} protons per bunch in a transverse emittance of 1.1 mm mrad and 48 bunches per extraction are delivered to the SPS since mid-2016. The LIU project aims at increasing the bunch intensity to 2.6×10^{11} protons per bunch for both beam types as given in Table 10.2.

For fixed-target experiments at the Super Proton Synchrotron (SPS), a 14 GeV/c proton beam (3×10^{13} per pulse) was spilled out over five turns by cutting it with an electro-static septum in horizontal phase space in order to fill by box-car stacking 5/11 parts of the SPS circumference being 11 times longer than the one of the CPS. A second CPS beam pulse fills further 5/11 parts, thus leaving 1/11 for the kicker rise-times. Since 2010 a novel scheme, using a fourth order betatron resonance for capturing the beam in five stable islands in the horizontal phase space, avoids the losses at the electrostatic extraction septum [9].

The slow-extraction based on a third-order betatron resonance is still in use to produce primary 24 GeV/c proton beams. Up to 5×10^{11} protons can be spilled out in up to 450 ms for the production of secondary beams for fixed-target experiments at the CPS, but also for primary protons to the IRRAD and CHARM irradiation facilities [10].

As soon as L2 was available, L1 was modified to provide Deuterium and α-particle beams for collisions in the ISR and, later, Sulphur beams for fixed target experiments in the SPS. In 1994, L1 was replaced by linac 3 (L3) providing 4.2 MeV/amu Pb^{+53} ions which, fully stripped after CPS extraction, are used for fixed-target experiments in the SPS. Since 2010 the lead ions are also fast-ejected to the SPS for lead-lead collisions in LHC. They no longer pass through the PSB but a small storage ring, the Low Energy Ion Ring (LEIR), acts as accumulator between the fast-cycling L3 and the slow-cycling CPS and is equipped with stochastic and electron cooling to decrease the transverse emittance of the accumulated beam. The CPS provides a total of 8×10^{10} ions per pulse in four bunches.

The acceleration cycles for the different users are grouped in a supercycle depending on the user requirements allowing for a quick, reproducible switching from one to another mode of operation [11].

10.1.2 Brookhaven Alternating Gradient Synchrotron (AGS)

After successful completion of the 3 GeV Cosmotron in 1952 the design study for a more powerful accelerator was launched coincident with the invention of the alternating-gradient principle [12, 13]. Construction led by G.K. Green and J.P. Blewett started in 1953 and commissioning was completed with the first proton beam accelerated to 31 GeV in July 1960 [14]. In order to test experimentally whether the beam would pass transition energy, an electron analogue had been built in 1954 and operated until 1957. This model had a circumference of 43.1 m accelerating electrons from 1 to 10 MeV with transition energy at 3.5 MeV. In order to reduce cost, the alternating-gradient, strong-focusing was provided by electrostatic lenses and bending by electrostatic fields [15]. The test showed that transition can be crossed without problems but the price was a delay in the AGS construction relative to the CERN PS, which however was at the end compensated by better preparation of the experimental programme compared to CERN.

The first injector was a 50 MeV Alvarez-type proton linear accelerator. The present 200 MeV linear accelerator began operation in 1970. In 1982 H^- charge exchange injection into the AGS was introduced [16] and in 1991 a 1.5 GeV booster synchrotron was commissioned [17]. The Booster can provide 1.5×10^{13} protons per pulse at 1.9 GeV at the design repetition frequency of 7.5 Hz. The acceleration harmonic schemes (Booster harmonic, AGS harmonic, transfers) evolved from (3, 4, 12) to (2, 4, 8) and finally to (1, 6) in pursuit of higher intensity [18] (Table 10.3).

Secondary beams from the AGS were initially provided from internal targets. This also creates high beam loss and activation in the accelerator not compatible with high-intensity operation. The first fast-extraction was installed in the mid-60s followed by slow-extraction in 1967 which served up to six target stations and spilling out protons with repetition periods from 1.8 s to 5.8 s. To cope with the intensity increases, the AGS underwent a series of upgrades including a new main magnet power supply, addition of transverse feed-back, special magnets to provide fast crossing of the transition energy, and a high power RF system [20]. In the early 2000s the AGS provided a slow-extracted beam of 7×10^{13} protons per pulse at 24 GeV [21].

With the appropriate source added to the 200 MeV linear accelerator, polarized protons have been produced by the injector chains from 1985 onward for fixed target experiments. To meet injector requirements for RHIC—intensity and polarization—the polarized source underwent a major upgrade [22]. The polarization transmission efficiency in the AGS has been substantially improved with the installation of two

Table 10.3 Basic parameters of the AGS [19]

Accelerated particles	Protons, polarized protons, heavy ions (up to Au)
Particle energy	30 GeV, 25 GeV, 14.5 GeV/n
Circumference [m]	256.9 π
Magnetic lattice	Alternating-gradient focusing, combined-function
Focusing order	(F/2)O(F/2)(D/2)O(D/2)
Magnetic field index	$n = 365$
Number of main magnets	240
Bending magnetic field	0.105 T at injection, 1.31 T at maximum particle momentum
Betatron oscillations/turn	8.75 (h), 8.75 (v)
Rise time/flat top time	0.6 s/0.5 to 2.5 s
Long straight sections	Number = 24, length = 3.15 m
RF system (tunable)	10 cavities, 1.8 to 4.5 MHz, 200 KeV/turn total maximum
Auxiliary RF system (fixed RF)	92 MHz
Vacuum chamber	Inconel, 173 × 78 mm^2 in the bending magnets

"partial Siberian" snakes [23, 24] and a system to rapidly cross weak resonances. Polarization at transfer to RHIC (24 GeV) is 70% (with 82% at 200 MeV) and with intensity 2×10^{11} protons per bunch [25, 26].

Since 1986 the Booster has also accelerated ions (d to Au) using a Tandem Van de Graaff as injector. For RHIC operation about 5×10^9 Au^{31+} ions at 41.6 MeV/n are injected over 60 turns into the Booster, accelerated to 101 MeV/n, stripped to Au^{77+} and injected into the AGS. The ions are fully stripped before injection into RHIC [27]. A new pre-injector is being commissioned based on an EBIS source followed by a new linear accelerator [28]. The new system increases the available ions for RHIC to include Uranium [29].

10.1.3 The 70 GeV Proton Synchrotron (U-70) of NRC "Kurchatov Institute": IHEP (Protvino)

The study of a powerful synchrotron started in the mid-60s in the then Soviet Union and focused onto the Protvino site since 1958 [30]. The project was led by V.V. Vladimirski from 1958 and A.A. Logunov from 1963 after the foundation of the Institute for High-Energy Physics. The construction started in 1961 and the commissioning [31] took place in 1967 culminating in a test run at 76 GeV in the same year, which was the world record at that time (Table 10.4).

The first proton injector has been the 100 MeV Alvarez-type DTL proton linear accelerator (I-100) providing a pulse current of 100 mA over five turns in U-70. It served until 1985 as injector [34] and is still in operation as light ion injector [33]. In order to increase the intensity, the 1.5 GeV booster synchrotron (U-1.5) cycling at $16^2/_3$ Hz came on line in 1985 [35]. Its injector is a 30 MeV RFQ-type DTL linear

Table 10.4 Basic parameters of the U-70 [32, 33]

Accelerated particles	Protons, carbon nuclei
Particle energy	70 GeV (34 GeV/n carbon nuclei)
Circumference	1483.7 m
Magnetic lattice	Alternating-gradient focusing, combined-function
Focusing order	FODO
Magnetic field index	$n = 443$
Number of main magnets	120
Bending magnetic field	0.035 T at injection, 1.2 T at maximum energy
Betatron oscillations/turn	9.9 (h), 9.8 (v)
Rise time/flat top time	2.8 s/2 s
Long straight sections	Number = 24, length = 4.87 m
RF system (tunable)	38 (+2 spare) cavities, 2.6–6.1 MHz, 150 keV/turn total maximum
Auxiliary RF system (fixed RF)	2 cavities at 200 MHz, 500 kV peak total voltage
Vacuum chamber	stainless steel, 200×100 mm^2 in the bending magnets

accelerator (URAL-30) injecting 1–4 turns into the booster with a pulse current up to 80 mA. Twenty nine single-bunch pulses of the booster with up to 8×10^{11} protons per pulse build up the beam in U-70 within 1.8 s using bunch-to-bucket transfer. The U-1.5 has a circumference 1/15 times the one of U-70. It operates now at 1.3 GeV limited by the power supply.

The U-70 accelerator has been equipped with a new vacuum chamber in 1997 and the control system has been modernized from 1998 onwards. Thanks to all these upgrades and, in particular, to the addition of the booster and the new linear accelerator, the beam intensity has reached 1.5×10^{13} protons per pulse with a repetition time of 9.8 s, which has to be compared with the initially planned 1×10^{12} protons per pulse [32]. In recent years, U-70 operates with 1.1×10^{13} protons per pulse at 50 GeV to save energy [36].

Initially, only internal targets have provided a large variety of secondary beams and some internal targets are still in use providing spills up to 1.8 s. A fast-extraction system has been added soon. A slow-extraction system based on a third order resonance came into operation in 1979 and it has been upgraded for higher intensity in 1989. Its spill-length could be extended up to 1.3 s. Now, a stochastic slow extraction system provides smooth spills over up to 3 s at top energy. Extraction, beam splitting and collimation using bent crystals have been achieved being under study since 1990 [37].

Deuterons have been accelerated from 2008 onwards in I-100 and U-1.5 to 16.7 MeV/n and 455 MeV/n, respectively. In 2009 deuterons were further accelerated in U-70 to 23.6 GeV/n corresponding to 50 GeV protons. The latter has the potential to accelerate them to 34 GeV/n. An intensity of 5×10^{10} deuterons per pulse has been achieved.

In 2011, carbon ions have been first accelerated in I-100, U-1.5 and U-70 to 34 GeV/n [38]. In 2013, validation tests of all the top-energy extractions available with carbon beam have been accomplished successfully. Carbon beam intensity is 3–5 × 10^9 ions per pulse (8.2 s), in a single bunch. At the DC flat-bottom, U-70 now also operates in a beam storage- and stretcher-ring mode for 455 MeV/n carbon ions enabling their square-wave slow stochastic extraction (0.5–1 s long) via a Piccioni-Wright technique for an applied fixed-target research [39, 40].

10.1.4 The CERN Intersecting Storage Rings (ISR)

The first ideas for a realistic proton-proton collider were publicly discussed in 1956 [41, 42]. The Accelerator Research Group set up by the CERN Council in 1956 formulated in 1960 the first proposal for a proton-proton collider attached to the CERN PS. In 1960 construction began on a small-proof-of-principle 1.9 MeV electron storage ring, the CERN Electron Storage and Accumulation Ring (CESAR) which experimentally proved the accumulation of particles by RF stacking in 1964, an essential technique [43] to build up intense beams, a prerequisite for getting the required luminosity. CESAR also was an important test bed for the for Ultra-High Vacuum (UHV) technology which had to developed to achieve a very low vacuum pressure, indispensable for a long lifetime of stored beams. The Design report [44] was issued in 1964, the project was approved in 1965, and construction lead by K. Johnsen started without delay. The first proton-proton collisions took place in 1971 with a beam momentum up to 26.5 GeV/c, the maximum momentum available from the CPS. The ISR consisted of two independent storage rings intersecting at eight points at an angle of 14.8°. To create space for long straight sections in the interaction regions, the circumference of the ring was 1.5 times of that of the CPS which supplied particles to the ISR through two long transfer lines. The ISR operated for physics as collider from 1971 to 1983. It was decommissioned after 1984 (Table 10.5).

The CPS was the injector for the ISR supplying mainly protons up to 26.5 GeV/c. Typical intensities were 3 × 10^{12} protons per pulse in 20 bunches every 2.4 s during the filling process. The filling of one ISR ring took less than 10 min. Later, also deuterons, alpha particles and antiprotons were accelerated for the ISR.

The ISR team had to tackle a number of technological challenges but the most important was to assure UHV imperative for a long beam lifetime. The stainless steel vacuum chamber was in situ bakeable eventually to 300 °C. Pumping was provided by sputter ion-pumps and, at critical places, Ti-sublimation pumps. All vacuum chambers had to be glow-discharge cleaned. The continuous effort eventually resulted in average pressure below 10^{-11} Torr (N_2 equivalent) reducing beam loss rates to typically around one part per million per minute during physics runs. Beams of physics quality could last 40–50 h. Beam currents of 10 A were achieved already after start-up. Later up to 57 A were stored per ring with 30–40 A as typical value. The proton-proton initial luminosity (design 4 × 10^{30} cm^{-2} s^{-1}) was increased from 1.6 × 10^{30} cm^{-2} s^{-1} in 1971 to 1.4 × 10^{32} cm^{-2} s^{-1} in the superconducting low-beta section installed in one of the interaction points in 1982, which stayed the

Table 10.5 Basic parameters of the ISR [45]

Colliding particles	pp, dd, pd, αα, αp, $p\bar{p}$
Particle momentum	3.5–31.4 GeV/c, typically 26 GeV/c
Circumference [m]	300 π
Magnetic lattice	Alternating-gradient focusing, combined-function
Focusing order	FODO
Number of main magnets	132 per ring
Magnetic dipole field	1.33 T at maximum momentum
Length of main magnets	4.88/2.44 m
Magnetic field index	$n = 248$
Betatron oscillations/turn	8.90 (h), 8.88 (v)
Long straight sections	Number = 8, length = 16.8 m
β_* (h/v)	21 m/12 m
β_* (h/v)	2.5 m/0.28 m in superconducting low-beta section
RF system per ring	7 cavities, 9.5 MHz, 16 kV RF peak voltage
Auxiliary RF system	3rd harmonic
Vacuum chamber	Stainless steel, 160 mm/52 mm full width (h/v)

world record luminosity until 1991. The superconducting low beta-section had been preceded by an insertion based on conventional magnets, operational from 1974. From 1973 onwards, the beams could be accelerated in the ISR to 31.4 GeV/c by phase displacement acceleration [46].

From 1976 onward deuterons were stored in the ISR so that dd and pd collisions became available. Alpha particles were stored in 1980 for αα and αp collisions. Initial dd luminosities reached 1.6×10^{30} cm^{-2} s^{-1} and were 4×10^{28} cm^{-2} s^{-1} in the αα case. Antiprotons were stored as soon as the antiproton injector complex had become operational in 1981 (see Sect. 10.1.6). This required a new transfer line from the CPS.

The ISR will be remembered for a number of breakthroughs in accelerator physic and technology: UHV-technology for a large scale facility, control of intense coasting beams, discovery of Schottky scans, experimental demonstration of stochastic cooling, and absolute luminosity measurement by van der Meer scans [46].

10.1.5 The CERN Super Proton Synchrotron (SPS)

Design studies of a powerful proton synchrotron started at CERN in 1961 when the Accelerator Research Division had been created, not long after the CPS had become operational in 1960. The design energy was 300 GeV and the specified flux 10^{13} protons/s. A study group lead by K. Johnsen presented a first design report in 1964 [47]. However, the idea that the facility should be constructed on a green field, i.e. not in Switzerland near CERN, made the choice of the site difficult and funding

Table 10.6 Basic parameter of the SPS in proton fixed-target mode [51]

Accelerated particles	Protons
Momentum protons	450 GeV/c
Circumference [m]	2200 π
Magnetic lattice	Alternating-gradient focusing, separated-function
Focusing order	FODO
Number of dipole magnets	744
Dipole magnetic field	0.056 T at injection, 1.8 T at maximum particle momentum
Betatron oscillations/turn	26.6 (h), 26.6 (v)
Rise time/flat top time	0.75 s/2.5s
Long straight sections	Number = 6, length = 128 m
RF system	4 traveling-wave structures at 200 MHz, 4 MeV/turn total
Vacuum chamber	Stainless steel, 150 × 50 mm^2

problems delayed the decision over many years. In 1969, J.B. Adams was appointed leader of the 300 GeV Programme and the project gathered new momentum with the suggestion to construct the facility close to the existing laboratory and using the CPS as injector, which had been advocated already in 1961 by C. Ramm [48]. This and a new design based on the alternating-gradient principle but with the function of bending and focusing separated instead of combined-function allowed a considerable cost reduction [49]. The project was approved in 1971 and a beam energy of 400 GeV, exceeding the design energy, was reached in 1976 after almost a decade of planning and decision making [50] (Table 10.6).

The CPS acts as proton injector having 1/11th circumference of the SPS. Initially, the SPS beam was created by peeling off the required beam from the CPS beam by an electrostatic septum over 10 turns of the CPS at 10 GeV/c, leaving 1/11th of the circumference for the SPS injection kicker fall-time and ejection kicker rise-time, in a process called continuous transfer. From 1978 onwards, after the CPS intensity had been increased by the new linac 2 and the booster synchrotron, two CPS pulses were consecutively sent to the SPS, each CPS pulse was peeled five times. The CPS acted also as positron-electron injector during lepton operation for LEP and as injector during the fixed-target ion runs. It provides also a 26 GeV/c proton beam and 5.9 GeV/u ion beams to the SPS when the latter is used to fill the CERN Large Hadron Collider (LHC).

The intensity of the proton beam for fixed-target experiments had been raised gradually. In 1984, the injection momentum was raised to 14 GeV/c providing a more stable, reproducible injection and a beam of lower emittance. After the kinetic beam energy of the PS-Booster (PSB) had been raised to 1.4 GeV in 1998 resulting in a further emittance reduction, the SPS delivered more than 4×10^{13} protons per pulse with a record value of 5.3×10^{13} ppp.

Since the start-up in 1976 three types of extraction modes have been available towards the two experimental areas: (i) fast extraction of part or the entire beam (spill 3–23 μs); (ii) slow-resonant extraction (0.5–2 s); (iii) fast-resonant extraction (<3 ms). A typical pattern was extraction to the West-Area at an energy of 200 GeV

(250 GeV maximum) during a short pause in the acceleration followed by extraction at top energy to the North Hall. After the upgrading to 450 GeV of the transfer line to the West Hall in 1983, simultaneous resonant extraction to both areas was implemented by an appropriate adjustment of the horizontal betatron phase advance in the ring. The sharing ratio between the two clients was fully adjustable. Extraction towards the West Hall was terminated in 2003 to free resources for LHC.

In 1986 and 1987 two exploratory fixed-target runs with ions took place after linac 2 had become operational freeing linac 1 which had in the meantime been equipped with the appropriate front-end for ion operation. In 1986, fully stripped oxygen ions were accelerated for the first time to 200 GeV/n after the SPS had been set up with a deuteron beam of more convenient higher intensity. In the second run, in 1987, sulphur ions were used. The sulphur runs were resumed from 1990 to 1992 providing 9×10^9 charges per pulse with four batches injected from the CPS. After the construction of the dedicated heavy-ion linac (linac 3) lead ions at 177 GeV/u were available for the experiments from 1994 to 2002 and indium at 158 GeV/u in 2003. No ion runs took place in 1997 and 2001. The SPS delivered up to 6×10^{10} fully-stripped ions in terms of charges per pulse [52].

It is worthwhile to mention that the SPS also accelerated electrons and positrons for injection into LEP in the years 1989 to 2000. This had required substantial modifications. Since the traveling-wave cavities could accelerate the beam only to 14 GeV, 32 copper-based standing-wave cavities operating at 200 MHz and providing a peak-voltage of 30 MV were added. Ejection and injection channels were equipped appropriately and a campaign of meticulously shielding the magnet coils against synchrotron radiation was conducted [53]. The LEP injection energy was first set to 20 GeV and, later, when the standing-wave cavities were replaced by two four-cell superconducting RF structures, the LEP injection energy could be raised to 22 GeV.

The SPS acts as injector into the LHC providing protons at 450 GeV since 2008 and ions at 176 GeV/u since 2010 [54]. Lead is the preferred ion species for the first years of operation. This new role required a number of hardware modifications and, in particular, the addition of a new extraction system for the anti-clockwise LHC ring which serves also the target for the new long-base line neutrino beam towards Gran Sasso in Italy. Two new beam lines towards the LHC had to be built.

10.1.6 The CERN Super Proton Synchrotron (SPS) as Proton-Antiproton Collider

The proposal to use the SPS as proton-antiproton collider was made in 1976 [55]. The required increase of phase space density of the secondary antiprotons was to be produced by stochastic cooling invented by S. van der Meer [56] which had been experimentally demonstrated in the ISR [57]. Simultaneous stochastic cooling in transverse and longitudinal phase space at cooling rates several orders higher than

Table 10.7 Basic parameter of the SPS in proton-antiproton collision mode [60]

Accelerated particles	Protons and antiprotons
Maximum particle energy	315 GeV
Circumference [m]	2200 π
Magnetic lattice	Alternating-gradient focusing, separated-function
Focusing order	FODO
Number of dipole magnets	744
Dipole magnetic field	0.12 T at injection, 1.4 T at maximum particle energy
Long straight sections	Number = 6, length = 128 m (including 2 low-β insertions)
β_* (h/v)	1.0 m/0.5 m
Filling time	29 s
RF system	4 traveling-wave structures at 200 MHz, 3.6 MeV/turn total
Auxiliary RF system	100 MHz, 2 MV/turn
Vacuum chamber	Stainless steel, 150 × 50 mm^2

those achieved in the ISR was proven to work in the Initial Cooling Experiment (ICE) in 1978. ICE was a small storage ring of 74 m circumference operating at 1.73 and 2.1 GeV/c and fed with protons from the CPS [58]. With the project decision taken in 1978, the construction of the Antiproton Accumulator Ring jointly led by R. Billinge and S. van der Meer started as well as the modifications of CPS and SPS. The first collision of protons and antiprotons occurred in 1981 and the data taking of the experiments took place from 1982 to 1991 except 1986 [59]. All modifications of the SPS for collider operation were removed after 1991 (Table 10.7).

The antiproton injector chain [59] consisted of the CPS with its proton injectors and of the Antiproton Accumulator storage ring (AA). The CPS produced an intense pulse of 1.2 × 10^{13} protons at 26 GeV every 4.8 s consisting of five bunches and having a pulse length matched to the circumference of the AA (157 m). The pulse impinged on a tungsten target followed by a magnetic horn or a lithium lens focusing the emerging antiprotons. The latter were collected in the AA operating at 3.5 GeV/c, the momentum where the production was close to the peak. Subsequently, the stochastic cooling systems increased the phase space density by a factor of more than 5 × 10^8. The best daily production was close to 2 × 10^{11} per day. For the SPS fill, the collected and cooled antiprotons were transferred to the CPS, accelerated to 26 GeV/c and injected into the SPS through a new transfer line built for anti-clockwise injection. The fill was terminated by acceleration of the protons in the CPS and their injection through the existing transfer line upgraded from 14 GeV/c to 26 GeV/c. Subsequently, both beams were simultaneously accelerated to collision energy and, after the β_* had been lowered, brought to collision by adjusting the separators. The duration of a coast was between 10 and 20 h.

In order to increase the transverse and longitudinal acceptance after the target, an additional storage ring of 187 m circumference was constructed around the AA in 1986, the Antiproton Collector (AC). This new ring featured an RF system providing 1.5 MV at 9.5 MHz rotating the antiproton bunches in longitudinal phase space to reduce their momentum spread. It took over from AA the stochastic pre-cooling so that AA could be simplified but had its cooling systems upgraded. With these measures the antiproton production rate could be raised to more than 1×10^{12} per day.

The SPS required a number of modifications: the vacuum ion pumps were doubled and Ti-sublimation pumps added resulting in a reduction of pressure from 2×10^{-7} to 6×10^{-9} Torr; an electrostatic deflector separated the beams horizontally at injection; the magnet lattice included two low-beta sections for focusing the beam in the two interaction points; the RF travelling wave structures at 200 MHz had to accelerate particles travelling in both directions, and two underground experimental areas had to be constructed [61]. The initial beam energy was raised from 273 to 315 GeV after an upgrade of the magnet cooling. Towards the end of the operation, the SPS was cycled between 100 and 450 GeV, thus providing collisions at 900 GeV c.m. The performance of the collider was steadily increased during its lifetime by the upgrade of the antiproton injectors, in particular after the commissioning of AC; by adding electrostatic deflectors allowing to raise the number of bunches per beam from three to six avoiding collisions except in the two experiments and in the mid-arc between them; by the installation of a 100 MHz RF system to increase the acceptance at injection. Hence, the average initial luminosity and the centre-of-mass energy rose from 0.05×10^{30} at 546 GeV in 1982 to $3 \times 10^{30} \mathrm{cm}^{-2}\ \mathrm{s}^{-1}$ at 630 GeV with β_* (h/v) at 0.6 m/0.15 m in 1991, the last year of collider operation. The collider collected an integrated luminosity of 17 pb^{-1} during its lifetime [59].

10.1.6.1 Acknowledgement

Thanks to Karel Cornelis (CERN) for his critical reading and useful suggestions.

10.1.7 Tevatron of Fermi National Laboratory (FNAL)

The study of superconducting magnets started in 1972 in view of doubling the proton energy available at FNAL by adding another accelerator in the tunnel of the Main Ring (MR). In the same year, MR equipped with conventional magnets had reached its design energy producing 200 GeV protons. Since a magnet study had shown the feasibility of ramped superconducting magnets of 4–5 T, the official design study of this new accelerator, the Tevatron, with the MR as injector was launched in 1974 under the leadership of R.R. Wilson. Project authorization was granted in 1979 and the accelerator reached 512 GeV during commissioning in 1983 with the MR operating at 150 GeV as injector [62]. The accelerator was then routinely used between 1983 and 2000 in eight runs for fixed-target physics in the three experimental areas dedicated to physics with mesons, neutrinos, and protons respectively. The beam energy was 800 GeV except in the first run (400 GeV) and

Table 10.8 Basic parameters of the Tevatron [63]

Accelerated/colliding particles	Protons/protons—antiprotons
Particle energy	800 GeV/980 GeV
Circumference [m]	2000 π
Magnetic lattice	Alternating-gradient focusing, separated-function
Focusing order	FODO
Number of main bending magnets	774
Bending magnetic field	0.67 T at injection, 4.35 T at maximum energy
Rise time	15 s
Bending magnet	Nb-Ti conductor at 4.3 K, cold-bore, iron at ambient temperature
Quadrupole field gradient	11.4 T/m at injection, 76 T/m at maximum energy
Betatron oscillations/turn	20.59 (h), 20.59 (v)
Long straight sections	Number = 6, length = 50 m (incl. 2 low-β insertions)
β_* (h/v)	0.28 m in low-β sections
RF system (tunable)	8 cavities, 53.1 MHz, 1.2 MeV/turn total maximum
Vacuum chamber	Stainless steel, rounded square 63 mm full aperture (h,v) (in dipoles)

the repetition rate of the order of one per minute. Slow-spills (20 s) and fast beam extraction (2 ms) were available (Table 10.8).

The potential of the accelerator for proton-antiproton collisions with a centre-of-mass energy close to 2 TeV had been realized very early [64]. The bunched proton beam and anti-proton beam counter-rotating in the same vacuum chamber would be simultaneously accelerated in the Tevatron and brought to collision at top energy. Initiated in 1977, a study of the anti-proton beam cooling methods revealed that an anti-proton beam of sufficient intensity and density in phase space could be produced. This led to the construction of the anti-proton source in the period 1982–1985. The MR would also accelerate antiprotons and protons for injection into the Tevatron. It was equipped with two overpasses to provide space for the large detectors located in the two crossing points of the Tevatron. In 1985, the conversion of the Tevatron itself to a collider was started by the installation of the first low-β section in the straight section housing later the CDF detector. First collisions were recorded at a centre-of-mass energy of 1.6 TeV in 1985 and this energy could be raised to 1.8 TeV in 1986. It was increased again to 1.96 TeV for Run II [65] which started in 2001 [66].

In order to increase the average proton beam power on the anti-proton target and to eliminate the perturbation of the experiments by the MR overpasses, a new injector synchrotron, the Main Injector (MI), was constructed between 1993 and 1999 and MR was decommissioned. MI is a synchrotron of 3.32 km circumference and a minimum repetition time of 1.4 s. It is located in a new tunnel and provides the protons at 120 GeV for anti-proton production. It also accelerates the antiprotons and the protons from 8 GeV to 150 GeV for injection into the collider.

For the antiproton production, the full complement of accelerators and storage rings in the injector chain is used. The first accelerator in the injector chain is a 400 MeV H^- linear accelerator (linac) operating at 200 MHz. Its initial energy was 200 MeV but half of the original drift-tube linac was replaced in 1992–1993 by side-coupled accelerating structures providing 300 MeV. The protons are transferred to the booster synchrotron operating with 15 Hz repetition rate and are accelerated to a kinetic energy of 8 GeV. The intensity is up to 5×10^{12} protons per booster pulse. Next the particles are accelerated to 120 GeV in the Main Injector (MI), and hit a solid metal target followed by a lithium lens.

From the emerging secondary particles anti-protons of 8 GeV kinetic energy are selected and their phase space density is stepwise increased in two rings in series, the Debuncher Ring and the Anti-Accumulator Ring; the latter also accumulates the particles. The anti-protons pass then to the latest addition, the Recycler Ring (RR), were they are cooled further and a second step of accumulation takes place. The RR could be fitted into the MI tunnel as its bending and focusing magnets have a small cross-section the magnetic fields being generated by permanent magnets (1.45 T field strength). This new storage ring has therefore the same circumference as MI; it is in operation since 2004. Due to this elaborate anti-proton source the antiproton production rate has reached nearly 3×10^{11} anti-protons/h. Still the production rate of the antiprotons is rather low compared to proton production. Hence, the injector chain produces and accumulates anti-protons all the time except during the period of filling the Tevatron with protons and antiprotons which takes about 1 h.

When the Tevatron needs a new fill, which happens about every 10–20 h, the linac, the booster and MI produce the 36 proton bunches for the Tevatron, then anti-protons collected in RR are transferred to MI for acceleration to 150 GeV and injection into the Tevatron. The 36 bunches are arranged in 3 trains of 12 bunches. Protons and antiprotons circulate on helical orbits produced by high-voltage electrostatic separators to prevent collisions during injection and acceleration. At top energy, after increasing the focusing in the two collision points equipped with detectors, the separator configuration is modified to bring the beams in collision in these two points [67].

A peak luminosity of 4×10^{32} cm^{-2} s^{-1} has been reached by continuously upgrading all the systems, introducing advanced accelerator technology and stream-lining the operational procedures. The Tevatron has delivered an integrated luminosity of more than 12 fb^{-1} to each of the experiments (CDF and D0) before it ceased operation at the end of September 2011. It was one of the most complex research instruments ever built and will be known for its advances in accelerator physics and technological breakthroughs [68].

10.1.7.1 Acknowledgement

Vladimir Shiltsev (FNAL) has contributed with useful suggestions and pertinent comments to this chapter. Sincere thanks are due to him.

10.2 RHIC[1]

T. Roser

10.2.1 The RHIC Facility

With its two independent superconducting rings RHIC is a highly flexible collider of hadron beams ranging from intense beams of polarized protons to fully stripped gold ions [69, 70]. The layout of the RHIC accelerator complex is shown in Fig. 10.1. The collision of 100 GeV/nucleon gold ions probes the conditions of the early universe by producing extreme conditions where quarks and gluons are forming a new state of matter, the strongly interacting quark-gluon plasma. Several runs of high luminosity gold-gold collisions as well as comparison runs using proton, deuteron and copper beams have demonstrated that indeed a new state of matter with extreme density is formed in the RHIC gold-gold collisions. (See also Sect. 11.5).

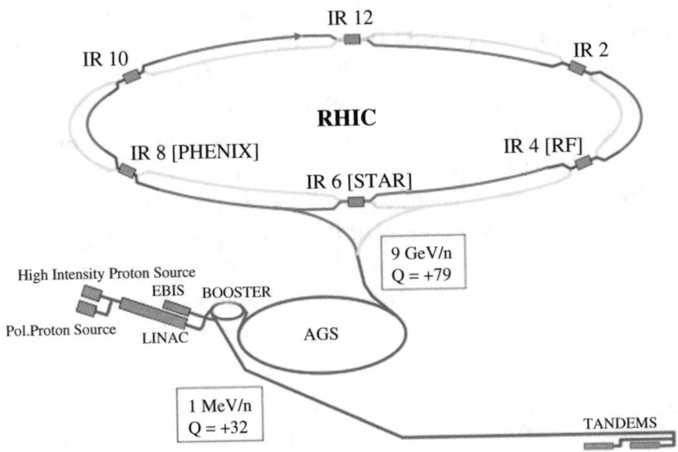

Fig. 10.1 Layout of RHIC and the injector accelerators. The gold ions are stepwise ionized as they are accelerated to RHIC injection energy

[1]This manuscript has been authored by Brookhaven Science Associates, LLC under Contract No. DE-SC0012704 with the U.S. Department of Energy. The United States Government retains and the publisher, by accepting the article for publication, acknowledges that the United States Government retains a non-exclusive, paid-up, irrevocable, world-wide license to publish or reproduce the published form of this manuscript, or allow others to do so, for United States Government purposes.

The RHIC polarized proton collider has opened up the completely unique physics opportunities of studying spin effects in hadronic reactions at high-luminosity high-energy proton-proton collisions. It allows the study of the spin structure of the proton, in particular the degree of polarization of the gluons and anti-quarks, and also verification of the many well-documented expectations of spin effects in perturbative QCD and parity violation in W and Z production. The RHIC center-of-mass energy range of 200–500 GeV is ideal in the sense that it is high enough for perturbative QCD to be applicable and low enough so that the typical momentum fraction of the valence quarks is about 0.1 or larger. This guarantees significant levels of parton polarization.

During its 10 years of operation RHIC has greatly exceeded the design parameters for gold-gold collisions, has successfully operated in an asymmetric mode of colliding deuteron on gold with both beams at the same energy per nucleon, and thereby at different rigidities, and successfully completed a comparison run of colliding copper beams with record luminosities. Operation at unequal rigidities of the two colliding beams is a unique feature of RHIC with its two independent rings. The interaction regions are designed to separate the two beams first before they go through their separate final focus triplets. This is shown in Fig. 10.2 for equal species and for the most unequal species of protons and gold beams with a rigidity ratio of 2.47 for equal energy per nucleon. The necessary increase in distance between the interaction point and the final focus triplet limits the achievable luminosity in RHIC.

In addition to heavy ions, RHIC successfully demonstrated its capabilities as a high luminosity polarized proton collider both at 100 and 250 GeV proton beam energy. For most of the heavy ion runs RHIC was operating with beam energies of 100 GeV/nucleon—the gold beam design energy. Additional running at lower beam energy was also accomplished again demonstrating the high level of flexibility of RHIC. Gold collisions at energies much below the RHIC injection energy of 10 GeV/nucleon is allowing the study of the critical point in the quark-gluon phase diagram. Figure 10.3 shows the achieved integrated nucleon-pair luminosities for the many modes of operation of RHIC since its start of operation in 2000. Using nucleon-pair luminosity, which is calculated as the ion-ion luminosity times the

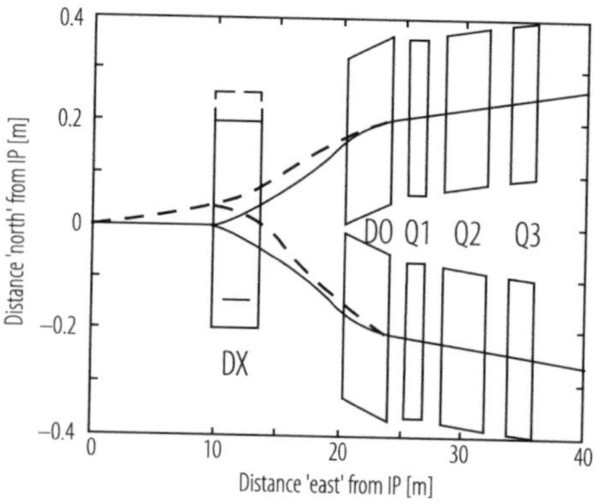

Fig. 10.2 Interaction region geometry in RHIC for equal species (solid line) and the most extreme example of dissimilar species—protons and gold (dashed line)

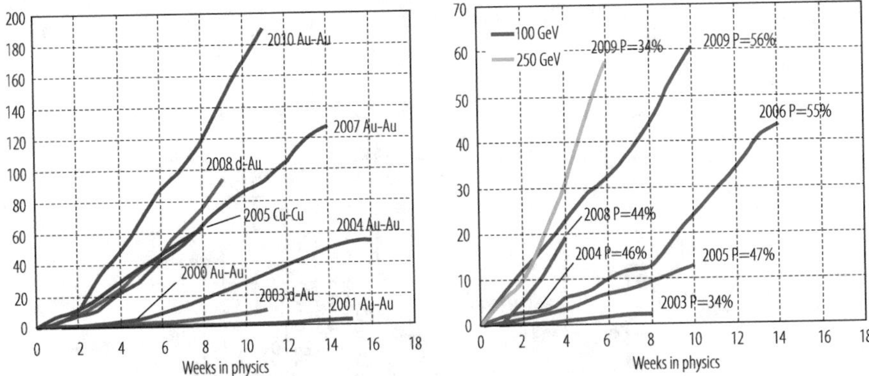

Fig. 10.3 Integrated nucleon-pair luminosity for the heavy ion (left) and the polarized proton (right) running modes since the start of RHIC operation

Table 10.9 Major parameters of RHIC

Parameter	Value
Circumference	3833.845 m
Number of interaction points	6
Harmonic number, acceleration	360
Harmonic number, storage	2520
Typical betatron tunes for Au	28.23/29.22
Transition energy γ_T	26.7 (Au), 22.8 (p)
Maximum magnetic rigidity	839.5 Tm
Total number of dipoles, both rings	396
Total number of quadrupoles, both rings	492
Dipole field at 100 GeV/nucleon, Au	3.458 T
Arc dipole effective length	9.45 m
Arc quadrupole gradient	71.2 T/m

number of nucleons in each of the two beam ions, allows the comparison of the different modes of operation properly reflecting the relative statistical relevance of the data samples and also the degree of difficulty in achieving high luminosity.

For RHIC's major parameters and achieved performance parameters see Tables 10.9 and 10.10, respectively.

10.2.2 Collider Operation

Gold-Gold Operation

Starting with Au^{1-} from a sputter source the gold ions are stepwise ionized as they are accelerated in the Tandem Van de Graaff, the AGS Booster and the AGS to RHIC injection energy. The electrostatic acceleration in the Tandem Van de Graff

Table 10.10 Achieved performance parameters of RHIC

Parameter	Unit	Au–Au operation	d–Au operation	Pol. proton–proton operation	
Total energy at store	GeV/nucleon	100	100	100	250
Total energy at injection	GeV/nucleon	9.8	9.9	23.8	23.8
Number of bunches		111	95	109	109
Bunch intensity	10^9	1.1	100d/1.0Au	135	110
IP beta-function	m	0.75	0.85	0.7	0.7
Normalized rms emittance	μm	2.8	2.6	2.5	3.0
Rms bunch length	m	0.3	0.3	0.85	0.6
Hourglass factor		0.93	0.95	0.70	0.80
Beam-beam parameter/IP		0.0015	0.0014	0.0045	0.0054
Peak luminosity	$cm^{-2}\ s^{-1}$	40×10^{26}	27×10^{28}	50×10^{30}	85×10^{30}
Average luminosity	$cm^{-2}\ s^{-1}$	20×10^{26}	14×10^{28}	28×10^{30}	55×10^{30}
Average polarization	%	–	–	55	34
Calendar time in store	%	53	58	54	53
Integrated luminosity per week	nb^{-1}	0.65	40	8300	1840

provides an extremely bright gold beam that can be captured and bunch-merged to provide the necessary bright bunches of 1×10^9 Au ions with a normalized transverse rms emittance of less than 2.5 μm and a total longitudinal emittance of less than 0.3 eVs/nucleon. The final stripping to bare Au^{79+} occurs on the way to RHIC. Recently the Tandem van de Graff pre-injector has been replaced by an Electron Beam Ion Source (EBIS) followed by an RFQ and IH-Linac. The EBIS can produce very bright high-charge state heavy ion beams by accumulating and stripping heavy ions with a 10 A space-charge neutralizing electron beam. The new source can produce beams of many species including uranium.

The two RHIC rings, labeled blue and yellow, are intersecting at six interaction regions (IRs). All IRs can operate at a betastar between 2 and 10 m. In two interaction regions, occupied by the two main detectors STAR and PHENIX, the quality of the triplet quadrupoles allows further reduction of betastar to less than 1 m. Typically betastar is about 10 m at injection energy for all IRs and is then squeezed during the acceleration ramp first to 5 m at the transition energy ($\gamma_T = 26.7$), which minimizes the momentum dependence of the transition energy crossing, and then to less than 1 m for PHENIX and STAR. A typical acceleration cycle consists of filling the blue ring with 111 bunches in groups of 4 bunches, filling the yellow ring in the same way and then simultaneous acceleration of both beams to storage energy. During acceleration the beam bunches are longitudinally aligned but are separated vertically by 10 mm in the interaction regions to avoid beam losses from beam-beam interaction.

The collision rate is measured using identical Zero Degree Calorimeters (ZDC) at all detectors. The ZDC counters detect at least one neutron on each side from mutual Coulomb and nuclear dissociation with a total cross section of about 10 barns. Typical stores in RHIC last about 4–5 h. Due to intra-beam scattering, which is particularly important for the fully stripped, highly charged gold beams, the initial normalized rms emittance of about 2.5 μm grows to about 5 μm at the end of the store and a significant amount of beam is lost from the RF bucket. The resulting luminosity lifetime is about 2.5 h.

The reduced number of gold ions compared to typical proton bunches makes it possible to contemplate stochastic cooling of the 100 GeV/nucleon bunched beam. A 6–9 GHz longitudinal and a 5–8 GHz vertical stochastic cooling system was installed using novel high power multi-cavity kickers and the high energy bunched beam was successfully cooled [71]. Figure 10.4 shows the successful cooling of the transverse emittance in the yellow ring compared to the emittance growth in the blue ring without cooling. Both the vertical and horizontal emittance are reduced due to x-y coupling. Figure 10.5 shows the vertex distribution at the end of the store with and without stochastic cooling. It clearly shows a significant increase in luminosity in the central region. The modulation of the vertex distribution is a result of the 200 MHz storage RF system that provides about 4 MV/turn to compress the bunch length and consequently reduce the length of the vertex distribution to better match the acceptance of the collider detectors. Again due to intra-beam scattering the initially narrow distribution widens during the store. A new 2 MV, 56 MHz superconducting cavity will provide improved RF focusing that, together

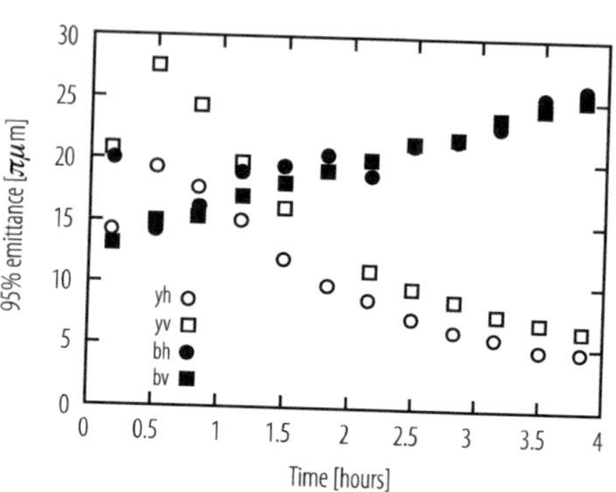

Fig. 10.4 Transverse emittance evolution in the blue (filled) and yellow (open) RHIC rings with vertical stochastic cooling in the yellow ring [71]

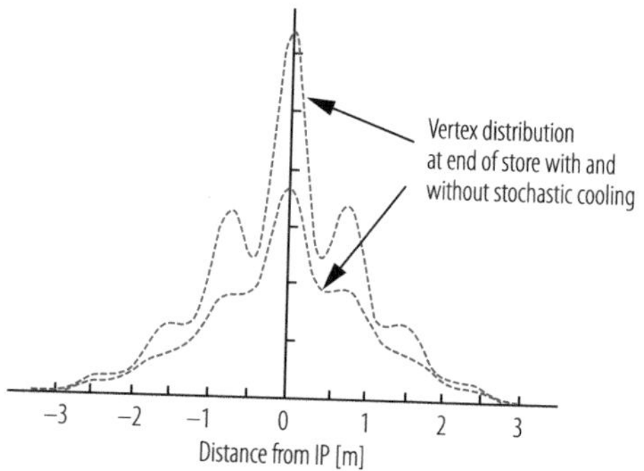

Fig. 10.5 Vertex distribution at the end of a 100 GeV/n Au-Au store with and without longitudinal stochastic cooling

with the longitudinal stochastic cooling, will maintain the short vertex distribution throughout the store.

With the two planes of stochastic cooling operational in both rings a peak luminosity at PHENIX and STAR of up to 40×10^{26} cm^{-2} s^{-1} (160×10^{30} cm^{-2} s^{-1} nucleon-pair luminosity) with an average store luminosity of 20×10^{26} cm^{-2} s^{-1}: ten times the original RHIC design average store luminosity. The full stochastic cooling system will include also the horizontal plane and should result in another doubling of the luminosity.

The total gold beam intensity in RHIC was initially limited by vacuum breakdowns in the room temperature sections of the RHIC rings [72]. This pressure rise is associated with the formation of electron clouds, which in turn appear when the bunch peak intensity is high around transition and after bunch compression, and when the bunch spacing is below about 200 ns. This situation was greatly improved by installing vacuum pipes with an internal coating of non-evaporative getter (NEG) that is properly activated. The resulting residual static pressure is about 10^{-11} Torr. The NEG coating acts as a very effective distributed pump and also suppresses electron cloud formation due to its low secondary electron yield.

Very fast single bunch transverse instabilities that develop near transition, where the chromaticity needs to cross zero, originally limited bunch intensity and is still responsible for occasional emittance growth, especially towards the end of bunch trains. The instability can be stabilized using octupoles [73] and the instability threshold can be increased by lowering the peak current during transition crossing. This instability has a growth rate faster than the synchrotron period and is similar to a beam break-up instability. The instability is also enhanced by the presence of electron clouds [74].

Deuteron-Gold Operation

Colliding 100 GeV/nucleon deuteron beam with 100 GeV/nucleon gold beam will not produce the required temperature to create a new state of matter and therefore serves as an important comparison measurement to the gold-gold collisions. The rigidity of the two beams is different by about 20%, which results in different deflection angles in the beam-combining dipoles on either side of the interaction region. This requires a non-zero angle at the collision point, which slightly reduces the available aperture.

The injection energy into RHIC was also the same for both beams requiring the injector to produce beams with different rigidity. With same energy beams throughout the acceleration cycle in RHIC the beams can remain cogged. Without this the effect of the time-modulated long-range beam-beam interaction in the IRs would lead to rapid beam loss. The typical bunch intensity of the deuteron beam was about 1×10^{11} with a transverse rms emittances of about 2 μm and a total longitudinal emittance of 0.3 eVs/nucleon. The gold beam parameters were similar to the gold-gold operation described above. Recently the ring with the gold beam was operated with increased transverse focusing to reduce the effect of intra-beam scattering on the transverse emittance. As a beneficial side effect the two beams crossed transition at different times, which reduced the development of the fast transverse instability. A peak luminosity of 27×10^{28} cm^{-2} s^{-1} (100×10^{30} cm^{-2} s^{-1} nucleon-pair luminosity) and store-averaged luminosity of 14×10^{28} cm^{-2} s^{-1} was reached at the IRs with a 0.85 m betastar.

Polarized Proton Operation

Figure 10.6 shows the layout of the RHIC accelerator complex highlighting the components required for polarized beam acceleration. The 'Optically Pumped Polarized Ion Source' [75] is producing about 10^{12} polarized protons per pulse. A single source pulse is captured into a single bunch, which is sufficient beam intensity to reach in RHIC the nominal bunch intensity of 2×10^{11} polarized protons.

In the AGS two partial Siberian snakes are installed, an iron-based helical dipole that rotates the spin around the longitudinal direction by 11° and a superconducting helical dipole that can reach a 3 T field and a spin rotation of up to 45°. A view down the magnet gap is shown in Fig. 10.7. With the two partial snakes placed with one third of the AGS ring between them all vertical spin resonances are avoided up to the required RHIC transfer energy of 23.8 GeV as long as the vertical betatron tune is placed at 8.98, very close to an integer [76]. With an 80% polarization from the source 65% polarization was reached at AGS extraction. The remaining polarization loss in the AGS comes from weak spin resonances driven by the horizontal motion of the beam [77]. It is planned to overcome them by quickly shifting the betatron tune during resonance crossing.

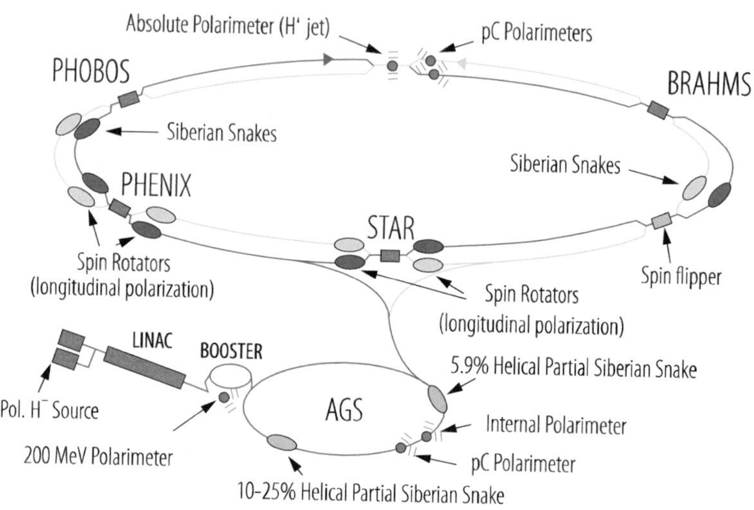

Fig. 10.6 The RHIC accelerator complex with the elements required for the acceleration and collision of polarized protons highlighted

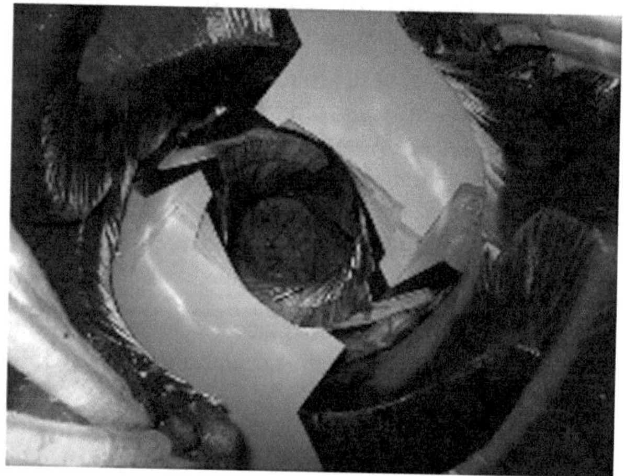

Fig. 10.7 View down the magnet gap of the warm, iron-based helical partial Siberian snake of the AGS

The full Siberian snakes [78], two for each ring, and the spin rotators, four for each collider experiment, in RHIC each consist of four 2.4 m long, 4 T superconducting helical dipole magnet modules each having a full 360° helical twist. Figure 10.8 shows the orbit and spin trajectory through a RHIC snake. The two Siberian snakes are installed in the location of the missing dipole of the dispersion suppression section and the orbit angle is exactly 180° in between the two snakes.

The spin rotation axis of the two snakes is pointing 45° in and out, respectively, relative to the direction of the beam resulting in the required 90° angle between the

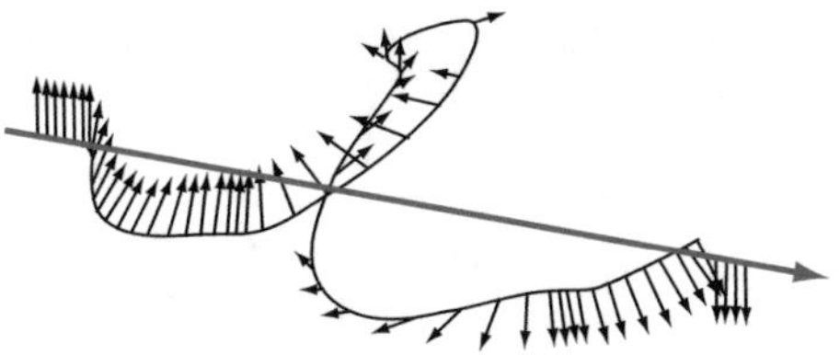

Fig. 10.8 Orbit and spin tracking through the four helical magnets of a Siberian Snake. The spin tracking shows the reversal of the vertical polarization. The spin rotation axis is in the horizontal plane and is pointing 45° away from the beam direction

spin rotation axes. This configuration yields a spin tune $Q_s = 0.5$ and a vertical stable spin direction around the ring. Here, spin tune is defined as the number of spin precessions in one orbital revolution. Since the betatron tunes in circular accelerators are kept away from half integer to keep the beam stable, both intrinsic and imperfection depolarizing spin resonances at $G\gamma = kP \pm Q_y$ and $G\gamma = k$ are avoided [79]. Here, G is the anomalous g factor, γ is the Lorentz factor, k is an integer, P is the super-periodicity of the accelerator, and Q_y is the vertical betatron tune.

The accurate measurement of the beam polarization is required for set-up and operation of the polarized proton collider. Very small angle elastic scattering in the Coulomb-Nuclear interference region offers the possibility for an analyzing reaction with a high figure-of-merit, which is not expected to be strongly energy dependent [80]. For polarized beam commissioning in RHIC an ultra-thin carbon ribbon is used as an internal target, and the recoil carbon nuclei are detected to measure both vertical and radial polarization components. The detection of the recoil carbon with silicon detectors using both energy and time-of-flight information shows excellent particle identification. It was demonstrated that this polarimeter can be used to monitor polarization of the high energy proton beams in an almost non-destructive manner and that the carbon fiber target could be scanned through the circulating beam to measure beam and polarization profiles. A polarized atomic hydrogen jet was also installed as an internal target for small angle proton-proton scattering which allows the absolute calibration of the beam polarization to better than 5%.

Figure 10.9 shows circulating beam current, luminosity and measured circulating beam polarization of a typical store with a beam energy of 100 GeV. The maximum peak luminosity achieved with 100 GeV proton beams is about $50 \times 10^{30}\ \text{cm}^{-2}\ \text{s}^{-1}$ and store beam polarization of about 55% calibrated at 100 GeV with the absolute polarimeter mentioned above. To preserve beam polarization in RHIC during acceleration and storage the vertical betatron tune has to be controlled to better than 0.005 and the orbit has to be corrected to better than 1 mm rms to avoid depolarizing "snake" resonances.

During acceleration from 100 GeV to the maximum RHIC energy of 250 GeV, several very strong depolarizing spin resonances need to be crossed [81]. In a first running period an average store polarization of 34% was measured at 250 GeV again calibrated with the absolute polarimeter. The peak luminosity

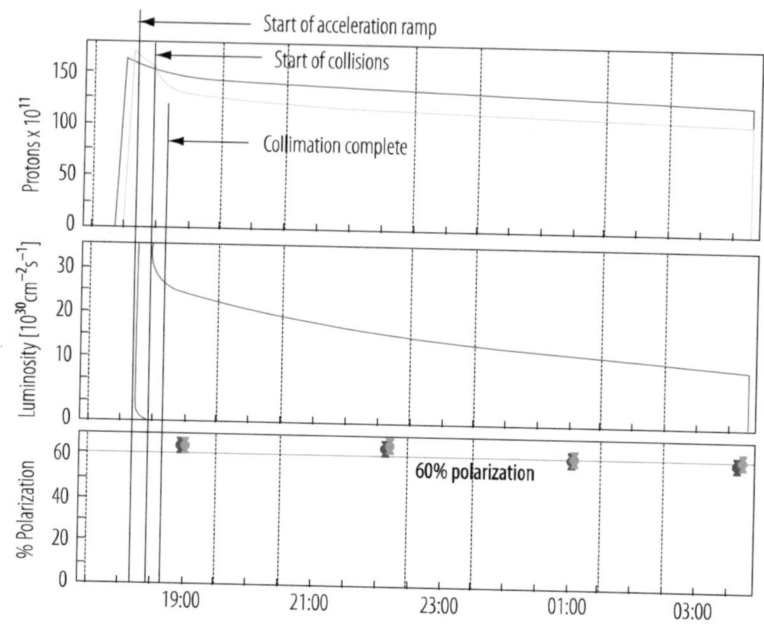

Fig. 10.9 Circulating beam in the blue and yellow ring, luminosity at STAR (red), as well as the measured circulating beam polarization in the blue and yellow RHIC ring (blue (dark) and yellow (light) lines and symbols, respectively) for a typical store with 100 GeV beam energy

reached was about 85×10^{30} cm^{-2} s^{-1} and the average store luminosity was about 55×10^{30} cm^{-2} s^{-1}. The goal for operation of RHIC with polarized 250 GeV proton beams is a peak luminosity of 200×10^{30} cm^{-2} s^{-1} with a beam polarization of 70%.

Head-on beam-beam effects are the main limitation for increasing the luminosity in proton-proton operation. The maximum beam-beam parameter reached was 0.0065 per IP with 100 GeV beams when operating with beams in collision at two detector IPs. This value is somewhat smaller than what was reached at the Tevatron proton-antiproton collider probably due to the smaller momentum aperture available at RHIC as a result of the dipole first IR design. Additional sextupole circuits are available to correct non-linear chromaticity and the beta-function momentum dependence. In the future it is planned to mitigate the beam-beam effects with the installation of two electron lenses in RHIC. Space is available to partially compensate the beam-beam effect with an opposite charge electron beam separated from the proton-proton collisions by the correct phase advance. With the electron lenses and also a more intense polarized proton source it is expected that the proton-proton luminosity could be increased by about a factor of two.

Recent Progress

The performance of RHIC has been continuously improved with store-averaged luminosity reaching 87×10^{26} cm^{-2} s^{-1}, exceeding the design luminosity by more than a factor of 40, and polarized proton operation at 510 GeV center-of-mass energy with peak luminosity of 250×10^{30} cm^{-2} s^{-1} with store-averaged beam polarization of up to 60%. Major upgrades to RHIC included electron lenses to compensate for head-on beam-beam interactions [82] and a superconducting storage RF system. In preparation for operation at very low center-of-mass energies with beam energies below the RHIC injection energy bunched beam electron cooling of both RHIC ion beams is being installed [83].

10.3 Electron Accelerators and Electron–Positron Colliders

J. Seeman

10.3.1 Cyclotrons

Cyclotrons are one of the first accelerators invented and use a fixed magnetic field and a radiofrequency (RF) cavity to accelerate electrons in ever increasing orbits [84]. The electron beam is injected into the center of a circular magnetic field region between magnet poles and then made to circulate. These electrons are accelerated twice on each turn and are extracted at high energy near the outer edge of the accelerator. The cyclotron can produce continuous beams leading to high power applications. The first cyclotron was built at Berkeley in 1931 with a diameter of 11 cm. Many more cyclotrons were to follow over the years. The TRIUMF cyclotron in use now is located in Vancouver, Canada, and has a diameter of 18 m and has a 4000 ton magnet. The beam energy is limited by the field strength and the diameter of the magnetic pole. Cyclotrons are mainly used to produce high beam power for science, for materials analysis, and industrial processing applications. Two cyclotrons are shown in Fig. 10.10.

10.3.2 Synchrotrons

A synchrotron is a fixed circumference accelerator that increases the electron beam energy using RF accelerating cavities but keeps the radius constant by increasing the magnetic fields. The field used to bend the beam in a circle is ramped in proportion to the beam energy [85]. The beam is injected at low energy and magnetic fields, accelerated, and then extracted at high energy and magnetic fields. The highest beam energy is limited by the strength of the magnetic field and the diameter of the accelerator which can be over 1 km and is usually limited by the site or its building. Synchrotrons are pulsed machines due to the cyclic ramping of the magnetic fields. The RF system accelerates the beam during the ramp and keeps the beam bunched.

Fig. 10.10 On the left is the first electron cyclotron constructed at the Berkeley in 1931 (courtesy LBNL Berkeley, CA). The right photograph is a modern cyclotron at PSI (courtesy PSI Lausanne, Switzerland)

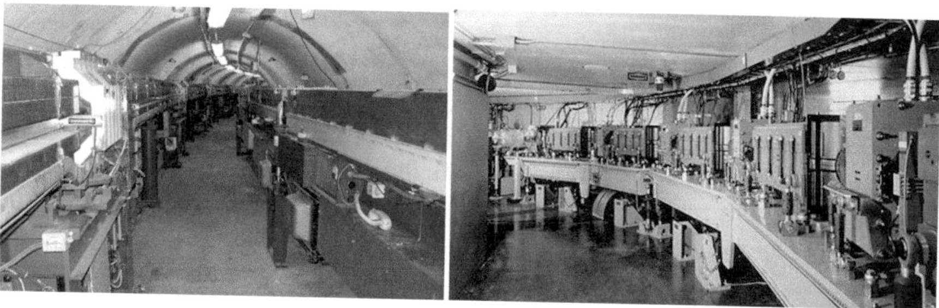

Fig. 10.11 The left photograph is the 12 GeV electron synchrotron at Cornell (courtesy Wilson Laboratory Ithaca, NY) used as an injector into the CESR storage ring collider, also shown. The right photograph is the Advanced Light Source injector synchrotron (courtesy LBNL Berkeley, CA)

The magnets are often divided into separated units to allow simplified construction including dipoles for bending, quadrupoles for beam focusing, sextupoles for chromatic corrections, and correction dipoles for trajectory tuning. The quadrupoles often form an alternating focus and defocusing magnetic lattice around the ring to produce "strong focusing" which reduces the beam's excursions. As a result, the cross sections of the magnet gaps can be made significantly smaller and less expensive. The radiofrequency RF system can have a large range of frequencies and cavity designs. A vacuum system in the milli-Torr to nano-Torr level is also needed. Synchrotrons are often used as injectors for higher energy accelerators and storage rings. In Fig. 10.11 are shown the electron synchrotrons at Cornell (12 GeV, ~756 m) and the injector at the Berkeley Advanced Light Source.

10.3.3 Electron Positron Circular Colliders

In the late 1950s the accelerator community realized that colliding beams head-on would increase the center of mass energy significantly over fixed target collisions and could significantly expand the reach for new particles and physics. The goal was to design an accelerator that could provide enough centre of mass energy to make new physics, high enough beam currents and small enough collision spot sizes to provide a sufficient data rate, and to provide a volume for the particle physics detector surrounding the collision point with backgrounds sufficiently low to allow clean data collection. A world wide effort was initiated to design these accelerators and has continued to today [86–88]. At least 25 electron-electron or electron-positron colliders have been built and operated for particle physics over the past 50 years with ever increasing beam currents, luminosity, energy, and sophistication. Several colliders are shown in Fig. 10.12. In Table 10.11 are listed in chronological order these 24 colliders along with several of their technical parameters. Following the table are descriptions of each of the colliders discussing several of their

Fig. 10.12 The ADA collider on the top left (courtesy INFN Frascati, Italy). Top right is the PEP-II B-Factory collider (courtesy SLAC Menlo Park, CA). The LEP collider is shown on the bottom left (courtesy CERN Geneva, Switzerland). SuperKEKB is shown on the bottom right (courtesy KEK Tsukuba, Japan)

Table 10.11 Historical listing of electron-electron and electron-positron colliders

Collider	Laboratory	Date (start–end)	Circumf. (type) [m]	Beams	*E* [GeV]	Luminosity [10^{30} cm^{-2} s^{-1}]
ADA	Frascati, Italy	1961–1964	3 (SR)	e^-/e^+	0.25	Measured
VEP-1	BINP, Russia	1962–1967	2.7 (DR)	e^-/e^-	0.13	0.003
CBX	Stanford, USA	1963–1967	12 (DR)	e^-/e^-	0.5	0.0017
VEPP-2	BINP, Russia	1967–1970	11.5 (SR)	e^-/e^-	0.13	0.02
ACO	Orsay, France	1967–1972	22 (SR)	e^-/e^+	0.5	0.1
VEPP-2M	BINP, Russia	1974–2000	17.8 (SR)	e^-/e^+	0.7	100
ADONE	Frascati, Italy	1969–1993	105 (SR)	e^-/e^+	1.5	0.6
CEA Bypass	Cambridge, USA	1971–1974	225 (SR)	e^-/e^+	3.0	10
SPEAR	SLAC, USA	1972–1988	234 (SR)	e^-/e^+	2.5	12
DORIS	DESY, Germany	1973–1993	288 (DR,SR)	e^-/e^+	6.0	33
DCI	Orsay, France	1977–1984	95 (DR)	e^-/e^+	1.8	1.4
PETRA	DESY, Germany	1978–1986	2304 (SR)	e^-/e^+	19	23
CESR	Cornell, USA	1979–2008	768 (SR)	e^-/e^+	6.0	1100
VEPP-4	BINP, Russia	1979–present	366 (SR)	e^-/e^+	7.0	50
PEP	SLAC, USA	1980–1988	2200 (SR)	e^-/e^+	15	59
Tristan	KEK, Japan	1986–1995	3016 (SR)	e^-/e^+	32	140
SLC	SLAC, USA	1989–1998	4000 (linear)	e^-/e^+	49	2.8
BEPC	IHEP, China	1989–2004	240 (SR)	e^-/e^+	2.8	8
LEP	CERN, Switzerland	1989–2000	26,659 (SR)	e^-/e^+	104	100
DAFNE	Frascati, Italy	1998–present	98 (DR)	e^-/e^+	0.7	453
PEP-II	SLAC, USA	1998–2008	2200 (DR)	e^-/e^+	9.0×3.1	12,069
KEKB	KEK, Japan	1999–2009	3016 (DR)	e^-/e^+	8.0×3.5	21,083
BEPC-II	IHEP, China	2008–present	240 (DR)	e^-/e^+	2.1	1000
VEPP-2000	BINP, Russia	2006–present	24.4 (SR)	e^-/e^+	1.0	120
SuperKEKB	KEK, Japan	2016–present	2200 (DR)	e^-/e^+	7.0×4.0	NA

The energy E shown is for each beam and is doubled for the center of mass energy
SR single ring, *DR* double ring

unique features and results. The intent is to illustrate the advancement of collider technology over the years. Several of the technical advances are high power RF systems, superconducting RF systems, high field magnets, superconducting magnets, high field permanent magnets, ultra-high vacuum systems, low emittance beam lattices, complex interaction region designs, and sophisticated diagnostics and controls. Many of the colliders in later life became synchrotron radiation sources and several serve in this capacity today.

10.3.3.1 ADA

ADA at Frascati was a pioneering electron-positron collider [89]. This collider was used to study stored beam parameters, injection, beam lifetime, and collisions. Injection of electrons and positrons was made by converting 1 GeV gamma rays in a small target inside the ring. The mean lifetime was consistent with gas bremsstrahlung. With the electrons stored, the whole magnet was rotated about a horizontal axis to inject positrons without losing electrons. A pulsed kicker was used to reduce the betatron oscillations of the injected particles. Luminosity was measured. The effects of Touschek scattering was discovered and studied.

10.3.3.2 VEP-1

VEP-1 at BINP was an early electron-electron collider used to study elastic scattering and double bremsstrahlung experiments [90]. This accelerator demonstrated the possibility to carry out colliding beam experiments for particle physics.

10.3.3.3 CBX

The Princeton-Stanford Colliding Beam Experiment CBX on the Stanford campus was an electron-electron collider used to study elastic scattering and double bremsstrahlung experiments [91]. G. O'Neill on this team proposed that radiation damping for a stored electron beam could help with injection and emittance reduction. The beam–beam tune shift limit was seen in this machine on the level of 0.02–0.05. CBX saw indications of the single beam resistive wall instability. Electron currents up to 1 A were stored. This accelerator also demonstrated the possibility to carry out colliding beam experiments for particle physics.

10.3.3.4 VEPP-2

VEPP-2 at BINP was one of the first electron-positron collider in the world [92, 93]. The rho and phi meson parameters were measured using annihilation into two pions followed later by omega and phi meson decay parameters in the vector meson region and the first observation of two-photon pair production. This collider was quickly rebuilt as VEPP-2M.

10.3.3.5 ACO

ACO at Orsay was an early electron–positron collider used to demonstrate collider principles [94]. This accelerator had six dipoles with quadrupoles in between. The ring was later used as a synchrotron light source demonstrating ring based FEL science. This accelerator gave way to the DCI collider a few years later.

10.3.3.6 ADONE

ADONE at Frascati was a one ring collider with a 50 MHz RF system [95]. ADONE had longitudinal phase feedback to damp beam instability modes, a 4 kG detector solenoidal field (MEA), magnet shunts to adjust the beta functions in the ring, and adjustable damping partition numbers generated by an RF frequency change. Injection was from a linac with an energy ramp after the fill. The measured luminosity increased a little over the fourth power of the energy. ADONE later confirmed the existence of the J/Psi.

10.3.3.7 CEA

The CEA Cambridge Electron Colliding Beam Facility used a linac for injection at 120 MeV (e^+) and 240 MeV (e^-) and then ramped both beams in the ring to full energy [96]. The accelerator technology advancements for the CEA are the first demonstration of a low beta insertion at the IP, single-turn pulsed orbit switching into an interaction point (IP) magnetic bypass, switching from energy cycling to dc operation at full energy and the addition of two damping magnets to redistribute the radial damping so that both betatron and synchrotron oscillations were damped. In particle physics the CEA measured that the R cross section ratio rose at higher center of mass energy, hinting at many discoveries to come later at other colliders.

10.3.3.8 SPEAR

SPEAR at SLAC was designed as a two ring collider but due to funding only one ring was built. Injection was at full energy from the SLAC 2-mile linac with a positron source in linac sector 11 [97]. The ring lattice was extremely flexible in the choice of operating tunes, dispersion, and beta values the IPs. Transverse horizontal and vertical instabilities (head-tail) were observed at about 0.5 mA per bunch which were cured by a positive chromaticity. The luminosity varied as the fourth power of the beam energy. SPEAR was very productive in particle physics with the discovery of the J/Psi and tau. SPEAR has been upgraded to a medium emittance light source.

10.3.3.9 VEPP-2M

VEPP-2M at BINP reached a luminosity 100 times VEPP-2. VEPP-2 served for a while as the injector [93, 98]. The ring operated with a superconducting wiggler with an 8 T field that was used to increase the radial emittance, decrease the damping time, increase the beam-beam tune shift for a higher luminosity, and also for suppression of intra-beam scattering. The collider could generate and use polarized beams. The particle physics results are many. Round beams have been studied at VEPP-2M to try to reduce beam-beam effects and, thus, increase the allowed beam-beam tune shift. Round beams required the use of solenoidal focusing at the IPs.

10.3.3.10 DORIS

DORIS at DESY started as a two ring collider with beams brought into collision with vertical dipoles in the IR and had a vertical crossing angle [99]. Injection was from a linac and a booster synchrotron. The single bunch currents in the collider were limited by higher parasitic modes in the RF cavities. The luminosity was limited by effects of the vertical crossing angle. DORIS was subsequently converted to a single ring collider with head-on collisions and went on to do many years of B meson physics.

10.3.3.11 DCI

DCI at Orsay was a two ring collider designed for high energy physics studies of charged and neutral particles and also of two photon physics [100]. The two rings were mounted one above the other and the beams were brought into collision with vertical bends near the IR. Both rings could store both e^- and e^+. The rings could operate with either two bunch collisions or four bunch collisions (opposite in both rings). The four bunch collisions aimed at charge cancelation of the beam-beam effects allowing higher beam-beam tune shifts and thus higher luminosity. The four beam scheme in reality partially worked but was strongly limited by incoherent and coherent beam-beam modes. This topic of four beam collisions has been theoretically studied for many years since. The DCI collider mostly operated in a one bunch mode in each ring (out of time) generating twice the luminosity. The peak luminosity scaled as the energy squared.

10.3.3.12 PETRA

PETRA at DESY had four interaction points and operated up to an energy of 24 GeV per beam with 2 × 2 bunches [101]. Additional seven cell RF cavities were installed over the years to achieve these energies. Second harmonic cavities (1 GHz) were installed to reduce the bunch length, cure a vertical single bunch instability and reduce several synchrotron-betatron resonances. The free space for the detector was reduced to ±4.45 m using mini-beta insertions allowing a vertical beta of 6 cm. The injected positrons were predamped in the accumulator storage ring PIA. PETRA is known for the discovery of the Gluon and for QCD studies. PETRA has been upgraded to a low emittance light source.

10.3.3.13 CESR

CESR at Cornell operated for B-meson studies for 20 years [102, 103]. The particle physics results included V_{ub} observations of "penguin" modes, b $\rightarrow$ sγ decays, CKM matrix constraining the unitarity triangle, and B mass and lifetime measurements. Injection was from a 200 MeV linac and a 12 GeV synchrotron. With electrostatic plates installed, CESR could collide up to 27 bunches separated in the accelerator arcs by what is now called a "Pretzel orbit" that was used to suppress parasitic beam-beam collisions and the related tune shifts. Several other colliders went on to use this technique to increase the number of bunches. It was discovered at CESR that a horizontal tune just above the half integer (<0.51) increased significantly the beam-beam limit allowing higher luminosity. Superconducting single-cell cavities with HOM damping were installed in CESR allowing up to 325 mA beams each of electrons and positron to the stored. The CESR interaction region used a combination of permanent-magnet and superconducting technologies for the vertically focusing quadrupoles. A ±2.5 mrad uncompensated IP crossing angle was ultimately used. Superconducting wigglers were later installed to allow operation at lower energy at the charm threshold. CESR is presently being used as a light source and as a test accelerator to study low emittance damping rings and electron cloud physics.

10.3.3.14 VEPP-4

VEPP-4 at BINP has been colliding beams for a long time including two modernizations [104]. Over the years, the vertical beta at the IP was reduced to 5 cm, superconducting wigglers were added to increase the luminosity, and a superconducting RF cavity was installed for bunch length reduction allowing the use of the lower IP beta. Bunch currents were limited by beam induced wakefields in the vacuum chambers at the level of 17 mA. Eight pairs of electrostatic separation plates allow two bunch operation in a Pretzel scheme. A transversely polarized beam could be injected into VEPP-4 from VEPP-3 for accurate measurements of the masses of the upsilon family particles. Other particle physics studied at VEPP-4M included precise tau and J/Psi mass measurements and two photon physics.

10.3.3.15 PEP

PEP at SLAC operated with three bunches per beam up to 14 GeV delivered to six interaction points [105]. PEP had aluminum vacuum chambers and aluminum RF cavities coated with TiN to suppress breakdowns. The head-tail microwave beam instability was studied extensively at PEP and cures investigated. Collisions with 1, 2, 4, and 6 IPs allowed studies of the beam-beam limits with different damping times per collision. The particle physics results at PEP included the first measurements of the tau lepton lifetime, the discovery that the B meson lifetime was unexpectedly long, analysis of jet structures, and the measurements of lifetimes and properties of charm and bottom hadrons.

10.3.3.16 Tristan

Tristan at KEK collided 2×2 bunches in four interaction points and was designed to search for high mass resonances [106]. The injector was a 2.5 GeV linac and a 377 m accumulator ring ramped to the 8 GeV injection energy. Tristan was the first large accelerator to use extensive superconducting RF technology.

10.3.3.17 SLC

The SLAC Linear Collider SLC was the first (and so far only) linear collider constructed. It was built to precisely measure the properties of the Z^0 meson. It was the first to measure the width of the Z^0 indicating only three families of light quarks and neutrinos. It also provided a precise indirect constraint on the Higgs mass [107]. The SLC collided single e^+ and e^- bunches at 120 Hz with 80% longitudinal e^- polarization at the IP coming from a polarized strained GaAs photo-gun. Other accelerator advances include BNS emittance damping in the linac, reduced emittances from e^- and e^+ damping rings, pulse-by-pulse IP position feedback, and a positron source and target with one-to-one e^- in to e^+ out conversion rate. About 10^{13} positrons were made and collided per second.

10.3.3.18 BEPC

BEPC at IHEP was built to produce tau and charm particle physics [108]. BEPC was a single ring collider with two collision points reaching 2.5 GeV per beam. The single bunch current was up to 22 mA and 140 mA in multi-bunch mode. Injection was from a full energy linac with two transport lines. A mini-beta optics at the IP using permanent magnet quadrupoles (0.5 m long) was installed reaching a vertical beta of 8.5 cm. Higher RF voltage was used to reduce the bunch length to match. BEPC measured precisely the tau lepton mass to 0.2 MeV out of a mass of 1777 MeV.

10.3.3.19 LEP

LEP at CERN has four collision points and is the largest (27 km) and highest energy (104.5 GeV per beam) collider built to date [109]. The particle physics completed at LEP included precise measurements of Z and W bosons, determination of the number of light neutrinos to be three, and exclusion of the Higgs mass below 114 GeV. To reduce power usage five cell RF cavities with attached spherical copper storage cavities were used to reach about 80 GeV. To reach 104.5 GeV additional superconducting RF cavities were added incorporating sputtered Nb on Cu surfaces. The bending dipoles in LEP needed only a low field so the steel laminations were spaced by concrete filler material. At the Z resonance the luminosity was limited by the beam-beam effect. At higher energies the luminosity limit was the available RF power. A Pretzel orbit scheme (8 and later 12 bunches) was used to increase the luminosity. At high energy, a low emittance optics was implemented. A beam-beam tune shift of about 0.083 as reached at 98 GeV.

10.3.3.20 DAFNE

DAFNE at Frascati was built to make precision measurements of K meson physics. Injection is made with a full energy linac and damping ring [110]. DAFNE operates with damping wigglers to decrease the emittances and shorten the damping time. A crab waist scheme was installed in the IP for increased luminosity and for tests of the SuperB IP concept [111]. Rolled permanent-magnet IP quadrupoles were used to compensate the detector solenoidal field.

10.3.3.21 PEP-II

PEP-II at SLAC was an asymmetric collider with two rings made to measure the properties of the b quark sector, the CP violation in the $B\overline{B}$ system, and to confirm the CKM matrix [112]. Injection was made at full energy at up to 30 Hz from the SLAC 3 km linac with the SLC e^- and e^+ damping rings and either e^- or e^+ accelerated on any given linac pulse. Accelerator advances at PEP-II include head-on asymmetric collisions at one IR, large bore permanent-magnet IP dipole and quadrupoles, local beta beats to correct chromaticity in the IP, fast IP position feedback, and bunch-by-bunch transverse and longitudinal feedbacks. The nearest final focus quadrupoles were inside the detector leaving a free space to the IP of about 0.5 m on each side. PEP-II holds the world's record of stored positrons at

3.2 A and for electrons at 2.1 A. PEP-II was the first collider to allow top-up injection (only a few Hz were needed) to keep the beam currents and luminosity constant, all with full continuous data collection by the particle physics detector.

10.3.3.22 KEKB

KEKB at KEK was an asymmetric two ring collider made to measure CP violation in the $B\overline{B}$ system and to confirm the CKM matrix [113]. Injection was made at full energy with the KEK J-shaped linac with either e^- or e^+ injected at 50 Hz with a several minute switch time between modes. KEKB had a lattice with a $5\pi/2$ phase advance to reduce the emittance well below that of a FODO cell. It had a crossing angle IP, used ARES RF copper cavities with an attached energy storage cell, and superconducting RF cavities to suppress longitudinal modes. Transverse bunch-by-bunch feedbacks were used to suppress instabilities. KEKB was the first collider to use superconducting crab cavities to reduce the effects of crossing angle collisions. KEKB also had top-up injection of both beams. KEKB holds the world's record for highest luminosity at 2.1×10^{34} cm^{-2} s^{-1}.

10.3.3.23 BEPC-II

BEPC-II at IHEP was built to provide tau and charm particle physics as a factory and also have synchrotron radiation production [114, 115]. BEPC-II is a double ring collider with one collision point reaching up to 2.1 GeV per beam with 93 bunches. The RF system has two superconducting single cavities at 500 MHz. The luminosity was recently limited by longitudinal instability from HOMs but was cured by a bunch-by-bunch longitudinal feedback system. BEPC-II has delivered several billion J/Psi events to the BES-III detector.

10.3.3.24 VEPP-2000

VEPP-2000 at BINP is a recent e+e− collider and has a single ring with two detectors and twofold symmetry [116, 117]. The particle physics being done at VEPP-2000 is tau and psi measurements and two gamma-physics. Injection comes from a 900 MeV booster synchrotron with one beam at a time. A round beam concept was applied in the ring design where a particle's angular momentum ($M = xz' - zx' =$ constant) is conserved yielding an enhancement of the beam's dynamic stability even with nonlinear effects of the beam-beam force included. This scheme requires equal emittances, equal small fractional tunes, equal betas at the IP, and no betatron coupling in the collider arcs. In practice with collisions for the detectors, only small adjustments in the tunes are needed to arrive at good luminosity conditions at the beginning of a fill. Observations show that a beam-beam parameter of 0.13 has been achieved and that round beams give a solid luminosity enhancement.

10.3.3.25 SUPERKEKB

SuperKEKB at KEK is an asymmetric two ring collider upgraded from KEKB and made to measure CP violation in the B$\overline{\text{B}}$ system to extend the understanding of the CKM matrix [118]. This collider has a 7.0 GeV High Energy Ring HER for e− and a 4.0 GeV Low Energy Ring LER for e+. The luminosity goal is 8×10^{35} cm^{-2} s^{-1}. The design beam currents will be doubled from KEK to 2.6 A on 3.6 A. The nano-beam scheme for increased luminosity decreases the overlapping length of colliding particles to about 0.25 mm with a 41.4 mrad half crossing angle [111]. The interaction region beta functions will be about 30/0.3 mm (h/v) with a bunch length of about 5.5 mm. The two rings without the interaction region were commissioning in 2016 where 1.01 A of e− were successfully stored in the HER and 0.87 A of e+ stored in the LER. The installation of the interaction region will be completed in early 2018 when luminosity commissioning will begin.

10.4 Asymmetric B-Factories

K. Oide

10.4.1 Physics Motivation

The idea of asymmetric B-factories was first introduced by P. Oddone in 1987 [119] to collide e^+e^- beams with different energies to measure the CP-asymmetry between the decay of B_0 and $\overline{B}_0$ mesons. The asymmetry of the energies of two beams boosts the generated particles longitudinally, then the difference of the decay time can be measured by the difference of the vertices, which was expected to be in about an order of 100 μm. The center-of-mass energy of the collision is set to the $\Upsilon(4S)$ resonance at 10.58 GeV. A very high luminosity around 10^{34} cm^{-2}s^{-1} was required, which was more than 100 times higher than what had been achieved in colliders by that time.

10.4.2 Double Ring Collider

There may be several ways to realize the asymmetric collision. One way is to build a linear–linear or a ring–linear collider. Such a linear machine needs a very strong focusing $\beta_y^* \sim 100\ \mu$m to achieve the luminosity, then the bunch length must be as short as β_y^* to avoid the hour-glass effect. The bunch length itself can be obtained by bunch compressors, but the associated energy spread degrades the effective luminosity, since the width of the resonance $\Upsilon(4S)$ is only 20 MeV (2×10^{-4}). A huge damping ring would be necessary to realize such a short bunch length and a small energy spread simultaneously. Thus linear collision schemes seemed difficult.

As for the double-ring collision, a question is the sizes of the rings. If one can collide a large high energy ring (HER), for instance at 25 GeV, with a small low energy ring (LER) at 1.2 GeV, the total cost will be saved, assuming an existing tunnel for such a high energy ring. It was pointed out [120] that the collision of rings with different circumferences has somewhat fundamental difficulty: if two rings have the ratio of circumferences $m : n (m > n)$, the periodicity of the system becomes very long, *i.e.*, LCM$(m, n)/m$ times the revolution period of the larger ring. Then both rings will have dense resonance lines in the tune space which reduces the operable area, especially with a certain amount of the beam-beam tune shift. Thus collision of rings with different circumferences seemed difficult. Therefore only the double ring collider scheme with equal cicumferences remained.

Two projects of the asymmetric B-factories, PEP–II [121] at SLAC and KEKB [122] at KEK, were approved and started the construction by 1994. Both projects utilized the components and facilities of their previous generation colliders, PEP and TRISTAN, and built the BaBar and Belle detectors, respectively. Both machines started the collision experiments in 1999 and stopped the operation in April 2008 (PEP–II) and June 2010 (KEKB). Table 10.12 lists the main machine parameters corresponding to their best records [123, 124]. Both colliders achieved higher performance than their designs, and experimentally verified the Kobayashi–Maskawa model to bring the 2008 Nobel Prize in Physics.

10.4.3 Luminosity

The luminosity $\mathscr{L}$ of an asymmetric ring collider can be expressed by the following expression:

$$\mathcal{L} = \frac{\gamma_\pm}{2er_e}\left(1 + \frac{\sigma_y^*}{\sigma_x^*}\right)\left(\frac{I\xi_y}{\beta_y^*}\right)_\pm \left(\frac{R_\mathcal{L}}{R_y}\right) , \tag{10.1}$$

where γ, e, r_e, $\sigma_{x,y}^*$, I, $\beta_{x,y}^*$ are the Lorentz factor, electron charge, classical electron radius, beam sizes at the interaction point (IP), stored beam current in the ring, and the β-function at the IP, respectively. The suffix $\pm$ denotes each beam. The expression (10.1) is obtained from the beam-beam tune-shift parameter

$$\xi_{\pm x,y} = \frac{r_e}{2\pi\gamma_\pm} \frac{N_\mp \beta_{\pm x,y}^*}{\sigma_{x,y}^* \left(\sigma_x^* + \sigma_y^*\right)} R_{x,y} \tag{10.2}$$

Table 10.12 Progress of machine parameters of the PEP–II and KEKB B-factories

	PEP–II		KEKB (no crab)		KEKB (crab)	
	8/16/2006		11/15/2006		6/17/2009	
Date	LER	HER	LER	HER	LER	HER
Circumference [m]	2200		3016			
Beam energy [GeV]	3.1	9.0	3.5	8.0	3.5	8.0
Effective crossing angle [mrad]	0		22		0 (crab)	
Beam current [A]	2.90	1.88	1.65	1.33	1.64	1.19
Bunches	1722		1389		1584	
Bunch current [mA]	4.02	1.09	1.19	0.96	1.03	0.71
Bunch spacing [m]	1.2		1.8–2.4		1.8	
Horizontal emittance ε_x [nm]	30	50	18	24	18	24
RF frequency [MHz]	476		509			
Bunch length σ_z [mm]	10	10	8	6	8	6
β_x^* [cm]	30	30	59	56	120	120
β_y^* [cm]	0.9	1.1	0.65	0.59	0.59	0.59
Horizontal size @ IP [μm]	95	158	103	116	147	170
Vertical size @ IP [μm]	4.7	4.7	1.9	1.9	0.94	0.94
Beam-beam ξ_x	0.072	0.064	0.115	0.075	0.125	0.100
Beam-beam ξ_y	0.064	0.053	0.104	0.058	0.130	0.090
Luminosity [$\text{nb}^{-1}\text{s}^{-1}$]	12.1		17.6		21.1	
Integrated luminosity/day [pb^{-1}]	858		1260		1479	
Integrated luminosity/7 days [fb^{-1}]	5.41		7.82		8.43	
Integrated luminosity/30 days [fb^{-1}]	19.8		30.2		23.0	
Total integrated luminosity [fb^{-1}]	557		1040			

The left, center, right correspond to the highest performance of PEP–II, KEKB (no crab) and KEKB (crab), respectively. The integrated luminosities are the delivered numbers for PEP–II, and recorded for KEKB. 1 $\text{nb}^{-1} = 10^{33}\ \text{cm}^{-2}\text{s}^{-1}$

and the definition of luminosity

$$\mathcal{L} = \frac{N_+ N_- f}{4\pi\sigma_x^*\sigma_y^*} R_{\mathcal{L}} \tag{10.3}$$

where N and f are the number of particles per bunch and the collision frequency ($I = Nef$), respectively, and we have assumed the beam sizes are common in two beams. The factors $R_{\mathcal{L},x,y}$ are the geometric reduction factors due to the hour-glass effect and the crossing angle.

While a round-beam scheme may have a merit of a factor of 2 on the luminosity according to Eq. (10.1), a flat beam scheme has been chosen in most e^+e^- colliders, as the round-beam focusing in both planes is more difficult for an extremely small β_*. For a flat beam, $\sigma_x^* \gg \sigma_y^*$, the luminosity is written as

$$\mathcal{L} \approx \frac{1}{2er_e}\left(\frac{\gamma I \xi_y}{\beta_y^*}\right)_{\pm}\left(\frac{R_{\mathcal{L}}}{R_y}\right) . \tag{10.4}$$

Then if there is no reason to differentiate ξ_y and β_y^* in the two rings,

$$\gamma_+ I_+ = \gamma_- I_- \tag{10.5}$$

is resulted. As the ratio of beam energies gets larger, the boost at the collision becomes larger, but the low energy ring must store higher beam current. Thus the energy ratio was a compromise between the physics merit and the accelerator difficulty. PEP–II chose 3.1 GeV and 9 GeV for positrons and electrons, while KEKB chose 3.5 GeV and 8 GeV. A larger ratio was more favored at PEP–II as it needs a magnetic separation of two beams at the IP as described later. The flavor of beams, the LER for positrons, was uniquely chosen at KEKB, where the positron acceleration for the HER was very difficult.

The actual operation of these machines, the condition (10.5) was not kept strictly, as shown in Table 10.12. One reason was that the natural size of each beam was not equal, for instance, the LER positron beam was relatively easy to be blown up due to the electron clouds at high current. Then there was a certain limit on the positron beam current and the HER current was increased beyond Eq. (10.5). This tendency was stronger in KEKB than PEP–II, as the former had stronger electron cloud effects than the latter as described later.

10.4.4 Crossing Angle

One of the design choices is the beam separation scheme near the IP. A crossing angle is a natural and easy solution of the separation, but the question was the experience at DORIS [125]. KEKB applied a horizontal crossing angle $2\theta_x = 22$ mrad, relying on simulation of the beam-beam effect. The corresponding Piwinski angle ($\equiv \theta_x \sigma_z / \sigma_x^*$) was 0.86. Their conclusion at the design stage was that the effect of the crossing angle on the beam-beam interaction would not be harmful up to their design beam-beam parameter 0.05, if the operating betatron tunes were carefully chosen. Their choice was right and verified the vertical beam-beam parameter of 0.06 in their luminosity marching. Crossing angles were also applied at CESR and DAΦNE colliders successfully in parallel with the KEKB operation. KEKB even prepared a crab-crossing scheme [126, 127] for the backup of the crossing angle. The ratio of the geometric reduction factors R_L/R_y in Eq. (10.1) does not drastically decrease for a large crossing angle as shown in [122].

PEP–II was much more nervous on the crossing angle, then they installed a magnetic separation scheme near the IP with permanent dipole magnets [128]. This scheme also worked, but their design around the IP had to be more complicated than with a crossing angle, and gave some limitations on the performance such as the detector background due to radiative Bhabha events [129], which was much less significant in Belle. As the space at the IP was limited, they could not install a compensation system for the detector solenoid field, which might have degraded the beam-optical performance. Another issue of the magnetic separation was the non-negligible effect due to the parasitic collision [130], which was never observed at KEKB.

10.4.5 Storing High Current

As described above, the luminosity is proportional to the stored current. To achieve the luminosity as high as $10^{34}\text{cm}^{-2}\text{s}^{-1}$, a stored current of near 3 A was required, which was one order higher than any high energy electron storage rings at that time. The first fundamental difficulty is to ensure the longitudinal stability of the beam.

The beam loading of the accelerating cavity is huge: a normal conducting cavity at the RF frequency $f_{\text{RF}} = 500$ MHz for the B-factories has a shunt impedance $R_s \approx 1.7\,\text{M}\Omega$. If the cavity is tuned at the harmonics, the 3 A beam generates 5.1 MV decelerating voltage at the cavity, which is even higher than the accelerating voltage V_c of the cavity, typically 0.5 MV. Thus the detuning of the cavity is necessary and the optimal amount of the detuning frequency is given by

$$\Delta f = -\frac{I \sin\phi_s}{2V_c}\frac{R_s}{Q} f_{\text{rf}} = -\frac{P_b \tan\phi_s}{4\pi U} , \qquad (10.6)$$

where ϕ_s, Q, P_b, U are the synchronous phase, the Q-value, the beam power, and the stored energy of the cavity, respectively. If the magnitude of the detuning frequency becomes higher than or comparable to the revolution frequency, the cavity impedance hits the side bands of synchrotron motion to excite strong longitudinal coupled-bunch instabilities.

This issue of the beam-loading instability was solved in two B-factories in different ways. KEKB developed two types cavities with large stored energy, as Eq. (10.6) is inversely proportional to the stored energy. Both ARES [131] and superconducting [132] cavities could store electromagnetic energy 10 times larger than a conventional cavity. Then together with the HOM damping mechanism of them, the RF system of KEKB did not induce any beam instability up to the design current without a help of bunch-by-much feedback system. On the other hand, PEP–II took a different strategy to develop a sophisticated feedback system to reduce the effective impedance seen by the beam [133]. PEP–II applied a direct RF feedback system with newly developed sideband klystrons combing a longitudinal bunch-by-bunch feedback [134]. Both KEKB and PEP–II systems basically worked as expected nearly up to or even beyond their design currents.

Storing high currents caused a number of issues on the beam pipes, bellows, collimators, and even on the detectors. Direct hit of the beam of an ampere caused by beam instability or anything else easily melted down such components. The wakes at transitions resulted in discharge and heating to drive the catastrophe. A number of models have been developed and tried for the collimators, bellows, and HOM absorbers. Also machine protection system, loss monitors, and beam abort system had to evolve as the stored current increased.

10.4.6 Electron Cloud

Electron cloud was the one of the toughest issues for the asymmetric B-factories, specifically on the accumulation of the positron beam. The electron cloud had been known as a possible cause of beam instability in positive-charged beams since a long time ago such as the ISR era. Its observation [135] had been made at the Photon Factory (PF) of KEK and a theoretical explanation [136] had been done well before the start of the B-factories. What was new at the B-factories was the single-bunch instability induced by electron clouds [137]. The previous instability observed at the PF had been interpreted as a coupled-bunch instability, which was supposed to be cured by a bunch-by-bunch feedback. Thus at least KEKB was not well prepared for the single-bunch phenomena which have much higher frequency than the available feedback. Actually possibility of such a single-bunch effect had been suggested [138] before the construction of the B-factories, it had not been, however, well recognized. The single-bunch effect was experimentally confirmed at KEKB [139] as well as at CesrTA.

The electron cloud blowed up the vertical beam size drastically, and the threshold beam current was 0.4 A with four-bucket (2.4 m) bunch spacing at KEKB. The electron cloud appeared more severely in KEKB than in PEP–II, as the former had a round Cu beam pipe while the latter an Al antechamber with TiN coating. Thus the initial startup of the luminosity at KEKB was slower than PEP–II.

By applying weak magnetic field at the beam pipe, the electron cloud was removed at least in the free space. Either permanent magnets or solenoids were installed at KEKB and PEP–II to cover almost all straight sections and inside of some magnets such as quadrupoles and weak dipoles by 2004. The mitigation worked as expected and the blowup became unnoticeable at least for three-bucket spacing in the case of KEKB [140]. Beside the magnetic field, various mitigation techniques have been developed and tested at the B-factories, against the formation of the electron cloud, including antechambers [141], TiN or Diamond-like carbon coatings, grooved surface pipes [142], and clearing electrodes [143]. Those techniques will be effective for future super B-factories and damping rings of linear colliders. Also several measurements of the cloud density have been carried out.

Although the density of the electron cloud became below the instability threshold by magnetic field, the betatron tune shift due to the cloud still remained in the LER at KEKB to make the tune variation along the bunch train. A mitigation for the tune variation was making use of pulsed quadrupoles as done at KEKB [144].

10.4.7 Beam Optics

The luminosity of a ring collider is inversely proportional to the vertical β-function at the IP as shown in Eq. (10.4). The B-factories have used the smallest β_y^* as a ring collider so far. Generally speaking, a small β_y^* means higher chromaticity and higher nonlinearity arisen from sextupoles for the chromaticity correction. Thus the design of the ring lattice needs special care to ensure the dynamic aperture. One technique applied for KEKB was non-interleaved sextupole pairs separated by a $-I$ transformation that cancel the geometric nonlinearity of the sextupoles up to the second order [145]. Although the idea was very old but the reason why the application to a real ring had to wait until the B-factories was probably the necessary computer power to optimize the sextupole setting, as it requires a large number of sextupole families to extend the momentum acceptance. For instance, KEKB had 54 families of sextupole pairs. The relative betatron phase advance between the pairs became adequate by using the 2.5π cell structure in the case of KEKB arc section [146].

Another technique to enlarge the dynamic aperture was to place a special chromaticity correction section near the IP. The beam optics became somewhat similar to that for linear colliders in this case. KEKB designed such a section for the vertical correction, while PEP–II horizontal for their LERs.

These schemes worked as expected for the B-factories, and expected to work for future super B-factories and light sources. Once the chromaticity correction is solved, the next source of the nonlinearity is the fringe field of the final quadrupole and the geometric nonlinearity at the IP [147], which may be mitigated by additional octupoles placed at the final quadrupoles.

The x-y coupling and the residual vertical dispersion all over the ring were one of the keys to achieve a high luminosity by reducing the vertical emittance. Various techniques have been applied for such optics measurements and corrections [148–150]. A counter solenoid to the detector solenoid was also effective to reduce the coupling source in the case of KEKB. This was also important in the case of crab crossing where the luminosity performance was sensitive to the chromatic x-y coupling as described later.

10.4.8 Beam Diagnostics and Control

A number of beam diagnostic methods were developed and applied for the B-factories:

- Beam position monitors (BPMs) with a resolution better than 1 μm in the average mode. In some cases turn-by-turn or bunch-by-bunch electronics were equipped [151]. In the case of KEKB, the gain imbalance of the electrodes through the electronics was calibrated using a beam-mapping technique [152]. The design of the electrodes and the electronics were carefully done for the high-current operation.

- Beam-based alignment of BPMs was regularly carried out. Displacements of BPMs near sextupoles, caused by heating from the stored current, were monitored at KEKB [153].
- Bunch-by-bunch feedback systems were installed both in PEP–II and KEKB. Only PEP–II had the longitudinal system to suppress the beam-loading instability as described above. A collaboration including the DAΦNE team has been developed on the system for the present and future applications [154].
- Betatron tune monitor: controlling the betatron tune was extremely important to maximize the luminosity. The basic idea at KEKB was to monitor the tunes of pilot bunches in each ring that did not collide to the other beam. Tune feedback with these bunches was also applied to control them within an accuracy of $\Delta\nu \approx 10^{-4}$.
- Synchrotron radiation beam profile monitors. For the visible light, an interferometer was used especially for the vertical size measurement [155]. Special gated cameras were also used to observe the beam size of individual bunch, esp. to diagnose the electron-cloud effects [156].
- Beam loss monitors and beam abort system: both machines were very nervous to protect the machine against accidental beam losses caused by instabilities, RF trips, wrong injection, or whatever. The most sensitive and expensive loss monitor was the BaBar and Belle detectors, which generated beam abort signals if necessary. A number of beam loss monitors such as ionization chambers and PiN diodes were distributed around the ring, esp. near the collimators. The beam abort system consists of an abort kicker and a beam damp. The abort kicker had a rise time of 0.5 μs in the case of KEKB.
- The injectors had developed their own diagnostics including BPMs, wire scanners, streak cameras, etc.
- All accelerator components were controlled by computer control systems either by EPICS at KEKB [157] or a legacy system at PEP–II. An online modeling such as SAD for KEKB [149] was also important to achieve the luminosity.

10.4.9 Collision Tuning

Starting up the colliders after a period of long shut down, the following procedures were necessary to recover the luminosity:

- Global coupling/dispersion/β-function correction all over the ring. The global orbit was then locked to the "golden" orbit that was resulted by the optics correction.
- Locking the betatron tunes of the pilot bunches.
- The beam steering at the IP looking at the beam-beam deflection.

- In the case of the crossing angle at KEKB, the horizontal offset at the IP was controlled by looking at the vertical beam size measured by the interferometer [158].
- Tuning of the local coupling and dispersion at the IP by making offsets of orbits at sextupoles near the IP [149].
- Dithering technique was used at PEP–II to maximize the luminosity against the beam offsets [159].
- Skew sextupoles were introduced at KEKB to correct the local chromatic x-y coupling terms at the IP [124].

The horizontal tunes were chosen as close to a half integer as possible, to maximize the luminosity using the dynamic-β effect and expecting the reduction of the degree-of-freedom of the beam-beam interaction [160]. In the case of KEKB, the LER and the HER was operated at $\nu_x \approx 0.506$ and $\nu_x \approx 0.510$, respectively. Both the optics correction and the tune feedback were necessary to maintain a collision near the stop band.

10.4.10 Injector

The electron-positron injector must provide enough number of charges to the collider rings. PEP–II could fully utilized the injection system for SLC, which had more than enough performance for PEP–II, in the intensity, repetition, and the emittance, especially with the damping rings for both beams. On the other hand, the injector for KEKB was upgraded from that for TRISTAN, having only the minimum performance to satisfy the requirements of the injection to KEKB as shown in Table 10.13. In early days, it was thought that the performances of the two machines would be eventually limited by the performance of each injector. Actually such a situation did not happen. The key was the top-up operation applied for the both machines since 2004. Then the necessary strength of the injected beam became much smaller than the maximum performance even at KEKB [161]. Both machines were the first to have utilized the top-up for high-energy colliders, even earlier than the most of light sources. The 2-bunch per pulse acceleration and installing a C-band section in the linac [162] also contributed to make the gap between PEP–II

Table 10.13 Comparison of positron injection

	KEKB		PEP–II
	1999	2010	
Production energy [GeV]	4		30
Particles per pulse [10^{10}]	0.4	1.0	2
Repetition rate [Hz]	50		≤120
Invariant emittances H/V [μm]	~3000/3000		3/0.3
e^+/e^- switching time [s]	300	0.02	0 (simultaneous)

and KEKB smaller. KEKB has solved the conflict with the injection to the light sources by introducing a pulse-to-pulse switching of the linac.

10.4.11 *Crab Crossing*

KEKB operated with crab crossing from 2007 through 2010 using superconducting crab cavities [163] installed one cavity per ring [164]. KEKB had already achieved a luminosity $1.76 \times 10^{34}\ \mathrm{cm}^{-2}\mathrm{s}^{-1}$ by then with the crossing angle, so the intension of the crab crossing was not simply the backup. Simulations of beam-beam effect indicated that a head-on or crab collision could increase the beam-beam parameter ξ_y even higher than 0.15 combining with a horizontal betatron tune close to a half integer [165]. Thus the hope was to experimentally verify the possibility of such a high beam-beam parameter, considering super B-factories.

The crab cavities were successfully installed and operated for more than 3 years. The luminosity was actually increased as shown in Table 10.12. The resulted beam-beam parameter was 0.09, which was indeed higher than the value with crossing angle, but much less than the simulation (0.15). It was not easy to single out the cause, but there were indications that remaining higher-order terms of the beam optics at the IP degraded the luminosity and ξ_y. One example was the chromatic x-y coupling terms at the IP [166]. By installing skew sextupoles in the arc as the tuning knobs, the luminosity was improved by up to 10% [124]. Then a speculation was that any higher order terms at the IP could degrade the performance. It was not easy to estimate how many terms were relevant and how to correct them, as there were almost no direct beam diagnostics at the IP except the luminosity. What was verified at KEKB was that the crab crossing itself should work for any colliders up to $\xi_{x,y} \lesssim 0.1$, including the Large Hadron Collider (LHC).

10.4.12 *SuperKEKB*

Although the crab crossing scheme tested at KEKB achieved then-highest luminosity, it also showed the difficulty to realize the very high beam-beam parameter $\gg$0.1. Thus the direction toward a higher luminosity asymmetric collider must change. An alternative idea, *Nano-beam scheme*, developed by P. Raimondi [167] saved the next generation of KEKB. The idea by Raimondi consists of:

- A large crossing angle, in terms of *Piwinski angle* $\theta_x\sigma_z/\sigma_x^* \gg 1$, where θ_x is the half horizontal crossing angle.
- A very short $\beta_y^* \sim \sigma_x^*/\theta_x \ll \sigma_z$.
- Small horizontal/vertical emittances.

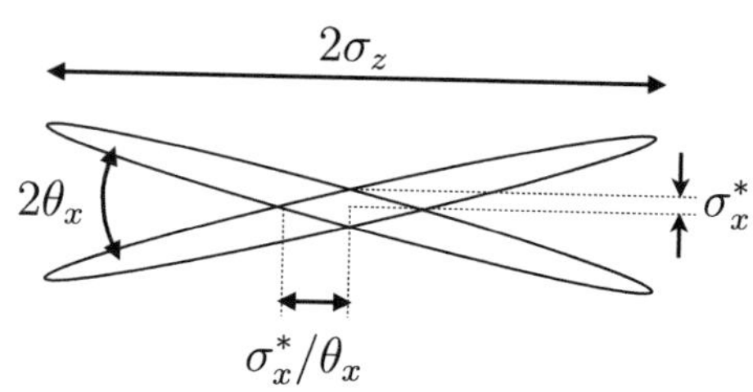

Fig. 10.13 Beam crossing in the nano-beam scheme

The formula for the luminosity, Eq. (10.1) is still valid in the nano-beam scheme. The major gain of the luminosity comes from a very short β_y^*. Unlike a head-on scheme, the bunches intersect one another only within σ_x^*/θ_x around the IP due to the large Piwinski angle as shown in Fig. 10.13.

As the length of the intersecting region is $\sigma_x^*/\theta_x \ll \sigma_z$, the condition to avoid an hour-glass effect becomes $\beta_y^* \lesssim \sigma_x^*/\theta_x \ll \sigma_z$. In the case of SuperKEKB, it is possible to choose $\beta_y^* \sim 0.3$ mm, which gives $\times 20$ gain compared to KEKB. Then the bunch length σ_z is not necessary to be very short, thus it is possible to avoid unfavorable effects such as the coherent synchrotron radiation. This scheme does not require a very high ξ_y, so $\xi_y \sim 0.09$, which has been achieved at KEKB, is assumed at SuperKEKB. The crossing angle itself can be larger than the previous machines, since it does not need crab crossing any more. As it does not require an operation very close to a half-integer tune, and also the horizontal beam-beam parameter is very small, the dynamic emittance effect is not an issue. The situation of the nano-beam crossing is similar to a collision with many micro-bunches which have a short bunch length $\sigma_x^*/\theta_x \ll \sigma_z$.

Actually Raimondi has proposed one more important idea, *crabbed waist scheme*, on top of the nano-beam scheme [167]. The crabbed waist aligns the vertical waist along the center line of the other beam, to reduce the dependence of the beam-beam effect on the horizontal displacement of a particle. This scheme should improve ξ_y by reducing the synchrotron-betatron coupling caused by the crossing angle. Although the crabbed waist scheme has merits on the collision itself, its realization needs further study. A simple way to introduce the crabbed waist is to install a pair of sextupoles in both sides of the IP. The nonlinearities of these sextupoles can be canceled by $-I$ or I transformation between the pair, but the unavoidable nonlinearities around the IP interfere the cancellation to reduce the dynamic aperture drastically in the case of SuperKEKB. Such nonlinearities include the fringe field of the final quadrupoles, geometric nonlinearities at the IP, and nonlinear fields in the quadrupoles and solenoids. As these terms increase for smaller β_y^*, the solution may be non-trivial and has not been found at least for SuperKEKB yet. Thus the crabbed waist scheme is not included in the base line design at SuperKEKB.

SuperKEKB started the beam operation in early 2016. The commissioning has been carried out in three phases: no collision (Phase 1, February–June 2016), collision without central vertex detector of Belle II (Phase 2, March–July 2018) [168], and collision with full Belle II (Phase 3, March 2019–). The design parameters [169]

Table 10.14 Parameters of SuperKEKB, design and typical values in phase 2, comparing to KEKB with crab crossing, where the effective crossing angle is zero

	SuperKEKB		SuperKEKB		KEKB (crab)		
	Design		June 2018		June 17, 2009		
Date	LER	HER	LER	HER	LER	HER	
Circumference	3016						m
Beam energy	4.0	7.0	4.0	7.0	3.5	8.0	GeV
Crossing angle	83				22/0 (crab)		mrad
Beam current	3.8	2.6	0.27	0.225	1.64	1.19	A
Bunches	2500		395		1584		
Bunch current	1.5	1.0	0.67	0.55	1.03	0.71	mA
Bunch spacing	1.2		~7.2		1.8		m
Hor. emittance	3.2	4.6	1.8	4.6	18	24	nm
RF frequency	509						MHz
Bunch length σ_z	6	5	4.6	5.3	8	6	mm
β_x^*	3.2	2.5	20	10	120	120	cm
β_y^*	0.27	0.3	3	3	5.9	5.9	mm
Hor. size @ IP	10	11	19	21	147	170	μm
Ver. size @ IP	0.048	0.048	0.56	0.56	0.94	0.94	μm
Piwinski Angle	24.9	18.9	10.0	10.5	(0.60)	(0.39)	
σ_x^*/θ_x	0.24	0.27	0.46	0.51	(13.3)	(15.4)	mm
Beam-beam ξ_x	0.0028	0.0012			0.125	0.100	
Beam-beam ξ_y	0.088	0.081	0.030	0.021	0.130	0.090	
Luminosity	800		2.3		21.1		/nb/s

1/nb = 10^{33} cm^{-2}s^{-1}

and a typical performance of Phase 2 operation [170] of SuperKEKB are listed in Table 10.14.

The commissioning in Phase 2 has achieved several milestones of the project:

- Verified the collision with nano beam scheme with Piwinski angle ~10. Although the luminosity was still much less than the design or achieved at KEKB, no fundamental limitations are found. The highest luminosity 56/nb/s was recorded during Phase 2 [170]. An experimental verification of coherent beam-beam instability [171] also assures the understanding of the beam-beam effect.
- The upgrade of the injector several new components: RF gun, positron target, and positron damping ring was basically successful [172].
- The mitigation of e-cloud in the LER has well suppressed the blowup of the vertical beam size up to 1 A of the stored beam [173].

The key toward the design luminosity is higher stored current with smaller $\beta*$s. Both the stored currents and $\beta*$s are still far from the design. Table 10.14 shows that the achieved β_y^* is much longer than the length of the interaction area, σ_x^*/θ_x. It means that the merit of nano-beam scheme has not been obtained yet. One possible obstacle for higher current and smaller $\beta*$s is the robustness of superconducting final focus quadrupoles [174] against quenches due to beam losses. Quenches of those magnets have been seen in the Phase 2 commissioning both for the stored and injected beams.

10.5 Tevatron—HERA—LHC

Karl-Hubert Mess · Peter Schmüser

10.5.1 Three Steps in the Evolution of Superconducting Accelerator Magnets

The first particle accelerator approaching the TeV energy range was the proton-antiproton collider Tevatron at the Fermi National Accelerator Laboratory. Much pioneering work was done at Fermilab, and many of the successful design and construction principles of the Tevatron dipole and quadrupole magnets have been adopted at the electron-proton collider HERA and the Large Hadron Collider LHC (CERN). The superconducting coils used in these accelerators show great similarities. They are wound with high precision from a multistrand cable containing many thousand fine niobium-titanium filaments in a copper matrix. The coils have a helium-transparent insulation and are confined by nonmagnetic clamps (often called *collars*). The clamps are assembled from precision-stamped stainless-steel or aluminum-alloy laminations and serve three purposes: they define the exact coil geometry, they exert a large prestress on the coils to prevent conductor motion during excitation of the magnet, and they take up the huge magnetic forces at large fields. Interesting differences, however, have evolved in the layout of the iron yoke. The Tevatron magnets are "warm-iron" magnets with a yoke at room temperature while the HERA and LHC magnets feature a "cold-iron" yoke inside the liquid helium cryostat.

In the beginning of the HERA project two different design lines were followed. Our group at DESY constructed and built 6 m long prototype dipoles of the "warm-iron" type which were basically copies of the Tevatron dipoles. These magnets were made with tooling suitable for series production and performed very well, the design field of 5.2 T was exceeded and excellent field quality was achieved. In parallel, an industrial company (Brown Bovery in Mannheim) designed and built two "cold-iron" prototype dipoles following a concept developed at Brookhaven National Laboratory. Here the coil was surrounded with epoxy-fibreglass form pieces and then directly mounted in an iron yoke which served as the clamping structure. Also the BBC dipoles showed remarkable performance. They exceeded the design field by an ample margin, however the field quality became poor for fields above 4 T owing to saturation effects in the nearby iron yoke.[2] With the aim in mind of combining the virtues of both design lines, while avoiding the relative drawbacks, the idea was conceived to take the well-proven aluminum-collared coil of the first design line and put an iron yoke immediately around the collars. Our main motivation was magnet safety in case on a quench (breakdown of superconductivity).

Quench protection is an important task in any large superconducting (sc) magnet system. Accelerator magnets have slim coils with a small copper-to-superconductor ratio. They are definitely not cryostable (cryostability means that the copper in the cable is able to carry the full current). If a HERA dipole quenches it is mandatory to ramp down the current of 5000–6000 A with a time constant of less than a second to prevent overheating and possible destruction of the coil. The protection of a single magnet is straightforward: if a quench is detected the coil is connected to a dump resistor via a thyristor switch and the current decays exponentially $I(t) = I_0 \exp(-t/\tau)$. The stored magnetic energy is dissipated in the dump resistor.

In an accelerator this simple solution is not possible since the magnets in a long string are connected in series. In HERA this string is a 45° arc comprising 52 dipoles and 26 quadrupoles. Current leads from the sc coils to the room-temperature environment are only installed at the ends of the string but not at individual magnets. Owing to the large inductivity of the 54-magnet string a long decay time constant $\tau \approx 20$ s must be chosen in order to limit the induced voltage against ground potential to less than 1000 V. The quenched coil would burn up during such a slow current decay, hence an electric bypass must be provided. Here the great advantage of a cold iron yoke comes in: it is easy to provide such a bypass by mounting a superconducting cable, reinforced with a copper bus bar, in a groove at the outer rim of the yoke. The bypass conductor is connected to the main current conductor via a “cold” silicon diode inside the liquid helium cryostat.[3] The diode has a threshold voltage of about 1 V at liquid helium temperature, hence the current flow through the bypass is zero as long as the magnet coil is in the superconducting state (during particle accceleration the ramp speed must be chosen so low that the inductive voltage stays well below the 1 V threshold). However, a quenched coil develops rapidly a resistive voltage exceeding 1 V, and then the main current switches automatically over to the bypass. The Tevatron possesses an active quench protection system which is more complicated and shall not be discussed here.

Potentially dangerous are quenches at localized spots in the coil. If the normal zone does not propagate fast enough along the coil it may happen that the stored magnetic energy is dissipated in the vicinity of the quench origin leading to local overheating. To avoid this, so-called “quench-heaters” are installed which are electrically heated when a quench has been detected. Their task is to spread the normal zone over the entire coil. Figure 10.14 shows that a warm-iron magnet may

---

[2] It has found out later in the LHC and RHIC projects that the detrimental effects of iron saturation can be alleviated by punching a suitable hole pattern into the iron yoke laminations.

[3] The cold diode concept was invented and thoroughly investigated at BNL.

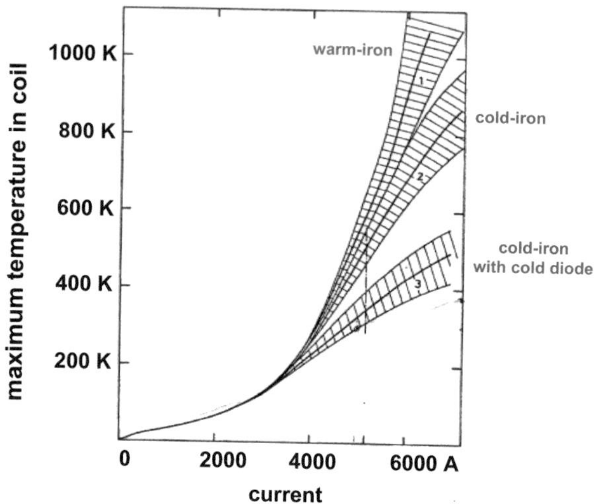

Fig. 10.14 Simulation of temperature rise in a dipole after a quench at a localized spot. Plotted is the maximum temperature in the coil as a function of coil current. It is assumed that the quench heaters fail and that the current is ramped down exponentially with a time constant of 17 s. Three magnets types are considered: the warm-iron HERA prototype dipole, the cold-iron BBC dipole without a cold diode, and the cold-iron BBC dipole with a cold diode. In the first two cases the bypass is provided by two curent leads from the superconducting cable in the liquid helium container to the room-temperature environment, a normal-conducting cable and a thyristor switch outside the cryostat. Adapted from a DESY Internal Note January 1984 by K.H. Mess and P. Schmüser

heat up to quite dangerous temperatures if the quench heaters fail, while the cold-iron magnet equipped with a cold diode and a superconducting bypass remains in a safe temperature regime.

The HERA-type magnet offers several more advantages. The cryostat can be a simple steel vessel which is easier to fabricate and less expensive than the very slim stainless-steel cryostat needed in the warm-iron magnet. The coil must be accurately centered with respect to the yoke to avoid field distortions and asymmetry forces. In the HERA design a precise centering is achieved by a tongue-groove combination in the collar and yoke laminations. In contrast to this, coil centering in a warm iron yoke requires many adjustable supports which constitute a considerable heat load on the cryogenic system.

The successful HERA magnet design has had a considerable impact on the layout of the LHC magnets.[4] The elegant twin-aperture magnet can be realized with this concept. The two counterrotating proton (or heavy ion) beams are guided and focused by two nearby sc coils of opposite polarity which are mounted in a common iron yoke. This yoke can be made rather slim since most the magnetic flux

[4]The evolution of the twin-aperture LHC magnets is described in an article by Lucio Rossi, CERN Courier October 2011.

of one coil returns through the aperture of the neigbouring coil and not through the iron. Another novel feature of the LHC magnets is the cooling by superfluid helium which extends the achievable field into the 9 T regime. A significant cost reduction is achieved by the fact that only one cryostat system is needed for both particle beams. The compactness of the magnets was very important for the installation in the tunnel, it would have been extremely demanding if not impossible to mount two separate proton storage rings in the narrow LEP tunnel.

The elegance and cost-effectiveness of LHC becomes obvious in comparison with the Superconducting Super Collider SSC project which was promoted in the USA in the 1980s. Here two counterrotating 20 TeV proton beams were planned to be accelerated and stored in two separate rings equipped with magnets of an improved Tevatron design. This "conservative" approach turned out far more costly than the innovative LHC solution. The SSC was cancelled in 1993 for budget reasons.

10.5.2 HERA Experience and the Design of Future Lepton-Hadron Colliders

The scientific results from HERA on quantum chromo-dynamics (QCD) are an essential pillar of our present understanding of the nature of strong interactions. These insights are of critical importance for the interpretation of the measurements of particle production processes at the Large Hadron Collider (LHC) at CERN. There are, however, very interesting phenomena and physics questions that were raised by the HERA experiments but remain unanswered as of today:

- How does QCD lead to the rich and complex structure we observe in nuclei?
- How are quarks and gluons and their spins distributed in nucleons and nuclei?
- Why do protons and neutrons have spin 1/2?
- How does the increase of gluon density with decreasing gluon fractional momentum x saturate?
- What is the distribution of partons that seed the new form of matter discovered at RHIC, the quark-gluon plasma?

The U.S. Nuclear physics community has formulated a physics program that addresses these questions in a White Paper [175]. This physics program is based on an Electron Ion Collider (EIC), which can provide collisions between polarized electrons and ions ranging from polarized proton to Uranium over a wide range of center of mass energies 29–140 GeV with a high luminosity in the order of 10^{34} cm^{-2} s^{-1}. There are two designs for an EIC under development in the U.S., one by Thomas Jefferson National Accelerator Facility, called JLEIC [176], and one by Brookhaven National Laboratory, called eRHIC [177]. Both designs are based on intersecting storage rings.

The experience gained at HERA with high luminosity lepton proton collisions, beam dynamics, various enabling technologies, the electron storage ring spin dynamics, and the control of collective effects have a tremendous impact on the design of these machines, in particular, the layout of the interaction regions.

HERA beam dynamics performance is used to benchmark computer codes needed to develop the EIC designs. HERA operational experience is extremely important for designing the complex interaction regions with unprecedented detector acceptance, assessing detector backgrounds induced by the beams and their mitigation. The design of the magnets which were developed for the HERA interaction regions have been carried over for the design of other colliders, such as BEPC-II, KEKB, ILC, LHeC, and eRHIC. Some beam dynamic phenomena observed in HERA directed the attention of the EIC designers to specific aspects of electron ion collisions.

The examples in the following sections illustrate the positive impact of the HERA design and operational experience on other colliders and in particular to the Electron Ion collider eRHIC.

10.5.2.1 Lepton-Hadron Beam-Beam Interactions

The concept of lepton proton collisions in HERA was to adjust the the strength of the beam-beam force for each beam such as this beam would collide with a beam of its own species. A typical beam-beam strength in terms of incoherent beam-beam tuneshift for $e^+ - e^-$ collisions was $x^e_{x,y} \simeq 0.03$. For Hadron, the strengths of beam beam interactions was typically below $x^p_{x,y} \simeq 0.01$. This succesful concept was used successfully in other colliders such as KEKB and is also adapted at the EIC design.

Occasionally, when the HERA beams were brought into collisions. an observation was made which is a concern for the electron ion collider. Under the influence of strong beam-beam interaction, the electron beam can develop coherent transverse oscillations. If this happens, the unstable motion of the electrons affects in turn the hadron beam via the beam-beam coupling. The hadrons will start to oscillate as well driven by the beam beam field of the unstable electron beam and the hadron beam emittance will then filament in phase space and become diluted that way. The effective beam size grows considerably and practically irreversely in the process which quickly renders the beam unusable.

This phenomenon was observed only occasionally at HERA and its rare occurrence made it impossible to perform systematic studies. Furthermore the impact on operations was very small and therefore, no resources have been invested to study the effect in detail.

The experience with lepton proton collisions in HERA, raised a concern that this effect would occur regularly in the EIC, because of significantly enhanced beam-beam parameters per collision in the EIC. This triggered realistic beam-beam interaction simulations for eRHIC. The beam-beam study was performed using the so-called strong, strong model, where each beam is described by a large number of super-particles and the thin lens beam-beam lenses are split into many beam-beam-lens slices. The individual particle's trajectories are impacted by the collective electro-magnetic forces induced by the particles of the opposite beam, respectively. This consitutes a fully dynamical model of the beam-beam interactions. The result of this study was that this effect indeed exists and would affect the performance of the EIC if not avoided by a careful choice of the machine parameters. The threshold in proton bunch intensity for this catastrophic instability is a factor of two above the chosen eRHIC design values of proton bunch intensity [177].

10.5.2.2 Beam-Gas Backgrounds of the Colliding Beam Detectors

The experience with beam induced backgrounds at HERA is very important for the design of the interaction regions and the collision detectors of the EIC or LHeC.

The combination of synchrotron radiation emitted by the electron beam in the separator- and the focusing fields when entering the interaction region and the corresponding strong sources of desorbed gases in the vicinity of the collision detectors caused initially strong hadron-beam-gas induced detector backgounds.

These backgrounds improved slowly by conditioning the surfaces of masks and absorbers by the presence of electron beam and scattered synchroton photons. Eventually a very high dynamical vacuum quality in the order of 1 nbar with full beam intensity was achieved. This required a installation of a large pumping speed and good vacuum conductance with integrated NEG pump in the IR quadrupole magnets and Titanium supplimation pumps. The cold beam pipe of the superconducting IR magnets (at a temperature of 80 K) acted as a cryopump and helped to improve the vacuum. But once the surface was saturated, the magnes needed to be warmed up and the accumulated gas gas molecules on the cold surfaces needed to be removed from the IR vacuum system using turbo pumps. Each warm-up and cooldown cycle marked a step in background improvement.

Another important lesson learned from HERA IR operations was that the vacuum system must be designed such that IR vacuum leaks are unlikely and that venting of the IR and detector vacuum for maintenance and repair purposes must be avoided by design.

The analysis of HERA backgrounds [178] shows the importance of extremely good vacuum quality in the beam vacuum chamber inside the central colliding beam detector. This makes high vacuum pumping speed and high vacuum conductance within the detector beam pipe mandatory. Techniques such as in situ bakeout and NEG coating of the detector beam chamber might be unavoidable to overcome this difficulty.

This experience is exploited in the design of the interaction regions of the EIC, in particular at eRHIC. First background simulations indicate that difficult initial background conditions as experienced in HERA can be avoided taking the lessons learned into account.

10.5.2.3 Hadron Beam Collimation

In lepton-hadron collisions with stong beam-beam interaction, the hadron beam will develop tails in the tranverse particle distribution. The controlled removal of these halo-particles is very important for low detector background and efficient data taking.

A sophisticated collimation system and collimator optimization scheme was implemented in HERA. It consisted of two stages of collimation, with one single primary collimator per oscillation plane and two secondary collimators per plane. A detailed and fully automated collimator adjustment procedure was developed which provided in general good background conditions routinely.

The HERA collimation system and its operational optimization techniques are being carried over to the eRHIC design version of the EIC.

Furthermore, high energy proton operation with the HERA-B [179] fixed target in HERA, revealed that the wire target could be integrated in the existing collimator system. It consitituted an excellent beam-edge spoiler target which significantly increased the efficiency of the two stage collimation system and led to a significant reduction of particle background in the detectors H1 and ZEUS.

Early simulations of halo particle backgrounds in HERA showed clearly, that the halo must be removed far away from the detectors and local shields or masks will only aggravate the background conditions for the colliding beam detectors [180].

Experience with HERA collimation system have been taken into account in the elaborated LHC collimation system and are an important input for the design of the collimation system of eRHIC.

10.5.2.4 Spin Polarization of the HERA Electron Beam

The good performance of the HERA electron spin in colliding beam operations at the North and the South interaction points with three pairs of spin rotators in the North, South, and East straight sections which provided a polarization of $\geq$50% [181] was vital for the HERMES experiment [182] and augmented the physics program of the H1 and ZEUS detectors [183]. This performance was achieved with uncompensated detector solenoids in the North (H1), South (ZEUS) interaction reagions. The spin tuning procedures developed for HERA which included the system of harmonic bumps, and spin-matching of the straigth sections between the rotator pairs, as well as the electron beam working point near the integer resonance were important factors in this success.

The HERA experience constitutes an important reference point for designing the EIC for high electron spin polarization. Spin physics is an important part of the EIC physics program. Collisions between highly polarized electron and Hadron beams with all combination of spin helicities enable an this physics program.

The spin matching techniques and the scheme of orbit optimization with harmonic bumps which contributed to the success of the HERA spin program, are being carried over to the eRHIC design to enable operation with highly polarized electron beams. As in HERA, an equilibrium polarization of at least 50% has been shown to be achievable in eRHIC at the highest electron beam operation energy of 18 GeV. This performance will enable to maintain a high level of average polarization of $\simeq$80% with the initial polization of the the electron beam of 85% if the stored electron bunches are replaced every 6 min on average [177].

10.5.2.5 Lessons Learned from HERA Dynamic Aperture

The dynamic aperture optimization in the HERA electron ring after the year 2000 was accomplished with only two sextupole families for the correction of chromaticity. Correction of higher order chromaticity which is due to the strong contribution of the interaction region quadrupole magnets to the chromaticity and the corresponding strong off-momentum distortion of the optical functions (beta-beat) is necessary to confine the beam tune foot print and to avoid destabilizing resonances.

The straight forward correction of off-momentum beta-beats with more families of sextupoles was avoided thereby avoiding large peak strengths of the sextupole magnets, which would have deteriorated the dynamic aperture. Instead, the horizontal and vertical betatron phase differences between the North and the South interaction points were chosen to be an odd integer of $\pi/2$. The off-momentum beta beat caused by the low beta quadrupoles of one IR was intrinsically canceled by the correponding beta beat of the other IR. This way, the nonlinear tuneshift with momentum was minimized.

The HERA chromatic correction scheme [184] also suppressed the generation of driving terms of the higher order nonlinear synchro-betatron resonances $Q_x + 3 \cot Q_s$=integer which otherwise would have affected the beam stability at the low tune values needed for high polarization operation.

Last but not least, the HERA chromatic correction scheme reduced the magnet currents of the sextupoles circuits which would have been required using the multi-family scheme. The HERA scheme is being carried over to eRHIC which is planned to be operated with two collidng beam detectors as well. The eRHIC chromatic correction scheme is an important constituent of the design which enables highest luminosity values.

10.5.2.6 HERA IR Magnet Design

In order to achieve hightest luminosity, the HERA interaction regions have been re-designed to allow for very small beta functions at the IP. The change was implemented in the year 2000 and HERA operated with about 3.5 times the design luminosity in the years 2004–2007.

This upgrade included accelerator magnets that were placed inside the colliding beam detectors H1 and ZEUS, and in particular inside the solenoidal detector fields of up to 5 T. Obviously, no magnetic steel was allowable inside the solenoid fields of the detectors. Furthermore, the accelerator magnets should have minimum mass to minimize the distorions of the trajectories of scattered particles, maximizong the detector acceptance this way and avoiding introduction of strong scattering targets insode the detector. Thus superconducting technology without steel collaring needed to be used. The available space for these magnets was very limited, as the detectors already existed and could not be modified. The magnets needed to be combinded function to separate the lepton and the hadron beam and to start focusing the electron beam as early as possible. The aperture of the magnet needed to be very large to allow the synchrotron beam to pass through the entire detector and low-beta quadrupole section. Furthermore, the aperture needed to provide space for the envelopes of the separated beams.

To meet these requirements, a new type of superconducting magnet was developed by a collaboration of scientists from Brookhaven National Laboratory and DESY [185]. The supercnducting magnet coil was produced by developing the 2-D direct wind method used for the RHIC corrector magnets to 3-D geometry such that the superconducting wire could be attached directly to the surface of a cylinder. The magnet coil was wound directly on the beam pipe using a seven-strand NbTi superconducting cable and the coil was stabilized with glass-fiber tape, a technique carried over from the HERA multipole corrector coils.

High field quality of superconducting magnets requires high precision in the manufacturing of the coil. For the HERA interaction region magnets, the precision was achieved by combining the RHIC and HERA corrector coil technologies.

The coil was wound on a precision-machined cylinder which was coated with epoxy. A computer controlled stylus placed superconducting wire with high precision on the cylinder. It also transmitted ultra-sound waves to the wire, which by means of friction melted the epoxy surface of the cylinder thereby fixing the superconducting wire in its position. After winding the first layer of the coil, it was complemented with GF spacers and fixed with glass-fiber tape under high tension before the coil was vacuum impregnated with epoxy. The field of the first layer was measured and compared with the calculated field for perfect wire position. Discrepancies were corrected by modifying the definition of the second layer wire positions correspondingly. After high precision machining the new surface, the magnet was ready for the second layer of coil to be wound. This way, the magnet achieved very high field quality of one unit of 10^{-4} at a radius of 25 mm with the center set off by up to 20 mm.

The HERA IR magnets were operated with 4.5 K supercritical He. The performance in routine operation was excellent.

The technique of the very successful HERA low beta quadrupoles was further developed by BNL scientists and the technique has been successfully applied to the low beta quadrupoles of the Beijing Electron-Positron Collider BEPC-II [186], the KEK b-factory SuperKEKB [187] and designs for the interaction regions of the International Linear Collider [188] have bee produced and prototyped. The Fig. 10.15 shows a photograph taken during the winding of the low-beta quadruole for superKEKB in September 2013. This technology is now quite mature and is used for many of the advanced eRHIC interaction region magnets [177].

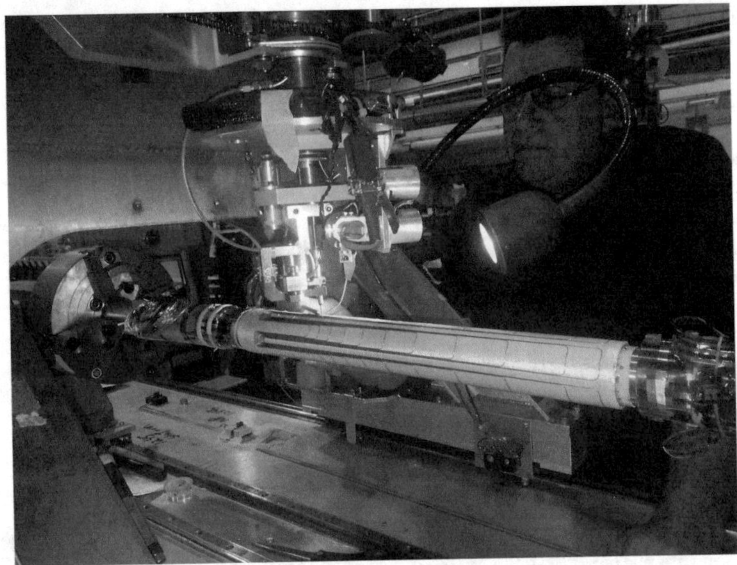

Fig. 10.15 Winding of the superKEKB coil in September 2013 (Courtesy Brett Parker, BNL)

10.5.2.7 Conclusion

The accelerator and technical solutions as well as the operational proceedures which enabled the successful operation of the HERA Lepton-Hadron collider are an important resource for the design of future colliders, in particular for the design of ring-based electron-ion colliders.

10.6 LHC Layout and Performance to Date

R. Bailey · J. Wenninger

10.6.1 Introduction

The Large Hadron Collider (LHC) [189] is a two-ring superconducting accelerator and collider installed in a 27 km underground tunnel at CERN, on the Swiss-French border near to Geneva. The tunnel was originally constructed for the Large Electron Positron collider (LEP), which operated from 1989 to 2000.

The LHC has two high luminosity experiments, ATLAS and CMS, designed for a peak luminosity of 10^{34} cm^{-2} s^{-1}, one dedicated ion experiment, ALICE, and one

Table 10.15 LHC basic beam parameters

Parameter	Unit	Injection	Collision
Beam data			
Proton energy	GeV	450	7000
Relativistic gamma γ		479.6	7461
Number of particles per bunch		1.15×10^{11}	
Number of bunches		2808	
Longitudinal emittance (4σ)	eVs	1.0	2.5[a]
Transverse normalized emittance	μm rad	3.5[b]	3.75
Circulating beam current	A	0.582	
Stored energy per beam	MJ	23.3	362
Peak luminosity related data			
RMS bunch length[c]	cm	11.24	7.55
RMS beam size at the IP1 and IP5[d]	μm	375.2	16.7
RMS beam size at the IP2 and IP8[e]	μm	279.6	70.9
Geometric luminosity reduction factor F[f]		–	0.836
Peak luminosity in IP1 and IP5	$\text{cm}^{-2}\ \text{s}^{-1}$	–	1.0×10^{34}
Peak luminosity per bunch in IP1 and IP5	$\text{cm}^{-2}\ \text{s}^{-1}$	–	3.5×10^{30}

[a]The base line machine operation assumes that the longitudinal emittance is deliberately blown up in the middle of the ramp in order to reduce the intra beam scattering growth rates
[b]The emittance at injection energy refers to the emittance delivered to the LHC by the SPS without any increase due to injection errors and optics mis-match. The RMS beam sizes at injection assume the nominal emittance value quoted for top energy (including emittance blowup due to injection oscillations and mismatch)
[c]Dimensions are given for Gaussian distributions. The real beam will not follow a Gaussian distribution but more realistic distributions do not allow analytic estimates for the IBS growth rates
[d]The RMS beam sizes in IP1 and IP5 assume a β-function of 0.55 m
[e]The RMS beam sizes in IP2 and IP8 assume a β-function of 10 m
[f]The geometric luminosity reduction factor depends on the total crossing angle at the IP. The quoted number assumes a total crossing angle of 285 μrad in IR1 and IR5

experiment for B-physics, LHCb. The main parameters required to reach the high luminosity for ATLAS and CMS are given in Table 10.15.

The high beam intensities required for a luminosity of $L = 10^{34}\ \text{cm}^{-2}\ \text{s}^{-1}$ exclude the use of anti-proton beams and a single common vacuum and magnet system for both circulating beams (as was done in the SPPbarS and the TEVATRON) and implies the use of two proton beams. To collide two beams of equally charged particles requires opposite magnet dipole fields in both beams. The LHC is therefore designed as a proton-proton collider with separate magnet fields and vacuum chambers in the main arcs and with common sections only at the insertion regions where the experimental detectors are located.

There is not enough room for two separate rings of magnets in the LEP tunnel. Therefore the LHC uses twin bore magnets which consist of two sets of coils and beam channels within the same mechanical structure and cryostat.

The peak beam energy in a storage ring depends on the integrated dipole field along the storage ring circumference. Aiming at peak beam energies of up to 7 TeV inside the existing LEP tunnel implies a peak dipole field of 8.33 T and the use of superconducting magnet technology.

This presented quite a challenge; to make the most profitable use of the existing tunnel and to obtain the highest possible bending strength by exploiting the well-proven technology based on Nb-Ti Rutherford cables. To meet this challenge and to find the cheapest solution compatible with the required performance needed a substantial R&D program on magnets and associated technology. This was carried out between 1988 and 2001 by CERN in close collaboration with other European laboratories and European industry.

10.6.2 Layout

The basic layout of the LHC follows the LEP tunnel geometry and is depicted in Fig. 10.16. The LHC has eight arcs and straight sections, with dispersion suppressors in between.

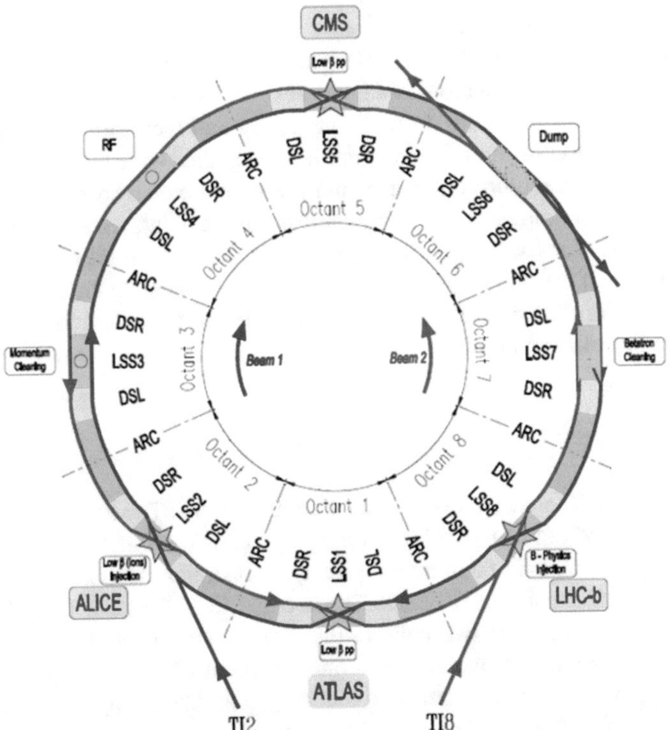

Fig. 10.16 Schematic layout of the LHC. Beam 1 (Blue) circulates clockwise. Beam 2 (Red) circulated counter-clockwise

10.6.2.1 The Straight Sections

Each straight section is approximately 528 m long and can serve as an experimental or utility insertion. The two high luminosity experimental insertions are located at diametrically opposite straight sections: the ATLAS experiment is located at point 1 and the CMS experiment at point 5. Two more experimental insertions, for ALICE and LHCb are located at point 2 and point 8 respectively. These insertion regions also contain the injection systems, for Beam 1 arriving from the SPS through the transfer line TI2, and for Beam 2 arriving from the SPS through the transfer line TI8. The injection kick occurs in the vertical plane with the two beams arriving at the LHC from below the LHC reference plane. The beams only cross from one magnet bore to the other at these four locations. The remaining four straight sections do not have beam crossings. Insertion 3 and 7 each contain two collimation systems, for momentum cleaning and betatron cleaning respectively. Insertion 4 contains two RF systems: one independent system for each LHC beam. The straight section at point 6 contains the beam dump insertion where the two beams are vertically extracted from the machine using a combination of horizontally deflecting fast-pulsed ('kicker') magnets and vertically-deflecting double steel septum magnets. Each beam features an independent abort system.

The two beams share an approximately 130 m long common beam pipe along the interaction regions (IR). The exact length is 126 m in IR2 and IR8 which feature superconducting separation dipole magnets next to the triplet assemblies and 140 m in IR1 and IR5 which feature normal conducting and therefore longer separation dipole magnets next to the triplet assemblies. Together with the large number of bunches (2808 for each proton beam), and a nominal bunch spacing of 25 ns, the long common beam pipe implies 34 parasitic collision points for each experimental insertion region (for four experimental IR's this implies a total of 136 unwanted collision points). Dedicated crossing angle orbit bumps separate the two LHC beams left and right from the central interaction point (IP) in order to avoid collisions at these parasitic collision points.

10.6.2.2 The Arcs

The arcs of the LHC are each made of 23 regular arc cells. The arc cells are 106.9 m long and are made out of two 53.45 m long half cells each of which contains one 5.355 m long cold mass (6.63 m long cryostat) short straight section (SSS) assembly and three 14.3 m long dipole magnets. The two apertures for Ring 1 and Ring 2 are separated by 194 mm. The two coils in the dipole magnets are powered in series and all dipole magnets of one arc form one electrical circuit. The quadrupoles of each arc form two electrical circuits: all focusing quadrupole magnets in Ring 1 and Ring 2 are powered in series and all defocusing quadrupole magnets of Beam 1 and Beam 2 are powered in series. The optics of Beam 1 and Beam 2 in the arc cells is therefore strictly coupled via the powering of the main magnetic elements.

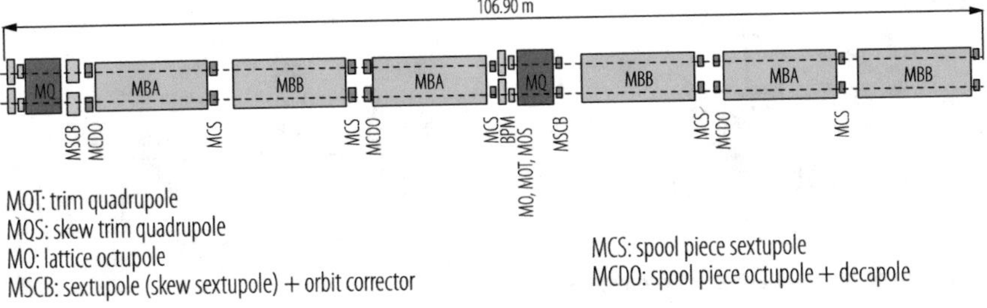

Fig. 10.17 The LHC arc cell layout

An LHC arc cell is also equipped with corrector magnets, which can be split into two distinct categories (see Fig. 10.17).

- The lattice corrector magnets attached on both sides of the main quadrupole magnets are installed in the Short Straight Section (SSS) cryostats.
- The spool-piece corrector magnets which are thin non-linear windings attached directly on the extremities of the main dipoles.

Contrary to the main dipole circuits and the two families (QF and QD) of lattice quadrupoles for which, in each sector, Ring 1 and Ring 2 are powered in series, the arc corrector magnets can be adjusted independently for the two beams.

10.6.2.3 The Dispersion Suppressors

A dispersion suppressor is located at the transition between an LHC arc and a straight section yielding a total of 16 dispersion suppressor sections. The aim of the dispersion suppressors is threefold:

- adapt the LHC reference orbit to the geometry of the LEP tunnel;
- cancel the horizontal dispersion arising in the arc and generated by the separation/recombination dipole magnets and the crossing angle bumps;
- help in matching the insertion optics to the periodic solution of the arc.

A generic design of a dispersion suppressor uses standard arc cells with missing dipole magnets. The LEP dispersion suppressor, which defines the geometry of the tunnel, was made of 3.5 cells with a 90° phase advance, optimized to suppress the dispersion. With the 2.5 times longer LHC dipole and quadrupole magnets, only two LHC cells can be fitted in the dispersion suppressor tunnel.

10.6.2.4 LSS1 and LSS5

IR1 and IR5 house the high luminosity experiments of the LHC and are identical in terms of hardware and optics (except for the crossing-angle scheme: the crossing

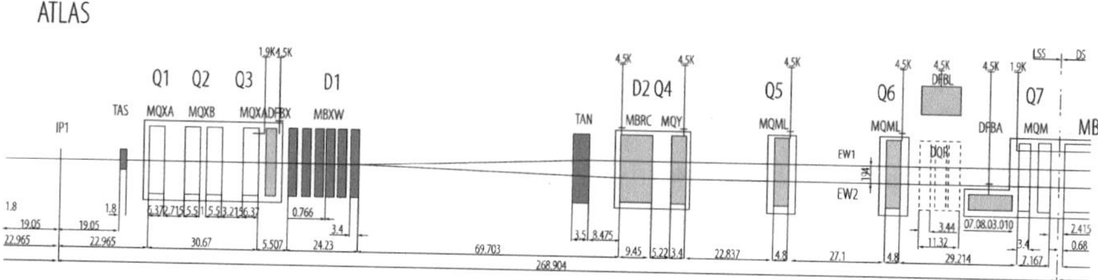

Fig. 10.18 Schematic layout of the right side of IR1

angle in IR1 is in the vertical plane and in IR5 in the horizontal plane). The small β-function values at the IP are generated with the help of a triplet quadrupole assembly. At the IP, the two rings share the same vacuum chamber, the same low-beta triplet magnets and the D1 separation dipole magnets. The remaining matching section (MS) and the dispersion suppressor (DS) consist of double-bore magnets with separate beam pipes for each ring.

Figure 10.18 shows the schematic layout of IR1.

Apart from the DS the insertions are comprised of the following sections, given in order from the interaction point:

- A 31 m long superconducting low-β triplet assembly operated at a temperature of 1.9 K and providing a nominal gradient of 205 T/m.
- A pair of separation/recombination dipoles separated by approximately 88 m. The D1 dipole located next to the triplet magnets has a single bore and consists of six 3.4 m long conventional warm magnet modules yielding a nominal field of 1.38 T. The following D2 dipole is a 9.45 m long, double bore, superconducting dipole magnet operating at a cryogenic temperature of 4.5 K with a nominal field of 3.8 T. The bore separation in the D2 magnet is 188 mm and is thus slightly smaller than the arc bore separation.
- Four matching quadrupole magnets. The first quadrupole following the separation dipole magnets, Q4, is a wide-aperture magnet operating at a cryogenic temperature of 4.5 K and yielding a nominal gradient of 160 T/m. The remaining three quadrupole magnets are normal-aperture quadrupole magnets operating at a cryogenic temperature of 1.9 K with a nominal gradient of 200 T/m.

The triplet assembly features two different quadrupole designs: the outer two quadrupole magnets are made by KEK and require a peak current of 6450 A to reach the nominal gradient of 205 T/m, whereas the inner quadrupole block consists of two quadrupole magnets made by FNAL and requires a peak current of 10,630 A.

The triplet quadrupoles are powered by two nested power converters: one 8 kA power converter powering all triplet quadrupole magnets in series and one 6 kA power converter supplying additional current only to the central two FNAL magnets. The Q1 quadrupole next to the IP features an additional 600 A trim power converter. The triplet quadrupoles are followed by the separation/recombination dipoles, D1 and D2, which guide the beams from the IP into two separated vacuum chambers.

Q4, Q5, Q6, Q7, Q8, Q9 and Q10 are individually powered magnets. The aperture of Q4 is larger to provide sufficient aperture for the crossing angle separation orbit. Two absorbers protect the cold magnets from particles leaving the IP. The TAS absorber protects the triplet quadrupole magnets and the TAN absorber, located in front of the D1 dipole magnet, protects the machine elements from neutral particles leaving the IP.

The matching section extends from Q4 to Q7 and the DS extends from Q8 to Q11. In addition to the DS, the first two trim quadrupoles of the first arc cell (QT12 and QT13) are also used for the matching procedure. All insertion and DS magnets are equipped with a beam screen [2]. The magnets left and right from the IP up to Q7 inclusive are placed symmetrically with respect to the IP. The positions of Q8, Q9 and Q10 left and right from the IP differ by approximately 0.5 m with respect to the IP due to the limited space in the DS.

10.6.2.5 LSS2

The straight section at IR2 houses the injection elements for Ring-1 as well as the ion beam experiment ALICE. During injection the optics must obey the special constraints imposed by the beam injection for Ring-1 and the geometrical acceptance in the interaction region (IR) must be large enough to accommodate both beams in the common part of the ring with a beam separation of at least 10 σ.

Figures 10.19 and 10.20 show the schematic layout of IR2.

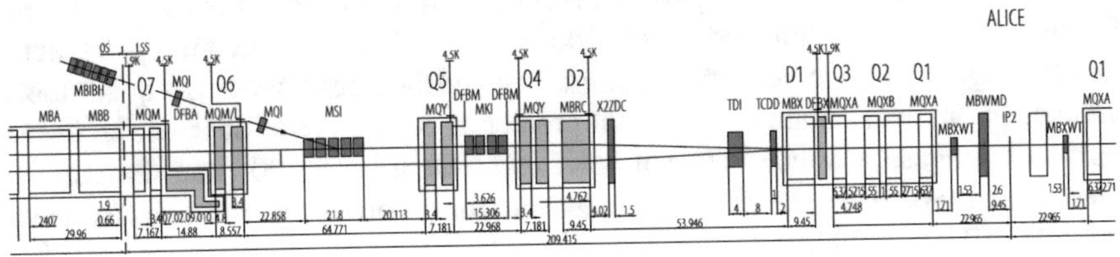

Fig. 10.19 Schematic layout of the left side of IR2

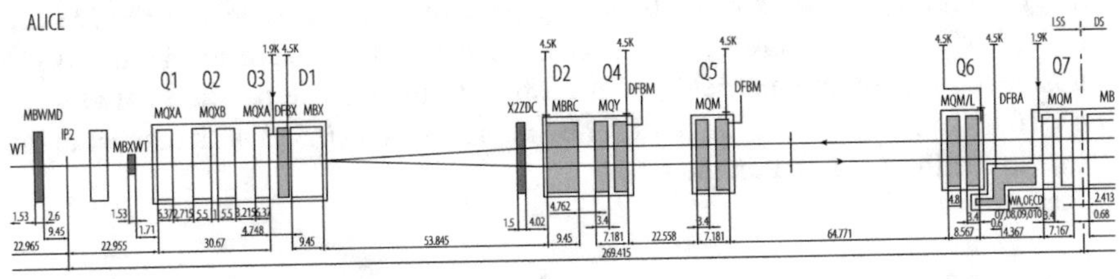

Fig. 10.20 Schematic layout of the right side of IR2

Apart from the DS the insertions comprise the following sections, given in order from the interaction point:

- A 31 m long superconducting low-β triplet assembly operated at 1.9 K and providing a nominal gradient of 215 T/m.
- A pair of 9.45 m long superconducting separation/recombination dipole magnets separated by approximately 66 m.
- Four matching quadrupole magnets. The first two quadrupole magnets following the separation dipole magnets, Q4 and Q5, are wide aperture magnets operating at 4.5 K and yielding a nominal gradient of 160 T/m. The remaining two quadrupole magnets are normal aperture quadrupole magnets operating at 1.9 K with a nominal gradient of 200 T/m.

The triplet quadrupoles are powered in series and are followed by the separation/recombination dipoles D1 and D2, which guide the beams from the IP into two separated vacuum chambers. Q4, Q5, Q6, Q7, Q8, Q9 and Q10 are individually powered magnets. The aperture of Q4 is increased to provide sufficient aperture for the crossing-angle separation orbit. The aperture of Q5 left of the IP is increased to provide sufficient aperture for the injected beam. The injection septum MSI is located between Q6 and Q5 on the left-side of the IP and kicks the injected beam in the horizontal plane towards the closed orbit of the circulating beam (positive deflection angle). The injection kicker MKI is located between Q5 and Q4 on the left-hand side of the IP and kicks the injected beam in the vertical plane towards the closed orbit of the circulating beam (negative deflection angle). In order to protect the cold elements in case of an injection failure a large absorber (TDI) is placed 15 m upstream from the D1 separation/recombination dipole left from the IP. The TDI absorber is complemented by an additional shielding element 3 m upstream of the D1 magnet and two additional collimators installed next to the Q6 quadrupole magnet. In order to obtain an optimum protection level in case of injection errors the vertical phase advance between MKI and TDI must be 90° and the vertical phase advance between the TDI and the two auxiliary collimators must be an integer multiple of $180° \pm 20°$.

The matching section extends from Q4 to Q7 and the DS extends from Q8 to Q11. In addition to the DS, the first two trim quadrupoles of the first arc cell (QT12 and QT13) are also used for the matching procedure. All magnets of the DS are equipped with a beam screen. The magnets left and right from the IP up to Q7 inclusive are placed symmetrically with respect to the IP. The positions of Q8, Q9 and Q10 left and right from the IP differ by approximately 0.5 m with respect to the IP due to the limited space in the DS.

10.6.2.6 LSS8

IR8 houses the LHCb experiment and the injection elements for Beam 2. The small β-function values at the IP are generated with the help of a triplet quadrupole assembly. At the IP, the two rings share the same vacuum chamber, the same low-

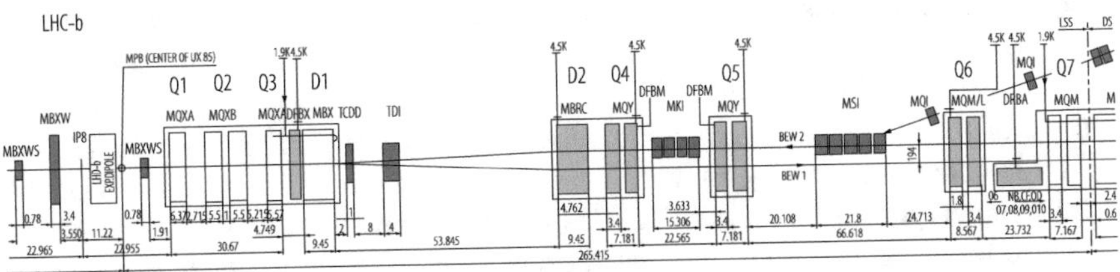

Fig. 10.21 Schematic layout of the right side of IR8

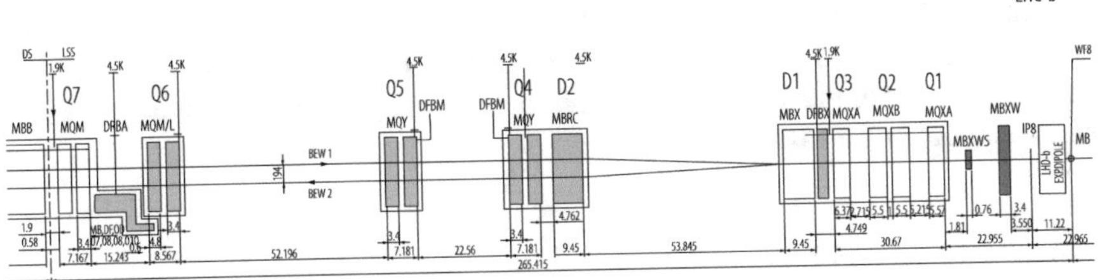

Fig. 10.22 Schematic layout of the left side of IR8

beta triplet magnets and the D1 separation dipole magnet. The remaining matching section (MS) and the DS consist of double-bore magnets with separate beam pipes for each ring.

Figures 10.21 and 10.22 show the schematic layout of IR8.

Apart from the DS the insertions contain the following sections, given in order from the interaction point:

Three warm dipole magnets compensate the deflection generated by the LHCb spectrometer magnet.

- A 31 m long superconducting low-β triplet assembly operated at 1.9 K and providing a nominal gradient of 205 T/m.
- A pair of separation/recombination dipole magnets separated by approximately 54 m. The D1 dipole located next to the triplet magnets is a 9.45 m long single-bore superconducting magnet. The following D2 dipole is a 9.45 m long, double bore, superconducting dipole magnet. Both magnets are operated at 4.5 K. The bore separation in the D2 magnet is 188 mm and is thus slightly smaller than the arc bore separation.
- Four matching quadrupole magnets. The first quadrupole following the separation dipole magnets, Q4, is a wide aperture magnet operating at 4.5 K and yielding a nominal gradient of 160 T/m. The remaining three matching section quadrupole magnets are normal aperture quadrupole magnets operating at 1.9 K with a nominal gradient of 200 T/m.
- The injection elements for Beam 2 on the right hand side of IP8. The 21.8 m long injection septum consists of five modules and is located between the Q6 and Q5 quadrupole magnets on the right-hand side of the IP. The 15 m long

injection kicker consists of four modules and is located between the Q5 and Q4 quadrupole magnets on the right-hand side of the IP. In order to protect the cold elements in case of injection failure a large absorber (TDI) is placed 15 m in front of the D1 separation/recombination dipole magnet right from the IP. The TDI is complemented by an additional shielding element between the TDI and D1 magnet (placed 3 m in front of D1) (TCDD) and by two additional collimators placed on the transition of the matching section left from the IP to the next DS section.

In order to provide sufficient space for the spectrometer magnet of the LHCb experiment, IP8 is shifted by 15 half RF wavelengths (3.5 times the nominal bunch spacing ~11.25 m) towards IR7. This shift of the IP has to be recuperated before the beam returns to the dispersion suppressor sections and implies a non-symmetric magnet layout in the matching section.

10.6.2.7 LSS3 and LSS7

The insertion IR3 houses the momentum cleaning systems of both beams, while IR7 houses the betatron cleaning systems of both beams. Particles with a large momentum offset are scattered by the primary jaw of IR3. Particles with a large H, V or combined H-V betatron amplitudes are scattered by the primary collimator jaws in IR7. In both cases the scattered particles are absorbed by secondary collimators.

Figure 10.23 shows the schematic layout of IR7.

The dispersion suppressor extends from Q8 to Q11. In addition to the DS, the first two trim quadrupoles of the first arc cell (QT12 and QT13) are also used for the matching procedure. All cryo-magnets are equipped with a beam screen. In IR3 and IR7, the underground galleries are not wide enough to house many high current power supplies. Therefore, contrary to the layout of the other IR's, the DS quadrupoles (Q7, Q8, Q9 and Q10) are made of a MQ+MQTL assembly (MQ + 2 MQTL at Q9) where the MQ's magnets are powered in series with the main arc quadrupoles. To avoid producing two kinds of MQ+MQTL assemblies, the dispersion suppressors left and right from the IP are not mirror symmetric with respect to each other. Instead, the DS quadrupole assemblies have the same orientation in the dispersion suppressors left and right from the IP and the MQ positions differ by approximately 0.5 m with respect to the IP in the two DS.

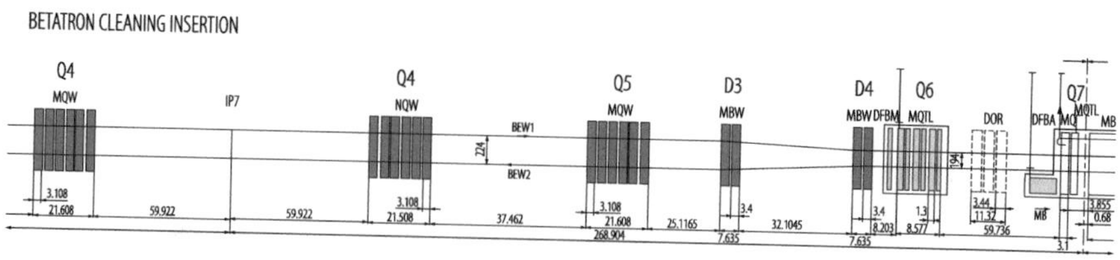

Fig. 10.23 Schematic layout of the right side of IR7

The layout of the Long Straight Section between Q7L and Q7R is mirror symmetric with respect to the IP. The right side of IR7 is shown in Fig. 10.23. This allows the symmetrical installation for the collimators of the two beams and minimizes the space conflicts in the insertion. Starting from Q7 left, the quadrupole Q6 (made of six superconducting MQTL modules) is followed by a dog-leg structure made of two sets of MBW warm single bore wide aperture dipole magnets (two warm modules each). The dogleg dipole magnets are labeled D3 and D4 in the LHC sequence with D3 being the dipole closer to the IP. The Primary Collimators are located between the D4 and D3 magnets, allowing neutral particles produced in the jaws to point out of the beam line, and most charged particles to be swept away. The inter-beam distance between the dogleg assemblies left and right from the IP is 224 mm, i.e. 30 mm larger than in the arc. This increased beam separation allows a substantially higher gradient in the Q4 and Q5 quadrupoles which are made out of six warm MQW modules. The space between Q5 left and right from the IP is used to house the secondary collimators at adequate phase advances with respect to the primary collimators.

The Q4 and Q5 quadrupoles left and right from the IP are powered in series. The warm dual-bore MQW quadrupole cannot be powered with different currents for each magnet aperture because the field quality is degraded to an unacceptable level even for a small imbalance in the field of the two apertures. The current must be equal or of opposite value in the bores to provide a good field quality. In order to obtain the required flexibility for the optics, two different kinds of powering schemes are used for the Q4 and Q5 quadrupole units. The magnets are identical, but in the MQWA type magnet the field is identical in both apertures, while in the MQWB type magnet, the field is opposite for both apertures. Each Q4 and Q5 assembly is made of five MQWA and one MQWB module. The nominal gradient of the MQWB unit is limited to 29.6 T/m while it can reach 35 T/m in the MQWA unit. This powering scheme breaks the exact antisymmetry by 29% providing enough flexibility to satisfy all the optics constraints. Again, Q5AL+Q5AR and Q5BL+Q5BR respectively are powered in series. As a by-product, this freedom in the straight section allows the trim strength needed in the DS to be limited so that regular MQTL's can be used.

In IR3, the most difficult constraint was to generate a large dispersion function in the straight section. Since the layout of the DS cannot be changed in IR3 this constraint means that the natural dispersion suppression generated in the DS is over compensated. To this end Q6 and Q5 were moved towards each other by a substantial amount, thus shrinking the space granted to the dog-leg structure D4-D3. It was therefore necessary to add a third MBW element to D3 and D4 in IR3. Apart from this IR3 and IR7 are identical.

10.6.2.8 LSS4

IR4 houses the RF and feed-back systems as well as some of the LHC beam instrumentation.

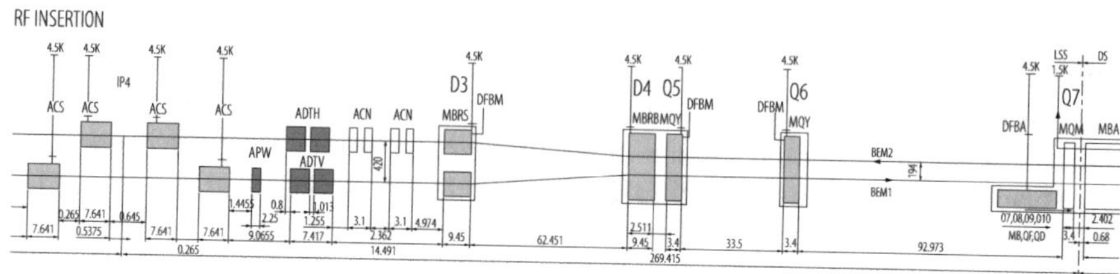

Fig. 10.24 Schematic layout of the right side of IR4

Figure 10.24 shows the schematic layout of IR4.

There are six superconducting quadrupole magnets in the straight section: Q5 Q6 Q7 on the left- and Q5 Q6 Q7 on the right-hand side of the IP. The outer dogleg dipoles, D4, sit next to the Q5 quadrupole magnets on each side of the IP. The RF cavities sit between the inner dogleg dipoles, D3. The layout of the DS, between Q7 and Q11, is identical to that in IR1 and IR5.

10.6.2.9 LSS6

IR6 houses the beam abort systems for Beam 1 and Beam 2. The beam extraction from the LHC is done by kicking the circulating beam horizontally into an iron septum magnet which deflects the beam in the vertical direction away from the machine components to absorbers in a separate tunnel. Each ring has its own system and both are installed in IR6.

Figure 10.25 shows the schematic layout of IR6.

In each of the dispersion suppressors up to six quadrupoles can be used for matching. The total of 16 quadrupoles is more than necessary to match both β-functions, the dispersion (both at the crossing point and in the arc) and adjust the phases. Although this number of parameters seems considerable, their variation is strongly limited by the aperture constraints which set limits on the β-functions and the dispersion inside the insertion. Special detection devices protect the extraction septum and the LHC machine against losses during the extraction process. The TCDS absorber is located in front of the extraction septum and the TCDQ in front of the Q4 quadrupole magnet downstream of the septum magnet.

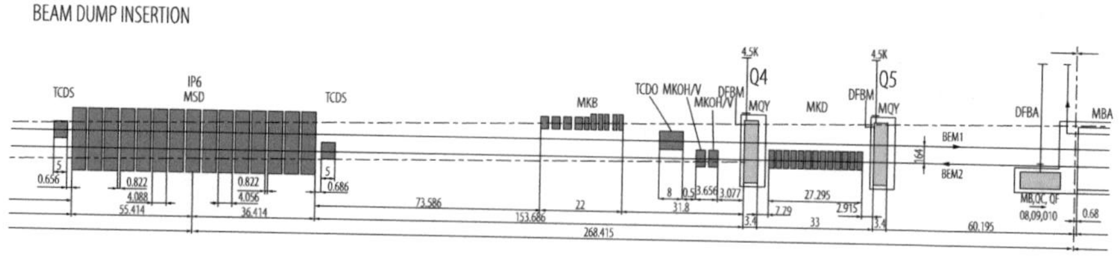

Fig. 10.25 Schematic layout of the right side of IR6

10.6.3 Performance

The LHC was first operated with beam for short periods in 2008 and 2009. In 2010 a first experience with the machine was gained at a beam energy of 3.5 TeV, and moderate beam intensity of up to around 200 bunches of 1.1×10^{11} protons per bunch (ppb). The reduced energy was chosen such as to minimize risks associated to the quality issue of the soldering of the busbar cables in the magnet interconnections. In 2011 the beam intensity was pushed to around 1400 bunches of 1.4×10^{11} ppb while 2012 was dedicated to luminosity production with higher bunch intensities 1.6×10^{11} ppb and a beam energy of 4 TeV. A bunch spacing of 50 ns was used in 2011 and 2012 to minimize the complication of electron clouds. The first operation period is commonly referred to as run 1. In early 2013 beam operation was stopped for a 2-year long shutdown (LS1) to consolidate the magnet interconnections in view of reaching the design beam energy. Many details on LHC operation during Run 1 may be found in [190].

Beam operation resumed in 2015 at 6.5 TeV following a dipole training campaign of 169 quenches at the end of Long Shutdown 1 (LS1). The LHC experiments expressed a strong preference for beams with 25 ns bunch spacing, as opposed to the 50 ns spacing used in 2011–2012, as this would result in a too high number of inelastic collisions per crossing (pile-up). On the machine side 25 ns beams pose additional challenges, e.g. the formation of electron clouds (e-clouds) in the vacuum chamber and a higher number of fast loss events, named Unidentified Falling Objects (UFOs). Given the number of new territories had to be explored, 2015 was considered a re-commissioning and a learning year, dedicated to preparing the machine for full luminosity production in 2016–2018. In 2016 the machine performance was pushed for the first time above the design luminosity of 10^{34} cm^{-2} s^{-1}, and by the end of that year the design had been exceeded by 40%. This excellent performance was possible thanks to beams of much lower emittance produced by the LHC injector chain, coupled to a reduced β^* of 40 cm as compared to the design value of 55 cm. In 2017 the performance could be pushed further with more than twice the design luminosity with a further reduction of the beam emittances and of β^* to 30 cm. The performance in 2017 was so exceptional that the luminosities of ATLAS and CMS had to levelled down to limit the event pile-up to 60. Figures 10.26 and 10.27 present the evolution of the peak and of the integrated luminosity between 2011 and 2017. A summary of the main parameters is presented in Table 10.16. Details on operation in LHC Run 2 up to and including 2016 can be found in [191].

The remarkable performance of LHC during run 2 between 2015 and 2017 was achieved despite unexpected limitations. In 2016 the SPS beam dump was damaged when high intensity LHC beams were dumped onto the block at 450 GeV, developing a vacuum leak. Because it was not practical to exchange the dump during

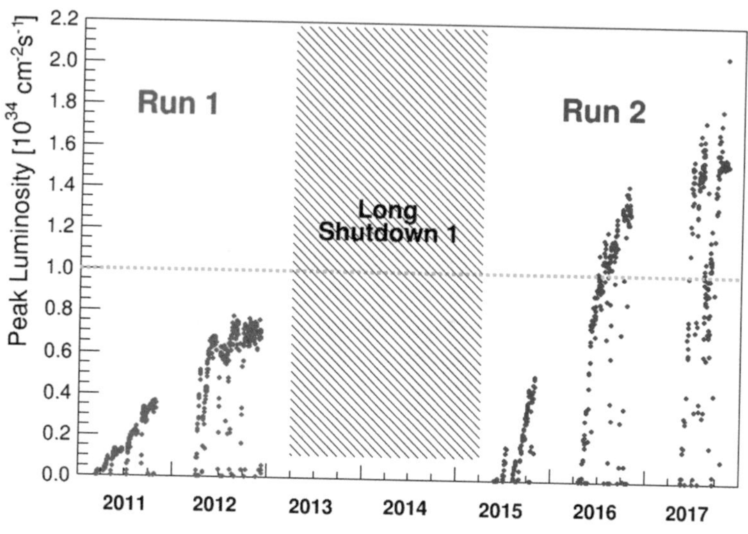

Fig. 10.26 Peak luminosity performance of the LHC in ATLAS and CMS between 2011 and 2017. In July 2016 the LHC reached the design luminosity of 1×10^{34} cm^{-2} s^{-1}

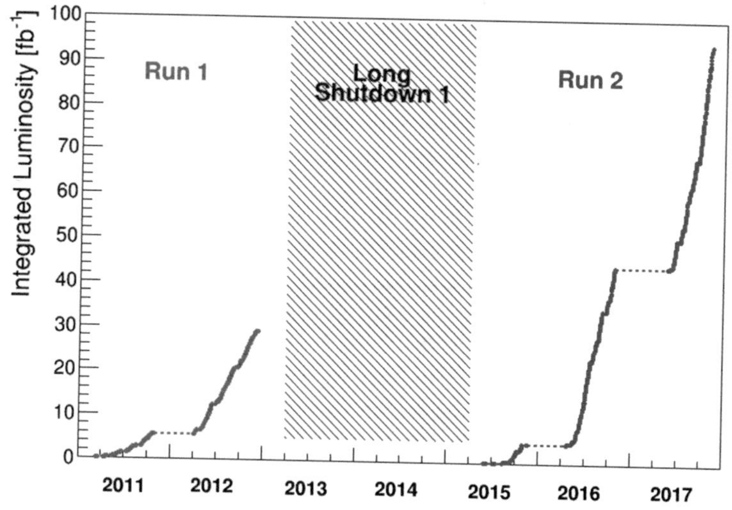

Fig. 10.27 Integrated luminosity delivered to the ATLAS and CMS experiments between 2011 and 2017

the run, the LHC beams were limited to 144 bunches during that year. This limitation was lifted in 2017 when a new dump was installed in the SPS. Unfortunately, an undetected vacuum pumping problem during the cool down of one LHC sector brought another limitation for the 2017 run. A few liters of Air introduced by a

Table 10.16 Performance reach in 2010 and 2011

Parameter	Unit	2012	2016	2017	Design
Energy	TeV	4	6.5	6.5	7
Bunch intensity	10^{10}	16	11	12.5	11.5
Bunches per beam		1380	2556	2556	2802
Emittance	μm	2.40	2.3	2.0	3.75
β_*	m	0.6	0.4	0.3	0.55
Luminosity 1 and 5	$cm^{-2}\ s^{-1}$	7×10^{33}	1.5×10^{34}	2×10^{34}	1×10^{34}

vacuum pump issue condensed as ice on the vacuum beam chamber. In the presence of very high intensity beams local losses would develop leading to beam dumps. Fortunately, it was possible to restore operation of the LHC with a low electron cloud variant of the 25 ns spacing LHC beam [192].

References

1. E. Regenstreif: Report CERN 62-03 (1962).
2. A. Hermann: History of CERN, Vol. I, North-Holland (1987).
3. U. Mersits, in: A. Hermann, et al.: History of CERN, Vol. II, North-Holland (1990)
4. S. Gilardone, D. Manglunki (eds.): Fifty Years of the CERN Proton Synchrotron, CERN-2011-004.
5. D. Simon, J.P. Riunaud: Proc. 17th Intern. Conf. High-Energy Accelerators, Accelerator Catalogue, I. Meshkov (ed.), Dubna (1998), p. 72
6. J. Coupard et al. (eds): LIU TDR Vol 1, CERN–ACC–2014-0337 15 December 2014
7. F. Gerigk, M. Vretenar (eds): LINAC4 TDR, CERN–AB–2006–084 ABP/RF
8. O. Brüning et al. (eds): LHC Design Report Vol. 1, CERN-2004-003
9. M. Giovanni (ed.): Multi Turn Eextraction design report, CERN-2006-011
10. J. Bernhard et al.: Proc. IPAC'18, IPAC2018, Vancouver, Canada doi:10.18429/JACoW-IPAC2018-TUPAF023
11. D. Simon: Proc. EPAC, Sitges (1996), p. 295.
12. N.C. Christofilos: unpublished manuscript (1950).
13. E.D. Courant, M.S. Livingston, H.S. Snyder: Phys. Rev. 88 (1952) 1190.
14. G.K. Green: Proc. Intern. Conf. High Energy Accelerators, M.H. Blewett (ed.), Brookhaven (1961), p. 39.
15. M. Plotkin: Brookhaven Nat. Lab. (1991) report-45058.
16. D.S. Barton: IEEE Trans. Nucl. Sci NS-30(4) (1983)2787.
17. L. Ahrens: HEACC'92, Intern. J. Mod. Phys. A (Proc. Suppl.) (1993) 109.
18. J.M. Brennan: IEEE Proc. Part. Accel. Conf., New York (1999), p. 614.
19. D.I. Lowenstein: Proc. 17th Intern. Conf. High-Energy Accelerators, Accelerator Catalogue, I. Meshkov (ed.), Dubna (1998), p. 72.
20. J.M. Brennan: IEEE Part. Accel. Conf., Dallas (1995), p. 1489.
21. K.A. Brown: Proc. IEEE Part. Accel. Conf., Portland (2003), p. 1545.
22. A. Zelenski: Proc. EPAC08, Genoa (2008), p. 1010.
23. Ya.S. Derbenev, A.M. Kondratenko: Sov. Phys. Dokl. 20 (1976) 562. Ya.S. Derbenev, et al.: Part. Accel. 8 (1978) 115.
24. T. Roser: 8th Intern. Symp. High-Energy Spin Physics, Particle and Fields Series 37, Minneapolis, MN (1988), p. 1442.

25. L.A. Huang: Proc. IEEE Part. Accel. Conf., Vancouver (2009), p. 4251.
26. V. Schoefer: Proc. IEEE Part. Accel. Conf., New York (2011), to be published.
27. C.J. Gardner: Proc. IEEE Part. Accel. Conf., Albuquerque (2007), p. 1862.
28. A. Pikin, et al.: Proc. Intern. Symp. Electron Beam Ion Sources and Traps, IOP Publishing, Stockholm (2010). http://iopscience.iop.org/1748-0221/5/09/C09003/pdf/jinst10_09_c09003.pdf
29. J. Alessi: Proc. IEEE Part. Accel. Conf., New York (2011), to be published.
30. V.V. Vladimirski, et al.: Proc. 4th Intern. Conf. High-Energy Accelerators, Dubna (1963), p. 233.
31. Institute of High Energy Physics, et al.: Proc. 6th Intern. Conf. High-Energy Accelerators, Cambridge (1967), p. 248.
32. E.F. Troyanov: Proc. 17th Intern. Conf. High-Energy Accelerators, Accelerator Catalogue, I. Meshkov (ed.), Dubna (1998), p. 29.
33. S. Ivanov: Proc. Russian Particle Accelerator Conf., Zvenigorod (2008), p. 130.
34. V.A. Teplyakov: Proc. 17th Intern. Conf. High-Energy Accelerators, Accelerator Catalogue, I. Meshkov (ed.), Dubna (1998), p. 31.
35. A.S. Gurevich: Proc. 17th Intern. Conf. High-Energy Accelerators, Accelerator Catalogue, I. Meshkov (ed.), Dubna (1998), p.30.
36. S. Ivanov. Proc. Russian Particle Accelerator Conf., Protvino (2010), p. 27.
37. A.G. Afonin: Proc. Russian Particle Accelerator Conf., Protvino (2010), session WECHX01.
38. S. Ivanov et al. Advances of Light-Ion Acceleration Program in the U70. Proc. 23rd Russian Particle Accelerators Conference RUPAC-2012, St.-Petersburg, 2012, pp. 100–102.
39. S. Ivanov, O. Lebedev. Transverse Noise Blow-up of the Beam in the U-70 Synchrotron. Instruments and Experimental Technique, Vol. 56, No. 3, 2013, pp. 249–255.
40. S. Ivanov, O. Lebedev. Attaining Square-Wave Stochastic Slow Extraction Spills from the U-70 Synchrotron. Instruments and Experimental Techniques, 2015, Vol. 58, No. 4, pp. 456–464.
41. D.W. Kerst: Symp. High-Energy Accel. and Pion Physics, CERN (1956), CERN Report 56-26, p. 36.
42. G.K. O'Neill: Symp. High-Energy Accel. and Pion Physics, CERN (1956), CERN Report 56-26, p. 64.
43. K.R. Symon, A.M. Sessler: Symp. High-Energy Accel. and Pion Physics, CERN (1956) CERN Report 56-26, p. 44.
44. CERN Study Group New Accelerators: Report CERN/542 (1964).
45. A. Hofmann: Proc. 11th Intern. Conf. High-Energy Accelerators, Accelerator Catalogue, J.H.B. Madsen, P.H. Standley (eds.), CERN (1980), p. 44.
46. K. Johnsen: CERN Report 84-13 (1984).
47. K. Johnsen: Proc. 5th Intern. Conf. High-Energy Accelerators, Frascati (1965), p. 3.
48. G. Plass: CERN Internal Report NPA/Int.61-8 (1961).
49. J.B. Adams: Proc. 8th Intern. Conf. High-Energy Accelerators, CERN (1971), p. 25.
50. J.B. Adams: Proc. 10th Intern. Conf. High-Energy Accelerators, Protvino (1977), p. 17.
51. K.-H. Kissler: Proc. 17th Intern. Conf. High-Energy Accelerators, Accelerator Catalogue, I. Meshkov (ed.), Dubna (1998), p. 8.
52. H. Haseroth: Phys. Reports 403-404 (2004) 27.
53. LEP Injector Study Group: Report CERN-LEP/TH/83-29 (1983).
54. M. Benedikt, et al. (eds.): LHC Design Report, Vol. III, CERN-2004-003.
55. C. Rubbia, et al.: Proc. Intern. Neutrino Conf., Aachen (1976), p. 683.
56. S. van der Meer: CERN Internal Report ISR-PO/72-31 (1972).
57. P. Bramham, et al.: Nucl. Instrum. Meth. 125 (1975) 201.
58. G. Carron, et al.: Proc. Part. Accel. Conf., San Francisco (1979), p. 3456.
59. H. Koziol, D. Möhl: Phys. Reports 403-404 (2004) 91.
60. G. Brianti: Proc. 14th Conf. High-Energy Accelerators, Accelerator Catalogue, S. Kurokawa (ed.), Tsukuba (1989), p. 56.
61. J. Gareyte: Proc. 11th Intern. Conf. High-Energy Accel., CERN (1980), p. 79.

62. H. Edwards: Annu. Rev. Nucl. Part. Sci. 35 (1985) 605.
63. M. Church: Proc. 17th Intern. Conf. High-Energy Accelerators, Accelerator Catalogue, I. Meshkov (ed.), Dubna (1998), p. 62.
64. F.R. Huson: Proc. 10th Intern. Conf. High-Energy Accelerators, Protvino (1977), p. 30.
65. S. Holmes (ed.): Report FNAL-TM-2484 (1998).
66. M. Church: Proc. Eur. Accelerator Conf., Paris (2002), p. 11.
67. A. Valishev, et al.: Proc. 23rd Particle Accelerator Conf., Vancouver (2009), p. 4230.
68. S.D. Holmes and V.D. Shiltsev, Annual Review of. Nuclear. and Particle Science, Vol.63, p.435.
69. M. Harrison, S. Peggs, T. Roser: Annu. Rev. Nucl. Part. Sci. 52 (2002) 425.
70. M. Harrison, et al.: Nucl. Instrum. Meth. A 499 (2003) 235.
71. M. Blaskiewicz, J.M. Brennan, K. Mernick: Phys. Rev. Lett. 105 (2010) 094801.
72. W. Fischer, et al.: Phys. Rev. ST Accel. Beams 11 (2008) 041002.
73. C. Montag, et al.: Phys. Rev. ST Accel. Beams 5 (2002) 084401.
74. W. Fischer, et al.: Proc. EPAC08, Genoa, Italy (2008), p. 1616.
75. A. Zelenski, J. Alessi, A. Kponou, D. Raparia: Proc. EPAC08, Genoa, Italy (2008) p. 1010.
76. H. Huang, et al.: Phys. Rev. ST Accel. Beams 7 (2004) 071001.
77. F. Lin, et al.: Phys. Rev. ST Accel. Beams 10 (2007) 044001.
78. Ya.S. Derbenev, A.M. Kondratenko: Part. Accel. 8 (1978) 115.
79. S.Y. Lee: Spin dynamcs and Snakes in Synchrotrons, World Scientific, Singapore (1997).
80. A. Bazilevsky, et al.: Proc. EPAC08, Genoa, Italy (2008), p. 1140.
81. M. Bai, et al.: Phys. Rev. Lett. 96 (2006) 174801.
82. W. Fischer, et al.: Phys. Rev. Lett. 115 (2015) 264801.
83. A. Fedotov et al.: Proc. NAPAC2016, Chicago, IL, USA (2016), p. 867
84. H. Blosser: Handbook of Accelerator Physics and Engineering, World Scientific (2006) 13.
85. E. Wilson: Handbook of Accelerator Physics and Engineering, World Scientific (2006) 57.
86. M. Tigner: Nuovo Cimento 37 (1965) 1228.
87. J. Seeman: *Nonlinear Dynamics Aspects of Particle Accelerators*, Springer-Verlag, Proc. 247 (1985) 121.
88. J. Rees: Handbook of Accelerator Physics and Engineering, World Scientific (2006) 11.
89. C. Bernardini, et al: High Energy Accelerator Conf. (1961) 256.
90. G. Budker, et al: J. Nucl. Energy C 8 (1966) 676.
91. G. O'Neill: High Energy Accelerator Conf. (1961) 247.
92. V. Auslender, et al: High Energy Accelerator Conf. (1965).
93. A. Skrinsky: Particle Accelerator Conf. (1995) 14.
94. G. Arzelier, et al: VIII High Energy Accelerator Conf. (1971) 127.
95. ADONE Group: Particle Accelerator Conf. (1971) 217.
96. J. Paterson, et al: Particle Accelerator Conf. (1971) 196.
97. J. Paterson, et al: Particle Accelerator Conf. (1975) 1366.
98. G. Tumaikin, et al: X High Energy Accelerator Conf. (1977) 443.
99. H. Nesemann, et al: Particle Accelerator Conf. (1983) 1998.
100. J. LeDuff, et al: XI High Energy Accelerator Conf. (1980) 566.
101. G. Voss, et al: XI High Energy Accelerator Conf. (1980) 748.
102. B. McDaniel, et al: Particle Accelerator Conf. (1981) 1984.
103. D. Rubin, et al: Particle Accelerator Conf. (1995) 481.
104. A. Blinoz, et al: XII High Energy Accelerator Conf. (1983) 183.
105. R. Helm, et al: Particle Accelerator Conf. (1983) 2001.
106. T. Nishikawa: XII High Energy Accelerator Conf. (1983) 143.
107. N. Phinney, et al: Particle Accelerator Conf. (1999) 3384.
108. J. Xu, et al: XII High Energy Accelerator Conf. (1983) 157.
109. S. Myers: Intern. Particle Accelerator Conf. (2010) 3663.
110. C. Milardi, et al: Particle Accelerator Conf. (2009) 80.
111. P. Raimondi: 2nd SuperB Meeting, Frascati (2006).
112. J. Seeman, et al: Eur. Particle Accelerator Conf. (2008) 946.

113. M. Tanaka, et al: Intern. Particle Accelerator Conf. (2011) 3735.
114. Q. Qin, et al: Intern. Particle Accelerator Conf. (2011) 3708.
115. Chinese Academy of Sciences: https://phys.org/news/2016-04-bepcii-luminosity-world-11033cm2s.html.
116. D. Shwartz, et al: Russ. Particle Accelerator Conf. (2010) 1.
117. Y. Rogovsky, et al: Physcis and Technique of Accelerators, Springer, 2014, Vol. 11, No. 5, pp. 651-655.
118. H. Koiso: Intern. Particle Accelerator Conf. (2017) 1275.
119. P. Oddone: Proc. UCLA Workshop Linear Collider BB Factory Conceptual Design, D. Stork (ed.), (1987), p. 243.
120. K. Hirata, E. Keil: Nucl. Instrum. Meth. A 292 (1990) 156.
121. *PEP-II: An Asymmetric B Factory. Conceptual Design Report*, SLAC–418, QCD183:S56:1993 (1993).
122. *KEKB B–Factory Design Report*, KEK Report 95–7 (1995).
123. J. Seeman: Conf. Proc. C 0806233 (2008) TUXG01.
124. Y. Funakoshi, T. Abe, K. Akai, Y. Cai, K. Ebihara, K. Egawa, A. Enomoto, J. Flanagan, et al.: Conf. Proc. C 100523 (2010) WEOAMH02.
125. A. Piwinski: IEEE Trans. Nucl. Sci. 24 (1977) 1408.
126. R.B. Palmer: Proc. 1988 DPF Summer Study on High-energy Physics in the 1990s (Snowmass 88), Snowmass, Colorado, 27 Jun – 15 Jul 1988, (1988), p. 613.
127. K. Oide, K. Yokoya: Phys. Rev. A 40 (1989) 315.
128. M. Sullivan, G. Bowden, H. DeStaebler, S. Ecklund, J. Hodgson, T. Mattison, M.E. Nordby, A. Ringwall, et al.: Conf. Proc. C 960610 (1996) 460.
129. B. Aubert, et al. (BABAR Collaboration): Nucl. Instrum. Meth. A 479 (2002) 1.
130. J. Seeman, M. Sullivan, M. Biagini, Y. Cai, F.J. Decker, M. Donald, S. Ecklund, A. Fisher, et al.: Proc. EPAC 2002, 3-7 Jun 2002, Paris, France, (2002), p. 434-436.
131. Y. Yamazaki, T. Kageyama: Part. Accel. 44 (1994) 107.
132. T. Furuya, et al.: Gif-sur-Yvette 1995, RF superconductivity, Vol. 2 (1995), p. 729.
133. H. Schwarz, R. Rimmer: Conf. Proc. C 940627 (1994) 1882.
134. J. Fox, T. Mastorides, C. Rivetta, D. Van Winkle, D. Teytelman: Phys. Rev. ST Accel. Beams 13 (2010) 052802.
135. M. Izawa, Y. Sato, T. Toyomasu: Phys. Rev. Lett. 74 (1995) 5044.
136. K. Ohmi: Phys. Rev. Lett. 75 (1995) 1526.
137. K. Ohmi, F. Zimmermann: Phys. Rev. Lett. 85 (2000) 3821.
138. T.O. Raubenheimer, F. Zimmermann (SLAC): Phys. Rev. E 52 (1995) 5487.
139. J.W. Flanagan, K. Ohmi, H. Fukuma, S. Hiramatsu, M. Tobiyama, E. Perevedentsev: Phys. Rev. Lett. 94 (2005) 054801.
140. H. Fukuma, J. Flanagan, K. Hosoyama, T. Ieiri, T. Kawamoto, T. Kubo, M. Suetake, S. Uno, et al.: AIP Conf. Proc. 642 (2003) 357.
141. Y. Suetsugu, K. Shibata, H. Hisamatsu, M. Shirai, K. Kanazawa: Vacuum 84 (2009) 694.
142. M.T.F. Pivi, F. King, R.E. Kirby, T. Markiewicz, T.O. Raubenheimer, J. Seeman, L. Wang: Conf. Proc. C 0806233 (2008) MOPP064.
143. Y. Suetsugu, H. Fukuma, L. Wang, M. Pivi, A. Morishige, Y. Suzuki, M. Tsukamoto, M. Tsuchiya: Nucl. Instrum. Meth. A 598 (2009) 372.
144. T. Mimashi, T. Ieiri, M. Kikuchi, A. Tokuchi, K. Tsuchida: Conf. Proc. C 0806233 (2008) TUPD011.
145. At least an application for a storage ring is seen in: R. Servranckx, K.L. Brown: IEEE Trans. Nucl. Sci. 26 (1979) 3598.
146. K. Oide, H. Koiso, K. Ohmi: AIP Conf. Proc. 391 (1997) 215.
147. K. Oide, H. Koiso: Phys. Rev. E 47 (1993) 2010.
148. J. Irwin, C.X. Wang, Y.T. Yan, K.L.F. Bane, Y. Cai, F.J. Decker, M.G. Minty, G.V. Stupakov, et al.: Phys. Rev. Lett. 82 (1999) 1684.
149. K. Akai, N. Akasaka, A. Enomoto, J. Flanagan, H. Fukuma, Y. Funakoshi, K. Furukawa, T. Furuya, et al.: Nucl. Instrum. Meth. A 499 (2003) 191.

150. Y.T. Yan, Y. Cai: Nucl. Instrum. Meth. A 558 (2006) 336.
151. T. Ieiri, K. Akai, H. Fukuma, M. Tobiyama: Nucl. Instrum. Meth. A 606 (2009) 248.
152. K. Satoh, M. Tejima: Conf. Proc. C 950501 (1995) 2482.
153. M. Tejima, M. Arinaga, T. Ieiri, H. Ishii, H. Fukuma, M. Tobiyama, S. Hiramatsu: Conf. Proc. C 0505161 (2005) 3253.
154. A. Drago, J.D. Fox, D. Teytelman, M. Tobiyama: Conf. Proc. C 0806233 (2008) THPC116.
155. T. Mitsuhashi, J.W. Flanagan, S. Hiramatsu: Proc. Seventh EPAC2000, 26-30 Jun 2000, Vienna, Austria, (2000), p. 1783-1785.
156. J.W. Flanagan, N. Akasaka, H. Fukuma, S. Hiramatsu, T. Mitsuhashi, T. Naito, K. Ohmi, K. Oide, et al.: Proc. Seventh EPAC2000, 26-30 Jun 2000, Vienna, Austria, (2000), p. 1119-1121.
157. N. Akasaka, A. Akiyama, S. Araki, K. Furukawa, T. Katoh, T. Kawamoto, I. Komada, K. Kudo, et al.: Nucl. Instrum. Meth. A 499 (2003) 138.
158. Y. Funakoshi, M. Masuzawa, K. Oide, J. Flanagan, M. Tawada, T. Ieiri, M. Tejima, M. Tobiyama, et al.: Phys. Rev. ST Accel. Beams 10 (2007) 101001.
159. L. Hendrickson, T. Gromme, P. Grossberg, T. Himel, D. Macnair, R. Sass, H. Smith, N. Spencer, et al.: Proc. Seventh EPAC2000, 26-30 Jun 2000, Vienna, Austria, (2000), p. 1897-1899.
160. K. Ohmi, K. Oide, E. Perevedentsev: Conf. Proc. C 060626 (2006) 616.
161. Y. Ogawa, A. Enomoto, K. Furukawa, T. Kamitani, M. Satoh, T. Sugimura, T. Suwada, Y. Yano, et al.: Conf. Proc. C 060626 (2006) 2700.
162. T. Kamitani, N. Delerue, M. Ikeda, K. Kakihara, S. Ohsawa, T. Oogoe, T. Sugimura, T. Takatomi, et al.: Conf. Proc. C 0505161 (2005) 1233.
163. K. Akai, J. Kirchgessner, D. Moffat, H. Padamsee, J. Sears, T. Stowe, M. Tigner: Proc. B factory workshop, 6-10 Apr 1992, Stanford, California, (1992).
164. K. Hosoyama, K. Akai, K. Ebihara, T. Furuya, K. Hara, T. Honma, A. Kabe, Y. Kojima, et al.: Conf. Proc. C 0806233 (2008) THXM02.
165. K. Ohmi, M. Tawada, Y. Cai, S. Kamada, K. Oide, J. Qiang: Phys. Rev. ST Accel. Beams 7 (2004) 104401.
166. Y. Ohnishi, K. Ohmi, H. Koiso, M. Masuzawa, A. Morita, K. Mori, K. Oide, Y. Seimiya, et al.: Phys. Rev. ST Accel. Beams 12 (2009) 091002.
167. P. Raimondi, D. Shatilov, M. Zobov, "Beam-Beam Issues for Colliding Schemes with Large Piwinski Angle and Crabbed Waist", arXiv:physics/0702033 [physics.acc-ph] (2007).
168. Y. Ohnishi, Proc.15th Annual Meeting of Particle Accelerator Society of Japan August 7-10, 2018, Nagaoka, Niigata, Japan (2018) WEOLP01.
169. *SuperKEKB Design Report*, https://kds.kek.jp/indico/event/15914/ (2014).
170. Y. Ohnishi, presentation at the 62nd ICFA Advanced Beam Dynamics Workshop on High Luminosity Circular e+e- Colliders (eeFACT2018), September 24-26 (2018), IAS, Hong Kong MOXAA02.
171. K. Ohmi, et al., "Coherent Beam-Beam Instability in Collisions with a Large Crossing Angle", Phys.Rev.Lett. 119 (2017) no.13, 134801.
172. K. Furukawa, et al., Proc. IPAC2018, Vancouver, BC, Canada, doi:10.18429/JACoW-IPAC2018-MOPMF073 (2018).
173. Y. Suetsugu, et al., "Mitigating the electron cloud effect in the SuperKEKB positron ring", Phys.Rev.Accel.Beams 22 (2019) no.2, 023201.
174. N. Ohuchi, et al., Proc. IPAC2018, Vancouver, BC, Canada, doi:10.18429/JACoW-IPAC2018-TUZGBE2 (2018).
175. A. Accardi et al, Electron Ion Collider: The Next QCD Frontier, arXiv:1212.1701 (2012)
176. https://www.jlab.org/jleic/index.html
177. J Beebe-Wang (Editor), eRHIC, Preconceptual Design Report, arXiv:xxxxxxxx
178. R. Yoshida, EIC Collaboration meeting Oct 2017 (BNL) unpublished, https://indico.bnl.gov/event/3492/contributions/10260/attachments/9195/11238/yoshida.pdf
179. Yu. Vassiliev et al, Multi-target operation at the HERA-B experiment, AIP Conference Proceedings 512, 359 (2000); https://doi.org/10.1063/1.1291460

180. R. Brinkmann. Simulation Of Background From Proton Losses In The Hera Straight Sections, Jul 1987. 22 pp. DESY-HERA-87-19 (1987) unpublished
181. http://www.desy.de/mpybar/psdump/eliana-cracow.paper.pdf
182. http://www-hermes.desy.de
183. Ziqing Zhang, Physics from Polarized ep Collisions at HERA, Proceeedings of the 25th Conference of Physics in Collisions, Prague (2005)
184. F. Willeke, Proceedings of the European Particle Accelerator Conference, Lausanne (2004).
185. B. Parker et al, TUOA02A, Proceedings of the European particle Conference 1998, Stockholm(1998); http://accelconf.web.cern.ch/AccelConf/e98/PAPERS/TUOA02A.PDF
186. B. Parker, Serpentine Coil Topology for BNL Direct Wind Superconducting Magnets, Particle Accelerator Conference Knoxville (2005) https://doi.org/10.1109/PAC.2005.1590546
187. N. Ouchi et al, DESIGN OF THE SUPERCONDUCTING MAGNET SYSTEM FOR THE SUPERKEKB INTERACTION REGION, Proceedings of PAC2013, Pasadena, CA USA WEODA1 (2013)
188. B. Parker, M. Anerella, J. Escallier, A. Ghosh, A. Jain, A. Marone, J. Muratore, P. Wanderer, BNL Direct Wind Superconducting Magnets, BNL-96547-2011-CP, Presented at the 22nd International Conference on Magnet Technology (MT-22) Marseille, France September 9-16, (2011) November 2011
189. The LHC Design Report, CERN-2004-003.
190. R. Alemany-Fernandez et al, Operation and Configuration of the LHC in Run 1, CERN-ACC-NOTE-2013-0041.
191. J. Wenninger, LHC towards nominal performance, Proceedings of IPAC17, Copenhagen, Dk (2017).
192. Proceedings of the 8th LHC operation workshop, Evian, France (2017), https://indico.cern.ch/event/663598.

Chapter 11
Application of Accelerators and Storage Rings

M. Dohlus, J. Rossbach, K. H. W. Bethge, J. Meijer, U. Amaldi, G. Magrin, M. Lindroos, S. Molloy, G. Rees, M. Seidel, N. Angert, and O. Boine-Frankenheim

11.1 Synchrotron Radiation and Free-Electron Lasers

M. Dohlus · J. Rossbach

11.1.1 Synchrotron Radiation

11.1.1.1 Basic Properties of Synchrotron Radiation

It is well known from Maxwell theory that electromagnetic radiation is emitted whenever electric charges are accelerated in free space. This radiation assumes quite extraordinary properties whenever the charged particles move at ultrarelativistic

---

M. Dohlus (✉) · J. Rossbach (✉)
DESY, Hamburg, Germany
e-mail: martin.dohlus@desy.de; joerg.rossbach@desy.de

K. H. W. Bethge · J. Meijer
University Leipzig, Leipzig, Germany
e-mail: jan.meijer@uni-leipzig.de

U. Amaldi · M. Lindroos
CERN (European Organization for Nuclear Research)Meyrin, Genève, Switzerland
e-mail: Ugo.Amaldi@cern.ch; Mats.Lindroos@cern.ch

G. Magrin
EBG MedAustron, Wiener Neustadt, Austria
e-mail: giulio.magrin@medaustron.at

S. Molloy
Accelerator Operations, MAX IV Laboratory, Lund, Sweden
e-mail: stephen.molloy@esss.se

S. Myers, H. Schopper (eds.), *Particle Physics Reference Library* ,
https://doi.org/10.1007/978-3-030-34245-6_11

speed: The radiation becomes very powerful and tightly collimated in space, and it may easily cover a rather wide spectrum ranging from the THz into the hard X-ray regime. When generation of such radiation is intended rather than being a side effect, the charged particles are normally electrons, thus kinetic energies are then typically in the multi-MeV range.

The theoretical treatment of synchrotron radiation starts traditionally from retarded Lienard-Wiechert potentials [1], allowing quantitative determination of radiation properties in detail:

$$\phi(0,t) = \frac{q}{4\pi\varepsilon_0}\left[\frac{1}{R\left(1-\vec{n}\cdot\vec{\beta}\right)}\right]_{t'=t-R/c}, \quad \vec{A}(0,t) = \frac{q}{4\pi\varepsilon_0 c}\left[\frac{\vec{\beta}}{R\left(1-\vec{n}\cdot\vec{\beta}\right)}\right]_{t'=t-R/c}. \tag{11.1}$$

Here, q is the accelerated point charge, ε_0 is the dielectric constant, and $\vec{R}$ is the distance vector from the charge to the observer, with the unit vector $\vec{n} = \vec{R}/R$ and $\vec{\beta} = \vec{v}/c$ the particle's velocity $\vec{v}$ normalized to the vacuum speed of light c, see Fig. 11.1. For simplicity, and without loss of generality, we assume the observer is located at the origin at $R = 0$.

All quantities in Eq. (11.1) must be taken at the retarded time $t' = t - R/c$, i.e. not at the time t of observation but at the time when a signal moving at speed c must

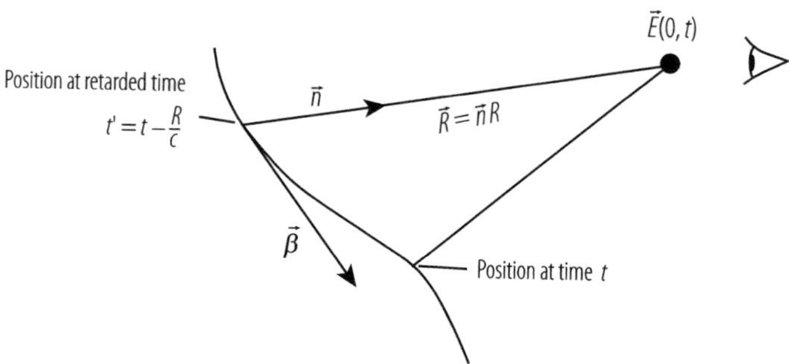

Fig. 11.1 Geometry of actual and retarded position of moving particle, and position of observer

G. Rees
Rutherford Appleton Laboratory, Didcot, UK

M. Seidel
Paul Scherrer Institut, Villigen, Switzerland
e-mail: mike.seidel@psi.ch

N. Angert · O. Boine-Frankenheim
GSI, Darmstadt, Germany
e-mail: N.Angert@gsi.de; O.Boine-Frankenheim@gsi.de

be emitted from the charge's position in order to reach the observer at time t. This latter fact is actually the origin for many peculiarities of synchrotron radiation.

The electric and magnetic fields are derived from Eq. (11.1) using $\gamma^2 = 1/(1-\beta^2)$, $\vec{E} = -\nabla\phi - \partial\vec{A}/\partial t$ and $\vec{B} = \nabla \times \vec{A}$:

$$\vec{E}\,(0,t) = \frac{q}{4\pi\varepsilon_0}\left[\frac{\vec{n}-\vec{\beta}}{\gamma^2\left(1-\vec{n}\cdot\vec{\beta}\right)^3 R^2}\right]_{t-R/c} + \frac{q}{4\pi\varepsilon_0 c}\left[\frac{\vec{n}\times\left\{\left(\vec{n}-\vec{\beta}\right)\times\dot{\vec{\beta}}\right\}}{\left(1-\vec{n}\cdot\vec{\beta}\right)^3 R}\right]_{t-R/c}. \tag{11.2}$$

and

$$\vec{B} = \frac{1}{c}\left[\vec{n}\times\vec{E}\right]_{t-R/c}. \tag{11.3}$$

While Eqs. (11.2) and (11.3) are well suited to calculate quantitatively the field of a charge moving on a well-known path, some general properties of the field are more conveniently determined from the equivalent expression Eq. (11.4) given in [2]:

$$\vec{E}\,(0,t) = \frac{q}{4\pi\varepsilon_0}\left[\frac{\vec{n}}{R^2} + \frac{R}{c}\frac{d}{dt}\left(\frac{\vec{n}}{R^2}\right) + \frac{1}{c^2}\frac{d^2}{dt^2}\vec{n}\right]. \tag{11.4}$$

The first term describes the static Coulomb field, scaling with R^{-2}. The second term modifies the field direction of the Coulomb field such that, in case of a charge moving at constant speed, the Coulomb field is NOT directed towards the retarded position of the particle (as it might be suggested by the first term) but rather to the instantaneous position of the charge just at the time t of observation. This will be shown below.

All radiation is described by the third term, and since it contains a contribution scaling with R^{-1}, it is the dominant field at large distances from the source. Thus, in order to consider properties of synchrotron radiation qualitatively, it is sufficient to understand the behaviour of

$$\vec{E}_{\text{rad}}\,(0,t) \propto \left[\frac{d^2}{dt^2}\vec{n}\right]. \tag{11.5}$$

Since the unit vector $\vec{n}$ cannot change its length, one can see [2] that $d^2\vec{n}/dt^2$ is always perpendicular to $\vec{n}$, i.e. to the retarded position of the particle if observed from a far distance. The fact that $\vec{E}_{\text{rad}}\,(0,t)$ is perpendicular to $\vec{n}$ is also seen from the second term of Eq. (11.2) (describing the radiation in far zone) which would yield zero if scalar multiplied by $\vec{n}$. Together with Eq. (11.3) it is seen that the Poynting vector $\vec{S} = \left(\vec{E} \times \vec{B}\right)/\mu_0$ is parallel to $\vec{n}$. It is thus immediately clear that the radiation detected by the observer always seems to be originating from the retarded position of the particle, i.e. the retarded position is the apparent origin of the radiation, as it is intuitively expected when the speed of light is taken into account.

Equation (11.5) suggests that one just has to inspect the acceleration of the charge transverse to its apparent line of sight in order to understand the behaviour of the electric field component. For example, it is thus easily seen that radiation from an electron moving on a circular orbit is linearly polarized in the plane of the circle if the observer is located in the same plane. If, however, the radiation is observed from a position elevated out of this (say, horizontal) plane, one expects also a vertical field component.

Since Eqs. (11.2) and (11.4) look so differently, it is useful to sketch how Eq. (11.2) can be derived from Eq. (11.4): a key point is that Eq. (11.2) requires all derivations to be taken with respect to the retarded time $t' = t - R/c$ while Eq. (11.5) contains derivatives with respect to the time of observation t. Explicitly, this reads

$$\vec{\beta} = -\frac{d\vec{R}}{cdt'} = -\frac{d\left(R\vec{n}\right)}{cdt'} = -\frac{dR}{cdt'}\vec{n} - R\frac{d\vec{n}}{cdt'}. \tag{11.6}$$

From $t' = t - R/c$ it follows: $dt'/dt = 1 - (dR(t)/cdt')(dt'/\mathrm{d}t)$. Solving for dt'/dt yields: $dt'/dt = 1/(1 + \mathrm{d}R/cdt')$.

Due to $\vec{n} \perp d\vec{n}/dt'$ we get from Eq. (11.6): $dR/cdt' = -\vec{n}\cdot\vec{\beta}$ and thus

$$\frac{dt'}{dt} = \frac{1}{1 - \vec{n}\cdot\vec{\beta}}. \tag{11.7}$$

For ultrarelativistic motion, and in particular in forward direction $\vec{n} \parallel \vec{\beta}$, *dt'* and *dt* differ by a really large factor

$$\frac{1}{1-\beta} = \frac{1+\beta}{1-\beta^2} \approx 2\gamma^2 \gg 1 \tag{11.8}$$

thus resulting in a huge compression of the time scale at which radiation properties are observed compared to the one at which electron motion takes place. It should be noted that this has nothing to do with any Lorentz transform but is rather a property of relativistic Doppler shift. The particle position at the retarded time is just a different point in space-time.

With Eq. (11.7) in mind, Eq. (11.4) can be rewritten (note we write $(n\beta)$ for $\left(\vec{n}\cdot\vec{\beta}\right)$ everywhere):

$$\vec{E}\,(0,t)=\frac{q}{4\pi\varepsilon_0}\frac{1}{R^2}\left\{\begin{array}{c}\vec{n}+\frac{1}{[1-(n\beta)]^3}\Big[-\vec{\beta}+2\vec{\beta}\,(n\beta)-\vec{\beta}\,(n\beta)^2+3\vec{n}\,(n\beta)\\ -6\vec{n}\,(n\beta)^2+3\vec{n}\,(n\beta)^3+3\vec{n}\,(n\beta)^2-\vec{n}\,\beta^2-2\vec{\beta}\,(n\beta)\\ -2\vec{n}\,(n\beta)^3+\vec{\beta}\,\beta^2+\vec{\beta}\,(n\beta)^2\Big]\end{array}\right\}$$
$$+\frac{q}{4\pi\varepsilon_0}\frac{1}{cR[1-(n\beta)]^3}\underbrace{\left[\vec{n}\,(n\dot{\beta})-\dot{\vec{\beta}}+\dot{\vec{\beta}}\,(n\beta)-\vec{\beta}\,(n\dot{\beta})\right]}_{\vec{n}\times\left[\left(\vec{n}-\vec{\beta}\right)\times\dot{\vec{\beta}}\right]}. \tag{11.9}$$

It is interesting noting that not only the last line but also all expressions after $3\vec{n}\,(n\beta)^3$ in the curly bracket stem from the "radiation term" $\frac{1}{c^2}\frac{d^2}{dt^2}\vec{n}$.

After expanding $\vec{n}$ by $\frac{[1-(n\beta)]^3}{[1-(n\beta)]^3}$, the curly bracket in Eq. (11.9) can be simplified into $\{\,\}=\frac{1}{[1-(n\beta)]^3}\frac{\vec{n}-\vec{\beta}}{\gamma^2}$, completing the proof that Eqs. (11.2) and (11.4) are equivalent.

In the following, we will restrict ourselves to properties of the radiation field $\vec{E}_{\text{rad}}$ described by the second term in Eq. (11.2) or by Eq. (11.5), which are equivalent if the observation is made at sufficiently large distance R from the charge ("far-zone approximation"), since in this case the contribution from the curly bracket in Eq. (11.9) to Eq. (11.5) can be neglected:

$$\vec{E}\,(0,t)_{\text{rad}}=\frac{q}{4\pi\varepsilon_0 c}\left[\frac{\vec{n}\times\left\{\left(\vec{n}-\vec{\beta}\right)\times\dot{\vec{\beta}}\right\}}{\left(1-\vec{n}\cdot\vec{\beta}\right)^3 R}\right]_{t-R/c}. \tag{11.10}$$

It should be emphasised that there are indeed practical cases where this approximation is not valid, e.g. if radiation from an undulator (see below) of length L_u is observed from a distance not much larger than L_u [3].

In most experimental cases the time evolution of the electric field vector is not observable but only the radiation power and its angular or spectral distribution.

In discussing properties of synchrotron radiation it is important to distinguish the instantaneously emitted power (and its angular distribution) from the time development of the power observed by an experimentalist fixed in the lab system. While the first is described in terms of the retarded time t', the latter is observed on the time scale t, which makes a big difference for ultrarelativistic motion, see Eq. (11.7). We point out again that this has nothing to do with a Lorentz transform.

Radiation Power and Its Angular Distribution

The radiation power density is described by the Poynting vector $\vec{S} = \left(\vec{E} \times \vec{B}\right)/\mu_0 = \varepsilon_0 c \left(\vec{E}_{\text{rad}}\right)^2 \vec{n}$, where $\vec{E}_{\text{rad}}$ is the far zone radiation field given in observer time t according to Eq. (11.10). In the present section, we want to refer the radiation power P in far zone to the motion of the electron (i.e. to an emission time interval dt'), so we need to consider

$$\frac{dP(t')}{d\Omega} = R^2\left(\vec{S}\cdot\vec{n}\right)\frac{dt}{dt'} = R^2\left(\vec{S}\cdot\vec{n}\right)\left(1-\vec{n}\cdot\vec{\beta}\right). \tag{11.11}$$

where $d\Omega$ is the solid angle element into which the power is emitted. Thus,

$$\frac{dP(t')}{d\Omega} = \frac{q^2}{(4\pi)^2\varepsilon_0 c}\frac{\left[\vec{n}\times\left\{\left(\vec{n}-\vec{\beta}\right)\times\dot{\vec{\beta}}\right\}\right]^2}{\left[1-\vec{n}\cdot\vec{\beta}\right]^5}. \tag{11.12}$$

This is the most general expression for the angular dependence of the energy loss into radiation of an accelerated point charge in far-field approximation.

There are essentially two different acceleration mechanisms which need to be distinguished: Acceleration $\dot{v}_\perp$ perpendicular to the direction of motion, usually provided by a magnetic field B_{ext}, and acceleration $\dot{v}_\parallel$ by an electric field E_{ext} parallel to the momentarily velocity.

The total radiation power due to $\dot{v}_\perp$ can be shown to be

$$P_{\text{rad},\perp} = \frac{q^2}{6\pi\varepsilon_0 c^3}\gamma^4\dot{v}_\perp^2 = \frac{q^2 c}{6\pi\varepsilon_0}\beta^4\frac{\gamma^4}{\rho^2}. \tag{11.13}$$

Here, $\varrho = \left|\vec{p}\right|/(qB_{\text{ext}})$ is the bending radius of the particle with momentum $\vec{p}$ and charge q in presence of a magnetic field B_{ext} directed perpendicular to $\vec{p}$.

The total radiation power due to $\dot{v}_\parallel$ is, on the other hand,

$$P_{\text{rad},\parallel} = \frac{q^2}{6\pi\varepsilon_0 c^3}\left(\gamma^3\dot{v}_\parallel\right)^2 = \frac{q^2 c}{6\pi\varepsilon_0}\left(\frac{d\gamma}{ds}\right)^2, \tag{11.14}$$

with $d\gamma/ds$ the acceleration due to the longitudinal electric field. In almost all practical cases this radiation power is much smaller than the one generated in a magnetic field of 1 T.

Thus, for the remainder of this paper, we restrict ourselves to acceleration due to magnetic fields.

For circular accelerators the total energy loss U_{rad} during one turn of the charged particle due to synchrotron radiation is an important quantity. From Eq. (11.13) one calculates

$$U_{\text{rad}} = \frac{1}{c}\int P_{\text{rad}} ds = \frac{2\pi\varrho}{c} P_{\text{rad}} = \frac{q^2}{3\varepsilon_0}\frac{\gamma^4}{\rho} \approx 88.5\,\text{keV}\,\frac{(E_0/\text{GeV})^4}{\rho/\text{m}}, \qquad (11.15)$$

where the integral extends over all bending magnets. The last expression represents the radiation loss in practical units in the case of electrons or positrons, with E_0 the particle energy. For simplicity is has been assumed that ρ is constant in all bending magnets.

According to the γ^4-scaling of Eqs. (11.13) and (11.15), synchrotron radiation constitutes a massive challenge on the construction of electron/positron synchrotrons in the multi-GeV range. As an extreme example, in the electron-positron storage ring LEP at CERN each particle lost approximately $U_{\text{rad}} = 2850$ MeV per turn when running at its maximum particle energy of $E_0 = 100$ GeV, even though the bending radius was as large as $\rho = 3100$ m.

On the contrary, emission of synchrotron radiation is a negligible effect for hadrons in most practical cases.

In a magnetic field, the acceleration is always perpendicular to the velocity: $\dot{\vec{\beta}} \perp \vec{\beta}$. Then, Eq. (11.12) can be expressed in more practical units, reflecting the geometry illustrated in Fig. 11.2:

$$\frac{dP(\varphi,\theta,t')}{d\Omega} = \frac{q^2}{(4\pi)^2\varepsilon_0 c}\frac{\left|\dot{\vec{\beta}}\right|^2}{(1-\beta\cos\theta)^3}\left(1-\frac{\sin^2\theta\cos^2\varphi}{\gamma^2(1-\beta\cos\theta)^2}\right). \qquad (11.16)$$

The direction of observation $\vec{n}$ is expressed here in terms of the angles φ,θ as illustrated in Fig. 11.2.

For ultrarelativistic particles, i.e. if $1-\beta \ll 1$, the denominators $1-\beta\cos\theta$ get very small around $\theta \approx 0$ such that the radiation is concentrated very much in the direction of $\vec{\beta}$.

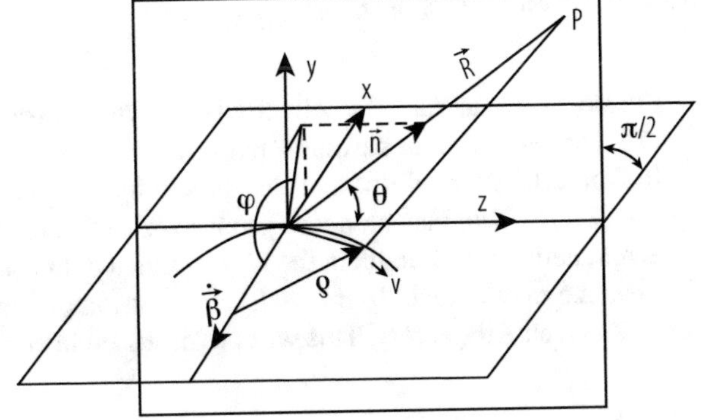

Fig. 11.2 Geometry and definition of parameters used in Eq. (11.16)

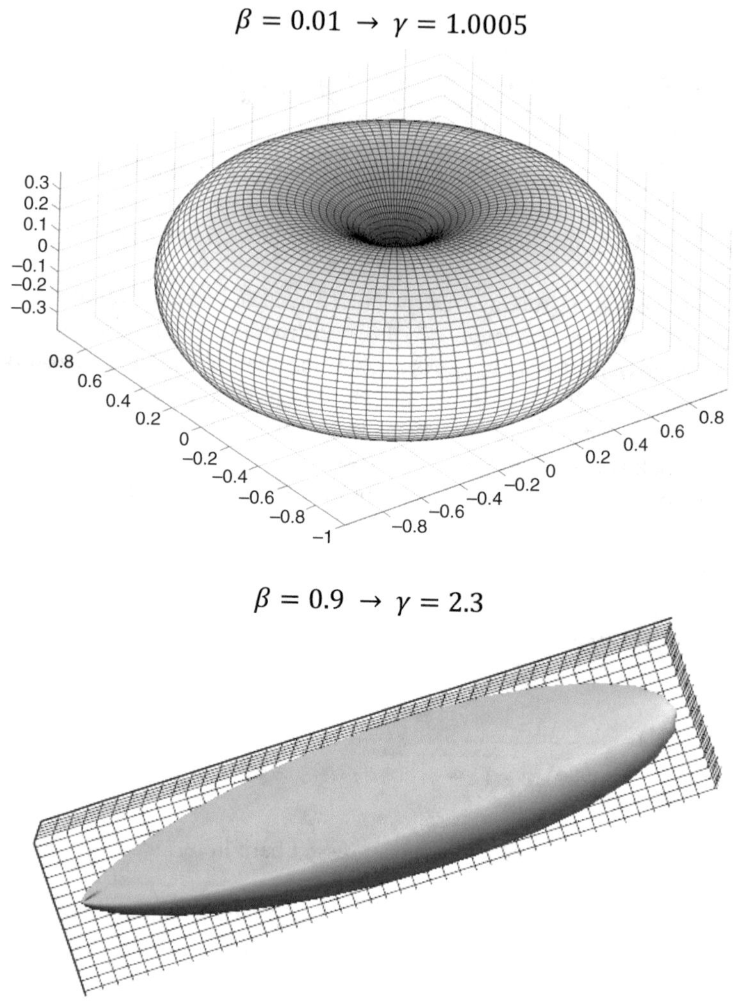

Fig. 11.3 Synchrotron radiation emitted by an ultrarelativistic charged particle is concentrated into a narrow cone with opening angle $1/\gamma$

Figure 11.3 illustrates the directivity of synchrotron radiation according to Eq. (11.16) for two different particle energies.

It should be pointed out that the directivity of synchrotron radiation described by Eq. (11.16) is an instantaneous property of emission, thus is depends only on the local magnetic field strength in the very moment of emission. An observer located in far distance may in fact observe field contributions stemming from several sections of the electron's trajectory. This will be discussed later.

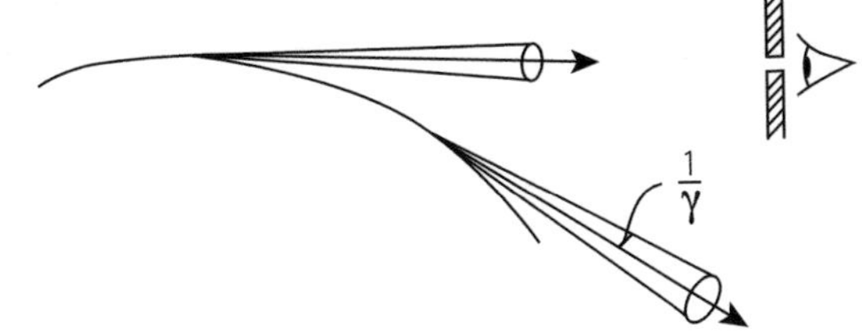

Fig. 11.4 Scenario of synchrotron radiation detection by a distant observer

For $\gamma \gg 1$ and $\theta \ll 1$ we get $1 - \beta \cos \theta \approx (1 + \gamma^2\theta^2)/(2\gamma^2)$ and thus:

$$\frac{dP\left(\varphi, \theta, t'\right)}{d\Omega} = \frac{q^2}{2\pi^2\varepsilon_0 c} \frac{\left|\dot{\vec{\beta}}\right|^2 \gamma^6}{\left(1 + \gamma^2\theta^2\right)^3} \left(1 - \frac{4\gamma^2\theta^2\cos^2\varphi}{\left(1 + \gamma^2\theta^2\right)^2}\right). \tag{11.17}$$

Equation (11.17) illustrates even more that the emission is concentrated into a cone of opening angle $\theta \approx 1/\gamma$ with respect to the forward direction. A rigorous calculation shows that the rms-opening angle is indeed exactly $1/\gamma$ [4].

For practical calculations it might be useful to replace $\left|\dot{\vec{\beta}}\right|$ by $\beta^2 c/\rho \approx c/\rho$ with $1/\rho = qB/p_0$ describing the bending radius ϱ.

An observer in far distance sees a radiation field only during the short time when the cone passes the observer's aperture, see Fig. 11.4.

For an observer in the plane of deflection, the radiation field has only a non-zero component $E_x(t)$ in this plane, as can be understood easily from Eq. (11.5). It is thus linearly polarized. Due to the strong retardation effect described by Eq. (11.7), this time duration is indeed much smaller than the time $\Delta t' \approx (2/\gamma)(\rho/c)$ the electron needs to cover an angle of $2/\gamma$ on its curved trajectory, namely shorter by a factor $2\gamma^2$. Thus one expects a radiation pulse duration $\Delta t \approx (2\rho)/(c\gamma^3)$. More precisely, the time profile (in the plane of deflection) is illustrated in Fig. 11.5 [4]. It consists essentially of a single spike with a characteristic duration $1/\omega_c$. The "critical frequency" ω_c is defined by

$$\omega_c = \frac{3}{2}\frac{\gamma^3 c}{\rho}. \tag{11.18}$$

11.1.1.2 Spectrum of Synchrotron Radiation from a Long Bending Magnet

In most practical cases, a user of synchrotron radiation is interested in the spectral properties rather than in time domain features. Since the user refers to radiation

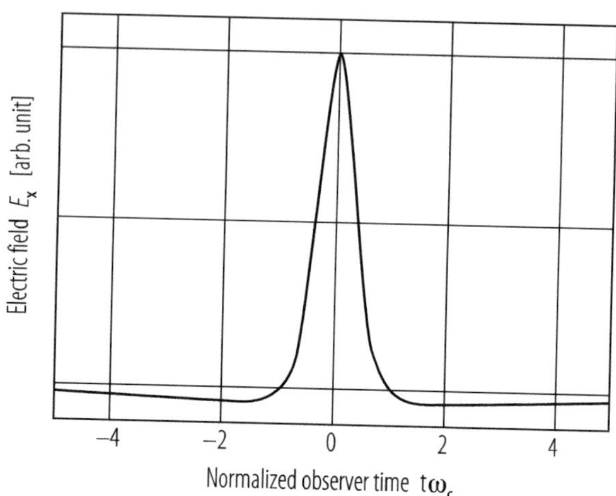

Fig. 11.5 Time profile of radiation field experienced by a distant observer

properties at the location of the observer, we have to investigate the expression

$$\frac{dP(\varphi,\theta,t)}{d\Omega} = R^2\left(\vec{S}\cdot\vec{n}\right) = R^2\varepsilon_0 c\left(\vec{E}\,(t)\right)^2 \tag{11.19}$$

instead of Eq. (11.11). $P(t) = \Delta W/\Delta t$ describes the amount of energy ΔW radiated within the observer's time interval Δt. The total energy radiated per passage into the solid angle $d\Omega$ is thus

$$\frac{dW}{d\Omega} = \varepsilon_0 c R^2 \int_{-\infty}^{\infty} \left|\vec{E}\,(t)\right|^2 dt. \tag{11.20}$$

With the help of Parseval's theorem, the r.h.s. of Eq. (11.20) can be turned into frequency domain: $\int_{-\infty}^{\infty}\left|\vec{E}\,(t)\right|^2 dt = \int_{-\infty}^{\infty}\left|\vec{E}\,(\omega)\right|^2 d\omega$. Here, the Fourier transform of the electric field $\vec{E}\,(t)$ is used:

$$\vec{E}\,(\omega) = \frac{1}{\sqrt{2\pi}}\int_{-\infty}^{\infty}\vec{E}\,(t)e^{i\omega t}dt. \tag{11.21}$$

The energy radiated into the solid angle $d\Omega$ and frequency interval $d\omega$ is thus given, in far field approximation, by

$$\frac{d^2W}{d\Omega d\omega} = 2\varepsilon_0 c R^2\left|\vec{E}\,(\omega)\right|^2. \tag{11.22}$$

The factor of 2 appears since $\vec{E}(t)$ is a real quantity, such that negative frequencies are not considered. According to Eq. (11.10), this means that the expression

$$\vec{E}(\omega) = \frac{q}{4\pi\varepsilon_0 c\sqrt{2\pi}} \int_{-\infty}^{\infty} e^{i\omega(t-R/c)} \frac{\vec{n} \times \left\{ \left(\vec{n} - \vec{\beta} \right) \times \dot{\vec{\beta}} \right\}}{\left(1 - \vec{n} \cdot \vec{\beta} \right)^3 R} dt \qquad (11.23)$$

needs to be evaluated. The term $(t - R/c)$ appears in the exponent since all quantities $\vec{n}$, $\vec{\beta}$, $\dot{\vec{\beta}}$, R must be evaluated at the retarded time. It should be noted that, in order the solid angle $d\Omega$ to be well defined in Eq. (11.20), the relevant part of the trajectory should remain for all times within a volume of diameter much smaller than R, e.g. in a circular accelerator.

For motion on a circle in a constant magnetic field, the result reads [1]:

$$\frac{d^2W(\theta)}{d\Omega d\omega} = \frac{3q^2}{16\pi^3\varepsilon_0 c}\gamma^2\left(\frac{\omega}{\omega_c}\right)^2\left(1+\gamma^2\theta^2\right)^2\left[K_{2/3}^2(\xi) + \frac{\gamma^2\theta^2}{1+\gamma^2\theta^2}K_{1/3}^2(\xi)\right], \qquad (11.24)$$

with the abbreviation

$$\xi = \frac{1}{2}\frac{\omega}{\omega_c}\left(1+\gamma^2\theta^2\right)^{3/2}. \qquad (11.25)$$

$K_{2/3}^2$ and $K_{1/3}^2$ are Bessel functions of fractional order.

Equation (11.24) is of course not any more an instantaneous property but refers to the average over the entire passage of the electron. The angle θ describes the elevation of the observer with respect to that tangent to the circle of motion where $\vec{n} \cdot \vec{\beta}$ becomes maximum in Fig. 11.2 (corresponding to $\varphi = 90°$).

It should also be noted that, although the integral in Eq. (11.23) extends from minus to plus infinity, it does not consider the fact that the electron (normally) performs multiple revolutions in the synchrotron, such that the radiation cone passes the observer many times per second. This fact can be accounted for by multiplying Eq. (11.23) by the revolution frequency which would turn the radiated energy per passage into an average radiation power, and the spectrum would become a discrete one.

The term with $K_{2/3}$ describes the radiation polarized in the (horizontal) plane of deflection (σ-polarization), while the $K_{1/3}$ contribution is vertically polarized (π-polarization). As mentioned before and inferred from Eq. (11.5), the π-polarization has no intensity in the plane of deflection ($\theta = 0$), while the horizontal polarization has its maximum there. Both components have a rather broad spectral distribution with a maximum close to ω_c.

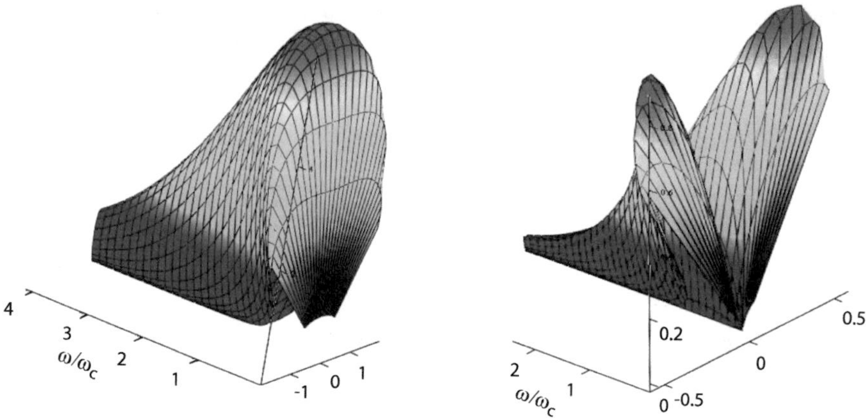

Fig. 11.6 σ-polarization (left) and π-polarization (right) component of synchrotron radiation from a bending magnet. The axis to the left describes the frequency normalized to the critical frequency ω_c, while in the plane perpendicular to ω/ω_c, the directivity of intensity is depicted, with the angle θ taken in normalized $\gamma\theta$ units. The intensity is in arbitrary units

According to Eq. (11.25), the functional dependence of $d^2W/(d\Omega d\omega)$ on ω and θ is a universal one if these quantities are normalized to ω_c and $1/\gamma$, respectively. Using such normalized variables, the frequency and angular dependence of the polarization components are illustrated separately in Fig. 11.6. The vertical opening angle decreases with rising frequency and is about $\pm 1/\gamma$ around $\omega = \omega_c$.

Since in most practical cases γ is very big and the opening angle is thus very small, the angular dependence is often not resolved by users. Integration of Eq. (11.24) over the vertical angle θ yields the spectral power density

$$\frac{dW}{d\omega} = \frac{\sqrt{3}q^2}{4\pi\varepsilon_0 c}\gamma \frac{\omega}{\omega_c} \int_{\omega/\omega_c}^{\infty} K_{5/3}(x)dx = \frac{\sqrt{3}q^2}{4\pi\varepsilon_0 c}\gamma \left(\frac{dW}{d\omega}\right)_{\text{norm}}. \tag{11.26}$$

Equation (11.26) summarizes both polarization components. Again, if the frequency is normalized to ω_c, the spectral power density can be expressed by the universal function

$$\left(\frac{dW}{d\omega}\right)_{\text{norm}} = \frac{\omega}{\omega_c} \int_{\omega/\omega_c}^{\infty} K_{5/3}(x)dx, \tag{11.27}$$

see Fig. 11.7 (note that, in contrast to Fig. 11.6, a double logarithmic scale is used). For a discussion of the individual σ- and π-components see, for instance, [4].

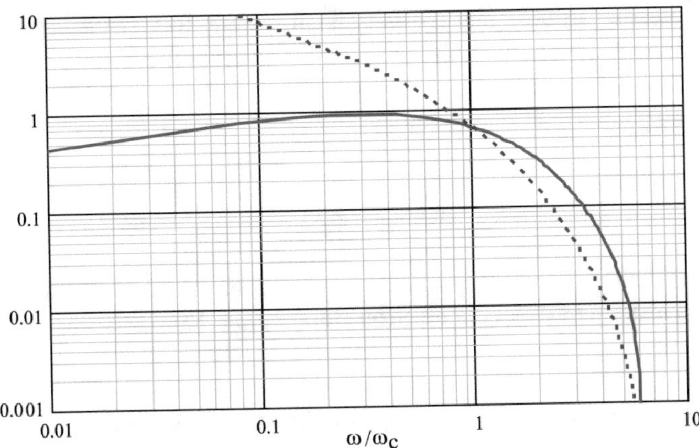

Fig. 11.7 Normalized power spectrum see Eq. (11.27) (solid line) and photon number spectrum (broken line) of synchrotron radiation

Photon Distribution

Sometimes it is necessary to pay attention to the fact that synchrotron radiation is emitted, as any electromagnetic radiation, in quanta (photons) of energy $\varepsilon_\gamma = \hbar\omega$, where $\hbar$ is Planck's constant. The spectral angular photon flux can be obtained from Eq. (11.24), dividing by $\hbar$:

$$\frac{d^2 N_\gamma}{d\Omega\, d\varepsilon_\gamma/\varepsilon_\gamma} = \frac{d^2 W}{d\Omega\, \hbar d\omega}. \tag{11.28}$$

Equation (11.28) calculates the number of photons N_γ emitted per unit solid angle into a relative photon energy interval $d\varepsilon_\gamma/\varepsilon_\gamma$. Again it is noted that this quantity refers to a single turn in the synchrotron.

The angular-integrated spectral photon spectrum corresponding to Eq. (11.26) can be expressed in the form

$$\frac{dN_\gamma}{d\omega} = \frac{\sqrt{3}\alpha\gamma}{\omega_c} \int\limits_{\omega/\omega_c}^{\infty} K_{5/3}(x)dx, \tag{11.29}$$

with the fine structure constant $\alpha = 1/137.036$. This spectrum is also depicted in Fig. 11.7.

Integrating Eq. (11.29) over all frequencies yields the total number N_γ of photons emitted per electron per turn in the synchrotron:

$$N_\gamma = \frac{5\pi}{\sqrt{3}}\alpha\gamma. \tag{11.30}$$

It is interesting to note that N_γ is typically about 100, i.e. it is a rather small number, although the photon number spectrum diverges for $\omega \to 0$, see Eq. (11.29) and Fig. 11.7.

The mean photon energy is given by

$$\langle \varepsilon_\gamma \rangle = \frac{8}{15\sqrt{3}} \hbar\omega_c. \tag{11.31}$$

The considerable granularity in the emission process has quite some impact on the electron beam parameters in electron storage rings [5].

11.1.1.3 Simple Means of Changing the Emission Spectrum

Users often don't appreciate the spectrum of synchrotron radiation, either because of its large frequency width or because it might not contain sufficiently high frequencies for the particular application. In the latter case, the most straight forward solution would be making use of the strong γ-dependence of Eqs. (11.24) and (11.26) and increasing the electron energy. However, there are often considerable technical limitations in this respect, in particular at electron storage rings operating in the GeV regime.

Wavelength Shifters

As the synchrotron radiation spectrum is normalized to the critical frequency ω_c, according to Eq. (11.18) the spectrum can also be hardened by increasing the magnetic field strength, thus reducing the bending radius ρ. To this end, often superconducting magnets are applied. In order to restrict the subsequent modification of the electron beam's design orbit to a small section of the storage ring, a sequence of dipole magnets is frequently used with zero net deflection angle. Such an arrangement is called *wavelength shifter*. The radiation properties are determined in the same way as for ordinary synchrotron radiation. Beyond hardening the spectrum and increasing the flux, there is also some advantage in terms of flexibility in the geometrical arrangement of radiation beam lines.

Short-Magnet Radiation, Edge Radiation

When discussing the characteristic time profile of synchrotron radiation it was assumed in the context of Figs. 11.4 and 11.5 that the electron would propagate in the magnetic field for sufficiently long time such that the radiation cone passes the observer's aperture in its entire angular extension of $\pm 1/\gamma$. To this end, the dipole magnet must have a length of at least

$$\Delta l_c \approx c\Delta t' \approx \frac{2\rho}{\gamma} \approx \frac{2m_0c}{eB}. \tag{11.32}$$

For electrons and magnetic fields in the $B \approx 1$ T range, this results in a few millimeters, which would be of little relevance in most cases. However, if B is much weaker, or if protons are considered (e.g. at LHC/CERN), the resulting time profile of the radiation field is shorter than shown in Fig. 11.5 and assumed for the calculation of the synchrotron radiation spectrum. As the time profile gets shorter, its Fourier transform extends to higher frequencies, thus the spectrum gets "harder". In contrast to increasing B or γ, this hardening is, however, not accompanied by increased flux as the instantaneous properties of emission depend only on the local magnetic field strength which does not change by shortening the magnet.

A spectrum hardening effect similar to short-magnet radiation takes place if the magnetic field rises at the entry face of the magnet (or drops at the exit, respectively) over a distance comparable to or smaller than given by Eq. (11.32). In such case, the time profile of the radiation field exhibits a rising (or falling, respectively) edge steeper than that seen in Fig. 11.5. As a consequence, the Fourier transform extends to rather high frequencies. At high-energy proton synchrotrons this is being used to extend the spectrum towards wavelengths which are easy to observe.

11.1.1.4 Wigglers and Undulators

Definitions

Undulators and wigglers provide a periodic magnetic field over a part of the synchrotron's circumference. In most cases, the magnetic field perpendicular to the electron beam's design orbit can be described by a pure sinusoidal—at least within the small spatial area where the electron beam is present. If the field acting on the electron beam has only one non-zero Cartesian component, the device is called a planar undulator (or wiggler, respectively). The field close to the axis can then be described by

$$B_y(z) = B_0 \sin(k_u z) \,. \tag{11.33}$$

Here, B_0 is the field amplitude and $k_u = 2\pi/\lambda_u$ is the undulator wave number, with λ_u the undulator's period. The z-axis is along the electron beam's initial design momentum, and we have chosen arbitrarily the magnetic field to be in the vertical y-direction.

Typically, the length L_u of these devices is a few meters, and the field integral

$$I_1 = \int_0^{L_u} B_y(s)ds \tag{11.34}$$

is made zero such that there is no over-all deflection of the electron beam's trajectory.

A key parameter is the undulator parameter

$$K = \frac{eB_0\lambda_u}{2\pi m_0 c} \approx (0.934 B_0/\mathrm{T}) \cdot (\lambda_u/\mathrm{cm})\,. \tag{11.35}$$

As this quantity refers to electrons (or positrons), we have assumed m_0 to be the electron rest mass m_e in the last part of the equation. A device with $K \leq 1$ is called undulator, while wigglers exhibit values $K > 1$. Sometimes devices with $K > 1$ are also called undulators if they are used in terms of radiation properties typical of undulators, e.g. when observing the line spectrum in forward direction (see below).

Undulators and wigglers are realized in basically three varieties of technology: iron-based electromagnets, permanent magnets and superconducting magnets. For details, see e.g. [6].

Particle Trajectory

The equation of motion of an electron (with elementary charge $-e$) moving in presence of the field of Eq. (11.33) reads in Cartesian components

$$\ddot{x} = \frac{e}{\gamma m_e} B_y(z) \cdot \dot{z}, \text{ and}$$

$$\ddot{z} = -\frac{e}{\gamma m_e} B_y(z) \cdot \dot{x}. \tag{11.36}$$

To first-order approximation, the periodic solution reads

$$x(t) \cong \frac{K}{\beta\gamma k_u} \sin(k_u \beta c t)\,, \text{ or } x(z) \cong \frac{K}{\beta\gamma k_u} \sin(k_u z)\,, \text{ and}$$

$$z(t) \cong \beta c t. \tag{11.37}$$

This motion is illustrated in Fig. 11.8.

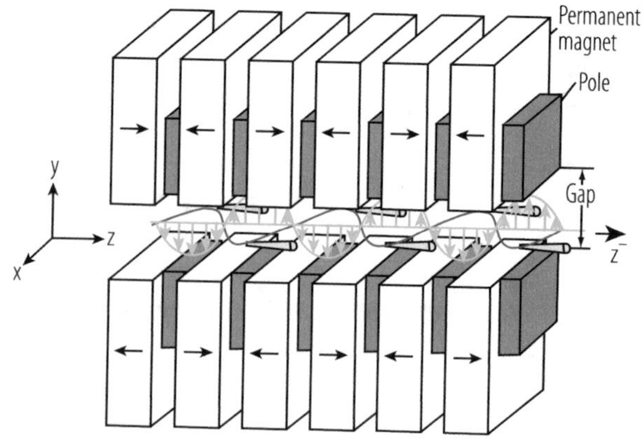

Fig. 11.8 Periodic electron motion (red) in a planar undulator fabricated in hybrid permanent magnet technology. The magnet field is indicated by green arrows

It should be noted that, in order this periodic trajectory to happen, the electron beam must enter the device on an orbit representing the appropriate initial conditions. In addition, the undulator field must begin and end with a quarter-period undulator section.

From Eq. (11.37) the maximum deflection angle can be calculated:

$$\vartheta_{\max} \approx \left(\frac{dx}{dz}\right)_{\max} = K \cdot \frac{1}{\gamma\beta} \approx \frac{K}{\gamma}. \tag{11.38}$$

The maximum orbit excursion of the electron is

$$x_{\max} \cong \frac{K}{\gamma k_u}. \tag{11.39}$$

Under many typical conditions, this results in only a few micrometers and is thus much smaller than the typical electron beam diameter.

In Eq. (11.37) the longitudinal motion $z(t)$was described only to first order, not taking into account its coupling to the transverse motion. A more precise, second order calculation results in

$$z(t) = \overline{v}_z t - \frac{K^2}{8\gamma^2 k_u} \sin\left(2\omega_u t\right), \tag{11.40}$$

where the abbreviation $\omega_u = \overline{\beta} c k_u$ and the average longitudinal velocity

$$\overline{v}_z \approx c\left[1 - \frac{1}{2\gamma^2}\left(1 + \frac{K^2}{2}\right)\right] \equiv \overline{\beta} c \tag{11.41}$$

have been introduced. According to Eq. (11.40), the oscillatory motion in the transverse plane translates into a small modulation of the longitudinal velocity around $\overline{v}_z$. This average longitudinal velocity differs, as described by Eq. (11.41), from c due to two effects: the factor $1 - 1/(2\gamma^2) \approx \beta$ describes by how much the electron's total speed differs from c. The factor $1 + K^2/2$ describes the *mean* additional longitudinal retardation due to the transverse velocity components.

Radiation Properties

Calculation of the radiation spectrum starts again from Eqs. (11.22) and (11.23), however, now the oscillatory motion in the undulator field must be considered when evaluating Eq. (11.23). Qualitatively, it should be clear from Figs. 11.4 and 11.9 that the time profile of the wiggler radiation field ($K > 1$) should resemble the standard dipole field case with the only difference that there will be a series of radiation spikes, see Fig. 11.10. The radiation spectrum differs thus not very much from the synchrotron radiation case but will be N times more intense with N being the number of wiggler periods (see Fig. 11.11).

On the contrary, in the undulator case $K \leq 1$, an observer will only see radiation field within a small angle of approximately $\pm 1/\gamma$ with respect to the forward

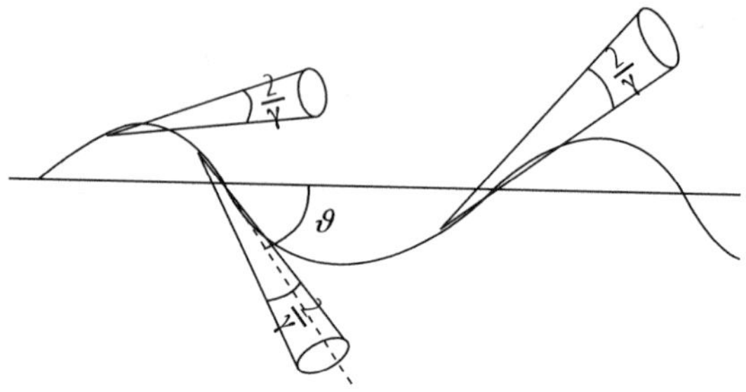

Fig. 11.9 Emission of radiation cones along the oscillatory trajectory in an undulator or wiggler. In this schematic, the wiggler case $K > 1$ is illustrated, where the deflection angles $\vartheta = K/\gamma$ are larger than $1/\gamma$

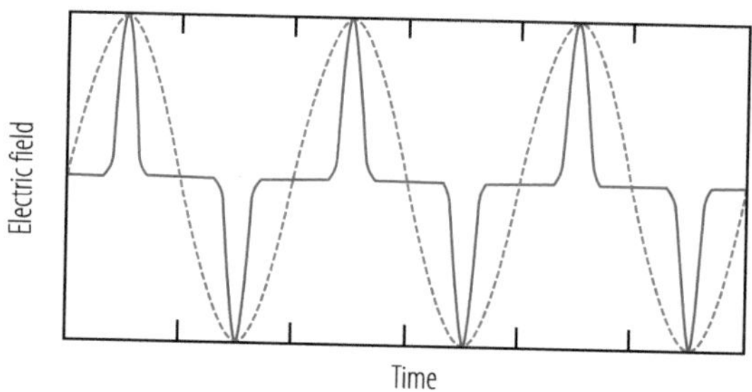

Fig. 11.10 Time profile of radiation field for wiggler (solid line) and undulator radiation (broken line). Note that these two time profiles do not belong to the same undulator period, although they have obviously the same fundamental period in time. The reason is that the slippage between the longitudinal electron velocity and c is larger in the wiggler than in the undulator case

direction, and within this angle he will observe a continuous oscillatory field with N periods. Fourier transform of this time profile results in a fundamental wavelength

$$\lambda_1 = \frac{\lambda_u}{2\gamma^2}\left(1 + \frac{K^2}{2} + \gamma^2\theta^2\right), \tag{11.42}$$

and higher harmonics $\lambda_k = \lambda_1/k$. From symmetry consideration one can immediately understand that in forward direction $\theta = 0$ there will be only odd harmonics.

The angular spectral energy distribution in forward direction (see Fig. 11.11) is given by [4]

$$\begin{aligned}\left(\frac{d^2W}{d\Omega d\omega}\right)_{\theta=0} &\cong \frac{q^2N^2K^2\gamma^2}{4\pi\varepsilon_0 c\left(1+\frac{K^2}{2}\right)^2}\sum\nolimits_{k=2n+1} k^2\left(\frac{\sin\left(\pi N\frac{\omega-k\omega_1}{\omega_1}\right)}{\pi N\frac{\omega-k\omega_1}{\omega_1}}\right)^2 \\ &\times\left|J_{\frac{k-1}{2}}\left(\frac{kK^2}{4\left(1+\frac{K^2}{2}\right)}\right) - J_{\frac{k+1}{2}}\left(\frac{kK^2}{4\left(1+\frac{K^2}{2}\right)}\right)\right|^2.\end{aligned} \tag{11.43}$$

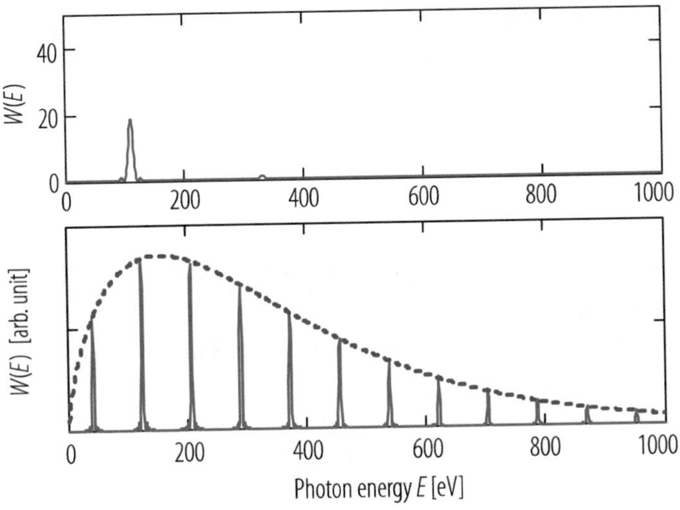

Fig. 11.11 Undulator spectrum in forward direction for $\gamma = 1000$ and $N = 10$ undulator periods. The undulator parameter is $K = 0.5$ (top) and $K = 2$ (bottom). The undulator period is $\lambda_u = 2$ cm. The vertical scales are in arbitrary units. The broken line in the lower diagram shows the spectrum of synchrotron radiation for a magnetic dipole field equal to the peak field of the undulator, with the vertical scale matched to the undulator spectrum arbitrarily

The sinc-function with argument $\pi N(\omega - k\omega_1)/\omega_1$ describes the spectral distribution of the individual undulator harmonics. The total line width of harmonics k is

$$\frac{\Delta\omega}{\omega_k} \approx \frac{1}{kN}. \tag{11.44}$$

The photon number spectrum radiated into the frequency interval $\Delta\omega/\omega_k$ in forward direction is given by

$$\frac{dN_\gamma}{d\Omega} = \alpha N^2 \gamma^2 \frac{\Delta\omega}{\omega_k} k^2 \frac{K^2}{\left(1+\frac{K^2}{2}\right)^2} \left| J_{\frac{k-1}{2}}\left(\frac{kK^2}{4\left(1+\frac{K^2}{2}\right)}\right) - J_{\frac{k+1}{2}}\left(\frac{kK^2}{4\left(1+\frac{K^2}{2}\right)}\right)\right|^2. \tag{11.45}$$

α is again the fine structure constant.

As compared to the synchrotron radiation spectrum Eq. (11.24), undulator radiation provides a spectral density in forward direction larger by some factor N^2. One factor N results from having a number of N undulator periods. The other factor N stems from the fact that, due to interference, the radiation spectrum is concentrated into narrow resonance lines. Thus, in forward direction, the field amplitudes add up coherently, not the intensities. The total radiation energy increases with N as expected.

Sometimes, e.g. for spectroscopy applications, only the spectral contribution of the undulator harmonics in forward direction is of interest. This radiation has a very narrow opening angle:

$$\sigma_\theta \approx \sqrt{\frac{1+K^2/2}{2\gamma^2 kN}} = \sqrt{\frac{\lambda_1}{\lambda_u kN}} = \sqrt{\frac{\lambda_k}{N\lambda_u}} = \sqrt{\frac{\lambda_k}{L_u}}. \tag{11.46}$$

It is some factor of $1/\sqrt{kN}$ narrower than with synchrotron radiation.

Helical Undulators

A helical undulator generates a magnetic field vector following a helical orientation like

$$\vec{B}_{\text{hel}} \cong B_0 \begin{pmatrix} -\sin k_u z \\ \cos k_u z \\ 0 \end{pmatrix}. \tag{11.47}$$

The resulting electron orbit is (appropriate initial conditions given) also a helix, with velocity components

$$\frac{v_\perp}{c} = \frac{1}{c}\sqrt{v_x^2 + v_y^2} = \frac{K}{\gamma}, \text{ and}$$

$$\beta_z = \frac{1}{c}\sqrt{v^2 - v_x^2 - v_y^2} = \sqrt{\beta^2 - \left(\frac{K}{\gamma}\right)^2} \approx 1 - \frac{1}{2\gamma^2}\left(1 + K^2\right), \tag{11.48}$$

which are both constant.

The major differences compared to planar undulators are:

(i) For calculation of the fundamental wavelength according to Eq. (11.42) one needs to replace $K^2/2$ by K^2.
(ii) According to Eq. (11.5), the electric field vector of undulator radiation in forward direction will also move on a perfect helix, thus this radiation will be circular polarized, and it will consist of only the fundamental harmonics, independent of the magnitude of K.

11.1.1.5 Radiation from Many Electrons

Brilliance

The emittance of a synchrotron radiation photon beam, i.e. the product of rms opening angle σ_γ' and rms source size σ_γ is limited due to diffraction. For a perfect

Gaussian optical mode, the minimum-emittance is achieved, given by [4, 7]

$$\varepsilon_\gamma = \sigma'_\gamma \sigma_\gamma = \frac{\lambda}{4\pi}. \tag{11.49}$$

with the wavelength of radiation λ.

As the emittance of the photon beam can never be smaller than indicated by Eq. (11.49), the very narrow opening angle of undulator radiation, Eq. (11.46), means that photons radiated from an undulator have a rather large apparent source size.

If a radiating electron beam has an emittance ε_e smaller than given by Eq. (11.49), the resulting radiation will be indistinguishable from radiation of a point source, if the opening angle σ'_e of the electron beam is matched to the opening angle of the radiation. A crude estimate for σ'_e is $\sigma'_e \approx \sqrt{\varepsilon_e/\beta}$, where β is the magnet optics beta function. Thus, a comparison with Eq. (11.46) would result in an estimated matched beta function $\approx L_u/4\pi$. However, such a small value cannot be kept constant within the undulator length if there are no further magnetic focusing elements. A more appropriate value is thus somewhat larger [4]:

$$\beta_{\text{match}} \approx \frac{L_u}{2}. \tag{11.50}$$

It should be noted that these considerations apply for both transverse directions *x/y* independently.

An electron storage ring providing an electron emittance

$$\varepsilon_e \leq \frac{\lambda}{4\pi} \tag{11.51}$$

in both x and y is said to work at the diffraction limit. Designing electron storage rings operating at (or close to) the diffraction limit is a difficult task, in particular if the wavelengths of interest are very short, e.g. in the hard X-ray regime. In tendency, this requires large ring diameters and sophisticated electron beam optics arrangements. Quite some efforts in this direction are under way at several big laboratories at the time of writing the present article.

In practice, Eq. (11.51) can often not be fulfilled. In this case, since both the photon and electron distributions are Gaussian in good approximation, the effective, combined distribution is given by a convolution and results in

$$\Sigma^2 \approx \sigma_\gamma^2 + \sigma_e^2. \tag{11.52}$$

This holds for both, transverse dimensions and opening angles, and applies again for both transverse directions *x/y* independently.

Many experiments using synchrotron (or undulator) radiation rely on the possibility to focus a small spectral fraction $\Delta\omega/\omega$ of the photon beam, selected e.g. by a monochromator, onto a small spot at the experiment. The figure of merit for such experiments is the brilliance B:

$$B = (dn/dt) / \left(4\pi^2 \Sigma_x \Sigma_y \Sigma_{x'} \Sigma_{y'} d\omega/\omega\right). \tag{11.53}$$

This quantity is also called brightness by some authors. dn/dt is the number of photons per unit time, and Σ_x, etc., are the rms photon beam extensions in x,y dimensions, convoluted in the spirit of Eq. (11.52).

In RF accelerators, electrons are arranged within bunches such that the peak beam current is much larger than the average current. Thus, the instantaneous value of brilliance during the very moment of bunch passage is much larger that its time average.

Undulator beam lines at modern storage rings reach peak brilliance values up to $B_{\text{peak}} \approx 10^{25}$ mm^{-2} mrad^{-2} s^{-1} (0.1%)$^{-1}$, see Fig. 11.12.

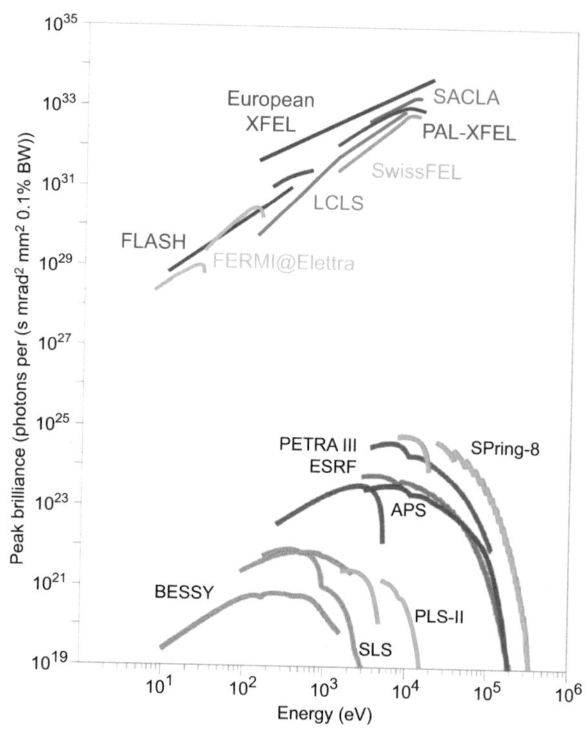

Fig. 11.12 Peak brilliance values delivered by typical undulator beamlines at some state-of-the-art synchrotron radiation storage rings. Values for FEL facilities are also shown. It should be noted that some of the FEL facilities (FLASH, European XFEL) are driven by a superconducting linear accelerator providing an electron bunch repetition rate larger by several orders of magnitude than normal conducting ones, resulting in a correspondingly larger rate of photon pulses. This difference is not visible in this figure of peak values but may be beneficial for several scientific applications

Average brilliance values can exceed values $B_{\mathrm{avg}} \approx 10^{21}$ mm^{-2} mrad^{-2} s^{-1} $(0.1\%)^{-1}$.

In case of a perfectly diffraction limited beam in both *x/y* directions, it is concluded from Eqs. (11.49), (11.51), and (11.52) that Eq. (11.53) simplifies to:

$$B = 4\,(dn/dt) \,/\, \left(\lambda^2 d\omega/\omega\right). \tag{11.54}$$

Coherent Synchrotron Radiation

In sections "Basic Properties of Synchrotron Radiation" through "Wigglers and Undulators", emission by a point charge has been considered. In this case, the intensity scales with the radiating charge q squared. In other words, if this quasi-point charge consists of N_e electrons, the total power would be N_e^2 larger than the power P_0 radiated by a single electron.

In general, if radiation of many electrons is considered, we need to add up first the electric field vectors $\vec{E}_k$ of all electrons coherently before calculating the intensity. In the following we restrict ourselves to the one-dimensional case, where all electrons follow the same path, but just at different times ("pencil beam"). In most cases, the transverse extension of the bunch does not alter the results significantly.

In case of a pencil beam, the electric field contributed by any of the electrons differs from the field of the others only in terms of the phase factor $\exp\{i\omega(t - R_k/c)\}$ in Eq. (11.23), since, at time t, the electron with index k is located at a retarded distance R_k while others will be located at a different distance. Calling these individual phases differences φ_k, the power radiated by the ensemble scales like

$$\begin{aligned} P(\omega) \propto \left|E^2\right| &= \left(\vec{E}_1 + \vec{E}_2 + \ldots\right)\left(\vec{E}_1 + \vec{E}_2 + \ldots\right)^* \\ &= \sum_{k,j}^{N_e} \vec{E}_k \vec{E}_j^{\,*} \propto \sum_{k,j}^{N_e} e^{-i(\varphi_k - \varphi_j)} \\ &\propto N_e + \sum_{k \neq j}^{N_e} e^{i(\varphi_k - \varphi_j)}. \end{aligned} \tag{11.55}$$

How much of phase difference is introduced by a difference in longitudinal position z_k depends, of course, on the wavelength λ considered:

$$\varphi_k - \varphi_j = \frac{2\pi}{\lambda}\left(z_k - z_j\right) = k\left(z_k - z_j\right) = \frac{\omega}{c}\left(z_k - z_j\right). \tag{11.56}$$

If the longitudinal positions of electrons are random, the last term in Eq. (11.55) cancels and the resulting power is given by

$$P_{\mathrm{inc}}(\omega) = N_e P_0(\omega), \tag{11.57}$$

with P_0 the power of a single electron. This contribution is called incoherent radiation.

If, however, the charge distribution is non-random, we describe the longitudinal charge distribution by a normalized density function $\rho(z)$. This function determines, when evaluating the double sum of phase factors in Eq. (11.55), how often each phase difference value appears, so the double sum can be translated into a Fourier transform of $\rho(z)$. The expectation value of the power radiated by the ensemble is then

$$P(\omega) = N_e P_0 + N_e (N_e - 1) \left| F_{\text{long}}(\omega) \right|^2 P_0(\omega) . \tag{11.58}$$

Here,

$$F_{\text{long}}(\omega) = \int_{-\infty}^{\infty} \rho(z) e^{i \frac{\omega}{c} z} dz \tag{11.59}$$

is the longitudinal form factor of the charge distribution inside the electron bunch. The contribution

$$P_{\text{coh}}(\omega) = N_e (N_e - 1) \left| F_{\text{long}}(\omega) \right|^2 P_0(\omega) . \tag{11.60}$$

to the total radiation power is called coherent synchrotron (or undulator, respectively) radiation. Due to the large number N_e of electrons in the bunch, P_{coh} can exceed P_{inc} by many orders of magnitude, even if the longitudinal charge profile has only a small Fourier content at the wavelength of interest.

At storage rings, where the charge distribution can be described very well by a Gaussian, with typical rms bunch length values of a few millimetres, P_{coh} becomes noticeable only at wavelengths in the far infrared. On the other hand, emission of radiation is suppressed very effectively by shielding due to the vaccum chamber. This happens, if in the vacuum chamber, which can be regarded as a curved waveguide, is no propagating mode whose phase velocity matches with the particle velocity. A typical parameter is the shielding wavelength $2\pi\sqrt{h^3/R}$ that can be calculated for a flat chamber of height h and curvature radius R [8]. Thus, coherent synchrotron radiation is not observed at most storage rings, but it has been provoked by arranging the storage ring parameters to achieve very short bunches [9].

The dramatic increase of radiation power can also be achieved if the longitudinal charge density is modulated at the wavelength of interest. This is the physical basics of the free-electron laser (FEL).

At linear accelerators much shorter bunches can be realized than at storage rings, and longitudinal bunch profiles may exhibit a very rich internal structure at scales down to the micrometer range, in particular at high-gain FEL facilities, where bunch lengths in the few micrometer range are needed, see Sect. 11.1.2. At such accelerators, infrared spectroscopy of coherent synchrotron radiation (or of optical transition radiation) represents a powerful tool for electron beam diagnostics [10].

11.1.2 Free-Electron Lasers

11.1.2.1 One Dimensional FEL Theory

The interaction of electrons with electromagnetic waves in an undulator of an FEL is sketched in Fig. 11.8. Important aspects of the FEL process can be described in a model, where electromagnetic fields do not depend on the transverse coordinates x,y and the trajectories of particles with different transverse initial conditions are just transversely shifted. This does not exclude transverse motion of particles.

For the description of the interaction of electrons with waves we need only three types of state quantities. Two of them are particle coordinates in longitudinal phase space: the ponderomotive phase ψ, see below and the relative energy offset η. A third quantity $\hat{E}_x$ stands for the complex amplitude of the electric field of the plane electromagnetic wave, and z, the length along the undulator axis, is the independent coordinate. To simplify the FEL equations, we are not interested in oscillations with the undulator period λ_u, but in the variation of our quantities from period to period or in average versus one period. Therefore the FEL equations for a bunch with N particles (of index ν) per period are in principle of the following type

$$\begin{aligned}\tfrac{d}{dz}\psi_\nu &= f_\psi\left(\psi_1\cdots\psi_N,\eta_1\cdots\eta_N,\hat{E}_x,z\right)\\ \tfrac{d}{dz}\eta_\nu &= f_\eta\left(\psi_1\cdots\psi_N,\eta_1\cdots\eta_N,\hat{E}_x,z\right)\\ \tfrac{d}{dz}\hat{E}_x &= f_E\left(\psi_1\cdots\psi_N,\eta_1\cdots\eta_N,\hat{E}_x,z\right).\end{aligned} \tag{11.61}$$

If the undulator parameters are z independent, and with some approximations they can be written as

$$\begin{aligned}\tfrac{d}{dz}\psi_\nu &\sim \eta_\nu\\ \tfrac{d}{dz}\eta_\nu &\sim \mathrm{Re}\left\{\hat{E}_x\exp\left(i\psi_\nu\right)\right\}\\ \tfrac{d}{dz}\hat{E}_x &\sim b = N^{-1}\sum\exp\left(-i\psi_\nu\right).\end{aligned} \tag{11.62}$$

The first equation relates the relative position in the bunch (ponderomotive phase) to the energy offset. This is just a linearized version of the equation of motion, that describes that a particle with more energy (higher η) is deflected less by the undulator, performs oscillations with smaller amplitude, has a shorter trajectory through one period and increases its phase (longitudinal position) relative to a particle without deviation from reference energy.

The second equation represents the change of particle energy caused by the electromagnetic wave. This depends on the phase of the particle ψ_ν relative to the phase of the wave $\arg\left(\hat{E}_x\right)$ and on the absolute amplitude $\left|\hat{E}_x\right|$. This clarifies the definition of the ponderomotive phase: ψ_ν is $\Delta z_\nu 2\pi/\lambda_l$ with Δz_ν the length shift relative to a reference particle of reference energy ($\eta_{ref} = 0$), and wavelength λ_l into forward direction according to Eq. (11.42).

The third equation assumes that only one spectral line (usually the fundamental wavelength) is excited and is characterized by the amplitude $\hat{E}_x$. This amplitude is driven by the microbunching of the particle distribution with the same wavelength. Therefore the Fourier coefficient of that wavelength, the bunching factor b, determines the increase of $\hat{E}_x$. This approach of a resonant and slowly varying amplitude (SVA) is used in most FEL programs (as in ALICE or GENESIS [11, 12]).

For a systematic presentation of FEL theory see [7, 13–16] and the literature quoted therein.

11.1.2.1.1 Low-Gain FEL Theory

The low-gain approximation assumes that the electromagnetic wave amplitude does not change during the passage of one electron bunch through the undulator. This assumption may sound like a contradiction to the purpose of the FEL of amplifying the intensity of a radiation field. The approximation makes it possible, however, to estimate in a simple way the amount of amplification as a cumulated effect during a single passage of the undulators. Therefore only the change of the ponderomotive phase and of the particle energy are considered. This causes microbunching and a net gain or loss of the particle energy. The microbunching can reach saturation if the electromagnetic wave is strong enough or the undulator sufficiently long. The net change of particle energy is an indirect method to calculate the change of field energy by utilizing energy conservation. The gain function $G = \Delta I/I_0$ is the ratio of the intensity change ΔI to I_0, the intensity of the electromagnetic wave assumed for low-gain theory. Strong bunching is not in contradiction to low-gain operation if G is small, but it requires presence of a very large initial intensity. This is normally accumulated over many round trips of the radiation within an optical cavity, with low-gain amplification at each round trip. The technical realization of such a FEL "oscillator" is thus based on the existence of an optical cavity consisting of low-loss mirrors.

The light wave co-propagating with the electron beam is taken as a plane wave $E_x(z,t) = E_0 \cos(k_l z - \omega_l t + \psi_0)$ with wavelength λ_l and wave number $k_l = \frac{2\pi}{\lambda_l}$. The motion of a particle in a planar undulator is described by Eqs. (11.37) and (11.40):

$$x(z(t)) \approx \frac{K}{\gamma k_u} \sin(k_u z(t))$$
$$z(t) \approx \overline{v}_z t - \frac{K^2}{8\gamma^2 k_u} \sin(2\omega_u t) .$$

The transverse velocity is

$$v_x(t) \approx \frac{K\overline{v}_z}{\gamma} \cos(k_u z(t)) , \tag{11.63}$$

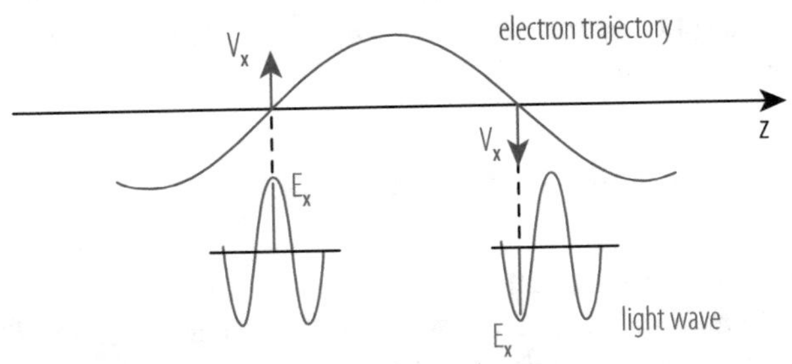

Fig. 11.13 Condition for sustained energy transfer from electron to light wave

and the time derivative of the electron energy is

$$m_e c^2 d\gamma/dt = -e\mathbf{v}\cdot\mathbf{E} = -eE_x\left(z(t),t\right)v_x(t). \tag{11.64}$$

Sustained energy transfer from electron to light wave requires that the light wave slips forward with respect to the electron by $\lambda_l/2$ per half period of the electron trajectory, see Fig. 11.13. This is fulfilled if $\overline{v}_z T_u + \lambda_l = cT_u$, with $T_u = 2\pi/\omega_u \approx \lambda_u/c$. This is a condition for the average longitudinal velocity Eq. (11.41) that relates the particle energy γ and undulator properties λ_u, K to the photon wavelength

$$\lambda_l = \frac{\lambda_u}{2\gamma^2}\left(1+\frac{K^2}{2}\right). \tag{11.65}$$

It is the same wavelength as for the radiation of a single electron in forward direction. (Compare Eq. (11.42) with $\theta = 0$.) Slippages by $3\lambda_l/2$, $5\lambda_l/2$ ... are also permitted, leading to odd higher harmonics ($\lambda_l/3$,$\lambda_l/5$ ...) of the FEL radiation. However slippages of $2\lambda_l/2$, $4\lambda_l/2$ ... yield zero net energy transfer, hence even harmonics are absent in FEL radiation.

To calculate the average derivative of electron energy, as it is required for the FEL Eq. (11.62), we have to calculate the mean value of Eq. (11.64) in a time interval of the length T_u. Supposed the resonance condition Eq. (11.65) is fulfilled, the derivative of energy is a periodic function in time (with period T_u) and the mean value is

$$\langle d\gamma/dt\rangle_{T_u} = -e\frac{E_0\overline{v}_z}{2\gamma m_e c^2}\hat{K}\cos\psi_0, \tag{11.66}$$

with

$$\hat{K} = K\left[J_0\left(\frac{K^2}{4+2K^2}\right) - J_1\left(\frac{K^2}{4+2K^2}\right)\right]. \tag{11.67}$$

We rewrite Eq. (11.66) for $\eta(z) = (\gamma - \gamma_r)/\gamma_r$ with $z = \overline{v}t$ as independent coordinate to get an equation of the required type:

$$\frac{d\eta}{dz} = -e\frac{\hat{K}}{2\gamma^2 m_e c^2}\,\mathrm{Re}\left\{E_0 \exp\left(i\psi_0\right)\exp\left(i\psi\right)\right\}. \tag{11.68}$$

Additional to the wave phase ψ_0 we consider the ponderomotive phase ψ which is individual per particle. This reflects that an arbitrary particle can be shifted in time relative to the "reference" particle with the trajectory $x(t)$, $z(t)$.

The slip $\Delta\psi = k_l\left(\overline{v}_z\left((1+\eta)\,\gamma_r\right) - \overline{v}_z\left(\gamma_r\right)\right)T_u$ of the ponderomotive phase in one undulator period due to η is caused by the energy dependency of the average longitudinal velocity Eq. (11.41). Therefore the longitudinal dispersion is in linear approximation

$$\frac{d\psi}{dz} = 2k_u\eta. \tag{11.69}$$

FEL Pendulum Equations and Gain Function

The particle dynamic in longitudinal phase space (ψ,η) is fully determined by Eqs. (11.68) and (11.69). They are formally equivalent to that of a mathematical pendulum. The phase space trajectories of few particles is illustrated in Fig. 11.14. The region of bounded motion is separated from the region of unbounded motion by a curve called the separatrix. Initially the particles are evenly distributed on the black line, but the endpoints are, in average, closer to the phase $\pi/2$. This illustrates microbunching. In the right diagram for the initial condition $\eta > 0$, we see a loss of the energy averaged over all particles. This net change of particle energy is an indirect method to calculate the change of field energy and light intensity by utilizing energy conservation.

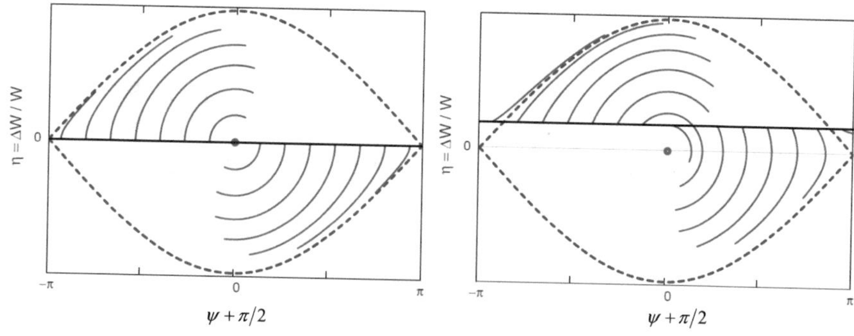

Fig. 11.14 Phase space (ψ,η) trajectories for 15 electrons of different initial phase (red) and separatrix (blue). Left picture: electrons are initially on resonance energy. Right picture: electron energy is initially above resonance energy

The FEL gain function is defined as the relative increase in light intensity during one passage of the undulator: $G = (I - I_0)/I_0 = \left|\tilde{E}_x/E_0\right|^2 - 1$. The gain is proportional to the negative derivative of the line-shape curve of undulator radiation (Madey theorem) [17]

$$G(\xi) = -\frac{\pi e^2 \hat{K}^2 N_u^2 \lambda_u^2 n_e}{4\varepsilon_0 m_e c^2 \gamma_r^3} \cdot \frac{d}{d\xi}\left(\frac{\sin^2 \xi}{\xi^2}\right), \tag{11.70}$$

with the detuning $\xi = \pi N_u \eta$, n_e the number of electrons per unit volume and N_u the number of undulator periods.

Low Gain FELs

Low gain FELs like the infrared FEL at JLAB consist of a short undulator in an optical cavity fed by a multi bunch source. Upon each bunch passage through the undulator the light intensity grows by only a few per cent, but after very many round trips a large average FEL beam power can be achieved, e.g. more than 10 kW in the infrared FEL at JLAB [16, 18].

11.1.2.1.2 High-Gain FEL Theory

For wavelengths far below the visible it is not possible to build optical cavities. An option is then making the undulator much longer, such that the gain during a single passage becomes attractive and an optical cavity arrangement becomes obsolete. In this case, the low-gain assumption $|G| < < 1$ cannot be made any more, meaning that the stimulation and propagation of electromagnetic waves must be taken into account. Results from LCLS (Linac Coherent Light Source, SLAC, Stanford, USA) are shown in Fig. 11.15 [19].

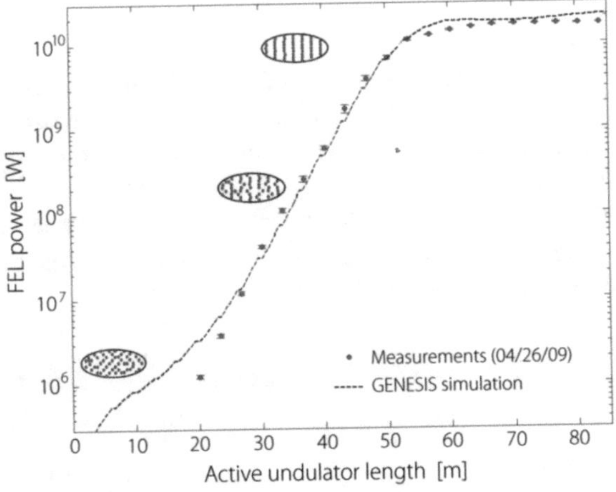

Fig. 11.15 Exponential growth and saturation of the FEL power in LCLS at $\lambda = 0.15$ nm as function of active undulator length [19]. The progressing microbunching is indicated schematically

Some important effects as microbunching, exponential growth and saturation of the FEL power and even the start-up-from-shot-noise can be studied with help of the one dimensional periodic model Eq. (11.62). The equations for $\frac{d}{dz}\psi_\nu$ and $\frac{d}{dz}\eta_\nu$ are the same as for the low gain case (Eqs. 11.68 and 11.69). The stimulation of the electric field amplitude is calculated with help of the wave equation.

Wave Equation

In one dimensional theory the electromagnetic field is described by a plane wave. According to this approach the finite beam cross-section is extended to infinity and each electron gets an infinite number of doubles in the expanded volume. The point particles are replaced by 1D charge sheets. All quantities as charge density, bunch current and electromagnetic fields are independent on the transverse coordinates x, y. The radiation field obeys the 1D inhomogeneous wave equation

$$\left[\frac{\partial^2}{\partial z^2} - \frac{1}{c^2}\frac{\partial^2}{\partial t^2}\right] E_x(z,t) = \mu_0 \frac{\partial j_x}{\partial t}, \tag{11.71}$$

where j_x is the transverse current density resulting from the sinusoidal motion. We make the ansatz

$$E_x(z,t) = \hat{E}_x(z,t)\exp\left[i\left(k_l z - \omega_l t\right)\right] \tag{11.72}$$

with a complex amplitude function $\hat{E}_x(z,t)$. The microbunching effect is anticipated by assuming a small periodic modulation $\hat{j}_1(z,t)$ of the longitudinal current density

$$j_z(z,t) = j_0(z - \overline{v}t) + \hat{j}_1(z,t)\cdot\exp\left(ik_l\left(\frac{z + \hat{z}\sin(2k_u z)}{\overline{\beta}} - ct\right)\right), \tag{11.73}$$

with $j_0(z - \overline{v}t)$ the current density without microbunching and $\hat{z} = K^2/\left(8\gamma^2 k_u\right)$. Note that the exponential term describes the fast oscillation in time and space, while $\hat{E}_x(z,t)$, $\hat{j}_1(z,t)$ and $j_0(\overline{v}t)$ are slowly varying amplitudes (SVA). The variations of SVA quantities in z and t are slowly compared to the λ_u respectively λ_l/c. Therefore the longitudinal oscillation of the bunch (compare Eq. (11.40)) can be neglected for SVA quantities but not for the exponential function. The transverse current density is

$$j_x(z,t) \approx \frac{v_x(z)}{\overline{v}_z} j_z(z,t), \tag{11.74}$$

with $v_x(z)$ from Eq. (11.63). Combining Eqs. (11.71), (11.72), and (11.74) and neglecting derivatives of SVA quantities compared to fast terms yields

$$\left(\frac{\partial}{\partial z}+\frac{1}{c}\frac{\partial}{\partial t}\right)\hat{E}_x(z,t)=-\frac{c\mu_0 K}{2\gamma_r}\hat{j}_1(z,t)\cos(k_u z)\exp(ik_u z)\exp\left(ik_l\hat{z}\sin(2k_u z)\right).$$

On the right hand side appears the product of three functions that are z periodic in λ_u respectively $\lambda_u/2$, but for the SVA approach we are only interested in variations large compared to the undulator period. Therefore we average this product along one undulator period and get

$$\left(\frac{\partial}{\partial z}+\frac{1}{c}\frac{\partial}{\partial t}\right)\hat{E}_x(z,t)=-\frac{c\mu_0\hat{K}}{4\gamma_r}\hat{j}_1(z,t). \tag{11.75}$$

Again the longitudinal oscillation is regarded by the modified undulator parameter $\hat{K}$, Eq. (11.67).

Periodic Approach

The periodic approach is applicable if the initial bunch and electromagnetic stimulation are sufficiently long in time, or more precisely if time variations are slowly compared to coherence time L_{coh}/c, with L_{coh} the coherence length defined below in Eq. (11.85). The time dependency in Eq. (11.75) is neglected $\hat{E}_x(z,t)\rightarrow\tilde{E}_x(z)$, $\hat{j}_1(z,t)\rightarrow\tilde{j}_1(z)$, $j_0(z-\overline{v}t)\rightarrow j_0$ and only particles in one micro-period are considered. We choose N electrons with start phases ψ_ν in the range $0\leq\psi<2\pi$. Then the modulation amplitude follows from Fourier series expansion

$$\tilde{j}_1=2j_0\sum_{\nu=1}^{N}\exp(-i\psi_\nu)/N. \tag{11.76}$$

The sum $\sum_{\nu=1}^{N}\exp(-i\psi_\nu)/N$ is called bunching factor.

Coupled First-Order Equations

Combining Eqs. (11.68), (11.69) and (11.75) one obtains a set of coupled first-order equations

$$\begin{aligned}\frac{d}{dz}\psi_\nu&=2k_u\eta_\nu\\ \frac{d}{dz}\eta_\nu&=-\frac{e\hat{K}}{2m_ec^2\gamma^2}\,\mathrm{Re}\left\{\tilde{E}_x\exp(i\psi_\nu)\right\}\\ \frac{d}{dz}\tilde{E}_x&=-\frac{\mu_0c\hat{K}}{4\gamma}\tilde{j}_1\end{aligned} \tag{11.77}$$

which, together with Eq. (11.76), describe the evolution of the phases ψ_n and energy deviations η_n of the N electrons, as well as the growth of $\tilde{E}_x(z)$ and $\tilde{j}_1(z)$. Longitudinal Coulomb forces ("space charge forces") are of minor importance in short-wavelength FELs [14] and are neglected here and in Eq. (11.78) below.

Third Order Equation
The main physics of the high-gain FEL is contained in the first-order Eq. (11.77) but these can only be solved numerically. If the modulation current $\tilde{j}_1$ remains small a linear third order differential equation for the electric field can be derived (see e.g. [7, 14, 20, 21]):

$$\tilde{E}_x''' + 4ik_u\eta\tilde{E}_x'' - 4k_u^2\eta^2\tilde{E}_x' - \frac{i}{\left(\sqrt{3}L_{g0}\right)^3}\tilde{E}_x = 0 \tag{11.78}$$

with the 1D power gain length

$$L_{g0} = \frac{1}{\sqrt{3}}\left[\frac{4\gamma_r^3 m_e}{\mu_0\hat{K}^2e^2k_un_e}\right]^{1/3}. \tag{11.79}$$

Exponential Gain and Saturation
The solution of Eq. (11.78) is of the form $\tilde{E}_x(z) = \sum_{j=1}^{3}A_j\exp\left(\alpha_j z\right)$. For the special case $\eta = 0$ (electrons are on resonance energy) one finds $\alpha_{1,2} = \left(\pm 1 + i/\sqrt{3}\right)/\left(2L_{g0}\right)$, $\alpha_3 = -i/\left(\sqrt{3}L_{g0}\right)$. In case of laser seeding (see below) with an initial field E_0, all amplitudes are equal, $A_j = E_0/3$. The light power stays almost constant in the "lethargy regime", $0 \leq z < {\sim}2L_{g0}$, but then it grows exponentially (see Fig. 11.16)

$$P(z) \propto \exp\left(2\,\mathrm{Re}\left[\alpha_1\right]z\right) \equiv \exp\left(z/L_{g0}\right). \tag{11.80}$$

The Eq. (11.77) yield the same result as Eq. (11.78) in the lethargy and exponential gain regimes but describe FEL saturation in addition. The saturation

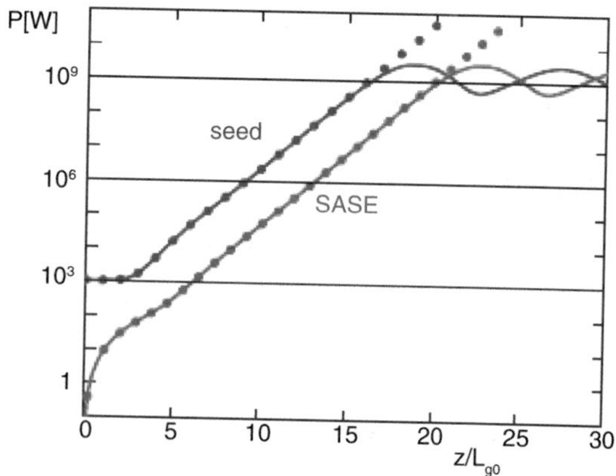

Fig. 11.16 FEL power as a function of z/L_{g0} in a seeded FEL (blue) and a SASE FEL (red). Solid curves: numerical integration of coupled first-order Eq. (11.77). Dots: analytic solution of third-order Eq. (11.78)

power is

$$P_{sat} \approx \rho P_b, \tag{11.81}$$

where P_b the electron beam power, and ρ is the dimensionless FEL (Pierce) parameter [21]

$$\rho = \frac{\lambda_u}{4\pi\sqrt{3}L_{g0}} = \left[\frac{\pi}{8}\frac{I_0}{I_A}\frac{\hat{K}^2}{\gamma_r^3 A_b k_u^2}\right]^{1/3}. \tag{11.82}$$

(I_0 peak current, $I_A \approx 17$ kA Alfven current, A_b beam cross section). For short-wavelength FELs ρ is typically of the order $10^{-4} \dots 10^{-3}$.

FEL Gain-Function and Bandwidth

For a short undulator (length $\leq L_{g0}$), the high-gain FEL theory agrees with the low-gain theory, but in long undulators strong differences are seen: the gain is much larger and the gain-function approaches a Gaussian (Fig. 11.17). The high-gain FEL acts as a narrow-band amplifier with an rms bandwidth [7]

$$\sigma_\omega/\omega = \sqrt{\frac{\rho\lambda_u}{z}\frac{9}{2\pi\sqrt{3}}}. \tag{11.83}$$

Self-Amplified Spontaneous Emission (SASE)

SASE [20, 21] permits the startup of lasing without seed radiation. Intuitively speaking, spontaneous undulator radiation produced in the first section of a long undulator serves as seed radiation in the remaining section. More precisely speaking, because of the random electron distribution, the current contains a noise term which has a

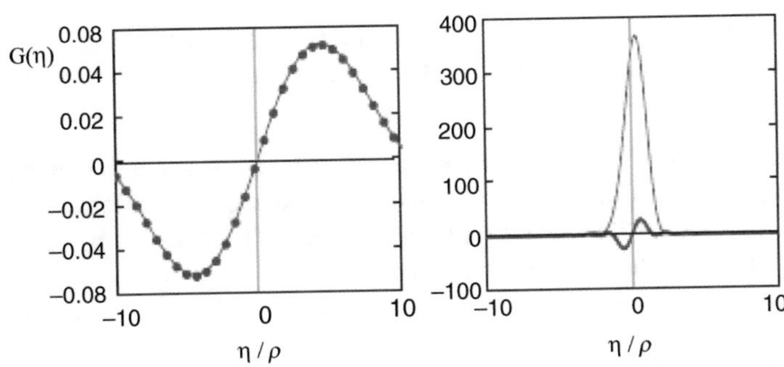

Fig. 11.17 FEL gain function $G(\eta)$ plotted vs. η/ρ at two positions in a long undulator: left $z = 1L_{g0}$, right $z = 8L_{g0}$. Red curves: high-gain theory. Blue dots/curve: low-gain theory (Madey theorem)

spectral component within the FEL bandwidth. The effective shot-noise power and modulated current density are [7, 22]

$$P_n = \rho \gamma m_e c^2 \sigma_\omega / (2\pi), \quad \tilde{j}_1(0) \approx \sqrt{e I_0 \sigma_\omega} / A_b. \tag{11.84}$$

The computed power rise for typical parameters of the soft x-ray FEL FLASH (see e.g. [7]) is shown in Fig. 11.16. Saturation is achieved at an undulator length $L_u \approx 20 L_{g0}$. The SASE bandwidth at saturation is $\sigma_\omega^s/\omega \approx \rho$. SASE radiation exhibits shot-to-shot fluctuations in its output spectrum. The coherence length at saturation is

$$L_{coh} \approx \sqrt{\pi} c / \sigma_\omega^s = \lambda_l / \left(2\sqrt{\pi}\rho\right) \approx 11 \frac{\lambda_l}{\lambda_u} L_{g0}. \tag{11.85}$$

For a bunch length $L_b > L_{coh}$, the average number of spikes in the wavelength spectra is $M = L_b / L_{coh}$ (assuming full transverse coherence). M can be interpreted as the number of coherent modes of the FEL pulse. In the exponential gain regime the normalized radiation pulse energy $u = U_{rad} / \langle U_{rad} \rangle$ fluctuates according to the gamma distribution [14]

$$\rho_M(u) = \frac{M^M u^{M-1}}{\Gamma(M)} e^{-Mu}, \quad \sigma_u^2 = 1/M. \tag{11.86}$$

Phase Space and Simulation of Microbunching
The FEL dynamics resembles the synchrotron oscillations of a proton in a synchrotron or storage ring. In the (ψ, η) phase space the particles rotate clockwise, hence particles in the right half of an FEL bucket transfer energy to the light wave, while those in the left half withdraw energy, see Fig. 11.18 and compare Fig. 11.14. Equation (11.77) are well suited for modelling the microbunching. For $z \geq 12 L_{g0}$ pronounced microbunches evolve in the right halves of the FEL buckets and increase the light intensity, while beyond $18 L_{g0}$ they move into the left halves and reduce it. The FEL power oscillations in Fig. 11.16 are caused by this rotation in phase space.

Higher Harmonics
Close to saturation, the periodic sequence of narrow microbunches (see Fig. 11.18) corresponds to a modulation current with rich harmonics contents. In a planar undulator, odd higher harmonics will be amplified. The third (fifth) harmonic can reach 1% (0.1%) of the fundamental power.

11.1.2.2 Three Dimensional Effects

The realistic description of high-gain FELs has to be based on a three-dimensional (3D) theory, taking into account electron beam emittance and energy spread, and optical diffraction. In idealized cases, e.g. a round beam with uniform longitudinal charge density, an FEL eigenmode equation including all these effects can be

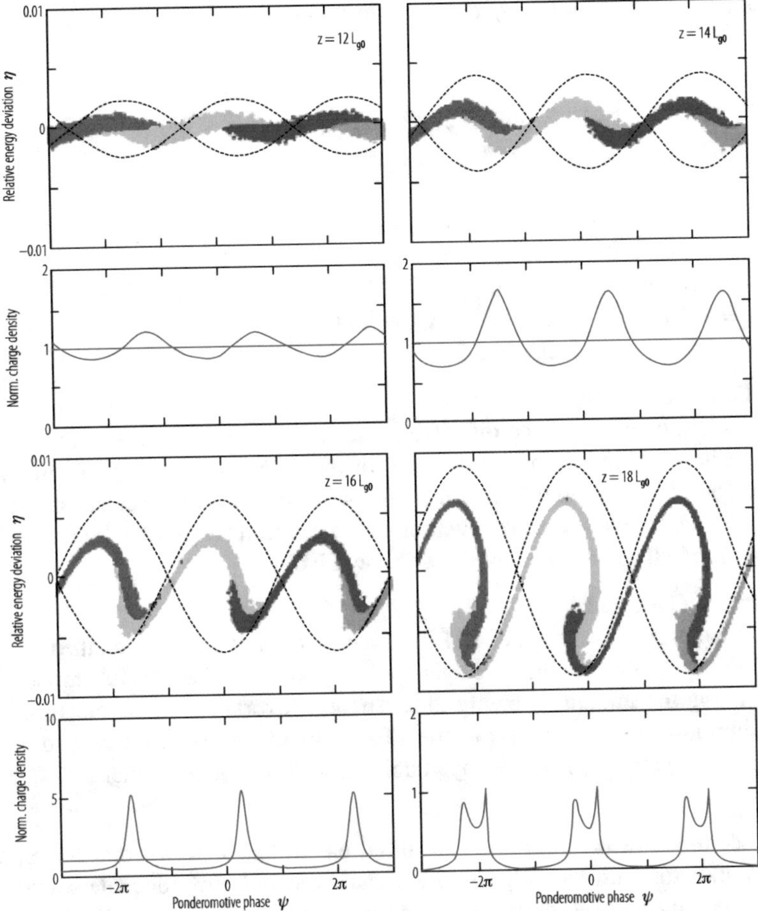

Fig. 11.18 Evolution of the microbunch structure at $z = 12L_{g0},\ 14L_{g0},\ 16L_{g0},\ 18L_{g0}$. Upper subplots: distribution of particles in (ψ, η) phase space. Three FEL buckets are indicated by dashed curves. Lower subplots: normalized charge density as function of ψ

developed [14, 23]. More realistic cases require sophisticated simulation codes such as FAST [24], GENESIS [11] or GINGER [25]. These are indispensable for the design of short-wavelength FELs.

3D Gain Length

The 3D gain length L_g is typically 30–50% longer than the 1D gain length L_{g0}. According to [23] L_g can be expressed in terms of three dimensionless parameters: $X_\gamma = L_{g0}4\pi\sigma_\eta/\lambda_u$ (energy spread parameter), $X_d = L_{g0}\lambda_l/\left(4\pi\sigma_r^2\right)$ (diffraction parameter, σ_r rms beam radius) and $X_\varepsilon = L_{g0}4\pi\varepsilon/(\beta_{av}\lambda_l)$ (angular spread parameter, ε emittance, β_{av} average beta function).

$$L_g = L_{g0}\,(1+\Lambda) \tag{11.87}$$

$$\begin{aligned}\Lambda = a_1 X_d^{a_2} + a_3 X_\varepsilon^{a_4} + a_5 X_\gamma^{a_6} + a_7 X_\varepsilon^{a_8} X_\gamma^{a_9} + \\ + a_{10} X_d^{a_{11}} X_\gamma^{a_{12}} + a_{13} X_d^{a_{14}} X_\varepsilon^{a_{15}} + a_{16} X_d^{a_{17}} X_\varepsilon^{a_{18}} X_\gamma^{a_{19}}\end{aligned}$$

$$\begin{aligned}&a_1 = 0.45, a_2 = 0.57, a_3 = 0.55, a_4 = 1, 6,\\ &a_5 = 3.0, a_6 = 2.0, a_7 = 0.35, a_8 = 2.9,\\ &a_9 = 2.4, a_{10} = 51, a_{11} = 0.95, a_{12} = 3.0,\\ &a_{13} = 5.4, a_{14} = 0.7, a_{15} = 1.9, a_{16} = 1140,\\ &a_{17} = 2.2, a_{18} = 2.9, a_{19} = 3.2.\end{aligned}$$

Gain Guiding

Gain guiding counteracts the diffractive widening of the FEL beam since most of the light is generated in the central core of the electron beam [26]. Gain guiding permits the FEL beam to follow slow, "adiabatic" motions of the electron beam and is thus crucial for the tolerable deviation of the electron beam orbit from a perfectly straight line in the long undulator of an x-ray FEL.

Transverse Coherence

The fundamental Gaussian mode TEM00 has its highest intensity on the beam axis while higher modes extend to larger radial distances. The TEM00 mode grows fastest along the undulator, owing to its superior overlap with the electron beam. Near saturation it dominates and the FEL radiation possesses a high degree of transverse coherence, as verified by double-slit diffraction experiments [27].

Seeding

FEL "Seeding" means to provide an initial electromagnetic wave at the entrance of the undulator with the help of an external laser pulse of adequate wavelength. Various seeding methods have been proposed to improve the longitudinal coherence properties of SASE radiation, and to reduce the relative timing jitter between pump and probe signals in time domain experiments aiming at femtosecond level resolution. Direct seeding requires a coherent signal well above the shot-noise level. In the VUV such signals may be obtained by high harmonic generation (HHG) in a gas [28, 29]. At shorter wavelengths, self-seeding [30] may be applied: a SASE signal, produced in a short undulator, is passed through a monochromator and serves as narrow-band seed radiation in the main undulators following further downstream. In a high-gain harmonic generation (HGHG) FEL [31], the electron beam is energy-modulated in an undulator by interaction with a powerful laser. A magnetic chicane converts the energy modulation to a density modulation. A second undulator causes the density-modulated beam to emit coherent radiation at a higher harmonic frequency. In an echo-enabled harmonic generation (EEHG) FEL [32], a second modulator followed by a second chicane are inserted before the radiator. The electron beam interacts twice with two laser pulses in the two modulators. The longitudinal phase space distribution becomes highly nonlinear, leading to density modulations at a very high harmonic number initiated by a modest energy modulation.

11.1.2.3 Technical Requirements

Very bright electron beams are required to drive ultraviolet and x-ray FELs. Higher peak current and smaller cross sections reduce the gain length (see Eq. (11.82)). High peak currents require longitudinal bunch compression, but the energy spread is increased by this process (which affects the gain length through X_γ in Eq. (11.87)). Very low-emittance beams can be generated by specially designed particle sources with photocathode or with thermionic emission. The beam cross section in the undulator can be reduced by stronger focusing (i.e., smaller β_{av}), but the increased angular spread will eventually degrade the FEL gain (through X_ε in Eq. (11.87)). The FEL design optimization is therefore multi-dimensional and beyond our scope here. Typical requirements on electron beams are

$$I_0 \geq 1\ \text{kA}, \quad \sigma_\eta < \rho/2, \quad \varepsilon \sim \lambda_l/(4\pi)\,. \tag{11.88}$$

These requirements apply to the "slice" beam qualities defined on the scale of the coherence length (see Eq. (11.85)). For harmonic generation FELs, the slice energy spread should be much smaller than the ρ-parameter of the final amplifier because the additional energy modulation imposed on the beam becomes the effective energy spread there. Beam current, slice emittance and energy spread should be "flat" along the bunch in order not to increase the final radiation bandwidth. However, there are proposals for introducing, on purpose, some longitudinal variation of the electron energy within the bunch to generate a controlled frequency chirp in the FEL radiation pulse [33].

High-quality electron beams as described can be produced with linear accelerators but not with storage rings, mainly due to synchrotron radiation effects.

11.2 Accelerators in Industry

K. H. W. Bethge · J. Meijer

11.2.1 Introduction

Accelerators in their earliest stages of development served exclusively as tools of fundamental research; today they find application over a broad range of technical, industrial and medical areas.

One driving force of Ernest Lawrence for the development of the cyclotron was a cancer illness of his mother. Thus the possibility of treating tumors by radiation stands at the beginning of accelerator development.

Initially the application of accelerators in industry [34, 35] was, a by-product of fundamental research. This was because it was not needed continuously and because radiation safety requirements were in many cases too high to make industrial application economical.

On the other hand the installation of accelerators in special industrial laboratories or production lines does not find much publicity and therefore information about these accelerators is hard to obtain.

Two main lines of machine installations have to be considered (a) electron accelerators and (b) ion accelerators.

Additionally, the accelerators can be sub-divided into machines that analyze or modify materials. Electron microscopes are a simple example related to analysis of materials. In addition, there is Accelerator Mass Spectrometry (AMS) which is used to analyse samples of specific species for their applicability particularly in the pharmaceutical industry.

Some applications belong to different fields like the production of radioactive probes for medical application, e.g., PET (positron emission tomography) or SPECT (single photon emission computer tomography). Such probes are industrially produced in close consultation with medical consumers.

The industrial application of synchrotron radiation for photolithographic processes is in general part of the cooperation between large research institutes with suitable installations and relevant industrial partners: the beamtime is paid for by the relevant industries.

11.2.2 Electron Accelerators

Electron accelerators for industrial applications [36] are classified by their energy, focusing capability as well as by the obtainable dose. The large number of possible applications in different industries are listed in Table 11.1 [35]. The electron microscope is the classical low energy electron accelerator for analyzing samples. Unfortunately, the Scherzer theorem forbids a defocussing symmetric lenses with constant voltage. Electron microscopes with achromatic and aspherical lenses are very challenging. In the case of electron beam processing, the incident energy determines the maximum material thickness and the electron beam current and power determine the maximum processing rate.

The purpose of electron irradiation is the transfer of energy doses to produce neutral or positive charged radicals which react very rapidly with other chemical compounds. Spin correlation plays the important role particularly for neutral radicals. Thus the processes listed in Table 11.1 become production processes in the different branches of industry [36]. The energy transfer rate is nearly constant over a large energy range between 500 keV and several MeV, thus a homogeneous

Table 11.1 Possible applications of electron accelerators in industries [35]

Industries	Processes	Products
Chemical	Crosslinking	Polyethylene
Petrochemical	Depolymerisation	Polypropylene
	Grafting	Copolymers
	Polymerisation	Lubricants
		Alcohol
Electrical	Crosslinking	Building
	Heat-shrink	Instrument
	Memory	Telephone wires
	Semiconductor	Power cables
	Modification	Insulating tapes
		Shielded cable splices
		Zener diodes
		ICs, SCRs
Coatings	Curing	Adhesive tapes
Adhesives	Grafting	Coating paper products
	Polymerization	Wood/plastic composites
		Veneered panels
		Thermal barriers
Plastics	Crosslinking	Food shrink wrap
Polymers	Foaming	Plastic tubing and pipes
	Heat shrink memory	Molded packing forms
		Flexible packing laminates
Rubber	Vulcanization	Tire components
	Green strength	Battery separators
	Graded cure	Roofing membrane
Health	Sterilization	Medical disposals
Pharmaceutical	Polymer modification	Membranes
		Powders and ointments
		Ethic drugs
Pollution	Disinfection	Agricultural fertilizers
Control	Precipitation	Safe stack gas emission
	Manomer entrapment	Ocean-life nutrients form
Sludge		OSHA and EPA compliances
Pulp	Depolymerization	Rayon
Textiles	Grafting	Permanent-press textiles
	Curing	Soil-release textiles
		Flocked and printed fabrics
Aerospace	Curing, repair	Composite structures

IC integrated circuit, *SCR* silicon controlled rectifier (thyristor), *OSHA* Occupational Safety & Health Administration, *EPA* Environmental Protection Agency production of defects like point defects in semiconductors is possible. In comparison to ion beam related defects, radiation damages produced byelectrons are loosely connected, thus avoiding extended defects.

The electron irradiation of polymers produces a substantial fraction of radicals (charged or neutral) which react strongly with other polymers, monomers or additional substances. The processing of polymers by electron bombardment fulfills the demands of modern industrial production e.g. the compatibility with environmental requirements. This is particularly the case for all processes which include the curing of adhesive coatings. Furniture, cloth production and paper industries have adopted electron beam processes. Electron beams of lateral dimensions of 1 m or more are incorporated into production lines [36]. In many of these processes the need to remove dangerous fumes can be avoided.

The vulcanization process in the rubber industry proceeds under electron beam treatment without the addition of sulfur. The applied doses range from 30 to 50 kGy.

For curing of coatings, adhesives, ink on paper, plastic and metal substrates low energy accelerators (75–300 keV) are used. Such materials consist of oligomers (acrylated urethane polyesters, acrylated epoxies and polyethers) and monomers (trimethylolpropane triacrylate) to provide fluidity before curing. The radiation technique avoids the use of volatile solvents, thus helping to reduce pollution. Comparatively low doses of less than 50 kGy are used.

High energy accelerators (up to 10 MeV) are used to cure fiber-reinforced composite materials reducing processing time and the costs. Doses of 150–250 kGy are needed to obtain a combination of polymerization and crosslinking. These composite parts are now being used in automobiles and aircrafts.

The insulation of electrical wires as well as the production of jackets on multi-conductor cables by radiation crosslinking was one of the first commercial applications. Materials used in this application are, e.g., polyethylene, polyvenylchloride, ethylene-propylene rubber, polyvenylidene fluoride and ethylene tetrafluoroethylene copolymer. Besides the creation of radicals, the electron beam can be used to create defects in crystals. One of the applications is the change or creation of the color of gemstones e.g. to create fancy diamonds. The electron beam in combination with a heat treatment changes the diamond color from yellow over green to red. These modified diamonds must be declared as "treated" to distinguish them from very rare colored natural diamonds with extreme high worth [37]. In the semiconductor industry, electron radiation is used to produce defects. These defects induce a life time reduction of charge carriers and allow a fast switching of electron high power devices [38]. This procedure is known as life-time killing and became very important for the production of high power devices e.g. for automotive or energy industrial products.

Radiation crosslinking stabilizes the initial dimensions of products and imparts the so-called "memory" effect. If the material is heated the original resistivity is maintained. Examples of commercial products using this effect are encapsulations for electronic components or exterior telephone cable connectors.

Low doses (30–50 kGy) of electron beam radiation are applied to automobile tires before assembling the different components.

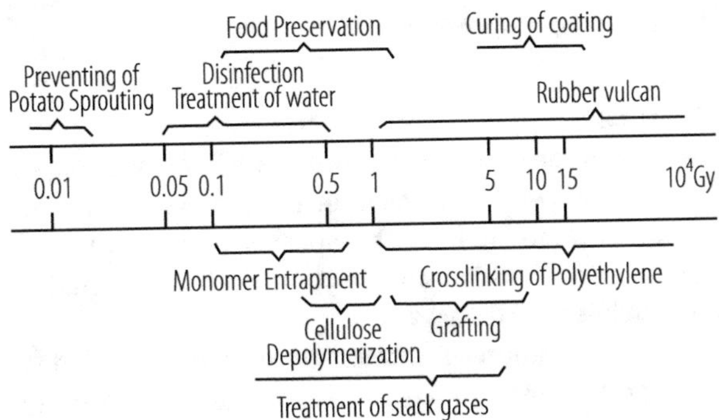

Fig. 11.19 Doses for electron irradiation [35]

Very high doses in the range 500–1000 kGy are needed to degrade polytetrafluoroethylene, which can be ground into fine particles or powder to be used as an additive to grease, engine oil, printing inks, coatings and thermoplastics.

Polypropylene and other polymers can be degraded by irradiation in air. This effect increases the melt flow and decreases the melt viscosity, which improves extrusion processes. By blending irradiated polymer with unirradiated material the desirable mechanical properties can be obtained. The doses for these processes range from 15 to 80 kGy.

An overview of applied doses is shown in Fig. 11.19.

The treatment of stack gases can be performed with electron beam radiation according to the reaction:

$$SO_2 + NO_x + 3/2 \cdot H_2O + O_2 + NH_3 \xrightarrow{\text{electron irradiation}} (NH_4)_2SO_4 + NH_4NO_3$$

The flue gas concentration of SO_2 and NO_x are reduced by adding oxygen and ammonia to a large extent depending on the applied energy dose. High-energy electron beams (few MeV) can penetrate in air through a thin foil and they are therefore applicable for medical applications. Low dose beams will be used to treat tumors by direct radiation or by production of high energetic X rays. With special structured filters a three dimensional irradiation is possible which reduces the damage of the healthy tissue.

High energetic electron beams (5–10 MeV) with high dose rates are used for sterilization of medical products and food e.g. spice. These kinds of machines produce a huge radiation level and must be installed in a locked safety radiation area. Typically, an endless treadmill is used to transport the items to the irradiation area. The electron beam is scanned in air over a stripe with a width of some cms.

The majority of electron accelerators for chemical processes are electron linacs in some cases as superconducting installations. Some special designed accelerators are the superconducting "Helios" compact electron synchrotron (Oxford Instruments) and the "Rhodotron" (Ion Beam Applications, Louvin-la-Neuve, Belgium).

11.2.3 Ion Accelerators

The structure of condensed matter materials can be modified by adding additional elements in order to achieve the special behavior required for a specific application. For this purpose, the application of accelerators is ideal, and enables high selectivity to be achieved in both the species of atoms to be implanted as well as their kinetic energy which allows control of the desired composition, including depth and thickness of the layers.

The application of accelerated ion beams in industry has two main directions, one for the analysis of materials and the other for the modification and production of materials. A new topic is the use of ion beams in medicine to treat cancer by irradiation with protons or carbon ions. In contrast to electron beams the energy transfer of heavy particles is concentrated at low kinetic energies (Bragg peak) and allows a three dimensional control of the desired cell damage.

11.2.3.1 Materials Modifications

The implantation of atoms into solid structures allows tailored changes in the properties of materials to be achieved. Targets are metals, semiconductors and insulators.

Tribological applications like the change of wear and friction are one of the domains in that field. Particularly the modification of surface hardness, ductility and lubrication effects are of interest. The implantation of nitrogen into metals and also alloys producing hardening of the surfaces has improved dramatically the efficiency of cutting tools.

Magnetic properties of materials have been modified by ion implantation [39]. The ion implantation into ceramics has started another industrial application [40].

In semiconducting materials, ion implantation is one of the most important processes for producing integrated electronic circuits. Mainly the basic semiconductors consisting of pure silicon and germanium are doped by elements of the third and fifth group to produce suitable acceptors and donors for transistor functions. Whereas for logic devices low energy implantation down to a few eV is important, the production of high power devices requires ion energies up to several MeV.

The profiles of implanted species are superior to those achieved by previously used diffusion processes. Ion implantation, however, produces radiation damage in the sample material which needs an annealing process in every case to produce functional electronic devices. Three dimensional ICs can be produced by a sequence of implantation and annealing if the necessary insulating layers can be provided. In many cases these layers are also produced by ion implantation.

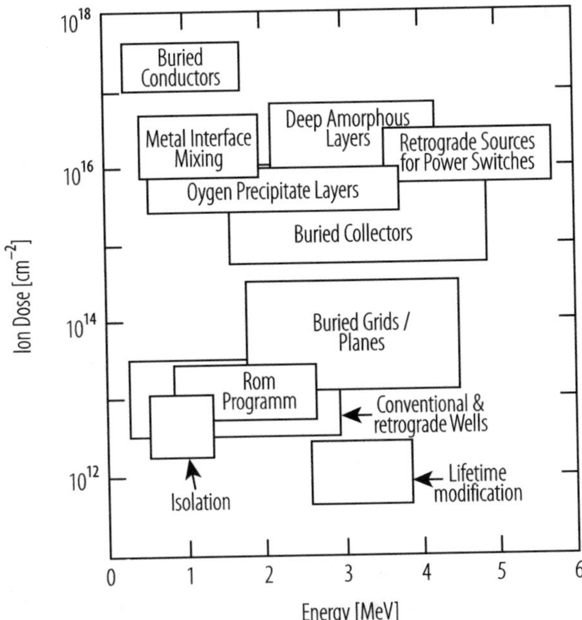

Fig. 11.20 Present and future high energy implantation [35]

The implantation can be used not only for elemental semiconductors, but also for compound semiconductors as GaAs, InP or superlattices. Figure 11.20 shows diagrammatically the whole field as dose dependent on ion energy [34]. To increase the performance of high power devices high energy ion beam irradiation is also used to reduce the charge carrier life time. In contracts to the common used life-time killing with electron beams, the damage area of ions is concentrated at the Bragg peak. This allows the reduction of the carrier concentration without a changing of the main electrical specification of the device.

Accelerators used for that work are van de Graaff accelerators, dynamitrons and in increasing number also RFQs. In all these types of accelerators the controllability of the ion energy is one of the major advantages.

11.2.3.2 Analysis of Materials

The analysis of materials uses many methods of the original nuclear research work, originally developed mainly for basic research. In many laboratories the RBS (Rutherford Back Scattering) is applied. The principle and one schematic spectrum are shown in Fig. 11.21. Ions impinge with energy E_o on the sample, which is composed of the species A_A and A_B. After scattering they leave the sample with energy E_1 and are than energy analyzed. A typical spectrum is shown in the lower part of Fig. 11.21. According to the kinematics for each element an upper energy is measured which indicate the elements on the surface of the sample. With

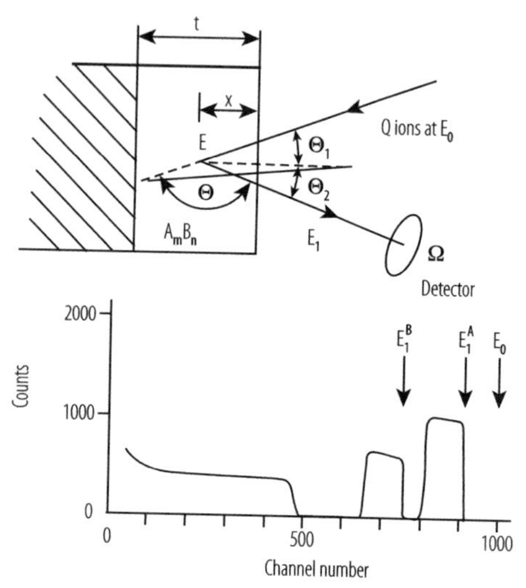

Fig. 11.21 Schematic view of a RBS spectrum [39]

increasing penetration the energy loss of the incoming and outgoing particles has to be considered. Important quantities are the sensitivity, the mass resolution and also the depth resolution. They all depend on the available parameters such as the mass and energy of the projectile, and the energy resolution of the detector as well as on the cross section of the scattering process. In most cases ^{4}He with energies 0.5 MeV $< E <$ 2.5 MeV are used as projectiles. The Coulomb repulsion is sufficient for the analysis as Rutherford scattering. At energies $E >$ 2.5 MeV Rutherford scattering is applicable for higher masses, however, if the energy of the α-particles is high enough Non-Rutherford scattering and the influence of nuclear reactions has to be taken into account. In these cases the cross sections have to be determined experimentally.

The sensitivity for low target masses is limited due to the underlying spectrum from the heavy substrate. The mass resolution is optimal for lower target masses, poor for heavier species and can also be limited by intrinsic detector resolution. For increasing beam energy the mass resolution improves linearly.

The depth profiling is dependent on the energy loss factor which incorporates the energy loss of the incoming and outgoing particles. Small energy loss factors result in large profiling depth, but limited resolution. The resolution is limited at high energy by the energy loss factor and at lower energies by an increasing energy straggling.

At high energies with broad resonances in the excitation functions the limitations are similar to the conventional backscattering.

Many other ions are suitable for RBS like ^{12}C, ^{16}O, ^{19}F or ^{35}Cl with energies >5 MeV. The beam energy has to be so chosen, that Rutherford cross sections can be applied. In some cases screening corrections are required particularly for heavy target species. Increased beam energy cancels potential increase in sensitivity arising from higher Z_1. It is less sensitive for lighter species than conventional backscattering due to increase in cross section. Mass separation improves with increasing beam energy. An increasing energy loss factor compared to light ions leads to shallower profiling depths and superior potential depth resolution. All target species heavier than the beam ion may be analyzed simultaneously.

The most frequently applied detector is the surface barrier detector. In some installations time-of-flight detectors are used.

A further well established method is ERD (Elastic Recoil Detection) which is also called Forward Recoil Spectrometry (FRES). In this process heavy projectiles are used to measure the content of hydrogen. For that method an accelerator with sufficiently high energy is required.

A general review of the methods of backscattering analysis is given e.g. by J.A. Leavitt et al. in [41].

Many nuclear reactions are used for the analysis of materials. In particular, those reactions which exhibit resonances in their excitation functions. By varying the energy of the projectiles to excite these resonances and exploit their enhanced cross sections, materials samples can be scanned with higher sensitivity.

A very special method of analysis is the charged particle activation analysis. In Table 11.2 [42] for a few particles, their energy and the reactions used are listed.

Table 11.2 Reactions for activation analysis [39]

Incident ion	Energy [MeV]	Reaction	Sampling depth
p	10–30	(p,n)	100 μm to few mm
		(p,2n)	
		(p,pn)	
		(p,α)	
d	3–20	(d,n)	10 μm to 2 mm
		(d,2n)	
		(d,p)	
		(d,α)	
t	3–15	(t,n)	10 μm to 100 μm
		(t,p)	
		(t,d)	
^{3}He	3–20	(^{3}He,n)	few μm to 100 μm
		(^{3}He,2n)	
		(^{3}He,p)	
		(^{3}He,α)	
α	15–45	(α,n)	10 μm to some 100 μm
		(α,2n)	
		(α,3n)	
		(α,p)	
		(α,pn)	
		(α,αn)	

11.2.4 Accelerator Mass Spectroscopy

Many methods and also facilities which were developed for solving basic scientific questions have also found their way into the applications. One of these methods is the Accelerator Mass Spectrometry (AMS). It was developed for the measuring of very small ratios of radioisotopes to stable isotope concentrations by detecting atoms, rather than by detecting their radioactive decay. The samples which contain the isotopes under investigation are incorporated in the ion source either in gaseous form or as solid material in a sputter source. One characteristic isotope is ^{14}C, which has a half life of 5760 years, and is used for age determination in many organic substances. Trees, e.g., incorporate also ^{14}C as long as they live. From the analysis of the amount of remaining ^{14}C the age of a piece of wood can be determined.

Tracing the radioactivity in organic substances can also determine e.g. metabolism or the paths of drugs and medicine. By measuring the concentration of the parent ions in beams, rather than their detecting radioactive decay products, AMS can determine values within the low 10^{-16} range for the ratio of ^{14}C/^{12}C. The accuracy is of the order of 0.3%. The measurements need much less time than detecting the radioactive decay products, an important factor in industrial application.

For industrial applications particularly compact instruments have been developed [42, 43].

As example—in Fig. 11.22 the installation of VERA (Vienna Environmental Research Accelerator) [42]—as a quite universal installation is shown. In this installation the measurement of negative as well as positive ions is possible. Particularly for the detection of ^{14}C the measurement of negative ions is important. The mass difference of ^{14}C and ^{14}N is so small that a magnetic separation of both is impossible, but since nitrogen forms no negatively charged ions ^{14}C appears as a very pure line in the spectrum.

Therefore the detection of ^{14}C allows the analysis of pharmaceuticals which is one of the dominant industrial applications of AMS.

11.2.5 Conclusion

The present experience has shown that accelerators with energies below 100 MeV fulfill the needs of industrial applications because electron and ion beams can steer fabrication processes in much smaller areas compared to conventional methods. Particularly the well-defined energies and doses are essential for tailoring layered structures in some cases also with isolating layers.

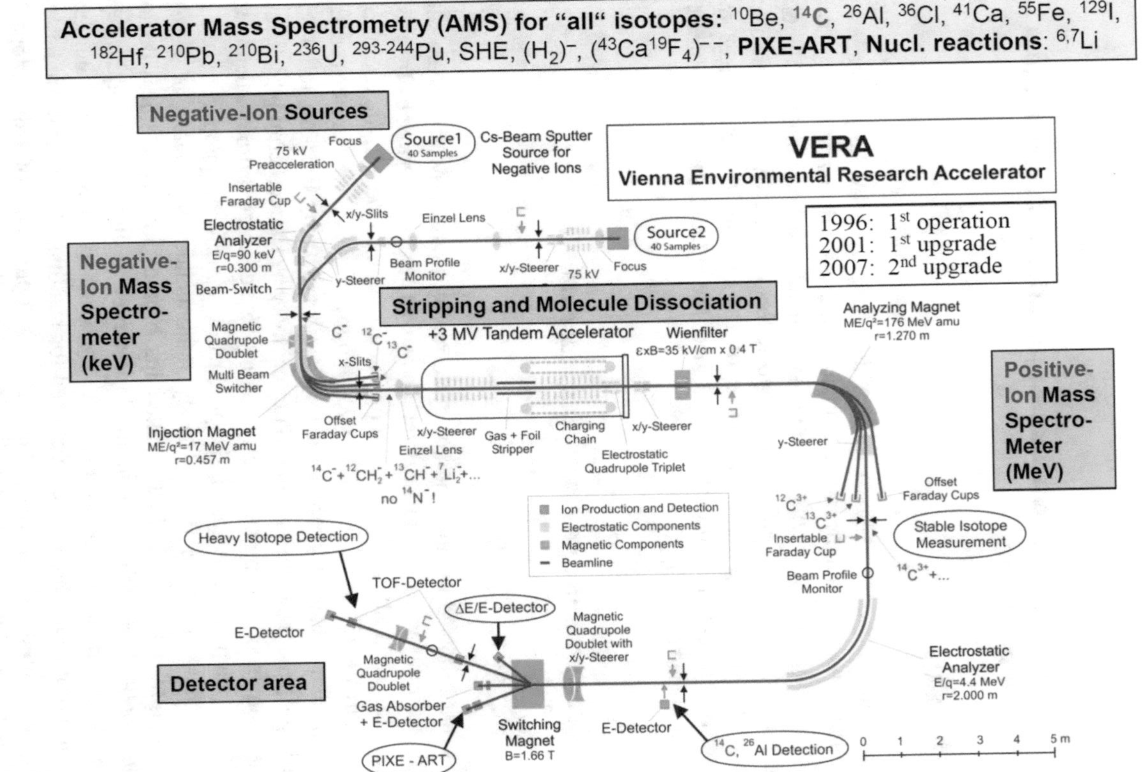

Fig. 11.22 Outlay of the VERA installation [42] (courtesy W. Kutschera)

11.2.6 Accelerator Suppliers

National Electrostatics Corp., Middleton, Wi, USA
High Voltage Engineering Europa B.V., Amersfoort, NL
Ion Beam Applications S.A., Louvain-la-Neuve, B

11.3 Accelerators in Medicine: Applications of Accelerators and Storage Rings

U. Amaldi · G. Magrin

Electron and hadron accelerators are widely employed in the production of radionuclides—used in diagnostic and brachytherapy—and in *teletherapy* i.e. in the formation of radiation beams directed from outside the body towards tumours and malformations. In this section these two topics are discussed in a historical perspective by describing both, the status of the art and the challenges that accelerator developers have to face today to meet the highest standards in nuclear diagnostic and cancer therapy.

11.3.1 Accelerators and Radiopharmaceuticals

11.3.1.1 History

The years that followed the invention of cyclotron by Ernest Lawrence were very prolific in defining what would be later known as nuclear medicine. In 1932, Lawrence, Stand and Sloan were able to produce, with their new 27-in. cyclotron, a proton beam of 4.8 MeV. In 1934 two important discoveries influenced the future use of that accelerator, alpha-induced radioactivity by Frédéric Joliot and Iréne Curie [44] and the neutron-induced radioactivity by Enrico Fermi [45]. Those discoveries convinced Lawrence to fully employ its accelerated proton, deuteron, neutron, and alpha particle beams for the production of artificial isotopes.

The medical exploitation of the newly radionuclides was clear in Ernest Lawrence's mind when in 1935 he called his brother John, a medical doctor from Yale school of Medicine, to join him in Berkeley to study the use of the new radioisotopes. Although the mainstream activity was nuclear research, a number of cyclotron-produced radionuclides was used for health applications in studies of physiology in animals and humans. The medical applications of radioisotope—radiotracing, endotherapy, and diagnostic—were defined at that time.

Radionuclides have the same chemical and physiological characteristics of the stable elements, so, when a radio-labeled organic compound is supplied to the organism, it follows the normal metabolism. From the mid 1930s, radionuclides were supplied to animals and humans and traced, with Geiger counters, in the internal organs following the physiologic uptake. Today the same concept is used in sophisticated diagnostics procedures as *Single Photon Emission Computer Tomography* (SPECT)—also called 'scintigraphy'—and *Positron Emission Tomography* (PET). Since the pioneering years, the challenge was to find the right radioisotope that, as part of a molecule, would enter in the physiological processes and would be preferably taken by a specific organ. The right isotopes must be chosen for their half-life and decay type, both compatible with the physiology of the processes investigated, the detection procedure, and the radiation protection of the patient.

Radio Phosphorus-32 (beta-emitter with half-life of 14.3 days), made from the 27-in. cyclotron was used as a tracer to study the absorption and metabolism of phosphorus so that the cellular regeneration activity could be followed over several weeks [46]. In 1936 John Lawrence gave for the first time Phosphorus-32 to a leukemia patient beginning the therapeutic use of artificial radionuclides [47].

In 1936 radioactive Sodium-24 (half-life of 15 h), obtained at the Berkeley cyclotron bombarding ordinary sodium with deuterium, was one of the first artificial radioisotopes applied in physiology to study transport and uptake in animal and humans and to determine the speed of absorption on the circulatory system. Radioactive sodium was first given to treat leukemia patients by Hamilton in 1937 [48].

In those years radionuclides were also used to study the transfer of traceable elements and compounds across the cell membrane. Today cytopathology and molecular biology are using radiotracers to study cellular diseases and physiology.

Larger cyclotrons were produced in the late 1930s at Berkeley, Harvard, and Leningrad. Berkley's 60-in. accelerator built in 1939 was used by Ernest Lawrence to produce radioactive gasses (nitrogen, argon krypton, and xenon) that were used to study the decompression sickness suffered by the aviators of the Second World War. The ^{81m}Kr gas is still used after 80 years in some studies of lung ventilation and functionality.

At the time of the first experiments in humans, the effects of X-rays on healthy biological tissues were known and avoiding overdose of ionizing radiation was a concern. It is important to mention that in 1936 John Lawrence, comparing identical doses of X-ray and neutrons [49], recognized different biological effects giving one of the earliest contribution to the knowledge of radiation qualities in correlation with internal and external radiotherapy and with radiation protection.

The isotope 131 of Iodine was selected by Joseph Gilbert Hamilton because its half-live of 8 days was considered to be the optimal time to treat the thyroid and avoid side effects to the patients. The radionuclide was isolated from a sample of tellurium bombarded with deuterons and neutrons from the 37-in. cyclotron. From 1941 Iodine-131 was used in diagnostic of thyroid functionality and, from 1946, in cancer therapy. In the 1950s Iodine-131, ingested as a liquid solution, became a common treatment. Commercially producer in nuclear reactors, it dominated for 40 years the market of radiopharmaceuticals for cancer treatment and it is still widely used today.

György Hevesy gave a decisive impulse to the studies with radiotracers of human metabolism and physiological processes. He carried on a program, supported by private funding, to introduce Sodium-24, Phosphorus-32, Cobalt-60, Technetium-99, and Iodine-131 into medical practice. In 1941 the first cyclotron fully dedicated to production of radionuclides for medical purposes was built at the Washington University of St Louis and used for the production of radionuclides for diagnostic and internal radiotherapy.

In the years that followed, the production of neutron-generated radioisotopes was moved to more efficient nuclear reactors. Nevertheless a number of cyclotron facilities, fully dedicated to medical applications, were built with the purpose of continuing the research of new products. In the 1950s, the use of Thallium-201 as tracer for cardiac flow revitalized the use of medical cyclotrons. Thallium-201 is produced via Lead-201 generator obtained in cyclotron from protons or deuterons beams.

In the late 1960s, a turning point that drastically transformed the medical application of cyclotrons was the idea of developing the radiochemistry labs inside the cyclotron facility, possibly close to the clinics where they were the patients were injected. Organic molecules containing short-living positron-emitting isotopes could be immediately transformed in pharmaceutical products and distributed in the facilities of the region to be used within few hours. The most important product was the Flourodeoxygloucose, FDG, made for the first time in 1976 with Fluorine-18, a positron emitter with a half-live of 110 min [50]. FDG was developed by a scientific collaboration of the Brookhaven National Laboratory, the US National Institute of Health and the University of Pennsylvania with the purpose of studying the brain metabolism.

Since then, FDG radiochemistry has evolved to fast and efficient production processes and today it is used in 75% of PET medical imaging [51]. The applications include diagnoses of brain diseases like epilepsy and dementia, examination of heart functionality, and detection of evolution of many different tumours.

New radiopharmaceuticals labeled with short half-lives positron emitters (in particular Fluorine-18 and Carbon-11) are still studied and developed today. Although Oxygen-15 is not commercially used for PET due to the fast decay time of 2 min, the Company Ion Beam Applications (IBA), built a 3 MeV deuterium cyclotron dedicated to the production of Oxygen-15 labeled gasses [52].

11.3.1.2 Accelerator for Radioisotope Production

Today the number of cyclotrons used for medical purposes, officially registered by the last IAEA survey published in 2006, is 262 [53]. The real number is certainly higher and it is increasing continuously by approximately 50 units per year, mainly because of new small PET accelerators. A survey made in 2010 combining inputs of production of the four major manufacturers already estimated in 671 the number of cyclotrons in operation [54].

The technology of cyclotrons has advanced in parallel to the development of PET. In the 1970s The Cyclotron Company (TCC) [55] studied the acceleration on negatively charged hydrogen (H^-). The extraction of the beam is in this way simplified since it can be obtained with a thin stripping foil conveniently positioned on the beam trajectory. In the foil all electrons are stripped leaving the protons, positively charged, deviate to the opposite direction under the effect of the magnetic field. The result is a clean extraction with efficiency very close to 100%. The elimination of beam loss (of the order of 25% with normal extraction) drastically reduces the activation of the cyclotron components simplifying the maintenance, the regular inspection, and the decommissioning. A remarkable characteristic of these cyclotrons is the possibility of obtaining simultaneous multiple beams making partial extraction with additional more foils.

In recent years several companies entered into the production of cyclotrons, machines that are stable, reliable, and can run continuously for days with limited supervision and maintenance requirements. The accelerators design has been adapted to the radiopharmaceutical request that is mainly concentrate on PET products. The cyclotrons dedicated to production of Fluorine-18, Carbon-11, Nitrogen-13, and Oxygen-15, have abandoned the versatile characteristics of the traditional machines and are made to accelerate only a specific ion (typically H^-), run at low energy (20 MeV and below), and at relative low current (approximately 100 μA).

Commercial cyclotrons characteristics are chosen taking into account the half-life of the radionuclides they produced. For the optimal logistics the cyclotrons should be homogeneously distributed in the territory with a large number of small cyclotrons, 10 MeV and 20 MeV, to fulfill the requests of FDG and other radiopharmaceuticals made on ^{18}F, ^{11}C, ^{13}N, ^{15}O, and a sparse distribution of cyclotrons with wider range of energy, 30 MeV or 70 MeV, able to produce all PET and SPECT radioisotopes used in research and in less common diagnostic (Fig. 11.23). Some of the higher-energy accelerators have the possibility of changing the

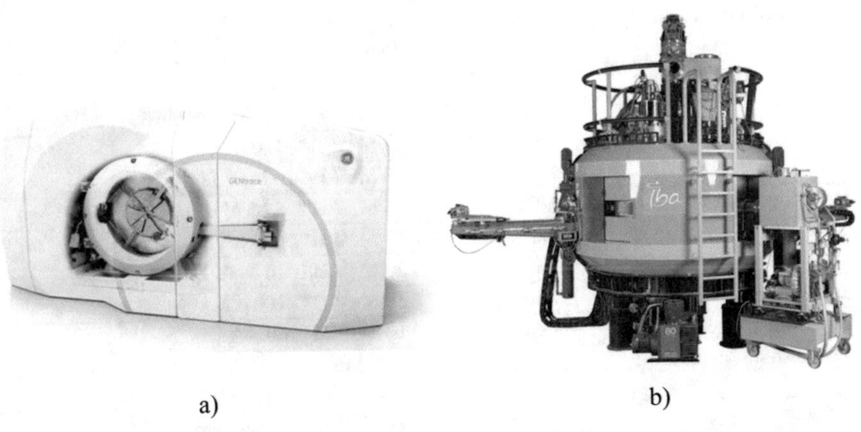

Fig. 11.23 Examples of two commercial cyclotrons. (**a**) The self-shielded 7.8 MeV cyclotron GENtrace produce by General Electric (courtesy of GE Healthcare) for common PET radioisotopes. (**b**) The 30 MeV cyclotron Cyclone 30 produced by IBA for a variety of PET and SPECT radioisotopes (courtesy of IBA SA)

energy moving the stripping foil along the cyclotron radius to adapt to the different nuclear reactions.

The commercial cyclotrons available today are presented here in two lists, Table 11.3 describes the cyclotrons in the range between 10 and 20 MeV, which correspond to the vast majority of machines today in operation for PET applications; Table 11.4 describes the cyclotrons with energies of 22 MeV and above for a more general use.

Besides cyclotrons, other particle accelerators have been considered for medical applications. Studies for low energy, high current accelerators have been financed in the 1980s, by the Star Wars program of the US Department of Defense. RFQ accelerators have been designed to obtain beam intensities of hundreds of milliampere at energies below 10 MeV. In the framework of the same program a 3.7 MeV, 750 μA Tandem Cascade Accelerator was designed, realized, and put into operation by

Table 11.3 Characteristics of commercial cyclotrons for range energies 10–20 MeV (from [53, 56, 57])

Company	Model	Description
Advanced Cyclotron Systems, Inc.	TR 14	11–14 MeV H^-, 100 μA
	TR 19	14–19 MeV H^-, 300 μA
Advanced Biomarkers Technology	BG-75	7.5 MeV H^+, 5 μA
Best Cyclotron System, Inc.	15	15 MeV H^-, 400 μA
China Inst. Atomic Energy	CYCCIAE14	14 MeV H^-, 400 μA
D.V. Efremov Institute	CC-18/9	18 MeV H^-, 9 MeV D^-, 100 μA
EuroMeV	Isotrace	12 MeV H^-, 100 μA
General Electric Healthcare	MiniTrace	9.6 MeV H^-, 50 μA
	PETrace	16.5 MeV H^-, 8.6 MeV D^-, 100 μA
Ion Beam Applications	Cyclone 3	3.8 MeV D^+, 60 μA
	Cyclone 10/5	10 MeV H^-, 150 μA, 5 MeV D^-,
	Cyclone 11	11 MeV H^+, 120 μA
	Cyclone 18/9	18 MeV H^-, 9 MeV D^-, 150 μA
Japan Steel Works	BC168	16 MeV H^+, 8 MeV D^+, 50 μA
	BC1710	17 MeV H^+, 10 MeV D^+, 60 μA
	BC2010N	20 MeV H^-, 10 MeV D^-, 60 μA
KIRAMS	Kirams-13	13 MeV H^+, 100 μA
Oxford Instrument Co.	OSCAR 12	12 MeV H^-, 60 μA
Scanditronix Medical AB	MC17	17.2 MeV H^+, 8.3 MeV D^+, 60 μA 12 MeV $3He^{++}$, 16.5 $4He^{++}$, 60 μA
Siemens	Eclipse	11 MeV H^-, 2 × 60 μA
Sumitomo Heavy Industries	HM 7	7.5 MeV H^-, 3.8 MeV D^-
	HM 10	9.6 MeV H^-, 4.8 MeV D^-
	HM 12	12 MeV H^-, 6 MeV D^-, 60 μA
	HM 18	18 MeV H^-, 10 MeV D^-, 90 μA

Table 11.4 Characteristics of commercial cyclotrons for energies of 22 MeV and above (from [53, 56, 57])

Company	Model	Description
Advanced Cyclotron Systems, Inc.	TR24	24 MeV H^-, 300 μA
	TR30/15	30 MeV H^-,1000 μA/15 MeV D^-, 160 μA
Best Cyclotron System, Inc.	35p	15–35 MeV H^-, 1000 μA
	70p	70 MeV H^-, 700 μA
China Inst. Atomic Energy	CYCCIAE70	70 MeV H^-, 750 μA
Ion Beam Applications	Cyclone 30	30 MeV H^-, 1500 μA/15 MeV D^-
	Cyclone 70	30–70 MeV H^-, 2 × 350 μA 35 MeV D^-, 17.5 MeV H_2^{++}, 70 MeV He^{++}, 50 μA
	Cyclone 235	240 MeV H^-
Japan Steel Works	BC2211	22 MeV H^+, 11 MeV D^+, 60 μA
	BC3015	30 MeV H^+, 15 MeV D^+, 60 μA
KIRAMS	Kirams-30	15–30 MeV H^-, 500 μA
Scanditronix Medical AB	MC30	30 MeV H^+, 15 MeV D^+, 60 μA
	MC32NI	15–32 MeV H^-; 8–16 MeV D^-, 11–23 MeV 3He^{++}, 15–31 4He^{++}, 60 μA
	MC40	10–40 MeV H^+, 5–20 MeV D^+, 13–53 MeV 3He^{++}, 10–40 4He^{++}, 60 μA
	MC50	18–52 MeV H^+, 9–25 MeV D^+, 24–67 MeV 3He^{++}, 18–50 4He^{++}, 60 μA
	MC60	50 MeV H^+, 60 μA
	K130	6–90 MeV H^-, 10–65 MeV D^-, 16–173 MeV 3He^{++}, 20–130 MeV 4He^{++}, 60 μA
Sumitomo Heavy Industries	AVF series	30, 40, 50, 70, 80, 90 MeV H^+, 60 μA
	Ring Cyclotron 400	400 MeV H^+ (K = 400), 60 μA
	Ring Cyclotron 540	240 MeV H^+ (K = 540), 60 μA
	C235	240 MeV H^-, 60 μA

Science Research Laboratory in Massachusetts for the production of Nitrogen-13, Oxygen-15, and Fluorine-18 for PET [58]. Today, AccSys Technologie proposes a linear proton accelerator for PET isotope productions. This RFQ accelerates protons to 3 MeV with a current of 150 μA. The energy can be upgraded to 10.5 MeV coupling the RFQ to a Drift Tube Linac (DTL). The system is suitable for supplying PET radioisotopes to a single diagnostic centre.

11.3.1.3 The Radionuclides Used in Nuclear Medicine

In the world the request of radionuclides for imaging is constantly increasing. Multimodality scanning systems have been favorably accepted by the medical doctors. The leading companies report that 35% of the SPECT scanners produced are sold combined with Computed Tomography (CT). In the same way the request of PET/CT and PET/MRI systems that combine metabolic and morphological imaging is stably increasing and so is the demand of positron emitter radionuclides. In the years to come many factors will contribute to define the role of accelerators in the commercial production of medical isotopes. Unforeseen events as the shortage of Molibdenum-99 described at the end of this section can weaken a situation that was considered stable. The primary role of reactors, to which it is ascribed approximately 80% of the medical radioisotope production, can be diminished by downsides as the rigidity of a centralized production, the risk of incident in nuclear power plant, and the problems connected to radioactive waste. These drawbacks are strongly reduced with radioisotope production based on accelerators and this can play in their favor.

Today PET imaging is certainly the largest medical use of the cyclotron radioisotopes. IAEA reports in its 2006 survey that 75% of the cyclotrons are dedicated to FDG [53]. Concerning SPECT, some radioisotopes for (gamma- and gamma/beta-emitters) are produced only with cyclotrons (Iodine-123, Indium-111, Gallium-67, Cobalt-57) some others, as Copper-67 and Rhenium-186, are produced both with cyclotrons or reactors.

Immuno-positron emission tomography is an imaging technique that, using radio-labeled monoclonal antibodies, allows tracking and quantifying their distribution in the body, and can be applied to imaging of human malignancies. Zirconium-89 has a primary role in immunoPET being used in most of the diagnoses based on radioactive antibodies [59]. Its favorable characteristics are a half-life of 3.3 days which allows a manageable production and distribution, and the favorable decay mode where the unavoidable gamma rays have energies well distinguishable from PET photons. Finally Zirconium-89 production, via proton-neutron reaction in Yttrium-89, is optimal at energies of 14 MeV, widely available in cyclotron facilities [see Table 11.3].

Radiopharmaceuticals for therapy represent only 5% of the total production [60], nevertheless the research in the field of cancer targeted radionuclide treatments is active and the role of cyclotrons is becoming more and more important. The isotopes that better conform to the therapy are those that decay transferring locally all their energy. The favorable decay products are alpha particles, beta particles and Auger electrons. The optimal radionuclide is selected based on essential characteristics, which are (i) the possibility to associate it to molecules or, in case of radio-immunotherapy, to monoclonal antibody, with affinity to tumour cells, (ii) the right decay time to allow the kinetics and avoid overdose, (iii) the availability of the product in reliable quantities, and (iv) the affordable cost. The research is trying to balance these demanding needs and the number of radionuclides under examination is getting large.

Alpha-particles emitters are Astatine-211, produced with the reaction ^{209}Bi($^{4}{}_{2}$He,2n) by 28 MeV alpha particle beams, and Bismuth-212, obtained from a

Actinium-225 generator produced with the reaction $^{226}Ra(p,2n)$ by 22 MeV protons. Bromine-77, an Auger-electron emitter produced with either alphas (27 MeV) or protons (in various possible reactions starting at the energy of 13 MeV), is a good candidate for its simple association with physiological molecules. Among the beta emitters, Renium-186 is favorably regarded since, it belongs to the same chemical family of Technetium and therefore it can follow similar the chemical process well establishes for ^{99m}Tc.

Brachytherapy with radioactive seeds is an established way of endoradiotherapy based almost exclusively on reactor-produced radionuclides. One exception is Palladium-103, an Auger-electron emitter produced in reasonable quantities by 18 MeV cyclotrons.

Interest is rising in endoradiotherapy based on pre-therapeutic PET dosimetry. Two radioisotopes of the same element can be used for complementary proposes. First the pre-therapeutic positron emitter is injected to trace the physiological uptake of the element and then the therapeutic isotope is administered to the patient according to the dose estimation. Isotope pairs of Copper and Scandium (β^+ emitted from ^{64}Cu or ^{44}Sc used for PET imaging, β^- emitted from ^{67}Cu or ^{47}Sc used for therapy) are considered for these procedures [61].

It is important to underline the steering role played by the medical community in determining the future of diagnostic, with SPECT and PET, and of endoradiotherapy, which is today only a marginal segment. Medical doctors—who today use preferably one diagnostic modality, SPECT, and one radionuclide, Technetium-99m—with their future orientations will influence the development on new tools than could complement or substitutes the existing one. A challenge is also the complexity of transforming an effective radioisotope to an approved pharmacological product. The most important factors to consider are the availability and cost of the raw material, the access to accelerators or reactors with appropriate energy and fluxes, the existence of fast radiochemistry, and the logistic to delivery the radioisotopes on time before they decay.

To address the needs of the research of new medical applications, and also to partially take over the aging accelerators, some dedicated facilities have been put in place worldwide. Among others (i) ARRONAX in Nantes, financed by French regional and National authorities and the European Union, that hosts a cyclotron for H^- (up to 70 MeV, 350 μA protons), $H_2{}^+$, D^-, and He^{++}; (ii) LANSCE in Los Alamos, USA, that produces from 2005 medical radioisotopes from a high-current 100-MeV proton linear accelerator; (iii) the PEFP center in South Korea, a facility based on a 20 mA proton linear accelerator (an RFQ followed by two Drift Tube Linacs to energies of 20 MeV and 100 MeV) for medical and industrial applications.

A motivation for studying new applications of accelerators came from the worldwide shortage of Molybdenum-99, the generator of Technetium-99m. The production dropped to about 50% of the market needs for 16 months between 2009 and 2010. Commercial Molybdenum is a fission product of highly enriched uranium target produced almost exclusively in five reactors that are old and need continuous maintenance. Each year 30 millions diagnostic studies, 80% of all medical scans that use radioisotopes, are made with Technetium-99m (6 h half-life) from a Molibdenum-99 generator (66 h of half-life).

The shortage of Molybdenum-99 stimulated several initiatives for finding alternative solutions and the use accelerator was considered for the production of either Molybdenum-99 or directly Technetium-99m. If, from one side, a single accelerator facility is unable to compete in production with large reactors, on the other side it allows a better distributed production with advantages for the logistics, the possibility of producing directly technetium-99m for locally distributed end users, and undisputable environmental benefits. The accelerators considered for different productions processes are electrostatic accelerators, electron linacs, and cyclotrons.

Two Canadian initiatives were established to produce Technetium-99m trough the reactions ^{100}Mo(p,2n)^{99m}Tc [62]. The projects use existing cyclotrons available for positron emission tomography (PET) from General Electric (130 μA, 16.5 MeV) and Advance Cyclotron Systems, Inc. (300 μA, 19 MeV) and the productions should start in 2017. The clinical trial which compared reactor- and cyclotron-produced Technetium-99m was concluded in 2017 receiving the approval from Health Canada.

Some project plan to substitute reactor-generated neutrons with accelerator-generated neutrons. To reach the capabilities of a reactor production, the neutron rate should exceed 10^{14} neutrons per second. The concept developed by the company Shine is based on the fusion of deuteron and tritium for producing neutrons (and helium). The electrostatic accelerated deuterons (300 keV, 60 mA) are directed toward a tritium target. The resulting neutrons ($5 \cdot 10^{13}$ neutrons per second) cross a natural uranium target for neutron multiplication and irradiate a low enriched uranium target producing, among other fission products, Molybdenum-99. The first facility is in construction phase.

The company NorthStar is developing a facility for the production of Molybdemum-99 from photo-nuclear reactions ^{100}Mo(γ,n)^{99}Mo using electron linacs. Two accelerators (40 MeV, 3 mA each) direct the electron beams from opposite sides to a Molybdenum-100 target where the bremsstrahlung photons are created. The production is expected to start in 2017 [63].

Another project linked to TRIUMF is the development of a high-current electron linac (50 MeV, 100 mA) for the generation of photons that, directed to a target of Uranium-238 produce Molybdenum-100 from photo-fission.

Electron linacs for the generation of neutrons are foreseen in different project from Shine (35 MeV, 0.6 mA), and Niowave (superconductive 40 MeV, 2.5 mA). The generated neutrons irradiate a target containing a solution of depleted uranium.

Several other ways of using accelerators for production of Molibdemun-100 or Technetium-99m have been proposed and are still under study and development [64].

11.3.2 Accelerators and Cancer Therapy

11.3.2.1 History

Conventional Therapy

The roots of brachytherapy (discussed in section 11.3.1.3) and teletherapy date back to the discoveries of X-rays and radium made by Roengten and the Curies in the years 1895–1897. X-ray tubes and gamma-ray sources were soon employed in medical or technological uses. Teletherapy with X-ray photons was performed to cure superficial tumours few years after the discovery.

The first electron linac was built in 1947 at Stanford by Bill Hansen and his group for research purposes [65] and was powered by a klystron produced by Varian. Soon after, this new tool superseded all other electron/photon sources. Also in 1947, in England, Fry and collaborators build a 40-cm linac that accelerated electrons from 45 to 538 keV [66].

In the years that followed, those two groups and others, in particular the group of John Slater at MIT, studied the parameters of the irises to improve the power efficiency of the structure and to adapt the traveling wave parameters of wavelength and phase velocity. Around 1950 megavoltage' tubes were built and, complementing the 'cobalt bombs' entered clinical practice. The positive surprise was that, due to the longer ranges of the electrons—put in motion by the gammas of the beam mainly through the Compton effect—these 'high-energy' radiations had a much better sparing of the skin.

In those years a process was initiated to adapt the complex machines developed for research purposes to the clinical environment and to the treatment needs. After a short season in which the X rays of energy larger than 5 MeV were produced with medical 'betatrons', electron linacs, running at the by-now standard 3 GHz frequency, became the instrument of choice. Varian in 1960 produced for the first time a linac that, mounted on a gantry, rotated around the patient and in 1968 the first medical machine based on standing-wave acceleration.

From the side of the traveling-wave accelerators, further studies have improved the efficiency and the start up time of the radiation so that today both designs are still used in medical linac: the two major manufacturers, Varian and Elekta (see Table 11.5) produce accelerators the first based on standing-wave and the second on travelling-wave. The average accelerating fields are in the range 10–15 MV/m.

In 2017 about 12,000 [67] linacs are installed in hospitals all over the world and in the developed countries there is one linac every 200,000–250,000 inhabitants. It is interesting to remark that all the electron linacs used to treat patients are close, as far as the overall length is concerned, to the 27 km circumference of LHC.

About 50% of all tumour patients are irradiated as exclusive treatment or combined with other modalities so that radiotherapy is used every year to treat about 20,000 patients on a Western population of 10 million inhabitants. On average, 40% of the patients who survive 5 years without symptoms have been irradiated.

Table 11.5 A selection of commercial electron linacs for radiotherapy

Company	Complete system	Acceleration characteristics	Maximum photon energies	Tumour conformation
AccuRay	Cyberknife®	Standing-wave, X-band	6 MV	Computer-controlled robotic arm
Elekta AB	Versa™	Traveling-wave, diode injection, S-band	18 MV	80 multileaf collimator
Siemens Medical Solutions	Artiste™	Standing-wave, triode injection, S-band	18 MV	82 multileaf collimator
AccuRay	TomoTherapy®	Standing-wave, S-band	6 MV	64 multileaf collimator
Varian Medical	HyperArc™	Standing-wave, triode injection, S-band	20 MV	120 multileaf collimator

This enormous development has been possible because of the advancements made in computer-assisted treatment systems and in imaging technologies, such as CT imaging, PET scans, MRIs. The most modern irradiation techniques are the *Intensity Modulated Radiation Therapy* (IMRT), which uses 6–12 no coplanar and non-uniform X-ray fields, and the *Image Guided Radiation Therapy* (IGRT), a technique capable of following tumour target which moves, mainly because of patient respiration.

Varian is the worldwide market leader with more than 50% of the share of electron linac for radiotherapy. One of its last development, HyperArc, is an irradiation system in which the uniform prescribed dose is delivered in a single revolution of the gantry. The times are drastically reduced thanks to the computer assisted continuous variation of the beam intensity and the use of a multileaf collimator that adapts the irradiation fields to the requirement of the treatment planning.

Cyberknife produced by Accuray (USA) is an original development in which a robotic arm substitutes the gantry to support the accelerator. To reduce dimension and weight, a X-band, 6 MV linac is chosen. The resulting pencil beam can penetrate and aim to the tumour with the optimal direction to spare the organs at risk.

In helical TomoTherapy a narrow x-ray beam constantly irradiates the tumour while rotating around the patient and, at same time, produces an image of the patient organs. The combination with the movement of the couch creates the helical irradiation which is modulated in intensity and collimated by a multileaf system to conform to the requirements of the panning. In 2011 the company TomoTherapy has been bought by Accuray.

The so-called MRI-guided radio therapy, i.e. the simultaneous combination of radiation therapy treatment and MRI scanning, has been performed for the first time by the company ViewRay. The system based on a 6 MeV linac, employs a MRI with 0.35 T magnets. A similar system which combines a 1.5 Tesla MRI and a

7 MeV linac, was developed and built by a collaboration between Elekta, Philips, and several clinics from Europe and North America [68]. The advantages of MRI scanning are the capability of identifying movements of soft tissue and organs, which are not detectable with other imaging media, and the absence of ionizing radiation exposure. Today the average frame rate during dynamic acquisitions is of the order of 10–20 frames per seconds with spatial resolution of the order of 2 mm.

The major companies producing electron linacs for photon therapy and the characteristics of some of the most advanced models are listed in Table 11.5.

It must be mentioned that, because of the increase risk due to the non-negligible production of photoneutrons, the use of the 16–20 MV photons is limited to the cases for which the size of the patient and the depth of the tumour require it. In normal conditions 6–8 MV energies are employed.

Neutron Therapy

At variance with the case of conventional radiotherapy, the use of atomic nuclei to treat cancer has benefited of a long series of accelerator developments, which is still continuing today. This is the reason for which we focus, in this second Part, on hadron accelerators.

'Hadron therapy' ('hadronthérapie' in French, 'hadronentherapie' in German, 'adroterapia' in Italian, 'hadroterapia' in Spanish) is a collective word which covers all forms of radiation therapy which use beams of particles made of quarks: neutrons, protons, pions, and ions of helium, carbon, neon, silicon, and argon have been used for treating cancer. Furthermore, lithium and boron ions, as well as antiprotons have been investigated as potential candidates. 'Hadron therapy', 'particle therapy', 'ion-beam therapy', 'heavy ion therapy' and 'light ion therapy' are other terms often used to indicate the same procedure, and every performing facility has its favorite term.

The first hadrons used for treatment purposes are the neutrons. The time progression in implementing neutron therapy was so fast that is astonishing today. The neutron was discovered by James Chadwick in 1932. Soon after, Ernest Lawrence and his brother John were experimenting with the effects of fast neutrons on biological systems. Following a paper by Gordon Locher [69], who in 1936 underlined the therapeutic potentialities of both, fast and slow neutrons, at the end of September 1938, the first patient was treated at the Berkeley 37-in. cyclotron. The first study on 24 patients, which used single fractions, was considered a success and led to the construction of the dedicated 60-in. Crocker Medical Cyclotron. Here Robert Stone and his collaborators (Fig. 11.24) treated patients with fractionated doses using fast neutrons. The tuning of the dose to control the tumour and prevent, at the same time, the complication of the normal tissue was still developing. The dose given to healthy tissues was too high, so that in 1948 the program was discontinued [70].

Neutron therapy was revived in 1965 by Mary Catterall at Hammersmith Hospital in London and later various centres were built where fast neutrons were used for many years. In particular in the 1970s radiation oncologists of Chicago worked with Robert Rathbun (Bob) Wilson, Fermilab Director, to build the 'Neutron Therapy

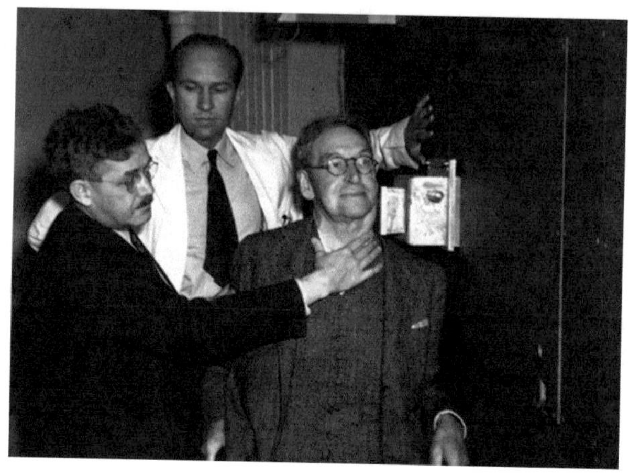

Fig. 11.24 Robert Stone is watched by John Lawrence while aligning a patient in the neutron beam produced by the 60-in. cyclotron

Facility' at Fermilab based on the injector linac [71]. At Michigan University, Henry Blosser built for the Harper Hospital a proton cyclotron rotating around the patient.

The worldwide effort for high-current 40–60 MeV proton cyclotrons was large. However, at present, this technique is rarely used. The reason is the poor depth-dose distribution of fast neutron and the fact that, all along the path, the biological effects, due mainly to highly ionizing protons put in motion by the neutrons, are difficult to determine and tissue-dependent. As discussed in the following, carbon ion beams are a much better solution to deliver doses being highly ionizing particles, i.e. particles which have energy losses (Linear Energy Transfers—or LET—in the medical parlance) much larger than the electrons put in motion by X rays.

Thermal and epithermal neutron are also used in Boron Neutron Capture Therapy (BNCT). This cancer treatment is based in two subsequent procedures, first the injection of the boron carrier, conceived to be absorbed preferably by the tumour, and second the irradiation of the patient with low energy neutrons. Alpha particles resulting from the reaction $^{10}B(n,^{4}{}_{2}He)^{7}Li$ release all their energy close to the point where they are originated affecting the tumour cells.

Proton Therapy

The use of protons in teletherapy was proposed in 1946 by Bob Wilson [72], who was asked by Lawrence to measure, at the Berkeley Cyclotron, proton depth-dose profiles. Describing the significant increase in dose at the end of particle range, the so-called Bragg peak, which had been observed 40 years before in the tracks of alpha particles by W. Bragg, Wilson recognized in his paper the advantages in treating tumours. The Bragg peak—which can be 'spread' with modulator wheels—can be used to concentrate the dose sparing healthy tissues better than with X-rays. It is interesting to remark that in his paper Wilson discussed mainly protons but mentions also alpha particles and carbon ions.

In 1954, the first patient was treated at Berkeley with protons, followed by helium treatment in 1957 and by neon ions in 1975 [73]. In these treatments—as in most of the following facilities—the beam was distributed over the target volume using 'passive' shaping systems, like scatterers, compensators, and collimators that were adapted from the conventional photon therapy. The first treatments consisted of irradiation to stop of the pituitary gland from producing hormones that stimulated some cancer cells to grow. Between 1954 and 1957 at Berkeley, under the leadership of Cornelius Tobias, 30 pituitary glands and pituitary tumours were treated with protons [74].

In 1957 the first tumour was irradiated with protons at the Uppsala cyclotron by Börje Larsson [75], but the facility that made the largest impact on the development of proton therapy is certainly the 160 MeV Harvard cyclotron, which was commissioned in 1949 (Fig. 11.25).

The Harvard staff got interested in using protons for medical treatments only after proton therapy was started at both, LBL and Uppsala. In 1961, Raymond Kjellberg, a young neurosurgeon at Massachusetts General Hospital in Boston, was the first to use the Harvard beam to treat a malignant brain tumour.

The results obtained for eye melanoma—by Ian Constable and Evangelos Gragoudas of the Massachusetts Eye and Ear Hospital—and for chordomas and chondosarcomas of the base of the skull [76]—by Herman Suit, Michael Goitein and colleagues of the Radiation Medicine Dept of Massachusetts General Hospital—convinced many radiation oncologists of the superiority of protons to X rays in particularly for tumours that are close to organs at risk. In 2002, when the cyclotron was definitely stopped, more than 9000 patients had been irradiated and the bases were put in place for the following development of the field. The competences developed in Boston were soon transferred to the new hospital-based facility of the Massachusetts General Hospital, now called "Francis H. Burr Proton Therapy

Fig. 11.25 Picture of the just completed Harvard cyclotron. Many years later Norman Ramsey was awarded the Nobel prize, together with Wolfang Paul

Table 11.6 The pioneers of proton therapy

Facility	Country	Years in operation
Lawrence Berkeley Laboratory	USA	1954–1957
Uppsala	Sweden	1957–1976
Harvard Cyclotron Laboratory	USA	1961–2002
Dubna	Russia	1967–1996
Moscow	Russia	1969–
St. Petersburg	Russia	1975–
Chiba	Japan	1979–2002
Tsukuba	Japan	1983–2000
Paul Scherrer Institute	Switzerland	1984–

These facilities, that today are out of operation except for the centres Moscow, St. Petersburg, and PSI, treated together 10,200 patients

Center", which is equipped with the first commercial accelerator for therapy—built by the market leader, IBA.

Soon after the start-up of the Harvard facility other nuclear physics laboratories in USSR and Japan assigned horizontal beams of their research facilities to proton therapy. As shown in Table 11.6 in 1984 the Paul Scherrer Institute, Switzerland, did the same. This facility has been fundamental to the progress of proton therapy with its pioneering realizations of the gantry to change the angle of the beam penetration and the active scanning [77], the technique that, avoiding passive elements, moves the pencil beam in three dimensions painting the tumour with Bragg spots (a short description of active scanning is provided in the following paragraph on dose distribution).

Not by chance the first hospital-based centre was built at the Loma Linda University (California), where the determination of James Slater lead to the agreement with Fermilab in 1986, to design and built the 7-m-diameter synchrotron which accelerates protons to 250 MeV.

A smooth conversion from a physics laboratory to a hospital facility took place in Japan. The University of Tsukuba started proton clinical studies in 1983 using a synchrotron constructed for physics studies at the High Energy Accelerator Research Organization (KEK). A total of 700 patients were treated at this facility from 1983 to 2000. In 2000, a new in-house facility, called *Proton Medical Research Center* (PMRC), was constructed adjacent to the University Hospital. Built by Hitachi, it is equipped with a 250 MeV synchrotron and two rotating gantries [78].

Light Ion Therapy

Ions heavier than protons, such as helium and argon, came into use at Berkeley in 1957 and 1975, respectively. At the old 184-in. cyclotron 2800 patients received treatments to the pituitary gland with helium beams: the lateral spread and range straggling being much smaller than in the proton case described before.

About 20 years later, argon beams were tried at the Bevalac in order to increase the effectiveness against hypoxic and otherwise radioresistant tumours, i.e. tumours that need deposited doses 2–3 times higher if they are to be controlled with either photons or protons (radiation oncologists speak of 'Oxygen effect'). But problems arose owing to non-tolerable side effects in the normal tissues. After the irradiations of some 20 patients, Cornelius Tobias and collaborators decided to use lighter ions, first silicon for 2 patients and then neon, for 433 patients. Only towards the end of the program it was found that the neon charge (Z = 10) is too large and undesirable effects were produced in the traversed and downstream healthy tissues [79]. The Bevalac stopped operation in 1993 and, at the conclusion of this first phase of hadrontherapy, almost 3600 patients have been treated worldwide with pions, helium ions or other ions.

The larger effects that ions have with respect to the ones produced by cobalt-60 gammas of identical dose is quantified by introducing the 'RBE' (*Relative Biological Effectiveness*). RBE depends upon the cell type, the radiation used (particle, energy, dose) and the chosen effect. For a given cell type and effect RBE varies with the LET of the ionizing particles. Only at the beginning of the 1990s carbon ions (Z = 6) were chosen as the optimal ion type. Indeed light ions produce a radiation field which is qualitatively different from the ones due to either photons or protons and succeed in controlling also radioresistant tumours.

The carbon choice was made in Japan by Yasuo Hirao and collaborators of the National Institute of Radiological Science, NIRS [80], who proposed and built HIMAC (Heavy Ion Medical Accelerator in Chiba) in the Chiba Prefecture (Fig. 11.26).

In 1994, under the leadership of Hirohiko Tsujii, the facility treated the first patient with a carbon-ion beam of energy smaller than 400 MeV/u, corresponding to a maximum range of 27 cm in water. By the end of 2015 about 10,500 patients have

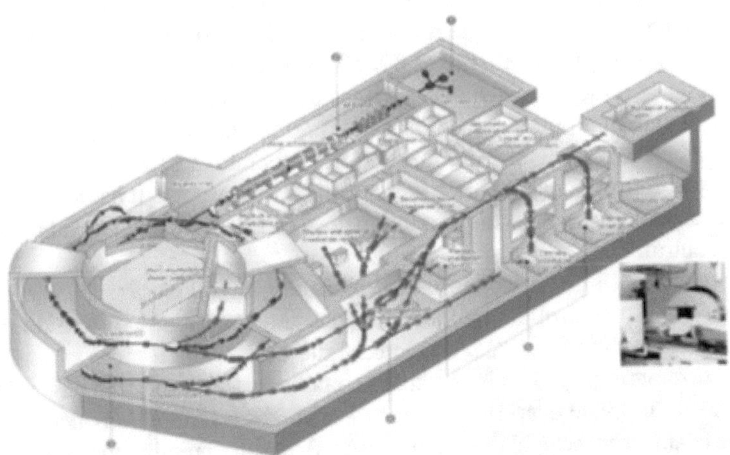

Fig. 11.26 HIMAC features two large synchrotrons, injected by a Alvarez linac, and three treatment rooms for a total of two horizontal and two vertical beams

been treated and it has been shown that many common tumours otherwise difficult to cure (e.g. lung and liver) can be controlled [81, 82].

In 1987 in Europe an important initiative was launched to create a full-fledged European light-ion therapy centre. The characteristics of the needed hadron beams were defined in a series of expert meetings. EULIMA, the European Light Ion Medical Accelerator project, financed by the European Commission, was led by Pierre Mandrillon and involved many European laboratories and centres. The European therapy synchrotron was never built but instead national-based projects in Germany and Italy were pushed through giving to the radiation oncologists facilities similar to the HIMAC at Chiba and the Hyogo Ion Beam Medical Center.

In 1993 Gerhard Kraft and colleagues obtained the approval for the construction of a carbon ion facility at GSI (Darmstadt), later called the "pilot project". Hadron therapy started in 1997 and, at the time of the closure in 2009, 440 patients [83] were treated with carbon-ion beams. The medical and technical competences were both used to build the Heidelberg Ion-Beam Therapy Center (HIT), which started operation in 2009 [84] and transferred to Siemens Medical, which built the HIT treatment rooms. The main new features of the GSI pilot project where: (i) the active 'raster' scanning system [85]; (ii) the sophisticated models and codes that take into account the biological effects of each irradiated sub-volume in the treatment planning system; (iii) the in-beam PET system which determined 'on-line' the location and shape of the irradiated volume by detecting the β^+ radioactivity produced by the incident carbon ions, mainly ^{11}C and ^{15}O [86–88].

11.3.2.2 The Bases of Cancer Radiation Therapy

The absorbed dose due to a conventional beam of photons has a roughly exponential absorption in matter after a slight initial increase. For instance, the highest dose for beams having a maximum energy of 8 MeV, is reached at a depth of about 2–3 cm of soft tissue and the dose is about one third of the maximum at a depth of 25 cm. Because of this non-optimal dose distribution, the unavoidable dose given to the healthy tissues represents the limiting factor to obtain the best local control of the pathology in conventional radiation therapy. In this connection, it has to be remarked that even a small increase of the maximum dose can be highly beneficial in terms of tumour control.

The radiation is given to the patient balancing two opposing goals, deliver a sufficient dose to control the tumour and spare the normal tissue from dose levels that could cause complications. To increase the dose to the tumour and not to the healthy surrounding organs it is essential to 'conform' the dose to the target. In order to selectively irradiate deep-seated tumours with X-rays, radiation oncologists use multiple beams from several directions (portals), usually pointing to the geometrical centre of the target. This is achieved by using a mechanical structure containing the linac which rotates around a horizontal axis passing through the isocentre ('isocentric gantry'). The already mentioned IMRT makes use of up to 6–12 X-

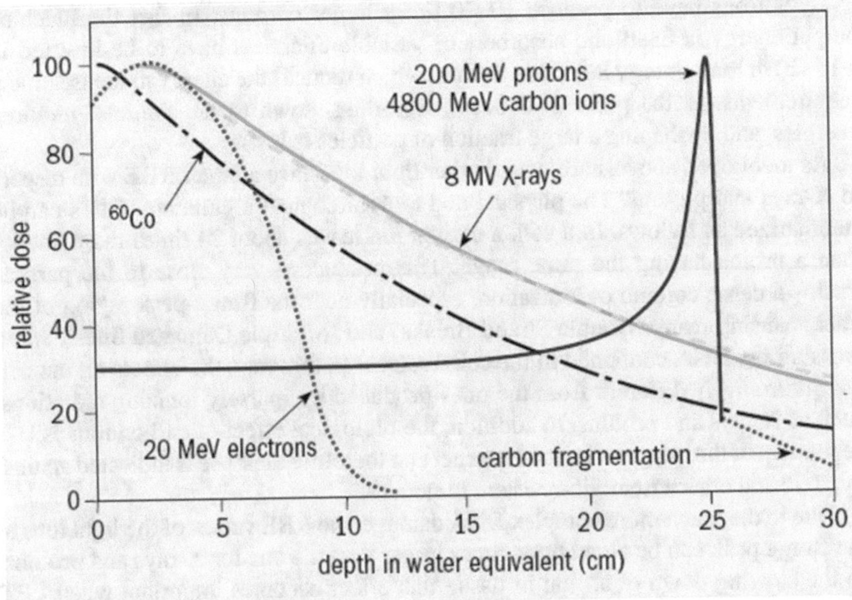

Fig. 11.27 Comparison of depth dependence of the deposited dose for each radiation type, with the narrow Bragg peak at the end. For monoenergetic beams, the carbon peak is three times narrower than the proton beam, a feature which is not represented in this schematic drawing

ray beams; the beams may be non-coplanar and their intensity is varied across the irradiation field by means of computer-controlled multi-leaf collimators [89].

The depth-dose curves of proton and light ion beams are completely different from those of X-rays, because these charged particles give the highest dose near the end of their range in the Bragg peak, just before coming to rest. Figure 11.27 illustrates how hadrons are more suitable to treat deep-seated tumours as they can be used to target profound regions preserving the more superficial healthy tissue.

In order to reach depths of more than 25 cm in soft tissues—necessary to treat *deep-seated tumours*—proton and carbon ion beams must have an initial energy not lower than 200 MeV and 4500 MeV (i.e. 375 MeV/u), respectively. In practice, the standard maximal energies are 230 MeV for protons and 400 MeV/u for carbon ions, which correspond to a 32 cm and 27 of water range, respectively. The minimal energies, used for shallow tumours, are 60 MeV and 100 MeV/u respectively.

For this reason sizeable particle accelerators are needed in hadron therapy. These machines are not demanding in terms of output current since 2 nA and 0.2 nA are sufficient for treating patients with protons and carbon ions, respectively, when active spreading systems are used. Higher currents are needed for 'passive' delivery systems in which the beam is reduced in energy with a variable absorber, spread by multiple scattering with two 'scatterers' and transversally shaped with collimators.

Cyclotrons have to produce 10–20 times larger currents. In fact the machine output energy is fixed and absorbers of variable thickness have to be inserted in a 15–20 m long Energy Selection System which reduces the energy to the treatment requirements. In the process of beam degrading, down to the minimal required energies, and re-shaping a large fraction of particles is lost.

As mentioned above, carbon and other light ions have a larger RBE with respect to X-rays and protons. The physical and radiobiological arguments of this can be summarized as follows. In a cell, a carbon ion leaves about 24 times more energy than a proton having the same range. This produces—very close to the particle track—a dense column of ionization, especially near the Bragg-peak region of the track, causing many 'Double Strand Breaks' and 'Multiple Damaged Sites', when crossing the DNA contained in the cell nucleus. In this way, the effects on the cell are *qualitatively* different from the ones produced by sparsely ionizing radiations, such as X-rays and protons. In addition, the biological effect of carbon ions is less dependent of the oxygenation of the target and therefore their use is indicated against hypoxic and otherwise radio-resistant tumours.

Due to the much more complex DNA damage, the RBE values of the light ions at the Bragg peak can be about three times larger than the one for X-rays and protons. In the slowing down of an ion in tissue this effect becomes important when LET becomes larger than about 20 keV/μm. For carbon ions this happens in the last 5 cm of their range in water.

11.3.2.3 Cyclotrons and Synchrotrons in Hadron Therapy

Proton Therapy

The first facilities that treated patients with protons and ions were based on existing accelerators built for fundamental research in nuclear and particle physics: LBL at Berkeley (USA); GWI in Uppsala (Sweden); JINR in Dubna (Russia); PMRC-1 in Tsukuba (Japan); UCL, in Louvain La Neuve (Belgium); MPRI-1 in Indiana (USA) 60 MeV protons at Chiba (Japan). Moreover pions where used at TRIUMPH and PSI (SIN at that time) in the periods 1979–1994 and 1980–1993, respectively. The clinical results have shown that pions are not superior to protons and light ions neither to obtain conformal dose volumes nor in the treatment of radioresistant tumours.

The proton therapy facilities running in the world are listed in Table 11.7. The data are extracted from PTCOG online database [90] and the data on the number of patients are updated to December 2015.

Table 11.7 includes three centres producing 60–70 MeV proton beams which are exclusively used for the treatment of eye tumours and malformations. The most recent facilities in the table are hospital-based, in the sense that they feature an accelerator built specifically for medical purposes and have more treatment rooms so that several hundred of patients can be treated every year. The companies which have built the accelerators and the high-tech parts of these centres are: Optivus (USA),

Table 11.7 Proton therapy facilities in operation [90]

Centre	Country	Acc.[a]	Max. energy [MeV]	Beam direct.[b] (del. method)[c]	Start date	Total patients
ITEP, Moscow	Russia	S	250	H(P)	1969	4368
St. Petersburg	Russia	S	100	H(P)	1975	1386
CPT, PSI, Villigen	Switzerland	C	250	2G(A),H(P)	1984	5458
Clatterbridge	England	C	62	H(P)	1989	2813
J. Slater PTC, Loma Linda	USA, CA	S	250	3G,H(P)	1990	18,362
CPO, Orsay	France	C	230	G,H(P)	1991	7560
CAL/IMPT, Nice	France	SC	230	G,H(P)	1991	5478
NRF—iThemba Labs	South Africa	C	200	H(P)	1993	524
UCSF-CNL, San Francisco	USA, CA	C	60	H(P)	1994	1839
TRIUMF, Vancouver	Canada	C	72	H(P)	1995	185
HZB, Berlin	Germany	C	250	H(P)	1998	2750
NCC, Kashiwa	Japan	C	235	2G(P,A)	1998	1560
JINR 2, Dubna	Russia	C	200	H(P)	1999	1122
PMRC 2, Tsukuba	Japan	S	250	2G(P,A)	2001	4502
MGH Francis H. Burr PTC, Boston	USA, MA	C	235	G,H(P,A)	2001	8358
INFN-LNS, Catania	Italy	C	60	H(P)	2002	350
Shizuoka Cancer Center	Japan	S	235	3G,H(P)	2003	1873
WPTC, Wanjie, Zi-Bo	China	C	230	2G,H(P)	2004	1078
UFHPTI, Jacksonville	USA, FL	C	230	3G,H(P,A)	2006	6107
MD Anderson Cancer Center, Houston	USA, TX	S	250	3G,H(P,A)	2006	6631
KNCC, Ilsan	South Korea	C	230	2G,H(P)	2007	1781
STPTC, Koriyama-City	Japan	S	235	2G,H(A)	2008	2797
RPTC, Munich	Germany	C	250	4G(A),H(P)	2009	2725
ProCure PTC, Oklahoma City	USA, OK	C	230	G,H,OB(P)	2009	2079
Chicago Proton Center, Warrenville	USA, IL	C	230	G(A),H(P),OB(P)	2010	2316

(continued)

Table 11.7 (continued)

Centre	Country	Acc.[a]	Max. energy [MeV]	Beam direct.[b] (del. method)[c]	Start date	Total patients
Roberts PTC, UPenn, Philadelphia	USA, PA	C	230	4G(P,A),H(P)	2010	3376
HUPTI, Hampton	USA, VA	C	230	4G(P),H(P)	2010	1399
MPTRC, Ibusuki	Japan	S	250	3G(A,P)	2011	1654
Fukui Prefectural Hospital PTC, Fukui City	Japan	S	235	2G(A,P),H(P)	2011	646
IFJ PAN, Krakow	Poland	C	230	G(A),H	2011	128
PTC Czech r.s.o., Prague	Czech Republic	C	230	3G(A),H	2012	780
ProCure Proton Therapy Center, Somerset	USA, NJ	C	230	4G(A,P)	2012	1892
WPE, Essen	Germany	C	230	4G(A),H	2013	366
Nagoya PTC, Nagoya City, Aichi	Japan	S	250	2G(A),H	2013	1095
S. Lee Kling PTC, Barnes Jewish Hospital, St. Louis	USA, MO	SC	250	G(P)	2013	270
SCCA ProCure Proton Therapy Center, Seattle	USA, WA	C	230	4G(A,P)	2013	844
UPTD, Dresden	Germany	C	230	G(A)	2014	106
APSS, Trento	Italy	C	230	2G(A),H	2014	92
Hokkaido Univ. Hospital PBTC, Hokkaido	Japan	S	220	G(P)	2014	
Aizawa Hospital PTC, Nagano	Japan	C	235	G(P)	2014	1
Scripps Proton Therapy Center, San Diego	USA, CA	C	250	3G(A),2H	2014	400
Willis Knighton PTCC, Shreveport	USA, LA	C	230	G(A)	2014	151
Provision Center for Proton Therapy Knoxville	USA, TN	C	230	3G(A)	2014	856
Samsung PTC, Seoul	South Korea	C	230	2G(P)	2015	4

(continued)

Table 11.7 (continued)

Centre	Country	Acc.[a]	Max. energy [MeV]	Beam direct.[b] (del. method)[c]	Start date	Total patients
The Skandion Clinic, Uppsala	Sweden	C	230	2G(A)	2015	1431
Chang Gung Memorial Hospital, Taipei	Taiwan	C	230	4G(A,P)	2015	
Ackerman Cancer Center, Jacksonville	USA, FL	SC	250	G(P)	2015	140
Mayo Clinic PBTC, Rochester	USA, MN	S	220	4G(A)	2015	186
Laurie Proton Center of Robert Wood Johnson Univ. Hospital, New Brunswick	USA, NJ	SC	250	G(P)	2015	50
St. Jude Red Frog Events PTC, Memphis	USA, TN	S	220	2G(A),H	2015	1
Texas Center for Proton Therapy, Irving	USA, TX	C	230	2G(A),H	2015	1
Tsuyama Chuo Hospital, Okayama	Japan	S	235	G(P)	2016	
Mayo Clinic Proton Therapy Center, Phoenix	USA, AZ	S	220	4G(A)	2016	
Orlando Health PTC, Orlando	USA, FL	SC	250	G(P)	2016	
Maryland PTC, Baltimore	USA, MD	C	250	4G(A),H(A)	2016	
UH Sideman CC, Cleveland	USA, OH	SC	250	G(P)	2016	
Cincinnati Children's PTC, Cincinnati	USA, OH	C	250	3G(A)	2016	
Beaumont Health Proton Therapy Center, Detroit	USA, MI	C	230	G(A)	2017	

[a]Cyclotron (C), Synchrotron (S), Synchrocyclotron (SC)
[b]Gantry (G), horizontal (H), oblique (OB)
[c]Active pencil-beam scanning (A), Passive beam spread (P)

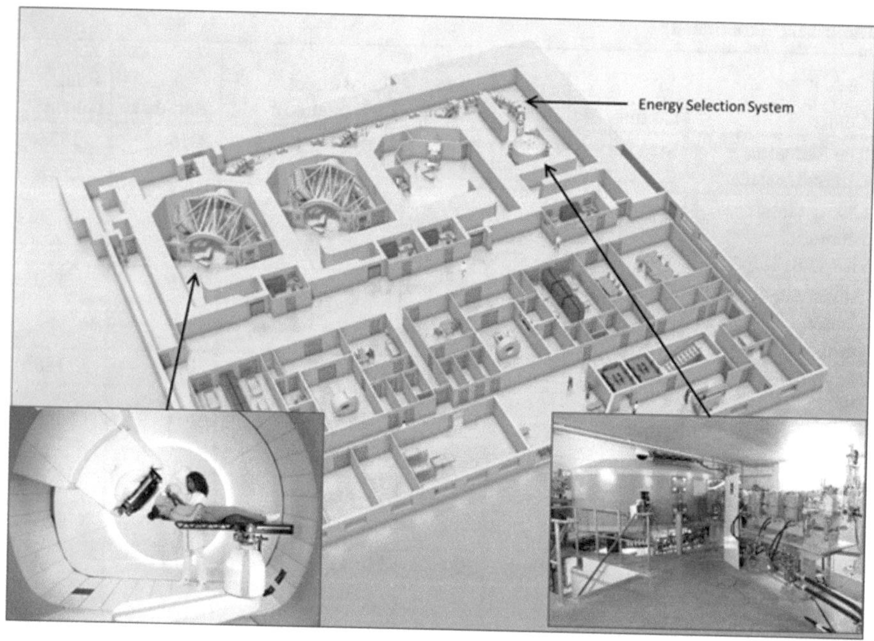

Fig. 11.28 An Ion Beam Applications (IBA) proton therapy centre featuring two gantries and a fix-beam room (insert at the bottom left). The magnetic channel, called ESS (Energy Selection System), is needed to reduce the 230 MeV proton energy produced by the cyclotron (insert at the bottom right) to lower energies, down to 70 MeV (with courtesy of IBA SA)

IBA (Belgium), Varian/Accel (USA/Germany), Hitachi (Japan), and Mitsubishi (Japan).

A centre by IBA, which is the market leader of protontherapy, is shown in Fig. 11.28.

The PTCOG is the organization, which associates experts from the technical and the clinical sectors in the field of hadron therapy. In its website [90] statistics on patients and the information about new proton and carbon-ion centres are updated with data from the companies and the clinics. By summing the numbers of all the centres in operation and out of operation one obtains that the number of patients treated with protons until the end of 2015 exceeded 131,000.

The most recent figures show that 39 new proton therapy centres are in construction and should start operation within 2020 while 24 more centres are planned. The geographical areas of major expansions are the USA with nine new centres under construction, China with seven, UK with six, and Japan with four. Proton therapy will be performed for the first time in Danemark, the Netherland, and Slovak Republic, and, outside Europe, in India, Saudi Arabia, Singapore, and Taiwan. Centres are planned to be built for the first time in South America (at the Instituto de Oncologia Angel Ruffo Hospital, Boenos Aires, Argentina), in Australia (at the Australian Bragg Centre for Proton Therapy and Research in Adelaide), and in North Africa (at the Children's Cancer Hospital Foundation, Cairo, Egypt).

Carbon-Ion Therapy

As far as carbon ions are concerned, the situation is summarized in Table 11.8. It is worth remarking that all these centres have the possibility of treating patient with both, carbon-ion and proton beams, so that it the name 'dual centres' are more appropriate than 'carbon-ion centres'. Several centres opted to use in clinic exclusively carbon ions in particular in Japan where the number of proton therapy facilities is large and well distributed in the territory.

NIRS promoting role has continued after the first experience of HIMAC and its activity was instrumental in the development and consolidation of carbon-ion therapy in Japan. Starting in 2004, NIRS was involved on the design of a new compact carbon-ion synchrotron. The new machine, that was employed for the first time in Gumna, is adapted to the medical use accelerating carbon ions up to 400 MeV/u and has a circumference of 63 m, more than two times smaller than HIMAC accelerator which was conceived for silicon ions up to 800 MeV/u [91]. Mitsubishi is commercializing the accelerator offering a complete turn-key solution as implemented in Saga-HIMAT. The centre Kanagawa i-ROCK is the result of the partnership of NIRS and Toshiba.

Table 11.8 Carbon-ion and dual facilities in operation

Centre	Country	Acc.[a]	Max. energy [MeV/u]	Beam direction[b] (scanning method)[c]	Treatment start	Total patients
HIMAC, Chiba	Japan	C	800/u	H,V,G (A,P)	1994	10,769
HIBMC, Hyogo	Japan	p, C	320/u	H,V (P)	2002	2366
IMP-CAS, Lanzhou	China	C	400/u	H (P)	2006	213
HIT, Heidelberg	Germany	p, C	430/u	2H,G (A)	2009	2086
GHMC, Gunma	Japan	C	400/u	3H (P)	2010	1909
CNAO, Pavia	Italy	p, C	480/u	3H,V (A)	2012	591
Saga-HIMAT, Tosu	Japan	C	400/u	3H,V,OB (P)	2013	1136
SPHIC, Shanghai	China	p, C	430/u	3H (A)	2014	1498
MIT, Marburg	Germany	p, C	430/u	3H,OB (A)	2015	0
Kanagawa i-ROCK, Yokohama	Japan	C	430/u	4H,2V (P)	2015	0
MedAustron, Wiener Neustadt	Austria	p, C	430/u	2H,V (A)	2016	67

The data are extracted from PTCOG online database [90] and the statistics on patients are compiled basing on update of December 2015 and form direct information. All accelerators are synchrotrons

[a] Accelerated protons (p), accelerated carbon ions (C)

[b] Horizontal (H), vertical (V), oblique (OB), gantry (G)

[c] Active pencil-beam scanning (A), Passive beam spread (P)

Not linked to NIRS developments is the synchrotron-based facility of HIBMC in Hyogo, which was developed by Mitsubishi and is in operation since 2001. This was the first clinical centre featuring dual modality, protons with energy between 70 MeV and 230 MeV and carbon ions between 70 MeV/u and 320 MeV/u.

Today the five carbon-ion therapy centres in operation in Japan treat, combined, more than 2000 patients each year, with a total at the end of 2016 of 18,000 treatments [92].

Built by Siemens Medical, the centre of HIT in Heidelberg applies the competences developed by GSI for its pilot project including the active scanning capabilities. This is the first dual centre featuring a gantry. Its construction has been a real challenge since it weighs 600 tons and consumes, at maximum energy, about 400 kW. The dual centre of Magburg was initiated as the second centre of Siemens Medical, and, when the company withdrew from hadron therapy it was bought by Marburger Ionenstrahl-Therapiezentrum which recommissioned and started treating patients in 2015.

In 1996 the *Proton Ion Medical Machine Study* (PIMMS) [93] was initiated at CERN with the participation of TERA Foundation (Italy), MedAustron (Austria), and Oncology 2000 (Czech Republic) to study a proton and carbon-ion accelerator for the therapy. The project of the high-tech part of the *Centro Nazionale di Adroterapia Oncologica* (CNAO)—shown in Fig. 11.29—is due to the TERA Foundation [94] which in 2002 obtained the financing of the centre by the Italian Health Minister. The synchrotron, features a unique 'betatron core' to obtain a very uniform extracted beam. The centre, built in Pavia, treated the first patient with protons in 2011, a year later, treated the first patient with carbon ions. The MedAustron centre built in Wiener Neustadt, uses the synchrotron design of CNAO and treated the first patient with proton beams in December 2016.

A new carbon-ion therapy facility is under construction in China (HITFil, Lanzhou). A dual center will be operational starting in 2018 in South Korea (KIRAMS, Busan). By the year 2020 two more carbon-ion centres will be operational in Japan, in the cities of Osaka and Yamagata. All these centers are based on synchrotrons.

Dose Distribution Systems

In order to maximize the outcomes of the therapy, the optimal hadron accelerators systems should deliver a biological equivalent dose defined with an uncertainty lower than 5% to tumour volumes that varies from few cubic milliliters up to 2 L preserving as much as possible the healthy surrounding tissue. The desirable dose rate is 2 grays/min/L, where 1 Gy = 1 J/kg.

Till the end of the last century, in all facilities, except PSI, the conformation of the dose to the tumour was realized with 'passive' dose delivery systems, using individually machined collimators, boluses, and passive absorbers to spread longitudinally the Bragg peak. With this technique, which is still widely used, the same spread out Bragg peak applies to the whole transversal section of the tumour, so that large volumes of healthy tissue, that surrounds the tumour, are exposed to the same dose as the tumour tissues.

Fig. 11.29 Above. CNAO has been built in Pavia by the CNAO Foundation in collaboration with INFN, the Italian National Institute for Nuclear Physics. To reduce the overall dimensions, the 7 MeV/u linac (built by GSI) is placed inside the 25 m diameter synchrotron. Below, MedAustron built in Wiener Neustadt (© Kästenbauer/Ettl.): the same synchrotron structure of CNAO hosts sources and the linac in a separate room, not visible in the picture

Modern system have been develop to minimize the over dosage of the healthy tissue. The spot scanning technique developed for protons by PSI and the raster scanning technique developed for carbon ions by GSI operate dividing the whole region to be treated in sub-volumes of the order of few cubic millimeters (named *voxel* from *volume cells*), directing beams with reduced transversal section (*pencil* beams) toward them, and delivering a dose that is varying from one position to the next. To further minimize the dose to healthy tissue the radiation is directed to the patients from several directions (typically three). These techniques, which have various practical realizations, are collectively called 'active dose spreading systems'.

The accelerating systems that realize such treatments require the highest beam stability, the possibility of changing promptly and frequently beam intensity and energy eider acting on the acceleration or using passive Energy Selection Systems with rapidly moving absorbers.

A recent development that resulted from the common effort of NIRS and HIMAC is the implementation of fast active spot scanning system. The fast beam scanning speed of 100 mm s^{-1} allows a high repetition rate of the scanning spots so that a 'volumetric multipainting' of the tumour target is achievable. One-liter tumour volume at a rate of 100 Hz can be repainted by the pencil beam several times in few minutes reducing the statistical uncertainty in the delivered dose by a factor $\sqrt{n}$ (where n is the number of re-paintings) and minimizing any local accidental under-dosage or over-dosage. This method also improves the synchronization of the irradiation with the breathing movements. A second development concerns the superconductive gantry. The structure, studied for carbon-ion beams up to 430 MeV/u, has a radius of 5.5 m and a wait of approximately 200 tons, comparable with normal-conductive proton therapy gantries and sensibly lighter than the 600 tons gantry built in Heidelberg.

The ultimate goal is the treatment of moving organs, a challenge that is more critical in hadron therapy than in conventional therapy because of the dose is concentrated in the Bragg peak. In this case the accelerating system and beam delivering system require a feedback that, based on on-line movement detection, adapt in few milliseconds the beam characteristics to the three dimensional movements of the target.

11.3.2.4 Present and Future Challenges

There is still space for improving the dose delivery, for instance, the active dose delivery systems for protons and carbon ions have been used for a small part of the patients treated with protons or ions. In spite of the fact that most new centres are featuring 'active' scanning systems and some of the existing centres are upgrading to them, the implementation in the clinical practice has been for many years quite slow.

As far as the accelerators and transfer lines are concerned, two main lines of research have to be intensively pursued:

- systems which actively scanned are able to follow the tumours which are subject to movements,
- accelerator and delivery systems which are lighter, smaller, more efficient, and less power consuming than the present cyclotrons

Fast Cycling Accelerators to Follow Moving Organs

In conventional radiotherapy various techniques have been introduced to determine in real time the position of an irradiated tumour which moves, for instance, because of the respiration cycle and to deliver the dose following these movement in IGRT. In this case computer controlled multileaf collimators are used. In hadron therapy the tumour position can be detected with the same methods and a better follow-up can be obtained by rapidly moving the Bragg spot so to compensate for *three-dimensional* movements.

In the transverse plane the active scanning of tumours with hadron pencil beams is obtained by adjusting *two* perpendicular magnetic fields located many meters upstream of the patient. The time needed to move the beam transversally, following the indication of a precise feedback system, is of the order of milliseconds.

Due to the respiration cycle, the target organs can also move longitudinally, i.e. in the direction of the beam. A good example of the problems concerning the movements is a rib which enters the irradiation field of a lung tumour. In cyclotrons and synchrotrons the time needed to change the energy, and thus the depth of the Bragg peak, is of the order of 1 s because of the delay necessary either to move energy absorbers (in cyclotrons) or to change the energy (in synchrotrons). An accelerator with an energy cycle of the order of a few milliseconds would allow a longitudinal follow-up of the tumour similar to the transversal one.

In synchrotrons, the beam is delivered to the treatment rooms only for fraction of the machine time. Ramping up and down the magnets after the extraction of the beam and waiting for the magnetic field to stabilize are time-consuming operations. Few seconds are needed in this procedure and to reduce this dead time the centre of HIT studied a system in which probes for the measurements of the field in the magnet were inserted on the feed-back loop of the regulation of the magnetic fields in the elements of synchrotron and beam lines. The result was an overall increase in efficiency of 24% and a reduction of the time between two successive extractions to approximately 1 s. If such time intervals are still not compatible with movement compensation, nevertheless the improved efficiency has important impacts on the facility, since a larger number of patients can be treated and the cost per treatment decreases [95].

NIRS and HIMAC optimized the efficiency of their extraction system allowing to extract of multiple energies, during a single machine cycle [96]. It is important to remind that, in a treatment based on active scanning, the first layer irradiated is the deepest, and then step by step all other layers. The number of particles accelerated to certain energy in general exceeds the total number needed for irradiating the corresponding layer at the planned dose. The beam not used is dumped and this corresponds also to a waste of time. At NIRS, a system was studied to avoid beam dumping using the synchrotron to decelerate and use the remaining beam to the next more superficial layer. The process takes approximately 100 ms and, since the procedure can be repeated several times within the same beam cycle, the optimization of the irradiation time is consistent. Similar techniques are feasible to compensate tumour movements.

Recently two new fast cycling accelerators—particularly suited for the movement compensation and repainting of moving organs—have been considered: high frequency linacs and Fixed Field Alterenating Gradient (FFAG) accelerators.

Since 1993 the linac approach is pursued by the TERA Foundation with the choice of using the same frequency (3 GHz) employed in the construction of electron medical linacs [97]. Since the acceleration of very low velocity protons (and carbon ions) is problematic, it was proposed to use as injector a commercial 30 MeV cyclotron. The cyclotron-linac complex has been dubbed 'cyclinac'. Between 1998 and 2003 a TERA, CERN, INFN collaboration built and tested a 3 GHz 1-m long side-coupled standing-wave structure which accelerated protons from 62 to 74 MeV [98] and could stand gradients as large as 27 MV/m, corresponding to surface

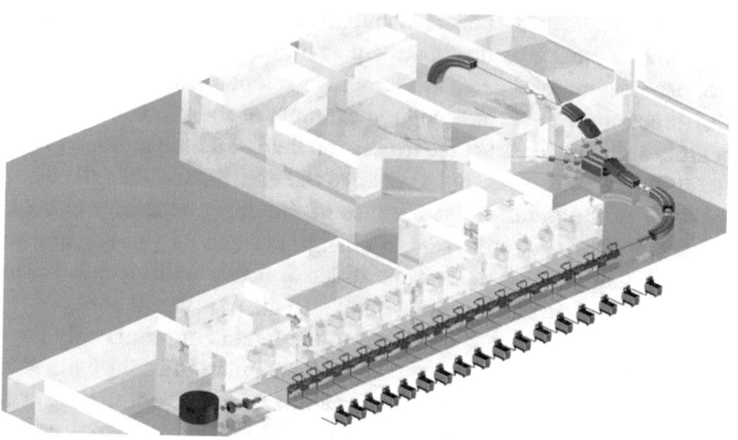

Fig. 11.30 This design of CABOTO (CArbon BOoster for the Therapy in Oncology) features a 150 MeV/u superconducting isochronous cyclotron followed by a 5.7 GHz standing-wave linac powered by seventeen 12 MW klystrons (courtesy of TERA Foundation)

electric fields larger than 150 MV/m. In parallel an *all-linac* solution was studied by L. Picardi and collaborators [99] which has been adopted for the TOP-IMPLART a proton-therapy facility under development at the Laboratories of Enea, in Frascati, Italy. The 150 MeV linac will be installed in the IFO Hospital, in Rome.

In the last years various cyclinacs have been designed: for proton therapy a 15-m long linac could be used to accelerate the protons between 30 MeV and a maximum of 230 MeV. A second cyclinac design is based on a superconducting cyclotron that accelerates carbon ions C^{+6} (and $H_2{}^+$ molecules) to 150 MeV/u (Fig. 11.30) with a 300–400 Hz repetition rate [100]. The linac is subdivided in modules, which interspaced with permanent magnetic quadruples are so short that the beam energy can be continuously varied in a couple of milliseconds between the cyclotron energy and the maximum by varying the output power of the klystrons.

Fixed Field Alternating Gradient (FFAG) accelerators have been proposed as fast cycling accelerators for hadron therapy since the magnetic fields of the ring are constant in time—as in a cyclotron—and the frequency of the accelerating electric field can vary up to 500–1000 Hz. The many bending magnets are arranged in triplets: the central magnet bends the circulating beam inwards while the two external ones bend it outwards. During acceleration the orbits maintain the same oscillating shape but increase in average radius so to remain inside the wide vacuum chamber and find stronger and stronger deflecting fields. In the recent 'non scaling' FFAGs the orbits of increasing radius do not maintain the same shape so that the full excursion requires bending magnets which are radially smaller than in classical 'scaling' FFAGs. It has to be noted that in every case beam extraction at different energies needs fast kickers [101].

A 150 MeV FFAG has been built in Japan for high-current applications [102] and low-current uses of non-scaling FFAGs in proton therapy have been proposed [103, 104].

In the case of carbon ions, due to the high magnetic rigidity of the beam, a solution with three concentric non-scaling FFAG machines has been studied [105]. In this case there is the added complication of the many injection and extraction systems which are needed in multi-ring facilities. This and other arguments are discussed in a recently published comparison between linacs and FFAGs for hadron therapy [106].

Compact Accelerators and 'Single Room' Facilities

An important line of development concerns what are now known as 'single room' proton facilities. The rational can be best appreciated by browsing Table 11.9 which has been constructed by using the results of the epidemiological studies performed in Austria, France, Italy and Germany in the framework of the EU funded network ENLIGHT [107]. They can be summarized by saying that in the medium-long term about 12% (3%) of the patients treated with high-energy photons would be better cured with fewer secondary effects if they could be irradiated with proton (carbon ion) beams. The table presents the number of treatment rooms needed for a population of 10 million people living in a developed country. The estimated numbers of rooms turn out to be in the easy to remember proportions 1:8:8^2. Since a typical hadron therapy centre has 3–4 rooms, the above figures tell that a proton (carbon ion) centre would be needed every about 5 (40) million people.

For proton therapy this shows the way to a flexible and patient-friendly solution: instead of a multi-room centre with its large building, one should develop single-room proton accelerator/gantry systems, constructed on a relatively small area (approximately 500 square metres) attached to existing hospital buildings homogeneously distributed in the territory. It is worth remarking that the overall cost of a full proton (dual) centre with three treatment rooms is of the order of 100 M€ (200 M€). Single-room proton therapy facilities, which cost approximately 25 M€ [108] are strategic for the future expansion of proton therapy in existing clinical facilities where cost and size is the most important factors.

Single room facilities are offered by Mevion, IBA, Varian, Sumitomo and other companies. Two lines of developments have to be considered: compact accelerators

Table 11.9 Estimate of the number of X ray and hadron treatment rooms

Radiation treatment	Patients per year in 10^7 inhabitants	Av. number of sessions per patient	Sessions/d in 1 room (d = 12 h)	Patients/y in 1 room (y = 230 d)	Rooms per 10 million people	Relative ratio
Photons	20, 000	30	48	370	54	8^2
Protons (12%)	2400	24	36	345	7.0	8
C ions (3%)	600	12	36	690	0.87	1

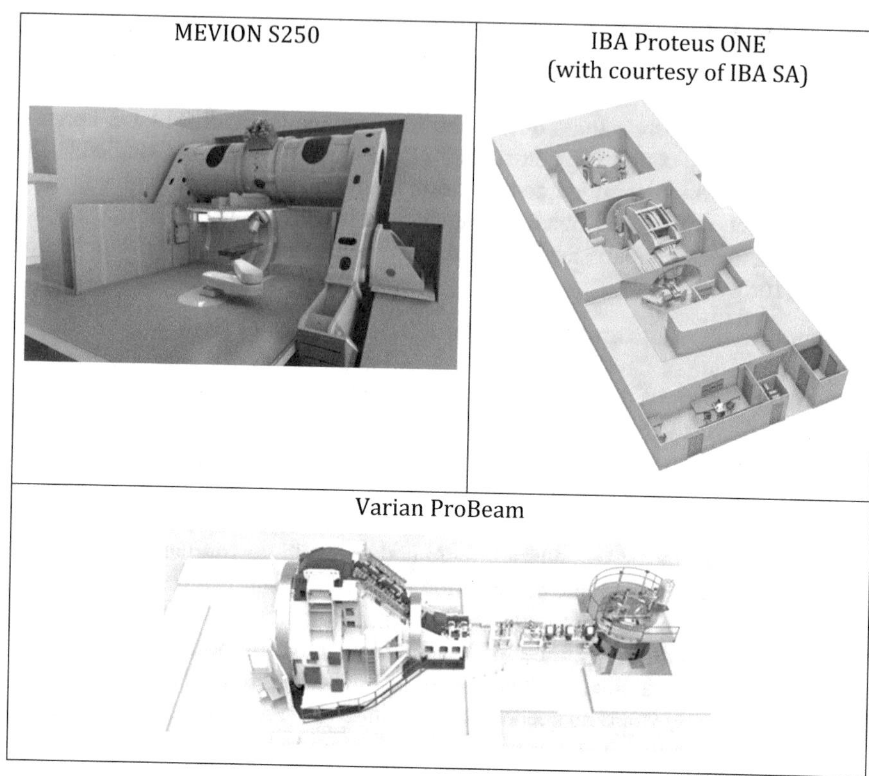

Fig. 11.31 Single-room facilities commercialized by MEVION, IBA, and Varian

followed by short magnetic channels and accelerators which rotate around the patient.

A very high field 15-tons niobium-tin superconducting 250 MeV synchrocyclotron was developed by Mevion Medical Systems Inc. in collaboration with the Massachusetts Institute of Technology (MIT) and the Massachusetts Medical Hospital. Mevion S250 is the firstly constructed system based on the concept of single-room facility which treated the first patient in December 2013. The accelerator is mounted directly on the 190° isocentric gantry (Fig. 11.31a) with a superconductive coils producing a 8.7 T central magnetic field. The first version of the system which featured a passive beam delivery system has been updated to a 'pencil beam' system for active scanning.

The Proteus ONE system of IBA consists of a compact 230 MeV superconducting synchrocyclotron, named S2C2. The energy selection system is mounted on the 220° isocentric gantry (Fig. 11.31b). In September 2016 the Lacassagne Centre in Nice treated the first patient with S2C2 accelerator.

The single-room system ProBeam of Varian system combines a 250 MeV cyclotron—the first commercial superconductive accelerator developed for proton therapy used at PSI since 2007—and the 360° gantry (Fig. 11.31c).

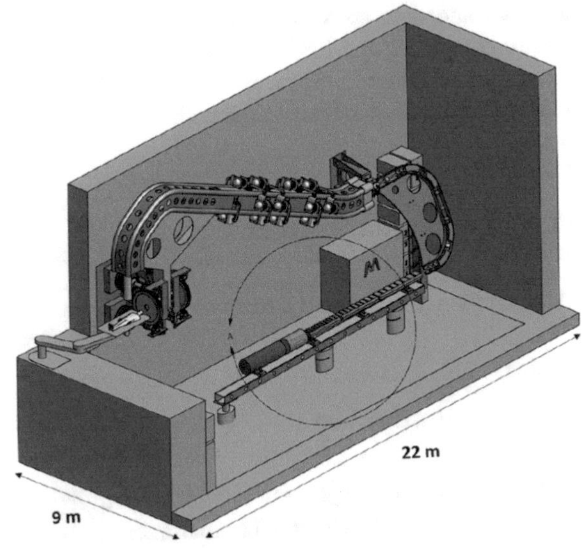

Fig. 11.32 TULIP (Turning LInac for Proton therapy) all-linac solution (courtesy of Mohammad Vaziri—TERA Foundation)

A proposed system is the high-frequency proton linac rotating around the patient—according to a scheme named TULIP and patented by TERA [109] which is based on the same 5.7 GHz linac designed for CABOTO (Fig. 11.32).

An original development concerns laser-driven ion beams and is based on protons accelerated by illuminating a thin target with powerful (10^{18}–10^{20} W/cm^2) and short (30–50 fs) laser pulses. When the laser pulse hits the target—with thickness of the few micrometers—electrons are violently accelerated producing an electric field of the order of teravolt per meter which draws behind the protons for the surface of the target itself. The phenomenon has been studied experimentally reaching proton energies of more than 10 MeV with promising results [110]. The energy spectrum is continuous and computations show that, using proper combination of target shapes and laser power, a 3% energy spread can be obtained [111]. The advantages of a laser-driven system are that the ion beams can be generated very close to the patient and that optical elements can substitute the cumbersome and heavy beamline elements including the gantries. The most relevant challenge is probably the short pulse duration which is linked to the unknown biological effects of the high dose rates, the impossibility of controlling the beam intensity during the irradiation, and the potential risk of overdose. An important issue is also linked to very low repetition rate of few hertz which would extend the irradiation to unacceptable time. The implementation process is complex and still several years are foreseen to create, around the laser-driven 200 MeV proton beam, the whole infrastructures for a therapy facility [112].

Acknowledgment
The authors are grateful to the International Atomic Energy Agency and in particular to Joao Alberto Osso and Amirreza Jalilian for the discussions and for providing

statistics and data on worldwide radioisotope production, to Eiichi Takada and Hikaru Souda for the information about the recent developments on carbon-ion therapy accelerators in Japan, and to Saverio Braccini for the precious contributions and comments he provided on the text.

11.4 Spallation Sources

M. Lindroos · S. Molloy · G. Rees · M. Seidel

11.4.1 Introduction

Spallation is a nuclear process in which neutrons at different energies are emitted in several stages following the bombardment of heavy nuclei with highly energetic particles. In this chapter we limit ourselves to the description of the accelerator that drives spallation sources. For a more detailed discussion of the spallation process itself, the target, the moderators and the physics at such facilities see (for example) [113]. However, it is worth noting that there are other ways to produce neutrons with accelerators, for example photo-fission induced by an intense electron beam. The spallation process is the most practical and feasible way of producing neutrons for a reasonable effort (or simply cost) of the neutron source cooling system, see Table 11.10. Research reactors also require fissile material handling, potentially a major constraint for both handling and licensing.

Spallation sources come in at least three types: short pulse sources (a few μs), long pulse sources (a few ms) and continuous sources. In general, synchrotrons or accumulator (compressor) rings provide short neutron pulses, linear accelerators provide long neutron pulses, and cyclotrons provide continuous beams of neutrons.

Pulsed neutron sources are more efficient in their use of neutrons than continuous sources, in a majority of all applications, according to a survey performed for the Autrans meeting in 1996 [115]. The time-of-flight of the neutrons—readily measured from a pulsed source—allows us to determine the neutron speed and energy. Additional tools are needed to select or determine the neutron speed in a

Table 11.10 The number of fast neutrons produced per joule of heat energy where the energy in joule is taken as heat produced over energy consumed [114]

Fission reactors	~10^9	in ~50 litre volume
Spallation	~10^{10}	in ~1 litre volume
Fusion	~2×10^{10}	in huge volume
Photo neutrons	~10^9	in ~0.01 litre volume
Nuclear reaction (p, Be):	~10^8	in ~0.001 litre volume
Laser induced fusion	~10^4	in ~10^{-9} litre volume

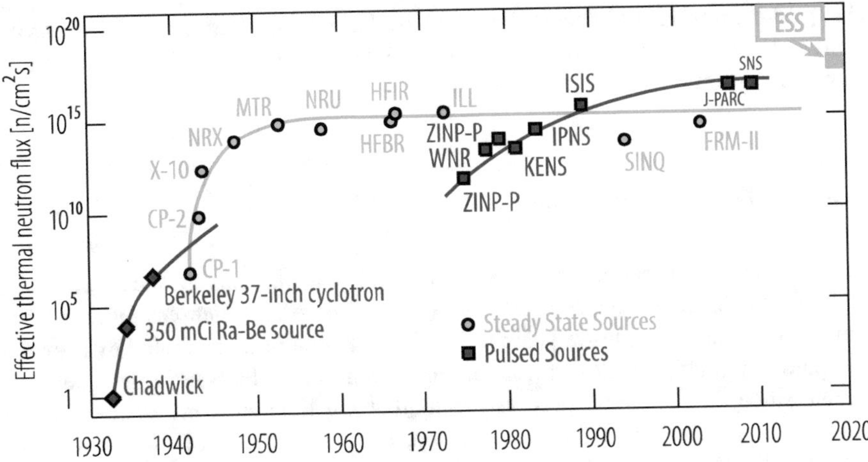

Fig. 11.33 The instant neutron flux at different research facilities plotted as a function of time

standard research reactor source. For example, a pulsed structure can be achieved by chopping a continuous beam with a shutter. Consequently many of the neutrons are then lost. Figure 11.33 plots the instantaneous neutron flux at different facilities as a function of the year of their completion. It is not possible to extract the average neutron flux from the figure without detailed knowledge of the time structure of the different facilities. However, the future European Spallation Source (ESS) will be the first spallation source with a time average neutron flux as high as that of the most intense research reactors.

The spallation cross section for protons on heavy nuclei increases as a function of proton energy up to several tens of GeV [116]. Nonetheless it is generally agreed that a kinetic proton energy between 1–3 GeV is optimal for practical target and moderator designs, and in order to keep the shielding requirements reasonable.

The first spallation source, ZING-P, based on a synchrotron, was built and operated at the Argonne National Laboratory. It provided a short pulse well suited for Time-Of-Flight (TOF) techniques at the "instruments" (experiments). Spallation sources at Argonne were superseded by the Intense Pulsed Neutron Source (INPS), by the KEK Neutron Science Center (KENS) in Japan, and by the ISIS neutron facility at Rutherford in UK. These synchrotron based spallation facilities were recently exceeded in power by J-PARC, at Tokai in Japan.

The first linear accelerator used to provide protons for a spallation source – the Los Alamos Meson Physics Facility (LAMPF) in the USA – later became the Los Alamos Neutron Science Center (LANCSE). The highest power Spallation Neutron Source (SNS) in Oak Ridge combines a full energy linear accelerator with an accumulator ring to provide very high intensity short pulses of neutrons to the instruments. The European Spallation (ESS) source will provide even higher intensities, but is developing instruments able to use longer linac pulses directly for spallation, avoiding the need for a costly and performance-limiting accumulator ring

[117]. Synchrotrons and accumulators, alike, have a intensity limits due to space charge effects that introduce tune spreads and cause beam instabilities, resulting in significant beam losses.

11.4.2 The Linear Accelerator

For many spallation-based sources, a good choice of technology for the acceleration of the proton beam is a linear accelerator (linac). These very complex machines are conceptually quite simple—a beam of H+ ions or protons is generated by a source and passed through a series of accelerating structures on its way to the spallation target. Since the spallation process means that the time structure of the neutron pulses will match that of the proton beam, the requirements of the experiments directly dictate the time structure of the proton current, and thus the high-level design of the linac.

11.4.2.1 High-Level Machine Design

Typically the specifications given by the experiments are the following.

- The desired repetition rate of the facility, usually less than 120 Hz.
- The duty factor or the length of the neutron pulses.
- The beam power on target; either average or maximum.

Given these parameters, the accelerator designers then begin sketching out the specifications of the design.

Kinetic Energy and Current

The maximum beam power on target constrains the maximum energy and current of the proton beam, and engineering capabilities will place additional limitations on these quantities. For example, the spallation target may have an energy threshold beyond which it will no longer survive.

As mentioned previously, the optimal beam energy for a spallation target system is in the range 1–3 GeV, and so an accelerator designer is likely to pick an energy in this range. From this, the required beam current is easily derived.

After this, these parameters will be optimised in an iterative way. Engineering limitations will first be reached in either the power transmission capabilities of the couplers on the accelerating structures, or the acceleration gradients required within the cavities. Once these have been determined, the designer can iteratively tweak the beam energy and current in order to balance the performance requirements of the cavities.

RF Frequency

Due to unavoidable losses in the generation and transmission of the RF, the installed capability will need to be more than a factor of two larger than the beam power. The high power nature of these machines means that one of the largest cost drivers will be the power sources for the accelerating RF. In order to save cost it is advisable to take advantage of already-designed components, and thus RF frequencies previously adopted by other facilities.

Typically this results in acceleration RF whose frequency is an integer multiple of 176.105 MHz or 201.5 MHz.

Since it is possible to achieve higher acceleration gradients with higher frequencies, the self-force (i.e. space charge) of the beam at the lower energy regions of the machine mean that the strong focusing that results from high frequencies can be problematic. It can therefore be optimal to use multiple acceleration frequencies, with that in later sections of the linac being a multiple of two or four of the low energy acceleration.

11.4.2.2 Linac Layout

A linac is laid out as a series of consecutive stages of acceleration. The following sections each provide some detail on the most common technologies.

Proton Source

The ion source for a long pulse spallation source delivers protons rather than H- ions, since there is no need to inject into an accumulator ring using charge exchange injection (see section on synchrotrons). Proton sources can be implemented using various techniques. For example, compact Electron Cyclotron Resonance (ECR) sources such as VIS [118] and SILHI [119] are well suited to the task, delivering continuous proton beams of up to 100 mA. The diverging beam leaving the source has to be transported and matched to RFQ, usually using Einzel lenses or solenoids in the Low Energy Beam Transport (LEBT). The LEBT has to be designed carefully because it defines the beam quality throughout the rest of the linac [120].

Radio Frequency Quadrupole

Modern proton sources typically output a continuous beam of protons, however it is not possible to directly accelerate such a CW current. The structure immediately following the proton source must be one that compresses the current into a series of bunches at the fundamental frequency of the accelerating RF. In the majority of linacs a Radio Frequency Quadrupole (RFQ) performs this job.

An RFQ is a resonant cavity loaded with four vanes or rods in such a way that the beam experiences a quadrupolar electrical field that strongly confines the size of the beam. These vanes or rods are modulated in such a way that the beam also experiences longitudinal forces. Initially these forces are tuned so that the continuous current will begin to coalesce into a series of bunches at the frequency of the accelerating RF.

As the beam proceeds further through the RFQ, the modulation of the vanes/rods will be adapted so that there will be an acceleration force in addition to the focusing terms. This will be increased throughout the length of the RFQ in a way that preserves the quality of the beam, while keeping the structure to a reasonable length.

For a given frequency the four-vane structure is the most power efficient, for low current applications 4-rod structures may be cheaper to build. Currents of up to 100 mA can be handled with a single RFQ in the frequency range of 80–400 MHz, while higher currents require beam from two parallel RFQs to be combined in a funnelling section that operates at half the frequency of the main linac.

Drift Tube Linac

Once the beam has reached an energy of approximately 3 MeV, the efficiency of the acceleration provided by an RFQ will drop to a level where it would be optimal to use an alternative method of acceleration. In high-power proton linacs, a Drift Tube Linac (DTL) is often chosen to succeed the RFQ due to its high efficiency, and well-understood design principles.

A DTL is a large RF cavity through which the beam passes, absorbing power from the accelerating field. The time taken for the beam to traverse this structure is normally many tens of RF periods, and so metallic drift tubes are added to shield the beam from the decelerating phase of the RF. The drift tubes must then be designed to have the correct length and location for the beam to see the correct amplitude and phase of the accelerating fields.

The DTL accelerates the bunched beam from the RFQ to energies between 50 and 100 MeV [121]. The accelerating efficiency of DTLs drops at energies above 50 MeV and therefore a change of structure is required. The problem is that the physics & engineering effort required to build a more efficient structure costs more than any potential operational savings while increasing the risks. The normal conducting alternatives are Separated DTLs (JPARC), Side Coupled DTLs (Los Alamos and SNS), and Cavity coupled DTLs (LINAC4). Transverse focusing is achieved by permanent or electromagnetic quadrupoles housed inside the DTL drift tubes and/or between RF cavities.

Intermediate Energy Acceleration

In the energy range between approximately 100 MeV and 500 MeV, there tends to be a split in the technologies chosen for beam acceleration.

For machines with a duty cycle of more than a few percent, the ohmic losses due to the finite resistance of the metallic cavities is a significant driver of the operational costs of the accelerator. For this reason, these machines tend to choose superconducting structures in order to increase the fraction of the RF power that accelerates the beam. Despite the running cost of the cryogenic facility, superconducting facilities with significant duty cycles will cost substantially less to operate than room-temperature machines. An emerging alternative to higher energy DTL-variants is the superconducting spoke cavity technology. Spoke resonators have a large transverse and longitudinal acceptance and are mechanically very stiff, reducing their sensitivity to microphonics and to Lorenz force detuning compared to elliptical resonators in this energy range.

For machines whose duty cycle is less than this cut-off, the long time taken for filling the superconducting cavities with RF before beam arrival becomes significant, thereby degrading the efficiency. For these machines, the higher RF power costs are easily outweighed by the removal of the large cryogenic facility.

These linacs tend to favour the alternatives mentioned in the previous section—separated DTL's, Side-Coupled DTL's, and Cavity-Coupled DTL's.

Superconducting Structures
In the high-energy part of the linac, superconducting technology tends to dominate, however care must be taken to set the energy at which the linac transitions to superconducting cavities at a point that properly balances costs and benefits.

Due to several decades of R&D by the ILC community, elliptical cavities are a very good choice for these structures. The complex manufacturing processes are very well understood, and the design is very robust to mechanical disturbances.

High-energy efficiency requires that each RF structure is designed to have an accelerating mode with a very high Quality Factor (Q). Consequently, each Higher Order Mode (HOM) also tends to have long damping time, with a significant risk that its amplitude will still be at a high level when the subsequent pulses arrives. Thus, it is necessary to consider a scheme to damp HOMs. In general two solutions are available to the accelerator designer.

One solution is to mount one or more HOM couplers in locations where it is expected that the more destructive parasitic fields will have large amplitudes. These ports couple HOM power into an external load, dramatically shortening the decay times of the modes. It is common for accelerating structures to join multiple resonant cells into a single cavity structure. In this case, it is possible (as shown in Fig. 11.34) that the amplitude of a particular HOM may be negligible in the region of the extraction coupler, while remaining high in the body of the cavity. This is an inherent weakness of the multiple-cell structure, and care should be taken in the design of the cavity to ensure that such "trapped modes" do not couple well to the beam.

An alternative HOM damping solution is to include low Q, lossy, material around the beam pipe between one cavity and the next, in order to induce losses in any field that extends into this volume. This is implemented in a region where the amplitude and losses of the fundamental accelerating field harmonic are negligible. Nonetheless, a reduced fundamental Q is an unavoidable consequence of these designs.

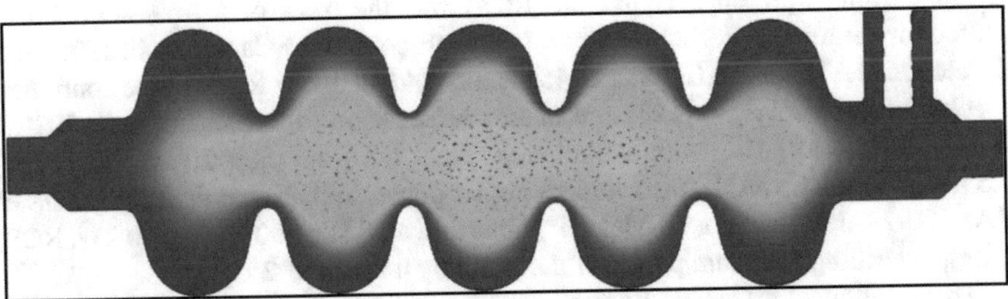

Fig. 11.34 Field magnitude for a particular dipole HOM oscillation within a five-cell accelerating cavity

HOM couplers typically have an inductive and a capacitive component, such that there is a very sharp rejection at the frequency of the accelerating mode. Although the addition of this filter successfully limits the impact on the accelerating efficiency of the cavity, the requirement for a non-zero capacitance between the central conductor and the cavity wall strongly increases the risk of a resonant cascade of electrons being field-emitted from the metallic surfaces. This would severely limit the successful operation of the cavity.

It has been shown [122] possible to instrument the signals excited on HOM couplers, to act as a diagnostic device measuring the 4D transverse location of the bunch (with position and angular resolutions of 4 μm and 140 μrad in FLASH), as well as to measure the phase of the bunch with respect to the accelerating RF (with a resolution of 0.08° at 1.3 GHz in FLASH). A determination of the cavity-cavity transverse alignment using the excited HOMs has also been demonstrated. The ability to instrument every accelerating cavity in this way is a major advantage to the HOM coupling scheme.

11.4.3 Rapid Cycling Accelerators for Short Pulse Spallation Neutron Sources

Accelerator schemes proposed for short pulse spallation neutron sources include:

1. Charge exchange injection from a pulsed H^- linac to a fast cycling synchrotron
2. Charge exchange injection from a pulsed H^- linac to a pulsed compressor ring
3. Direct proton injection from a pulsed proton linac to a fast cycling synchrotron
4. Direct proton injection from a pulsed proton linac to a pulsed compressor ring
5. Direct proton injection from a pulsed proton linac to a fast cycling FFAG ring

11.4.3.1 Charge Exchange Injection from a Pulsed H^- Linac to a Fast Cycling Synchrotron

A rapid cycling proton synchrotron (RCS) was the basis of a first, short pulse source for neutron scattering research. It was proposed by J Carpenter in 1972, and developed at ANL as the IPNS. The 450 MeV, 6.4 kW 30 Hz RCS source continued successfully for 26 years [123] and other sources soon followed. In Japan, KENS was developed at KEK, with a spallation target fed from their 20 Hz, 500 MeV, 3.5 kW proton RCS. ANL then studied higher power sources, but next came ISIS at RAL, (1979–1984) with a 70 MeV H^- linac and a 50 Hz, 800 MeV, 160 kW, RCS, [124], exceeding the beam power of the IPNS by a factor of 25.

There followed a Central European Initiative for a spallation source in 1993–1994, involving detailed feasibility studies [125] at both Vienna and CERN. Austron proposed to use a 0.13 GeV, H^- linac with a 1.6 GeV, 213 m circumference, proton

RCS to deliver neutron target beam powers of 205 (410) kW, at 25 (50) Hz. Despite the extent of the study, the project was not approved.

Now, Japan's Proton Accelerator Research Complex, J-PARC [126] has a 3 GeV RCS for a spallation neutron source as a part of a larger facility. The initial phase had a 25 Hz, 0.18 GeV, H^- linac with a 25 Hz, 3 GeV RCS, for a 0.5 MW beam power, but the H^- linac energy has been raised to 0.4 GeV to double the beam power. The 348.3 m circumference RCS has three superperiods, and uses novel, MA (magnetic alloy) loaded cavities for its harmonic number 2, RF system.

A RCS based, spallation source, the CSNS, is being built in China. It has a 81 MeV, H^- injector linac and a 25 Hz, 120 kW, 1.6 GeV, proton synchrotron [127]. Later linac energies of 134 and 230 MeV will allow beam powers of 240 and 500 kW, respectively. The 41.5 m long H^- linac has a 324 MHz, RFQ and DTL. An initial hybrid lattice of doublet straights and FBDB arc cells has been replaced by an all triplet lattice.

ISIS experience showed the importance of reliable 50 Hz, ion sources, long-life H^- stripping foils, RF shields for ceramic vacuum chambers, glass bonding for ceramic sections, low loss fast extraction, and quick release for flanges and water fittings. H^- charge exchange injection is a key issue and includes painting smooth beam distributions, reducing proton foil traverses (to limit foil temperatures), collecting foil stripped electrons, and locating a momentum and a betatron collimation system, allowing hands-on and "active" maintenance, for improved reliability and availability. Anti-correlated is preferred to correlated painting, as the latter produces some large amplitude protons in both transverse planes, which may be lost via coupling. Every fifth, RCS pulse is now diverted, at 10 Hz, to a new target station, which has become a major development.

A sub-tunnel within a main ring tunnel may reduce air activation near the collectors. Fractional loss design levels are set at $\approx 10^{-3}$ at collimators and 10^{-4} elsewhere in the ring, with the latter leading to a beam loss of <1 W/m. A RCS needs rectangular vessels to avoid vertical loss of any un-trapped beam, which spirals to protective momentum collectors, set in a high, normalized dispersion region. Rectangular vessels also aid transverse halo collection. Betatron collectors may be in three adjacent straights, with three long secondary units placed at 17°, 90° and 163°, transverse, betatron phase shifts after a primary scatterer.

Beam density limitations due to Liouville's Theorem are circumvented in charge exchange injection. Two designs are available for low-loss H^- injection systems for RCS rings. In one, the merging of the H^- ions and the circulating protons occurs at a ring dipole and, in the other, at the centre of an orbit chicane of four fixed and eight variable dipoles, in a very low dispersion region (SNS scheme) [128]. The former does not need an injection chicane, or an injection septum unit, or horizontal painting magnets, but it is not as flexible as the latter for longitudinal painting.

Both injection schemes aim to scrape the H^- beam (~$\pm 10\sigma$) in the input line at $\pm 5\sigma$ and set the beam centre at 5σ (~5 mm) from the adjacent, stripping foil edges. Input steering errors (of one polarity) result in some H^- ions missing the foil. Partially stripped H^o atoms appear in a range of quantum states [129] and, together with unstripped H^- ions, diverge from the protons in the region's bend field, and

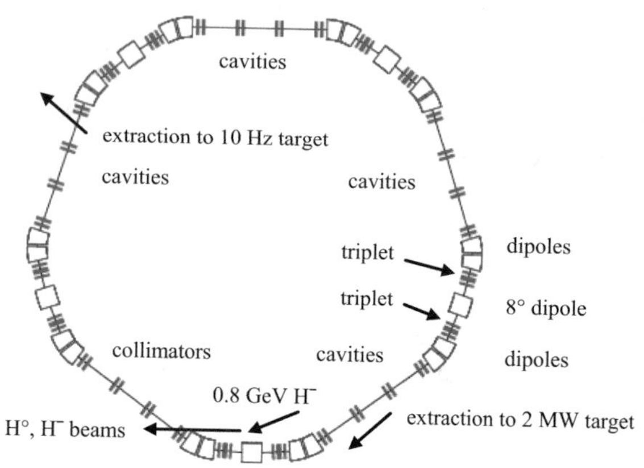

Fig. 11.35 Lattice for a 0.8–3.2GeV RCS, ISIS upgrade

traverse a second stripper and septum extraction unit(s) to exit the ring. Atoms with low principal quantum number, n, remain as H^o up to the second foil, while atoms of high n strip rapidly and are accepted as protons. Graphs of H^o lifetime versus field show "gaps" between n = 4 and 5 and between n = 5 and 6 states. A "gap" may be chosen by optimum choice of the region's field.

An upgrade study for ISIS [130] has considered a 0.8–3.2 GeV RCS, fed either from the ISIS ring for a 1 MW source at 50 Hz, or from a new 800 MeV H^- linac for a 2 (or 5) MW source at 30 (or 50) Hz. The injection is in an arc dipole as shown schematically in Fig. 11.35. The ring has been designed with five superperiods, each with arcs of triplets and straight sections of doublets (one pair of which is back to back). The lattice has been designed to minimize peak stored energy in the bend magnets and provide sufficient straight section space for economic cavity and RF system designs. In comparison with an all triplet lattice, it has 20 quadrupoles less, 20 m more of straight sections and ~67% less, total dipole stored energy.

11.4.3.2 Charge Exchange Injection from a Pulsed H^- Linac to a Pulsed Compressor Ring

Charge exchange injection from a pulsed H^- linac to a pulsed compressor ring was considered for a short pulse, 5 MW, European Spallation Source (ESS) shortly after the success of ISIS [131]. The study was overtaken in the USA, where it was decided to build a similar source, the SNS, at Oak Ridge. Injection was similar to that proposed for the RCS rings. SNS parameters were set at a 1.4 MW beam power at 60 Hz and 1 GeV, with a later upgrade to 3 MW at 1.3 GeV. Operating experience at the SNS [132] soon identified problem areas for future high power short pulse sources. These included the beam losses in the H^- linac following intra-beam scattering, and during H^- charge exchange injection in the compressor ring. Use of a H^- linac injector for a compressor ring or RCS now needs to be re-evaluated and other schemes also considered.

11.4.3.3 Direct Proton Injection from a Pulsed Proton Linac to a Fast Cycling Synchrotron

In a new scheme, a proton linac replaces the H^- linac and a ring with only one type of lattice cell is assumed, as a H^- injection insertion isn't needed. A more stable, higher current, lower emittance linac beam is realised and intra-beam and injection stripper losses are avoided. The RCS, compressor or FFAG ring is fed directly from the proton linac, using a two plane, multi-turn injection scheme, as proposed for a heavy ion fusion research project in 1998 [133]. Direct proton injection is similar for all three rings. A tilted, corner, electrostatic septum unit is set at a long drift section centre point. The septum wires have an effective 1 mm thickness and are at a 40–60° angle, θ, to the horizontal. Control of closed orbits at the septum is via vertical and horizontal bump magnets. Correlated painting is used, with closed orbits reduced during injection to reach the design emittances.

Basic parameters are input beam and ring emittances, angle θ, ring betatron tunes, Twiss parameters, space charge levels, and positions and directions, at the septum output, of the incoming beam centre and the ring's closed orbit. The ring lattice parameters may be kept constant during the injection or varied, to minimise injection beam losses. Evolution of beam distributions under the non-linear space charge forces, during and after injection, was simulated in the study [133]. Emittances were found to continue to grow after injection and evolve towards a stationary state so, to avoid loss, the late beam-septum separations were increased. A maximum number of turns injected without loss, for $F = 10–20$, and $(\varepsilon h\ \varepsilon v)_{ring}$ and $(\varepsilon h\ \varepsilon v)_{inj}$ the respective products of ring and injected emittances, was found to be approximately: $N_{inj} = (1/F)\ (\varepsilon h\ \varepsilon v)_{ring}/(\varepsilon h\ \varepsilon v)_{inj}$.

Ring circumference and straight sections must be large enough to achieve the required injected beam. RAL studies [134] have compared combined function, DFD and FDF triplet, and dFDFd and fDFDf, pumplet lattice cells. Magnets d and D provide vertical focusing (horizontal defocusing) and f and F give opposite focusing. Doublet cells are avoided as beam waists aren't at an injection straight centre. RCS pumplet cells are similar to those proposed for FFAGs [135], but their fields are linear and their magnets all have positive bends. The bending radii are thus larger and magnetic fields much reduced. For equal bending and corrector lengths, triplet and pumplet cells have the same length. Pumplets are preferred as the Twiss parameters are lower and hence the magnet apertures smaller. Ring designs use only one type of pumplet lattice cell (dFDFd or fDFDf), and a low dispersion in the injection straight section is advantageous.

An ISIS upgrade considers replacing the present ring by a large emittance, 0.6–1.8 GeV pumplet RCS. A fDFDf cell is preferred to a dFDFd as it has lower β(h) values and dispersion. The 12 cells have a 26.0 m mean and a 12.5 m bend radius, and a 0.686 T maximum on-axis magnetic field. Assumed is a 70% chopped, 0.1 A linac proton beam, of 3 (π) mm mr, full un-normalised (y, x) emittances, F = 20, two-plane painting, a 450 (π) mm mr, ring vertical acceptance, and 150 and 300 (π) mm mr, respective, full un-normalised (y, x) ring emittances. These allow 75 10^{+12} protons to be injected in 172 μs over 250 turns, at a peak space charge tune shift ≈0.23 and a 1.08 MW beam power. ISIS second harmonic cavities may be used at a h = 2, frequency range of 2.91–3.45 MHz.

11.4.3.4 Direct Proton Injection from a Pulsed Proton Linac to a Pulsed Compressor Ring

A compressor ring and a RCS may be compared for the case of direct proton injection and equal beam powers. The compressor ring requires a higher energy, longer, more costly proton linac injector, a little larger ring circumference, slightly smaller ring acceptances, and a simpler, lower power radio frequency system. A detailed cost estimate for both linacs and rings is needed in order to determine the choice.

11.4.3.5 Direct Proton Injection from a Pulsed Proton Linac to a Fast Cycling FFAG Ring

FFAG accelerators have non-linear fields and focusing, defined by $B = Bo\,(1 + x/r_o)^K$, with x and r the distances from the ring centre and $K = \rho\ r\ (B'/B\ \rho) = \rho\ r\ Kv$. They may be non-scaling or scaling (NSFFAG or SCFFAG) [134, 135], with the former more common for higher beam energies. Two types have been considered for a 26.0 m radius ISIS upgrade. One, proposed by S. Machida [136], has a novel type of DF spiral lattice cell and construction of a model magnet cell is under study.

The second, by G.H. Rees [134], has 12 [O f(+) o D(–) o F(+) o D(–) o f(+) O] pumplet cells, where the non-linear magnets, d & D, give vertical focusing (horizontal defocusing), the f & F provide opposite polarity focusing, straight sections are O and o and bend directions are (+) or (–), with (–) the reverse bends.

Ring closed orbits needed for direct, multi-turn, proton injection into the FFAG are influenced by the non-linear main fields. Injection orbit bumps reduce the ring periodicity below that set by the number of lattice cells, and the design must ensure the beam dynamic aperture is not reduced during the injection.

FFAG-RCS differences include the former's lower chromaticities and beam-size dependent tune shifts and the latter's lower injection dispersion. Errors in FFAG's non-linear magnets are harder to correct than those in RCS's linear magnets. A RCS has the advantage of no reverse bends so, in the case of equal outer radii, room temperature magnets, it may have a higher energy and beam power than a FFAG. It may also include trim quadrupoles, sextupoles and octupoles to adjust the respective tunes, chromaticities and tune spreads, and so aid in obtaining beam stability.

11.4.4 High Intensity Cyclotrons

Cyclotrons have a long history in accelerator physics and are used for a wide range of medical, industrial and research applications [137]. First cyclotrons were designed and built by Lawrence and Livingston [138] back in 1931. The cyclotron is a resonant accelerator concept with several properties that make it well suited for the acceleration of hadron beams with high average intensity.

As a circular accelerator the cyclotron repetitively uses the same accelerating resonators and radio frequency (RF) sources, thereby utilizing these expensive components in an effective and economical way. Cyclotrons allow continuous injection, acceleration and extraction of a beam. Neither RF frequency nor magnetic bending field must be cycled. The continuous wave (CW) operation results in low bunch charges, which leads to moderate space charge effects in comparison to pulsed accelerator concepts. Basic components of a separated sector cyclotron, suited for high intensity operation, are the bending magnets in sector shape, RF resonators for beam acceleration and the injection/extraction elements. The Fig. 11.36 shows the layout of the PSI Ring cyclotron [139, 140] as an example of a high intensity, separated sector cyclotron.

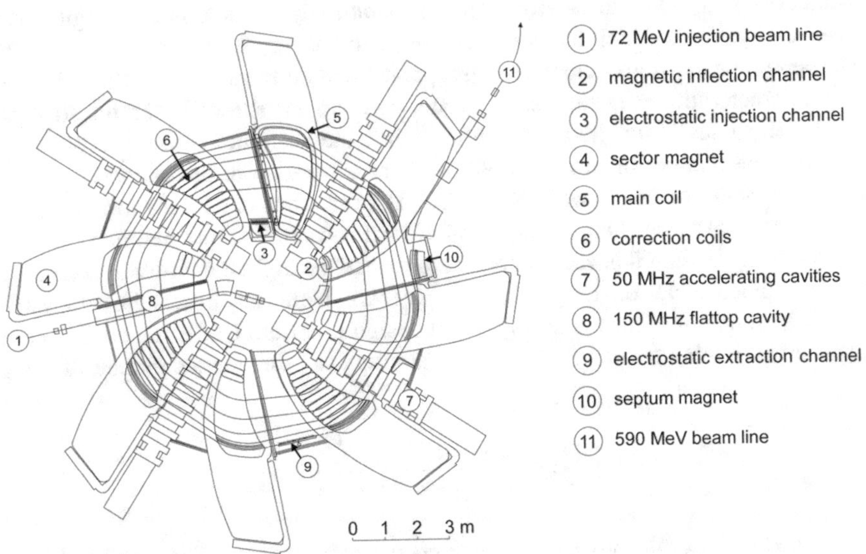

Fig. 11.36 The PSI Ring cyclotron generates a 1.4 MW proton beam at 590 MeV kinetic energy

During the course of acceleration the revolution time of the beam must be kept constant in order to ensure synchronicity with the accelerating RF voltage. At relativistic energies this is achieved by raising the average magnetic bending field in proportion to the relativistic γ factor towards larger orbit radii. This positive slope of the bending field as a function of radius would result in a loss of vertical focusing if the field was uniform azimuthally. Sufficient focusing is achieved due to the azimuthally varying field strength in sector magnets and intermittent gaps (Thomas focusing [141]), and by the introduction of spiral magnet shapes that enhance the edge focusing effect of the bending magnets.

In practice the strength of vertical focusing is typically just sufficient for operation, but weak in comparison with alternate gradient focusing. In the PSI Ring cyclotron the number of vertical betatron oscillations per turn is slightly smaller than 1 which reflects the weak focusing properties of the cyclotron magnet lattice.

The key for the acceleration of a high intensity beam in a cyclotron is a clean extraction with small losses. In the PSI cyclotron the extraction is realized by deflecting the beam with an electrostatic element whose 50 μm tungsten electrode has to be placed between last and second last turn. Particles in the beam halo may hit this electrode, are scattered out of the beam and are lost in the extraction beamline. According to practical experience losses of the order of 100 W can be accepted with respect to this major loss mechanism in the PSI facility. The highest activation levels in the extraction beamline typically amount to 10 mSv/h. All efforts to optimize the accelerator towards higher beam powers are focused on lowering the particle density in the beam tails and maximizing the turn separation at the extraction point. One dominating effect generating beam tails is the longitudinal space charge force which increases the correlated energy spread inside the bunch. The bending fields transform the energy spread finally into transverse beam tails that may interact with the extraction electrode. As Joho has shown [142], the strength of this effect scales quadratically with the time the beam needs to be accelerated, i.e. with the number of turns. In addition turn separation at the extraction of the cyclotron scales inversely with the number of turns. Thus in total the losses caused by longitudinal space charge scale with the third power of the turn number. Because of this strong scaling it is very beneficial to realize the highest possible accelerating voltages. For a cyclotron designed purposely for high intensity beams the achievable radius increment per turn is an important design criterion. Under the condition of constant revolution frequency the turn separation can be expressed in the following way:

$$\Delta R = \frac{R}{\gamma\left(\gamma^2 - 1\right)} \frac{U_t}{m_0 c^2} \,. \tag{11.89}$$

Here R is the orbit radius and U_t the energy gain per turn. Thus the ideal high intensity cyclotron exhibits a large diameter, provides a large energy gain per turn and accelerates to moderate energies $\leq$1 GeV, since for large γ the turn separation decreases very fast.

An alternative way to realize a clean extraction is charge stripping with a thin foil. In this case incompletely stripped ions are accelerated in the cyclotron, for example H^- or H_2^+. At the extraction foil the ions change their charge and are separated from the circulating beam by the action of the bending field. This scheme is employed in the TRIUMF cyclotron to extract proton beams up to 520 MeV [143]. A disadvantage of this method is the non-negligible rate of stripping that occurs in the bending field, causing losses and activation. The H_2^+ molecule has a higher binding energy in the ground state, resulting in significantly lower dissociation probability. However, disadvantages are the lower charge to mass ratio requiring stronger fields, and the complex extraction path across the centre of the cyclotron, resulting from a reduction of the bending radius by a factor 2 after stripping.

Transverse space charge forces present a limitation for the maximum feasible beam intensity. Above a certain charge density the repelling space charge forces exceed the relatively weak vertical focusing forces in a cyclotron. With the PSI concept the realization of cyclotrons for beam powers up to 10 MW seems feasible.

Classical cyclotron magnets represent rather heavy and complicated devices since they need to cover a wide radius variation of the beam orbit. To ensure the condition of constant revolution time the radial field shape must be precisely controlled. Radially distributed trim coil circuits are used for fine tuning on the basis of beam phase measurements. Modern cyclotrons like the RIKEN SRC employ superconducting magnets [144, 145] to obtain higher bending strength. In view of turn separation and efficient extraction it is generally not desirable to design very compact high intensity cyclotrons with the help of superconducting magnets. However, with a large radius the gained space can be used to maximize the installed RF voltage.

For the production of MW-class beams the energy efficiency of the accelerator systems is important. Cyclotrons with continuous RF generated by tetrode amplifiers and utilizing copper resonators can reach a relatively high efficiency. A comparison of different accelerator concepts suited for producing high intensity proton beams is presented in [146]. At the PSI facility a fraction of 18% of the total grid power is converted to beam power.

In Table 11.11 selected properties of three cyclotrons are listed.

In summary cyclotrons present an effective and relatively compact solution to generate high intensity proton beams at energies below 1 GeV and beam powers of a few MW.

Acknowledgements
Mohammad Eshraqi, ESS
Steve Peggs, ESS and BNL
Romuald Duperrier, CEA
Jim Stowall, Los Alamos
John Galambos, SNS
Ference Mezei, ESS

Table 11.11 Parameters of existing cyclotrons [147]

Cyclotron	K [MeV]	$N_{mag.}$	Harmonic number	R_{inj} [m]	R_{extr} [m]	Extraction method	Overall transmission	P_{max} [kW]
TRIUMF	520	6 (sect.)	5	0.25	3.8..7.9	H^- stripping	0.70	110
PSI-Ring	592	8	6	2.1	4.5	Electrostatic channel	0.9997	1400
RIKEN s.c. Ring	2600	6	6	3.6	5.4	Electrostatic channel	0.63	1 (^{86}Kr)

11.5 Heavy Ion Accelerators for Nuclear Physics

N. Angert · O. Boine-Frankenheim

11.5.1 Accelerator Facilities for Heavy Ion Nuclear Physics: Background and Aims

Nuclear spectroscopy, reaction studies, and nuclear astrophysics using beams of accelerated heavy ions have been research fields in nuclear physics since the middle of last century. An especially interesting topic is the search for new super-heavy elements (SHE) using fusion reactions between medium-mass ions, with energies of 4–5 MeV/u (around the Coulomb barrier), and very heavy target nuclei. Such studies have been conducted for many years now at LBL, Berkeley; GSI, Darmstadt; JINR, Dubna; RIKEN, Saitama, and other laboratories [148]. Using the cold fusion technique to produce super-heavy compound nuclei, the element chart has been expanded up to element $Z = 118$ in recent years using actinide target nuclei [149]. Today, a growing research community is conducting experiments in nuclear spectroscopy, mass and life time measurements as well as in the nuclear chemistry of super-heavy elements [150].

Since the mid-1980s the availability of medium-energy ion beams up to 100 MeV/u in a new generation of normal conducting sector cyclotron [151] and superconducting cyclotron facilities [152] has provided access to radioactive isotopes in unexplored regions of the nuclear chart, far from stable isotopes. Fragmentation and fission of heavy projectile nuclei in thin light element targets were used to produce fast exotic nuclei using the in-flight method. Light exotic nuclei produced in this way such as ^{11}Li have exhibited the new phenomenon of neutron halos [153, 154]. In other exotic nuclei produced using this method, changes have been detected in the neutron shell structure for large N/Z-ratios shedding new light on nucleon-nucleon interaction. Precision studies of masses, life times, and new decay channels have revealed interesting phenomena, extended the understanding of nuclear forces, and raised new questions. In addition, experiments with unstable nuclei have contributed to a better understanding of an increasing number of observations in astronomy.

A new era in nuclear astrophysics has opened up with the rare-ion beam facilities dedicated to the measurement of nuclear reactions involving post-accelerated short-lived nuclides of particular relevance to astrophysics. Whereas at large facilities such as RHIC and LHC the aim is to simulate the early state of the universe in the laboratory, exotic beam facilities will increase our understanding of the development of stars and the creation of heavy elements. It is estimated that more than 8000 isotopes may remain bound, with only about a third of them having already been identified. Nuclear behaviour is expected to change significantly in the yet unexplored regions. Figure 11.37 shows, as an example, the landmark results

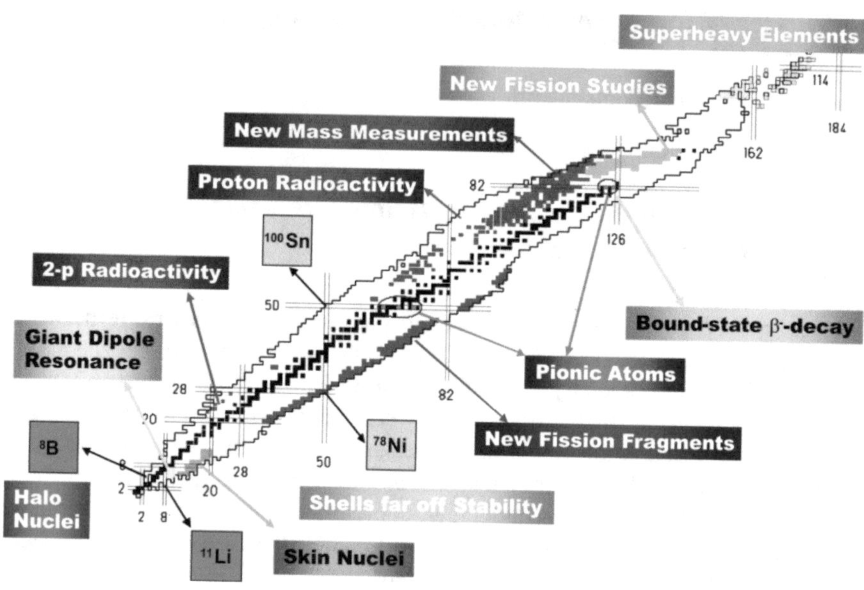

Fig. 11.37 Landmark results from experiments with in-flight produced super heavy elements (SHE), with exotic beams at the Synchrotron-FRagment-Separator facility (SIS-FRS), and at the Experimental Storage Ring (ESR) at GSI Darmstadt. [Reprinted from Nucl. Phys. A 701 (2002) 259, H. Geissel, G. Muenzenberg, and H. Weick, Copyright (2002), with permission from Elsevier]

from experiments with in-flight produced super-heavy elements, exotic beams at the Synchrotron-FRagment-Separator facility (SIS-FRS) and at the Experimental Storage Ring (ESR) at GSI Darmstadt [155].

Another method of producing isotopes far off stability and which can deliver higher phase-space density exotic ion beams is the ISotope On-Line technique (ISOL) [156]. This method is to some degree complementary to the in-flight method. ISOL uses intense proton or light-ion beams in the energy range from 20 to 1400 MeV to generate radioactive nuclei using spallation, fission, or fragmentation reactions induced by the projectile in thick targets of heavy elements. Unstable isotopes produced in this way have thermal kinetic energy. After effusing out of the hot target into an ion source they are singly ionized and separated in a high-resolution magnetic separator. Then they are either trapped for precision mass measurements or nuclear spectroscopy in a Penning trap, for example, or further ionised in a charge breeder and post-accelerated to low energies in the keV/u to MeV/u range. The principles of the in-flight and the ISOL methods are shown in the diagrams in Fig. 11.38.

To summarize: rare isotopes either stopped ones or those with very low energy (0–100 keV) are used for precision measurements of masses, moments, and symmetries. Re-accelerated isotopes (0.2–20 MeV/u) are used for detailed nuclear structure studies, high-spin studies, and measurements of astrophysical reaction

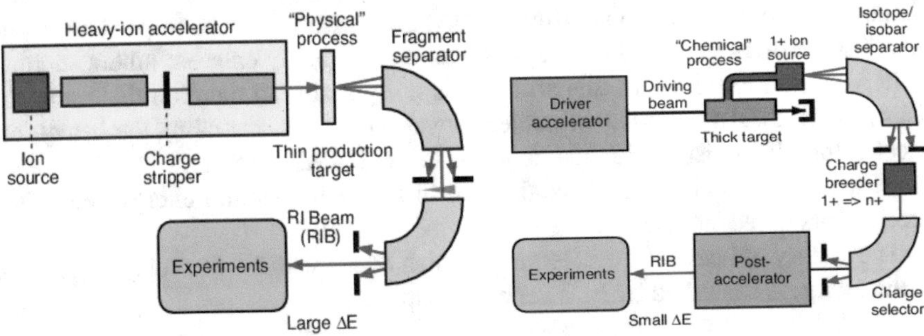

Fig. 11.38 Diagram of the in-flight production method for radioactive isotopes (left) and of the ISotope On-Line (ISOL) production method (right). [Copyright ©, O. Kamigaito, IPAC'10, Kyoto (2010), MOYBMH01, jacow.org (2010)]

rates. In-flight produced fast isotopes (>100 MeV/u) make possible the farthest reach from stability to the limits of existence and to the shortest life times. Generally speaking, with the expansion of nuclear physics research to areas far off stability in the nuclear chart, the atomic nucleus has become a laboratory for studies of fundamental interactions and symmetries at the interface of particle, nuclear, atomic, and astrophysics. This program requires the availability of both stable beam and radioactive beam facilities, along with the development of new experimental techniques and instrumentation.

Also, new facilities dedicated to delivering high-intensity heavy ion beams are needed to further extend the possibilities for synthesis and experiments involving super-heavy elements. In addition, smaller accelerator facilities are needed to develop and test new instruments, and to educate the next generation of scientists.

11.5.2 Accelerators

11.5.2.1 Introduction

In the last two decades exciting challenges in physics described above have stimulated many laboratories to begin research with radioactive isotope beams (RIB) in existing or expanded accelerator facilities. Progress in ion source and superconducting accelerator technology has facilitated this development.

At present there are a number of small and large in-flight and ISOL facilities. The larger ones are NSCL at Michigan State University (MSU); ISOLDE at CERN; ISAC (I and II) at TRIUMF, Vancouver; SPIRAL1 at GANIL, Caen; RIKEN RIBF, Saitama; and SIS/ESR at GSI, Darmstadt. Large facilities planned for the near future are GSI-FAIR at GSI and FRIB at MSU [157].

SPES in Legnaro and SPIRAL2 at GANIL are facilities on the way to EURISOL, which is still in the planning stage. The quest for and the study of super-heavy elements has been an ongoing effort at JINR, GSI, and RIKEN, in particular, but other laboratories such as JYFL, Jyväskylä, have recently entered; others such as SPIRAL2 plan to do so. In this article it is not possible to mention the complete list of facilities. An overview of nuclear physics facilities, including the heavy ion accelerators that exist worldwide, is given in the IUPAP Report 41 [157]. Indeed, there has been something of a renaissance in low- and medium-energy heavy ion accelerators in recent years.

The history of heavy ion accelerators, which began with the Berkeley cyclotrons in the 1950s, is described by E. Wilson in Chap. 1.

11.5.2.2 Special Issues of Heavy Ion Accelerators and Storage Rings

Some specific issues must be considered when designing accelerators and storage rings for heavy ions. These issues are not relevant for proton or light ion machines. Ion sources are needed which can generate highly charged, intense ion beams of accelerator-compatible beam quality for the full spectrum of elements, with their varying physical and chemical properties. During the acceleration of partially ionized, highly charged heavy ions, charge-changing processes occur, whether by accident or design. Knowledge of the equilibrium ion beam charge after it passes targets (strippers), and of the yield in the desired charge state after stripping, is important when designing and optimising an accelerator layout. The high specific energy loss and irradiation effects are much more important issues for heavy ions than for light ones. Knowledge of charge-changing cross sections is important when estimating beam transmission in synchrotrons and beam lifetimes in storage rings for given vacuum conditions, as well as recombination processes during electron cooling. In synchrotrons and storage rings, heavy-ion-induced gas desorption processes are crucial with respect to intensity limits and lifetimes of ion beams. The reference atom for the design of the large in-flight driver accelerators to be described later is ^{238}U. Therefore, the explanation of aspects specific to heavy ions will mainly be confined to that element.

11.5.2.2.1 Ion Sources

The first ion sources used for multiply charged ions in accelerators were been Penning ion sources. In the 1950s, the cold-cathode Penning type was developed at Berkeley for ion beam production [158]. Later, the hot-cathode Penning ion source, which provided higher ion-beam currents for medium-charged ions, was the favourite choice both for linear and cyclotron heavy ion accelerators up to the mid-1980s [159]. It was now possible to deliver beams of up to several 100 μA for U^{10+} [160]. However, the source lifetime was limited by erosion of the discharge electrodes. In addition, in view of the fact that beams from metallic elements were produced by using the sputtering technique, the lifetime of sputter electrodes was a limiting factor, too, along with rather high material consumption. This was a serious drawback when beams from expensive isotope-enriched materials were needed in super-heavy element research.

For the production of short-pulsed high-charge-state beams of light ions Laser Ion Sources (LIS) were used, for purposes such as injection into the Synchrophasotron in Dubna beginning in the mid-1970s; now they are used for injection into the Nuclotron [161]. Pulsed Electron Beam Ion Sources (EBIS) were used there for medium-mass ion beams up to krypton [162]. Both low duty cycle ion source types (LIS and superconducting EBIS) are now in use at the Nucletron based collider facility NICA for high charge state light and heavy ions, respectively [163]. A powerful superconducting EBIS is delivering intense beam pulses of highly charged heavy ions up to uranium for the RHIC injector at BNL [164, 165]. EBIS-type devices are also used as charge breeders at various ISOL facilities [166].

The most influential development on the field of heavy ion sources has been the Electron Cyclotron Resonance Ion Source (ECRIS), which has replaced the Penning type ion sources at most heavy ion accelerators over the years and has substantially increased the capabilities of cyclotron-, linac-, and synchrotron-facilities for research on the field of nuclear physics over the last decades. The ECRIS concept evolved from plasma devices for fusion research and has been investigated as a source for highly charged ions at several laboratories, such as Grenoble, Jülich, Karlsruhe, Louvin-la-Neuve, Marburg, and Oak Ridge since the 1970s. In the ECR ion source, a plasma-confining magnetic configuration is generated by the superposition of an axial solenoid mirror and a radial hexapole field. Energetic electrons are generated for the ionization process using electron cyclotron heating by GHz microwaves in magnet field regions which meet the electron-cyclotron-resonance condition for the plasma electrons.

R. Geller's group in Grenoble developed this device by introducing hexapole magnets to improve ion confinement to a compact efficient generator for highly charged ions (Minimafios) [167, 168]. An ECRIS was used for the first time at an accelerator for ion injection in CYCLONE at Louvain-la-Neuve begin of the 1980s [169]. The first ECRIS used at a large-scale accelerator was installed at the CERN facility. An ECRIS, operated with a 10 GHz klystron, delivered 40 μA oxygen 6+ in pulsed mode for injection into the CERN PSB-PS-SPS complex in the mid-1980s [167]. Other accelerator laboratories followed CERN's lead. Figure 11.39 shows LBL's superconducting Versatile ECR ion source for NUclear Science (VENUS), as an example of the third generation of ECR ion sources which are used at various laboratories today [170]. These third generation ECR sources, such as SECRAL and SECRALII at IMP (Lanzhou) [171], SuSI at NSCL/MSU (East Lansing) [172], the RIKEN-SCECR at Nishina-Center (Wako) [173], and the KBSI-SCECR for RAON (Deajeon) [174], are essential for the capabilities of existing and next generation heavy ion accelerator facilities for Nuclear Physics.

Third-generation ECRIS devices with superconducting magnets and operated with microwave generators for electron heating up to 28 GHz are the standard now. Developments in the various laboratories and operating experience over many years have led to steady improvements of performance. These ion sources can deliver now beams of several mA of oxygen 6+. Record beam currents of more

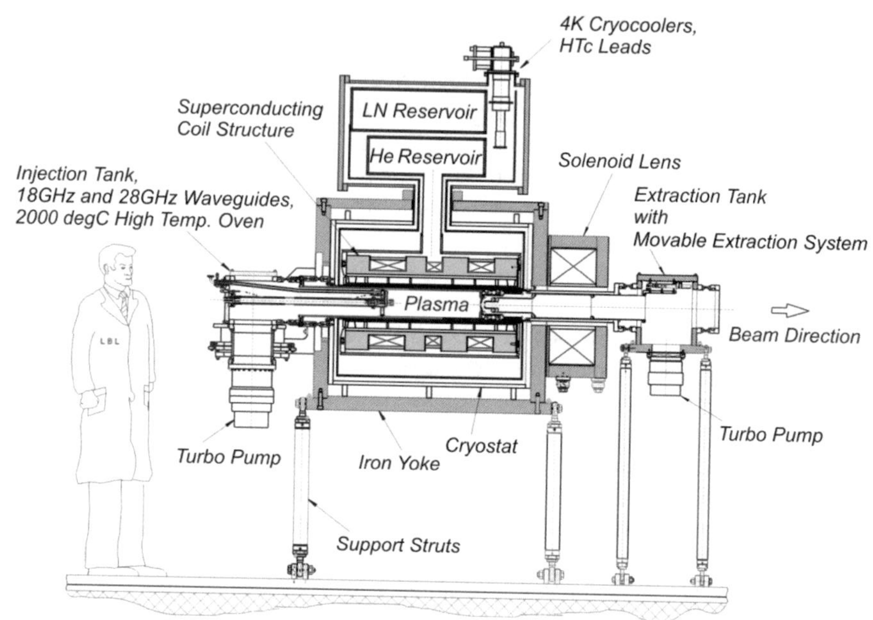

Fig. 11.39 Mechanical layout of the LBL VENUS ion source and cryogenic system. [The Physics and Technology of Ion Sources, 2nd ed., ECR Ion Sources, 203. D. Leitner, C. Lyneis, Copyright © (2004), Wiley-VHC Verlag]

than 400 μA have been reached for U^{31+} with VENUS [175], 680 μA for Bi^{31+} and 10 μA for Bi^{50+}, respectively, with SECRAL [176]. In long term accelerator operation typically about half of these record beam currents are used in order to have stable operating conditions over days or weeks. Figure 11.40 shows a charge-state spectrum for uranium achieved with VENUS, when operated in the high-current mode for medium charge states.

For cyclotrons and superconducting linacs, the ECRIS is the ideal choice. It can be operated at a duty cycle of 100%. It has no consumable electrodes; therefore, it can run for weeks when operated with gaseous elements. For metal-ion production, furnaces have been developed at many laboratories for a variety of elements up to uranium (e.g. [177–180]) [175]. Material consumption is at least an order of magnitude lower than for Penning ion sources, in the range of 10 mg/h or below, an important issue if beams from enriched isotope materials are needed. For injection into synchrotrons the pulsed afterglow mode can be used which provides higher peak currents of very highly charged ions than the cw operation [181, 182]. A survey on the progress, challenges, and experiences in intense highly charged ion beam production and long term operation with the third generation ECRISs is given in [176].

The impact of ECRIS on the ion-accelerator field is best illustrated by the following example. For the Berkeley 88-in. cyclotron (K = 140) it was possible to

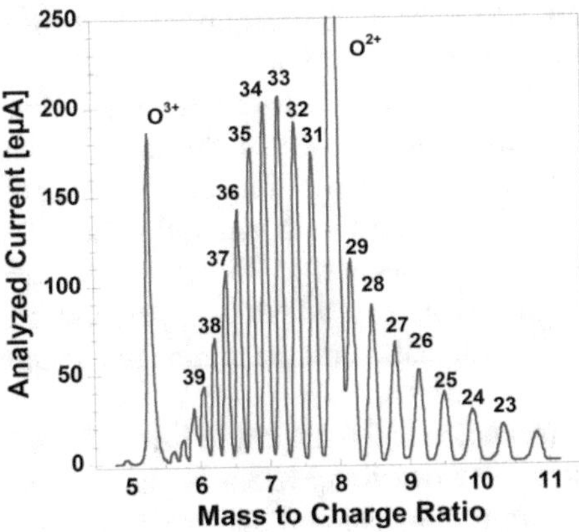

Fig. 11.40 Spectrum of uranium charge states from the LBL VENUS in the high-current mode. [Copyright ©, D. Leitner, S. Caspi, P. Ferracin, C.M. Lyneis, S. Prestemon, G. Sabbi, D. Todd, F. Trillaud, HIAT'09, Venice, Italy, WE-10 jacow.org (2009)]

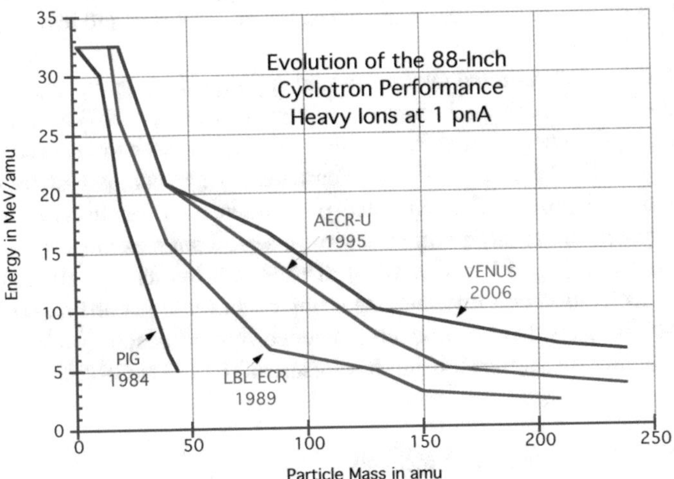

Fig. 11.41 The evolution of beam energies vs. element mass at the 88-in. (K = 140) cyclotron at LBL from the mid-1980s, using the best Penning ion sources, up to the present using the third-generation superconducting ECRIS of the VENUS type. [Courtesy of LBL Berkeley, C.M. Lyneis]

expand the mass range, for which energies of more than 5 MeV/u can be provided, from argon ($A = 40$) in the 1980s using Penning ion sources to medium-mass ions (xenon) using the normal conducting AECR-U ion source, and finally to uranium ($A = 238$) using the superconducting VENUS (Fig. 11.41) [183]. There are many cyclotrons worldwide with $K \geq 100$, which can take advantage of the high charge-states delivered by the ECRIS to reach beam energies in order to perform nuclear physics experiments with heavy ions.

ECRIS can also be used for charge breeding of ISOL-produced or stopped in-flight-produced rare isotopes before post- or re-acceleration [184, 185].

Based on the experience and progress with third generation ECRIS in increasing beam intensities of highly charged ions in the last decade, proposals to build next generation ECRIS for microwave frequencies beyond 28 GHz have been discussed since some years. The 1st fourth generation ECRIS is under construction now at IMP, Lanzhou, to be operated at 45 GHz, with correspondingly higher magnetic fields, based on Nb3Sn- instead NbTi- superconducting technology [186]. Expectations are to reach with that fourth generation two to three times higher beam currents in the future.

With respect to the formation and transport of beams from ECR ion sources, one has to bear in mind that the extraction system is placed in a region with a superposition of stray fields from the axial coil and the radial hexapole field, affecting beam formation and beam density distribution. In addition, beams with different ion charge states have different emittances due to the magnet-field structure, the production and confinement processes inside the source. The ion-current density is not uniform within the beam [187–189]. In third generation ECRIS, with higher magnetic fields and microwave frequencies, beam emittance and low energy beam transport is an issue [176].

In contrast to cw-operated cyclotrons and superconducting linacs, synchrotrons require injection at high peak currents within a brief period. Those peak beam currents of highly charged ions can be delivered from laser ion sources, or by an EBIS, or by an ECRIS along with an accumulator ring such as LEIR, which uses fast extraction to provide the requested peak intensities for the LHC [190]. For the FAIR synchrotrons, high-current short-pulse ion sources are used, which can deliver many mA (e.g. 15 mA of U^{4+}) of ions in a low charge state [191, 192]. After pre-acceleration and intermediate stripping in the injector linac (Unilac) the charge state is reached (U^{28+}) which is needed in the booster synchrotron SIS18 (section “Facility for Antiproton and Ion Research (FAIR) at GSI”).

11.5.2.2.2 Charge-Changing Processes

The design of the first stage of a heavy ion accelerator depends on the charge state delivered from the ion source for the heaviest ion species to be accelerated (highest A/q-ratio). In accelerator facilities for medium to relativistic energies (some 10–1000 MeV/u), usually one or two strippers are needed to increase the charge state of the ions for an efficient acceleration scheme. In a stripper target, the initially charge-homogeneous beam splits up into several charge components. One gets a charge state distribution centred on the equilibrium charge state. To optimize the overall accelerator layout, it is important to know the ion and energy dependence of the equilibrium charge states and the yield of the charge states as well.

Pioneering theoretical work on these processes was done by Bohr [193]. Bohr assumed that a fast heavy ion passing a gas stripper retains the electrons which have orbital velocities which are greater than the ion velocity (Bohr’s Criterion). This physically reasonable criterion has been proven to be a good first-order approximation for estimating the charge states which can be reached. Semi-empirical approximation formulas were developed later to describe the energy and target dependence of the equilibrium charge and the width of charge-state distributions. For the low energy range, that means for energies from about 1 to some 10 MeV/u, the equilibrium charge state for heavy ions can be approximately described by a simple parameterized formula of the type

$$1 - (\overline{q}/Z_P) = C\ \exp(-\delta\beta/\alpha)\,, \tag{11.90}$$

with $\overline{q}$ = average charge state, Z_P = nuclear charge of the ion, $\beta = v/c$, $\alpha = 1/137$, $\delta = Z_P^{-\gamma}$, see among others [194, 195]. For the stripping data gathered for example in the operation of the Unilac (GSI) over the years, this formula is used with $C = C^* + 140/Z_P^2$, $C^* = 1.0285$. Averaging the experimental data, γ-values of 0.56 for foil and 0.65 for nitrogen-gas-jet data have been calculated [196]. Foil strippers deliver higher charge-states, because the higher collision frequency in solid materials leads to higher electron loss probability [195], but also suffer from degradation due to beam intensity. For the range of $0.2 \leq \overline{q}/Z_P \geq 0.8$, which means if the ion has a sufficient number of electrons with comparable binding energy, the charge state distribution is roughly Gaussian in shape. The width of the Gaussian curve is proportional to $Z_P^{1/2}$ (see e.g. [195]).

Simple exponential relations for the average charge begin to deviate from measured data, and charge state distributions are no longer Gaussian-like in shape, when the electron configuration of the ion approaches inner shells with large steps in ionization energies. This was discovered in the case of bromine ions by C.D. Moak et al. in 1967 when approaching the L-shell by the stripping process in the energy range between 1 and 2 MeV/u [197]. Therefore, after estimations with simple semi-empirical formulas, experimental data should be taken into account whenever available.

For beams of energies in the range from 10 MeV/u to 80 MeV/u, the computer program ETACHA was developed by Rozet et al. to calculate charge-exchange cross-sections, charge-state evolutions in targets, and equilibrium charge-state distributions for highly charged ions, taking into account the electronic structure of inner shells [198]. Experimental results in this energy range can be found in [199], for example.

An overview of theoretical and experimental results for Xe, Au, and U projectiles impinging with kinetic energies from 80 to 1000 MeV/u on solid and gaseous targets ranging from Be to U is given in [200]. The calculations are compared to data from experiments carried out at the BEVALAC (LBL) and at the heavy ion synchrotron SIS (GSI) and associated facilities. Measured equilibrium charge state distributions for uranium in the energy range from 1 to 1000 MeV/u are shown in Fig. 11.42 [200].

For high intensity beams, gas or liquid strippers have advantages compared to foil strippers concerning their durability. Gas strippers lead to much lower equilibrium

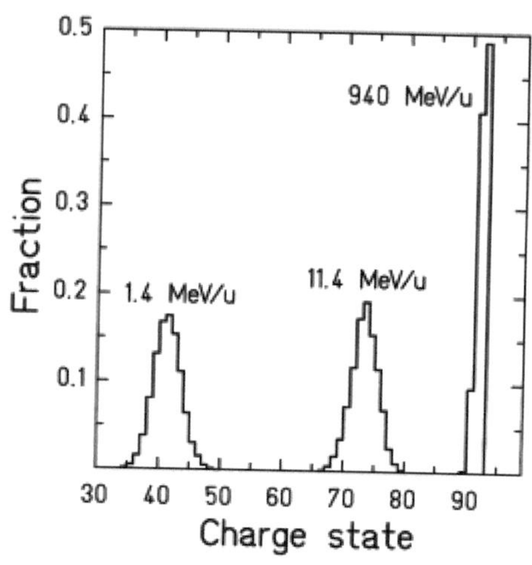

Fig. 11.42 Measured equilibrium charge state distributions at different energies for ^{238}U behind foil strippers. [Reprinted from Nucl. Instrum. Meth. B 142 (1998) 441, C. Scheidenberger, Th. Stöhlker, W.E. Meyerhof, H. Geissel, P.H. Mokler, B. Blank, Copyright (1998) with permission from Elsevier]

charge states due to the strongly reduced influence of density effects compared to solid strippers [195]. Since electron capture cross sections of the heavy ions in the low-Z gases are considerably suppressed, in particular hydrogen promises higher equilibrium charge states as compared to nitrogen which is routinely used at the UNILAC gas stripper [201].

In synchrotrons and storage rings, charge-changing reactions between beam ions and the residual gas in the vacuum chamber can lead to the immediate loss of the ion. These processes are not significant in linear and cyclotron accelerators, because of the short acceleration path length there. However, in synchrotrons for partially stripped ions and in storage rings, these processes can limit the intensity and life time of the beam. In order to determine the lifetime, it is necessary to know the charge-changing cross-sections for both the capture and the loss of electrons during the interaction of the ions with the residual gases.

For practical purposes, the one-electron-capture cross-section σ_c, for example, is usually estimated by applying a semi-empirical formula of Schlachter et al., derived from experimental data in gases from H_2 to Xe [202]. It describes most of the experimental data for σ_c in the low- and medium-energy range within a factor of two. For the electron-loss cross-section σ_l, simple semi-formulas cannot be found for a comparable wide energy range.

In recent years methods calculating charge-changing cross-sections have been developed taking into account the electronic structure of projectile ions and target atoms. As an example calculated cross-sections of U^{39+} + Ar as a function of the ion energy are shown in Fig. 11.43 in comparison with experimental data [203–206]. Figure 11.43 shows quite well the general energy dependence of cross-sections of partially ionized highly charged heavy ions: The probability of electron capture saturates at very low velocities and decreases steeply at energies above a few 100 keV/u. Loss cross-sections increase with increasing energy, pass through

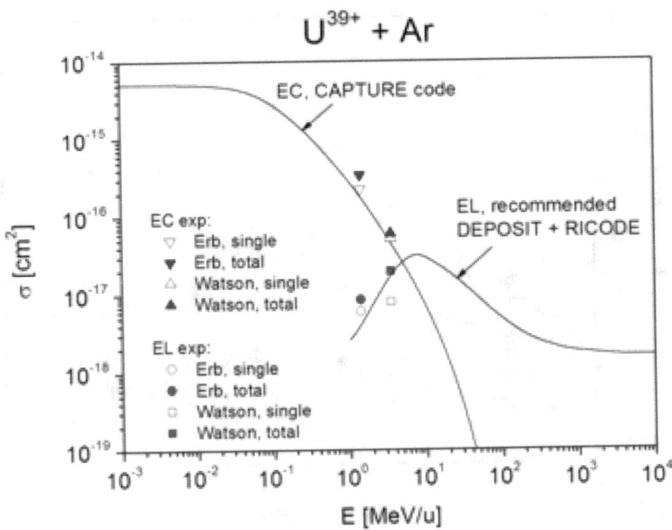

Fig. 11.43 Calculated cross sections (solid line) for electron capture (EC) and loss (EL) for uranium 39+ ions in argon gas as a function of the collision energy [190, 191]. Experimental single-electron capture and loss cross-sections (open symbols) are given to show their contribution to the total cross-sections (solid symbols) [192, 193]. Solid curves: EC calculations using the CAPTURE code, and EL calculations using the RICODE and DEPOSIT codes, respectively. The DEPOSIT code is described in [190]. [Reprinted from Nucl. Instrum. Meth. B 269(12) (2011) 1455, V.P. Shevelko, I.L. Beigman, M.S. Litsarev, H. Tawaras, I.Yu. Tolstikhina, G. Weber, Copyright (2011) with permission from Elsevier]

a maximum at intermediate energies, and reach a lower nearly constant value at relativistic energies.

For heavy gases, the cross-sections can be orders of magnitudes larger than for helium or hydrogen, especially the capture cross-sections. This is important with respect to the desorption of heavy gas molecules from vacuum chamber walls induced by impinging heavy ions (see section "Heavy Ion-Induced Desorption"). Cross-sections for electron capture and loss in the medium-energy range have been measured at RIKEN [207].

Charge changing processes can also occur due to ion-electron recombination during electron cooling of an ion beam. Significant recombination processes under these conditions, that means near zero relative energy $E_{\text{rel}} = 0$, are Radiative Recombination (RR) and Dielectric Recombination (DR). Unexpectedly high recombination rates have been observed in merged-beam experiments using a cold dense electron target ($n_e = 4 \times 10^8$ cm^{-3}) in an experiment with 6.3 MeV/u U^{28+} at the GSI Unilac [208], indicating that the lifetime of a stored electron-cooled U^{28+} beam would be only seconds. Later, in a series of experiments at CERN's Low Energy Antiproton Ring (LEAR) Pb^{53+} ions were stored and cooled

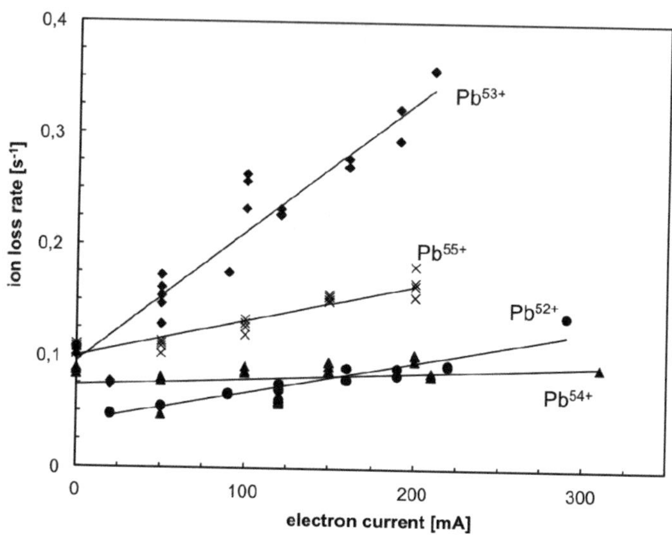

Fig. 11.44 Beam decay rates $1/\tau$ as a function of the electron cooler current for 4.2 MeV/u lead ions with charge states 52+ to 55+ in LEAR [209]. The decay rate for an electron cooler current of zero is determined by the residual gas pressure in the ring, which varies for the different runs. [Data for the graph have been provided from CERN, C. Carli]

in preparation for further acceleration [209]. The recombination rates obtained from the observed lifetimes of a few seconds were by far larger (by a factor of about 50) than the calculated RR. Neighbouring charge states Pb^{52+} and Pb^{54+}/Pb^{55+} behaved quite differently and provided sufficient time for cooling without extensive beam losses (Fig. 11.44). Experiments at the Test Storage Ring (TSR) at the Max-Planck-Institut für Kernphysik (MPIK) in Heidelberg investigated $Au^{49+,50+,51+}$ ions with an energy of 3.6 MeV/u; these are isoelectronic with $Pb^{52+,53+,54+}$. In the TSR-experiments recombination rates were not determined indirectly from lifetime measurements; rather, they were measured as a function of the relative energy in the electron-ion-center-of-mass frame [210]. An extremely sharp recombination peak was found for Au^{50+} at $E_{\mathrm{rel}} = 0$. The enhancement factor was about 60 for Au^{50+} in comparison to RR theory. Obviously, for this ion DR resonances dominate the recombination at relative energy zero. For Au^{49+}, the maximum recombination rate at $E_{\mathrm{rel}} = 0$ is lower by roughly a factor of 10.

Apparently, the huge recombination rates observed for Au^{50+} and Pb^{53+} are caused by the individual electronic structure of these ions, which happens to support dielectronic recombination resonances at very low relative energies, leading to much shorter lifetimes under cooling conditions. Unfortunately, so far theory is not able to predict DR rates for these complex multi-electron systems.

11.5.2.2.3 Heavy Ion-Induced Desorption

During high-intensity, heavy-ion operation of several particle accelerators worldwide, dynamic pressure build-ups of several orders of magnitude have been observed. The pressure increase is caused by lost beam ions that impact the vacuum chamber walls at a grazing angle. Ion-induced desorption, which has been observed at BNL, CERN, and GSI, can seriously limit beam intensity and lifetime in synchrotrons or storage rings, because mainly heavy gas molecules are desorbed, resulting in large cross-sections for projectile charge changing processes. In the past several years, experiments have been performed at several laboratories to study the observed dynamic vacuum degradations; it is important to understand and overcome this problem for present and future heavy ion accelerators [211]. The ion-induced desorption yield is defined as

$$\eta = \frac{\#\text{desorped molecules}}{\#\text{incident ions}}. \tag{11.91}$$

Experimental studies of the desorption yield for energetic ions on stainless steel at room temperature indicate a scaling law $\eta \sim (dE/dx)^n_{\mathrm{el}}$, where $(dE/dx)_{\mathrm{el}}$~Z^2/A is the electronic energy loss, Z is the charge number and A the mass number. In the 5–100 MeV/u energy range and for perpendicular incidence, $n \approx 3$ was obtained for uranium ions. Unfortunately, the measured desorption yields at 100 MeV/u were $\eta \approx 100$ for uranium and $\eta \approx 4$ for argon (Fig. 11.45) [212]. As a consequence, uncontrolled beam loss must be minimized to avoid pressure bumps caused by lost heavy ions. Ions lost due to collisions with the residual gas should hit only low-desorption materials or dedicated collimators and catchers [213]. Vacuum pressures in the range of 10^{-11} mbar and lower are required for highly charged and partially ionized heavy ion beams in order to control these processes. Therefore, the vacuum requirements and beam-loss collimation issues in high-intensity heavy ion synchrotrons and storage rings differ greatly from those in proton facilities.

11.5.2.3 Ion Accelerator Facilities

In this article is not possible to describe all existing accelerator facilities for stable or radioactive ion beams, or those under construction. A few examples have been selected and described. The focus will be on facilities that are considered by the scientific community especially important for heavy ion nuclear physics in the long term, as described in the IUPAP Report 41 and the NuPECC Long Range Plan 2010 [157, 214]. The facilities described here have been categorized on the basis of the main production methods for radioactive ions: In-flight and ISOL facilities.

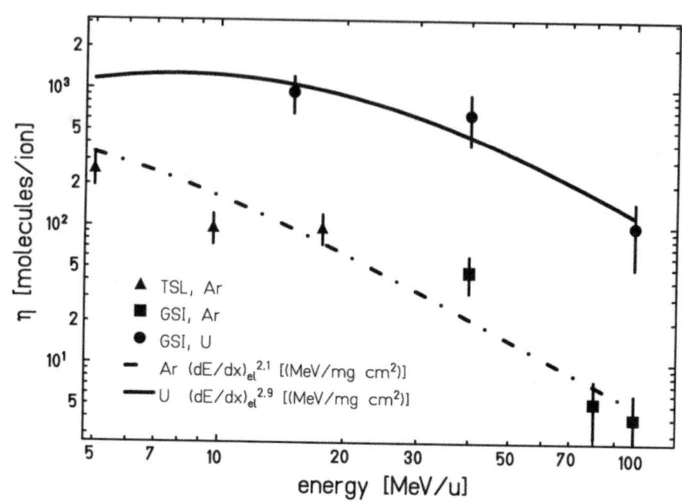

Fig. 11.45 Desorption yields for argon and uranium ions impacting stainless steel at a perpendicular angle, shown here as a function of the projectile energy. The solid lines represent the calculated electronic energy loss $(dE/dx)_{\text{el}}^{n}$ for argon and uranium to the power of $n = 2.1$ and $n = 2.9$, respectively, adapted to the experimental values (left) [212]. [Reprinted with permission from J. Vac. Sci. Technol. A 27(2) (2009) 245, H. Kolmus, A. Krämer, M. Bender, M.C. Bellachioma, H. Reich-Sprenger, E. Mahner, E. Hedlund, L. Westerberg, O.B. Malyshev, M. Leandersson, E. Edqvist, Copyright (2009), American Vacuum Society]

11.5.2.3.1 In-Flight Facilities

The in-flight method takes advantage of reaction kinematics to efficiently separate short-lived nuclei, from the limits of stability to lifetimes in the μs range and even to the level of a single ion. At medium and high energies (approximately several 10 to several 100 MeV/u), the radioactive isotope beams produced are forward-directed at a velocity approaching beam velocity. Due to interaction with the target, the emittance is larger than that of the projectile beam. Therefore, a large acceptance separator system is needed for the fragment beam separation. The advantage of this rare isotope-production process is its independence of chemical properties; it can be used with isotopes from all elements.

In a new generation of in-flight facilities, powerful and versatile heavy ion accelerators, as projectile sources for the production of exotic nuclei, are combined with large-acceptance electromagnetic separators and different high-resolution systems, such as high-resolution spectrometers, storage rings, and ion traps. Post-acceleration up to more than 10 MeV/u for stopped radioactive isotopes is also included to obtain high-quality exotic beams at low energies. The different in-flight scenarios are described in [215].

In the following sections, examples of large scale cyclotron-, synchrotron-, and linear-accelerator facilities as the primary projectile source for in-flight-produced nuclei are briefly described and discussed. The production targets, electromagnetic separators, and experimental set-ups (ion traps, high-resolution spectrometers, etc.) are not addressed here. However, storage rings and re- or post-accelerators are included where they are part of the overall facility.

RIKEN Radioactive Isotope Beam Facility (RIBF)

The history of cyclotron accelerators at RIKEN Nishina Center, Wako, Japan, began before World War II [216]. The era of heavy ion beams began in 1966 with a 160-cm cyclotron. Work with radioactive ion beams began in 1986 when the RIKEN K540 Ring Cyclotron (RRC) [217] went into operation along with the Projectile Fragment Separator (RIPS). A K70 AVF cyclotron was built for light ion injection [218] and the existing variable-frequency linac (RILAC) [219] was used as injector for heavy ions. The RILAC has been in operation since 1980 and has successfully accelerated light ions to energies of 4 MeV/u and heavy ions to 0.8 MeV/u.

The RIKEN accelerator facility was expanded beginning in 1997 and became the Radioactive Isotope Beam Facility (RIKEN RIBF). The aim of the expansion was to improve experimental conditions for research with radioactive isotopes [220]. In the first phase of the expansion project, a new multi-stage acceleration system was built based on the existing RIKEN accelerators RILAC/AVF and RRC. To these were added three booster cyclotrons, the fixed-frequency Ring Cyclotron fRC/K = 570 [221], the Intermediate Ring Cyclotron IRC/K = 980 [209], and the Superconducting Ring Cyclotron SRC/K = 2600 [222]. Figure 11.46 shows a schematic diagram of the facility, including the key parameters of the accelerator stages. The main accelerator is the SRC, with six sector magnets. Figure 11.47

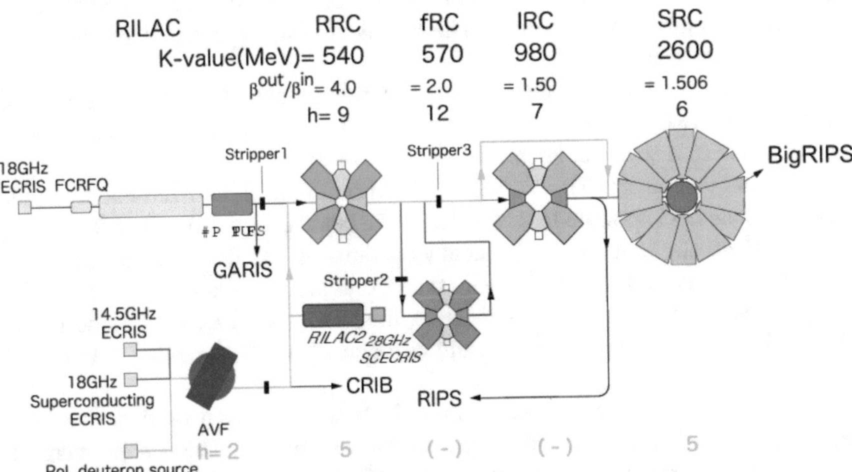

Fig. 11.46 Schematic diagram of the RIKEN RIBF showing the key parameters of the accelerator stages (see article) [220]. The injectors RILAC, RILAC2, and AVF are used for different acceleration modes of the four booster cyclotrons: RRC, fRC, IRC, and SRC. [Copyright ©, Y. Yano, Cyclotrons'04, Tokyo, Japan, 18A1, jacow.org (2004)]

Fig. 11.47 Picture of the RIKEN Superconducting Ring Cyclotron SRC [Copyright © RIKEN]

shows a picture of this facility. When these accelerators are combined in a cascade, with strippers in between, all ions of all elements can be accelerated up to at least 70% of the speed of light. The accelerated stable ion beams pass a thin low-Z target. Radioactive ion beams are produced there using projectile fragmentation and sometimes in-flight fission as well. The beams are then selected in the following BigRIPS spectrometer [223].

RIBF is operated in three different acceleration modes [224]. The first one uses RILAC, RRC, IRC, and SRC to accelerate medium-mass ions. The beam energy can be varied in a wide range below 400 MeV/u by varying the RF frequency of the linac and cyclotrons. The second acceleration mode is the fixed-energy mode, which uses the fRC between RRC and IRC. The beam energy from the SRC is fixed at 345 MeV/u, due to the fixed frequency operation of the fRC. This mode is used to accelerate heavy ions such as uranium and xenon. The third mode uses the AVF cyclotron as the injector and two boosters, the RRC and SRC. This mode is only used for light ions such as deuterons and nitrogen, also with variable RF frequency in the SRC.

The RIKEN RIBF started operation in 2006 and has been used to accelerate beams from the full mass range from deuteron to uranium. The design energy of 345 MeV/u for uranium was achieved in 2007. Since then, experimental studies on rare isotopes have been carried out. For example, 45 new neutron-rich isotopes were created in 4 days [225], halo structure and large deformation were found in the medium-mass nuclei far from the stability line [226, 227], and decay half-lives of very neutron-rich isotopes were measured to study cosmic r-process [228].

Recently the main focus of accelerator development has been on improving the heavy ion beam intensities by optimizing the accelerator tuning and by reducing matching and transmission losses in the individual stages. Overall transmission for light ions is now approaching 100%. Construction of a second linac for the RRC (RILAC2) began in 2008, fed from a new superconducting 28-GHz ECR ion source, to provide an independent choice of ion beams for rare isotope physics and super-heavy element research [224].

GSI Accelerator Facility (Unilac, SIS18, ESR)

The accelerator facility at the GSI Helmholtzzentrum for heavy ion research, Darmstadt, Germany, (see Fig. 11.52 below), consists of the universal heavy ion linear accelerator (Unilac), the heavy ion synchrotron SIS18 (Schwer-Ionen-Synchrotron, 18 Tm), and the Experimental Storage Ring (ESR, 10 Tm). Construction of the Unilac started in 1970 [229, 230]. The first heavy ions were accelerated in 1975, and uranium ions in 1976 [231]. The Unilac can deliver beams with energies >12.5 MeV/u for all elements up to uranium. Upgrade measures were performed for beam energy in the 1980s and for beam intensity in the 1990s [232]. The present layout of the Unilac is shown in Fig. 11.48. The Unilac pre-stripper linac consists of a 36-MHz IH-type RFQ and an IH-drift tube linac [235], followed by a nitrogen gas-jet stripper and an achromatic charge-analysis system at 1.4 MeV/u [233]. The high-current pre-stripper linac is fed from a Penning ion source for high-duty-cycle (~25%) operation or from high current sources of the MEVVA or CORDIS type for synchrotron injection (low-duty-cycle ~ 1%) [223]. In the low-duty-cycle mode, ions with *A*/*q*-ratio up to 60 (U^{4+}) can be accelerated there [236].

Switching between the different ion sources with up to 50 Hz enables the acceleration of up to three different ion species; this way, experiments at the Unilac, the SIS18, and in ESR can be performed with different ion species in parallel [237].

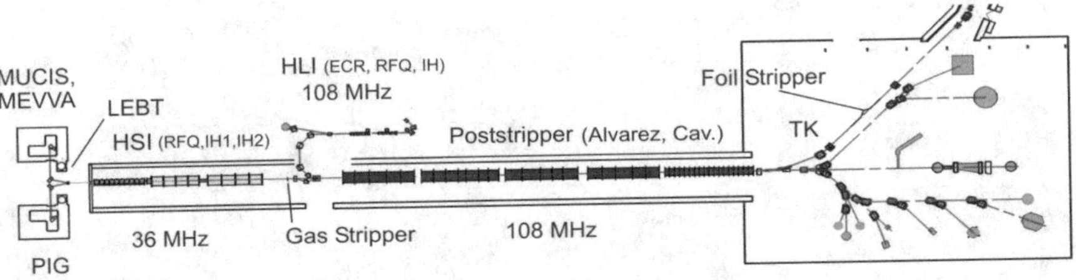

Fig. 11.48 Layout of the Unilac: The pre-stripper linac consists of a 36-MHz high-current IH-type RFQ and an IH-drift tube linac, followed by a nitrogen gas-jet stripper and a charge-analysis system at 1.4 MeV/u [233]. The post-stripper linac consists of four 10-m 108-MHz Alvarez drift-tube sections (up to 11.4 MeV/u) and a chain of single gap cavities either for post-acceleration (up to about 12.5 MeV/u for heavy ions) or for interpolation of the energy steps (from 2.6 to 11.4 MeV/u) of the Alvarez sections. In addition to the high current IH-pre-stripper linac, there is a high-charge-state 1.4 MeV/u injector linac (108 MHz), equipped with an ECR ion source which is used for direct beam delivery (e.g. of U^{28+}-ions) without stripping into the Alvarez post-stripper linac [234]. [Copyright © L. Dahl, HIAT'09, Venice, Italy, FR-01, jacow.org (2009)]

Construction of the synchrotron storage ring facility [225] started in 1985; it went into operation in 1990 [226, 227]. The synchrotron SIS18 delivers U^{72+} beams (foil stripped after the Unilac at 11.4 MeV/u) at energies of up to 1 GeV/u; U^{92+}-ions (using stripping at 1 GeV/u and re-injection into the SIS18 via the ESR) are delivered at energies of up to 1.4 GeV/u. Light ion beams ($q/A = 1/2$) can be accelerated at energies of up to 2 GeV/u. Beam intensities currently range from 10^{10} ions (for heavy elements) to 10^{11} ions (for light elements) per synchrotron cycle (1 Hz). The Unilac and the SIS18 will be used as the injector system in the Facility for Antiproton and Ion Research (FAIR) (Fig. 11.52 below). Intensities are being increased using a current upgrade program for injection in FAIR (see section "Facility for Antiproton and Ion Research (FAIR) at GSI").

Experiments with in-flight-produced isotopes are performed both at Unilac and SIS18 energies. Super-heavy elements, produced with low-energy Unilac beams using fusion reactions in very thin heavy targets, move forward with the center-of-mass energy. They are analyzed in an electromagnetic separator (SHIP) before being stopped in a detector, or investigated at rest in an ion trap, or separated using chemical methods [148]. Fast radioactive isotopes produced by means of the projectile fragmentation of SIS18 beams in a thin target are analyzed in the FRagment Separator (FRS) [238]. The isotopes thus analyzed can either be transferred to the ESR (Fig. 11.49) or used for investigations of nuclei with lifetimes as short as microseconds in high-resolution spectrometers.

The nuclear fragments emerge from the fragmentation target with increased momentum-spread in all phase spaces, but with comparable average velocities. Therefore, many isotopes with the same magnetic rigidity may be injected into the ESR, where all components of the mixed beam are phase-space-cooled to the

Fig. 11.49 Photo of the experimental storage ring (ESR) at GSI, a unique tool for studying in-flight-produced radioactive ion beams with or without beam cooling, using stochastic and electron beam cooling or the isochronous operation mode (see text). [Photo from GSI Helmholtzzentrum für Schwerionenforschung, A. Zschau]

same velocity, with extremely small velocity spread [239]. Therefore, the frequency spectrum of the beam noise (Schottky spectrum) from the multi-component beam stored in the ESR consists of clearly separated lines. The frequency differences between the lines are determined only by the mass-to-charge ratio A/q. The difference in the revolution frequency (or revolution time) between ions of varying A/q ratios is given by

$$\frac{\Delta f}{f} = -\frac{\Delta T}{T} = -\frac{1}{\gamma_t^2}\frac{\Delta(A/q)}{A/q} + \left(1 - \frac{\gamma^2}{\gamma_t^2}\right)\frac{\Delta v}{v}. \tag{11.92}$$

where γ_t is the transition energy and $\Delta v/v$ is the velocity spread in the ion beam; the velocity spread can be reduced by stochastic or electron cooling. The best results in terms of accuracy and resolution are achieved with electron cooling at very low beam intensities, and when the number of total stored ions within a small A/q-interval is below 1000 (this minimizes the counteracting effect of intra-beam scattering ($\sim q^4/A^2$) on the cooling process). Figure 11.50 shows the equilibrium-momentum spread as a function of the number of fully stripped uranium ions in the ESR storage ring at GSI. Above a threshold of roughly 1000 ions, the equilibrium-momentum spread is defined by the balance of intra-beam scattering and electron cooling. Below this threshold, intra-beam scattering is suppressed because the beam ions form a longitudinal string, in which the ions are no longer able to pass each other. In this ordered state, the equilibrium is determined by the balance of cooling and machine noise [240, 241]. For electron cooling, the precision of mass measurements performed using the Schottky Mass Spectroscopy (SMS) is around 1×10^{-6} or better. The resolution power of this method is illustrated by Fig. 11.51, which shows the separation of mass-resolved ^{52}Mn isomers [155, 242].

The cooling times at high energies restrict the application of electron cooling to nuclei with lifetimes of about 10 s. For shorter lifetimes, stochastic pre-cooling must be applied to the "hot" beams, with typical time constants of roughly 1 s [243].

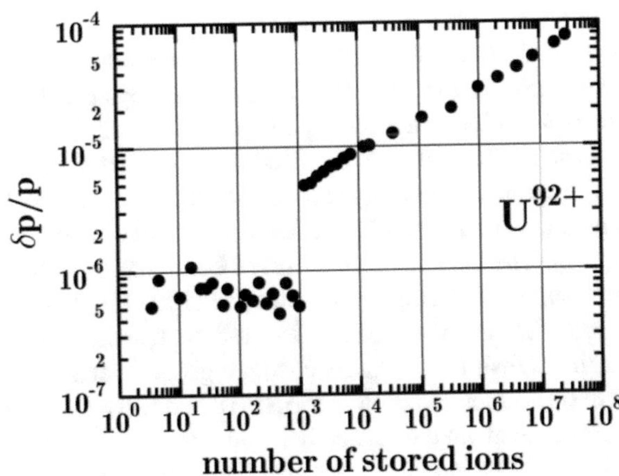

Fig. 11.50 Experimental momentum spread plotted against the number of stored ions in the ESR for fully stripped electron cooled uranium ions at 360 MeV/u. [Copyright © R. W. Hasse, M. Steck, EPAC'00, Vienna, Austria, TUOBF201, jacow.org (2000)]

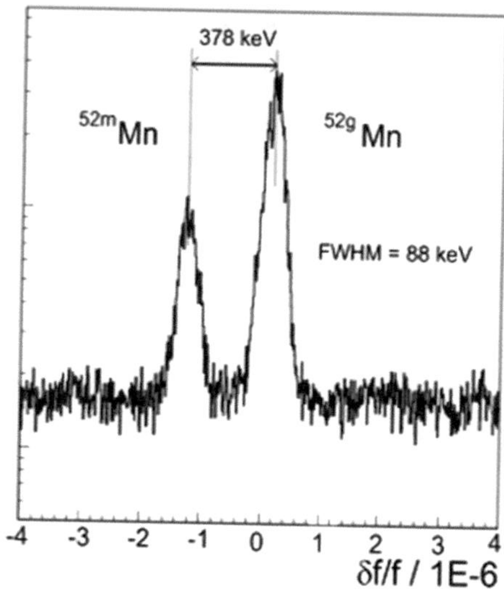

Fig. 11.51 High-resolution mass spectra of electron cooled ^{52}Mn ions at the ground (right) and isomeric (left) states in the ESR. [Reprinted from Nucl. Phys. A 701 (2002) 259, H. Geissel, G. Muenzenberg, and H. Weick, Copyright (2002), with permission from Elsevier]

Operation in the isochronous mode with $\gamma = \gamma_t$ makes possible the investigation of short-lived nuclei with lifetimes of as little as few milliseconds [244].

For atomic and nuclear physics experiments with low energy, electron cooled ion beams in 2012 the CRYRING storage ring was delivered from Stockholm to GSI. Since 2017 the modified CRYRING is operational at GSI. In combination with the ESR and later also with the new FAIR facility CRYRING is the only facility world-wide that provides low-energy highly charged stable beams and beams of rare isotopes with a free choice of the charge state, including bare ions [245].

Facility for Antiproton and Ion Research (FAIR) at GSI

With the new international Facility for Antiproton and Ion Research (FAIR), shown in Fig. 11.52, the research possibilities in nuclear physics with radioactive ions will be expanded considerably at GSI, and the range of accessible isotopes further extended [246, 247]. However, this is only one part of the physics research to be performed at the new facility. Hadron physics, the study of highly compressed nuclear matter, and atomic physics will be additional fields of research [248].

The FAIR facility in the Modularized Start Version (MSV) [247] will consist of six circular accelerators (SIS18, SIS100, CR, HESR, ESR and CRYRING), of two linear accelerators (p-Linac, UNILAC) and of about 1.5 km of beam lines (see Fig. 11.52). The existing Unilac-SIS18 combination will be used as the injector for the new superconducting synchrotron SIS100 (100 Tm), which will have five times the circumference of the SIS18. For radioactive-isotope research much higher primary beam intensities will be reached. To give one example, in the case of uranium U^{28+} will be accelerated both in the SIS18 and SIS100, without stripping after the Unilac in order to avoid the related beam losses. In addition, the ramping and cycling rate of SIS18 will be increased (2.7 Hz). The achievable energy for U^{28+} will be 2.7 GeV/u

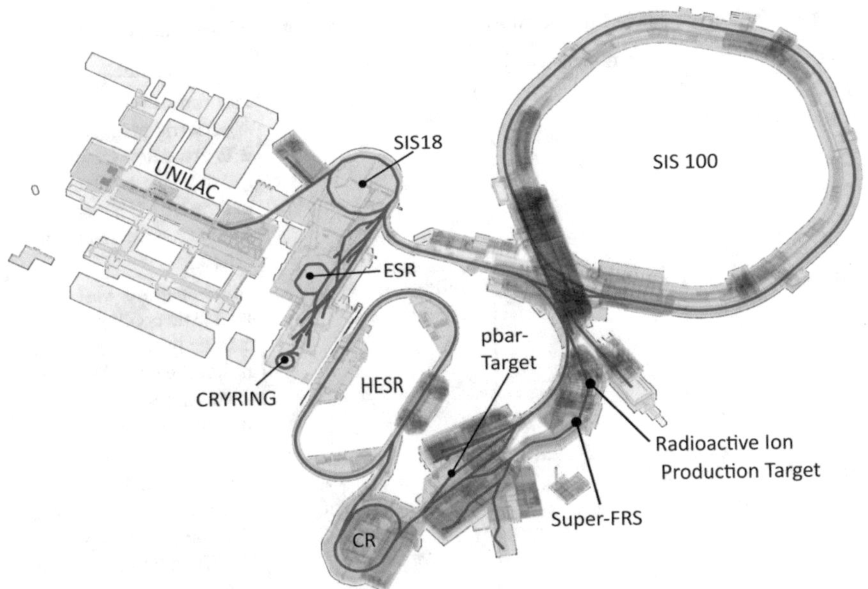

Fig. 11.52 Left: Layout of the existing GSI facilities with the accelerators Unilac, SIS18, the Experimental Storage Ring ESR and the CRYRING storage ring. Right: Layout of the planned new FAIR facilities, with the existing accelerators (Unilac and SIS18) acting as the injector system: the diagram shows the new Superconducting Synchrotrons SIS100 (100 Tm), the Collector Ring (CR), the Superconducting FRagment Separator (Super-FRS), the antiproton production target, the additional proton-injector linac, and the High Energy Storage Ring (HESR) (see text). [Courtesy of GSI Helmholtzzentrum für Schwerionenforschung, C. Pomplun]

and up to 11 GeV/u for fully stripped U^{92+} beams in the SIS100 [249]. With a new 70 MeV proton linac injector for the SIS18, intense proton beams with energies of up to 29 GeV will be provided from the SIS100 for antiproton production [250].

The ion beam from SIS100 can either be transferred to different experimental set-ups or be sent through a production target for fast radioactive-isotope beams. Together with a new large acceptance fragment separator (SuperFRS) behind the SIS100, it is expected that radioactive-isotope intensities will be reached that are 10,000 times the current level [251]. The SIS100 RF-system is also designed to compress the accelerated heavy ion or proton beams into short bunches (to ~60 ns in the case of heavy ions and to ~25 ns in the case of protons). That is required for the production and subsequent storage and efficient cooling of "hot" rare isotope and antiproton beams in the following CR cooler-storage ring [252]. The main task of the collector ring (CR) is stochastic cooling of radioactive ions or antiproton beams from the production targets. In addition, this ring offers the possibility for mass measurements of short-lived ions, by operating in isochronous mode. For research with high-energy antiprotons up to 14 GeV and with heavy-ions, a high-energy storage ring (HESR) will be available, equipped with a high-energy electron cooler (up to 8 MeV), a stochastic cooling system, internal target, and an associated

Table 11.12 Key parameters and features of the synchrotrons and cooler/storage rings [247]

Ring	Circumference [m]	Energy [GeV/u]	Specific features
SIS-100 Synchrotron	1083	2.7 for U^{28+}, 29 for p	Fast ramped (up to 4 T/s) superferric magnets up to 2 T, extraction of short (60 ns), single pulses of up to 5×10^{11} U^{28+} and 4×10^{13} p (25 ns) or slow extraction.
CR Collector Ring	215	0.740 for A/Z = 2.7, 3 for antiprotons	Large aperture. Fast stochastic cooling of radioactive ion beams and antiprotons. Isochronous mode for mass measurements of short-lived nuclei.
HESR High Energy Storage Ring	575	14 for anti-protons, heavy ions (50 Tm)	Stochastic cooling of antiprotons up to 14 GeV. Electron cooling up to 14 GeV. Internal pellet or cluster target.

detector set-up [253]. The key parameters and features of the synchrotrons and cooler/storage rings are given in Table 11.12.

The FAIR facility is presently under construction. The CRYRING storage ring has already started its operation. The upgraded SIS18 will be available 2018. Start of commissioning of the SIS100, the production targets and the storage rings CR and HESR is expected for 2024/2025.

National Superconducting Cyclotron Laboratory (NSCL) at Michigan State University (MSU)

The first superconducting cyclotron, the K500 (K = 500) was built and went into operation at MSU in 1982 [254, 255]. A second cyclotron, the K1200, with greater bending power and hence higher energy beams (K = 1200) was commissioned in 1988 [256]. It was capable accelerating fully stripped light ions with $N = Z$ to 200 MeV/u and heavy ions to approximately 50 MeV/u, depending on the charge state. At these energies it was possible to convert the primary beam into radioactive secondary (sometimes called rare isotope) beams using projectile fragmentation or projectile fission. The A1200 separator was constructed for cyclotron beam analysis and for production and separation of radioactive fragment beams [257]. This device allowed radioactive beams to be delivered to all experimental set-ups and made possible a successful research program with radioactive-isotopes. The Coupled Cyclotron Facility (CCF) was built to increase the intensities (by several orders of magnitude) and energies for heavy ion beams. The project included an upgrade to the K500 cyclotron and its use as an injector for the K1200 cyclotron, along with a

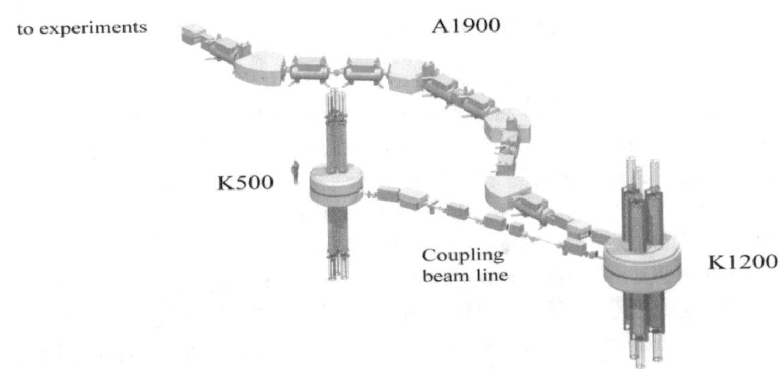

Fig. 11.53 The NSCL coupled cyclotron K500+K1200 facility with the superconducting large-acceptance A1900 fragment separator [246]. To show scale, the size of a person is shown near the K500 cyclotron (Courtesy of NSCL, Michigan State University, B. Sherrill)

new A1900 fragment separator. The CCF went into operation in 2001 [258, 259]. The program is centered on experimentation with radioactive ion beams produced by the A1900. Nearly 1000 different radioactive beams have been produced and used for experiments at the CCF since its inception.

Figure 11.53 shows the MSU-NSCL accelerator facility with the superconducting A1900 fragment separator with a collection efficiency near to 100% as compared to a few % for the A1200 system [260]. The maximum energy of this coupled cyclotron facility is limited to 200 MeV/u for $q/A = ½$ by the focussing in the K1200. Energies up to 80 MeV/u can be delivered for the heaviest ions. High transmission efficiencies allow 0.7–1.0 kW beams to be routinely delivered for experiments at the NSCL. Net beam transmission measured from just before the K500 to extracted beam from the K1200 can be about 30% depending on the ion used (factoring out the unavoidable loss due to the charge stripping foil in the K1200) [261].

The facility is currently being expanded for the investigation of radioactive isotopes at low energies; this will be done by stopping the in-flight-produced isotopes in a cryogenic gas stopping system and re-accelerating them in a compact linac (ReA3) [261, 262]. The ReA3 linear accelerator will provide low-energy radioactive isotope beams of high beam quality. The stopped ions will be re-ionized in an Electron Beam Ion Trap (EBIT), then re-accelerated in a room-temperature RFQ and a superconducting linac built of $\lambda/4$-resonators (QWR). ReA3 beam energies range from 0.3 to 6 MeV/u for $A < 50$ and up to 3 MeV/u for uranium. A later upgrade to 12 MeV/u (ReA12) will be done by expanding the ReA3 linac with QWRs. That will be carried out prior to completion of the FRIB project described in the following section.

MSU Facility for Rare Isotope Beams (FRIB)

The major new initiative in the US in radioactive beams is the Facility for Rare Isotope Beams (FRIB). FRIB is a research facility for the creation and utilization of radioactive isotope beams based on the concept of a high-intensity (400 kW) and high-energy (200 MeV/u) heavy-ion linac. It will include the capability to provide radioactive ion beams at all energies from thermal to nearly 200 MeV/u. Originally, the Rare Isotope Accelerator (RIA) with 100 kW and 400 MeV/u beams was proposed in 2003. In the years that followed an alternative design, FRIB, based on a lower energy (200 MeV/u) but higher intensity (400 kW) heavy-ion driver linac, was developed. Michigan State University was selected near the end of 2008 to build FRIB and the project attained DOE Critical Decision 1, CD-1, in 2010 [263].

Figure 11.54 shows the overall layout and the topology of the FRIB facility [264]. The heavy ion driver linac is fed by ECR ion sources. The low-energy beam transport (LEBT), the RFQ structure, and the medium-energy beam transport (MEBT) to the first section of the main linac are capable of transporting and accelerating heavy ion beams containing more than one charge state, such as 33+ and 34+ for uranium. The linac segment 1 consists of two types of superconducting quarter-wave resonators (QWR) operating at 80.5 MHz with $\beta_{\text{opt}} = 0.041$ and 0.085 to increase uranium-beam energy to 17.5 MeV/u and higher for lighter ions [265].

A stripping section will be installed after linac section 1 to increase the charge states of the ions and hence the acceleration efficiency. An isochronous charge

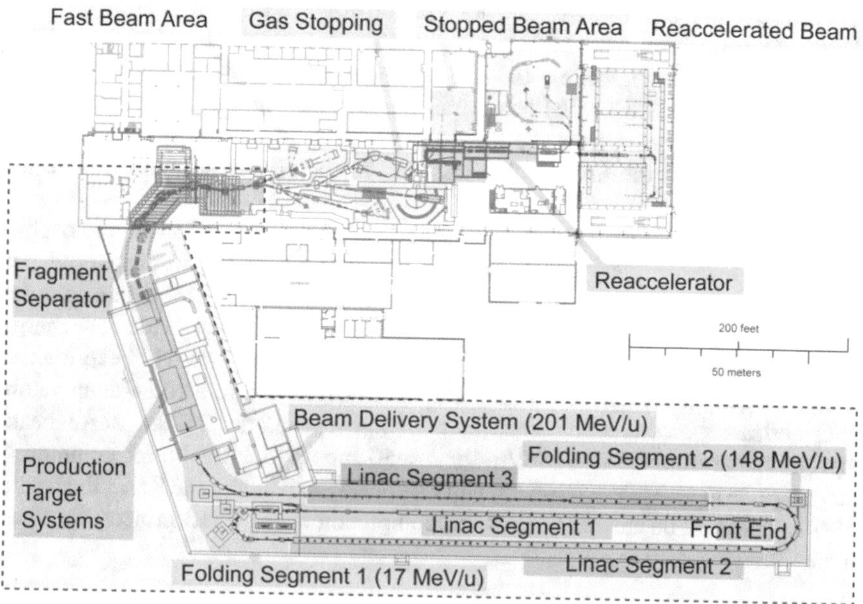

Fig. 11.54 Layout of the MSU FRIB project (Courtesy of FRIB, Michigan State University, B. Sherrill)

analysis system will make possible the selection of several charge states for further acceleration. Additional RF-cavities will help to optimize the longitudinal matching to linac section 2. This system will provide efficient acceleration of all beams.

The post-stripper linac will be built with two types of Half-Wave Resonators (HWR) with $\beta_{opt} = 0.285$ and 0.53 operating at a frequency of 322 MHz. The four cavity types will be able to efficiently cover the full velocity range from $\beta \sim 0.025$ to $\beta \sim 0.57$ ($E/A \geq 200$ MeV/u). Transverse focusing in the linac structures will be performed using superconducting 9T solenoids. With this condition, linac sections 2 and 3 will be able to accept several charge states (e.g. five charge states from 77+ to 81+ for uranium), and hence most of the intensity of the stripped beam charge state distribution for further acceleration. For purposes of hands-on machine maintenance, the specification for uncontrolled beam loss has been set to 1 W/m.

The high-power beam will be transferred to RIB production target systems, which are followed by a large-acceptance fragment separator with three stages of separation. The experimental area will reuse the existing NSCL experimental facilities. The fragments will be either transported to experimental set-ups for fast radioactive beams or slowed down using a gas-stopping system. Stopped isotopes can be investigated in traps or transferred to a charge breeder and then re-accelerated to low or medium energies up to 12 MeV/u (ReA12) [262].

The cryogenic facility, ECR, and RF support facilities are located above the linac tunnel, which according to plans will be about 13 m underground.

The project includes several upgrade options. Space will be left in the linac tunnel to make possible the addition of cryomodules; this will allow the beam energy to be upgraded to 400 MeV/u. In addition, the facility will include a path for a beam line to an optional ISOL production area that might be added in the future. There will also be space to add a light-ion injector and fast beam-switching system to make simultaneous multiple use of the FRIB possible. Construction of FRIB started in 2014. The project should be completed by 2022, with a possible early completion in 2021.

Cyclotron Facility at the Flerov Laboratory for Nuclear Reactions (FLNR) Dubna

The Laboratory for Nuclear Reactions, now named Flerov Laboratory for Nuclear Reactions (FLNR) in Dubna, Russia, was founded in the Joint Institute for Nuclear Research (JINR) in Dubna in 1957 [266]. For more than 40 years, classical and isochronous cyclotrons have been used there for the acceleration of heavy ions for nuclear physics research and heavy ion beam applications. Two isochronous cyclotrons (U400 and U400M) are used for heavy ion nuclear physics. The U400 (K = 625) has been in operation since 1978. In 1996 it was upgraded; the Penning ion source was replaced by an ECR ion source [267]. The synthesis of super heavy elements, predominantly using ^{48}Ca-beams, was and is still the main research approach there [268]. Steady improvement of ^{48}Ca-beams allowed to successfully synthesise new elements from Z = 114 to 118. A review of the discovery of super-heavy nuclei at FLNR, JINR is presented in [149].

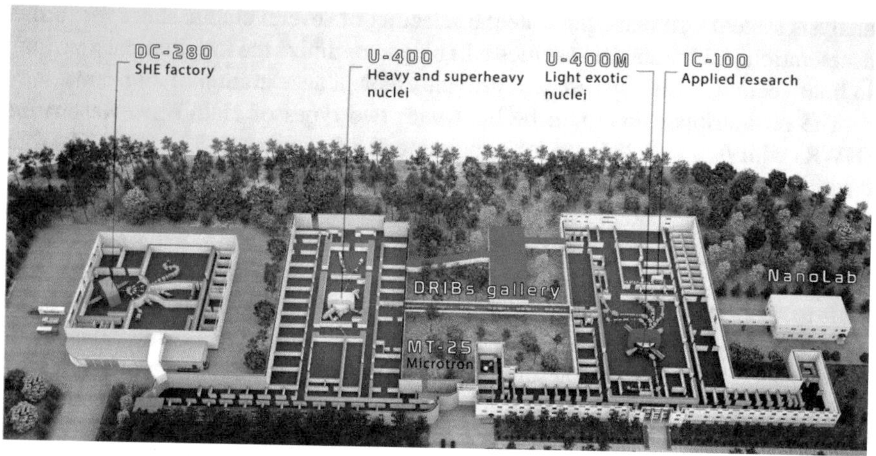

Fig. 11.55 Layout of the Dubna Radioactive Ion Beam (DRIBsIII) facility at the Flerov Laboratory for Nuclear Research (FLNR). It shows from left to right the new DC280 cyclotron, the U400 cyclotron, the U400M, and the beam line (DRIBs gallery) for transport of RIBs from the U400M to the U400 cyclotron, used for the acceleration of RIBS (see text). [Copyright © FLNR, JINR, Dubna]

The U400M (K = 550) has delivered light ion beams in the energy range of up to 50 MeV/u since the beginning of the 1990s. Originally, it has mainly been used for the production of light radioactive isotopes such as ^{6}He or ^{8}He using the ISOL method. These ions are transferred to an ECRIS for charge breeding, and then transported with keV/u energies via a 120-m beam transport system to the U400, where they are post-accelerated to MeV/u energies [269]. Upgrades of the U400M in the past years allowed accelerating ions of very heavy elements e. g. bismuth up to energies of 15 MeV/u [270]. A new cyclotron DC280 (K = 280) has been built, which will be the main accelerator of a Super Heavy Element Factory in the future, expected delivering first beams for experiments in 2018. Fed by normal and superconducting (18 GHz) ECR-ion sources, it should provide ten times higher beam intensities than the cyclotrons used so far for masses up to A = 50. The layout of the FLNR accelerator complex is shown in Figs. 11.55 and 11.56.

Heavy Ion Research Facility in Lanzhou (HIRFL)

One new and rather versatile accelerator combination of cyclotrons and synchrotron-storage rings is the Heavy Ion Research Facility in Lanzhou (HIRFL), China [271]. Its first stage consists of a superconducting ECR ion source, a compact sector-focusing injector cyclotron (K = 69), and a main sector-cyclotron accelerator SSC (K = 450). The SSC went into operation in 1988. The construction of a new heavy ion synchrotron (CSRm) (with the SSC as injector), combined with a fragment separator, and an experimental storage ring (CSRe) for maximum uranium beam energies of 500 MeV/u, started in 1999; the facility went into operation in 2007.

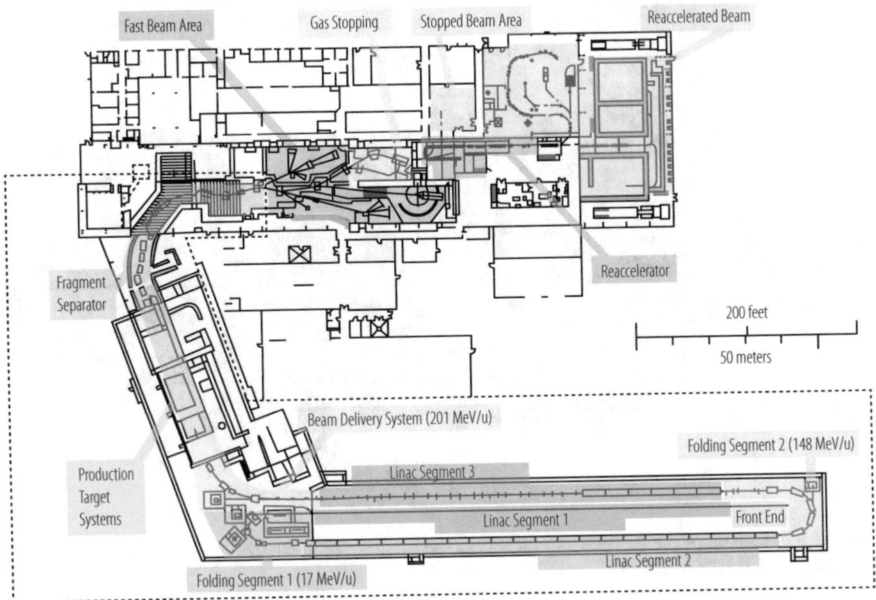

Fig. 11.56 Picture of the U400 cyclotron at the Flerov Laboratory for Nuclear Research (FLNR) in Dubna, used for super-heavy element synthesis. [Copyright © FLNR, JINR, Dubna]

The synchrotron and storage rings are both equipped with electron-cooler devices. The HIRFL facility is used for in-flight production of exotic isotopes at medium and high energies, and other basic and applied science, as well (Fig. 11.57).

Other In-Flight Facilities

A number of other facilities exist which can or could produce radioactive isotopes using the in-flight method. As can be concluded from Fig. 11.41, cyclotrons with K-values $\geq$150 (this includes all superconducting cyclotrons) can, using modern ECR ion sources, deliver sufficient beam energy for heavy ions. Again we refer to the IUPAP Report 41 and the NuPECC Long Range Plan 2010 [157, 214].

11.5.2.3.2 ISOL Facilities

Two examples of ISOL facilities have been selected and will be described here; both involve the post-acceleration of radioactive isotope beams. The first is the ISAC facility at TRIUMF, Vancouver, which like ISOLDE at CERN uses a high-energy proton beam to produce radioactive isotopes in a thick target. It has recently been expanded, the existing linac post-accelerator being upgraded to achieve higher energies. The second one, SPIRAL2 at GANIL, is an extension of the existing in-flight/ISOL facility GANIL-SPIRAL. It involves an additional new powerful

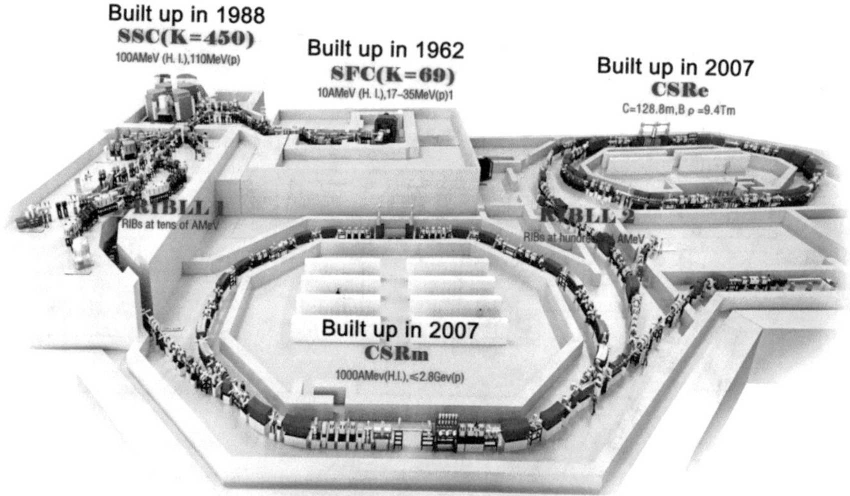

Fig. 11.57 Layout of the Accelerator facility at the Institute of Modern Physics (IMP) in Lanzhou. [Copyright © IMP, Lanzhou]

medium-energy light ion linac driver to produce much higher intensities of exotic isotopes than is currently possible, and which will make accessible new regions of exotic nuclei.

ISAC at TRIUMF

TRIUMF has operated a 500 MeV H^- cyclotron since 1974. The TRIUMF facility (Fig. 11.58) was expanded in 1995 with the addition of a radioactive beam facility, ISAC [272]. The radioactive species at ISAC are produced using the ISOL method with a 500 MeV proton beam of up to 100 μA bombarding a thick target. After production the species are ionized, mass-separated and sent to either a low-energy area or pass through a string of linear accelerators to feed experiments at higher energies. The first beams from ISAC, now named ISAC-I, were available in 1998, while the first accelerated beams were delivered in 2001 to a medium-energy area and used chiefly for nuclear astrophysics [273]. The TRIUMF ISAC-II superconducting linac, proposed in 1999, was designed to raise the energy of radioactive ion beams above the Coulomb barrier to support nuclear physics at TRIUMF. The first stage of this project, Phase I, commissioned in 2006, involved the addition of 20 MV of superconducting linac [274]. Phase II of the project consisting of an additional 20 MV of superconducting linac was installed and commissioned in 2010 [275].

Typically, 1+ beams are produced in the on-line source, but an ECR ion source was installed in 2009 to act as a charge-state booster (CSB) to raise the q/A ratio of low-energy high-mass beams so that they could be accelerated through the ISAC

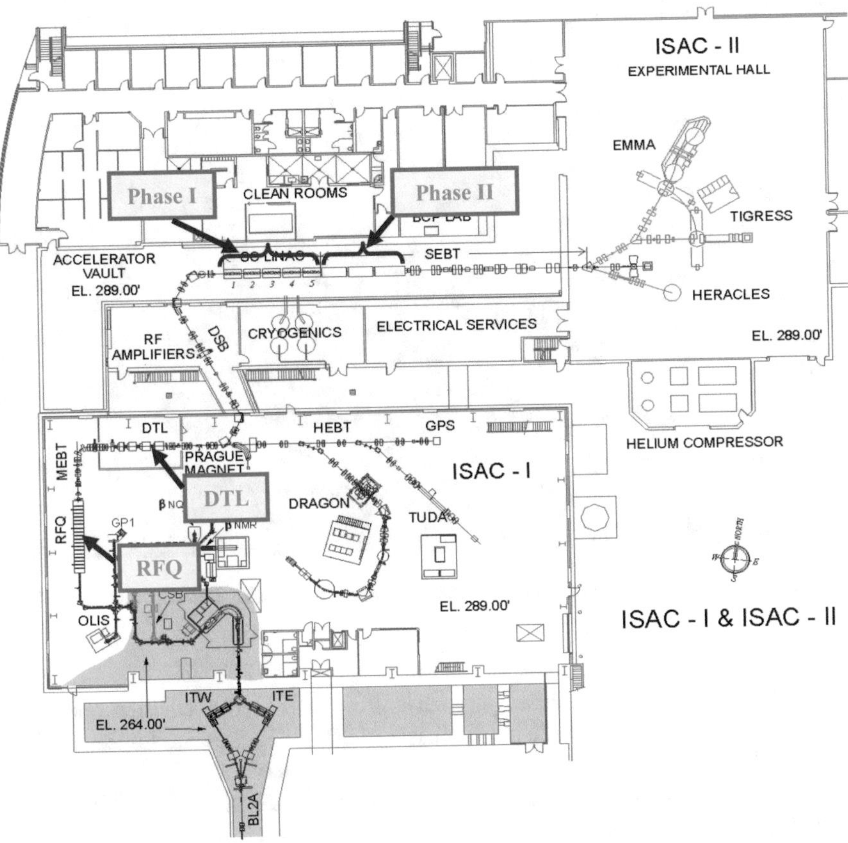

Fig. 11.58 The ISAC facility at TRIUMF showing the RFQ, DTL in ISAC-I and the two installation phases of the superconducting heavy ion linac in ISAC-II. [Courtesy of ISAC/TRIUMF, Vancouver, R. E. Laxdal]

accelerators. An off-line ECR ion source provides stable beams for accelerator commissioning and tuning as well as for the science program.

The ISAC-I accelerator chain includes a four-vane split-ring structure RFQ (35.4 MHz), which accelerates beams of $A/q \leq 30$ from 2 keV/u to 153 keV/u [276]. The post-stripper, a 106 MHz variable-energy drift tube linac (DTL), accelerates ions of $2 \leq A/q \leq 6$ to a final energy between 0.153 MeV/u and 1.53 MeV/u. The variable-energy DTL is based on a unique separated-function approach with five independent interdigital H-mode (IH) structures [277]. Both the RFQ and DTL have been used since 2001 to reliably provide a variety of radioactive and stable ions.

The ISAC-II superconducting linac is composed of bulk niobium, quarter wave resonators (QWR) for acceleration, and superconducting solenoids for periodic transverse focusing, housed in several cryomodules. The Phase-I linac consists of 20 QWR housed in five cryomodules. The first eight cavities have a geometric

Fig. 11.59 The ISAC-II Superconducting linac. [Courtesy of ISAC/TRIUMF, Vancouver, R. E. Laxdal]

$\beta = 0.057$ and the remainder a geometric $\beta = 0.071$. The cavities operate at 106 MHz. The Phase-II upgrade also consists of 20 QWR; they are housed in three cryomodules. These bulk niobium cavities have a geometric $\beta = 0.11$ and resonate at 141.44 MHz. One 9T superconducting solenoid is installed in the middle of each of the eight cryomodules in close proximity to the cavities. The ISAC-II SC-linac is shown in Fig. 11.59 [278].

The performance of the SC-linac is quoted in terms of the peak surface field achieved at a cavity RF power of 7 W. The Phase-I section has operated at Ep = 33 ± 1 MV/m since 2006 with little sign of degradation. The Phase-II section averaged Ep = 26 MV/m in the first 6 months of operation [279].

GANIL-SPIRAL2 Accelerator Facility

Construction of the cyclotron facility at the French national heavy-ion laboratory (Grand Accelerateur National d'Ion Lourd GANIL) in Caen, France, started at the end of the 1970s. It consisted of two compact cyclotrons C01 and C02, (K = 30 each) and two separated-sector cyclotrons CSS1and CSS2, (K = 380 each). The facility went into operation in 1982. The first beams for experiments were delivered in 1983 [280]. Using one of the compact cyclotrons as the injector for the CSS1 and a stripper before CSS2, it can achieve ion energies of up to 95 MeV/u for light ions and up to 24 MeV/u for uranium beams. Stable beams of intermediate energies in the range of ≤1 MeV/u to 13 MeV/u are used after C01 or C02, and CSS1 for atomic physics, biology, and solid state physics. Upgrades to improve beam energies and intensities have been performed since mid-1980s, with the special aim of improving the situation for exotic-beam experiments. This has been done by introducing ECR ion sources, by modifying the injection systems into the C01/C02 injector cyclotrons, and by improving performance of components for the operation with beam powers of up to several kW in and after the CSS2 [281, 283]. In-flight technology was used to produce exotic beams [283].

Since 2001, when a fifth cyclotron, CIME, went into operation, the Isotope Separation On-Line (ISOL) technique was also used to produce rare isotopes [284]. The exotic isotopes, which were produced in thick targets using high power ion beams of up to 3 kW from the CSS2 cyclotron, were transferred into a charge breeding ECR ion source, and post-accelerated in the compact cyclotron CIME (K = 265). Depending on the isotope, energies in the range from 1.2 to 25 MeV/u are delivered from CIME. These beams are transferred to the existing experimental facilities. An overview of the existing GANIL-SPIRAL1 accelerator and experiment facility displays Fig. 11.60 (right) [285]. A review on stable and radioactive beams produced there is given in [286].

To improve the possibilities for heavy-ion nuclear physics research both for stable and for radioactive ions an expansion of the GANIL facilities has been proposed named SPIRAL2. The SPIRAL2 project at GANIL was approved in 2005 [287, 288]. This new facility will provide intense beams of neutron rich exotic nuclei (10^{+6} to 10^{+11}) by the ISOL method in the mass range from $A = 60$ to $A = 140$. The facility will be composed of a flexible high power superconducting

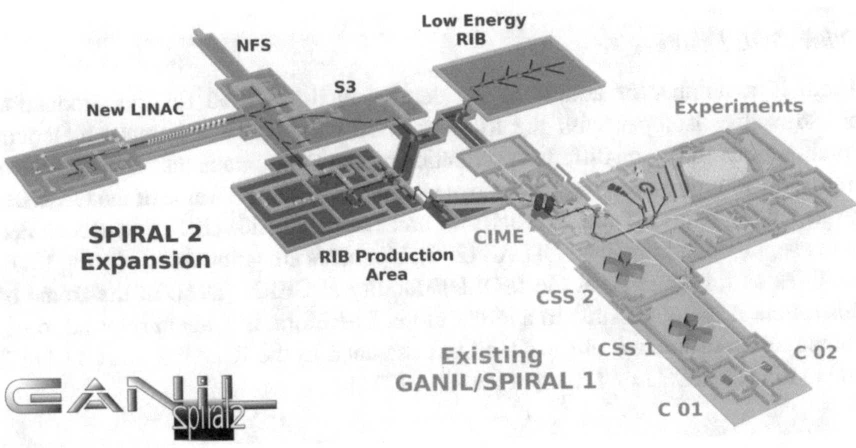

Fig. 11.60 Layout of the existing GANIL-SPIRAL1 accelerators and the experiment facilities (right), and of the new accelerator and experiment facilities under construction for SPIRAL2 (left): the existing GANIL-SPIRAL1 facility builds on the two injector cyclotrons C01/C02 the two sector cyclotrons CSS1 and CSS2 with stripper in between and the cyclotron CIME for acceleration of exotic beams. SPIRAL2 is based on a new superconducting linear ion accelerator as particle source for different production methods of radioactive isotopes (RIB) and stable ion beams. [Courtesy of GANIL/SPIRAL2, Caen, T. Junquera]

linear driver accelerator (5 mA/40 MeV deuterons, 5 mA/33 MeV protons, and 1 mA/14.5 MeV/u heavy ions with $q/A = 1/3$), a dedicated building for the production of radioactive isotopes (RI), the existing cyclotron CIME for the re-acceleration, and new experimental facilities. The high power deuteron beam is directed on a carbon converter target, producing fast neutrons. The main process for the production of the RI is based on the fast neutron induced fission in uranium carbide targets. Figure 11.60 (left) [285] shows an overview of the GANIL-SPIRAL2 facility [287, 289].

The new linac is composed of two families of 88 MHz superconducting QWR ($\beta = 0.07$, $\beta = 0.12$), which permits the acceleration of the different ions and energies mentioned before. The basic principle of this design is to install the resonators in separate cryomodules (one QWR per cryomodule in the $\beta = 0.07$ section and two QWR per cryomodule in the $\beta = 0.12$ section). Between each cryomodule beam focusing is performed by means of 2 room temperature quadrupoles with short vacuum/diagnostics boxes in between to allow for flexible beam optics and tuning.

Phase 1 of SPIRAL2 includes construction of buildings, which started in 2011, installation of the driver linac, the high energy beam lines and first experimental equipment in 2012. Proton beams were accelerated in the RFQ in 2015. The phase 2 contains the radioactive isotope production cave, low energy experiments for RIB and the connection to GANIL/SPIRAL1 facilities. Construction is expected to start in 2014. As a long term option a heavy ion source with $q/A = 1/6$ will be added to the linac. It will accelerate heavy ions up to 8.5 MeV/u.

Other ISOL-Facilities

There is a number of additional accelerator facilities used for the production of radioactive isotopes with the ISOL method building on different accelerator combinations and using different production methods. Spontaneous fission products from californium are used as radioactive isotope source in the frame of the CARIBU project at Argonne, in order to reach new areas of the nuclide chart, and accelerated in the superconducting linac ATLAS [290]. Upgrades are going on at existing ISOL facilities as for example at the ISOLDE facility at CERN [291]. In the frame of this article it is not possible to address more. Therefore, it is again referred to the compendium of nuclear physics facilities included in the IUPAP Report 41 [157] and in the NuPECC Long Range Plan 2010 [214].

11.5.2.3.3 Conclusions

Since about 20 years there is a growing worldwide demand for in-flight and in ISOL facilities for the production of radioactive isotopes. Developments in the fields of ion sources and superconducting accelerator technologies have facilitated access of many nuclear physics laboratories to exotic beams. The new facilities will further increase the production rates and the measurement precision for exotic ions.

In-flight facilities have advantages in the field of nuclear physics, whereas ISOL facilities have their strengths in the field of nuclear astrophysics [292]. A new generation of in-flight facilities builds on very powerful driver accelerators covering a broad range of light to heavy primary ion beams. For new and expanding ISOL facilities with superconducting post-accelerators, target handling, charge breeding, acceleration, and beam handling of radioactive ions are important issues. ISOL-post-accelerators are operated typically in the same energy range as machines for super-heavy element synthesis.

References

1. J.D. Jackson: *Classical Electrodynamics*, Wiley, 1962.
2. R. Feynman: *Feynman Lectures on Physics*, Vol. 2, eq. (21.1) ff.
3. G. Geloni, et al.: *Undulator radiation in a waveguide*, Nucl. Instrum. Meth. A 584 (2008) 219.
4. A. Hofmann: *The Physics of Synchrotron Radiation*, Cambridge Univ. Press, 2004.
5. M. Sands: *The Physics of Electron Storage Rings*, SLAC Report 121, SLAC, Stanford, 1970.
6. J.A. Clarke: *The Science and Technology of Undulators and Wigglers*, Oxford Publ., 2004.
7. Schmüser, P., Dohlus, M., Rossbach, J., Behrens, C.: *Free-Electron Lasers in the Ultraviolet and X-Ray Regime*, Springer, 2014.
8. G. Stupakov, D. Zhou: *Analytical theory of coherent synchrotron radiation wakefield of short bunches shielded by conducting parallel plates*, Phys. Rev. Accel Beams 19 (2016) 044402.
9. M. Abo-Bakr, et al.: *Steady-State Far-Infrared Coherent Synchrotron Radiation detected at BESSY II*, Phys. Rev. Lett. 88(25) (2002) 254801.
10. S. Wesch, et al.: *A Multi-Channel THz and Infrared Spectrometer for Femtosecond Electron Bunch Diagnostics by Single-Shot Spectroscopy of Coherent Radiation*, Nucl. Instrum. Meth. A 665 (2011) 40.
11. S. Reiche: *GENESIS 1.3: a fully 3D time-dependent FEL simulation code*, Nucl. Instrum. Meth. A 429 (1999) 243.
12. I. Zagorodnov, *Numerical modeling of collective effects in free electron laser, in Proceedings of 11th International Computational Accelerator Physics Conference, Rostock-Warnemünde, Germany* (JACoW, Geneva, Switzerland, 2012) p. 81.
13. J.B. Murphy, C. Pellegrini: *Introduction to the physics of free electron lasers*, Laser Handbook, Vol. 6, (1990), p. 115.
14. E.L. Saldin, E.A. Schneidmiller, M.V. Yurkov: *The Physics of Free Electron Lasers*, Springer (2000).
15. Z. Huang, K.-J. Kim: *A Review of X-Ray Free-Electron Laser*, Phys. Rev. ST Accel. Beams 30 (2007) 034801.
16. Z. Huang, P. Schmüser: *Handbook of Accelerator Physics and Engineering*, A.W. Chao, M. Tigner (eds.), World Scientific, 1999.
17. J.M.J. Madey: *Relationship between mean radiated energy, mean squared radiated energy and spontaneous power spectrum in a power series expansion of the equations of motion in a free-electron laser*, Nuovo Cimento B 50 (1979) 64.
18. FEL at JLAB: see www.jlab.org/FEL
19. P. Emma, et al.: *First Lasing of the LCLS X-Ray FEL at 1.5 Å*, PAC2009 Proc., 2009.
20. A.M. Kondratenko, E.L. Saldin: *Generation of Coherent Radiation by a Relativistic Electron Beam in an Undulator*, Part. Accel. 19 (1980) 207.
21. R. Bonifacio, C. Pellegrini, M. Narducci: *Collective instabilities and high-gain regime in a free electron laser*, Opt. Comm. 50 (1984) 373.

22. K.-J. Kim: *Three-Dimensional Analysis of Coherent Amplification and Self-Amplified Spontaneous Emission in Free Electron Lasers*, Phys. Rev. Lett. 57 (1986) 1871.
23. M. Xie: *Exact and variational solutions of 3D eigenmodes in high gain FELs*, Nucl. Instrum. Meth. A 445 (2000) 59.
24. E.L. Saldin, et al.: *FAST a three-dimensional time-dependent FEL simulation code*, Nucl. Instrum. Meth. A 429 (1999) 233.
25. W. Fawlay: *A User Manual for GINGER and Its Post-Processor XPLOTGIN*, Report LBNL-49625, Lawrence Berkeley Laboratory (2002).
26. G. Moore: *The high-gain regime of the free electron laser*, Nucl. Instrum. Meth. A 239 (1985) 19.
27. R. Ischebeck, et al.: *Study of the transverse coherence at the TTF free electron laser*, Nucl. Instrum. Meth. A 507 (2000) 175.
28. P.B. Corkum: *Plasma perspective on strong field multiphoton ionization*, Phys. Rev. Lett. 71 (1993) 1994.
29. K.J. Schafer, et.al.: *Above threshold ionization beyond the high harmonic cutoff*, Phys. Rev. Lett. 70 (1993) 1599.
30. J. Feldhaus, et al.: *Possible application of X-ray optical elements for reducing the spectral bandwidth of an X-ray SASE FEL*, Opt. Comm. 140 (1997). 341.
31. L.H. Yu: *Generation of intense uv radiation by subharmonically seeded single-pass free-electron lasers*, Phys. Rev. A 44 (1991) 5178.
32. G. Stupakov: *Using the Beam-Echo Effect for Generation of Short-Wavelength Radiation*, Phys. Rev. Lett 102 (2009) 074801.
33. E.L. Saldin, et al.: *Self-amplified spontaneous emission FEL with energy-chirped electron beam and its application for generation of attosecond x-ray pulses*, Phys. Rev. ST Accel. Beams 9 (2006) 050702.
34. W. Henning, C. Shanks: *Accelerators for Americas Future*, US. Dep. Energy, 2010.
35. K. Bethge, in: Advances of accelerator physics and technologies, H. Schopper (ed.), World Scientific, 1993.
36. M.R. Cleland: CAS-CERN Accelerator School and KVI specialized CAS course on small accelerators, 2005, p. 383.
37. A. Zaitsev, I.A. Dobrinets, G. Vins; HPHT-Treared Daimonds, Springer-Verlag, ISBN: 9783662506462, 2016.
38. J. Lutz et al.: Semiconductor Power Devices: Springer Verlag, ISBN 978-3-642-11125-9, 2011.
39. J.R. Tesmer, et al.: Handbook of modern ion beam materials analysis, Materials Research Society, 1995.
40. T. Schulze-König, et al.: Nucl. Instrum. Meth. B 268 (2010) 891.
41. U. Zoppi, et al.: Radiocarbon 49 (2007) 171.
42. W. Kutschera: private communication.
43. J.J. Nelson, D.J. Muehlner, in: Magnetic bubbles, H. Jouve (ed.), Acad. Press, London, 1986.
44. F. Joliot, I. Curie, *Artificial Production of a New Kind of Radio-Element,* Nature, Volume 133, Issue 3354, (1934) 201-202
45. E. Fermi, *Radioactivity Induced by Neutron Bombardment*, Nature 133, (1934) 757-757
46. O. Chievitz, G. Hevesy, *Radioactive indicators in the study of phosphorous in the metabolism in rats*, Nature 136 (1935) 754
47. J. H. Lawrence, K. G. Scott, and L. W. Tuttle, *Studies on leukemia with the aid of radioactive phosphorus*. Internat. Clin., 3, (1939) 33
48. J. G. Hamilton, R. S. Stone, *The Intravenous and Intraduodenal Administration of Radio-Sodium,* Radiobiology, 28, (1937) 178
49. J. H. Lawrence and R. Tennant, *The comparative effects of neutrons and x-rays on the whole body*, J Exp Med 66, (1937) 667-688
50. T. Ido, C.N. Wan, J.S. Fowler, A.P. Wolf, *Fluorination with F_2: a convenient synthesis of 2-deoxy-2-fluoro-D-glucose*. J Org Chem 42 (1977) 2341-2342

51. J.S. Fowler, T. Ido, *Initial and subsequent approach for the synthesis of 18FDG*, Semin Nucl Med. 32(1) (2002) 6-12
52. D. B. Mackay, C. J. Steel, K. Poole, S. McKnight, F. Schmitz, M. Ghyoot, R. Verbruggen, F. Vamecq and Y. Jongen, *Quality assurance for PET gas production using the Cyclone 3D oxygen-15 generator*, Applied Radiation and Isotopes, Volume 51, Issue 4 (1999) 403-409
53. IAEA-DCRP/2006, *Directory of Cyclotrons used for Radionuclide Production in Member States - 2006 Update*, October 2006
54. N. Ramamoorthy, *Production of radioisotopes for medical applications*, Oral presentation, Workshop Physics For Health in Europe, 2-4 February 2010, CERN, Geneva
55. G.O. Hendry et al., "Design and Performance of a H-Cyclotron," Proc. of the 9th Int. Conference on Cyclotrons and their Applications (1981) 125.
56. P. W. Schmor, Review of cyclotrons used in the production of radioisotopes for biomedical applications, Proceedings of CYCLOTRONS 2010, Lanzhou, China
57. C. Oliver, Compact and efficient accelerators for radioisotope production, IPAC2017 - Proceedings, June 2017, ISBN 978-3-95450-182-3
58. R. E. Shefer, R. E. Klinkowstein, M. J. Welch and J. W. Brodack, *The Production of Short Lived PET Isotopes at Low Bombarding Energy with a High Current Electrostatic Accelerator*. Proc. Third Workshop on Targetry and Target Chemistry, T. J. Ruth, ed., Vancouver, B.C., (1989)
59. A.R. Jalilian, J.A. Osso, Production, applications and status of zirconium-89 immunoPET agents, J Radioanal Nucl Chem (2017) 314: 7. https://doi.org/10.1007
60. IAEA, *Cyclotron Produced Radionuclides: Principles and Practice*, Technical Reports Series No. 465, International Atomic Energy Agency, Vienna, 2008
61. *Therapeutic Nuclear Medicine*, Richard P Baum, (Ed.), Springer-Verlag Berlin Heidelberg, 2014, ISBN 978-3-540-36719-2. DOI https://doi.org/10.1007/978-3-540-36719-2
62. Making Medical Isotopes, Report of the Task Force on Alternatives for Medical-Isotope Production, TRIUMF University of British Columbia Advanced Applied Physics Solutions, Inc. (2008)
63. Molybdenum-99 for Medical Imaging, The National Academies Press, Washington, DC, ISBN 978-0-309-44531-3, DOI 10.17226/23563
64. IAEA, Non-HEU production technologies for molybdenum-99 and technetium-99m, Nuclear Energy Series No. NF-T-5.4, Vienna: IAEA, 2013, ISBN 978–92–0–137710–4
65. W. K. H. Panofsky, *Big Physics and Small Physics at Stanford*, Stanford Historical Society, Sandstone and Tile, Volume 14, no. 3 (1990) 1-10
66. D.W. Fry, R.B.R.S. Harvie, L.B Mullett, W. Walkinshaw, *Travelling-wave linear accelerator for electrons*, Nature 160 (1947) 351-353
67. IAEA, DIrectory of RAdiotherapy Centres https://dirac.iaea.org/Query/Countries, as of 30 November 2017 [25] http://iopscience.iop.org/article/10.1088/1361-6560/aa9517
68. B W Raaymakers et al. *First patients treated with a 1.5 T MRI-Linac: clinical proof of concept of a high-precision, high-field MRI guided radiotherapy treatment*, Phys. Med. Biol. (2017) 62 L41
69. G. L. Locher, Biological effects and therapeutic possibilities of neutrons, Am. J. Roentgenol. 36, (1936) 1-13.
70. R.S. Stone, *Neutron therapy and specific ionization*, Am. J. Roentgenol. 59 (1948) 771-785
71. M. Awschalom, et al. *The Fermilab Neutron Radiotherapy Facility*, 1977 PAC, Chicago, IEEE Trans. Nuc. Sci., Vol. NS-24, No. 3, (1977) 1055.
72. R.R. Wilson, *Radiological use of fast protons*, Radiobiology 47, (1946) 487-491
73. R.P. Levy, J.I. Fabrikant, K.A. Frankel, M.H. Phillips, J.T. Lyman, J.H. Lawrence, C.A. Tobias, *Heavy-charged-particle radiosurgery of the pituitary gland: clinical results of 840 patients*, Stereotact. Funct. Neurosurg. 57 (1-2), (1991) 22-35
74. C. A. Tobias, J. H. Lawrence, J. L. Born, R. K. Mccombs, J. E. Roberts, H. O. Anger, B. V. A. Low-Beer, and C. B. Huggins, *Pituitary Irradiation with High-Energy Proton Beams - A Preliminary Report*, Cancer Research 18, No 2, (1958) 121

75. Börje Larsson, Lars Leksell, Bror Rexed, Patrick Sourander, William Mair, and Bengt Andersson, *The High-Energy Proton Beam as a Neurosurgical Tool*, Nature 182, (1958) 1222-1223
76. H.D. Suit, M. Goitein, J. Munzenrider, L. Verhey, K.R. Davis, A. Koehler, R. Linggood, R.G. Ojemann, *Definitive radiation therapy for chordoma and chondrosarcoma of base of skull and cervical spine*, J Neurosurg, 56(3) (1982) 377-385
77. E. Pedroni, R. Bacher, H. Blattmann, T. Böhringer, A. Coray, A. Lomax, S. Lin, G. Munkel, S. Scheib, U. Schneider and A. Tourosvsky, The 200-MeV proton therapy project at the Paul Scherrer Institute: conceptual design and practical realization, Med. Phys. 22 (1995) 37-53.
78. K. Umegaki, K. Hiramoto, N. Kosugi, K. Moriyama, H. Akiyama, S. Kakiuchi, *Development of Advanced Proton Beam Therapy System for Cancer Treatment,* Hitachi Review Vol. 52 No. 4, (2003) 196-201
79. J. R. Castro, *Heavy ion therapy: the BEVALAC epoch. In: "Hadron therapy in oncology"*, U. Amaldi and B. Larsson eds., B., Elsevier, Amsterdam-Lausanne-New York-Oxford-Shannon-Tokyo, (1994) 208-216
80. Y. Hirao et al., Heavy ion synchrotron for medical use, Nucl. Phys. A 538 (1992) 541c.
81. H. Tsujii et al., *Clinical Results of Carbon Ion Radiotherapy at NIRS*, Journal of Radiation Research, Vol. 48, Suppl. A, A1-A13.
82. H. Tsujii et al., *Clinical advantages of carbon-ion radiotherapy*, New J. Phys. 10 (2008)
83. http://www.klinikum.uni-heidelberg.de/
84. S.E. Combs, M. Ellerbrock, T. Haberer, D. Habermehl, A. Hoess, O. Jäkel, A. Jensen, S. Klemm, M. Münter, J. Naumann, A. Nikoghosyan, S. Oertel, K. Parodi, S. Rieken, J. Debus, *Heidelberg Ion Therapy Center (HIT): Initial clinical experience in the first 80 patients*, Acta Oncol. 49(7), (2010) 1132-40
85. T. Haberer, W. Becher, D. Schardt and G. Kraft, Magnetic scanning system for heavy ion therapy, Nuclear Instruments and Methods A 330 (1993) 296.
86. W. Enghardt et al., Charged hadron tumour therapy monitoring by means of PET, Nucl. Instr. Methods A525 (2004) 284;
87. P. Crespo, Optimization of In-Beam Positron Emission Tomography for Monitoring Heavy Ion Tumour Therapy, Ph. D. Thesis, Technische Universitaet Darmstadt (2005).
88. K. Parodi et al., The feasibility of in-beam PET for accurate monitoring of proton therapy: results of a comprehensive experimental study, IEEE Trans. Nucl. Sci. 52 (2005) 778.
89. S. Webb, Intensity-Modulated Radiation Therapy, Institute of Physics Publishing, Bristol and Philadelphia, 2001.
90. Particle Therapy CoOperative Group (PTCOG), www.ptcog.com and ptcog.web.psi.ch.
91. H. Souda, *Facility Set-up and operation*, International Training Course on Carbon-ion Radiotherapy 2016, Chiba/Maebashi, Japan
92. H. Souda, *Carbon Ion Therapy Facilities in Japan*, 2017, Asian Forum for Accelerators and Detectors (AFAD) 2017, Lanzhou, China
93. U. Amaldi, *Particle Accelerators: From Big Bang Physics to Hadron Therapy*, Springer, 2012, ISBN 978-3-319-08870-9, DOI 10.1007/978-3-319-08870-9_4
94. U. Amaldi, G. Magrin (Eds), *The Path to the Italian National Centre for Ion Therapy*, Ed. Mercurio, Vercelli, (2005)
95. E. Feldmeier, T. Haberer, M. Galonska, R. Cee, S. Scheloske, A. Peters, *The First Magnetic Field Control (B-train) to Optimize the Duty Cycle of a Synchrotron in Clinical Operation*, Proceedings of IPAC 2012, New Orleans (Louisiana, USA), THPPD002, pp. 3503-3505
96. Y. Iwata et al. *Multiple-energy operation with extended flattops at HIMAC*, NIM A, 624 (1), 2010, pp. 33-38.
97. K. Crandall, M. Weiss, *Preliminary design of compact linac for TERA, TERA 94/34 ACC 20, September 1994*
98. U. Amaldi et al., *LIBO—a linac-booster for protontherapy: construction and tests of a prototype*, Nuclear Instruments and Methods in Physics Research A, 512 (2004) 521
99. L. Picardi et al., Progetto del TOP Linac, ENEA-CR, Frascati 1997, RT/INN/97-17.

100. A. Garonna, U. Amaldi, R. Bonomi, D. Campo, A. Degiovanni, M. Garlasché, I. Mondino, V. Rizzoglio and S. Verdú Andrés, *Cyclinac medical accelerators using pulsed* C6+/H2+ *ion sources*, J. Inst. 5 C09004 (2010)
101. J. Fourrier et al, Variable energy proton therapy FFAG accelerator, proceedings of EPACS08, 1791-1793.
102. Y. Yonemura et al, Development of RF acceleration system for 150 MeV FFAG accelerator, NIM A 576 (2007) 294-300.
103. D. Trbojevic et al., Design of a non-scaling FFAG accelerator for proton therapy, Proc. Cycl. 2004 (2005) 246–248;
104. S. Antoine et al, Principle design of a protontherapy, rapid-cycling, variable energy spiral FFAG, Nuclear Instruments and Methods A 602 (2009) 293-305.
105. E. Keil, A. M. Sessler and D. Trbojevic, Hadron cancer therapy complex using non-scaling fixed field alternating gradient accelerator and gantry design, Phys. Rev. ST Accel. Beams 10 (2007) 054701.
106. S. Vérdu Andrés, U. Amaldi and A. Faus-Golfe, *Literature review on Linacs and FFAGs for hadron therapy,* Int J Mod Phys. A26, (2011) 1659–1689
107. Carbon ion therapy, Proceedings of the HPCBM and ENLIGHT meetings held in Baden and in Lyon', Radiotherapy and Oncology 73/2 (2004) 1-217.
108. T.R. Bortfeld & J.S. Loeffler *Three ways to make proton therapy affordable*, 28 September 2017, Nature, V 549 (2017) 451-453
109. U. Amaldi, S. Braccini, G. Magrin, P. Pearce and R. Zennaro, patent WO 2008/081480 A1.
110. J. Fuchs et al, Laser-driven proton scaling laws and new paths towards energy increase, Nature Physics 2 (2005) 48-54.
111. S. V. Bulanov, G. A. Mourou and T. Tajima, Optics in the relativistic regime, Rev. Mod. Phys. 78 (2006) 309-372.
112. U. Linz and J. Alonso, What will it take for laser driven proton accelerators to be applied to tumour therapy?, Phys. Rev. ST Acc. Beams 10 (2007) 094801.
113. B.T.M. Willis, C.J. Carlile: Experimental neutron scattering, Oxford University Press, 2009, ISBN 978-0-19-851970-6.
114. F. Mezei: 2011, private communication.
115. Scientific Prospects for Neutron Scattering with support from EC/TMR and ILL, in Autrans, France, 11-13 January 1996.
116. K. van der Meer, et al.: Nucl. Instrum. Meth. B 217 (2004) 202–220.
117. F. Mezei: *Comparison of neutron efficiency of reactor and pulsed source instruments*, Proc. ICANS-XII (Abingdon 1993) (RAL Report No. 94-025), I-137; and F. Mezei: *The raison d'être of long pulse spallation sources*, J. Neutron Research 6 (1997) 3-32.
118. S. Gammino, L. Celona, R. Miracoli, D. Mascali, G. Castro, G. Ciavola, F. Maimone, R. Gobin, O. Delferrière, G. Adroit, F. Senèe: Proc. 19th Workshop ECR Ion Sources (MOPOT012), Grenoble, August 2010, to be published on Jacow.
119. R. Gobin, et al.: *High intensity ECR ion source (H^+, D^+, H^-) developments at CEA/Saclay*, Rev. Sci. Instrum. 73 (2002) 922.
120. M. Eshraqi, G. Franchetti, A.M. Lombardi: *Emittance control in rf cavities and solenoids*, Phys. Rev. ST Accel. Beams 12 (2009) 024201.
121. J. Stovall: *Low and medium energy beam acceleration in high intensity linacs*, Eur. Particle Accelerator Conf., 2004, Lucerne, Switzerland.
122. S. Molloy, et al.: *High precision superconducting cavity diagnostics with higher order mode measurements*, Phys. Rev. ST Accel. Beams 9 (2006) 112802.
123. G.E. McMichael; et al.: *Accelerator research on the RCS at IPNS*, Proc. EPAC'06, MOPCH126 (2006).
124. G. H. Rees: *Status of the SNS (now ISIS)*, Proc. PAC83, IEEE Trans. Nucl. Sci. NS-30(4) (1983) 3044-3048.
125. P. Bryant, M. Regler and M. Schuster, (eds), *The AUSTRON Feasibility Study*, Vienna (1994).
126. JAERI-KEK Project Team, *Accelerator Technical Design Report for J-PARC*, J-PARC 03-01 (2003).

127. Shinian Fu et al, *Accelerator design for China Spallation Neutron Source*, ICFA Beam Dynamics Newsletter, pp. 120-123, April, (2011).
128. J. Wei et al, Injection choice for Spallation Neutron Source ring, Proc. of PAC01, (2001).
129. R. Damburg et al, Chapter 3, in: Rydberg states of atoms and molecules, Cambridge University Press, pp 31-71, (2003).
130. G.H. Rees, *Linac, beam line and ring studies for an upgrading of ISIS*, Internal RAL note GHR1/ASTeC/ December (2009).
131. The ESS (European Spallation Source) Project, Volume III, Technical Report, pp.2-4 to 2-56, (2002).
132. M.A. Plum et al, *SNS ring commissioning results*, Proceedings of EPAC'06, MOPCH131 (2006).
133. C.R. Prior and G.H. Rees, *Multi-turn injection and lattice design for HIDIF,* Nuclear Instruments and Methods, Physics Research A, 415 pages 357-362, (1998).
134. G.H. Rees, *Direct proton injection for high power, short pulse, spallation source rings,* Internal RA note GHR1/ASTeC/ November (2016).
135. G. H. Rees, Non-isochronous & isochronous, non-scaling FFAG designs, Proc. of 18th International Conf. on Cyclotrons & their Applications, Sicily, MOP1197, p.189-192 (2007).
136. S. Machida, Parameters of a DF spiral ring for ISIS Hall, RAL note smb://ISIS/Shares/Accelerator R&D/IBIS Meetings/ ISIS II FFAG Parameters.
137. L.M. Onishchenko: *Cyclotrons*, Phys. Part. Nucl. 39 (2008) 950.
138. E.O. Lawrence, N.E. Edlefsen: *On the production of high speed protons*, Science 72 (1930) 376.
139. H. Willax: *Proposal for a 500 MeV Isochronous Cyclotron with Ring Magnet*, Proc. Intern. Conf. Sector-Focused Cyclotrons (1963) 386.
140. M. Seidel, et al.: *Production of a 1.3 Megawatt Proton Beam at PSI*, Proc. IPAC10, Kyoto, Japan (2010) 1309-1313.
141. L.H. Thomas: *The Path of Ions in the Cyclotron*, Phys. Rev. 54 (1938) 580-598.
142. W. Joho: *High Intensity Problems in Cyclotrons*, Proc. 5th Intern. Conf. Cyclotrons and their Applications, Caen (1981).
143. G. Dutto, et al.: *TRIUMF High Intensity Cyclotron Development for ISAC*, Proc. 17th Intern. Conf. Cyclotrons and Their Applications, Tokyo (2004) 82–88.
144. M. Kase, et al.: *Present Status of the RIKEN Ring Cyclotron*, Proc. 17th Intern. Conf. Cyclotrons and their Applications, Tokyo (2004) 160–162.
145. H. Okuno, et al.: *Magnets for the RIKEN Superconducting RING Cyclotron*, Proc. 17th Intern. Conf. Cyclotrons and their Applications, Tokyo (2004) 373–377.
146. V. Yakovlev et al.: *The Energy Efficiency of High Intensity Proton Driver Concepts*, Proc. IPAC'17, Copenhagen (2017) 4842-4847
147. *List of Cyclotrons*, Proc. 18th Intern. Conf. Cyclotrons and their Applications, Giardini Naxos (2007).
148. S. Hofmann, G. Münzenberg: Rev. Mod. Phys. 72(3) (2000) 733.
149. Yu.Ts. Oganessian, V.K. Utyonkov, Rep. Prog. Phys. 78, (2015) 036301
150. S. Hofmann, *J. Phys. G: Nucl. Part. Phys. 42 (2015) 114001*
151. J.R. Alonso: Proc. EPAC'90, Nice (1990) 95.
152. H.W. Schreuder: Proc. EPAC'90, Nice (1990) 82.
153. I. Tanihata, et al.: Phys. Rev. Lett. 55(24) (1985) 2676.
154. P.G. Hansen, B. Jonson: Europhys. Lett. 4(4) (1987) 409.
155. H. Geissel, G. Muenzenberg, H. Weick: Nucl. Phys. A 701 (2002) 259.
156. U. Koester: Eur. Phys. J. A 15 (2002) 255.
157. http://www.iupap.org/wg/
158. C.E. Anderson, K.W. Ehlers: Rev. Sci. Instrum. 27 (1956) 809.
159. A.S. Pasuyk, Y.P. Tretiakov, S.K. Gorbacher: Dubna-Report 3370 (1967).
160. P. Spaedtke, et al.: Proc. LINAC'96, Geneva (1996) 163.
161. V.A. Monchinsky, L.V. Kalagin, A.I. Govorov: Laser Part. Beams 14 (1996) 439.
162. E.D. Donets: Rev. Sci. Instrum. 69(2) (1998) 614.

162. A. V. Butenko et al., Proc. IPAC'14, Dresden (2014) 2103
164. J. Alessi, et al.: Proc. HIAT'09, Venice (2009) 138.
165. C. J. Gardner et al., Proc. IPAC'15, Richmond (2015) 3805
166. F. Wenander: EBIST2010, J. Instrum. 5 (2010) C10004.
167. P. Briand, R. Geller, B. Jacquot, C. Jacquot: Nucl. Instrum. Meth. 131 (1975) 407.
168. R. Geller, B. Jacquot, M. Pontonnier: Phys. Rev. 56(8) (1985) 1505.
169. Y. Jongen and G. Ryckewaert, IEEE Trans. Nucl. Sci. 30(4):2685 (1983)
170. D. Leitner, C. Lyneis, in: Physics and Technology of Ion Sources, I.G. Brown (ed.), Wiley-VCH (2004) 203.
171. L. Sun et al., Proc. LINAC'16, East Lansing (2016) 1028
172. G. Machicaone et al., Proc. ECRIS'14, Nizhny Novgorod (2014) 1
173. Y. Higurashi et al., Proc. ECRIS'16, Busan (2016) 10
174. S. Jeong, Proc. IPAC'16 Busan (2016) 4261
175. J. Benitez et al., Proc. ECRIS'12, Sydney (2012) 153
176. H. W. Zhao et al., Phys. Rev. Accel. Beams **20** 094801 (2017)
177. K. Tinschert et al., Proc. ECRIS'08, Chicago (2008) 97
178. H. Koivisto et al., Proc. HIAT'09, Venice, (2009) 128
179. T. Loew et al., Proc. PAC'07, Albuquerque (2007) 1742
180. S. L. Bogomolov et al., Proc. EPAC'98, Stockholm (1998) 1391
181. S. Gammino et al., Proc. Cyclotrons'01, East Lansing (2001) 223
182. P. Sortais et al., Rev. Sci. Instrum. **75** 1610 (2004)
183. D. Leitner et al., Proc. ECRIS'08, Chicago (2008) 2.
184. T. Lamy, J. Angot, C. Fourel, Proc. HIAT'09, Venice (2009) 114
185. L. Maunoury et al., Proc. ECRIS'16, Busan (2016) 35
186. L. Sun et al., Proc. LINAC'16, East Lansing (2016) 1027
187. S. Gammino, ISIBHI collaboration, Proc. Cyclotrons'07, Giardini-Naxos, Sicily (2007) 256
188. D. Leitner, C. Lyneis, Physics and Technology of Ion sources, 223, Ed. I. G. Brown, Wiley-VCH
189. P. Spädtke et al., Rev. Sci. Instrum. 83 02B720 (2012)
190. M. Chanel, Nucl. Instrum. Meth. A 532 (2004) 137
191. R. Hollinger et al., Rev. Sci. Instrum. 79 (2008) 02C703
192. R. Hollinger et al., Proc. RUPAC'12, Saint-Petersburg (2012) 436
193. N. Bohr: Kgl. Danske. Videnskab. Selskab. Mat.- Fys. Medd. 18(8) (1948).
194. H.H. Heckmann, E.L. Hubbard, W.G. Simon: Phys. Rev. 129(3) (1963) 1240.
195. H.-D. Betz: Rev. Mod. Phys. 44 (1972) 465.
196. P. Strehl, in: Handbook of Accel. Phys. Eng., A.W. Chao, M. Tigner (eds.), World Sci. (2006) 603.
197. C.D. Moak, et al.: Phys. Rev. Lett. 18(2) (1967) 41.
198. J.P. Rozet, C. Stéphan, D. Vernhet: Nucl. Instrum. Meth. B 107 (1996) 67.
199. A. Leon, et al.: Atomic Data & Nuclear Data Tables 69 (1998) 217.
200. C. Scheidenberger, et al.: Nucl. Instrum. Meth. B 142 (1998) 441.
201. W. Bath et al., Phys. Rev. Accel. Beams 18 (2015) 040101
202. A.S. Schlachter, et al.: Phys. Rev. A 27(11) (1983) 3372.
203. V.P. Shevelko, et al.: J. Phys. B 37 (2004) 201.
204. V.P. Shevelko, et al.: Nucl. Instrum. Meth. B 269 (2011) 1455.
205. W. Erb: GSI-Report GSI-P-7-78, (1978).
206. A.N. Perumal, et al.: Nucl. Instrum. Meth. B 227 (2005) 251.
207. H. Okuno, et al.: Phys. Rev. ST Accel. Beams 14 (2011) 033503.
208. A. Mueller, et al.: Phys. Ser. T 37 (1991) 62.
209. J. Bosser, et al.: Part. Accel. 63 (1999) 171; S. Baird, et al.: Phys. Lett. B 361 (1995) 184.
210. O. Uwira, et al.: Hyp. Interact. 108, (1997) 149.
211. E. Mahner: Phys. Rev. ST Accel. Beams 11 (2008) 104801.
212. H. Kolmus, et al.: J. Vac. Sci. Technol. A 27(2) (2009) 245.
213. C. Omet, H. Kollmus, H. Reich-Sprenger, P. Spiller: Proc. EPAC'08 Genoa (2008) 295.

214. http://www.nupecc.org
215. H. Geissel, G. Münzenberg, H. Weick: Nucl. Phys. A 701 (2002) 259.
216. http://www.rarf.riken.go.jp/rarf/acc/history.html
217. H. Kamitsubo: Proc. Cyclotrons'84, East Lansing (1984) 257.
218. A. Goto, et al.: Proc. Cyclotrons'89, Berlin (1989) 51.
219. M. Odera, et al.: Nucl. Instrum. Meth. 227 (1984) 187.
220. Y. Yano: Proc. Cyclotrons'04, Tokyo (2004) 18A1; Y. Yano: Nucl. Instrum. Meth. B 261 (2007) 1009.
221. N. Inabe, et al.: Proc. Cyclotrons'04, Tokyo (2004) 200; T. Mitsumoto, et al.: Proc. Cyclotrons'04, Tokyo (2004) 384.
222. H. Okuno, et al.: IEEE Trans. Appl. Supercond. 17 (2007) 1063.
223. T. Kubo: Nucl. Instrum. Meth. B 204 (2003) 97.
224. O. Kamigaito, et al.: Proc. Cyclotrons'10, Lanzhou (2010) TUM2CIO01.
225. T. Ohnishi, et al.: J. Phys. Soc. Jpn. 79 (2010) 073201.
226. T. Nakamura, et al.: Phys. Rev. Lett. 103 (2009) 262501.
227. P. Doornenbal, et al.: Phys. Rev. Lett. 103 (2009) 032501.
228. S. Nishimura, et al.: Phys. Rev. Lett. 106 (2011) 052502.
229. Ch. Schmelzer, D. Böhne: Proc. Prot. Lin. Accel. Conf. NAL (1970) 981.
230. D. Böhne: Proc. Prot. Lin. Accel. Conf. Los Alamos (1972) 25.
231. D. Böhne: Proc. PAC'77, IEEE Trans. Nucl. Sci. 14(3) (1977) 1070.
232. N. Angert: Proc. PAC'83, IEEE Trans. Nucl. Sci. 30(4) (1983) 2980.
233. L. Dahl: Proc. HIAT'09, Venice (2009) 193.
234. N. Angert, et al.: Proc. EPAC'92, Berlin (1992) 167.
235. U. Ratzinger: Proc. LINAC'96, Geneva (1996) 288.
236. J. Ohnishi, et al.: Proc. Cyclotrons'04, Tokyo (2004) 197.
237. J. Glatz: Proc. LINAC'86, SLAC-Rep. 303, Stanford (1986) 302.
238. H. Geissel, et al.: Nucl. Instrum. Meth. B 70 (1992) 286.
239. B. Franzke, et al.: Proc. EPAC'98, Stockholm (1998) 256.
240. R. Hasse: Phys. Rev. Lett. 83, 3430 (1999).
241. M. Steck, et al.: Phys. Rev. Lett. 77 (1996) 3803; M. Steck, et al.: Proc. PAC'01, Chicago (2001) 137.
242. H. Irnich, et al.: Phys. Rev. Lett. 75(23) (1995) 4182.
243. F. Nolden, et al.: Proc. EPAC'00 Vienna (2000) 1262.
244. M. Hausmann, et al.: Nucl. Instrum. Meth. A 446 (2000) 569.
245. M. Lestinsky et al.: Physics book: CRYRING@ESR, Eur. Phys. J. Spec. Top. 225, 797 (2016).
246. FAIR Technical Design Reports, GSI, Darmstadt (2008).
247. O. Kester, et al.: Proc. IPAC'15, Richmond (2015) 1343.
248. W. Henning: Proc. EPAC'04 Lucerne (2004).
249. P. Spiller, et al.: Proc. EPAC'08, Genoa (2008) 298.
250. U. Ratzinger, et al.: Proc. LINAC'06, Knoxville (2006) 526.
251. H. Geissel, et al.: Nucl. Instrum. Meth. B 204 (2003) 71.
252. M. Steck, et al.: Proc. PAC'09, Vancouver (2009) 4246.
253. R. Toelle, et al.: Proc. PAC'07, Albuquerque (2007) 1482.
254. H.G. Blosser: Proc. Cyclotrons'78, Bloomington (1978), Nucl. Sci. NS-26(2) (1979) 2040.
255. H. Blosser, et al.: Proc. Cyclotrons'86, Tokyo (1986) 157.
256. J.A. Nolen, et al.: Proc. Cyclotrons'89, Berlin (1989) 5.
257. B.M. Sherrill, et al.: Nucl. Instrum. Meth. 56-57(2) (1991) 1106.
258. F. Marti, et al.: Proc. Cyclotrons'01, East Lansing (2001) 64.
259. P. Miller, et al.: Proc. Cyclotrons'04, Tokyo (2004) 62.
260. D.J. Morrissey, et al.: Nucl. Instrum. Meth. B 126 (1997) 316.
261. J. Stetson, et al.: Proc. Cyclotrons'10, Lanzhou (2010) MOA1CIO01.
262. O. Kester, et al.: Proc. SRF'09, Berlin (2009) 57.
263. R.C. York, Proc. PAC'09, Vancouver (2009) 70.

264. http://frib.msu.edu/
265. X. Wu, et al.: Proc. PAC'09, Vancouver (2009) 4947.
266. http://flerovlab.jinr.ru/flnr
267. G. Gulbekyan, et al.: Proc. Cyclotrons'95, Cape Town (1995) 95.
268. Yu. Oganessian: Eur. Phys. J. A 42 (2009) 361.
269. V.V. Bashevoy, et al.: Proc. Cyclotrons'01, East Lansing (2001) 387.
270. G. Gulbekyan et al. Proc. Cyclotrons'16, Zurich (2016), 278
271. W. Zhan, et al.: Proc. Cyclotrons'07, Catania (2007) 110.
272. P.W. Schmor, et al.: Proc. LINAC'04, Lübeck (2004) 251.
273. R.E. Laxdal, et al.: Proc. PAC'01, Chicago (2001) 3942.
274. R.E. Laxdal: Proc. LINAC'06, Knoxville (2006) 521.
275. V. Zvyagintsev, et al.: Proc. RuPAC'10, Protvino (2010) 292.
276. R. Poirier, et al.: Proc. LINAC'00, Monterey (2000) 1023.
277. R.E. Laxdal, et al.: Proc. LINAC'00, Monterey (2000) 97.
278. R.E. Laxdal, et al.: Proc. PAC'05, Knoxville (2005) 3191.
279. R.E. Laxdal: priv. commun.
280. A. Joubert, et al.: Proc. Cyclotrons'84, East Lansing (1984) 3.
281. J. Ferme: Proc. Cyclotrons'86, Tokyo (1986) 24.
282. E. Baron, et al.: Proc. Cyclotrons'95, Cape Town (1995) 39.
283. E. Baron, et al.: Nucl. Instrum. Meth. A 362 (1995) 90.
284. M. Lieuvin, et al.: Proc. Cyclotrons'01, East Lansing (2001) 59.
285. By courtesy of GANIL/SPIRAL2, T. Junquera.
286. F. Chautard, et al.: Proc. EPAC'04, Lucerne (2004) 1270.
287. F. Chautard: Proc.Cyclotrons'10, Lanzhou (2010) MOM2CIO02.
288. M. Lewitowicz: *Acta Physica Polonica B 40* (*2009*) 811.
289. T. Junquera (SPIRAL2 Team): Proc. Linac'08, Victoria (2008) 348.
290. R.C. Pardo, et al.: Proc. PAC'09, Vancouver (2009) 65.
291. M. Pasini (HIE-ISOLDE design team): Proc. SRF'09, Berlin (2009) 924.
292. I. Tanihata: Nucl. Instrum. Meth. B 266 (2008) 4067.

Chapter 12
Outlook for the Future

C. Joshi, A. Caldwell, P. Muggli, S. D. Holmes, and V. D. Shiltsev

12.1 Plasma Accelerators

C. Joshi · A. Caldwell · P. Muggli

12.1.1 Introduction

The charge separation between electrons and ions that exists within an electron plasma density wave can create large electric fields. In 1979 Tajima and Dawson first recognized that the longitudinal component of the field of a so-called "relativistic" wave (one propagating with a phase velocity close to c), could be used to accelerate charged particles to high energies in a short distance [1]. The accelerating gradient of such a plasma wave, E_0, can be approximated—assuming a total separation of

Coordinated by C. Joshi, A. Caldwell

C. Joshi
Los Angeles, CA, USA
e-mail: joshi@ee.ucla.edu

A. Caldwell (✉) · P. Muggli
Max-Planck-Institut, Munich, Germany
e-mail: caldwell@mpp.mpg.de; muggli@mpp.mpg.de

S. D. Holmes · V. D. Shiltsev
Fermi National Accelerator Laboratory, Batavia, IL, USA
e-mail: shiltsev@fnal.gov

S. Myers, H. Schopper (eds.), *Particle Physics Reference Library*,
https://doi.org/10.1007/978-3-030-34245-6_12

electrons and ions in such a wave with wavelength $\lambda_p = 2\pi c/\omega_p$—as

$$E_o \approx \frac{m_e c \omega_p}{e},$$
$$E_o\,(\text{eV/m}) \approx 96.2\sqrt{n_o\,(\text{cm}^{-3})}. \qquad (12.1)$$

Here λ_p is the plasma wavelength, $\omega_p = \sqrt{4\pi e^2 n_o/m_e}$ is the plasma frequency and n_o is the background plasma density, c is the speed of light, m_e and e are the mass and charge of an electron, respectively. The majority of plasma accelerators have been operated at plasma densities $10^{14}\ \text{cm}^{-3} < n_o < 10^{20}\ \text{cm}^{-3}$ giving $10^9 < E_o(\text{eV/m}) < 10^{12}$. Such ultrahigh accelerating gradients are the principal attraction of plasma accelerators.

In a laser-plasma accelerator (LPA), an accelerating structure is created as an intense laser pulse propagates through the plasma driving an electron plasma wave also known as a *wake*. Specifically it is the ponderomotive force, F_p, of the intense laser that displaces the electrons from the heavier ions creating the charge separation and accelerating *wakefield*. The ponderomotive force is related to E_l or A_l, the electric field and vector potential of the laser respectively via $F_p \propto \nabla E_l^2 \propto \nabla A_l^2$. Often the normalized vector potential of the laser driver, a_o, is used to characterize the strength or magnitude of the laser driver and subsequent wake that is driven. The normalized vector potential of the laser is given by

$$a_o = \frac{eA_l}{m_e c^2} = \frac{eE_l}{\omega_o m_e c} \simeq 8.6 \times 10^{-10} \lambda_o\,(\mu\text{m})\sqrt{I_o\,(\text{W/cm}^2)}, \qquad (12.2)$$

where I_o is the focused intensity of the laser pulse, and λ_o and ω_o are the vacuum laser wavelength and frequency respectively. In laser wakefield accelerators, laser intensities of $10^{17} < I_o(\text{W/cm}^2) < 10^{20}$ are commonly used. For $a_o \approx 1$ and laser pulse duration on the order of half-a-plasma wavelength (typically 50–100 fs), wakes with accelerating fields of approximately 100 GeV/m are produced in a $10^{18}/\text{cm}^3$ density plasma. This gradient is more than three orders of magnitude larger than the accelerating gradient in a conventional RF driven accelerator.

It was recognized in the 1980s that wakes in plasmas could also be driven by a relativistic beam of electrons [2]. Given an intense enough beam of electrons, the plasma is both created [3] and excited by the passage of the bunch. Recently, driving the wake with a proton bunch has also been suggested [4].

In a beam-driven plasma wakefield accelerator (PWA), it is the space charge force of a highly relativistic beam of particles which excites the wakefield. For such a bunch, the electric field seen by the plasma electrons is in the transverse direction, and the plasma electrons initially move away from the beam axis, while the more massive ions remain effectively frozen. These transversely expelled electrons are attracted back towards the beam axis by the space charge force of the ions, creating a cavity with very strong electric fields. An appropriately timed witness bunch can be placed in a region of very strong longitudinal component of the electric field of this cavity and accelerated. See Fig. 12.1. The cavity also provides a radial force that

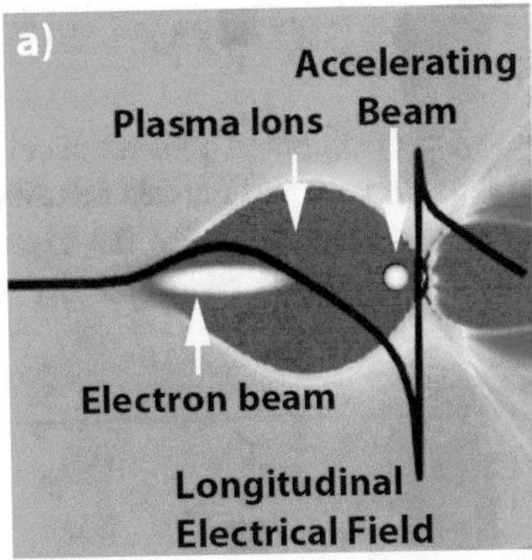

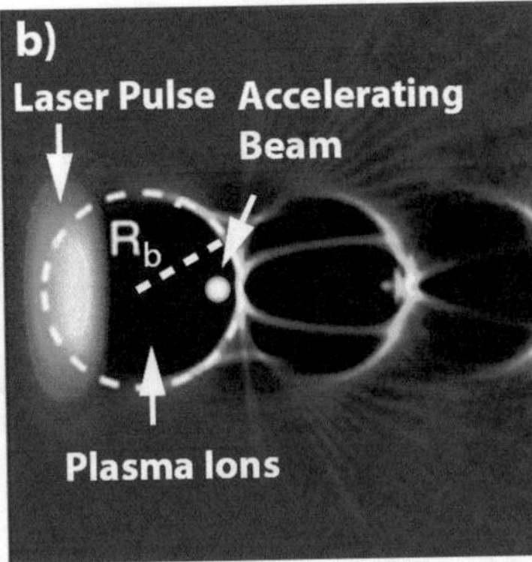

Fig. 12.1 Schematic of the accelerating structures produced in a plasma in the case of (**a**) electron beam-driven plasma wakefield accelerator and (**b**) laser wakefield accelerator (LWFA), both operating in the blow-out regime. In this regime all the plasma electrons are expelled by the electron beam or the laser pulse thereby creating a "cavity" containing only plasma ions. The extremely nonlinear longitudinal electric field generated in both cases is shown as a black curve in (**a**). In the LWFA case the laser spot size is matched to the plasma to produce a near spherical ion cavity with radius R_b equal to the laser spot size. The blow-out regime is attractive because the transverse focusing field increases linearly and the longitudinal accelerating field is constant with radius from the axis

Table 12.1 This table summarizes the scaling laws for the three distinct LWFA regimes characterized by the laser a_o, normalized spot-size $k_p w_o$ and longitudinal electric field amplitude ε_{LW} $(m_e c\omega_p/e)$

Regime	a_o	$k_p w_o$	$\epsilon_{LW}(E_o)$	$k_p L_d$	$k_p L_{pd}$	λ_W	γ_ϕ	$\Delta W(mc^2)$
Linear	<1	2π	a_o^2	$\frac{\omega_o^2}{\omega_p^2}$	$\frac{\omega_o^2}{\omega_p^2}\frac{\omega_p\tau'}{a_o^2}$	$\frac{2\pi}{k_p}$	$\frac{\omega_o}{\omega_p}$	$a_o^2\frac{\omega_o^2}{\omega_p^2}$
1D Nonlinear	>1	2π	a_o	$4a_o^2\frac{\omega_o^2}{\omega_p^2}$	$\frac{1}{3}\frac{\omega_o^2}{\omega_p^2}\omega_p\tau'$	$\frac{4a_o}{k_p}$	$\sqrt{a_o}\frac{\omega_o}{\omega_p}$	$4a_o^2\frac{\omega_o}{\omega_p^2}$
3D Nonlinear	>2	$2\sqrt{a_o}$	$\frac{1}{2}\sqrt{a_o}$	$\frac{4}{3}\frac{\omega_o^2}{\omega_p^2}\sqrt{a_o}$	$\frac{\omega_o^2}{\omega_p^2}\omega_p\tau'$	$\sqrt{a_o}\frac{2\pi}{k_p}$	$\frac{1}{\sqrt{3}}\frac{\omega_o}{\omega_p}$	$\frac{2}{3}\frac{\omega_o^2}{\omega_p^2}a_o$

Here $k_p L_d$ and $k_p L_{pd}$ are the normalized dephasing and pump depletion lengths respectively, λ_W is the plasma wavelength, γ_ϕ is the Lorentz factor associated with the phase velocity of the wake, and $\Delta W(mc^2)$ is the dephasing limited net energy gain in a single stage of a LWFA. τ' is the characteristic laser pulse duration given approximately by $\tau' = \frac{4}{3}\frac{1}{\omega_p}\sqrt{a_o}$. This table has been reproduced from [70]

keeps the witness bunch from expanding radially. A conceptual e^-e^+ linear collider based on plasma acceleration envisions multiple plasma wakefield stages powered by either a laser or a particle beam driver. Each stage adds typically 10 GeV to the accelerating beam. Using a multi-TeV class proton beam as a driver, it may be possible to accelerate an electron beam to energies of interest (TeV) in a single stage.

12.1.2 Physical Concepts

We now discuss some of the important physical concepts in plasma accelerators. These concepts will determine the necessary laser/beam and plasma conditions for the efficient acceleration of electrons. Unless specified otherwise the expressions given in this section are in the limit of small amplitude or sinusoidally varying (linear) wakes.

12.1.2.1 Phase Velocity v_ϕ

The phase velocity, v_ϕ, of the accelerating wake structure is equal to the group velocity of the driver. In the case of a LPA, v_ϕ of the wakefield is equal to $v_g = c\left(1 - \omega_p^2/\omega_0^2\right)^{1/2}$ of the laser pulse as it propagates through the plasma. Often it is also useful to define a relativistic Lorentz factor of the wake $\gamma_\phi = \left(1 - v_\phi^2/c^2\right)^{-1/2} \simeq \omega_0/\omega_p$. For a PWA driven by a highly relativistic electron bunch $v_\phi \approx c$.

12.1.2.2 Dephasing Length L_d

In a LPA, electrons that are accelerated by the wakefield can gain enough energy as to move faster than the phase velocity of the accelerating wake structure. The difference in velocity causes a relative slip in position between the electron beam and the accelerating phase of the wakefield. Eventually the electron beam will slip out of the accelerating phase and into the decelerating phase of the wakefield and begin to lose energy. The length over which the electron beam can gain energy from the accelerating phase of the wakefield is known as the dephasing length, L_d. Within the wake, the accelerating phase extends over approximately half a plasma wavelength and the dephasing length is approximated by assuming that the velocity of the relativistic electrons equals c. The dephasing length is then given by $L_d = c\lambda_p/\{2(c - v_\phi)\}$. This length limits the maximum energy gain that an electron experiences in a LPA.

In a PWA driven by an electron beam, where $v_\phi \approx c$ energy gain is not limited by dephasing. However, in a proton driven PWA the dephasing between the drive and witness bunch must be considered. The phase slippage between the drive and witness bunch has been estimated as [5]

$$\delta \approx \frac{\pi L}{\lambda_p}\left[\frac{1}{\gamma_f \gamma_i}\right], \tag{12.3}$$

where L is the distance traveled, and γ_i and γ_f are the initial and final Lorentz factors of the particles in the driving bunch. With an initial proton energy of 1 TeV and final energy of 0.5 TeV, it is nevertheless possible to have dephasing lengths of many hundreds of meters.

In either LPA or PWA cases, it is conceivable to control the plasma wavelength by adjusting the density of the plasma, thus in part or fully compensating for the phase slippage.

12.1.2.3 Pump Depletion Length L_{pd}

In a LPA, the laser pulse is depleted of its energy as it excites the wake. Additionally, laser light that is not coupled into driving the wakefield is lost due to diffraction/refraction within the plasma. The depletion of energy from the laser pulse limits the distance over which the laser remains intense enough to drive a wakefield. The characteristic distance over which the laser pulse is depleted of it energy is defined as the pump depletion length L_{pd}. In order to gain further energy, two or more laser-plasma accelerator stages must be combined in series. This is known as staging. In a PWA if $\gamma_f \ll \gamma_i$, then the pump depletion length is approximately $\gamma_i mc^2/E_-$, where E_- is the decelerating field of the wake.

12.1.2.4 Injection and Trapping

Once a plasma accelerator has been created, electrons can be *injected* into the accelerator to gain energy. In a LPA, the background plasma typically provides the source for injected electrons. In order to gain a significant amount of energy these injected plasma electrons must be *trapped* by the wakefield. An electron is considered to be trapped once it has gained enough energy to have a longitudinal velocity equal to the phase velocity of the wakefield. After an electron is trapped it can remain in the accelerating phase of the wakefield and continue to gain energy until it travels a dephasing length. Next different methods of electron injection are introduced.

In a PWA, a relativistic witness beam is injected at the appropriate phase with respect to the drive beam in order to experience the accelerating phase of the wakefield. Since both the witness beam and the wake propagate at c, there is no relative phase slippage.

Self-Injection
In a thermal plasma, (such as those typically created using lasers with joules of energy and pulse durations on the ps-ns time scale), some electrons in the tail of the electron distribution function may have sufficiently high momentum so as to reside on a "trapped-orbit". This occurs when the electron velocity is equal to v_ϕ in the plasma wake potential.

In a cold plasma, plasma electrons from outside the wake can cross into or be self-injected into the accelerating phase of the wake. If the wake amplitude is large enough, these injected electrons can quickly gain enough momentum from the accelerating field to move with the wake and become trapped [6, 7]. Self-injection process can be facilitated in a density downramp [8] or at a sudden transition between high and low density plasma regimes [9].

Downramp Injection
If the drive laser pulse or the particle bunch traverses across a plasma downramp [9] some of the plasma electrons forming the sheath around the plasma ions of a 3D nonlinear wake (discussed later) can be injected into the wake. This happens because the wavelength of the wake increase as the drive bunch goes from higher to lower density causing some electrons to cross the sheath and gain sufficient longitudinal velocity to move synchronously with the wake as they reach the back of the wake. These electrons can also have a very small transverse emittance because their transverse velocity is reduced to near zero by the strong defocusing field they feel as they converge on the axis of the wake.

Colliding-Pulse Injection
In a LPA, if a second laser pulse is collided with the laser pulse that induces the wake, then the ponderomotive force associated with the beating of the two pulses can impart enough longitudinal momentum to the initially untrapped electrons so that they are injected into the separatrix of the wakefield to be trapped [10, 11].

Ionization Injection
Electrons can be injected into a fully formed wake via ionization-trapping [12, 13]. A binary mixture of atoms is ionized by the field of the laser or space charge of the driving bunch. The minority atoms in this mixture typically have a step in the ionization potential such that the inner electrons are ionized close to the wake axis, near the peak field of the laser pulse or beam driver. These electrons experience a much greater wake potential so as to be trapped by the wake.

External Injection
In analogy to a conventional traveling wave accelerating structure, a witness beam of electrons or other charged particles may be pre-accelerated and injected into a plasma accelerating structure to increase their energy [14]. Within the wake the magnitude of accelerating field varies with space. Therefore, to minimize the difference in the accelerating gradient that is experienced, the injected electron beam should ideally have a longitudinal length that is much less than the plasma wavelength of the accelerator. For a plasma density of 10^{18} cm^{-3}, $\lambda_p \sim 30$ μm ($\sim$100 fs). Additionally, the external electron beam must be synchronized to the accelerating structure such that it is injected into the correct accelerating and focusing phases of the wakefield.

12.1.2.5 Net Energy Gain ΔW

For a LPA, The net energy gain of an electron ΔW, can be estimated as

$$\Delta W = eE_{LW}L_{acc} = \left(\frac{E_{LW}}{E_o}\right)\left(\frac{\omega_p}{c}L_{acc}\right)mc^2. \tag{12.4}$$

Here E_{LW} is the local longitudinal accelerating field seen by the electron of the wake, and L_{acc} is the distance over which the electron interacts with the accelerating field.

12.1.2.6 Beam Loading

The accelerated particles can modify the wakefield in a process known as beam loading [15, 16]. It can be understood as destructive interference between the fields of the wake and fields of the accelerated particles. Beam loading places limitations on the number of particles, the peak current, the energy spread and duration of the accelerated particles. The plasma-accelerator can be optimized for one or more of these parameters. The maximum number N_o of electrons that can be loaded into the accelerating phase of the wake is [15]

$$N_o = 5 \times 10^5\,(E_1/E_o)\,n_o^{1/2}\left[\mathrm{cm}^{-3}\right]S\left[\mathrm{cm}^2\right], \tag{12.5}$$

where $S \gg \pi/k_p^2$ is the cross sectional area of the wake. As $N \rightarrow N_o$ the energy spread $\Delta\gamma/\gamma \rightarrow 1$. From energy conservation arguments the beam loading efficiency scales as $(N/N_o)(2 - N/N_o)$.

12.1.2.7 Drive Pulse Evolution

In a LPA, plasma can act as a lens to focus the spot size of the laser in a process known as relativistic self-focusing [17]. Relativistic self-focusing of the laser pulse occurs when the power of the laser pulse exceeds P_c, the critical power for self-focusing given by

$$P_c(GW) \simeq 17.4\frac{\omega_o^2}{\omega_p^2}. \tag{12.6}$$

The laser pulse will continue to focus until all the electrons are expelled from within the spot size of the laser and there is no longer a gradient in the index of refraction in the radial direction. The ratio of the laser power, P, to P_c gives a good indication of how strongly the laser pulse will be focused by the plasma.

The laser pulse can also be compressed longitudinally in time by the density gradient of a wake via photon acceleration and deceleration [18, 19]. If the laser pulse width is on the order of λ_p the entire pulse will be compressed and this can lead to an increase the laser intensity and a_o [20]. However, if the laser pulse width is much greater than a plasma wavelength, the laser pulse will be compressed within each wake driving a laser-plasma accelerator in the self-modulated regime.

In a PWA, the particles in a symmetric drive pulse lose their energy to the wake at a different rate. For ultra-relativistic electrons, the pulse shape of the drive beam does not evolve longitudinally since there is insignificant relative motion between the particles. But for a proton drive beam the spatial spread of particles d due to momentum spread will induce a lengthening of the bunch. The effect can be evaluated for vacuum propagation as

$$d \approx \frac{L}{2\Delta\gamma^2} \approx \left(\frac{\sigma_p}{p}\right)\frac{M_p^2c^4}{p^2c^2}L, \tag{12.7}$$

where M_p is the proton mass, and p is the proton momentum. Given a 1 TeV proton beam, a 10% momentum spread leads to a growth of ~0.1 μm/m. Large relative momentum spreads will still allow for long plasma acceleration stages provided the drive beam is relativistic.

12.1.2.8 Guiding

The length of the accelerating structure can limit the energy gain of the particles. For a LPA the length of the accelerator (assuming Gaussian optics) is approximately equal to π times the Rayleigh length of the focused laser pulse, $z_R = \pi w_o^2/\lambda_o$, where w_o is the spot size of the laser. Often, in order to reach the intensities necessary to drive large amplitude wakefields, the laser must be focused to a spot sizes on the order of ~10 μm. This limits the accelerating structure length to ~1 mm and subsequently severely limits the energy gain of electrons if the L_d is longer than this.

To overcome this limitation, a channel can be created that has an appropriate radial plasma density profile such that the diffraction of the laser pulse is minimized. This allows the spot size of the laser pulse to remain small, or 'guided', over the length of the channel that is much greater than a Rayleigh length. For a channel with a radial parabolic electron density profile, a change in electron density of $\Delta n = 1/\left(\pi r_e w_o^2\right)$ is required to optical guide the laser pulse, where here $r_e = e^2/m_e c^2$ is the classical electron radius. Two main methods have been used to guide short laser pulses.

In the first method, an electrical discharge is struck across a tube, or capillary containing hydrogen gas creating a plasma. The electron temperature close to the capillary wall is lower than that near the center of the discharge. Therefore over a short amount of time (~100 ns), hotter electrons located on axis of the discharge will diffuse radially, creating an appropriate density channel that can guide a laser pulse [21].

A second method, known as self-guiding, relies on matching the ponderomotive force of the laser, which expels the plasma electrons and drives the wake, to the space-charge attraction force that the plasma ions exert on the expelled electrons. When this is achieved, the radial plasma density profile of the wake that is created has the appropriate shape and depth to minimize the diffraction of the laser pulse [22, 23].

Both of these techniques have been used to extend the length of a laser-plasma accelerator from hundreds of microns to centimeter scale lengths [24–26].

In a PWA, if the density-length product of the plasma is large enough, the electron drive pulse can be guided by the transverse focusing force provided by the plasma ions. If the beam density $n_b > n_o$, the beam electrons can blow-out all the plasma electrons creating an ion channel. In this blow-out regime, the beam envelope is described by the differential equation [27]:

$$\sigma_r'' + \left[K^2 - \epsilon_{Nr}^2/\gamma^2\sigma_r^4(z)\right]\sigma_r(z) = 1, \tag{12.8}$$

where $K = \omega_p/(2\gamma)^{1/2}c$ is the restoring force provided by the plasma ions or equivalently the betatron wavenumber $k_\beta = \omega_\beta/c$ where ω_β is the betatron frequency. The beam is said to be matched to the plasma if $\beta_{beam} = 1/K = \beta_{plasma}$. In this case the beam propagates through the plasma with a constant radius [28]. The matched beam radius r_b is found by letting $\sigma_r''(z) = 0$ in this equation giving $r_b = (\epsilon_N/\gamma k_\beta)^{1/2}$. The beam now propagates through the plasma until it is either pump depleted or its front is slowly eroded away by finite emittance of the particles.

12.1.2.9 Head Erosion

In a LPA, the front portion of the laser pulse will diffract at a rate of some fraction of c/ω_p per z_R unless the pulse is guided in a pre-formed plasma channel [23, 29]. This head erosion will eventually limit the distance over which a wake can be excited. Similarly, in a PWA, the head of the drive bunch propagates in a region of no wakefield, so that the beam emittance will eventually erode the drive bunch [30]. A lower emittance drive beam can reduce head erosion as an obstacle limiting the energy gain, thus extracting more energy from the bunch. Proton beams have larger emittances than electron beams, therefore longer beams, lower plasma densities and thus long plasma cells are envisaged for efficient energy extraction. Therefore strong magnetic focusing of the proton drive bunch along the length of the plasma channel is likely necessary.

12.1.2.10 Instabilities, Scattering, and Radiation Loss

Generally short laser/beam drivers are relatively immune to laser-plasma instabilities because both the driver and the wake are continuously entering undisturbed plasma at $\sim c$. Nevertheless, there are two instabilities that can grow the emittance of the beam being accelerated. These are laser and electron beam hosing instability [31] and plasma ion motion [32].

In the hosing instability any transverse offset of the electron beam from the axis of propagation of the wake or any head-to-tail tilt can grow. The growth rate depends on the initial offset of a particular slice, how far the slice is from the front of the bunch and how far the bunch has propagated into the plasma. In the nonlinear 3D regime the instability involves a coupling of the centroid of the wake cavity and the bunch. In plasma accelerators the ions are usually assumed to be immobile because they are far more massive than the electrons. However if the accelerating bunch density exceeds the plasma density by a factor (mi/me), the focusing force exerted by the electrons on the plasma ions becomes so large that the plasma ions implode inward toward the axis. This can locally alter the linear focusing force provided by the plasma ions and affect the emittance of the accelerating electrons.

Coulomb scattering in the plasma can increase the emittance of the witness bunch. In plasma accelerators, typical values of n_o will be in the range of $10^{14} - 10^{17}$ cm^{-3}, and both the radiation length and the mean-free-path are orders of magnitude larger than the expected plasma length on the order of 100 m. A 1 TeV proton beam in Li vapor of density 1×10^{15} atoms/cm^3 gives a transverse growth rate of the proton beam of less than 0.01 μm/m due to multiple scattering, which is small compared to the size of the drive bunch. Multiple scattering for high energy electrons will also be small.

In addition to these instabilities, as particles are accelerated they will oscillate transversely emitting betratron radiation. The energy lost to radiation per unit is given by [33, 34]:

$$\frac{dW_{loss}}{dz} = \frac{1}{3} r_e \gamma_b^2 \omega_\beta^2 K_w^2. \tag{12.9}$$

Here $K_w^2 = \gamma_b \omega_\beta r_o / c$ is the effective wiggler strength of the ion column. At extremely high energies, the betatron radiation loss rate will equal the rate particles gain energy. This ultimately limits the maximum energy gain in a plasma accelerator.

12.1.3 Beam Driven Plasma Wakefield Accelerators

12.1.3.1 Electron Beam Driven PWA

The basic concept of the plasma wakefield accelerator involves the passage of an ultra-relativistic charged particle bunch through a stationary plasma. The plasma may be formed by ionizing a gas with a laser or through field-ionization by the Coulomb field of the relativistic bunch itself. This second method allows the production of meter-long, dense ($10^{16} - 10^{17}$cm^{-3}) plasmas suitable for the PWA and greatly simplifies the experimental set-up. In single bunch experiments [35] carried out with ultra-short electron bunches, the head of the bunch creates the plasma and drives the wake. The wake produces a high-gradient longitudinal field that in turn accelerates particles in the back of the bunch. The system effectively operates as a transformer, where the energy from the particles in the bulk of the bunch is transferred to those in the back, via the plasma wake. The physics is unchanged if there are two bunches rather than a single bunch; energy from the leading drive bunch is transferred to a trailing witness bunch. The maximum energy which can be given to a particle in the witness bunch in a PWA is limited by the transformer ratio, defined as

$$R = \frac{\Delta W_{max}^{witness}}{\Delta W_{max}^{drive}} \leq 2 - \frac{N_{witness}}{N_{drive}} \,, \tag{12.10}$$

which is at most two for longitudinally symmetric drive bunches and an unloaded wake ($N_{witness} \ll N_{drive}$) [36]. Here ΔW_{max} is the change in energy of the particles in the drive/witness beam and N is the number of particles in the witness/drive beam. This upper limit can in principle be overcome by nonsymmetric bunches [37]. Another option currently under study is to use an appropriately phased train of bunches so that the wakefield is increased with each bunch [38]. According to linear plasma theory [39], maximum accelerating field of the wake is given by [40]:

$$eE_{linear} \simeq 100 \frac{N}{2 \times 10^{10}} \frac{20}{\sigma_z\ (\mu\mathrm{m})} \ln \sqrt{\frac{2.5 \times 10^{17}}{n_o\ (\mathrm{cm}^{-3})} \frac{10}{\sigma_r\ (\mu\mathrm{m})}}\ \mathrm{GeV/m}, \tag{12.11}$$

where N is the number of particles in the electron bunch and σ_z is the bunch length, and σ_r is the spot size. Linear theory is valid when the normalized charge per unit length of the beam, $\Lambda = \frac{n_b}{n_o}\left(k_p \sigma_z\right) \simeq 2.5 \left(\frac{N}{2\times 10^{10}}\right)\left(\frac{20}{\sigma_z(\mu\mathrm{m})}\right)$ is less than 1. Equation (12.11) indicates that generating large gradient wakefields requires short, high density electron bunches. For high Λ the nonlinear equivalent of Eq. (12.11) should be used [41].

In a PWA operating in the so-called blowout regime, a short but high-current electron bunch, with density n_b larger than the plasma density n_p, expels all the plasma electrons from a region surrounding the beam as shown in Fig. 12.1. The expelled plasma electrons rush back in because of the restoring force of the relatively immobile plasma ions and thus generate a large plasma wakefield. This wakefield has a phase velocity equal to the beam velocity ($\approx c$ for ultra-relativistic beams). Since the electrons in the bunch are ultra-relativistic, there is no relative phase slippage between the electrons and the wake over meter-scale plasmas.

When the wakes are in the blowout regime, shaped bunches can be used to transfer a substantial amount of energy from the drive bunch to the wake and optimize the energy extraction efficiency from the wake to the trailing bunch while not further increasing its energy spread. It can be shown that particles in a trapezoid shape drive bunch—with beam current increasing from the front to the back—loose energy at a near constant rate. A trailing bunch with a reverse shape can load the wake in such a way that nearly all the particles gain energy at the same rate. Significant beam loading will occur [42] when the loaded charge

$$Q(nC) > [(0.047mcp)\,/eEs]\,(1016/np\,(cm - 3))\;1/2(kpRb)4 \qquad (12.12)$$

Here, we take the normalized electric field eEs/mcp = kpRb/2 as the electric field seen by the accelerating electrons, kp−1 is the plasma skin depth, and Rb is the blowout radius of the wake.

12.1.3.2 Short Proton and Positron Beam Driven PWA

In contrast to plasma wakes driven by bunches of electrons, only limited investigations of the plasma wave excitation by a positively charged driver exist [42–46]. For a positively charged driver (such as a proton bunch or a positron bunch), plasma electrons are attracted towards the axis of the driver, overshoot and setup the accelerating wake structure [47, 48]. For positively charged drivers with $n_b/n_o \ll 1$ the electric field distribution of the wake is the same as that for the negative driver but shifted in phase. However, for a positively charged driver driver with $n_b/n_o \gtrsim 1$ the wake behavior is significantly different than for a negatively charged driver with the same beam density. Due to the radial symmetry, for the wakefield driven by a strong positively charged driver ($n_b/n_o \geq 1$), there is an electron density enhancement on-axis and effective increase of the local plasma frequency. As a result, the proton or positron driver beams must be even shorter in order to excite the plasma wake resonantly. However, given that protons can be accelerated to the TeV regime in conventional accelerators it is conceivable to accelerate electron bunches in the wake of the proton bunch up to several TeV (e.g., in the wake of a Large Hadron Collider (LHC) proton beam) in a single plasma stage.

12.1.3.3 Long Proton Beam Driven PWA

Proton bunches available today are in principle very interesting to drive wakefields because, besides being relativistic, they also have a large population and thus carry large amounts of energy (10s to 100s of kJ). A proton bunch can therefore drive wakefields over a very long single plasma and lead to very large energy gain of a witness bunch. It was shown through simulations [4] that a 10 GeV electron bunch injected into wakefields driven by a 100 μm-long, 10 kA, 1 TeV proton bunch can reach an energy of ~0.5 TeV in a single plasma, only ~300 m-long. This corresponds to an average accelerating gradient larger than 1 GeV/m in a plasma with density of 6×10^{14} cm^{-3}. We note here that in these simulations the proton bunch length σ_z is shorter and the width σ_r is narrower than the wakefields period λ_p, a condition necessary to effectively drive wakefields. These two conditions are usually expressed as $k_p\sigma_z \leq 1$ and $k_p\sigma_r \leq 1$.

Proton bunches produced for example by the CERN Super Proton Synchrotron (SPS) or Large Hadron Collider (LHC) are long: σ_z~6–12 cm.

Compression of proton bunches produced by these synchrotrons is in principle possible [49]. However, it would first require means to impart a %-level correlated energy chirp along the already high-energy bunch. Second, the magnetic compressor (chicane) would have to be rather long (km), again because of the high energy of the particles. Therefore, producing short proton bunches such as the one used in the simulations of [4] remains a challenge.

Matching the plasma density to the long SPS proton bunch ($k_p\sigma_z \cong 1$) would mean using a low density plasma (~10^{11}cm^{-3}) and driving low amplitude wakefields (~10s of MV/m range from E_0 in Eq. 12.11).

However, proton bunches can be focused to small transverse sizes (e.g., $\sigma_r = 200$ μm). Choosing the plasma density to reach $k_p\sigma_r \cong 1$, leads to a much larger plasma density (7×10^{14}cm^{-3}) and also much larger possible accelerating field (E_0 ~ 2.5 GV/m, Eq. 12.11). This choice of $k_p\sigma_r \cong 1$ means $k_p\sigma_z \gg 1$ since $\sigma_z \gg \sigma_r$. Such bunches are subject to a symmetric transverse self-modulation process [50]. The self-modulation process transforms the long continuous bunch into a train of short bunches separated by the wakefields period. The short bunches are formed by the action of the periodically focusing/defocusing transverse wakefields. The train can then resonantly drive wakefields to amplitudes on the order of the 2.5 GV/m predicted by Eq. 12.11. Furthermore, the process can be seeded, so that it becomes a well-controlled, seeded self-modulation (SSM) process [51]. In this case the beam-plasma system is turned into an amplifier of seed wakefields, with well determined final fields amplitude *and phase*. The process is also weakly sensitive to variations of the input bunch parameters (±5%) [52]. The SSM process makes it possible to deterministically inject an electron bunch where wakefields reach their peak amplitude and into the range of accelerating and focusing phase of the wakefields (corresponding to region ~$\lambda_p/4$ long), a large number of periods (~σ_z/λ_p) behind the seed point [53]. Simulation results show that with this scheme, electrons can be accelerated to multi-TeV energies in a plasma a few km-long [54].

12.1.4 *Laser-Driven Plasma Accelerators*

12.1.4.1 Plasma Beat Wave Accelerator (PBWA)

Two co-propagating long ($c\tau \gg \lambda_p$) but moderate intensity ($a_o < 1$) laser beams with frequencies ω_1 and ω_2 can resonantly excite a relativistic plasma wave when $\omega_1 - \omega_2 \cong \omega_p$ [55]. Here τ is the pulse width of the laser pulse. For a square pulse, the amplitude of the plasma wave grows linearly in time and eventually saturates due to change in plasma frequency caused by relativistic mass increase of the plasma electrons. For laser pulses with $a_o \ll 1$, the maximum wave amplitude is given by $E_1/E_o = (4a_{o1}a_{o2}/3)^{1/3}$ and occurs when $\omega_1 - \omega_2 = \omega_p(1 - (a_{o1}a_{o2})^{2/3}/8)$. The relativistic plasma wave is prone to modulational instability [56], and mode coupling

[57]. For linear wakes, electrons need to be externally injected for acceleration [58, 59].

12.1.4.2 Self-Modulated Laser-Wakefield accelerator (SM-LWFA)

An accelerating wakefield can be driven by a single long laser pulse ($c\tau > \lambda_p$) via the Raman forward-scattering (RFS) instability [60] if $\omega_o > 2\omega_p$ and the laser power $P \cong P_c$ the critical power for relativistic self-focusing. In this process the laser pulse amplitude is modulated at ω_p by the creation of frequency sidebands at $\omega_o \pm \omega_p$. The spatiotemporal gain for RFS instability [61] has been estimated as $G = e^g(2\pi g)^{-1/2}$, where g is given by

$$g = \frac{a_o}{\sqrt{2}}\left(1 + \frac{a_o^2}{2}\right)\left(\frac{\omega_p^2}{\omega_o^2}\right)\frac{\omega_o}{c}\sqrt{x\phi}, \tag{12.13}$$

where x is the length of the plasma, and ϕ/c is the time over which the plasma experiences a constant laser intensity. The normalized plasma wave amplitude is $\delta n/n_o = \alpha_n G$, where α_n is the initial wave amplitude associated with the thermal noise of the plasma. When $\alpha_n G \approx 1$ the wave amplitude becomes large enough to self-trap and accelerate electrons [62, 63].

12.1.4.3 Laser Wakefield Accelerator (LWFA)

A laser wakefield accelerator (LWFA) is a laser-plasma accelerator which is created directly by the ponderomotive force of a single short ($c\tau \sim \lambda_p$), intense laser pulse. With the advent of Ti:sapphire laser systems, such ultra-short, ultra-intense laser pulses are readily available. An advantage of using such short laser pulses to drive an accelerating wake is that they are relatively immune to most laser-plasma instabilities. Additionally, these short, intense laser pulses can drive wakefields with large amplitudes, that are capable of self-trapping and accelerating electrons to high energies [64–67]. As a result of this, laser wakefield accelerators are the focus of the majority of current laser-plasma accelerator research. Next, details on a few specific regimes in LWFA research are presented.

12.1.5 LWFA Regimes

12.1.5.1 Linear Regime

The wake induced by a short laser pulse ($c\tau \sim \lambda_p$) is said to be in the linear regime if $\delta n/n_o < 1$, where δn is the change in electron density associated with the wake. Laser drivers with modest intensities $a_o < 1$ and $P/P_c \leq 1$, excite wakefields in the linear regime. In the linear regime the electron density response and the longitudinal electric field of the wake vary sinusoidally. The transverse and longitudinal fields are $\pi/2$ out of phase with one another. Thus there is a quarter wavelength region where the fields are both accelerating and focusing for the particles.

12.1.5.2 Nonlinear Regime

A nonlinear wake is created when electron density perturbation of the wake is on the order of or greater than the background plasma density ($\delta n/n_o \gtrsim 1$) and the longitudinal accelerating field of the wake becomes larger than E_o. Such nonlinear wakes are typically created by using a laser driver with an $a_o \gtrsim 2$ or with a beam driver with $n_b/n_o \gg 1$ and a $k_p\sigma_r \ll 1$. A characteristic of a nonlinear wake is a "saw toothed" shape longitudinal electric field profile. Additionally, the plasma wavelength of a nonlinear wake increases as the a_o of the laser (wake amplitude) is increased. A nonlinear wake is said to be in the 1D regime if the normalized laser pulse spot is broad, i.e., $k_p w_o \gg 1$. However, when $k_p w_o \sim 1$, and $a_o > 1$ one approaches the 3D-nonlinear regime.

The electron density response and corresponding fields of nonlinear wakes driven by laser pulses in 3-D regime are difficult to study analytically. To gain a better understanding in this regime, particle in cell (PIC) codes have been used to model the LWFA in this regime. Early work in this regime focused on simulating and developing scalings to describe the dynamics of a singular nonlinear wake driven in the so called bubble regime by a laser pulse with an $a_o \geq 2\omega_o/\omega_p$ [68, 69]. More recently a nonlinear regime in which a laser pulse with an $2 \leq a_o \leq 2\omega_o/\omega_p$ is used to drive a periodic nonlinear wake, in the so called blowout regime, has been studied with simulations, a phenomenological theory and experimental investigation [23, 25, 26, 41, 70–73]. In both the bubble and blowout regimes, the ponderomotive force of the laser pushes out all the electrons from within the first period of the wake creating an ion bubble into which background plasma electrons are self-injected, trapped and accelerated.

An advantage of operating in the blowout regime is that there exits a matched self-guiding condition, in which the intensity, spot size and pulse width of the laser pulse are matched to the plasma density, such that the driven wake serves to minimize the diffraction of the laser pulse [70]. This allows a strong and stable wake to be sustained over tens of Rayleigh lengths and can increase the interaction length between trapped electrons and the accelerating field of the wake. The matched self-guiding condition is valid when the laser $a_o \gtrsim 2$ and $c\tau \sim \lambda_p/2$. When these conditions are met, the matching condition is given by

$$k_p w_o \simeq k_p R_b = 2\sqrt{a_o}, \tag{12.14}$$

where R_b is the blowout radius of the spherically shaped wake. Remarkably, even when the spot size and or pulse width are not precisely matched to the plasma density, the laser pulse evolves within the plasma towards the matched spot size and pulse width, via relativistic self-focusing and longitudinal pulse compression [67, 72, 73].

Simulations indicate that in the blowout regime self-trapping of background electrons will occur for large amplitude wakefields when the normalized blowout radius $k_p R_b \approx 4 - 5$.

12.1.5.3 Scaling Laws

The scaling laws for a LWFA in the regimes which were discussed above are now presented.

12.1.6 Status

12.1.6.1 LWFA and PWA

The development of laser-plasma accelerators has closely followed the progress of high-power, short pulse lasers [74]. Prior to 1990, laser-plasma accelerator experiments focused primarily on the PBWA [14] or SM-LWFA concepts [54, 56]. These demonstrated the existence of greater than many GeV/m gradients and acceleration of self-injected as well as externally injected electrons [14, 69]. The accelerated particles typically had an exponentially falling energy spectrum [57]. In the mid 1990s, the chirped pulse amplification (CPA) scheme was implemented using titanium-sapphire as a gain medium and enabled the development of terawatt-class, sub-100 fs laser systems. Such short pulses (cp) were ideal for exciting wakes in underdense plasmas. Through better control of plasma density and length, and laser pulse parameters, electron bunches with a significant charge and relatively narrow (¡20%) energy spread were observed [58–60]. The electron beam energy spread and charge were controlled further by employing the colliding pulse injection technique [10, 11]. The CPA technique has extended the peak power reach of the Ti-sapphire lasers to multi-petawatt (PW) levels. For instance at Lawrence Berkeley National Laboratory (LBNL) a petawatt laser facility became operational as part of the Berkeley Lab Laser Accelerator (BELLA) project [74]. Using a plasma-channel to extend the acceleration distance, narrow energy electron bunches with energy up to 4.3 GeV have been observed [24] at LBNL. Similar peak power laser facilities have either come on line (for instance at GIST (Korea)) or expected to soon be operational in several institutes worldwide. Most of the LPA experiments are being conducted using the highly nonlinear 3D (sometimes called the blow-out or bubble) regime using either self-injected or ionization injected charge into the plasma wake. For example electrons were accelerated to beyond 1.4 GeV using ionization injection into a LPA operating in the nearly matched, self-guided blowout regime [70]. At present, several groups are exploring this regime to obtain electron beams for the generation of betatron and undulator radiation [71–73]. In past few years much effort has been devoted to controlling the charge, minimizing the transverse emittance and energy spread of electrons created from laser wakefield accelerators [74]. New techniques for beam injection such as downramp injection [74] and transverse collide pulse injection [74] have been proposed for generating electron beams with sub 100 nm transverse emittances. Possible applications for such beams might be fifth generation light sources (ultra compact XUV and X-FELs) and high-energy electron therapy of cancerous tumors. As in the past, continued progress in this field will be tied to progress in making lasers more reliable, efficient and cheaper. Additionally, to extend the interaction between the electron beam and the LPA progress must be made in making reproducible, meter-scale plasma sources.

Experiments on beam-driven PWA have been going on at Argonne National Laboratory (ANL) [75], Fermilab [76], KEK [77], Brookhaven National Laboratory (BNL) [78] and at SLAC [79]. New facilities such as FLASHForward at GSI, Germany are being commissioned. With the exception of the SLAC experiments, all other work has hitherto used modest energy electron beam (¡100 MeV) to study both acceleration and focusing effect of plasmas on beams. Experiments at SLACs FFTB facility were initially carried out using 28 GeV electron and positron beams. These used 4 ps long beams containing more than 10 kA current to systematically study beam propagation, focusing, betatron radiation emission, and acceleration in meter-scale, laser-ionized plasmas [80, 81]. These were followed by experiments that used a sub-50 fs beam to both produce the plasma via tunnel ionization and excite the wakefield in much denser, meter-scale vapor columns. The front portion of the bunch was used to excite the field while the particles in the back of the same bunch were used to sample to accelerating wake. An example of data on changes in electrons energy in the above mentioned PWA experiment at SLAC is shown in Fig. 12.2. Most of the initially 28.5 GeV electrons in the pulse are seen to lose energy but electrons in the back of the same pulse are accelerated by the wake as its electric field changes sign. The energy gain is seen to increase with distance. When the electron beam energy was increased to 42 GeV some electrons were seen to gain more than 42 GeV in just 85 cm, implying that a gradient of 50 GeV/m was sustained throughout the plasma [82].

In 2006 the FFTB facility was replaced by a new experimental facility called FACET for Advanced Acceleration research [83]. Between 2019 and 2016 FACET provided electron and positron capability with a driver beam-witness beam structure so that high gradient acceleration of a significant number of charged particles with a narrow energy spread could be demonstrated. The drive and witness beams were only about 50 fs long and were separated by about 300 fs. In order to preserve its narrow energy spread the witness beam had to beam-load the wake so that particles in the witness beam flattened the electric field of the wake thereby experiencing a nearly uniform accelerating field. With this loading, the energy contained in the wake is efficiently transferred to the accelerating particles. In a landmark experiment carried out at FACET, it was shown that the efficiency of transferring drive bunch energy to the core of the accelerated bunch was up to 30% [84]. In a follow up campaign an energy gain of 9 GeV for a bunch containing 30 pC of charge with a 5% energy spread in a 1.2 m long plasma was observed [85]. Using the energy and

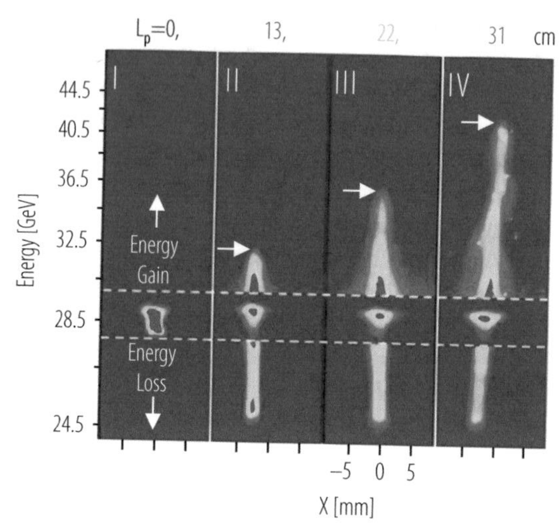

Fig. 12.2 Changes in electron energy in the PWA experiment at SLAC

the spot size changes experienced by different slices of the drive bunch itself, the PWA cavity in the nonlinear blowout regime was shown to have a longitudinal and transverse field structure that in principle will accelerate electrons without emittance growth, as long as the electrons (to be accelerated) are matched in and out of the plasma with minimal energy spread [86].

The plasma wake produced by an electron bunch cannot be used to accelerate a positron beam when the wake is in the nonlinear blow out regime because the plasma ions strongly defocus the positrons. Until recently it was not very clear how efficient positron acceleration at a high gradient could be carried out using highly nonlinear plasma wakes. Experiments at FACET showed that a certain positron beam current profile can lead to a loaded wake where the longitudinal electric field reverses sign (from decelerating to accelerating) in the middle of the single drive bunch [87]. This happens because the presence of the positrons pulls in the plasma electrons towards the axis. These plasma electrons can cross the axis in the middle of the drive bunch. Most of the electrons overshoot and set up a bubble like wake cavity but a significant fraction of the electrons are confined by the back of the positron beam close to the axis. This flattens the wake shape by beam loading [87]. A significant amount of positron charge can now be accelerated at the same electric field gradient producing a well-defined narrow energy spectrum. The energy extraction efficiency is similar to the electron bunch acceleration case described above.

In 2016, FACET ceased operation to make way for the LCLS II facility that will occupy the first 1 km space of the original SLAC linac tunnel. A new facility for advanced accelerator research, known as FACET II, is being constructed between the LCLS II linac and the LCLS linac. Together with the BELLA laser the FACET II facility [88] will arguably be the backbone of research facilities for short drive pulse driven plasma accelerator research for the next decade. The foremost physics challenge is the generation of collider quality transverse and longitudinal emittance bunches and identifying the factors that cause emittance growth in plasma accelerators [88]. Transverse emittance growth may occur as a result of the hosing instability and also ion motion [31, 32]. Even if the electrons have a very small emittance inside the wake, such a beam must be extracted and matched either to another plasma acceleration stage or to conventional magnetic optics [89–91] to avoid emittance growth. Longitudinal emittance will depend on how small the energy spread of the beam can be, which in turn depends on optimizing the beam loading to give a constant accelerating field. Continued development of diagnostic techniques to visualize the fields of the highly nonlinear wake and the injected electrons is also needed. Issues of generating asymmetric emittance flat beams and spin polarized beams are still open questions. Creative solutions for the generation and acceleration of collider quality positron bunches using plasmas are needed. In the near future it is highly likely that a single stage of LWFA and a PWA will produce 10 GeV bunches with a percent level energy spread and a high efficiency.

12.1.6.2 Proton-Driven Plasma Wakeeld Acceleration

The AWAKE experiment at CERN studies the driving of plasma wakefields with proton bunches [53, 90, 91]. AWAKE uses the (6–12)cm-long bunch delivered by the SPS with (1–3)$\times 10^{11}$, 400 GeV protons focused to a transverse rms size of $\sigma_r = 200$ μm. The plasma density is chosen so that the accelerating field can reach ~1 GV/m, i.e.: $n_0 > 10^{14}$ cm^{-3} corresponding to $\lambda_p < 3$ mm. Since the proton bunch is long when compared to the wakefield period, the experiment relies on the seeded self-modulation (SSM) process [51] to reach these wakefield amplitudes.

A 10 m-long rubidium source was developed to produce a column of vapor with a very uniform density ($\delta n_{Rb}/n_{Rb} < 0.2\%$) and with sharp density ramps (<10 cm) at the entrance and exit [92]. A 450 mJ, 100 fs laser pulse propagating within with the proton bunch creates a sharp (< 100fs), relativistic ionization front that seeds the SSM process. With full ionization of the only electron of the rubidium atom outer shell, the plasma density longitudinal profile is identical to that of the rubidium vapor. The plasma radius is on the order of 1 mm. The plasma density is adjustable from 1×10^{14} to 10×10^{14} cm^{-3}.

The occurrence of the SSM on the proton bunch is observed with three diagnostics [51]: a two-screen method [93], the time resolved emission of the optical transition radiation (OTR) emitted by the protons [94], and spectral analysis of the coherent transition radiation (CTR) emitted by the bunch train [95].

Preliminary experimental results obtained recently show clear evidence of the occurrence of SSM [51]. These results also show that the SSM leads to stable excitation of wakefields and that the process corresponds to the amplification of the seed wakefields with a final phase that is very weakly dependent on the variations of the bunch initial parameters. This was also shown in numerical simulations [52]. Excitation of seed wakefields was demonstrated with a low energy electron bunch [96].

First acceleration experiments of low energy electrons (~15 MeV) externally injected into the wakefields are currently underway. In these experiments the electron bunch is purposely made longer than the plasma period in order to ease the temporal synchronization requirements between the witness electron bunch and the wakefields. Numerical simulation results show that a fraction of these electrons could emerge from the plasma with an energy in the GeV range and with a finite final energy spread ($\delta E/E \sim 10\%$) [90].

Future experiments will use a short witness electron bunch ($\sigma_z < \lambda_p$) that will load the wakefields in order to minimize the final the energy spread and at the same time preserve the emittance of the a large fraction of the bunch population [97] while gaining a few GeVs of energy.

The application of the proton-driven plasma wakefield accelerator is to fixed target experiments and to a possible very high energy electron/proton collider [98]. Electron/proton collision applications ease the requirements on the accelerated electron parameters since proton beams are not focused as tightly as beams of an electron/positron collider and do not require production of a high quality positron beam. Also, electron/proton collisions are used to study QCD physics in which interaction cross-sections tend to increase and not decrease with collision energy.

Self-modulation experiments with low energy electron bunches are also performed at DESY-Zeuthen [99].

12.2 Muon Collider

S. D. Holmes · V. D. Shiltsev

Both e^+e^- and $\mu^+\mu^-$ colliders have been proposed as possible candidates for a lepton collider to complement and extend the reach of the Large Hadron Collider (LHC) at CERN. The physics program that could be pursued by a new lepton collider (e^+e^- or $\mu^+\mu^-$) with sufficient luminosity would include understanding the mechanism behind mass generation and electroweak symmetry breaking; searching for, and possibly discovering, super symmetric particles; and hunting for signs of extra space-time dimensions and quantum gravity. However, the appropriate energy reach for such a collider is currently unknown, and will only be determined following initial physics results at the LHC. It is entirely possible that such results will indicate that a lepton collider with a collision energy well in excess of 1 TeV will be required to illuminate the physics uncovered at LHC. Such a requirement would require consideration of muons as the lepton of choice for such a collider.

The lifetime of the muon, 2 μs in the muon rest frame, is just long enough to allow acceleration to high energy before the muon decays into an electron, a muon-type neutrino and an electron-type antineutrino ($\mu^- \to e^- \nu_\mu \overline{\nu}_e$). However, constructing and operating a muon based collider with useable luminosity requires surmounting significant technical challenges associated with the production, capture, cooling, acceleration, and storage of muons in the required quantities and with appropriate phase space densities. Over the last decade there has been significant progress in developing the concepts and technologies needed to produce, capture, cool, and accelerate muon beams with high intensities of the order of $O(10^{21})$ muons/year. These developments have established a multi-TeV Muon Collider (MC) in which μ^+ and μ^- are brought to collision at high luminosity in a storage ring as a viable option for the next generation lepton-lepton collider for the full exploration of high energy physics in the era following the LHC discoveries.

Muon colliders were proposed by Budker [100] in 1969 and later conceptually developed by a number of authors and collaborations (see comprehensive list of references in [101]). Figure 12.3 presents a possible layout on the Fermilab

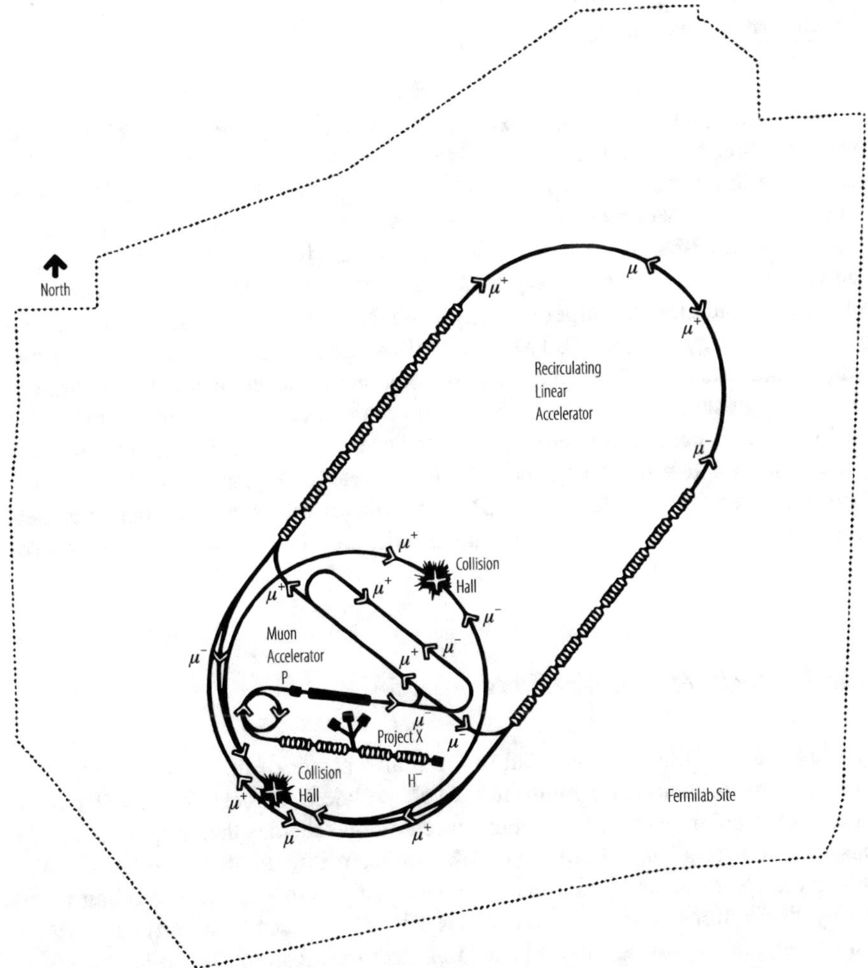

Fig. 12.3 Schematic of a 4 TeV Muon Collider on the 6 × 7 km FNAL site

Table 12.2 The parameters of the low- and high-energy Muon Collider options

Parameter	Higgs factory	Low E	High E
Center-of-mass energy [TeV]	0.126	1.5	6
Luminosity [cm^{-2} s^{-1}]	$0.005{\cdot}10^{34}$	$4.5{\cdot}10^{34}$	$7{\cdot}10^{34}$
Number of bunches	1	1	1
Muons/bunch [10^{12}]	2	2	2
Circumference [km]	0.3	2.8	6.3
Focusing at IP β_*/σ_z [mm]	25/5	10/10	10/5
Beam energy spread dp/p (rms) [%]	0.003	0.1	0.10
Ring depth [m]	~10	13	~150
Proton driver pulse rate [Hz]	30	12	15
Proton driver power [MW]	≈4	≈4	≈2
Transverse emittance ε_T [π μmrad]	300	25	25
Longitudinal emittance ε_L [π mmrad]	1	72	72

site of a MC that would fully explore the physics responsible for electroweak symmetry breaking. Such a MC requires a center-of-mass energy ($\sqrt{s}$) of a few TeV and a luminosity in the 10^{34} cm^{-2} s^{-1} range (see Table 12.2 for the list of parameters). The MC consists of a high power proton driver based, e.g., on the "Project X" SRF-based 8 GeV 2–4 MW H^- linac [102]; pre-target accumulation and compressor rings where very high intensity 1–3 ns long proton bunches are formed; a liquid mercury target for converting the proton beam into a tertiary muon beam with energy of about 200 MeV; a multi-stage ionization cooling section that reduces the transverse and longitudinal emittances and creates a low emittance beam; a multistage acceleration (initial and main) system—the latter employing Recirculating Linear Accelerators (RLA) to accelerate muons in a modest number of turns up to 2 TeV using superconducting RF technology; and, finally, a roughly 2-km diameter Collider Ring located some 100 m underground where counter-propagating muon beams are stored and collide over the roughly 1000–2000 turns corresponding to the muon lifetime.

12.2.1 *Technical Motivations*

Synchrotron radiation (proportional to the fourth power of the Lorentz factor γ^4) poses severe limitations on multi-TeV e^+e^- colliders, namely they must have a linear, not circular, geometry. Practical acceleration schemes then require a facility tens of kilometers long. Furthermore, beam-beam effects at the collision point induce the electrons and positrons to radiate, which broadens the colliding beam energy distributions. Since $(m_\mu/m_e)^4 = 2 \times 10^9$, all of these radiation-related effects can be mitigated by using muons instead of electrons. A multi-TeV $\mu^+\mu^-$ collider can be circular and therefore have a compact geometry that will fit on existing accelerator sites, and may be significantly less expensive than alternative machines.

The center-of-mass energy spread for a 3-TeV $\mu^+\mu^-$ collider, $dE/E < 0.1\%$, is an order of magnitude smaller than for an e^+e^- collider of the same energy. Additionally, the MC needs lower wall plug power and has a smaller number of elements requiring high reliability and individual control for effective operation [103].

An additional attraction of a MC is its possible synergy with the Neutrino Factory concept [104]. The front-end of a MC, up to and including the initial cooling channel, is similar (perhaps identical) to the corresponding Neutrino Factory (NF) front-end [105]. However, in a NF the cooling channel must reduce the transverse emittances (ε_x, ε_y) by only factors of a few, whereas to produce the desired luminosity, a MC cooling channel must reduce the transverse emittances (vertical and horizontal) by factors of a few hundred and reduce the longitudinal emittance ε_L by a factor $O(10)$. Thus, a Neutrino Factory could offer the opportunity of a staged approach to a Muon Collider, and also the opportunity of shared R&D.

12.2.2 Design Concepts

Since muons decay quickly, large numbers of them must be produced to operate a muon collider at high luminosity. Collection of muons from the decay of pions produced in proton-nucleus interactions results in a large initial phase volume for the muons, which must be reduced (cooled) by a factor of 10^6 for a practical collider. Without such a cooling, the luminosity reach will not exceed $O(10^{31}\ \text{cm}^{-2}\ \text{s}^{-1})$, a substantial limitation on the discovery reach of the MC. The technique of ionization cooling [106] is proposed for the $\mu^+\mu^-$ collider [107, 108]. This technique is uniquely applicable to muons because of their minimal interaction with matter.

Ionization cooling involves passing the beam through some material absorber in which the muons lose momentum essentially along the direction of motion via ionization energy loss, commonly referred to as dE/dx. Both transverse and longitudinal momentum are reduced via this mechanism, but only the longitudinal momentum is then restored by reacceleration, leaving a net loss of transverse momentum (transverse cooling). The process is repeated many times to achieve a large cooling factor. The energy spread can be reduced by introducing a transverse variation in the absorber density or thickness (e.g., a wedge) at a location where there is dispersion (a correlation between transverse position and energy). This method results in a corresponding increase of transverse phase space and represents in an exchange of longitudinal and transverse emittances. With transverse cooling, this allows cooling in all dimensions. The cooling effect on the emittance is balanced against stochastic multiple scattering and Landau straggling, leading to an equilibrium emittance.

Theoretical studies have shown that, assuming realistic parameters for the cooling hardware, ionization cooling can be expected to reduce the phase space volume occupied by the initial muon beam by a factor of 10^5–10^6. A complete

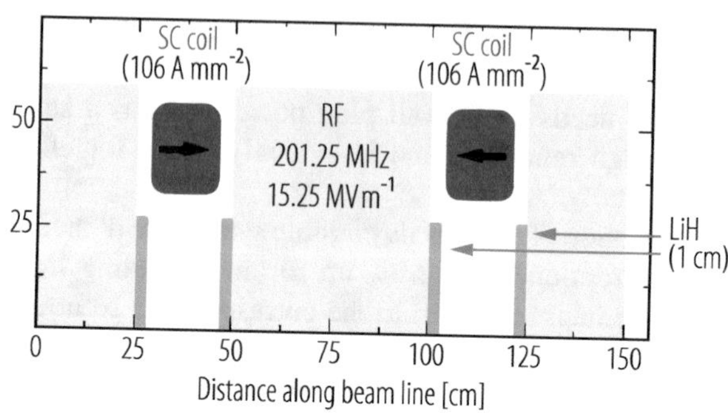

Fig. 12.4 Cooling-channel section. Muons lose energy in lithium hydride (LiH) absorbers (*blue*) that is replaced when the muons are reaccelerated in the longitudinal direction in radio frequency (RF) cavities (*green*). The few-Tesla superconducting (SC) solenoids (*red*) confine the beam within the channel and radially focus the beam at the absorbers. Some representative component parameters are also shown (from [101])

cooling channel would consist of 20–30 cooling stages, each stage yielding about a factor of 2 in 6D phase space reduction—see Fig. 12.4.

Such a cooling method seems relatively straightforward in principle, but has proven quite challenging to implement in practice. One of the main issues is breakdown suppression and attainment of high accelerating gradients in normal-conducting RF cavities immersed in strong magnetic fields. The International Muon Ionization Cooling Experiment (MICE [109]) at RAL (UK) was set to test an ionization cooling channel cell consisting of a sequence of LiH absorbers and 201 MHz RF cavities within a lattice of solenoids that provide the required focusing in a 200 MeV muon beam [110]. The initial results indicate anticipated significant emittance reduction $O(10\%)$ in the "no re-acceleration" configuration [111] and, therefore, can be considered as the first experimental proof of the ionization cooling concept.

12.2.3 Technology Development

Multi-MW target R&D has greatly advanced in recent years, and has culminated in the Mercury Intense Target experiment (MERIT [112]) which has successfully demonstrated a Hg-jet injected into a 15 T solenoid and hit by an intense proton beam from the CERN PS. A high-Z target is chosen to maximize $\pi^{\pm}$ production. The solenoid radially confines essentially all the $\pi^{\pm}$ coming from the target. The Hg-jet choice avoids the shock and radiation damage related target-lifetime issues that arise in a solid target. The jet was viewed by high speed cameras which enabled measurement of the jet dynamics. MERIT results suggest this technology could support beam powers in excess of 4 MW. More advanced solutions for multi-MW targets are under considerations, too, such as granular waterfall targets [113].

Significant efforts are presently focused on high gradient normal conducting RF cavities operating in multi-Tesla magnetic fields as required in the bunching, phase rotation, and cooling channel designs. Closed 805 MHz RF cells with thin Be windows have initially shown significant reduction of maximum RF gradient in a 3 T field—12 MV/m vs. 17 MV/m specified. Further R&D as part of the U.S. based Muon Accelerator Program (MAP) has experimentally demonstrated some 50 MV/m gradients in the RF cavities with high pressure hydrogen gas [114] and in the Be-coated vacuum cavities [115].

Several self-consistent concepts based on different technologies have recently emerged for the MC six-dimensional cooling channel which plays a central role in reaching high luminosity. To achieve the desired mixing of transverse and longitudinal degrees of freedom, the muons must either pass through a series of wedge absorbers in a ring [116] or be put onto a helical trajectory, e.g., as in a "Helical Cooling Channel" [117] or a "FOFO-snake" [118]. The design simulations of the channels are not yet complete and the main challenges are attainment of sufficiently large dynamic apertures, taking into account realistic magnetic fields, RF cavities and absorbers, optimization of the B-fields in RF cavities and technological complexity. The design of the final cooling stages is particularly challenging as it requires very high solenoid fields (up to ~30 T have been considered [119]). The final MC luminosity is proportional to this field. High-field superconducting magnets for the collider ring and for the cooling have been actively developed [120], including feasibility studies of a high temperature superconductor (HTS) option for the 25–50 T final cooling solenoids [121].

A Recirculating Linac with SC RF cavities (e.g. 1.3 GHz ILC-like cavities) is a very attractive option for acceleration of muons from the low energies emerging from the cooling sections to the energy of the experiments. The recirculating linac offers small lengths and low wall plug power consumption but requires small beam emittances.

Recently, realistic collider ring beam optics has been designed which boasts a very good dynamic aperture for about $dP/P = \pm 0.5\%$ and small momentum compaction [122, 123]. The distortions due to the beam-beam interaction will need to be studied as well as practical issues of the machine-detector interface.

Representative performance parameters for a multi-TeV Muon Collider are given in Table 12.2. These parameters are based on the design concepts described above and represent reasonable extrapolations of technologies currently under development. The luminosities displayed are appropriate for the physics research programs that would be undertaken at such a facility [124–127].

12.2.4 Advanced Muon Collider Concepts

In the last few years several advanced muon collider concepts were proposed. An alternative low emittance muon source based on near-threshold production of muons in the reaction $e+e- \rightarrow \mu+\mu-$ was considered in [128]. The scheme relies on availability of high intensity beam of 45 GeV positrons hitting solid, liquid or crystal target of Be, C or diamond. The resulting emittance of the muon beam is very small and allows direct acceleration with extensive ionization cooling. Synchrotron radiation of high-energy muons channelling in between crystal planes results in very small emittances, too, and opens opportunities for crystal-based muon colliders. Given natural advantages of muons, such as absence of nuclear interaction characteristic of protons and greatly reduced synchrotron radiation compared to electrons, the muons are particle of choice for ultra-high gradient acceleration in crystals, originally proposed in [129]. Such colliders with gradients $O(0.1–1\ \text{TeV/m})$ can potentially reach c.o.m energies hundreds of times higher than in the LHC collisions, though, by necessity, with lower luminosities due to practical limits on the facility total electrical power consumption $O(100\ \text{MW})$ [130]. Of course, significant R&D is needed to demonstrate feasibility of the channelling acceleration in crystals or, as a first step, in carbon nanotubes [131].

References

1. Tajima, T., Dawson, J.M.: Phys. Rev. Lett. 43 (1979) 267.
2. Chen, P., et al.: Phys. Rev. Lett. 54 (1985) 693.
3. O'Connell, C.L., et al.: Phys. Rev. ST Accel. Beams 9 (2006) 101301.
4. Caldwell, A., et al.: Nature Phys. 5 (2009) 363.
5. Ruth, R., et al.: Particle Accelerators 17 (1985) 171.
6. Tsung, F., Narang, R., Mori, W.B., Joshi, C., Fonseca, R.A., Silva, L.O.: Phys. Rev. Lett. 93 (2004) 185002.
7. Mangles, S.P.D., et al.: IEEE Trans. Plasma Sci. 36 (2008) 1715.
8. Geddes, C.G.R., Nakamura, K., Plateau, G.R., Toth, Cs., Cormier-Michel, E., Esarey, E., Schroeder, C.B., Cary, J.R., Leemans, W.P.: Phys. Rev. Lett. 100 (2008) 215004.
9. Suk, H., Barov, N., Rosenzweig, J.B., Esarey, E.: Phys. Rev. Lett. 86 (2001) 1011.
10. Esarey, E., Hubbard, R.F., Leemans, W.P., Ting, A., Sprangle, P.: Phys. Rev. Lett. 76 (1997) 2682.
11. Faure, J., et al.: Nature 444 (2006) 737.
12. Pak, A., Marsh, K.A., Martins, S.F., Lu, W., Mori, W.B., Joshi, C.: Phys. Rev. Lett. 104 (2010) 025003.
13. Oz, E., et al.: Phys. Rev. Lett. 98 (2007) 084801.
14. Clayton, C.E., et al.: Phys. Rev. Lett. 70 (1993) 37.
15. Katsouleas, T., Wilks, S., Chen, P., Dawson, J.M., Su, J.J., et al.: Part. Accel. 22 (1987).
16. Tzoufraz, M., et al.: Phys. Plasmas 16 (2009) 056705.
17. Guo-Zheng Sun, Ott, E., Lee, Y.C., Guzdar, P.: Phys. Fluids 30 (1987) 526.
18. Wilks, S.C., Dawson, J.M., Mori, W.B.: Phys. Rev. Lett. 61 (1988) 337.
19. Esarey, E., Ting, A., Sprangle, P.: Phys. Rev. A 42 (1990) 3526.
20. Faure, J., et al.: Phys. Rev. Lett. 95 (2005) 205003.
21. Butler, A., Spence, D.J., Hooker, S.M.: Phys. Rev. Lett. 89 (2002) 185003.
22. Lu, W., Huang, C., Zhou, M., Mori, W.B., Katsouleas, T.: Phys. Rev. Lett. 96 (2006) 165002.
23. Ralph, J.E., Marsh, K.A., Pak, A.E., Lu, W., Clayton, C.E., Fang, F., Mori, W.B., Joshi, C.: Phys. Rev. Lett. 102 (2009) 175003.
24. Leemans, W.P., et al.: Nature Phys. 2 (2006) 696.
25. Ralph, J.E., et al.: Phys. Plasmas 17 (2010) 056709.
26. Kneip, S., et al.: Phys. Rev. Lett. 103 (2009) 035002.

27. Clayton, C.E., et al.: Phys. Rev. Lett. 88 (2002) 154801.
28. Muggli, P., et al.: Phys. Rev. Lett. 93 (2004) 014802.
29. Esarey, E., et al.: IEEE Trans. Plasma Sci. 24 (1996) 252.
30. Zhou, M.: UCLA Ph.D. Thesis (2008).
31. Huang, C., et al.: Phys. Rev. Lett. 99 (2007) 255001.
32. Rosenweig, J.B., et al.: Phys. Rev. Lett. 95 (2005) 195002.
33. Wang, S., et al.: Phys. Rev. Lett. 88 (2002) 135004.
34. Johnson, D., et al.: Phys. Rev. Lett. 97 (2006) 175003.
35. Hogan, M., et al.: Phys. Rev. Lett. 95 (2005) 054802.
36. Chen, P., et al.: Phys. Rev. Lett. 56 (1986) 1252.
37. Bane, K., et al.: IEEE Trans. Nucl. Sci. NS-32 (1985) 3524.
38. Muggli, P., et al.: Phys. Rev. ST Accel. Beams 13 (2010) 052803.
39. Lu, W., et al.: Phys. Plasmas 12 (2005) 63101.
40. Lu, W., et al.: Phys. Rev. Lett. 96 (2006) 165002.
41. Lu, W., et al.: Phys. Plasmas 13 (2006) 56709.
42. M. Tzoufras et al.: Phys. Rev. Lett. 101(2008) 145002.
43. Blue, B., et al.: Phys. Rev. Lett. 90 (2003) 214801.
44. Blue, B., et al.: Laser Part. Beams 21 (2003) 497.
45. Lee, S., et al.: Phys. Rev. E 64 (2001) 04550.
46. Lotov, K.V., et al.: Phys. Plasmas 14 (2007) 023101.
47. Muggli, P., et al.: Phys. Rev. Lett. 101 (2008) 055001.
48. Wang, X., et al.: Phys. Rev. Lett. 101 (2008) 124801.
49. G. Xia et al.: Proceedings IPAC2010, Kyoto, Japan, June 2010, p. 4395
50. Kumar, N., Pukhov, A., Lotov, K.: Phys. Rev. Lett. 104 (2010) 255003
51. P. Muggli et al., Plasma Physics and Controlled Fusion, 60(1) 014046 (2017).
52. N. Savard et al.: in Proc. North American Particle Accelerator Conf. (NAPAC'16), Chicago, IL, USA, Oct. 2016, paper WEPOA01, pp. 684, 2017, M. Moreira, Phys. Rev. Accel. Beams 22, 031301 (2019)
53. AWAKE Collaboration: *Plasma Phys. Control. Fusion* **56** (2014) 084013
54. A. Caldwell et al.: Physics of Plasmas 18 (2011) 103101
55. Joshi, C., et al.: Nature 311 (1994) 525.
56. Amiranoff, F., Bernard, D., Cros, B., Jacquet, F., Matthieussent, G., Mine, P., Mora, P., Morillo, J., Moulin, F., Specka, A.E., Stenz, C.: Phys. Rev. Lett. 74 (1995) 5220.
57. Darrow, C., et al.: Phys. Rev. Lett. 56 (1986) 2629.
58. Everett, M., et al.: Nature 368 (1994) 527.
59. Tochitsky, S., et al.: Phys. Rev. Lett. 92 (2004) 095004.
60. Joshi, C., et al.: Phys. Rev. Lett. 47 (1981) 1285.
61. Mori, W.B., Decker, C.D., Hinkel, D.E., Katsouleas, T.: Phys. Rev. Lett. 72 (1994) 1482.
62. Coverdale, C.A., Darrow, C.B., Decker, C.D., Mori, W.B., Tzeng, K.-C., Marsh, K.A., Clayton, C.E., Joshi, C.: Phys. Rev. Lett. 74 (1995) 4659.
63. Modena, A., Najmudin, Z., Dangor, A.E., Clayton, C.E., Marsh, K.A., Joshi, C., Malka, V., Darrow, C.B., Danson, C., Neely, D., Walsh, F.N.: Nature 377 (1995) 606.
64. Faure, J., Glinec, Y., Pukhov, A., Kisetev, S., Gordienko, S., Lefebvre, E., Rousseau, J.-P., Burgy, F., Malka, V.: Nature 431 (2004) 541.
65. Geddes, C.G.R., Toths, C., van Tilborg, J., Esarey, E., Schroeder, C.B., Bruhwiler, D., Nieter, C., Cary, J., Leemans, W.P.: Nature 431 (2004) 538.
66. Mangles, S.P.D., Murphy, C.D., Najmudin, Z., Thomas, A.G.R., Collier, J.L., Dangor, A.E., Divall, E.J., Foster, P.S., Gallacher, J.G., Hooker, C.J., Jaroszynsk, D.A., Langley, A.J., Mori, W.B., Norreys, P.A., Tsung, F.S., Viskup, R., Walton, B.R., Krushelnick, K.: Nature 431 (2004) 535.
67. Tsung, F.S., Lu, W., Tzoufras, M., Mori, W.B., Joshi, C., Vieira, J.M., Silva, L.O., Fonseca, R.A.: Phys. Plasma 13 (2006) 56708.
68. Pukhov, A., Meyer-Ter-Vehn, J.: Appl. Phys. B 74 (2002) 355.
69. Gordienko, S., Pukhov, A.: Phys. Plasmas 12 (2005) 043109.

70. Lu, W., Tzoufras, M., Joshi, C., Tsung, F.S., Mori, W.B., Vieira, J., Fonseca, R.A., Silva, L.O. Phys. Rev. ST Accel. Beams 10 (2007) 061301.
71. Martins, S.F., Fonseca, R.A., Lu, W., Mori, W.B., Silva, L.O.: Nature Phys. 6 (2010) 311.
72. Froula, D.H., Clayton, C.E., Döppner, T., Marsh, K.A., Barty, C.P.J., Divol, L., Fonseca, R.A., Glenzer, S.H., Joshi, C., Lu, W., Martins, S.F., Michel, P., Mori, W.B., Palastro, J.P., Pollock, B.B., Pak, A., Ralph, J.E., Ross, J.S., Siders, C.W., Silva, L.O., Wang, T.: Phys. Rev. Lett. 103 (2009) 215006.
73. Osterhoff, J., Popp, A., Major, Zs., Marx, B., Rowlands-Rees, T.P., Fuchs, M., Geissler, M., Hörlein, R., Hidding, B., Becker, S., Peralta, E.A., Schramm, U., Grüner, F., Habs, D., Krausz, F., Hooker, S. M., Karsch, S.: Phys. Rev. Lett. 101 (2008) 085002.
74. Perry, M.D., Mourou, G.: Science 264 (1994) 917.
75. Umstadter, D., Chen, S.-Y., Maksimchuk, A., Mourou, G., Wagner, R.: Science 273 (1996) 472.
76. Clayton, C.E., Ralph, J.E., Albert, F., Fonseca, R.A., Glenzer, S.H., Joshi, C., Lu, W., Marsh, K.A., Martins, S.F., Mori, W.B., Pak, A., Tsung, F.S., Pollock, B.B., Rosse, J.S., Silva, L.O., Froula, D.H.: Phys. Rev. Lett. 105 (2010) 105003.
77. Rousse, A., et al.: Phys. Rev. Lett. 93 (2004) 135005.
78. Fuchs, M., et al.: Nature Phys. 5 (2009) 826.
79. Schenvoigz, H.P., et al.: Nature Phys. 4 (2008) 130.
80. Leemans, W.P., Esarey, E. Physics Today 62 (2009) 44.
81. Rosenzweig, J.B., et al.: Phys. Rev. Lett. 61 (1988) 98.
82. Barov, N., et al.: Phys. Rev. Lett. 80 (1998) 81.
83. Nakanishi, H., et al.: Phys. Rev. Lett. 66 (1991) 1870.
84. Hogan, M., et al.: Phys. Plasmas 7 (2000) 2241.
85. Joshi, C., et al.: Phys. Plasmas 14 (2007) 055501.
86. Muggli, P., et al.: IEEE Trans. Plasma Sci. 27 (1999) 791.
87. Blumenfeld, I., et al.: Nature 445 (2007) 741.
88. Hogan, M., et al.: New J. Phys. 12 (2010) 055030.
89. Muggli, P., et al.: New J. Phys. 12 (2010) 045022.
90. E. Gschwendtner et al.: *Nucl. Instr. and Meth. in Phys. Res. A***829** (2016) 76.
91. A. Caldwell et al.: *Nucl. Instr. and Meth. in Phys. Res. A***829** (2016) 3.
92. E. Öz et al.: *Nucl. Instr. and Meth. in Phys. Res. A***740** (2014) 197, E. Öz et al.: *Nucl. Instr. and Meth. in Phys. Res. A***829** (2016) 321.
93. M. Turner et al.: submitted to Nucl. Inst. Meth. Phys. Res. A, (2017), M. Turner et al.: *Nucl. Instr. and Meth. in Phys. Res. A***829** (2016) 314, M. Turner et al.: *Nucl. Instr. and Meth. in Phys. Res. A***854** (2017) 100.
94. K. Rieger et al.: *Review of Scientific Instruments***88** (2017) 025110.
95. M. Martyanov et al.: in preparation, F. Braunmueller et al., Nucl. Instr. and Meth. in Phys. Res. A, 909, 76 (2018).
96. Y. Fang et al.: *Phys. Rev. Lett.***112** (2014) 045001.
97. V. K. Berglyd Olsen et al.: accepted for publication in Phys. Rev. Accelerators and Beams (2017).
98. A. Caldwell et al.: *Eur. Phys. J. C***76** (2016) 463.
99. A. Martinez de la Ossa et al.: AIP Conference Proceedings 1507 (2012) 588, O.Lishilin et al.: *Nucl. Instr. and Meth. in Phys. Res. A***829** (2016) 37.
100. G. Budker: Proc. 7th Intern. Conf. High Energy Accel., Yerevan, (1969) 33.
101. S. Geer: Annu. Rev. Nucl. Part. Sci. 59 (2009) 347–365.
102. S.D. Holmes, in: Proc. 2010 Intern. Part. Accel. Conf., Kyoto, Japan, (2010) 1299.
103. V. Shiltsev: Mod. Phys. Lett. A 25 (2010) 567-577.
104. S. Geer: Phys. Rev. D 57 (1998) 6989.
105. *The Neutrino Factory Intern. Scoping Study Accelerator Working Group Report*, J. Instrum. 4 (2009)P07001.
106. Yu. Ado, V. Balbekov: Atomnaya Energiya, 31 (1971) 40; transl. in: Sov. Atomic Energy 31 (1971) 731.

107. A. Skrinsky, V. Parkhomchuk: Sov. J. Nucl. Phys. 12 (1981) 3.
108. D. Neuffer: Part. Accel. 14 (1983) 75.
109. R. Sandstrom (MICE Collab.): AIP Conf. Proc. 981 (2008) 107.
110. M. Bogomilov et al. (The MICE collaboration), Phys. Rev. Accel. Beams 20 (2017) 063501
111. Rogers, C. T. et al, in Proc. 2017 IPAC (Copenhagen, Denmark), (2017) 2874.
112. H. Kirk, et al. (MERIT Collab.), in: Proc. 2007 IEEE Part. Accel. Conf. (Albuquerque, NM, USA), (2007) 646.
113. H.J. Cai, et al.: Phys. Rev. Accel. Beams 20 (2017) 023401.
114. M. Chung, et al, Phys. Rev. Lett. 111 (2013) 184802.
115. D. Bowring, et al, in: Proc. 2016 IPAC (Busan, Korea), (2016) 444.
116. R. Palmer, et al: Phys. Rev. ST Accel. Beams 8 (2005) 061003.
117. Ya. Derbenev, R. Johnson: Phys. Rev. ST Accel. Beams 8 (2005) 041002.
118. Y. Alexahin, AIP Conference Proceedings 1222, no.1 (2010) 313.
119. D. Neuffer, et al, JINST 12 T07003 (2017)
120. G. Apollinari, S. Prestemon, A. Zlobin, Annu. Rev. Nucl. Part. Sci. 2015.65:355-377
121. V. Kashikhin, et al, IEEE Trans. Appl. Superconductivity, 18, no. 2 (2008) 938
122. Y. Alexahin, E. Gianfelice-Wendt, in: Proc. 2009 IEEE Part. Accel. Conf. (Vancouver, Canada), (2009) 3817.
123. M.H. Wang, et al, JINST 11 P09003 (2016).
124. E. Eichten, A. Martin, Physics Letters B 728 (2014) 125
125. R. Brock, et al, arXiv:1401.6081
126. A. Conway, et al, arXiv:1405.5910.
127. N. Chakrabarty, et al, Physical Review D, 91(1), 015008 (2015).
128. M. Antonelli, et al, NIM-A 807 (2016) 101.
129. T. Tajima, M. Cavenago Phys. Rev. Lett. 59 (1987) 1440.
130. V. Shiltsev, Physics-Uspekhi 55.10 (2012) 965.
131. X. Zhang, et al, Phys. Rev. Accel. Beams 19, 101004 (2016)

Chapter 13
Cosmic Particle Accelerators

W. Hofmann and J. A. Hinton

13.1 Introduction

In the century since the measurements of Victor Hess [1]—considered as the discovery of cosmic rays—the properties of cosmic rays, as they arrive on Earth, have been studied in remarkable detail; we know their energy spectrum, extending to 10^{20} eV, their elemental composition, their angular distribution, and we understand the basic energetic requirements of cosmic ray production in the Galaxy. The energy density of cosmic rays in the Galaxy is known to be comparable to the energy density in Galactic magnetic fields and in the thermal energy of interstellar gas, hence cosmic rays play a non-negligible role in shaping the evolution of galaxies. Charged cosmic-ray particles are strongly deflected in the few μG interstellar magnetic fields—the radius of curvature of a $Z = 1$ particle in astronomical distance units of parsec is given by $R_{\mathrm{gyro,pc}} \sim E_{\mathrm{PeV}}/B_{\mu\mathrm{G}}$ (1 parsec (pc) = 3.1×10^{16} m = 3.26 light years). Therefore, except for energies in the 10^{20} eV range, cosmic rays cannot be traced back to their sources, the cosmic particle accelerators. Much of the current effort goes into identifying and quantitatively describing cosmic accelerators, with supernova remnant shocks as the likely dominant source of Galactic cosmic rays. Properties of cosmic particle accelerators can be studied and inferred in two ways: on the one hand based on the characteristics of local cosmic rays, as they arrive—more or less isotropically—on Earth; this is subject of Sect. 13.2, with a discussion of acceleration mechanisms given in Sect. 13.3. On the other hand, cosmic rays propagating through the Galaxy and in particular cosmic particle accelerators can be imaged using neutral particles and radiation created during the acceleration process and during propagation through the interstellar medium (ISM),

W. Hofmann (✉) · J. A. Hinton (✉)
Max-Planck-Institut für Kernphysik, Heidelberg, Germany
e-mail: wh@mpi-hd.mpg.de; Jim.Hinton@mpi-hd.mpg.de

S. Myers, H. Schopper (eds.), *Particle Physics Reference Library*,
https://doi.org/10.1007/978-3-030-34245-6_13

with synchrotron radiation in the radio and X-ray regimes, high-energy gamma rays, high-energy neutrinos, and neutrons. So far, most information is obtained from electromagnetic probes, as discussed in Sect. 13.4. The small interaction cross section makes detection of high-energy cosmic neutrino sources challenging; only recently, the first detection was reported. Neutron decay limits their range. The flux on Earth of various high-energy cosmic messengers used to explore cosmic particle accelerators is illustrated in Fig. 13.1. The following discussion will mostly concentrate on Galactic particle populations and those particle accelerators which plausibly contribute to cosmic rays observed on Earth. For further details, we refer to reviews such as [2] on cosmic rays in the Galaxy, [3, 4] on supernovae as Galactic cosmic ray sources, [5] on cosmic rays above the knee, [6, 7] on ultra-high-energy cosmic rays, and [8, 9] on very high energy gamma ray astronomy to explore cosmic particle accelerators.

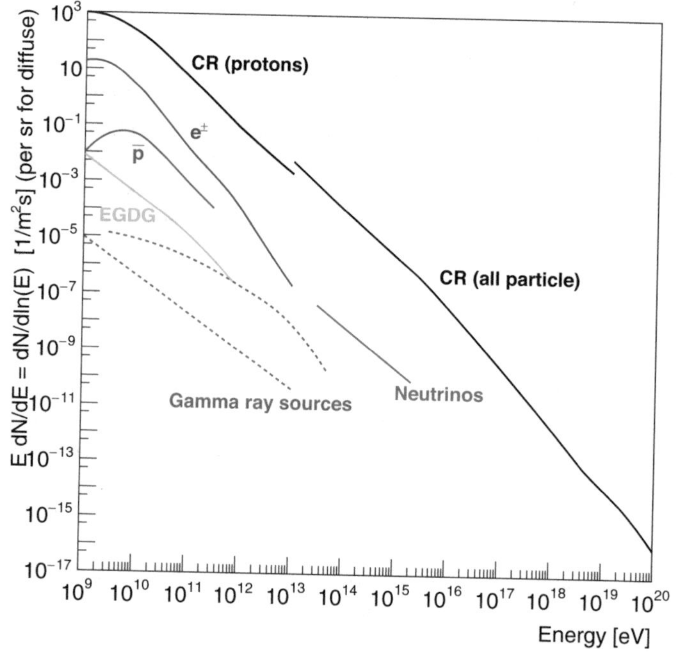

Fig. 13.1 Illustration of the flux of various cosmic messengers expressed as the rate per logarithmic energy interval $EdN/dE = dN/d\ln(E)$. For diffuse and close to isotropic fluxes, a solid angle of 1 sr is used. Indicated are the diffuse all-particle cosmic-ray flux resp. the proton flux (black), the antiproton flux (red), the flux of cosmic $e^{\pm}$ (blue), the flux of extragalactic diffuse gamma rays (EGDG, green), and the high-energy neutrino flux (magenta). Also shown are the gamma-ray fluxes from a localized strong source (RX J1713.7−3946) and from a faint source (NGC 253) (red dashed). Event numbers per spectral interval $\Delta\ln(E) = 1$ are obtained by multiplying the flux with the exposure (effective detection area × time × solid angle of the detector (for diffuse fluxes))

13.2 Cosmic Ray Properties and Implications for Cosmic Ray Sources

Certain properties of cosmic particle accelerators can be inferred from the cosmic ray spectrum measured on Earth, from the elemental composition of cosmic rays, from any anisotropies in their arrival directions, and also from the yield of electrons and antiparticles—positrons and antiprotons–among cosmic rays. The interpretation of data, however, assumes that cosmic rays measured on Earth are reasonably representative for cosmic rays in the Galaxy, and that one is able to disentangle effects reflecting properties of cosmic ray sources from those arising from cosmic ray propagation from their sources the Earth, over kpc distances.[1]

13.2.1 Cosmic Ray Spectrum

At lower energies, up to at most to 10^{14} eV, cosmic ray properties have been studied by space-based instruments, or instruments carried by giant balloons into the upper atmosphere, equipped with magnetic spectrometers or calorimeters for momentum/energy determination, with time-of-flight systems, transition radiation detectors and/or Cherenkov detectors for velocity determination, and with ionisation measurements for determination of the charge state. Examples of such instruments, illustrating the different techniques employed, include PAMELA [10], AMS [11], TRACER [12], CALET [13] and DAMPE [14]. While direct detection in space frequently provides superior performance, in particular regarding the measurement of elemental and isotopic composition, the steeply-falling energy spectrum of cosmic rays, coupled with at most m^2-scale detection areas and—in case of magnetic spectrometers—limited bending power, restricts the energy range of these detection systems. Starting at 10^{12} to 10^{13} eV, ground-based instruments take over, using the Earth's atmosphere as an absorbing medium and detecting the particle cascade created when a high-energy particle interacts in the atmosphere. Instruments detect either the shower particles reaching the ground—with an effective detection area determined by the size of the detector array, which can be as large as 3000 km^2 for the Pierre Auger Observatory [15]—or image the cascade by focusing onto a photosensor array the light emitted by shower particles in the atmosphere. The forward-beamed Cherenkov light illuminates areas of $\sim 10^5$ m^2 on the ground and is readily detectable for showers beyond 10^{11} eV; the isotropically emitted air fluorescence light—imaged e.g. by the Auger fluorescence telescopes [16] and the Telescope Array instrument [17]—allows detection over multi-km distances and provides detection areas in excess of 10^7 m^2, but only in the energy range above

[1] 1 kpc = 1000 pc $\approx$ 3300 light years; the distance from the Sun to the centre of the Galaxy is about 8 kpc.

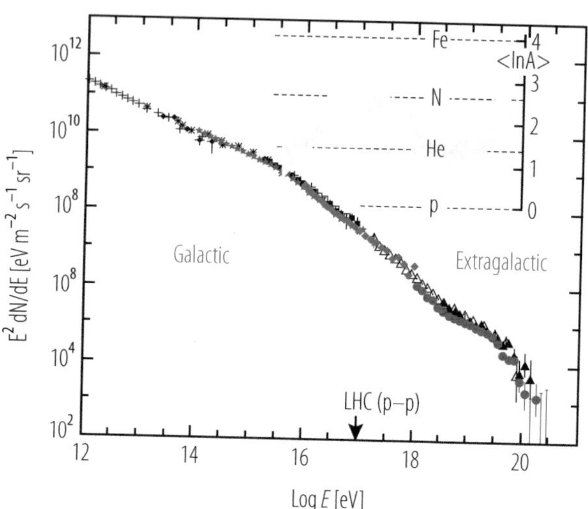

Fig. 13.2 Cosmic ray energy spectrum above 10^{12} eV, presented as a spectral energy distribution $E^2\,dN/dE = E\,dN/d\log(E)$ representing the energy contained per logarithmic energy interval (from [6]). The inset shows the variation of elemental composition—$\langle \ln A \rangle$—with energy, from [31]. While the absolute value of $\langle \ln A \rangle$ is depends on the model used to interpret the air shower data, the pattern of change—from 'light composition' to 'heavy' then back to 'light' and again 'heavy' persists

10^{17} to 10^{18} eV. Identification of primary particles is achieved via the longitudinal and transverse development of the air shower, the radial distribution of shower particles on the ground, and the ratio of nucleons, electrons and muons among shower particles. Compared to detection in space, cosmic-ray identification power is dramatically reduced, allowing essentially to distinguish light from heavy elements, and nuclei from electrons. Examples of air-shower arrays include KASKADE [18], optimised for measurements of cosmic ray composition, the Tibet array [19], optimised for low energy threshold, and the large LHAASO array under construction in China [20].

The cosmic ray energy spectrum shown in Fig. 13.2, compiled from numerous measurements using different techniques, is a mostly featureless power-law spectrum, $dN/dE \sim E^{-\Gamma}$, ranging from GeV energies to 10^{20} eV, covering over 30 orders of magnitude in flux, with the power-law index Γ varying at the "knee" from about 2.7 below 10^{15} eV to about 3 for energies from 10^{16} eV to 10^{18} eV, to flatten again at the "ankle" between 10^{18} eV to 10^{19} eV, and cutting off around 10^{20} eV. Below about 10^{10} eV, the spectrum is modulated by the solar wind; at these energies particles do not penetrate into the inner Solar system. Newer measurements indicate another spectral break in the 10^{11} eV to 10^{12} eV range, with a change in proton and Helium spectral indices [21–23]. The power-law shape, with its lack of features and characteristic energies in the spectrum, indicates that cosmic rays are of non-thermal origin. The local energy density in cosmic rays is dominated by lower-energy particles and amounts to about 1 eV/cm^3. Different mechanisms are discussed for the origin of the slight changes in spectral index: a harding of the spectrum can indicate the emergence of a new component/a new source of cosmic rays, a steepening the peak energy of accelerators, in particular if accompanied by a shift towards heavier composition, since the peak energy of accelerators will usually scale with the charge Z of the accelerated nuclei. Changes in spectral slope can, however, also indicate a change of cosmic ray propagation, i.e. of the energy-dependent diffusion coefficient. The spectral hardening observed at the ankle by 10^{19} eV is most easily explained as the emergence of a new component of cosmic rays, with a harder spectrum. Most models aiming at explaining the overall cosmic ray spectrum assume such a transition from Galactic sources to extragalactic sources in the 10^{17} to 10^{19} eV range (e.g. [24]). The steepening of spectra at the "knee" could be caused by a cutoff in the acceleration mechanism, characterising the upper end of the energy range of Galactic cosmic particle accelerators. A cutoff in the cosmic ray spectrum around 10^{20} eV—the Greisen-Zatsepin-Kuzmin (GZK) cutoff [25, 26]—has been predicted as a result of pion production by ultra-high-energy protons interacting with the cosmic microwave background radiation, limiting the range of protons beyond 10^{20} eV to about 100 Mpc or less (e.g. [27]). While early results questioned the existence of the GZK cutoff, both Auger [28, 29] and Telescope Array (TA) [17] have with much improved statistics established a cutoff in the spectrum of ultra-high-energy cosmic rays (UHECR). The energy of the cutoff is consistent with the GZK cutoff, indicating that most sources of UHECR are located at distances beyond 100 Mpc, but data are also well reproduced in terms of more nearby sources with a rigidity-dependent cutoff in the acceleration mechanism [30].

The cosmic-ray spectrum measured on Earth reflects the source spectrum (averaged over the lifetime of each source and over the population of sources), modified by factors arising from the energy-dependent propagation from sources throughout the Galaxy to Earth. Cosmic ray composition measurements discussed below have proven particularly useful in disentangling the two contributions.

13.2.2 *Cosmic Ray Composition, Cosmic Ray Propagation, and Cosmic Ray Energetics*

Lacking directional information, clues regarding the origin of cosmic rays (beyond those from shape of the energy spectrum) have been drawn mainly from the elemental composition (see e.g., reviews [2, 32]). In the GeV energy range, cosmic-ray chemical composition resembles the composition of solar-system material, with hydrogen and helium nuclei as dominant components (see Fig. 13.3). For some elements and isotopes, there are, however, marked differences between cosmic-ray composition and solar-system composition. This concerns e.g. boron (Z=5), which is suppressed by more than five orders of magnitude relative to carbon (Z=6) in the solar system, but is (within a factor of a few) equally represented among cosmic rays. The reason for this is that boron is not produced directly in stellar nucleosynthesis, but can easily be produced by spallation of heavier cosmic-ray nuclei interacting with interstellar gas. From the boron to carbon ratio, it is inferred that GeV cosmic rays arriving at the Earth have traversed about 10 g/cm^2 of interstellar medium (ISM) (e.g. [33]), with the amount of target material traversed

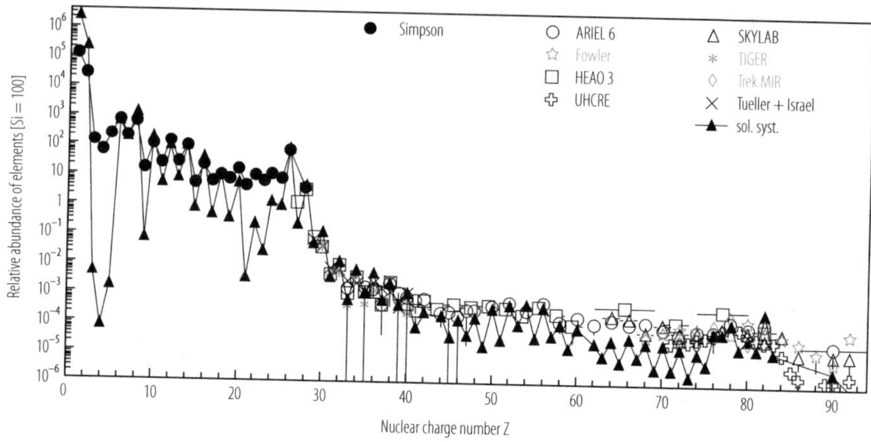

Fig. 13.3 Abundance of elements in cosmic rays as function of their nuclear charge Z at momenta around 1 GeV/c. Solar-system abundances are shown in grey for comparison. From [5]

decreasing roughly as $R^{-0.5}$ for higher rigidities R (defined as momentum/unit charge). Radioactive secondary nuclei, such as ^{10}Be with its half-life of about a million years, are also detected among cosmic rays and can be used to estimate the residence time of cosmic rays in the Galaxy: the observed ^{10}Be/^{9}Be ratio implies a typical age of GeV cosmic rays reaching the Earth in the 10^7 year range, with a trend towards lower ages at higher energy. Both the grammage traversed by cosmic rays and in particular the residence time, which is much larger than the time of about 10^5 years required to cross the entire Galaxy on a straight path, indicate that cosmic rays are trapped inside the Galaxy and propagate diffusively away from their sources, due to frequent deflection in interstellar magnetic fields. However, the combination of the age and the grammage traversed implies that cosmic rays spend a significant fraction of their lifetime outside the Galactic disc with its typical gas density of 1 atom/cm^3. This has lead to the development of "leaky box" models or diffusion models (see Fig. 13.4), where cosmic rays do not escape the Galaxy when leaving the disc with its scale height of $\mathcal{O}(100\ \text{pc})$, but continue diffusive motion out to distances of several kpc from the disc, with the possibility of returning into the disc (for overviews see e.g. [2, 34]). Measurements of Galactic magnetic fields, via synchrotron emission or rotation measures (e.g. [35]), indeed indicate that particle-confining fields extend well beyond the height of the Galactic disc. For diffusive propagation, $<r^2> = 2Dt$. The diffusion coefficient D depends on the scale and degree of turbulence of Galactic magnetic fields, both poorly known. Assuming particles escape the system when reaching a halo height h above the disc, a particle will escape after a time of order $T \sim h^2/2D$. With h much larger than the disc scale height b, and a gas density in the halo which is negligible compared to the density ρ in the disc, the grammage traversed is $X \sim cT\rho(b/h) \sim h/D$. From the residence time and the grammage, or by directly comparing element ratios with model calculations, the halo height h and the diffusion coefficient D can hence be inferred;

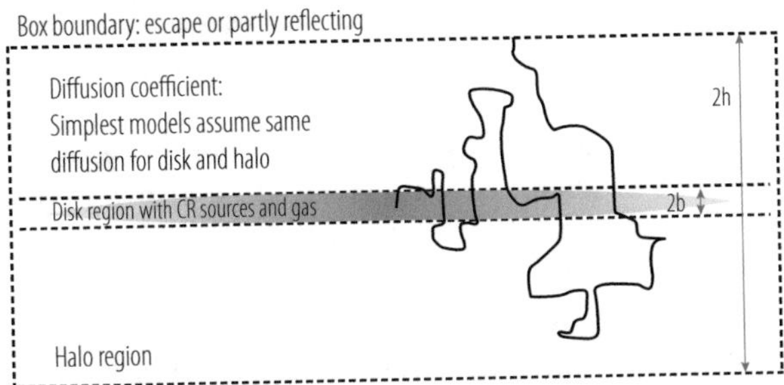

Fig. 13.4 "Leaky Box" cosmic-ray propagation models [34]. Cosmic-ray sources are located in the gaseous disc of the Galaxy, but diffuse propagation continues into the kpc-sized halo. Depending on the specific model, particles are assumed lost when reaching the halo boundary, or are partially reflected back

depending on model details, halo heights of a few kpc, up to 10 kpc, and energy-dependent diffusion coefficients of $D(E)$ of order $10^{28} E_{GeV}^{\alpha}$ cm^2/s are obtained, with α in the range from 0.3 to 0.6; $\alpha = 0.5$ is often used as a representative value. This value of D corresponds to an rms propagation distance of cosmic rays of $d_{pc} \sim 0.3 t_{\rm year}^{1/2}$ at GeV energies; the diffusive approximation, however, holds only for reasonable large time and spatial scales, corresponding to many gyro radii and distances much larger than the coherence length of magnetic fields. Since the average spectrum of cosmic rays in the Galaxy is given by the source spectrum multiplied by the average residence time $T(E) \sim 1/D(E)$, the average source spectrum has to be harder than the observed spectrum, $\Gamma_{\rm source} \approx \Gamma_{\rm CR} + 0.5$. For the interpretation of cosmic rays spectra and composition, initial analytical models (e.g. [34]) have increasingly been replaced by numerical simulations such as GALPROP, allowing the inclusion of specific assumptions regarding the distribution of cosmic ray sources, of material in the Milky Way, and of energy loss and re-acceleration processes, see e.g. [2] for an overview and references. Beyond the effects of cosmic ray spallation products and radioactive decay, differences between solar system composition and cosmic ray composition are observed which seem to depend on the ionization potential or the volatility of elements (e.g. [36]), and which are attributed to the efficiency with which elements are injected into the acceleration process.

Assuming that cosmic rays more or less uniformly permeate the Galaxy, as confirmed by gamma-ray observations (see Sect. 13.4) and that they are not a temporary or local phenomenon, leaky box-models allow constraints to be placed on the energy requirements of Galactic cosmic-rays: to sustain the flux of cosmic rays in the Galaxy and its halo with a volume V of a few 10^{67} cm^3, a typical energy density $\rho \approx 1$ eV/cm^3 and a typical escape time $T \approx 10^7$ year, an energy input of $\rho V/T \approx 10^{41}$ erg/s needs to be provided by Galactic cosmic particle accelerators [37]. Only a few percent of this energy is dissipated in form of ionization and radiative or adiabatic losses; over 95% of the energy leaves the Galaxy into intergalactic space [38].

At higher energies, cosmic-ray composition is determined from ground-based measurements of the electron and muon content of air showers—heavy primaries have a higher fraction of muons—or from the depth of the shower maximum in the atmosphere, which is sensitive to the interaction cross section of the primary, which scales with mass number A roughly as the $A^{2/3}$, see e.g. [31]. As they are not sensitive to individual species, results are often presented in terms of the mean of $\log(A)$ and tend to be rather sensitive to the algorithms used to model the air shower (see e.g. [18, 31]), and inconsistencies between models and data are seen [29, 39]. Nearly all measurements, however, show a change from a dominantly light (H, He) composition up to the knee, to a heavy composition above the knee (see Fig. 13.2, and also Fig. 13.11 below), and while the knee is clearly seen in showers initiated by light elements, no change in the slope of spectra of heavy primaries is evident up to 10^{17} eV. Well beyond the knee, at energies of few 10^{18} eV, studies of the height of shower maximum again suggest a light composition, with a transition to a heavy composition at a few 10^{19} eV [40, 41], although details remain under discussion [42]. A common interpretation is in terms of two components, a Galactic component and an extragalactic component, the latter dominating above 10^{17} to 10^{18} eV, each component with a cutoff in the acceleration mechanism at fixed particle gyro radius (i.e. rigidity), corresponding to a peak energy which scales with the nuclear charge Z. In this scenario typical Galactic accelerators are required to reach an energy in the range of a few $Z \times 10^{15}$ eV (e.g. [43]).

13.2.3 Cosmic Ray Anisotropy

The diffusive propagation of all but the very highest energy cosmic rays, with a gyro-radius much smaller than the scale of the Galaxy, destroys almost all directional information in cosmic rays; nevertheless, cosmic rays will on average flow away from their sources, resulting in small anisotropies.

The arrival directions of cosmic rays are indeed almost uniform on the sky, with (dipole) anisotropies at the 10^{-3} level at energies below 0.1 PeV, an indication of a minimum in anisotropy in the 0.1–1 PeV range, and an increase up to 10^{-2} at higher energies (see e.g. [2, 44]). The phase of the anisotropy—i.e. the direction of maximum intensity—varies with energy. Effects which might cause anisotropies include [34]: (a) a diffusive flow of cosmic rays governed by gradients in cosmic ray density ρ, with a resulting anisotropy of order $3D\nabla\rho/c\rho$; since the cosmic ray density likely decreases with galactocentric radius, an outward flow results at the location of the Solar System. A density gradient and hence flow might also be caused by single nearby cosmic-ray sources, in which case the magnitude and direction cannot be predicted a priori. Given that the diffusion coefficient $D(E)$ increases with energy, density-gradient related anisotropies are expected to increase with energy. (b) Anisotropies of order $(\Gamma + 2)(v/c)$ caused by the motion of the Earth relative to the cosmic-ray rest frame (the Compton-Getting effect [45]). Cosmic-ray energies are also Doppler-shifted, for power-law spectra with index Γ the anisotropy hence depends on the spectral index, but not on the energy. Compton-Getting anisotropies could be caused by the 220 km/s motion of the Sun around the centre of the Galaxy, and/or—an order of magnitude smaller—by the orbital motion of the Earth around the sun. (c) Anisotropies arising from special magnetic field configurations near the Earth or sun, or from modulation by the solar wind. Such anisotropies should decrease with increasing rigidity (i.e. energy) of particles, but large-scale structures such as the heliotail caused by the motion of the sun relative to the local interstellar medium, possible extending beyond 1000s of AU,[2] could influence particles beyond multi-TeV energies. Decomposition of anisotropies into the different components is difficult also because most ground-based detectors measure the variation of rates along right ascension, for a fixed viewing direction, i.e., whereas the declination dependence is usually not measured directly. The observed energy dependence is non-trivial to explain and may include partial compensation between different contributions. In a study of a sample of simulated spiral galaxies and their cosmic-ray sources, most realisations show larger anisotropies than measured, and a uniform increase with energy [46]. The minimal anisotropy in the sub-PeV range can also be interpreted as evidence that the cosmic-ray "gas" co-rotates with the Galaxy [47]—quite plausible given that cosmic rays couple via magnetic fields to the interstellar plasma.

At TeV energies, anisotropies of a few 10^{-4} on smaller angular scales (few 10°) are observed [48, 49]. The origin of these small-scale anisotropies is not well understood; explanations e.g. assume field lines connecting a cosmic ray source and the solar system [50], but the effect may also simply reflect the local concrete realisation of the turbulent magnetic field within the cosmic ray scattering length [51].

13.2.4 Electrons and Antiparticles Among Cosmic Rays

Electrons and antiparticles among cosmic rays play a special role, since their "natural" yields are quite low and hence their fluxes are most sensitive to contributions from "exotic" sources such as annihilation of Dark Matter particles. Contrary to cosmic ray nuclei, electrons and positrons suffer significant energy losses, mostly from synchrotron radiation, with an energy loss timescale of $\mathcal{O}(10^5\,\text{year}/E_{\text{TeV}})$. Combined with typical diffusion coefficients $D(E)$, this implies that electron and positron sources need to be within a distance of $d_{pc} \approx 500/E_{\text{TeV}}^{1/4}$ from Earth. Unlike cosmic ray nuclei, electrons and positrons therefore act as probes of local sources. While electrons are assumed to be accelerated together with nuclei in cosmic ray sources, antiparticle yields—antiprotons and positrons—were traditionally modeled as arising exclusively from nuclear interactions during cosmic ray propagation, resulting in highly suppressed yields.

[2] AU = Astronomical Unit, the mean Sun-Earth distance of $\approx 1.5 \times 10^{11}$ m.

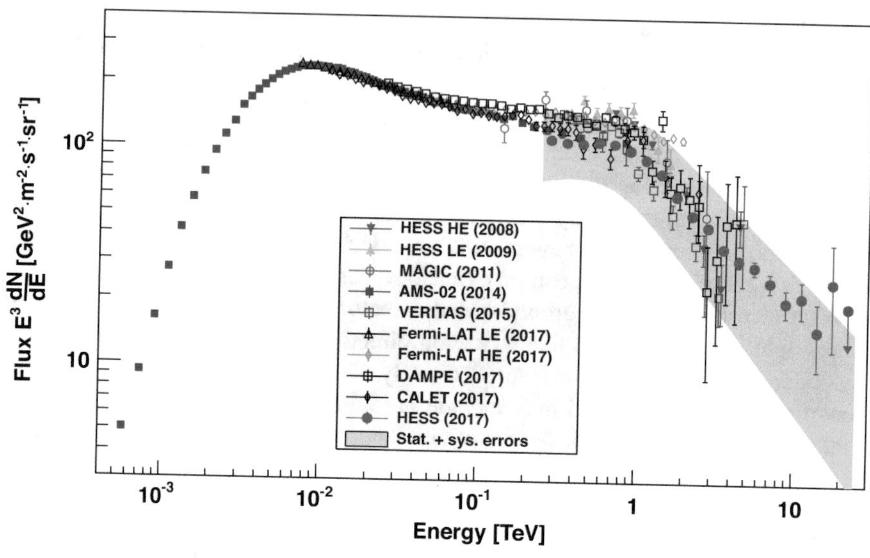

Fig. 13.5 Cosmic-ray electron spectrum, multiplied by E^3, as measured by H.E.S.S., MAGIC, VERITAS, AMS-02, Fermi, DAMPE and CALET ([13, 14, 53–55] and refs. given there)

Antiproton yields have been measured up to energies of 10^{11} eV. The antiproton to proton ratio rises with energy up to about 2×10^{-4} around 10 GeV, and then levels off, in good agreement with expectations for secondary antiprotons (e.g. [52]).

The cosmic-ray electron spectrum is illustrated in Fig. 13.5. The spectrum falls steeper than the spectrum of nuclei, with an index $\Gamma \approx 3$, and steepens further at around 10^{12} eV. The flux of cosmic-ray electrons at 10^{12} eV is about 0.1% of that of cosmic ray nuclei. The spectrum can be modeled by assuming that cosmic-ray sources accelerate electrons and nuclei in a ratio of about 1:100 at a given energy; with increasing energy, and hence shorter electron range, fewer and fewer sources contribute, resulting in both a steeper spectrum and increasing uncertainty in the predicted flux due to the stochastic distribution of sources (see e.g. [57]).

Recent measurements, however, reveal deviations from this picture for the electron (plus positron) flux (see Fig. 13.5), and in particular for the positron/electron ratio (Fig. 13.6). Beyond the range of solar modulation, the electron flux is predicted to decrease slightly faster than E^{-3}; the data suggest an additional component appears in the energy range between about 100 GeV and 1 TeV, before the flux cuts off. A similar, but much more dramatic effect is seen in the electron/positron ratio, which increases beyond 10 GeV, rather than continue to drop as predicted for secondary production of positrons. This feature, first detected by PAMELA [58] and Fermi [59], and studied with high statistics using AMS-02 [56], gives rise to considerable speculation regarding its origin (e.g. [60]). Both the electron yield and the positron/electron ratio can be described by assuming an additional, charge-symmetric source of electrons and positrons, with a (propagation-modified)

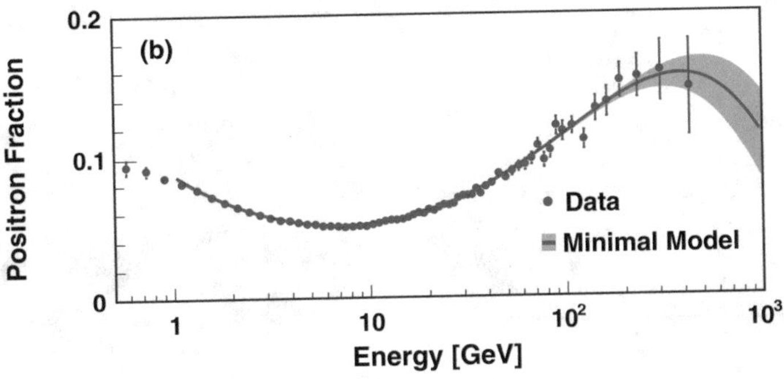

Fig. 13.6 Positron fraction, $N_{e^+}/(N_{e^-} + N_{e^+})$, as a function of energy, as measured by AMS-02 (from [56])

spectrum which rises faster than E^3 up to a few 100 GeV, and cuts off at about one TeV [60]. Dark matter annihilation of particles in the TeV mass range can account for the shape of the spectra, but would require enhanced annihilation rates compared to typical models, and annihilation modes which produce only leptons but no baryons; most conventional dark matter annihilation models predict the positron excess to be associated with an excess in antiprotons, which is not seen. An alternative explanation is provided by electrons escaping from nearby pulsars, or, more specifically, pulsar wind nebulae [61] (see also Sect. 13.4), although results regarding very low diffusion coefficients for cosmic-ray electrons disfavor this interpretation somewhat, making it difficult for electrons from known pulsars to reach the Earth [62].

13.2.5 Astronomy with Ultra High Energy Cosmic Rays

Due to the strong deflection of cosmic rays in Galactic and extragalactic magnetic fields, astronomical imaging of their sources is impossible over most of their energy range. Only at energies of several 10^{19} eV deflections for $Z = 1$ ultra high energy cosmic ray (UHECR) particles are predicted to become small enough—a few degrees—that particles can be traced back to their sources. The arrival directions of the highest-energy cosmic rays are essentially isotropic (see Fig. 13.7), but indications of correlations with astrophysical objects start to emerge. In the Auger data, the most significant over-density of UHECR lies roughly in the direction of Centaurus A, the closest AGN at 3.8 Mpc distance; the statistical significance of this fact is about 3σ after accounting for the number of trials [29]. Arrival directions also correlate with the distributions of starbust galaxies, and of gamma-ray AGN, at the $3-4\sigma$ level [29]. The 7-year Telescope Array data [17] exhibit at energies above $10^{19.2}$ eV a 3.7σ post-trial enhancement around RA = 9 h 16 min, Dec = 45°. The interpretation of these data is not obvious. One possibility is that few sources are

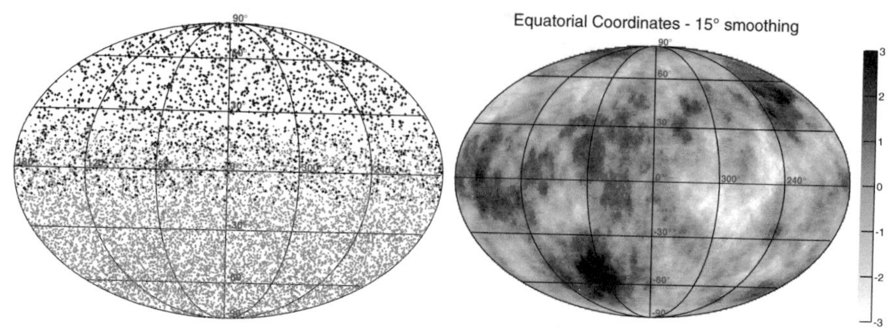

Fig. 13.7 Left: Arrival directions of Auger events (red points in the South hemisphere) and Telescope Array ones (black crosses in the Northern hemisphere) above 10^{19} eV in equatorial coordinates. Right: Excess significance sky map smoothed out at a 15° angular scale. From [63]

responsible for UHECR, emitting a mixture of light and heavy nuclei; the protons among those particles create the detected directional enhancements whereas the heavier nuclei are strongly deflected and are responsible for a uniform background, with similar spectra in both hemispheres. Verifying or disproving such a scenario requires particle-by-particle mass identification and increased statistics, both goals of next-generation UHECR experiments.

13.3 Particle Acceleration Mechanisms and Supernova Shocks as Cosmic Accelerators

A few general conditions can be imposed regarding sources of Galactic cosmic rays. To sustain the flux of cosmic rays in the Galaxy, an energy input of $\approx 10^{41}$ erg/s by Galactic cosmic ray sources is required (see Sect. 13.2), and a spectrum extending at least up to $\approx Z \times 10^{15}$ eV with an average source spectral index in the range $\Gamma \approx 2$ to 2.4. In addition, if cosmic accelerators use regular or turbulent magnetic fields to confine particles, the acceleration region has to have a size at least equal to the gyro radius of particles, $R_{gyro,pc} \sim E_{\mathrm{PeV}}/B_{\mu G}$ [7]. Figure 13.8 illustrates that there are astrophysical objects which fulfil this condition up to energies of 10^{20} eV, but not much beyond.

In Sect. 13.3.1 below we focus on the most established mechanism, diffusive shock acceleration, and in particular the well-studied case of supernova remnants. Note however, that acceleration associated with magnetic reconnection is now increasingly discussed, in particular in the cases of objects with relativistic bulk motions such as in the jets of active galaxies.

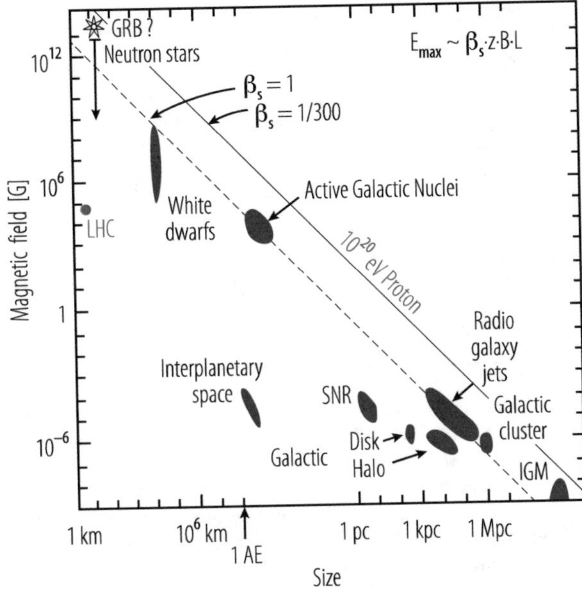

Fig. 13.8 The maximum energy attainable in a cosmic accelerator depends on the product of magnetic field strength B and size L [7]. The condition that the gyro radius of a particle of charge z is contained in the acceleration region yields $E_{max} \sim zBL$. For shock acceleration, see Sect. 13.3.1, one finds that the actually reachable E_{max} is lower by a factor of $\mathcal{O}(\beta_s)$, the shock speed in units of the speed of light. The lines corresponds to a maximum energy of 10^{20} eV, for $z = 1$ and $\beta_s = 1$ (dashed) or $\beta_s = 1/300$ (full). Potential sources range from very compact, high-field objects such as gamma-ray bursts (GRB) or neutron stars via the parsec-size supernova remnants (SNR) to the extended low-field intergalactic medium (IGM). From [5] and [7]

13.3.1 *Shock Acceleration in Supernova Remnants*

As sources of Galactic cosmic rays, supernova explosions were suggested very early as a suitable source [64], providing both sufficient energy—10^{51} erg kinetic energy per explosion, or 10^{42} erg/s for a supernova rate in the Galaxy of 1/30 year—as well as a plausible acceleration mechanism—first order Fermi acceleration (see e.g. [65, 66] and further references given in [3, 4]), the appropriate spectral index $\Gamma \approx 2$ and a peak energy around 10^{15} eV (see Fig. 13.8).

In a supernova explosion, stellar material of up to a few solar masses is ejected with initial speeds of up to 10^4 km/s or $\beta_{sh} = v_{sh}/c \sim 0.03$, and creates a shock where the ambient interstellar medium is compressed and piled up. While the piled-up ambient material slows down the ejecta, speeds remain supersonic for time scales of order 10^4 years (e.g. [67]). Shocks with high Mach number are characterized by a compression ratio $r = (\gamma + 1)/(\gamma - 1) = 4$, governed by the adiabatic index $\gamma = 5/3$ of the compressed monatomic gas. Charged particles which cross the shock in either direction find themselves in a medium moving with velocity $\approx v_{sh}$ relative

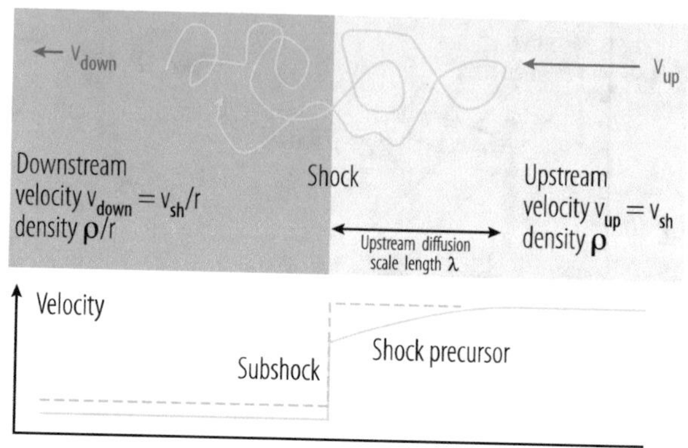

Fig. 13.9 Cartoon of cosmic-ray acceleration at shock fronts, viewed in the rest frame of the shock. In this frame, interstellar medium material streams into the shock with the shock propagation speed $v_{up} = v_{sh}$, is compressed by a factor r and streams out with $v_{down} = v_{sh}/r$. Charged particles can be scattered across the shock front and back. Particles will on average propagate a distance $\lambda \approx D/v_{sh}$ upstream before they are swept back through the shock. The lower part of the diagram illustrates how flow velocity varies across the shock for the case that the energy in accelerated particles is negligible ("test particle case", dashed line) and for the case where the shock is modified by the cosmic-ray pressure, see text for details

to the medium on the other side of the shock (Fig. 13.9). Particles are scattered off anisotropies in the magnetic field and isotropised in the new medium, with an average increase in energy $\Delta E/E$ of $\mathcal{O}(\beta_{sh})$. In their diffusive motion, particles can cross the shock front multiple times, their energy growing as $E \sim (1+k)^n$, where n is the number of crossing cycles and $k = (4/3)(1 - 1/r)(v_{sh}/c)$; here $(1 - 1/r)v_{sh}$ is the difference in flow speed before and after the shock and the factor 4/3 arises from averaging over shock crossing angles. Since, viewed in the rest frame of the shock, material is inflowing ("upstream") with speed v_{sh} and outflowing ("downstream") after shock compression with speed v_{sh}/r, in the long run particles tend to be carried into the downstream region, with a probability of order $4\beta_{sh}/r$ not to return to the shock at each cycle ($c\beta_{sh}/r$ is the downstream flow speed and hence loss rate, $c/4$ the angle-averaged shock crossing speed and hence rate of initiating another cycle). Given the gain per cycle and the loss probability per cycle, the energy spectrum of accelerated particles can be calculated as $dN/dE \sim E^{-\Gamma}$ with $\Gamma = (r+2)/(r-1)$. The time per acceleration cycle is governed by the diffusion coefficient D. Maximum magnetic field turbulence $\Delta B/B \sim 1$ implies that the mean free path of relativistic particles, between scattering off field inhomogeneities, is of order of the gyro radius. In this so-called 'Bohm diffusion' regime, a more detailed calculation gives $D = R_{gyro}c/3$. The cycle time ΔT can be estimated to be $4D(1+r)/(v_{sh}c) \approx (20/3)(R_{gyro}/v_{sh})$ (for $r = 4$). With a gain per cycle of order $\Delta E \approx \beta_{sh}E$, the resulting acceleration rate is hence $dE/dt \approx \Delta E/\Delta T \sim E\beta_{sh}^2/R_{gyro} \sim \beta_{sh}^2 B$, of order $0.05\beta^2 B_{\mu G}$ PeV/year. The maximum achievable energy is governed by the age of the system, by energy losses (in particular radiation losses of electrons), and by the scale λ of the upstream diffusion length of particles, $\lambda \approx R_{gyro}/3\beta_{sh}$ (determined from the balance between the upstream medium flow speed v_{sh} and the diffusion speed $v_{Diff} = D\nabla\rho/\rho \approx D/\lambda$), which has to be small compared to the size of the supernova remnant to give particles a chance to return to the shock. The age-limited peak energy is given by [4] as $E_{max}(age) \approx 0.5T_3 v_{sh,8}^2 B_{\mu G} f^{-1}$ TeV where T_3 is the remnant age in kyr, $v_{sh,3}$ is the shock speed in units of 10^3 km/s and f parametrizes diffusion effects, with $f \approx 1$ for Bohm diffusion. Injection efficiency, peak energy and particle acceleration rate also depend on the angle of the average magnetic field relative to the shock front. For heavier nuclei, the acceleration rate and peak energy scale with their charge Z, reflecting the reduced gyro radius for a given energy.

The acceleration process is predicted to be highly efficient, converting as much as 50% of the kinetic energy of the ejecta into non-thermal particles. This high efficiency makes the process non-linear (e.g. [68, 69] and further references in [4]); the accelerated cosmic ray currents induce turbulent magnetic fields—determined under certain assumptions as high as 300 μG (e.g. [70])—which in turn reduce the gyro radius and the diffusion coefficient D and increase the speed of particle acceleration. A shock precursor of scale λ develops since the in-streaming gas reacts to the upstream cosmic-ray pressure, reducing the compression ratio at the subshock, see Fig. 13.9. On the other hand, since the shock now compresses a mixture of normal gas with $\gamma = 5/3$ and relativistic cosmic ray gas with $\gamma = 4/3$ upstream of the shock, the total compression ratio increases up to $r = 7$ and the spectrum of accelerated particles becomes harder, up to $dN/dE \sim E^{-3/2}$. This hardening of spectra affects mainly the highest-energy particles, with gyro radii of order λ, which probe both the shock precursor and the subshock (see Fig. 13.10).

Strictly speaking, the calculated spectral index $\Gamma = (r+2)/(r-1)$ applies to the particles swept downstream and confined inside the remnant. The exact mechanism of cosmic-ray escape from supernova shocks into the upstream region is not well understood; it is usually assumed that the time-integrated spectrum of particles released from the remnant reflects the spectrum of accelerated particles, but that particles of highest energy are released early, and those of low energy late in the lifecycle of the remnant [71]. The escaping cosmic rays represent a current that generates turbulent magnetic fields outside the remnant, reducing the diffusion coefficient in the upstream region, thereby enhancing the rate of particle crossing of the shock front, and speeding up the acceleration process [72].

Cosmic ray spectra measured at the Earth result from the superposition of many sources, up to distances comparable to the scale height of the halo, and are essentially stationary, despite the stochastic nature of the sources. At any given time thousands of supernovae will contribute to the cosmic rays on Earth; an individual supernova will cause a change in cosmic ray intensity over a volume of only about 100 pc in radius; beyond that the energy density of it's cosmic rays falls below the 1 eV/cm^3 level of the cosmic ray sea. The spectrum of cosmic rays on Earth is given by the source spectra multiplied by the average residence time $T(E) \sim 1/D(E)$ in the Galaxy, qualitatively explaining the difference between the source spectral index $\Gamma \approx 2$ and the observed cosmic ray spectral index $\Gamma \approx 2.7$. The knee

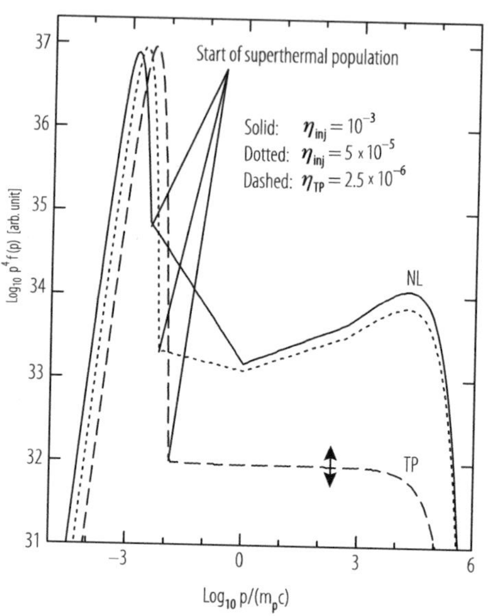

Fig. 13.10 Spectra of particles accelerated in a simulated supernova remnant shock, at a given time during the evolution of the remnant, as a function of momentum in units of the proton mass × c. The density in momentum space, $p^4 f(p)$, is shown, for relativistic particles equivalent to $E^2 dN/dE$. The 'TP' curve refers to the test particle case, where the energy carried by cosmic rays is modest compared to the kinetic energy of ejecta; in this case, the thermal distribution of unaccelerated protons extends into a power law with index $\Gamma = 2$. For efficient acceleration ('NL'), the shock is modified due to cosmic-ray pressure and the local spectral index of accelerated particles varies from $\Gamma > 2$ at low energy to $\Gamma < 2$ at high energy, with details depending on the efficiency η governing the rate of particle injection into the acceleration process. From [69]

in the cosmic ray spectrum is then associated with the peak energy of particles in the acceleration process; heavier nuclei of charge Z can be accelerated to Z-times higher energies than protons and dominate beyond the knee. With plausible parameters, supernova remnant-based models can reproduce the observed spectrum and variation of composition across the knee, see Fig. 13.11.

Shocks, and particle acceleration in shocks, can occur in all situations where non-relativistic or relativistic outflows exist with Mach numbers greater than unity; examples include stellar winds, Galactic outflows, winds driven by pulsars, or jets emerging from the vicinity of black holes driven by matter accretion. Shocks can also arise in collisions or from the infall of matter, e.g. during structure formation in galaxies and galaxy clusters. Less well understood than supernova shocks are relativistic shocks with $\beta_{sh} \approx 1$, here particles can gain significant energy in one or few shock crossings, but crossing the shock becomes increasingly difficult (e.g. [74]).

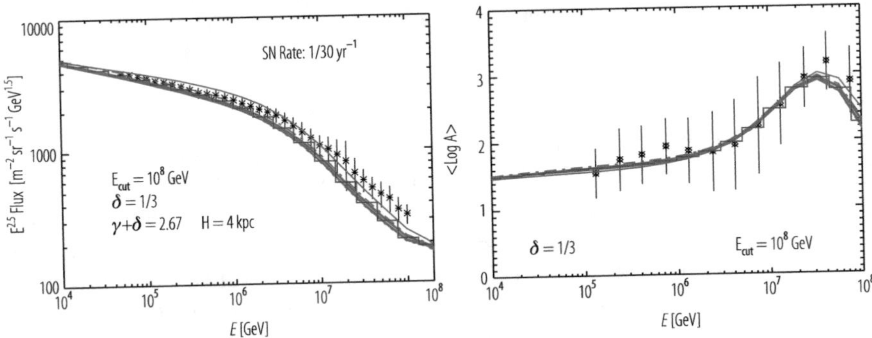

Fig. 13.11 Predicted cosmic ray spectra and change of composition—$\langle \log(A) \rangle$—with energy for ten simulated spiral galaxies populated with supernova remnants, with plausible acceleration parameters. The model includes an assumed extragalactic component which takes over at highest energies. From [73]

13.3.2 Pulsars as Particle Sources

An alternative mechanism for particle acceleration are neutron stars (or generally, rotating compact objects) with large magnetic fields, acting as unipolar inductors building up large electric fields and radiating Poynting flux.

A pulsar (a rotating magnetised neutron star) created in a supernova explosion radiates energy because its magnetic moment is misaligned by an angle α with respect to its rotation axis. The pulsar rotational energy $E = I\Omega^2/2 \approx 2 \times 10^{52}/P_{ms}^2$ erg serves as an energy source; here $I \approx 10^{45}$ cm^{2}g is the pulsar moment of inertia and P_{ms} is the rotation period in ms, with typical pulsar birth periods of order tens of ms. Assuming energy loss through dipole radiation, energy is radiated at a rate $\dot{E} = I\dot{\Omega}\Omega \approx (8\pi/3\mu_0 c^3)B_s^2 R^6 \Omega^4 \sin^2\alpha \approx 4 \times 10^{43} B_{12}^2 P_{ms}^{-4} \sin^2\alpha$ erg/s, causing the pulsar to spin down, with $\dot{\Omega} \sim \Omega^3$. Here, $R \approx 10$ km is the pulsar radius and B_{12} is the surface field in units of 10^{12} G. In case other dissipation mechanisms contribute, such as outflowing winds of particles, $\dot{\Omega}$ is parametrized as $\dot{\Omega} = -K\Omega^n$, where n is the braking index, $n = 3$ for dipole radiation. Measured values for n range from 2 to 3. Integrating the $\dot{\Omega}(\Omega)$ relation, one obtains $\dot{E}(t) = \dot{E}_0/(1+t/\tau)^{-(n+1)/(n-1)}$ for the general case, or $\dot{E}(t) = \dot{E}_0/(1+t/\tau)^{-2}$ for dipole radiation, with τ as the characteristic spin-down time. For dipole radiation, $\tau \approx (3\mu_0 c^3 I)/(16\pi R^6 B_s^2 \sin^2\alpha\ \Omega_0^2) \approx 15 P_{0,ms}^2/(B_{12}^2 \sin^2\alpha)$ years; in the time up to $t = \tau$, the pulsar loses half of its rotational energy, with the energy output diminishing like $1/t^2$ for $t >> \tau$.

In the near field, the rotating pulsar magnetic field with a surface strength in the 10^{12} G range creates electric fields with voltage drops of order $10^{17} B_{12}/P_{ms}^2$ volt. Electrons and positrons are generated by pair cascades near the pulsar surface and are accelerated until currents short-circuit the fields. Inside the light cylinder, $r < \Omega/c$, where the magnetic field and the currents co-rotate with the neutron star, particles either near the polar cap of the pulsar or in the 'outer gap' close to the light cylinder create beams of radiation swept across the sky by the pulsar rotation, see [75, 76]. In the far field, magnetic fields spiral up and cause a Poynting flux of electromagnetic energy. By a mechanism, which is yet to be understood in detail (e.g. [76]), this Poynting flux drives a particle wind, effectively converting much of the radiated energy into particle kinetic energy. It is usually assumed that the particle wind is dominated by electrons and positrons, but a component of nuclei extracted from the pulsar surface cannot be excluded. In the pulsar wind, reconnection of opposite magnetic field lines can provide a mechanism for energy release and particle acceleration [77]. The particle wind, initially assumed to be spherical [78, 79] but in more recent models concentrated in the equatorial plane, ends in a standing wind termination shock where the pressure of the wind is balanced by the ambient pressure; the termination shock is visible in high-resolution X-ray images of pulsars (see e.g. [80]). In the termination shock, particle velocities are randomised, particles are accelerated and emerge in a subsonic flow, creating a large and expanding magnetised bubble filled with high-energy electrons and positrons, see e.g. [78, 79, 81]. Pulsars are hence cosmic sources of high-energy electrons and positrons, possibly of nuclei, and of a complex mix of pulsed and beamed as well as of more or less steady and isotropic radiation spanning the range from radio to gamma rays, as elaborated in Sect. 13.4.3. However, since not all supernova explosions result in pulsars and since their initial rotational energy is usually much smaller than the kinetic energy released in a supernova explosion, the contribution of pulsars to overall cosmic-ray energetics should be modest.

13.4 Probing Cosmic-Ray Sources and Propagation Using Gamma-Rays and Neutrinos

For energies below $E \sim hZeB_{\rm ISM} \sim 10^{18}$ eV to 10^{19} eV (see e.g. [82]) cosmic ray protons are strongly deflected when propagating in the Galaxy on scales of the Galactic halo height h in typical interstellar fields $B_{\rm ISM}$; their arrival directions therefore carry almost no information on their source locations. Directional messengers are therefore required to study Galactic cosmic-ray sources. Strong interactions of cosmic ray protons and nuclei with target protons and nuclei in the interstellar medium (ISM) lead to pion production and hence gamma-ray, neutrino and secondary electron and positron signatures (for a detailed description, see e.g. [83]). The following discussion will focus largely on the well-explored gamma ray signatures. Protons or nuclei with power-law spectra of index Γ_p generate gamma-ray spectra with $\Gamma_\gamma \approx \Gamma_p$, and a cutoff in proton spectra translates into a (smoother) cutoff in gamma-ray spectra about a decade in energy below the cutoff in primary spectra. For a typical ISM density, n, of 1 hydrogen atom per cm^3 the energy loss timescale for relativistic protons is $(f\,\sigma_{pp}\,n\,c)^{-1}$ or a few 10^7 years, comparable to or—in particular at high energy—longer than their residence time in the Galaxy. The fraction f of the primary energy lost in a typical collision ("inelasticity") is

$\sim$0.5 of which $\approx 1/3$ goes into the gamma-ray channel. Overall, about 1% of the energy injected into relativistic hadrons in the Galaxy in the end emerges in photons [38]. Inside, or in the vicinity of cosmic accelerators, particle density is strongly enhanced and the objects are visible as gamma-ray sources, assuming that sufficient amounts of target material (interstellar gas) are present. The intensity and extent of the gamma-ray or neutrino emission depends the distribution of target material, on whether accelerated particles are efficiently confined within the accelerator, and on how quickly particles diffuse away after escaping from the acceleration region (see Sect. 13.2.2) and merge into the cosmic-ray "sea".

For cosmic electrons and positrons, ionisation, bremsstrahlung, synchrotron radiation and Inverse Compton (IC) scattering of ambient radiation fields compete as energy-loss processes [84, 85]. For the highest energy electrons, at TeV energies and above, synchrotron and IC emission dominate and synchrotron X-rays and IC gamma-rays can be used as effective tracers of electron acceleration and propagation. The targets for IC scattering are typically the cosmic microwave background radiation (CMBR), starlight, and reprocessed starlight remitted in the far infrared, with typical energy densities of order 1 eV/cm^3. The typical lifetime of a high energy electron in the ISM is $5 \times 10^5 (B/5\,\mu\text{G} + U_{\text{rad}}/(\text{eVcm}^{-3}))^{-1}(E/\text{TeV})^{-1}$ years, much shorter than propagation time scales in the Galaxy. Radiative losses modify the energy spectra of very-high-energy electrons, introducing—for burst-like injection—an age-dependent cutoff in the electron spectra at the energy where the lifetime corresponds to the source age, at $E_{\text{cut,TeV}} \approx 3 \times 10^5/T_{\text{year}}$, or—for continuous injection—increasing the spectral index Γ_e by one unit, since at high energy only electrons injected within a period corresponding to the electron lifetime survive [86]. Electron spectra with power-law index Γ_e generate power-law IC and synchrotron spectra with index $(\Gamma_e + 1)/2$, and a cutoff energy E_c in electron spectra translates into cutoffs $E_{\gamma,\text{TeV}} \approx 10 E^2_{c,\text{TeV}} E_{ph,\text{eV}}$ for IC gamma rays (in the Thomson regime where $E_{c,\text{TeV}} E_{ph,\text{eV}} \sim< 1$), where E_{ph} is the typical energy of the target photons, and $E_{X,\text{eV}} \approx 0.01 E_{c,\text{TeV}} B_{\mu G}$ for X-rays. With the rapid energy loss, emission by electrons is usually concentrated relatively close to the sites of acceleration. Figure 13.12 gives an example spectral energy distribution for emission dominated by energetic electrons.

Gamma-ray detection at high energies is based on pair-production and subsequent electromagnetic cascading. The most sensitive satellite-based gamma-ray detector currently operating is the Fermi Large Area Telescope (LAT), which has $\approx$1 m^2 detection area and $\approx$2.5 steradian field-of-view (FoV). The LAT combines a silicon-strip tracker for directional reconstruction and a 8.6 radiation-length thick calorimeter for energy determination [90]. The angular resolution achievable is strongly energy dependent: improving from 5° at 100 MeV to 0.25° at 10 GeV, where photon statistics become very limited for most sources. The most sensitive ground-based approach (see e.g. [9] for a review) is the Imaging Atmospheric Cherenkov Technique (IACT), which uses the Cherenkov light produced by electromagnetic cascade electrons and positrons in the atmosphere to establish the properties of the primary gamma-ray; the gamma-ray direction is determined by imaging the cascade, the gamma-ray energy is derived from the Cherenkov

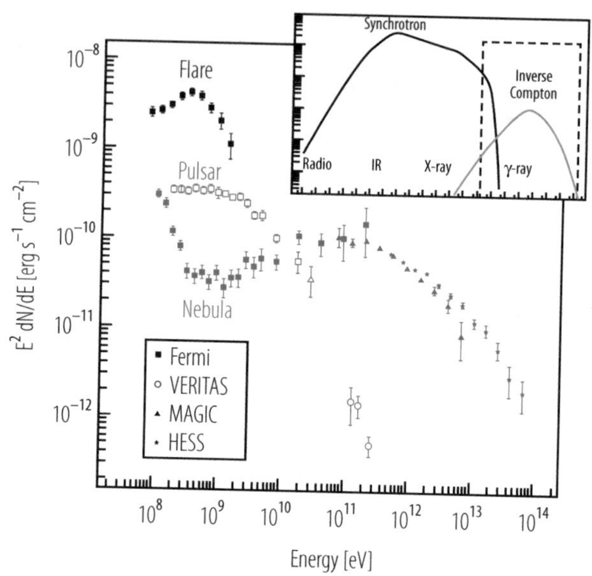

Fig. 13.12 The spectral energy distribution $E^2 dN/dE$ of the Crab Nebula, with its synchrotron and Inverse Compton (IC) components. The inset illustrates the full spectral range from radio to gamma rays [87], the main figure shows the steady-state gamma-ray flux from the nebula (blue solid symbols), the pulsed emission from the vicinity of the pulsar (red open symbols), and the brightest flare observed from the nebula so far (black solid symbols). Data from Fermi (squares), MAGIC (triangles), VERITAS (circles) and HESS (stars) are shown. Adapted from [88, 89]

light yield. The technique is in principle applicable for photon energies above $\sim$5 GeV, where the Cherenkov yield becomes significant. Current instruments HESS, MAGIC and VERITAS are sensitive from $\sim$20–50 GeV to $\sim$50 TeV, have $\sim 4^\circ$ field-of-view and collection areas at TeV energies of $\sim 10^5\,\mathrm{m}^2$. The directional precision achievable from the ground is limited by shower fluctuations to $\approx 0.01^\circ/(E/1\,\mathrm{TeV})^{-0.6}$ [91], with about 0.1° (and $\approx$15% energy resolution) achieved for current instruments [8]. Compared to Cherenkov telescopes, ground-level detection of shower particles allows large field-of-view and duty cycle, at the expense of higher energy threshold and reduced sensitivity and energy resolution [9]. The HAWC instrument, combining a high-altitude location at 4200 m asl. with the calorimetric detection of shower particle energy flow using large water Cherenkov detectors, instrumenting 60% of its 22,000 m^2 array area, has for the first achieved gamma-ray detection performance competitive with current IACTs [92].

At this time, over 3000 sources of GeV gamma rays have been detected using the space-based instruments Fermi [93] and AGILE, and well over 200 sources of TeV gamma rays [94] are seen with ground-based instruments, showing the abundance and indeed ubiquity of cosmic particle accelerators. Around one half of the TeV sources are of extragalactic nature, half or slightly more are associated with our

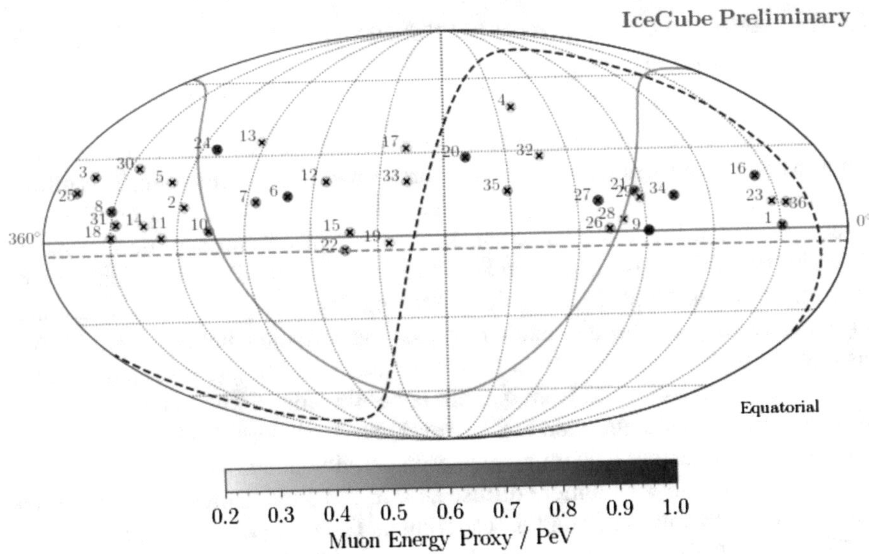

Fig. 13.13 Reconstructed arrival directions of observed IceCube neutrino events with estimated muon energies above 200 TeV, in equatorial coordinates. The marker color indicates the energy. The solid gray line indicates the galactic plane and the dashed black line the supergalactic plane. Reproduced from [96]

Galaxy. Of the Galactic TeV gamma ray sources, a handful are SNR where shells are resolved, a similar number are clearly associated with SNR but not resolved, about two dozen are associated with pulsars and their pulsar wind nebulae, and for the remaining 15–20 sources the identification is unclear, either because of a lack of counterparts or because multiple potential counterparts exist. Despite the much larger number of GeV sources, the number of well-identified Galactic objects is similar, with pulsars, identified by their pulsed emission, as the dominant class of Galactic GeV emitters.

For hadronic sources similar fluxes are generated in gamma-rays and neutrinos (see e.g. [83]). Neutrino telescopes primarily sensitive above $\sim$1 TeV now exist, with the best sensitivity reached by the IceCube detector beneath the South Pole. IceCube has for the first time detected astrophysical neutrinos [95, 99], emerging at energies beyond 100 TeV, from the strong background of atmospheric neutrinos. The neutrino flux appears to be of diffuse nature (see Fig. 13.13), and no consensus exists regarding its exact origin. No localized very-high-energy cosmic neutrino source has yet been detected [97]; a detection would provide completely unambiguous identification of hadronic accelerators and allow high-density environments, from which TeV photons may not emerge, to be probed. The KM3Net collaboration [98] is building a larger detector in the Mediterranean sea with greater sensitivity and the potential for detection of neutrino sources in the inner parts of the Galaxy, and IceCube is studying options for a tenfold increase of detection volume [96].

13.4.1 Diffuse Gamma Ray Emission: Tracing Cosmic Rays in the Galaxy

The Galactic cosmic rays interact with the material, radiation fields and magnetic fields in and around the Galaxy to produce broad-band diffuse emission. This diffuse emission peaks in the gamma-ray band due to strong contributions from π^0 decay, inverse Compton scattering and bremsstrahlung. Diffuse emission dominates the GeV sky (see Fig. 13.14) and provides a means to test the distribution and energy spectrum of cosmic rays in the Galaxy. Probing cosmic rays in this way requires an understanding of the distribution of gas and radiation fields in the Galaxy; historically the opposite has often been the case, with the gas distribution in the Milky Way being estimated from gamma-ray observations. The main components of the Galactic diffuse emission are, however, clear: π^0 decay and bremsstrahlung emission from interactions with molecular material with a scale-height in the disc of $\sim$50 parsecs, and with a more diffuse atomic component, plus a very extensive halo generated by inverse Compton scattering on Galactic radiation fields.

The diffuse GeV γ-ray emission suggests that the ISM is permeated with cosmic ray electrons, protons and nuclei, occupying a much greater volume than the Galactic disc (consistent with the picture presented in Sect. 13.2.2). In the outer parts of the Galaxy, where it is easiest to measure, this "sea" of cosmic rays has spectral properties similar to those measured at the Earth. The emissivity (gamma-ray flux per hydrogen atom, proportional to the cosmic-ray density) is seen to be roughly constant from the location of the sun up to $\sim$14 kpc from the Galactic Centre [101]. This relative uniformity suggests that the radial distribution of acceleration sites in the Galaxy is flatter than that of identified SNRs or pulsars.

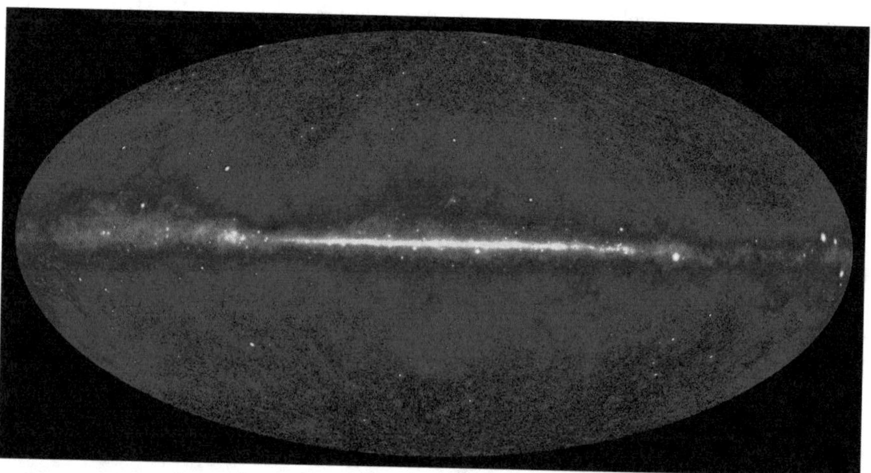

Fig. 13.14 The gamma-ray sky above 1 GeV energies as measured using the Fermi-LAT, plotted in Galactic coordinates [100]

Diffuse Galactic gamma-ray emission has also been detected at TeV energies: from the outer parts of the Galaxy, with the wide field-of-view shower-particle detector Milagro [102] and in the Galactic Centre region with HESS [103]. The level of emission seen is well above the extrapolation of the cosmic ray "sea" emission at lower energies and suggests the existence of recently injected cosmic rays and/or unresolved cosmic ray sources. The diffuse emission seen with HESS from the $\sim$100 pc radius Central Molecular Zone is correlated with the distribution of target material, illuminated by cosmic rays diffusing away from the Galactic Centre and hence suggesting a π^0-decay origin of the emission, with the central supermassive ($3 \times 10^6 M_\odot$) black hole Sgr A* as likely candidate for the origin of these cosmic rays. The local density of TeV cosmic rays at the centre of the Galaxy is enhanced by an order of magnitude compared to local cosmic rays.

Diffuse emission is seen with Fermi from beyond the Milky way, in satellite galaxies such as the Large and Small Magellanic Clouds and Local Group galaxies such as Andromeda. The gamma-ray fluxes seen from these objects support the idea that the energy input into cosmic ray acceleration is proportional to the star formation rate, with this trend continuing to more distant starburst galaxies, which are undergoing phases of enhanced star-formation [104–106].

A completely unexpected discovery were the Fermi bubbles—kpc-size emission regions of GeV gamma rays above and below the Galactic Centre, with relatively sharp boundaries, indicating confined populations of high-energy particles well beyond the Galactic disc and near halo [107–109]. Similar-shaped radio features hint at the presence of (primary or secondary) electrons. The origin of the Fermi bubbles is under discussion; possibilities discussed include (a) a Gyr-old stellar wind driven by star formation in the Galactic Centre region, similar to the outflows seen in starburst galaxies such as M82 or NGC 253, carrying cosmic rays along which then interact in the thin medium above and below the disc, (b) earlier jet activity of the supermassive hole at the Galactic Centre, or (c) in-situ acceleration of electrons by plasma-wave turbulence [108, 110, 111].

13.4.2 *Supernova Remnants Viewed in Gamma Rays*

As described in Sect. 13.3.1, the idea that supernova remnants (SNR) accelerate the majority of the Galactic cosmic rays has been with us for half a century, until relatively recently, however, the level of experimental support for this idea was relatively modest. Radio emission from the shells of SNRs was attributed to GeV electrons, but evidence for the acceleration of very-high-energy particles, in particular protons and nuclei, was essentially absent. Observations in the X-ray band from the 1990s onwards have established the presence of synchrotron emission from $>$100 TeV electrons in these objects. These observations also indicate the presence of enhanced magnetic fields [70], and for some objects "missing energy" is evident when comparing expected shock heating of the gas to measured thermal X-ray emission [70, 112]. A likely form for this missing energy is a population of cosmic-ray protons and nuclei, which may also be responsible for the amplification of magnetic fields in the SNR shell, via cosmic ray driven instabilities—the positive feedback process that leads to an increase in the efficiency and maximum energy of cosmic ray acceleration, provided that conversion of shock kinetic energy to cosmic rays occurs with a significant efficiency [113].

Detection of very-high-energy gamma rays from SNR was proposed in [114] as a test of cosmic-ray origin; the gamma-ray flux from interacting protons was predicted to be $F(>E)_{\mathrm{cm}^{-2}\,\mathrm{s}^{-1}} \approx 9 \times 10^{-11} \theta E_{\mathrm{TeV}}^{-1.1} E_{SNR,51} d_{kpc}^{-2} n_{\mathrm{cm}^{-3}}$, where θ is the efficiency of energy conversion, E_{SNR} the kinetic energy in the explosion in units of 10^{51} ergs, d the distance in kpc, and n the ambient density in hydrogen atoms per cm^3. With θ assumed as 0.1 or larger and E_{SNR} and n typically of order unity, supernova remnants within a few kpc were predicted to be bright enough for detection with air-Cherenkov instruments.

The subsequent discovery of resolved TeV emission from the shells of SNRs with H.E.S.S.—RX J1713.7−3946, RX J0852.0−4622, RCW 86 and SN 1006—(e.g. [8]) can be seen as the direct and definitive proof that very-high-energy particles are accelerated in the shells of SNR.

Figure 13.15 shows keV X-ray and TeV gamma-ray images of the nearby Galactic SNR RX J1713.7−3946 [117], the best-studied gamma-ray remnant. However,

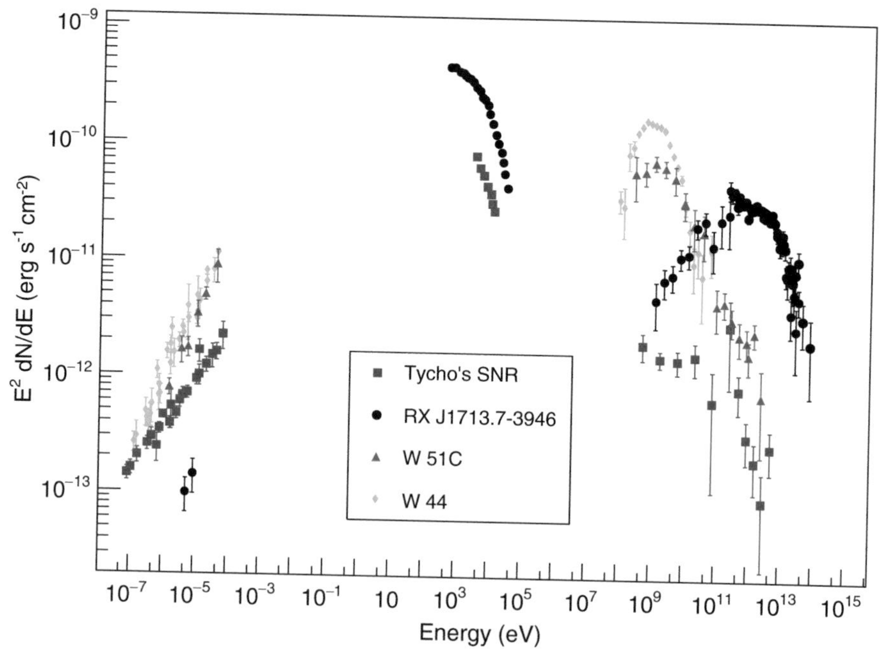

Fig. 13.15 The radio to gamma-ray spectral energy distributions E^2dN/dE of three archetypal Galactic supernova remnants of increasing age: Tycho's SNR [115, 116], RX J1713.7−3946 [117], W 51 [118, 119] and W 44 [120]

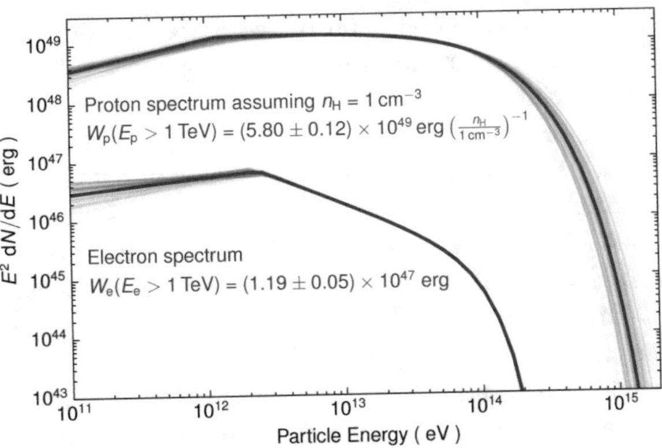

Fig. 13.16 Spectrum of primary particles required to reproduce the wide-band gamma ray and X-ray spectra of RX J1713.7−3946, assuming either dominant emission by a population of accelerated protons, or of electrons. From [117]

there is still significant debate regarding the nature of the parent particle populations. The close resemblance of the X-ray and gamma-ray images, taken a factor 10^9 apart in energy, has been taken as implying a common parent population, i.e. electrons with spectra extending to ~100 TeV, with the observed TeV emission arising from Inverse Compton scattering rather than from proton interactions. This picture is supported by the spectral shape of the gamma-ray emission, see Fig. 13.15; in the GeV energy range, the spectral index of gamma rays is $\Gamma \approx 1.5$, as expected from an electron population with the canonical E^{-2} energy spectrum resulting from shock acceleration, whereas proton interactions should result in a gamma-ray index $\Gamma \approx 2$. These data do not exclude efficient proton acceleration in this SNR; in the low-density environment, protons may simply not find enough targets to create a gamma-ray flux which is competitive with the flux produced by an energetically sub-dominant, but much more efficiently-radiating population of accelerated electrons. This is illustrated in Fig. 13.16, which shows the spectrum of primary particles that reproduce the measured wide-band gamma ray spectrum: either about 6×10^{49} ergs are required in protons above 1 TeV (assuming a target density of 1/cm^3), or only about 10^{47} ergs in electrons.

Other gamma-ray detected SNRs have dramatically different spectra; this may be due to a combination of the environment and evolutionary stage of the system. TeV emission is for example seen also from extremely young Galactic SNRs such as Tycho's SNR and Cassiopeia A, although the angular resolution of TeV instruments is insufficient to resolve their shells. These objects are firmly associated with historically observed supernova and are bright X-ray and radio sources. Figure 13.15 shows the spectral energy distribution of Tycho's SNR [116], which exhibits a spectral softening at a few hundred GeV. The maximum energy to which a particle can be accelerated inside an SNR shell is very likely time-dependent, and determined by the shock velocity and magnetic field strength. Very young SNRs with fast shocks and strong magnetic fields may accelerate extremely high energy particles (up to PeV) which can escape ahead of the shock at later times [71], leading to a decrease in the maximum or break energy with time. In addition, the presence or absence of dense target material (in the form of molecular gas) in the neighbourhood of the SNR will affect the balance between inverse Compton emission and Bremsstrahlung and π^0-decay emission. The nature of the environment depends strongly on the nature of the explosion. Type Ia supernova occur in evolved systems, with the explosion occurring far from the birth place of the star and typically far from molecular gas clouds. The majority of core-collapse supernova explosions occur inside massive stellar clusters, born from molecular clouds, but where powerful stellar winds have already redistributed the molecular material to produce a highly non-uniform density environment.

Emission correlated with the distribution of target material, rather than following the emission seen in radio or X-rays, would be a clear sign for a hadronic origin of gamma rays, and is seen for several Galactic SNRs (see e.g. [121]), all of which are significantly older than the ~1000 year age of RX J1713.7−3946. W 51 is an example of this class of object: as for RX J1713.7−3946 it is thought to be the remnant of the core-collapse of a very massive ($>8\,M_{\odot}$) star but its age is estimated to about 10^4 y and it is clearly interacting with dense molecular material. The spectrum of W 51 is rather steep at TeV energies (Fig. 13.15), with the peak energy output in the GeV domain [118], in marked contrast to RX J1713.7−3946. The centroid of the GeV and TeV emission is also consistent with the point of interaction of the SNR with gas clouds rather than the centre of the remnant. Both these facts point to a hadronic origin of the gamma-ray emission, with 5×10^{50} ergs of cosmic rays present in the SNR [118], consistent with the picture that a significant fraction of the energy of a typical SNR goes into the acceleration of cosmic ray protons and nuclei. The steep spectrum of the emission suggests that particles with TeV energies may already have escaped the SNR. Hadronic origin is also demonstrated for remnants such as W 44 and IC 443, where spectra show a clear pion-decay feature—a break in spectral index at about the pion mass [120], which is not expected in spectra from the inverse Compton process.

There is now compelling evidence that SNRs are effective accelerators of both electrons and nuclei, but due to uncertainties on the population and evolution of particle-accelerating SNRs one cannot yet confidently conclude that they are the dominant sources of the Galactic cosmic rays.

13.4.3 Pulsars and Pulsar Wind Nebulae

The newly-formed neutron stars left behind in some types of supernova explosion are generally rapidly rotating and highly magnetised. The rotational energy of such pulsars is converted into pulsed emission (primarily and gamma-ray energies) and into an ultra-relativistic outflow of electron-positron pairs, see Sect. 13.3.2. The termination shock where this wind is halted by external pressure is thought be a site of acceleration up to very high energies. After the shock the direction of particle motions become randomised and synchrotron and Inverse Compton emission is produced in a pulsar wind nebula (PWN). There is as of yet no clear physical picture of how particle acceleration operates in these systems. None the less, there is a huge body of evidence that acceleration to PeV energies occurs in these systems. The Crab Nebula, the most prominent example of a PWN, is a unique object which is bright and well-studied in every waveband of the electromagnetic spectrum, with synchrotron emission dominating from the radio up to 1 GeV and inverse Compton emission seen above, up to almost 100 TeV (see Fig. 13.12). The Crab pulsar was born in a historically observed supernova explosion in 1054 and has the most extreme rate of conversion of rotational energy ($\dot{E} \approx 5 \times 10^{38}$ erg/s) of any Galactic pulsar. In 1989 it became the first source to be detected at TeV energies [122]. Until very recently the gamma-ray emission from the Nebula was thought to be steady in time, but recent dramatic flaring activity has been seen at GeV energies, apparently corresponding to a rapid increase in the number of synchrotron-emitting >PeV electrons (Fig. 13.12) [89, 123, 124]. It seems very difficult to explain these flares in terms of diffusive shock acceleration, as this process seems to be too slow to offset catastrophic synchrotron energy losses. A single shot acceleration mechanism such as magnetic reconnection at the termination shock is an attractive option.

The gamma-ray emission from the Crab Nebula appears almost point-like with the $\sim$0.1° resolution of current instruments, but this situation is atypical. In the last decade, about two dozen TeV gamma-ray sources were discovered and firmly or tentatively identified as PWN, including objects such as Vela X, MSH 15–52, G 21.5−0.9, G 0.9+0.1, N 175B, and the nebula surrounding PSR B1706–44 (for summaries, see e.g. [8, 125–127]). Most of these PWN are extended TeV gamma-ray sources, with size of a fraction of a degree, often accompanied by a significantly smaller-scale X-ray nebula surrounding the pulsar. While X-ray luminosities tend to correlate quite well with the instantaneous spin-down energy loss $\dot{E}$ of the pulsar, no clear correlation is observed between the gamma-ray luminosity and $\dot{E}$ [126]. Gamma-ray luminosities range from a fraction of a percent of the pulsar spin-down energy loss $\dot{E}$ to tens of percent, indicating a relatively efficient conversion of (rotational) kinetic energy into high-energy particles. A likely explanation for the difference in size between X-ray and gamma-ray PWN and for the different $\dot{E}$-dependence of luminosities is that in the typical μG fields derived for extended PWN, keV X-ray emitting electrons have energies of 100s of TeV and cooling times of order 1000 years, whereas TeV gamma ray emitting electrons have energies in the 10 TeV range and cooling times beyond a few 10,000 years. For pulsars with ages between a 1000 and a few 10,000 years (most of the gamma-ray PWN population), only "recently" accelerated electrons with number $\sim\dot{E}$ therefore contribute to the X-ray emission, whereas all electrons ever accelerated contribute to the gamma-ray emission, reflecting essentially the initial rotational energy E_0 of the pulsar (half of which is lost during the initial few 100 years of spin-down history, see Sect. 13.3.2), rather than $\dot{E}$. The sizes of gamma-ray PWN tend to increase with the age of the pulsar, saturating at sizes of a few tens of pc. The gamma-ray PWN are frequently displaced from the pulsar, locating the pulsar (and the associated X-ray PWN) at the edge of the gamma-ray PWN (e.g. Fig. 13.15). One explanation is that PWN are often crushed and/or displaced by the supernova reverse shock [128], another that—as discussed above—the gamma-ray PWN reflects relic electrons abundantly created in the early history of the pulsar, whereas now the pulsar may have moved away from its birth place due to a kick from the explosion. However, in the few cases where pulsar motion is known, it does not line-up well with the vector connecting the pulsar and the centroid of the VHE PWN. In a few cases, such as for the source HESS J1825−137 [129] shown in Fig. 13.15, gamma-ray PWN (as well as X-ray PWN) show energy-dependent morphology, with the nebula shrinking towards the pulsar with increasing gamma-ray energy, presumably due to radiative cooling of electrons as the propagate away from the pulsar. GeV-TeV gamma-ray emission can therefore be used to measure the time-integrated particle injection of the pulsar and to understand the propagation of relativistic particles away from their sources.

PWN represent the bulk of Galactic TeV gamma-ray sources, by far outnumbering emission traced to SNR shock-accelerated protons. At first, this seems surprising. However, contrary to supernova remnant shocks, where acceleration of very-high-energy particles stalls after a few 10^3 to at most 10^4 years, a pulsar can supply the nebula with energy for many 10^4 years. In addition, under typical conditions, electrons and positrons are more efficient TeV gamma-ray emitters than are SNR-accelerated protons—radiative energy loss timescales are smaller by one to two orders of magnitude. Therefore, PWN dominate the Galactic population of TeV gamma-ray sources even though their energy reservoir—the rotational energy of the pulsar—is typically an order of magnitude smaller than the $\approx 10^{51}$ ergs released in a supernova explosion. In fact, it seems likely that a sizeable fraction of the currently unidentified TeV gamma-ray sources—where no counterpart is seen in other wavebands—are PWN where the pulsar is not detected (due to beaming effects or simply a lack of sensitive observations) and where radiation-cooled electrons no longer have sufficient energy to produce keV X-ray synchrotron photons.

PWN also occur inside binary systems, where the huge radiation fields and stellar wind/outflow of the companion star dramatically modify the PWN properties. PSR B1259−63 is young and powerful pulsar in an eccentric 3.4 year orbit around a $\approx$10 solar mass companion. Variable and point-like TeV and GeV emission is seen around the periastron passage of the neutron star when radiation densities are highest but the time-profile of the emission and the spectral energy distribution are complex and very poorly understood (e.g. [130]). Extended radio emission is seen on milliarcsecond scales supporting the idea that this system is a PWN "compactified" by the high pressure environment and rapid radiative losses. In other well-established TeV binaries systems, LS 5039, LS I +61 303 and HESS J0632+057, the nature of the compact object is not certain and the systems may be accretion-, rather than rotation-powered.

PWN naturally accelerate positrons and electrons in equal number. An increase in the proportion of cosmic ray electrons and positrons contributed by PWN (and/or the presence of a small number of dominant local/recent accelerators) may explain the increase in the fraction of positrons seen in the locally measured cosmic rays at high energies, as discussed in Sect. 13.2.4. Given a suitable pulsar age T, energy-dependent diffusion $D(E)$ of electrons causes the detected spectrum to harden compared to the source spectrum—high energy particles reach Earth faster—and radiative energy losses during propagation over a time T cause the spectrum to cut off at a certain energy. The two effects combine to produce an excess in an E^3

weighted spectrum, the peak position and peak level being adjustable via pulsar age, distance, and energy output, and matching the detected excess contribution for plausible parameters. For example, positrons released $T \approx 10^5$ years ago exhibit a cutoff at $E_{\mathrm{cut,TeV}} \approx 3 \times 10^5 / T_{\mathrm{year}} \approx 3\,\mathrm{TeV}$ and at TeV energies travel over a distance $d \approx (2DT)^{1/2} \approx 500\,\mathrm{pc}$ (for $D \approx 10^{28} E_{\mathrm{GeV}}^{0.5}\ \mathrm{cm^2/s}$, see Sect. 13.2.2). Assuming that about 10% of the rotational kinetic energy of $\approx 10^{50}$ erg of a 10 ms pulsar is released in electrons and positrons, the resulting average electron energy density in the 500 pc volume is of order $10^{-3}\ \mathrm{eV/cm^3}$, comparable to the density of secondary electrons from nuclear interactions of cosmic rays. In an energy range where such a source dominates the flux of electrons and positrons, the positron fraction is 1/2. While many details of this scheme remain to be clarified, such an explanation of the effect in terms of conventional astrophysics would need to be ruled out first, before more exotic schemes such as Dark Matter annihilation are invoked. Local PWN, for which the signature of electron escape and subsequent propagation to the Earth may be apparent in the cosmic ray electron spectrum at very high energies, include Vela-X [131] and Geminga. A recent measurement of the cosmic-ray diffusion around from Geminga [62], however, revealed unusually small diffusion coefficients, which make it difficult for those electrons to reach Earth during the relevant time scale.

In addition to the unpulsed emission from nebulae, pulsed GeV emission from many pulsars is observed, with cutoffs at a few GeV e.g. [88], as expected since higher-energy gamma rays have difficulty escaping from the pulsar magnetosphere with its huge magnetic fields. In this context the recent detection of pulsed emission up to energies of a few 100 GeV, following a steep power law rather than the expected super-exponential cutoff [132, 133] was very surprising. Cascade processes may be responsible for transporting the pulsed signal away from the pulsar, reducing the suppression at higher energies.

13.4.4 *Other Galactic Systems as Sources of High-Energy Radiation*

Several additional classes of cosmic particle accelerators have recently been identified in our galaxy. These objects are generally stellar binary systems of various types, or related to the collective effects of clusters of stars. Stellar binaries containing a normal star and a black-hole are known to (episodically) host accretion-powered jets which can be relativistic. These systems are the Galactic analogs of the active galactic nuclei described in Sect. 13.4.5, and have been dubbed "micro-quasars". Cygnus X-3 appears to be black hole with a massive stellar companion, and periodic emission has been detected using Fermi. The emission is correlated with the appearance of radio features and seems to be associated with the formation of a jet in the system. The only TeV detection of emission from a well-established black-hole binary is that of a single flare from Cygnus X-1 with the MAGIC telescope [134]. Whilst intriguing, further TeV detections will be required to confirm Cygnus X-1 as a TeV source.

The well-studied stellar binary Eta Carina contains two very massive ($M > 30M_{\odot}$) stars which both produce powerful (radiatively driven) winds. The collision of these stellar winds results in strong shocks and a situation akin to a supernova explosion, except in a much denser (in terms of both matter and radiation) and higher magnetic field environment (e.g. [135]). Gamma-ray emission is seen from this system up to $\sim$100 GeV [136], with two distinct components, and variability observed in the higher energy component which emerges above 20 GeV [137]. Whilst many questions remain, it now seems clear that Eta Carinae is a cosmic particle accelerator and a hadronic origin of one of the components seems plausible.

Evidence for acceleration associated with the collective effects of stellar winds is provided by the detection of very extended emission from the massive stellar cluster Westerlund 1 [138]. Degree scale emission is seen stretching well beyond the stellar cluster, which is one of the most massive in the Galaxy. However, a supernova remnant, unseen at other wavelengths due to the unusual environment, is not excluded as the origin of the TeV emission. Systems such as Westerlund 1 can be seen as the Galactic analogues of the starburst galaxies described in Sect. 13.4.1.

A single Nova (powered by a thermonuclear explosion on the surface of a degenerate white dwarf star) has been detected in high energy gamma-rays [139]. In an $\sim$10 day flare from the white dwarf V407 Cyg in 2010 emission was seen up to $\sim$5 GeV.

13.4.5 *Particle Acceleration Driven by Supermassive Black Holes*

The accretion of material onto supermassive ($M \sim 10^4 M_{sol} - 10^{10} M_{sol}$) black holes leads to the formation of oppositely-directed and highly-collimated jets of material. The mechanism for launching of these jets is still hotly debated, but magnetic fields in the accretion disc around the black hole look to be playing a crucial role [140, 141]. Radio emission associated with these jets has been know about for a long time, and is most dramatic in the case of Radio Galaxies where the jets are seen at a large angle to the observer and their true scale of (often) 100s of kiloparsecs can be seen, often dwarfing the host galaxy. Particle acceleration is clearly taking place in these objects, at several different locations, for example inside the inner (relativistic) jets, at the termination shocks of powerful jets and at the edges of the so-called radio "lobes" which result when jets are decelerated and spread out. Inverse Compton X-ray emission has been detected from several such systems, tracing the same population of electrons as seen in radio and allowing magnetic field strength and total energy densities to be estimated. X-ray Synchrotron emission, indicating the presence of very high energy electrons, is also seen in several places within systems powered by an active galactic nucleus (AGN). Despite the well established relativistic particle populations in these objects the nature of the acceleration mechanism is still not known. Diffusive shock acceleration does not seem to be the whole story in these objects, with a more distributed acceleration mechanism apparently needed to explain the X-ray synchrotron emission from the inner jets of nearby AGN [142].

Active galaxies are prime candidates for the acceleration of UHECRs, as can be seen from Fig. 13.8, with both the nucleus and the extended jets as possible acceleration sites for 10^{20} eV particles. As discussed in Sect. 13.2.5 there are hints of UHECR anisotropy correlated with the distribution of matter in the nearby universe (within the GZK horizon) and in particular an excess in the direction of the very nearby active galaxy Cen A (and also the direction of the more distant Centaurus cluster of galaxies). Gamma-ray emission his been detected from Cen A on a wide range of spatial scales. Figure 13.17 shows emission from the giant (10°) lobes of Cen A in radio synchrotron emission and (very likely) inverse Compton gamma-ray emission in the Fermi band [143]. Non-thermal emission is also seen in Cen A from the nucleus, inner jets and inner lobes. The synchrotron X-ray emission seen from the termination shock of the inner lobe closely resembles the situation for supernova remnants, but on a very much larger scale [144]. TeV emission is seen from the inner parts of Cen A (see the right-hand panel of Fig. 13.17) with a position consistent with the inner parts of the jet or the nucleus itself [145]. The giant lobes, jets and nucleus are all candidate acceleration sites for UHECR [146–148].

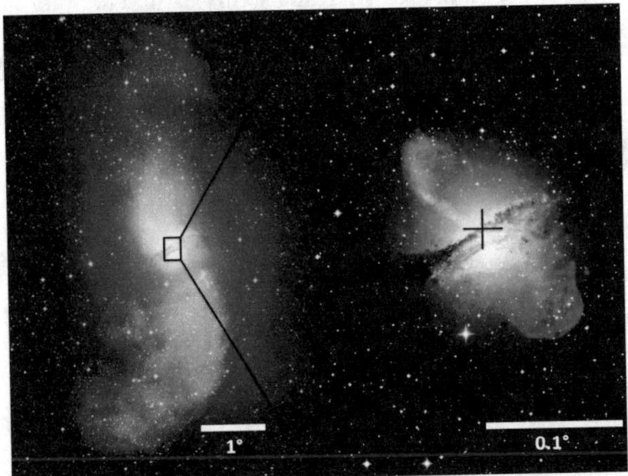

Fig. 13.17 The nearby active galaxy Centaurus A on scales of many degrees (left) and fractions of a degree (right). The left-hand plot is a multi-wavelength composite showing visible light overlaid with radio data in orange and gamma-ray data from Fermi in purple. The right-hand plot indicates the centroid of the TeV emission (cross) on a MWL composite of optical, X-ray (Chandra: blue) and microwave (orange) data. (Credit—left hand—NASA/DOE/Fermi LAT Collaboration, Capella Observatory, and Ilana Feain, Tim Cornwell, and Ron Ekers (CSIRO/ATNF), R. Morganti (ASTRON), and N. Junkes (MPIfR); right hand—ESO/WFI (visible); MPIfR/ESO/APEX/A.Weiss et al. (microwave); NASA/CXC/CfA/R.Kraft et al. (X-ray))

The bulk of the ~60 know extragalactic TeV gamma ray sources is of blazer type, as is the bulk of the extragalactic GeV sources. In blazers, jets are oriented close (within ~10°) to the line-of-sight to the observer, presenting a dramatically different perspective compared to radio galaxies. Apparently super-luminal motions observed in very-long-baseline-interferometer (VLBI) images of the central regions of these objects indicate bulk relativistic motion. Particle acceleration is presumably powered by this bulk motion, related to inhomogenities and shocks arising in the flow. Blazers are characterised by rapidly variable emission and complete non-thermal dominance of their spectral energy distributions, with the peak energy output usually in the gamma-ray domain. Markarian 421 was the first extragalactic object to be discovered in TeV gamma-rays [150] and is a dramatic example of a high-energy-peaked blazar. The measured energy spectra of these objects are significantly influenced by gamma-ray absorption by pair production with infrared or optical extragalactic background light (EBL) (e.g. [151]); vice versa, with assumptions regarding the intrinsic blazar spectra the absorption features can be used to constrain the level of EBL (e.g. [152]), which traces the history of star formation in the Universe and which, in certain spectral ranges, is difficult to measure directly due to overwhelming foregrounds [153]. TeV Blazars exhibit extreme variability, down to minute time scale [154]. This extreme variability places tight constraints on the size r of the emission region, from causality arguments: $r < \delta c t_{var}$, where δ is a Doppler boost factor reflecting the bulk motion of the particle acceleration region in the blazer jet. Blazars usually exhibit double-humped spectral energy distributions (Fig. 13.18), most likely reflecting a synchrotron component at

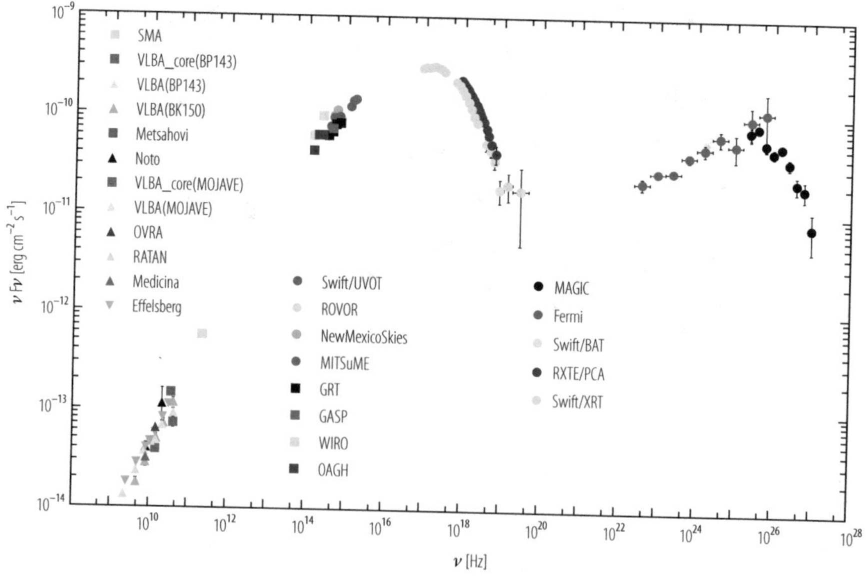

Fig. 13.18 The spectral energy distribution $\nu F_\nu \sim E^2 dN/dE$ of Mrk 421 as a function of frequency ν (where 1 TeV $\approx 2 \times 10^{26}$ Hz), measured simultaneously with a variety of instruments. F_ν is the energy flux per unit frequency. Reproduced from [149]

radio and X-ray energies, and an IC component at GeV/TeV energies, hinting at electrons as the parent particles. Often, the synchrotron X-ray intensity is such that X-rays form the dominant target for IC scattering. The parameters of the system—the electron spectrum, size, boost factor and magnetic field—can be determined from the location of the peaks in the radiation spectra, the exact spectral shape, the variability timescale and the additional condition that the photon density in the source region must be low enough to allow gamma rays to escape without pair-producing. Combined with short variability timescales this approach usually enforces large Doppler factors well beyond $\delta = 10$; how such bulk motions are achieved is not fully understood. While blazers obviously act as cosmic particle accelerators, it is unclear if they accelerate significant amounts of protons in addition to the electrons evident from their spectra, and most likely they have little impact on the cosmic rays detected on Earth.

13.5 Outlook

Significant progress has been achieved in the study of cosmic particle accelerators, in particular through the improved detection and imaging of high energy (GeV) and very high energy (TeV) gamma rays emitted in interactions of accelerated charged particles; also the first detection of very high energy cosmic neutrinos was reported. Particle acceleration is a ubiquitous feature in the Universe, associated especially with the evolution of massive stars and with black holes. Over 3000 multi-GeV accelerators are detected, and over 200 multi-TeV accelerators, with maximum energies well beyond 100 TeV. A variety of acceleration mechanisms seems to be realised in nature, including acceleration in non-relativistic SNR shocks and colliding-wind shocks, in relativistic pulsar wind shocks, in the unipolar inductors of pulsars, and in the compact and extended jets of AGN. In many systems, the acceleration process is highly efficient in converting bulk kinetic energy. The observed rapid variability in systems such as AGN or in the Crab Nebula represent a challenge to models and may require additional acceleration schemes. Despite this significant progress, a quantitative confirmation that SNR represent the sources of the bulk of nucleonic cosmic rays in our own galaxy and a first-principle calculation of their yield and spectrum is still lacking; other open questions concern the details of diffusive propagation of cosmic rays and of the resulting cosmic-ray anisotropies, the origin of the high-energy cosmic-ray positrons, and the origin of the highest-energy cosmic rays. A variety of new instruments aim to address these issues, including next-generation air-Cherenkov instruments such as CTA [155], ground-based cosmic-ray detectors like LHAASO [20], ultra-high-energy cosmic ray detectors such as JEM-EUSO [156] and a larger version of the Pierre Auger detector, or the very-high-energy neutrino detectors KM3Net [98] and the planned IceCube Gen2 instrument [157].

References

1. V.F. Hess: Phys. Z. 21 (1912) 1084.
2. A.W. Strong, I.V. Moskalenko, V.S. Ptuskin: Annu. Rev. Nucl. Part. Sci. 57 (2007) 285.
3. A.M. Hillas: J. Phys. G 31 (2005) 95.
4. S.P. Reynolds: Annu. Rev. Astron. Astrophys. 46 (2008) 89.
5. J. Blümer, R. Engel, J.R. Hörandel: Prog. Part. Nucl. Phys. 63 (2009) 293.
6. K. Kotera, A.V. Olinto: Annu. Rev. Astron. Astrophys. 49 (2011) 119.
7. A.M. Hillas: Annu. Rev. Astron. Astrophys. 22 (1984) 425.
8. J.A. Hinton, W. Hofmann: Annu. Rev. Astron. Astrophys. 47 (2009) 523.
9. F.A. Aharonian, et al.: Rep. Prog. Phys. 71 (2008) 096901.
10. P. Picozza, et al.: Astropart. Phys. 27 (2007) 296.
11. M. Aguilar, et al. (AMS Coll.): Phys. Rep. 366 (2002) 331.
12. S. Wissel, et al.: arXiv:1107.3272 (2011).
13. O. Adriani et al.: Phys. Rev. Lett. 119 (2017) 181101.
14. G. Ambrosi et al.: Nature 552 (2017) 63.
15. I. Allekotte, et al. (Pierre Auger Coll.): Nucl. Instrum. Meth. A 586 (2008) 409.
16. J. Abraham, et al. (Pierre Auger Coll.): Nucl. Instrum. Meth. A 620 (2010) 227.
17. U. Abbasi et al.: Astropart. Phys. 80 (2016) 131.
18. K.H. Kampert, et al.: Nucl. Phys. B Proc. Suppl. 136 (2004) 273.
19. M. Amenomori, et al.: Astrophys. J. 678 (2008) 1165.
20. G. Di Sciascio: Nucl. Part. Phys. Proc. 279–281 (2016) 166.
21. O. Adriani, et al.: Science 332 (2011) 69.
22. M. Aguilar et al.: Phys. Rev. Lett. 114 (2015) 171103.
23. M. Aguilar et al.: Phys. Rev. Lett. 115 (2015) 211101.
24. D. Allard, E. Parizot, A. V. Olinto: Astropart. Phys. 27 (2007) 61.
25. K. Greisen: Phys. Rev. Lett. 16 (1066) 747.
26. G.T. Zatsepin, V.A. Kuz'min: JETP Lett. 4 (1966) 78.
27. V. Berezinsky, A.Z. Gazizov, S.I. Grigorieva: Phys. Rev. D 74 (2006) 043005.
28. J. Abraham et al.: Phys. Lett. B685 (2010) 239.
29. A. Aab et al.: arXiv:1708.06592 and Proceedings, 35th International Cosmic Ray Conference (ICRC 2017): Bexco, Busan, Korea, July 12–20, 2017.
30. A. Aab et al.: JCAP 1704 (2017) 038.
31. K.H. Kampert, M. Unger: Astropart.Phys. 35 (2012) 660.
32. J.A. Simpson: Annu. Rev. Nucl. Part. Sci. 33 (1983) 323.
33. M. Simon, A. Molnar, S. Roesler: Astrophys. J. 499 (1998) 250.
34. V.L. Ginzburg, V.S. Ptuskin: Rev. Mod. Phys. 48 (1976) 161, (Erratum) Rev. Mod. Phys. 48 (1976) 675.
35. R. Jansson, G.R. Farrar: Astrophys. J. 761 (2012) L11.
36. L.O'C. Drury, D.C. Ellisson, J.-P. Meyer: Nucl. Phys. A 663 (2000) 843–843.
37. B.D. Fields, et al.: Astron. Astrophys. 370 (2001) 623.
38. A.W. Strong, et al.: Astrophys. J. Lett. 722 (2010) L58.
39. A. Aab et al.: Phys. Rev. Lett. 117 (2016) 192001
40. J. Abraham, et al.: Phys. Rev. Lett. 104 (2010) 091101.
41. A.Aab et al: Phys. Rev. D90 (2014) 122006.
42. R.U. Abbasi, et al.: Astropart. Phys. 64 (2015) 49.
43. J.R. Hörandel: Adv. Space Res. 41 (2008) 442.
44. M. Amenomori, et al.: Astrophys. J. 626 (2005) L29.
45. A.H. Compton, I.A. Getting: Phys. Rev. 47 (1935) 817.
46. P. Blasi, E. Amato: arXiv:1105.4529 (2011).
47. M. Amenomori, et al.: Science 314 (2006) 439.
48. A.A. Abdo, et al.: Phys. Rev. Lett. 101 (2008) 221101.
49. R.U. Abbasi, et al.: Astrophys. J. 740 (2011) 16.

50. L.O'C. Drury, F.A. Aharonian: Astropart. Phys. 29 (2008) 420.
51. G. Ciacinti, G. Sigl: arXiv:1111.2536 (2011).
52. O. Adriani, et al.: Phys. Rev. Lett. 102 (2009) 051101.
53. M. Aguilar et al.: Phys. Rev. Lett. 113 (2014) 121102.
54. S. Abdollahi et al.: Phys. Rev. D95 (2017) 082007.
55. D. Kerszberg: 5th International Cosmic Ray Conference (ICRC 2017): Bexco, Busan, Korea, July 12–20, 2017.
56. L. Accardo et al.: Phys. Rev. Lett. 113 (2014) 121101.
57. M. Pohl, et al.: Astron. Astrophys. 409 (2003) 581.
58. O. Adriani, et al.: Nature 458 (2009) 607.
59. M. Ackermann, et al.: arXiv:1109.0521.
60. D. Grasso, et al.: Astroparticle Phys. 32 (2009) 140.
61. F.A. Aharonian, A. Atoyan, H.J. Völk: Astron. Astrophys. 294 (1995) L41.
62. A.U. Abeysekara et al.: Science 358 (2017) 911.
63. A. Aab et al.: Astrophys. J. 794 (2014) 172.
64. F. Zwicky: Phys. Rev. 55 (1939) 986.
65. R.D. Blandford, J.P. Ostriker: Astrophys. J. Lett. 221 (1978) L29; Astrophys. J. 237 (1980) 793.
66. A.R. Bell: Mon. Not. R. Astron. Soc. 182 (178) 147; Mon. Not. R. Astron. Soc. 182 (1978) 443.
67. J.K. Truelove, C.F. McKee: Astrophys. J. Suppl. 120 (1999) 299.
68. L.O'C. Drury, H.J. Völk: Astrophys. J. 248 (1981) 344.
69. D.C. Ellison, A. Decourchelle, J. Ballet: Astron. Astrophys. 413 (2004) 189.
70. J. Vink: arXiv:1112.0576 (2011).
71. V.S. Ptuskin, V.N. Zirakashvili: Astron. Astrophys. 429 (2005) 755.
72. M.A. Malkov et al.: Astrophys. J. 768 (2013) 73.
73. P. Blasi, E. Amato: arXiv:1105.4521 (2011).
74. G. Pelletier, M. Lemoine, A. Marcowith: arXiv:0811.1506 (2008).
75. A.K. Harding, arXiv:0710.3517 (2007).
76. J. Arons: arXiv:0708.1050 (2007).
77. Y. Lyubarsky, J.G. Kirk: Astrophys. J. 547 (2001) 437.
78. C.F. Kennel, F.V. Coroniti: Astrophys. J. 283 (1984) 694.
79. C.F. Kennel, F.V. Coroniti: Astrophys. J. 283 (1984) 710.
80. C.-Y. Ng, R.W. Romani: Astrophys. J. 601 (2004) 479; Astrophys. J. 673 (2008) 411.
81. M.J. Rees, J.E. Gunn: Mon. Not. R. Astron. Soc. 167 (1974)1.
82. M. Nagano, A.A. Watson: Rev. Mod. Phys. 72 (2000) 689.
83. S.R. Kelner, F.A. Aharonian: Phys. Rev. D 78 (2008) 034013.
84. G.R. Blumenthal, R.J. Gould: Rev. Mod. Phys. 42 (1970) 237.
85. M.S. Longair: High Energy Astrophysics, Cambridge University Press.
86. N.S. Kardashev: Sov. Astron. 6 (1962) 317.
87. A.M. Atoyan, F.A. Aharonian: Mon. Not. R. Astron. Soc. 287 (1996) 525.
88. A.A. Abdo, et al.: Astrophys. J. 708 (2010) 1254.
89. R. Bühler, et al.: arXiv:1112.1979 (2011).
90. W.B. Atwood, et al.: Astrophys. J. 697 (2009) 1071.
91. W. Hofmann: arXiv:astro-ph/0603076 (2006).
92. U. Abeysekara et al.: Astrophys. J. 843 (2017) 40.
93. F. Acero et al.: Astrophys. J. Suppl. 218 (2015) 23.
94. TeVCat, http://tevcat.uchicago.edu/
95. M.G. Aartsen et al.: Phys. Rev. Lett. 115 (2015) 081102
96. M.G. Aartsen et al., arXiv:1710.01191.
97. M.G. Aartsen et al.: Astrophys. J. 835 (2017) 151
98. P. Bagley, et al.: KM3NeT Technical Design Report (ISBN 978-90-6488-033-9).
99. M. Ackermann et al.: arXiv:1710.01207 and Proceedings, 35th International Cosmic Ray Conference (ICRC 2017): Bexco, Busan, Korea, July 12–20, 2017

100. https://fermi.gsfc.nasa.gov/ssc/observations/types/allsky/
101. M. Ackermann, et al.: Astrophys. J. 726 (2011) 81.
102. A.A. Abdo, et al.: Astrophys. J. 688 (2008) 1078.
103. A. Abramowski et al.: Nature 531 (2016) 476.
104. A.A. Abdo, et al.: Astrophys. J. Lett. 709 (2010) L152.
105. F. Acero, et al.: Science 326 (2009) 1080.
106. V.A. Acciari, et al.: Nature 462 (2009) 7274.
107. G. Dobler, et al.: Astrophys. J. 717 (2010) 825.
108. M. Su, T.R. Slatyer, D.P. Finkbeiner: Astrophys. J. 724 (2010) 1044.
109. M. Ackermann et al.: Astrophys.J. 793 (2014) 64
110. R.M. Crocker, F.A. Aharonian: Phys. Rev. Lett. 106 (2011) 101102.
111. P. Mertsch, S. Sarkar: Phys. Rev. Lett. 107 (2011) 091101.
112. E.A. Helder, et al.: Science 325 (2009) 719.
113. S.G. Lucek, A.R. Bell: Mon. Not. R. Astron. Soc. 314 (2000) 65.
114. L.O'C. Drury, F.A. Aharonian, H.J. Völk: Astron. Astrophys. 287 (1994) 959.
115. H.J. Völk, E.G. Berezhko, L.T. Ksenofontov: Astron. Astrophys. 483 (2008) 529.
116. S. Archambault et al.: Astrophys. J. 836 (2017) 23.
117. H. Abdalla et al.: Astron. Astrophys. 612 (2018)
118. A.A. Abdo, et al.: Astrophys. J. 706 (2009) L1.
119. J. Aleksić, et al.: Astron. Astrophys. 541 (2012) 11.
120. M. Ackermann et al.: Science 339 (2013) 807.
121. Y. Uchiyama, et al.: Astrophys. J. Lett. 723 (2010) L122.
122. T.C. Weekes, et al.: Astrophys. J. 342 (1989) 379.
123. A.A. Abdo, et al.: Science 331 (2011) 739.
124. M. Tavani, et al.: Science 331 (2011) 736.
125. P.M. Gaensler, P.O. Slane: Annu. Rev. Astron. Astrophys. 44 (2006) 17.
126. F. Mattana, et al.: Astrophys. J. 694 (2009) 12.
127. O. Kargaltsev, G. Pavlov: arXiv:1002.0885 (2010).
128. J.M. Blondin, R.A. Chevalier, D.M. Frierson: Astrophys. J. 563 (2001) 806.
129. F.A. Aharonian, et al.: Astron. Astrophys. 460 (2006) 365.
130. S.W. Kong, et al.: Mon. Not. R. Astron. Soc. 416 (2011) 1067.
131. J.A. Hinton, et al.: Astrophys. J. Lett. 743 (2011) L7.
132. E. Aliu, et al.: Science 334 (2011) 69.
133. J. Aleksić, et al.: arXiv:1109.6124 (2011).
134. J. Albert, et al.: Astrophys. J. 665 (2007) L51.
135. E.R. Parkin, et al.: Astrophys. J. 726 (2011) 105.
136. A.A. Abdo, et al.: Astrophys. J. 723 (2010) 649.
137. C. Farnier, R. Walter: Mem. Soc. Astron. Italiana 82 (2011) 796.
138. A. Abramowski, et al.: arXiv:1111.2043 (2011).
139. A.A. Abdo, et al.: Science 329 (2010) 817.
140. R.D. Blandford, R.L. Znajek: Mon. Not. R. Astron. Soc. 179 (1977) 433.
141. R.D. Blandford, D.G. Payne: Mon. Not. R. Astron. Soc. 199 (1982) 883.
142. M.J. Hardcastle, et al.: Astrophys. J. 670 (2007) L81.
143. A.A. Abdo, et al.: Science 328 (2010) 725.
144. J.H. Croston, et al.: Mon. Not. R. Astron. Soc. 395 (2009) 1999.
145. F.A. Aharonian, et al.: Astrophys. J. Lett. 695 (2009) L44.
146. M.J. Hardcastle, et al.: Mon. Not. R. Astron. Soc. 393 (2009) 1041.
147. M. Honda, Astrophys. J. 706 (2009) 1517.
148. F.M. Rieger, F.A. Aharonian: Astron. Astrophys. 506 (2009) L41.
149. A.A. Abdo, et al.: Astrophys. J. 736 (2011) 131.
150. M. Punch, et al.: Nature 358 (1992) 477.
151. A. Franceschini, G. Rodighiero, M. Vaccari: Astron. Astrophys. 487 (2008) 837.
152. D. Mazin, M. Raue: Astron. Astrophys. 471 (2007) 439.
153. M.G. Hauser, E. Dwek: Annu. Rev. Astron. Astrophys. 39 (2001) 249.

154. F.A. Aharonian, et al.: Astrophys. J. 664 (2007) L71.
155. M. Actis, et al.: Exp. Astron. 32 (2011) 193.
156. T. Ebisuzaki, et al.: AIP Conf. Proc. 1367 (2011) 120–125.
157. M. Ackermann, et al.: Intern. Cosmic Ray Conf. 2017, arXiv:1710.01207

Printed in the United States
By Bookmasters